Encyclopedia of
Applied Electrochemistry

Gerhard Kreysa • Ken-ichiro Ota
Robert F. Savinell
Editors

Encyclopedia of Applied Electrochemistry

Volume 2

F–O

With 1250 Figures and 122 Tables

Editors
Gerhard Kreysa
Eppstein, Germany

Robert F. Savinell
Case Western Reserve University
Cleveland, OH, USA

Ken-ichiro Ota
Yokohama National University, Fac. Engineering
Yokohama, Japan

ISBN 978-1-4419-6995-8 978-1-4419-6996-5 (eBook)
ISBN Bundle 978-1-4419-6997-2 (print and electronic bundle)
DOI 10.1007/978-1-4419-6996-5
Springer New York Heidelberg Dordrecht London

Library of Congress Control Number: 2014934571

Printed on acid-free paper

Springer is part of Springer Science+Business Media (www.springer.com)

Preface

Electrochemistry provides the opportunity to run chemical redox reactions directly with electrons as reaction partners and against the free energy gradient via external electrical energy input into the reaction system. It also serves as a way to efficiently generate energy from stored energy in chemical bonds. Applied electrochemistry has been the basis for many industrial processes ranging from metals recovery and purification to chemical synthesis and separations since the latter half of the 1800s when large-scale electricity generation became possible. Electrochemical processes have a large impact on energy as it has been estimated that these processes consume about 6–10 % of the world's electricity generation capacity. Applied electrochemistry is now impacting industry and society more and more with technologies for waste water treatment, efficient chemical separation, and environmental sensing and remediation. Electrochemistry is the foundation for electrochemical energy storage by batteries and electrochemical capacitors and energy conversion by fuel cells and solar cells. In fact, applied electrochemistry will play a major role in the world's ability to harness and use renewable energy sources. Electrochemistry is also fundamental to biological cell transport and many aspects of living systems and their activities. It is exploited for use in medical diagnostics to detect abnormalities and in biomedical engineering to relieve pain and deliver function.

The application of electrochemistry involves not just a fundamental understanding of the sciences, but also applying engineering principles to device and technology design by considering mass and energy balances, transport processes in the electrolyte and at electrode interfaces, and multi-scale modeling and simulation for predicting and optimizing performance. The interaction of the interfacial reactions, the transport driving forces, and the electric field defines the field of electrochemical engineering. The understanding of the scientific and engineering principles of electrochemical systems has driven advances in the application of electrochemistry especially during the last half century.

The purpose of this collection is to summarize the A–Z of the application of electrochemistry and electrochemical engineering for use by electrochemists and electrochemical engineers as well as nonspecialists such as engineers and scientists of all disciplines, economists, students, and even politicians. Electrochemical fundamentals, electrochemical processes and technologies, and electrochemical techniques are described by many experts in their fields from around the world, many from industry. Each entry is meant to be an introduction and also gives references for further study. With this collection, we hope that current technology and operating practices can be made available for future generations to learn from. We hope that this encyclopedia will stimulate understanding of the current state of the art and lead to advances in new and more efficient technologies with breakthroughs from new theory and materials. We hope you find it of value to your work.

Gerhard Kreysa
Ken-ichiro Ota
Robert F. Savinell
Editors-in-Chief

Acknowledgments

The Editors-in-Chief would like to acknowledge the backing of their institutions in supporting this work. Specifically we want to thank the Technical University of Dortmund (GK), Yokohama National University (KO), and Case Western Reserve University (RFS). We also want to thank the topical editors and authors; it is because of their dedication to their fields and hard work that this collection was possible. We thank our families for their understanding of the importance of this project to us and our profession. Finally, we want to thank the editorial staff of Springer, especially Barbara Wolf who worked long and hard on this project and Kenneth Howell specifically for encouraging us to embark on this project and nudging us along.

About the Editors

Prof. Dr.rer.nat. Dr.-Ing. E.h. Dr.tekn.h.c. Gerhard Kreysa Retired Chief Executive of DECHEMA e.V.

Gerhard Kreysa was born in 1945 in Dresden. He studied chemistry at the University of Dresden and received his Ph.D. in 1970. In 1973, he joined the Karl Winnacker Institute of DECHEMA in Frankfurt am Main. He developed new concepts for the utilization of three-dimensional electrodes, which became prominent for electrochemical waste water treatment in the process industry. He also played a leading role in the clarification of the "cold fusion" affaire in 1989. In 1985, he was appointed as professor in the Chemical Engineering Department at the University of Dortmund. In 1993, he was appointed as honorary professor at the University of Regensburg. From 1985 to 1995, he served as executive editorial board member of the *Journal of Applied Electrochemistry*. He was a recipient of the Chemviron Award in 1980, the Max-Buchner-Research-Award of DECHEMA and the Castner Medal of the Society of Chemical Industry in 1994, and the Wilhelm Ostwald Medal of the Saxon Academy of Sciences in Leipzig in 2006.

From 1992 to 2009, Dr. Kreysa served as chief executive of DECHEMA Society of Chemical Engineering and Biotechnology in Frankfurt am Main, Germany. During this time he also served as general secretary of the European Federation of Chemical Engineering and the European Federation of Biotechnology. He obtained many distinctions: Honorary doctor degrees of Technical University of Clausthal and of the Royal Institute of Technology in Stockholm, Foreign Member of the Royal Swedish Academy of Engineering Sciences, elected honorary fellow of the Institution of Chemical Engineers, and honorary member of the Czech Society of Chemical Engineering. In 2007 he was awarded with the Order of Merit of the Free State of Saxony, and in 2008 he became a member of the German Academy of Technical Sciences (acetech). He has 196 scientific publications and books and has given 312 scientific and public lectures. Despite the numerous duties and responsibilities of his former senior management position, he continues to have a lively interest in the further development of science and engineering and is highly regarded as an advisor on national and international issues.

Prof. Dr. G. Kreysa
Weingasse 22
D-65817 Eppstein

Ken-ichiro Ota

Ken-ichiro Ota is a professor at and the chairman of Green Hydrogen Research Center at Graduate School of Engineering, Yokohama National University, Japan. He received his B.S.E. in applied chemistry in 1968 and Ph.D. in engineering in 1973, both from the University of Tokyo. After graduation, he became a research associate at the university until 1979. In the same year, he became an associate professor at the Yokohama National University, and a professor in 1995. He has worked on hydrogen energy and fuel cell since 1974, focusing on materials science for fuel cells and hydrogen energy system including water electrolysis. In the fuel cell field he has worked on direct methanol fuel cell, molten carbonate fuel cell, and polymer electrolyte fuel cell. Recently, he is developing transition metal oxide–based cathode for polymer electrolyte fuel cell. He is also working on storage and transport of renewable energies by hydrogen technology. He has published more than 190 original papers, 80 review papers, and 50 scientific books. He received the Molten Salt Award in 1998 and the Industrial Electrolysis Award in 2002 from the Electrochemical Society of Japan. He received the Canadian Hydrogen Society Award in 2004 and the Society Award of the Electrochemical

Society of Japan in 2011. He is now the chairman of the National Committee for Standardization of the Stationary Fuel Cells. He was the president of the Hydrogen Energy Systems Society of Japan from 2000 to 2008 and also the president of the Electrochemical Society of Japan from 2008 to 2009. He is now the chairman of the Fuel Cell Development Information Center of Japan.

Dr. Robert F. Savinell Distinguish University Professor, George S. Dively Professor of Engineering, Department of Chemical Engineering, Case Western Reserve University, Cleveland, OH, USA

Dr. Robert F. Savinell received his B.Che. from Cleveland State University in 1973, and his M.S. (1974) and Ph.D. (1977), both in chemical engineering from University of Pittsburgh. He worked as a research engineer for Diamond Shamrock Corporation, then as a faculty member at the University of Akron before joining the faculty at Case Western Reserve University (CWRU) in 1986. Professor Savinell was the Director of the Ernest B. Yeager Center for Electrochemical Sciences at CWRU for ten years and served as Dean of Engineering at CWRU for seven years. Professor Savinell has been engaged in electrochemical engineering research and development for 40 years. Savinell's research is directed at fundamental science and mechanistic issues of electrochemical processes; and at electrochemical technology systems and device design, development, modeling and optimization. His research has addressed applications for energy conversion, energy storage, sensing, and electrochemical materials extraction and synthesis. Savinell has over 120 peer-reviewed and over 168 other publications, eight patents, and has been an invited and keynote speaker at hundreds of national and international conferences and workshops in the electrochemical field. He has supervised over 50 Ph.D./M.S. student projects.

Professor Savinell is the former North American editor of the *Journal of Applied Electrochemistry* and currently is the editor of the *Journal of the Electrochemical Society*. He is a Fellow of the Electrochemical Society, Fellow of the American Institute of Chemical Engineers, and Fellow of the International Society of Electrochemistry.

Section Editors

Electrocatalysis

Radoslav R. Adzic Chemistry Department, Brookhaven National Laboratory, Upton, NY, USA

Primary Batteries

George Blomgren Imara Corporation, Menlo Park, CA, USA

Environmental Electrochemistry

Christos Comninellis EPFL, Lausanne, Switzerland

Organic Electrochemistry

Toshio Fuchigami Department of Electrochemistry, Tokyo Institute of Technology, Midori-ku, Yokohama, Japan

Solid State Electrochemistry

Ulrich Guth Kurt-Schwabe-Institut für Mess- und Sensortechnik e.V. Meinsberg, Ziegra-Knobelsdorf, Germany

High-Temperature Molten Salts

Rika Hagiwara Graduate School of Energy Science, Kyoto University, Sakyo-ku, Kyoto, Japan

Inorganic Electrochemical Synthesis (Including Chlorine/Caustic, Chlorates, Hypochlorite)

Kenneth L. Hardee Research & Development Division, De Nora, Fairport Harbor, OH, USA

Bioelectrochemistry

Dirk Holtmann DECHEMA e.V., Frankfurt am Main, Germany

Electrochemical Instrumentation and Laboratory Techniques

Rudolf Holze Institut für Chemie, Technische Universität Chemnitz, Chemnitz, Germany

Fuel Cells

Minoru Inaba Department of Molecular Chemistry and Biochemistry, Doshisha University, Kyotanabe, Kyoto, Japan

Supercapacitors

Hiroshi Inoue Department of Applied Chemistry, Graduate School of Engineering, Osaka Prefecture University, Sakai, Osaka, Japan

High-Temperature Electrochemistry

Tatsumi Ishihara Department of Applied Chemistry, Kyushu University, Nishi-ku, Fukuoka, Japan

Secondary Batteries

Kiyoshi Kanamura Applied Chemistry Graduate School of Engineering, Tokyo Metropolitan University, Hachioji, Tokyo, Japan

Semiconductor Synthesis and Electrochemistry

Paul A. Kohl Georgia Institute of Technology, School of Chemical and Biomolecular Engineering, Atlanta, GA, USA

Electrolytes

Werner Kunz Institute of Physical and Theoretical Chemistry, Regensburg University, Regensburg, Germany

Electrodeposition (Electrochemical Metal Deposition and Plating)

Uziel Landau Chemical Engineering Department, Case Western Reserve University, Cleveland, OH, USA

Electrochemical Analysis and Sensors

Chung-Chiun Liu Case Western Reserve University, Cleveland, OH, USA

Electrocatalysis

Nebojsa Marinkovic Synchrotron Catalysis Consortium, University of Delaware, Newark, DE, USA

Environmental Electrochemistry

Yunny Meas Vong CIDETEQ, Parque Tecnológico Querétaro, Sanfandila, Pedro Escobedo, CP, México

Photoelectrochemistry

Tsutomu Miyasaka Graduate School of Engineering, Toin University of Yokohama, Kanagawa, Japan

Electrochemical Engineering

Trung Van Nguyen Department of Chemical and Petroleum Engineering, The University of Kansas, Lawrence, KS, USA

Fuel Cells

Thomas Schmidt Electrochemistry Laboratory, Paul Scherrer Institut, Villigen, Switzerland

Advisory Board

Contributors

Abd El Aziz Abd-El-Latif Institute of Physical and Theoretical Chemistry, University of Bonn, Bonn, Germany

Luisa M. Abrantes Departamento de Química e Bioquímica, FCUL, Lisbon, Portugal

Radoslav R. Adzic Chemistry Department, Brookhaven National Laboratory, Upton, NY, USA

Sheikh A. Akbar The Ohio State University, Columbus, OH, USA

Francisco Alcaide Energy Department, Fundación CIDETEC, San Sebastián, Spain

Antonio Aldaz Instituto Universitario de Electroquímica, University of Alicante, Alicante, Spain

Leonardo S. Andrade Universidade Federal de Goiás, Catalão, Brazil

Juan Manuel Artés Institució Catalana de Recerca i Estudis Avançats (ICREA), Institute for Bioengineering of Catalonia (IBEC), Barcelona, Spain

Electrical and Computer Engineering, University of California Davis, Davis, CA, USA

Mahito Atobe Graduate School of Environment and Information Sciences, Yokohama National University, Yokohama, Japan

Nicola Aust BASF SE, Ludwigshafen, Germany

Arseto Bagastyo Advanced Water Management Centre (AWMC), The University of Queensland, Brisbane, QLD, Australia

Helmut Baltruschat Institute of Physical and Theoretical Chemistry, University of Bonn, Bonn, Germany

Cesar Alfredo Barbero Department of Chemistry, National University of Rio Cuarto, Rio Cuarto, Cordoba, Argentina

Scott Barnett Department of Materials Science and Engineering, Northwestern University, Evanston, IL, USA

Romas Baronas Faculty of Mathematics and Informatics, Vilnius University, Vilnius, Lithuania

Damien Batstone Advanced Water Management Centre (AWMC), The University of Queensland, Brisbane, QLD, Australia

Pierre Bauduin Institut de Chimie Separative de Marcoule, UMR 5257 – ICSM Site de Marcoule, CEA/CNRS/UM2/ENSCM, Bagnols sur Ceze, France

Dorin Bejan Department of Chemistry, Electrochemical Technology Centre, University of Guelph, Guelph, ON, Canada

Luc Belloni CEA Saclay, Gif-sur-Yvette, France

Henry Bergman Anhalt University, Anhalt, Germany

Sonia R. Biaggio Universidade Federal de Goiás, Catalão, Brazil

Salma Bilal National Centre of Excellence in Physical Chemistry, University of Peshawar, Peshawar, Pakistan

José M. Bisang Programa de Electroquímica Aplicada e Ingeniería Electroquímica (PRELINE), Facultad de Ingeniería Química, Universidad Nacional del Litoral, Santa Fe, Santa Fe, Argentina

George Blomgren Blomgren Consulting Services, Lakewood, OH, USA

Nerilso Bocchi Universidade Federal de Goiás, Catalão, Brazil

Pierre Boillat Paul Scherrer Institut, Villigen PSI, Switzerland

Nikolaos Bonanos Department of Energy Conversion and Storage, Technical University of Denmark, Roskilde, DK

Antoine Bonnefont Institut de Chimie de Strasbourg, CNRS-Université de Strasbourg, Strasbourg, France

Oleg Borodin Electrochemistry Branch, Sensor and Electron Devices Directorate, U.S. Army Research Laboratory, Adelphi, MD, USA

Stanko Brankovic Electrical and Computer Engineering Department, Chemical and Bimolecular Engineering Department, and Chemistry Department, University of Houston, Houston, TX, USA

Enric Brillas Laboratory of Electrochemistry of Materials and Environment, Department of Physical Chemistry, Faculty of Chemistry, University of Barcelona, Barcelona, Spain

Ralph Brodd Broddarp of Nevada, Inc., Henderson, NV, USA

Michael Bron Institut für Chemie, Technische Chemie, Martin-Luther-Universität Halle-Wittenberg, Halle, Germany

Nigel W. Brown Daresbury Innovation Centre, Arvia Technology Ltd., Daresbury, UK

Felix N. Büchi Paul Scherrer Institut, Villigen PSI, Switzerland

Richard Buchner Institute of Physical and Theoretical Chemistry, University of Regensburg, Regensburg, Germany

Ratnakumar V. Bugga Jet Propulsion Laboratory, Pasdena, CA, USA

Nigel J. Bunce Department of Chemistry, Electrochemical Technology Centre, University of Guelph, Guelph, ON, Canada

Andreas Bund FG Elektrochemie und Galvanotechnik, Institut für Werkstofftechnik, Technische Universitüt Ilmenau, Ilmenau, Germany

Erika Bustos Centro de Investigacióny Desarrollo Tecnológico en Electroquímica, S. C., Sanfandila, Pedro Escobedo, Querétaro, México

Julea Butt School of Chemistry, University of East Anglia, Norwich, UK

Yun Cai Material Science, Joint Center for Artificial Photosynthesis, Lawrence Berkeley National Laboratory, Berkeley, CA, USA

Claudio Cameselle Department of Chemical Engineering, University of Vigo, Vigo, Spain

Maja Cemazar Department of Experimental Oncology, Institute of Oncology Ljubljana, Ljubljana, Slovenia

Vidhya Chakrapani Department of Chemical and Biological Engineering, Rensselaer Polytechnic Institute, Troy, NY, USA

François Chellé Universite Catholique de Louvain, Louvain-la-Neuve, Belgium

Kaimin Chen MECC, Medtronic Inc., Minneapolis, MN, USA

Po-Yu Chen Department of Medicinal and Applied Chemistry, Kaohsiung Medical University, Kaohsiung, Taiwan

Kazuhiro Chiba Tokyo University of Agriculture and Technology, Fuchu, Tokyo, Japan

Masanobu Chiku Department of Applied Chemistry, Graduate School of Engineering, Osaka Prefecture University, Osaka, Japan

YongMan Choi Chemistry Department, Brookhaven National Laboratory, Upton, NY, USA

David E. Cliffel Department of Chemistry, Vanderbilt University, Nashville, TN, USA

Christos Comninellis Institute of Chemical Sciences and Engineering, Ecole Polytechnique Fédérale de Lausanne (EPFL), Lausanne, Switzerland

Ann Cornell School of Chemical Science and Engineering, Applied Electrochemistry, KTH Royal Institute of Technology, Stockholm, Sweden

Serge Cosnier Department of Molecular Chemistry, CNRS UMR 5250 CNRS-University of Grenoble, Grenoble, France

Vincent S. Craig Department of Applied Mathematics, Research School of Physical Sciences and Engineering, Australian National University, Canberra, ACT, Australia

Hideo Daimon Advanced Research and Education, Doshisha University, Kyotanabe, Kyoto, Japan

Manfred Decker Kurt-Schwabe-Institut fuer Mess- und Sensortechnik e.V. Meinsberg, Waldheim, Germany

Dario Dekel CellEra Inc., Caesarea, Israel

I. M. Dharmadasa Electronic Materials and Sensors Group, Materials and Engineering Research Institute, Sheffield Hallam University, Sheffield, UK

Petros Dimitriou-Christidis Environmental Chemistry Modeling Laboratory, Ecole Polytechnique Fédérale de Lausanne (EPFL), Lausanne, Switzerland

Pablo Docampo Clarendon Laboratory, Department of Physics, Oxford University, Oxford, UK

Deepak Dubal AG Elektrochemie, Institut für Chemie, Technische Universität Chemnitz, Chemnitz, Germany

Laurie Dudik Case Western Reserve University, Cleveland, OH, USA

Jean François Dufreche Institut de Chimie Séparative de Marcoule and Université Montpellier, Marcoule, France

Christian Durante Department of Chemical Sciences, University of Padova, Padova, Italy

Prabir K. Dutta The Ohio State University, Columbus, OH, USA

Ulrich Eberle Government Programs and Research Strategy, GM Alternative Propulsion Center, Adam Opel AG, Rüsselsheim, Germany

Obi Kingsley Echendu Electronic Materials and Sensors Group, Materials and Engineering Research Institute, Sheffield Hallam University, Sheffield, UK

Minato Egashira College of Bioresource Sciences, Nihon University, Fujisawa, Kanagawa, Japan

Takashi Eguro Frukawa Battery, Iwaki, Fukushima, Japan

Martin Eichler AG Elektrochemie, Institut für Chemie, Technische Universität Chemnitz, Chemnitz, Germany

Robert Eisenberg Department of Molecular Biophysics and Physiology, Rush University Medical Center, Chicago, IL, USA

Bernd Elsler Johannes Gutenberg-University Mainz, Mainz, Germany

Eduardo Expósito Instituto Universitario de Electroquímica, University of Alicante, Alicante, Spain

Emiliana Fabbri Electrochemistry Laboratory, Paul Scherrer Institute, Villigen, Switzerland

Yujie Feng Harbin Institute of Technology, Harbin, China

Rui Ferreira Instituto de Tecnologia Química e Biológica, Universidade Nova de Lisboa, Oeiras, Portugal

Sergio Ferro Department of Chemical and Pharmaceutical Sciences, University of Ferrara, Ferrara, Italy

Stéphane Fierro Institute of Chemical Sciences and Engineering, Ecole Polytechnique Fédérale de Lausanne (EPFL), Lausanne, Switzerland

Michael A. Filler School of Chemical and Biomolecular Engineering, Georgia Institute of Technology, Atlanta, GA, USA

Alanah Fitch Department of Chemistry, Loyola University, Chicago, IL, USA

Robert Forster School of Chemical Sciences National Center for Sensor Research, Dublin City University, Dublin, Ireland

György Fóti Institute of Chemical Sciences and Engineering, Ecole Polytechnique Fédérale de Lausanne (EPFL), Lausanne, Switzerland

Alejandro A. Franco Laboratoire de Réactivité et de Chimie des Solides (LRCS) - UMR 7314, Université de Picardie Jules Verne, CNRS and Réseau sur le Stockage Electrochimique de l'Energie (RS2E), Amiens, France

Matthias Franzreb Institute of Functional Interfaces, Karlsruhe Institute of Technology, Eggenstein-Leopoldshafen, Germany

Stefano Freguia Advanced Water Management Centre (AWMC), The University of Queensland, Brisbane, QLD, Australia

Bernardo A. Frontana-Uribe Centro Conjunto de Investigación en Química Sustentable, UAEMéx–UNAM, Toluca, Estado de México, Mexico

Instituto de Química UNAM, Mexico, Mexico

Albert J. Fry Weslayan University, Middletown, CT, USA

Toshio Fuchigami Department of Electrochemistry, Tokyo Institute of Technology, Midori-ku, Yokohama, Japan

Akira Fujishima Kanagawa Academy of Science and Technology, Takatsu–ku, Kawasaki, Kanagawa, Japan

Photocatalysis International Research Center, Tokyo University of Science, Noda, Chiba, Japan

Klaus Funke Institute of Physical Chemistry, University of Muenster, Muenster, Germany

Ping Gao AG Elektrochemie, Institut für Chemie, Technische Universität Chemnitz, Chemnitz, Germany

Vicente García-García Instituto Universitario de Electroquímica, University of Alicante, Alicante, Spain

Helga Garcia Instituto de Tecnologia Química e Biológica, Universidade Nova de Lisboa, Oeiras, Portugal

Darlene G. Garey CIDETEQ, Centro de Investigación y Desarrollo Tecnológico en Electroquímica Parque Tecnológico Querétaro, Pedro Escobedo, Edo. Querétaro, México

Armando Gennaro Department of Chemical Sciences, University of Padova, Padova, Italy

Abhijit Ghosh Advanced Ceramics Section Glass and Advanced Materials Division, Bhabha Atomic Research Centre, Mumbai, India

M. Mar Gil-Diaz IMIDRA, Alcalá de Henares, Madrid, Spain

Luc Girard Institut de Chimie Separative de Marcoule, UMR 5257 – ICSM Site de Marcoule, CEA/CNRS/UM2/ENSCM, Bagnols sur Ceze, France

Jean Gobet Adamant-Technologies, La Chaux-de-Fonds, Switzerland

Luis Godinez Centro de Investigación y Desarrollo Tecnológico en Electroquímica S.C., Querétaro, Mexico

Alan Le Goff Department of Molecular Chemistry, CNRS UMR 5250, CNRS-University of Grenoble, Grenoble, France

Muriel Golzio CNRS; IPBS (Institut de Pharmacologie et de Biologie Structurale), Toulouse, France

Ignacio Gonzalez Department of Chemistry, Universidad Autónoma Metropolitana-Iztapalapa, México, Mexico

Heiner Jakob Gores Institute of Physical Chemistry, Münster Electrochemical Energy Technology (MEET), Westfälische Wilhelms-Universität Münster (WWU), Münster, Germany

Pau Gorostiza Institució Catalana de Recerca i Estudis Avançats (ICREA), Barcelona, Spain

Lars Gundlach Department of Chemistry and Biochemistry and Department of Physics and Astronomy, University of Delaware, Newark, DE, USA

Ulrich Guth Kurt-Schwabe-Institut für Mess- und Sensortechnik e.V. Meinsberg, Waldheim, Germany

FB Chemie und Lebensmittelchemie, Technische Universität Dresden, Dresden, Germany

Geir Martin Haarberg Department of Materials Science and Engineering, Norwegian University of Science and Technology (NTNU), Trondheim, Norway

Jonathan E. Halls Department of Chemistry, The University of Bath, Bath, UK

Ahmad Hammad Research and Development Center, Saudi Aramco, Dhahran, Saudi Arabia

Achim Hannappel DECHEMA Research Institute of Biochemical Engineering, Frankfurt am Main, Germany

Falk Harnisch Institute of Environmental and Sustainable Chemistry, Technical University Braunschweig, Braunschweig, Germany

Akitoshi Hayashi Department of Applied Chemistry, Osaka Prefecture University, Sakai, Osaka, Japan

Christoph Held Department of Biochemical and Chemical Engineering, Technische Universität Dortmund, Dortmund, Germany

Wesley A. Henderson Department of Chemical and Biomolecular Engineering, North Carolina State University, Raleigh, NC, USA

Peter J. Hesketh School of Mechanical Engineering, Georgia Institute of Technology, Atlanta, GA, USA

Michael Heyrovsky J. Heyrovsky Institute of Physical Chemistry of the ASCR, Prague, Czech Republic

Takashi Hibino Graduate School of Environmental Studies, Nagoya University, Nagoya, Japan

Yoshio Hisaeda Department of Chemistry and Biochemistry, Kyushu University, Graduate School of Engineering, Fukuoka, Japan

Tuan Hoang University of Southern California, Los Angeles, CA, USA

Dirk Holtmann DECHEMA Research Institute of Biochemical Engineering, Frankfurt am Main, Germany

Rudolf Holze AG Elektrochemie, Institut für Chemie, Technische Universität Chemnitz, Chemnitz, Germany

Michael Holzinger Department of Molecular Chemistry, CNRS UMR 5250, CNRS-University of Grenoble, Grenoble, France

Dominik Horinek Institute of Physical and Theoretical Chemistry, University of Regensburg, Regensburg, Germany

Teruhisa Horita Fuel Cell Materials Group, Energy Technology Research Institute, National Institute of Advanced Industrial Science and Technology (AIST), Tsukuba, Ibaraki, Japan

Barbara Hribar-Lee Faculty of Chemistry and Chemical Technology, University of Ljubljana, Ljubljana, Slovenia

Chang-Jung Hsueh Electronics Design Center, and Chemical Engineering Department, Case Western Reserve University, Cleveland, OH, USA

Chi-Chang Hu Chemical Engineering Department, National Tsing Hua University, Hsinchu, Taiwan

Gary W. Hunter NASA Glenn Research Center, Cleveland, OH, USA

Jorge G. Ibanez Department of Chemical Engineering and Sciences, Universidad Iberoamericana, México, Mexico

Munehisa Ikoma Panasonic, Moriguchi, Japan

Nobuhito Imanaka Department of Applied Chemistry, Faculty of Engineering, Osaka University, Osaka, Japan

Shinsuke Inagi Tokyo Institute of Technology, Midori-ku, Yokohama, Japan

Hiroshi Inoue Osaka Prefecture University, Sakai, Osaka, Japan

György Inzelt Department of Physical Chemistry, Eötvös Loránd University, Budapest, Hungary

Tsutomu Ioroi AIST, Ikeda, Japan

Hiroshi Irie Yamanashi University, Yamanashi Prefecture, Japan

John Thomas Sirr Irvine School of Chemistry, University of St Andrews, St Andrews, UK

Manabu Ishifune Kinki University, Higashi-Osaka, Osaka, Japan

Akimitsu Ishihara Yokohama National University, Hodogaya-ku, Yokohama, Japan

Tatsumi Ishihara Department of Applied Chemistry, Faculty of Engineering, International Institute for Carbon Neutral Energy Research (WPI-I2CNER), Kyushu University, Nishi ku, Fukuoka, Japan

Masashi Ishikawa Kansai University, Suita, Osaka, Japan

Adriana Ispas FG Elektrochemie und Galvanotechnik, Institut für Werkstofftechnik, Technische Universitüt Ilmenau, Ilmenau, Germany

Gaurav Jain MECC, Medtronic Inc., Minneapolis, MN, USA

Metini Janyasupab Electronics Design Center, and Chemical Engineering Department, Case Western Reserve University, Cleveland, OH, USA

Fengjing Jiang Institute of Fuel Cells, School of Mechanical Engineering, Shanghai Jiao Tong University, Shanghai, People's Republic of China

Maria Jitaru Research Institute for Organic Auxiliary Products (ICPAO), Medias, Romania

Jakob Jörissen Chair of Technical Chemistry, Technical University of Dortmund, Germany

Pavel Jungwirth Institute of Organic Chemistry and Biochemistry, Academy of Sciences of the Czech Republic, Prague, Czech Republic

Yoshifumi Kado Asahi Kasei Chemicals Corporation, Tokyo, Japan

Heike Kahlert Institut für Biochemie, Universität Greifswald, Greifswald, Germany

Yijin Kang University of Pennsylvania, Philadelphia, PA, USA

Agnieszka Kapałka Institute of Chemical Sciences and Engineering, Ecole Polytechnique Fédérale de Lausanne (EPFL), Lausanne, Switzerland

Shigenori Kashimura Kinki University, Higashi-Osaka, Japan

Alexandros Katsaounis Department of Chemical Engineering, University of Patras, Patras, Greece

Jurg Keller Advanced Water Management Centre (AWMC), The University of Queensland, Brisbane, QLD, Australia

Geoffrey H. Kelsall Department of Chemical Engineering, Imperial College London, London, UK

Sangtae Kim Department of Chemical Engineering and Materials Science, University of California, Davis, CA, USA

Woong-Ki Kim Faculty of Electrical Engineering and Computer Science, Ingolstadt University of Applied Sciences, Ingolstadt, Germany

Axel Kirste BASF SE, Ludwigshafen, Germany

Naoki Kise Department of Chemistry and Biotechnology, Graduate School of Engineering, Tottori University, Tottori, Japan

Norihisa Kobayashi Chiba University, Chiba, Japan

Svenja Kochius DECHEMA Research Institute, Frankfurt am Main, Germany

Paul A. Kohl Georgia Institute of Technology, School of Chemical and Biomolecular Engineering, Atlanta, GA, USA

Ulrike I. Kramm Technical University Cottbus, Cottbus, Germany

Mario Krička AG Elektrochemie, Institut für Chemie, Technische Universität Chemnitz, Chemnitz, Germany

Nedeljko Krstajic Faculty of Technology and Metallurgy, University of Belgrade, Belgrade, Serbia

Akihiko Kudo Tokyo University of Science, Tokyo, Japan

Andrzej Kuklinski Fakultät Chemie, Biofilm Centre/Aquatische Biotechnologie, Universität Duisburg-Essen, Essen, Germany

Juozas Kulys Department of Chemistry and Bioengineering, Vilnius Gediminas Technical University, Vilnius, Lithuania

Werner Kunz Institut für Biophysik, Fachbereich Physik, Johann Wolfgang Goethe-Universität Frankfurt am Main, Frankfurt am Main, Germany

Manabu Kuroboshi Okayama University, Okayama, Japan

Jan Labuda Institute of Analytical Chemistry, Faculty of Chemical and Food Technology, Slovak University of Technology, Bratislava, Slovakia

Claude Lamy Institut Européen des Membranes, Université Montpellier 2, UMR CNRS n° 5635, Montpellier, France

Ying-Hui Lee Chemical Engineering Department, National Tsing Hua University, Hsinchu, Taiwan

Carlos A. Ponce de Leon Electrochemical Engineering Laboratory, University of Southampton, Faculty of Engineering and the Environment, Southampton, Hampshire, UK

Jean Lessard Universite de Sherbrooke, Quebec, Canada

Hans J. Lewerenz Joint Center for Artificial Photosynthesis, California Institute of Technology, Pasadena, CA, USA

Claudia Ley DECHEMA Research Institute, Frankfurt am Main, Germany

Meng Li Chemistry Department, Brookhaven National Laboratory, Upton, NY, USA

R. Daniel Little University of California, Santa Barbara, CA, USA

Chen-Wei Liu Institute for Material Science and Engineering, National Central University, Jhongli City, Taoyuan County, Taiwan

Chung-Chiun Liu Electronics Design Center, and Chemical Engineering Department, Case Western Reserve University, Cleveland, OH, USA

Ping Liu Brookhaven National Laboratory, Upton, NY, USA

Yoav D. Livney Faculty of Biotechnology and Food Engineering, The Technion, Israel Institute of Technology, Haifa, Israel

Leonardo Lizarraga Université Lyon 1, CNRS, UMR 5256, IRCELYON, Institut de recherches sur la catalyse et l'environnement de Lyon, Villeurbanne, France

M. Carmen Lobo IMIDRA, Alcalá de Henares, Madrid, Spain

Svenja Lohner Stanford University, Stanford, CA, USA

Manuel Lohrengel University of Düsseldorf, Düsseldorf, Germany

Reiner Lomoth Department of Chemistry - Ångström Laboratory, Uppsala University, Uppsala, Sweden

Daniel Lowy FlexEl, LLC, College Park, MD, USA

Roland Ludwig Department of Food Science and Technology, Vienna Institute of Biotechnology BOKU-University of Natural Resources and Life Sciences, Vienna, Austria

Dirk Lützenkirchen-Hecht Fachbereich C- Abteilung Physik, Wuppertal, Germany

Vadim F. Lvovich NASA Glenn Research Center, Electrochemistry Branch, Power and In-Space Propulsion Division, Cleveland, OH, USA

Johannes Lyklema Department of Physical Chemistry and Colloid Science, Wageningen University, Wageningen, The Netherlands

Hirofumi Maekawa Department of Materials Science and Technology, Nagaoka University of Technology, Nagaoka, Japan

Anders O. Magnusson DECHEMA Research Institute, Frankfurt am Main, Germany

J. Maier Max Planck Institute for Solid State Research, Stuttgart, Germany

Frédéric Maillard Laboratoire d'Electrochimie et de Physico-chimie des Matériaux et des Interfaces, Saint Martin d'Héres, France

Daniel Mandler Institute of Chemistry, The Hebrew University, Jerusalem, Israel

Klaus-Michael Mangold DECHEMA-Forschungsinstitut, Frankfurt am Main, Germany

Yizhak Marcus Department of Inorganic and Analytical Chemistry, The Hebrew University, Jerusalem, Israel

Nebojsa Marinkovic Synchrotron Catalysis Consortium, University of Delaware, Newark, DE, USA

Frank Marken University of Bath, Bath, UK

István Markó Universite Catholique de Louvain, Louvain-la-Neuve, Belgium

Marko S. Markov Research International, Williamsville, NY, USA

Jack Marple Research and Development, Energizer Battery Manufacturing Inc, Westlake, OH, USA

Virginie Marry Laboratoire Physicochimie des Electrolytes, Colloïdes et Sciences Analytiques, CNRS, ESPCI, Université Pierre et Marie Curie, Paris, France

Guillermo Marshall Laboratorio de Sistemas Complejos, Departamento de Ciencias de la Computación, Facultad de Ciencias Exactas y Naturales, Universidad de Buenos Aires, Buenos Aires, Argentina

Carlos Alberto Martinez-Huitle Institute of Chemistry, Federal University of Rio Grande do Norte, Lagoa Nova, Natal, RN - CEP, Brazil

Marco Mascini Dipartimento di Chimica "Ugo Schiff", Università degli Studi di Firenze, Sesto Fiorentino, Firenze, Italy

Rudy Matousek Severn Trent Services, Sugar Land, TX, USA

Hiroshige Matsumoto Kyushu University, Fukuoka, Japan

Kouichi Matsumoto Faculty of Science and Engineering, Kinki University, Higashi-osaka, Japan

Werner Mäntele Johann Wolfgang Goethe-Universität Frankfurt am Main, Institut für Biophysik, Fachbereich Physik, Frankfurt am Main, Germany

Steven McIntosh Department of Chemical Engineering, Lehigh University, Bethlehem, PA, USA

Jennifer R. McKenzie Department of Chemistry, Vanderbilt University, Nashville, TN, USA

Ellis Meng University of Southern California, Los Angeles, CA, USA

Pierre-Alain Michaud Institute of Chemical Sciences and Engineering, Ecole Polytechnique Fédérale de Lausanne (EPFL), Lausanne, Switzerland

Richard L. Middaugh Independent Consultant, formerly with Energizer Battery Co, Rocky River, OH, USA

Alessandro Minguzzi Dipartimento di Chimica, Università degli Studi di Milano, Milan, Italy

Shigenori Mitsushima Yokohama National University, Yokohama, Kanagawa Prefecture, Japan

Tsutomu Miyasaka Graduate School of Engineering, Toin University of Yokohama, Yokohama, Kanagawa, Japan

Kenji Miyatake University of Yamanashi, Kofu, Yamanashi, Japan

Junichiro Mizusaki Tohoku University, Funabashi, Chiba, Japan

Mogens Bjerg Mogensen Department of Energy Conversion and Storage, Technical University of Denmark, Roskilde, Denmark

Charles W. Monroe Department of Chemical Engineering, University of Michigan, Ann Arbor, MI, USA

Vicente Montiel Instituto Universitario de Electroquímica, University of Alicante, Alicante, Spain

Somayeh Moradi AG Elektrochemie, Institut für Chemie, Technische Universität Chemnitz, Chemnitz, Germany

J. Thomas Mortimer Case Western Reserve University, Cleveland, OH, USA

Hubert Motschmann Institute of Physical and Theoretical Chemistry, University of Regensburg, Regensburg, Germany

Christopher B. Murray University of Pennsylvania, Philadelphia, PA, USA

Katsuhiko Naoi Institute of Symbiotic Science and Technology, Tokyo University of Agriculture and Technology, Koganei, Tokyo, Japan

Hiroki Nara Faculty of Science and Engineering, Waseda University, Okubo, Shinjuku-ku, Tokyo, Japan

George Neophytides Electrochemistry Laboratory, Paul Scherrer Institut, Villigen, Switzerland

Roland Neueder Institute of Physical and Theoretical Chemistry, University of Regensburg, Regensburg, Germany

Jinren Ni Peking University, Beijing, China

Ernst Niebur Mind/Brain Institute, Johns Hopkins University, Baltimore, MD, USA

Branislav Ž. Nikolić Department of Physical Chemistry and Electrochemistry, University of Belgrade, Faculty of Technology and Metallurgy, Belgrade, Serbia

Yoshinori Nishiki Development Department, Permelec Electrode Ltd, Kanagawa, Japan

Shigeru Nishiyama Department of Chemistry, Keio University, Hiyoshi, Yokohama, Japan

Toshiyuki Nohira Graduate School of Energy Science, Kyoto University, Kyoto, Japan

Atusko Nosaka Department Materials Science and Technology, Nagaoka University of Technology, Nagaoka, Niigata, Japan

Naoyoshi Nunotani Department of Applied Chemistry, Faculty of Engineering, Osaka University, Osaka, Japan

Tsuyoshi Ochiai Kanagawa Academy of Science and Technology, Takatsu–ku, Kawasaki, Kanagawa, Japan

Photocatalysis International Research Center, Tokyo University of Science, Noda, Chiba, Japan

Wolfram Oelßner Kurt-Schwabe-Institut für Mess- und Sensortechnik e.V. Meinsberg, Kurt-Schwabe-Straße, Waldheim, Germany

Andreas Offenhäusser Institute of Complex Systems, Peter Grünberg Institute: Bioelectronics, Jülich, Germany

Ulker Bakir Ogutveren Anadolu University, Eskişehir, Turkey

Bunsho Ohtani Catalysis Research Center, Hokkaido University, Sapporo, Hokkaido, Japan

Yohei Okada Tokyo University of Agriculture and Technology, Fuchu, Tokyo, Japan

Osamu Onomura Nagasaki University, Nagasaki, Japan

Immaculada Ortiz Department of Chemical Engineering and Inorganic Chemistry, University of Cantabria, Santander, Cantabria, Spain

Juan Manuel Ortiz Instituto Universitario de Electroquímica, University of Alicante, Alicante, Spain

Tetsuya Osaka Faculty of Science and Engineering, Waseda University, Okubo, Shinjuku-ku, Tokyo, Japan

Ken-ichiro Ota Yokohama National University, Fac. Engineering, Yokohama, Japan

Lisbeth M. Ottosen Department of Civil Engineering, Technical University of Denmark, Lyngby, Denmark

Ilaria Palchetti Dipartimento di Chimica "Ugo Schiff", Università degli Studi di Firenze, Sesto Fiorentino, Firenze, Italy

Vladimir Panić University of Belgrade, Institute of Chemistry, Technology and Metallurgy, Belgrade, Serbia

Marco Panizza University of Genoa, Genoa, Italy

Juan Manuel Peralta-Hernández Centro de Innovación Aplicada en Tecnologías Competitivas, Guanajuato, Mexico

Cristina Pereira Instituto de Tecnologia Química e Biológica, Universidade Nova de Lisboa, Oeiras, Portugal

Araceli Pérez-Sanz IMIDRA, Alcalá de Henares, Madrid, Spain

Laurence (Laurie) Peter Department of Chemistry, University of Bath, Bath, UK

Marija Petkovic Instituto de Tecnologia Química e Biológica, Universidade Nova de Lisboa, Oeiras, Portugal

Ilje Pikaar Advanced Water Management Centre (AWMC), The University of Queensland, Brisbane, QLD, Australia

Antonio Plaza IMIDRA, Alcalá de Henares, Madrid, Spain

Dmitry E. Polyansky Chemistry Department, Brookhaven National Laboratory, Upton, NY, USA

Jinyi Qin Department of Environmental Microbiology, Helmholtz Centre for Environmental Research - UFZ, Leipzig, Germany

Jelena Radjenovic Advanced Water Management Centre (AWMC), The University of Queensland, Brisbane, QLD, Australia

Krishnan Rajeshwar The University of Texas at Arlington, Arlington, TX, USA

Nayif A. Rasheedi Research and Development Center, Saudi Aramco, Dhahran, Saudi Arabia

David Rauh EIC Laboratories, Inc, Norwood, MA, USA

Thomas B. Reddy Department of Materials Science and Engineering, Rutgers, The State University of New Jersey, Piscataway, NJ, USA

David Reyter INRS Energie, Matériaux et Télécommunications, Varennes, Quebec, Canada

Marcel Risch MIT, Cambridge, MA, USA

Vivian Robinson ETP Semra Pty Ltd, Canterbury, NSW, Australia

Romeu C. Rocha-Filho Universidade Federal de Goiás, Catalão, Brazil

Manuel A. Rodrigo Department of Chemical Engineering, Faculty of Chemical Sciences and Technology, Universided de Castille la Mandne, Ciudad Real, Spain

Paramaconi Rodriguez School of Chemistry, The University of Birmingham, Birmingham, UK

Alberto Rojas-Hernández Department of Chemistry, Universidad Autónoma Metropolitana-Iztapalapa, México, Mexico

Sandra Rondinini Dipartimento di Chimica, Università degli Studi di Milano, Milan, Italy

Benjamin Rotenberg Laboratoire Physicochimie des Electrolytes, Colloïdes et Sciences Analytiques, CNRS, ESPCI, Université Pierre et Marie Curie, Paris, France

Anna Joëlle Ruff Lehrstuhl für Biotechnologie, RWTH Aachen University, Aachen, Germany

Luís Augusto M. Ruotolo Department of Chemical Engineering, Federal University of São Carlos, São Carlos, SP, Brazil

Jennifer L. M. Rupp Electrochemical Materials, ETH Zurich, Zurich, Switzerland

Yoshihiko Sadaoka Ehime University, Matsuyama, Japan

Gabriele Sadowski Department of Biochemical and Chemical Engineering, Technische Universität Dortmund, Dortmund, Germany

Hikari Sakaebe Research Institute for Ubiquitous Energy Devices, National Institute of Advanced Industrial Science and Technology (AIST), Ikeda, Osaka, Japan

Hikari Sakaebe National Institute of Advanced Industrial Science and Technology (AIST), Ikeda, Osaka, Japan

Mathieu Salanne Laboratoire PECSA, UMR 7195, Université Pierre et Marie Curie, Paris, France

Wolfgang Sand Fakultät Chemie, Biofilm Centre/Aquatische Biotechnologie, Universität Duisburg-Essen, Essen, Germany

Shriram Santhanagopalan National Renewable Energy Laboratory, Golden, CO, USA

Hamidreza Sardary AG Elektrochemie, Institut für Chemie, Technische Universität Chemnitz, Chemnitz, Germany

Kotaro Sasaki Chemistry Department, Brookhaven National Laboratory, Upton, NY, USA

Richard Sass DECHEMA e.V., Informationssysteme und Datenbanken, Frankfurt, Germany

Shunsuke Sato Toyota Central Research and Development Laboratories, Inc., Nagakute, Aichi, Japan

André Savall Laboratoire de Génie Chimique, CNRS, Université Paul Sabatier, Toulouse, France

Elena Savinova Institut de Chimie et Procédés pour l'Energie, l'Environnement et la Santé, UMR 7515 CNRS, Université de Strasbourg-ECPM, Strasbourg, France

Natascha Schelero Stranski-Laboratorium, Institut für Chemie, Fakultät II, Technical University Berlin, Berlin, Germany

Günther G. Scherer Electrochemistry Laboratory, Paul Scherrer Institute, Villigen, Switzerland

Thomas J. Schmidt Electrochemistry Laboratory, Paul Scherrer Institut, Villigen, Switzerland

Jens Schrader DECHEMA Research Institute of Biochemical Engineering, Frankfurt am Main, Germany

Uwe Schroeder Institute of Environmental and Sustainable Chemistry, Technical University Braunschweig, Braunschweig, Germany

Brooke Schumm Eagle Cliffs, Inc., Bay Village, OH, USA

Ulrich Schwaneberg Lehrstuhl für Biotechnologie, RWTH Aachen University, Aachen, Germany

Hans-Georg Schweiger Faculty of Electrical Engineering and Computer Science, Ingolstadt University of Applied Sciences, Ingolstadt, Germany

Hisanori Senboku Hokkaido University, Sapporo, Hokkaido, Japan

Daniel Seo Department of Chemical and Biomolecular Engineering, North Carolina State University, Raleigh, NC, USA

Karine Groenen Serrano Laboratory of Chemical Engineering, University of Paul Sabatier, Toulouse, France

Anwar-ul-Haq Ali Shah Institute of Chemical Sciences, University of Peshawar, Peshawar, Pakistan

Yang Shao-Horn MIT, Cambridge, MA, USA

Hisashi Shimakoshi Department of Chemistry and Biochemistry, Kyushu University, Graduate School of Engineering, Fukuoka, Japan

Yasuhiro Shimizu Graduate School of Engineering, Nagasaki University, Nagasaki, Japan

Youichi Shimizu Department of Applied Chemistry, Graduate School of Engineering, Kyushu Institute of Technology, Kitakyushu, Fukuoka, Japan

Komaba Shinichi Department of Applied Chemistry, Tokyo University of Science, Shinjuku, Tokyo, Japan

Soshi Shiraishi Division of Molecular Science, Faculty of Science and Technology, Gunma University, Kiryu, Gunma, Japan

Pavel Shuk Rosemount Analytical Inc. Emerson Process Management, Solon, OH, USA

Jean-Pierre Simonin Laboratoire PECSA, UMR CNRS 7195, Université Paris, Paris, France

Subhash C. Singhal Pacific Northwest National Laboratory, Richland, WA, USA

Ignasi Sirés Laboratory of Electrochemistry of Materials and Environment, Department of Physical Chemistry, Faculty of Chemistry, University of Barcelona, Barcelona, Spain

Stephen Skinner Department of Materials, Imperial College London, London, UK

Marshall C. Smart Jet Propulsion Laboratory, Pasdena, CA, USA

Henry Snaith Clarendon Laboratory, Department of Physics, Oxford University, Oxford, UK

Stamatios Souentie Department of Chemical Engineering, University of Patras, Patras, Greece

Bernd Speiser Institut für Organische Chemie, Universität Tübingen, Tübingen, Germany

Jacob Spendelow Los Alamos National Laboratory, Los Alamos, NM, USA

Daniel Steingart Department of Mechanical and Aerospace Engineering, Andlinger Center for Energy, the Environment Princeton University, Princeton, NJ, USA

John Stickney University of Georgia, Athens, GA, USA

Margarita Stoytcheva Instituto de Ingenieria, Universidad Autonoma de Baja California, Mexicali, Baja California, Mexico

Svetlana B. Strbac ICTM-Institute of Electrochemistry, University of Belgrade, Belgrade, Serbia

Eric M. Stuve Department of Chemical Engineering, University of Washington, Seattle, WA, USA

Stenbjörn Styring Department of Chemistry - Ångström Laboratory, Uppsala University, Uppsala, Sweden

Seiji Suga Okayama University, Okayama, Japan

Wataru Sugimoto Faculty of Textile Science and Technology, Shinshu University, Nagano, Japan

I-Wen Sun Department of Chemistry, National Cheng kung University, Tainan, Taiwan

Jin Suntivich Cornell University, Ithaca, NY, USA

Hitoshi Takamura Department of Materials Science, Graduate School of Engineering, Tohoku University, Sendai, Japan

Prabhakar A. Tamirisa MECC, Medtronic Inc., Minneapolis, MN, USA

Shinji Tamura Department of Applied Chemistry, Faculty of Engineering, Osaka University, Osaka, Japan

Hideo Tanaka Okayama University, Okayama, Japan

Tadaaki Tani Society of Photography and Imaging of Japan, Tokyo, Japan

Akimasa Tasaka Doshisha University, Kyotanabe, Kyoto, Japan

Tetsu Tatsuma Institute of Industrial Science, University of Tokyo, Tokyo, Japan

Masahiro Tatsumisago Department of Applied Chemistry, Osaka Prefecture University, Sakai, Osaka, Japan

Pierre Taxil Laboratoire de Génie Chimique, Université de Toulouse, Toulouse, France

Justin Teissie CNRS; IPBS (Institut de Pharmacologie et de Biologie Structurale), Toulouse, France

Ingrid Tessmer Rudolf-Virchow-Zentrum, Experimentelle Biomedizin, University of Würzburg, Würzburg, Germany

Anders Thapper Department of Chemistry - Ångström Laboratory, Uppsala University, Uppsala, Sweden

Masataka Tomita Department of Applied Chemistry, Tokyo University of Science, Shinjuku, Tokyo, Japan

Marc Tornow Institut für Halbleitertechnik, Technische Universität München, München, Germany

Taro Toyoda Department of Engineering Science, The University of Electro-Communications, Chofu, Tokyo, Japan

Dimitrios Tsiplakides Department of Chemistry, Aristotle University of Thessaloniki, Thessaloniki, Greece

Pierre Turq Laboratoire Physicochimie des Electrolytes, Colloïdes et Sciences Analytiques, CNRS, ESPCI, Université Pierre et Marie Curie, Paris, France

Makoto Uchida Fuel Cell Nanomaterials Center, Yamanashi University, 6-43 Miyamaecho, Kofu, Yamanashi, Japan

Kai M. Udert Eawag, Dübendorf, Switzerland

Helmut Ullmann FB Chemie und Lebensmittelchemie, Technische Universität Dresden, Dresden, Germany

Soichiro Uno Nissan Research Center/EV System Laboratory, NISSAN MOTOR CO., LTD., Yokosuka-shi, Kanagawa-ken, Japan

Kohei Uosaki International Center for Materials Nanoarchitectonics (WPI-MANA), National Institute for Materials Science (NIMS), Tsukuba, Japan

Ane Urtiaga Department of Chemical Engineering and Inorganic Chemistry, University of Cantabria, Santander, Cantabria, Spain

Francisco J. Rodriguez Valadez Centro de Investigación y Desarrollo Tecnológico en Electroquímica S.C., Querétaro, Mexico

S. C. Sanfandila, Research Branch, Center for Research and Technological Development in Electrochemistry, Querétaro, Mexico

Vladimir Vashook Kurt-Schwabe-Institut für Mess- und Sensortechnik e.V. Meinsberg, Waldheim, Germany

FB Chemie und Lebensmittelchemie, Technische Universität Dresden, Dresden, Germany

Ruben Vasquez-Medrano Department of Chemical Engineering and Sciences, Universidad Iberoamericana, México, Mexico

Nicolaos Vatistas DICI, Università di Pisa, Pisa, Italy

Constantinos G. Vayenas Department of Chemical Engineering, University of Patras, Patras, Achaia, Greece

Danae Venieri Department of Environmental Engineering, Technical University of Crete, Chania, Greece

Philippe Vernoux Université Lyon 1, CNRS, UMR 5256, IRCELYON, Institut de recherches sur la catalyse et l'environnement de Lyon, Villeurbanne, France

Alberto Vertova Dipartimento di Chimica, Università degli Studi di Milano, Milan, Italy

Bernardino Virdis Advanced Water Management Centre (AWMC), The University of Queensland, Brisbane, QLD, Australia

Centre for Microbial Electrosynthesis (CEMES), The University of Queensland, Brisbane, QLD, Australia

Vojko Vlachy Faculty of Chemistry and Chemical Technology, University of Ljubljana, Ljubljana, Slovenia

Rittmar von Helmolt Government Programs and Research Strategy, GM Alternative Propulsion Center, Adam Opel AG, Rüsselsheim, Germany

Regine von Klitzing Stranski-Laboratorium, Institut für Chemie, Fakultät II, Technical University Berlin, Berlin, Germany

Winfried Vonau Kurt-Schwabe-Institut fuer Mess- und Sensortechnik e.V. Meinsberg, Waldheim, Germany

Yunny Meas Vong CIDETEQ, Parque Tecnológico Querétaro, México, Estado de Querétaro, México

Lj Vracar Faculty of Technology and Metallurgy University of Belgrade, Belgrade, Serbia

Miomir B. Vukmirovic Chemistry Department, Brookhaven National Laboratory, Upton, NY, USA

Vlastimil Vyskocil UNESCO Laboratory of Environmental Electrochemistry, Faculty of Science, Department of Analytical Chemistry, Charles University in Prague, Prague, Czech Republic

Jay D. Wadhawan Department of Chemistry, The University of Hull, Hull, UK

Siegfried R. Waldvogel Johannes Gutenberg-University Mainz, Mainz, Germany

Frank C. Walsh Electrochemical Engineering Laboratory, University of Southampton, Faculty of Engineering and the Environment, Southampton, Hampshire, UK

Jia X. Wang Brookhaven National Laboratory, Upton, NY, USA

Masahiro Watanabe University of Yamanashi, Kofu, Yamanashi, Japan

Takao Watanabe Central Research Institute of Electric Power Industry, Yokosuka, Kanagawa, Japan

Andrew Webber Technology, Energizer Battery Manufacturing Inc., Westlake, OH, USA

Adam Z. Weber Lawrence Berkeley National Laboratory, Berkeley, CA, USA

Hermann Weingärtner Lehrstuhl für Physikalische Chemie II, Ruhr-Universität Bochum, Bochum, Germany

Nina Welschoff Johannes Gutenberg-University Mainz, Mainz, Germany

Alan West Department of Chemical Engineering Columbia University, New York, NY, USA

Reiner Westermeier SERVA Electrophoresis GmbH, Heidelberg, Germany

Lukas Y. Wick Department of Environmental Microbiology, Helmholtz Centre for Environmental Research - UFZ, Leipzig, Germany

Andrzej Wieckowski University of Illinois, Urbana, IL, USA

Alexander Wiek AG Elektrochemie, Institut für Chemie, Technische Universität Chemnitz, Chemnitz, Germany

John P. Wikswo Department of Physics and Astronomy, Vanderbilt University, Nashville, TN, USA

Frank Willig Fritz-Haber-Institut der Max-Planck-Gesellschaft, Berlin, Germany

Yuping Wu Department of Chemistry, Fudan University, Shanghai, China

Naoaki Yabuuchi Department of Applied Chemistry, Tokyo University of Science, Shinjuku, Tokyo, Japan

Kohta Yamada Asahi Glass Co. Ltd., Research Center, Kanagawa-ku, Yokohama, Japan

Yoshiaki Yamaguchi Technical Development Division, Global Technical Headquarters, GS Yuasa International Ltd., Kyoto, Japan

Ichiro Yamanaka Department of Applied Chemistry, Tokyo Institute of Technology, Tokyo, Japan

Shigeaki Yamazaki Kansai University, Suita, Osaka, Japan

Harumi Yokokawa The University of Tokyo, Institute of Industrial Science, Tokyo, Japan

H.-I. Yoo Department of Materials Science and Engineering, Seoul National University, Seoul, Korea

Nobuko Yoshimoto Graduate School of Science and Engineering, Yamaguchi University, Yamaguchi, Japan

Akira Yoshino Yoshino Laboratory, Asahi Kasei Corporation, Fuji, Shizuoka, Japan

Jun-ichi Yosida Kyoto University, Kyoto, Japan

Zaki Yusuf Research and Development Center, Saudi Aramco, Dhahran, Saudi Arabia

Junliang Zhang Institute of Fuel Cells, School of Mechanical Engineering, Shanghai Jiao Tong University, Shanghai, People's Republic of China

Fengjuan Zhu Institute of Fuel Cells, School of Mechanical Engineering, Shanghai Jiao Tong University, Shanghai, People's Republic of China

Hongmin Zhu School of Metallurgical and Ecological Engineering, University of Science and Technology Beijing, Beijing, China

Roumen Zlatev Instituto de Ingenieria, Universidad Autonoma de Baja California, Mexicali, Baja California, Mexico

Jens Zosel Kurt-Schwabe-Institut für Mess- und Sensortechnik e.V. Meinsberg, Waldheim, Germany

Sandra Zugmann EVA Fahrzeugtechnik, Munich, Germany

Institució Catalana de Recerca i Estudis Avançats (ICREA), Institute for Bioengineering of Catalonia (IBEC), Barcelona, Spain

Electrical and Computer Engineering, University of California Davis, Davis, CA, USA

Andreas Züttel Empa Materials Science and Technology, Hydrogen and Energy, Dübendorf, Switzerland

Faculty of Applied Science, DelftChemTech, Delft, The Netherlands

F

Formic Acid Oxidation

Yijin Kang and Christopher B. Murray
University of Pennsylvania, Philadelphia, PA, USA

Introduction

Electrocatalysis of formic acid (FA) oxidation reactions has been intensively studied for two main reasons: (1) FA is an attractive chemical fuel for fuel cell applications due to its high energy density (1,740 Wh/kg, 2,086 Wh/L) and easy storage [1], and (2) FA is the smallest molecule that has four most common chemical bonds in organic compounds (C–H, C=O, C–O, O–H), making FA an ideal model molecule for studying electrooxidation reactions.

Three possible reaction pathways of FA oxidation have been proposed [2–4]:

(i) $HCOOH^* \rightarrow COOH^* + H^+ + e^- \rightarrow CO_2 + 2H^+ + 2e^-$

(ii) $HCOOH^* \rightarrow HCOO^* + H^+ + e^- \rightarrow CO_2 + 2H^+ + 2e^-$

(iii) $HCOOH^* \rightarrow CO^* + H_2O \rightarrow CO_2 + 2H^+ + 2e^-$

Here, * represents an adsorbed state. (i) is referred to as the direct pathway, where the removal of the first hydrogen atom (dehydrogenation) from the C–H bond occurs to produce a hydroxy carbonyl, and then a second dehydrogenation step occurs at the O–H bond to produce CO_2. (ii) is called the formate pathway, in which the dehydrogenation first occurs at the O–H bond to produce formate, and then at the C–H bond to generate CO_2. (iii) is known as the indirect pathway, in which CO^* is produced by non-Faradaic dehydration of FA, then is further oxidized to CO_2. Both (i) and (ii) are accomplished by dehydrogenation (via different intermediates) to form CO_2 directly, so (i) and (ii) are sometimes jointly referred to as the "direct pathway."

The electrocatalysis of FA oxidation has been studied on various metal surfaces. Among all the pure metals, platinum (Pt) exhibits the highest activity. Figure 1b shows a typical voltammetry curve of FA oxidation on Pt catalyst. The first peak (peak **I**) in the anodic scan appears at $\approx$0.5 V, corresponding to the direct pathway (including the formate pathway here). The current of peak **I** is usually very low, because the inactive "poisoning" intermediates are accumulated long before the potential reaches 0.5 V. Taking the example of reactions at 0 V, as shown in Fig. 1c, the first steps of the formate pathway (ii) and the indirect pathway (iii) are favorable due to their low activation barriers while the generated CO^* species are strongly adsorbed on the Pt surface. In addition, the large activation barriers of the second steps of (ii) and (iii) make the poisoning intermediates particularly stable on the Pt surfaces. The second anodic peak (peak **II**) shows up at $\approx$0.9 V, with higher current than peak **I**, because at high potential (Fig. 1d, e) the reaction barrier of the second

G. Kreysa et al. (eds.), *Encyclopedia of Applied Electrochemistry*, DOI 10.1007/978-1-4419-6996-5,

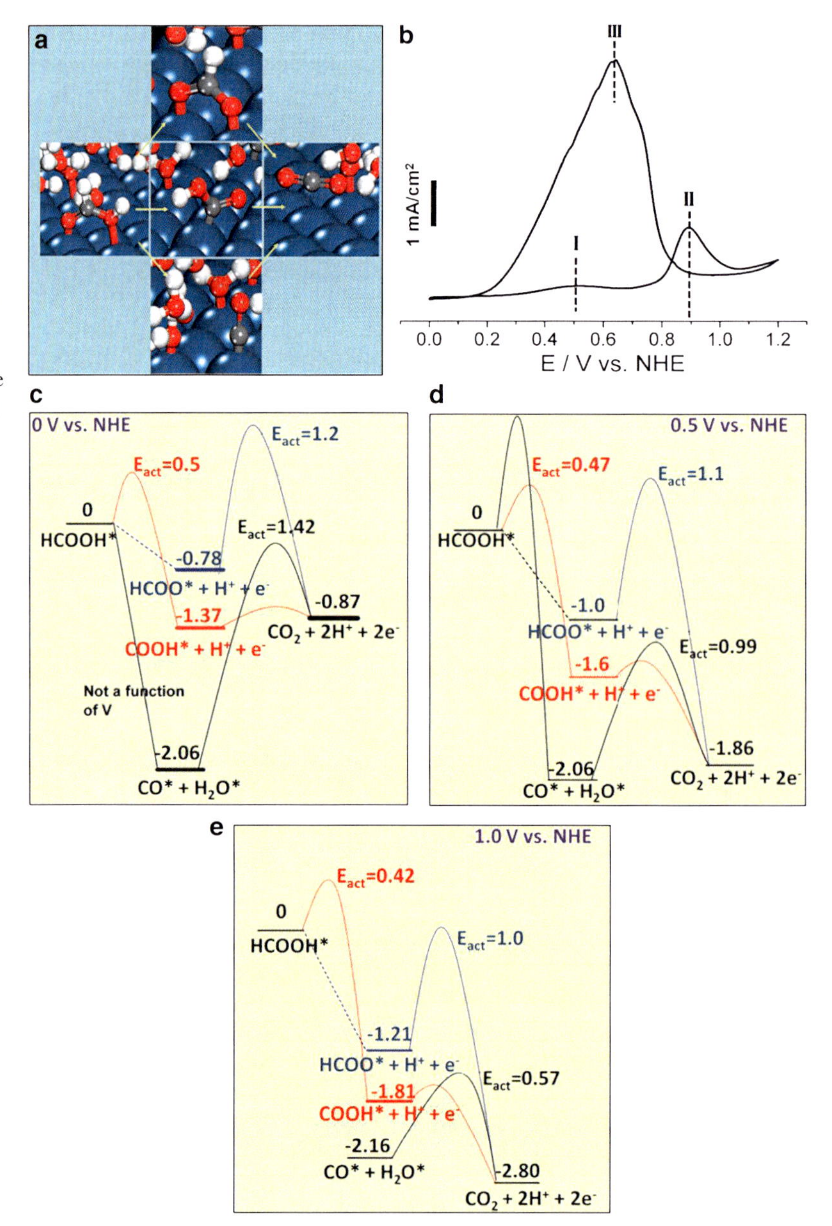

Formic Acid Oxidation, Fig. 1 (**a**) Models representing formate pathway (*top*), direct pathway (*center*), and indirect pathway (*bottom*); (**b**) a typical cyclic voltammetry curve for formic acid electrooxidation on (poly-oriented) Pt catalyst; (**c–e**) DFT-calculated potential energy surface for the direct, indirect, and formate pathways for the oxidation of formic acid over Pt (111), held at constant potentials (Reproduced by permission of The Royal Society of Chemistry [4])

step is significantly reduced. In the cathodic scan, the current (peak **III**) is much higher than that of peak **I** and **II** due to the removal of the poisoning intermediate at high potential.

The FA oxidation is structure sensitive to the surface of Pt. Experiments done on the well-defined single crystals of Pt (Fig. 2) have demonstrated that the poisoning effect is strongest on Pt(100) and weakest on Pt(111), and the total activity follows the trend of Pt(100) > Pt(110) > Pt(111) [5–7]. The studies on preferentially oriented Pt nanoparticles also suggest the same trends as observed on single crystals [8]. In the real catalysts, Pt polycrystals inevitably bring the poisoning problem. For this reason, existing Pt-based electrocatalysts can accomplish FA oxidation only at extremely positive potentials (i.e., at potentials without technological interest). Therefore, the development of direct formic acid fuel cells (DFAFCs) has been impeded by the

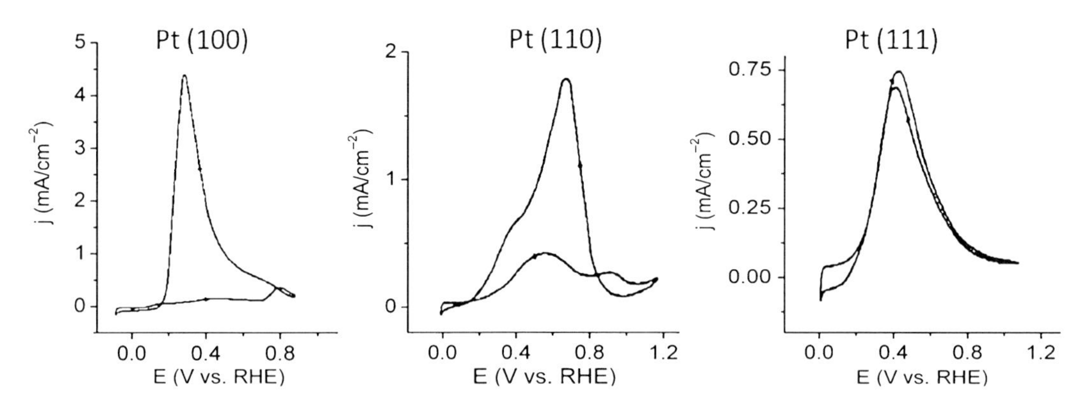

Formic Acid Oxidation, Fig. 2 Electrooxidation of formic acid (0.26 M) on single crystal Pt electrodes with (100), (110), and (111) orientations in 1 M $HClO_4$, sweep rate: 50 mV/s (Reproduced by permission of Nature Publishing Group [6])

low activity of existing electrocatalysts at desirable operating voltage (i.e., at low oxidation potential).

Various efforts have been made to improve the Pt-based electrocatalysts [9–24]. One direction is the use of foreign metal adatoms. As early as in 1975, Adzic et al. demonstrated that monolayers of a foreign metal (i.e., Cd, Tl, Pb, Bi) underpotential deposited on Pt exhibit enhanced activity and pronounced catalytic effects on electrocatalysis of FA oxidation [25]. By recognizing that a foreign metal may promote the performance of Pt toward FA electrooxidation [25–27], scientists have attempted to employ Pt-based alloys for the new generation of electrocatalysts. For instance, cubic Mn-Pt alloy nanocrystals (NCs) [28] and Pt-Ru alloy surface [29] can effectively reduce the poisoning effect on the electrocatalyst. However, the maximum current is still obtained at fairly high oxidation potential. Ideally, an electrocatalyst should have both good poisoning tolerance and high activity at low oxidation potential. This end, Pb has been targeted as promising adatom or alloying element to enhance FA oxidation activity and CO-poisoning tolerance. Adding Pb into Pt electrocatalysts to improve their performance has not only been demonstrated by underpotential deposited Pb, but has also been reinforced by Pt-Pb alloys made under high temperature annealing [10]. Pt-Pb nanoparticle electrocatalysts in various compositions and phases have also been developed subsequently [30–32]. Taking Pt_3Pb NCs synthesized via solution-phase-synthesis as an example, Fig. 3a shows the superior electrochemical performance of Pt_3Pb NCs, compared to pure Pt (Pt black, Aldrich). As described earlier, FA oxidation on electrocatalysts follows both direct (including the formate pathway) and indirect pathways, corresponding to peak **I** and peak **II** in Fig. 3d, respectively. The indirect pathway dominates the reactions on Pt catalysts, making the Pt unsuitable for DFAFC. However, on Pt_3Pb the direct pathway is more favorable, with little reaction occurring via the indirect pathway. At 0.3 V (typical anodic working voltage in DFAFC), the FA oxidation activity on Pt_3Pb is 33 times greater than on Pt. No peak for CO oxidation is observed for both Pt_3Pb and other Pt-Pb nanoparticles, because Pb alloying can relieve the CO-poisoning effect by weakening the adsorption strength of CO, as described in previous DFT calculations [33].

Not only the composition, but also the structure of electrocatalysts is important for a good performance [30, 34, 35]. Core-shell structured Pt-Pb NCs are made by growing Pt shells on the Pt_3Pb NCs. At lower temperature, slower growth allows Pt to grow uniformly onto Pt_3Pb seeds at a low yield, forming epitaxial layers; while high temperature reaction facilitates the Pt precursor decomposition inducing growth on any possible seed sites, usually forming islands as well as layers.

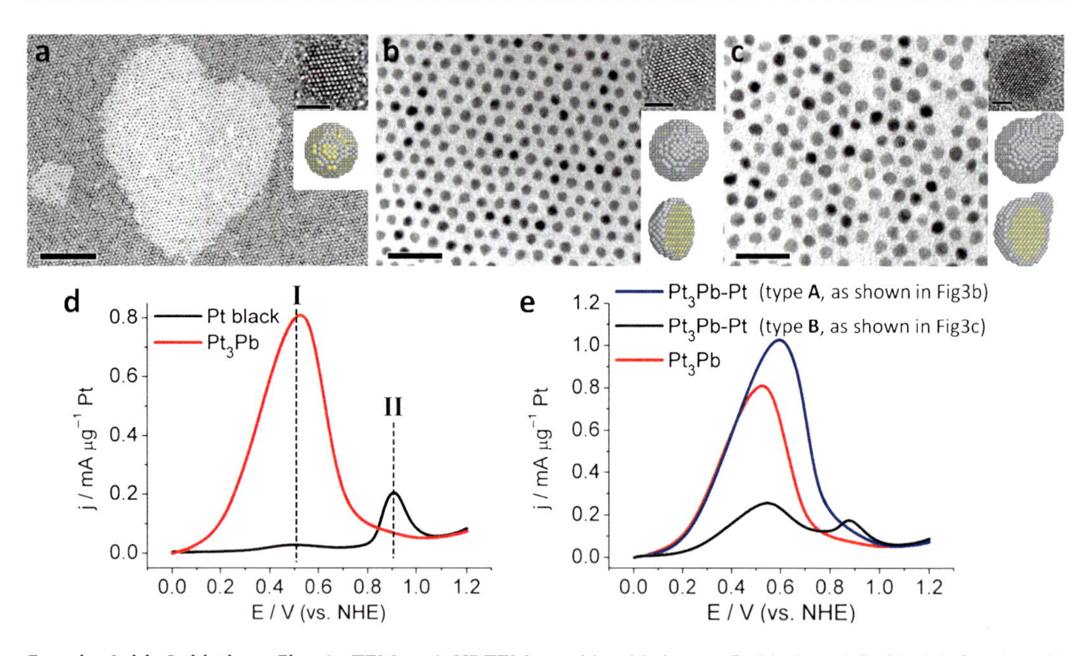

Formic Acid Oxidation, Fig. 3 TEM and HRTEM (insets) images of (**a**) Pt_3Pb NCs, (**b**) type **A**, and (**c**) type **B** Pt-Pb core-shell NCs. Scale bars: (**a**)100 nm, (**b**, **c**) 20 nm, (insets) 2 nm. Polarization curves of (**d**) formic acid oxidation on Pt black and Pt_3Pb, (**e**) formic acid oxidation on Pt_3Pb and Pt-Pb core-shell NCs, in the solution of 0.5 M formic acid and 0.1 M H_2SO_4, sweep rate: 20 mV/s

Figure 3b shows the Pt-Pb NCs with a monolayer of Pt on the Pt_3Pb NC seeds (we call this type **A** core-shell Pt-Pb NCs); Fig. 3c shows the NCs with much thicker Pt shell, as well as some Pt islands (we call this type **B** core-shell Pt-Pb NCs). Interestingly, the type **A** NCs show a higher activity toward FA oxidation than pure Pt_3Pb; the activity on type **B** NCs is much lower than that on the other two Pt-Pb nanostructures, while a peak at higher potential is observed, indicating the FA oxidation on type **B** NCs does occur via the indirect pathway. As described in all Pt-Pb cases, the FA oxidation activity increases at low potential due to the reduced CO poisoning. On the other hand, the low activity of type **B** NCs can be explained by the structural relaxation of the Pt atoms deposited on Pt_3Pb NCs. The type **B** NCs have a lattice constant close to pure Pt, smaller than lattice constants for Pt_3Pb NCs and type **A** NCs. Excess Pt atoms form thick Pt layers and Pt cluster islands on the Pt_3Pb NCs. These Pt thick layers and islands have similar electrocatalytic properties to those of pure Pt, which generates CO* from FA dehydration and decreases the overall activity. However, this mechanism cannot explain the further increase of total activity for type **A** NCs relative to Pt_3Pb NCs, since both have almost no CO-poisoning effect. Thus, there must be other factors contributing to the increase of the activity beside that the Pb alloys tend to avoid CO poisoning. Subsequent DFT studies suggest that the further increase of activity for type **A** NCs relative to Pt_3Pb NCs may result from continuous decreases of dehydrogenation barriers, strong adsorption of reactant molecules and, most importantly, the suppression of the formate path and avoidance of stable formate intermediates. Thus, both the existence and thickness of the Pt shell surrounding Pt_3Pb NCs are critical to achieve a high reaction rate, reinforcing the observation that even materials with very similar chemical compositions can be manipulated by control of nanoscale structure to obtain high catalytic activities.

Metals other than Pt have also been extensively investigated. Nevertheless, noble metals are still the focus. Pd has comparable activity to Pt in nanocrystalline form [36]. Additionally, a Pd thin film deposited on Au shows a high CO-poisoning tolerance and an improved activity in the low oxidation potential region. The FA electrooxidation on Au/Pd is structure sensitive, as Au(100)/Pd exhibits the highest activity among Au/Pd electrocatalysts [37]. Besides the metal (or alloy) electrocatalysts themselves, the supporting materials and their interaction with an electrocatalyst are receiving more and more attention in recent years, especially since materials such as carbon nanotubes and graphene have emerged as potentially useful support materials [12, 38]. Nonetheless, there is still a long way to go before the discovery of either "super cheap" or "super active" electrocatalysts suitable for the large-scale adoption of DFAFC.

Future Directions

Though many improved electrocatalysts have been reported, the development of practical electrocatalysts for FA oxidation remains challenging due to the following issues:

1. Insufficient activities: existing electrocatalysts for FA oxidation are still not active enough for the practical application of DFAFC at large scale.
2. Poor stabilities: Pt-Pb electrocatalysts have already achieved high activities, especially at low oxidation potentials; however, their stabilities are still less than projected. Pt-based alloys have a variety of advantages, but all such alloys face the problem of leaching, especially in acid media. None of the existing electrocatalysts meets the standard of stability for industrial implementation (e.g., 5,000 h lifetime).
3. High cost: considering the cost for large-scale applications (e.g., automobile), Pt-based materials and other noble metals are far too costly for wide adoption.

Therefore, the future research on FA electrooxidation should address the issues stated above. Possible future directions are:

1. Further understanding of the electrocatalytic processes on electrode surfaces: For this purpose, Pt and Pt-based alloys are still appropriate to be the research subject. The knowledge learned from noble metals and their alloys are to be generalized and then applied as guidelines for developing inexpensive electrocatalysts.
2. Improvement of stability: In the discussed Pt-Pb case, Pb leaches out from electrocatalysts in acid electrolyte. It is believed that a complete Pt shell can better protect the Pt_3Pb core. On the other hand, if the Pt shell is too thick, it lowers the overall performance. Such a dilemma exists in many cases of electrocatalyst development. Improving the stability, while maintaining the activity, will be a major challenge in the future research.
3. Development of inexpensive electrocatalysts: For instance, Pb nanoparticles covered with a monolayer of Pt could be an option of an "inexpensive" electrocatalyst, although it is very challenging to make such a structure (e.g., the large lattice mismatch between Pt and Pb prohibits the epitaxial growth of Pt on Pb). Ultimately, non-noble-metal electrocatalyst is expected to be the main research direction for DFAFC applications.

Cross-References

- ► Anodic Reactions in Electrocatalysis - Methanol Oxidation
- ► Anodic Reactions in Electrocatalysis - Oxidation of Carbon Monoxide
- ► Direct Alcohol Fuel Cells (DAFCs)
- ► Polymer Electrolyte Fuel Cells (PEFCs), Introduction

References

1. Rice C et al (2002) Direct formic acid fuel cells. J Power Sources 111:83–89
2. Capon A, Parsons R (1973) Oxidation of formic-acid at noble-metal electrodes part. 3. Intermediates and mechanism on platinum-electrodes. J Electroanal Chem 45:205–231
3. Samjeske G, Miki A, Ye S, Osawa M (2006) Mechanistic study of electrocatalytic oxidation of formic acid at platinum in acidic solution by time-resolved surface-enhanced infrared absorption spectroscopy. J Phys Chem B 110:16559–16566

4. Neurock M, Janik M, Wieckowski A (2008) A first principles comparison of the mechanism and site requirements for the electrocatalytic oxidation of methanol and formic acid over Pt. Faraday Discuss 140:363–378
5. Clavilier J, Parsons R, Durand R, Lamy C, Leger JM (1981) Formic-acid oxidation on single-crystal platinum-electrodes – comparison with polycrystalline platinum. J Electroanal Chem 124:321–326
6. Adzic RR, Tripkovic AV, Ogrady WE (1982) Structural effects in electrocatalysis. Nature 296:137–138
7. Sun SG, Clavilier J, Bewick A (1988) The mechanism of electrocatalytic oxidation of formic acid on Pt (100) and Pt (111) in sulfuric acid solution – an EMIRS study. J Electroanal Chem 240:147–159
8. Solla-Gullon J et al (2008) Shape-dependent electrocatalysis: methanol and formic acid electrooxidation on preferentially oriented Pt nanoparticles. Phys Chem Chem Phys 10:3689–3698
9. Tian N, Zhou ZY, Sun SG, Ding Y, Wang ZL (2007) Synthesis of tetrahexahedral platinum nanocrystals with high-index facets and high electro-oxidation activity. Science 316:732–735
10. Casado-Rivera E et al (2004) Electrocatalytic activity of ordered intermetallic phases for fuel cell applications. J Am Chem Soc 126:4043–4049
11. Zhang J, Sasaki K, Sutter E, Adzic RR (2007) Stabilization of platinum oxygen-reduction electrocatalysts using gold clusters. Science 315:220–222
12. Ji XL et al (2010) Nanocrystalline intermetallics on mesoporous carbon for direct formic acid fuel cell anodes. Nat Chem 2:286–293
13. Steele BCH, Heinzel A (2001) Materials for fuel-cell technologies. Nature 414:345–352
14. Stamenkovic VR et al (2007) Trends in electrocatalysis on extended and nanoscale Pt-bimetallic alloy surfaces. Nat Mater 6:241–247
15. Lim B et al (2009) Pd-Pt bimetallic nanodendrites with high activity for oxygen reduction. Science 324:1302–1305
16. Stamenkovic VR et al (2007) Improved oxygen reduction activity on Pt3Ni(111) via increased surface site availability. Science 315:493–497
17. Wang C et al (2011) Multimetallic Au/FePt(3) nanoparticles as highly durable electrocatalyst. Nano Lett 11:919–926
18. Mazumder V, Lee Y, Sun SH (2010) Recent development of active nanoparticle catalysts for fuel cell reactions. Adv Funct Mater 20:1224–1231
19. Wang HF, Liu ZP (2009) Formic acid oxidation at Pt/H_2O interface from periodic DFT calculations integrated with a continuum solvation model. J Phys Chem C 113:17502–17508
20. Rice C, Ha S, Masel RI, Wieckowski A (2003) Catalysts for direct formic acid fuel cells. J Power Sources 115:229–235
21. Lee HJ, Habas SE, Somorjai GA, Yang PD (2008) Localized Pd overgrowth on cubic Pt nanocrystals for enhanced electrocatalytic oxidation of formic acid. J Am Chem Soc 130:5406–5407
22. Stamenkovic V et al (2006) Changing the activity of electrocatalysts for oxygen reduction by tuning the surface electronic structure. Angew Chem Int Ed 45:2897–2901
23. Zhang JL, Vukmirovic MB, Xu Y, Mavrikakis M, Adzic RR (2005) Controlling the catalytic activity of platinum-monolayer electrocatalysts for oxygen reduction with different substrates. Angew Chem Int Ed 44:2132–2135
24. Bauer JC, Chen X, Liu QS, Phan TH, Schaak RE (2008) Converting nanocrystalline metals into alloys and intermetallic compounds for applications in catalysis. J Mater Chem 18:275–282
25. Adzic RR, Simic DN, Despic AR, Drazic DM (1975) Electrocatalysis by foreign metal monolayers – oxidation of formic-acid on platinum. J Electroanal Chem 65:587–601
26. Adzic RR, Tripkovic AV, Markovic NM (1983) Structural effects in electrocatalysis – oxidation of formic-acid and oxygen reduction on single-crystal electrodes and the effects of foreign metal adatoms. J Electroanal Chem 150:79–88
27. Xia XH, Iwasita T (1993) Influence of underpotential deposited lead upon the oxidation of HCOOH in HClO4 at platinum-electrodes. J Electrochem Soc 140:2559–2565
28. Kang YJ, Murray CB (2010) Synthesis and electrocatalytic properties of cubic Mn-Pt nanocrystals (nanocubes). J Am Chem Soc 132:7568–7569
29. Markovic NM et al (1995) Electrooxidation mechanisms of methanol and formic-acid on Pt-Ru alloys surfaces. Electrochim Acta 40:91–98
30. Kang YJ et al (2012) Highly active Pt_3Pb and core-shell Pt_3Pb-Pt electrocatalysts for formic acid oxidation. ACS Nano 6(3):2818–2825
31. Maksimuk S, Yang SC, Peng ZM, Yang H (2007) Synthesis and characterization of ordered intermetallic PtPb nanorods. J Am Chem Soc 129:8684–8685
32. Alden LR, Han DK, Matsumoto F, Abruna HD, DiSalvo FJ (2006) Intermetallic PtPb nanoparticles prepared by sodium naphthalide reduction of metal-organic precursors: electrocatalytic oxidation of formic acid. Chem Mater 18:5591–5596
33. Wang LL, Johnson DD (2008) Electrocatalytic properties of PtBi and PtPb intermetallic line compounds via DFT: CO and H adsorption. J Phys Chem C 112:8266–8275
34. Alayoglu S, Nilekar AU, Mavrikakis M, Eichhorn B (2008) Ru-Pt core-shell nanoparticles for preferential oxidation of carbon monoxide in hydrogen. Nat Mater 7:333–338
35. Sasaki K et al (2010) Core-protected platinum monolayer shell high-stability electrocatalysts for fuel-cell cathodes. Angew Chem Int Ed 49:8602–8607

36. Mazumder V, Sun SH (2009) Oleylamine-mediated synthesis of Pd nanoparticles for catalytic formic acid oxidation. J Am Chem Soc 131:4588–4589
37. Naohara H, Ye S, Uosaki K (2001) Thickness dependent electrochemical reactivity of epitaxially electrodeposited palladium thin layers on Au(111) and Au(100) surfaces. J Electroanal Chem 500:435–445
38. Liang Y et al (2011) Co_3O_4 nanocrystals on graphene as a synergistic catalyst for oxygen reduction reaction. Nat Mater 10:780–786

Fuel Cell Vehicles

Soichiro Uno
Nissan Research Center/EV System Laboratory, NISSAN MOTOR CO., LTD., Yokosuka-shi, Kanagawa-ken, Japan

Introduction

A fuel cell vehicle (hereinafter "FCV") is an electric vehicle (hereinafter "EV") that travels by powering its onboard electric motor by the electricity generated in its onboard fuel cell. Figure 1 shows examples of FCVs [1–7] as of 2012.

A fuel cell is a device that produces electricity through an electrochemical reaction between hydrogen and oxygen contained in the air. The fuel cells installed in FCVs are mainly polymer electrolyte fuel cells whose major benefit is their short start-up time. Pure hydrogen is used as the fuel for FCVs, and in most cases, the hydrogen is stored in a gaseous state in a compressed hydrogen storage system.

FCVs are completely free of hazardous emissions (CO, HC, NOx, etc.) and greenhouse gases (CO_2, CH_4, etc.) during operation and are therefore called ultimate zero-emission vehicles. Hydrogen to fuel FCVs can be produced from various primary energy sources (Fig. 2). Furthermore, FCVs have a high energy efficiency compared with conventional internal combustion engine vehicles (hereinafter "ICEVs") (Fig. 3) [8] and thus contribute also to energy security from the viewpoint of reducing energy consumption.

Thanks to these features, FCVs are, parallel to battery electric vehicles (hereinafter "BEVs"), drawing attention as examples of next-generation vehicles that can solve environmental and energy issues.

The pioneer of the global FCV development trend was the NECAR 1 (New Electric Car) that Mercedes-Benz developed in 1994 by modifying the MB 100 van [9]. Figure 4 shows the appearance of the NECAR 1. The drive system occupied the entire load compartment of the MB 100 van.

The world's first FCV, on the other hand, was a converted GMC Handi-Van called the "Electrovan," which GM scientists and engineers presented in 1966 [10]. Figure 5 shows the appearance and a transparent view of the Electrovan. The Electrovan was equipped with cryogenic liquid hydrogen and liquid oxygen tanks.

As of 2012, the development of FCVs is mainly driven by the world's major automobile manufacturers, and demonstration tests are being conducted in North America, Europe, Japan, Korea, etc., to enable the smooth introduction of FCVs to the market. Furthermore, regulations related to FCVs are being reviewed and developed in various countries, and the Global Technical Regulations (GTR) and the International Organization for Standardization (ISO) are also being promoted.

Configuration

Figure 6 shows an example of FCV configuration.

The hydrogen that fuels FCVs is stored in an onboard compressed hydrogen storage container that is filled up at hydrogen stations. There are two pressure levels of hydrogen filling: 35 MPa and 70 MPa. Recently, the 70 MPa-specification has been the mainstream. When an FCV is started up, the pressure of its high-pressure hydrogen gas is reduced to an appropriate level and the gas is led to a fuel cell stack consisting of multiple layers of fuel cells. At the same time, air is

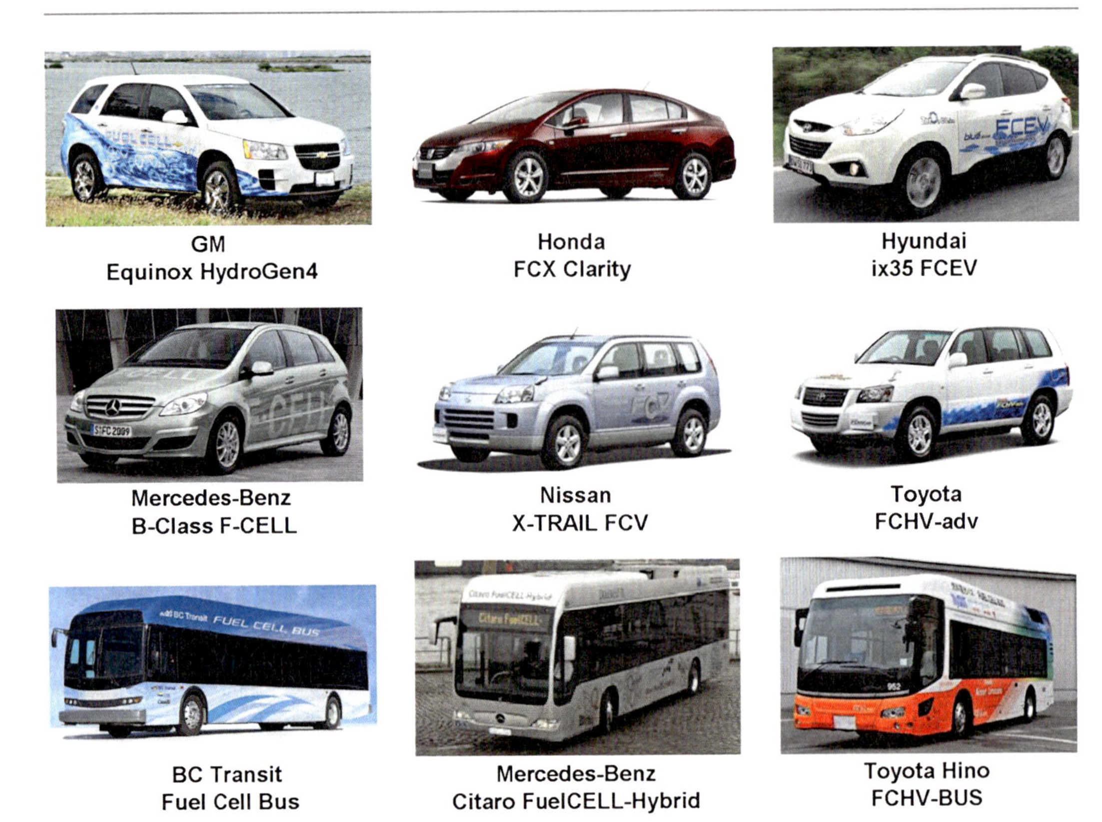

Fuel Cell Vehicles, Fig. 1 Examples of FCVs in 2012

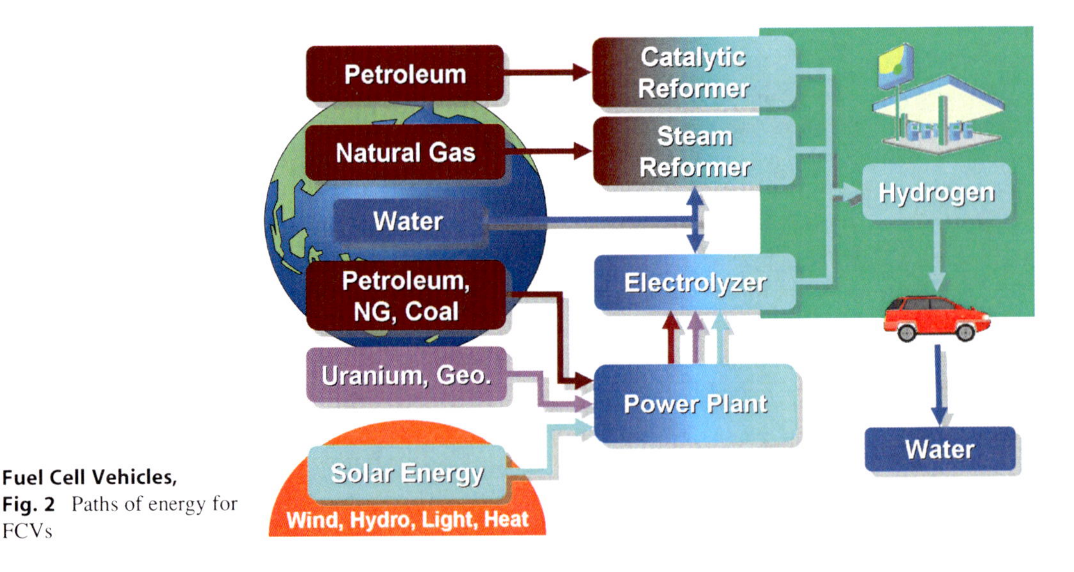

Fuel Cell Vehicles, Fig. 2 Paths of energy for FCVs

supplied to the fuel cell stack by an air compressor. The direct current generated in the fuel cell stack is converted into alternating current by an inverter and supplied to the driving motor of the vehicle. Generally, the motor can create a strong torque at a low rotation speed near zero and can easily be switched into a reverse direction of rotation. Therefore, there is no need of a torque converter, a clutch, or a transmission as required in an ICEV. Normally, the configuration of an

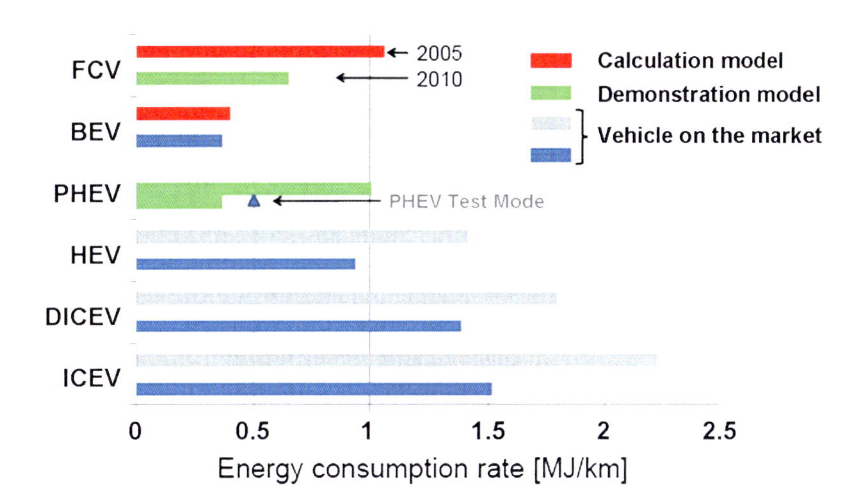

Fuel Cell Vehicles, Fig. 3 TtW calculation result < efficiency > (10-15Mode in Japan)

Fuel Cell Vehicles, Fig. 4 Mercedes-Benz "NECAR 1 (New Electric Car)" in 1994

FCV is such that only a reduction gear and a differential gear are installed between the driving motor and the drive shaft.

Many FCVs use their driving motor as a generator during deceleration and convert the deceleration energy into electric energy that is then stored in a secondary battery such as a lithium-ion rechargeable battery. When accelerating the vehicle, the electric energy stored in the secondary battery is reused as energy for driving. In FCVs, such a regeneration control is carried out for the efficient use of hydrogen energy.

Moreover, the motor control is configured so as to simulate a "creep condition" (a situation in which the vehicle starts to run at a low speed if the driver releases the brake pedal and the acceleration pedal at the same time) in which a driver who is well used to automatic transmission vehicles can drive FCVs without any discomfort. Various other measures like this have been taken for a smooth transition to FCVs.

Additionally, as an EV-specific arrangement, the air-conditioning compressor which is driven by the engine in an ICEV is driven by a motor specific to that type. The hydraulic power steering system is replaced with an electric power steering system. The conventional 12 V battery that supplies power to the headlights and

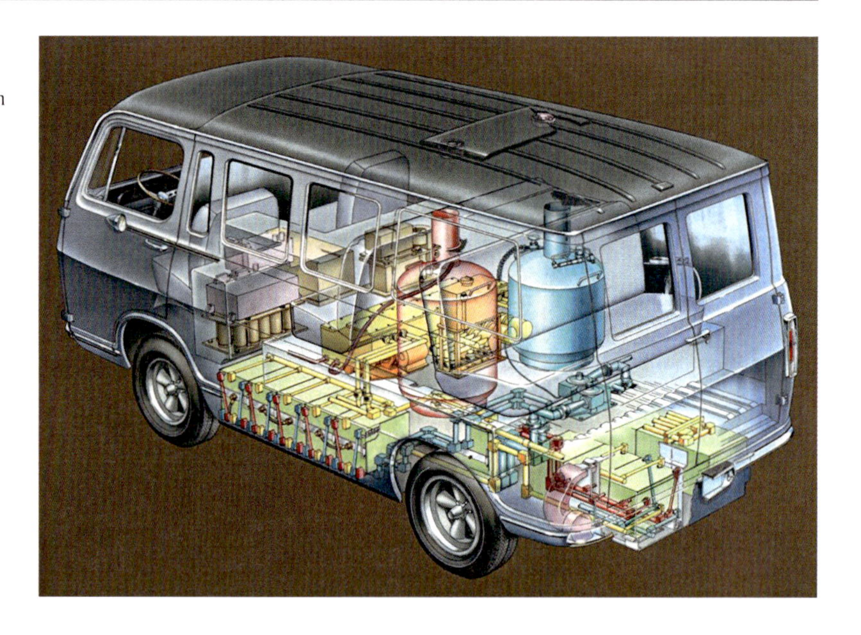

Fuel Cell Vehicles, Fig. 5 The world's first FCV, GM "Electrovan" in 1966

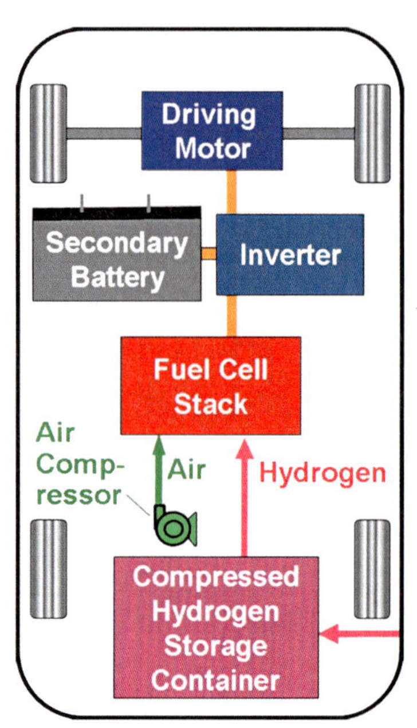

Fuel Cell Vehicles, Fig. 6 Example of FCV mechanism

other electric instruments is fed with a power of 12 V from the fuel cell stack via a DC-DC converter.

In terms of safety, FCVs are different from conventional ICEVs in that high-pressure hydrogen gas is stored and a high voltage is used. For this reason, international regulations have been reviewed and revised to ensure the same or better safety in FCVs as compared with ICEVs. For compressed hydrogen storage containers, not only sturdy structure but also a regular inspection and replacement are required. To cope with hydrogen leakage in FCVs, hydrogen sensors are provided to constantly monitor hydrogen leakage, and the compressed hydrogen storage system shuts off the passage of hydrogen after the key has been turned off in FCVs. In the event of a collision, the compressed hydrogen storage system is designed so as to shut off the passage of hydrogen and the high voltage system. If the vehicle is involved in a collision and the compressed hydrogen storage container reaches a high temperature by fire, in a similar manner as in CNG vehicles and LPG vehicles, a relief fusible plug that melts at high temperatures releases hydrogen to prevent the compressed hydrogen storage container from exploding.

Development Status as of 2012

As of 2012, FCVs are released only on a limited leasing basis and are not available freely for general customers. However, the development is taking place globally so that the general public will begin to be able to use them around 2015.

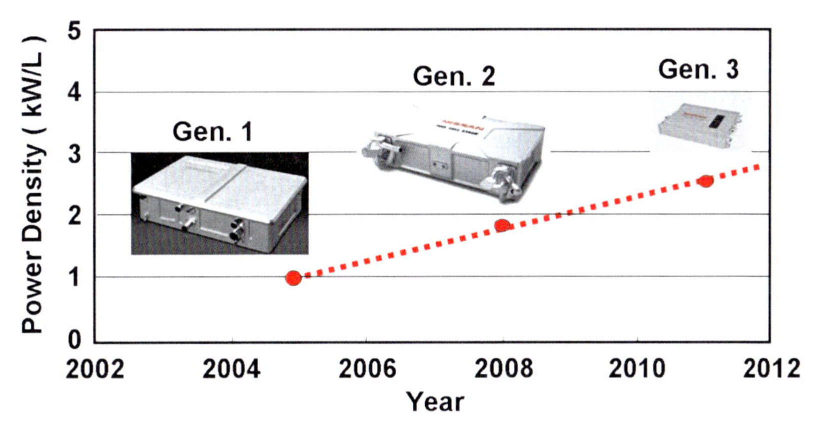

Fuel Cell Vehicles, Fig. 7 Improvement of FC stack power density (Reprinted with permission from SAE paper 2011-01-1344 © 2011 SAE International)

The issues raised for FCVs around 2005 were concerned with power density of fuel cell stack which related power train size, cruising range, start-up performance at temperatures below the freezing point, durability, and cost.

Figure 7 shows the development of fuel cell stacks in terms of power density from 2005 through 2011 reported by Nissan [11].

In these 5 years, the volume of a fuel cell stack generating the same amount electric power was reduced to 1/2.5.

In terms of the cruising range, the Honda FCX Clarity has reached 620 km, and the Toyota FCHV-adv of the 70Mpa specification has reached 830 km according to the 10–15 mode of measurement in Japan. In November 2009, the Nissan X-TRAIL FCV and the abovementioned two vehicles made a demonstration to travel over a distance of 1,137 km from Tokyo to Fukuoka by refilling hydrogen only twice [12] and proved that their cruising range was almost comparable with that of gasoline engine vehicles.

Concerning the FCV start-up performance below the freezing point, however, there were cases in which water inside the fuel cell turned to ice and blocked the passage of air so that the vehicle could not be started smoothly. Such phenomena have been analyzed since then, and a confirmation result of start-up performance in a temperature range of minus 25–30 °C has been reported [12–14], indicating that the FCV technology is now reaching the phase of practical market's application.

In terms of the durability of fuel cell stacks, analyses have been made to clarify reactions on the level of chemical elements and their mechanisms as well as deterioration modes in the field. As Fig. 8 shows, it has been reported by Nissan [14] that the level of deterioration in real vehicles driven in Japan and the United States coincides with the level of deterioration estimated from their travel measurement data, thus establishing a method of predicting deterioration.

And Toyota informed that the fuel cell stack on Toyota FCHV-adv maintains about 70 % of power after 500,000 km equivalent traveling as Fig. 9 [12].

On June 1, 2011, three Mercedes-Benz FCVs completed their driving program in 14 countries on four continents, thus completing the first around-the-world FCV travel to prove the high durability of FCVs [15].

In order to lower the cost of FCVs, technical developments have been promoted through all possible approaches, starting naturally with a reduction in the amount of platinum used in fuel cell stacks, and by replacing materials, improving processes, and simplifying and eliminating auxiliary machines. Figure 10 illustrates the cost development of Nissan fuel cell stacks shown in Fig. 7. The 2010 model has reached a cost level that can almost achieve the 2010 target defined by the DOE [11].

And from 2011, developments for cost reduction are still going on towards practical realization.

With regard to the compressed hydrogen storage systems, the development may be getting closer to the stage of practical market application, but the systems are still large, heavy, and costly compared with gasoline tanks. Therefore, there is strong need

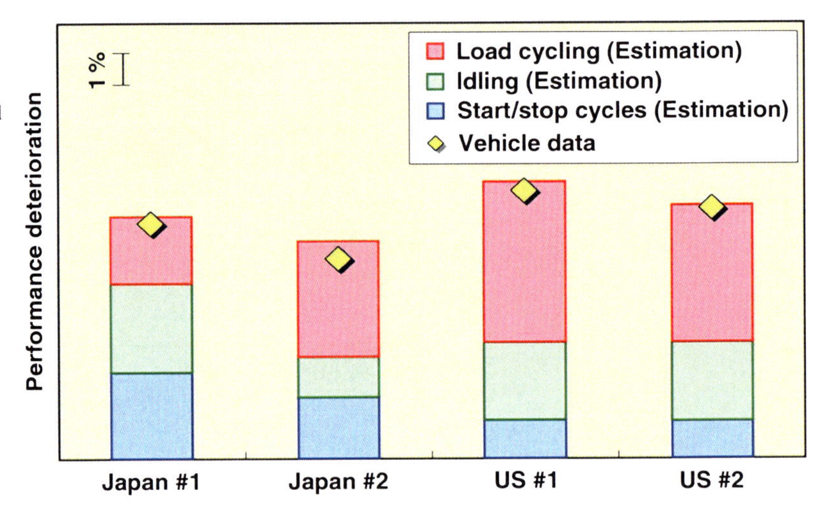

Fuel Cell Vehicles, Fig. 8 Deterioration of fuel cell stack (estimation vs. vehicle data) (Reprinted with permission from SAE paper 2009-01-1014 © 2009 SAE International)

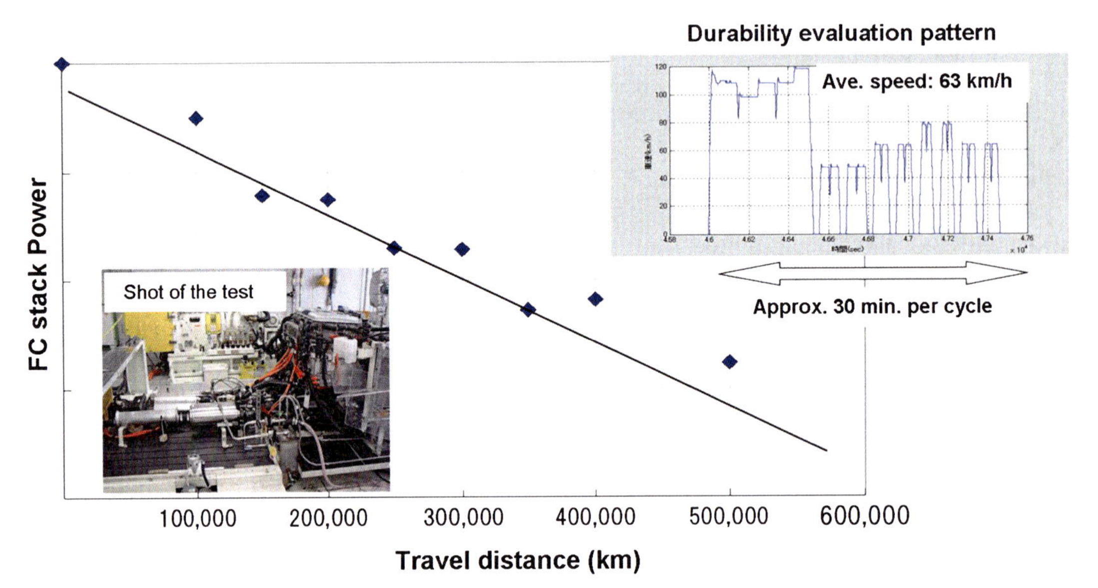

Fuel Cell Vehicles, Fig. 9 Toyota FCHV-adv fuel cell stack power maintenance ratio (Source: September 2009 JSAE Kanto branch lecture meeting)

for the development of a new technology such as hydrogen storage materials that can replace compressed hydrogen storage systems.

Implementation of Hydrogen Stations as of 2012

As of 2012, hydrogen stations are operated as case studies in North America, Europe, Japan, Korea, etc. Figure 11 shows examples of hydrogen stations as of 2012 [16–19].

The examples include a mobile hydrogen station on a truck, a hydrogen station that is packaged in a container for easy transportation and can change its specification by adding containers, a combined station in which a hydrogen dispenser is installed next to existing gas station dispensers, and a hydrogen-specific station.

Hydrogen stations can largely be classified into two groups: on-site type stations that produce hydrogen from other materials on the site and off-site type stations that are supplied with hydrogen by large-scale hydrogen plants in ironworks, refineries, etc.

The methods of producing hydrogen in on-site type stations include steam reforming of hydrocarbon-based materials such as city gas (CNG), naphtha, desulfurized gasoline, and methanol as well as alkaline water electrolysis.

The methods of supplying hydrogen to off-site type hydrogen stations include high-pressure hydrogen tank trailers that are suitable for a relatively small hydrogen supply, liquid hydrogen trailers, and hydrogen pipelines that are suitable for a large supply. In the world especially North America and Europe, hydrogen pipelines reaching a total length of about 3,000 km have already been installed.

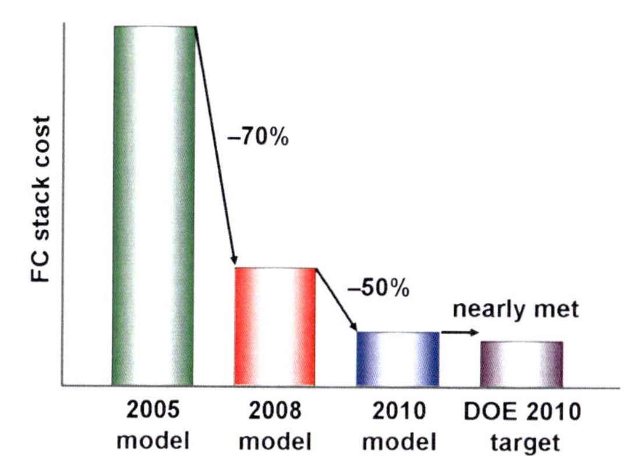

Fuel Cell Vehicles, Fig. 10 Comparison of estimated FC stack cost with the 2010 target of DOE's hydrogen program (Reprinted with permission from SAE paper 2011-01-1344 © 2011 SAE International)

There are three main issues associated with hydrogen stations: the number of hydrogen stations, the cost of hydrogen, and the hydrogen-filling speed.

With regard to the number of hydrogen stations, many countries are considering how to install hydrogen stations because FCV users always need one close to them to supply their vehicles with hydrogen. The general idea is to install hydrogen stations first in cities where the density of FCV users is expected to rise and on highways that connect such cities with one another and then to increase the number of hydrogen stations as the sales of FCVs increase.

In order to enhance the sales of FCVs, from the viewpoint of FCV users, it is important to reduce the cost of FCVs and to lower the cost of hydrogen to the level of fuel price of HEVs and diesel ICEVs when covering the same mileage. To lower the price of hydrogen, it is considered effective to reduce the costs of equipment and personnel.

On a small trailer

On a truck

In a container

Combined with existing gasoline station

Hydrogen is produced from NG at the station

Hydrogen is delivered by liquid hydrogen tankers

Fuel Cell Vehicles, Fig. 11 Examples of hydrogen stations in 2012

As a method of cutting the cost of equipment, container-type hydrogen stations with a unified specification suitable for mass production as well as hydrogen stations that are installed in the existing gasoline stations are both under investigation. Demonstration tests have been going on in North America and Europe to reduce the labor cost by introducing self-service systems in which hydrogen is filled up by the users as well as unmanned hydrogen stations (systems that do not have an administrator in each station and is instead monitored and operated remotely by the control center).

A short hydrogen fill-up time will also be an attractive feature for the users. In a gas-filling process in general, the shorter the gas-filling time and the higher the filling pressure, the higher the gas temperature inside the storage container to be filled becomes. Since the operating temperature of compressed hydrogen containers is limited to 85 °C or lower, it will be necessary to provide a precooling system to fill cool hydrogen gas in hydrogen stations in advance as well as a communication system to exchange information between FCVs and hydrogen stations about the material of storage containers as well as the pressure and temperature in the storage containers in order to fill hydrogen quickly in summer while maintaining the temperature of hydrogen inside the compressed hydrogen storage containers at 85 °C or lower. For this purpose, their specifications are now being examined and demonstration test data is also being collected.

Future Directions

In general, currently, it is impossible to separate, capture, and store CO_2 emitted from the internal combustion engines that are installed in vehicles as moving bodies. In order to eliminate CO_2 emissions from vehicles, the only effective way is to replace ICEVs with CO_2-free FCVs and BEVs.

As mentioned earlier, hydrogen and electricity as the fuel for FCVs and BEVs can be produced from various primary energies sources including CO_2-free sources. Even if fossil fuel is used as the primary energy, its manufacturing plants are stationary and can use the technology called "Carbon Dioxide Capture and Storage" ("CCS") to separate, recover, and store CO_2, thus making it possible to produce hydrogen and electricity without CO_2 emissions.

Furthermore, the price of fossil fuel which is unevenly distributed throughout the world is subject to significant fluctuations and is expected to rise with the demand grows in the world. Hydrogen and electricity, in contrast, can be produced from thermal, hydraulic, nuclear power, and many other energy sources such as solar, wind, geothermal, and wave energy and are therefore an attractive alternative in terms of energy security, particularly for energy-importing countries.

On the other hand, however, hydrogen and electricity are costly compared with primary energy, and also FCVs and BEVs are more expensive than ICEVs.

In order to address the issues of FCVs mentioned above, including the need for cost reduction, large-scale research facilities are necessary, and a further industry-government-academia collaboration is expected to promote research and development activities and to leverage analyses and investigations.

Particularly in the early stages of FCV introduction into the market, it is important to have scenarios that can be shared by the industry, government, and municipalities to bring FCVs into wider use.

Cross-References

▶ Compressed and Liquid Hydrogen for Fuel Cell Vehicles
▶ Polymer Electrolyte Fuel Cells (PEFCs), Introduction

References

1. GM HP, GM news, photos, new pilot program to create fuel cell stations in Hawaii2010-05-11: http://gm.wieck.com/forms/gm/X10CO-FC002.jpg?download = 053583
2. JHFCHP, Participating vehicles: http://www.jari.or.jp/jhfc/e/fcv/index.html

3. Hyundai HP, Hyundai News Room: http://worldwide.hyundai.com/company-overview/news-view.aspx?idx = 365&&nCurPage = 1&strSearchColumn = Title&strSearchWord = FCEV&ListNum = 3
4. Daimler HP, Global media site: http://media.daimler.com/dcmedia/0-921-656547-1-829622-1-0-2-0-0-1-11701-614316-0-3842-0-0-0-0-0.html?TS = 1324891911326
5. BC Transit HP: http://www.bctransit.com/> Corporate: http://www.bctransit.com/corporate/> NEWS & MEDIA Hydrogen Fuel Cell: http://busonline.ca/fuelcell/, ISE Corporation Photo Gallery: http://www.isecorp.com/gallery/Fuel-Cell-Hybrid-Vehicles > BC Transit Fuel Cell Bus: http://www.isecorp.com/gallery/BC-Transit-Fuel-Cell-Bus/BCTransit_fuel_cell_bus
6. Daimler HP, Global media site: http://media.daimler.com/dcmedia/0-921-656547-1-829622-1-0-1-0-0-1-11701-614316-0-3842-0-0-0-0-0.html?TS = 1324891521095
7. HySUT HP: http://hysut.or.jp/business/2011/bus/fchvbus.html
8. Hisashi Ishitani (2011) Analysis of total efficiency and greenhouse gas emission, FY2010 JHFC international seminar: http://www.jari.or.jp/jhfc/data/seminor/fy2010/pdf/day1_E_20.pdf
9. Daimler HP, Global media site: http://media.daimler.com/dcmedia/0-921-656547-1-829622-1-0-3-0-0-1-11701-614316-0-3842-0-0-0-0-0.html?TS = 1324964014306
10. GM HP, GM News, Photos, 1966 The world's first fuel cell vehicle (Electrovan) 2008-08-25: http://gm.wieck.com/forms/gm/W66HV-CO001.jpg?download = 048825
11. Mitsutaka Abe,Takanori Oku, Yasuhiro Numao, Satoshi Takaichi, Masanari Yanagisawa (2011) Low-cost FC stack concept with increased power density and simplified configuration utilizing an advanced MEA,SAE 2011-01-1344
12. Yoichi Hori, Hidemi Onaka (2010) WG2 fuel cell vehicles WG, FY2010 JHFC international seminar: http://www.jari.or.jp/jhfc/data/seminor/fy2010/pdf/day1_E_10.pdf
13. Nobuyuki Kitamura, Kota Manabe, Yasuhiro Nonobe, Mikio Kizaki (2010) Development of water content control system for fuel cell hybrid vehicles based on AC impedance, SAE 2010-01-1088
14. Ryoichi Shimoi, Takashi Aoyama, Akihiro Iiyama (2009) Development of fuel cell stack d based on actual vehicle test data: current status and future work, SAE 2009-01-1014
15. Daimler HP, Successful finish: F-CELL World Drive reaches Stuttgart after circling the globe, F-CELL World Drive, WEV SPECIALS, Company: http://www.cms.daimler.com/dccom/0-5-1367004-1-1367069-1-0-0-0-0-1-8-7145-0-0-0-0-0-0-0.html
16. Powertech HP, Areas of Focus, Clean Transportation, Fueling Stations,Mobile Hydrogen Fueling Stations: http://www.powertechlabs.com/areas-of-focus/clean-transportation/fueling-stations/mobile-hydrogen-filling-stations/
17. JHFC HP: JHFC Hydrogen Station, JHFC Kasumigaseki Hydrogen Station: http://www.jari.or.jp/jhfc/e/station/kanto/kasumi.html
18. Shell Hydrogen, Alternative Fuel Projects: http://www.thelivingmoon.com/41pegasus/02files/Alternate_Fuel_Shell_Oil_Hydrogen.html
19. JHFC HP: JHFC Hydrogen Station, JHFC Senju Hydrogen Station: http://www.jari.or.jp/jhfc/e/station/kanto/senju.html

Fuel Cells, Non-Precious Metal Catalysts for Oxygen Reduction Reaction

Ulrike I. Kramm
Technical University Cottbus, Cottbus, Germany

Introduction

Today, platinum-based catalysts are the state-of-the-art material in fuel cell applications. The costs of these catalysts, however, contribute by 33 % to the overall costs of a fuel cell stack [1]. This makes it reasonable to search for cheap alternatives, especially non-precious metal catalysts (NPMC). Some metal nitrides (Me = W, Mo) and oxynitrides (Me = Ta, Zr, Nb) are promising regarding the observed onset potentials (up to 0.8 V vs. NHE) and could be of interest for further investigation. In this respect, however, the readers are referred to the original contributions (W_2N [2], Mo_2N [3], ZrO_xN_y [4–6], TaO_xN_y [7], NbO_xN_y [8, 9]).

This essay will focus on Me–N–C catalysts that are currently the best-performing NPMC for the ORR.

Recently, researchers from INRS-EMT and LANL published impressive volumetric current densities (230 A cm^{-3} at 0.8 V) for μ-Fe–N–C and long-term stability for (Fe,Co)–PANI–C (PANI or polyaniline), respectively [10, 11]. Unfortunately, the combination of high currents and sufficient durability was not reached, yet. Therefore, for fuel cell application, these catalysts still have to be improved in both respects.

The *durability* benchmarks are defined by potentiostatic long-term stability but even more important by the stability of a certain catalyst under changing load conditions. It was found that especially this issue has a large impact on the performance decrease, whereas the operation under constant conditions seems less challenging.

The *volumetric current density* J_V gives the current in relation to the catalyst volume (A cm^{-3}). In general, you might be able to increase the power density of a FC by adding a larger amount of catalyst material on the electrode. However, above a certain limit, the transport properties of this electrode layer will break down. A further increase of the catalyst-layer thickness above this critical value is not useful. Thus, the definition of the current per volume takes care of this fact.

$$J_V = S_D \cdot TOF \cdot e, \quad (1)$$

with [A cm^{-3}] = [sites cm^{-3}] · [e $sites^{-1}$ s^{-1}]. As defined in Eq. 1, the volumetric current density J_V is the product of *the turnover frequency TOF(U)* of the catalyst's active sites, its *site density* S_D, and the elementary charge e.

With respect to literature, some preparation routes are beneficial for reaching high S_D (A) or TOF (B) and some others for achieving a promising long-term stability (C). Therefore, three preparation routes that are beneficial with respect to either S_D (A), TOF (B), or durability (C) will be discussed. This will, of course, only reflect a small part of the beneficial achievements in the development of Me–N–C catalysts. For the optimization process, however, it might give some hints about how to enhance the performance of Me–N–C catalysts. In this respect, the paper of Jaouen et al. might be of interest as it compares several NPMC under the same experimental conditions [12]. Before I will introduce these preparation routes, you will find a short historical overview regarding the most important steps in the development of Me–N–C catalysts and a small contribution regarding the beneficial role of ^{57}Fe Mössbauer spectroscopy in the identification of iron species. In order to get a more complete overview on Me–N–C catalysts, the readers are referred to the review articles [13–15] and book chapters [16, 17].

Overview of the Preparation Improvements During the Decades

In 1964, Jasinski found out that different metallophthalocyanines were able to reduce oxygen in alkaline media [18]. Soon after, this finding was expanded to a variety of MeN_4 macrocycles that were applied for the oxygen reduction at low and high pH values. In 1976, Jahnke et al. published an improvement of the ORR activity and stability when carbon-supported complexes were heat treated in inert atmosphere at temperatures ranging from 400 °C to 1,000 °C [19].

To that time, the outcome was quite surprising as during the heat treatment, some of the MeN_4 centers (that are the active sites in the macrocycles) were destroyed [20]. Although only a smaller number of MeN_4 centers were present, higher activities were gained. This observation was the starting point of a still ongoing debate regarding the nature of the catalytic sites in these heat-treated materials. Van Veen proposed that the change in the electronic environment of the MeN_4 centers enables a higher turnover frequency [20, 21]. Indeed, recently, it was shown that with increasing electron density on the metal center (measured by the isomer shift of the active FeN_4 center in Mössbauer spectroscopy), the turnover frequency increases by a factor of about 100 comparing the non-pyrolyzed carbon-supported macrocycle with its best-performing heat treatment product [22].

The controversy was further igniting when it became clear that instead of the macrocycles, much simpler molecules might be used as precursors [23, 24]. Nowadays, we understand that for the preparation of a Me–N–C catalyst, one might mix nearly any kind of metal, nitrogen, and carbon source; after the heat treatment at appropriate temperatures, the material will have some activity for the ORR. Some metal, nitrogen, and carbon sources that were successfully applied for the preparation of highly active Me–N–C are summarized in Fig. 1.

Depending on the ratio of the precursor components used for the synthesis as sketched in Fig. 1 and the heat treatment conditions, your catalyst might contain active MeN_4 centers as

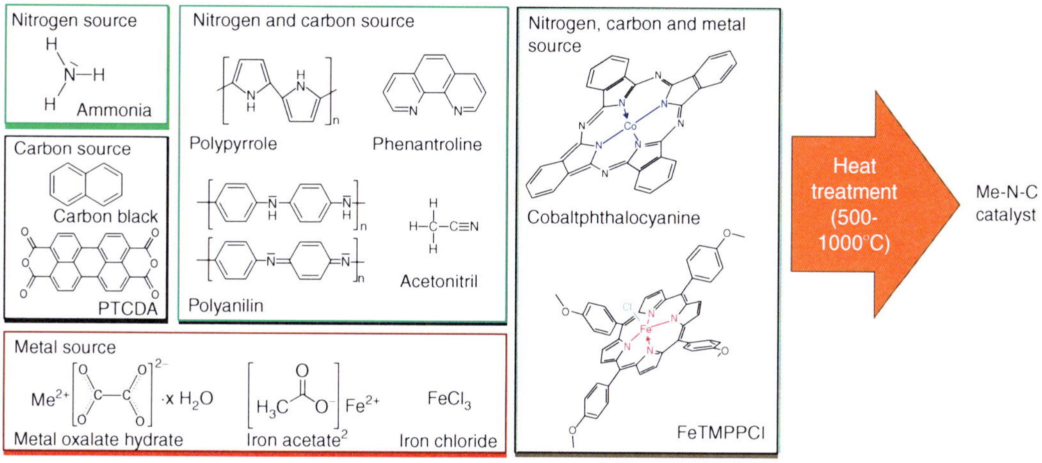

Fuel Cells, Non-Precious Metal Catalysts for Oxygen Reduction Reaction, Fig. 1 Different nitrogen, carbon, and metal sources that might be used as precursors in the synthesis of Me–N–C catalysts. *PTCDA*, perylenetetracarboxylic dianhydride; *FeTMPPCl*, chloroirontetramethoxyphenylporphyrin

well as inorganic metal phases in different ratios. For structural characterization and for the ORR activity, it can be helpful to remove the inorganic metal phases by an acid leaching subsequent to this first heat treatment step [22, 25–30]. The benefit of such acid leaching for the structural characterization by Mössbauer spectroscopy (a) and extended X-ray absorption fine structure (EXAFS) analysis is illustrated in Fig. 2.

In 2008, it was demonstrated that a second heat treatment of the (acid leached) catalysts can further enhance the kinetic current density for ORR [32–34]. This second heat treatment (SHT) might be performed in inert gas atmosphere but will lead to higher activity increases when performed in a reactive gas atmosphere like CO_2 or NH_3 [12, 34, 35]. Today, most of the groups use two (or more) heat treatment steps in order to prepare the final catalyst.

For some catalysts, it was found that besides the contents of iron and nitrogen, the catalytic activity was dependent on the micropore content of the final catalyst [36, 37]. In order to illustrate the importance of the micropores, these catalysts are labeled as μ–Fe–N–C in this essay. So far, these are the Me–N–C catalysts that reach the highest volumetric current densities [10] which are basically determined by their very high turnover frequency [31] (see the text on TOF, below).

Identification of Iron Sites

In Fig. 2, the structural results of ^{57}Fe Mössbauer spectroscopy and EXAFS are compared. It becomes apparent that EXAFS gives an average overview of all compounds that contribute to the Fe–N distance and Fe–Fe distance. On the other hand, the ^{57}Fe Mössbauer spectra indicate that beside iron nitride (contributing to Fe–N and Fe–Fe), three FeN_4 centers can be distinguished. In all cases, the iron atoms have the same chemical environment (planar N_4 coordination) but different electronic structures.

Thus, ^{57}Fe Mössbauer spectroscopy allows distinguishing iron sites by their chemical environment **and/or** by their electronic states.

It is reasonable that the electronic signature of a FeN_4 center will affect its probability to reduce oxygen. Thus, for structure-activity correlations, it is mandatory distinguishing the different FeN_4 centers and determining their absolute contents in a catalyst. This might be done by the combination of the Mössbauer data (A_{Abs}) and the absolute iron content c_{Fe} using Eq. 2.

$$c_{Fe}(\mathrm{SiteX}) = A_{Abs}(\mathrm{SiteX}) \cdot c_{Fe} \qquad (2)$$

Indeed, this "active-site-only approach" enabled a direct correlation of the iron content of one specific center with the kinetic current density,

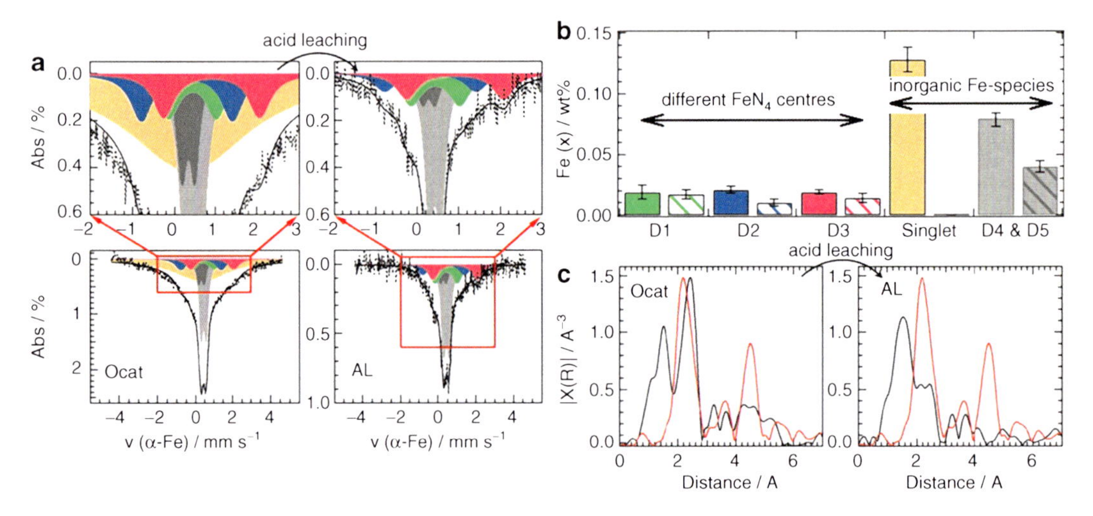

Fuel Cells, Non-Precious Metal Catalysts for Oxygen Reduction Reaction, Fig. 2 ^{57}Fe Mössbauer spectra (**a**) and EXAFS analysis (**c**) of an as-prepared catalyst (*Ocat*) and after acid leaching (*AL*). For better visualization in (**b**), the iron contents assigned to the different iron sites are given for the *Ocat* and *AL* (*filled bars*, *Ocat*; *striped bars*, *AL*). It is evidenced that induced by the acid leaching, a large fraction of inorganic iron species was removed, whereas only a minor decrease of the iron contents assigned to FeN_4 centers was observed. The *black* and *red curves* in (**c**) give the distribution function of the catalysts and an iron foil (for reasons of comparison), respectively. The first peak at 1.6 Å is related to the Fe–N distance, whereas the second and third peaks (2.2 and 4.4 Å) are assigned to the first and second shell of the Fe–Fe distribution function. In the present case, the Fe–Fe contribution is caused by iron nitride particles (*gray doublets* D4+D5) that contribute minor to the Fe–N distance and major to the Fe–Fe distance (Adapted from Fig. 5 of [31] with permission from the PCCP owner societies, RSC)

for the first time [34]. The related Mössbauer site was a ferrous low-spin FeN_4 center (compare Fig. 3a, $\delta_{Iso} \approx 0.3$ mm s^{-1}, $\Delta E_Q = 0.8 \ldots 1.1$ mm s^{-1}). This iron site is found in high concentrations in porphyrin-based catalysts but in addition to this it is also found in alternatively prepared Fe–N–C catalysts [22, 31, 32, 34, 35, 38].

A second active FeN_4 center seems to be unique for μ–Fe–N–C catalysts [31]. It is a five-fold coordinated ferrous high-spin $NFeN_4$ center with a Mössbauer signature similar to picket fence porphyrins ($\delta_{Iso} \approx 1.0$ mm s^{-1}, $\Delta E_Q = 2.1 \ldots 2.4$ mm s^{-1}, compare Fig. 3b). This site reaches its best performance as a composite center in combination with a protonated nitrogen atom in its vicinity [39]. Sketches that might give an idea about how both FeN_4 centers look like are given in Fig. 3.

Preparation Strategy to Reach High Densities of Active Sites

Up to the year 2000, in most of the various preparation strategies, different kinds of carbon black were used. These were impregnated either with metallomacrocycles (like porphyrins or phthalocyanines) or with different iron sources in combination with various nitrogen sources. Often, the achieved current density was dependent on (i) the surface area of the carbon support and the (ii) content of macrocycle that was utilized (or the contents of iron and nitrogen). However, all these efforts did not lead to a breakthrough in the improvement of the catalysts. In 2004, Bogdanoff et al. published an oxalate-supported pyrolysis of porphyrins [25]. In contrast to previous attempts, no carbon black was added. The precursor of these catalysts was just a mixture of porphyrin plus oxalate, whereas the oxalate worked as a template and foaming agent during the heat treatment. After the heating, an acid leaching was performed in order to remove the decomposition products of the oxalate template. As a consequence, the carbon present in the final catalyst was only related to the carbon provided by the macrocycle. The in situ formed carbon matrix contained a high density of active sites.

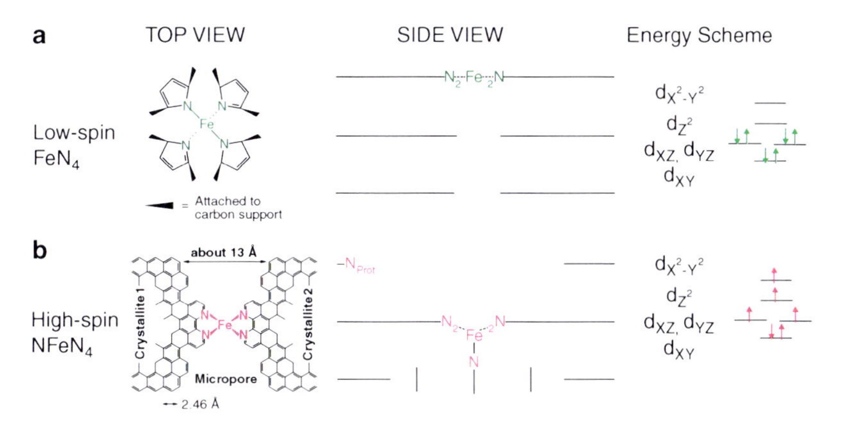

Fuel Cells, Non-Precious Metal Catalysts for Oxygen Reduction Reaction, Fig. 3 Sketches of the deduced structures of active FeN_4 centers in Fe–N–C catalysts. On the basis of the given results, the low-spin $Fe^{2+}N_4$ center seems to enable a direct reduction of oxygen as monomolecular active site, but only with low turnover frequencies of less than 0.15 electron $site^{-1}$ s^{-1} at 0.8 V (Refs. [22, 31, 34, 35]). In contrast to this, the high-spin $NFe^{2+}N_4$ center can reach very large turnover frequencies when it is working as binary site in combination with a protonatable nitrogen atom ($NFe^{2+}N_4 \ldots NH^+$), with TOF(0.8 V) = 11 electron $site^{-1}$ s^{-1} [31, 39] (Adapted from Fig. 2 of [31] with permission of the PCCP owner societies)

In the following years, it was found that even higher kinetic current densities were reached when, in addition to the iron oxalate, sulfur was added to the precursor [27]. Structural investigations showed that the beneficial effect of sulfur was related to the hindered formation of iron carbide, as this process was at the origin of the disintegration of MeN_4 centers [26]. In general, in this preparation technique, several different porphyrins and oxalates can be utilized yielding a variety of mono- and bimetallic Me–N–C catalysts [28, 40]. Induced by a second heat treatment, the ORR activity of these Me–N–C catalysts can be increased by a factor of up to 20 [34, 35]. This increase was assigned to a larger number of active sites [34] and improved TOF when NH_3 was utilized [35]. As a consequence, for these catalysts, site densities of $S_D > 10^{20}$ *sites* cm^3 and TOF(0.8 V) < 0.15 electron $site^{-1}$ s^{-1} are reached. Under PEM–FC conditions, these catalysts achieve volumetric current densities of about 35 A cm^{-3}. While potentiostatic long-term stability tests at RT revealed a stable performance, activity losses of 87 % were observed during 100 h of FC run (at 0.5 V, 80 °C) [41]. Therefore, this type of Me–N–C catalysts should be improved with respect to the TOF and long-term stability.

It should be noted that today, several other groups use different types of template-assisted preparations (see Refs. in [12]). A comparison of the FC performance of Fe-N-C catalysts prepared by different preparation strategies can be found in Fig. 4. The overall morphology of the final catalyst is strongly related to the structure of the employed template.

Preparation Strategy that Leads to a High Turnover Frequency

Today, the highest volumetric current density (at 0.8 $V_{iR-free}$) of any kind of NPMC catalysts is 230 A cm^{-3}. This value is reached by a μ–Fe–N–C catalyst of Dodelet's group that – again – was prepared by a template-assisted method [10]. In this case, a metal organic framework (MOF) with a high nitrogen content and high microporous surface area was used as the template. In order to obtain the precursor, the MOF was mixed with iron acetate and phenanthroline. Even after a first heat treatment in Argon at 1,050 °C, this catalyst reached impressive current densities (about one fifth of the above-reported value). The highest value was achieved after 15 min of a second heat treatment in ammonia at 950 °C. In contrast to other template-assisted methods, the used MOF has the

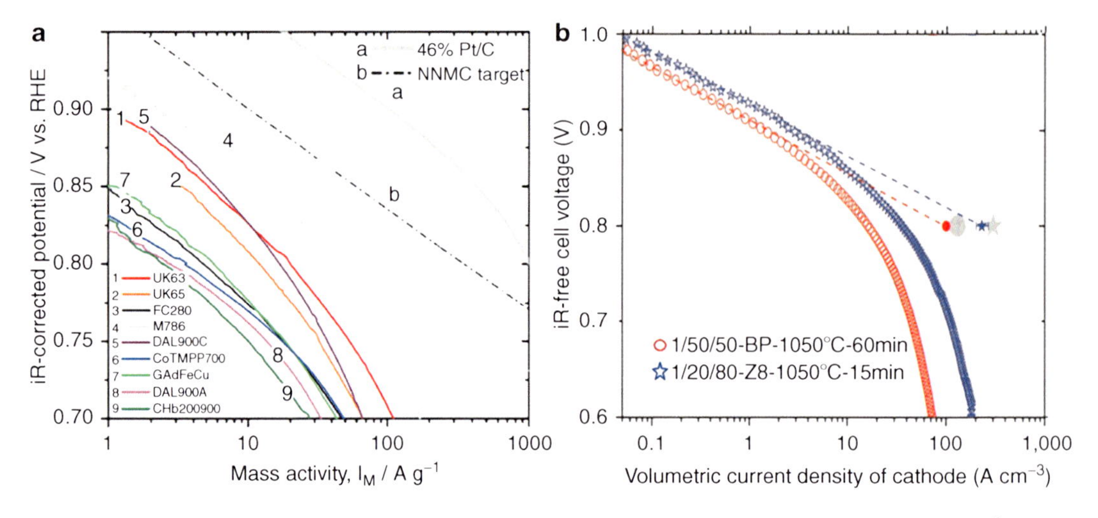

Fuel Cells, Non-Precious Metal Catalysts for Oxygen Reduction Reaction, Fig. 4 (**a**) Polarization curves of several Me–N–C catalysts all examined under the same experimental conditions (1.5 bar H_2/O_2, 1 mg cm^{-2}; see Ref. [12] et al. for details), (**b**) today's most active NPMC catalysts (μ–Fe–N–C) as prepared by the pore-filling method (*PFM*) described in Lefèvre et al. [44] and by the method described by Proietti et al. [10]. The values were obtained at 1.5 bar H_2/O_2 with 1 mg cm^{-2} but are already converted to DOE standard conditions of 1 bar H_2/O_2; the catalyst's density was determined to be 0.37 g cm^{-3}. (Part **a** reprinted with permission from Fig. 3, Jaouen et al., Appl. Mater & Interf [12]. Copyright 2009 American Chemical Society. Part **b** reprinted from Fig. 3 by permission from Macmillan Publishers Ltd: Nature Communications [10], Copyright 2011)

advantage that it decomposes without any metal-containing residuals at the applied heat treatment temperatures. As a consequence, an acid leaching after the first heat treatment is not mandatory as in the case of other template-assisted methods.

The preparation technique as described above was the logical outcome of several previous investigations on differently prepared μ–Fe–N–C catalysts. These were well understood for the factors affecting their ORR activity: (i) the iron content [42], (ii) the nitrogen content [43], and (iii) the microporous surface area [36, 37]. In the preparation, iron acetate was impregnated on a nonporous carbon black (72 m^2 g^{-1}) followed by a heat treatment at 950 °C in NH_3 until a weight loss of 30–40 % was observed. For these catalysts, Jaouen et al. [42] investigated the effect of increasing iron content (0.005 wt% Fe up to 2 wt% Fe in the precursor) on the ORR activity. It was found that with increasing iron content, a linear increase of the kinetic current density was observed up to a concentration of 0.2 wt% Fe. Above this value, the slope became smaller, and for iron contents exceeding 1 wt%, a breakdown of the current density was observed. Therefore, with this old preparation strategy, only small site densities were achieved. In order to identify the reason for this behavior, Mössbauer spectroscopy, EXAFS (see Fig. 2), and transmission electron microscopy were performed [31]. On the basis of these results, the structural model for the (most active) catalytic site in these μ–Fe–N–C catalysts was refined to what is sketched in Fig. 3b. In combination with a protonated nitrogen atom, this high-spin $NFe^{2+}N_4$ site (thus as composite site $NFe^{2+}N_4 \ldots NH^+$) can achieve turnover frequencies (at 0.8 V) of up to 11 electron $site^{-1}$ s^{-1} which is nearly 50 % of what is reached with commercial Pt/C. (Please note: This value was determined from RDE experiments at RT; under FC conditions, this value might even be larger!) The major problem of this center, however, is given in the low stability that is connected to this composite site [39]. The protonated nitrogen atoms become readily neutralized by anions. This effect was found to be reversible in acid [39]; however, under FC conditions, a way to prevent the anion binding has to be developed.

A model for the formation of iron sites in this preparation route was deduced (see Fig. 9 in Ref. [31]). On the basis of this, it was possible to explain the iron content ↔ ORR activity behavior observed by Jaouen et al. [42]. At higher iron concentrations, the formation of iron nitride particles was enhanced instead of active site formation. The reason for this was found in the occupation of the iron acetate molecules on the carbon support: in order to enable the active site formation, only less than a monolayer of iron acetate should be present on the carbon surface. In this case, the formation of active sites is faster than the agglomeration of iron atoms. For higher occupations, however, the agglomeration of iron atoms becomes dominant, yielding higher concentrations of iron nitride instead of active sites [31].

Thus, the new preparation technique as published by Proietti et al. [10] exhibits several advantages in comparison to the old approach: a much higher initial (microporous) surface area is utilized so that larger concentrations of iron can be used. Furthermore, iron will already form complexes with phenanthroline that is used as nitrogen and carbon precursor. Doing so, it might be more stable against agglomeration so that higher concentrations of active sites can be achieved.

In addition to this, the nitrogen content of the MOF might be beneficial.

Preparation Strategies that Enable a Good Durability

Today's most stable Me–N–C catalysts are the Me–PANI–C catalysts of Zelenay's group [11]. In order to obtain the nitrogen precursor, aniline is impregnated on a carbon support and then an oxidative polymerization is initiated in the presence of iron chlorine and/or cobalt nitride (both in various concentrations). As oxidant, either hydrogen peroxide or ammonium peroxydisulfate can be used [38]. In order to obtain the final catalysts, two heat treatment steps are performed, both at the same temperature and both followed by an acid leaching. Best results were obtained for catalysts prepared at 900 °C [29]. These catalysts were stable for 4 weeks of FC run under potentiostatic conditions (0.4 V) and even showed increasing currents (in the mass transport-related range) under cycling conditions [11]. It should be noted that also for Pt/C catalysts a stabilization by modification with PANI was observed [45]. The reason why in the presence of PANI a better long-term stability is reached can only be suspected. It might be possible that it stabilizes dangling bonds within the carbon framework enhancing the catalyst's durability. Further work on this will be needed to get a better understanding of the processes involved. While these catalysts are promising regarding the stability question, however, they reach only volumetric current densities of 32 and 8 A cm^{-3}, respectively, for the best-performing (Co, Fe)–PANI–C and Fe–PANI–C (as calculated by linear regression of the current density values in Fig. 3 of Ref. [11]. Volumetric current density was calculated assuming a catalyst density of 0.4 g cm^{-3}). The reason why the bimetallic catalyst exhibits a better performance compared to the iron-only catalyst is not known, yet. However, similar observations were made by others [27, 28, 40, 46].

Mössbauer spectroscopy was performed in order to identify the iron species [38]. The catalysts contained to some degree the low-spin FeN_4 center that is shown in Fig. 3a. However, there were large fractions of inorganic iron sites. This might indicate that with further optimization of the preparation route, the number of active sites might be enhanced further in order to increase the volumetric current density of these catalysts.

Future Directions

The different preparation strategies introduced in this essay might give further ideas improving Me–N–C catalysts. It was shown that template-assisted methods working without additional carbon are beneficial in reaching high densities of active sites.

Furthermore, the composite site structure of $NFe^{2+}N_4 \ldots NH^+$ reaches TOF(0.8 V) of 11 electron $site^{-1}$ s^{-1}. For the formation of this center, the integration of FeN_4 centers in micropores as well as the presence of protonatable nitrogen is needed to gain the high TOF values. These conditions might be reached by performing a heat treatment at high temperature in ammonia. This composite structure, however, has to be improved with respect to its stability as it gets deactivated by anion binding.

A stabilization effect can be attributed to PANI. It is not yet clarified if this is due to hindered carbon corrosion or to another effect. In addition to these findings, the utilization of two (or maybe more) different kinds of metal centers seems to enable higher current densities.

It can be assumed that the development of a stabilization mechanism for the composite site $NFe^{2+}N_4 \ldots NH^+$ and/or the enhancement of the site density of Me–PANI–C catalysts will lead to a further breakthrough in the development of NPMC. Both strategies might be successful in the next few years.

Cross-References

► Platinum-Based Cathode Catalysts for Polymer Electrolyte Fuel Cells
► Polymer Electrolyte Fuel Cells, Oxide-Based Cathode Catalysts

References

1. James BD, Kalinoski JA, Baum KN (2010) Mass production cost estimation for direct H2 PEM fuel cell systems for automotive applications: 2010 update. Washington
2. Zhong H, Zhang H, Liang Y, Zhang J, Wang M, Wang X (2007) A novel non-noble electrocatalyst for oxygen reduction in proton exchange membrane fuel cells. J Power Sources 164(2):572–577
3. Zhong H, Zhang H, Liu G, Liang Y, Hu J, Yi B (2006) A novel non-noble electrocatalyst for PEM fuel cell based on molybdenum nitride. Electrochem Commun 8:707–712
4. Liu G, Zhang HM, Wang MR, Zhong HX, Chen J (2007) Preparation, characterization of ZrOxNy/C and its application in PEMFC as an electrocatalyst for oxygen reduction. J Power Sources 172:503–510
5. Doi S, Ishihara A, Mitsushima S, Kamiya N, Ota K (2007) Zirconium-based compounds for cathode of polymer electrolyte fuel cell. J Electrochem Soc 154(3):B362–B369
6. Ohgi Y, Ishihara A, Matsuzawa K, Mitsushima S, Ota K (2010) Zirconium oxide-based compound as new cathode without platinum group metals for PEFC. J Electrochem Soc 157(6):B885–B891
7. Ishihara A, Lee K, Doi S, Mitsushima S, Kamiya N, Hara M, Domen K, Fukuda K, Ota K (2005) Tantalum oxynitride for a novel cathode of PEFC. Electrochem Solid State Lett 8(4):A201–A203
8. Ohnishi R, Katayama M, Takanabe K, Kubota J, Domen K (2010) Niobium-based catalysts prepared by reactive radio-frequency magnetron sputtering and arc plasma methods as non-noble metal cathode catalysts for polymer electrolyte fuel cells. Electrochim Acta 55(19):5393–5400
9. Takagaki A, Takahashi Y, Yin F, Takanabe K, Kubota J, Domen K (2009) Highly dispersed niobium catalyst on carbon black by polymerized complex method as PEFC cathode catalyst. J Electrochem Soc 156(7):B811–B815
10. Proietti E, Jaouen F, Lefèvre M, Larouche N, Tian J, Herranz J, Dodelet J-P (2011) Iron-based cathode catalyst with enhanced power density in polymer electrolyte membrane fuel cells. Nat Commun 2:416–424
11. Wu G, More KL, Johnston CM, Zelenay P (2011) High-performance electrocatalysts for oxygen reduction derived from polyaniline, iron and cobalt. Science 332:443–447
12. Jaouen F, Herranz J, Lefèvre M, Dodelet J-P, Kramm UI, Herrmann I, Bogdanoff P, Maruyama J, Nagaoka T, Garsuch A, Dahn JR, Olson T, Pylypenko S, Atanassov P, Ustinov EA (2009) Cross-laboratory experimental study of non-noble-metal electrocatalysts for the oxygen reduction reaction. ACS Appl Mater Interfaces 1(8):1623–1639
13. Jaouen F, Proietti E, Lefèvre M, Chenitz R, Dodelet J-P, Wu G, Chung HT, Johnston CM, Zelenay P (2011) Recent advances in non-precious metal catalysis for oxygen-reduction reaction in polymer electrolyte fuel cells. Energy Environ Sci 4:114–130
14. Chen Z, Higgins D, Yu A, Zhang L, Zhang J (2011) A review on non-precious metal electrocatalysts for PEM fuel cells. Energy Environ Sci 4:3167–3192
15. Matter PH, Biddinger EJ, Ozkan US (2007) Non-precious metal oxygen reduction catalysts for PEM fuel cells. Catalysis 20:338–366
16. Dodelet J-P (2006) Oxygen reduction in PEM fuel cell conditions: heat-treated non-precious metal-N4 macrocycles and beyond. In: N4 macrocyclic metal complexes. Springer, New York, pp 83–148
17. Kramm UI, Bogdanoff P, Fiechter S (2013) Non-noble metal catalysts for the oxygen reduction in polymer electrolyte membrane fuel cells (PEM-FC). In: Encyclopedia of sustainable science and technology. Springer, New York, pp. 8265–8307
18. Jasinski R (1964) A new fuel cell cathode catalyst. Nature 201:1212–1213

19. Jahnke H, Schönborn M, Zimmermann G (1976) Organic dye stuffs as catalysts for fuel cells. Topics Curr Chem 61:133–182
20. van Veen JAR, van Baar JF, Kroese KJ (1981) Effect of heat treatment on the performance of carbon-supported transition-metal chelates in the electrochemical reduction of oxygen. J Chem Soc Faraday Trans 77:2827–2843
21. van Veen JAR, Colijn HA, van Baar JF (1988) On the effect of a heat treatment on the structure of carbon-supported metalloporphyrins and phthalocyanines. Electrochim Acta 33(6):801–804
22. Kramm UI, Abs-Wurmbach I, Herrmann-Geppert I, Radnik J, Fiechter S, Bogdanoff P (2011) Influence of the electron-density of FeN4-centers towards the catalytic activity of pyrolyzed FeTMPPCl-based ORR-electrocatalysts. J Electrochem Soc 158(1): B69–B78
23. Johansson LY, Larsson R (1986) Electrochemical reduction of oxygen in sulphuric acid catalyzed by porphyrin-like complexes. J Mol Catal 38:61–70
24. Gupta S, Trzk D, Bae I, Aldred W, Yeager E (1989) Heat-treated polyacrylonitrile-based catalysts for oxygen electroreduction. J Appl Electrochem 19:19–27
25. Bogdanoff P, Herrmann I, Hilgendorff M, Dorbandt I, Fiechter S, Tributsch H (2004) Probing structural effects of pyrolysed CoTMPP-based electrocatalysts for oxygen reduction via new preparation strategies. J New Mater Electrochem Syst 7:85–92
26. Kramm UI (née Koslowski), Herrmann I, Fiechter S, Zehl G, Zizak I, Abs-Wurmbach I, Radnik J, Dorbandt I, Bogdanoff P (2009) On the influence of sulphur on the pyrolysis process of FeTMPP-Cl-based electro-catalysts with respect to oxygen reduction reaction (ORR) in acidic media. ECS Trans 25:659–670
27. Herrmann I, Kramm UI, Radnik J, Fiechter S, Bogdanoff P (2009) Influence of sulfur on the pyrolysis of CoTMPP as electrocatalyst for the oxygen reduction reaction. J Electrochem Soc 156(10): B1283–B1292
28. Herrmann I, Kramm UI, Fiechter S, Bogdanoff P (2009) Oxalate supported pyrolysis of CoTMPP as electrocatalysts for the oxygen reduction reaction. Electrochim Acta 54(18):4275–4287
29. Wu G, Johnston CM, Mack NH, Artyushkova K, Ferrandon M, Nelson M, Lezama-Pacheco JS, Conradson SD, More KL, Myers DJ, Zelenay P (2011) Synthesis – structure – performance correlation for polyaniline–Me–C non-precious metal cathode catalysts for oxygen reduction in fuel cells. J Mater Chem 21:11392–11405
30. Bouwkamp-Wijnoltz AL, Visscher W, van Veen JAR, Boellaard E, van der Kraan AM, Tang SC (2002) On active-site heterogeneity in pyrolyzed carbon-supported Iron porphyrin catalysts for the electrochemical reduction of oxygen: an in situ Mössbauer study. J Phys Chem B 106:12993–13001
31. Kramm UI, Herranz J, Larouche N, Arruda TM, Lefèvre M, Bogdanoff P, Fiechter S (2012) Structure of the catalytic sites in Fe/N/C-catalysts for O_2-reduction in PEM fuel cells. Phys Chem Chem Phys 14:11673–11688
32. Koslowski UI, Herrmann I, Bogdanoff P, Barkschat C, Fiechter S, Iwata N, Takahashi H, Nishikori H (2008) Evaluation and analysis of PEM-FC performance using non-platinum cathode catalysts based on pyrolysed Fe- and Co-porphyrins – influence of a secondary heat-treatment. ECS Trans 13(17):125–141
33. Olson TS, Chapman K, Atanassov P (2008) Non-platinum cathode catalyst layer composition for single membrane electrode assembly proton exchange membrane fuel cell. J Power Density 183:557–563
34. Koslowski UI, Abs-Wurmbach I, Fiechter S, Bogdanoff P (2008) Nature of the catalytic centers of porphyrin-based electrocatalysts for the ORR: a correlation of kinetic current density with the site density of Fe–N4 centers. J Phys Chem C 112(39):15356–15366
35. Kramm UI, Herrmann-Geppert I, Bogdanoff P, Fiechter S (2011) Effect of an ammonia treatment on structure, composition, and oxygen reduction reaction activity of Fe-N-C catalysts. J Phys Chem C 115:23417–23427
36. Jaouen F, Lefèvre M, Dodelet J-P, Cai M (2006) Heat-treated Fe/N/C catalysts for O_2 electroreduction: are active sites hosted in micropores? J Phys Chem B 110:5553–5558
37. Charreteur F, Jaouen F, Ruggeri S, Dodelet J-P (2008) Fe/N/C non-precious catalysts for PEM fuel cells: influence of the structural parameters of pristine commercial carbon blacks on their activity for oxygen reduction. Electrochim Acta 53:2925–2938
38. Ferrandon M, Kropf AJ, Myers DJ, Artyushkova K, Kramm U, Bogdanoff P, Wu G, Johnston CM, Zelenay P (2012) Multitechnique characterization of a polyaniline-iron-carbon oxygen reduction catalyst. J Phys Chem C 116(30):16001–16013
39. Herranz J, Jaouen F, Lefèvre M, Kramm UI, Proietti E, Dodelet J-P, Bogdanoff P, Fiechter S, Abs-Wurmbach I, Bertrand P, Arruda TM, Mukerjee S (2011) Unveiling N-protonation and anion-binding effects on Fe/N/C catalysts for O_2 reduction in proton-exchange-membrane fuel cells. J Phys Chem C 115:16087–16097
40. Tributsch H, Koslowski UI, Dorbandt I (2008) Experimental and theoretical modeling of Fe-, Co-, Cu-, Mn-based electrocatalysts for oxygen reduction. Electrochim Acta 53:2198–2209
41. Kramm UI, Lefèvre M, Herrmann-Geppert I, Bogdanoff P, Dodelet J-P, Fiechter S (2012) Fe-N-C catalysts – investigating the degradation induced by PEM fuel cell vs. room temperature conditions. In: ISE conference, Prague, no 2009

42. Jaouen F, Dodelet J-P (2007) Average turn-over frequency of O_2 electro-reduction for Fe/N/C and Co/N/C catalysts in PEFCs. Electrochim Acta 52:5975–5984
43. Lalande G, Côte R, Guay D, Dodelet JP, Weng LT, Bertrand P (1997) Is nitrogen important in the formulation of Fe-based catalysts for oxygen reduction in solid polymer fuel cells? Electrochim Acta 42(9):1379–1388
44. Lefèvre M, Proietti E, Jaouen F, Dodelet J-P (2009) Iron-based catalysts with improved oxygen reduction activity in polymer electrolyte fuel cells. Science 324:71–74
45. Ettingshausen F, Weidner A, Zils S, Wolz A, Suffner J, Michel M, Roth C (2009) Alternative support materials for fuel cell catalysts. ECS Trans 25(1):1883–1892
46. Herrmann I, Koslowski UI, Radnik J, Fiechter S, Bogdanoff P (2008) Preparation and structural analysis of heat treated Co-and Fe-porphyrins as cathode catalysts for the oxygen reduction reaction. ECS Trans 13(17):143–160

Fuel Cells, Principles and Thermodynamics

Ken-ichiro Ota
Yokohama National University, Fac. Engineering, Yokohama, Japan

Introduction

Through fuel cells we can convert chemical energy to electrical energy directly where fuel and oxidant are supplied from the outside of a cell. Fuel cells are the energy conversion systems rather than the energy storage devices such as primary or secondary batteries. Fuel cell was invented by Schoenbein [1] or Sir William Grove [2] in 1939. This invention was before those of a lead acid battery and a manganese dry cell.

From the early stage of the fuel cell development, several types have been developed. A polymer electrolyte fuel cell (PEFC or PEMFC), a phosphoric acid fuel cell (PAFC), a molten carbonate fuel cell (MCFC), a solid oxide fuel cell (SOFC), and an alkaline fuel cell (AFC) are the fuel cells which use principally hydrogen as a fuel. The difference is the electrolyte. Alcohols, ethers, and hydrides, including hydrazine besides hydrogen, can be used directly for a fuel cell system. A methanol fuel cell is called a direct methanol fuel cell (DMFC).

After the invention of a novel membrane electrolyte, the output power density of a PEFC has improved, and fuel cells can be applied to electric vehicles, stationary power sources, and mobile applications such as laptop computers and cellular phones. The high efficiency of the energy conversion of stationary fuel cells would be an important feature considering the limited resources of fossil fuels and their global warming effect.

In this paper the principle and characteristics of fuel cells will be stated from the thermochemical point of view. The future prospects of fuel cells will be also considered especially related to the sustainable growth of human beings.

Principle of Fuel Cell

Fuel cells can produce electricity and heat simultaneously through an electrochemical reaction using a fuel and an oxidant. Fuel cells can produce electric energy through an electrochemical system composed of two electrodes (anode and cathode) and an electrolyte. Electrodes are electronic conductors and should have a good electrochemical catalytic activity for the oxidation of fuels and/or the reduction of oxidants. An electrolyte is an ionic conductor.

Normally, hydrogen is used for a fuel and air (oxygen) is used for an oxidant. When a fossil fuel is used as a primary fuel, it has to be converted to hydrogen through a steam reforming or a partial oxidation. Since hydrocarbons are not electrochemically active at room temperature, they have to change to hydrogen for fuel cells. In a fuel cell system, an oxidation reaction of a fuel (hydrogen) takes place at the anode and a reduction reaction of an oxidant (oxygen, air) takes place at the cathode that is shown in Fig. 1.

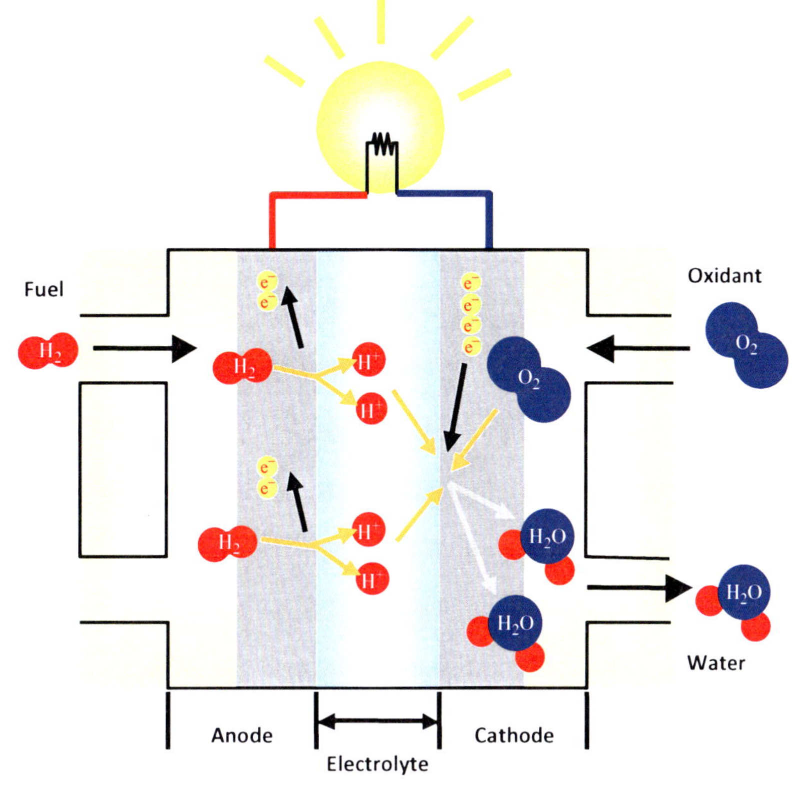

Fuel Cells, Principles and Thermodynamics, Fig. 1 Schematic drawing of fuel cell

The electrochemical reactions which take place at an anode and a cathode for hydrogen–oxygen can be expressed as follows:

Anode (hydrogen): $H_2 = > 2H^+ + 2e^-$

Cathode (oxygen): $1/2O_2 + 2H^+ + 2e^- = > H_2O$

Total: $H_2 + 1/2O_2 = > H_2O$

The electrons produced through the oxidation reaction of hydrogen (fuel) at an anode can transfer to a cathode via the outside circuit. The oxidation reaction takes place at the anode and the reduction reaction takes place at the cathode. These are the important features of an electrochemical system.

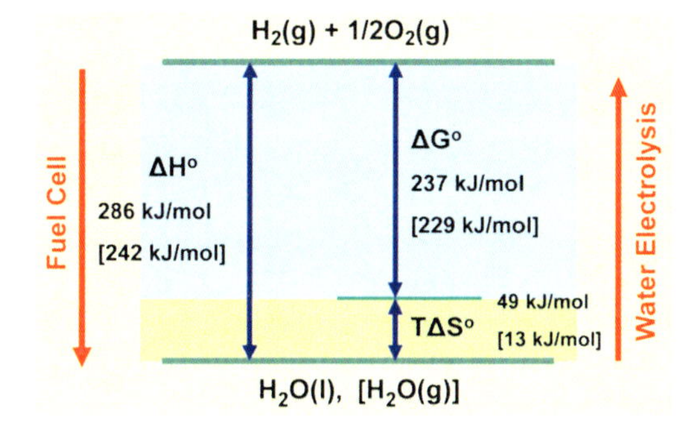

Fuel Cells, Principles and Thermodynamics, Fig. 2 Energy conversion of fuel cell reaction

Figure 2 shows the energy change of the water formation reaction at 25 °C. The total energy change is expressed in ΔH which is –286 kJ/mol when the formed H_2O is liquid water (HHV). If water vapor is formed, the total energy changed to −242 kJ/mol (LHV). In this paper the HHV value is used unless otherwise expressed.

The total energy can be divided into work and heat. The work which is expressed in ΔG is −237 kJ/mol. This value is the theoretical amount of the electrical energy that can be obtained through the water formation reaction at 25 °C. The theoretical energy conversion efficiency of a fuel cell can be calculated by ΔG/ΔH. This value becomes 83 % for hydrogen–oxygen fuel cell at 25 °C. No mechanical energy conversion systems such as an internal

Fuel Cells, Principles and Thermodynamics, Table 1 Oxidation reaction of fuels (25 °C)

Fuel	Reaction	ΔH° (kJ/mol)	ΔG° (kJ/mol)	EMF (V)	Eff. (%)
Hydrogen	$H_2(g) + 1/2O_2(g) => H_2O(I)$	−286	−237	1.23	83
Methane	$CH_4 (g) + 2 O_2(g) => CO_2(g) + 2 H_2O(I)$	−890	−817	1.06	92
CO	$CO(g) + 1/2 O_2(g) => CO_2(g)$	−283	−257	1.33	91
Carbon	$C(s) + O_2(g) => CO_2(g)$	−394	−394	1.02	100
Methanol	$CH_3OH(I) + 1/2O_2(g) => CO_2(g) + 2H_2O(I)$	−727	−703	1.21	97
Ethanol	$C_2H_5OH(I) + 3O_2(g) => 2CO_2(g) + 3H_2O(I)$	−1,367	−1,325	1.18	96
Hydrazine	$N_2H_4(I) + O_2(g) => N_2(g) + 2H_2O(I)$	−622	−623	1.61	100
Ammonia	$NH_3(g) + 4/3O_2(g) => 3/2H_2O(I) + 1/2 N_2(g)$	−383	−339	1.17	89
DME	$CH_3OCH_3(g) + 3O_2(g) => 2CO_2(g) + 3H_2O(I)$	−1,460	−1,390	1.20	95

combustion engine can reach this high efficiency. Furthermore the efficiency of a mechanical system is zero at room temperature because of the Carnot limit.

The water formation reaction is an exothermic reaction where the reversible heat of 49 kJ/mol is formed at 25 °C. Considering the exothermic reaction of the water formation reaction, the higher temperature is not suitable theoretically in order to get a high efficiency of electrical energy.

Hydrocarbons and alcohols including carbon can be fuels for a fuel cell system theoretically. Table 1 shows the characteristics of the oxidation reactions of these fuels at 25 °C.

The theoretical voltages are also shown. The voltages of all these reactions are a little bit more than 1 V and not much different among these reactions. This is a nature of chemical reactions. The theoretical efficiency of hydrogen is the lowest among these fuels. The theoretical efficiency of hydrazine is over 100 %. The efficiency exceeds over 100 %, if the reaction is endothermic (or $T\Delta S > 0$).

These efficiencies are theoretical and obtained from the thermochemical data. In the actual applications, the reaction rates of fuels are important to get usable energy. In this point hydrogen is far more reactive compared to other fuels. It is very difficult to get a detectable reaction rate for methane, carbon monoxide, and carbon at room temperature. Although methanol can be used for a fuel of a fuel cell, the oxidation rate is far smaller compared to that of hydrogen.

Ethanol is considered to be a CO_2-free fuel when it is obtained from a biomass. The electrochemical activity of ethanol is almost the same as that of methanol. However, ethanol has a C–C bond and the electrochemical breaking of the C–C bond is not easy at room temperature. Ethanol can be oxidized theoretically at high efficiency, but the reaction rate is very slow.

In order to utilize the high conversion efficiency of these carbon-containing fuels, the development of good catalysts is most important. At present platinum-based catalysts are used for the oxidations of these fuels in an acidic electrolyte. However, the catalytic activity is not enough except for hydrogen. Steam reforming of carbon-containing fuels (e.g., natural gas and alcohols) and their conversion into hydrogen helps to improve the system efficiency.

The theoretical efficiency of a fuel cell depends on the operating temperature. Since the water formation reaction is an exothermic reaction, the ΔG value decreased at higher temperatures. In other words the theoretical efficiency of hydrogen–oxygen fuel cell decreased at higher temperatures.

Figure 3 shows the temperature dependence of the theoretical efficiency of a hydrogen–oxygen fuel cell. The Carnot efficiency which is a theoretical efficiency of a mechanical energy conversion is also shown for comparison. The efficiency of a fuel cell is much higher than the Carnot efficiency at low temperature. However, the Carnot efficiency increased very much at higher temperature. At the temperature of higher than 1,000 K, the Carnot efficiency is higher than the theoretical efficiency of a fuel cell. In another words, a fuel

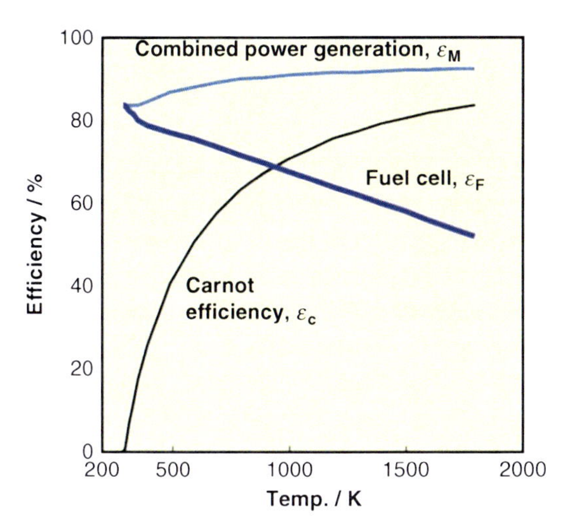

Fuel Cells, Principles and Thermodynamics, Fig. 3 Theoretical efficiency of hydrogen–oxygen fuel cell and Carnot efficiency

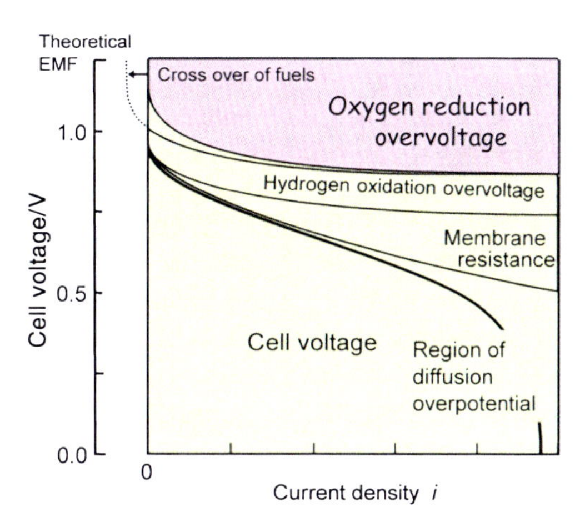

Fuel Cells, Principles and Thermodynamics, Fig. 4 Typical current–voltage relation of fuel cell

cell should be operated at low temperatures in order to get a high energy efficiency.

The electric power of a fuel cell can be expressed as follows:

[Electrical Power] = [Voltage] [Current]

[Electrical Energy] = [Voltage] [Electric Quantity]

[Electrical Energy] = [Voltage] [Current] [Time]

Since the output voltage of a fuel cell is at most 1 V, a high output current is required to get a high power. Figure 4 shows a typical current–voltage relation of a fuel cell.

The output voltage decreased as the current increased owing to the voltage loss due to the internal resistance. The internal resistance is mainly from the reaction resistance of electrodes and the membrane resistance. The resistance of the oxygen electrode causes the largest voltage loss in general. A PEFC and a DMFC do not show a theoretical voltage at the open-circuit potential. This is because a platinum electrode does not have an enough catalytic activity at room temperature for the oxygen reduction and the methanol oxidation. If a suitable catalyst is developed for these reactions, the efficiency of the fuel cell would be improved further.

A fuel cell has the following characteristics:

1. Theoretical efficiency is very high especially at low temperatures.
2. The output voltage of a unit cell is 1 V or less. In order to get a high power, a large current is needed. In another words, the large quantity of reactants should react at the electrode/electrolyte interface smoothly.
3. The system efficiency does not depend on the size of the system. The fuel cell system is suitable for the smaller size power generation system compared to engines and turbine systems.
4. A fuel cell is an environmentally friendly system with no NOx emission, no noise, and no vibration.

Types of Fuel Cells

Several types of fuel cells are developing. AFC, PEFC, and PAFC are the low-temperature fuel cells using hydrogen as a fuel. [3–5]

An AFC has good characteristics for a fuel cell. It can operate without platinum and its components are flexible. Many metallic materials can be used in alkaline media. However, it has to be protected against CO_2. Even CO_2 in air causes the damage of electrolyte. In a potential hydrogen economy, where pure hydrogen will be got more easily, an AFC will have a more interest. This fuel cell is the cheapest system at present.

A PAFC is commercially available in 100–400 kW size at this time. It can operate at

200 °C with the efficiency of 38 % in electricity. The proved lifetime is more than 8 years that is the longest among the fuel cells.

An MCFC operates at 650 °C using alkaline carbonate eutectic melt (Li/K or Li/Na) as electrolyte. The internal reforming is possible by using the high operating temperature. A special feature of an MCFC is the separation or condensation of CO_2 during its operation, since the migration species in the electrolyte is the carbonate ion. It can operate at more than 0.85 V which is the highest operation voltage among fuel cells. It might have the highest efficiency.

There are two types of SOFC: planar type and tubular type. An SOFC operates at 800 °C or higher temperatures. Although it can operate with high power density due to its high temperature, the stability of the component materials should be checked for a longtime operation.

A PEFC is developing in many countries for many applications. It was originally developed for the space use (Gemini Program in the 1960s). The hydrocarbon membrane which was used for the Gemini, however, was not stable. A stable fluoride resin membrane (Nafion®) was invented soon after this. In1980s this membrane proved a high power density for a PEFC system that showed the possibility of a vehicle application. Once a fuel cell is applied to commercial vehicles, its market is huge. PEFCs would be applied for vehicles, stationary uses, and power sources for small mobile applications.

Future Directions

Fuel cells can be used for stationary applications and mobile applications including vehicles.

Fuel cells for stationary distributed power source might be important for the future energy system where very high conversion efficiency would need to be achieved using electric energy and heat simultaneously. MCFCs might be suitable for the size from 200 kW to 2,000 kW since they have proved the highest efficiency and the high-quality heat. For smaller size fuel cells (less than 10 kW) PEFCs might be suitable since they can operate easily at room temperature.

Fuel cell vehicles which can be operated with hydrogen are the most suitable vehicles for carbon dioxide-free transportation considering the future problem of global warming. A high power density has been achieved for PEFC vehicles. The durability has improved in recent years. The cost and the amounts of resources including platinum which are used in PEFCs might be problems for the wide application of fuel cells.

The applications of fuel cells to small portable devices such as laptop computers and cellular phones are rather new applications considering the long history of fuel cell development. The high energy density of a fuel cell system is an important feature. Table 2 shows the electrochemical energy density of fuels at 25 °C. Hydrogen has high weight density. However, its volume density is small. Metal hydrides have relatively high volume density.

Fuel Cells, Principles and Thermodynamics, Table 2 Electromechanical energy density of fuels[a]

Fuel	Volume density [kJ/ml]	Weight density [kJ/g]
H_2(liquid)	8.3	117.6
$LaNi_5H_6$	10.6	1.6
CH_3OH	17.5	22.1
C_2H_5OH	24.9	31.5
N_2H_4	19.9	19.7
Li	21.8	40.8
Zn	34.3	4.8

[a]Calculated from ΔG of oxidation by O_2

Conclusion

Considering the world climate change due to a large consumption of fossil fuels, we have to move to a clean energy system as soon as possible. The hydrogen energy is a clean energy system [6]. If hydrogen is obtained from water using renewable energies, we can reach an ultimate clean energy system. Fuel cells would be the most suitable device to use hydrogen since it has high conversion efficiency. In order to create an ultimate clean energy system for the sustainable growth of human beings, the hydrogen-fuel cell system is inevitable and should be promoted.

However, its weight density is small. Alcohols, namely, methanol and ethanol, have relatively high volume density and weight density. Considering their reaction rate, methanol as well as hydrogen can be used for micro fuel cell applications. In future, ethanol might be used directly for these applications if a good catalyst for ethanol oxidation had been developed.

Cross-References

- ▶ Alkaline Membrane Fuel Cells
- ▶ Direct Alcohol Fuel Cells (DAFCs)
- ▶ Molten Carbonate Fuel Cells, Overview
- ▶ Polymer Electrolyte Fuel Cells (PEFCs), Introduction
- ▶ Solid Oxide Fuel Cells, Introduction

References

1. Schoenbein CF (1839) Philosophical magazine. p 43
2. Grove WR (1939) Philosophical magazine. p 129
3. Vielstich W, Yokokawa H, Gasteiger H (eds) (2009) Handbook of fuel cells, vol 1–5. Wiley, New York
4. Srinivasan S (ed) (2006) Fuel cells. Springer, Boston
5. Blomen LJMJ, Mugerwa MN (eds) (1993) Fuel cell systems. Plenum Press, New York
6. Winter C, Nitsch J (eds) (1988) Hydrogen as an energy carrier. Springer, New York

G

Galvanostat

Manuel Lohrengel
University of Düsseldorf, Düsseldorf, Germany

Introduction

Before 1900, electrochemical experiments focused on thermodynamic aspects and, thus, were reduced to potential measurements of systems, which were stationary or in equilibrium. But with time, scientists became interested in kinetics, i.e., systems away from equilibrium [1]. This was often realized by constant current (galvanostatic) experiments, which were for two reasons advantageous: they were easily realized and guaranteed a constant reaction rate, which was relevant in some cases. Moreover, time dependent reactions could be monitored, if the potential was recorded versus time. These "charging curves" [2] were the main technique to follow electrode kinetics up to the 60s of the last century.

Potentiostats were much later developed. The first completely electronic device was presented by Hickling in 1942 [3], and from 1960 most galvanostatic experiments were substituted by potentiostatic sweep techniques.

Basic Circuits

Modern concepts of galvanostats or potentiostats are based on special amplifiers, so-called operational amplifiers [4, 5], which became available as compact devices or integrated circuits after 1950. Galvanostatic experiments, however, could be easily realized by a number of batteries, connected in series, or by a valve tube amplifier (Fig. 1). These power sources delivered constant output voltages U between 50 and 300 V. The cell and adjustable resistances were connected in series. The resulting current I_c is given by

$$I_c = \frac{U - U_{\text{cell}}}{R_c} \approx \frac{U}{R_c} \qquad (1)$$

as long as the cell voltage U_{cell} is much smaller than U. This is true for elements such as Fe, Cr, Ni, or noble metals, where the cell voltage remains smaller than 3 V. Disadvantageous is the low power efficiency. Most of the power is lost in the resistors resulting in an intense heat production at larger current densities.

An impedance buffer (IB) is added in common experiments. It is necessary to monitor the potential, as reference electrodes are sensitive to current flow due to their large source resistance (typically 10 kΩ–100 MΩ).

Potentiostats [cross-reference] can be converted into galvanostats by adding a resistor between reference electrode input of the potentiostat and ground (Fig. 2). The cell is connected between counter electrode output and reference electrode input. Potential monitoring requires a special differential amplifier (IA in Fig. 2, instrumentation amplifier; see Fig. 3).

G. Kreysa et al. (eds.), *Encyclopedia of Applied Electrochemistry*, DOI 10.1007/978-1-4419-6996-5,

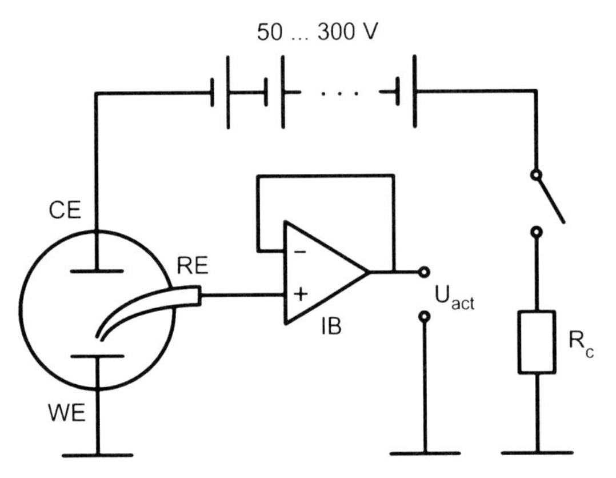

Galvanostat, Fig. 1 Galvanostatic circuit, based on a constant voltage source. An impedance buffer (*IB*) is added to monitor the potential

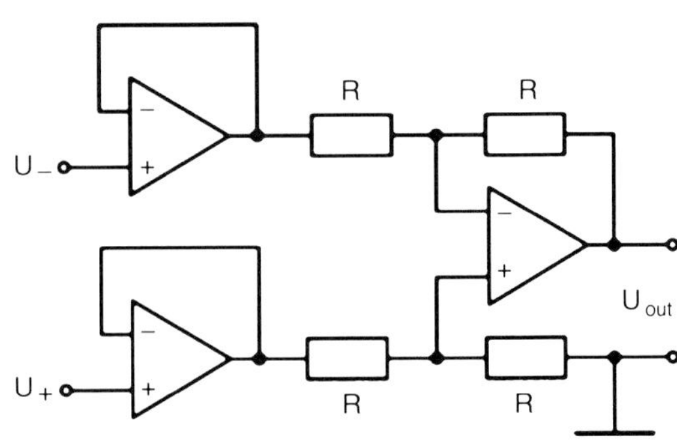

Galvanostat, Fig. 3 Instrumentation amplifier consisting of a differential amplifier and two impedance buffers to minimize input currents

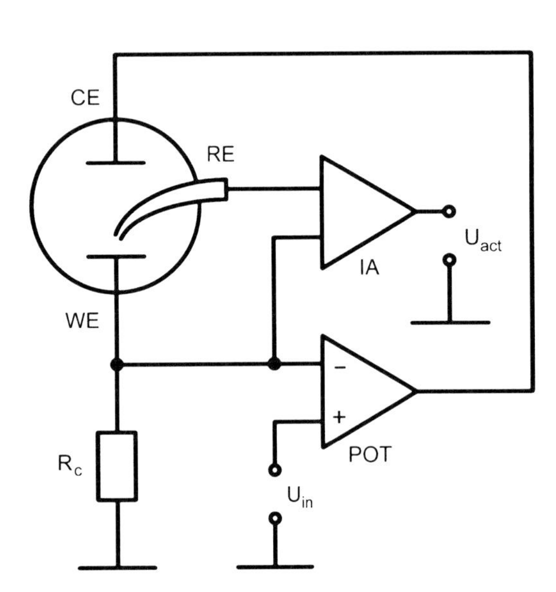

Galvanostat, Fig. 2 Galvanostat concept using a potentiostat

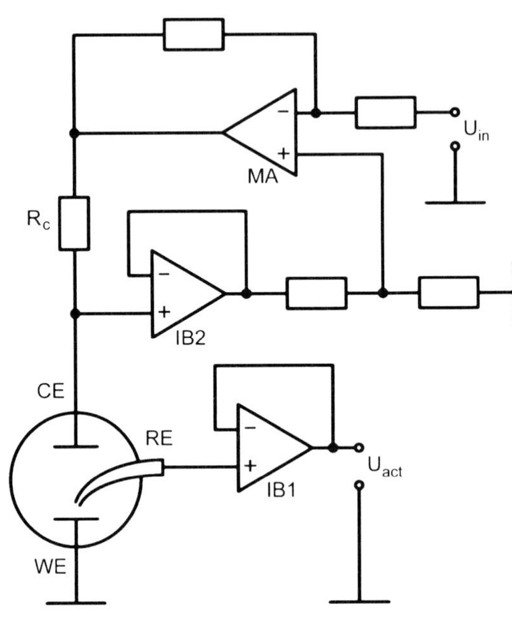

Galvanostat, Fig. 4 Galvanostat with main amplifier MA. The working electrode is connected to ground

The resulting cell current I_c is given by

$$I_c = \frac{U_{in}}{R_c} \quad (2)$$

The sign of I_c depends on the concept of the potentiostat, as some devices internally invert U_{in}.

A more convenient galvanostat concept, where the working electrode is connected to ground, is presented in Fig. 4. The resulting cell current I_c is delivered by the main amplifier MA and given by

$$I_c = \frac{U_{in}}{R_c} \quad (3)$$

Again, an impedance buffer (IB1) is used to monitor the potential. A second impedance buffer (IB2) is necessary to avoid errors by current flow across the resistors *R*.

Applications

Up to 1970, galvanostatic charging curves were the most common dynamic technique. In case of reversible processes, such as oxide formation on noble metals, it was possible to start with an anodic current and then switch to a cathodic current, somewhat similar to a cyclovoltammogram. In fact, a cyclovoltammogram looks like the first derivative of a galvanostatic charging curve, provided that the processes are less time dependant, as

$$I = \frac{\partial Q}{\partial t} \qquad (4)$$

Moreover, derivatives are more sensitive to noise.

The charge is simply given by the time interval in charging curves:

$$\Delta Q = \int I dt = I_{\text{const}} \cdot \Delta t \qquad (5)$$

Charge determination in cyclovoltammograms means more efforts such as graphic, numeric, or electronic integration. After 1960, however, when potentiostats, potential sweep generators, electronic integrators, and laboratory computers became available, the number of galvanostatic experiments decreased rapidly.

Future Directions

The quality of galvanostat systems is nowadays almost perfect and allows, supported by computer and software, effective and convenient experiments. A dominance of potentiostatic experiments must be stated, but galvanostatic experiments may be advantageous in fast pulse applications, as constant current control requires no reference electrode. The inner resistance of reference electrodes R_{RE} may be large (>1 MΩ), which means, together with an input capacity of the potentiostat C_{in}, a time constant

$$\tau = R_{\text{RE}} \cdot C_{\text{in}} \qquad (6)$$

and, therefore, time delays of some 10 μs. Modern galvanostats have settling times down to some ns. Moreover, current control is not affected by the iRdrop (potential drop across the electrolyte resistance between reference and working electrode) in experiments with large currents or electrolytes of low conductivity.

Cross-References

► iR-Drop Elimination
► Potentiostat
► Reference Electrodes

References

1. Tafel J (1905) Polarisation during cathodic Hydrogen development. Z physik Chem 50:641–712
2. Vetter KJ (1967) Electrochemical kinetics theoretical and experimental aspects. Academic, New York
3. Hickling A (1942) Studies in electrode polarisation Part IV-The automatic control of the potential of a working electrode. Trans Faraday Soc 38:27–33
4. Bard AJ, Faulkner LR (1980) Electrochemical methods: fundamentals and applications. Wiley, New York
5. Macdonald DD (1977) Transient techniques in electrochemistry. Plenum, New York

Gas Solubility of Electrolytes

Vincent S. Craig
Department of Applied Mathematics, Research School of Physical Sciences and Engineering, Australian National University, Canberra, ACT, Australia

Introduction

It is well known that dissolved electrolytes influence the solubility of gases in aqueous solution. This is generally called the salting-out effect if the solubility of the gas is decreased relative to pure water as is most often the case.

Gas Solubility in Water

Gas solubility has been measured by a variety of means, for a wide variety of purposes. As a consequence of this, there exist at least ten expressions for the solubility of a gas in a liquid [5]. This is problematic as it is not always simple to convert between units. The units that are most relevant to expressing the influence of electrolytes on gas solubility are the Bunsen Coefficient (α), the mole fraction (X_2), and the Henry's law constant (K_H,).

The Bunsen coefficient is unitless and is defined as the volume of pure gas (adjusted to 1 atm. and 273.15 K) V_g, adsorbed by a unit volume of the pure solvent, V_s, measured at a given temperature and a total pressure equivalent to the sum of the partial pressure of the solvent and the gas at that temperature.

$$a = (V_g/V_s) \quad (1)$$

For oxygen in water at 298.15 K and $p_{O2} = 101.325$ kPa, $\alpha = 0.02847$.

The mole fraction (in the absence of electrolyte) is defined as the ratio of the number of molecules of the dissolved gas (n_2) to the total number of molecules of dissolved gas and solvent (n_1).

$$X_2 = n_2/(n_2 + n_1) \quad (2)$$

For oxygen in water at 298.15 K and $p_{O2} = 101.325$ kPa, $X_2 = 2.295 \times 10^{-5}$.

The most common form of Henry's law is

$$p = k_H c \quad (3)$$

where c is the concentration of gas in solution in mol/L. Note that a variety of Henry's law constants are in use, and it is therefore necessary to check that the appropriate units are employed. Further, Henry's law is a limiting law that only applies to sufficiently dilute solutions. This condition is usually met when dealing with gas solubilities in water at atmospheric pressures.

For oxygen in water at 298.15 K and $p_{O2} = 101.325$ kPa,

$$k_H = 769.23\,\text{L . atm . mol}^{-1}$$

When oxygen is the gas of interest, the preferred method for determining the solubility is known as the Winkler titration (or modifications thereof). KI and $MnCl_2$ is added to the solution which is made strongly basic, whereby $Mn(OH)_2$ is formed as a white precipitate. The O_2 in solution oxidizes this precipitate and forms a pinkish-brown manganese precipitate containing Mn^{3+} ions, which consists of either ($MnO(OH)_2$) or $Mn(OH)_3$. The precise nature of this precipitate is not important, rather the stoichiometry between the dissolved oxygen and the production of the Mn^{3+} ions. On acidification, the pinkish-brown precipitate dissolves and reduction of Mn^{3+} to Mn^{2+}converts iodide (I^-) to iodine (I_2), the amount of which is equivalent to the concentration of dissolved oxygen. The concentration of iodine is determined by titration with $Na_2S_2O_3$ and the end point is determined spectrophotometrically. The error in this determination is better than 2 % and can be as low as 0.1 %. This technique is most easily applied to a batch process and is useful in field studies as the oxygen in the sample can be fixed by completing the steps up to and including the acidification in the field and the final titration performed later. Many other methods have been developed, some of which give rise to continuous determination of O_2 concentrations. The relative unreactive nature of N_2 demands that N_2 gas concentrations are determined by physical methods rather than chemical methods as is the case with the noble gases.

The solubility of gases in electrolyte solutions is usually expressed in a form that is relative to the solubility of the same gas in pure water under the same conditions of temperature and pressure. For oxygen and nitrogen, there exist useful empirical equations for calculating the solubility of gas, expressed as a mole fraction, at atmospheric pressure over a given temperature range.

For oxygen in water [5] and $273 < T < 373$

$$-\ln X_2 = 3.71814 + 5.59617 \times 10^3/T - 1.049668 \times 10^6/T^2 \quad (4)$$

For nitrogen in water [2] at $T < 350$ K where $\tau = T/100$ [2]

$$\ln X_2 = -67.38765 + 86.32129/\tau + 24.79808 \ln \tau \quad (5)$$

Solubility of Gases in Electrolyte Solutions

The Sechenov Coefficient

The first quantitative investigation of the influence of electrolytes on the solubility of dissolved gas was presented by Sechenov, (also known as Setschenow or Setchenov, see entry on "Salting-In and Salting-Out") in the late nineteenth century. He proposed an empirical relation between the concentration of electrolyte and the gas solubility:

$$\log(\alpha_0/\alpha) = K\alpha\, c, \quad (6)$$

where α_0 and α are the solubility of the gas expressed as Bunsen coefficients, c is the concentration of electrolyte, and $K\alpha$ is the Sechenov coefficient which is a function of the electrolyte, the temperature, and the gas type. For low electrolyte concentrations, this equation is very accurate. For appreciable electrolyte concentrations, the equation is best modified to the form [9]:

$$\log(\alpha_0/\alpha) = K_I I, \quad (7)$$

where the concentration of the solution has been replaced by the ionic strength I.

$$I = \frac{1}{2}\sum c_i Z_i^2 \quad (8)$$

This equation holds for most systems [12] and is accurate at least up to 2 mol L^{-1}. The Sechenov coefficients for a wide range of electrolytes at a range of temperatures for different gases have been determined and tabulated. Therefore, the solubility of a gas in single electrolyte solutions can be found easily by referring to the appropriate coefficient. For mixtures of gases, the solubility can be determined to within ~1 % by considering the partial pressures of the gases and their solubilities. Acids are found to have significantly lower salting-out coefficients than salts, indicating that cations have a much stronger influence on gas solubility than the hydrogen ion. The lithium ion is found to be particularly effective. This is attributed to the strong hydration of lithium ions. It is found that pH and pressure up to many atmospheres have minimal effects on the Sechenov coefficients. Therefore, if the Sechenov coefficient is known for a given pressure, the solubility of gas at other pressures can be determined using Henry's law.

Finally, we note that the solubility of nitrogen at a partial pressure of 1 atm. in deuterated water has been found to increase by about 7–10 % relative to normal water [1].

Gas Solubility in Salt Mixtures and Sea Water

In practice, many solutions of interest contain more than one electrolyte and the salinity and composition may change with time. Therefore, it is desirable to have a scheme whereby the gas solubility can be predicted for these more complex electrolyte solutions. The most common approach employed is to evaluate a salting-out coefficient not for each salt present but for each ion present. Then by combining these coefficients appropriately, the gas solubility can be determined for any electrolyte solution.

This approach leads to a modification of the Sechenov equation to the form [9]

$$\log\left(\frac{\alpha_0}{\alpha}\right) = \sum_i H_i I_i = \frac{1}{2} c_s \sum_i H_i x_i z_i^2 \quad (9)$$

where I is the ionic strength of the individual ionic species present in solution and the H_i's are specific with respect to the different cations and anions and to the gas and depend upon temperature. In practice, these have to be determined from experimental data and it is not possible to uniquely determine these values in an absolute sense due to the requirement of electroneutrality. Therefore, a single ion needs to be assigned a reference value and this allows the value of H_i to be determined for other ions.

For a strong electrolyte that completely dissociates, this means that the Sechenov coefficient can be calculated from the single ion parameters H_i and the number, x_i, and the charge z_i of the ions of type i in the electrolyte. Thus:

$$k_\alpha = \frac{1}{2}\sum_i H_i x_i z_i^2 \tag{10}$$

Good agreement has been found between experimental and predicted oxygen solubilities using this approach for a large number of electrolytes, with the deviation between predicted and experimental Sechenov coefficients being $\pm$ 3.5 %.

The above approach is empirical. Thermodynamic models for describing solution behavior can also be employed to determine gas solubilities, and these models are amenable to the estimation of gas solubilities in multicomponent systems from sets of single salt data. The thermodynamic approach employed is known as the Pitzer species interaction model, and it is used to determine the activity coefficient of the gas from a summation of interaction terms with anions, cations, and neutral species [3, 10, 11]. These interaction parameters are determined empirically from solubility data in a range of electrolyte solutions and have been tabulated for a wide range of salts, permitting the solubility of oxygen to be determined in mixed electrolyte solutions over a wide range of temperature and concentrations.

An alternative theoretical approach to determining gas solubility utilizes the reference interaction site model theory. This theory has been applied for the noble gases and is found to give reasonable agreement to experimental data without any adjustable parameters or empirical fits to solubility data [8]. These calculations work best for Helium, and the error increases as the gas molecule increases in size. The theory is yet to be applied to diatomic gases.

Solubility Data

For the reader seeking useful compilations of solubility data, the following manuscripts are suggested. For oxygen, there are the reviews of Groisman et al. [5] and Battino [1] and the work of Lang et al. [9] and Millero et al. [10, 11], for Nitrogen the review of Battino [2], for Helium the work of Gerth [4] as well as studies of methane [7], carbon dioxide [6], and the noble gases [8]. This list is by no means exhaustive.

Future Directions

As theoretical understanding of electrolyte solutions improves, we can expect that the ability to predict gas solubilities over a range of concentrations and temperatures will improve and be less dependent upon empirical data.

Cross-References

- ▶ Ion Properties
- ▶ Salting-In and Salting-Out
- ▶ Thermodynamic Properties of Ionic Solutions - MSA and NRTL Models

References

1. Battino R, Rettich TR, Tominaga T (1983) The solubility of oxygen and ozone in liquids. J Phys Chem Ref Data 12:163–178
2. Battino R, Rettich TR, Tominaga T (1984) The solubility of nitrogen and air in liquids. J Phys Chem Ref Data 13:563–600
3. Clegg SL, Brimblecombe P (1990) The solubility and activity-coefficient of oxygen in salt-solutions and brines. Geochim Cosmochim Acta 54:3315–3328
4. Gerth WA (1983) Effects of dissolved electrolytes on the solubility and partial molar volume of helim in

water from 50 to 400 atmospheres at 25-degrees-C. J Solut Chem 12:655–669
5. Groisman AS, Khomutov NE (1990) Solubility of oxygen in electrolyte solutions. Uspekhi Khimii 59:1217–1250
6. Kiepe J, Horstmann S, Fischer K, Gmehling J (2002) Experimental determination and prediction of gas solubility data for CO_2 + H_2O mixtures containing NaCl or KCl at temperatures between 313 and 393 K and pressures up to 10 MPa. Ind Eng Chem Res 41:4393–4398
7. Kiepe J, Horstmann S, Fischer K, Gmehling J (2004) Experimental determination and prediction of gas solubility data for methane plus water solutions containing different monovalent electrolytes. Ind Eng Chem Res 43:3216–3216 (42:5392, 2004)
8. Kinoshita M, Hirata F (1997) Analysis of salt effects on solubility of noble gases in water using the reference interaction site model theory. J Chem Phys 106:5202–5215
9. Lang W, Zander R (1986) Salting-out of oxygen from aqueous-electrolyte solutions – prediction and measurement. Ind Eng Chem Fundam 25:775–782
10. Millero FJ, Huang F, Laferiere AL (2002) The solubility of oxygen in the major sea salts and their mixtures at 25 degrees C. Geochim Cosmochim Acta 66:2349–2359
11. Millero FJ, Huang F, Laferiere AL (2002) Solubility of oxygen in the major sea salts as a function of concentration and temperature. Mar Chem 78:217–230
12. Ueyama K, Hatanaka J (1982) Salt effect on solubility of non-electrolyte gases and liquids. Chem Eng Sci 37:790–792

Gas Titration with Solid Electrolytes

Jens Zosel[1] and Ulrich Guth[1,2]
[1]Kurt-Schwabe-Institut für Mess- und Sensortechnik e.V. Meinsberg, Waldheim, Germany
[2]FB Chemie und Lebensmittelchemie, Technische Universität Dresden, Dresden, Germany

Introduction

Definition

The physicochemical principles of the titration of gases with solid electrolytes (SE) are similar to those used in cells with liquid electrolytes. The gaseous reactant (*titrant*) is added to the analyte (*titrand*) in form of a gaseous mixture with known amount or flow or alternatively by SE pumping cells dosing primarily oxygen or hydrogen and changing the amount of given species in the titrand according to *Faraday's law*, also called *coulometry*. The titration end point can be detected principally with *potentiometric*, *amperometric*, or *conductometric* devices.

History

Early gas titrations based on solid electrolyte devices for end point detection were carried out by Möbius [1], using air as the gaseous titrant and a potentiometric cell made of *stabilized zirconia* for the end point detection. The setup is schematically outlined in Fig. 1.

The analyte gas (titrand) is guided through a heated and temperature-controlled SE tube at a controlled flow rate. Before entering the SE tube, air (titrant) is added to the titrand also at controlled flow rate. The equilibrium is established immediately at the catalytically highly active inner Pt electrode of the cell, and its potential is measured versus a Pt/air reference electrode. The accuracy of this method depends exclusively on the accuracy of the flow rate control. Some results of measurements on the combustibles city gas, hydrogen, and CO/CO_2-mixtures are given in Fig. 2. The deviation of measured potentials from the calculated curve of the mixture with 47.8 vol.-% CO below and above the stoichiometric point is related to the temperature drift powered by the exothermic reaction at the platinum electrode surface. An important precondition of this method consists in the establishment of a perfect mixture between titrant and titrand before the passage of the SE cell. The stoichiometric point as the target value of flow rate control is indicated as a filled circle with a horizontal bar in three of the curves in Fig. 2.

One of the first gas titrations by pumping ions through SE was carried out by Yuan and Kröger [2], who utilized stabilized zirconia as an oxygen pump. This device was combined later by Ullmann et al. with a potentiometric cell and applied inter alia for safety monitoring in nuclear

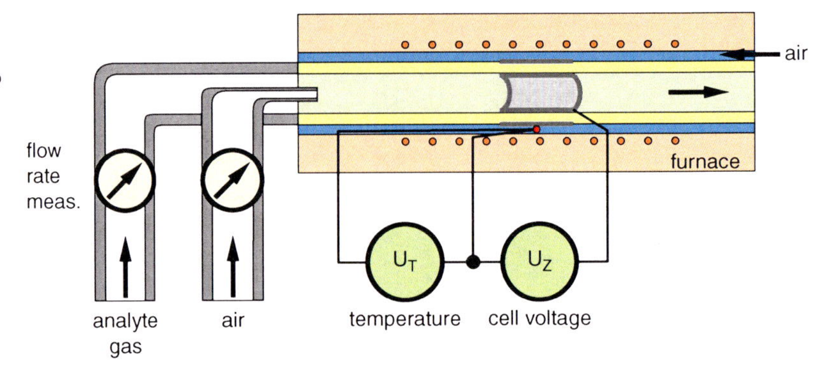

Gas Titration with Solid Electrolytes, Fig. 1 Experimental setup for potentiometric gas titration of combustibles with air [1]

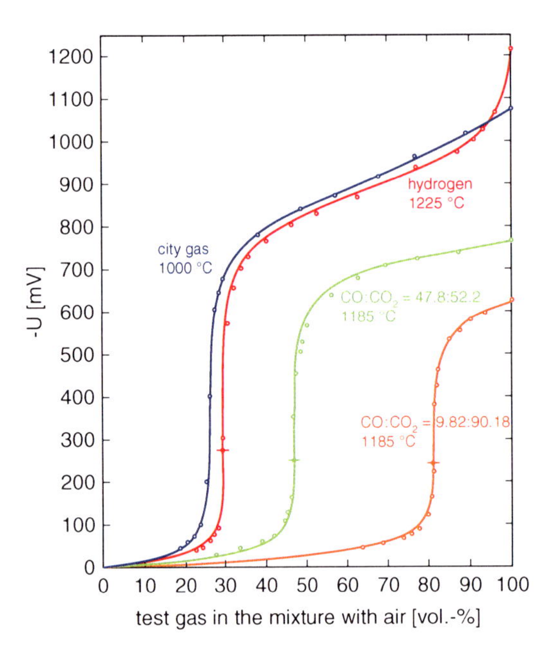

Gas Titration with Solid Electrolytes, Fig. 2 Results of the potentiometric gas titration with the setup given in Fig. 1, according to [1]

power plants [3]. A detailed review on gas titration cells using solid electrolytes is given in [4].

Fundamentals

The fundamentals of gas titration with solid electrolyte cells are described in detail within the entries *titration* and *coulometry*, and special aspects are also treated within *solid electrolyte*, *potentiometry*, and *amperometry*. Therefore, the focus is set here to the most important errors of gas titration with SE cells. These errors are related mainly to the peripheral parameters as well as to the solid electrolyte itself. If gas titration is used for concentration measurements for instance, it is essential to know the volume of the titrand containing gas or in case of continuous titration its actual flow rate. The measurement errors of these parameters are directly related to the resulting accuracy of the concentration measurement. If the gas titration is used to measure the amount of given species by using SE for pumping the titrant in form of ions, the electronic conductivity of the SE might be taken into account as another source of measurement error. According to [5], the electronic conductivities of electrons σ_e and defect electrons σ_h of yttria-stabilized zirconia (YSZ) can be calculated from temperature and oxygen partial pressure by Eqs. 1 and 2:

$$\sigma_e\left[\frac{\text{S}}{\text{cm}}\right] = 1.31 \cdot 10^7 \cdot \exp\left(-3.88eV/kT\right) \cdot [p(O_2)/\text{atm}]^{-1/4} \quad (1)$$

$$\sigma_h\left[\frac{\text{S}}{\text{cm}}\right] = 2.35 \cdot 10^2 \cdot \exp\left(-1.67eV/kT\right) \cdot [p(O_2)/\text{atm}]^{1/4} \quad (2)$$

with k as the Boltzmann constant ($1.38065 \cdot 10^{-23}$ J · K^{-1}), T as the absolute temperature, and $p(O_2)$ as the oxygen partial pressure of the surrounding gas atmosphere.

Therefore, pumping the titrant oxygen through a YSZ cell with the area A and the thickness d from the environment with the oxygen partial pressure $p(O_2)^I$ to the titrand containing gas flow with $p(O_2)^{II}$ results in an additional current I_e according to Eq. 3:

$$I_e = \frac{ART}{Fd}\left[\sigma_h\left(p(O_2)^I\right)\left\{\left(\frac{p(O_2)^{II}}{p(O_2)^I}\right)^{1/4} - 1\right\}\right] \quad (3)$$

with R as the gas constant ($R = 8.31446\ \mathrm{J} \cdot \mathrm{mol}^{-1} \cdot \mathrm{K}^{-1}$) and F as the Faraday constant ($F = 96485.3\ \mathrm{As} \cdot \mathrm{mol}^{-1}$). Equation 3 is valid for oxygen partial pressures in the vicinity of the stoichiometric point where $\sigma_h \gg \sigma_e$.

This electronic current I_e can be quantified within the temperature range by measurement in pure nitrogen [6] and is strictly related to the cell. Since the temperature of the pumping cell varies due to control instabilities, changes in gas composition and exothermic reactions at the electrode in contact with the titrand, the electronic current has also a random component which determines the limit of detection [7]. If these errors are taken correctly into account, the continuous gas titration with controlled end point by means of SE cells can be used as a calibration-free method strictly based on Faraday's law.

The third important source of error comes with the end point detection. In case of potentiometric detection, these errors are related mostly to cell temperature measurement and temperature differences between the two electrodes [8]. If amperometric end point detection is used, the temperature control and the aging of the diffusion barrier can be serious sources of error [9].

Examples of Applications

Titrant Pumping with Controlled SE Cells

The gas titration with coulometric SE cells can be subdivided into two different applications: the batch-like titration into nearly stationary titrands [10] and the continuous titration into flowing titrants [11]. As the batch-like titration needs a certain amount of time depending on the volume to be titrated and the applicable current, it does not belong to the real-time methods. In contrast to that, the continuous titration delivers an uninterrupted current as a constant measure for the analyte flow rate. An important example for batch-like gas titration comprises coulometric SE gas sensors. One of the most popular designs is schematically illustrated in Fig. 3.

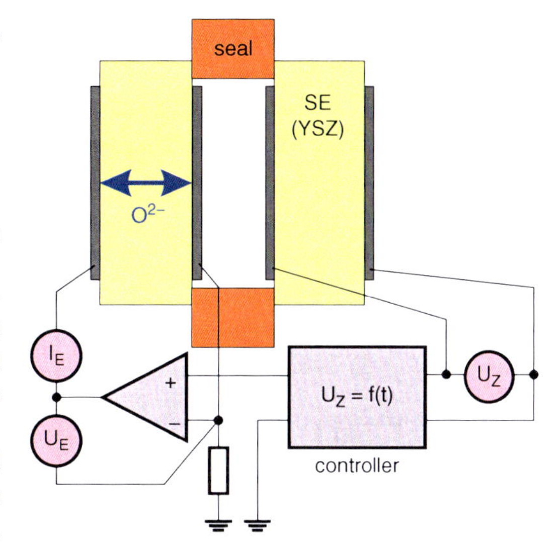

Gas Titration with Solid Electrolytes, Fig. 3 Schematic drawing of a coulometric SE sensor based on batch-like oxygen titration into a closed chamber [9]

This design contains four Pt electrodes, enabling optimization of both the potentiometric and the pumping cell for their specific purposes. To simplify the design, nowadays also coulometric sensors with two electrodes were developed, which pump and measure oxygen consecutively [12]. Due to the advantage that no reference gas is needed for this kind of devices, these sensors were developed and marketed on a large scale for the measurement of oxygen. Other developments are based on proton conducting SE for hydrogen detection [13]. A review on the specific characteristics of these coulometric gas sensors was published in [14].

A schematic drawing of a possible setup for continuous titration is given in Fig. 4. It contains the electrolysis cell, which can be controlled galvanostatically [15] or potentiometrically [7] by the end point potential of a downstream potentiometric cell.

The advantages of controlling the titration galvanostatically consist in a high signal to

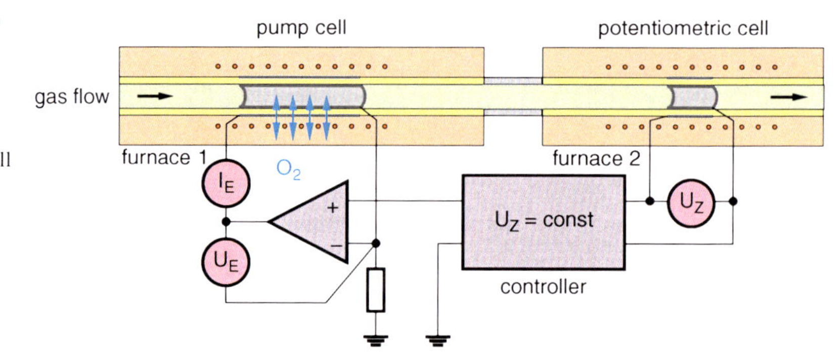

Gas Titration with Solid Electrolytes, Fig. 4 Continuous gas titration using a galvanostatically controlled SE pumping cell

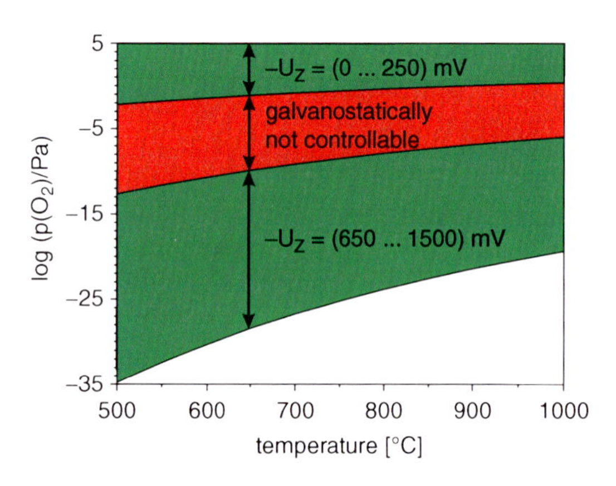

Gas Titration with Solid Electrolytes, Fig. 5 Oxygen partial pressure range adjustable with galvanostatic control of the SE pumping cell by means of a downstream potentiometric cell

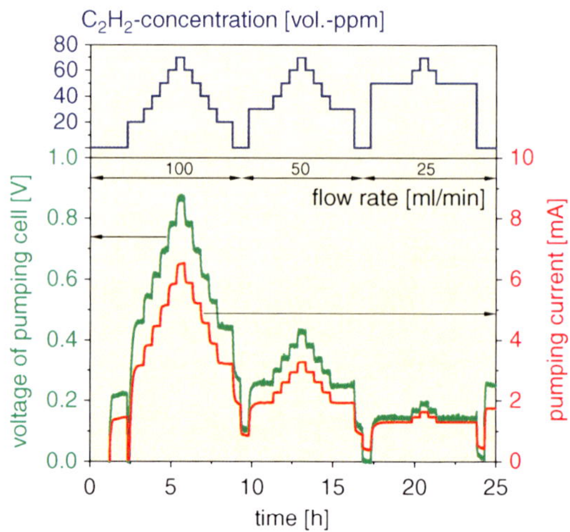

Gas Titration with Solid Electrolytes, Fig. 6 Gas titration of ethine in N_2 with a galvanostatically controlled YSZ cell, end point voltage of the potentiometric cell: $U_z = -180$ mV

noise ratio (SNR) and comparably short response times. On the other hand, this control method is connected with the disadvantage of a temperature dependent range of oxygen partial pressures, which cannot be established as end point values due to control oscillation in the vicinity of the stoichiometric point, as it is shown in Fig. 5. This disadvantage can be diminished significantly by controlling the titration cell potentiometrically, but this is often connected with decreasing SNR and elevated response time. As an example the galvanostatically controlled titration of ethine-containing mixtures is illustrated in Fig. 6.

Titrant Pumping with Noncontrolled SE Cells

In case of low titrand concentrations and subsequently low current densities at the inner electrode of the pumping cell, complete turnover can be established without control already by fixing the pumping voltage [16]. This means that the downstream potentiometric cell shown in Fig. 4 is not necessary. The flow rate and concentration limits of this simplified design can be quantified for a given cell design and every component to be measured easily by determining the oxygen partial pressure in the exhaust of the pumping cell.

This requirement of small analyte amounts for the utilization of fixed voltage titration cells is fulfilled, e.g., in chromatography applications. With the setup given in Fig. 7, it could be shown that the application of simple and miniaturized SE titration cells can be a successful way to calibration-free and easy to use chromatographic detectors for the measurement of oxygen as well as a huge

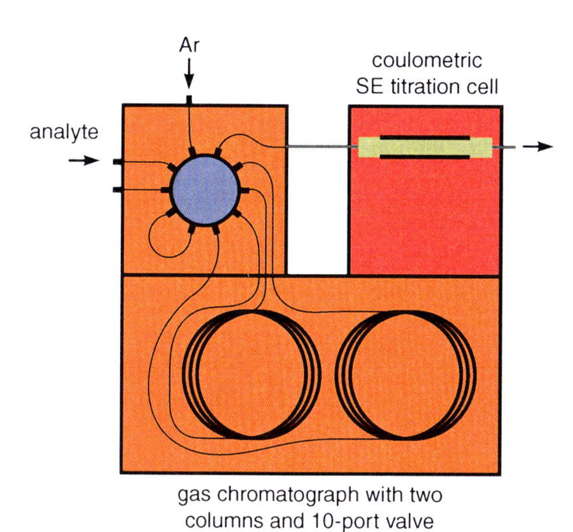

Gas Titration with Solid Electrolytes, Fig. 7 Utilization of a noncontrolled SE pumping cell as a multi-gas detector for gas chromatography

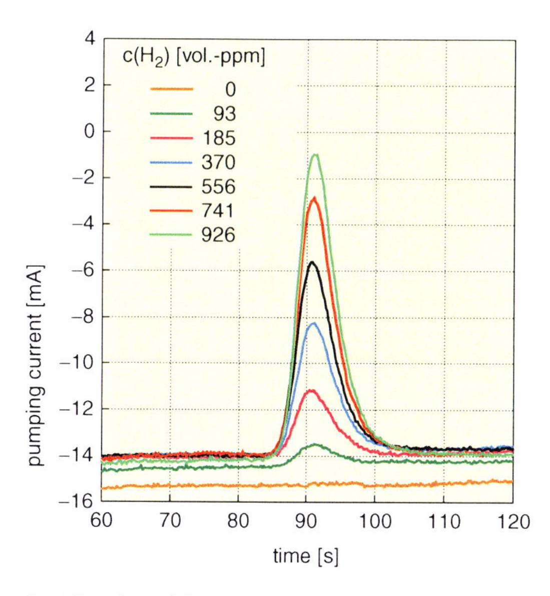

Gas Titration with Solid Electrolytes, Fig. 8 Titration of the hydrogen peak from 1 ml injected sample with the setup according to Fig. 7

variety of oxidizable gases. The titration of hydrogen shown in Fig. 8 proofs that the measured amount of charge as the peak area is consistent with the amount of hydrogen in the injected probe volume. This is also valid for many other analytes as the curves in Fig. 8 of the entry ▶ Combustion Control Sensors, Electrochemical indicate.

Future Directions

Some future directions concerning gas titration with solid electrolytes will be:

- Broader utilization of coulometric sensors for in situ analytics
- Miniaturization of coulometric titration cells
- Expansion of the method in the field of material research and
- New applications in the field of ultra-pure gas supply.

Cross-References

▶ Amperometry
▶ Combustion Control Sensors, Electrochemical
▶ Coulometry
▶ Potentiometry
▶ Solid Electrolytes

References

1. Möbius H-H (1966) Sauerstoffionenleitende Festelektrolyte und ihre Anwendungsmöglichkeiten. Z Phys Chem 231:209–214
2. Yuan D, Kröger FA (1969) Stabilized Zirconia as an oxygen pump. J Electrochem Soc 116:594–600
3. Ullmann H, Teske K, Reetz T (1973) Die Entwicklung elektrochemischer Meßverfahren für die Kontrolle schädlicher Verunreinigungen in Natriumkreisläufen. Kernenergie 16:291–297
4. Vashook V, Zosel J, Guth U (2012) Oxygen solid electrolyte coulometry (OSEC). J Solid State Electrochem 16:3401–3421
5. Park J-H, Blumenthal RN (1989) Electronic transport in 8 mole percent Y_20_3-$Zr0_2$. J Electrochem Soc 136:2867–2876
6. Rickert H (1982) Electrochemistry of solids. Springer, Heidelberg, p 101
7. Schelter M, Zosel J, Oelßner W, Guth U, Mertig M (2012) A solid electrolyte sensor for trace gas analysis. Sens Actuators B. doi:10.1016/j.snb.2012.10.111
8. Hartung R (1968) Über galvanische Sauerstoffketten mit Zirkondioxyd-Festelektrolyten und deren gasanalytische Anwendung. Dissertation, Greifswald
9. Zosel J, Gerlach F, Ahlborn K, Guth U, Solbach A, Tuchtenhagen D, Treu C Heelemann H (2011) Characterization of ageing of solid electrolyte sensors by impedance spectroscopy. Proceedings Sensor + Test Conference Nuremberg, Vol. I, 527–531
10. Besson J, Déportes C, Kleitz M (1969) French Patent 1 580 819

11. Haaland DM (1977) Internal-reference solid-electrolyte oxygen sensor. Anal Chem 49:1813–1817
12. Turwitt M, Wienand K, Ullrich K, Asmus T, Dittrich J, Schönauer U, Guth U (2010) European Patent 2226629A1
13. Katahira K, Matsumoto H, Iwahara H, Koidea K, Iwamoto T (2001) A solid electrolyte hydrogen sensor with an electrochemically-supplied hydrogen standard. Sens Actuators B 73:130–134
14. Mascell WC (1987) Inorganic solid state chemically sensitive devices: electrochemical oxygen gas sensors. J Phys E Sci Instrum 20:1156–1168
15. Ullmann H, Teske K (1991) Determination of oxygen activities in melts and solid materials by solid electrolyte cells. Sens Actuators B4:417–423
16. Zosel J, Schelter M, Vashook V,Guth U (2012) Coulometrische Festelektrolyt-Gassensoren für Konzentrationen im ppb-Bereich. 16. GMA/ITG-Fachtagung Sensoren und Messsysteme, Nürnberg 22./23.5.2012, Proceedings, pp 324–331. doi:10.5162/sensoren2012/3.2.4

Gel Electrolyte for Supercapacitors, Nonaqueous

Nobuko Yoshimoto
Graduate School of Science and Engineering, Yamaguchi University, Yamaguchi, Japan

Introduction

The utilization of a solid ion conductor as the electrolyte component has been attempted in order to fabricate an all-solid supercapacitor (also referred to as electrochemical capacitor) device, similar to rechargeable battery systems. The merits of the solidification of the electrolyte in such energy devices would be to avoid the risk of electrolyte leakage from the cell that would cause various problems on the reliability and safety. A solid-state supercapacitor using electric double-layer capacitor (EDLC) using polymeric electrolyte was first reported in the early 1990s. High specific capacitance was observed under a limited charge–discharge cycling condition. However, further improvements in the system were required to obtain practical capacitor performances. In designing such all-solid electrochemical devices and in order to maintain the contact between solid electrode and solid electrolyte phases, a kind of "flexible interface" is needed, where the fast charge (or ion) transfer process can occur. The ionic mobility in the solid phase is generally too low to be used in capacitor devices at ambient temperature. From these standpoints, a concept of polymeric gel electrolyte has been introduced and examined extensively, especially for the supercapacitor system.

Definition of Polymeric Gel Electrolytes

Organic polymeric gels are defined as polymeric composite in which low molecular weight solvents are included in three-dimensional network structures formed by chemical cross-linking (covalent, ionic, or coordinate bonding) or physical intermolecular forces [1, 2]. Those are sometimes classified as chemically and physically cross-linked gels, respectively. The polymeric gels containing ionic components are generally called "polymeric gel electrolytes." Ions move through solvent-rich domains in the gels while the polymeric components have roles of keeping the mechanical strength and holding the liquid components in the whole system as shown in Fig. 1. This structural characteristic leads to high ionic conductance of the gel electrolyte compared with that observed in usual liquid electrolytes. The morphological classification of the gel electrolyte is ranged from soft and fluid form to mechanically hard one as solid. As the gel electrolytes contain some amounts of liquid components, compatibility with solid electrodes becomes much better than the case of intrinsic solid polymer electrolytes, and the flexibility of the cell design is also improved by using soft and flexible electrolyte layer.

Constitution of Gel Electrolytes

According to the definition [1, 2], gel electrolytes are classified by the combination of constituting components. Most of the gel electrolytes that can be utilized in energy storage devices are

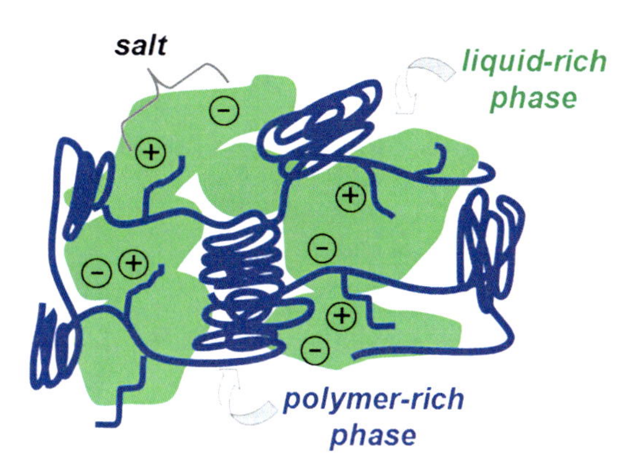

Gel Electrolyte for Supercapacitors, Nonaqueous, Fig. 1 Schematic image of polymeric gel electrolyte

composed of solid polymers and liquid electrolyte components where the latter are physically held in the former matrices. Up to the present, many sorts of gel electrolytes have been proposed with various polymer-liquid combinations. Chemical structure of typical polymers used as the matrices for gel electrolytes of capacitor devices are shown in Fig. 2. These polymers have polar units in their structures, which assist to dissociate the salts dissolved in liquid components and to move ionic species in the gel systems. Usually, a quaternary salt-based organic electrolyte or a lithium salt-based electrolyte, an ionic liquid, is for the liquid component of gel electrolytes.

In some gel systems, chemical or physical cross-linking is made to form network structures of the matrices. Composition and characteristics of typical gel electrolyte systems for suparcapacitors are described in the following paragraphs.

Poly(ethylene oxide)-Based Gel Electrolyte

Polymers with repeated units of ethylene oxide (EO: $-CH_2CH_2-O-$) group in their main chain are commonly called poly(ethylene oxide) (PEO) although they are correctly named poly (oxyethylene) (POE). The EO unit in the polymer takes a role of a donor (or basic ligand) for cationic species that work as the charge carrier in the solid gel. There have so far been proposed many variations of PEO family.

A research group at Yamaguchi University [3] has first reported an all-solid EDLC system using a polymeric gel electrolyte composed of poly (ethylene oxide)-modified polymethacrylate (PEO-PMA) swollen with propylene carbonate (PC) that dissolves tetraethylammonium tetrafluoroborate ($TEABF_4$) as the ionic charge carrier. The ionic conductivity of (PEO-PMA)-PC-$TEABF_4$ gel, depending on its composition, was measured to be around 0.4 mS cm^{-1} at room temperature, being lower than those of conventional organic electrolyte solutions.

In recent years, IL-based polymeric gel electrolytes have been examined for their applications promising electrolytes in supercapacitors [4]. The addition of IL to PEO-based electrolytes may improve not only the ionic conductivity of the electrolytes but also the interfacial stability of electrode/electrolyte compared to the addition of such conventional organic solvents as PC and ethylene carbonate (EC).

Poly(acrylonitrile)-Based Gel Electrolyte

Gel electrolytes consisting of poly(acrylonitrile) (PAN) as the polymer matrix have high mechanical strength even when they contain large amounts of liquid components. This property mainly comes from the –CN functionality in the chemical structure of the matrix. The PAN-based gels also show wide potential window and good compatibility with electrode materials.

In 1995, the gel electrolyte containing PC solution of $TEABF_4$ as the plasticizing and ion conducting component has proposed as the electrolyte of an EDLC [5]. This gel electrolyte has wide potential window of ca. 5.0 V, and the ionic conductivity of the gel electrolytes comparable to those of liquid organic electrolytes. The EDLC device using this type of PAN electrolyte showed good charge–discharge cycle performances and long voltage-retention

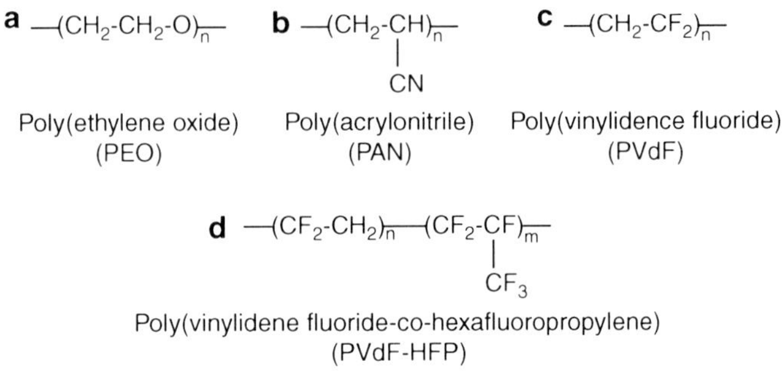

Gel Electrolyte for Supercapacitors, Nonaqueous, Fig. 2 Chemical structures of typical polymer matrices for polymeric gel electrolytes

a $-(CH_2\text{-}CH_2\text{-}O)_n-$ Poly(ethylene oxide) (PEO)

b $-(CH_2\text{-}CH(CN))_n-$ Poly(acrylonitrile) (PAN)

c $-(CH_2\text{-}CF_2)_n-$ Poly(vinylidence fluoride) (PVdF)

d $-(CF_2\text{-}CH_2)_n-(CF_2\text{-}CF(CF_3))_m-$ Poly(vinylidene fluoride-co-hexafluoropropylene) (PVdF-HFP)

characteristics. The gel electrolyte containing mixture of EC and PC solution (or dimethyl carbonate (DMC)) of lithium perchlorate ($LiClO_4$) as the plasticizing and ion conducting component has also proposed as the electrolyte of supercapacitor.

Poly(vinylidene fluoride)-Based Gel Electrolyte

The polymeric composite of poly(vinylidene fluoride) (PVdF) swollen with an ester solvent tends to form physical gel with phase-separation structure. The gel has microporous structure, so that the ion transport is not hindered by the polymer matrix itself. Thus, it is easy to form a gel electrolyte with high ionic conductivity. Copolymers of vinylidene fluoride and hexafluoropropylene (PVdF-HFP) have also been used as excellent polymer matrices of gel electrolytes containing alkylammonium salts for EDLC [6]. PVdF-HFP-based gel electrolyte has a heterogeneous structure of crystalline and amorphous phases. The PVdF unit in the copolymer forms a crystalline phase that contributes to mechanical strength of the gel, and the HFP unit holds the liquid component to form an amorphous phase that takes charge in ionic conduction. With respect to the electrolytic salt in polymeric gel electrolyte, alkylammonium salts have been used widely for capacitor devices because of their high solubility in such organic solvents as PC and high ionic conductivity of the resulting gel electrolyte. Effects of the ions in gel electrolyte on the capacitor performances of EDLC are significant. For example, higher specific capacitance is observed for activated carbon electrode in the gel containing trimethylammonium (TEMA) salt than that in the gel dissolving TEA salt [7]. It is worthy of note that higher specific capacitance is often observed in both nonaqueous polymeric gel electrolyte and aqueous one than in the original liquid electrolyte consisting of the same electrolytic salt [8]. That is, the polymeric component itself can enhance the electric double-layer capacitance of the high-surface-area carbon electrode. Although the mechanisms of such capacitance enhancements are still unknown in detail, some pseudocapacitance or changes in the double-layer structure can possibly contribute to the additional capacitance on polymeric gel electrolytes.

To the rate capability of the capacitor device, proton as the charge carrier will be preferable because of the higher mobility in polymeric gel electrolyte than that of Li^+. A polymeric gel electrolyte made of PVdF-HFP swollen with DMF dissolving CF_3SO_3H as the proton source was examined for an asymmetric cell that had activated carbon (AC) as the negative and hydrous RuO_2 as the positive electrode [9]. This type of gel electrolyte had high proton conductivity (10^{-3} S cm^{-1}) and showed good thermal stability. As the system contains no free-water, the operating voltage of the asymmetric AC/RuO_2 capacitor was maximized to 1.6 V or higher [9].

Future Directions

The energy and power capabilities of supercapacitors have been incrementally increased in the past several years. New concepts for capacitive energy storage processes and new functional materials have been proposed. In the recent years, there has been renewed interest in new polymeric gel electrolytes for high-energy supercapacitors. For example, a research group of Kansai University first reports an alginate-based gel electrolyte containing an ionic liquid for nonaqueous EDLC [10]. Alginate (Alg) is one of the common polysaccharides derived from brown seaweeds. It is a nature-friendly material with high abundance in nature. This transparent gel electrolyte consisting of Alg and 1-ethyl-3-methylimidazolium tetrafluoroborate ($EMImBF_4$) has the high levels of mechanical strength and retentively of ionic liquid needed to construct EDLC.

Hereafter, safety and reliability issues will become more and more important for developing the energy storage devices for large-scale utilization. However, improvement of the capacitor performances will be needed as well as reliability and safety issues. Such problems as dry out of liquid components from the gel and disposal/recycle of waste materials are still remaining. As the supercapacitor devices have to guarantee higher-rate capability and longer cycle life than the battery system, higher target performances will be needed for the gel electrolytes in capacitor devices. Thus, wisdom not only in electrochemistry but also in wide area of materials science should be concentrated to solve the above mentioned problems.

Cross-References

- ▶ Electrolytes for Electrochemical Double Layer Capacitors
- ▶ Ionic Liquids
- ▶ Super Capacitors

References

1. Ross-Murphy S. B (1991) Physical gelation of Synthetic and biological macromolecules, 23–25. In: De Rossi D et al (eds) Polymer gels. Plenum, New York
2. Osada Y (1997) Polymer gel - About the formation of cross-linking (in Japanese) In: Kajiwara K et al (eds) Gel handbook (in Japanese). NTS, Tokyo
3. Matsuda Y, Morita M, Ishikawa M, Ihara M (1993) New electric double-layer capacitors using polymer solid electrolytes containing tetraalkylammonium salts J Electrochem Soc 140:L109–L110; Ishikwa M, Morita M, Ihara M, Matsuda Y (1994) Electric double-layer capacitor composed of activated carbon fiber cloth electrodes and solid polymer electrolytes containing alkylammonium salts ibid: 1730–1734
4. Pandey GP, Kumar Y, Hasami SA (2010) Ionic liquid incorporated polymer electrolytes for supercapacitor application. Indian J Chem 49A:743–751
5. Ishikawa M, Ihara M, Morita M, Matsuda Y (1995) Electric double layer capacitors with new gel electrolytes. Electrochim Acta 40:2217–2222
6. Osaka T, Liu X, Nojima M, Momma T (1999) An electrochemical double layer capacitor using an activated carbon electrode with gel electrolyte binder. J Electrochem Soc 146:1724–1729
7. Ishikawa M, Yamamoto L, Morita M, Ando Y (2001) Performance of electric double layer capacitors with gel electrolytes containing an asymmetric ammonium salt. Electrochemistry 69:437–439
8. Iwakura C, Wada H, Nohara S, Furukawa N, Inoue H, Morita M (2003) New electric double layer capacitor with polymer hydrogel electrolyte. Electrochem Solid-State Lett 6: A37–A39; Morita M, kaigaishi T, Yoshimoto N, Egashira M, Aida T (2006) Effects of the electrolyte composition on the electric double-layer capacitance at carbon electrodes. Electrochem Solid-State Lett 9: 386–389
9. Morita M, Ohsumi N, Yoshimoto N, Egashira M (2007) Proton-conducting non-aqueous gel electrolyte for a redox capacitor system. Electrochemistry 75:641–644
10. Yamagata M, Soeda K, Yamazaki S, Ishikawa M (2011) Alginate gel containing an ionic liquid and its application to non-aqueous electric double layer capacitors. Electrochem Solid-State Lett 14: A165–A169

Gene Electrotransfer for Clinical Use

Maja Cemazar
Department of Experimental Oncology, Institute of Oncology Ljubljana, Ljubljana, Slovenia

Introduction

Gene therapy refers to the introduction of nucleic acids, e.g., DNAs or RNAs, into the cells of target tissues with a therapeutic purpose. Gene therapy can be applied as a gene replacement

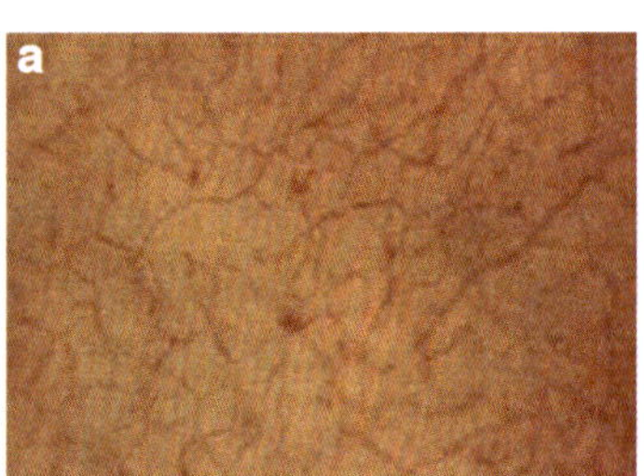

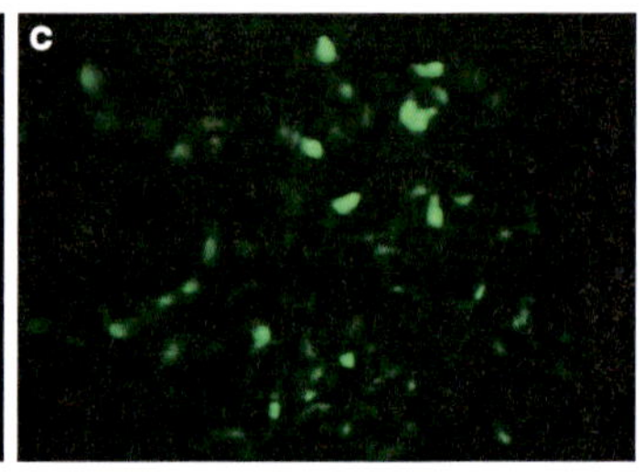

Gene Electrotransfer for Clinical Use, Fig. 1 Electrotransfer of plasmid DNA encoding green fluorescent protein into P22 rat carcinosarcoma growing in dorsal skin window chamber preparation. 50 mg pEGFP-N1 was placed on the top of the tumor that was exposed to 8 square wave pulses of 5 micro second duration at 600 V/cm ratio immediately thereafter. (**a**) tumor under visible light, (**b**) 5 h after gene electrotransfer few cells expressing green fluorescent protein are already visible, (**c**) 24 h after gene electrotransfer, a large number of fluorescent cells are visible

therapy or mutation compensation (e.g., replacement of mutated oncogene), gene immunopotentiation (e.g., introduction of cytokines and co-stimulatory molecules), a molecular chemotherapy (e.g., gene directed enzyme/prodrug therapy) in cancer gene therapy, and gene vaccination (e.g., introduction of specific antigen).

The main faltering block to effective and broader use of gene therapy remains the effective delivery of therapeutic gene to target tissue. In general, gene delivery systems can be divided in viral and nonviral delivery. Among viral delivery vectors, adenoviral and retroviral vectors are the most advanced [1, 2]. Nonviral delivery systems are based on chemical vectors (e.g., liposomes, dendrimers, nanoparticles) or physical methods of delivery (e.g., electroporation, ultrasound, gene gun) [2–5]. In electroporation, by using short intense electric pulses, the plasma membrane becomes permeable to molecules otherwise deprived of membrane transport mechanisms. When a cell in vitro or in vivo in tissue is exposed to an electric field, transmembrane potential is induced across the cell membrane. If the induced transmembrane potential is sufficiently high, structural changes leading to increased membrane permeability are induced. Although the exact mechanism operating on the molecular level and the various structures has not yet been fully elucidated, a flow of molecules was demonstrated through areas of membranes in regions where the highest absolute value of the induced transmembrane potential was observed after exposure of cells to electric pulses. Depending on the size of molecules to be introduced, i.e., small molecules such as chemotherapeutic drugs and siRNA molecules or plasmid DNA, different mechanism of transport across the cell membrane is observed. In the case of small molecules, diffusion across the membrane is observed, being the highest at the cathode site, but occurs also at the anode site [6, 7]. When larger molecules, such as plasmid DNA, are introduced by electroporation, first the interaction between DNA and cell membrane occurs during application of electric pulses, being observed only at the cathode site. After that translocation and intracellular migration take place. The exact mechanism of translocation is not known, but one of the hypotheses is to be endocytotic process [8]. The expression of introduced genes in vivo can be detected already 5 h after transfection [9] (Fig. 1). Successful introduction of nucleic acids into the cells of different tissues depends on many variables. Firstly, parameters of electric pulses should be carefully chosen for introduction of nucleic acid molecules into cells of different tissues, such as skin, muscle, and tumor. Besides the size of the nucleic molecules, also tissue properties, such as size and shape of the cells and content of extracellular matrix, influence the choice of parameters of electric pulses. In addition, the timing of the injection of nucleic acids and application of electric pulses is also crucial for the successful electrotransfection [3]. Therefore, a lot of studies on gene electrotransfer are devoted to optimization of electrical parameters and design of electrodes for the use in different tissues as well as studies elucidating basic mechanisms of nucleic acids electrotransfer. However, successful use of electroporation in electrochemotherapy in clinical setting and studies with plasmids DNA encoding different therapeutic genes demonstrating high efficiency in preclinical studies lead to rapid growth of clinical studies on gene electrotransfer [7, 10, 11].

The use of electroporation for transfection of cells dates back to 1982, when Neumann et al. demonstrated increased uptake of plasmid DNA into mouse lyoma cells using electroporation [12]. After this first demonstration that electroporation can be used for delivery of plasmid DNA, the method gained a lot of attention since it represented a promising alternative to viral and chemical methods for introduction of genes of interest into the cells in vitro. Gene electrotransfer to tissues was introduced by Titomirov et al. in 1991. The skin of newborn mice was transfected with plasmid DNA. The skin was then excised and NEO-resistant colonies were found in primary cell cultures obtained from the treated skin [13]. After this first study, the use of gene electrotransfer to tissues grew rapidly. Optimization studies aiming to optimize the parameters of electric pulses for application of gene electrotransfer to different tissues, as well as studies dealing with the therapeutic effect of gene electrotransfer, were performed.

Therapeutic applications of gene electrotransfer are focused mainly on two fields: cancer gene therapy and DNA vaccination against infectious disease. The first use of electrogene therapy (a term used to describe gene therapy in which transfection of cells is achieved by means of electroporation) for treatment of cancer was published in 1999. Niu et al. demonstrated that electrotransfer of a *Stat3* variant with dominant-negative properties to melanoma subcutaneous tumors induced in C57Bl/6 mice suppressed the growth of transfected tumors by inducing apoptotic cell death [14]. A year later, a protection against the lethal influenza virus was achieved by intramuscular electrotransfer of plasmid DNA encoding for neuraminidase from different subtype-A influenza viruses [15]. Clinical studies with electroporation in cancer patients were mainly focused on electrochemotherapy, as this approach was highly successful in preclinical trials, resulting in up to 100 % complete tumor regression. In addition, both chemotherapeutic drugs used in electrochemotherapy, bleomycin and cisplatin, were already used in clinical routine [7, 16]. The first published human clinical trial in cancer patients with gene electrotransfer was performed on cutaneous metastases of melanoma patients, who were treated with plasmid DNA encoding interleukin-12 (IL-12), which was delivered directly into the tumor nodules by application of electric pulses [10]. The study was a phase I dose escalation study carried out in 24 patients. Patients were treated three times on days 1, 5, and 8. The maximum dose of plasmid per tumor nodule was 1.6 mg, which resulted in a cumulative dose of 3.8 and 5.8 mg in two patients with multiple nodules. Fine-needle aspiration biopsy was performed before, while excisional biopsy was performed after the treatment to assess histology of tumors, immune cell infiltration and to determine the levels of IL-12 protein in tumors. Reponses to treatment were evaluated by the modified Response Evaluation Criteria in Solid Tumors (RECIST). Of major importance is the fact that the response to therapy was observed in treated as well as in distant non-treated tumor nodules. In 53 % of patients, a systemic response was observed resulting in either stable disease or an objective response. The major adverse side effect was transient pain after application of electric pulses. The systemic response to IL-12 is due to its diverse biological properties, which are still not fully elucidated. Briefly, they consist of induction of interferon–ϒ(IFN-ϒ) production that induces infiltration of CD8+ T lymphocytes and NK cells into tumors, which exhibit cytolytic activities. In addition, IL-12 has antiangiogenic action through activation of IFN-ϒ induction of interferon inducible protein-10 (IP10) and monokine Mig. Furthermore, IL-12 augments the CD4+ Th1 response, leading to activation of a specific B cell response. Recently, new studies of electrogene therapy with IL-12 were launched on subcutaneous nodules of Merkel cell carcinoma and melanoma [17]. Besides electrogene therapy based on augmentation on immune system, a phase I clinical trial was initiated on safety and efficacy of antiangiogenic gene therapy with plasmid encoding AMEP also on subcutaneous melanoma nodules [17]. However, most of the clinical trials are devoted to gene electrotransfer for gene vaccination, both for treatment of cancer and infectious diseases. Currently 10 clinical trials are recruiting patients mainly for vaccination against HIV-1, but also for cancer, influenza virus, hepatitis B and C virus, Hantaan, and Puumala virus [17]. In addition to more than 10 clinical trials that are already published, the first one being published in 2009, used electrotransfer of p.DOM-PSMA, a plasmid encoding a domain of fragment C of tetanus toxin to induce CD4(+) T helper cells, fused to a tumor-derived epitope from prostate-specific membrane antigen (PSMA) for use in HLA-A2(+) patients with recurrent prostate cancer. The results of the study demonstrated safety and tolerability of vaccine deliver by electrotransfer to the muscle [18]. The reason for such success of gene electrotransfer for DNA vaccination lays in the fact that plasmid DNA is an excellent candidate for vaccination. Plasmid DNA is namely based on bacterial backbone that contains a gene encoding the antigen(s) of interest under the control of eukaryotic promoter. Once delivered to cells, the antigen is produced endogenously; therefore, plasmid DNA can elicit both cell-mediated and humoral response, which is one of the important advantages. In addition, the manufacturing of plasmid DNA can be tightly controlled, and the stability of DNA vaccine at ambient temperature makes it perfect candidate for the veterinary use and the use in developing countries. However, after many successful pre-clinical trials in small laboratory animals, the injection of plasmid DNA alone in large target species did not result in the desired magnitude of immune response. Compared to other vaccination types, especially humoral, response by DNA vaccine was lower. The main reason for that was low transfection efficiency of simple injection of plasmid DNA. Thus, the use of electroporation for enhanced delivery of plasmid DNA was an obvious track to proceed. Electroporation besides enhancing plasmid DNA delivery by several orders of magnitude and prolonging antigen expression also induce inflammation, which enhance the infiltration of antigen-presenting cells at the site of gene electrotransfer and thus increase immune response. This is especially important in muscles, as they do not contain many resident antigen-presenting cells. Although most of the first studies of gene electrotransfer were performed in muscles, nowadays the researches are more oriented towards the skin for several reasons. The skin contains antigen-presenting cells and is more easily accessible, and anesthesia to ameliorate pain due to the application of electric pulses can be topical [19, 20]. However, technical problems, mainly with production of suitable electrodes for skin electroporation, have to be solved before wider use of skin electroporation for DNA vaccination.

Future Directions

In conclusion, the future for gene electrotransfer for clinical use is bright. Based on many clinical trials in human and veterinary medicine that showed positive results and safety of the approaches, especially two fields, prophylactic vaccination for infectious disease and curative vaccination combined with standard treatments for cancer will benefit in the future. Further technical developments of electrodes and generators of electric pulses on one hand and further optimization of plasmids DNA regarding the controlled expression and safety on the other hand will bring gene electrotransfer into wider use.

References

1. Warnock JN, Daigre C, Al-Rubeai M (2011) Introduction to viral vectors. Methods Mol Biol 737:1–25
2. Mauro G (2010) Methods for gene delivery. In: Giacca M (ed) Gene therapy. Springer, Berlin

3. Cemazar M, Golzio M, Rols MP, Sersa G, Teissie J (2006) Electrically-assisted nucleic acid delivery in vivo: Where do we stand? Review. Curr Pharm Des 12(29):3817–3825
4. Kay MA (2011) State-of-the-art gene-based therapies: the road ahead. Nat Rev Genet 12:316–328
5. Heller LC, Heller R (2010) Electroporation gene therapy preclinical and clinical trials for melanoma. Curr Gene Ther 10:312–317
6. Escoffre JM, Teissié J, Rols MP (2010) Gene transfer: how can the biological barriers be overcome? J Membr Biol 236:61–74
7. Sersa G, Miklavcic D, Cemazar M, Rudolf Z, Pucihar G, Snoj M (2008) Electrochemotherapy in treatment of tumours. EJSO 34:232–240
8. Rosazza C, Phez E, Escoffre JM, Cézanne L, Zumbusch A, Rols MP (2012) Cholesterol implications in plasmid DNA electrotransfer: evidence for the involvement of endocytotic pathways. Int J Pharm 423:134–143
9. Cemazar M, Wilson I, Dachs GU, Tozer G, Sersa G (2004) Direct visualization of electroporation-assisted in vivo gene delivery to tumors using intravital microscopy – spatial and time dependent distribution. BMC Cancer 4:81
10. Daud AI, DeConti RC, Andrews S, Urbas P, Riker AI, Sondak VK, Munster PN, Sullivan DM, Ugen KE, Messina JL, Heller R (2008) Phase I trial of interleukin-12 plasmid electroporation in patients with metastatic melanoma. J Clin Oncol 26:5896–5903
11. Pavlin D, Cemazar M, Coer A, Sersa G, Pogacnik A, Tozon N (2011) Electrogene therapy with interleukin-12 in canine mast cell tumors. Radiol Oncol 45:31–39
12. Neumann E, Schaeferridder M, Wang Y, Hofschneider PH (1982) Gene transfer into mouse lypoma cellls by electroporation in high electric fields. EMBO J 1:841–845
13. Titomirov AV, Sukharev S, Kistanova E (1991) In vivo electroporation and stable transformation of skin cells of newborn mice by plasmid DNA. Biochim Biophys Acta 1088:131–134
14. Niu GL, Heller R, Catlett-Falcone R, Coppola D, Jaroszeski M, Dalton W et al (1999) Gene therapy with dominant-negative Stat3 suppresses growth of the murine melanoma B16 tumor in vivo. Cancer Res 59:5059–5063
15. Chen Z, Kadowaki S, Hagiwara Y, Yoshikawa T, Matsuo K, Kurata T et al (2000) Cross-protection against a lethal influenza virus infection by DNA vaccine to neuraminidase. Vaccine 18:3214–3222
16. Marty M, Sersa G, Garbay JR, Gehl J, Collins CG, Snoj M, Billard V, Geertsen PF, Larkin JO, Miklavcic D, Pavlovic I et al (2006) Electrochemotherapy – an easy, highly effective and safe treatment of cutaneous and subcutaneous metastases: results of ESOPE (European Standard Operating Procedures of Electrochemotherapy) study. EJC Suppl 4:3–13
17. International Clinical Trials registry Platform (2012) Search Portal. http://apps.who.int/trialsearch/default.aspx. Accessed 17 Apr 2012
18. Low L, Mander A, McCann K, Dearnaley D, Tjelle T, Mathiesen I, Stevenson F, Ottensmeier CH (2009) DNA vaccination with electroporation induces increased antibody responses in patients with prostate cancer. Hum Gene Ther 20:1269–1278
19. van Drunen Littel-van den Hurk S, Hannaman D (2010) Electroporation for DNA immunization: clinical application. Expert Rev Vaccines 9:503–517
20. Sardesai NY, Weiner DB (2011) Electroporation delivery of DNA vaccines: prospects for success. Curr Opin Immunol 23:421–429

General Concepts and Global Parameters (EOD, COD, O_x)

Christos Comninellis[1], Agnieszka Kapałka[1], Stéphane Fierro[1], György Fóti[1], Pierre-Alain Michaud[1] and Petros Dimitriou-Christidis[2]
[1]Institute of Chemical Sciences and Engineering, Ecole Polytechnique Fédérale de Lausanne (EPFL), Lausanne, Switzerland
[2]Environmental Chemistry Modeling Laboratory, Ecole Polytechnique Fédérale de Lausanne (EPFL), Lausanne, Switzerland

Introduction

Wastewater contains in general a variety of organic pollutants; analysis of these pollutants and their degradation products during treatment is a complex matter. In order to avoid this problem, global parameters are frequently used.

Depending on the wastewater treatment technique used, different global parameters have been employed. In chemical treatment using strong oxidants like O_3, H_2O_2/Fe^{+2}, etc., TOC (Total Organic Carbon) and COD (Chemical Oxygen Demand) are frequently used.

In biological wastewater treatment, the 5-day Biological Oxygen Demand (BOD_5) is the main parameter used.

These parameters are not specific for the chemical nature of the organic pollutants present in the wastewater. They can, however, allow estimation of the extent (concentration) of pollution and evaluation of treatment efficiency. Furthermore, the ratio between these parameters can give very useful information. In fact the BOD_5/COD ratio can give information on the biodegradability of the wastewater under investigation.

In this work, the parameters Electrochemical Oxidation Index (EOI) and Electrochemical Oxygen Demand (EOD), involved in the electrochemical treatment of organic pollutants, are presented [1–6]. From these parameters, both the degree of oxidation (X) and the average oxidation state of carbon (Ox) in the wastewater can be calculated during treatment [6].

Both EOI and EOD values depend on anode material, electrolysis conditions (T, pH, current density, organics concentration), and nature of the organic species [1–6]. These parameters are very useful for the optimization of electrochemical treatment of the wastewater under investigation.

Electrochemical Oxidation Index (EOI)

The Electrochemical Oxidation Index (EOI) expresses the average current efficiency for the oxidation of organics and is a measure of the ease of electrochemical oxidation of the target organic compound (or the wastewater under investigation) on the investigated anode material under the experimental conditions used.

EOI is dimensionless and can take values between zero and one. EOI = 0 means that the organic compound is not oxidized on the selected anode material under investigation conditions, in which case the main reaction is oxygen evolution. EOI = 1 indicates that there are no side reactions, and all of the current is used for the oxidation of organics. It is worthwhile to notice that EOI depends on electrolysis time.

EOI for the system under investigation, after a given time t of electrolysis (EOI_t), can be calculated from the relationship:

$$EOI_\tau = \frac{\int_0^\tau ICEdt}{\tau} \quad \text{with } 0 \leq EOI \leq 1 \qquad (1)$$

where:

ICE: instantaneous current efficiency for organics oxidation (–)

τ: duration of the electrochemical treatment (s).

Two techniques have been proposed for estimation of the instantaneous current efficiency (ICE) for organics oxidation during electrolysis: Chemical Oxygen Demand (ICE_{COD}) and Oxygen Flow Rate (ICE_{OFR}) [1–6].

In the Chemical Oxygen demand technique (ICE_{COD}), the COD of the electrolyte (wastewater) is measured at regular intervals (Δt) during constant current (galvanostatic) electrolysis, and the instantaneous current efficiency is calculated using the relationship [1–6]:

$$ICE_{COD} = \frac{FV}{8I}\frac{\left[(COD)_t - (COD)_{t+\Delta t}\right]}{\Delta t} \qquad (2)$$

where $(COD)_t$ and $(COD)_{t+\Delta t}$ are the chemical oxygen demand (g m^{-3}) at times t and t + Δt (s) respectively, I is the applied current (A), F is the Faraday constant (96487 Cmol^{-1}), and V is the volume of electrolyte (m^3).

In the Oxygen Flow Rate technique (ICE_{OFR}), a two compartment electrochemical cell is used, and the oxygen flow rate is measured continuously in the anodic compartment during constant current (galvanostatic) electrolysis. The instantaneous current efficiency is then calculated using the relationship:

$$ICE_{OFR} = \frac{\dot{V}_o - (\dot{V}_t)_{org}}{\dot{V}_o} \qquad (3)$$

where $\dot{V}_o(m^3s^{-1})$ is the theoretical oxygen flow rate calculated from Faraday's law (or measured in a blank experiment in the absence of organic compounds) and $\dot{V}_t(m^3s^{-1})$ the oxygen flow rate measured at regular intervals (or continuously) during electrochemical treatment of the wastewater.

Both the (ICE_{COD}) and (ICE_{OFR}) techniques have their limitations, typical ones being:

- If volatile organic compounds (VOC) are formed during treatment, only the OFR technique will give reliable measurements.

- If, for example, $Cl_2(g)$ is evolved during treatment (due to the oxidation of Cl^- present in the wastewater), only the COD technique will give reliable measurements.
- If insoluble organic products are formed during treatment (for example, polymeric material), only the OFR technique will give reliable measurements.

Furthermore, using both the COD and OFR techniques simultaneously during the electrochemical process will allow a better control of the side reactions involved in the electrochemical treatment.

The Electrochemical Oxygen Demand (EOD)

The Electrochemical Oxygen Demand after a given time t of electrolysis (EOD_τ) expresses the amount of "electrochemically" formed oxygen (g m^{-3}) used for the oxidation of the organic pollutants. EOD_t can be calculated using the relationship [1–6]:

$$EOD_\tau = EOI_\tau \frac{8I\tau}{VF} \tag{4}$$

where τ (s) is the electrolysis time, I is the applied current (A), F is the Faraday constant (96487 $Cmol^{-1}$), and V is the volume of electrolyte (m^3).

Degree of Oxidation (X_t)

From the EOD value, the degree of oxidation after a given time τ of electrolysis (X_τ) can be calculated using the relationship [1–6]:

$$X_\tau = \frac{EOD\tau}{COD^\circ} \quad \text{with } 0 \leq X_\tau \leq 1 \tag{5}$$

where COD° is the initial COD value.

Values for the degree of oxidation close to one ($X_\tau \approx 1$) will indicate a quasi-complete mineralization of the organic pollutants initially present in the wastewater.

The Average Oxidation State of Carbon (Ox_τ)

The average oxidation state of carbon after a given time τ of electrolysis (Ox_τ) can be calculated from the relationship [6]:

$$Ox_\tau = 4\left(1 - \frac{12EOD_\tau}{32TOC_\tau}\right) \quad \text{with } -4 \leq OX_\tau \leq +4 \tag{6}$$

Where, TOC_τ (g m^{-3}) is the Total Organic Carbon of the electrolyte (wastewater) after a given time τ of electrolysis.

The average oxidation state of carbon (Ox_τ) can have values between −4 and + 4. Deviation from this domain (i.e., in case of $Ox_\tau > 4$ and/or $Ox_\tau < -4$) will indicate that the measured global parameters TOC_τ and EOD_τ values are not reliable.

In fact, the presence of inorganic pollutants (Cl^-, S^{-2}, Fe^{+2}, etc.) in the wastewater or the formation of polymeric material and/or volatile organics during electrolysis will result in non-reliable TOC_τ and EOD_τ values.

Future Directions

The measurement of the global parameters TOC_t and EOD_t during electrochemical treatment of wastewater containing organic pollutants can allow a fast optimization of the electrochemical process. However, new in situ techniques need to be developed for the analysis of evolved oxygen, CO_2, volatile organics (adsorption on an adsorbent column), and active oxidants (chlorine, persulfate, H_2O_2, O_3, etc.) formed during electrochemical treatment. This can allow more reliable values of these important parameters.

Cross-References

▶ Electrochemical Sensors for Environmental Analysis
▶ Organic Pollutants in Water Using DSA Electrodes, In-Cell Mediated (via Active Chlorine) Electrochemical Oxidation

- ► Organic Pollutants in Water Using SnO_2, Direct Electrochemical Oxidation
- ► Organic Pollutants in Water, Direct Electrochemical Oxidation Using PbO_2
- ► Organic Pollutants, Direct and Mediated Anodic Oxidation
- ► Organic Pollutants, Direct Electrochemical Oxidation
- ► Organic Pollutants, Oxidation on Active and Non-Active Anodes
- ► Wastewater Treatment, Electrochemical Design Concepts
- ► Water Treatment with Electrogenerated Fe(VI)

References

1. Comninellis C, Chen G (eds) (2010) Electrochemistry for the environment. Springer, New York/Dordrecht/Heidelberg/London
2. Comninellis C, Plattner E (1988) Electrochemical wastewater treatment. Chimia 42:250–252
3. Comninellis C, Pulgarin C (1991) Anodic oxidation of phenol for wastewater treatment. J Appl Electrochem 21:703–708
4. Comninellis C, Pulgarin C (1993) Electrochemical oxidation of phenol for wastewater treatment using tin dioxide anodes. J Appl Electrochem 23:108–112
5. Comninellis C (1992) Electrochemical treatment of wastewater containing phenol. Trans IChemE 70(B):219–224
6. Comninellis C (1994) Traitement Electrochimique des eaux Résiduaires. Informat Chimie 357:109–112

Glass Ion Conductors for Solid-State Lithium Batteries

Masahiro Tatsumisago and Akitoshi Hayashi
Department of Applied Chemistry, Osaka Prefecture University, Sakai, Osaka, Japan

Introduction

Inorganic solid electrolytes have been studied in two types of solids: crystal and glass [1, 2]. Glass ion conductors have several advantages: high conductivity based on so-called open structure is achieved at wide chemical compositions and single ion conduction where only targeted ions such as Li^+ ions are mobile is realized. Glass ion conductors are a promising solid electrolyte for highly safe all-solid-state batteries on the basis of a free of hazards of leakage and flammability of the electrolytes [3]. In this chapter, features of glass electrolytes with Li^+ ion conductivity are demonstrated and recent development of all-solid-state rechargeable lithium batteries with glass electrolytes are reported.

Conductivity of Glass Ion Conductors

A principal strategy to develop the conductivity of glass electrolytes is to increase the number and mobility of lithium ions. Figure 1 shows the composition dependence of electrical conductivity at 25 °C for oxide and sulfide glasses in the systems Li_2O-SiO_2, Li_2O-P_2O_5, Li_2S-SiS_2, and Li_2S-P_2S_5. Rapid quenching using a twin-roller apparatus and mechanical milling using a planetary ball mill apparatus were used for preparation of the glasses [4, 5]. By increasing lithium ion concentration in the glasses, the conductivity of the glasses in all the systems monotonously increased. The conductivity drastically increased by changing the glass matrix from oxides to sulfides. The conductivity of the Li_2S-SiS_2 sulfide glass at the composition of $[Li]/([Li] + [M]) = 0.75$ was in the order of 10^{-4} S cm^{-1}, which was two orders of magnitude higher than that of the Li_2O-SiO_2 oxide glass with the same lithium ion concentration. The activation energy for conduction of the sulfide glass (33 kJ mol^{-1}) was lower than that of the oxide glass (48 kJ mol^{-1}), suggesting that the mobility of lithium ions increased by replacing oxide matrix with sulfide one.

The addition of third components is effective in enhancing conductivity of sulfide glasses; the Li_2S-SiS_2 glasses added with LiI or Li_3PO_4 exhibited a high conductivity of 10^{-3} S cm^{-1} at 25 °C [6–8]. Another effective technique to improve conductivity of glasses is the precipitation of a metastable or high-temperature phase by crystallization. In general, crystallization of glass

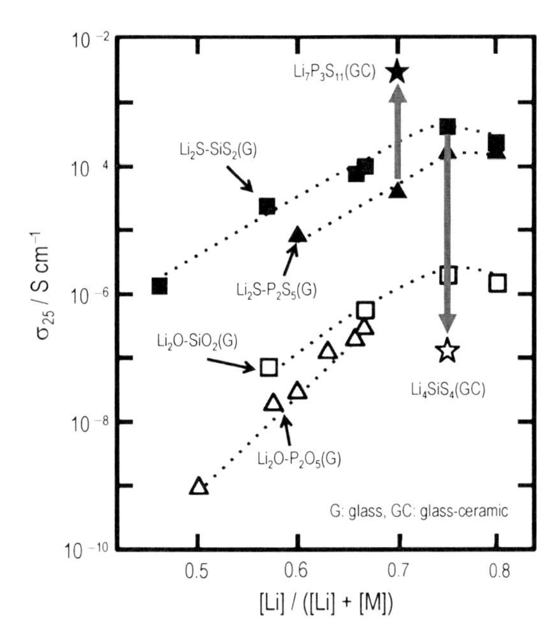

Glass Ion Conductors for Solid-State Lithium Batteries, Fig. 1 Composition dependence of conductivity at 25 °C for oxide and sulfide glass-based electrolytes

electrolytes is known to decrease ionic conductivity. As shown in Fig. 1, the crystallization of the $60Li_2S \cdot 40SiS_2$ glass ([Li]/([Li] + [M]) = 0.75) decreased the conductivity by three orders of magnitude [9]. On the other hand, the enhancement of conductivity by crystallization was found in the $70Li_2S \cdot 30P_2S_5$ glass ([Li]/([Li] + [M]) = 0.70) [10]. A high-temperature phase of $Li_7P_3S_{11}$ was crystallized and the obtained glass-ceramic exhibited an extremely high conductivity of 5.4×10^{-3} S cm^{-1} [11]. A metastable or high-temperature phase with high conductivity tends to precipitate as a primary crystal by heating a supercooled liquid of glass electrolytes. Thus, glass electrolytes have an advantage as a precursor for stabilizing at room temperature a high-temperature phase such as $Li_7P_3S_{11}$, which is difficult to synthesize by a conventional solid-state reaction [12, 13]. The sulfide glass-based materials mentioned above also have a wide electrochemical window of over 5 V, which is a good feature as a solid electrolyte for solid-state batteries.

Application of Glass Electrolytes to Solid-State Rechargeable Lithium Batteries

The electrochemical performance of bulk-type solid-state In//$LiCoO_2$ cells with the Li_2S-SiS_2-Li_3PO_4 oxysulfide glass electrolyte was firstly reported in the early 1990s [14], and then these cells with sulfide solid electrolytes were developed. A schematic of a typical bulk-type solid-state electrochemical cell is shown in Fig. 2. The cell consists of a three-layered powder compressed pellet. The first layer is an indium foil as a negative electrode. The second layer is a Li_2S-P_2S_5 glass-ceramic powder as a solid electrolyte (SE) with a room-temperature conductivity of over 10^{-3} S cm^{-1}. The third layer is a composite powder as a positive electrode. In order to form continuous lithium ion and electron conducting paths to an active material, a composite positive electrode composed of the active material, the SE, and a conductive additive such as acetylene black (AB) and vapor grown carbon fiber (VGCF) is commonly used; a typical weight ratio of those three components is 20 : 30 : 3.

Figure 3 shows the charge–discharge curves of the all-solid-state cell In//Li_2S-P_2S_5 glass-ceramic//$LiCoO_2$ at the 500th cycle under constant current density of 0.064 mA cm^{-2} at 25 °C. The inset shows cycle performance of the cell. Although an irreversible capacity was initially observed at the first several cycles, the all-solid-state cell maintained the reversible capacity of about 100 mAh g^{-1} and the charge–discharge efficiency of 100 % (no irreversible capacity) for 500 cycles, suggesting that the cell operated as a lithium rechargeable battery without decomposition of the solid electrolyte [15, 16].

High temperature operation at 100 °C brings about high rate performance of solid-state cells because of decreasing cell resistance. Figure 4 shows the charge–discharge curve at the 100th cycle (inset) and cycle performance of the all-solid-state cell Li-In//Li_2S-P_2S_5 glass-ceramic//$Li_4Ti_5O_{12}$ at 100 °C [17]. The cell operated reversibly for 700 cycles with the charge–discharge efficiency of 100 % even at

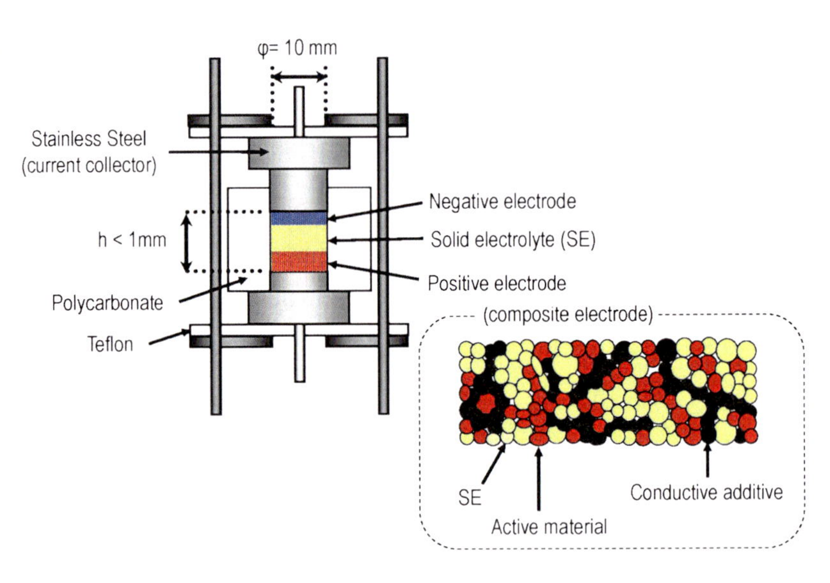

Glass Ion Conductors for Solid-State Lithium Batteries, Fig. 2 Schematic of a bulk-type solid-state cell

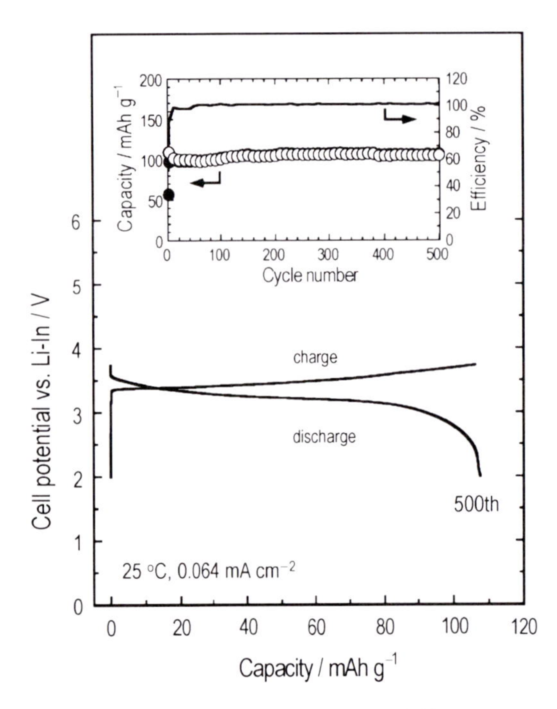

Glass Ion Conductors for Solid-State Lithium Batteries, Fig. 3 Charge–discharge curves at the 500th cycle of the all-solid-state cell In/Li_2S-P_2S5 glass-ceramic/$LiCoO_2$. The inset shows cycling performance of the rechargeable capacities and charge–discharge efficiencies for the cell at 25 °C under the current density of 0.064 mA cm^{-2}

a high current density of 12.7 mA cm^{-2}. The cell showed the discharge and charge capacity of about 140 mAh g^{-1} and exhibited no capacity fading for 700 cycles. There are few reports about the high temperature operation of lithium rechargeable batteries with long cycle life. On the contrary, it is beneficial to use all-solid-state batteries using glass-ceramic electrolyte for high temperature application.

Concluding Remarks

Sulfide glass ion conductors have several advantages as solid electrolytes for bulk-type solid-state batteries because of high conductivity, single Li^+ ion conduction, and a wide electrochemical window. Stabilization of a high temperature or metastable phase by crystallization is another merit of glass materials. Glass-ceramic electrolytes with the $Li_7P_3S_{11}$ phase exhibited the conductivity of 5.4×10^{-3} S cm^{-1}, which is almost the same Li^+ ion conductivity as organic liquid electrolytes in light of the Li^+ ion transference number of liquid electrolytes. Bulk-type solid-state batteries with glass electrolytes have

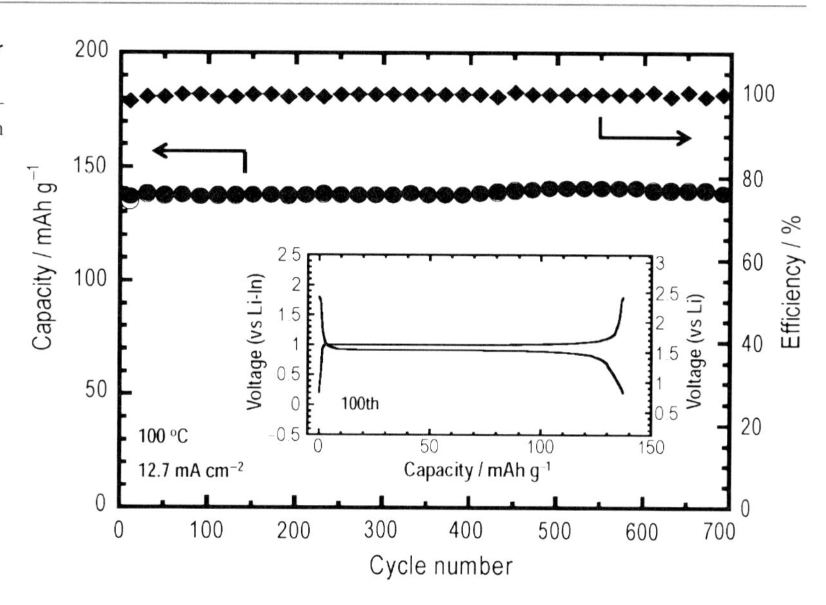

Glass Ion Conductors for Solid-State Lithium Batteries, Fig. 4 Charge–discharge curve at the 100th cycle (inset) and cycle performance of the all-solid-state cell Li-In/ Li_2S-P_2S_5glass-ceramic/ $Li_4Ti_5O_{12}$ at 100 °C under the current density of 12.7 mA cm^{-2}

a lot of merits of high safety, high reliability and simple design because of the elimination of leakage and flammability of cell components. High temperature operation at 100 °C, where a conventional lithium-ion battery with an organic liquid electrolyte is difficult to operate, is effective in increasing rate performance of solid-state batteries with sulfide electrolytes.

Cross-References

- ▶ Lithium Solid Cathode Batteries
- ▶ Lithium-Ion Batteries
- ▶ Lithium-Sulfur Batteries

References

1. Julien C, Nazri GA (1994) Solid state batteries: materials design and optimization. Kluwer Academic, Boston
2. Fergus JW (2010) Ceramic and polymeric solid electrolytes for lithium-ion batteries. J Power Sources 195:4554–4569
3. Minami T, Tatsumisago M, Wakihara M, Iwakura C, Kohjiya S, Tanaka I (2005) Solid state ionics for batteries. Springer, Tokyo
4. Tatsumisago M, Minami T (1987) Lithium ion conducting glasses prepared by rapid quenching. Mater Chem Phys 18:1–17
5. Hayashi A, Hama S, Morimoto H, Tatsumisago M, Minami T (2001) Preparation of Li_2S-P_2S_5 amorphous solid electrolytes by mechanical milling. J Am Ceram Soc 84:477–479
6. Kennedy JH, Zhang Z, Eckert H (1990) Ionically conductive sulfide-based lithium glasses. J Non-Cryst Solids 123:328–338
7. Aotani N, Iwamoto K, Takada K, Kondo S (1994) Synthesis and electrochemical properties of lithium ion conductive glass, Li_3PO_4-Li_2S-SiS_2. Solid State Ion 68:35–39
8. Minami T, Hayashi A, Tatsumisago M (2000) Preparation and characterization of lithium ion-conducting oxysulfide glasses. Solid State Ion 136–137:1015–1023
9. Hayashi A, Tatsumisago M, Minami T (1999) Electrochemical properties for the lithium ion conductive (100-x)(0.6Li_2S · 0.4SiS_2) · xLi_4SiO_4 oxysulfide glasses. J Electrochem Soc 146:3472–3475
10. Mizuno F, Hayashi A, Tadanaga K, Tatsumisago M (2005) New, highly ion-conductive crystals precipitated from Li_2S-P_2S_5 glasses. Adv Mater 17:918–921
11. Hayashi A, Minami K, Ujiie S, Tatsumisago M (2010) Preparation and ionic conductivty of $Li_7P_3S_{11-z}$ glass-ceramic electrolytes. J Non-Cryst Solids 356:2670–2673
12. Yamane H, Shibata M, Shimane Y, Junke T, Seino Y, Adams S, Minami K, Hayashi A, Tatsumisago M (2007) Crystal structure of a superionic conductor, $Li_7P_3S_{11}$. Solid State Ion 178:1163–1167
13. Minami K, Hayashi A, Tatsumisago M (2010) Preparation and characterization of superionic conducting Li7P3S11 crystal from glassy liquids. J Ceramic Soc Jpn 118:305–308
14. Iwamoto K, Aotani N, Takada K, Kondo S (1994) Rechargeable solid state battery with lithium

conductive glass, Li_3PO_4-Li_2S-SiS_2. Solid State Ion 70–71:658–661
15. Minami T, Hayashi A, Tatsumisago M (2006) Recent progress of glass and glass-ceramics as solid electrolytes for lithium secondary batteries. Solid State Ion 177:2715–2720
16. Tatsumisago M, Hayashi A (2008) All-solid-state lithium secondary batteries using sulfide-based glass ceramic electrolytes. Funct Mater Lett 1:1–4
17. Minami K, Hayashi A, Ujiie S, Tatsumisago M (2011) Electrical and electrochemical properties of glass-ceramic electrolytes in the systems Li_2S-P_2S_5-P_2S_3 and Li_2S-P_2S_5-P_2O_5. Solid State Ion 192:122–125

Grain-Boundary Conductivity

Sangtae Kim
Department of Chemical Engineering and Materials Science, University of California, Davis, CA, USA

Introduction and Description

A grain boundary (Fig. 1) is the internal interface between two arbitrarily orientated crystallites in a polycrystalline solid, which comprises several atomic layers. The transitional symmetry of the crystallite breaks at the grain boundary at which the atomic layers are reconstructed to respond to its local environmental changes. Consequently, the grain boundary is anticipated to have an atomic structure that deviates from the crystal interior (namely, the bulk), implying that the physical (and/or mechanical) properties of the grain boundary inherently differ from those of the bulk. The grain boundary also serves as a sink and/or a source for point defects (i.e., irregular atoms in the lattice structure) such as impurity atoms and vacancies which tend to segregate to the grain boundary.

Most engineering materials are polycrystalline in nature. The solid electrolytes (SEs) [1–3] used in solid oxide fuel cells (SOFCs) are made of crystallites with a size of typically few micrometers. The SEs conduct ions exclusively by definition, and there are several different types of oxide materials currently being considered for such applications. Acceptor-doped ZrO_2 and CeO_2, conventional SEs, have a fluorite structure with oxide-ion conductivity of $\sim$ 0.1 S/cm at around 900 °C [1]. $LaGaO_3$-based SEs [4], a relatively recent discovery, on the other hand, has a perovskite structure and presents the conductivity higher than that of the fluorites. Some other oxides with perovskite structure such as acceptor-doped $BaZrO_3$ [5] exhibit the highest conductivity among these in the temperature range of 300–600 °C under wet atmosphere. Such high conductivity is attributed to proton transport in this material, in contrast to those oxygen-ion conducting SEs mentioned above. Regardless of the difference in their bulk conductivity, one of the features they have in common is that the effective conductivities of polycrystalline ceramics of the SEs is noticeably lower than the corresponding bulk conductivities, implying that the grain boundaries further impede the ion transport in the materials.

One of the earlier understandings of the possible origins of the very high resistance to the ion transport across the grain boundary was that impurities introduced to the SEs during processing precipitate in the grain boundaries to form amorphous secondary phases which constrict the ionic current (Fig. 2) [6–8]. So do pores in not fully dense materials since they tend to segregate in the grain boundaries as well. On the other hand, numerous studies [9, 10] have demonstrated later that the resistivity of even a pristine and pore-free grain boundary is higher than the bulk counterpart by orders of magnitude, suggesting that there must be intrinsic causes rather than such extrinsic ones.

As mentioned above, grain boundaries serve as a sink for the point defects. Since these defects are electrically charged, it is inevitable for a grain boundary to acquire excess charge. This excess charge will repel/attract the nearby mobile ions (e.g., oxygen vacancies or protons in the SEs) with the same/opposite charge so that the ions should deplete/accumulate to build, namely, the space charge in the vicinity of the grain boundary (i.e., the space-charge zones) for the charge compensation. As a result, there is substantial

Grain-Boundary Conductivity, Fig. 1 A Tunneling Electron Microscopy (TEM) image of a high angle grain boundary of a $SrTiO_3$ (Courtesy http://ipg.ucdavis.edu/research.html)

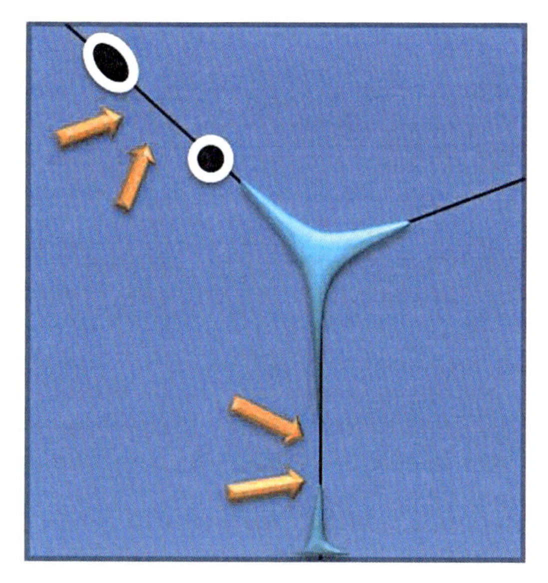

Grain-Boundary Conductivity, Fig. 2 A schematic description of current constriction (*yellow arrows*) at grain boundaries (*solid lines*) due to either impurity phase (*colored light blue*) or pores (*black circles*)

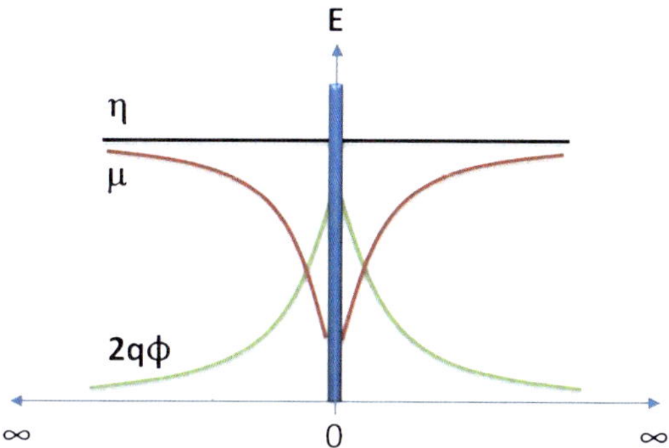

Grain-Boundary Conductivity, Fig. 3 A schematic energy diagram for the oxygen vacancy in the vicinity of a positively charged grain boundary in an SE. It should be noted that E is in a logarithmic scale

modification in the number of the ions in the space-charge zones in which electroneutrality no longer holds [11, 12]. For instance, oxygen vacancies (with a charge of + 2) in an oxide-ion-conducting SE would deplete in the space-charge zone if the excess charge in neighboring grain boundary were positive. A thermodynamic explanation of the oxygen-vacancy depletion in the space-charge zone is the following. The electrochemical potential (η) of the oxygen vacancy should be kept unchanged throughout the SE at an electrochemical equilibrium at a given temperature. On the other hand, the potential (ϕ) built in the grain boundary decays almost exponentially towards the bulk value at a location sufficiently far from the grain boundary. In light of the fact that the electrochemical potential is the sum of chemical potential (μ) and electrical potential ($\eta = \mu + zq\phi$, z and q being the charge number and the elementary charge, respectively.), the chemical potential of the oxygen vacancy must be adjusted to keep the electrochemical potential constant in the space-charge zone, meaning that the concentration of oxygen vacancy in the space-charge zone should be lower than that in the bulk, as shown in Fig. 3.

Figure 3 can also be understood as oxygen vacancies in the vicinity of a grain boundary move over into the grain boundary due to the difference in its chemical potential between in the bulk and at the grain boundary, resulting in the positively charged grain boundary with excess oxygen vacancies contacting with the negatively charged space-charge zone due to depletion of oxygen vacancy. One may then argue that the charge-carrier depletion in the space-charge

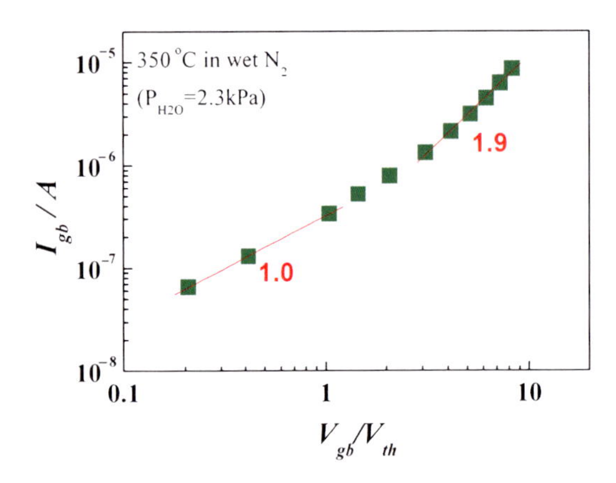

Grain-Boundary Conductivity, Fig. 4 A log-log plot of the current–voltage curve measured for a single grain boundary in 2 mol% Y-doped $BaZrO_3$ at 350 °C [15]

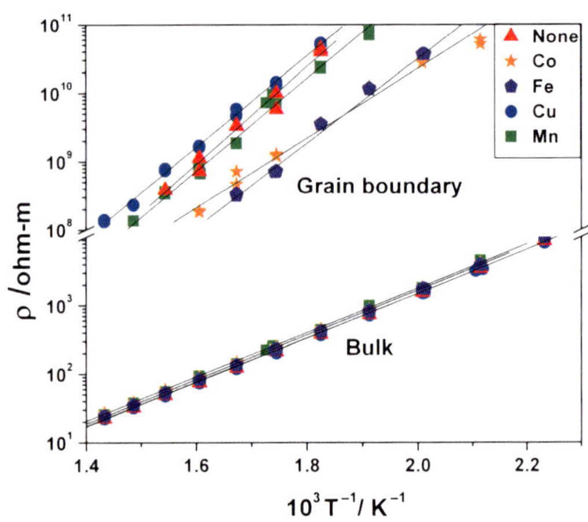

Grain-Boundary Conductivity, Fig. 5 Arrhenius plots of the bulk and grain boundary resistivities of 0.5 mol% transition metal doped $Ce_{0.985}Gd_{0.01}O_{2-x}$ measured under O_2 [21]

zone may primarily be responsible for the high grain-boundary resistance in the SEs.

Extensive investigations have been conducted over the past decade to experimentally verify such a hypothesis, and in fact, numerous results [13–22] have demonstrated that a potential barrier such as shown in Fig. 3 indeed exists at the grain boundary in the SEs mentioned above to limit the ionic current.

One of the most straightforward and widely accepted ways to experimentally confirm the existence of such a potential barrier is to measure the current–voltage (*I–V*) characteristics of the grain boundary. The current across a potential barrier is expected to increase with the increasing applied voltage in a nonlinear manner to display the transition from ohmic to superohmic at a temperature sufficiently higher than the thermal voltage, $V_{th} = \frac{k_B T}{q}$ (with k_B being the Boltzmann constant), at a given temperature. For instance, Fig. 4 exhibits a log-log plot of the *I–V* relation measured for a single grain boundary in 2 mol% Y-doped $BaZrO_3$ at 350 °C [15]. It is apparent that the relation is ohmic (with the slope of 1) at $V/V_{th} = 1$ or below while superohmic (with the slope of ~2) at $V > 3V_{th}$.

This material conducts protons under a wet environment, as mentioned above, and is known to have very high grain-boundary resistivity. The displacement of atoms in the grain boundary was previously attributed as the cause because, according to theoretical calculations, the proton hopping in this material is highly sensitive to local environments [5]. However, the primary cause of the high resistivity of the grain boundary manifests itself in Fig. 4.

To enhance the effective conductivity of those SEs, it is imperative to reduce, if not eliminate, the excess charge in the grain boundary. Namely, the grain-boundary engineering with the intention of manipulating the grain-boundary conductivity is clearly one the future directions of the research foci in this field. The result of one of few attempts with such an idea previously reported is shown in Fig. 5 [21]. In this study, very small amount (<0.5 mol%) of transition metal cations were introduced in 1 mol% Gd-doped CeO_2, in which the grain-boundary conduction is known to primarily be governed by the space charge. The metal cations segregate into the grain boundaries almost exclusively so that their introduction hardly affects the bulk resistivity at all as seen in Fig. 5. On the other hand, the specific grain-boundary resistivity (i.e., the grain-boundary resistivity of a single grain boundary in the material) is reduced by an order of magnitude in some cases (e.g., Co, Fe cations). Such reduction in the grain-boundary resistivity upon introducing cations into the grain boundary was attributed to the following reason:

The cations with lower valence states (< + 4) segregated in the grain boundary serve as acceptors to reduce the net charge of the grain boundary (i.e., reduction in the height of the potential barrier), leading to increase in the ionic current across the grain boundary.

These results demonstrate that grain-boundary engineering to enhance the effective conductivity of the SEs is an achievable goal and yet raise other interrelated intriguing questions to be addressed in the future: (1) What is the source (s) of the net positive charge in the grain boundary; (2) whether the grain boundary can be engineered to serve as a source of the extra charge carriers rather than a sink; and finally (3) whether the grain-boundary itself can serve as a fast conduction path for those ions? The answers to these questions will determine the future success of grain-boundary engineering and ultimately nanostructuring [23, 24] of the SEs (i.e., fabrication of the SEs with very high density of the grain boundary). Currently intensive studies are being conducted to address these questions and the knowledge to be obtained from such efforts will provide great opportunities to achieve grain boundary engineering in the near future.

Cross-References

▸ Conductivity of Electrolytes
▸ Defects in Solids
▸ Solid Electrolytes

References

1. Goodenough JB (2003) Oxide-ion electrolytes. Annu Rev Mater Res 33:91–128
2. Goodenough JB (1997) Ceramic solid electrolytes. Solid State Ion 94:17–25
3. Goodenough JB (1995) Solid electrolytes. Pure Appl Chem 67:931–938
4. Ishihara T, Matsuda H, Takita Y (1994) Doped LaGao3 perovskite-type oxide as a new oxide ionic conductor. J Am Chem Soc 116:3801–3803
5. Kreuer KD (2003) Proton-conducting oxides. Annu Rev Mater Res 33:333–359
6. Badwal SPS (1995) Grain-boundary resistivity in zirconia-based materials - effect of sintering temperatures and impurities. Solid State Ion 76:67–80
7. Badwal SPS, Rajendran S (1994) Effect of microstructures and nanostructures on the properties of ionic conductors. Solid State Ion 70:83–95
8. Badwal SPS (1992) Zirconia-based solid electrolytes - microstructure, stability and ionic-conductivity. Solid State Ion 52:23–32
9. Aoki M, Chiang YM, Kosacki I, Lee IJR, Tuller H, Liu YP (1996) Solute segregation and grain-boundary impedance in high-purity stabilized zirconia. J Am Ceram Soc 79:1169–1180
10. Christie GM, vanBerkel FPF (1996) Microstructure - ionic conductivity relationships in ceria-gadolinia electrolytes. Solid State Ion 83:17–27
11. Maier J (1995) Ionic-conduction in-space charge regions. Prog Solid State Ch 23:171–263
12. Kim S, Fleig J, Maier J (2003) Space charge conduction: simple analytical solutions for ionic and mixed conductors and application to nanocrystalline ceria. Phys Chem Chem Phys 5:2268–2273
13. Guo X, Maier J (2001) Grain boundary blocking effect in zirconia: a Schottky barrier analysis. J Electrochem Soc 148:E121–E126
14. Guo X, Waser R (2006) Electrical properties of the grain boundaries of oxygen ion conductors: acceptor-doped zirconia and ceria. Prog Mater Sci 51:151–210
15. Iguchi F, Chen CT, Yugami H, Kim S (2011) Direct evidence of potential barriers at grain boundaries in Y-doped BaZrO(3) from dc-bias dependence measurements. J Mater Chem 21:16517–16523
16. Chen CT, Danel CE, Kim S (2011) On the origin of the blocking effect of grain-boundaries on proton transport in yttrium-doped barium zirconates. J Mater Chem 21:5435–5442
17. Avila-Paredes HJ, Choi K, Chen CT, Kim S (2009) Dopant-concentration dependence of grain-boundary conductivity in ceria: a space-charge analysis. J Mater Chem 19:4837–4842
18. Park HJ, Kim S (2008) The enhanced electronic transference number at the grain boundaries in Sr-doped LaGaO3. Solid State Ion 179:1329–1332
19. Park HJ, Kim S (2007) Space charge effects on the interfacial conduction in Sr-doped lanthanum gallates: a quantitative analysis. J Phys Chem C 111:14903–14910
20. Lee JS, Anselmi-Tamburini U, Munir ZA, Kim S (2006) Direct evidence of electron accumulation in the grain boundary of yttria- doped nanocrystalline zirconia ceramics. Electrochem Solid State Lett 9:J34–J36
21. Avila-Paredes HJ, Kim S (2006) The effect of segregated transition metal ions on the grain boundary resistivity of gadolinium doped ceria: alteration of the space charge potential. Solid State Ion 177:3075–3080
22. Kim S, Maier J (2002) On the conductivity mechanism of nanocrystalline ceria. J Electrochem Soc 149:J73–J83

23. Maier J (2005) Nanoionics: ion transport and electrochemical storage in confined systems. Nat Mater 4:805–815
24. Tuller HL, Bishop SR (2010) Tailoring material properties through defect engineering. Chem Lett 39:1226–1231

Graphene (or Reduced Graphite Oxide Nanosheets)

Wataru Sugimoto
Faculty of Textile Science and Technology, Shinshu University, Nagano, Japan

Introduction and Background

As a nanoscopic two-dimensional crystallite with an ultimate thickness of one sp^2-bonded carbon, graphene has attracted great interest from many researchers as the newest item in the nanocarbon family. The high theoretical surface area and good electronic conductivity of graphene have allowed this new material to be projected as a potential electrode material for many applications. Graphene has been known for a while; the term "graphene layer" was first proposed in 1986 as the name for a single carbon layer occurring in graphite intercalation compounds [1, 2]. It can be regarded as the final member of fused polycyclic aromatic hydrocarbons with infinite size. Thus, a freestanding slice of graphene will be composed of all surface atoms, with a theoretical surface area of 2,630 $m^2\ g^{-1}$. Ever since graphene was first experimentally isolated in 2004 by micromechanical cleavage of highly oriented pyrolytic graphite with scotch tape [3], researchers have excitedly turned to the material to discover its potential applications. One potential application is electrochemical double-layer capacitors, which presently use activated carbon as the electrode material. Research on graphene-based electrodes for electrochemical double-layer capacitors has already shown a remarkable increase, with already over 10 review articles appearing in different journals [4–18]. Close to 200 papers regarding EDLC performance of graphene-based material have been published in the past 5 years. About 60 % of these papers deal with composites of graphene with oxides, polymers, metals, and other carbonaceous materials (carbon nanotubes, carbon black, etc.). In order to clarify the properties of graphene with respect to EDLC performance, this section will concentrate on research conducted on "single-component" graphene electrodes, that is, material without any second material.

Single- to multilayer graphene can be obtained by micromechanical cleavage of graphite [3, 19, 20] or by epitaxially growth via controlled thermal decomposition of SiC [21–25] and chemical vapor deposition (CVD) of hydrocarbons [26, 27]. Graphene produced in such ways are sometimes called physical graphene in contrast to chemical graphene which is prepared by reduction of chemically exfoliated graphite oxide nanosheets. Micromechanical cleavage of graphite crystals affords multilayered graphene (or more precisely, thin small flakes of graphite). Single-layer graphene can be isolated by repeated peeling, a rather strenuous process unfit for large-scale production. Epitaxial graphene allows production of large area, high-quality samples free of impurities and defects, beneficial for fundamental studies on the physical properties of graphene and has been used as an integral part of 3-D structures [28]. The properties of such physically grown graphene often do not represent the intrinsic properties of pure freestanding graphene, as the products are mainly multilayered and significant interaction between the substrate cannot be avoided. Nonetheless, electrical double-layer capacitors based on physical graphene electrodes have a potential to replace aluminum electrolytic capacitors for ac line filtering [29, 30]. The challenge for energy storage applications is the preparation of high surface area graphene in microgram to milligram scale for fundamental studies and kilogram scale for practical application.

Graphene-Based Electrodes for Supercapacitor Application

Researchers have recently circumvented the problem of limited-scale production by switching

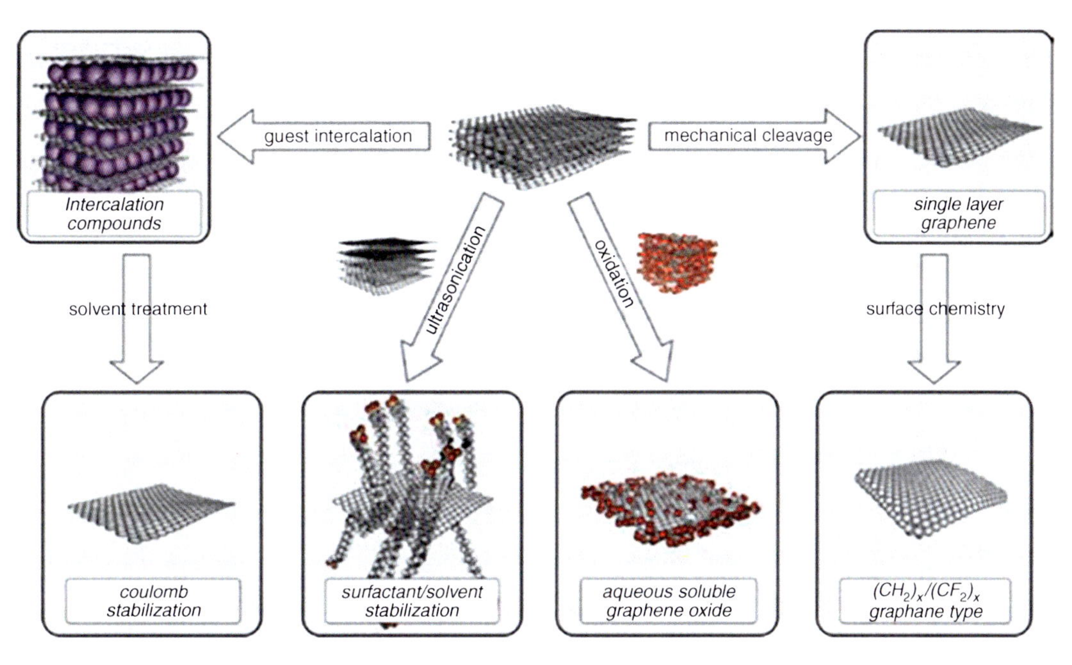

Graphene (or Reduced Graphite Oxide Nanosheets), Fig. 1 Selected pathways towards graphene [31]

from physical to chemical processing methods (Fig. 1). For example, exfoliation of graphite oxide (GO) in solvents affords single layers of graphite oxide nanosheets stabilized in the form of a colloid. GO itself is not a good electric conductor, thus the sp^2 network must be restored by chemical or gas-phase reduction. Various nomenclatures have appeared in the literature for such reduced graphite oxide (RGO) nanosheets, including chemical graphene (an antonym to physical graphene) and even simply graphene. Scalability is perhaps the largest advantage of such chemical processing techniques, leading to agglomerates of single- (or multi-) layered graphene with a re-stacked structure. Conversely, drawbacks of such preparation techniques include defects induced during the processing, incomplete reduction of graphite oxide nanosheets, and loss of accessible surface area due to re-stacking and aggregation. Incomplete reduction leaves functional groups on the surface which is normally considered as unsolicited impurities for electrical double-layer capacitor applications. Leakage current increases proportionally with increasing amount of surface functional groups [32] and thus, the amount of such "surface impurities" is decreased as much as possible for commercial capacitor grade activated carbon. It should be noted that partial reduction may in some cases have a positive impact on the electrode structure, as surface functional groups may obstruct re-stacking of the nanosheets leading to possible higher utilization of the surface area.

Table 1 summarizes a survey of the synthetic approaches and resulting capacitance values of graphene-based electrodes (see also Refs. [5–7, 11, 14, 31, 76–80] for a review of synthetic approaches). Natural graphite is the most commonly used starting material, while more atypical materials such as CNTs [46] and platelet carbon nanofibers [55] have also been used. The (modified) Hummers method [81] is the most mutual method for oxidation of carbon, while the Staudenmaier [82] and Brodie [83] methods have also been applied. Reduction of the oxidized graphite oxide nanosheets is often conducted with hydrazine.

Figure 2 illustrates the reported specific capacitance values, categorized by various electrolytes. Capacitance values for graphene-based electrodes range from <100 to 350 F g^{-1}. Neutral

Graphene (or Reduced Graphite Oxide Nanosheets), Table 1 Summary of specific capacitance of graphene-based electrodes prepared via various methods

Electrolyte	Capacitance/F g^{-1}	Electrode preparation	Surface area	Starting material	Oxidation process	Reduction/ post treatment	Treatment temperature /°C	Ref
$TEABF_4$	150	Sheet (50 μm)	3,100	–	Hummers	Microwave + KOH activation	800	[33]
$TEABF_4$	72	Drop cast	–	Sigma-Aldrich	Hummers	Wet-thermal (in PC)	150	[34]
$TEABF_4$	147	Drop cast	–	SNO-10; SEC carbon	Hummers	Electrochemical	–	[35]
$TEABF_4$	110	Sheet (50 μm)	–	–	Hummers	Wet-thermal	150	[36]
$TEABF_4$	220	Sheet (500 μm)	–	200 mesh; Alfa Aesar or Timrex SFG44 (synthetic graphite)	Brodie	Dry(Ar)-thermal	>200	[37]
$TEABF_4$	140	Bulk	–	Natural graphite 325 meh; Qingdao HuaTai Lubricant Sealing	Hummers	Hydrothermal	180	[38]
$TEABF_4$	94	Sheet	705	SP-1; Bay carbon	Hummers	Hydrazine	100	[39]
$TEABF_4$	30	–	–	Needle coke	Chemical ox	None	–	[40]
$BMIBF_4$	166	Sheet (50 μm)	2,400	–	Hummers	Microwave + KOH activation	800	[33]
$BMIBF_4$	155	Sheet (50 μm)	2,700	–	Hummers + thermal exfoliation	KOH activated	800	[33]
$BMIBF_4$	91	–	–	–	Hummers	Hydrazine	95	[41]
$BMIBF_4$	144	Drop cast	–	300 mesh; Alfa Aesar	Hummers	Hydrazine	95	[42]
$BMIPF_6$	132	Drop cast	617	Natural graphite; Alfa Aesar	Hummers	Hydrazine	–	[43]
$BMIPF_4$	158	Sheet	–	–	Hummers	HBr	110	[44]
$EMIBF_4$	91	–	–	–	Hummers	Hydrazine	95	[41]
$EMIBF_4$	154	Sheet	–	Natural graphite	Hummers	Hydrazine	95	[45]
Et_3MeBF_4	131	Sheet	356	Graphite; Wako	Staudenmaier	Thermal	1,050	[46]
Et_3MeBF_4	131	Sheet	871	SWCNT	Staudenmaier	Thermal	1,050	[46]
Et_3MeBF_4	122	Sheet	368	–	Hummers	Thermal (vacuum)	200	[47]
EMITFSI	200	Sheet (50 μm)	3,100	–	Hummers	Microwave + KOH activation	800	[33]

$MMIBF_4$	96	–	–	–	Hummers	Hydrazine	95	[41]
$PYR_{14}TFSI$	75	–	925	2–15 μm graphite; Alfa Aesar	Staudenmaier	Thermal	1,050	[48]
KCl	128	–	–	Aldrich	Hummers	Electrochemical	–	[49]
KCl	135	Sheet	–	–	Hummers	$NaBH_4$	80	[50]
Li_2SO_4	238	Sheet	353	Natural graphite 5 μm	Hummers	None	–	[51]
Na_2SO_4	99	Sheet	353	Natural graphite 5 μm	Hummers	None	–	[51]
Na_2SO_4	160	Drop cast	–	<45 μm; Sigma-Aldrich	Hummers	Electrochemical	–	[52]
Na_2SO_4	176	Layer-by-Layer	–	Crude flake; Qingdao Aoke	Hummers	None	–	[53]
Na_2SO_4	156	EPD	–	–	Hummers	Thermal-Electrochemical	1,050	[54]
H_2SO_4	82	Drop cast	–	3 μm graphite; Z-5 F, Ito Kokuen	Hummers	Hydrazine	60	[55]
H_2SO_4	132	Drop cast	–	PCNF	Hummers	Hydrazine	60	[55]
H_2SO_4	117	–	925	2–15 μm graphite; Alfa Aesar	Staudenmaier	Thermal	1,050	[48]
H_2SO_4	123	Sheet	871	Graphite; Wako	Staudenmaier	Thermal	1,050	[46]
H_2SO_4	199	Sheet	871	SWCNT	Staudenmaier	Thermal	1,050	[46]
H_2SO_4	132	Ink jet print	–	Commercial; Cheap tubes	–	N_2	200	[56]
H_2SO_4	193	Bulk	–	Natural graphite; 325 mesh Qingdao HuaTai Lubricant Sealing	Hummers	Sodium ascorbate	80	[57]
H_2SO_4	266	Drop cast	–	300 mesh; Alfa Aesar	Hummers	Hydrazine	95	[42]
H_2SO_4	240	Bulk	–	325 mesh natural graphite	Hummers	Sodium ascorbate	90	[58]
H_2SO_4	348	Sheet	–	–	Hummers	HBr	110	[44]
H_2SO_4	276	Drop cast	–	230U; Asbury	Hummers	Hydrothermal	150	[59]
H_3PO_4	247	Layer-by-Layer	–	SP-1; Bay carbon	Hummers	Hydrazine	50	[60]
KOH	164	EPD	–	Natural graphite	Hummers	*p*-Phenylenediamine	400	[61]
KOH	130	Bulk	–	325 mesh	Hummers	Hydrazine	95	[62]
KOH	255	Sheet	–	–	CVD	None	–	[63]
KOH	150	Sheet	524	Natural graphite	Staudenmaier	Thermal	1,000	[64]
KOH	211	Layer-by-Layer	–	230U; Asbury Carbons	Hummers	Hydrazine + thermal	500	[65]
KOH	136	Sheet	492	Commercial; XG sciences	–	None	–	[66]
KOH	264	Sheet	368	–	Hummers	Thermal(vacuum)	200	[47]

(continued)

Graphene (or Reduced Graphite Oxide Nanosheets), Table 1 (continued)

Electrolyte	Capacitance/F g^{-1}	Electrode preparation	Surface area	Starting material	Oxidation process	Reduction/ post treatment	Treatment temperature /°C	Ref
KOH	191	Sheet	463	–	Hummers	Microwave	–	[67]
KOH	80	Interfacial self-assembly	–	–	–	–	–	[68]
KOH	230	Sheet	404	Natural graphite universal grade	Hummers	Thermal	–	[69]
KOH	203	Bulk	964	Graphite	Hummers	Hydrothermal	180	[70]
KOH	184	Bulk	778	Graphite	Hummers	HI	100 (3 h)	[70]
KOH	205	Bulk	818	Graphite	Hummers	HI	100 (8 h)	[70]
KOH	187	Bulk	911	Graphite	Hummers	Hydrazine	95 (3 h)	[70]
KOH	222	Bulk	951	Graphite	Hummers	Hydrazine	95 (8 h)	[70]
KOH	120	Sheet	–	<45 μm Aldrich	Pyrenecarboxylic acid	None	–	[71]
KOH	205	–	320	Flake graphite; Qingdao Tianhe Graphite	Hummers	Hydrazine	72 h, vapor	[72]
KOH	90	–	–	Flake graphite; Qingdao Tianhe Graphite	Hummers	Hydrazine	24 h, vapor	[72]
KOH	68	–	–	Flake graphite; Qingdao Tianhe Graphite	Hummers	Hydrazine + Ar	24 h, vapor +400 °C (Ar)	[72]
KOH	260	Sheet	72	Flake graphite; Qingdao Tianhe Graphite	Hummers	N_2	200	[73]
KOH	135	Sheet	705	SP-1; Bay carbon	Hummers	Hydrazine	100	[39]
KOH	100	Bulk	542	–	CVD	–	–	[74]
NaOH	127	Dip-coat	464	Graphite	Hummers	Hydrazine	60	[75]
NaOH	165	Dip-coat	391	MWCNT	Hummers	Hydrazine	60	[75]

electrolytes tend to give lower capacitance values than acidic and basic electrolytes; nonaqueous systems are slightly lower than aqueous systems. Such trends are not astonishing and follow the general rules for other carbonaceous materials. One of the reasons for the wide range in reported capacitance can be attributed to the contribution from pseudocapacitance originating from surface functional groups [59]. This is particularly true for the case of acidic electrolytes, where a large portion of the capacitance is due to pseudocapacitance, as shown in Fig. 3. The porosity of the graphene-based material is another important property that must be considered, as electrochemical double-layer capacitors store charge at the electrode/electrolyte interface. The specific surface area measured by N_2 adsorption/desorption studies ranges from ~400 to 1,000 $m^2\ g^{-1}$ (Table 1), which are much smaller than the theoretical value for a freestanding graphene as well as most typical activated carbons. The theoretical surface area is of course a hypothetical case assuming that both sides of the nanomaterial are 100 % accessible, which naturally is not possible for electrodes. In a real electrode system, a significant amount of the surface area is not available for electrolyte penetration because overlaping or re-stacking of the individual graphene will always exist.

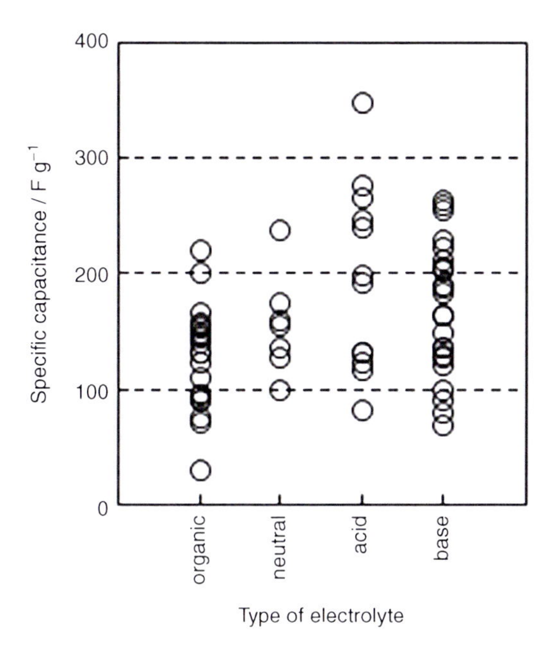

Graphene (or Reduced Graphite Oxide Nanosheets), Fig. 2 Reported specific capacitance in various electrolytes. See Table 1 for details

The type of starting material can also have a large impact on both lateral size and thickness of the graphite oxide nanosheet, which in turn impacts the specific capacitance of graphene-based electrodes. For example, it has been shown that the mass specific capacitance of commercially available graphene increases with

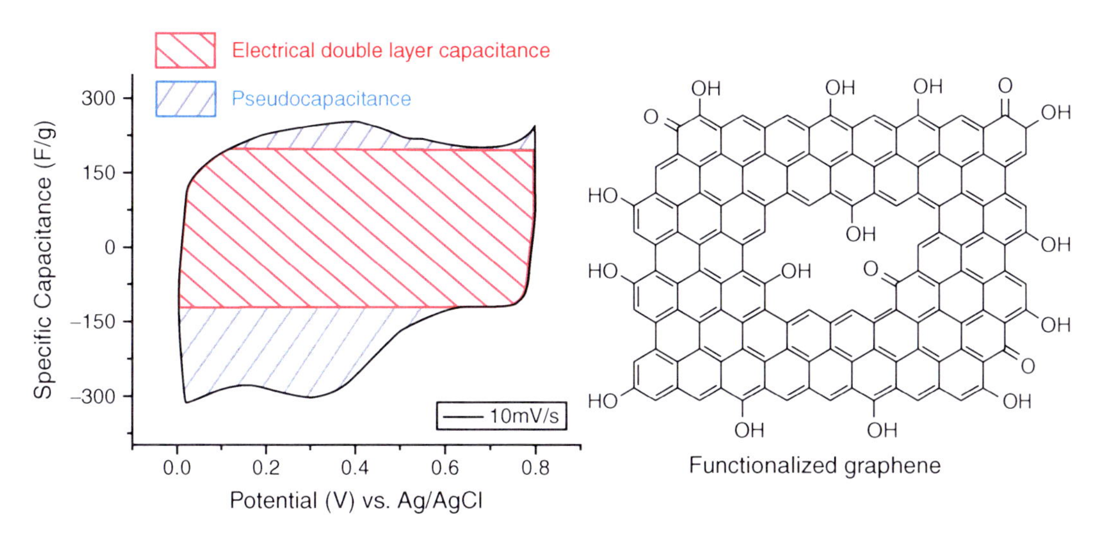

Graphene (or Reduced Graphite Oxide Nanosheets), Fig. 3 Cyclic voltammogram of functionalized graphene (partially reduced graphene) in 1 M H_2SO_4 [59]

increasing number of graphene layers from single, few, to multilayers [84], and the area specific capacitance has been reported to scale with increasing number of layers [85–87]. The size of the starting material is also another factor that influences the specific capacitance [55, 68]. The effect of reducing conditions have been pursued by various groups [15, 49, 64, 88].

Comparison with Graphene-Like Materials

While graphene has potentially high surface area, its morphology may make it more difficult to produce porous electrodes, as the graphene sheets are likely to stack on top of one another without intervening pore volume for the electrolyte. So far, reported specific surface area values of graphene-based materials are considerably lower than most capacitor grade activated carbons. However, activated carbons are primarily microporous, and not all of the surface area measured by gas adsorption/desorption studies can be utilized for electrochemical charge storage. Thus, despite the lower surface areas measured, the specific capacitance of graphene materials is generally comparable to most commercial activated carbons. As a result, the area specific capacitance of graphene-based electrodes is generally higher ($\sim$5–40 μF cm^{-2}) than activated carbons ($\sim$5 μF cm^{-2}). This suggests that graphene-based materials are potentially in favor for higher energy and power density per volume.

Activated carbon used for electrical double-layer capacitors is typically constituted by few layers of graphene with lateral dimension in the order of <50 nm. Thus one design principle for graphene-based materials would be to synthesize porous carbonaceous materials with similar dimensions. Activated carbon prepared from graphite oxide is a good example of such material, with well-developed micro/mesoporosity, specific surface area of 3,100 m^2 g^{-1}, and low content of surface functional groups [33]. Microporous carbons prepared by using zeolite as templates (ZTC) also have a graphene-like framework [89]. ZTC is comprised by the assembly of single, non-stacked nanometer-sized graphene fragments curved like buckybowls. ZTCs can have specific surface area as high as 4,000 m^2 g^{-1} [90], with good volumetric and gravimetric capacitance [91, 92].

A number of studies report capacitive behavior of graphene-based electrodes without any reducing process (thus, graphite oxide nanosheet electrodes). In such cases, there will be contribution of surface redox properties originating from the surface functional groups [59], and such material would be expected to be poor electronic conductors. However, relatively high capacitance values have been reported. These studies are consistent with earlier work on the capacitive behavior of expanded graphitic carbon fibers where the specific capacitance can be as high as 555 F g^{-1} with specific surface area of 300 m^2 g^{-1}.

Future Directions

In summary, perhaps it is too early in the stage of research to identify or isolate which variable affects the electrochemical double-layer capacitance most. In many cases, the material is not sufficiently or fully characterized, lacking identification on the purity of the material in terms of number of layers (distribution in thickness of the exfoliated graphite oxide) and lateral size of individual graphene, quantitative and qualitative analysis of surface functional groups, porous structure of the aggregate, etc. Such and other properties have been identified as to have an influence on the capacitive behavior. Chemical graphene would be a strong candidate for electrochemical capacitors if one can obtain enhanced performance in terms of either gravimetric or volumetric capacitance compared to conventional activated carbon, while minimizing the amount of surface functional groups.

References

1. Boehm HP, Setton R, Stumpp E (1986) Nomenclature and terminology of graphite intercalation compounds. Carbon 24:241–245
2. Boehm H-P, Setton R, Stumpp E (1994) Nomenclature terminology of graphite intercalation compounds

(IUPAC recommendations 1994). Pure Appl Chem 66:1893–1901
3. Novoselov KS, Geim AK, Morozov SV, Jiang D, Zhang Y, Dubonos SV, Grigorieva IV, Firsov AA (2004) Electric field effect in atomically thin carbon films. Science 306:666–669
4. Jang BZ, Zhamu A (2008) Processing of nanographene platelets (NGPs) and NGP nanocomposites: a review. J Mater Sci 43:5092–5101
5. Rao CNR, Sood AK, Subrahmanyam KS, Govindaraj A (2009) Graphene: the new two-dimensional nanomaterial. Angew Chem Int Ed Engl 48:7752–7777
6. Pumera M (2009) Electrochemistry of graphene: new horizons for sensing and energy storage. Chem Rec 9:211–223
7. Zhu Y, Murali S, Cai W, Li X, Suk JW, Potts JR, Ruoff RS (2010) Graphene and graphene oxide: synthesis, properties, and applications. Adv Mater 22:3906–3924
8. Zhang LL, Zhou R, Zhao XS (2010) Graphene-based materials as supercapacitor electrodes. J Mater Chem 20:5983
9. Rao CNR, Sood AK, Voggu R, Subrahmanyam KS (2010) Some novel attributes of graphene. J Phys Chem Lett 1:572–580
10. Zhai Y, Dou Y, Zhao D, Fulvio PF, Mayes RT, Dai S (2011) Carbon materials for chemical capacitive energy storage. Adv Mater 23:4828–4850
11. Chen XM, Wu GH, Jiang YQ, Wang YR, Chen X (2011) Graphene and graphene-based nanomaterials: the promising materials for bright future of electroanalytical chemistry. Analyst 136:4631–4640
12. Radich JG, Mcginn PJ, Kamat PV (2011) Graphene-based composites for electrochemical energy storage. Electrochem Soc Interface 20:63–66
13. Li C, Shi G (2011) Synthesis and electrochemical applications of the composites of conducting polymers and chemically converted graphene. Electrochim Acta 56:10737–10743
14. Inagaki M, Kim YA, Endo M (2011) Graphene: preparation and structural perfection. J Mater Chem 21:3280
15. Brownson DAC, Kampouris DK, Banks CE (2011) An overview of graphene in energy production and storage applications. J Power Sources 196:4873–4885
16. Davies A, Yu A (2011) Material advancements in supercapacitors: from activated carbon to carbon nanotube and graphene. Can J Chem Eng 89:1342–1357
17. Kamat PV (2011) Graphene-based nanoassemblies for energy conversion. J Phys Chem Lett 2:242–251
18. Bose S, Kuila T, Mishra AK, Rajasekar R, Kim NH, Lee JH (2012) Carbon-based nanostructured materials and their composites as supercapacitor electrodes. J Mater Chem 22:767
19. Lu X, Huang H, Nemchuk N, Ruoff RS (1999) Patterning of highly oriented pyrolytic graphite by oxygen plasma etching. Appl Phys Lett 75:193–195
20. Lu X, Yu M, Huang H, Ruoff RS (1999) Tailoring graphite with the goal of achieving single sheets. Nanotechnology 10:269–272
21. Van Bommel AJ, Crombeen JE, Van Tooren A (1975) LEED and auger electron observations of the SiC (0001) surface. Surf Sci 48:463–472
22. Forbeaux I, Themlin J-M, Debever J-M (1998) Heteroepitaxial graphite on 6H-SiC(0001): interface formation through conduction-band electronic structure. Phys Rev B 58:16396–16406
23. Berger C, Song Z, Li T, Li X, Ogbazghi AY, Feng R, Dai Z, Marchenkov AN, Conrad EH, First PN, de Heer WA (2004) Ultrathin epitaxial graphite: 2D electron gas properties and a route toward graphene-based nanoelectronics. J Phys Chem B 108:19912–19916
24. Berger C, Song Z, Li X, Wu X, Brown N, Naud C, Mayou D, Li T, Hass J, Marchenkov AN, Conrad EH, First PN, de Heer WA (2006) Electronic confinement and coherence in patterned epitaxial graphene. Science 312:1191–1196
25. Ohta T, Bostwick A, Seyller T, Horn K, Rotenberg E (2006) Controlling the electronic structure of bilayer graphene. Science 313:951–954
26. Land TA, Michely T, Behm RJ, Hemminger JC, Comsa G (1992) STM investigation of single layer graphite structures produced on Pt(111) by hydrocarbon decomposition. Surf Sci 264:261–270
27. Nagashima A, Nuka K, Itoh H, Ichinokawa T, Oshima C, Otani S (1993) Electronic states of monolayer graphite formed on TiC(111) surface. Surf Sci 291:93–98
28. Lin Y-M, Valdes-Garcia A, Han S-J, Farmer DB, Meric I, Sun Y, Wu Y, Dimitrakopoulos C, Grill A, Avouris P, Jenkins KA (2011) Wafer-scale graphene integrated circuit. Science 332:1294–1297
29. Miller JR, Outlaw RA, Holloway BC (2010) Graphene double-layer capacitor with ac line-filtering performance. Science 329:1637–1639
30. Miller JR, Outlaw RA, Holloway BC (2011) Graphene electric double layer capacitor with ultra-high-power performance. Electrochim Acta 56:10443–10449
31. Dreyer DR, Park S, Bielawski CW, Ruoff RS (2011) The chemistry of graphene oxide. Electrochem Soc Interface 20:53–56
32. Yoshida A, Tanahashi I, Nishino A (1990) Effect of concentration of surface acidic functional groups on electric double-layer properties of activated carbon fibers. Carbon 28:611–615
33. Zhu Y, Murali S, Stoller MD, Ganesh KJ, Cai W, Ferreira PJ, Pirkle A, Wallace RM, Cychosz KA, Thommes M, Su D, Stach EA, Ruoff RS (2011) Carbon-based supercapacitors produced by activation of graphene. Science 332:1537–1541
34. Ku K, Kim B, Chung H, Kim W (2010) Characterization of graphene-based supercapacitors fabricated on Al foils using Au or Pd thin films as interlayers. Synth Met 160:2613–2617
35. Harima Y, Setodoi S, Imae I, Komaguchi K, Ooyama Y, Ohshita J, Mizota H, Yano J (2011)

Electrochemical reduction of graphene oxide in organic solvents. Electrochim Acta 56:5363–5368
36. Zhu Y, Stoller MD, Cai W, Velamakanni A, Piner RD, Chen D, Ruoff RS (2010) Exfoliation of graphite oxide in propylene carbonate and thermal reduction of the resulting graphene oxide platelets. ACS Nano 4:1227–1233
37. Hantel MM, Kaspar T, Nesper R, Wokaun A, Kötz R (2011) Partially reduced graphite oxide for supercapacitor electrodes: effect of graphene layer spacing and huge specific capacitance. Electrochem Commun 13:90–92
38. Sun Y, Wu Q, Shi G (2011) Supercapacitors based on self-assembled graphene organogel. Phys Chem Chem Phys 13:17249–17254
39. Stoller MD, Park S, Zhu Y, An J, Ruoff RS (2008) Graphene-based ultracapacitors. Nano Lett 8:3498–3502
40. Yang S, Kim I, Jeon M, Kim K, Moon S, Kim H, An K (2008) Preparation of graphite oxide and its electrochemical performance for electric double layer capacitor. J Ind Eng Chem 14:365–370
41. Liu W, Yan X, Lang J, Xue Q (2011) Electrochemical behavior of graphene nanosheets in alkylimidazolium tetrafluoroborate ionic liquid electrolytes: influences of organic solvents and the alkyl chains. J Mater Chem 21:13205
42. Zhang K, Mao L, Zhang LL, On Chan HS, Zhao XS, Wu J (2011) Surfactant-intercalated, chemically reduced graphene oxide for high performance supercapacitor electrodes. J Mater Chem 21:7302
43. Fu C, Kuang Y, Huang Z, Wang X, Yin Y, Chen J, Zhou H (2010) Supercapacitor based on graphene and ionic liquid electrolyte. J Solid State Electrochem 15:2581–2585
44. Chen Y, Zhang X, Zhang D, Yu P, Ma Y (2011) High performance supercapacitors based on reduced graphene oxide in aqueous and ionic liquid electrolytes. Carbon 49:573–580
45. Liu C, Yu Z, Neff D, Zhamu A, Jang BZ (2010) Graphene-based supercapacitor with an ultrahigh energy density. Nano Lett 10:4863–4868
46. Inoue T, Mori S, Kawasaki S (2011) Electric double layer capacitance of graphene-like materials derived from single-walled carbon nanotubes. Jpn J Appl Phys 50:01AF07
47. Lv W, Tang D-M, He Y-B, You C-H, Shi Z-Q, Chen X-C, Chen C-M, Hou P-X, Liu C, Yang Q-H (2009) Low-temperature exfoliated graphenes: vacuum-promoted exfoliation and electrochemical energy storage. ACS Nano 3:3730–3736
48. Subrahmanyam KS, Vivekchand SRC, Govindaraj A, Rao CNR (2008) A study of graphenes prepared by different methods: characterization, properties and solubilization. J Mater Chem 18:1517
49. Peng X-Y, Liu X-X, Diamond D, Lau KT (2011) Synthesis of electrochemically-reduced graphene oxide film with controllable size and thickness and its use in supercapacitor. Carbon 49:3488–3496
50. Yu A, Roes I, Davies A, Chen Z (2010) Ultrathin, transparent, and flexible graphene films for supercapacitor application. Appl Phys Lett 96:253105
51. Hsieh C-T, Hsu S-M, Lin J-Y, Teng H (2011) Electrochemical capacitors based on graphene oxide sheets using different aqueous electrolytes. J Phys Chem C 115:12367–12374
52. Shao Y, Wang J, Engelhard M, Wang C, Lin Y (2010) Facile and controllable electrochemical reduction of graphene oxide and its applications. J Mater Chem 20:743
53. Ni P, Li H, Yang M, He X, Li Y, Liu Z-H (2010) Study on the assembling reaction of graphite oxide nanosheets and polycations. Carbon 48:2100–2105
54. Liu S, Ou J, Wang J, Liu X, Yang S (2011) A simple two-step electrochemical synthesis of graphene sheets film on the ITO electrode as supercapacitors. J Appl Electrochem 41:881–884
55. Sato J, Takasu Y, Fukuda K, Sugimoto W (2011) Graphene nanoplatelets via exfoliation of platelet carbon nanofibers and its electric double layer capacitance. Chem Lett 40:44–45
56. Le LT, Ervin MH, Qiu H, Fuchs BE, Lee WY (2011) Graphene supercapacitor electrodes fabricated by inkjet printing and thermal reduction of graphene oxide. Electrochem Commun 13:355–358
57. Wu Q, Sun Y, Bai H, Shi G (2011) High-performance supercapacitor electrodes based on graphene hydrogels modified with 2-aminoanthraquinone moieties. Phys Chem Chem Phys 13:11193–11198
58. Xuan SK, Xi XY, Li C, Quan SG (2011) High-performance self-assembled graphene hydrogels prepared by chemical reduction of graphene oxide. New Carbon Mater 26:9–15
59. Lin Z, Liu Y, Yao Y, Hildreth OJ, Li Z, Moon K, Ping WC (2011) Superior capacitance of functionalized graphene. J Phys Chem C 115:7120–7125
60. Yoo JJ, Balakrishnan K, Huang J, Meunier V, Sumpter BG, Srivastava A, Conway M, Reddy ALM, Yu J, Vajtai R, Ajayan PM (2011) Ultrathin planar graphene supercapacitors. Nano Lett 11:1423–1427
61. Chen Y, Zhang X, Yu P, Ma Y (2010) Electrophoretic deposition of graphene nanosheets on nickel foams for electrochemical capacitors. J Power Sources 195:3031–3035
62. Tai Z, Yan X, Lang J, Xue Q (2012) Enhancement of capacitance performance of flexible carbon nanofiber paper by adding graphene nanosheets. J Power Sources 199:373–378
63. Ning G, Fan Z, Wang G, Gao J, Qian W, Wei F (2011) Gram-scale synthesis of nanomesh graphene with

high surface area and its application in supercapacitor electrodes. Chem Commun 47:5976–5978
64. Du X, Guo P, Song H, Chen X (2010) Graphene nanosheets as electrode material for electric double-layer capacitors. Electrochim Acta 55:4812–4819
65. Tang LAL, Lee WC, Shi H, Wong EYL, Sadovoy A, Gorelik S, Hobley J, Lim CT, Loh KP (2012) Highly wrinkled cross-linked graphene oxide membranes for biological and charge-storage applications. Small 8:423–431
66. Li Y, van Zijll M, Chiang S, Pan N (2011) KOH modified graphene nanosheets for supercapacitor electrodes. J Power Sources 196:6003–6006
67. Zhu Y, Murali S, Stoller MD, Velamakanni A, Piner RD, Ruoff RS (2010) Microwave assisted exfoliation and reduction of graphite oxide for ultracapacitors. Carbon 48:2118–2122
68. Biswas S, Drzal LT (2010) Multilayered nano-architecture of variable sized graphene nanosheets for enhanced supercapacitor electrode performance. ACS Appl Mater Interfaces 2:2293–2300
69. Du Q, Zheng M, Zhang L, Wang Y, Chen J, Xue L, Dai W, Ji G, Cao J (2010) Preparation of functionalized graphene sheets by a low-temperature thermal exfoliation approach and their electrochemical supercapacitive behaviors. Electrochim Acta 55:3897–3903
70. Zhang L, Shi G (2011) Preparation of highly conductive graphene hydrogels for fabricating supercapacitors with high rate capability. J Phys Chem C 115:17206–17212
71. An X, Simmons T, Shah R, Wolfe C, Lewis KM, Washington M, Nayak SK, Talapatra S, Kar S (2010) Stable aqueous dispersions of noncovalently functionalized graphene from graphite and their multifunctional high-performance applications. Nano Lett 10:4295–4301
72. Wang Y, Shi Z, Huang Y, Ma Y, Wang C, Chen M, Chen Y (2009) Supercapacitor devices based on graphene materials. J Phys Chem C 113:13103–13107
73. Zhao B, Liu P, Jiang Y, Pan D, Tao H, Song J, Fang T, Xu W (2012) Supercapacitor performances of thermally reduced graphene oxide. J Power Sources 198:423–427
74. Jiang F, Fang Y, Xue Q, Chen L, Lu Y (2010) Graphene-based carbon nano-fibers grown on thin-sheet sinter-locked Ni-fiber as self-supported electrodes for supercapacitors. Mater Lett 64:199–202
75. Wang G, Ling Y, Qian F, Yang X, Liu X-X, Li Y (2011) Enhanced capacitance in partially exfoliated multi-walled carbon nanotubes. J Power Sources 196:5209–5214
76. Allen MJ, Tung VC, Kaner RB (2010) Honeycomb carbon: a review of graphene. Chem Rev 110:132–145
77. Green AA, Hersam MC (2010) Emerging methods for producing monodisperse graphene dispersions. J Phys Chem Lett 1:544–549
78. Park S, Ruoff RS (2009) Chemical methods for the production of graphenes. Nat Nanotechnol 4:217–224
79. Singh V, Joung D, Zhai L, Das S, Khondaker SI, Seal S (2011) Graphene based materials: past, present and future. Prog Mater Sci 56:1178–1271
80. Choi W, Lahiri I, Seelaboyina R, Kang YS (2010) Synthesis of graphene and its applications: a review. Crit Rev Solid State Mater Sci 35:52–71
81. Hummers WS Jr, Offeman RE (1958) Preparation of graphitic oxide. J Am Chem Soc 80:1339
82. Staudenmaier L (1898) Verfahren zur darstellung der graphitsäure. Ber Dtsch Chem Ges 31:1481–1487
83. Brodie B (1860) Sur le poids atomique du graphite. Ann Chim Phys 59:466–472
84. Goh MS, Pumera M (2010) Multilayer graphene nanoribbons exhibit larger capacitance than their few-layer and single-layer graphene counterparts. Electrochem Commun 12:1375–1377
85. Wei WD, Li F, Shuai WZ, Ren W, Ming CH (2009) Electrochemical interfacial capacitance in multilayer graphene sheets: dependence on number of stacking layers. Electrochem Commun 11:1729–1732
86. Stoller MD, Magnuson CW, Zhu Y, Murali S, Suk JW, Piner R, Ruoff RS (2011) Interfacial capacitance of single layer graphene. Energy Environ Sci 4:4685–4689
87. Ye J, Craciun MF, Koshino M, Russo S, Inoue S, Yuan H, Shimotani H, Morpurgo AF, Iwasa Y (2011) Accessing the transport properties of graphene and its multilayers at high carrier density. Proc Natl Acad Sci 108:13002–13006
88. Luo D, Zhang G, Liu J, Sun X (2011) Evaluation criteria for reduced graphene oxide. J Phys Chem C 115:11327–11335
89. Nishihara H, Yang Q-H, Hou P-X, Unno M, Yamauchi S, Saito R, Paredes JI, Martínez-Alonso A, Tascón JMD, Sato Y (2009) A possible buckybowl-like structure of zeolite templated carbon. Carbon 47:1220–1230
90. Matsuoka K, Yamagishi Y, Yamazaki T, Setoyama N, Tomita A, Kyotani T (2005) Extremely high microporosity and sharp pore size distribution of a large surface area carbon prepared in the nanochannels of zeolite Y. Carbon 43:876–879
91. Itoi H, Nishihara H, Kogure T, Kyotani T (2011) Three-dimensionally arrayed and mutually connected 1.2-nm nanopores for high-performance electric double layer capacitor. J Am Chem Soc 133:1165–1167
92. Nishihara H, Itoi H, Kogure T, Hou P-X, Touhara H, Okino F, Kyotani T (2009) Investigation of the ion storage/transfer behavior in an electrical double-layer capacitor by using ordered microporous carbons as model materials. Chem Eur J 15:5355–5363

Green Electrochemistry

Jorge G. Ibanez[1], Alanah Fitch[2], Bernardo A. Frontana-Uribe[3,4] and Ruben Vasquez-Medrano[1]
[1]Department of Chemical Engineering and Sciences, Universidad Iberoamericana, México, Mexico
[2]Department of Chemistry, Loyola University, Chicago, IL, USA
[3]Centro Conjunto de Investigación en Química Sustentable, UAEMéx-UNAM, Toluca, Estado de México, Mexico
[4]Instituto de Química UNAM, Mexico, Mexico

Introduction

Green chemistry (GC) has been defined as the utilization of a series of principles that *reduce or eliminate* the use or generation of dangerous substances during the *design, fabrication, or application* of chemical products [1].

Electrochemistry is naturally suited to conform to most of the principles involved in green chemistry. There are several environmentally favorable features of electrochemical transformations including [2–4] (a) electrons are intrinsically clean reagents; (b) most of the reactions may take place at room temperature which reduces energy consumption, the risk of corrosion, material failure, and the cost associated to temperature controls; (c) reactions may occur in low or null volatility reaction media (e.g., the use of ionic liquids), and this reduces accidental solvent releases to the atmosphere; (d) electrodes function as heterogeneous catalysts (they are easily separated from the products); (e) when the heterogeneous electron transfer is naturally slow, electrochemically active mediator species may help and later be electrochemically recovered; (f) electrochemistry is flexible in the sense that it can be used to treat neutral, positive, or negatively charged species and induce the production of precipitates, gaseous species, pH changes, or charge neutralization; it can deal with solids, liquids, or gases and with inorganic, organic, or biochemical substances; (g) cost-effectiveness, since the required equipment and operations are normally simple and if properly designed, they can also be made less expensive than other techniques.

Electrochemical Techniques/Operations

Green chemistry principles are now enunciated and exemplified with different electrochemical techniques/operations: [5]

1. *GC Principle: Use and generate non-dangerous substances with low or no toxicity*

   Example 1: The production of a nitrating agent friendlier than nitric acid [6, 7]

   N_2O_5 is a powerful nitrating agent that can be obtained easily by the electrochemical reduction of the nitrite ion to N_2O_4 with its subsequent oxidation at the anode to N_2O_5 (Fig. 1). It has been used in aromatic nitrations in liquid CO_2.

   Example 2: Synthesis of organic carbamates via O_2^- and CO_2

   The traditional route implies the use of toxic phosgene for the production of biochemicals and insecticides. The proposed greener route (Fig. 2) does not require $COCl_2$ and the yields are high (ca. 90 %, for the linear compounds) [8].

2. *GC Principle: Maximize the incorporation of all the process materials and elements in the final products.* For example, $AB + CD \rightarrow ABC + D$, *vs.* $AB + C \rightarrow ABC$

   Example 3: Electrochemical activation to replace chlorine

   The traditional route for the production of imidazoles, carboxamides, and sulfonamides has several disadvantages including the fact that dilute HCl is difficult to recover and that the process uses Cl_2. In the green route no HCl is produced and no Cl_2 is required (Fig. 3). If the process is coupled to the cathodic reduction of the phthalic acid dimethyl ester, phthalide can be efficiently obtained and the overall process has a 100 % atom economy [9].

3. *GC Principle: Minimize the use of auxiliary substances (e.g., solvents); use alternative, cleaner, and safer solvents.*

Electrochemical methods are uniquely suited to adopt ionic liquid technology. Ionic liquids are green replacements for traditional solvents because they are recyclable and have low p_{vap} (see an example in Fig. 4). The third advantage is that they are intrinsically ionic conductors, and therefore no additional electrolyte is required to obtain conductivity [10, 11].

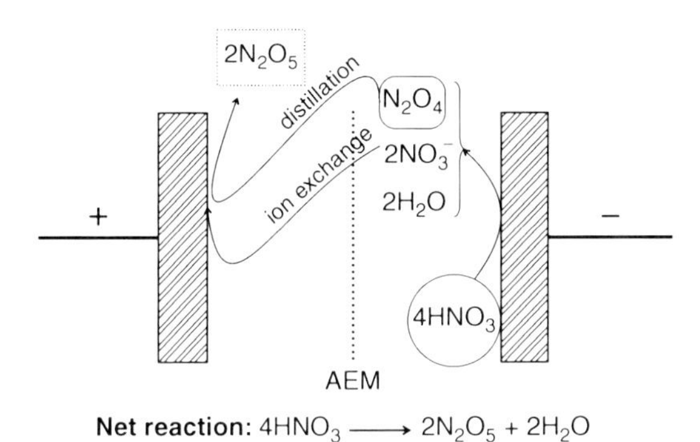

Green Electrochemistry, Fig. 1 Electrochemical production of N_2O_5 from the nitrite ion (*AEM* anionic exchange membrane)

Example 4: Electrogenerated organic cation radicals in room-temperature ionic liquids

Electrochemical mechanisms appear to be conserved when using ionic liquids as compared to conventional organic media [12]. Minor changes are noted. For example, the electron-transfer kinetics from aromatic molecules to the electrode becomes slower, indicating a higher solvent reorganization during charge transfer. The bimolecular reaction rates may decrease, partly due to a lowering of the limiting diffusion-controlled kinetics rate constant together with a specific solvation effect of reactants in these special media.

Example 5: Supercritical fluids

Many liquids and solids are highly soluble in supercritical fluids like CO_2 (see Fig. 5). The use of such solvents often vintages a great ease of separation [13].

Example 6: The electrochemical synthesis of CH_3OH in supercritical CO_2

The reaction is

$$CO_2 + 6H^+ + 6e^- \rightarrow CH_3OH + H_2O \quad (1)$$

(Cu cathode, 80 °C, 68 bar, $\eta = 40\,\%$) [14]. This process has several advantages such as (a) the CO_2 solvent system is environmentally friendly, (b) the CO_2 solvent can also serve as

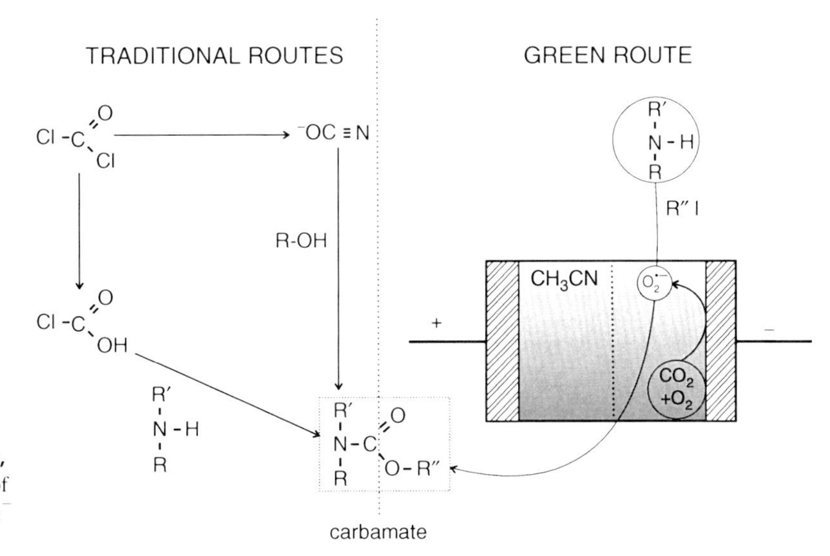

Green Electrochemistry, Fig. 2 Electrosynthesis of organic carbamates via O_2^- and CO_2

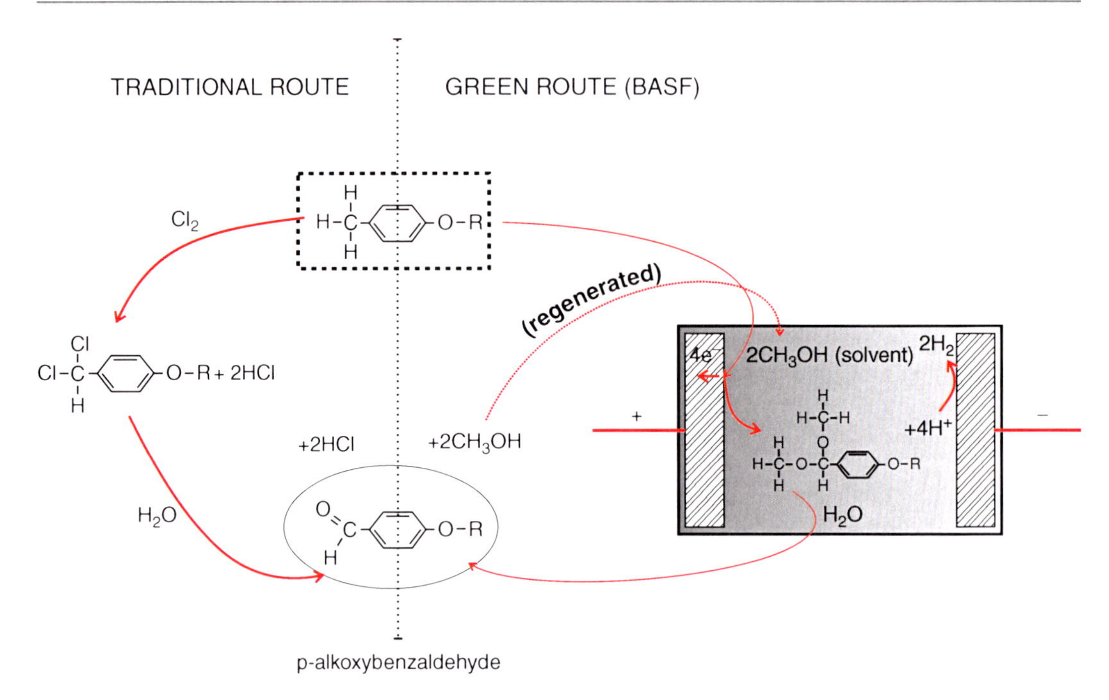

Green Electrochemistry, Fig. 3 Paired electrosynthesis of 4-(alkoxy)benzaldehyde dimethylacetal at the anode and of phthalide at the cathode by reduction of the dimethyl ester of the phthalic acid (not shown in the figure). This is the BASF-AG process

Green Electrochemistry, Fig. 4 1-Butyl-3-methyl imidazolium hexafluorophosphate, [bmim][PF_6]. m. p. = 10 °C

the raw material for the synthesis itself, (c) mass transfer limitations (which plague electrochemistry in ordinary liquids) are eliminated since the solvent is plentiful in the system and readily available, (d) the electrochemical reduction of CO_2 to methanol allows subsequent synthesis of higher-value materials, and (e) this is a possible route for a partial mitigation of CO_2 emissions.

4. *GC Principle: Prefer catalysts to reagents that need to be added in stoichiometric amounts.*

   This strategy has several advantages: (a) minimizes the use of energy by lowering the activation energy, (b) reduces the amount of reagents and waste products, and (c) enhances selectivity.

Example 7: The use of highly reactive metal complexes as electrocatalysts

Highly reactive metal complexes serve as electrocatalysts for a number of detoxification reactions. An example is the use of Co(II) N, N'-bis(salicylidene)ethylene (CoSalen) to facilitate the electrochemical reductive elimination of chlorine from hexachlorobenzene (HCB) [15]. Even though more environmentally friendly conditions have yet to be developed to scale up this process, the CV voltammogram in Fig. 6 shows separate dehalogenation electrochemical steps from the sequence below.

$$C_6Cl_6 \xrightarrow[-Cl^-]{2e^-,\,H^+} C_6HCl_5 \xrightarrow[-Cl^-]{2e^-,\,H^+} C_6H_2Cl_4 \xrightarrow[-Cl^-]{2e^-,\,H^+} C_6H_3Cl_3 \xrightarrow[-Cl^-]{2e^-,\,H^+} C_6H_4Cl_2 \xrightarrow[-Cl^-]{2e^-,\,H^+} C_6H_5Cl \xrightarrow[-Cl^-]{2e^-,\,H^+} C_6H_6$$

5. *GC Principle: Utilize real-time controls to prevent the formation of dangerous substances*

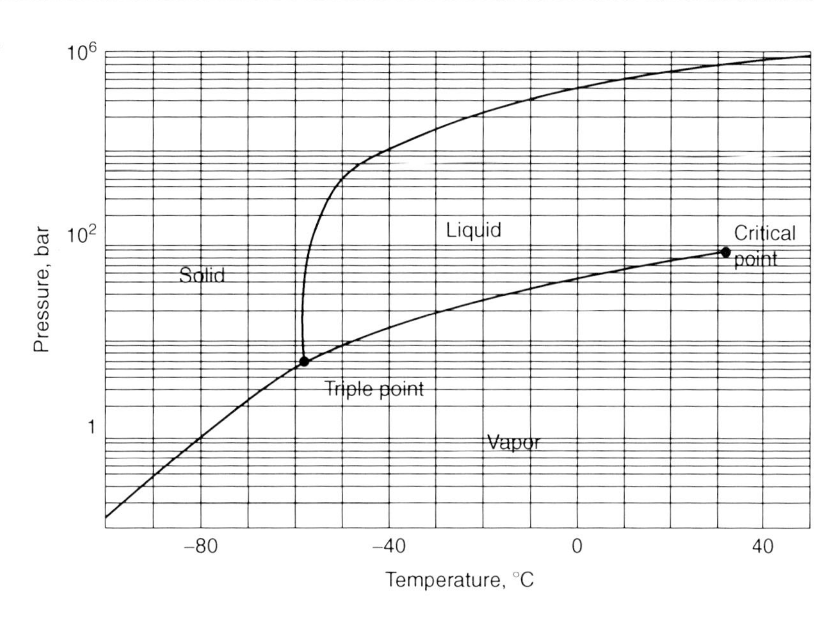

Green Electrochemistry, Fig. 5 Phase diagram of CO_2

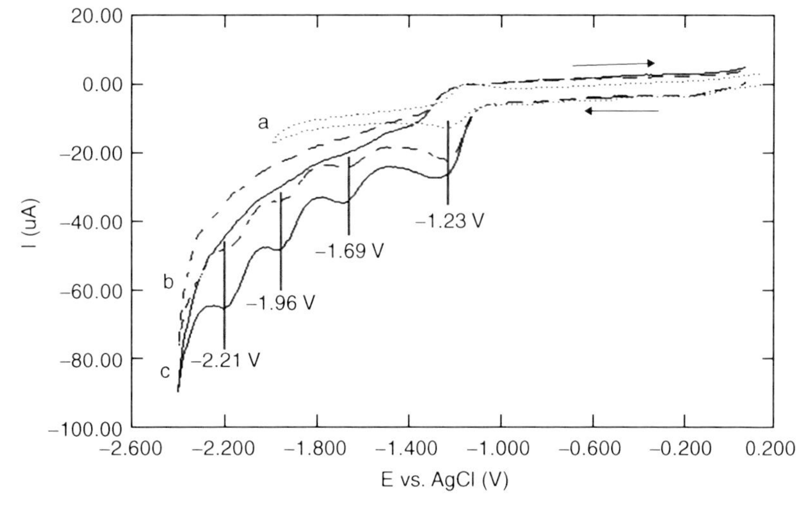

Green Electrochemistry, Fig. 6 Cyclic voltammogram of showing consecutive chlorine reductions from hexachlorobenzene (HCB) in the presence of CoSalen. HCB in 0.1 M TBA-BF4 in ACN + 0.5 mM Co(II) Salen: (**a**) supporting electrolyte, (**b**) 25 ppm, and (**c**) 50 ppm of HCB [v = 100 mV/s] [15]

The electrical variables inherent to the electrochemical processes (i, E) are particularly suited for facilitating data acquisition, automation, and process control. In addition, electroanalytical methods tend to be less costly than other more sophisticated analytical methods [4].

6. *GC Principle: Minimize energy requirements*

Lower temperatures and pressures are typically required in electrochemical operations compared to those of equivalent non-electrochemical counterparts (e.g., incineration, supercritical oxidation). In addition, the applied potentials can be controlled, and the electrodes and cells can be designed to minimize power losses due to poor current distribution and voltage drops [4].

Example 8: The design of simultaneous or paired electrochemical processes

This promotes the use of electrical energy on both sides of the cell. Classical examples include the production of chlorine–sodium hydroxide from brine and the electrorefining of Cu. More recent proposals include

(a) The divergent production (i.e., the production of two different substances from the

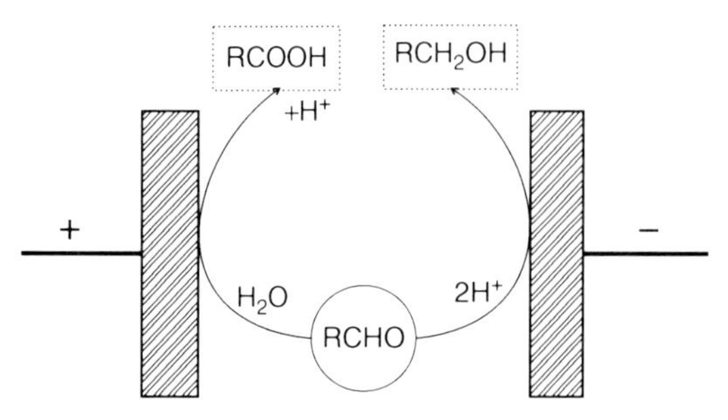

Green Electrochemistry, Fig. 7 Paired production of gluconic acid and sorbitol

same starting material) of gluconic acid and sorbitol from the same aldehyde (Fig. 7) [6, 16].

(b) The production of H_2SO_4 and NaOH by membrane salt splitting (Fig. 8) [4].

(c) The convergent simultaneous synthesis of ClO_2 (the Ibero process, Fig. 9) [17].

(d) The simultaneous removal of a metal pollutant and destruction of its accompanying ligand (Fig. 10) [4]. The metal here is deposited at the cathode and the organic ligand is oxidatively destroyed at the anode, as shown in the following reactions. Some anodes like the boron-doped diamond electrodes catalyze the total mineralization of the organic, but others merely degrade the ligand into carboxylic acids of low molecular weight.

Cathode:

$$M^{n+}(aq) + ne^- \rightarrow M^0(s) \quad (2)$$

Anode:

$$L(aq) \rightarrow nCO_2(g) + nH_2O(l) + ne^- \quad (3)$$

(e) The removal of insoluble metal salts, MX(s), by chelation and regeneration of M + L [18].

Here, thermodynamic diagrams greatly aid in the prediction of predominance zones for the electroactive compounds

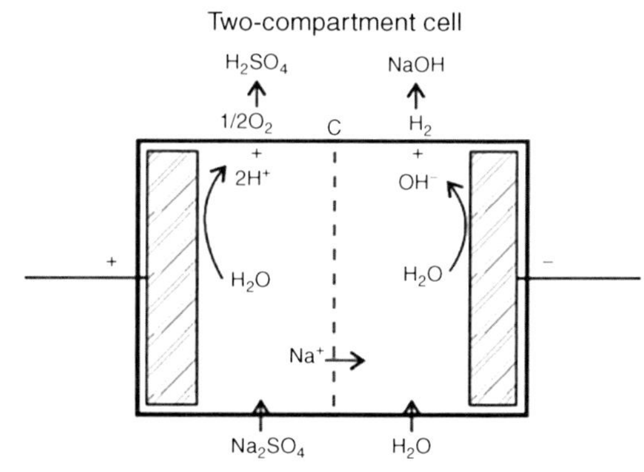

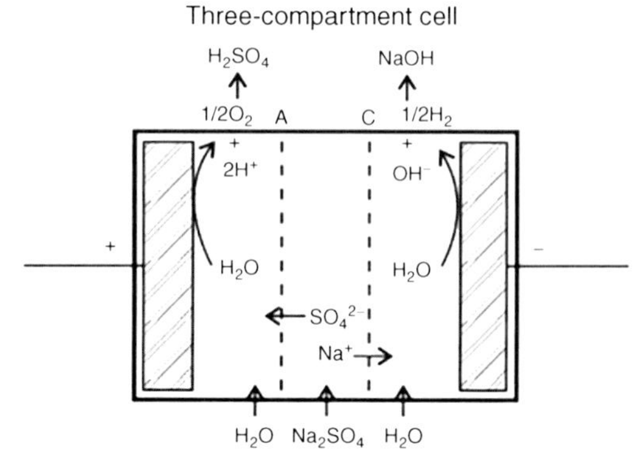

Green Electrochemistry, Fig. 8 Production of H_2SO_4 and NaOH by membrane salt splitting (*A* anionic exchange membrane, *C* cationic exchange membrane)

[19, 20]. For example, when a residual metal complex, CuL (where L = EDTA), is treated electrochemically, a Pourbaix-type diagram (Fig. 11) predicts the conditions under which each species predominates. Since the most electroactive species is $CuHEDTA^-$, its E-pH region of stability shown in the figure is suited for the recovery of Cu^0.

The crucial electrochemical reactions are thought to be as follows: [21]

Anode:

$$2H_2O(l) \rightarrow O_2(g) + 4H^+(aq) + 4e^- \quad (4)$$

$$EDTA^{4-}(aq) + 4H^+(aq) \rightarrow H_4EDTA(s) \quad (5)$$

Cathode:

$$\mathrm{CuHEDTA^-(aq) + 3H^+(aq) + 2e^- \rightarrow Cu^0(s) + H_4EDTA(s)} \quad (6)$$

and both the ligand and the metal are recovered as solids.

Other uses of electrochemistry in the treatment and destruction of pollutants arena abound [22, 23]. Due to its heterogeneous liquid–solid electrode operation, it has been mainly used to treat wastewaters, but it also has been applied to the electrorecovery of soils [24, 25]. Another popular application involves electrocoagulation, where adsorption of water pollutants on solid particles with their subsequent removal is a very effective method for the cleanup of various types of wastewaters. Colloidal particles are especially ubiquitous as pollutants, and their wide-ranging sizes preclude a single type of treatment based either on physical or chemical principles. Coagulation is commonly a fitting choice for this purpose. In order to prevent the concomitant addition of anions in the coagulation step, the electrical route offers the *electrocoagulation* process (EC) as an alternative [26, 27]. Here, Fe or Al anodes are used in an electrochemical cell to furnish the corresponding Fe^{n+} or Al^{3+} ions required for the coagulation process. When such ions encounter the hydroxyl ions produced by water electrolysis at the cathode of the cell, iron hydroxide/oxyhydroxide precipitates form that can adsorb a myriad of polluting species.

Many other examples exist which show that electrochemistry is a vibrant option for the greening of synthetic processes [28, 29] and that it can collaborate in the preservation of our environment.

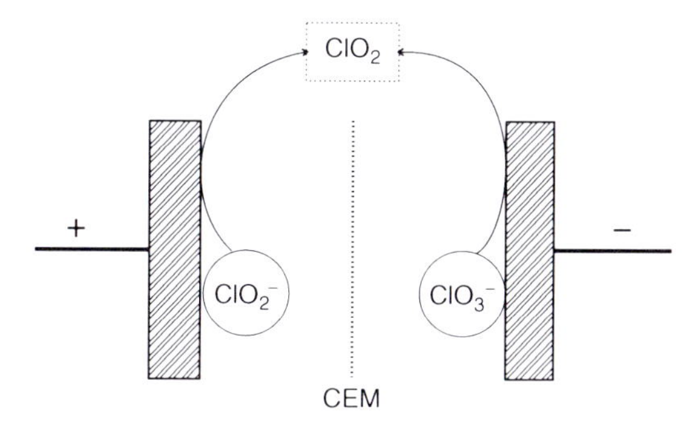

Green Electrochemistry, Fig. 9 The convergent simultaneous synthesis of ClO_2 (*CEM* cationic exchange membrane)

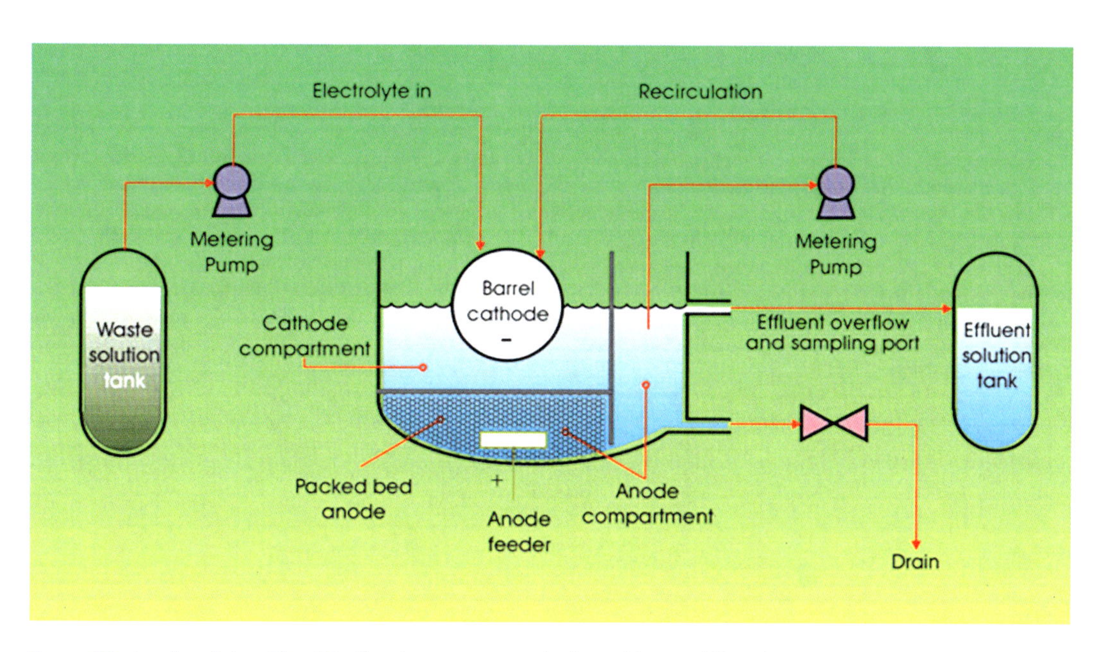

Green Electrochemistry, Fig. 10 Simultaneous removal of metal ions and ligands

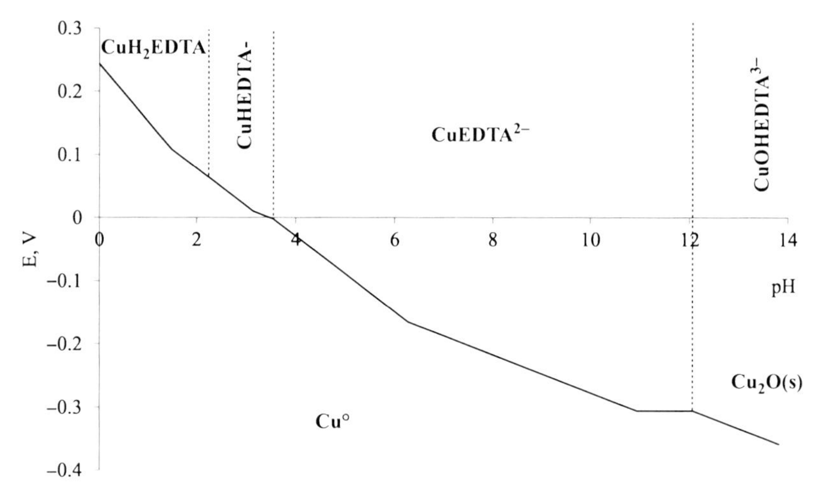

Green Electrochemistry, Fig. 11 Simultaneous cathodic recovery of a metal ion and anodic regeneration of its ligand [18]

Cross-References

▶ Electrocatalytic Synthesis

References

1. Anastas JWP (1998) Green chemistry: theory and practice. Oxford University Press, New York
2. Matthews MA (2001) Green electrochemistry. Examples and challenges. Pure Appl Chem (IUPAC) 73(8):1305–1308
3. Steckhan E, Arns T, Heineman WR, Hilt G, Hoorman D, Jorissen J, Krone L, Lewall B, Putter H (2001) Environmental protection and economization of resources by electroorganic and electroenzymatic syntheses. Chemosphere 43:63–73
4. Rajeshwar K, Ibanez JG (1997) Environmental electrochemistry: fundamentals and applications in pollution abatement. San Diego, Academic Press
5. Frontana-Uribe BA, Little RD, Ibanez JG, Palma A, Vasquez-Medrano R (2010) Organic electrosynthesis: a promising green methodology in organic chemistry. Green Chem 12:2099–2119
6. Walsh F (1993) A first course in electrochemical engineering. The Electrochemical Consultancy, Hants (England)
7. Bakke JM, Hegbom I, Ovreeide E, Aaby K (1994) Nitration of aromatic and heteroaromatic compounds by dinitrogen pentaoxide. Acta Chem Scand 48:1001–1006
8. Casadei MA, Moracci FM, Zappia G, Inesi A, Rossi L (1997) Electrogenerated superoxide-activated carbon dioxide. A new mild and safe approach to organic carbamates. J Org Chem 62:6754–6759
9. Degner D (1985) In: Weinberg N, Tilak BV (eds.) Techniques of electroorganic synthesis. Part III. Wiley: New York
10. Tsuda T, Hussey CL (2007) Electrochemical applications of room-temperature ionic liquids. The Electrochemical Society Interface 16(1):42–49
11. Ohno H (2006) Electrochemical aspects of ionic liquids. New York, John Wiley
12. Lagrost C, Carrié D, Vaultier M, Hapiot P (2003) Reactivities of some electrogenerated organic cation radicals in room-temperature ionic liquids: toward an alternative to volatile organic solvents? J Phys Chem A 107(5):745–752
13. Ibanez JG, Hernandez-Esparza M, Doria-Serrano C, Fregoso-Infante A, Singh MM (2007) Environmental chemistry: fundamentals. New York, Springer
14. Li J, Prentice G (1997) Electrochemical synthesis of methanol from CO_2 in high-pressure electrolyte. TECHNICAL PAPERS - electrochemical science and technology. J Electrochem Soc 144(12): 4284–4288
15. Paramo-Garcia U, Avila-Rodriguez M, Garcia-Jimenez MG, Gutierrez-Granados S, Ibanez-Cornejo JG (2006) Electrochemical reduction of hexachlorobenzene in organic and aquo-organic media with Co(II)salen as catalyst. Electroanal 18(9):904–910
16. Paddon CA, Atobe M, Fuchigami T, He P, Watts P, Haswell SJ, Pritchard GJ, Bull SD, Marken F (2006) Towards paired and coupled electrode reactions for clean organic microreactor electrosyntheses. J Appl Electrochem 36:617–634
17. Gomez-Gonzalez A, Ibanez JG, Vasquez-Medrano R, Zavala-Araiza D, Paramo-Garcia U (2009) Electrochemical paired convergent production of ClO_2 from $NaClO_2$ and $NaClO_3$. In: electrochemical applications to biology, nanotechnology, and environmental engineering and materials. ECS Transactions 20(1):91–101
18. Ibanez JG, Balderas-Hernández P, Garcia-Pintor E, Barba-Gonzalez S, Doria-Serrano MC, Hernaiz-Arce L, Diaz-Perez A, Lozano-Cusi A (2011) Laboratory experiments on the electrochemical remediation of

the environment. Part 9: microscale recovery of a soil metal pollutant and its extractant. J Chem Educ 88:1123–1125.
19. Rojas-Hernandez A, Rodriguez-Laguna N, Ramirez-Silva MT, Moya-Hernandez R (2012) Distribution diagrams and graphical methods to determine or to use the stoichiometric coefficients of acid–base and complexation reactions. Chap. 13. In: Alessio Innocenti. stoichiometry and research–the importance of quantity in biomedicine. InTech Open Access, Rijeka, Croatia, pp 287–310
20. Rojas A, Gonzalez I (1986) Relationship of Two-dimensional predominance-zone diagrams with conditional constants for complexation equilibria. Anal Chim Acta 187:279–285
21. Allen JS, Fenton SS, Fenton JM, Sundstrom DW (1996) 51st Purdue Ind. Waste conference proceeding. Ann Arbor Press, Chelsea, pp 601–612
22. Martinez-Huitle CA, Brillas E (2008) Electrochemical alternatives for drinking water disinfection. Angew Chem Int Ed 47:1998–2005
23. Cominellis C, Chen G (eds) (2010) Electrochemistry for the environment. New York, Springer
24. Alcántara MT, Gómez J, Pazos M, Sanromán MA (2008) Combined treatment of PAHs contaminated soils using the sequence extraction with surfactant–electrochemical degradation. Chemosphere 70:1438–1444
25. Pociecha M, Lestan D (2009) EDTA leaching of Cu contaminated soil using electrochemical treatment of the washing solution. J Hazard Mat 165:533–539
26. Ibanez JG, Vazquez-Olavarrieta JL, Hernández-Rivera L, García-Sánchez MA, Garcia-Pintor E (2012) A novel combined electrochemical-magnetic method for water treatment. Water Sci Technol 65(11):2079–2083
27. Cardenas-Peña AM, Ibanez JG, Vasquez-Medrano R (2012) Determination of the point of zero charge for electrocoagulation precipitates from an iron anode. Int J Electrochem Sci 7:6142–6153

28. Genders D, Weinberg N (eds) (1992) Electrochemistry for a cleaner environment. The Electrosynthesis Co, East Amherst, New York
29. Palomar M (ed) (2005) Applications of analytical chemistry in environmental research. Research Signpost, Kerala, India

H

High-Temperature Ceramic Electrochemical Sensors

Prabir K. Dutta and Sheikh A. Akbar
The Ohio State University, Columbus, OH, USA

Introduction

Chemical sensors are widely used for health and safety (e.g., medical diagnostics, air quality monitoring, and detection of toxic, flammable, and explosive gases), energy efficiency, and emission control in combustion processes and industrial process control for improved productivity. There is a continuing need for the development of sensitive, selective, and low-cost sensors for applications in automotive, aerospace, food processing, heat treating, metal processing and casting, glass, ceramic, pulp and paper, utility and power, and chemical and petrochemical processing industries. Over the last 15 years, our focus at the Center for Industrial Sensors and Measurements (CISM) has been on the development of a series of high-temperature gas sensors specifically for combustion processes [1–10]. The underlying theme in our sensor development has been the use of materials science and chemistry to promote high-temperature performance with selectivity. We have developed both the resistive as well as the electrochemical sensors. In a resistive sensor one detects a change in the resistance due to the interaction of the target gas with surface adsorbed oxygen on the sensor film that involves charge transfer. An electrochemical sensor, on the other hand, uses a Galvanic cell that detects the difference in the chemical potential/activity of the target gas between the sensing and the reference electrodes. The sensor response in terms of emf signal is created by the electrochemical reactions that occur at the triple phase boundary (TPB) of the electrolyte, the electrode, and the gas phase. This article focuses on the electrochemical type outlining critical developments in NO_x, O_2, and CO_2 gas sensors over the last decade.

NO_x Sensor

Nitrogen oxides contribute both to ground level ozone formation and acid deposition in the form of acidic particles, fog, and rain [11]. The major source of NO_x is from high-temperature combustion processes of fossil fuels in power plants, vehicles, and airplanes. A potentially very large field of application for high-temperature NO_x sensors is for emissions control for next-generation lean-burn engines, where catalytic converters are ineffective [12].

For transportation industry applications, solid-state NO_x sensing devices and materials operational at high temperatures are necessary [13–15]. While electrochemical sensors provide a promising approach for NO_x detection in harsh environments, they can exhibit response to many different gases, thus minimizing selectivity. The two main components of nitrogen oxides in

G. Kreysa et al. (eds.), *Encyclopedia of Applied Electrochemistry*, DOI 10.1007/978-1-4419-6996-5,

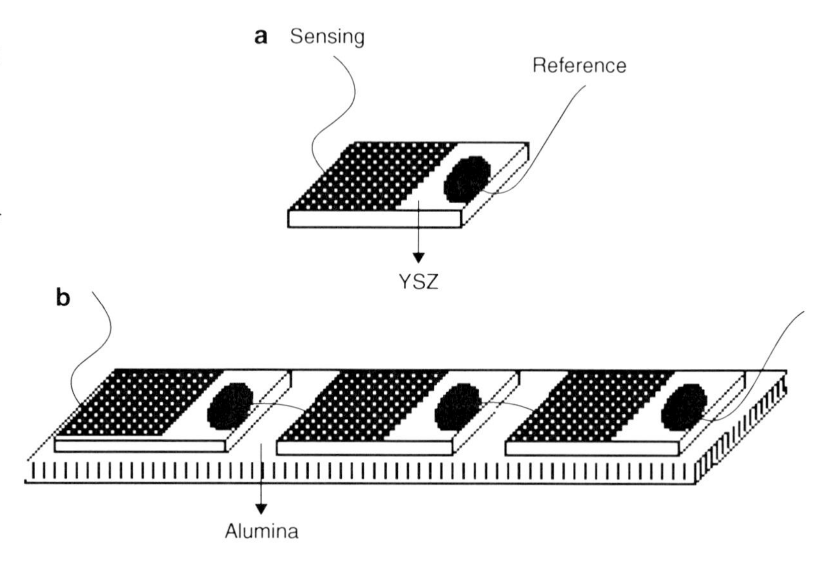

High-Temperature Ceramic Electrochemical Sensors, Fig. 1 Potentiometric sensors composed of YSZ, WO_3 sensing electrodes and PtY/Pt reference electrodes: (**a**) single sensor and (**b**) three-sensor array on an alumina substrate (Adapted from Ref. [17])

combustion environments are NO and NO_2, which generate opposite signals in an electrochemical sensor. Many NO_x sensors focus on NO since it is the major component of NO_x at high temperatures [16]. However, in lean-burn conditions, NO_2 is also present in significant amounts. Thus, sensors that can discriminate between the two gases or provide total NO_x ($NO + NO_2$) are required. Moreover, selectivity needs to be considered since typical lean-burn engine exhausts contain 1–10 ppm NO_x along with 20 % CO_2, 10 % H_2O, 3 % O_2, 10 ppm NH_3, 1,000 ppm hydrocarbons, and 2,000 ppm CO [12].

A typical electrochemical NO_x sensor design involves the use of two electrodes on an oxygen-ion conducting ceramic, such as yttria-stabilized zirconia (YSZ), as shown in Fig. 1a. Both chemical and electrochemical reactivity at each electrode is critical to sensor performance [8, 18–20]. We have obtained optimal results with a Pt electrode covered with Pt-containing zeolite Y (PtY) as the reference electrode and WO_3 as the sensing electrode [17, 21, 22]. These electrodes were identified by temperature programmed desorption of NO from NO_x/O_2-exposed PtY and WO_3, and the ability of PtY and WO_3 to equilibrate a mixture of NO and O_2. Significant reactivity differences were found between the PtY and WO_3, with the latter being largely inactive toward NO_x equilibration. Transmission electron microscopy of PtY showed that the Pt clusters were highly dispersed and no clusters larger than ~10 nm were found on the exterior of zeolite crystals. Thus, with PtY as the reference electrode, NO and NO_2 reach equilibrium ($2NO + O_2 \Leftrightarrow 2NO_2$) upon passing through the PtY before reaching the TPB and thus do not contribute to an electrochemical signal. On the other hand, because of the poor chemical reactivity of NO_x on WO_3, NO_x species reach the TPB chemically unmodified and undergo the electrochemical reaction $2NO + 2O^{2-} \Leftrightarrow 2NO_2 + 4e^-$, making this electrode primarily responsible for the sensor signal. The measured potential is referred to as a mixed-potential and is the steady-state potential where the added partial currents generated by electrode reactions ($i_{cathodic} + i_{anodic}$) equal to zero.

In order to obtain a total NO_x sensor ($NO + NO_2$), a second optimization step of passing the gases through the PtY filter prior to the sensor is necessary. It is necessary to maintain a temperature difference between the filter (typically at 400 °C) and the sensor at 600 °C, to obtain a signal. When NO or NO_2 passes through the PtY filter in the presence of oxygen, an equilibrium mixture of NO and NO_2 is formed. The NO/NO_2 ratio depends only on the filter temperature when the oxygen level is fixed, e.g.,

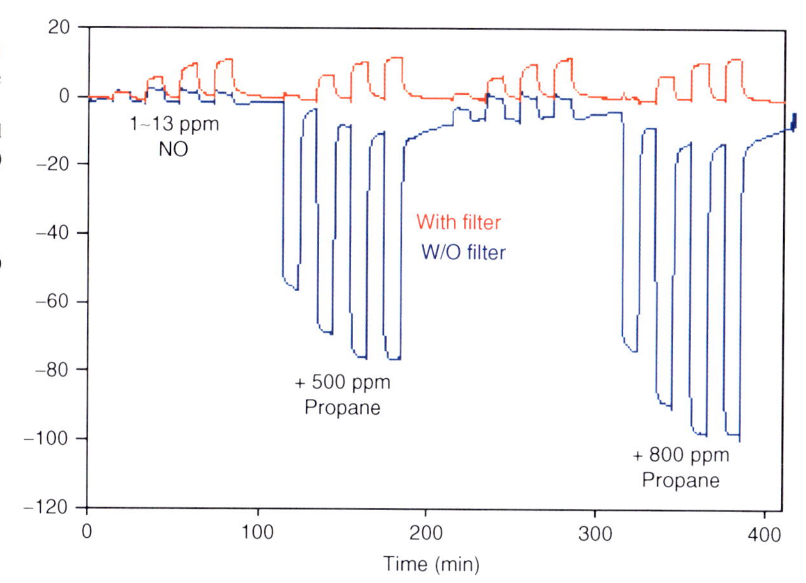

High-Temperature Ceramic Electrochemical Sensors, Fig. 2 Response curves of 1–13 ppm NO in the presence of 3 % O_2 and propane (500 and 800 ppm) with a PtY filter at 400 °C, 500, and 800 ppm propane without a filter (first and third sets of data are for NO only, while second and fourth sets are in presence of propane and NO) (Adapted from Ref. [17])

in 3 % oxygen, NO_2 is 37.7 % of total NO_x at 400 °C and 5.3 % at 600 °C. Thus, a NO/NO_2 equilibrated mixture emerging from the PtY filter at 400 °C will upon contact with a sensor at 600 °C generate a new equilibrium (NO_2 converting to NO) and is the basis for the total NO_x sensing.

This filter/sensor combination also reduces the interference to CO, CO_2, NH_3, propane, O_2, and H_2O. Figure 2 shows that 500–800 ppm propane generates a signal of about −80 to −100 mV, significantly higher than NO. After switching the gases through a PtY filter at 400 °C, propane does not cause significant interference (<0.5–1 % for 10 ppm NO). It was also possible to remove interferences from 2,000 ppm CO, 10 ppm NH_3, as well as minimize effects of 1 ~ 13 % O_2, CO_2, and H_2O.

There are two strategies to increase sensitivity in the present filter/sensor design. The first is to increase the temperature difference between the sensor and the filter. A free energy calculation for a device with the filter at 400 °C and sensor at 600 °C produces a ΔG value of 34.68 kJ/mol [17]. The positive free energy value indicates that the signal on the sensor is being produced due to the reduction of NO_2 on the sensing electrode, and this value would rise as the filter temperature is lowered, leading to a greater driving force for the reaction. A second method is by connecting sensors in series, as shown in Fig. 1b. The EMF is additive by connecting the sensors in series. It has been demonstrated that a ten-sensor array can detect NO concentrations as low as a few ppb [23, 24].

O_2 Sensor

Sensor-enabled feedback control makes it possible for combustion processes to operate at peak efficiency, maximizing energy output and minimizing pollutant emissions. It is estimated that yearly savings of $409 million could be enabled through combustion optimization within coal-fired power plants alone [25]. For optimal combustion, the levels of oxygen need to be carefully controlled and thus oxygen sensors are used extensively in the power and transportation industries.

For high-temperature applications, potentiometric tube-type yttria-stabilized zirconia sensors are most commonly used [26, 27]. However, such sensors require a source of reference oxygen (usually air) which implies the need for plumbing to get the air in or placement close to an air

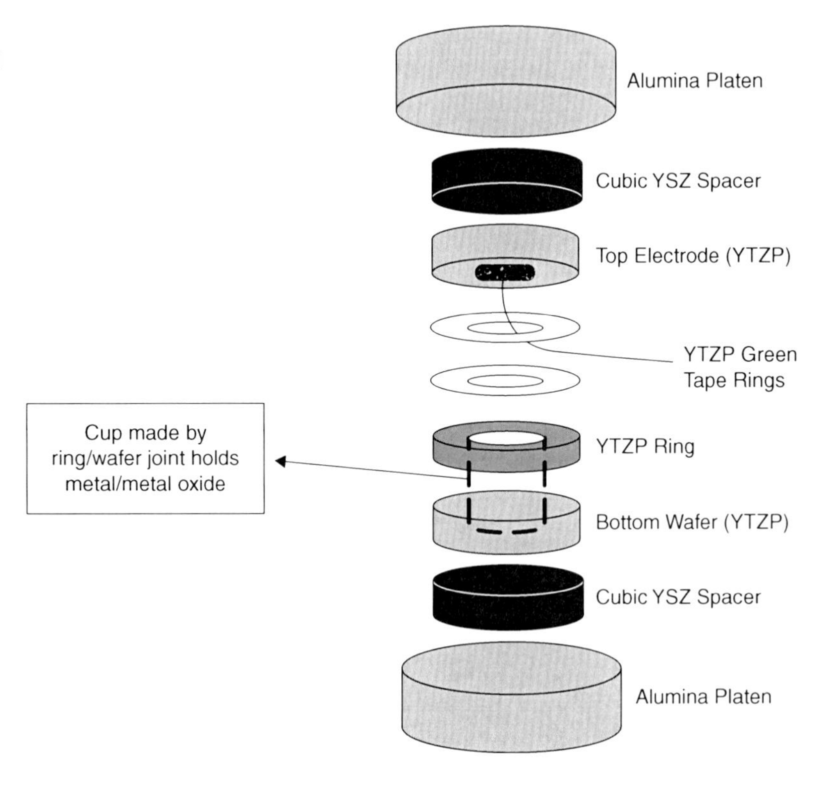

High-Temperature Ceramic Electrochemical Sensors, Fig. 3 Components of the sensor package. The cubic YSZ spacers were necessary as bonding occurred between YTZP and alumina during initial joining (Adapted from Ref. [28])

source. Isolating and sealing an internal oxygen reference would make it possible to eliminate the need for an air reference required by state-of-the-art commercial oxygen sensors, thus allowing for unrestricted placement of the sensors and significant miniaturization. It is known that a metal/metal oxide mixture encapsulated within a ceramic superstructure and placed in intimate contact with a Pt electrode would generate a stable oxygen pressure [5]. These sensors were never successfully commercialized because ceramic seals that contain intermediary bonding agents strain against mismatched thermal expansions in the course of long-term high-temperature operation. It is still a challenge to create a high-temperature seal that prevents the reference gas from equilibrating with the external environment.

We have used grain boundary sliding for sealing, which involves heating the materials under a load [28, 29]. Plastic deformation resulting from grain boundary sliding is strongly dependent on the grain size, since grains of the material diffuse into their neighbors while maintaining a constant shape [30–33]. This interpenetration produces an essentially perfect junction without an interlayer that leads to jointless structures.

The procedure to make the oxygen sensor package using grain boundary sliding is demonstrated in Fig. 3. The electrolyte, ring, and bottom wafer were made of 3 mol% yttria-stabilized tetragonal zirconia polycrystals (YTZP). The devices were constructed by sealing 100 % metal oxide into the reference chamber (PdO, RuO_2, NiO).

To assemble the sensor, the "sandwich" in Fig. 3 was compressed in an argon atmosphere at temperatures ranging from 1,250 °C to 1,290 °C in a universal testing machine (Instron, Model 1125) at crosshead speeds of 0.01–0.02 mm/min and a strain rate of $4 \times 10^{-5}\ s^{-1}$. At temperatures above 1,250 °C, this strain rate is expected to yield a total stress on a 1 cm^2 sample of less than 40 MPa [24]. During the heating cycle, the load on the sample was balanced as not to exceed 5 N. Upon reaching the target temperature, the system was left under a 5 N load for 30 min to attain thermal

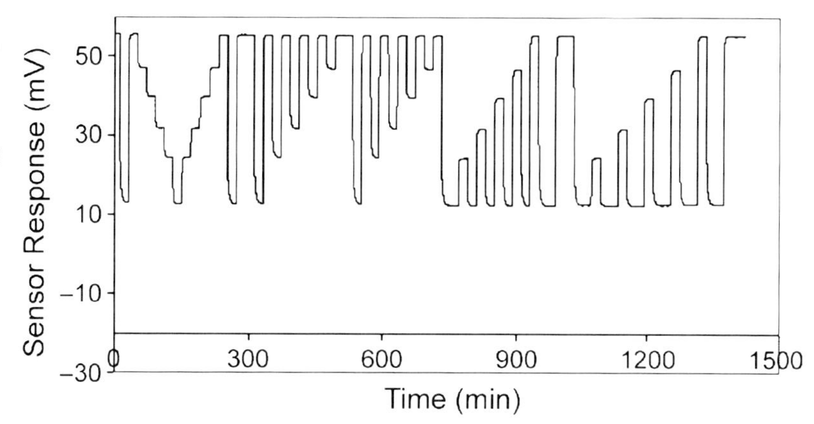

High-Temperature Ceramic Electrochemical Sensors, Fig. 4 Performance of the Pd/PdO-based sensor at 700 °C to changing external oxygen concentrations (Adapted from Ref. [28])

equilibrium. In order to complete the sensor fabrication, a glass plug was applied to the region of the sensor package where the Pt wire breached the inner-to-outer environment. Figure 4 shows the sensing of a Pd/PdO-containing reference electrode to changes in external oxygen concentration (from bottom 3, 5, 7, 10, 14, 21 % O_2) over 24 h at 700 °C. Over an 8-day test cycle, there was no baseline drift and no loss of sensitivity, and the sensor exhibited near-Nernstian behavior.

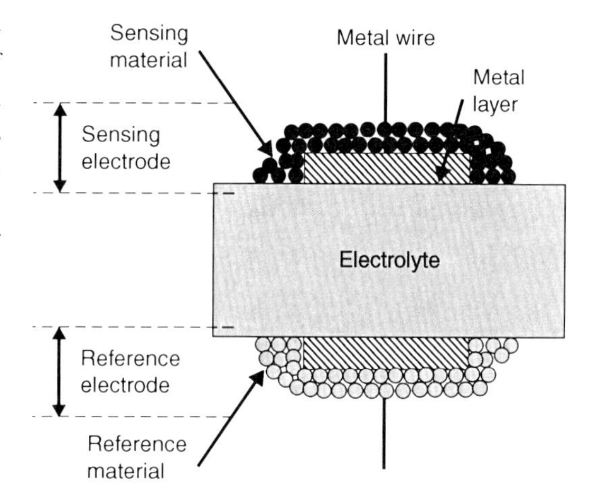

High-Temperature Ceramic Electrochemical Sensors, Fig. 5 Schematic of the sensor with open reference electrode

CO_2 Sensor

In addition to being a green house gas, CO_2 is an important component for metabolism process of plant and many living creatures [34]. Thus, reliable and selective CO_2 detectors are needed for a variety of applications including environmental and health monitoring [35, 36], fire detection [37], and controlling of fermentation [38]. There are several types of commercially available CO_2 gas sensors and most of them are based on nondispersed infrared (NDIR) and electrochemical methods.

NDIR type CO_2 sensors allow highly specific detection via the absorption of CO_2 in the infrared region [39–41]. However, because of bulk size, limited operation temperature range (<328 K) and high cost, their applications are not widespread. The use of light emitting diode (LED) has allowed dramatic reduction in the size of the optics system allowing NDIR sensor dimension comparable to that of the electrochemical type. Moreover, their power consumption, response time (t_{90}), and warm-up time are superior to the electrochemical type. However, their limited detection temperature range and high cost are still impediments for many applications requiring in situ monitoring.

A solid-state electrochemical CO_2 sensor with Li_3PO_4 electrolyte was developed in our laboratory in early 2000 [9]. For the sensor structure, we adopted the open reference system with a biphase mixture of Li_2TiO_3 and TiO_2 as the reference electrode and Li_2CO_3 as the sensing electrode, as shown in Fig. 5. While the

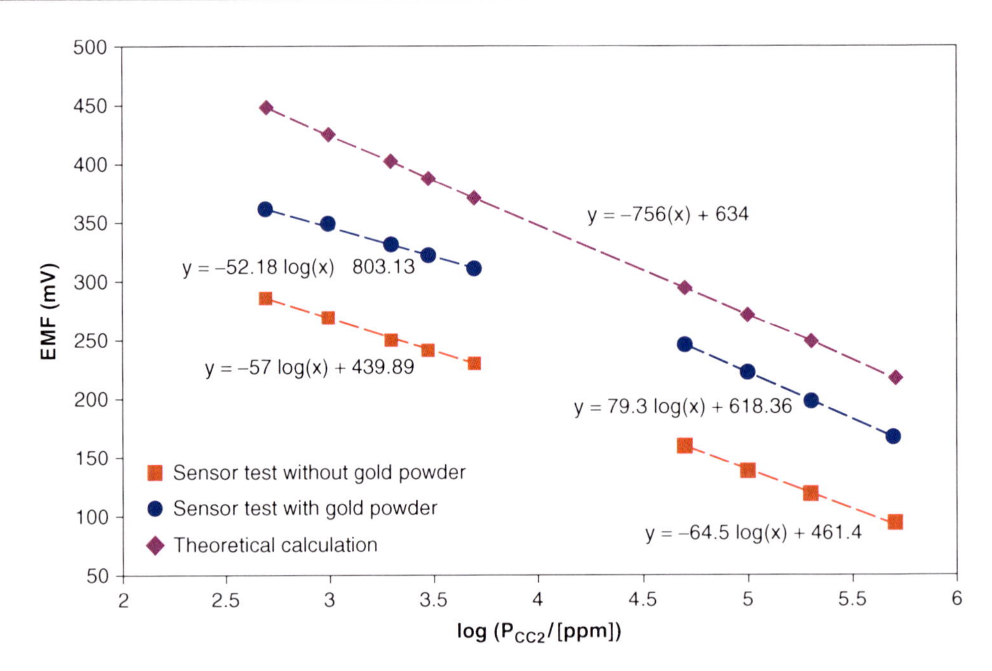

High-Temperature Ceramic Electrochemical Sensors, Fig. 6 Comparison of EMF between sensors with and without gold powder at 773 K along with theoretical Nernstian values (Adapted from Ref. [9])

interaction of the CO_2 gas on the sensing electrode allows its detection, the reference electrode fixes Li^+ activity as illustrated by the electrochemical reactions.

$$2Li^+ + CO_2(s) + \frac{1}{2}O_2(g) + 2e^- \leftrightarrow Li_2CO_3(s) \quad \text{(sensing electrode)}$$

$$2Li^+ + TiO_2(s) + \frac{1}{2}O_2(g) + 2e^- \leftrightarrow Li_2TiO_3(s) \quad \text{(reference electrode)}$$

The use of the bi-phase mixture as the reference electrode is a unique feature of this design allowing a simpler device than the closed reference system and associated gas-tight sealing challenges, particularly for high-temperature applications.

This sensor showed good sensitivity, selectivity, and linear response in the temperature range between 823 K and 873 K [9], though its response systematically deviated from ideal Nerstian values, especially at high temperatures and low CO_2 concentrations. Lee et al. [10] investigated the reasons for the non-Nernstian behavior and proposed two possibilities: (1) lack of reversibility of electrode reaction and (2) partial electronic conductivity of the electrolyte. Based on electrochemical impedance spectroscopy (EIS) study, the reaction on $Li_2TiO_3 + TiO_2$ mixture was found to be sluggish, while on Li_2CO_3 was relatively fast. To solve the kinetic problem, addition of gold powder or porous sputtered gold electrode was investigated that improved the sensitivity closer to the Nernstian value as shown in Fig. 6. However, since this could not explain the temperature dependence, they suspected mixed ionic and electronic conduction of Li_3PO_4 electrolyte. Based on EMF measurements and Hebb-Wagner (HB) DC polarization measurement, an n-type electronic conduction of Li_3PO_4 electrolyte was confirmed, particularly in low CO_2 concentration ranges (~ppm levels) and at higher temperatures.

This sensor also showed humidity interference and Lee et al. [42] used Li_2CO_3-$BaCO_3$ bi-phase as the sensing electrode to eliminate the problem. Unlike other binary carbonate studies, Li_2CO_3 was coated by $Ba(NO_3)_2$ via a wet chemical method and then $BaCO_3$ outer layer was formed on Li_2CO_3 in the presence of CO_2. The heat treatment allowed eutectic reaction promoting adhesion of the sensing electrode as well. The sensitivity of the sensor for 5–20 % CO_2 was nearly Nernstian at 773 K and the humidity interference was practically eliminated [42].

While this sensor works well at high temperatures (above 673 K), for many other applications there is a need for the development of a low-temperature CO_2 sensor. For such a sensor, the selection of the electrolyte and the electrode materials is critical. Lithium-lanthanum-titanate ($Li_{3x}La_{((2/3)-x)}TiO_3$; LLTO) is known as the highest lithium (Li)-ion conducting material at room temperature [43]. A sensor based on this electrolyte is currently being investigated for low-temperature applications (<473 K) [44]. A major challenge for such a sensor remains to be overcoming the slow CO_2 reaction kinetics at low temperatures.

Future Directions

In spite of the recent developments of solid-state electrochemical gas sensors reported in this paper, further research and development are needed for their commercialization. First, the operation temperature of the sensor should cover a wider range and clearly there are challenges specifically for low-temperature (from room temperature to 200 °C) devices. The enhancement of electrochemical reactions at the TPBs particularly in a dry condition will require multidisciplinary effort in the development of heterogeneous catalysts, new electrode and electrolyte materials, optimized electrode morphology, as well as study of electrochemical reaction kinetics. Second, novel fabrication processes should be developed for mass-scale production of devices. Traditional ceramic processing and wet chemical methods may not be practical for maintaining quality control from one device to another. Like the semiconductor industry, thin film technology can be a promising solution. The development of new processes will obviously open opportunities and challenges for the synthesis and characterization of novel materials, as well as fundamental studies on a variety of surface and interface problems. What makes this area exciting for future R&D is that the solution to the engineering problems is closely linked with the fundamental understanding yet to be uncovered.

Cross-References

▶ Electrochemical Sensor of Gaseous Contaminants
▶ Sensors

References

1. Birkefeld LD, Azad AM, Akbar SA (1992) Carbon monoxide and hydrogen detection by anatase modification of titanium dioxide. J Am Ceram Soc 75:2694–2698
2. Savage N, Chwieroth B, Ginwalla A, Patton BR, Akbar SA, Dutta PK (2001) Composite n-p semiconducting titanium oxides as gas sensors. Sens Actuators B 79:17–27
3. Szabo NF, Du H, Akbar SA, Soliman AA, Dutta PK (2002) Microporous zeolite modified Yttria Stabilized Zirconia (YSZ) sensors for Nitric Oxide (NO) determination in harsh environments. Sens Actuators B 82:142–149
4. Kohli A, Wang CC, Akbar SA (1999) Niobium pentoxide as a lean-range oxygen sensor. Sens Actuators B 56:121–128
5. Chowdhury AKMS, Akbar SA, Kapileshwar S, Schorr JR (2001) A rugged oxygen gas sensor with solid reference for high temperature applications. J Electrochem Soc 148:G91–G94
6. Narayanan B, Akbar SA, Dutta PK (2002) A phosphate-based proton conducting solid electrolyte hydrocarbon gas sensor. Sens Actuators B 87(3):480–486
7. Reddy C, Dutta PK, Akbar SA (2003) Detection of CO in a reducing, hydrous environment using CuBr as electrolyte. Sens Actuators B 92:351–355
8. Szabo NF, Dutta PK (2003) Strategies for total NO_x measurement with minimal CO interference utilizing a microporous zeolitic catalytic filter. Sens Actuators B 88(2):168–177
9. Lee C, Akbar SA, Park CO (2001) Potentiometric type CO_2 gas sensor with lithium phosphorous oxynitride electrolyte. Sens Actuators B 80:234–242
10. Lee C, Dutta PK, Ramamoorthy R, Akbar SA (2006) Study of mixed Ionic and electronic conduction in Li_3PO_4 electrolyte for a CO_2 gas sensor. J Electrochem Soc 153:H4–H14

11. Seinfeld JH, Pandis S (1998) Atmospheric chemistry and physics: from air pollution to climate change. Wiley, New York
12. Menil F, Coillard V, Lucat C (2000) Critical review of nitrogen monoxide sensors for exhaust gases of lean burn engines. Sens Actuators B 67:1–23
13. Raj ES, Pratt KFE, Skinner SJ, Parkin IP, Kilner JA (2006) High conductivity $La_{2-x}Sr_xCu_{1-y}(Mg, Al)_yO_4$ solid state metal oxide gas sensors with the K_2NiF_4 structure. Chem Mater 18:3351–3355
14. Gurlo A, Barsan N, Weimar U, Ivanovskaya M, Taurino A, Siciliano P (2003) Polycrystalline well-shaped blocks of indium oxide obtained by the sol–gel method and their gas-sensing properties. Chem Mater 15:4377–4383
15. Yoo J, Van Assche FM, Wachsman ED (2006) - Temperature-programmed reaction and desorption of the sensor elements of a WO_3/YSZ/Pt potentiometric sensor. J Electrochem Soc 153:H115–H121
16. Miura N, Lu G, Yamazoe N (2000) Progress in mixed-potential type devices based on solid electrolyte for sensing redox gases. Solid State Ion 136:533–542
17. Yang J-C, Dutta PK (2007) Promoting selectivity and sensitivity for a high temperature YSZ-based electrochemical total NOx sensor by using a Pt-loaded zeolite Y filter. Sens Actuators B 125(1):30–39
18. Dutta A, Kaabbuathong N, Grilli ML, Di Bartolomeo E, Traversa E (2003) Study of YSZ-based electrochemical sensors with WO_3 electrodes in NO_2 and CO environments. J Electrochem Soc 150:H33–H37
19. Szabo NF, Dutta PK (2004) Correlation of sensing behavior of mixed potential sensors with chemical and electrochemical properties of electrodes. Solid State Ion 171:183–190
20. Yang J-C, Dutta PK (2010) High temperature potentiometric NO_2 sensor with asymmetric sensing and reference Pt electrodes. Sens Actuators B 143(2):459–463
21. Yang J-C, Dutta PK (2007) Influence of solid-state reactions at the electrode-electrolyte interface on high-temperature potentiometric NOx-gas sensors. J Phys Chem C 111(23):8307–8313
22. Yang J-C, Dutta PK (2009) Solution-based synthesis of efficient WO_3 sensing electrodes for high temperature potentiometric NOx sensors. Sens Actuators B 136(2):523–529
23. Hunter GW, Xu JC, Biaggi-Labiosa AM, Laskowski D, Dutta PK, Mondal SP, Ward BJ, Makel DB, Liu CC, Chang CW, Dweik RA (2011) Smart sensor systems for human health breath monitoring applications. J Breath Res 5:037111
24. Mondal SP, Dutta PK, Hunter GW, Ward BJ, Laskowski D, Dweik RA (2011) Development of high sensitivity potentiometric NOx sensor and its application to breath analysis. Sens Actuators B 158:292–298
25. Linkins D, Lewis D, Frank R, Weiss J (1997) TVA's Kingston Unit 9 Distributed Control System (DCS) retrofit benefit documentation. TVA Technol Adv 1(2):69–75
26. Ramamoorthy R, Dutta PK, Akbar SA (2003) Oxygen sensors: materials, methods, designs and applications. J Mater Sci 38(21):4271–4282
27. Maskell WC, Steele BCH (1986) Solid state potentiometric oxygen gas sensors. J Appl Electrochem 16(4):475–489
28. Spirig JV, Ramamoorthy R, Akbar SA, Routbort JL, Singh D, Dutta PK (2007) High temperature zirconia oxygen sensor with sealed metal/metal oxide internal reference. Sens Actuators B 124(1):192–201
29. Spirig JV, Routbort JL, Singh D, King G, Woodward PM, Dutta PK (2008) Joining of highly aluminum-doped lanthanum strontium manganese oxide with tetragonal zirconia by plastic deformation. Solid State Ion 179(15–16):550–557
30. Mohamed FA, Li Y (2001) Creep and superplasticity in nanocrystalline materials: current understanding and future prospects. Mater Sci Eng A 298(1–2):1–15
31. Gutierrez-Mora F, Goretta KC, Majumdar S, Routbort JL, Grimdisch M, Dominguez-Rodriguez A (2002) Influence of internal stresses on superplastic joining of zirconia-toughened alumina. Acta Mater 50:3475–3486
32. Gutierrez-Mora F, Dominguez-Rodriguez A, Routbort JL, Chaim R, Guiberteau F (1999) Joining of yttria-tetragonal stabilized zirconia polycrystals using nanocrystals. Scr Mater 41(5):455–460
33. Dominguez-Rodriguez A, Gutierrez-Mora F, Jimenez-Melendo M, Routbort JL, Chaim R (2001) Current understanding of superplastic deformation of Y-TZP and its application to joining. Mater Sci Eng A 302(1):154–161
34. Zosel J et al (2011) Topical review-the measurement of dissolved and gaseous carbon dioxide concentration. Meas Sci Technol 22:072001
35. Lee C, Szabo N, Ramamoorthy R, Dutta P, Akbar S (2006) Solid-state electrochemical sensors: opportunities and challenges. In: Grimes CA, Dickey EC, Pishko MV (eds) Encyclopedia of sensors, Stevenson Ranch, CA: American Scientific Publisher, vol 10. pp 1–20
36. Hunter GW, Dweik RA (2008) Applied breath analysis: an overview of the challenges and opportunities in developing and testing sensor technology for human health monitoring in aerospace and clinical applications. J Breath Res 2:037020
37. Chen S-J et al (2007) Fire detection using smoke and gas sensors. Fire Saf J 42:507–515
38. Dixon NM, Kell DB (1989) Review article-the control and measurement of CO_2 during fermentations. J Microbiol Methods 10:155–176
39. Auble DL, Meyers TP (1992) An open path, fast response infrared absorption gas analyzer for H_2O and CO_2. Bound-Layer Meterol 59(3):243–256

40. Holzinger M, Maier J, Sitte W (1997) Potentiometric detection of complex gases: application to CO_2. Solid State Ion 94:217–225
41. Mulier M et al (2009) Development of a compact CO_2 sensor based on near-infrared laser technology for enological application. Appl Phys B-Lasers Opt 94:725–733
42. Lee I, Akbar SA, Dutta PK (2009) High temperature potentiometric carbon dioxide sensor with minimal interference to humidity. Sens Actuators B 142:337–341
43. Inaguma Y et al (1993) High ionic conductivity in lithium lanthanum titanate. Solid State Commun 86(10):689–693
44. Yoon J, Hunter G, Akbar SA, Dutta PK (2013) Interface reaction and its effect on performance of a CO2 sensor based on $Li_{0.35}La_{0.55}TiO_3$ electrolyte and Li_2CO_3 sensing electrode. Sensors and Actuators B 182: 95–103

High-Temperature CO_2 Electrolysis

Mogens Bjerg Mogensen
Department of Energy Conversion and Storage, Technical University of Denmark, Roskilde, Denmark

Introduction

High temperature is here defined as temperatures of 500 °C or higher. There are three different electrochemical cells that can be used for CO_2 electrolysis, and there are three main arguments in favor of high-temperature electrolysis of CO_2 into CO or CH_4, and O_2. Details and electrochemical reactions are given below.

Cell Types:

1. Solid oxide electrolyzer cells (SOEC) have a solid oxide ion conductor as electrolyte, often yttria-stabilized zirconia (YSZ). The cathode (CO evolution, negative) is often a Ni-YSZ composite called a cermet. The anode (O_2 evolution, positive) most often consists of a composite of YSZ electrolyte and an electron-conducting perovskite-structured oxide, e.g., $(La_{0.75}Sr_{0.25})_{0.95}MnO_3$ [1].
2. Solid proton-conducting electrolyzer cell (SPCEC) has a proton-conducting solid oxide electrolyte, e.g., yttria-doped barium zirconate, $BaZr_{0.85}Y_{0.15}H_{0.15}O_3$, i.e., ceramics that can take up H_2O and become proton conducting [2]. Electrodes may be similar to those used for SOEC.
3. Molten carbonate electrolyzer cell (MCEC) has molten Li_2CO_3 electrolyte, Ti-Al metal alloy as cathode (CO evolution), and graphite as anode (O_2 evolution) [3].

Arguments in Favor:

1. Electrolysis of CO_2 is an endothermic process that utilizes the unavoidable Joule heat of the electrochemical cell, which may be operated as a self-cooling (thermoneutral) system in which the Joule heat compensates the heat loss associated with the entropy loss.
2. High temperature in the range of 500–900 °C yields fast kinetics.
3. High temperature enables co-electrolysis of CO_2 and steam, H_2O, into $H_2 + CO$, also called synthesis gas or just syngas, which can be converted into hydrocarbon using commercially available catalytic reactors.

Argument 1 may also be expressed as the Joule heat contributes to the splitting of the water and CO_2 molecules. The higher the temperature, the less electrical energy is needed for the splitting, i.e., the higher is the electrical efficiency. Therefore, SOEC may be operated self-cooling if the used cell voltage is equal to the thermoneutral voltage, E_{tn}, which is equivalent to ΔH_f, the enthalpy of formation. E_{tn} is ca. 1.5 V for CO_2 electrolysis. This is illustrated in Fig. 1. If the cell is operated at E_{tn}, then there is no need for either heating or cooling. To compensate for unavoidable heat loss to the surroundings, it is necessary to operate the cell slightly above E_{tn}. How much above is dependent of the circumstances. In case of a large and well-insulated system, only a few mV may be necessary.

Argument 2 expresses that the rate of the electrochemical process, the CO (or syngas) production, is much faster at high temperature than at

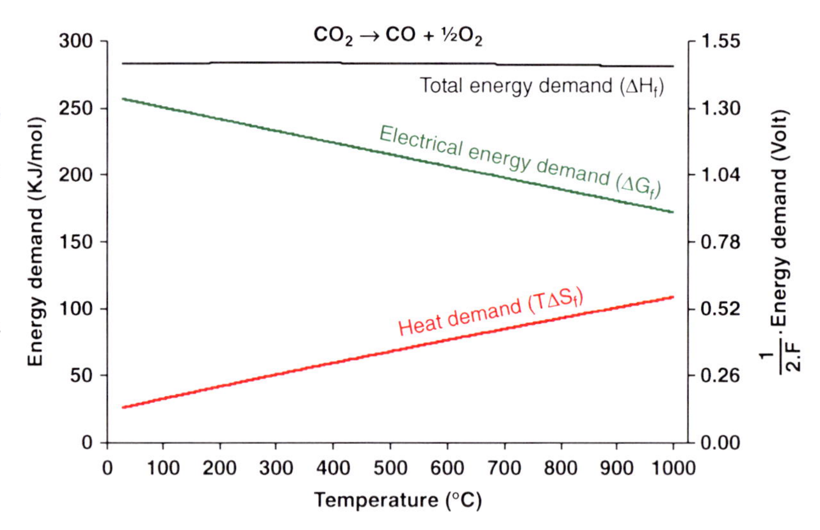

High-Temperature CO_2 Electrolysis, Fig. 1 Thermodynamic diagram of splitting CO_2 to CO and O_2. ΔH_f is the enthalpy of formation, ΔG_f is Gibb's free energy of formation, ΔS_f is the entropy of formation, and T is the temperature in K. Data are form Ref. [4] (By courtesy of Dr. Sune D. Ebbesen, Technical University of Denmark)

low temperature. More m^3 of H_2 and CO per m^2 of cell area per minute gives lower capital costs per m^3 syngas.

Argument 3 implies that it is possible to use renewable electric energy to convert steam and CO_2 into CO_2 neutral methane, also called green natural gas (GNG) or into green liquid fuels. Here "green" means CO_2 neutral, i.e., as much CO_2 has been used for the manufacturing of the fuel as will be released by using the fuel. Principles illustrated by experimental data may be found in ref. [1].

All three types of high-temperature CO_2 electrolysis cells are under development, and none of them are commercial yet. The SOEC is, however, by far the most explored.

Solid Oxide Electrolysis Cell

Principle

Physically, a solid oxide electrolyzer cells (SOEC) may be exactly the same cell as solid oxide fuel cell (SOFC) Solid Oxide Fuel Cells, Introduction as this cell type is fully reversible [1]. It can be used for CO_2 electrolysis only but will most often be intended for electrolysis of a mixture CO_2 and steam, H_2O. This is called co-electrolysis of steam and CO_2. The principle of both SOEC and SOFC is illustrated in Fig. 2.

The reactions at the negative electrode (the cathode in electrolyzer mode) are

$$CO_2(g) + 2e^- \rightarrow CO(g) + O^{2-}(s) \quad (1)$$

if only CO_2 electrolysis, and in case of co-electrolysis in parallel with this,

$$H_2O(g) + 2e^- \rightarrow H_2(g) + O^{2-}(s) \quad (2)$$

The reaction at the positive electrode (the anode in electrolyzer mode) is

$$2O^{2-}(s) \rightarrow O_2(g) + 4e^- \quad (3)$$

The overall reaction in case of co-electrolysis is

$$CO_2(g) + H_2O(g) \rightarrow CO(g) + H_2(g) + O_2(g) \quad (4)$$

Materials

The SOEC consists of relatively inexpensive materials and may be produced using low-cost processes. Electrolytes are most often yttria-stabilized zirconia (YSZ) of composition $Zr_{0.84}Y_{0.16}O_{1.92}$, often called 8YSZ, but recently scandia doping or co-doping of the zirconia has become popular in both SOFC and SOEC

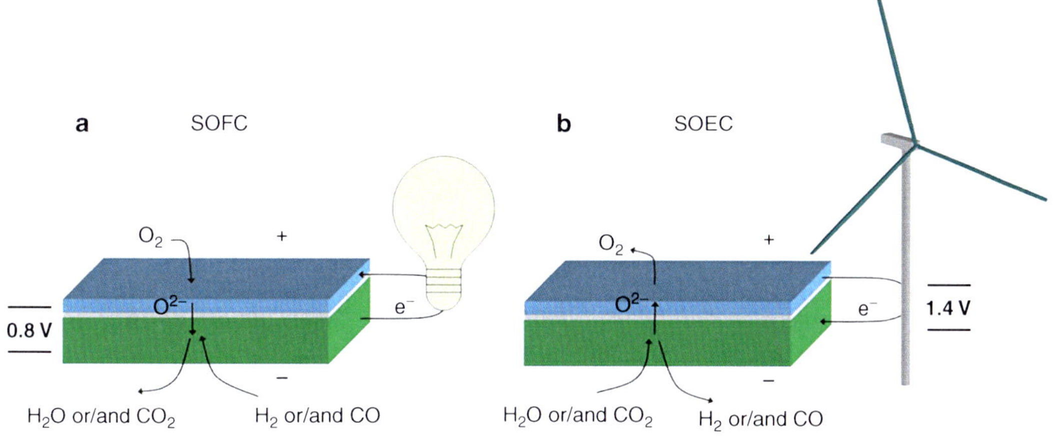

High-Temperature CO_2 Electrolysis, Fig. 2 Working principle of a reversible solid oxide cell (SOC). The cell can be operated as a SOFC (**a**) and as a SOEC (**b**). The operation temperature is typically in the range of 650–850 °C. The blue-colored layer is the oxygen electrode, the thin white layer is the solid oxide electrolyte, and the green is the fuel electrode with its support [1]. Other designs than this have been constructed and tested (see ► Solid Oxide Fuel Cells, Introduction)

applications due to lower electrical resistance. Typical electrolyte composition is $Zr_{0.79}Ce_{0.01}Sc_{0.20}O_{1.9}$ [5].

History and State of the Art

The high temperature (typically 650–850 °C) SOEC has been reported practical for co-electrolysis of H_2O and CO_2 even though it is not yet commercialized. The technology has been studied and developed during a long period. The SOC was invented by Bauer and Preis in 1937 [6], and Weissbart et al. demonstrated CO_2 electrolysis using SOEC in 1969 [7]. The purpose was to produce oxygen in a space craft. The reversibility of the SOFC, i.e., that it works equally well in SOEC mode, i.e., as electrolyzer, was published by Isenberg in 1981 [8]. Later, a significant R&D work on CO_2 electrolysis for space application was carried out using SOEC [9]. Today, the technology has been demonstrated with cell stacks of up to 15 kW size and tested over more than 1 year [10–13]. Initial cell performance is acceptably high in many reports – above 1 A cm^{-2} at 1.3 V – whereas cell and stack durability, which has been studied over prolonged periods, is less acceptable, and most reports indicate degradation rates typically around 5 ± 3 % per 1,000 h [10–14]. It seems to depend a lot on how the cell is manufactured and on type and concentration of impurities in the feed gas and steam.

At high current densities above 1 A cm^{-2}, state-of-the-art SOEC degrades relatively fast due to damage of the YSZ electrolyte [15]. Such current density (1 A cm^{-2} gives ca. 1 $Nm^3\ h^{-1}\ m^{-2}$, in words 1 normal cubic meter of SNG per hour and square meter of cell) may be required for economical reasons. Below ca. 0.7 A cm^{-2}, the degradation of the SOEC seems very low (less than 1 % per 1,000 h) in case the gases are cleaned using a newly invented method [16]. Apart from production of GNG based on renewable electricity sources, SOEC may also be used for upgrading of digester biogas [17].

Solid Proton-Conducting Electrolyzer Cell

Reduction of CO_2 may also be done in a cell with a proton-conducting ceramic such as the perovskite of the type $BaZr_{1-x}Y_xO_3$ as electrolyte [18]. Oxygen electrode is usually perovskites, and CO electrode is usually Ni-YSZ in similarity to SOEC, but recently ceramic fuel electrodes have been demonstrated [19]. The proton conduction is possible in a material like yttria-doped barium zirconate, $BaZr_{0.85}Y_{0.15}O_{2.925}$, because the substitution of Zr^{4+} with Y^{3+} causes oxygen vacancies in the crystal structure and these vacancies may take up water. If the material is fully saturated with H_2O then it becomes $BaZr_{0.85}Y_{0.15}H_{0.15}O_3$. The protons in the water then distribute themselves homogenously with only one proton (H^+) on each oxygen, i.e., OH^- ions are formed in the crystals. The protons are "free" to hop by thermal activation from one O^{2-} ion to another.

The electrochemical reactions in such cell may be either

$$\text{Anode}: \quad H_2O \rightleftarrows \tfrac{1}{2}O_2 + 2H^+ + 2e^- \tag{5}$$

$$\text{Cathode}: \quad CO_2 + 2H^+ + 2e^- \rightleftarrows CO + H_2O \tag{6}$$

$$\text{Total}: \quad CO_2 \rightleftarrows CO + \tfrac{1}{2}O_2 \tag{7}$$

or (most pronounced at relative low temperature)

$$\text{Anode}: \quad 4H_2O \rightleftarrows 2O_2 + 8H^+ + 8e^- \tag{8}$$

$$\text{Cathode}: \quad CO_2 + 8H^+ + 8e^- \rightleftarrows CH_4 + 2\,H_2O \tag{9}$$

$$\text{Total}: \quad CO_2 + 2H_2O \rightleftarrows 2O_2 + CH_4 \tag{10}$$

Molten Carbonate Electrolyzer Cell

Recently, a breakthrough in electrolysis of CO_2 into CO using molten carbonate electrolytes has occurred [3, 20]. The cell types used for molten carbonate fuel cells are not practically usable for electrolysis. Basically, the breakthrough is that it has shown feasible to use pure molten Li_2CO_3 above 800 °C as electrolyte, Ti-Al metal alloy as cathode (CO evolution), and graphite as anode (O_2 evolution). Surprisingly, the graphite electrode anode seems stable in this very high temperature and strongly oxidizing environment.

As a basis for the reaction, it is assumed that the chemical equilibrium exists in the molten Li_2CO_3:

$$CO_3^{2-} \rightleftarrows CO_2 + O^{2-} \tag{11}$$

Then the electrode reactions are

$$CO_2(l) + 2e^- \rightarrow CO(l) + O^{2-}(l) \tag{12}$$

$$O^{2-}(l) \rightarrow \tfrac{1}{2}O_2(l) + 2e^- \tag{13}$$

As CO and O_2 gases are very little soluble in molten Li_2CO_3, they will escape to the gas phase at each of their electrodes, and the total reaction is

$$CO_2(l) \rightarrow CO(g) + \tfrac{1}{2}O_2(g) \tag{14}$$

This cell may produce pure CO rather than a mixture of CO and CO_2 that is characteristic of the other two CO_2 electrolysis cells.

Cross-References

► Electrocatalysts for Carbon Dioxide Reduction
► Electrochemical Fixation of Carbon Dioxide (Cathodic Reduction in the Presence of Carbon Dioxide)
► Molten Carbonate Fuel Cells, Overview

References

1. Jensen SH, Larsen PH, Mogensen M (2007) Hydrogen and synthetic fuel production from renewable energy sources. Int J Hydrogen Energy 32:3253
2. Bonanos N (2001) Oxide-based protonic conductors: point defects and transport properties. Solid State Ionics 145:265
3. Kaplan V, Wachtel E, Gartsman K, Feldman Y, Lubomirsky I (2010) Conversion of CO_2 to CO by electrolysis of molten lithium carbonate. J Electrochem Soc 157:B552
4. National Institute of Standards and Technology (NIST) (2008) NIST Chemistry WebBook. http://webbook.nist.gov/chemistry/
5. Mogensen M, Lybye D, Kammer K, Bonanos N (2005) Ceria revisited: electrolyte or electrode material? In: Singhal SC, Mizusaki J (eds) Proceedings 9th international symposium on solid oxide fuel cells (SOFC IX), vol PV 2005–07, The Electrochemical Society, Pennington, p 1068
6. Baur E, Preis H (1937) Uber brennstoff-ketten mit festleitern. Zeitschrift für Elektrochemie 43:727
7. Weissbart J, Smart W, Wydeven T (1969) Oxygen reclamation from carbon dioxide using a solid oxide electrolyte. Aerospace Med 40:136

8. Isenberg AO (1981) Energy conversion via solid oxide electrolyte electrochemical cells at high temperatures. Solid State Ionics 3–4:431
9. Sridhar KR, Vaniman BT (1997) Oxygen production on Mars using solid oxide electrolysis. Solid State Ionics 93:321
10. Stoots CM, O'Brien JE, Condie KG, Hartvigsen JJ (2010) High-temperature electrolysis for large-scale hydrogen production from nuclear energy – Experimental investigations. Int J Hydrogen Energy 35:4861
11. Ebbesen SD, Høgh J, Nielsen KA, Nielsen JU, Mogensen M (2011) Durable SOC stacks for production of hydrogen and synthesis gas by high temperature electrolysis. Int J Hydrogen Energy 36:7363
12. Schefold J, Brisse A, Zahid M, Ouweltjes JP, Nielsen JU (2011) Long term testing of short stacks with solid oxide cells for water electrolysis. ECS Trans 35:2915
13. Schefold J, Brisse A, Tietz F (2012) Nine thousand hours of operation of a solid oxide cell in steam electrolysis mode. J Electrochem Soc 159:A137
14. Hauch A, Ebbesen SD, Jensen SH, Mogensen M (2008) Highly efficient high temperature electrolysis. J Mater Chem 18:2331
15. Knibbe R, Traulsen ML, Hauch A, Ebbesen SD, Mogensen M (2010) Solid oxide electrolysis cells: degradation at high current densities. J Electrochem Soc 157:B1209
16. Ebbesen SD, Mogensen M (2011) Method and system for purification gas streams for solid oxide cells. EPO patent EP2362475A1
17. Hansen JB (2012) Process for converting biogas to a gas rich in methane. Patent WO/2012/003849, published 12 Jan 2012
18. Stuart PA, Unno T, Kilner JA, Skinner SJ (2008) Solid oxide proton conducting steam electrolysers. Solid State Ionics 179:1120
19. Xie K, Zhang Y, Meng G, Irvine JTS (2011) Electrochemical reduction of CO2 in a proton conducting solid oxide electrolyser. J Mater Chem 21:195
20. Licht S Advanced materials (2011) Efficient solar-driven synthesis, carbon capture, and desalinization, STEP: solar thermal electrochemical production of fuels, metals, Bleach doi:10.1002/adma.201103198

High-Temperature Molten Salts

Mathieu Salanne
Laboratoire PECSA, UMR 7195, Université Pierre et Marie Curie, Paris, France

Introduction

Molten salts, together with their low-temperature analogues, the room-temperature ionic liquids, constitute a particular class of electrolytes which do not contain any solvent. They are composed of inorganic ions only: The cations (M^{p+}) are metallic species while the anions (X^{q-}) can either be halides (F^-, Cl^-, Br^-, I^-) or polyatomic species (NO_3^- and CO_3^{2-} mainly); the archetypal molten salt is sodium chloride NaCl. Being solid at ambient conditions, these salts are generally used in high-temperature applications (it is worth noting that some compounds have relatively low melting points, e.g., $NaAlCl_4$ melts at 152 °C).

Despite the fact that the most important industrial applications of molten salts have started at the beginning of the twentieth century with the large-scale production of aluminum, their structure has long remained largely unknown. Most of the short-range structural properties can be attributed to the competition between packing effects and electrostatic screening, which leads to the existence of charge ordering: Around a central ion, successive coordination shells of alternating charges are formed. Nevertheless, the details of the structure adopted by these solvation shells, which are usually expressed in terms of interatomic distance and coordination numbers, differ a lot depending on the ionic radii, valence charge, and polarizability of the involved species.

From the experimental point of view, the first picture of the structure of molten salts was provided by X-Ray diffraction and neutron scattering techniques. Due to the intrinsic multicomponent character of molten salts, the combination of several diffraction data is compulsory in order to gain insight into the structural informations for all the pairs of atoms. For example, in the simple case of sodium chloride, the first report of the whole set of radial distribution functions (Na^+–Na^+, Na^+–Cl^-, Cl–Cl^-) was obtained by J. Enderby and co-workers in the 1970s through the use of the technique of isotopic substitution in neutron diffraction [1]. Since then, many other spectroscopic techniques have been used to study the structure of molten salts. Raman spectroscopy brings important informations on the vibrational properties of the complexes $[MX_n]$ present in the melt [2, 3]. In many systems, one or more nuclei can be probed by nuclear magnetic resonance (NMR), also allowing for the identification of the nature of

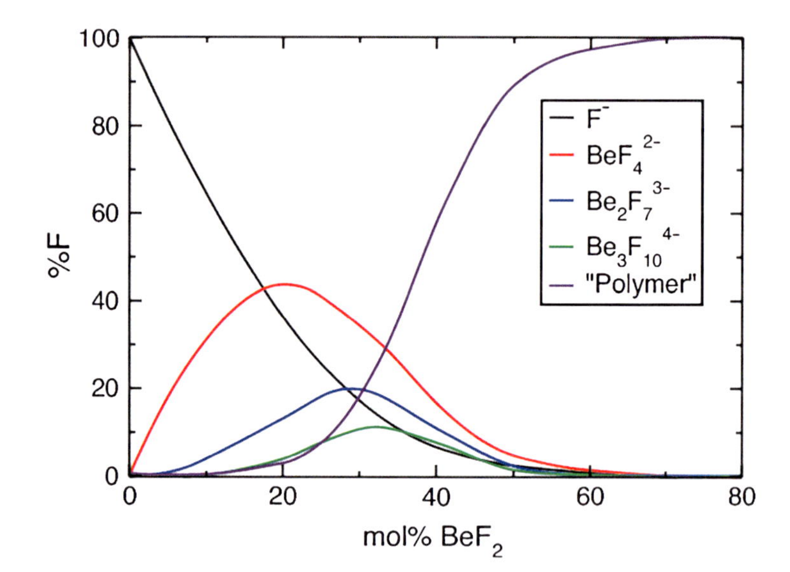

High-Temperature Molten Salts, Fig. 1 Speciation diagram for LiF-BeF_2 molten salts. "Polymer" means a fluroberyllate ion with more than 3 Be^{2+} cations, whereas F^- correspond to free fluoride ions

these complexes [4]. Recently, the design of cells allowing to perform high-temperature extended X-ray absorption fine structure (EXAFS) measurements [5] has opened the door for the determination of the first solvation shell structure in many molten salts, including rare earth halides.

In parallel, numerical simulation techniques have also been developed for the study of ionic compounds, based on the seminal work of Fumi and Tosi, who derived parameters for describing the interactions between the various ions in alkali halides [6]. This model includes three important interactions, namely, the Coulombic interaction, the short-range repulsion (which is a consequence of the Pauli principle), and the London dispersion term. Although the Fumi-Tosi potentials have successfully been used in many molecular dynamics studies concerning simple monovalent molten salts, they suffer from an important limitation as soon as more complex ionic species are involved: They do not include the effects of electronic polarization [7]. Those affect importantly the structure of the melts: Not only do they allow for the formation of highly coordinated multivalent cations, but the induced dipole on anionic species also screens the repulsive cation-cation Coulombic interaction, thus allowing the formation of bent M-X-M angles and thereby influencing the arrangements which are formed through the linkage of cation-centered coordination complexes [8]. In recent works, the combination of several spectroscopic techniques with molecular dynamics simulations employing polarizable interaction potentials has been favored; for example, it has allowed to understand the particular speciation occurring in mixtures of molten alkali fluorides with tetravalent zirconium fluoride [9].

In order to illustrate the abundant chemistry of molten salts, we discuss here the example of LiF-BeF_2 mixtures. Due to their use as coolant in molten salt nuclear reactor concepts, their structure has been investigated using a wide range of diffraction, spectroscopic, and simulation tools. From the first solvation-shell point of view, the beryllium cation binds with four fluoride anions at all the compositions, forming tetrahedral structural units. At low concentrations of BeF_2 in LiF, the mixture behaves as a well-dissociated ionic melt consisting of Li^+, BeF_4^{2-}, and F^- ions. As the BeF_2 concentration increases, some medium-range ordering is observed though the formation of Be-F-Be bonds; in other words, the [BeF_4] tetrahedra start to be connected through a corner-sharing mechanism [10]. The speciation diagram of these fluoroberyllate species is shown in Fig. 1. At intermediate concentrations, first $Be_2F_7^{3-}$ appears, comprising two [BeF_4] units sharing a common F^- ion, followed by $Be_3F_{10}^{4-}$, and so forth. As the BeF_2 concentration increases further, more polymeric species are formed, which arrange as a tridimensional network.

The important changes in the structure of molten salts depending on the chemical composition are at the origin of strong variations in their thermodynamic and transport properties. Concerning the latter, the most important effects are observed for the viscosity, which can increase (at constant temperature) by several order of magnitudes when polymeric species are formed. The details of such a decrease in the fluidity of the melts can now be probed thanks to the use of pulse field gradient NMR combined with laser heating, which has made possible the in situ determination of self-diffusion coefficient in molten salts [11]. Indeed, highly coordinated, multivalent species have been shown to diffuse much more slowly than the alkali ions, while the fluoride adopts an intermediate behavior depending on the degree of polymerization of the system. Interestingly, a much less pronounced electrical conductivity drop is generally observed [12], due to the decoupling of the dynamics of the mobile alkali ions with the one of the viscous network [10]. This situation is somewhat comparable to the case of solid-state superionic electrolytes such as alkali silicate glasses [13].

Many industrial applications of molten salts, such as aluminum, alkaline and alkaline earth, or refractory metal production, take profit of the thermodynamic properties of molten salts. Their ability to dissolve many materials is of particular importance, and a lot of work has therefore been devoted to the building of accurate phase diagrams for each application. For example, the knowledge of the melting points for each particular composition is compulsory for the assessment of nuclear reactor coolants performances [14]. Due to their excellent thermal storage capabilities, molten salts are also used in the most recent concentrating solar power plants such as Andasol, which was built in 2008 in Spain and has a maximal electric output of 50 MWe (this number is expected to be multiplied by 3 in the next years). In this complex, the sunlight is concentrated through parabolic mirrors in order to heat up some synthetic oil. The later is then used to generate steam, from which electricity is produced. The role of the molten salt (a mixture of potassium nitrate and sodium nitrate) consists in storing part of the heat generated during the day and give it back to the oil at night or when the weather is rainy [15]. This setup allows the plant to operate for an extra 7.5 h, ensuring a much better continuity of the electricity production; this number could even be enhanced by finding molten salts compositions with lower melting points [16]. Recently, Sadoway and co-workers have proposed to take profit from the excellent solvating properties of molten salts to use them as electrolytes for performing the recycling of a series of semiconductors by ambipolar electrolysis [17].

As for more conventional solvents, the properties of molten salts may vary a lot depending on their acidity. According to the definition of Lewis, an acid molecule is one which is capable of receiving an electron pair while a basic molecule is a species that may provide such an electron pair. A first approach consists in defining acidity scales for each kind of Lewis base: For example, chloroacidity refers to the acidity with regard to the base Cl^- [18]. To do so, one needs to be able to measure the activity of the base, a difficulty that still has to be overcome in many systems like molten fluorides [19]. An alternative approach is to build basicity scales by measuring the optical activity of specific probes [20]. In principle, this concept could be applied to any liquid, thus leading to a generalized Lewis acidity scale [21].

Future Directions

In order to increase the share of renewables in the world's energy consumption, an important challenge for the nearest years is the development of cheap electrochemical devices for stationary electricity storage. Recent work has shown that molten salt electrolytes provide an attractive option. Sadoway and co-workers have indeed proposed a battery comprising a negative electrode of Mg, a molten salt electrolyte ($MgCl_2$–KCl–NaCl), and a positive electrode of Sb, which operates at 700 °C [22]. Its big advantage over other promising technologies such as Li-ion batteries is the low-cost and abundance of the three components. Such an application would also benefit from the experience accumulated in running the aluminum production process over more than a century as well as renew the interest for the use of high-temperature molten salts as electrolytes in large-scale applications.

Cross-References

- ▶ Alkali and Alkaline Earth Metal Production by Molten Salt Electrolysis
- ▶ Aluminum Smelter Technology
- ▶ Refractory Metal Production by Molten Salt Electrolysis

References

1. Edwards FG, Enderby JE, Howe RA, Page DI (1975) The structure of molten sodium chloride. J Phys C Solid State Phys 8:3483–3490
2. Papatheodorou GN (1977) Raman spectroscopic studies of yttrium (III) chloride-alkali metal chloride melts and of Cs_2NaYCl_6 and YCl_3 solid compounds. J Chem Phys 66:2893–2900
3. Robert E, Olsen JE, Danek V, Tixhon E, Ostvold T, Gilbert B (1997) Structure and thermodynamics of alkali fluoride-aluminum fluoride – alumina melts. Vapor pressure, solubility, and Raman spectroscopic studies. J Phys Chem B 101:9447–9457
4. Lacassagne V, Bessada C, Florian P, Bouvet S, Ollivier B, Coutures JP, Massiot D (2002) Structure of high-temperature NaF-AlF_3-Al_2O_3 melts: a multinuclear NMR study. J Phys Chem B 106:1862–1868
5. Rollet A-L, Bessada C, Auger Y, Melin P, Gailhanou M, Thiaudière D (2004) A new cell for high temperature EXAFS measurements in molten rare earth fluorides. Nucl Instrum Methods Phys Res, Sect B 226:447–452
6. Fumi FG, Tosi MP (1964) Ionic sizes + born repulsive parameters in NaCl-type alkali halides. I. Huggins-Mayer + Pauling form. J Phys Chem Solids 25:31–44
7. Madden PA, Wilson M (1996) "Covalent" effects in "ionic" systems. Chem Soc Rev 25:339–350
8. Salanne M, Madden PA (2011) Polarization effects in ionic solids and melts. Mol Phys 109:2299–2315
9. Pauvert O, Salanne M, Zanghi D, Simon C, Reguer S, Thiaudière D, Okamoto Y, Matsuura H, Bessada C (2011) Ion specific effects on the structure of molten AF-ZrF_4 systems ($A^+ = Li^+$, Na^+, and K^+). J Phys Chem B 115:9160–9167
10. Salanne M, Simon C, Turq P, Madden PA (2007) Conductivity-viscosity-structure: unpicking the relationship in an ionic liquid. J Phys Chem B 111:4678–4684
11. Rollet A-L, Sarou-Kanian V, Bessada C (2009) Measuring self-diffusion coefficients up to 1500 K: a powerful tool to investigate the dynamics and the local structure of inorganic melts. Inorg Chem 48:10972–10975
12. Salanne M, Simon C, Groult H, Lantelme F, Goto T, Barhoun A (2009) Transport in molten LiF-NaF-ZrF_4 mixtures: a combined computational and experimental approach. J Fluorine Chem 130:61–66
13. Angell CA (1992) Mobile ions in amorphous solids. Annu Rev Phys Chem 43:693–717
14. Benes O, Beilmann M, Konings RJM (2010) Thermodynamic assessment of the LiF-NaF-ThF_4-UF_4 system. J Nucl Mater 405:186–198
15. Cartlidge E (2011) Saving for a rainy day. Science 334:922–924
16. Jarayaman S, Thompson AP, von Lilienfeld OA (2011) Molten salt eutectics from atomistic simulations. Phys Rev E 84:030201
17. Bradwell DJ, Osswald S, Wei WF, Barriga SA, Ceder G, Sadoway DR (2011) Recycling ZnTe, CdTe, and other compound semiconductors by ambipolar electrolysis. J Am Chem Soc 133:19971–19975
18. Tremillon BL (1987) Acid base effects in molten electrolytes. In: Mamantov G, Marassi R (eds) Molten salt chemistry, NATO ASI series, series C: mathematical and physical sciences, vol 202. Springer: Dordrecht
19. Bieber AL, Massot L, Gibilaro M, Cassayre L, Chamelot P, Taxil P (2011) Fluoroacidity evaluation in molten salts. Electrochim Acta 56:5022–5027
20. Duffy JA, Ingram MD (1971) Establishment of an optical scale for Lewis basicity in inorganic oxyacids, molten salts, and glasses. J Am Chem Soc 93:6448–6454
21. Salanne M, Simon C, Madden PA (2011) Optical basicity scales in protic solvents: water, hydrogen fluoride, ammonia and their mixtures. Phys Chem Chem Phys 13:6305–6308
22. Bradwell DJ, Kim H, Sirk AHC, Sadoway DR (2012) Magnesium–antimony liquid metal battery for stationary energy storage. J Am Chem Soc 134:1895–1897

High-Temperature Oxygen Sensor

Yoshihiko Sadaoka
Ehime University, Matsuyama, Japan

Introduction

In 1957, K. Kiukkola and C. Wagner discovered that zirconia- or thoria-base solid solution with

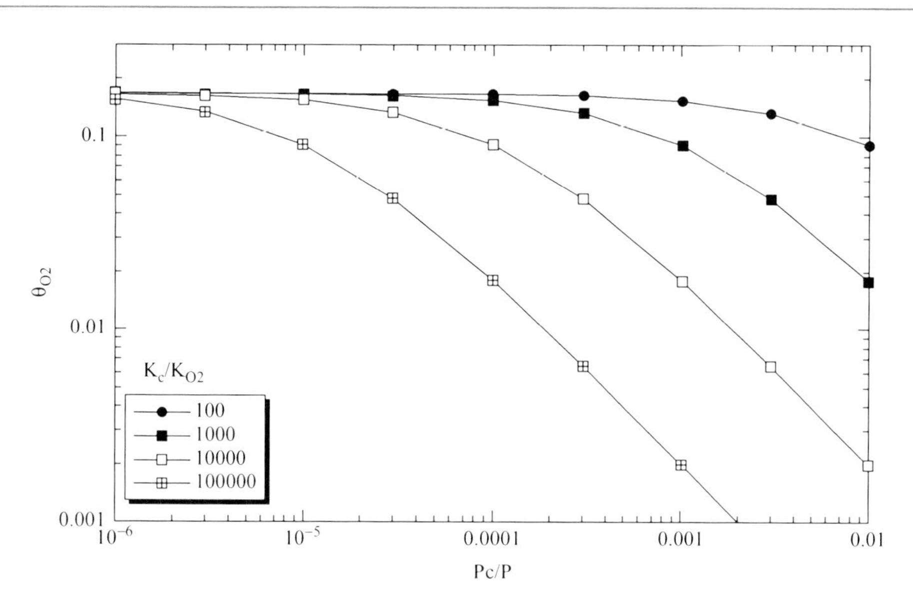

High-Temperature Oxygen Sensor, Fig. 1 The correlation between θ_{O_2} and P_C/P as a function of K $(= K_C/K_{O_2})$

CaO, MgO, or Y_2O_3 can be used as solid electrolyte for galvanic cells operating at high temperatures above 870 K [1]. In recent years, solid-state oxygen sensors have found extensive applications in many areas such as monitoring and control of industrial processes and of automotive engine operation. In the most important application of engine control, oxygen sensors are used to optimize the vehicle performance with respect to exhaust emissions and the air-to-fuel ratio (A/F) of the engine.

An electrochemical sensor is essentially an electrochemical cell consisting of two or more electrodes in contact with a solid electrolyte. They can be classified according to their operation mode, e.g., conductivity/impedance sensors, potentiometric sensors, and amperometric sensors.

Interaction of Oxygen and Solid-State Materials

For the mixed gas with N_2, O_2, and contaminant, C, and based on the adsorption isotherm-based Langmuir-type isotherm for the mixture with oxygen, contaminant, and nitrogen, the adsorption behavior could be expressed as follows [2]:

$$N_{O_2} = K_{O_2}P_{O_2} \tag{1}$$

$$N_C = K_C P_C \tag{2}$$

$$\theta_C = N_C/(N_C + N_{O_2}) \tag{3}$$

$$\theta_{O_2} = 1 - \theta_C \tag{4}$$

$$P_{O_2} + P_c = P \tag{5}$$

$$1/\theta_C = 1 - K_{O_2}/K_C + (K_{O_2}/K_C)/(P_C/P) \tag{6}$$

where P $(= P_{O2} + P_C)$ is the total pressure (concentration) and K_{O2} and K_C are the equivalent constants of oxygen and the contaminant, respectively.

Figure 1 shows the correlation between θ_{O_2} and P_C/P as a function of K $(= K_C/K_{O_2})$. It is obvious that the concentration of adsorbed oxygen decreased with an increase in the equivalent constant, K_C. A physisorption model should be introduced. It is well known that the surface collision number, Z, is given by the following Herz-Knudsen equation,

$$Z = P/(2\pi MRT)^{1/2} \tag{7}$$

where P is the pressure and T and M are the absolute temperature and molecular weight, respectively. The sticking probability (condensation coefficient), α, is defined by

$$\alpha = N_S(d\theta/dt)/Z = N_S(d\theta/dt)\,(2\pi MRT)^{1/2}/P \qquad (8)$$

Polanyi proposed the adsorption potential model in 1914. The adsorption potential, A, is expressed as follows:

$$A = RT\,\ln\left(P_0/P\right) = b \cdot f(y) \qquad (9)$$

where y is the adsorbed molecular volume. The characteristic, b, of organic molecules was estimated by Dubinin and Timofeev in 1948. These characteristics are well related to the molecular volume, V, of a compound in the liquid phase.

$$A_L/A_{benzene} = V_L/V_{benzene} \qquad (10)$$

Based on the classic models for the physisorption phenomena on solid surfaces, the coverage of the physisorbed molecules is a function of the sticking probability and/or adsorption potential and depends on the molecular weight and molecular volume in the liquid phase. It is obvious that these characteristics are related to the equivalent constant in the adsorption isotherm of the contaminants.

When the oxidation reaction of the contaminant expressed as follows:

$$C + O_2 \rightarrow \text{products}\uparrow$$

adsorbed site of the contaminant decreases to very small amount. It is considered that the various species of oxygen relevant to solid surface reactions. The energy difference between O_2^- adsorbed ($O_2^-{}_{ad}$) and O^{2-} bound to a lattice site ($O^{2-}{}_{lattice}$) is estimated to be about 20 eV and between $1/2O_2\,{}_{gas}$ and $1/2O_2^-{}_{ad}$ to be about 1.5 eV [3]. It seems that the equilibrium of the $O_2^-{}_{ad}$ with gaseous O_2 is approached slowly, and the reaction $O_2^-{}_{ad} + e = 2O^-{}_{ad}$ takes place with increasing temperature. Above 450 K $O^-{}_{ad}$ are found as the prevailing species. $O^{2-}{}_{ad}$ should not be stable if it does not react immediately or is trapped by an oxygen vacancy and stabilized by the Madelung potential of the lattice. The adsorbed oxygen species come not only from the gas phase but can also emerge from lattice sites. This process can be understood as an intermediate step in the thermal decomposition of the materials. A certain "coexistence" of vacancies and adsorbed species should be accepted.

High-Temperature Oxygen Sensor, Fig. 2 Illustration of oxide surface

The interactions of the oxide surface with water are important. The exposure of water to the oxides induces the formation of –OH group, and two equations can be proposed:

$$H_2O + M_{lattice} + O_{lattice} \rightarrow HO\text{-}M_{lattice} + O_{lattice}H + e^-$$

$$H_2O + 2M_{lattice} + O_{lattice} \rightarrow 2(HO\text{-}M_{lattice}) + Vo^{\bullet}$$

In thermal desorption behavior, desorption of water is mainly observed in two temperature ranges, around 400 K and 650–800 K. The low temperature desorption is attributed to molecularly adsorbed water, whereas the high temperature process is due to OH groups recombining to form H_2O. Larger hydrogen-containing molecules can also act as a source of hydrogen atoms. These mechanisms are shown in Fig. 2.

Conductivity/Impedance Sensors

The response of electrical-conducting oxygen sensors is generated by the equilibration of oxygen in the gas phase with ionic and electronic point defects in the oxide semiconductor. Several kinds of oxides are used in such sensors. The focus of this section is on the use of perovskite-type oxides. Perovskite-type oxides are particularly attractive for high-temperature applications. Often having high melting and/or decomposition temperature up to 1,200 K, they can provide reliability and long-term stability. The perovskite-type oxide most commonly used in oxygen sensors is $SrTiO_3$. The intrinsic ionic point defects of which are strontium and oxygen vacancies (Schottky defects) [4–6].

At high oxygen partial pressures, the predominant point defects are strontium vacancies and holes, which can form by oxygen occupying a vacant SrO unit on the surface according to Eq. (11)

$$1/2O_2 + \left[V''_{Sr} + Vo^{\bullet\bullet}\right] = V''_{Sr} + O_o^{x} + 2h^{\bullet} \quad (11)$$

and also by removing a strontium ion to form a separate SrO phase:

$$1/2O_2 + Sr^{x}_{Sr} = V''_{Sr} + 2h^{\bullet} + SrO \quad (12)$$

When combined with the charge neutrality requirement, either of these reactions,

$$p \propto 2[V''_{Sr}]$$

results in the same oxygen pressure dependence for the electronic conductivity as given by Eq. (13).

$$p = (2Kp)^{1/3} P_{O_2}{}^{1/6} \quad (13)$$

At low oxygen partial pressure, the predominant point defects are oxygen vacancies and electrons. The electronic conductivity is given by Eq. (14).

$$n = (2Kn)^{1/3} P_{O_2}{}^{-1/6} \quad (14)$$

In practically, some deviations of the observed oxygen partial pressure dependence of the electronic conductivity from that expected for intrinsic Schottky defects have been attributed to the presence of additional ions defects. One possibility is that, even in nominally undoped $SrTiO_3$, impurities are present in amounts large enough to establish the point defect concentration. For example, the aluminum doped $SrTiO_3$, Al_2O_3 would replace TiO_2 [7],

$$Al_2O_3\ (-2TiO_2) = 2Al'_{Ti} + 3O_O{}^{x} + Vo^{\bullet\bullet} \quad (15)$$

which result to create oxygen vacancies. If the charge balance is given by Eq. (16):

$$[Vo^{\bullet\bullet}] \sim \left[V''_{Sr}\right] + 1/2\left[Al'_{Ti}\right] \quad (16)$$

The conductivity in low oxygen partial pressure is given by

$$n = \left\{Kn/\left(\left[V''_{Sr}\right] + 1/2\left[Al'_{Ti}\right]\right)\right\}^{1/2} P_{O_2}{}^{-1/4} \quad (17)$$

The oxygen partial pressure dependence of the conductivity will be similarly changed at high oxygen partial pressures, and then the oxygen partial pressure dependence of the hole concentration is given by

$$p = \left(2Kp/\left[V''_{Sr}\right]\right)^{1/2} P_{O_2}{}^{1/4} \quad (18)$$

The conductivity of undoped $SrTiO_3$ at 800 °C is shown in Fig. 3.

Potentiometric Sensors

In a potentiometric oxygen sensor, the measuring electrode is generally made of a porous metallic coating, and the electrode reaction

$$O_2(gas) + 4e^-(electrode) \Longleftrightarrow 2O^{2-}(electrolyte) \quad (19)$$

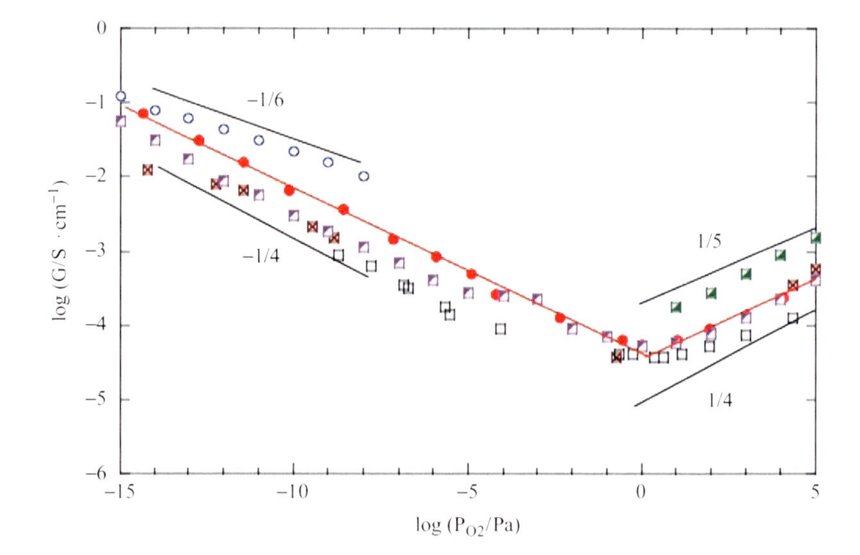

High-Temperature Oxygen Sensor, Fig. 3 Conductivity of undoped $SrTiO_3$ at 800 °C [8]

is viewed as occurring at the triple gas-electrolyte-metal contact (triple phase boundary). Many possibilities exist and have been envisaged. Some of the models emphasize a dissolving of oxygen in the metal or its adsorption on the metal or on the electrolyte surface a necessary intermediate step.

A solid-state oxide with the pure ionic conductivity of oxide ion can be used as the electrolyte in the galvanic cell.

$$P_{O_2}(I), \text{electrode}(I) \mid \text{solid electrolyte} \mid \text{electrode}(II), P_{O_2}(II) \quad (20)$$

This cell is sometimes called the oxygen concentration cell.

A typical device was the oxygen sensor patented by Weissbart et al. [8] and by Moebius et al. [9] It is based on stabilized zirconia, a predominant oxide ion conductor with minority electron–hole carriers in a free-oxygen environment. Platinum is typically used as an electronic sensing coating on the surface of this electrolyte. The basic principle of the sensor, within the framework of Wagner's theory, is a probing of the relevant electron–hole electrochemical potential variations. This probing is done with the coating of porous platinum. In this case the devices were operated at a higher temperature, and the basic assumptions were that

1. full equilibria are rapidly reached at the surfaces.
2. possible slow responses, due to the equilibrium process of the bulk of the solid electrolyte, are generally governed by diffusion-controlled adjustments of the relevant electrochemical potential profiles in the solid electrolyte.

At the contact between platinum and stabilized zirconia, an equilibrated electron exchange is assumed to prevail.

$$e^-(Pt) + h^{\bullet}(\text{zirconia}) \Longleftrightarrow 0 \quad (21)$$

The sensing electrode thus can be sketched by the triple contact (three phase boundary) as shown in Fig. 4a.

For the galvanic cell, the Gibbs free energy change caused by supplying n Faradays of electrons is related to the reversible electrical work by the following Nernst equation.

$$\Delta G = -nFE \quad (22)$$

Where G id Gibbs free energy, E is the reversible equilibrium electrode potential of the cell, and F is the Faraday constant. This equation is

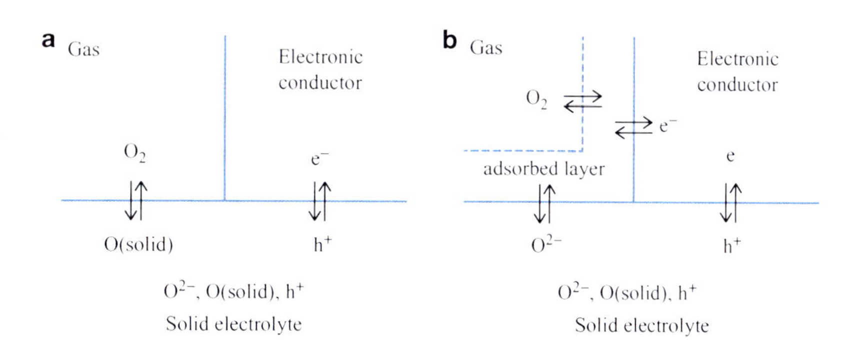

High-Temperature Oxygen Sensor, Fig. 4 (**a**) Triple contact according to Wagner's theory. (**b**) Triple contact with adsorption layer

consistent as long as temperature and total pressure remain a constant. When 4 Faradays of electrons are externally fed to the electrode II and taken out from the electrode I, the oxide ion will migrate to the electrode I. Under the condition, the following electrochemical reactions take place at the two electrode interfaces:

$$O_2(P_{O_2}(II)) + 4e^- \rightarrow 2O^{2-} \quad \text{at electrode II} \tag{23}$$

$$2O^{2-} \rightarrow O_2(P_{O_2}(I)) + 4e^- \quad \text{at electrode I} \tag{24}$$

Summarizing the above two equations, the cell reaction can be obtained as follows:

$$O_2(P_{O_2}(II)) \rightarrow O_2(P_{O_2}(I)) + 4e^- \tag{25}$$

Thus, the Gibbs free energy change of the cell reaction is equal to that caused by the change in partial pressure of oxygen.

$$\Delta G = RT\ ln[P_{O_2}(I)/P_{O_2}(II)] \tag{26}$$

The electrode potential, E, is expressed as follows:

$$E = (RT/4F)\ ln[P_{O_2}(I)/P_{O_2}(II)] \tag{27}$$

By a great many experiments using several stabilized zirconia, the validity of this equation has been verified above about 850 K, while now this type of the oxygen sensor is operable at around 600 K for the optimized sensor.

In many cases, the three-phase boundary system is formed with the layer of oxygen adsorbed on the solid electrolyte surface. Then, the triple contact diagram is that shown in Fig. 4b. It is assumed that oxygen adsorbs and dissociates at a rate proportional to its pressure in the gas and desorbs proportionally to the concentration of atomic species in the adsorbed layer. The response curves of the Pt/YSZ/Pt sensor are shown in Fig. 5.

This very simple model having the adsorbed layer formed on the surfaces of the electrode and the electrolyte has allowed us to successfully interpret the response time of oxygen sensors systematically measured under various conditions and the deviations from ideal behavior due to the oxygen semipermeability flux [10, 11]. The existence of the adsorbed layer becomes the most important factor to determine the response characteristics of the device to the oxygen pressure.

$$O_2(\text{gas}) \Longleftrightarrow 2O(\text{adsorbed layer}) \tag{28}$$

$$2O(\text{adsorbed layer}) + 4e^-(\text{metal}) \Longleftrightarrow 2O^{2-}(\text{electrolyte}) \tag{29}$$

The obvious interest of this diagram is that it expands the list of materials available to develop many kinds of gas sensing devices based on the potentiometric oxygen sensor, since adsorbed layer can be also controlled by the modification of the surface of the three-phase boundary with the additional materials having molecular sieve effects and catalytic activities [12–14].

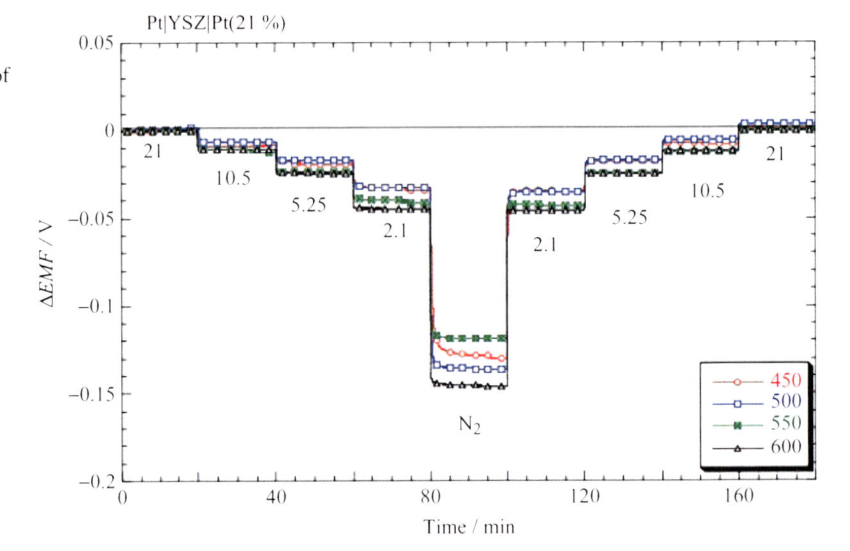

High-Temperature Oxygen Sensor, Fig. 5 Response curves of Pt/YSZ/Pt sensor

Amperometric Sensors

Oxygen sensors based on the principle of electrochemical pumping are interesting and useful. A limiting current type oxygen sensor, which is based on electrochemical pumping, was developed and used in a lean burn engine control system an in alarm and monitoring systems. In the early stages of the development of limiting current type oxygen sensors, the limiting current characteristic of the electrode of a stabilized zirconia electrolyte cell was utilized to measure oxygen concentration. Later, trials utilizing a gas diffusion limiting devices (pinhole type and porous layer type) attached to the outside of the electrode of a zirconia cell were made to improve the stability of the sensing characteristics. When voltage is applied to a zirconia electrolyte cell, oxygen is pumped through the zirconia electrolyte from the cathode to the anode side. At low voltages, the current is nearly proportional to the voltage. As the voltage is increased, the current shows saturation due to the rate-determining step in the transfer of oxygen at the cathode. The saturation current is called limiting current and is also nearly proportional to the ambient oxygen concentration. However, the limiting current of this type of sensor is not stable because it is directly affected by the degradation of the cathode and the electrolysis of the body. To overcome the disadvantages, a gas diffusion limiting device was attached on the surface of the cathode. For the sensor with the pinhole and/or porous layer, it is small enough for the oxygen diffusion in the pinhole/porous layer to be a rate-determining step, and the effective diameter of the pinhole/porous layer is so much larger than the mean path of oxygen that ordinary diffusion dominates and Knudsen diffusion can be neglected. In this case, the limiting current, I_p, is given with the following relation.

$$Ip = (4FDS\mathrm{Ct}/l)\ ln[1/(1 - \mathrm{C}/\mathrm{Ct})] \qquad (30)$$

where F is the Faraday constant, l, the length of the pinhole, Ct, the total molar concentration of gases, D and C are the diffusion coefficient and molar concentration of oxygen, respectively. This equation shows that the limiting current varies with oxygen concentration logarithmically. At a low concentration of oxygen, the current varies almost linearly with the concentration of oxygen.

$$Ip = (4FDS/l)\mathrm{C} \qquad (31)$$

In the case where the diameter of the hole in the diffusion layer is much larger than the mean free path of gases, ordinary diffusion dominates and the limiting current is expressed by Eq. (31).

The condition of a much larger diameter than the mean free path of gases at atmosphere pressure corresponds to a diameter larger than the order of 1 mm. On the other hand, in the case where the diameter of the hole is much smaller than the mean free path of gases, the Knudsen diffusion dominates. In this case, the diffusion of gases is not governed by the interaction between gas molecules themselves but by the interaction between gas molecules and the wall of the holes. The limiting current for the Knudsen diffusion is given by the following equation and is linear to oxygen concentration even at high oxygen concentrations [15].

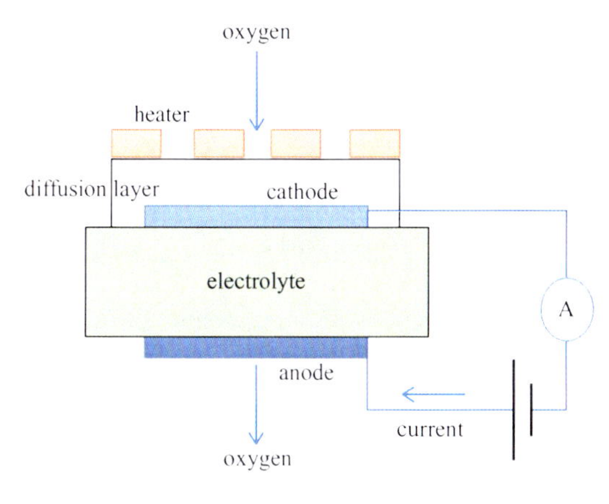

High-Temperature Oxygen Sensor, Fig. 6 The structure of the sensor

$$Ip = (4FD_K S/l)C \qquad (32)$$

where D_K is the Knudsen diffusion coefficient. Knudsen diffusion is dominant for a very small hole, usually smaller than the order of 10 nm.

For the practical applications, thin-film type sensors are fabricated. The structure of the sensor is shown in Fig. 6.

For this type of the oxygen sensor, the dependence of the limiting current on the total pressure becomes important. For ordinary diffusion, the limiting current is independent of total pressure, whereas the limiting current due to the Knudsen diffusion is proportional to the total pressure [16]. This type sensor usually operates at 770–970 K. In general, the limiting current is measured at a constant applied voltage and the measurable range of oxygen limited with the operating temperature as shown in Fig. 7.

Future Directions

Oxide ion conductive ceramics, such as stabilized zirconia or thoria, have high thermal and chemical stability. Oxide ion activity on the surface of ceramics is strongly influenced by the chemical components present in the environment. Many researchers have suggested that this changing property of oxide ion on the surface is useful to

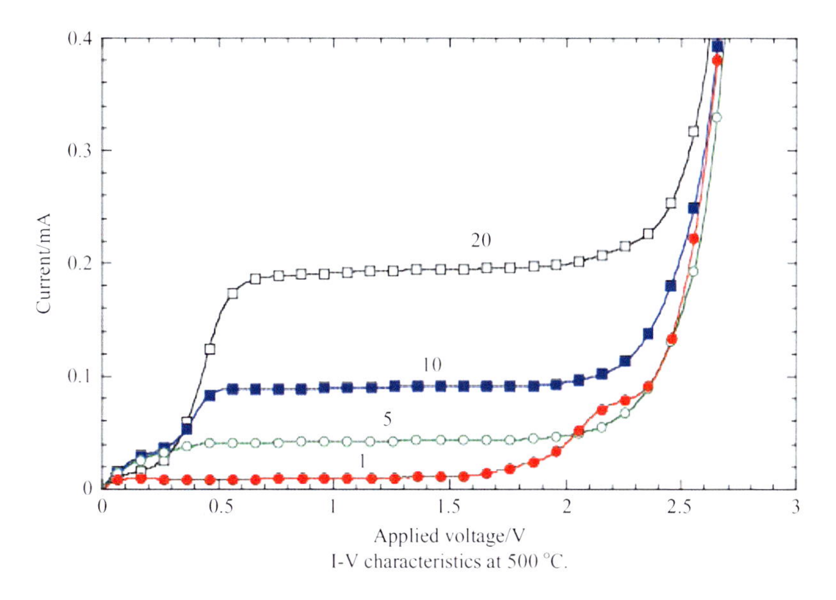

High-Temperature Oxygen Sensor, Fig. 7 The measurable range of oxygen limited with the operating temperature

develop gas monitoring devices with greater stability, accuracy, and repeatability. Oxide ion conductive materials operating at a lower temperature, below 670 K, used in sensing devices are helpful in many fields, not only in fuel/air monitoring but also in environment ambient analysis, health care, medical diagnosis, and so on.

Cross-References

- ► Sensors
- ► Solid Electrolytes
- ► Solid Electrolytes Cells, Electrochemical Cells with Solid Electrolytes in Equilibrium
- ► Oxide Ion Conductor

References

1. Kiukkola K, Wagner C (1957) Measurements of galvanic cells involving solid electrolytes. J Electrochem Soc 104:308–379
2. de Boer JH (1953) The dynamical character of adsorption. Clarendon, Oxford
3. Biela'nski A, Haber J (1979) Oxygen in catalysis on transition metal oxides. Catal Rev-Sci Eng 19(1):1–41
4. Meyer R, Waser R (2004) Resistive donor-doped SrTiO 3 sensors: I, basic model for a fast sensor response. Sens Actuators B 101(3):335–345
5. Chan NH, Sharma RK, Smyth DM (1981) Nonstoichiometry in StTiO//3. J Electrochem Soc 128(8):1762–1769
6. Moseley PT (1992) Materials selection for semiconductor gas sensors. Sens Actuators B Chem 6(1–3):149–156
7. Fergus JW (2007) Perovskite oxides for semiconductor-based gas sensors. Sens Actuators B Chem 123(2):1169–1179
8. Weissbart J, Ruka R (1961) Oxygen gauge. Rev Sci Instrum 32(5):593–595
9. Peters H, Moebius HH (1961) East Germany Patent. 21:673
10. Fouletier J, Fabry P, Kleitz M (1976) Electrochemical semipermeability and the electrode microsystem in solid oxide electrolyte cells. J Electrochem Soc 123(2):204–213
11. Fouletier J, Senera H, Kleitz M (1974) Measurement and regulation of oxygen content in selected gases using solid electrolyte cells. I. Discontinuous use of gauges. J Appl Electrochem 4(4):305–315
12. Mori M, Nishimura H, Itagaki Y, Sadaoka Y (2009) Potentiometric VOC detection in air using 8YSZ-based oxygen sensor modified with $SmFeO_3$catalytic layer. Sens Actuators B142(1):141–146
13. Mori M, Sadaoka Y (2010) Potentiometric VOC detection at sub-ppm levels based on YSZ electrolyte and platinum electrode covered with gold. Sens Actuators B 146(1):46–52
14. Mori M, Nishimura H, Yahiro H, Sadaoka Y (2008) Potentiometric VOCs detection using 8YSZ based oxygen sensor. J Ceram Soc Jpn 116(1335):777–780
15. Ruka RJ, Panson AJ (1972) US Patent 3/691023
16. Diets H (1982) Gas-diffusion-controlled solid-electrolyte oxygen sensors. Solid State Ion 6(2):175–183

High-Temperature Polymer Electrolyte Fuel Cells

Thomas J. Schmidt and George Neophytides
Electrochemistry Laboratory, Paul Scherrer Institut, Villigen, Switzerland

Introduction

Fuel cells have been known as power sources since the times of Sir William R. Grove and Christian F. Schönbein around 1840. Since then huge leaps in technology occurred leading to different kinds of fuel cells classified upon the physicochemical properties of their electrolyte and their operating conditions. Polymer electrolyte fuel cells (PEFC) utilize the ionic conductivity, the gas separation ability, and the electronic insulation provided by polymeric materials. These properties led to solutions of various engineering problems in liquid electrolyte fuel cells (e.g., phosphoric acid FC, alkaline FC).

Different technologies have been developed for the PEFCs classifying them into low- and high-temperature PEFCs. In low-temperature applications (−20 °C to 90 °C), the predominant polymer membrane material is from the group of perfluorsulfonic acids (PFSA) with the most prominent example being Nafion©, a polymer membrane discovered in the late 1960s by Walther Grot of DuPont [1]. The polymer becomes conductive for protons (H^+) in the presence of water therefore demanding humidification of the reactants for optimum fuel-cell performance [2]. Due to the properties of the material and the need for the presence of liquid

High-Temperature Polymer Electrolyte Fuel Cells, Fig. 1 Chemical structures of PBI and TPS variants: (**a**) m-PBI, (**b**) AB-PBI, (**c**) p-PBI, (**d**) TPS, (**e**) TPS 2 [6–8]

water, the operating temperature should not significantly exceed 80 °C. This rather low fuel-cell operating temperature leads to the necessity of very pure hydrogen fuel feed, since the tolerance of the PEFC anode catalysts towards typical contaminants (e.g., CO and H_2S) is very low.

A new type of polymer electrolyte fuel cell was proposed by Savinell in 1994 comprising of a Nafion© membrane imbibed in phosphoric acid to increase the operation temperature to 200 °C [3]. Later on Savinell et al. proposed the use of polybenzimidazole (PBI) films – a material used as fibers in textiles for high performance, nonflammable applications – imbibed in phosphoric acid as electrolytes for fuel cells operating in elevated temperatures up to 200 °C and giving a good performance without the need for humidifying the reactants [4]. These so-called high-temperature (HT) PEFCs are related to both the LT-PEFCs and phosphoric acid FCs, respectively. The increased operation temperature gave rise to high tolerance to poisoning species present in the reactant gases. Furthermore, these fuel cells can be used in combined heat and power (CHP) applications. Due to the elevated temperature waste heat can be effectively exploited since the operating temperature between 150 °C and 200 °C favors the heat transfer from the system to the environment.

Membranes for High-Temperature PEFCs

Several materials have been proposed and commercialized as electrolytes for HT-PEFCs. As introduced before, the main polymers used materials from the PBI family and the Advent tetramethyl pyridine sulfone (TPS) family, both being basic polymers allowing chemical interaction with mineral acids (e.g., phosphoric acid); see Fig. 1. Differences can be found both in the chemistry and in the synthetic process especially among the different PBIs, yielding different physicochemical properties such as glass transition temperatures, mechanical stabilities, proton conductivities, and achievable phosphoric acid doping levels (defined as either the ratio of phosphoric acid molecules per polymer repeat unit or the weight ratio of polymer and included phosphoric acid); for a summary, see Table 1.

Doping of the polymer film with H_3PO_4 is an important process which creates the final proton conductive membrane. The method used to dope the material can yield different doping levels, even when the same polymer structure is used, making it thus important in order to achieve good cell performance. For the PBI family, two approaches have been used by different manufacturers and researchers.

High-Temperature Polymer Electrolyte Fuel Cells, Table 1 Physicochemical properties of different HT polymer membranes for fuel cells

Polymer	p-PBI [7]	m-PBI (PPA) [7]	m-PBI [7]	AB-PB [7]I	TPS [9]	TPS 2 [6]
M_w	–	–	–	–	155000	75000
IV ($dLcm^{-1}$)	3	1.49	0.6–1.2	1.5–2.4	–	–
PA Content (n_{H3PO4}/p.r.u)	32	14.4	2–16	1.6–5	8.9	17
Conductivity (Scm^{-1}) at 160°C	0.26	0.127	0.059	0.062	0.013	0.065

The most common method and first approach used is the so-called dip-and-soak process. After the casting of the polymer film, it is dipped in H_3PO_4 and soaked until the desired doping level has been reached. This process was initially proposed by Savinell et al. and by Bjerrum et al. later on, leading to the commercialization of PBI-based MEAs [4, 5]. Membranes are prepared the same way in the case of Advent TPS despite the different membrane chemistry [6, 9].

A different approach was developed by Benicewicz and co-workers in collaboration with BASF fuel cells (formerly PEMEAS), respectively. In the so-called PPA (polyphosphoric acid) process, the polycondensation of the monomers (e.g., tetra-amino-biphenyl and an organic diacid, e.g., iso-phtalic acid) takes place in PPA as solvent [7]. The polymer-PPA solution is casted as a film followed by the defined hydrolysis of the PPA to orthophosphoric acid. During this hydrolysis, a sol–gel transition occurs solidifying the casted film to form a self-standing membrane to be used in HT-PEFCs. This novel method is leading to fuel-cell membranes with unique physicochemical properties, very high molecular weights, or inherent viscosities as compared to other PBI synthesis routes and high PA-doping levels of the formed PBI membranes (Table 1) [7].

Catalysts for High-Temperature PEFCs

One of the most important parts of the membrane-electrode assemblies (MEAs) is the catalyst. The typical catalyst used in both electrodes consists of supported nanoparticles on high surface area carbons, and depending on the manufacturer, it could comprise of either pure Pt or Pt alloys. The catalysts used in HT-PEFCs are usually quite common to the catalysts used in LT-PEFC and in PAFCs. The main differences between the anodic and cathodic catalysts are the Pt loading on each electrode and, in the case that an alloy is used, the alloying component.

Beginning with the anode, the most commonly used catalyst is Pt nanoparticles on high surface area carbon. Due to the high operating temperature of up to 180 °C for oxidation of typical reformate-based feed gases with high impurity levels (e.g., CO concentration of up to 3 %), Pt catalysts can be used providing high impurity tolerance. There are only small indications in the literature that PtRu catalysts would improve CO tolerance even in HT-PEFCs [10, 11]. As was demonstrated by Schmidt and Baurmeister [12], even at CO partial pressures of 3 % in hydrogen, the anode voltage losses at 180 °C and 0.5 A/cm^2 are tolerated by a pure Pt–Vulcan catalyst, and the associated voltage losses amount only to less than 20 mV; see Fig. 2. Pt nanoparticles and Pt alloys on carbon support are used in HT-PEFCs cathodes, basically mimicking catalyst technology from earlier PAFCs developments.

The most commonly used catalyst has been Pt supported on both ungraphitized Vulcan X72 and proprietary graphitized carbon blacks. Due to kinetic limitations of pure Pt catalysts at the

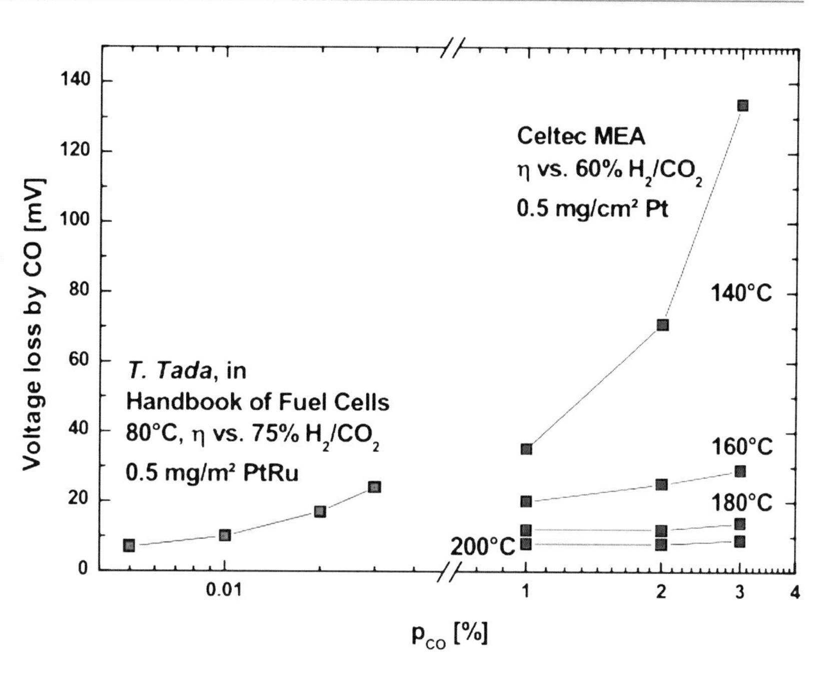

High-Temperature Polymer Electrolyte Fuel Cells, Fig. 2 CO-induced voltage loss for low- and high-temperature PEFCs. The LT-PEFC uses 0.5 mg/cm^2 PtRu/C, the HT-PEFC uses 0.5 mg/cm^2 Pt/C as catalyst. LT data are taken from ref. [16]. Figure is reprinted with permission from *ECS Transactions 16(2) (2008) 263–270*. Copyright 2008, The Electrochemical Society

interface to concentrated phosphoric acid in HT-PEFCs, currently mainly Pt alloys are used in order to increase the intrinsic activity of the catalyst. Several metals have been used such as Co, Cr, V, W, and Ti in binary and ternary alloys showing a vast increase in catalytic activity [13]. State-of-the-art HT MEAs are basically operating at the kinetic limits of Pt or Pt alloy catalysts, and the overall fuel-cell performance is limited by the strong interaction of the catalyst surface with phosphoric acid. The operation of a HT-PEFC imposes significant requirements on the MEA materials, especially on the cathode catalyst in the oxidizing atmosphere at high electrode potentials. As mentioned before, the most common support used in fuel cells catalysts is high surface carbon such as Vulcan X72, which can provide high specific surfaces but also suffers from corrosion in the extreme environment of a fuel-cell cathode. Therefore, new supports had to be investigated, being more resistant against oxidation leading to the use of graphitized carbon and carbon nanotubes. This further has been thoroughly tested for PAFC applications [13] and by BASF fuel cells as a support for Pt alloy nanoparticles in HT-PEFCs under start–stop cycling [14, 15] with its detrimental high potential excursions (up to 1.3 V). The materials showed a fivefold decrease in the voltage loss versus beginning-of-life (BOL) values [14].

Properties of High-Temperature PEFCs

The elevated operation temperatures give rise to unique characteristics and advantages and disadvantages in the design, operation, and durability of high-temperature PEFCs.

Impurity Tolerance

To begin with, a major aspect is the tolerance to impurities of both the anode and the cathode. An important property is the ability of a fuel cell to operate with reformate as fuel, which comprises of H_2, water, CO, and other impurities such as H_2S, since it is produced through steam reforming of natural gas. Usually realistic reformate mixtures contain up to 2 % CO and 20 ppm H_2S along with hydrogen and water, with their partial pressures varying depending on the reformer and the initial composition of the natural gas. Extensive durability of BASF Celtec P1000 MEAs of more than 3,000 h of continuous operation has been reported while using reformate gas mixtures (60 %H_2, 5 ppm H_2S, 2 % CO, 17 %CO_2, 21 %H_2O) [17].

Also the cathode can be affected by sulfur, an impurity among hydrocarbon fuels, air and from the catalyst carbon support. A study by Garsany et al. showed that PBI-based systems exhibit an excellent performance while poisoning with common air contaminants (e.g., H_2S, SO_2) in comparison to Nafion, and the recovery was done easily by simply purging with neat air [18].

Durability

Certain features of HT-PEFCs induce limitations in their durability, comprising mainly by the degradation of the membrane and the catalyst layer. Several degradation modes have been identified, such as membrane thinning and pinhole formation, PA evaporation (from membrane and catalyst layer), Pt dissolution, and carbon corrosion from the catalyst layer [7, 19]. These effects are caused by the realistic operating conditions which include high operating temperature and temperature cycles as well as start–stop cycling combined with operation on reformates produced by fuel processors. Especially for stationary CHP HT fuel-cell systems, durability of more than 40,000 h operation time is requested in order to become competitive to traditional heating devices. It is claimed that PBI-based systems can perform well with an acceptable performance degradation rate of only 6 μV/h for more than 20,000 h in continuous operation [20]. This long-term stability was proven to be a result of the low acid evaporation rates of membrane and MEAs prepared based on the aforementioned PPA process. Also systems based on other high-temperature polymer electrolytes have demonstrated promising durability, such as Advent TPS-based MEAs, where the materials endure for more than 4,000 h of continuous operation with a degradation rate of 9 μV/h [21]. Carbon corrosion of the catalyst support can be severe during start–stop cycling and local fuel starvation due to local potential excursions up to 1.3 V [15]. Carbon corrosion to CO_2 combined with Pt particle growth and dissolution can lead to significant losses of electrochemically active surface area and subsequent reduction of the cathode kinetics [14, 15]. One important degradation aspect is related to acid redistribution within the fuel cell, as it was shown that corrosion of binders used in bipolar plate materials leads to increased porosity of the plates and some acid wicking from the MEA into the pores of the plates and to severe cell failure [22] (Fig. 3).

Applications

(a) **Cogeneration of Heat and Power (CHP)**
The elevated operation temperature of high-temperature PEFCs offers simplifications in stack and system engineering in comparison to low-temperature PEFCs. In particular, humidification of the reactant gases is not needed due to the intrinsic characteristics of the electrolyte, and omission of the external humidification system is possible. Moreover the elevated operation temperature provides better heat transfer from the system to the environment, and therefore, cooling is simplified. For the same reason, the waste heat not only can be easily used in combined heat and power applications (CHP) increasing the overall efficiency of the fuel-cell system but also makes the thermal integration of fuel processors(e.g., methanol steam reformer) to the stack easier [20]. All the above have a great influence on the balance of plant and can lead to a threefold increase in power to volume ratio in comparison to low-temperature applications. Several examples of commercialized systems exist from manufacturers such as Serenergy, Plug Power, and ClearEdge Power targeting mainly the USA and the Japanese market. An example is shown in Fig. 4a.

(b) **Small and Micro Fuel Cells (sub-1kW)**
Micro fuel cells also make use of the advantages of HT-PEFCs. These systems, comprising of methanol reformers and HT-PEFCs, are used to power portable applications such as laptops, communication, and global positioning systems, respectively, with a power range of 5–50 W. A well-known application is the one proposed by Ultracell under the brand name XX25, which can provide 20 W of continuous maximum power for timescales ranging from 9 h to 25 days [20]. Compared to batteries these applications are lighter and more rugged and easily operated under extreme conditions (temperature, humidity). Another example of a reformed methanol HT-PEFC system is the off-grid battery charger unit

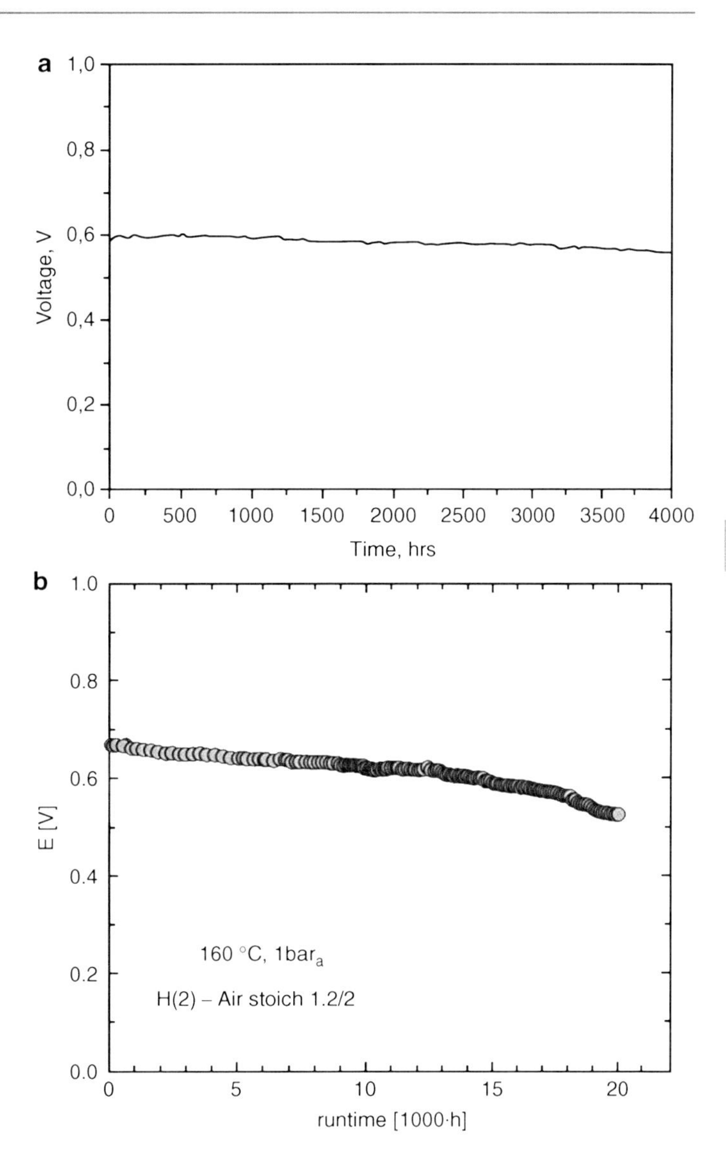

High-Temperature Polymer Electrolyte Fuel Cells, Fig. 3 Long-term durability of H2-Air HT-PEFCs (**a**) MEA using Advent TPS at 180 °C, 1 bar$_a$, H_2-Air stoich. 1.2/2 and (**b**) BASF Celtec P 1,000 (PBI) MEA at 160 °C, 1 bar$_a$, H_2-Air stoich. 1.2/2

Serenus H3-350W from Serenergy AS (Fig. 4b) with triple volumetric and double gravimetric power density, respectively, as compared to state-of-the-art direct methanol fuel-cell systems in the same power class, to be used either stand-alone or in 24 V hybrid battery–fuel-cell systems.

(c) **Aerospace Industry**

A significant effort is taking place in incorporating fuel cells in airplanes. A pioneer in this area is a consortium between DLR and Lange Aviation which has developed a new fuel-cell testing bed on the basis of a high-tech motor glider under the name DLR-H2. The aircraft is powered by a 25 kW HT-PEFC system using PBI membranes providing an electrical efficiency of 52 % and a total system efficiency of 44 %, presenting a significant advantage over conventional propulsion technologies [23].

(d) **H_2 Pump**

A very interesting application for high-temperature membrane electrolytes is their use in hydrogen pumps to purify hydrogen from gas mixtures. The electrochemical hydrogen pump was first explored in the 1960s along with the first development of low-temperature polymer electrolytes, but the purity of the hydrogen-containing gas stream needed to meet the same specifications that apply in the low-temperature

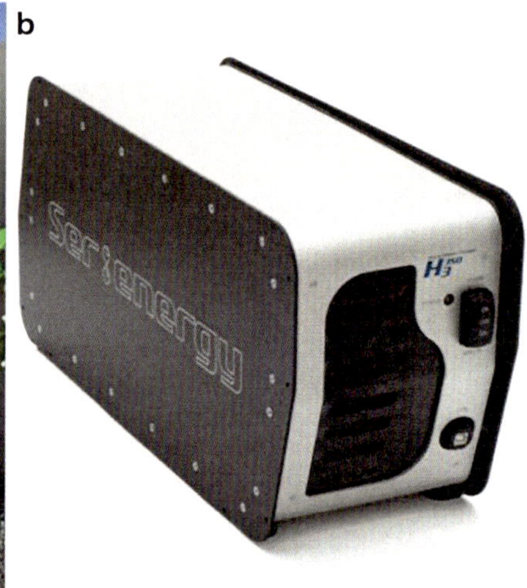

High-Temperature Polymer Electrolyte Fuel Cells, Fig. 4 (**a**) House installation of a ClearEdge Power CE5 HT-PEFC 5 kW commercial unit; natural gas reformer system (with kind permission from Clearedge Power, Hillsboro, OR, USA) and (**b**) Stand-alone Serenus H3 350 off-grid battery charger module (350 W, methanol reformer system) (With kind permission from Serenergy AS, Hobro, Denmark)

PEFCs, such as the absence of catalyst poisoning species like CO. Therefore, the high-temperature hydrogen pump can be proven a very efficient and low energy process to purify hydrogen in comparison to traditional methods. The quality of the purification varies, depending on the permeability of the gas stream components through the membrane. Perry et al. demonstrated that a high-temperature hydrogen pump based on PBI can be operated on relatively low power requirements with various gas feeds, showing a long-term durability of nearly 4,000 h [24].

Future Directions

Although fuel cells have been known for quite a while, the HT-PEFC branch has been only developed recently. The transfer of knowledge from PAFCs was a significant contributor to the rapid improvement of their performance and durability, but still further improvements need to take place in order to meet specifications and targets for their commercialization. The U.S. Department of Energy, for instance, has set minimum requirements for stationary applications in terms of durability, cost, degradation, and other various technical targets [25]. Therefore, future work should focus on solving several problems which inhibit the performance and durability of HT-PEFC.

Improving the corrosion stability would vastly favor the durability of HT-PEFC stacks since all of the components contribute to the degradation of performance over time. Corrosion of the catalytic layers lead to loss of active surface area, while in the case of the bipolar plates, great amounts of acid can be removed from the MEA [22]. Another significant reason for system failure is the polymer electrolyte membrane degradation. Although the material used (e.g., PBI, Advent TPS) are proven to be thermally and chemically stable, the harsh environment found at elevated temperatures can cause thinning and pinhole formation, both leading to adverse effects on the integrity of the material. Although the aforementioned improvements would have beneficial effects on durability, acid evaporation and acid management inside the fuel cell will remain an important factor for reaching lifetimes of more than 40,000 h. Finally, overcoming the kinetic limitation on the cathode catalysts imposed by the strong phosphoric acid–Pt interaction will remain the holy grail in HT-PEFC (and also PAFC) R&D. Although Yamamoto et al. showed possibilities to increase the catalytic activity of Pt for oxygen reduction, it remains elusive if this approach can be realized in a sustainable way inside an operating fuel cell [26]. Nevertheless, HT-PEFC systems proved to be a potentially suitable technology for applications ranging from several Watts to multiple Kilowatts in micro, mobile, and stationary applications.

Acknowledgement This work was carried out within the JTI-FCH Project DEMMEA of the European Community. We thank ClearEdge Power and Serenergy AS for providing the pictures in Fig. 4.

Cross-References

- ▶ Fuel Cells, Non-Precious Metal Catalysts for Oxygen Reduction Reaction
- ▶ Fuel Cells, Principles and Thermodynamics
- ▶ Polymer Electrolyte Fuel Cells (PEFCs), Introduction

References

1. Sandstede G, Cairns EJ, Bagotsky VS, Wiesener K (2003) History of low temperature fuel cells. In: Vielstich W, Lamm A, Gasteiger HA (eds) Handbook of fuel cells. Wiley, Chichester
2. Doyle M, Rajendran G (2003) Perfluorinated membranes. In: Vielstich W, Lamm A, Gasteiger HA (eds) Handbook of fuel cells. Wiley, Chichester, pp 351–395
3. Savinell R, Yeager E, Tryk D, Landau U, Wainright J, Weng D, Lux K, Litt M, Rogers C (1994) A polymer electrolyte for operation at temperatures up to 200-degrees-C. J Electrochem Soc 141(4):L46–L48
4. Wang JT, Wainright J, Yu H, Litt M, Savinell RF (1995) A H_2/O_2 fuel cell using acid doped polybenzimidazole as polymer electrolyte. In: Proceedings for the first international symposium on proton conducting membrane fuel cells 1. Electrochemical society, Chicago, IL
5. Li QF, Hjuler HA, Bjerrum NJ (2000) Oxygen reduction on carbon supported platinum catalysts in high temperature polymer electrolytes. Electrochim Acta 45(25–26):4219–4226
6. Kallitsis JK, Geormezi M, Neophytides SG (2009) Polymer electrolyte membranes for high-temperature fuel cells based on aromatic polyethers bearing pyridine units. Polym Int 58(11):1226–1233
7. Mader J, Xiao L, Schmidt TJ, Benicewicz BC (2008) Polybenzimidazole/acid complexes as high-temperature membranes, in fuel cells II. Adv Polym Sci 216:63–124
8. Geormezi M, Chochos CL, Gourdoupi N, Neophytides SG, Kallitsis JK (2011) High performance polymer electrolytes based on main and side chain pyridine aromatic polyethers for high and medium temperature proton exchange membrane fuel cells. J Power Sources 196(22):9382–9390
9. Daletou MK, Kallitsis JK, Voyiatzis G, Neophytides SG (2009) The interaction of water vapors with H(3) PO(4) imbibed electrolyte based on PBI/polysulfone copolymer blends. J Memb Sci 326(1):76–83
10. Kwon K, Yoo DY, Park JO (2008) Experimental factors that influence carbon monoxide tolerance of high-temperature proton-exchange membrane fuel cells. J Power Sources 185(1):202–206
11. Modestov AD, Tarasevich MR, Filimonov VY, Davydova ES (2010) CO tolerance and CO oxidation at Pt and Pt-Ru anode catalysts in fuel cell with polybenzimidazole-H(3)PO(4) membrane. Electrochim Acta 55(20):6073–6080
12. Schmidt TJ, Baurmeister J (2008) Development status of high temperature PBI based membrane electrode assemblies. ECS Trans 16:263–270
13. Landsman DA, Luczak FJ (2003) Catalyst studies and coating technologies. In: Vielstich W, Lamm A, Gasteiger HA (eds) Handbook of fuel cells. Wiley, Chichester, pp 811–831
14. Hartnig C, Schmidt TJ (2011) Simulated start-stop as a rapid aging tool for polymer electrolyte fuel cell electrodes. J Power Sources 196(13):5564–5572
15. Schmidt TJ, Baurmeister J (2008) Properties of high-temperature PEFC Celtec®-P 1000 MEAs in start/stop operation mode. J Power Sources 176(2):428–434
16. Tada T (2003) High dispersion catalysts including novel carbon supports. In: Vielstich W, Lamm A, Gasteiger HA (eds) Handbook of fuel cells. Wiley, Chichester, pp 481–488
17. Schmidt TJ, Baurmeister J (2006) Durability and reliability in high-temperature reformed hydrogen PEFCs. ECS Trans 3(1):861–869
18. Garsany Y, Gould BD, Baturina OA, Swider-Lyons KE (2009) Comparison of the sulfur poisoning of PBI and nafion PEMFC cathodes. Electrochem Solid State Lett 12(9):B138–B140
19. Yu S, Xiao L, Benicewicz BC (2008) Durability studies of PBI-based high temperature PEMFCs. Fuel Cells 8(3–4):165–174
20. Molleo M, Schmidt TJ, Benicewicz BC (2011) Polybenzimidazole fuel cell technology: theory, performance and applications. In: Meyers RA (ed) Encyclopedia of sustainability, science and technology. Springer Science + Business Media LLC, New York, DOI 10.1007/SpringerReference_226339 (2012)
21. Advent. Available from: http://www.adventech.gr/products.html
22. Hartnig C, Schmidt TJ (2011) On a new degradation mode for high-temperature polymer electrolyte fuel cells: how bipolar plate degradation affects cell performance. Electrochim Acta 56(11):4237–4242
23. Renouard-Vallet G, Saballus M, Schmithals G, Schirmer J, Kallo J, Friedrich KA (2010) Improving the environmental impact of civil aircraft by fuel cell technology: concepts and technological progress. Energ Environ Sci 3(10):1458–1468
24. Perry KA, Eisman GA, Benicewicz BC (2008) Electrochemical hydrogen pumping using

a high-temperature polybenzimidazole (PBI) membrane. J Power Sources 177(2):478–484
25. Department of Energy, U.S. Available from: http://www1.eere.energy.gov/hydrogenandfuelcells/fuelcells/systems.html
26. Yamamoto O, Kikuchi J, Ono T, Sounai A (2011) Membrane electrode assembly and fuel cells with increased performance. WO2011054503

High-Temperature Reactions, Anode

John Thomas Sirr Irvine[1] and Abhijit Ghosh[2]
[1]School of Chemistry, University of St Andrews, St Andrews, UK
[2]Advanced Ceramics Section Glass and Advanced Materials Division, Bhabha Atomic Research Centre, Mumbai, India

Introduction

A Solid Oxide Fuel Cell (SOFC) is capable of directly converting chemical energy to electrical energy at elevated temperatures. Like any other fuel cells, a SOFC consists of three major components: two electrodes – the anode and the cathode, separated by an electrolyte. Reduction of oxidant takes place in the cathode. Fuel is fed to the anode where it is oxidized releasing electrons, which flows to the cathode through the external circuit to complete the reduction process. A micrograph for the electrode-electrolyte interfaces with a schematic of the electrochemical processes is shown in Fig. 1. The movement of electrons through the outer circuit generates an electromotive force, emf, thus resulting in generation of direct-current electricity. Like other fuel cells, air is the most common oxidant for SOFC, since it is readily and economically available from the atmosphere. Unlike other fuel cells, a SOFC is not restricted to use hydrogen as the fuel. In fact, one of the advantages of SOFC is its ability to utilize "CO" as well as different hydrocarbon gases as fuels. Development of a suitable anode is a major challenge to exploit the readily available fuels to make SOFC as an efficient, economic, and environmental friendly device [1].

Ni-YSZ and Hydrogen-Based Fuel

Ni-YSZ (yttria stabilized zirconia) is the state-of-the-art anode for SOFC devices owing to its high electronic conductivity, physical and chemical compatibility with most of the common electrolyte materials, and good performance in H_2 and reformed fuel environments. Generally, a mixture of NiO and YSZ is used for the fabrication of the anode. In situ reduction of NiO to Ni (Eq. 1) during SOFC operation forms the Ni-YSZ cermet anode:

$$NiO + H_2 = Ni + H_2O \quad (1)$$

The anodic microstructure has a pronounced effect on the performance of the cell, because the reactions at the anode are influenced by the developed microstructure. Generally, moist hydrogen (with or without CO) is fed into the anode chamber. The oxidization of fuel takes place around fuel/anode/electrolyte interfaces (commonly known as triple phase boundary or TPB). The probable reaction at the anode involving charge transfer is shown below (Eqs. 2a and 2b):

$$H_2 + O^{2-} \rightarrow H_2O + 2e^- \quad (2a)$$

$$CO + O^{2-} \rightarrow CO_2 + 2e^- \quad (2b)$$

The whole process can be divided into several steps. For example, in case of H_2 as a fuel, the steps may be (a) transfer of fuel constituents to the anode chamber in gas phase, (b) adsorption of H_2 on the Ni surface, (c) transfer of reaction species towards the TPB, (d) charge transfer at the reaction site (i.e., TPB), and formation of H_2O. Among these various steps, the rate-determining step is very much influenced by the developed Ni-YSZ microstructure. A schematic of the whole process is shown in Fig. 2.

Presence of YSZ in the anode helps in extending the TPB away from the electrolyte-electrode interface. Most probably, YSZ itself directly contributes in the electrochemical reaction at anode [2]. The possible reactions involving charge transfer and the possible hydroxyl ion formation on the YSZ surface are shown below [3, 4]:

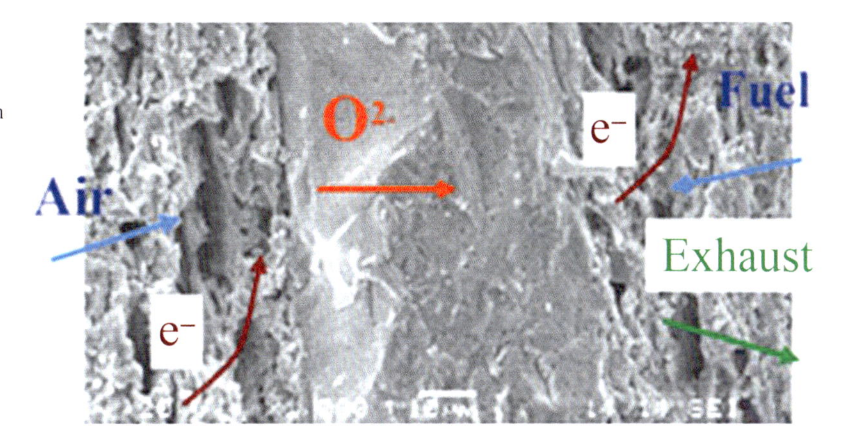

High-Temperature Reactions, Anode, Fig. 1 Schematic of processes occurring within the air electrode, electrolyte, and fuel electrode and at their interfaces (Ref. [1])

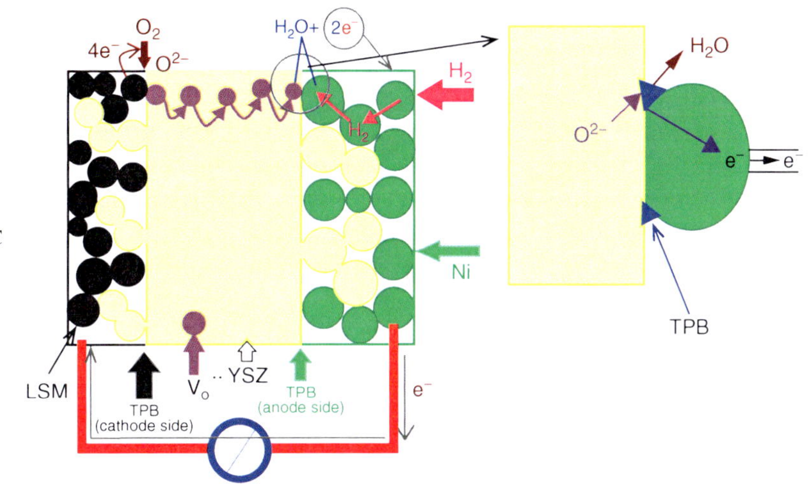

High-Temperature Reactions, Anode, Fig. 2 Schematic of SOFC operation presenting the four-step anodic. H_2O formation and electron transfer were shown in the insert. *LSM* strontium-doped lanthanum manganite (cathode for SOFC), $V_O^{\bullet\bullet}$ vacancy of oxygen

$$H_2 = 2H_{ad(Ni)} \tag{2c}$$

$$H_{ad(Ni)} = \left(H_{ad(Ni)}\right)^+ + e^- \tag{2d}$$

$$\left(H_{ad(Ni)}\right)^+ + O^{2-} = (OH)^- \tag{2e}$$

$$2(OH)^- = H_2O + O^{2-} \tag{2f}$$

Reactions mentioned in Eqs. 2e and 2f are postulated to take place on YSZ surface, and the oxygen ion involved in the reaction is available from the electrolyte structure. The abovementioned reactions were further reformulated to incorporate proton in the bulk of "Ni" and "YSZ" as well as on the surface [5] and those are shown below:

$$H_2(g) = 2H_{Ni}^{\bullet} + 2e^- \tag{2g}$$

$$H_{Ni}^{\bullet} = H_{i,YSZ}^{\bullet} \tag{2h}$$

$$2H_{i,YSZ}^{\bullet} + O_O^{x} = H_2O_{ad,YSZ} + V_O^{\bullet\bullet} \tag{2i}$$

$$H_2O_{ad,YSZ} = H_2O(g) \tag{2j}$$

Ni-YSZ and Hydrocarbon Fuel

Grain growth and sintering of Ni during operation and susceptibility of Ni towards sulfur poisoning and carbon deposition constrain the application of Ni-based cermet anode in certain applications. Generally, to use hydrocarbons as fuel in Ni cermet-based anode, a considerable amount of water vapor is to be fed into the anode along with hydrocarbon to facilitate partial steam reformation of the fuel and to avoid coking. Steam reformation of hydrocarbons consists of several reactions. For example, there are three reversible reactions involved in steam reformation of CH_4, as shown below (Eqs. 3a–3c):

$$CH_4 + H_2O = CO + 3H_2 \quad (3a)$$

$$CO + H_2O = CO_2 + H_2 \quad (3b)$$

$$CH_4 + 2H_2O = CO_2 + 4H_2 \quad (3c)$$

The reactions shown in Eqs. 3a and 3c are endothermic in nature, whereas the water-gas shift (WGS) reaction shown in Eq. 3b is moderately exothermic. If these reactions are carried out externally (external reformation), the efficiency and operation of the cell are significantly affected. Hence, internal reforming is preferred; however, Ni acts as an excellent coking catalyst. As a consequence, in the presence of carbonaceous fuels (and in the absence of sufficient water vapor), there is always a possibility of deposition of carbon filament on the surface of "Ni." The mechanism involves carbon formation on the metal surface followed by dissolution of the carbon into the bulk of the metal and finally precipitation of graphitic carbon at some surface of the metal particles after it becomes supersaturated with carbon [6]. It not only reduces the active sites for reactions mentioned in Eqs. 2c–j and 3a–c but also destroys the whole anode over a period of time. The following three reactions are the most probable catalytic reactions that lead to carbon formation in high-temperature systems:

$$2CO = CO_2 + C \quad (4a)$$

$$CH_4 = 2H_2 + C \quad (4b)$$

$$CO + H_2 = H_2O + C \quad (4c)$$

In addition, sulfur present in the fuel, often in form of H_2S, especially at SOFC temperature, gets chemisorbed on the Ni surface at elevated temperature and forms two-dimensional sulfide (Eq. 5). This causes blockage of Ni surface available for carrying out hydrogen adsorption and subsequent oxidation. Hence, overall cell performance degrades drastically:

$$Ni + H_2S = Ni\text{-}S + H_2 \quad (5)$$

Alternative Anode Materials

Various alternate materials have emerged, viz., doped or undoped ceria, Ni-$BaZr_{0.1}Ce_{0.7}Y_{0.2-x}Yb_xO_{3-\delta}$(BZCYYb), and Ni-free oxide anode materials including $La_{0.75}Sr_{0.25}Cr_{0.5}Mn_{0.5}O_{3-\delta}$(LSCM) and doped $(La,Sr)(Ti)O_3$ (LST) to counteract the problem related to redox instability, microstructural degradation, coking, and sulfur poisoning [7–10].

In case of Ni-BZCYYb, H_2O adsorbed on the surface of the BZCYYb facilitates oxidation of "S" chemisorbed on "Ni" surface around TPB. The most probable reaction is shown in Eq. 6:

$$Ni\text{-}S + 2H_2O = Ni + SO_2 + 2H_2 \quad (6)$$

Since, "SO_2" is readily desorbed from the "Ni" surface, the active area would be available for the catalytic activity.

Alternative Anode Materials-Oxide Anode and Hydrocarbon Fuels: Future Directions

Ni-free anodes, viz., perovskite oxides, are gaining importance for the fabrication of all ceramic SOFC. The candidate materials should have desirable electrochemical performance, electronic conductivity; oxygen diffusivity (ionic conductivity); oxygen surface exchange; chemical stability and compatibility; compatible thermal expansion; and mechanical strength and dimensional stability under redox cycling. As of now, the materials of interest are ceria (doped and undoped) and transition metal perovskite and fluorite-related structures [1]. Perovskite oxides can accommodate a large content of oxygen

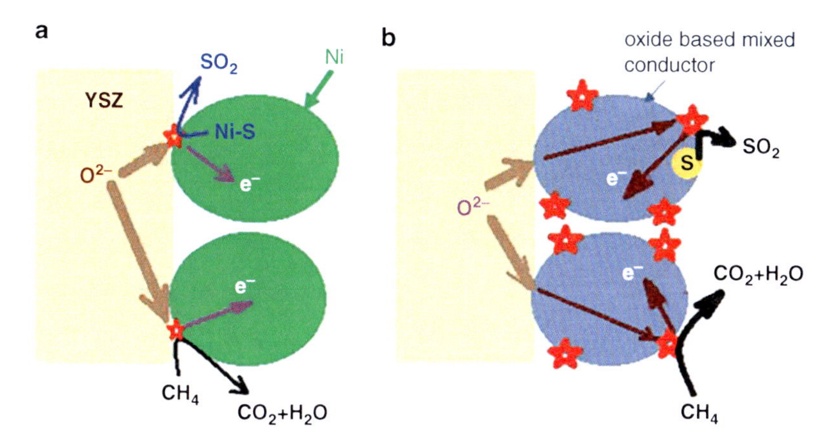

High-Temperature Reactions, Anode, Fig. 3 Enhancement of electrode reactive surface area for fuel oxidation in an SOFC anode by mixed conducting materials as compared to Ni-YSZ anode considering both carbon fuel oxidation and sulfur impurity reactions. (**a**) indicates the confinement of reaction area at the junction of YSZ and Ni particles, whereas (**b**) depicts the whole of mixed conducting oxide anode surface acting as reaction area

vacancies; hence, some perovskites are good oxygen ion conductors. The small "B" site in the perovskite allows transition elements to be introduced in the lattice. These elements exhibit multivalence under different conditions, which may be a source of electronic conductivity. Adequate ionic and electronic conductivities (mixed conductivity) are thus found in several perovskite oxides. Materials like LSCM are less likely to allow carbon and sulfur buildup, due to the higher availability of oxygen throughout the anode (Fig. 3) and less likely to suffer from coking and sulfur poisoning.

The charge compensation mechanism in LSCM is shown below:

$$LaBO_3 \xrightarrow{SrO} La^{3+}_{1-x}Sr^{2+}_{x}B^{3+}_{1-x}B^{4+}_{x}O_3 \tag{7}$$

where B represents both Cr and Mn. The electronic conductivity in this system is due to hopping of electron from one valance state to another. However, during operation, when LSCM is subjected to reducing atmosphere, Cr mostly remains in +3 state, whereas Mn attains ~ +2.7 valency state. The charge neutrality in this case is attained by forming oxygen vacancies:

$$LaBO_3 \xrightarrow{SrO} La^{3+}_{1-x}Sr^{2+}_{x}Cr^{3+}_{0.5}Mn^{3+}_{0.5-x}Mn^{2+}_{x}O_{3-\delta} \tag{8}$$

Thus, a combination of electronic and ionic conductivity is obtained in this system (both in moderate amount). The loss of oxygen during reduction or gain of oxygen during oxidation of reduced sample is reported in the literature [11, 12].

Similar to Ni-YSZ anode, the reactions desirable in the oxide-based anodes are shown in Eqs. 2a–b and 3a–c. However, undesirable reactions mentioned in Eq. 4a–c also occur in this system. Hydrocarbon-based fuel also contains "H_2S," which undergoes different reactions in the anodic compartment as shown below:

$$H_2S + O^{2-} = H_2O + 0.5S_2 + 2\ e^- \tag{9a}$$

$$H_2S + 3O^{2-} = H_2O + SO_2 + 6\ e^- \tag{9b}$$

$$H_2S = 0.5S_2 + H_2 \tag{9c}$$

$$2H_2S + SO_2 = 2H_2O + 3/nS_n\ (\text{where } n = 2-8) \tag{9d}$$

Complete oxidation of hydrogen sulfide to sulfur dioxide is most desirable. However, Eq. 9a–d indicate the inevitability of the formation of "S," which mostly covers the anode surface. Equation 4 also indicates the formation of C on the anode surface. Mixed ionic conduction in the oxide anode helps in the oxidation of C and S to their respective oxides as shown in Fig. 3b. This can be represented by the equation below:

$$C + 2O^{2-} = 2CO_2 + 4e^- \quad (10a)$$

$$1/2S_2 + 2O^{2-} = SO_2 + 4e^- \quad (10b)$$

Cross-References

- ► High-Temperature Reactions, Cathode
- ► Solid Oxide Fuel Cells, Direct Hydrocarbon Type
- ► Solid Oxide Fuel Cells, Introduction

References

1. Atkinson A, Barnett S, Gorte RJ, Irvine JTS, Mcevoy AJ, Mogensen M, Singhal SC, Vohs J (2004) Advanced anodes for high temperature fuel cells. Nat Mater 3:17–27
2. Dees DW, Balachandran U, Dorris SE, Heiberger JJ, McPheeters CC, Picciolo JJ (1989) Interfacial effects in monolithic solid oxide fuel cells. In: Singhal SC (ed) Proceedings of the first international symposium on solid oxide fuel cells, Hollywood, Electrochemical society, Pennington, October 16–18, 1989, p 317
3. Mogensen M, Lindegaard T (1993) The kinetics of hydrogen oxidation on a Ni-YSZ SOFC electrode at 1000°C. In: Singhal SC, Iwahara H (ed) Solid oxide fuel cells, Electrochemical society (Proceedings vol. 93-4), Pennington, NJ, 1993, p 484
4. Holtappels P, Vinke IC, de Haart LGJ, Stimming U (1999) Reaction of hydrogen/water mixtures on nickel-zirconia cermet electrodes. 2. AC polarization characteristics. J Electrochem Soc 146:2976–2982
5. de Boer B, (1998) SOFC Anode: Hydrogen oxidation at porous nickel and Nickel/yttria stabilised zirconia cermet electrodes. PhD thesis, University of Twente, The Netherlands
6. Toebes ML, Bitter JH, van Dillen AJ, de Jong KP (2002) Impact of the structure and reactivity of nickel particles on the catalytic growth of carbon nanofibers. Catal Today 76:33–42
7. Marina AO, Mogensen M (1999) High-temperature conversion of methane on a composite gadolinia-doped ceria-gold electrode. Appl Catal A 189:117–126
8. Yang L, Wang S, Blinn K, Liu M, Liu Z, Cheng Z, Liu M (2009) Enhanced sulfur and coking tolerance of a mixed ion conductor for SOFCs: $BaZr_{0.1}Ce_{0.7}Y_{0.2-x}Yb_xO_{3-\delta}$. Science 326:126–129
9. Tao S, Irvine JTS (2003) A redox-stable efficient anode for solid-oxide fuel cells. Nat Mater 2:320–323
10. Ruiz-Morales JC, Canales-Vázquez J, Savaniu C, Marrero-López D, Zhou W, Irvine JTS (2006) Disruption of extended defects in solid oxide fuel cell anodes for methane oxidation. Nature 439:568–571
11. Plint SM, Connor PA, Tao S, Irvine JTS (2006) Electron transport in novel SOFC anode material $La_{1-x}Sr_xCr_{0.5}Mn_{0.5}O_{3\pm\delta}$. Solid State Ionics 177:2005–2008
12. Tao SW, Irvine JTS (2006) Phase transition in perovskite oxide $La_{0.75}Sr_{0.25}Cr_{0.5}Mn_{0.5}O_{3-\delta}$ observed by in situ high temperature neutron powder diffraction. Chem Mater 18:5453–5460

High-Temperature Reactions, Cathode

Stephen Skinner
Department of Materials, Imperial College London, London, UK

Introduction

Solid oxide fuel cells (SOFCs), typically based on an oxide ion conducting electrolyte as previously discussed in earlier chapters, have a fundamental requirement for the high-temperature reduction of the oxidizing species. In general this reaction is viewed as the simple oxygen reduction reaction, as summarized in Eq. 1:

$$O_{2(g)} + 4e^- \rightarrow 2O^{2-} \quad (1)$$

Despite the apparent simplicity of this reaction, the process by which the oxygen reduction occurs followed by incorporation of the ionic species into the electrolyte is the subject of some debate and is dependent on the mode of operation of the cathode material. Two typical cathode types are currently utilized in SOFCs – electronic conductors and mixed ionic-electronic conductors (MIECs). The cathode reactions, while nominally the same in both types of materials, occur at different locations, and hence, the active region varies, leading to differences in the operating regime and ultimately performance. In the case of a single phase electronic conductor,

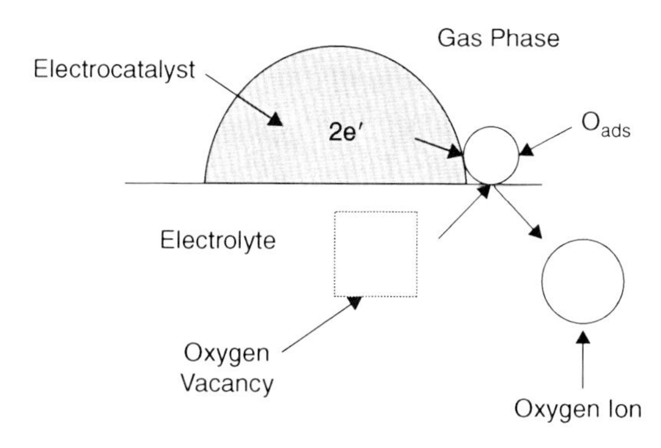

High-Temperature Reactions, Cathode, Fig. 1 Schematic of the triple phase boundary in a SOFC cathode (With permission from Ref. [1])

this is limited to the triple phase boundary (TPB) where the gaseous species (oxygen), electronic conductor, and ionic conductor meet, as illustrated schematically in Fig. 1.

An alternative strategy for high-temperature operation that extends the active zone of the cathode has been proposed by producing *composite* cathodes, where an electrolyte phase and an electronic phase are intimately mixed. This cathode structure involves optimization of several parameters such as the particle size, homogeneity of the composite, and the interpenetrating nature of the two components to ensure both ionic and electronic pathways exist through the entire cathode. On lowering the operating temperature an MIEC is potentially a more appropriate choice that has been demonstrated to enhance the catalytic performance of the cathode. With an MIEC the active zone extends over the cathode surface, utilizing both the ionic transport capabilities of the MIEC phase and the catalytically active surface. Currently all three of these strategies are employed in SOFC device development, with the MIEC used primarily to optimize cathode performance at lower temperatures.

The simple reaction represented in Eq. 1 is a summary of the complex incorporation reaction for ionic oxygen which may be more accurately represented by the following sequence of steps:

$1/2O_2 + V_O^{\bullet\bullet} + 2e^- \leftrightarrow O_O^x$ – overall reaction consisting of:

(i) $1/2O_{2(g)} \rightarrow 1/2O_{2(ad)}$

(ii) $1/2O_{2(ad)} \rightarrow O_{ad}$

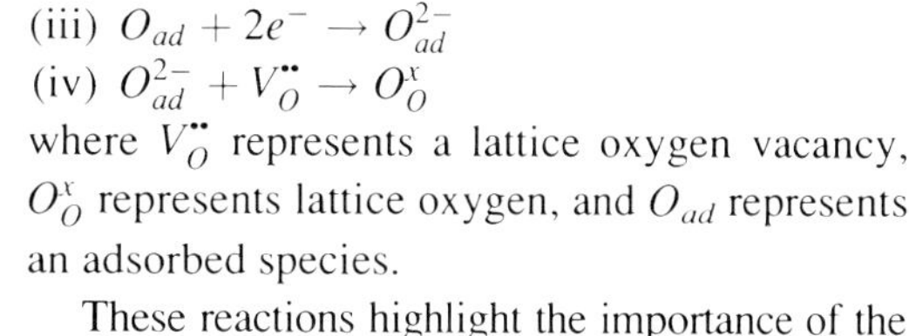

(iii) $O_{ad} + 2e^- \rightarrow O_{ad}^{2-}$

(iv) $O_{ad}^{2-} + V_O^{\bullet\bullet} \rightarrow O_O^x$

where $V_O^{\bullet\bullet}$ represents a lattice oxygen vacancy, O_O^x represents lattice oxygen, and O_{ad} represents an adsorbed species.

These reactions highlight the importance of the surface of the cathode to the overall reaction and indeed the composition of the cathode itself, in that the defect chemistry of the cathode material is such that the reduced oxygen has a vacant lattice site to occupy. This then leads to further discussion of the electrochemical processes occurring at the surface of the cathode, the near surface, and indeed through the bulk of the material to the interface with the electrolyte. Critical parameters governing the cathode reactions are therefore the kinetics of the reduction process, the surface exchange, and the diffusion of the ionic species through the cathode, in the case of a mixed conductor. However, these apparently simple processes do not account for a number of potential processes occurring elsewhere in the cathode structure. These are illustrated in Fig. 2, where it is clear that surface diffusion, grain boundary diffusion, surface conductivity of the electrolyte, and gas diffusion are all critical parameters in SOFC cathodes. Considering these factors and linking the cathode performance to standard electrochemical principles, Steele [2] outlined that the overall fuel cell stack performance is represented by the low-field Butler-Volmer equation, assuming the interfacial electrode kinetics are suitably fast. Through manipulation of Ohm's law Steele highlights that the Wagner equation derived for metal oxidation kinetics is equally applicable to SOFCs. From this basis a key parameter guiding the design of fuel cell electrodes is developed – the area-specific resistance of the electrolytes and electrodes, which is calculated as 0.15 Ω cm^2. This allowed corresponding values of the electrolyte thickness, conductivity, and oxygen flux to be determined.

A full discussion of the complexity of all potential cathode reactions is beyond the scope of this article, and the reader is therefore referred to the excellent discussion of oxygen reduction by Adler [3] and to the later work of Merkle and Maier in which oxygen incorporation into model oxide systems is discussed [4].

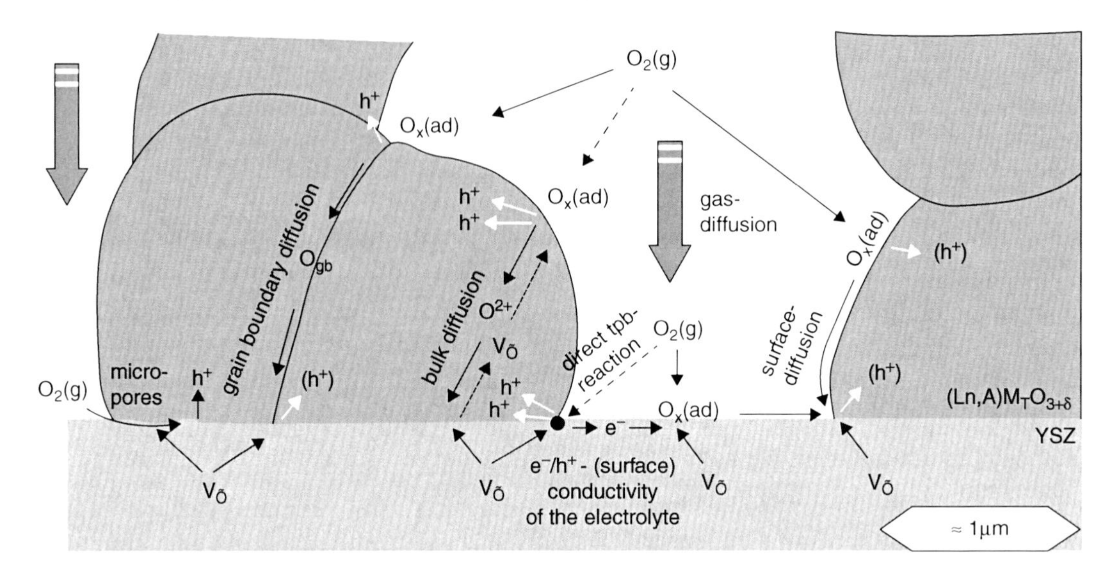

High-Temperature Reactions, Cathode, Fig. 2 Illustration of the complex series of processes governing the oxygen reduction and incorporation processes occurring in a SOFC cathode (With permission from Ref. [1])

Evidently the performance of the cathode is determined not only by the fundamental kinetic parameters discussed above but a number of further factors that are of equal significance, such as the microstructure of the cathode. In the case of a composite cathode the percolation of the two phases can have a significant effect on the overall cell performance, while in all cases the extent of open and interconnected porosity of the electrode is essential, allowing gas diffusion to the reaction sites. Each of these factors leads to polarization processes at the cathode, which limit the operation of the fuel cell. At low current densities the combination of both anode and cathode reactions leads to activation polarization, originating from reaction rate losses while at high current densities concentration polarization results from gas transport losses. These processes are a result of the electrochemical parameters and microstructure and are termed the overpotential of the electrode, leading to cathode activation and concentration polarization terms. Kim et al. [5] give a detailed account of the factors affecting the electrochemical performance of an anode supported SOFC, considering the impact of microstructural parameters such as tortuosity, combined with the electrochemical parameters, flux of gaseous species, Nernst potential, etc., and derive a model based on binary diffusion and gas transport which adequately describes the experimental performance of a composite cathode. They clearly demonstrate the importance of understanding both microstructure and electrochemistry in these high-temperature cells. A further example of the complex nature of electrode reactions is given by Adler et al. [6] where continuum modelling is used to analyze the oxygen reduction process in mixed conducting cathodes, rather than composite materials. Again, using binary diffusion theory and considering the porosity and tortuosity of the cathode, a relationship linking the cell resistance to the chemical contributions to the cell was derived:

$$R_{chem} = \left(RT/2F^2\right) \times \sqrt{\tau/(1-\varepsilon)c_v D_v a r_o(\alpha_f + \alpha_b)} \quad (2)$$

where R = gas constant, T = temperature, F = Faraday's constant, τ = tortuosity, ε = porosity, and all other terms are as defined in [6]. For a typical mixed conducting cathode, $La_{1-x}Sr_xCo_{1-y}Fe_yO_{3-d}$, this model was found to accurately reproduce the experimentally derived impedance spectra obtained from a model cell using data for the diffusion coefficient obtained from isotopic labelling experiments.

Perovskite-Type Cathode Materials

While the basic principles of cathode function are outlined above, it is instructive to now focus on specific examples of SOFC cathodes: single phase materials, composite cathodes, and MIECs. At high temperatures, in excess of 800 °C, a regime in which cells based on yttria-stabilized zirconia (YSZ) electrolytes function well, the cathodes are typically electronic conductors, often composites, which rely on the TPB mode of operation. Of these the most prevalent cathode oxide is $La_{1-x}Sr_xMnO_3$ (LSM). While the LSM, typically with a Sr content of $x = 0.2$, is an attractive electronic conductor [7], it is clear that oxygen diffusion in the single phase is limited [8], and therefore, mixed conductivity is poor. Thus, any single phase cathode based on LSM will be limited by the active triple phase boundary length. In order to increase the TPB length the production of composites is required, with several authors adopting different strategies to optimize this parameter. Further optimization of the LSM performance is gained by modifying the A-site composition, with Jiang et al. noting enhanced performance for the $(La_{0.8}Sr_{0.2})_{0.9}MnO_3$ material [9]. Understanding of the effect of the composition on performance leads to the consideration of the effect of the electrolyte content on the overall cathode performance. Kim et al. [10] systematically varied the LSM:YSZ ratio and considered the percolation of the two phases and the optimal YSZ content in terms of the cathode resistance, finding that irrespective of particle size 40 wt% YSZ produced the lowest polarization resistance of $0.05\ \Omega\ cm^2$ at 950 °C.

Further work by Steele et al. [7] suggested that the likely cathode performance of cathodes could be predicted using the oxygen diffusion and surface exchange data and the Adler Lane Steele (ALS) model [6] to calculate the R_{chem} values. This approach leads to calculated values for cathode resistance for the LSM single phase cathode of $>1,000\ \Omega\ cm^2$ and clearly indicates the requirement for composite materials or alternative cathode compositions. A thorough review of the state of development of LSM and LSM/YSZ cathodes has been produced by Jiang [11] in which a comprehensive data set of electrochemical parameters is presented and to which the reader is directed for further detail.

Efforts to optimize the electrochemical performance of the cathode at reduced temperatures (500–800 °C) have led to the exploration of alternative perovskite compositions, most notably the mixed conducting $La_{1-x}Sr_xCo_{1-y}Fe_yO_{3-d}$ materials (LSCF), both as single phases and as composites. Notably Dusastre and Kilner [12] studied the percolation effects of LSCF cathodes with the $Ce_{1-x}Gd_xO_{2-d}$ (CGO) electrolytes, finding that the experimental data accorded with effective medium percolation theory, with optimal CGO content of 36 vol%, with Murray et al. supporting this work by demonstrating a factor of 10 reduction in polarization resistance when CGO content was in the range 30–50 wt% [13]. Evidently a combined approach of investigating both microstructure and electrochemistry is required to fully optimize a cathode material.

Finally, in considering the nature of the cathode materials, whether single phase or composite, one has to carefully monitor the reactivity between the cathodes and electrolytes, and clearly the macroscopic mechanical and chemical behavior of these oxide ceramics has a dramatic impact on the device operation. It is well known that LSM and YSZ, under certain conditions will form insulating phases, and that $Ce_{1-x}Gd_xO_{2-d}$ will reduce at lower oxygen partial pressures. Also, $La_{1-x}Sr_xCoO_3$ has fast ionic conduction and good electronic conductivity but poor thermal expansion behavior and stability. Choosing a suitable cathode material is therefore a compromise between the electrochemical optimization and the structural chemistry. These choices remain a challenge for high-temperature devices.

Future Directions

SOFC cathodes are typically perovskite-type oxides formed as a composite with a suitable electrolyte (either YSZ for high-temperature operation or CGO for lower operating temperatures) and deposited as thin layers by conventional techniques such as screen printing. There are several possible avenues for future cathode developments. Firstly there is the drive towards new materials, with layered perovskites and double perovskites showing attractive kinetic parameters with good cell performance data. Demonstrating that these materials have acceptable degradation rates and durability is a key challenge. Of course to reduce the degradation of the electrochemical cell, one obvious strategy is to reduce the operating temperature which presents challenges and opportunities, either in the development of novel materials or in the development of novel design concepts. As discussed by Wachsman and Lee [14] understanding the kinetic parameters and polarization data of mixed and ionic conductors offers the opportunity to rationally design composition and microstructure of electrodes for reduced operating temperatures. Perhaps fully understanding these processes at any temperature is a larger challenge, and producing predictive models of electrochemical performance encompassing mass transport, charge transfer, and interface reactivity/processes for all structure types is the ultimate challenge.

Cross-References

- ► Defect Chemistry in Solid State Ionic Materials
- ► Defects in Solids
- ► Fuel Cells, Principles and Thermodynamics
- ► Grain-Boundary Conductivity
- ► High-Temperature Reactions, Anode
- ► Kröger-Vinks Notation of Point Defects
- ► Mixed Conductors, Determination of Electronic and Ionic Conductivity (Transport Numbers)
- ► Oxide Ion Conductor
- ► Oxygen Anion Transport in Solid Oxides
- ► Oxygen Nonstoichiometry of Oxide
- ► Solid Electrolytes
- ► Solid Electrolytes Cells, Electrochemical Cells with Solid Electrolytes in Equilibrium
- ► Solid Oxide Fuel Cells, History
- ► Solid Oxide Fuel Cells, Introduction
- ► Solid Oxide Fuel Cells, Thermodynamics
- ► Solid State Electrochemistry, Electrochemistry Using Solid Electrolytes
- ► X-Ray Absorption and Scattering Methods
- ► X-Ray Diffraction Methods

References

1. Singhal SC, Kendall K (eds) (2003) High temperature solid oxide fuel cells: fundamentals, design and applications. Elsevier, Oxford
2. Steele BCH (1994) Oxygen transport and exchange in oxide ceramics. J Power Sources 49:1–14
3. Adler SB (2004) Factors governing oxygen reduction in solid oxide fuel cell cathodes. Chem Rev 104:4791–4844. doi:10.1021/cr020724o
4. Merkle R, Maier J (2008) How is oxygen incorporated into oxides? A comprehensive kinetic study of a simple solid state reaction with $SrTiO_3$ as a model material. Angew Chem Int Ed 47:3874–3894. doi:10.1002/anie.200700987
5. Kim J-W, Virkar AV, Fung K-Z, Mehta K, Singhal SC (1999) Polarization effects in intermediate temperature anode supported solid oxide fuel cells. J Electrochem Soc 146:69–78
6. Adler SB, Lane JA, Steele BCH (1996) Electrode kinetics of porous mixed conducting oxygen electrodes. J Electrochem Soc 143:3554–3564
7. Steele BCH, Hori KM, Uchino S (2000) Kinetic parameters influencing the performance of IT-SOFC composite electrodes. Solid State Ionics 135:445–450
8. De Souza RA, Kilner JA (1998) Oxygen transport in $La_{1-x}Sr_xMn_{1-y}Co_yO_{3+/-d}$ perovskites Part I Oxygen tracer diffusion. Solid State Ionics 106:175–187
9. Jiang SP, Love JG, Zhang JP, Hoang M, Ramprakash Y, Hughes AE, Badwal SPS (1999) The electrochemical performance of LSM/zirconia-yttria interface as a function of a-site stoichiometry and cathodic current treatment. Solid State Ionics 121:1–10
10. Kim J-D, Kim G-D, Moon J-W, Lee H-W, Lee K-T, Kim C-E (2000) The effect of percolation on electrochemical performance. Solid State Ionics 133:67–77
11. Jiang SP (2008) Development of lanthanum strontium manganite perovskite cathode materials of solid oxide fuel cells: a review. J Mater Sci 43:6799–6833
12. Dusastre V, Kilner JA (1999) Optimisation of composite cathodes for intermediate temperature SOFC applications. Solid State Ionics 126:163–174
13. Murray EP, Sever MJ, Barnett SA (2002) Electrochemical performance of (La, Sr)(Co, Fe)O3-(Ce, Gd)O2 composite cathodes. Solid State Ionics 148:27–34
14. Wachsman ED, Lee KT (2011) Lowering the temperature of solid oxide fuel cells. Science 334:936–939. doi:10.1126/Science.1204090

Hybrid Li-Ion Based Supercapacitor Systems in Organic Media

Katsuhiko Naoi
Institute of Symbiotic Science and Technology, Tokyo University of Agriculture and Technology, Koganei, Tokyo, Japan

Introduction

Environmental protection and new energy development have been recently becoming a growing industry, in which energy storage devices are such important electronic components which play crucial roles. Energy storage devices include batteries, e.g., lithium-ion batteries, nickel-hydrogen batteries, and lead batteries, which have high energy densities. The batteries are now under vigorous study for further increasing their energy densities so that they can serve as power sources of electric vehicles. On the other hand, electric double-layer capacitors (EDLCs), which are another type of high-power energy storage devices capable of effectively utilizing energy, have been also studied and are practically used in trucks, buses, elevators, and in heavy industrial equipments such as forklifts and yard cranes[1].

However, since the EDLCs generally have low energy densities, their uses are limited and they cannot fully meet various performance demands required by the recent markets as shown in Fig. 1. Particularly in the field of automobiles, new energy devices are strongly desired to have hybrid characteristics between lithium-ion batteries and EDLCs and thereby which can be suitably employed in idle reduction systems. Accordingly, it is expected for them to form a large market [2]. In order to satisfy the performance demands, it is often said to be necessary for the EDLCs to enhance their energy density to 20–30 Wh L^{-1}, which is approximately twice or more than present EDLCs, namely, 5–10 Wh L^{-1}. For realizing the above high energy density, hybrid capacitor systems comprising nonaqueous redox materials are being dynamically researched and developed in recent years [3–10]. The present entry deals with the recent contributions to get this high energy density by focusing on two major hybridized cell configurations in organic media.

Limitation of Withstanding Voltage for Conventional EDLCs

As is described above, increasing energy density is one of the most crucial matters. For conventional EDLC systems, designed with two symmetrical activated carbon electrodes, increasing voltage is more effective because the energy density increases in proportion to the squared voltage. Thus, it is essential to develop higher electrochemical durability at the electrode/electrolyte interface. Currently, the maximum voltage of an AC/AC capacitor is limited to 2.5–2.7 V (Fig. 2).

At applied voltage over 2.7 V, there occurs a significant decrease in capacitance and an increase in internal resistance. In fact the float test revealed that the AC/AC capacitor loses its capacitance ($\Delta C < -13$ % up to 30 days) at 2.5 V, whereas at 2.9 V the capacitance loss became more significant ($\Delta C = -28$ %) for the same duration. The threshold cell voltages between 2.5 and 2.9 V certainly trigger a consecutive and fatal degradation. The undesired faradaic process (failure modes) that leads to a capacitance fade is the most critical factor that determines the life of conventional EDLCs and needs to be further investigated as in the recent review article by Simon et al. [2]. Anyway this withstanding voltage limitation (2.5–2.7 V) certainly exists because of gas evolutions at higher voltages and is a barrier for further enhancement of the energy density of the conventional activated carbon-based EDLCs.

Hybrid Capacitor Systems

There is presently a major effort to increase the energy density of EDLCs up to a target value in the vicinity of 20–30 Wh kg^{-1} [1].

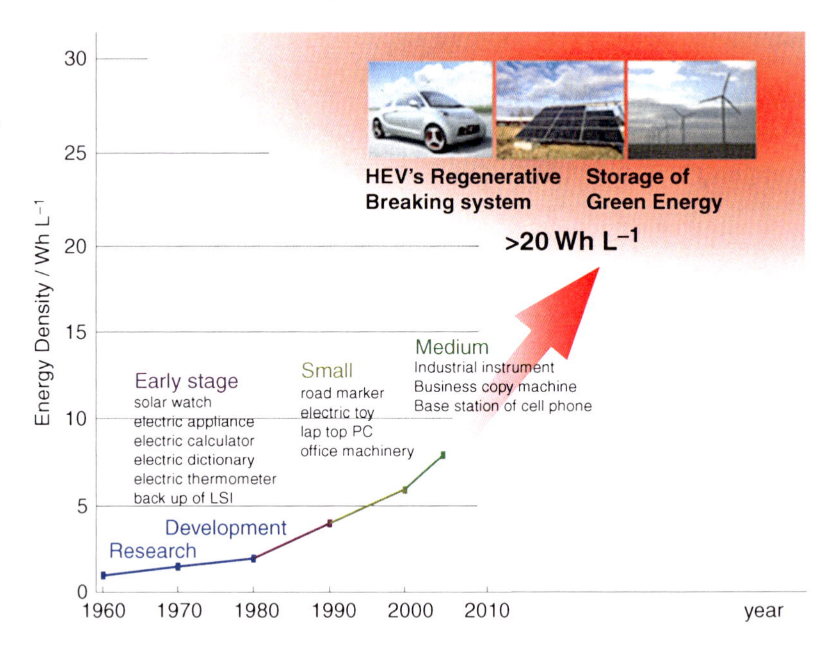

Hybrid Li-Ion Based Supercapacitor Systems in Organic Media, Fig. 1 Target value of the next-generation electrochemical supercapacitor

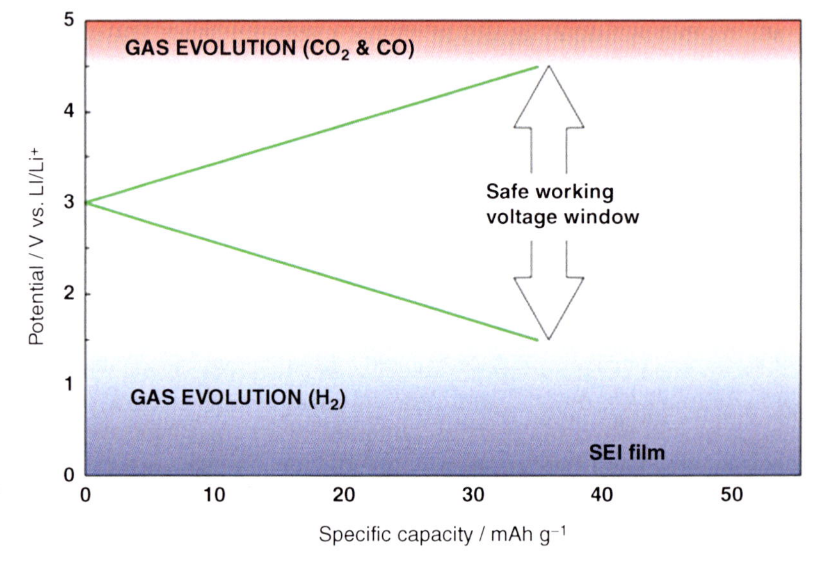

Hybrid Li-Ion Based Supercapacitor Systems in Organic Media, Fig. 2 Limitation of withstanding voltage or the safe working voltage window of EDLCs

Many studies have been undertaken as shown in Fig. 3. There are mainly three approaches; the first is to change the electrode (by higher-capacitance carbons or redox), and the second is to change electrolyte (by durable new electrolyte or ionic liquids).

The third and the most important approach to meet the goal that is under serious investigation is to develop asymmetric (hybrid) capacitors. Various hybrid capacitor systems are possible by coupling redox-active materials (e.g., graphite [3, 4], metal oxides [9–11], conducting polymers [12, 13]) and an activated carbon (AC). This approach can overcome the energy density limitation of the conventional EDLCs because it employs a hybrid system of a battery-like (faradic) electrode and a capacitor-like (non-faradic) electrode, producing higher working

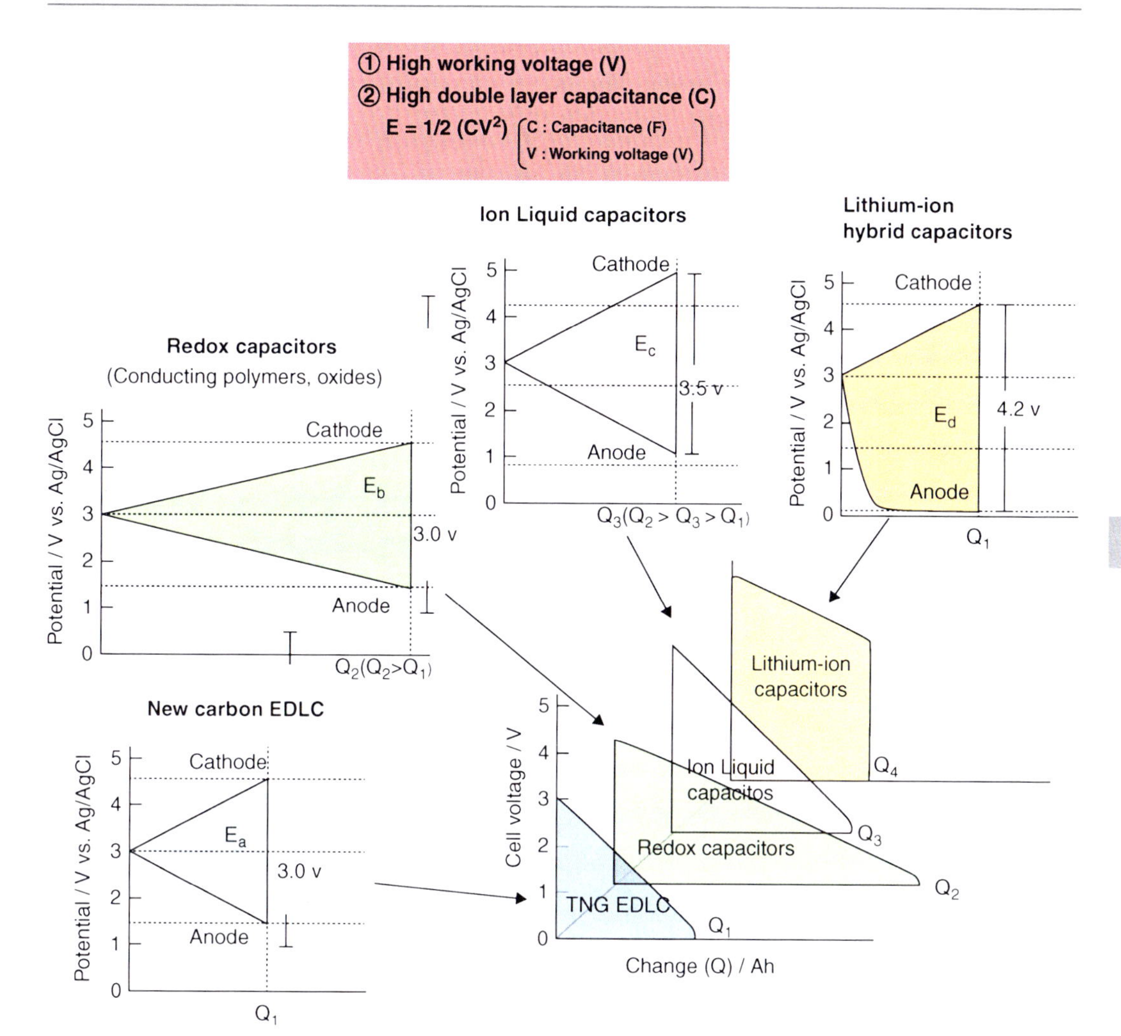

Hybrid Li-Ion Based Supercapacitor Systems in Organic Media, Fig. 3 Various attempts to get higher capacitance and voltage capacitors

voltage and capacitance. With these systems, one can certainly achieve twice or triple enhancements in energy density compared to the conventional EDLCs.

Figure 4 represents the voltage profiles of the conventional EDLC system, Nanogate, and LIC whereby one could have a good scenario attaining higher voltage and higher energy density.

Lithium-Ion Capacitor (LIC)

Among high-energy hybrid capacitors comprising nonaqueous redox materials, a hybrid system called "lithium-ion capacitor (abbreviated as LIC)" has particularly attracted the attention in these days [3–5]. The LIC is a hybrid capacitor in which the positive and negative electrodes are made of activated carbon and of graphite pre-doped with lithium ions, respectively (Fig. 5).

Since the negative electrode of graphite undergoes the reaction at a potential a little over 0 V versus Li/Li^+, the LIC has a high working voltage of 3.8 –4.0 V. This high working voltage enables the lithium-ion capacitor to realize both a high power density of approximately 5 kW kg^{-1} and an energy density of approx. 20–30 Wh kg^{-1}. The LIC thus exhibits favorable performance and

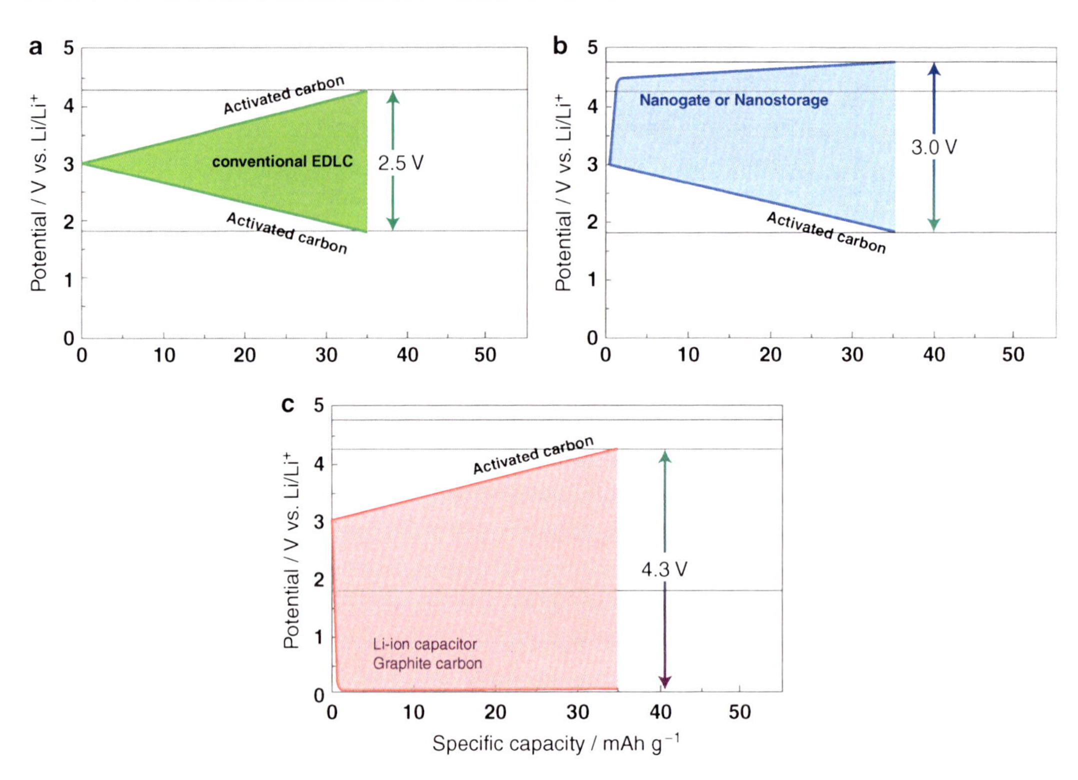

Hybrid Li-Ion Based Supercapacitor Systems in Organic Media, Fig. 4 Voltage profiles of a conventional (**a**) EDLC, (**b**) Nanogate, and (**c**) LIC systems

is regarded as a promising candidate of the next-generation electrochemical capacitor. Hence, some Japanese companies have already started commercializing LIC [5] (Fig. 6).

The limited charging rate and the low temperature performance are the possible drawbacks of the LIC because it has a lithium intercalation negative electrode. The process of pre-lithiation of the graphite electrode may lead to poor cost-effectiveness or instability in the quality of the LIC device when they are mass-produced (see Fig. 7). Generally speaking, instead of achieving energy density improvement, the high working voltage causes an electrolyte decomposition problem especially at negative graphite electrode [14]. This "pre-doing of Li^+" may bring about the same risks relevant to a long-term stability to keep low ESR. This is an issue that is specifically important for an electrochemical capacitor as a "power device" because it leads to create a high impedance electrode/electrolyte interface and thus eventually leads to a deteriorated power performance for longer cycles.

Nanohybrid Capacitor (NHC)

Very recently, Naoi's group developed a hybrid capacitor system that certainly achieves a high energy density, high stability, and high safety. This is called "Nanohybrid capacitor" (abbreviated as NHC) using a super-high-rate nanostructured lithium titanate ($Li_4Ti_5O_{12}$)/carbon composite negative electrode. The authors have kept their eyes on lithium titanate ($Li_4Ti_5O_{12}$) as a stable and safe redox material capable of increasing the energy densities of hybrid capacitors without sacrificing interfacial characteristics. The $Li_4Ti_5O_{12}$ operates at a potential (1.55 V vs. Li/Li^+) out of the range where the electrolyte solution may be decomposed and hence play key roles for providing capacitor systems as stable and safe as

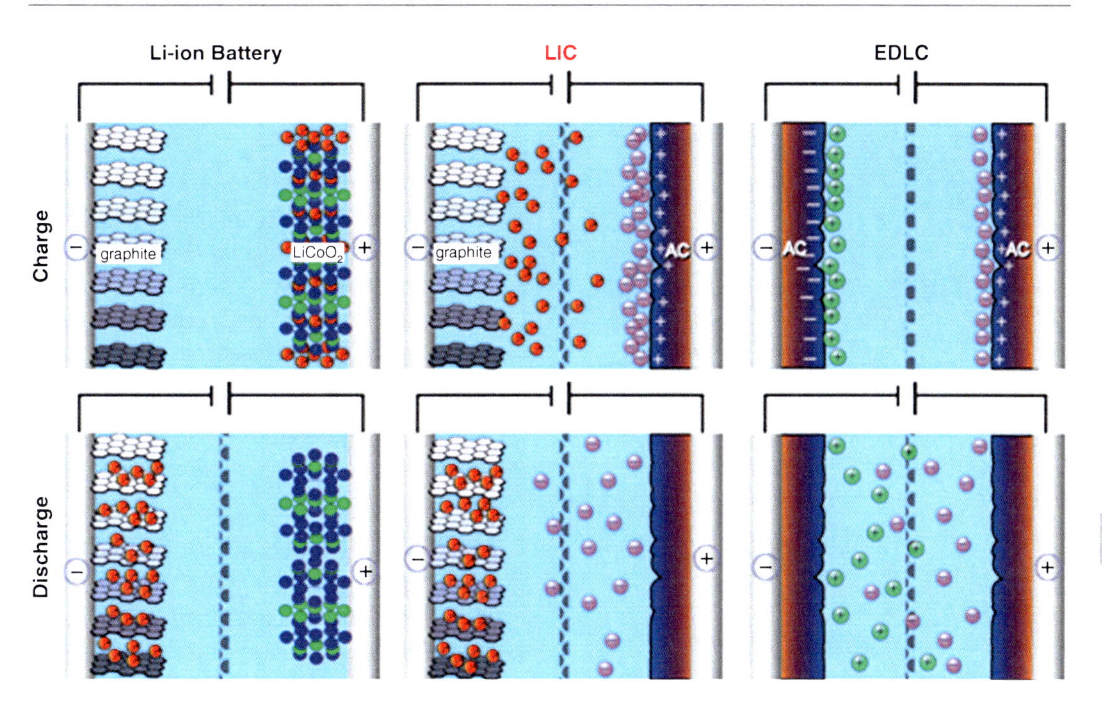

Hybrid Li-Ion Based Supercapacitor Systems in Organic Media, Fig. 5 Cell configurations of Li-ion battery (*LIB*), EDLC, and LIC

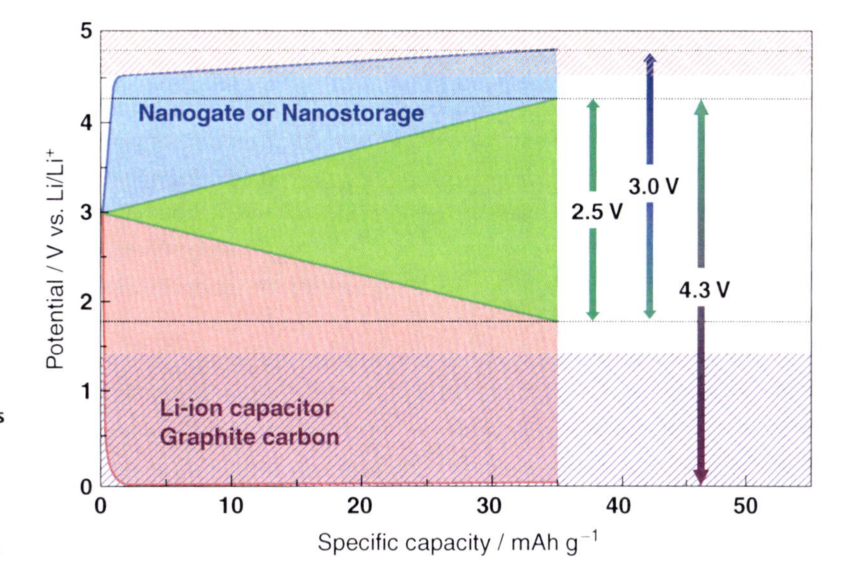

Hybrid Li-Ion Based Supercapacitor Systems in Organic Media, Fig. 6 Reliability and electrolyte decomposition at positive and negative electrodes of LIC systems

EDLCs (see Fig. 8). Amatucci et al. firstly introduced the $Li_4Ti_5O_{12}$/AC system as safer battery systems [15, 16]. However, the conventional $Li_4Ti_5O_{12}$ has the greatest problem of low power characteristics that stem from inherent poor Li^+ diffusion coefficient ($<10^{-6}$ cm^2 s^{-1}) [17] and poor electronic conductivity ($<10^{-13}$ W^{-1} cm^{-1}) [18]. Such slow output characteristics were not fully developed for the application in electrochemical capacitors at that time. For the purpose of solving the problem of poor output performance, some measures can be

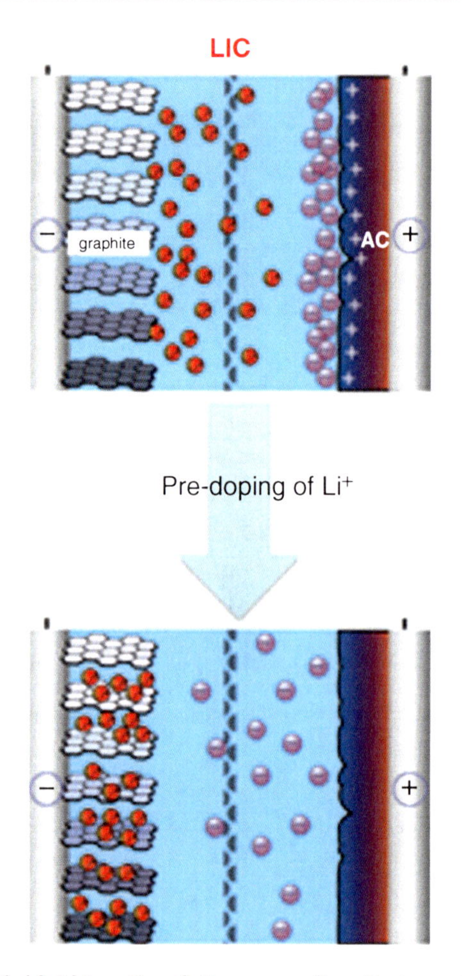

Hybrid Li-Ion Based Supercapacitor Systems in Organic Media, Fig. 7 Necessary pre-doing of Li^+ for LIC systems

thought of. For example, the $Li_4Ti_5O_{12}$ may be ground down to nanosized particles and may be combined with an electroconductive material to prepare a composite.

The $Li_4Ti_5O_{12}$ as a redox material, as described in Fig. 9, for hybrid capacitors has the following essential advantages in energy density, in stability, and in safety:

1. It exhibits a high coulombic efficiency close to 100 % during charge–discharge cycle [11–13].
2. It has a theoretical capacity (175 mAh g^{-1}) four times higher than as activated carbon [11–13].
3. It undergoes charging and discharging at a constant potential of 1.55 V versus Li/Li^+, where the electrolyte solution is free from the fear of decomposition (little SEI formation and little gas evolution) [12, 14, 19].
4. It changes the volume to a very small degree (0.2 %) in charging or discharging in intercalating or deintercalating Li^+ ions (zero-strain insertion) [11, 12].
5. Inexpensive raw material.

As summarized in Fig. 10, a capacitor cell comprising $Li_4Ti_5O_{12}$ is, therefore, expected to be free from necessity for pre-doping of lithium ions (because of the above advantage (1)), to be capable of having a high energy density (because of the advantage (2)), and to be excellent in stability and in safety (because of the advantages (3) and (4)). There is larger selection of

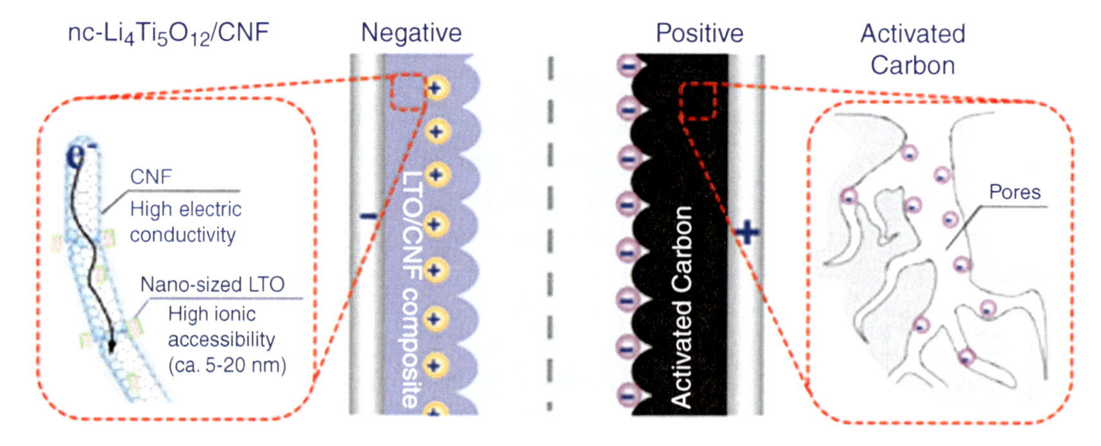

Hybrid Li-Ion Based Supercapacitor Systems in Organic Media, Fig. 8 The configuration of Nanohybrid capacitor (*NHC*) and its features

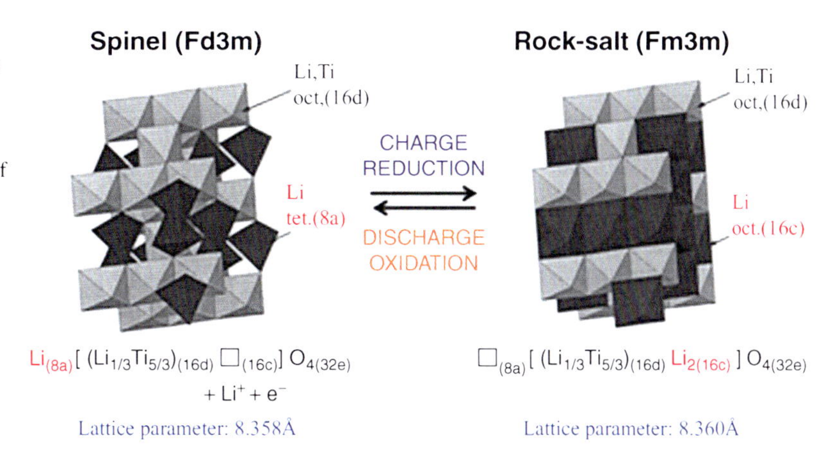

Hybrid Li-Ion Based Supercapacitor Systems in Organic Media, Fig. 9 Structural and volumetric change during charging and discharging of $Li_4Ti_5O_{12}$

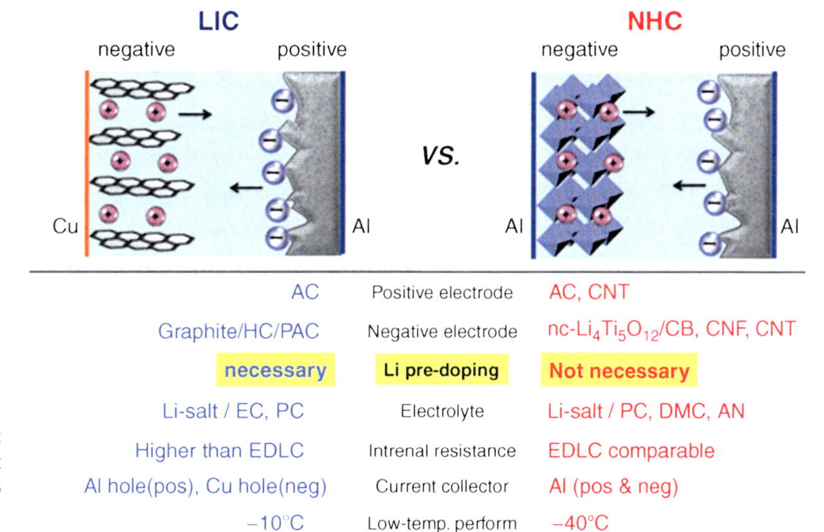

LIC		NHC
AC	Positive electrode	AC, CNT
Graphite/HC/PAC	Negative electrode	nc-$Li_4Ti_5O_{12}$/CB, CNF, CNT
necessary	Li pre-doping	Not necessary
Li-salt / EC, PC	Electrolyte	Li-salt / PC, DMC, AN
Higher than EDLC	Intrenal resistance	EDLC comparable
Al hole(pos), Cu hole(neg)	Current collector	Al (pos & neg)
−10°C	Low-temp. perform	−40°C

Hybrid Li-Ion Based Supercapacitor Systems in Organic Media, Fig. 10 Comparison of new-generation capacitors: LIC and NHC with respect to various items relating to cell configurations and performances

electrolyte for NHC because of the modest voltage window compared with LIC system. AN, ionic liquid, and linear carbonate (such as DMC and DEC) mixture systems can be utilized for NHC. The selection of electrolyte is very important for getting even better performances. In fact, an AN-based NHC exhibited extremely (9 times) higher performances compared with the conventional PC-based EDLCs. The low temperature performance is excellent down to −40 °C for NHC due to the nature of $Li_4Ti_5O_{12}$. Therefore the internal resistance can be minimized for NHC which is comparable to EDLCs in wide range of temperature.

Material Design for NHC

In the nanosized $Li_4Ti_5O_{12}$ particles, it is expected that Li^+ ions diffuse and electrons migrate in distances reduced by a factor of 1/1,000 or less as compared with the distances in normal micron-order $Li_4Ti_5O_{12}$ particles. Further, in the composite, electron paths are effectively formed in the electrode of nanosized $Li_4Ti_5O_{12}$ particles. On the basis of these expectations, the authors synthesized a composite (nc-$Li_4Ti_5O_{12}$/CNF) by means of a new method referred to as "ultracentrifugal force (UC) method" [20]. Specifically, the composite

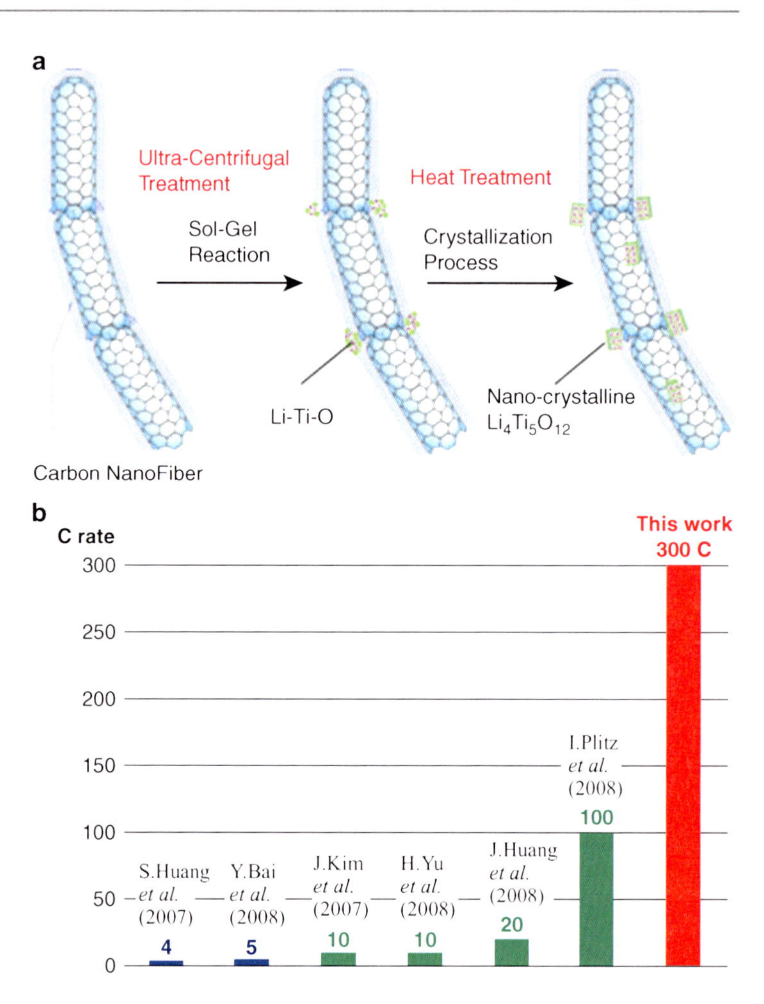

Hybrid Li-Ion Based Supercapacitor Systems in Organic Media, Fig. 11 Nanostructure model and high power performance of super-high-rate nanocrystalline $Li_4Ti_5O_{12}$ nested and grafted onto carbon nanofiber. (**a**) Schematic illustration for the two-step formation procedure of the nc-$Li_4Ti_5O_{12}$/CNF composite. (**b**) Maximum C-rate values of the nc-$Li_4Ti_5O_{12}$/CNF composite and various $Li_4Ti_5O_{12}$ materials in literatures reported so far

(nc-$Li_4Ti_5O_{12}$/CNF) comprises highly dispersed nanocrystalline $Li_4Ti_5O_{12}$ particles hyper-dispersed on carbon nanofibers (CNFs) [21] having high electro-conductivity (25 Ω^{-1} cm^{-1}) (see Fig. 11). The authors utilized the composite as the negative electrode active material and thereby succeeded in producing a novel hybrid capacitor (nanohybrid capacitor) realizing both high power and high energy density.

The nc-$Li_4Ti_5O_{12}$ negative electrode was developed to have a unique nanostructure that can operate at unusually high current densities. Nanocrystalline s attached onto carbon nanofibers were prepared by a unique technique (UC method) of a mechanochemical sol–gel reaction under ultracentrifugal force field [22], followed by an instantaneous heat treatment under vacuum for very short duration (see Fig. 11a). The UC method is induced to form, anchor, and graft the nano-Li-Ti-O precursors on the CNF matrices. The subsequent instantaneous heat treatment is of prime importance to precisely achieve all of the following: (1) high crystallization of $Li_4Ti_5O_{12}$, (2) inhibition of the CNF oxidation decomposition during the annealing at high temperature, and (3) suppression of the agglomeration of the $Li_4Ti_5O_{12}$ particles. These processes are quite simple and require only a few minutes. Actually, the power characteristic of the prepared composite (nc-$Li_4Ti_5O_{12}$/CNF) made a new bench mark (12 s) [9, 23–27] which exceeds greatly the shortest charge–discharge time that has ever been attained (see Fig. 11b).

XRD analysis was performed to confirm the formation of the nc-$Li_4Ti_5O_{12}$ and the presence of the CNF in the nc-$Li_4Ti_5O_{12}$/CNF composites.

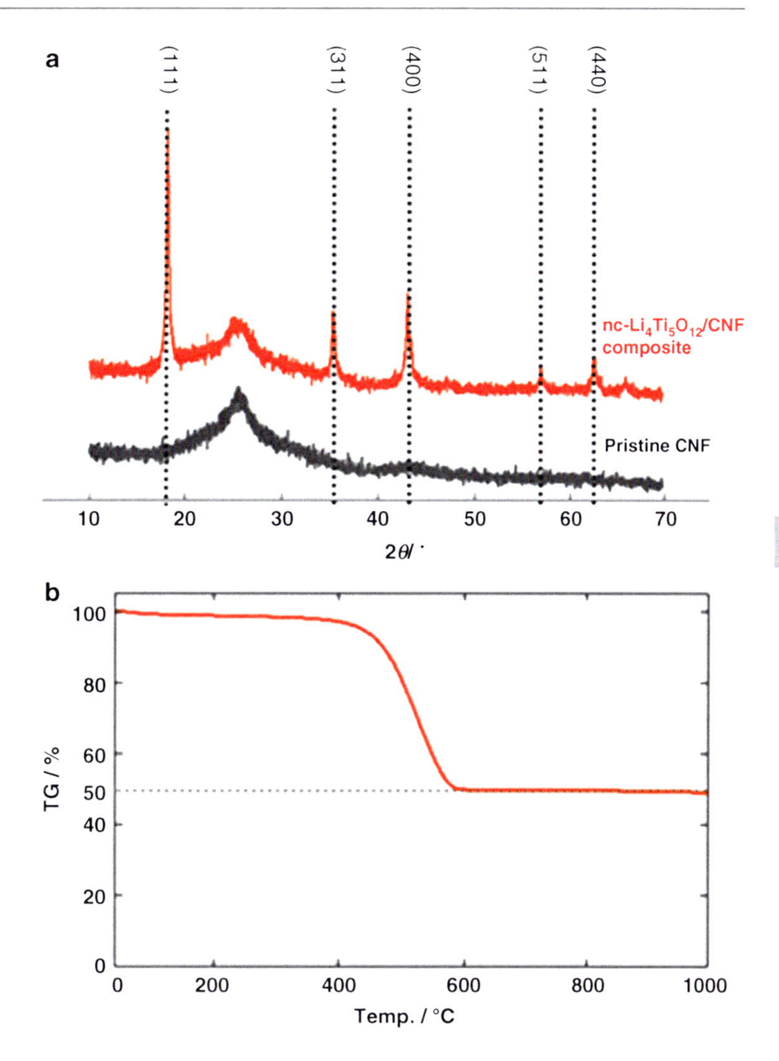

Hybrid Li-Ion Based Supercapacitor Systems in Organic Media, Fig. 12 Crystallinity and content of nc-$Li_4Ti_5O_{12}$ in the composite. (**a**) XRD patterns of the nc-$Li_4Ti_5O_{12}$/CNF composite and pristine CNF. (**b**) TG curve of the nc-$Li_4Ti_5O_{12}$/CNF composite at 1 °C min^{-1} under air. The residual weight ratio corresponds to the content of nc-$Li_4Ti_5O_{12}$ in the composite

Figure 12a shows the XRD patterns of the prepared nc-$Li_4Ti_5O_{12}$/CNF composite and pristine CNF. The composite has several sharp diffraction peaks at $2q = 18, 35, 42, 57$, and $63°$. These peaks correspond to (111), (311), (400), (511), and (440) planes of a face-centered cubic spinel structure with $Fd\bar{3}m$ space group [28, 29], respectively, indicative of the formation of crystalline $Li_4Ti_5O_{12}$. Annealing at 900 °C under vacuum well crystallizes the nanoparticles of $Li_4Ti_5O_{12}$. A broad peak at around $2q = 24.5°$ is observed which corresponds to the (002) plane of the pristine CNF [30]. This means that the CNF exists in the annealed composite and preserves its graphene layer structure. The fact that there are no other peaks observed corresponding to some possible impurities such as TiO_2, Li_2CO_3, and Li_2TiO_3 [31, 32] suggests that there are only two species, crystalline $Li_4Ti_5O_{12}$ and CNF.

Thermogravimetric (TG) measurement of the nc-$Li_4Ti_5O_{12}$/CNF composite was performed under air to estimate the residual weight ratio of the CNF. The obtained TG curve is shown in Fig. 12. The weight loss at 400–600 °C resulted by the oxidative decomposition of the CNF, and exactly 50 wt% of the nc-$Li_4Ti_5O_{12}$ remained. The 50wt% is well consistent with the weight ratio of the $Li_4Ti_5O_{12}$ to the CNF calculated on the basis of dosed Ti alkoxide weight before UC method. This fact implies that sol–gel reaction in the UC method stoichiometrically proceeds, and the optimized (very short duration) annealing

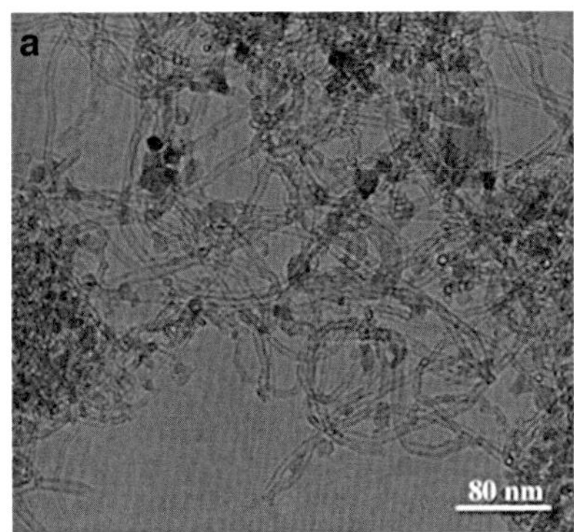

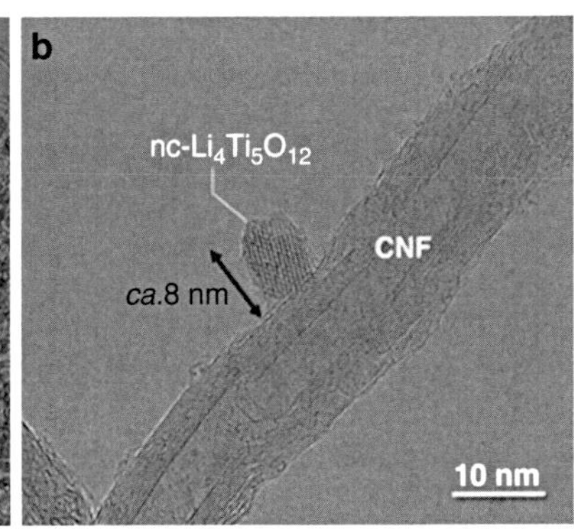

Hybrid Li-Ion Based Supercapacitor Systems in Organic Media, Fig. 13 Nanostructure of the nc-$Li_4Ti_5O_{12}$ and CNF. (**a**) A bird's eye view of a HR-TEM image of the nc-$Li_4Ti_5O_{12}$/CNF composite and (**b**) a worm's eye view focusing on the junction of the nc-$Li_4Ti_5O_{12}$ particle on CNF graphene surface

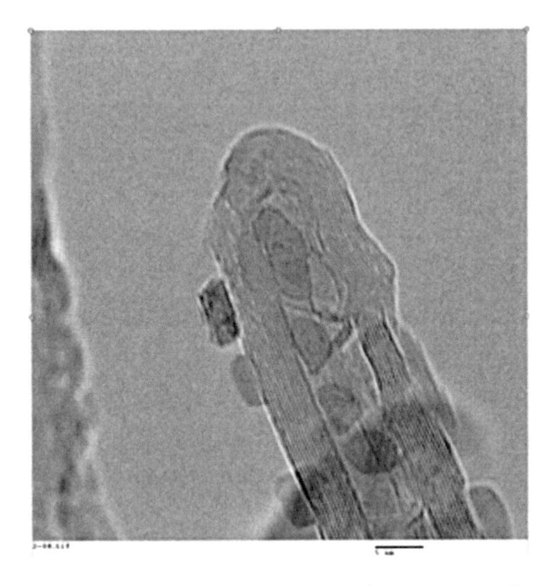

Hybrid Li-Ion Based Supercapacitor Systems in Organic Media, Fig. 14 Closer view of a HR-TEM image of multiple attachments of nc-$Li_4Ti_5O_{12}$ inside and outside wall of the CNF composite

does not cause oxidative decomposition of the CNF. Such a stoichiometric preparation process (UC method and instantaneous annealing) is one of the important factors for the cost-effectiveness of capacitor production.

The nanostructures and crystallinity of nc-$Li_4Ti_5O_{12}$ and CNF in the composite are observed by HR-TEM (see Fig. 13). The image indicates that the edge or defect sites of CNF graphenes accommodate and graft nc-$Li_4Ti_5O_{12}$ particles. The results support that the $Li_4Ti_5O_{12}$ particles are formed through a nucleation process onto the edge and defect CNF sites. The clear facet of $Li_4Ti_5O_{12}$ reflects the high crystallinity that is consistent with the sharp XRD spectrum (see Fig. 11a) despite such a nanosize. Such a high crystallinity resulted in a reversible, smooth Li^+ insertion performance and ca. 100 % of coulombic efficiency. Also, HR-TEM images of CNF show clear graphene layers indicative of crystalline structure. Thus, this composite is considered to be the junction material of two crystalline species, $Li_4Ti_5O_{12}$ and CNF. Of particular interest is that the lattice matching of the nc-$Li_4Ti_5O_{12}$ particles and CNFs is perfect and they are firmly tied at the atomic level. This could bring about an establishment of good electronic paths between the two species.

Another closer view of a HR-TEM (see Fig. 14) indicates the multiple attachments of the nc-$Li_4Ti_5O_{12}$ particles on the inside wall of the CNF graphene substrate as well as outside wall of the CNF matrices. This may indicate that the crystallization would be such efficient on CNF, and more importantly one can enhance both the energy density and the specific gravity of the nc-$Li_4Ti_5O_{12}$.

Figure 15a shows the charge–discharge curve of the half-cell Li/(nc-$Li_4Ti_5O_{12}$/CNF) at 1 °C. The horizontal axis represents specific capacity per unit weight of $Li_4Ti_5O_{12}$. The dominant plateau was observed at ca. 1.5 V versus Li/Li^+ that corresponds to the Li^+ intercalation-deintercalation process of the crystalline $Li_4Ti_5O_{12}$ [14, 29], indicating that the capacity

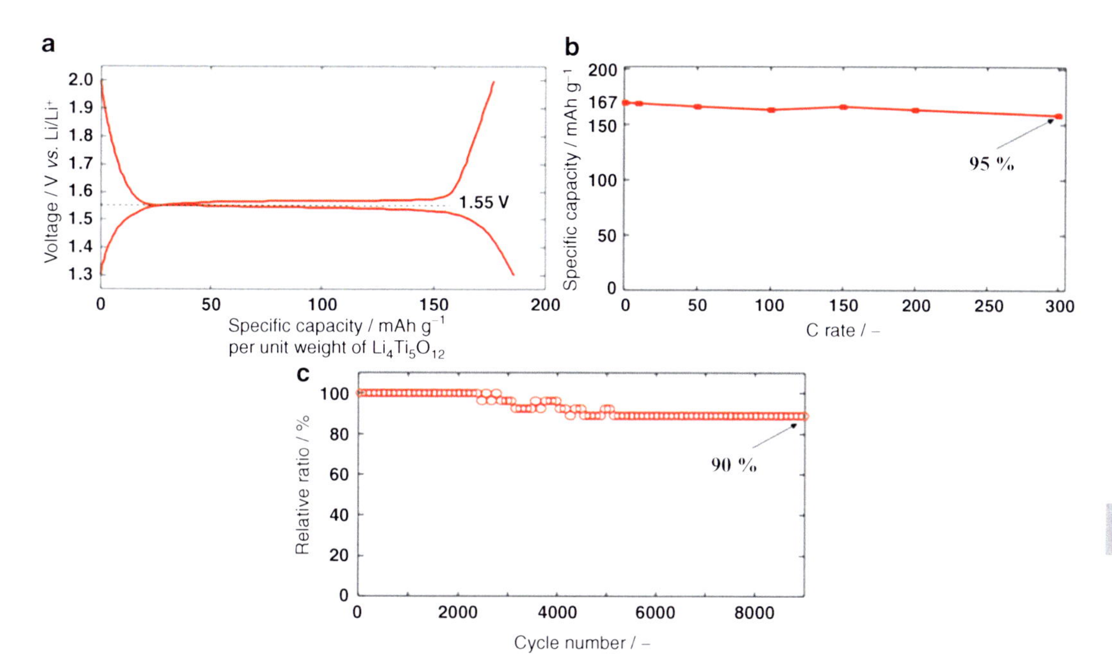

Hybrid Li-Ion Based Supercapacitor Systems in Organic Media, Fig. 15 Electrochemical properties of the nc-$Li_4Ti_5O_{12}$/CNF composite. (**a**) Charge–discharge curve of nc-$Li_4Ti_5O_{12}$/CNF composite between 1.0 and 3.0 V versus Li/Li^+ at 1 C. (**b**) Rate capability of the nc-$Li_4Ti_5O_{12}$/CNF composite in a C-rate range from 1 °C to 300 °C. (**c**) Cycleability of nc-$Li_4Ti_5O_{12}$/CNF composite between 1.0 and 3.0 V versus Li/Li^+ at 20 °C. 1 M $LiBF_4$/EC + DEC(1:1) was used as the electrolyte

of the composite is determined by the redox capacity of nc-$Li_4Ti_5O_{12}$ in the composite. The obtained capacity was 167 mAh g^{-1} per $Li_4Ti_5O_{12}$, which is 95 % of the theoretical capacity. It is noted here that the value of 167 mAh g^{-1} is obtained after subtracting the double-layer capacity of the CNF (8 mAh g^{-1}). This result indicates that almost all of the nc-$Li_4Ti_5O_{12}$ particles in the composite are electrochemically active, meaning that ionic and electric paths are fully established in the composite. The rate capability of the obtained composite is shown in Fig. 15b.

Even at a high rate of 300 °C, the composite shows reversible capacity of 158 mAh g^{-1} per $Li_4Ti_5O_{12}$ which corresponds to 95 % of the capacity obtained at 1 °C. Such an excellent rate capability indicates that the optimized nanostructure of the nc-$Li_4Ti_5O_{12}$/CNF composites as observed in HR-TEM image well overcomes the inherent problems of $Li_4Ti_5O_{12}$ materials like poor Li^+ diffusivity and poor electronic conductivity. Probably, this is because the nanocrystallized $Li_4Ti_5O_{12}$ and $Li_4Ti_5O_{12}$/CNF junctions lead to facile ionic diffusion and electronic conduction, respectively. The cycleability of the composite is shown in Fig. 15c. Even after 9,000 cycles, 90 % of the initial capacity is maintained, showing that the composite is electrochemically stable. The result strongly suggests that the aggregation and detachment of the nc-$Li_4Ti_5O_{12}$ particles hardly happen when they are operated at high-rate charge–discharge for a prolonged cycling.

Figure 16 shows Ragone plots obtained from charge–discharge measurements as laminate-type cell of the hybrid capacitor system ((nc-$Li_4Ti_5O_{12}$/CNF)/$LiBF_4$-PC/AC). The charge–discharge was performed between 1.5 and 3.0 V at various current densities ranging 0.2–30 mA cm^{-2} (0.18–26.8 A g^{-1}). For comparison, a conventional EDLC system (AC/$TEABF_4$-PC/AC) was

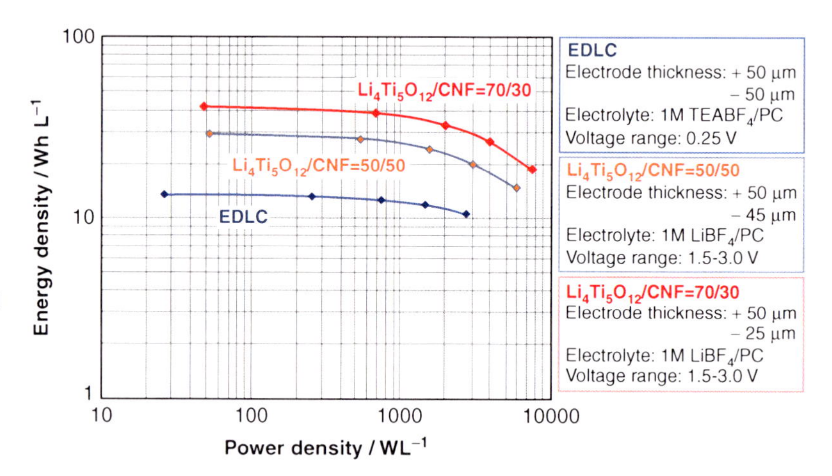

Hybrid Li-Ion Based Supercapacitor Systems in Organic Media, Fig. 16 Ragone plots of Nanohybrid capacitor systems ((nc-$Li_4Ti_5O_{12}$/CNF)/AC) with two different $Li_4Ti_5O_{12}$ loadings (50 % and 70 %) and conventional EDLC system (AC/AC) all in 1 M $LiBF_4$/PC

also assembled and measured between 0 and 2.5 V. In a low power density range of 0.1–1 kW L^{-1}, the hybrid capacitor shows the energy density as high as 28–30 Wh L^{-1}, which is a value comparable to that of the Li-ion capacitors [3]. Even at a high power of 6 kW L^{-1}, the energy density of the hybrid capacitor remains at 15 Wh L^{-1} which is double that of the conventional EDLC system (AC/AC). The result reveals that our capacitor system can provide higher energy as compared with conventional EDLCs not only in the low power density region (0.1–1 kW L^{-1}) but also in the high power density region (1–6 kW L^{-1}). Accordingly, this configuration of capacitor system is anticipated as an energy device utilizable for both high energy and high power applications.

Outlook

Hybrid capacitor systems are the promising approach to meet the goal to effectively increase the energy density. The investigation to develop hybrid capacitors has been initiated by "Li-ion capacitors (LIC)." And now "Nanohybrid capacitor (NHC)" certainly achieves as high energy density as Li-ion capacitors with higher stability, higher safety, and higher productivity. Both LIC and NHC attain three times higher energy density and regarded as the next-generation supercapacitor systems.

Cross-References

► Ionic Liquids
► Lithium-Ion Batteries
► Super Capacitors

References

1. Burke A (2007) Electrochim Acta 53:1083
2. Simon P, Gogotsi Y (2008) Nat Mater 7:845
3. Yoshino A, Tsubata T, Shimoyamada M, Satake H, Okano Y, Mori S, Yata S (2004) J Electrochem Soc 151:A2180
4. Hatozaki O (2008) Proceeding of advanced capacitor world summit 2008, San Diego
5. Azaïs P, Tetrais F, Caumont O, Depond JD, Lejosne J (2009) Abstract of ISEE'Cap 09, 19
6. Burke A (2000) J Power Sources 91:37
7. Pandolfo AG, Hollenkamp AF (2006) J Power Sources 157:11
8. Koetz R, Carlen M (2000) Electrochim Acta 45:2483
9. Plitz I, Dupasquier A, Badway F, Gural J, Pereira N, Gmitter A, Amatucci GG (2006) Appl Phys A 82:615
10. Baldsing WG, Puffy NW, Newnham RH, Pandolfo AG (2007) Proceeding of the advanced automotive battery and ultracapacitor conference, Long Beach
11. Kazaryan SA, Kharsov GG, Litvinrnko SV, Kogan VI (2007) J Electrochem Soc 154:A751
12. Laforgue A, Simon P, Fauvarque JF, Mastrangostino M, Soavi F, Sarrau JF, Lailler P, Conte M, Rossi E, Saguatti S (2003) J Electrochem Soc 150:A645
13. Machida K, Suematsu S, Ishimoto S, Tamamitsu K (2008) J Electrochem Soc 155:A970
14. (2008) Nikkei electronics 991:77
15. Amatucci GG, Badway F, DuPasquier A, Zheng T (1999) Abstract of 196th meeting of the Electrochemical Society, vol 122

16. Amatucci GG, Badway F, Pasquier AD, Zheng T (2001) J Electrochem Soc 148:A930
17. Takai S, Kamata M, Fujiine S, Yoneda K, Kanda K, Esaka T (1999) Solid State Ionics 123:165
18. Chen CH, Vaughey JT, Jansen AN, Dees DW, Kahaian AJ, Goacher T, Thackeray MM (2001) J Electrochem Soc 148:A102
19. Naoi K, Simon P (2008) Interface 17:34
20. Scharner S, Weppner W, Schmind-Beurmann P (1999) J Electrochem Soc 146:857
21. Jansen AN, Kahaian AJ, Kepler KD, Nelson PA, Amine K, Dees DW, Vissers DR (1999) J Power Sources 81–82:902
22. Naoi K, Ishimoto S, Ogihara N, Nakagawa Y, Hatta S (2009) J Electrochem Soc 156:A52
23. Huang S, Wen Z, Zhu X, Gu Z (2004) Electrochem Commun 6:1093
24. Kim J, Cho J (2007) Electrochem Solid-State Lett 10: A81
25. Yu H, Zhang X, Jalbout AF, Yan X, Pan X, Xie H, Wang R (2008) Electrochim Acta 53:4200
26. Huang J, Jiang Z (2008) Electrochim Acta 53:7756
27. Bai Y, Wang F, Wu F, Wu C, Bao L (2008) Electrochim Acta 54:322
28. Ohzuku T, Ueda A, Yamamoto N (1995) J Electrochem Soc 142:1431
29. Thackeray MM (1995) J Electrochem Soc 142:2558
30. Zhou JH, Sui ZJ, Li P, Chen D, Dai YC, Yuan WK (2006) Carbon 44:3255
31. Shen CM, Zhang XG, Zhou YK, Li HL (2002) Mater Chem Phys 78:437
32. Hao Y, Lai Q, Xu Z, Liu X, Ji X (2005) Solid State Ionics 176:1201
33. Shu J (2008) Electrochem Solid-State Lett 11:A238

Hydrocarbon Membranes for Polymer Electrolyte Fuel Cells

Kenji Miyatake and Masahiro Watanabe
University of Yamanashi, Kofu, Yamanashi, Japan

Introduction

In the preceding chapter, recent progress on perfluorinated ionomer membranes is reviewed. There is no doubt that the perfluorinated ionomer membranes will dominate in the area of polymer electrolyte fuel cells (PEFCs) for the years to come; however, drawbacks peculiar to perfluorinated materials such as high production cost, high gas permeability, and environmental incompatibility have to be addressed possibly by non-fluorinated materials.

Most studied alternative membranes are hydrocarbon ionomers. Hydrocarbon ionomers have a rather longer history (ca. a century) as cation exchange resins than the perfluorinated ionomers. Original hydrocarbon ionomers were based on sulfonated polystyrenes and phenol resins. In the earliest stage of the PEFC research (1960s), such hydrocarbon ionomer membranes were applied only for short-time operation because of insufficient durability of the membranes. Hydrocarbon ionomers have been reexamined in more detail in the last decade due to their possible low production cost, freedom in molecular design and chemical modifications, low gas permeability, and recyclability [1–3]. Hydrocarbon ionomers can be roughly classified into two classes, aliphatic and aromatic ionomers, depending on their main chain structures. Typical aliphatic ionomers are vinyl polymers with acidic functions. The acid-functionalized vinyl polymers often suffer from instability to oxidation commenced by hydrogen abstraction reactions with highly active radicals. Aromatic ionomers seem to have attracted more attention due to their chemical robustness. In this chapter, some of these emerging non-fluorinated aromatic ionomer membranes are reviewed.

Sulfonated Poly(arylene ether) Ionomers

Poly(arylene ether)s are one of the most studied aromatic polymers for the sulfonation reactions. Poly(arylene ether)s include poly(phenylene ether), poly(arylene ether ketone), poly(arylene ether ether ketone), and poly(arylene ether sulfone). These polymers contain electron-rich phenylene rings linked by ether bonds, which are susceptible for the sulfonation reactions with typical sulfonating reagents such as sulfuric acid, chlorosulfonic acid, or sulfur trioxide. The first attempt was the sulfonation of poly(2,6-dimethyl-phenylene ether) (PPO) by General Electric in 1960s [4]. Since then, a number of

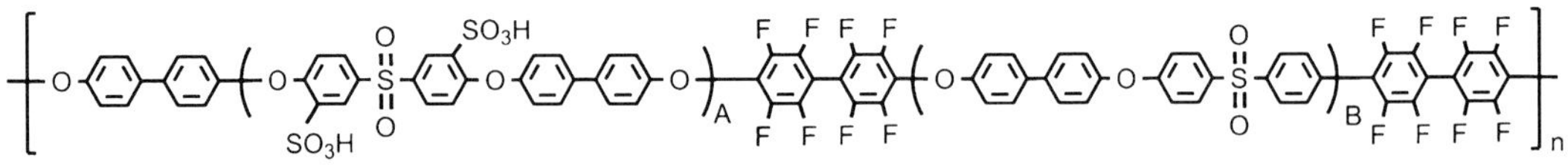

Hydrocarbon Membranes for Polymer Electrolyte Fuel Cells, Fig. 1 BPSH/BPS block copolymer (*A* and *B* are the numbers of repeat unit of hydrophilic and hydrophobic blocks, respectively. *n* is the degree of polymerization)

sulfonated poly(arylene ether)s have been reported. Advantages of the ionomers are reasonable solubility in polar organic solvents, good film-forming capability, high mechanical strength, and thermal stability. The sulfonated poly(arylene ether) membranes require much higher IEC to have proton conductivity comparable to that of the perfluorinated ionomer membranes. High IEC often causes excessive swelling and mechanical failure of the membranes during fuel-cell operation.

More recently, block copolymers with sequenced hydrophilic and hydrophobic units have been developed for the sulfonated poly (arylene ether)s with rather low IEC to show high proton conductivity [5, 6]. They have better-developed hydrophilic/hydrophobic nanophase separation and, accordingly, better-connected ionic channels compared to the conventional random copolymers of the same chemical structure and composition. A typical and well-examined example is block copolymers composed of biphenol-based disulfonated arylene ether sulfone (the so-called BPSH) units and the unsulfonated equivalents (BPS) (Fig. 1) by McGrath's group [7–9]. They have investigated detailed properties of the BPSH block copolymer membranes to obtain the following informative conclusions:

1. High molecular weight BPSH block copolymers were obtained when reactive perfluorinated linkage (nonafluorobiphenylene or pentafluorophenylene) groups were attached at the both ends on unsulfonated hydrophobic oligomers.
2. The linkage groups had some effect on the membrane properties. The fluorinated biphenylene groups seemed to promote nanophase separation and thus water uptake and proton conductivity at low humidity than the fluorinated phenylene groups.
3. The BPSH block copolymer membranes performed much better as proton exchange membranes for fuel cells than the random copolymers with similar IEC, especially in terms of proton conductivity at low humidity (on the order of mS/cm at 80 °C and 30 % RH, IEC = 1.5 meq/g).
4. The block length rather than the IEC was more important to dominate water uptake and proton conductivity, where longer block length lead to higher water uptake and higher proton conductivity. So were the nanophase separation (or connection of hydrophilic domains) and water diffusion coefficient.
5. In the hydrated BPSH block copolymers, more freezing water (free and loosely bound water to sulfonic acid groups) existed than in the random copolymers due to the developed morphology. The molar mass of the block should be higher than 10,000 in order to have noticeable improvement on the morphological order and proton conductivity.
6. In addition to the block length and IEC values, casting conditions (such as the solvent and drying temperature or solvent removal rate) did have significant impact on membrane morphology and properties.
7. The BPSH block copolymer membrane performed in an H_2/air fuel cell at 100 °C and 40 % RH comparable to Nafion membrane.

These findings are, more or less, applicable to the other aromatic hydrocarbon ionomers and useful to design the higher-order structure and properties of ionomer membranes.

Miyatake et al. have proposed an advanced block copolymers of their sulfonated poly (arylene ether)s, block copolymers containing highly dense sulfonic acid groups in the hydrophilic blocks. They have successfully synthesized a series of block poly(arylene ether sulfone

Hydrocarbon Membranes for Polymer Electrolyte Fuel Cells, Fig. 2 SPESK block copolymer (*A* and *B* are the numbers of repeat unit of hydrophilic and hydrophobic blocks, respectively. *n* is the degree of polymerization)

ketone)s (SPESKs) containing fully sulfonated fluorenylidene biphenylene units (Fig. 2) [10, 11]. The well-controlled post-sulfonation reaction of the precursor polymers enabled preferential sulfonation on each aromatic ring of the fluorenylidene biphenylene groups with 100 % degree of sulfonation. Pre-sulfonation method via oligomeric sulfonation reactions has also been developed [12]. The ionomer membranes showed unique morphology with well-developed hydrophilic/hydrophobic nanophase separation, depending on the block length of each segment. It has been concluded that longer block length and/or higher IEC resulted in larger and better-connected hydrophilic clusters under dry conditions, while the morphology was less dependent on these factors under fully hydrated conditions as confirmed by small angle X-ray-scattering (SAXS) analyses. The SPESK block copolymer membranes showed much higher proton conductivity than that of the random copolymer membranes with similar chemical structure and IEC. The proton conductivity was similar or even higher compared to that of Nafion over a wide range of humidity (Fig. 3). The SPESK membrane retained high proton conductivity at 110 °C. The high conductivity resulted from the high proton diffusion coefficient. Longer block length seemed effective in increasing proton diffusion coefficient, which coincided with the morphological observations. The SPESK membranes were highly stable to hydrolysis in water at 140 °C for 24 h or at 100 °C for 1,000 h. The membranes degraded to some extent under harsh oxidative conditions (in Fenton's reagent), which is the fate of hydrocarbon ionomer compounds. Oxidative degradation is likely to occur at phenylene carbon atoms ortho to the ether bonds

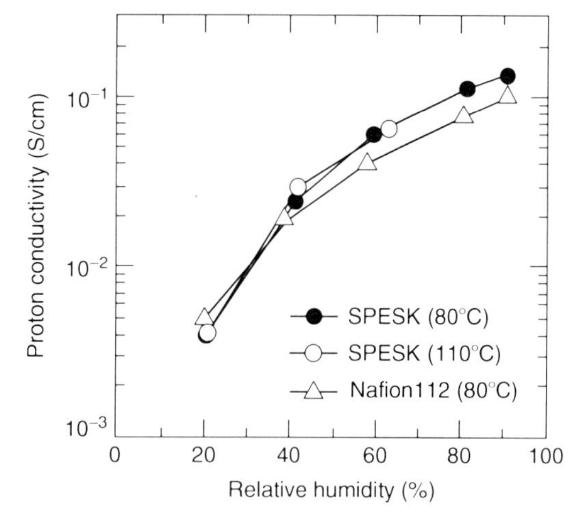

Hydrocarbon Membranes for Polymer Electrolyte Fuel Cells, Fig. 3 Humidity dependence of the proton conductivity of SPESK block copolymer membrane (IEC = 1.62 meq/g) at 80 °C and 110 °C

by the attack of highly oxidative hydroxyl radicals. The SPESK membrane showed much lower gas permeability than that of Nafion. The low gas permeability could mitigate their oxidative instability since hydrogen peroxide as a by-product is potentially less produced when the gas permeation through the membranes is low. A fuel cell was successfully operated with the SPESK membranes at 30 % and 53 % RH and 100 °C (Fig. 4) [13]. The current density was 250 mA/cm^2 at 30 % RH and 410 mA/cm^2 at 53 % RH at a cell voltage of 0.6 V. The high proton conductivity of the membrane at low RH and high temperature was well confirmed in the practical fuel-cell operation. The long-term stability of the SPESK membrane was tested under fuel-cell operation at a current density 200 mA/cm^2 at 53 % RH and 80 °C for 2,000 h. While the membrane maintained the original

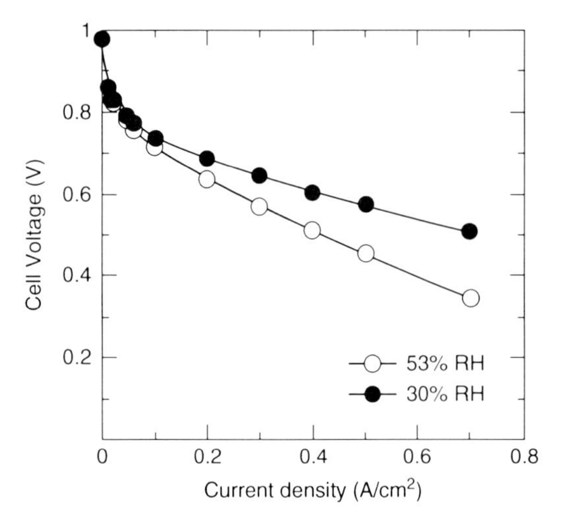

Hydrocarbon Membranes for Polymer Electrolyte Fuel Cells, Fig. 4 Hydrogen/air fuel-cell performance of SPESK block copolymer membrane (IEC = 1.62 meq/g) at 100 °C, 53 % and 30 % RH for both electrode

IEC, water affinity, and proton conductivity, minor oxidative degradation occurred in the hydrophilic blocks. The major decomposition products were acetate, formate, and sulfate anions, which were adsorbed onto the catalysts to cause degradation in fuel-cell performance [14].

Sulfonated Poly(phenylene) Ionomers

Poly(*p*-phenylene) (PP) provides advantages as fuel-cell membrane due to the rigid and robust polymer backbone. The wholly aromatic polymer main chains without hetero linkages such as ether and sulfone groups are highly stable to oxidation, which is a great advantage over the above-mentioned sulfonated poly(arylene ether)s. Disadvantages include its low solubility in most organic solvents to limit the formation of high molecular weight polymers and the molecular design for functionalization. The challenge is how to synthesize soluble high molecular weight poly(*p*-phenylene) ionomers (high enough to provide self-standing and bendable membranes) and how to introduce ionic groups. There are not many synthetic options available for poly(*p*-phenylene) derivatives compared to the other aromatic hydrocarbon polymers with hetero linkages. Therefore, there have been limited number of reports for poly(*p*-phenylene) ionomers. Among them, poly(*p*-phenylene) ionomers developed by JSR Corporation, Japan, are one of the most successful examples [15]. They have developed a new synthetic route for the sulfonated poly(*p*-phenylene) ionomers via Nickel-catalyzed coupling reaction of dichlorobenzophenone-containing neopentylsulfonate ester groups. Similar to the sulfonated poly (arylene ether)s, sequenced structure caused improved performance in the sulfonated poly(*p*-phenylene) ionomers. Block copolymerization with more flexible polymers such as poly (arylene ether ketone)s have provided flexible membranes with better handling capability. Chemical structure of a typical sulfonated poly(*p*-phenylene) block copolymer is shown in Fig. 5. In the literature, several other poly (phenylene) derivatives with acidic functions can also be found [16–18].

The advantage of the sulfonated poly(*p*-phenylene) ionomers is that the membranes can carry very high IEC values, higher than 3 meq/g, without sacrificing mechanical and chemical stability of the membranes. Such high IEC contributes to high proton conductivity under high-temperature and low humidity conditions. The membranes show well-developed hydrophilic/hydrophobic nanophase separation, which is also beneficial for proton-conducting properties as described above. The phase-separated morphology could be controlled by copolymer composition, chemical structure of hydrophobic block, sequenced structure and length of hydrophilic and hydrophobic blocks, and membrane fabrication (or casting) conditions. In Table 1 are summarized properties of a typical JSR membrane. According to the disclosed data, the JSR membranes absorb more water and show higher proton conductivity, better thermal and mechanical stability, comparable hydrolytic stability, and much lower gas permeability compared to those of the conventional perfluorinated ionomer membranes. It has been recently claimed that the introduction of basic

Hydrocarbon Membranes for Polymer Electrolyte Fuel Cells, Fig. 5 Sulfonated poly(*p*-phenylene)/poly(arylene ether ketone) block copolymer

Hydrocarbon Membranes for Polymer Electrolyte Fuel Cells, Table 1 Properties of the sulfonated poly(*p*-phenylene) copolymer (JSR) membrane

Proton conductivity	70 °C, 80 % RH	0.1 S cm^{-1}
	95 °C, 80 % RH	0.16 S cm^{-1}
Mechanical strength	Elongation at break at 23 °C, 50 % RH	100 %
	Stress at break at 23 °C, 50 % RH	130 MPa
Gas permeability	H_2 at 80 °C, dry	9×10^{-10} cm^3 cm cm^{-2} s^{-1} $cmHg^{-1}$
	O_2 at 40 °C, 90 % RH	6×10^{-10} cm^3 cm cm^{-2} s^{-1} $cmHg^{-1}$
Chemical stability	IEC change after 1,000 h in 95 °C water	0 %
	Weight change after 1,000 h in 95 °C water	0 %

groups such as pyridine and imidazole as a third comonomer component could improve the durability of the block poly(*p*-phenylene) ionomer membranes. The JSR membranes have been successfully installed on Honda FCX Clarity fuel-cell electric vehicles.

Future Directions

There is no doubt that the perfluorinated ionomer membranes take initiative in this field and contribute a great deal in the commercialization and wide diffusion of fuel cells in the early stage. In terms of environmental compatibility (recyclability or disposability) and production cost, the perfluorinated ionomer membranes should be replaced with non-fluorinated alternative materials within the next decade. Challenge is how to achieve comparable conductivity and durability with the non-fluorinated membranes. Currently, no alternative materials have overcome the trade-off relationship between these two conflicting properties. In addition to the abovementioned two representative ionomers, significant advancement has also been achieved with the other types of hydrocarbon ionomer membranes. It would probably need some more investment (time, effort, and money) to develop alternative membranes suitable for practical fuel-cell applications.

Cross-References

▸ Polymer Electrolyte Fuel Cells, Perfluorinated Membranes

References

1. Rikukawa M, Sanui K (2000) Proton-conducting polymer electrolyte membranes based on hydrocarbon polymers. Prog Polym Sci 25:1463
2. Hickner MA, Ghassemi H, Kim YS, Einsla BR, McGrath JE (2004) Alternative polymer systems for proton exchange membranes (PEMs). Chem Rev (Washington, DC, US) 104:4587

3. Miyatake K, Watanabe M (2005) Recent progress in proton conducting membranes for PEFCs. Electrochem (Tokyo, Jpn) 73:12
4. Fox DW, Shenian P (1966) Sulfonated polyphenylene ether cation-exchange resins. US3259592
5. Peckham TJ, Holdcroft S (2010) Structure-morphology-property relationships of non-perfluorinated proton-conducting membranes. Adv Mater (Weinheim, Germany) 22:4667
6. Elabd YA, Hickner MA (2011) Block copolymers for fuel cells. Macromolecules 44:1
7. Yu X, Roy A, Dunn S, Yang J, McGrath JE (2006) Synthesis and characterization of sulfonated-fluorinated, hydrophilic-hydrophobic multiblock copolymers for proton exchange membranes. Macromol Symp 245/246:439
8. Roy A, Hickner MA, Yu X, Li Y, Glass TE, McGrath JE (2006) Influence of chemical composition and sequence length on the transport properties of proton exchange membranes. J Polym Sci Part B Polym Phys 44:2226
9. Li Y, Roy A, Badami AS, Hill M, Yang J, Dunn S, McGrath JE (2007) Synthesis and characterization of partially fluorinated hydrophobic-hydrophilic multiblock copolymers containing sulfonate groups for proton exchange membrane. J Power Sources 172:30
10. Bae B, Yoda T, Miyatake K, Uchida H, Watanabe M (2010) Proton-conductive aromatic ionomers containing highly sulfonated blocks for high-temperature-operable fuel cells. Angew Chem Int Ed 49:317
11. Bae B, Miyatake K, Watanabe M (2010) Sulfonated poly(arylene ether sulfone ketone) multiblock copolymers with highly sulfonated block. Synthesis and Properties. Macromolecules 43:2684
12. Bae B, Hoshi T, Miyatake K, Watanabe M (2011) Sulfonated block poly(arylene ether sulfone) membranes for fuel cell applications via oligomeric sulfonation. Macromolecules 44:3884
13. Bae B, Yoda T, Miyatake K, Uchida M, Uchida H, Watanabe M (2010) Sulfonated poly(arylene ether sulfone ketone) multiblock copolymers with highly sulfonated block: fuel cell performance. J Phys Chem B 114:10481
14. Bae B, Miyatake K, Uchida M, Uchida H, Sakiyama Y, Okanishi T, Watanabe M (2011) Sulfonated poly(arylene ether sulfone ketone) multiblock copolymers with highly sulfonated blocks. Long-term fuel cell operation and post-test analyses. ACS Appl Mater Interfaces 3:2786
15. Goto K, Rozhanskii I, Yamakawa Y, Otsuki T, Naito Y (2008) Development of aromatic polymer electrolyte membrane with high conductivity and durability for fuel cell. Polym J 41:95
16. Bae JM, Honma I, Murata M, Yamamoto T, Rikukawa M, Ogata N (2002) Properties of selected sulfonated polymers as proton-conducting electrolytes for polymer electrolyte fuel cells. Solid State Ionics 147:189
17. Ghassemi H, McGrath JE (2004) Synthesis and properties of new sulfonated poly(p-phenylene) derivatives for proton exchange membranes. Polymer 45:5847
18. Fujimoto CH, Hickner MA, Cornelius CJ, Loy DA (2005) Ionomeric poly(phenylene) prepared by diels-alder polymerization: synthesis and physical properties of a novel polyelectrolyte. Macromolecules 38:5010

Hydrochloric Acid Electrolysis

Jakob Jörissen
Chair of Technical Chemistry, Technical University of Dortmund, Germany

Introduction

Hydrogen chloride is a co-product of numerous processes in chemical industry which use chlorine due to its high reactivity for selective formation of intermediates (See entry "▸ Chlorine and Caustic Technology, Overview and Traditional Processes"). In many cases, chlorine is subsequently removed and the final products are chlorine-free. The elimination of chlorine frequently yields hydrogen chloride. However, a worthwhile utilization of hydrogen chloride is not everywhere possible and thus, its usage for chlorine recycling is desired. This is possible in principle by several chemical processes which mostly use air (oxygen) as reactant (Deacon reaction). The formation of chlorine is an equilibrium reaction and requires sophisticated catalysts and procedures for sufficient yield [1–3]. Such processes are relatively uncomplicated if no free chlorine has to be produced because it is immediately consumed by a consecutive reaction as in case of oxychlorination, e.g., of ethylene to 1,2-dichloroethane for vinylchloride production.

Alternatively, hydrogen chloride can be absorbed in water as hydrochloric acid which is electrolyzed for chlorine recovery. A typical example is the polyurethane production: Chlorine reacts with carbon monoxide into phosgene which subsequently reacts with primary amines to give isocyanates under elimination of hydrogen chloride [3]. A closed loop of chlorine recycling is possible using hydrochloric acid electrolysis. Its products are very pure without further treatment; the process design is simple, and the operation is easy and flexible for fluctuating production rates [1, 4].

Hydrochloric acid electrolysis was developed in 1942. Since 1964, it is operated in industry and continuously optimized until today [4–6] (See section ► Hydrochloric Acid Electrolysis). A far-reaching modification starting in 1999 was the implementation of "Oxygen Depolarized Cathodes" (ODCs) which decrease substantially the energy consumption (See section Hydrochloric acid electrolysis using "Oxygen Depolarized Cathode" (ODC)). An overview about hydrochloric acid electrolysis is given in [1, 2], detailed information about both processes is available in [4].

Diaphragm Process of Hydrochloric Acid Electrolysis

The electrochemical reactions of hydrochloric acid electrolysis are simple because only two gaseous products are formed from the same electrolyte, less complicated than chlor-alkali electrolysis with two very different electrolytes (See entry "► Chlorine and Caustic Technology, Overview and Traditional Processes"). The overall electrolysis reaction is:

$$2\,HCl + 2\,F \rightarrow Cl_2 + H_2 \qquad (1)$$

Hydrochloric acid is delivered to the anode where chlorine gas is evolved:

$$2\,Cl^- \leftrightarrow Cl_2 + 2\,e^- \quad E_0 = 1.36\,V \qquad (2)$$

At standard conditions, water decomposition forming oxygen should be preferred, according to thermodynamics, because its standard potential is lower:

$$H_2O \leftrightarrow 1/2\,O_2 + 2\,H^+ + 2\,e^- \quad E_0 = 1.23\,V \qquad (3)$$

However, the very strong acidic conditions increase the oxygen potential and additionally oxygen formation is kinetically hindered at usual anode materials, i.e., it needs a high charge transfer over-potential. Therefore – in contrast to chlor-alkali electrolysis – no oxygen is detectable in the anode gas and no corrosion of graphite anodes occurs (service life 10 years [4]).

H^+ ions are reduced to hydrogen at the cathode:

$$2\,H^+ + 2\,e^- \leftrightarrow H_2 \qquad E_0 = \pm 0.00\,V \qquad (4)$$

Figure 1 shows a scheme of the electrolyzer in the traditional Bayer-Hoechst-Uhde process (for details see [4]). Graphite plates (a) are operating as bipolar electrodes. For optimal gas release, both surfaces are provided with grooves. The electrodes (a) in the dimension of 2.5 m^2 are embedded into frames (b) of synthetic resin which is resistant against chlorine and acid. Diaphragms (c) are mounted with gaskets between the electrodes/frames. About 36 frames are attached together with endplates and current distributors to one electrolyzer (filter press type) by spring-loaded tension rods. The electrolyzer includes inlet (d) and outlet (e) channels which are penetrating through frames and diaphragms and are connected by holes with all anode or cathode chambers respectively. Two separated small inlet channels at the bottom (d) supply the electrode chambers with electrolytes. The foamy mixtures of depleted electrolytes and produced gases are removed through six separated large outlet channels (e) on the top of the electrolyzer.

Anolyte and catholyte are circulating in two separated loops. Feed concentration is 21–23 wt% HCl and electrolytes are depleted in the electrolyzer to 17 wt% HCl. A part of depleted catholyte is reconcentrated to 28 wt% in an absorption column from a hydrogen chloride containing gas stream and is after filtering again mixed into the circular flows [4].

The most important function of the diaphragm (c) is the separation of the gases which is sufficiently attained (a mixture of chlorine and hydrogen would be explosive). A separation of anolyte and catholyte seems to be not necessary because

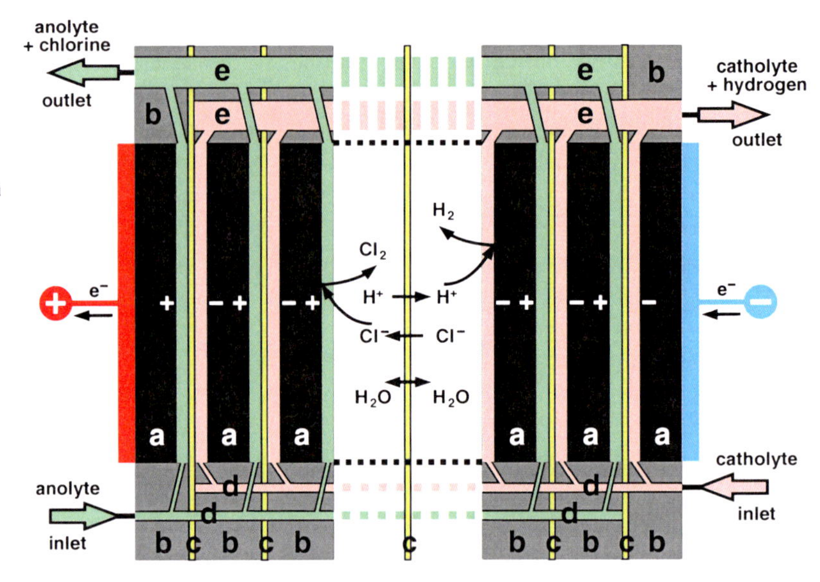

Hydrochloric Acid Electrolysis, Fig. 1 Scheme of diaphragm cell for hydrochloric acid electrolysis (**a**) bipolar graphite electrode, (**b**) resin frame, (**c**) diaphragm with gaskets, (**d**) inlet channels, (**e**) outlet channels

on both sides, hydrochloric acid of about 17 wt % HCl is used. The migration of H^+ and Cl^- ions and the diffusion of water is required (See Fig. 1) and provides a low voltage drop. However, in the anolyte, chlorine is dissolved and diffuses into the catholyte. It is partially reduced at the cathode and decreases the current efficiency by about 2.0–2.5 % [1]. A residue of chlorine remains in the produced hydrogen gas which has to be scrubbed with caustic soda solution.

The diaphragm (c) is made of a special PVC or PVC/PVDF (polyvinylchloride and polyvinylidenfluoride) cloth. Its service life (up to 4 years) is very much dependent on organic impurities in the feed. Therefore, the hydrochloric acid frequently is additionally purified by adsorption with activated carbon.

The diaphragm may be replaced by a cation exchange membrane which is suitable for high concentrated hydrochloric acid like Nafion® (DuPont, perfluorosulfonic acid polymer, PFSA, see entry "▶ Chlorine and Caustic Technology, Membrane Cell Process"). This membrane has almost no usual porosity and is nearly exclusively permeable for H^+ ions including a hydration shell of some water molecules. Thus, product quality is significantly increased, process operation can be simplified, and cell voltage is reduced by about 0.3 V [1, 6]. However, the mechanical durability of the membrane in the described electrolyzer is insufficient.

The cell voltage at standard conditions should be 1.36 V (difference of standard potentials of reaction (2) and (4)). At the usual hydrochloric acid concentration > 15 wt % and 70 °C, it should be decreased to 1.16 V [1]. In reality, the voltage is higher due to electrolyte and diaphragm voltage drop and over-potentials. The reliable state-of-the-art diaphragm process operates with 2.2 V at 5 kA m^{-2} [4]. Several measures were successfully tested to reduce the cell voltage: addition of depolarizing agents or catalytically active compounds into the catholyte or a catalytic coating of the cathode in order to decrease the cathodic hydrogen over-potential, optimization of the diaphragm material, or changing to a cation exchange membrane for a lower ohmic voltage drop. A cell voltage of 1.9 V at 85 °C and 4.8 kA m^{-2} is noted in [1]. However, this is not realized for long-term and safe operation.

Hydrochloric Acid Electrolysis Using "Oxygen Depolarized Cathode" (ODC)

A significant lower cell voltage can be achieved if the cathode reaction is changed to the reversed reaction of anodic oxygen evolution (Reaction (3)):

$$4\ H^+ + O_2 + 4\ e^- \leftrightarrow 2\ H_2O \qquad E_0 = 1.23\ V \qquad (5)$$

The overall electrolysis reaction is changed to:

$$4\ HCl + O_2 + 4\ F \rightarrow 2\ Cl_2 + 2\ H_2O \qquad (6)$$

The "Oxygen Depolarized Cathode" (ODC) needs oxygen as reactant and the usual co-product hydrogen is no longer available. On the other hand, the cell voltage is decreased: Theoretically by 1.23 V, realizable is a reduction by about 0.8 V in consequence of over-potentials. A cell voltage of nearly 1.4 V is achieved in industrial ODC electrolyzers, i.e., an energy saving of approximately 36 % is possible in comparison with 2.2 V of the diaphragm process as mentioned above [4].

The economic aspects of ODC application in hydrochloric acid electrolysis are similar as discussed for chlor-alkali electrolysis (See entry "▸ Chlorine and Caustic Technology, Using Oxygen Depolarized Cathode"). Especially the value of hydrogen as a co-product has to be balanced with costs and availability of electrical energy and oxygen.

Oxygen gas conversion in an ODC requires an electrode design as "gas diffusion electrode" (GDE). The reaction is possible only in "three-phase boundaries" where the gas oxygen, the liquid electrolyte hydrochloric acid, and the solid electro-catalyst – electrically connected with the current distributor – are in optimal contact and any transport hindrance is minimized.

The functional principle of an ODC/GDE is elucidated for chlor-alkali electrolysis in the entry "▸ Chlorine and Caustic Technology, Using Oxygen Depolarized Cathode." There are problems discussed which result from handling of the liquid product caustic soda solution and the pressure difference between gas phase and electrolyte (hydrostatic pressure). Such problems are irrelevant in case of hydrochloric acid electrolysis which is in consequence easier to operate.

The principle of operation is shown in Fig. 2. Chlorine gas is produced at the anode (especially optimized dimensionally stable anode) with an anolyte feed concentration of 14 wt % HCl. Anode and cathode are separated by a cation exchange membrane (perfluorosulfonic acid polymer, PFSA, e.g., Nafion® of DuPont). The ODC is based on a conductive carbon cloth which operates simultaneously as a gas diffusion layer because a suitable material is incorporated. The oxygen reduction reaction (5) takes place in three-phase boundaries of a thin, porous catalyst layer on the surface.

The ODC requires a catalyst which is suitable for the strongly acidic medium of hydrochloric acid. It has to be highly resistant toward corrosion and poisoning by organic species, which may be included as contaminants in technical grade hydrochloric acid. A novel rhodium sulfide catalyst was developed for this purpose [8] (See also entry "▸ Oxygen Reduction Reaction in Acid Solution").

An excess pressure in the anolyte presses the membrane and the ODC onto the current distributor for optimal contact. Thus, the membrane is in a fixed position and no problem of mechanical stability occurs. The membrane has almost no usual porosity and no pressurized liquid can arrive at the ODC. In the cathode compartment behind the ODC (right side in the scheme of Fig. 2), oxygen is supplied from a circulating loop and has free access to the ODC without hindrance by a liquid. H^+ ions, which are needed for the oxygen reduction reaction (5), predominantly migrate through the membrane. Additionally, together with each H^+ ion, a hydration shell of some water molecules is transferred. This water, including the produced water, mainly is vaporized into the oxygen gas phase and has to be condensed outside of the electrolyzer. The permeation selectivity of the perfluorosulfonic acid membrane is not very high (See entry "▸ Chlorine and Caustic Technology, Membrane Cell Process") so that several Cl^- ions can diffuse to the cathode. Additionally, chlorine can diffuse through the membrane and is reduced at the cathode to Cl^- ions. All Cl^- ions generate together with H^+ ions and water a small amount of diluted

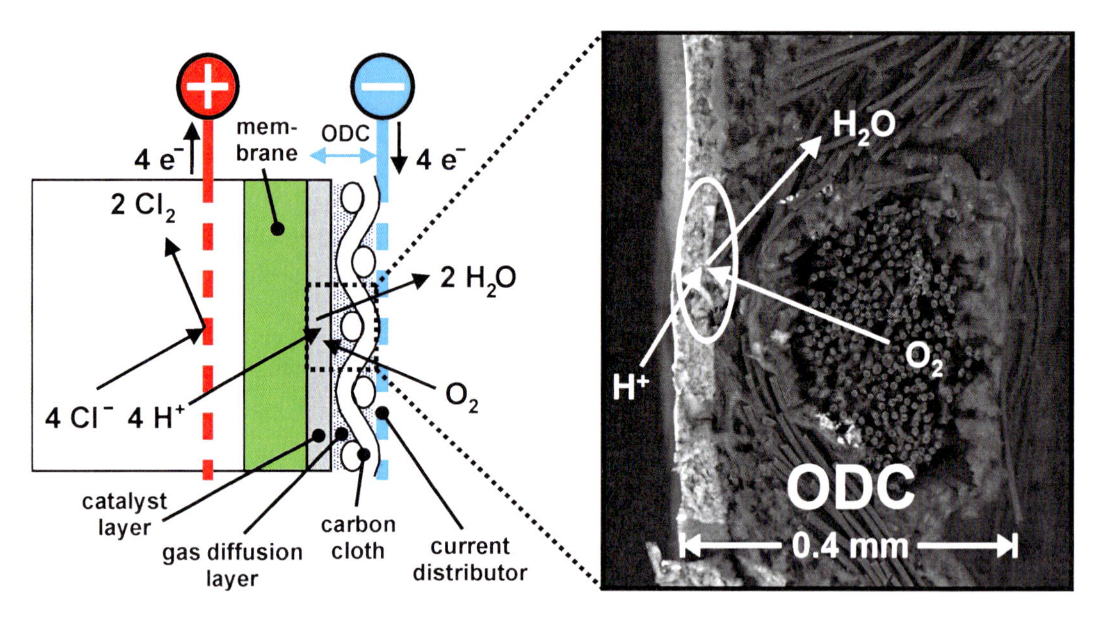

Hydrochloric Acid Electrolysis, Fig. 2 Scheme of hydrochloric acid electrolysis using Oxygen Depolarized Cathode (*ODC*) and cross-section photograph of the ODC (Based on Bulan et al. [7])

hydrochloric acid which is advantageous for a high ionic conductivity between membrane and catalyst layer. This liquid can trickle down through the ODC without blocking of oxygen gas supply and is discharged outside of the electrolyzer. This small amount of liquid is not a problem in the ODC.

A first ODC hydrochloric acid electrolysis plant in an industrial scale (10,000 and later 20,000 t/a chlorine capacity) is operated since 2003 and a first world scale plant (215,000 t/a) started in 2008. Further plants are under construction [4].

Future Directions

There are developments which could be interesting for realization in industry. Chlorine can be evolved from gaseous hydrogen chloride at an anode which is prepared as a coating of electro-catalytic active material directly on the surface of a Nafion® cation exchange membrane [9]. Also the cathode is a suitable coating on the opposite side of the membrane. In this case, hydrogen evolution is possible as the cathodic reaction from diluted hydrochloric acid. Alternatively, a cathode coating can be applied which operates as an ODC. This is in principle an electrochemical Deacon process according to the overall reaction (6). The cell is comparable with a "Polymer Electrolyte Fuel Cell" (PEFC, see entry "▶ Polymer Electrolyte Fuel Cells (PEFCs), Introduction").

Further possibilities could be opened using a chlorine-resistant anion exchange membrane, especially in case of hydrochloric acid with too low concentration for the above discussed processes [10]. The anion exchange membrane enables the transfer of Cl^- ions to the anode even though at low hydrochloric acid concentrations in the catholyte. Two variants of chlorine evolution at the anode under these conditions were experimentally proved:

- High chlorine purity can be achieved if the anolyte contains a high concentration of an additional chloride, e.g., calcium chloride.
- An electro-catalytic coating on the anode side of the membrane – comparable with a gas diffusion electrode – enables chlorine evolution without any anolyte.

Cross-References

► Chlorine and Caustic Technology, Membrane Cell Process
► Chlorine and Caustic Technology, Overview and Traditional Processes
► Chlorine and Caustic Technology, Using Oxygen Depolarized Cathode
► Oxygen Reduction Reaction in Acid Solution
► Polymer Electrolyte Fuel Cells (PEFCs), Introduction

References

1. Schmittinger P, Florkiewicz T, Curlin L, Lüke B, Scannell R, Navin P, Zelfel E, Bartsch R (2011) Chlorine. In: Ullmann's encyclopedia of industrial chemistry. Wiley-VCH, Weinheim, doi: 10.1002/14356007.a06_399.pub3
2. Bommaraju TV, Lüke B, O'Brien TF, Blackburn MC (2002) Chlorine. In: Kirk-Othmer encyclopedia of chemical technology. Wiley, Hoboken, NJ, doi: 10.1002/0471238961.0308121503211812.a01.pub2
3. Ando H, Uchida Y, Seki K, Knapp C, Omoto N, Kinoshita M (2010) Trends and views in the development of technologies for chlorine production from hydrogen chloride. Download: http://www.sumitomo-chem.co.jp/english/rd/report/theses/docs/01_chlorine_production_e.pdf
4. Thyssen Krupp Uhde (2013) Hydrochloric acid electrolysis sustainable chlorine production. Download: http://www.thyssenkrupp-uhde.de/fileadmin/documents/brochures/Hydrochloric_Acid_Electrolysis.pdf
5. Isfort H, Stockmans WJ (1984) Twenty years of experience, recent developments and economic aspects of hydrochloric acid electrolysis. Proc Electrochem Soc 84:259–275
6. Minz FR (1992) Hydrochloric acid electrolysis – recycling technique (Language: German). DECHEMA Monogr 125:195–203
7. Bulan A, Gestermann F, Turek T, Weber R, Weuta P (2004) Chlorine electrolysis using gas diffusion electrodes. GVC-DECHEMA annual conference 2004
8. Allen RJ, Giallombardo JR, Czerwiec D, De Castro ES, Shaikh K, Gestermann F, Pinter HD, Speer G (2002) Process for the electrolysis of technical-grade hydrochloric acid contaminated with organic substances using oxygen-consuming cathodes. US 6402930 B1 De Nora Elettrodi S.p.A., Bayer AG
9. Trainham JA, Law CG, Newman JS, Keating KB, Eames DJ (1995) Electrochemical conversion of anhydrous hydrogen halide to halogen gas using a cation-transporting membrane. US 5411641 A E.I. Du Pont de Nemours
10. Barmashenko V, Jörissen J (2005) Recovery of chlorine from dilute hydrochloric acid by electrolysis using a chlorine resistant anion exchange membrane. J Appl Electrochem 35:1311–1319. doi:10.1007/s10800-005-9063-1

Hydrogel Electrolyte

Hiroshi Inoue
Osaka Prefecture University, Sakai, Osaka, Japan

Introduction

Rechargeable alkaline batteries such as nickel-cadmium battery and nickel-metal hydride (Ni-MH) battery include concentrated alkaline aqueous solutions as an electrolyte. The electromotive force of the Ni-MH battery is ca. 1.2 V which is a third of that of lithium ion batteries (LIBs) using nonaqueous electrolyte solutions because the former is restricted by the decomposition voltage of water. On the other hand, electrical conductivity of aqueous electrolyte solutions is more than two digits higher than that of nonaqueous solutions, which enable the rechargeable alkaline batteries to charge and discharge with high current density. Moreover, they are superior in safety to the LIBs. The rechargeable alkaline batteries have been applied to mobile electronic devices, electric power tools, hybrid electric vehicles, etc.

Due to no leakage and no freezing of electrolyte solutions and thinness and compactness of batteries, all-solid-state batteries with proton- or hydroxide-conductive solid electrolytes have been developed. However, they are not practically used yet. The most serious issue was that the solid electrolytes had much lower electrical conductivity than alkaline electrolyte solutions. Since the second half of 1990s, new types of solid electrolytes, hydrogel electrolytes, have been developed by various research groups and applied to all-solid-state electrochemical devices like batteries and capacitors. In this section, the hydrogel electrolytes can be put into five categories.

Polyethylene Oxide-Based Hydrogel Electrolytes

Polyethylene oxide (PEO)-based hydrogel electrolytes in which electrolyte solutions are incorporated in PEO matrix usually have strong mechanical strength, high flexibility, easy formation of thin film, no leakage for battery applications, etc. Hydrogel electrolytes with high electrical conductivity were prepared by mixing various concentrations and volumes of potassium hydroxide aqueous solutions with PEO [1–3]. For example, a PEO-based hydrogel electrolyte which was composed of 60 wt % PEO, 30 wt % KOH, and 10 wt % H_2O showed electrical conductivity of $\sim 10^{-3}$ S cm^{-1} [1] which was much higher than any other conventional proton- or hydroxide-conductive solid electrolytes. In addition, the hydrogel electrolytes were applied to all-solid-state rechargeable alkaline batteries such as Ni–Cd, Ni–Zn, and Ni-MH batteries. Consequently, they worked well in the temperature range between −20 °C and 40 °C and are rechargeable over 500 cycles although their discharge capacity was significantly scattered every cycles. In the self-discharge test of an all-solid-state Ni-MH battery, a half of charge capacity was remained even after a month although a conventional Ni-MH battery with alkaline solution discharged out. A hydrogel electrolyte film could be formed with copolymer of epichlorohydrin and ethylene oxide, and it had higher electrical conductivity than that with PEO [4].

Polyvinyl Alcohol-Based Hydrogel Electrolytes

Polyvinyl alcohol (PVA) was also used for preparing a hydrogel electrolyte film [5–10]. The electrical conductivity of PVA-based hydrogel electrolyte films at room temperature was $4.7 \times 10^{-2} \sim 8.5 \times 10^{-4}$ S cm^{-1} and it depended on some factors like degree of polymerization. All-solid-state Ni-MH and Ni-Zn batteries were assembled with the PVA-based hydrogel electrolyte films, and their

Hydrogel Electrolyte, Fig. 1 A PVA-based hydrogel electrolyte film

charge–discharge cycle life was about several dozens of cycles. A blend of PVA with PEO was used for preparing a hydrogel electrolyte film [11]. Its electrical conductivity was around 10^{-2} S cm^{-1} and used for an electric double-layer capacitor (EDLC) application. A hydrogel electrolyte prepared by blending PVA and carboxymethyl cellulose exhibited electrical conductivity of 10^{-2} S cm^{-1} and are used for Ni-MH battery application [12]. The electrical conductivity of the PVA-based hydrogel electrolyte was improved by adding bentonite [13].

A new PVA-based hydrogel electrolyte film in which PVA was cross-linked with glutaraldehyde was prepared (Fig. 1). A great amount of 4 M sulfuric acid was absorbed and held in the hydrogel electrolyte, and it showed high conductivity of 0.6 S cm^{-1} [14] which was comparable to that of 4 M sulfuric acid solution. Moreover, an all-solid-state EDLC was assembled with the polymer hydrogel electrolyte film and exhibited good capacitor performance [14, 15].

Biopolymer-Based Hydrogel Electrolytes

Chitosan is a natural low-cost biopolymer and has high chemical, thermal, and mechanical stability and high hydrophilicity. But chitosan is almost nonconductive, but electrical conductivity was increased to 10^{-2} S cm^{-1} by cross-linking [16, 17]. The resultant cross-linked chitosan-based hydrogel electrolyte membrane was used for alkaline fuel cell applications. Gelatin-based hydrogel electrolytes with different concentrations of NaCl were prepared by cross-linking with glutaraldehyde and applied to electrochemical capacitors [18]. The electrical conductivity of the resultant hydrogel electrolytes was $10^{-3} \sim 10^{-1}$ S cm^{-1}.

Cross-Linked Potassium Polyacrylate-Based Hydrogel Electrolytes

The PEO- and PVA-based hydrogel electrolytes including KOH aqueous solutions could work as an electrolyte for all-solid-state rechargeable alkaline batteries, but their conductivity was not so high, leading to lower rate capability. Cross-linked potassium polyacrylate (PAAK) is well known to be a water-absorbing polymer. The cross-linked PAAK can absorb not only water but also electrolyte solutions such as KOH aqueous solution to some extent to form a hydrogel electrolyte. The hydrogel electrolyte, which was prepared by mixing cross-linked PAAK with 6 mol L^{-1} KOH aqueous solution, exhibited high electrical conductivity which was comparable to the 6 mol L^{-1} KOH aqueous solution (Fig. 2) [19]. In addition, the potential window of the cross-linked PAAK hydrogel electrolyte was the same as that of the KOH aqueous solution [20], suggesting that the polyacrylate backbone was stable in this potential window. All-solid-state Ni-MH battery with the cross-linked PAAK hydrogel electrolyte was quite similar to the conventional Ni-MH battery with the KOH aqueous solution in terms of discharge capacity, charge–discharge cycle durability, high-rate chargeability, and dischargeability [20, 21]. Moreover, as for self-discharge property, the former was superior to the latter [22]. The cross-linked PAAK was also used for assembling all-solid-state EDLC [23, 24].

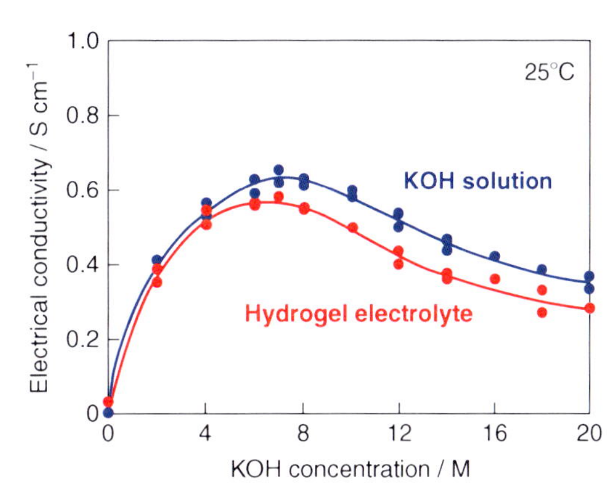

Hydrogel Electrolyte, Fig. 2 Electrical conductivity of polymer hydrogel electrolyte and 6 mol L^{-1} KOH aqueous solution at 25 °C

In terms of handling, a self-standing film is desirable for an electrolyte in electrochemical devices, but the cross-linked PAAK is insufficient in mechanical strength. For forming a self-standing hydrogel electrolyte film, the polymerization of acrylic acid monomer was directly carried out in a KOH aqueous solution containing an initiator and a cross-linking agent [22]. The resultant hydrogel electrolyte film showed conductivity of ca. 0.3 S cm^{-1}, and it was used for all-solid-state Zn-air, Zn-MnO_2, and Ni-Cd cell applications.

Inorganic Hydrogel Electrolyte and Hybrid Hydrogel Electrolyte

Hydrogel electrolytes with the inorganic polymer backbone were prepared [23]. Hydrotalcite, which is a clay with a layered double hydroxide structure, could absorb and hold a great amount of KOH aqueous solution to form a new hydrogel electrolyte. The inorganic hydrogel electrolyte showed electrical conductivity of 0.56 S cm^{-1} at 30 °C which was comparable to those of the KOH aqueous solutions. The diffusion coefficient of zinc ion in the inorganic hydrogel electrolyte was 3.1×10^{-6} cm^2 s^{-1}, which was close to those of the KOH aqueous solutions. Thus, the inorganic electrolyte hydrogel is expected to be a candidate for an electrolyte in all-solid-state rechargeable alkaline batteries.

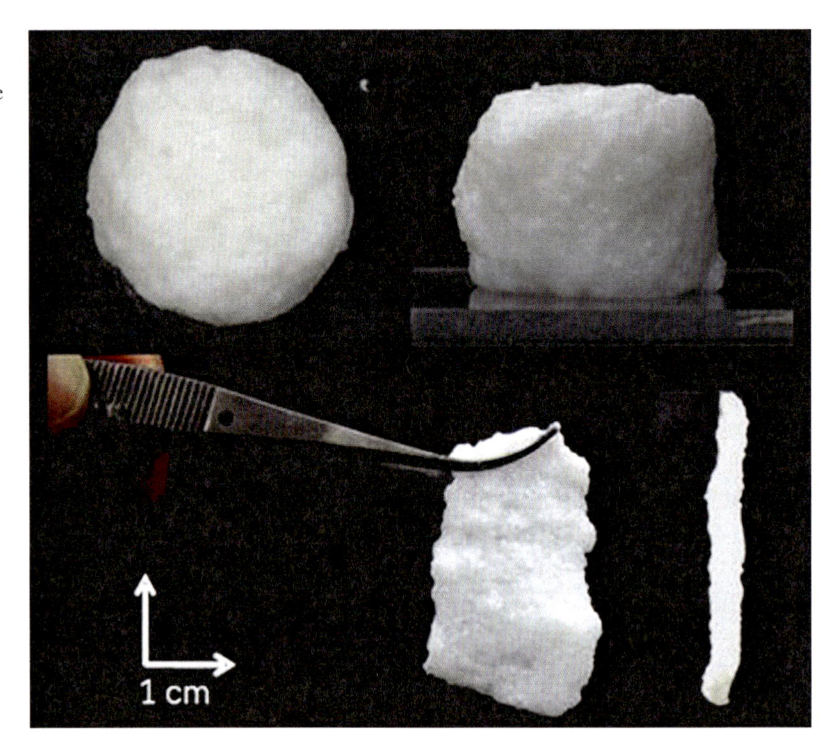

Hydrogel Electrolyte, Fig. 3 Photographs of hybrid hydrogel electrolyte and its membrane

Quite recently, organic–inorganic hybrid hydrogel electrolyte was prepared by mixing hydrotalcite, cross-linked potassium polyacrylate, and 6 M KOH solution [24]. The resultant hybrid hydrogel electrolyte also had high ionic conductivity (0.46–0.54 S cm^{-1}) at 30 °C. The hybrid hydrogel electrolyte has mechanical strength which was high enough to form a 2–3-mm-thick self-standing membrane (Fig. 3) because of the reinforcement with hydrotalcite, suggesting that the hybrid hydrogel electrolyte can be used for assembling all-solid-state rechargeable alkaline batteries.

Future Directions

Hydrogel electrolytes which include water-absorbable cross-linked polymers can show high electrical conductivity like liquid electrolytes. Thus they are clearly promising materials for all-solid-state rechargeable alkaline batteries. The rechargeable alkaline batteries have a peculiar safety mechanism in overcharging and overdischarging. For example, in Ni-MH batteries, overcharging oxygen was evolved at the cathode, and it moved to the anode through the electrolyte solution. And it is converted to water by the reaction with MH, which prevents the internal pressure from rising. In overdischarging hydrogen was evolved at the cathode, and it was converted to MH at the anode by the reaction with M which had been formed during discharging. If gas permeation property of the hydrogel electrolytes is improved, all-solid-state batteries should be commercialized in the near future.

Cross-References

► Electrical Double-Layer Capacitors (EDLC)
► Lithium-Ion Batteries
► Ni-Metal Hydride Batteries

References

1. Inoue H, Okuda S, Higuchi E, Nohara S (2009) Inorganic hydrogel electrolyte with liquidlike ionic conductivity. Electrochem Solid-State Lett 12:A58
2. Chiku M, Tomita S, Higuchi E, Inoue H (2011) Preparation and characterization of organic-inorganic

hybrid hydrogel electrolyte using alkaline solution. Polymers 3:1600
3. Fauvarquet JF, Guinot S, Bouzir N, Salmon E, Penneau JF (1995) Alkaline poly(ethylene oxide) solid polymer electrolytes. Application to nickel secondary batteries. Electrochim Acta 40:2449
4. Guinot S, Salmon E, Penneau JF, Fauvarque JF (1998) A new class of PEO-based SPEs: structure, conductivity and application to alkaline secondary batteries. Electrochim Acta 43:1163
5. Vassal N, Salmon E, Fauvarque JF (1999) Nickel/metal hydride secondary batteries using an alkaline solid polymer electrolyte. J Electrochem Soc 146:20
6. Vassal N, Salmon E, Fauvarque JF (2000) Electrochemical properties of an alkaline solid polymer electrolyte based on P(ECH-co-EO). Electrochim Acta 45:1527
7. Lewandowski A, Skorupska K, Malinska J (2000) Novel poly(vinyl alcohol)–KOH–H_2O alkaline polymer electrolyte. Solid State Ionics 133:265
8. Yang C-C, Lin S-J (2003) Preparation of alkaline PVA-based polymer electrolytes for Ni–MH and Zn–air batteries. J Appl Electrochem 33:777
9. Mohamad AA, Mohamed NS, Yahya MZA, Othman R, Ramesh S, Alias Y, Arof AK (2003) Ionic conductivity studies of poly(vinyl alcohol) alkaline solid polymer electrolyte and its use in nickel–zinc cells. Solid State Ionics 156:171
10. Kalpana D, Renganathan NG, Pitchumani S (2006) A new class of alkaline polymer gel electrolyte for carbon aerogel supercapacitors. J Power Sources 157:621
11. Yuan A, Zhao J (2006) Composite alkaline polymer electrolytes and its application to nickel–metal hydride batteries. Electrochim Acta 51:2454
12. Sang S, Zhang J, Wu Q, Liao Y (2007) Influences of Bentonite on conductivity of composite solid alkaline polymer electrolyte PVA-Bentonite-KOH-H_2O. Electrochim Acta 52:7315
13. Nohara S, Miura T, Iwakura C, Inoue H (2007) Electric double layer capacitor using polymer hydrogel electrolyte with 4 M H_2SO_4 aqueous solution. Electrochemistry 75:579
14. Wada H, Yoshikawa K, Furukawa N, Inoue H, Sugoh N, Iwasaki H, Iwakura C (2006) Electrochemical characteristics of new electric double layer capacitor with acidic polymer hydrogel electrolyte. J Power Sources 159:1464
15. Wan Y, Pepley B, Creber KAM, Tam Bui V, Halliop E (2006) Preliminary evaluation of an alkaline chitosan-based membrane fuel cell. J Power Sources 162:105
16. Wan Y, Creber KAM, Peppley B, Tam Bui V (2006) Chitosan-based electrolyte composite membranes: II. Mechanical properties and ionic conductivity. J Membrane Sci 284:331
17. Choudhury NA, Sampath S, Shukla AK (2008) Gelatin hydrogel electrolytes and their application to electrochemical supercapacitors. J Electrochem Soc 155:A74
18. Iwakura C, Furukawa N, Ohishi T, Sakamoto K, Nohara S, Inoue H (2001) Nickel/metal hydride cells using an alkaline polymer gel electrolyte based on potassium salt of crosslinked poly(acrylic acid). Electrochemistry 69:659
19. Iwakura C, Nohara S, Furukawa N, Inoue H (2002) The possible use of polymer gel electrolytes in nickel/metal hydride battery. Solid State Ionics 148:487
20. Iwakura K, Ikoma S, Nohara N, Furukawa H, Inoue J (2003) Charge-discharge and capacity retention characteristics of new type Ni/MH batteries using polymer hydrogel electrolyte. Electrochem Soc 150:A1623
21. Iwakura C, Ikoma K, Nohara S, Furukawa N, Inoe H (2005) Capacity retention characteristics of nickel/metal hydride batteries with polymer hydrogel electrolyte. Electrochem Solid-State Lett 8:A45
22. Nohara S, Wada H, Furukawa N, Inoue H, Morita M, Iwakura C (2003) Electrochemical characterization of new electric double layer capacitor with polymer hydrogel electrolyte. Electrochim Acta 48:749
23. Wada H, Nohara S, Furukawa N, Inoue H, Sugoh N, Iwasaki H, Morita M, Iwakura C (2004) Electrochemical characteristics of electric double layer capacitor using sulfonated polypropylene separator impregnated with polymer hydrogel electrolyte. Electrochim Acta 49:4871
24. Zhu X, Yang H, Cao Y, Ai X (2004) Preparation and electrochemical characterization of the alkaline polymer gel electrolyte polymerized from acrylic acid and KOH solution. Electrochim Acta 49:2533

Hydrogen Evolution Reaction

Nedeljko Krstajic
Faculty of Technology and Metallurgy,
University of Belgrade, Belgrade, Serbia

Introduction

Hydrogen evolution reaction (her) is one of the most frequently occurring cathodic reactions in industrial cell processes. Hydrogen is formed as a by-product in the chlor-alkali process and in the chlorate production, which are the major industrial electrochemical processes [1], but it is the desired reaction in water electrolysis. Traditional cathode materials for industrial applications have long been iron and mild steel. But, with the increasing cost of electrical power and especially with the advent of membrane technology in the chlor-alkali industry, the relative inefficiency of these cathode materials has remained one of the most outstanding factors of energy dissipation.

Hydrogen production calls for intensive improvement by working at higher temperatures with more efficient electrocatalysts, because mild steel and iron [2] do not withstand the more severe conditions of high caustic concentrations and higher temperatures. Under similar conditions, Ni is mostly used to replace above materials in spite of the fact that Ni is not more active than iron but its use is required by the need of stability. The competition of activity and stability, taking into account the economy of the whole process, dictates the application of electrode materials in industry.

With ever increasing energy costs owing to the dwindling availability of oil reserves, production and supply [3], and concerns with global warming and climate change blamed on man-made carbon dioxide (CO_2) emissions associated with fossil fuel use [4], particularly coal use [5], hydrogen has in recent years become very popular for a number of reasons: (1) it is perceived as a clean fuel and emits almost nothing other than water at the point of use; (2) it can be produced using any energy sources, with renewable energy being most attractive; and (3) it works with fuel cells [6–8], and together, they may serve as one of the solutions to the sustainable energy supply and use puzzle in the long run, in so-called hydrogen economy [9, 10].

In a conceptual distributed energy production, conversion, storage, and use system for remote communities, water electrolysis may play an important role in this system as it produces hydrogen using renewable energy as a fuel gas for heating applications and as an energy storage mechanism [11]. When abundant renewable energy is available, excessive energy may be stored in the form of hydrogen by water electrolysis. The stored hydrogen can then be used in fuel cells to generate electricity or used as a fuel gas. While possessing these advantages of availability, flexibility, and high purity, to achieve widespread applications, hydrogen production using water electrolysis needs improvements in energy efficiency, safety, durability, operability, and portability and, above all, reduction in costs of installation and operation.

Numerous investigations have been carried out during the last 30 years in the direction of cathode activation and hundreds of different materials have been tested with the purpose of finding the best one.

Reaction Fundamentals

Reaction Mechanism

The general mechanism of hydrogen evolution reaction is based [12, 13] on the following three steps:

$$M + H^+ + e \rightarrow MH_{ad} \text{ (Volmer step, V)} \quad (1)$$

$$MH_{ad} + H^+ + e \rightarrow H_2 + M \text{ (Heyrovsky step, H)} \quad (2)$$

$$MH_{ad} + MH_{ad} \rightarrow H_2 + 2M \text{ (Tafel step, T)} \quad (3)$$

Reaction starts with proton discharge electrosorption (Volmer reaction, Eq. 1), and follows either or both electrodesorption step (Heyrovsky reaction, Eq. 2), and/or H_{ads} recombination step (Tafel reaction, Eq. 3). If the steps are consecutive and one is the rate determining step (*rds*), theory predicts various values for Tafel slopes which are summarized in Table 1.

The nature of intermediate has been rarely disputed. The suggestion [14] that reaction (2) might involve $(H_2^+)_{ad}$, has not received any experimental confirmation. The possibility that the various steps proceed at comparable rates [15], rather than with a single *rds*, has also been suggested and theoretical calculated values of Tafel slopes for these cases are also presented in Table 1. Hydrogen evolution with barrierless discharge – step (1) or barrierless electrochemical desorption step – (3) as the *rds* [16] is expected to occur with a Tafel slope of −60 and −30 mV, respectively. This behavior has been observed with Au [17] and Ag. It has been also observed that hydrogen evolution could occur at Ni

Hydrogen Evolution Reaction, Table 1 Theoretical values of Tafel slopes for the hydrogen evolution reaction at 25 °C

Mechanism	Rate determining step	Tafel slope, b/mV dec^{-1} Langmuir adsorption condition $Q_H \to 0$	$Q_H \to 1$	Temkin adsorption conditions
V-H	Volmer	−120	−40	−60
	Heyrovsky	−40	−120	−60
V-T	Volmer	−120	−120	−120
	Tafel	−30	∞	−60
V-T	Volmer-Tafel	−120	∞	−180
V-H	Volmer-Heyrovsky	−120	−120	−120

electrode as an activationless process [18], at highly negative potentials when its energy of activation is no longer affected by the electrode potential.

Electrocatalytic Activity

A complete theory of electrocatalysis has been developed for hydrogen evolution reaction [12] because the reaction proceeds through a limited number of steps with possibly only one type of reaction intermediate. The theory predicts that the electrocatalytic activity depends on the heat of adsorption of the intermediate on electrode surface in a way giving rise to the well-known "volcano" curve [19], and prediction has been verified experimentally (Fig. 1).

According to Fig. 1, the most active catalysts for *her* are those of the platinum group metals; the platinum group metals are also special because they are generally stable in corrosive environments.

The heat of adsorption of the intermediate remains the most straightforward parameter on which development of new cathode catalysts can be based. It has been suggested that a combination of two metals from the two branches of the volcano curve could result in enhanced activity [20], which implies that a direct correlation exists between composition and heat of adsorption. Unfortunately, predictions based on the volcano curve do not show any general validity since only a few combinations give more active materials.

Catalysts

Classes of cathode materials which have been used more often in practical applications are considered.

Metallic Alloys

Mo-based alloys either electrodeposited [21], or thermally prepared [22], or added in situ [23] have constituted the main objective of research during the past 20 years. The basic catalyst is an alloy of Ni and Mo [24, 25], consisting in addition of small amounts of Re, W V, or, as a third component, Co, Cr, and Fe. Among this group of materials, Ni-Fe [21] and Co-Mo [26] have also been investigated. The activity enhancement has been found to be only due to larger surface area because Tafel slope [27] for *her* at Ni-Mo co-deposit or bulk alloys remains the same as for pure Ni.

However, for Ni-Mo-mixed layer prepared by thermal decomposition of suitable precursors [28], appreciable synergetic effects are observed. The Tafel slope decreases to −40 mV and extends to very high current densities. Part of the activity can be related to the large surface area, but intrinsic activity is manifested by the change in the reaction mechanism.

Investigations of several co-deposits on mild steel as support have shown [29] that the electrocatalytic sequence is Ni-Mo > Raney Ni > Ni-Co > Ni-Fe > Ni-Cr > Ni.

Intermetallic Compounds

Among the intermetallics of Ni, $LaNi_5$ pressed as a coating [30] on a support is the most promising, hence the most investigated. It has been shown that this catalyst exhibited high activity in spite of a number of current interruptions. However, it has been reported that some erosion of the surface may occur with roughening and scaling, probably related to hydride formation [31].

Ewe et al. [32] were among the first to report investigations of hydrogen evolution on

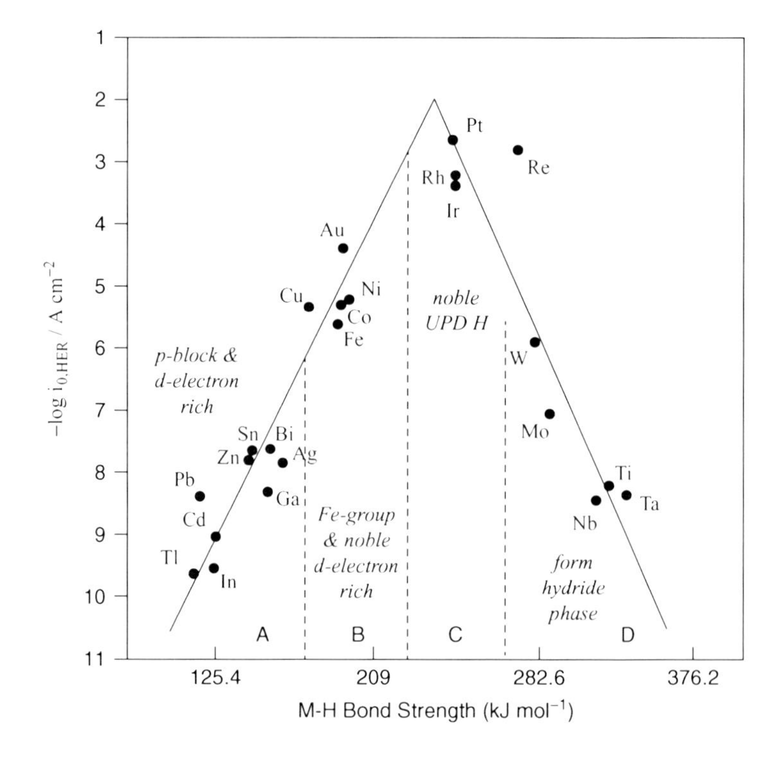

Hydrogen Evolution Reaction, Fig. 1 Volcano–shaped curve for hydrogen evolution on metals [19]

intermetallic compounds such as Ti_2Ni and TiNi in strongly alkaline solution. Later, Miles [33] tested a number of intermetallic compounds and found that TiNi exhibited the highest activity.

Amorphous Materials

Amorphous materials are usually highly corrosion resistant, and several amorphous [34] alloys have been investigated as cathode catalysts such as CuTi, CuZr, NiTi, NiZr, FeCo, and NiCo. Amorphous phases are obtained by rapidly quenching a melt, but mechanical alloying by compaction is also possible. The metallic components are usually alloyed with nonmetallic components such as B, Si, and P which stabilize the metastable noncrystalline structure.

Two general characteristics of amorphous alloys are as follows: (a) presence of inactive surface [35] oxides which impart low catalytic activity for her and (b) they usually absorb hydrogen [36] and this must be taken into account in evaluating the electrocatalytic activity. Activation is achieved by removing these surface oxides either by prepolarization or most frequently by leaching in HF.

Oxides

Several studies have discussed the mechanism of hydrogen evolution on passive metals or bulk semiconducting oxides. Some of them such as RuO_2 [37] and IrO_2 [38] exhibit high catalytic activity. It has been shown that $RuO_2$and also IrO_2 electrodes are among the most active materials for hydrogen evolution. Although thermodynamics predicts that RuO_2 should not be stable under hydrogen evolution, no bulk reduction [39] can be detected.

Oxide electrodes have been observed to be almost immune from poisoning effects due to traces of metallic impurities in solution. This is primarily due to the extended surface area but also in a different mechanism of electrodeposition. Hemisorption on wet oxides is weak due [40] to presence of chemisorbed OH groups.

Raney Nickel

In order to enhance the apparent current density of Ni cathodes, several surface treatments have been proposed to increase the surface area. Sandblasting is the common operation also prior to proceeding to the further activation of cathodes. Sintered, porous, or Teflon-bonded Ni electrodes [41] have been proposed. A mere increase in surface area is observed without any change in Tafel slope. A decisive modification of the kinetic pattern is obtained as Raney Ni is used [42]. Raney Ni is obtained by alloying Ni with a component (Al or Zn) which is then leached away in caustic solution. The dissolution of the soluble component leaves a porous structure with a very large surface area and particularly active metal sites. However, under long-term cathodic load, the structure of the electrocatalyst may change with extensive sintering and recrystallization phenomena [43] leading to progressive deactivation.

Valve metals are added to Raney Ni [44] in order to stabilize the catalyst structure, thus decreasing the recrystallization and sintering which take place as the solution temperature is raised.

Composite Cathodes

A new class of cathodes has been developed by Tavares and Trasatti [45] and Iwakura et al. [46] for hydrogen evolution. In these electrodes, RuO_2 particles were co-deposited with Ni onto smooth or rough Ni supports from Ni baths of different compositions, expediently named Watts, chloride, and thiosulfate baths. The cathodes were electrochemically characterized, and the results showed that the enhanced electrocatalytic activity of the cathodes was mainly ascribable to the increase of active sites or the content of RuO_2 particles. The performance of these new developed electrodes was evaluated for the use in a chlor-alkali electrolysis in a laboratory and industrial scale.

It has been recently shown [47] that composite Ni-MoO_2 electrode under conditions of industrial application in the "zero-gap" membrane cells for the chlorine production exhibited the similar activity for *her*, as commercial Ni-RuO_2 electrode.

Cross-References

- ▶ Electrocatalysis, Fundamentals - Electron Transfer Process; Current-Potential Relationship; Volcano Plots
- ▶ Electrocatalysis of Anodic Reactions

References

1. Sconce JS (1962) Chlorine, its manufacture, properties and uses. Reinhold Publishing Corporation/Chapman & Hall, New York/London
2. Trasatti S (1992) Electrolysis of hydrogen evolution: progress in cathode activation. In: Gerischer H, Tobias CW (eds) Advances in electrochemical science and engineering. Wiley-VCH, Weinheim, pp 2–85
3. Bockris JOM, Conway BE, Yeager E, White RE (1981) Comprehensive treatise of electrochemistry. Plenum Press, New York
4. Turner JA (1999) A realizable renewable energy future. Science 285:687–894
5. Mueller-Langer F, Tzimas E, Kaltschmitt M, Peteves S (2007) Techno-economic assessment of hydrogen production processes for the hydrogen economy for the short and medium term. Int J Hydrogen Energy 32:3797–3810
6. Grigoriev SA, Porembsky VI, Fateev VN (2006) Pure hydrogen production by PEM electrolysis for hydrogen energy. Int J Hydrogen Energy 31:171–175
7. Granovskii M, Dincer I, Rosen MA (2006) Environmental and economic aspects of hydrogen production and utilization in fuel cell vehicles. J Power Sources 157:411–421
8. Kreuter W, Hofmann H (1998) Electrolysis: the important energy transformer in a world of sustainable energy. Int J Hydrogen Energy 23:661–666
9. Bockris JOM (2002) The origin of ideas on a hydrogen economy and its solution to the decay of the environment. Int J Hydrogen Energy 27:731–740
10. Bockris JOM, Veziroglu TN (2007) Estimates of the price of hydrogen as a medium for wind and solar sources. Int J Hydrogen Energy 32:1605–1610
11. Zeng K, Zhang D (2010) Recent progress in alkaline water electrolysis for hydrogen production and applications. Prog Energy Combust Sci 36:307–326
12. Parsons R (1958) The rate of electrolytic hydrogen evolution and the heat of adsorption of hydrogen. Trans Faraday Soc 54:1053–1063
13. Bockris JOM (1954) Electrode Kinetics. In: Bockris JOM, Conway BE (eds) Modern aspects of electrochemistry. Butterworths, London, p 180
14. Horiuti J, Keii T, Hirota K (1951) Mechanism of hydrogen electrode of mercury. "Electrochemical mechanism". J Res Inst Catalysis, Hokkaido Univ 2:1–72
15. Divisek J (1986) Determination of the kinetics of hydrogen evolution by analysis of the potential-current and potential-coverage curves. J Electroanal Chem 214:615–632
16. Krishtalik LI (1969) On the conditions favourable to the detection of barrierless electrode processes. J Electroanal Chem 21:421–424
17. Parsons R, Picq G, Vennereau P (1984) Comparison between the protonation of atomic hydrogen resulting in hydrogen evolution and in the protonic trapping of

photoemitted electrons on gold electrodes. J Electroanal Chem 181:281–287
18. Krstajić N, Popović M, Grgur B, Vojnović M, Šepa D (2001) On the kinetics of the hydrogen evolution reaction on nickel in alkaline solution. Part I. The mechanism. J Electroanal Chem 512:16–26
19. Trasatti S (1972) Work function, electronegativity, and electrochemical behaviour of metals: III. Electrolytic hydrogen evolution in acid solutions. J Electroanal Chem 39:163–184
20. Srinivasan S, Salzano FJ (1977) Prospects for hydrogen production by water electrolysis to be competitive with conventional methods. Int J Hydrogen Energy 2:53–59
21. Elezović NR, Jović VD, Krstajić NN (2008) Kinetics of the hydrogen evolution reaction on Fe–Mo film deposited on mild steel support in alkaline solution. Electrochim Acta 50:5594–5601
22. Brown DE, Mahmood MN, Turner AK, Hall SM, Fogarty PO (1982) Low overvoltage electrocatalysts for hydrogen evolving electrodes. Int J Hydrogen Energy 7:405–410
23. Huot JY, Brossard L (1988) In situ activation of cobalt cathodes in alkaline water electrolysis. J Appl Electrochem 18:815–822
24. Imarisio G (1981) Progress in water electrolysis at the conclusion of the first hydrogen programme of the European communities. Int J Hydrogen Energy 6:153–158
25. Bailleux C, Damien A, Montet A (1983) Alkaline electrolysis of water-EGF activity in electrochemical engineering from 1975 to 1982. Int J Hydrogen Energy 8:529–538
26. Spasojević M, Krstajić N, Despotov P, Atanasoski R, Popov K (1984) The evolution of hydrogen on cobalt-molybdenum coating: polarization characteristics. J Appl Electrochem 14:265–266
27. Conway BE, Bai L (1985) Determination of the adsorption behaviour of 'overpotential-deposited' hydrogen-atom species in the cathodic hydrogen-evolution reaction by analysis of potential-relaxation transients. J Chem Soc Faraday Trans 81:1841–1862
28. Jakšić MM (1989) Brewer intermetallic phases as synergetic electrocatalysts for hydrogen evolution. Mater Chem Phys 22:1–26
29. Raj IA, Vasu KI (1990) Transition metal-based hydrogen electrodes in alkaline solution – electrocatalysis on nickel based binary alloy coatings. J Appl Electrochem 20:32–38
30. Machida K, Enyo M, Adachi G, Shiokawa J (1984) The hydrogen electrode reaction characteristics of thin film electrodes of Ni-based hydrogen storage. Electrochim Acta 29:807–815
31. Hall DE, Shepard VR (1984) AB5-catalyzed hydrogen evolution cathodes. Int J Hydrogen Energy 9:1005–1009
32. Ewe H, Justi EW, Stephan K (1973) Elektrochimiche Speicherung und Oxidation vonWasserstoff mit der intermetallichen Verbindung $LaNi_5$. Energy Conversion 13:109–113
33. Miles MH (1975) Evaluation of electrocatalysts for water electrolysis in alkaline solutions. J Electroanal Chem 60:89–96
34. Shibata M, Masumoto T (1987) Amorphous alloys as catalysts or catalyst precursors. In: Delmon B, Grange P, Jacobs PA, Poncelet G (eds) Preparation of catalysts IV. Elsevier, Amsterdam, p 353
35. Enyo M, Yamazaki T, Kai K, Suzuki K (1983) Amorphous Pd-Zr alloys for water electrolysis cathode materials. Electrochim Acta 28:1573–1579
36. Guojin L, Evans P, Zangari G (2003) Electrocatalytic properties of Ni-based alloys toward hydrogen evolution reaction in acid media. J Electrochem Soc 150: A551–A557
37. Galizzioli D, Tantardini F, Trasatti S (1975) Ruthenium dioxide: a new electrode material. II. Non-stoichiometry and energetics of electrode reactions in acid solutions. J Appl Electrochem 5:203–214
38. Baronetto D, Krstajić N, Trasatti S (1994) Reply to "note on a method to interrelate inner and outer electrode areas" by H. Vogt. Electrochim Acta 39:2359–2362
39. Kötz ER, Stucki S (1987) Ruthenium dioxide as a hydrogen-evolving cathode. J Appl Electrochem 17:1190–1197
40. Daghetti A, Lodi G, Trasatti S (1983) Interfacial properties of oxides used as anodes in the electrochemical technology. Mater Chem Phys 8:1–90
41. Brossard L (1988) Electrocatalytic performance of a cathode material for industrial water electrolysers. Int J Hydrogen Energy 13:315–317
42. Pshenichnikov AG, Chernyshov SF, Kryukov YI, Altentaller LI, Tumasova EI, Dudin VN (1982) Mekhanizm vydeniya vodoroda na katodakh s poverkhnostym skeletnym nikelovym katalizatorom. Elektrokhimiya 18:1011–1015
43. Korovin NV, Kumenko MV, Kozlova NI (1987) Ispolzovanie polozhitelnoi obratnoi svazi v peremennotokovoi inversionnoi voltamperometrii. Elektrokhimiya 23:408–412
44. Shalyyukhin VG, Padyukova GL, Kuanyshev AS, Fasman AB (1989) Sostoyanie povekhnosti katalizatorov reneya iz splavov Pt-Me-Al. Elektrokhimiya 25:134–137
45. Tavares AC, Trasatti S (2000) Ni + RuO_2 co-deposited electrodes for hydrogen evolution. Electrochim Acta 45:4195–4202
46. Iwakura C, Tanaka M, Nakamatsu S, Inoue H, Matsuoka M, Furukawa N (1995) Electrochemical properties of Ni(Ni + RuO_2) active cathodes for hydrogen evolution in chlor-alkali electrolysis. Electrochim Acta 40:977–982
47. Krstajić NV, Lačnjevac U, Jović BM, Mora S, Jović VD (2011) Non-noble metal composite cathodes for hydrogen evolution. Part II: the Ni-MoO_2 coatings electrodeposited from nickel chloride-ammonium chloride bath containing MoO_2 powder particles. Int J Hydrogen Energy 36(2011):6450–6461

Hydrogen Oxidation and Evolution on Platinum in Acids

Jia X. Wang
Brookhaven National Laboratory, Upton, NY, USA

Introduction

The hydrogen oxidation reaction (HOR) and the hydrogen evolution reaction (HER): $H_2(gas) \leftrightarrow 2H^+(aq) + 2e^-$ are a pair of reversible reactions frequently used for developing and illustrating the basic concepts in electrochemistry. The two reactions are highly reversible and facile on platinum (Pt). Their equilibrium potential follows the equation, $E = -\frac{RT}{2F}\ln(P_{H_2}) - \frac{2.303RT}{F}pH$, and is used as the primary reference electrode in electrochemistry. The reaction mechanism and kinetic behavior are actively discussed in recent years, partly due to the increasing need for green and renewable energy technologies.

In industrial applications, the HOR and HER reactions are involved in one of the two most promising technical routes to cut down carbon emissions and lower petroleum use in transportations [1]. As Fig. 1 illustrates, hydrogen is used as fuel in a fuel cell that converts chemical energy to electricity, through the HOR at the anode ($H_2 \rightarrow 2H^+ + 2e^-$) and via the oxygen reduction reaction (ORR) at the cathode ($O_2 + 4H^+ + 4e^- \rightarrow 2H_2O$). Inversely, in water electrolyzers, hydrogen is generated through the HER ($2H^+ + 2e^- \rightarrow H_2$) at the cathode, while the oxygen evolution reaction (OER) occurs at the anode. Combined, they function like a battery while allowing chemical energy to be stored outside of the devices. This is advantageous because refueling can be less frequent and time-consuming than battery recharging.

The reaction rates of electrocatalytic reactions often are measured by polarization curves using the rotating disk electrode (RDE) method. As Fig. 2 shows, the HER and HOR currents (blue lines) on Pt, respectively, rise sharply with potentials below and above the reversible hydrogen electrode (RHE). The HOR current reaches a maximum value above 0.05 V, where the reaction rate is determined by the slower mass-transfer rate of hydrogen, which increases linearly with the square root of electrode's rotating rate. Under the same electrode-rotation rate, the mass-transfer limiting current for the HOR is about half that for the ORR (red lines), because two electrons are transferred per H_2 molecule in the HOR rather than the four per O_2 molecule in the ORR. In kinetic studies, an overpotential (η) refers to the potential difference from where a reaction current is zero, which is the RHE for the HER and HOR. The kinetic current (j_k), defined as the reaction current at an overpotential without the mass-transport effect, often is extracted from the measured current, j, by $j_k = j/(1 - j/j_L)$ with a sufficiently large mass-transport limiting current, j_L. In the absence of reactants, the voltammetry curve (black line) shows current peaks below 0.4 V corresponding to H adsorption (negative sweep) and desorption (positive sweep) and above 0.6 V due to OH or O adsorption (positive sweep) and desorption (negative sweep).

Reaction Mechanism and Kinetic Equation

For the highly facile HER and HOR on Pt in acids, it is difficult to quantify the kinetic currents using the RDE method because of the insufficient mass-transport limiting current ($\sim$4 mA cm^{-2}) [2]. Significant progress was made in recent years. By reducing the radius of the Pt microelectrodes to the sub-micrometer region, the HOR polarization curves were obtained with the mass-transport limiting currents up to two orders of magnitude higher than that in RDE measurements [3]. As shown in Fig. 3, the kinetic current rapidly rises to 400 mA cm^{-2} (on right side axis) in about 60 mV and there is a plateau before further rising with increasing overpotential, η. This curve can be reproduced using a dual-pathway kinetic equation [4]:

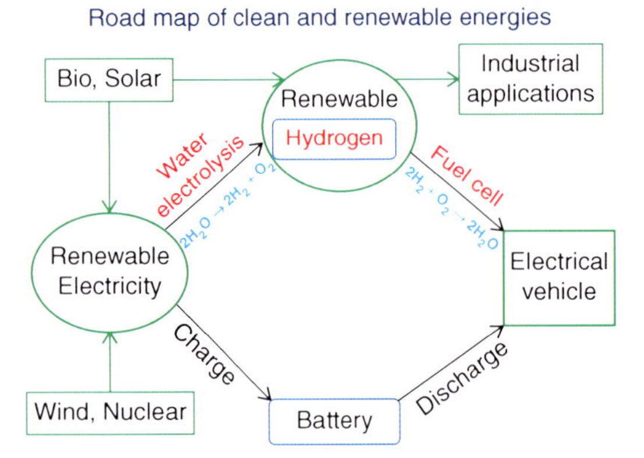

Hydrogen Oxidation and Evolution on Platinum in Acids, Fig. 1 Hydrogen electrochemical reactions in energy conversion

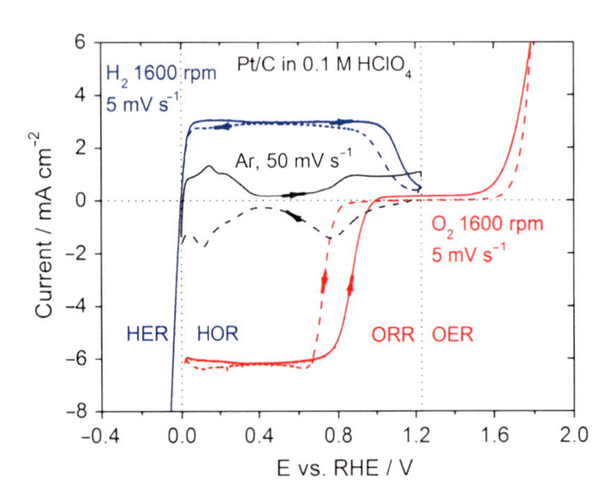

Hydrogen Oxidation and Evolution on Platinum in Acids, Fig. 2 Typical voltammetry (*black*) and polarization curves for the HER/HOR (*blue*) and ORR/OER (*red*) reactions measured, respectively, on carbon-supported Pt nanoparticles in Ar-, H_2-, and O_2-saturated 0.1 M $HClO_4$ solutions. The *vertical dotted lines* show the reversible potentials for the HER/HOR and ORR/OER

$$j_k = \left[j_{0T}\left(1 - e^{-2F\eta/\gamma RT}\right) + j_{0H}\left(e^{F\eta/2RT} - e^{-F\eta/\gamma RT} e^{-F\eta/2RT}\right)\right].$$

Based on the Tafel-Heyrovsky-Volmer mechanism [5–7], there are three elementary reaction steps for the reaction, $H_2 \leftrightarrows 2H^+ + 2e^-$, on Pt catalysts:

$$H_2 \rightleftarrows 2H_{ad} \quad \text{Tafel reaction (T)Dissociative adsorption (DA)} \tag{1}$$

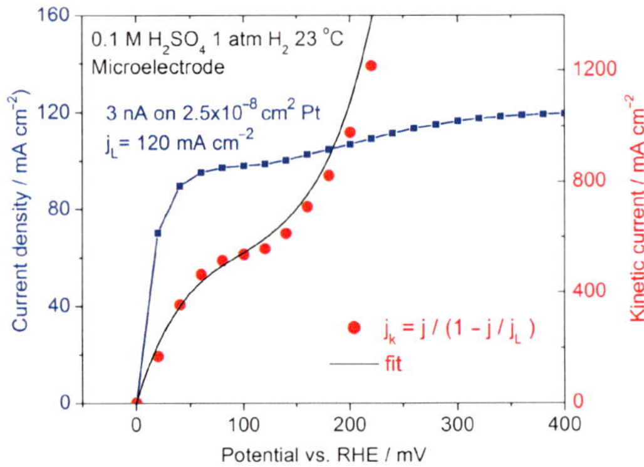

Hydrogen Oxidation and Evolution on Platinum in Acids, Fig. 3 HOR polarization curve on a single-particle microelectrode (*blue dots* and *line*) obtained by Chen and Kucernak [3] on the left axis, mass-transport-corrected kinetic current on the right axis (*red circles*) with fitted curve (*black line*) using the dual-path kinetic equation [4, 8]

$$H_2 \rightleftarrows H_{ad} + H^+ + e^- \quad \text{Heyrovsky reaction (H) Oxidative adsorption (OA)} \tag{2}$$

$$H_{ad} \rightleftarrows H^+ + e^- \quad \text{Volmer reaction (V) Oxidative desorption (OD)} \tag{3}$$

In the Tafel-Volmer (TV) pathway, the dissociative adsorption of a hydrogen molecule occurs without electron transfer and is followed by two separate one-electron oxidations of the H adatoms. By contrast, in the Heyrovsky-Volmer (HV) pathway, chemisorption and one-electron oxidation occur simultaneously, followed by another one-electron oxidation of the H adatom.

Furthermore, a dual-path kinetic equation was derived using the adsorption and activation free energies as the intrinsic kinetic parameters, wherein the adsorption isotherm, $\theta(\eta)$, was expressed using the exact solution for the stead-state rate equations [8]. The results demonstrated how the adsorption isotherm is determined by the standard ($\eta = 0$) free energies of adsorption (ΔG^0_{ad}) and activations (ΔG^{*0}_{DA}, ΔG^{*0}_{OA}, ΔG^{*0}_{OD}) for the elementary

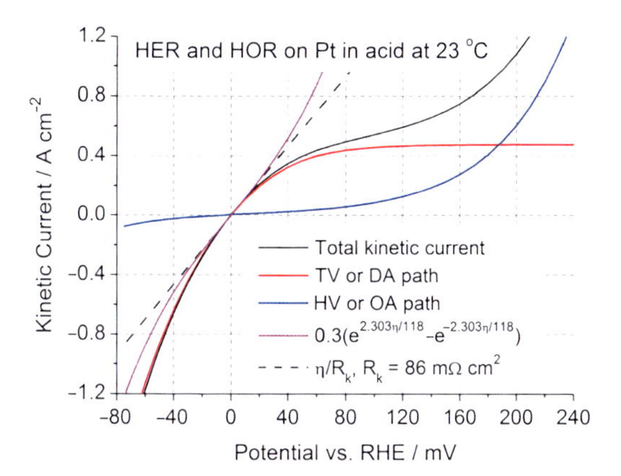

Hydrogen Oxidation and Evolution on Platinum in Acids, Fig. 4 Calculated kinetic current, with the contributions from two reaction pathways along with the exponential and linear approximations near 0 V

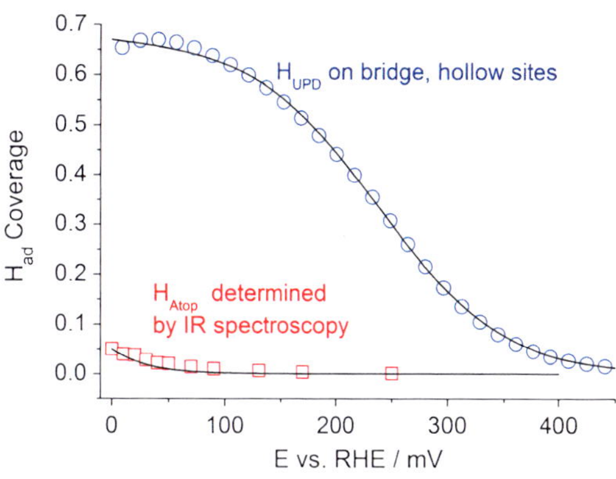

Hydrogen Oxidation and Evolution on Platinum in Acids, Fig. 5 Adsorption isotherms of H on Pt/C in 0.1 M $HClO_4$ obtained from electrochemical (*circle*) and IR (*square*) measurements

reactions (Eqs. 1–3), and how the $\theta(\eta)$ affects $j_k(\eta)$. By using the exact solution for the adsorption isotherm and free energies as the parameters, the kinetic equation applies for both the HOR and HER. As Fig. 4 shows, the Tafel pathway (or the dissociative adsorption pathway) is responsible for the high activities on Pt at small overpotentials ($-40 < \eta < 40$ mV) for both the HER and HOR (red); in contrast, the contribution from the Heyrovsky pathway (blue) becomes significant only at $\eta > 100$ mV.

Within ±10 mV, the HER and HOR kinetic current can be expressed approximately with an exchange current of 0.3 A cm^{-2} and a Tafel slope of 118 mV/decade (purple line) or by a kinetic resistance of 86 mΩ cm^2 (dashed line). The value of exchange current is two orders of magnitude higher than that of the few mA cm^{-2} measured by the RDE method for single crystals [2], polycrystals [9], and nanoparticles [10] of Pt, demonstrating a need of caution in using RDE method to compare HER and HOR activities of different catalysts. More in-depth understanding of the fast reaction rate on Pt was obtained by connecting the experimentally measured adsorption isotherm of the reaction intermediate and the free energies of adsorption calculated using the density functional theory (DFT).

H Adsorption and the Origin of High HOR and HER Activities on Pt

For hydrogen reactions on Pt, two types of H adatoms differing in adsorption sites and adsorption energies firstly were recognized as underpotentially and overpotentially deposited hydrogen, i.e., H_{UPD} and H_{OPD} [11, 12]. Infrared (IR) spectroscopy confirmed that the active HOR intermediate adsorbs at atop sites [13] as does that of the HER intermediate found earlier [14]. In contrast, H_{UPD} resides in hollow/bridge sites, embedded in the Pt surface's lattice, as inferred from IR and ultraviolet–visible reflectance measurements [11]. Furthermore, the continuing hydrogen evolution, even when manifold sites are completely blocked by chemisorbed S, was observed, supporting that H_{UPD} is not a precursor for HER and HOR [15]. This distinction is confirmed by the consistency between the IR-determined adsorption isotherm for H_{Atop} (Fig. 5) and that deduced from kinetic analysis for the reaction intermediate; both showed very low coverage at $\eta = 0$, and nearly vanished above 50 mV [8].

While the H_{UPD} at hollow or bridge sites is not much involved in the HOR, their lateral repulsion significantly weakens the adsorption of reaction intermediate at atop sites, as demonstrated by the DFT calculated adsorption energy as a function of H coverage [8]. This effect facilitates the

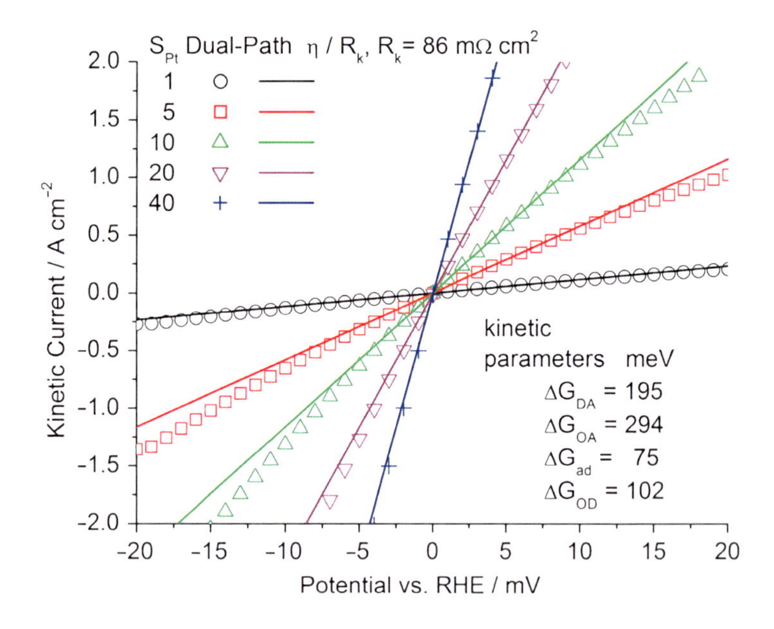

Hydrogen Oxidation and Evolution on Platinum in Acids, Fig. 6 Polarization curves with increasing Pt area ratio, S_{Pt}, (*symbols*) calculated using the dual-path kinetic equation (parameters determined from microelectrode polarization curve in Fig. 3) and their corresponding linear approximation, $j = S_{Pt}/R_k\ \eta$ (*lines*)

desorption of the reaction intermediate, H_{Atop}, so contributing to the sharply rising current near the equilibrium potential. In comparing HER and HOR activities on different metals, the exchange currents were plotted as a function of the free energy of hydrogen adsorption, which shows a volcano curve with the Pt on the top [16]. While this plot well explains the trend for metals that bind hydrogen too strongly or too weakly, predicting the maximum activity is difficult, because other factors may have a large impact when the activation barrier is minimized. The exceptionally high activity on Pt greatly benefits from hydrogen dissociative adsorption on atop sites that the more strongly bounded H_{UPD} does not block.

Nanocatalysts for Applications in Fuel Cells and Electrolyzers

For practical applications, a major goal of research is to minimize the overpotentials for HOR and HER at current density up to 2 A cm^{-2} in proton exchange membrane (PEM) fuel cells and water electrolyzer. An effective way to do it is to increase the ratio of Pt surface area to electrode area, S_{Pt}. Assuming full utilization of Pt nanocatalysts, the current density is proportional to S_{Pt} and thus, the overpotential for reaching 2 A cm^{-2} current decreases with increasing S_{Pt}. As shown in Fig. 6, the linear approximation holds for high current densities when increased S_{Pt} reduces the overpotential range. At 23 °C, the HER and HOR overpotential can be negligibly small (<5 mV) at 2 A cm^{-2} with $S_{Pt} = 40$. For 2.5-nm Pt particles having about 1 cm^2 μg^{-1} specific surface area, it means that 0.04 mg cm^{-2} loading is sufficient. Operating at 50 °C or 80 °C may lower the value to 0.02 mg cm^{-2}. Experimentally, a Pt loading of 0.05 mg cm^{-2} was demonstrated to be sufficient for hydrogen PEM fuel cell operating at 80 °C [17, 18]. Similar performance has not yet been demonstrated for water electrolysis, but should be achievable.

Concluding Remarks

For fundamental studies of electrocatalysis, the HOR and HER will continue to serve as the archetypal case, where the elementary reactions and the reaction intermediates are simple and well defined. Microelectrode method contributed to determining the HOR-HER kinetic currents over a large potential region, and IR spectroscopic technique determined the adsorption isotherm of active reaction intermediate. Fitting these data with the rigorously derived kinetic equation (without assuming the adsorption isotherm) generated a quantitative kinetic description of the facile HOR and HER, with the DFT calculations providing further physical insights. These basic concepts are applicable to studying other electrocatalytic reactions.

Cross-References

- ▶ Electrocatalysis - Basic Concepts, Theoretical Treatments in Electrocatalysis via DFT-Based Simulations
- ▶ Electrocatalysis of Anodic Reactions
- ▶ Electrocatalysis, Fundamentals - Electron Transfer Process; Current-Potential Relationship; Volcano Plots
- ▶ Platinum-Based Anode Catalysts for Polymer Electrolyte Fuel Cells

References

1. Service RF (2009) Transportation research hydrogen cars: fad or the future? Science 324:1257–1259
2. Markovic NM, Grgur BN, Ross PN (1997) Temperature-dependent hydrogen electrochemistry on platinum low-index single-crystal surfaces in acid solutions. J Phys Chem B 101:5405–5413
3. Chen SL, Kucernak A (2004) Electrocatalysis under conditions of high mass transport: investigation of hydrogen oxidation on single submicron Pt particles supported on carbon. J Phys Chem B 108:13984–13994
4. Wang JX, Springer TE, Adzic RR (2006) Dual-pathway kinetic equation for the hydrogen oxidation reaction on Pt electrodes. J Electrochem Soc 153: A1732–A1740
5. Tafel J (1905) The polarisation of cathodic hydrogen development. Z Phys Chem Stoch Verwandt 50:641–712
6. Heyrovsky J (1927) A theory of overpotential. Recl Trav Chim Des Pays-Bas 46:582–585
7. Erdey-Gruz T, Volmer M (1930) The theory of hydrogen high tension. Z Phys Chem Abt Chem Thermodyn Kin Elektrochem Eig 150:203–213
8. Wang JX, Springer TE, Liu P et al (2007) Hydrogen oxidation reaction on Pt in acidic media: adsorption isotherm and activation free energies. J Phys Chem C 111:12425–12433
9. Maruyama J, Inaba M, Katakura K et al (1998) Influence of Nafion (R) film on the kinetics of anodic hydrogen oxidation. J Electroanal Chem 447:201–209
10. Wang JX, Brankovic SR, Zhu Y et al (2003) Kinetic characterization of PtRu fuel cell anode catalysts made by spontaneous Pt deposition on Ru nanoparticles. J Electrochem Soc 150:A1108–A1117
11. Jerkiewicz G (1998) Hydrogen sorption at/in electrodes. Prog Surf Sci 57:137–186
12. Conway BE, Jerkiewicz G (2000) Relation of energies and coverages of underpotential and overpotential deposited H at Pt and other metals to the 'volcano curve' for cathodic H-2 evolution kinetics. Electrochim Acta 45:4075–4083
13. Kunimatsu K, Uchida H, Osawa M et al (2006) In situ infrared spectroscopic and electrochemical study of hydrogen electro-oxidation on Pt electrode in sulfric acid. J Electroanal Chem 587:299–307
14. Nichols RJ, Bewick A (1988) Spectroscopic identification of the adsorbed intermediate in hydrogen evolution on platinum. J Electroanal Chem 243:445–453
15. Protopopoff E, Marcus P (1988) HER on S-chemisobed Pt(110). J Electrochem Soc 135:3073
16. Skulason E, Tripkovic V, Bjorketun ME et al (2010) Modeling the electrochemical hydrogen oxidation and evolution reactions on the basis of density functional theory calculations. J Phys Chem C 114:22374
17. Gasteiger HA, Panels JE, Yan SG (2004) Dependence of PEM fuel cell performance on catalyst loading. J Power Sources 127:162–171
18. Neyerlin KC, Gu WB, Jorne J et al (2007) Study of the exchange current density for the hydrogen oxidation and evolution reactions. J Electrochem Soc 154: B631–B635

Hydrogen Storage Materials (Solid) for Fuel Cell Vehicles

Andreas Züttel
Empa Materials Science and Technology, Hydrogen and Energy, Dübendorf, Switzerland
Faculty of Applied Science, DelftChemTech, Delft, The Netherlands

Introduction

Due to the physical properties of hydrogen, i.e., the triple point of hydrogen is at T = 13.803 K and 7.0 kPa, and the low critical temperature of hydrogen of 32 K, liquefaction by compression at room temperature is not possible. This is the main difference between hydrogen and camping gas. Camping gas can be stored under pressure <300 kPa at 20 °C as a liquid. Hydrogen does not exist in the liquid phase at room

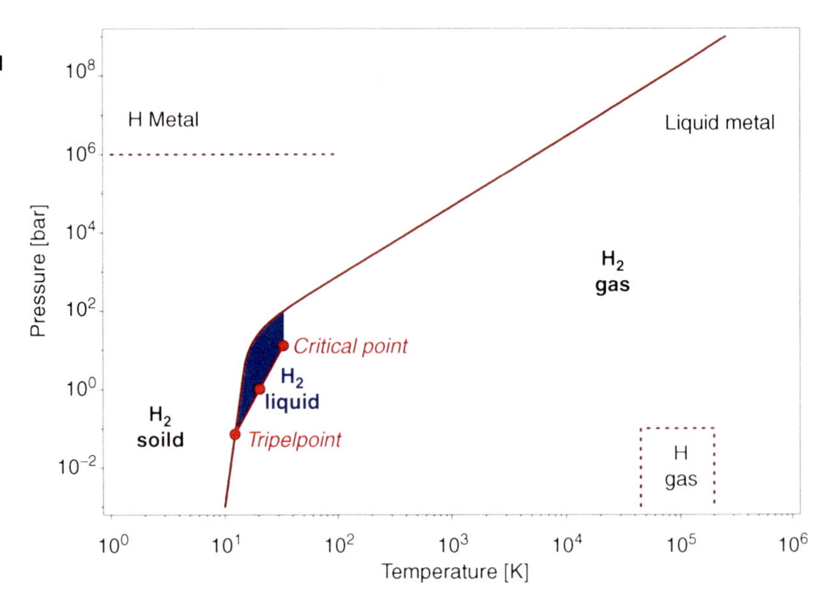

Hydrogen Storage Materials (Solid) for Fuel Cell Vehicles, Fig. 1 Primitive phase diagram for hydrogen [1]

temperature (20 °C). The density of solid and liquid hydrogen at the triple point is 86.5 kg·m^{-3} and 77.2 kg·m^{-3}, respectively. The boiling point at normal pressure (p = 101.3 kPa) is at 20.3 K, and the critical point is at T_c = 33 K and p_c = 1,293 kPa (Fig. 1).

Hydrogen storage implies the reduction of the large volume of the hydrogen gas. 1 kg of hydrogen at ambient temperature (20 °C) and atmospheric pressure (101.3 kPa) takes a volume of 11 m^3. Four parameters allow increasing the density of hydrogen:

(i) increase pressure,
(ii) lower temperature
(iii) reduction of the oscillation amplitude (dynamic volume) of the hydrogen molecules by the interaction with other materials, and
(iv) dissociating the hydrogen molecule and binding the atoms to another element.

There are basically six methods in order to reversibly store hydrogen with a high volumetric and gravimetric density. Reversibility in the context of hydrogen storage applications refers to the technical feasibility more than to a thermodynamic reversibility of the reaction. However, the energy demand to produce the hydrogen storage has to be considered in relation to the amount of stored energy. Furthermore, some storage systems are not in thermodynamic equilibrium at the storage conditions (ambient temperature and pressure), which may imply a time limitation or an additional energy demand of the storage (Fig. 2).

The volumetric hydrogen density describes the mass of hydrogen in a material or a system divided by the volume of the material or storage system $\rho_V = m_H/V$ [kg · m^{-3}]. The gravimetric hydrogen density describes the ratio of the mass of hydrogen to the mass of the material or storage system $\rho_m = m_H/m$ [mass%].

It is essential for the quality of the application to distinguish between the volumetric and gravimetric hydrogen density of the materials and of the storage system. In case of a metal hydride, the hydrogen density of the system is approximately 50 % of the hydrogen density of the metal hydride itself. Furthermore, the energy density of the storage system defines the range of the vehicle (Fig. 3).

The conventional vehicles today release [2] CO_2 [g/km] = 0.15 m[kg] – 25 and CO_2 [g/km] = 0.113 m[kg] – 13 for gasoline and diesel, respectively. Based on the CO_2 emission, the energy consumption and finally the work (equal to the energy times the efficiency) needed per distance can be estimated and is W[Wh/km] = 0.2 ·m[kg] – 20. Based on the work needed per distance, the maximum driving range (s_{max}) of a vehicle is estimated as a function of the energy density in the storage:

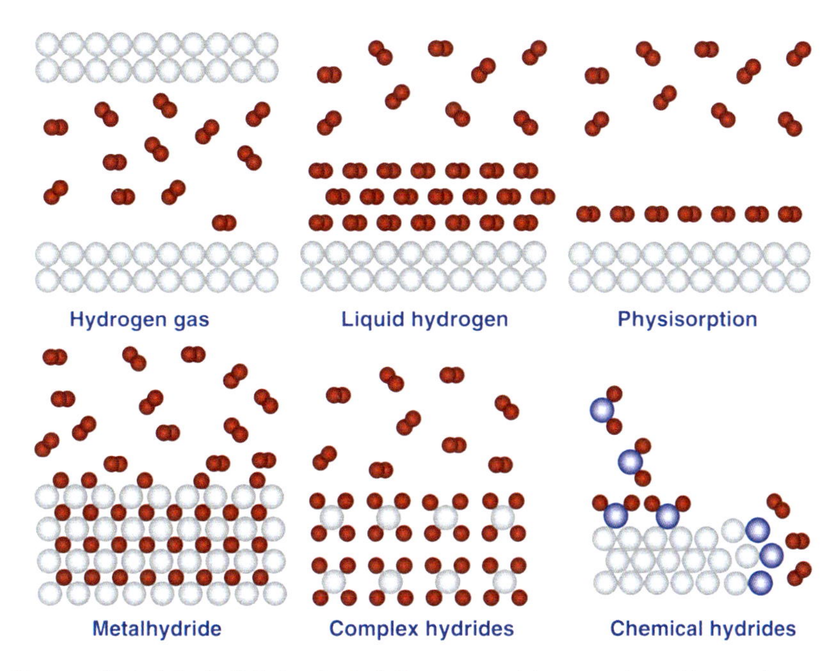

Hydrogen Storage Materials (Solid) for Fuel Cell Vehicles, Fig. 2 The six basic hydrogen storage methods and phenomena. The gravimetric density ρ_m, the volumetric density ρ_v, the working temperature T, and pressure p are listed. RT stands for room temperature (25 °C). From *top* to *bottom*: compressed gas (molecular H_2); liquid hydrogen (molecular H_2); physisorption (molecular H_2) on materials, e.g., carbon with a very large specific surface area; hydrogen (atomic H) intercalation in host metals and metallic hydrides working at RT are fully reversible; complex compounds ($[AlH_4]^-$ or $[BH_4]^-$), desorption at elevated temperature, and adsorption at high pressures; and chemical oxidation of metals with water and liberation of hydrogen

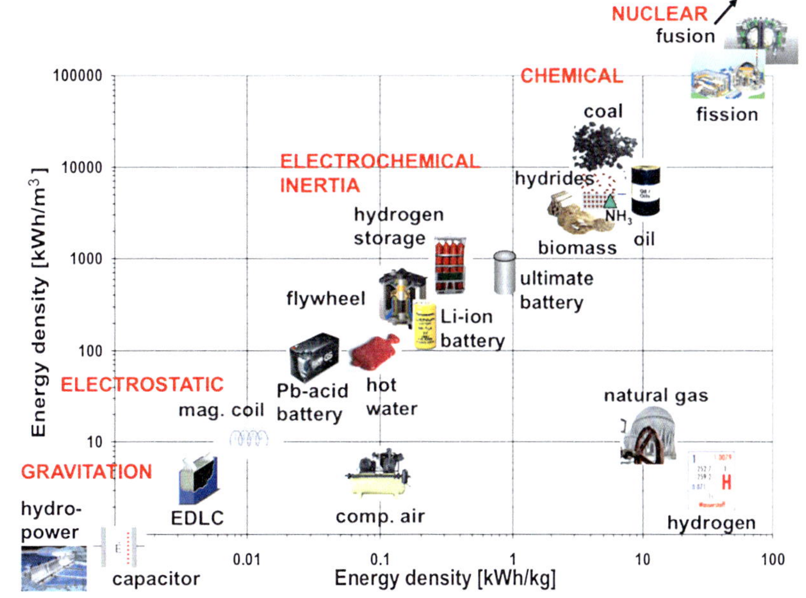

Hydrogen Storage Materials (Solid) for Fuel Cell Vehicles, Fig. 3 Volumetric versus gravimetric energy storage density on a double logarithmic scale (for visualization reasons only) for the most common storage systems used today. Close to oil and coal, but less than half of the energy density as compared to oil, we find ammonia, hydrides, and biomass

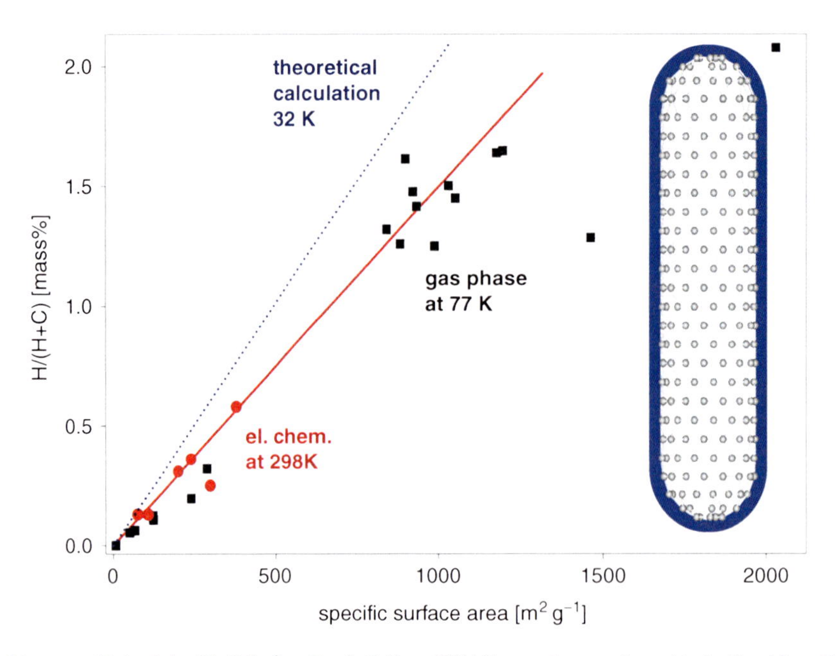

Hydrogen Storage Materials (Solid) for Fuel Cell Vehicles, Fig. 4 Reversible amount of hydrogen (electrochemical measurement at 298 K) versus the B.E.T. surface area (*round markers*) of a few carbon nanotube samples including two measurements on high-surface-area graphite (*HSAG*) samples together with the fitted line. Hydrogen gas adsorption measurements at 77 K from Nijkamp et al. [6] (*square markers*) are included. The *dotted line* represents the calculated amount of hydrogen in a monolayer at the surface of the substrate

s_{max} [km] = $C_{storage}$ [kWh/kg] · $\eta/(2 \cdot 10^{-4}$ [kWh/kg·km]), where η stands for the conversion efficiency. A car, for example, with a battery with an energy density of 0.1 kWh/kg and an efficiency of 90 % has a maximum driving range of 450 km. This is the limit and only considers the amount of work necessary and no use of thermal energy. Therefore, the energy density of a storage is the crucial parameter for any mobile application.

Physisorption

The adsorption of a gas on a surface is a consequence of the field force at the surface of the solid, called the adsorbent, which attracts the molecules of the gas or vapor, called adsorbate. In the physisorption process, a gas molecule interacts with several atoms at the surface of the solid. The potential energy of the molecule shows a minimum at a distance of approximately one molecular radius of the adsorbate. The energy minimum is of the order of 0.04–0.7 eV (4–7 kJ mol^{-1}) [3, 4]. Due to the weak interaction, a significant physisorption is only observed at low temperatures (<100 K). The amount of adsorbed hydrogen is proportional to the specific surface area of the adsorbent with $m_{ads}/S_{spec} = 2.27\ 10^{-3}$ mass % $\cdot m^{-2}$ g and can only be observed at very low temperatures [5] (Fig. 4).

Metal Hydrides

Metals, intermetallic compounds and alloys generally react with hydrogen and form mainly solid metal–hydrogen compounds. Hydrides exist as ionic, polymeric covalent, volatile covalent, and metallic hydrides [7–9].

Hydrogen reacts at elevated temperature with many transition metals and their alloys to form hydrides. The electropositive elements are the most reactive, i.e., scandium, yttrium, the lanthanides, the actinides, and the members of the titanium and vanadium groups (Fig. 5).

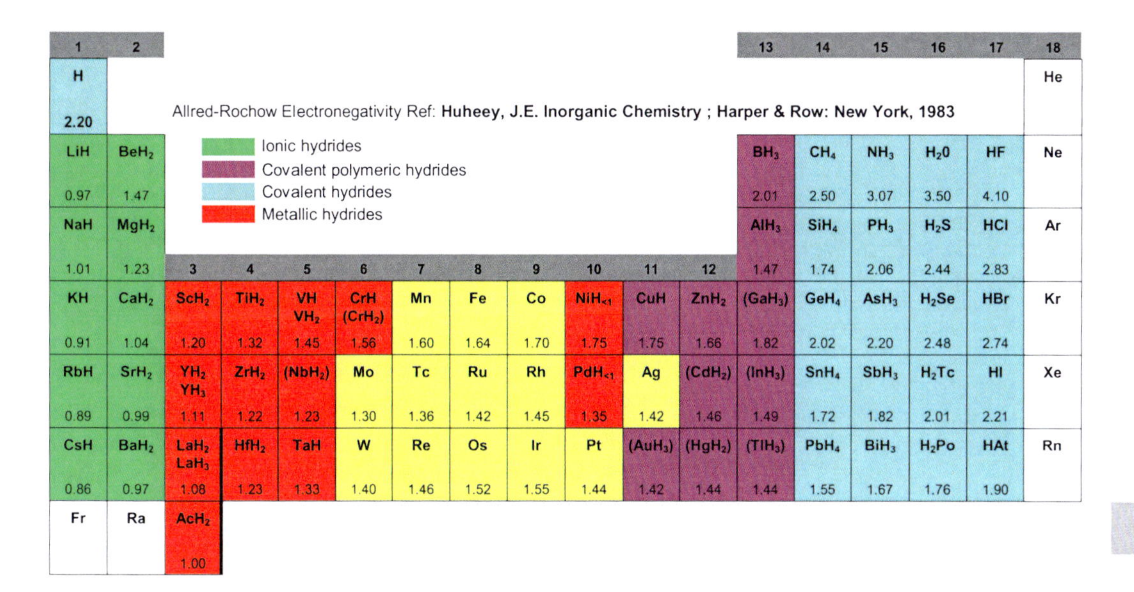

Hydrogen Storage Materials (Solid) for Fuel Cell Vehicles, Fig. 5 Table of the binary hydrides and the Allred–Rochow electronegativity [10]. Most elements react with hydrogen to form ionic, covalent, or metallic binary hydrides

The binary hydrides of the transition metals are predominantly metallic in character and are usually referred to as **metallic hydrides**. They are good conductors of electricity and possess a metallic or graphite-like appearance.

Many of these compounds (MH_n) show large deviations from ideal stoichiometry (n = 1, 2, 3) and can exist as multiphase systems. The lattice structure is that of a typical metal with atoms of hydrogen on the *interstitial sites*; for this reason they are also called interstitial hydrides. This type of structure has the limiting compositions MH, MH_2, and MH_3; the hydrogen atoms fit into *octahedral* or *tetrahedral* holes in the metal lattice, or a combination of the two types. In fcc and hcp structures, we find 1 octahedral site (r = 0.414) and 2 tetrahedral sites (r = 0.255), and in bcc (body-centered-cubic) structures we find 3 octahedral site (r = 0.155) and 6 tetrahedral sites (r = 0.291).

Especially interesting are the metallic hydrides of intermetallic compounds, in the simplest case the ternary system AB_xH_n, because the variation of the elements allows to tailor the properties of the hydrides (Table 1).

The A element is usually a rare earth or an alkaline earth metal and tends to form a stable hydride. The B element is often a transition metal and forms only unstable hydrides. Some well-defined ratios of B to A in the intermetallic compound x = 0.5, 1, 2, 5 have been found to form hydrides with a hydrogen-to-metal ratio of up to two.

The hydrogen is at small hydrogen-to-metal ratio (H/M < 0.1) exothermically dissolved (solid-solution, α-phase) in the metal. The metal lattice expands proportional to the hydrogen concentration by approximately 2 to 3Å^3 per hydrogen atom [11]. At greater hydrogen concentrations in the host metal (H/M > 0.1), a strong H–H interaction due to the lattice expansion becomes important, and the hydride phase (β-phase) nucleates and grows. The hydrogen concentration in the hydride phase is often found to be H/M = 1. The volume expansion between the coexisting α- and the β-phase corresponds in many cases to 10–20 % of the metal lattice. Therefore, at the phase boundary, large stress is built up and often leads to a decrepitation of brittle host metals such as intermetallic compounds. The final hydride is a powder with a typical particle size of 10–100 μm (Fig. 6).

Hydrogen Storage Materials (Solid) for Fuel Cell Vehicles, Table 1 The most important families of hydride forming intermetallic compounds including the prototype and the structure. A is an element with a high affinity to hydrogen, and B is an element with a low affinity to hydrogen

Intermetallic compound	Prototype	Hydrides	Structure
AB_5	$LaNi_5$	$LaNiH_6$	Haucke phases, hexagonal
AB_2	ZrV_2, $ZrMn_2$, $TiMn_2$	$ZrV_2H_{5.5}$	Laves phase, hexagonal or cubic
AB_3	$CeNi_3$, YFe_3	$CeNi_3H_4$	Hexagonal, $PuNi_3$-type
A_2B_7	Y_2Ni_7, Th_2Fe_7	$Y_2Ni_7H_3$	Hexagonal, Ce_2Ni_7-type
A_6B_{23}	Y_6Fe_{23}	$Ho_6Fe_{23}H_{12}$	Cubic, Th_6Mn_{23}-type
AB	TiFe, ZrNi	$TiFeH_2$	Cubic, CsCl- or CrB-type
A_2B	Mg_2Ni, Ti_2Ni	Mg_2NiH_4	Cubic, $MoSi_2$- or Ti_2Ni-type

The thermodynamic aspects of the hydride formation from gaseous hydrogen are described by means of pressure–composition isotherms in equilibrium (ΔG = 0). While the solid solution and hydride phase coexist, the isotherms show a flat plateau, the length of which determines the amount of H_2 stored. In the pure β-phase, the H_2 pressure rises steeply with the concentration. The two-phase region ends in a critical point T_C, above which the transition from α- to β-phase is continuous. The equilibrium pressure p_{eq} as a function of temperature is related to the changes ΔH^0 and ΔS^0 of enthalpy and entropy:

$$\ln\left(\frac{p_{eq}}{p_{eq}^{o}}\right) = \frac{\Delta H^{o}}{R}\cdot\frac{1}{T} - \frac{\Delta S^{o}}{R} \quad \text{Van't Hoff equation}$$

As the entropy change corresponds mostly to the change from molecular hydrogen gas to dissolved solid hydrogen, it amounts approximately to the standard entropy of

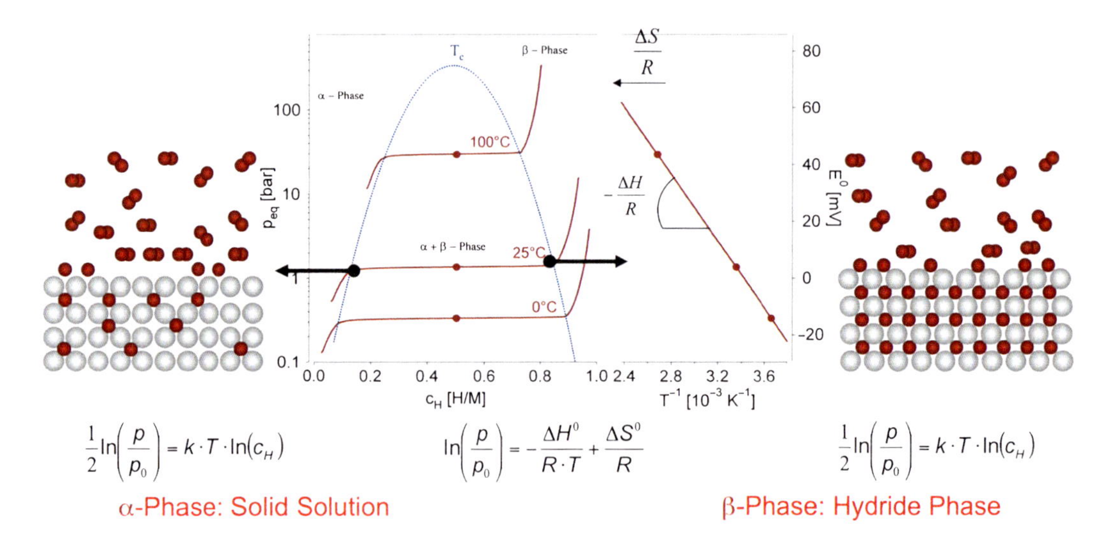

Hydrogen Storage Materials (Solid) for Fuel Cell Vehicles, Fig. 6 Pressure–composition isotherms for the hydrogen absorption in a typical metal on the left-hand side. The solid solution (α-phase), the hydride phase (β-phase), and the region of the coexistence of the two phases. The coexistence region is characterized by the flat plateau and ends at the critical temperature T_c. The construction of the van't Hoff plot is shown on the right-hand side. The slope of the line is equal to the enthalpy of formation divided by the gas constant, and the interception with the axis is equal to the entropy of formation divided by the gas constant

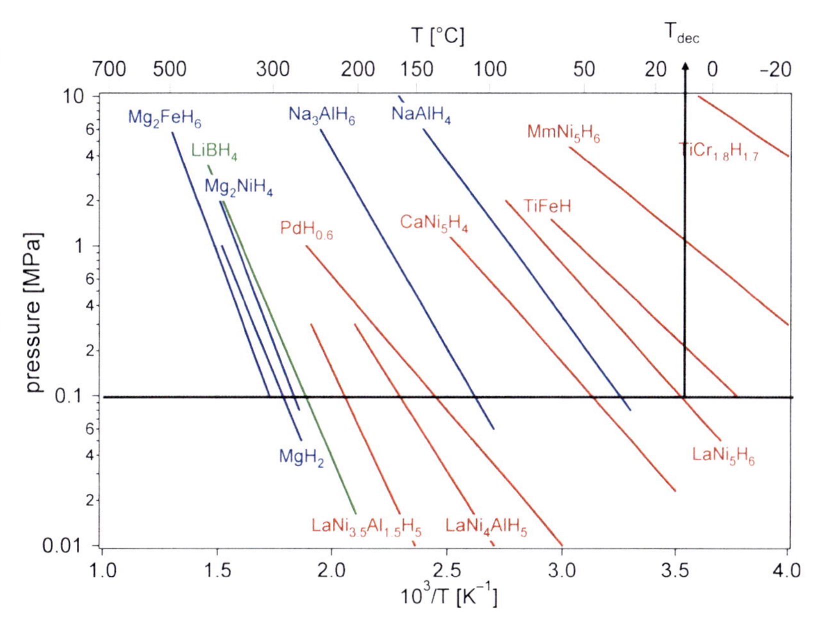

Hydrogen Storage Materials (Solid) for Fuel Cell Vehicles, Fig. 7 van't Hoff plots of some selected hydrides. The stabilization of the hydride of $LaNi_5$ by the partial substitution of nickel with aluminum in $LaNi_5$ is shown as well as the substitution of lanthanum with mischmetal (e.g., 51 % La, 33 % Ce, 12 % Nd, 4 % Pr)

hydrogen ($S^0 = 130\ J{\cdot}K^{-1}mol^{-1}$) and is therefore $\Delta S_f \approx -130\ J{\cdot}K^{-1}mol^{-1}H_2$ for all metal–hydrogen systems. The enthalpy term characterizes the stability of the metal–hydrogen bond. The decomposition temperature for $P = p^0 = 1$ bar (usually): $T_{dec} = \Delta H^0/\Delta S^0$ for $p = p^0$.

To reach an equilibrium pressure of 1 bar at 300 K, ΔH should amount to 39.2 $kJmol^{-1}H_2$. The entropy of formation term of metal hydrides (change of the hydrogen gas to a solid) leads to a significant heat evolution $\Delta Q = T{\cdot}\Delta S^0$ (exothermal reaction) during the hydrogen absorption. The same heat has to be provided to the metal hydride to desorb the hydrogen (endothermal reaction). If the hydrogen desorbs below room temperature this heat can be delivered by the environment. However, if the desorption is carried out above room temperature, the heat has to be delivered at the necessary temperature, from an external source which may be the combustion of the hydrogen. The ratio of the hydriding heat (ΔQ) to the upper heating value (ΔH_V) is approximately constant: $\Delta Q/\Delta H_V = \Delta S^0/\Delta H_V{\cdot}T = 4.6{\cdot}10^{-4}{\cdot}T$. The stability of metal hydrides is usually presented in the form of van't Hoff plots. The most stable binary hydrides have enthalpies of formation of $\Delta H_f = -226\ kJ{\cdot}mol^{-1}H_2$, e.g., HoH_2. The least stable hydrides are $FeH_{0.5}$, $NiH_{0.5}$ and $MoH_{0.5}$ with enthalpies of formation of $\Delta H_f = +20\ kJ{\cdot}mol^{-1}H_2$, $\Delta H_f = +20\ kJ{\cdot}mol^{-1}H_2$, and $\Delta H_f = +92\ kJ{\cdot}mol^{-1}H_2$, respectively [12] (Fig. 7).

Due to the phase transition upon hydrogen absorption, metal hydrides have the property of absorbing large amounts of hydrogen at a constant pressure, i.e., the pressure does not increase with the amount of hydrogen absorbed as long as the phase transition takes place. The characteristics of the hydrogen absorption and desorption can be tailored by partial substitution of the constituent elements in the host lattice. Some metal hydrides absorb and desorb hydrogen at ambient temperature and close to atmospheric pressure.

One of the most interesting features of the metallic hydrides is the extremely high volumetric density of the hydrogen atoms present in the host lattice. The highest volumetric hydrogen density know today is 150 $kg{\cdot}m^{-3}$, found in Mg_2FeH_6 and $Al(BH_4)_3$. Both hydrides belong to the complex hydrides and will be discussed in the next chapter. Metallic hydrides reach a volumetric hydrogen density of 115 $kg{\cdot}m^{-3}$ e.g., $LaNi_5$. Most metallic hydrides absorb hydrogen up to a hydrogen-to-metal ratio of H/M = 2.

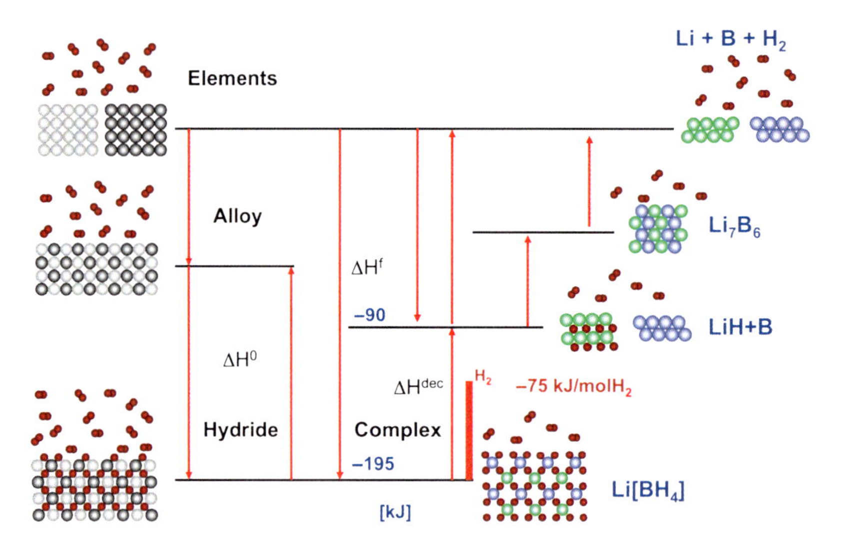

Hydrogen Storage Materials (Solid) for Fuel Cell Vehicles, Fig. 8 Comparison of the energy levels of metal hydrides and complex hydrides [17]. Metal hydrides consist of an almost unchanged metal lattice with the interstitial sites filled with hydrogen. From the elements the stable alloy or intermetallic compound is formed and from this level the hydride forms reversibly with a well-defined enthalpy of formation (ΔH^0). The complex hydrides exhibit structures where the hydrogen builds an anion with Al or B and a cation compensates the charge. Complex hydrides form several phases upon hydrogen desorption

Greater ratios up to H/M = 4.5, e.g., $BaReH_9$, have been found [13]; however, all hydrides with a hydrogen to metal ratio of more than two are ionic or covalent compounds and belong to the complex hydrides.

Complex Hydrides

All the elements of group 13 (boron group) form polymeric hydrides $(MH_3)_x$. The monomers MH_3 are strong Lewis acids and are unstable. Borane (BH_3) achieves electronic saturation by dimerization to form diborane (B_2H_6). All other hydrides in this group attain closed electron shells by polymerization. Aluminum hydride, alane $(AlH_3)_x$, has been extensively investigated [14], the hydrides of gallium indium and thallium much less so [15, 16] (Fig. 8).

The hydrogen in the p-element complex hydrides is located in the corners of a tetrahedron with boron or aluminum in the center. The bonding character and the properties of the complexes $M^+[AlH_4]^-$ and $M^+[BH_4]^-$ are largely determined by the localization of the negative charge on the central atom (Al or B), i.e., the difference in electronegativity between the cation and the aluminum or boron atom respectively. The IUPAC has recommended the names tetrahydroborate for $[BH_4]^-$ and tetrahydroaluminate for $[AlH_4]^-$ (Fig. 9).

The alkali metal tetrahydroborates are ionic, white, crystalline, high-melting solids that are sensitive to moisture but not to oxygen. Group 3 and transition metal tetrahydroborates are covalent bonded and are either liquids or sublimable solids. The alkaline earth tetrahydroborates are intermediate between ionic and covalent. The tetrahydroaluminates are very much less stable than the tetrahydroborates and therefore considerably more reactive. The difference between the stability of the tetrahydroaluminate and the tetrahydroborates is due to the different Pauling electronegativity of B and Al which is 2.04 and 1.61, respectively.

In contrast to the interstitial hydrides, where the metal lattice hosts the hydrogen atoms on interstitial sites, the desorption of the hydrogen from the complex hydride leads to a complete decomposition of the complex hydride, and a mixture of at least two phases is formed.

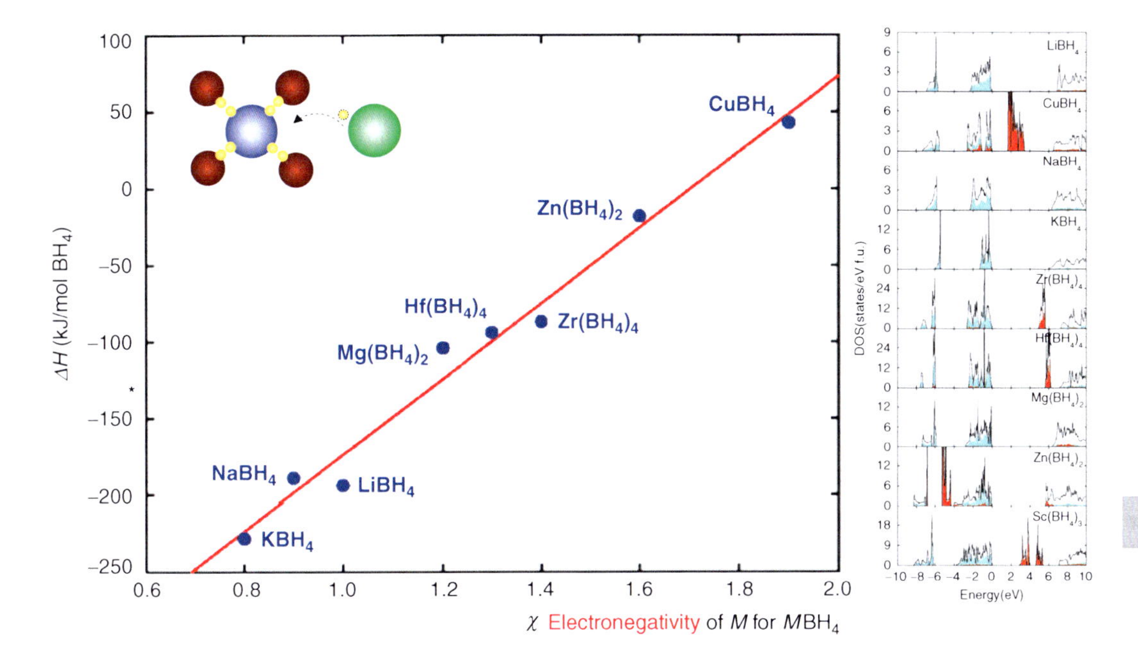

Hydrogen Storage Materials (Solid) for Fuel Cell Vehicles, Fig. 9 The enthalpy of formation of tetrahydroborates $M + nB + 2nH_2 \rightleftharpoons M[BH_4]$ as a function of the electronegativity of the cation forming atom was found based on DFT calculations by Nakamory et al. [18] to be ΔH [kJ/mol BH_4] = 247.4 · EN – 421.2

For alkali metal tetrahydroborates and tetrahydroaluminates, the decomposition reaction is generally described according to the following equations:

$$A(BH_4) \rightarrow AH + BH_3 \rightarrow AH + B + {}^3/_2H_2$$

and

$$3A(AlH_4) \rightarrow A_3AlH_6 + 2AlH_3 \rightarrow A_3AlH_6 + 2Al + 3H_2$$

$$A_3AlH_6 \rightarrow 3AH + AlH_3 \rightarrow 3AH + Al + {}^3/_2H_2$$

where the occurrence of the intermediate products depends on their stability at the thermodynamic conditions of the reaction.

The most industrially important complex hydrides are sodium tetrahydridoborate $NaBH_4$ and lithium tetrahydridoaluminate $LiAlH_4$. These are produced in tonnage quantities and used mainly as reducing agents in organic chemistry. Many other hydridoborates and hydridoaluminates are produced using these as starting materials [19].

Tetrahydroalanates

The $NaAlH_4$ and Na_3AlH_6 mixed ionic–covalent complex hydrides have been known for many years. $NaAlH_4$ consists of a Na^{1+} cation and a covalently bonded $[AlH_4]^{1-}$ complex. In the case of Na_3AlH_6, there is a related $[AlH_6]^{3-}$ complex (Fig. 10).

These alanates have been synthesized by both indirect and direct methods and used as chemical reagents [20]. However, the practical key to using the Na alanates for hydrogen storage is to be able to easily accomplish the following reversible two-step dry-gas reaction:

1. Reaction: 3 $NaAlH_4 \rightarrow Na_3AlH_6 + 2\ Al + 3\ H_2$ (3.7 wt% H)
2. Reaction: $Na_3AlH_6 \rightarrow 3NaH + Al + 3/2\ H_2$ (3.0 wt% H)

Stoichiometrically, the first step consists of 3.7 wt% H_2 release, and the second step 1.9 wt% H_2 release, for a theoretical net reaction of 5.6 wt% reversible gravimetric H storage.

Although Dymova et al. showed that the reversibility of reaction 1 was possible [21, 22],

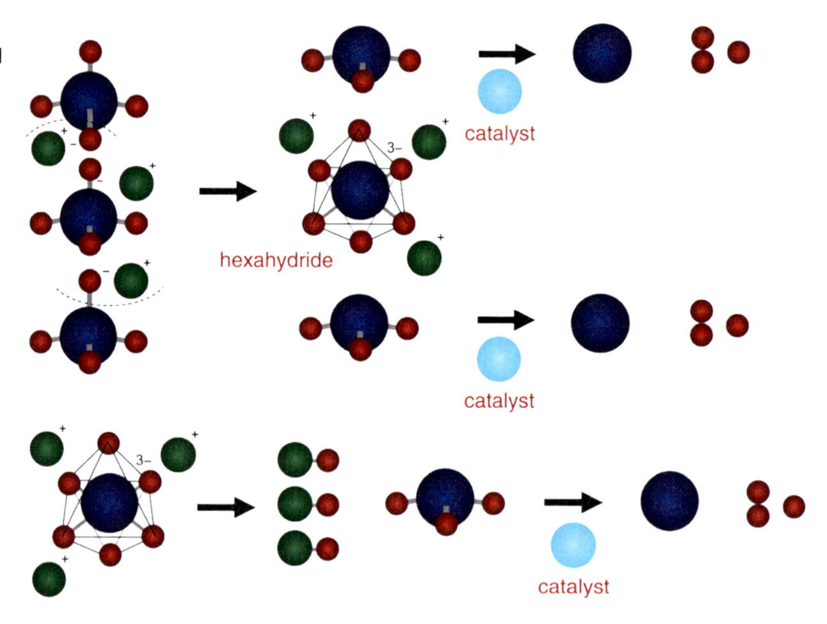

Hydrogen Storage Materials (Solid) for Fuel Cell Vehicles, Fig. 10 Schematic hydrogen desorption reaction for $NaAlH_4$

the conditions required were impractical in their severity. For example, the formation of $NaAlH_4$ from the elements required temperatures of 200–400 °C (i.e., above the 183 °C melting temperature of the tetrahydride) and H_2 pressures of 100–400 bar.

The practical use of the desorption reaction requires a catalyst for the improvement of the kinetics. The first work on catalyzed alanates by Bogdanovic at MPI – Mülheim – was derived from studies that used transition metal catalysts for the preparation of MgH_2. The $NaAlH_4$ was doped with Ti by solution chemistry techniques whereby nonaqueous liquid solutions or suspensions of $NaAlH_4$ and either $TiCl_3$ or the alkoxide Ti $(OBu\textit{n})_4$ [titanium(IV) *n*-butoxide] catalyst precursors were decomposed to precipitate solid Ti-doped $NaAlH_4$ (Fig. 11).

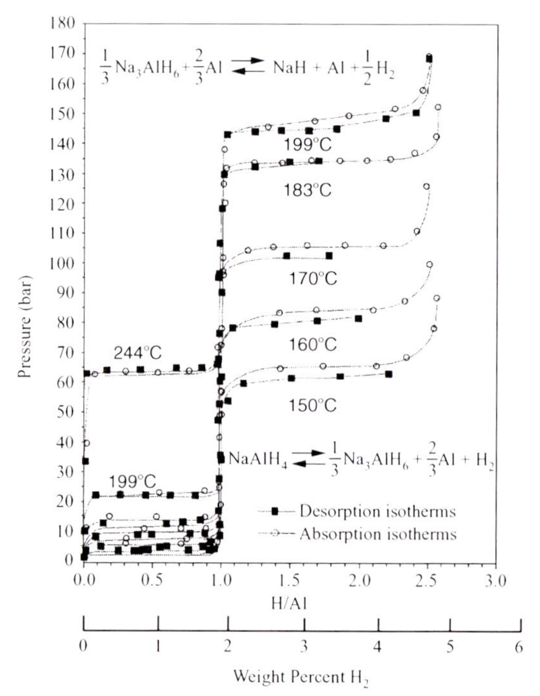

Hydrogen Storage Materials (Solid) for Fuel Cell Vehicles, Fig. 11 Pressure–composition isotherms for catalyzed $NaAlH_4$ and $Na_3AlH_{6.2}$ [23]

The thermodynamics of reaction 1 are comparable with other metallic, ionic, and covalent hydrides. $NaAlH_4$ exhibits typical low hysteresis and a two-plateau absorption and desorption isotherm.

The enthalpy changes ΔH for the $NaAlH_4$ and Na_3AlH_6 decompositions are about 37 and 47 $kJ{\cdot}mol^{-1}$ H_2, respectively. Therefore, $NaAlH_4$ is thermodynamically comparable to those of classic low-temperature hydrides, in the range useful for a near-ambient-temperature hydrogen store. Na_3AlH_6 requires about 110 °C for H_2 liberation at atmospheric pressure (Fig. 12).

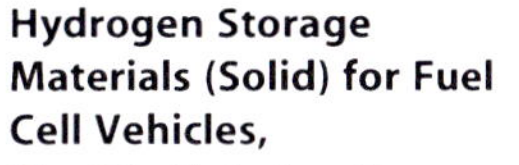

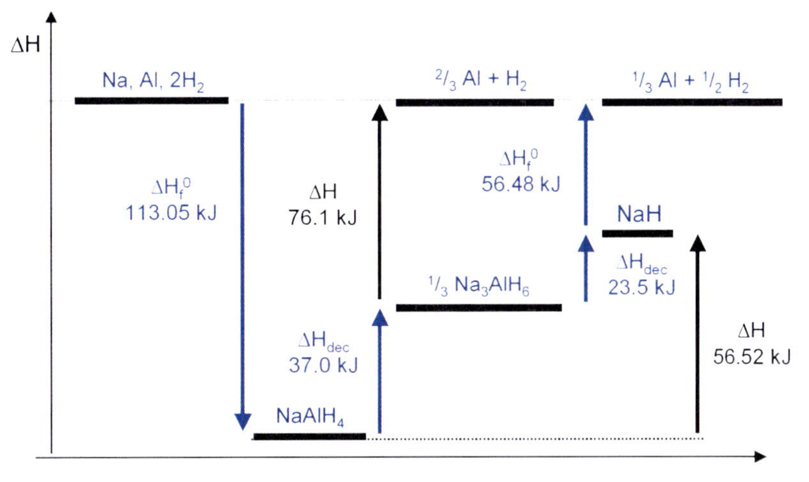

Hydrogen Storage Materials (Solid) for Fuel Cell Vehicles, Fig. 12 Enthalpy diagram for $NaAlH_4$

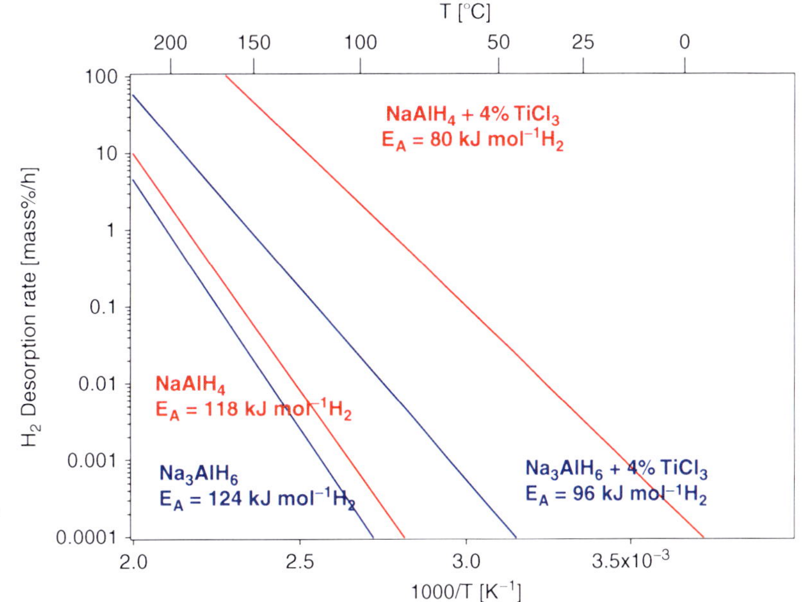

Hydrogen Storage Materials (Solid) for Fuel Cell Vehicles, Fig. 13 Arrhenius plots of the catalyzed (4 % $TiCl_3$) and pure $NaAlH_4$ and Na_3AlH_6

According to Sandrock et al. [24] both the $NaAlH_4$ and Na_3AlH_6 decomposition reactions obey thermally activated behavior consistent with the Arrhenius equation. Undoped and 4 mol% Ti-doped Na alanate samples were investigated (Fig. 13).

The history of intermetallic hydrides is deeply laced with thermodynamic tailoring by means of partial substitution of secondary and higher-order components. Li can be partially substituted for Na in $NaAlH_4$.

The alanates have numerous isostructural counterparts in the borohydrides [25] (e.g., $NaBH_4$). The borohydrides tend to be thermodynamically more stable than the alanates, yet less water reactive.

Tetrahydroborates

The first report of a pure alkali metal tetrahydroboride appeared in 1940 by

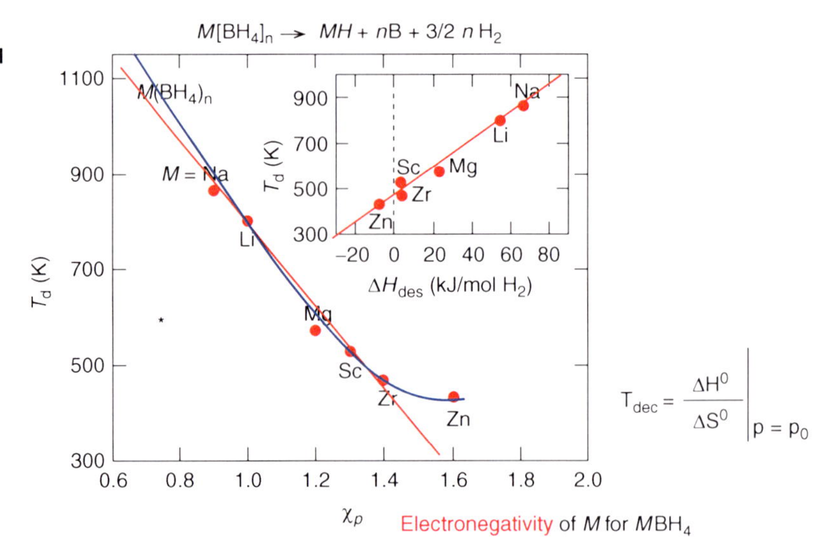

Hydrogen Storage Materials (Solid) for Fuel Cell Vehicles, Fig. 14 Temperature of decomposition of some tetrahydroborides as a function of the electronegativity [33]

Schlesinger and Brown [26] who synthesized the lithiumtetrahydroboride (lithiumborohydride) ($LiBH_4$) by the reaction of ethyllithium with diborane (B_2H_6). The direct reaction of the corresponding metal with diborane in ethereal solvents under suitable conditions produces high yields of the tetrahydroborides [27] $2\ MH + B_2H_6 \rightarrow 2\ MBH_4$ where M = Li, Na, K, etc. Direct synthesis from the metal, boron, and hydrogen at 550–700 °C and 30–150 bar H_2 has been reported to yield the lithium salt, and it has been claimed that such a method is generally applicable to groups IA and IIA metals [28]. The reaction of the elemental hydride with diborane in a ball mill leads to the tetrahydroborate with a high yield [29]. The formation of the product leads to a passivation layer; therefore, the tetrahydroborate has to be removed from the surface of the elemental hydride by dissolution in a solvent or by mechanical abrasion in a ball mill.

The stability of metal tetrahydroborides has been discussed in relation to their percentage ionic character, and those compounds with less ionic character than diborane are expected to be highly unstable [30]. Steric effects have also been suggested to be important in some compounds [31, 32]. The special feature exhibited by the covalent metal hydroborides is that the hydroboride group is bonded to the metal atom by bridging hydrogen atoms similar to the bonding in diborane, which may be regarded as the simplest of the so-called "electron-deficient" molecules. Such molecules possess fewer electrons than those apparently required to fill all the bonding orbitals, based on the criterion that a normal bonding orbital involving two atoms contains two electrons. The molecular orbital bonding scheme for diborane has been discussed extensively (Fig. 14).

The complex hydrides with the highest gravimetric hydrogen density at room temperature known today are $BeBH_4$ and $LiBH_4$ with 20.6 mass% and 18.4 mass%, respectively.

The enthalpy of formation of tetrahydroborides depends linearly on the electronegativity of the cation forming atom. The temperature of decomposition, which is the ratio of the enthalpy of decomposition and the corresponding entropy, is a parabolic function of the electronegativity of the cation. The enthalpy of decomposition, i.e., the enthalpy for the hydrogen desorption, is the difference of the enthalpy of formation and the Pauling enthalpy of the elemental hydrides, and both are functions of the electronegativity. Therefore, the enthalpy of hydrogen desorption can be calculated (Fig. 15).

The hydrogen desorption reaction involves the transfer of an H^- from the BH_4^- to the cation, leaving a BH_3 behind. Depending on the

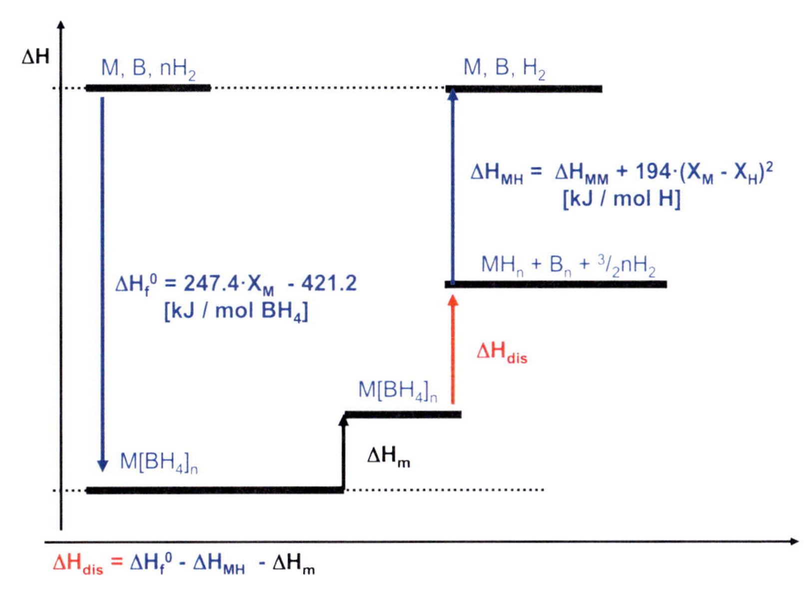

Hydrogen Storage Materials (Solid) for Fuel Cell Vehicles, Fig. 15 Enthalpy diagram for tetrahydroborates. The enthalpy for the hydrogen desorption is the difference between the enthalpy of formation and the enthalpy of the elemental hydride. In case of the desorption above the melting point, the heat of crystallization has to be considered [34]

thermodynamic conditions, i.e., temperature and hydrogen partial pressure, the BH_3 subsequently decomposes into hydrogen and boron. Alternatively, two BH_3 may form B_2H_6 and desorb into the gas phase or decompose into boron and hydrogen. Furthermore, the B_2H_6 may react with the tetrahydroboride to form higher boron–hydrogen compounds (Fig. 16).

It has been shown that the hydrogen desorption reaction is reversible and the end products lithium hydride and boron absorb hydrogen at 690 °C and 200 bar to form $LiBH_4$ [37]. Also the reaction of the elemental hydride with diborane forms the tetrahydroboride, and the reaction proceeds via an asymmetric splitting of the diborane due to the overlap of the molecular orbitals of diborane with the orbitals from the H^- at the surface of the elemental hydride. The diborane splits into an BH_4^- and a BH_3, formed with the hydrogen from the elemental hydride [38].

A large number of unstable and liquid tetrahydroborides are not yet investigated and have the potential to release large amounts (>15 mass%) of hydrogen around ambient temperature. The formation of diborane upon the hydrogen release at temperatures below 300 °C is a challenge for future applications.

Chemical Hydrides (Hydrolysis)

Hydrogen can be generated from metals and chemical compounds reacting with water. The common experiment – shown in many chemistry classes – where a piece of sodium floating on water produces hydrogen demonstrates such a process. The sodium is transformed to sodium hydroxide in this reaction. The reaction is not directly reversible but the sodium hydroxide could later be removed and reduced in a solar furnace back to metallic sodium. Two sodium atoms react with two water molecules and produce one hydrogen molecule. The hydrogen molecule produces again a water molecule in the combustion, which can be recycled to generate more hydrogen gas. However, the second water molecule necessary for the oxidation of the two sodium atoms has to be added. Therefore, sodium has a gravimetric hydrogen density of 4.4 mass%. The same process carried out with lithium leads to a gravimetric hydrogen density of 6.3 mass%. The major challenge with this storage method is the reversibility and the control of the exothermal reduction process in order to produce the metal in a solar furnace.

Complex hydrides reacted with water to hydrogen, a metal hydroxiyde and borax [39].

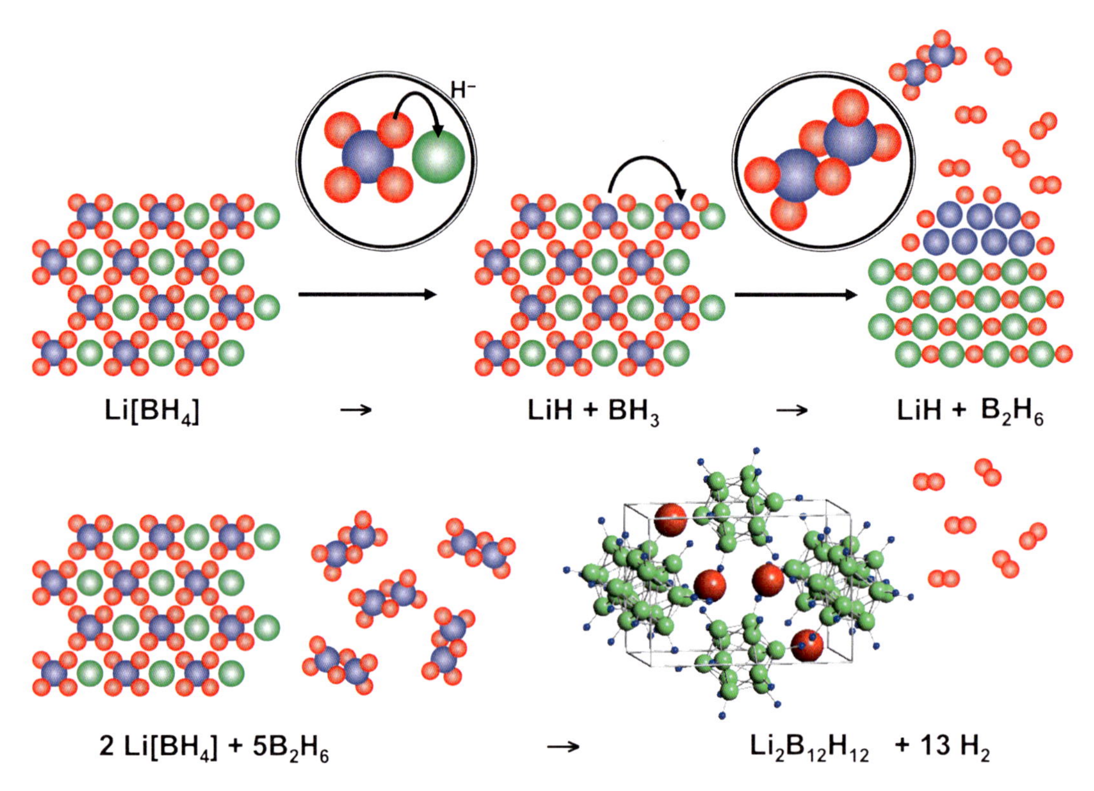

Hydrogen Storage Materials (Solid) for Fuel Cell Vehicles, Fig. 16 Hypothetical desorption mechanism for $LiBH_4$ [35]

Hydrogen Storage Materials (Solid) for Fuel Cell Vehicles, Table 2 Some reactions with chemical hydrides. The hydrogen density is calculated based on the reuse of the water from the combustion

Reaction	H mass%	H mass%
$NaAlH_4 + 4\ H_2O \rightarrow NaOH + Al(OH)_3 + 4\ H_2$	6.3	14.8
$NaBH_4 + 3\ H_2O \rightarrow NaOH + HBO_2 + 4\ H_2$	8.7	21.3
$NaBH_4 + 4\ H_2O \rightarrow NaOH + HBO_3 + 5\ H_2$	9.1	26.6
$LiBH_4 + 3\ H_2O \rightarrow LiOH + HBO_2 + 4\ H_2$	10.6	37.0
$LiBH_4 + 4\ H_2O \rightarrow LiOH + HBO_3 + 5\ H_2$	10.7	46.2

Very high hydrogen densities are reached if the water from the combustion of the hydrogen is reused (Table 2).

Hydrolysis reactions involve the oxidation reaction of chemical hydrides with water to produce hydrogen. The reaction of sodium borohydride has been the most studied to date. In the first embodiment, a slurry of an inert stabilizing liquid protects the hydride from contact with moisture and makes the hydride pumpable. At the point of use, the slurry is mixed with water and the consequent reaction produces high-purity hydrogen.

The reaction can be controlled in an aqueous medium via pH and the use of a catalyst [40]. While the material hydrogen capacity can be high and the hydrogen release kinetics fast, the borohydride regeneration reaction must take place off-board. Regeneration energy requirements, cost, and life cycle impacts are key issues currently being investigated (Fig. 17).

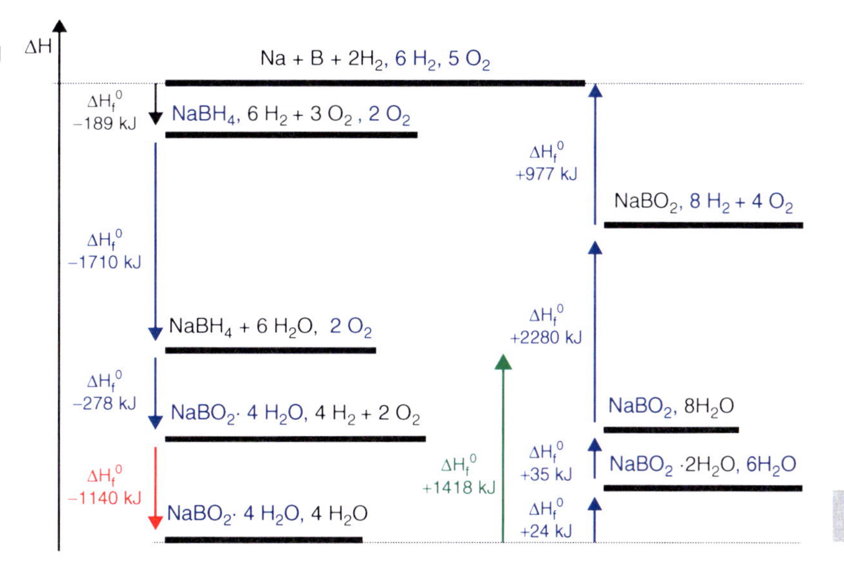

Hydrogen Storage Materials (Solid) for Fuel Cell Vehicles, Fig. 17 Energy diagram for the reaction of $NaBH_4$ with water

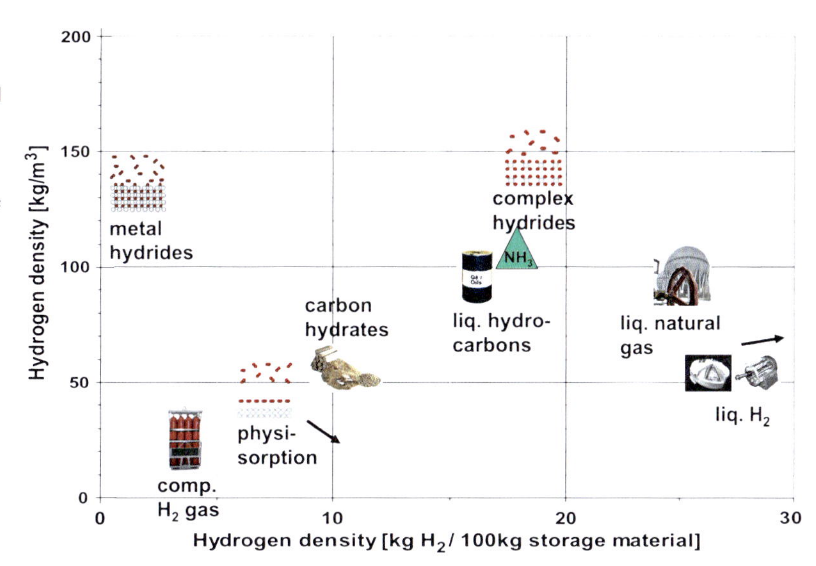

Hydrogen Storage Materials (Solid) for Fuel Cell Vehicles, Fig. 18 Max. volumetric and gravimetric hydrogen density of hydrogen storage systems. Mg_2FeH_6 shows the highest known volumetric hydrogen density of 150 kgm^{-3}, which is more than the double of liquid hydrogen. $BaReH_9$ has the largest H/M ratio of 4.5, i.e., 4.5 hydrogen atoms per metal atom. $BeBH_4$ exhibits the highest gravimetric hydrogen density of 20 mass%

Alternative Hydrogen Storage Materials

Hydrogen is stored in various ways depending on the application, e.g., mobile or stationary. Today we know about several efficient and safe ways to store hydrogen; however, there are many other new potential materials and methods possible to increase the hydrogen density significantly (Fig. 18).

The material science challenge is to better understand the electronic behavior of the interaction of hydrogen with other elements and especially metals. Unstable complex compounds like $Al(BH_4)_3$ have to be further investigated, and new compounds from the lightweight metals and hydrogen will be discovered.

New material systems have been investigated recently which are based on new effects and interactions.

Amides and Imides ($-NH_2$, $=NH$): The thermal hydrogen desorption from lithium amide ($LiNH_2$) was investigated by Chen et al. [41, 42]. The thermal desorption of pure lithium amide mainly evolves NH_3 at elevated

Hydrogen Storage Materials (Solid) for Fuel Cell Vehicles, Table 3 Combustion and explosion properties of hydrogen, methane, propane, and gasoline

	Hydrogen	Methane	Propane	Gasoline
Density of gas at standard conditions [kg/m^3(STP)]	0.084	0.65	2.42	4.4[a]
Heat of vaporization [kJ·kg^{-1}]	445.6	509.9		250–400
Lower heating value [kJ·kg^{-1}]	119.93·10^3	50.02·10^3	46.35·10^3	44.5·10^3
Higher heating value[kJ·kg^{-1}]	141.8·10^3	55.3·10^3	50.41·10^3	48·10^3
Thermal cond. of gas at stand. cond. [mW·cm^{-1} K^{-1}]	1.897	0.33	0.18	0.112
Diffusion coefficient in air at stand. cond. [cm^2·s^{-1}]	0.61	0.16	0.12	0.05
Flammability limits in air [vol%]	4.0–75	5.3–15	2.1–9.5	1–7.6
Detonability limits in air [vol%]	18.3–59	6.3–13.5		1.1–3.3
Limiting oxygen index [vol%]	5	12.1		11.6[b]
Stoichiometric composition in air [vol%]	29.53	9.48	4.03	1.76
Minimum energy for ignition in air [mJ]	0.02	0.29	0.26	0.24
Autoignition temperature [K]	858	813	760	500–744
Flame temperature in air [K]	2,318	2,148	2,385	2,470
Maximum burning velocity in air at stand. cond. [m·s^{-1}]	3.46	0.45	0.47	1.76
Detonation velocity in air at stand. cond. [km·s^{-1}]	1.48–2.15	1.4–1.64	1.85	1.4–1.7[c]
Energy[d] of explosion, mass-related [gTNT/g]	24	11	10	10
Energy[d] of explosion, volume-related [gTNT·m^3(STP)]	2.02	7.03	20.5	44.2

[a]100 kPa and 15.5 °C
[b]Average value for a mixture of C_1–C_4 and higher hydrocarbons including benzene
[c]Based on the properties of n-pentane and benzene
[d]Theoretical explosive yields

temperatures following the reaction of 2 $LiNH_2 \rightarrow Li_2NH + NH_3$. When $LiNH_2 + LiH$ is used, H_2 is released at a much lower temperature and imide is formed: $LiNH_2 + LiH \rightarrow Li_2NH + H_2$. H in $LiNH_2$ is partially positively charged, but in LiH the H is negatively charged. The redox of $H^{\delta+}$ + and $H^{\delta-}$ to H_2 and simultaneously combination of $N^{\delta-}$ and $Li^{\delta+}$ + (in LiH) are energetically favorable.

BCC Alloys: Hydrogen-absorbing alloys with bcc (body-centered-cubic) structures, such as Ti–V–Mn, Ti–V–Cr, Ti–V–Cr–Mn, and Ti–Cr–(Mo, Ru), have been developed since 1993 [43]. These alloys offer more interstitial sites than fcc and hcp structures and have a higher hydrogen capacity (about 3.0 mass%) than conventional intermetallic hydrogen-absorbing alloys.

AlH_3: The potential for using aluminum hydride, AlH_3, for vehicular hydrogen storage was explored by Sandrock [44]. It is shown that particle-size control and doping of AlH_3 with small levels of alkali metal hydrides (e.g., LiH) results in accelerated desorption rates.

Ammonia Storage: Metal ammine complexes represent a new promising solid form of storage. Using $Mg(NH_3)_6Cl_2$ as the example, it was show by Christensen [45] that it can store up to 9.1 % hydrogen by weight in the form of ammonia. Borazane: chemical hydrides with empirical formula BH_2NH_2 and $B_xN_xH_y$, have been probed [46] as high-capacity hydrogen storage materials for a long time. Ammonia borane (H_3BNH_3), the simplest stable AB, has been used in these gas generators and, more recently, in fuel cell applications. Another stable AB used in the gas generators was diborane diammoniate, $H_2B(NH_3)_2BH_4$ [47].

Safety

Hydrogen reacts, when the ignition energy (thermal activation energy) of ≈0.02 mJ is provided, violently with oxidizing agents such as oxygen (air), fluorine or chlorine, and N_2O. Combustion, deflagration, or detonation may occur depending on the conditions. The ignition and detonation properties of hydrogen–air mixtures are particularly important from the safety aspect. The flammability limits (i.e., the minimum and the maximum concentration of hydrogen in air) are exceptionally wide for hydrogen (Table 3).

The flammability limits of hydrogen in pure oxygen at room temperature are 4.65–93.9 vol.% hydrogen and those of deuterium in pure oxygen 5.0–95 vol.%. In general, the chemical properties of H, D, and T are essentially identical. Small differences can be found in processes where the atomic mass has a significant influence, e.g., the thermal dissociation of a D-X bond requires a greater energy than for the comparable H–X bond [48].

Cross-References

► Compressed and Liquid Hydrogen for Fuel Cell Vehicles
► Fuel Cell Vehicles

References

1. Leung WB, March NH, Motz H (1976) Primitive phase diagramm for hydrogen. Phys Lett 56A(6):425–426
2. Bach C (2012) Empa materials science & technology. Private communication. Switzerland
3. London F (1930) Z Physik 63:245; (1930) Z Physik Chem 11:222
4. Schlichtenmayer M, Hirscher M (2012) Nanosponges for hydrogen storage. J Mater Chem 22:10134
5. Züttel A, Sudan P, Mauron P, Kyiobaiashi T, Emmenegger C, Schlapbach L (2002) Int J Hydrogen Energy 27:203
6. Nijkamp MG, Raaymakers JEMJ, van Dillen AJ, de Jong KP (2001) Appl Phys A 72:619
7. Rittmeyer P, Wietelmann U (2012) Ullmann's encyclopedia of industrial chemistry, Chap. Hydrides. Peter Rittmeyer, Ulrich Wietelmann, 2012 Wiley-VCH, Weinheim
8. Louis S (ed) (1992) Hydrogen in intermetallic compounds II. Surface and dynamic properties, applications. Series: topics in applied physics, vol 67. Springer
9. Zuettel A, Borgschulte A, Schlapbach L (2008) Hydrogen as a future energy carrier. Wiley-VCH, Weinheim
10. Huheey JE (1983) Inorganic chemistry. Harper & Row, New York
11. Fukai Y (1989) Z Phys Chem 164:165
12. Griessen R, Riesterer T (1988) Heat of formation models. In: Schlapbach L (ed) Hydrogen in intermetallic compounds I electronic, thermodynamic, and crystallographic properties, preparation, vol 63, Series topics in applied physics. Springer, Berlin, pp 219–284
13. Yvon K (1998) Complex transition metal hydrides. Chimia 52(10):613–619
14. Fauroux JC, Teichner SJ (1966) N° 507. – Nouvelle méthode de préparation et identification du borohydrure d'aluminium; new methods of preparation and identification of Al(BH4)3. Bull de la Soc Chim Fr 9:3014–3016
15. Wiberg E, Amberger E (1971) Hydrides of the elements of main groups I–IV. Elsevier, Amsterdam
16. BD James, MGH Wallbridge (1970) Metal tetrahydroborates. Prog Inorg Chem ll:99–231
17. Miwa K, Ohba N, Towata S, Nakamori Y, Orimo S (2004) Phys Rev B 69:245120
18. Nakamori Y, Miwa K, Ninomiya A, Li H, Ohba N, Towata S-I, Züttel A, Shin-ichi O (2006) Phys Rev B 74:045126
19. Ashby EC (1966) The chemistry of complex aluminohydrides. Adv Inorg Chem Radiochem 8:283–338
20. Bogdanovic' B, Brand RA, Marjanovic' A, Schwickardi M, Tölle J (2000) J Alloys Compd 302:36
21. Dymova TN, Eliseeva NG, Bakum SI, Dergachev YM (1974) Dokl Akad Nauk SSSR 215:1369, Engl. p. 256
22. Dymova TN, Dergachev YM, Sokolov VA, Grechanaya NA (1975) Dokl Akad Nauk SSSR 224:591, Engl. p. 556
23. Bogdanovic' B, Schwickardi MJ (1997) J Alloys Compd 253:1
24. Sandrock G, Gross K, Thomas G (2002) J Alloys Compd 339:299
25. Sullivan EA (1995) Kirk-Othmer encyclopedia of chemical technology, vol 13. Wiley, New York, p 606
26. Schlesinger HJ, Brown HC (1940) J Am Chem Soc 62:3429–3435
27. Schlesinger HJ, Brown HC, Hoekstra HR, Rapp LR (1953) J Am Chem Soc 75:199–204
28. D Goerrig (1958) Ger Pat 1,077,644 (Dec. 27, 1958)
29. Friedrichs O, Buchter F, Borgschulte A, Remhof A, Zwicky CN, Mauron Ph, Bielmann M, Züttel A (2008) Direct synthesis of Li[BH_4] and Li[BD_4] from the elements. Acta Mater 56(5):949–954
30. Schrauzer GN (1955) Naturwissenschaften 42:438
31. Lippard SJ, Ucko DA (1968) Inorg Chem 7:1051
32. Lipscomb WN (1963) Boron hydrides. Benjamin, New York
33. Nakamori Y, Miwa K, Ninomiya A, Li H, Ohba N, Towata S-I, Züttel A, Shin-ichi O (2006) Phys Rev B 74:045126
34. Pauling L (1929) The principles determining the structure of complex ionic crystals. J Am Chem Soc 51:1010–1026 – pubs.acs.org
35. Friedrichs O, Remhof A, Hwang S-J, Züttel A (2010) Formation of intermediate compound Li2B12H12

during the dehydrogenation process of the LiBH4–MgH2 system. Chem Mater 22:3265–3268
36. Yigang Yan, Hai-Wen Li, Hideki Maekawa, Kazutoshi Miwa, Shin-ichi Towata, Shin-ichi Orimo (2011) Formation of intermediate compound $Li_2B_{12}H_{12}$ during the dehydrogenation process of the $LiBH_4$–MgH_2 system. J Phys Chem C 115(39):19419–19423
37. Orimo S, Nakamori Y, Kitahara G, Miwa K, Ohba N, Towata S, Züttel A (2005) Dehydriding and rehydriding reactions of LiBH4. J Alloys Comp 404–406:427–430
38. Gremaud R, Borgschulte A, Friedrichs O, Züttel A (2011) Synthesis mechanism of alkali borohydrides by heterolytic diborane splitting. J Phys Chem C 115(5):2489–2496
39. Aiellot R, Matthews MA, Reger DL, Collins JE (1998) Int J Hydrogen Energy 23(12):1103–1108
40. Li ZP, Liu BH, Zhu JK, Morigasaki N, Suda S (2007) NaBH4 formation mechanism by reaction of sodium borate with Mg and H_2. J Alloys Comp 437(1–2):311–316
41. Chen P, Xiong ZR, Luo J, Lin J, Tan KL (2003) Interaction between lithium amide and lithium hydride. J Phys Chem B 107:10967–10970
42. Chen P, Xiong ZR, Luo J, Lin J, Tan KL (2002) Interaction of hydrogen with metal nitrides and imides. Nature 420:302–304
43. Akiba E, Okada M (2002) Metallic hydrides III: body-centered-cubic solid-solution alloys. Mrs Bull 27:699–703
44. Sandrock G, Reilly J, Graetz J, Zhou W-M, Johnson J, Wegrzyn J (2005) Appl Phys A 80:687–690
45. Christensen CH, Sørensen RZ, Johannessen T, Quaade UJ, Honkala K, Elmøe TD, Køhler R, Nørskov JK (2005) J Mater Chem 7
46. Mohajeri N, Robertson T, Raissi AT (2003) Hydrogen storage in amine borane complexes. FSEC final report for task III-B
47. Grant LR, Flanagan JE (1983) Advanced solid reactants for hydrogen/deuterium generation. U.S. patent no. 4,381,206
48. Wiberg KB (1955) Chem Rev 55:713

Hypochlorite Synthesis Cells and Technology, Sea Water

Rudy Matousek
Severn Trent Services, Sugar Land, TX, USA

Introduction

Seawater (normally 15–35 g/l, 10–30 °C, and a pH around 7 or slightly higher) or other water containing NaCl may be used to generate a disinfecting solution containing chlorine by passing a direct electrical current through the solution. On-site generation of hypochlorite from seawater has been used for over 35 years. These systems can be purchased as completely skid mounted systems that generate sodium hypochlorite from seawater. These systems are used in refining, petrochemical power plants, offshore drilling production, and marine applications around the world. Systems can be scaled to the appropriate size depending on the quantity of hypochlorite required.

The type of electrolytic cell commonly used in these marine and offshore applications is a "tube within a tube", (but there are a variety of configurations.) A tube within a tube type cell consists of one anode, one cathode, and one bipolar tube with the necessary ancillary hardware to facilitate assembly-see Fig. 1. The outer anode and cathode are manufactured from seamless titanium pipe. The anode surface is coated with proprietary precious metal oxides, primarily ruthenium and iridium. Seawater enters one end of the cell and passes between the cathode, the anode and bipolar tube annular spaces. When direct current is applied to the cell, sodium hypochlorite results. One cell can produce up to 5.5 kg/day and a maximum of 12 cells can be connected in series for a capacity of 65 kg/day per train. Multiple trains can operate in parallel to produce the required capacity.

Chemistry

The process is based on the partial electrolysis of NaCl present in seawater as it flows through an unseparated electrolytic cell. The resulting solution exiting the cell is a mixture of seawater, sodium hypochlorite (hypo), hydrogen gas and Hypochlorous acid. Electrolysis of sodium chloride solution (seawater in this case) is the passage of direct current between an anode (positive pole) and a cathode (negative pole) to separate salt and water into their basic elements. Chlorine generated at the anode immediately goes through chemical reactions to form sodium hypochlorite

Hypochlorite Synthesis Cells and Technology, Sea Water, Fig. 1 Salt water hypochlorite "tube withn a tube" hypochlorite generator, SANILECR

and Hypochlorous acid. Reactions are shown below:

$$Cl^- \rightarrow Cl_2(aq) + 2e^- \quad Eo = 1.396\ V \tag{1}$$

Which is hydrolyzed in solution to form hypochlorous acid:

$$Cl_2 + 2H_2O \rightarrow 2HOCl + 2H^+ \tag{2}$$

$$HCLO \leftrightarrow OCl^- + H^+ \tag{2A}$$

In typical seawater hypochlorite cells, the solution pH typically ranges from neutral up to pH 9. Hypochlorous acid dissociates to hypochlorite at alkaline pH levels:

$$HOCl \rightarrow OCl^- + H^+ \quad pKa = 7.5 \tag{3}$$

Hydrogen and hydroxides are formed at the cathode, the hydrogen forms a gas and is vented and the hydroxide aids in the formation of sodium hypochlorite and increases the exit stream pH to approximately 8.5. This reaction is shown below:

$$2H_2O + 2e^- \rightarrow H_2(g) + 2OH^- \quad Eo = -0.828\ V \tag{4}$$

The hydrolysis of chlorine is rapid and complete at the pH value of the solution in the cell. The dissociation of hypochlorous acid is not complete; the relationship between pH and the degree of dissociation is shown in Fig. 2. Therefore the solution produced from a typical seawater cell contains a mixture of hypochlorite ions CLO^- and Hypochlorous acid HOCl and the ratio between the two depends on the pH of the solution and temperature.

The above reactions describe the ideal functions of a typical unseparated hypochlorite cell. There are, however, several competing reactions which take place in the cell and contribute to cell inefficiencies.

Chlorate Formation

Chlorate may be formed by either anodic oxidation of hypochlorite as shown in Eq. 5 or by the chemical reaction between hypochlorite and hypochlorous acid as shown in Eq. 6.

$$6ClO^- + 3H2O \rightarrow 2ClO3^- + 4Cl^- + 6H^+ + 3/2O2 + 6e^- \tag{5}$$

$$2HClO + ClO^- \rightarrow ClO_3^- + 2Cl2 + 2H^+ \tag{6}$$

The chemical chlorate formation (Eq. 6) proceeds very slowly at temperatures below 40 °C and basic pH. Therefore this reaction is not significant since the cells do not operate under these conditions.

Oxygen Formation

Some current efficiency is lost due to anodic discharge of hydroxyl ions to form oxygen.

$$2OH^- \rightarrow 2H^+ + O_2 + 4e^- \tag{7}$$

This reaction competes with anodic chlorine evolution since the feed to hypochlorite cells contains only 10–30 gpl NaCl.

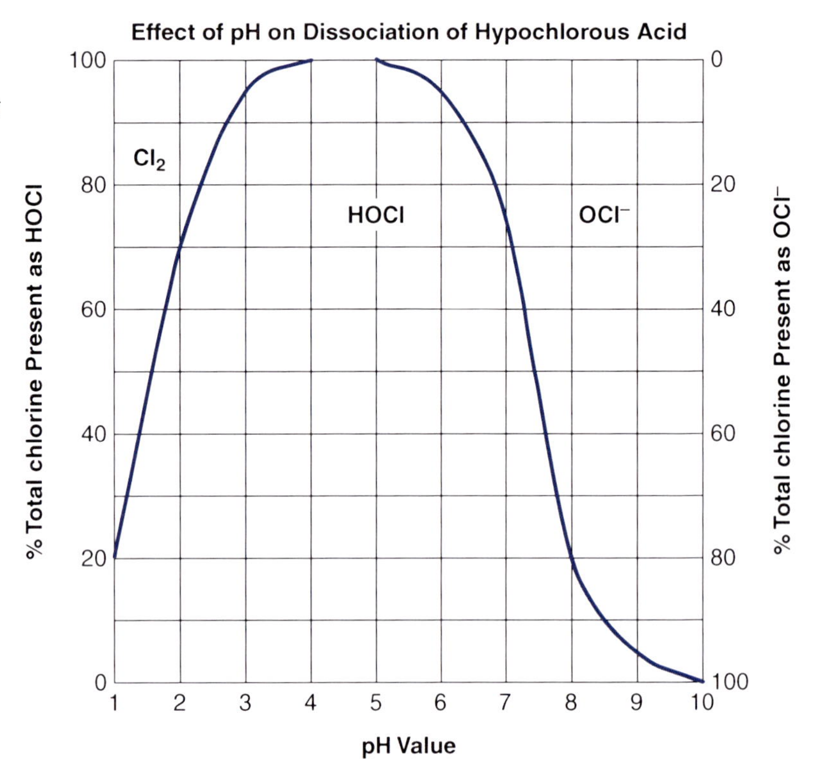

Hypochlorite Synthesis Cells and Technology, Sea Water, Fig. 2 Effect of pH on the dissociation of hypochlorous acid

Breakdown of Formed Hypochlorite

The largest and most important efficiency loss is due to the cathodic reduction of hypochlorite to chloride and hydroxyl.

$$ClO^- + H_2O + 2e \rightarrow Cl^- + 2OH^- \quad (8)$$

Operating Theory

In a seawater source operating on-site hypochlorite cell there typically is an abundance of seawater normally with a salt concentration above 15 gpl NaCl. The only significant cost in operating these on-site seawater systems is electricity to drive the electrolytic cells and pumping cost; feed stock seawater is considered free. Therefore seawater cells are operated to maximize production (minimize inefficiency reactions) and reduce operating cell voltage to minimize electricity consumption.

To maximize production one must consider the side reactions listed as Eqs. 5, 6, 7 and 8. The relative effects of operating parameters such as current density, temperature, salt content, cell geometry, have been studied extensively. Accordingly, seawater cells that operate in a region of lower available chlorine content (0.2–2.0 gpl) where cell efficiency is the highest. Also lower concentrations of available chlorine assist with the equilibrium in Eq. 8 to reduce breakdown of hypochlorite and reduces the formation of chlorates. The electrode current density is comparatively low to minimize loss by reaction 7 as well as ohmic losses. Impact of operating current density on current efficiency is well known and is inversely proportional (driving the inefficiency reactions). Typically in the range of 500–1500 Å/m^2 the bulk efficiency is reduced about 3 % for an increase in 250 Å/m^2.

Cell voltage results from the summation of electrode reversible potentials, overpotentials, ohmic voltage drops in the electrolyte

Hypochlorite Synthesis Cells and Technology, Sea Water, Table 1 Hypochlorite cell typical voltage breakdown for a current density of 1,200 A/m^2

1. Anode reversible potential plus overpotential	1.45 V
2. Anode concentration overvoltage	0.10 V
3. Solution ohmic drop	0.82 V
4. Cathode concentration overvoltage	0.15 V
5. Cathode reversible potential plus overpotential	1.27 V
6. Ohmic drop in cell hardware plus intercell	0.25 V
Typical Total Cell Voltage Drop	4.04 V

(conductivity), and cell hardware. An example for contribution to cell voltage of the above parameters is listed in Table 1. Typical current density is between 800 and 1,200 Å/m^2. This current density as compared to a higher operating current density as used in brine systems reduces the impact of almost all the components on increasing cell voltage (energy cost) in Table 1.

Seawater quality (composition, temperature, contamination) also has an impact on performance of the hypochlorite cell operation. Salinity has already been discussed. There are two significant reactions in which substances other than chloride take part. These are listed below.

Precipitation of magnesium hydroxide

$$Mg^{++} + 2OH^- \rightarrow Mg(OH)_2$$

Precipitation of calcium carbonate

$$Ca^{++} + HCO3^- + OH^- \rightarrow CaCO_3 + H2O$$

The precipitation of these two compounds starts with pH values above 8.5. These values may not be reached by the bulk solution pH value, but are exceeded by the pH of the alkaline film that forms at the cathode during electrolysis, which is approximately pH 11. Accordingly $Mg(OH)_2$ and $CaCO_3$ precipitate on the alkaline film. Most of the precipitates are carried out of the cell by the seawater stream due to the velocity (which is a design consideration for seawater hypochlorite cells), but a small amount may stick to the cathode. These precipitates in very small quantities depress the reduction of hypochlorite ion at the cathode due to a blinding effect (good thing). However, as the deposits continue to build the positive effect is overturned as the build-up prevents good circulation of the seawater and the electrodes are locally "starved" (local high depletion or not present) for NaCl. The inefficiencies discussed then come into the picture and if the deposit is not removed, eventually cause passivation of the electrodes. To prevent significant loss in efficiency and shortened electrode life, this temporary situation may be remedied by periodically dissolving the deposits with a solution containing 5–10 % hydrochloric acid.

Cell Configuration

Cell design often is specific to a given vendor but typically there are on-site hypochlorite generation units from a few grams per hour to hundreds of kilograms per hour. The discussed information in previous sections focused on what goes on at the "cell" level which is designated as a single anode and a single cathode. This activity between a single anode and a single cathode can be multiplied in parallel and in series. This accumulation of individual single anodes and cathodes in commercial practice is designated as an electrolyzer or generator. The design of the electrolyzer depends basically on capacity and required service, but most electrolyzers have some common features:

1. The use of Dimensionally Stable Anodes – commonly referred to as DSA and titanium cathodes
2. An interelectrodic gap (distance between facing anode and cathode) in the range of 1–8 mm. Such distances keep reasonably low ohmic drop through the electrolyte.
3. Minimum number of electric and hydraulic connections between electrolyzers.
4. Non-corroding durable enclosures (bodies) for the electrode packs.
5. Most electrolyzers have multiple electrodes (mostly bipolar).

The arrangement of the electrodes is based on the specific vendor design. The tube in a tube design discussed earlier has some limitation in production capacity, complexity for both electrical and hydraulic connections, and footprint associated with the unit. Therefore most vendors use the stacked plate cell arrangement for moderate and large production units.

Cross-References

- ► Chlorate Cathodes and Electrode Design
- ► Chlorate Synthesis Cells and Technology
- ► Chlorine and Caustic Technology, Overview and Traditional Processes

References

1. Bennett JE (1974) Non-diaphragm electrolytic hypochlorite generators. In: Chemical engineering progress
2. White GC (1999) Handbook of chlorination and alternative disinfectants, 4th edn. John Wiley, New York
3. DeNora Permelec PTE (1991) Seaclor systems for on-site electrolytic generation of hypochlorite solutions
4. Pletcher D (1984) Industrial electrochemistry. Chapman and Hall, New York, Chapter 5
5. O'M Bockris J, Conway BE, Yeager E, White RE (eds) (1981) Comprehensive treatise of electrochemistry-volume 2: electrochemical processing. Plenum, New York, Chapter 3

I

Infrared Spectroelectrochemistry

Michael Bron
Institut für Chemie, Technische Chemie,
Martin-Luther-Universität Halle-Wittenberg,
Halle, Germany

Introduction

The term "spectroelectrochemistry" describes experimental techniques in which an electrochemical experiment is combined with a spectroscopic technique. While electrochemical experiments typically yield information on macroscopic properties like reaction rates, spectroscopic techniques usually are applied to yield information on a molecular level, i.e., the structure of molecules, their electronic configuration, etc. The combination of both electrochemical and spectroscopic approaches will thus unravel a more complete picture of the system under study. A broad variety of spectroscopic techniques has been coupled with electrochemistry, including UV-vis, Raman, EPR, and NMR [1–4]. The application of infrared spectroscopy to study electrochemical systems is labelled infrared spectroelectrochemistry and will be described in the following.

General Principles

In infrared (IR) spectroscopy, a beam of electromagnetic radiation in the range of 400–4,000 cm^{-1} is directed through a sample. The electromagnetic radiation may excite molecular vibrations (and rotations) which match in their energy with the radiation and where a change in the dipole moment occurs by excitation to a higher energy level. During this process, the electromagnetic radiation is absorbed. Thus, the spectral intensity distribution of the electromagnetic radiation leaving the sample reflects the vibrational and rotational states of the sample. It hence contains chemical information on the structure of molecules, the presence of functional groups, the bond strengths, etc. and may be used to identify known substances or unravel the structure of unknown ones. For more details on the basics of IR spectroscopy, the reader is referred to textbooks on physical or analytical chemistry. IR spectroscopy is routinely applied to study solid, liquid, or gaseous samples employing comparably simple experimental setups where the IR light is transmitted through the sample. However, it may also be successfully used to investigate chemical processes occurring, for instance, in catalysis or electrochemistry, where more sophisticated setups are required. In the following, the principles and applications of IR spectroelectrochemistry are described. In electrochemistry, IR spectroscopy is usually applied to study processes in the electrochemical interphase (i.e., those occurring on or close to an electrode surfaces); however, solution processes may be examined as well. More generally, it refers to the study of electrified interfaces.

G. Kreysa et al. (eds.), *Encyclopedia of Applied Electrochemistry*, DOI 10.1007/978-1-4419-6996-5,

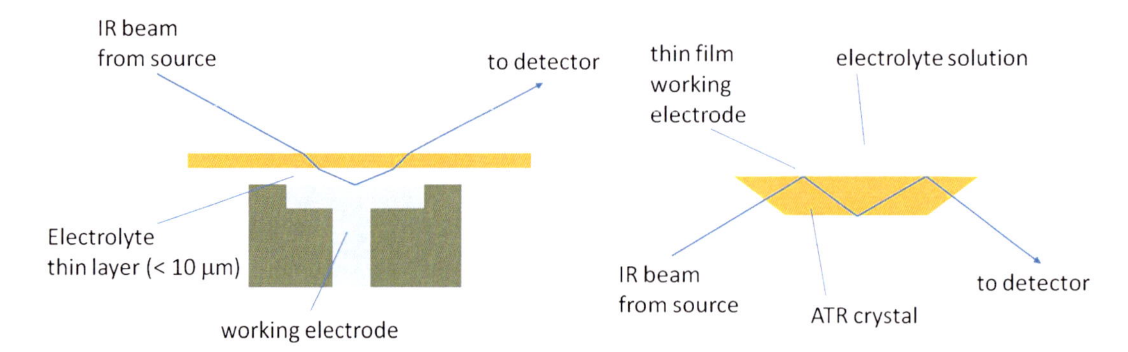

Infrared Spectroelectrochemistry, Fig. 1 Schematic representation of external reflection (*left*) and internal reflection (*right*). Note that electrolyte gap and thin film electrode are drawn disproportionately thick. The electrochemical cell would be completed by reference and counter electrode, electrolyte in- and outlet, and combined with a system of mirrors directing the IR beam to the electrode and further to the detector

Experimental Details

IR spectroelectrochemistry has been the subject of a sizeable amount of early reviews, where the experimental details and applications have been described [5–7]. Regardless the fact that electrochemistry is an extremely broad field, the following discussion will be restricted to "classical" electrochemical systems where a solid electrode is in contact with a liquid electrolyte solution which may contain electroactive species. Since the typically used electrolyte solutions (mostly aqueous solutions) are strongly IR absorbing, it is not possible to use a standard laboratory electrochemical cell, but for spectroelectrochemical experiments, special cell designs and beam paths have to be employed. There are two general principles on how the IR beam is directed to the electrode surface called internal reflection and external reflection, respectively.

In external reflection, a thin layer configuration is employed which is formed by pushing a planar electrode (the working electrode) against an IR transparent window, typically a single crystal of, e.g., ZnSe or CaF_2 [5–7]. Vertical designs, where an upright electrode is pressed to the window which forms the bottom of the cell, as well as horizontal designs have been described. The electrolyte gap between electrode and window should be smaller than 10 μm in order to allow for a good signal to noise ratio. Counter and reference electrode complete the three-electrode arrangement. Such an arrangement is advantageous since the typical bulk electrodes may be used; however, it may suffer from restricted diffusion in the thin layer.

The other approach, internal reflection, takes advantage of the fact that if an electromagnetic wave is totally reflected at the interface between an optically thicker and an optically thinner medium (refraction index of the optical thicker medium is higher), a part of the electromagnetic wave penetrates into the optical thinner medium (the so-called evanescent wave) where it may interact with any species present and may be absorbed. The spectral intensity distribution of the reflected light thus differs from that of the incident beam and contains information about species present in the interphase of the optical thinner medium. In practical setups ATR (attenuated total reflection) single crystals are used, onto which the electrode under investigation is deposited. The beam baths of internal and external reflection are visualized in Fig. 1.

Spectra Acquisition

As in standard infrared spectroscopy, the infrared radiation finally detected is influenced by the intensity distribution of the radiation source, any substance/material absorbing IR light which resides in the beam path (e.g., gas atmosphere in the spectrometer, but also water in the thin layer of an external reflection cell) as well as the spectral sensitivity of the detector and is usually called "single beam spectrum." Thus, a reference spectrum has to be recorded in order to elucidate the spectral features of the sample. In standard IR spectroscopy, the reference spectrum is simply obtained using an empty beam path either parallel to the sample path as typically found in dispersive IR spectroscopy or measuring sample spectrum and reference spectrum sequentially, as in FTIR spectroscopy. This approach does not work in infrared spectroelectrochemistry due to the complex beam path, and the reference spectrum has to be recorded in the same configuration as the sample spectrum, implying that in both the reference and the sample single beam information on the species under consideration are present. If however reference and sample single beam are recorded at different electrode potentials, any changes induced by this change in potential will be visible in the final spectrum. Initial developments in this field used dispersive/grating spectrometers. The potential was modulated at high frequency (e.g., 10 Hz) and the spectrum was recorded using phase-sensitive lock-in amplification. This approach is called EMIRS (electrochemically modulated IR spectroscopy) [8].

Modern FTIR spectrometers led to the development of different approaches to monitor changes at electrode surfaces occurring between two different potentials. If these changes are reversible (i.e., adsorption-desorption phenomena, changes in bonding strength), a repetitive switching between "sample potential" and "reference potential" has proven to be beneficial and is called "SNIFTIRS" (substractively normalized interfacial FTIR spectroscopy) [6]. The acronym SW-FTIRS (square-wave FTIR spectroscopy) has however been suggested by the respective IUPAC commission [1]. Irreversible changes could be analyzed by single potential steps in "SPAIRS," single potential alteration IR spectroscopy [1]. It would also be possible to stepwise alter the potential in multiple steps in one direction (potential difference FTIR spectroscopy, PDFTIRS [1]). In both SPAIRS and SNIFTIRS, the final IR spectrum is calculated from the single beam spectra using the normalized differential reflectance $\Delta R/R_0$, with $\Delta R = R - R_0$ (or $R_0 - R$) and R = reflectance at sample potential and R_0 = reflectance at reference potential, where in PDFTIRS any of the consecutively applied potentials could be used as reference [1, 6].

Another approach to acquire infrared spectra of the electrochemical interface takes advantage of the so-called surface selection rule, which states that only p-polarized light may interact with molecules at a metal electrode while s-polarized may not. P-polarized light is light in which the electrical field vector is parallel to the plane formed by the incident and the reflected beam, while s-polarized light is perpendicular to the plane of incidence. Furthermore, only molecules having a dipole perpendicular to the surface may interact with p-polarized light [5].

Thus, another straightforward approach would be to modulate the polarization of the light, employing s-polarized light ("blind" to surface molecules) for the reference and p-polarized light (interacting with surface molecules) for the sample single beam, and labelled PM-IRRAS (polarization-modulated infrared reflection absorption spectroscopy) [9]. This approach has been developed both for dispersive and FTIR spectrometers, where modulation between both polarizations has been achieved with a high-frequency photoelastic modulator [9]. In this case, the detected intensities I_P of the p-polarized light and I_S of the s-polarized light are used to calculate the quantity $(I_P - I_S)/(I_P + I_S)$ which is equivalent to absorbance [6, 9, 10].

Electrodes

From the description of the setup, it is clear that techniques working with external reflection require reflecting electrodes. The major type used is bulk metal electrodes, either as polycrystalline material or as single crystals. However, other reflecting electrode materials like glassy carbon may be used as well. These electrodes may either serve as system under study or simply as substrates, onto which the sample of interest (e.g., a conducting polymer) is deposited. Approaches where nanoparticles or carbon supported catalysts are deposited onto such electrodes and investigated with IR spectroelectrochemistry have been described as well (e.g., [11]).

Major Applications

Infrared spectroelectrochemistry has been established as a versatile tool to obtain detailed information about processes occurring in the electrochemical interphase and found application in many subfields of electrochemistry. Following shortly, important applications are summarized; for more details, the reader is referred to the above cited reviews.

One major application of IR-EC is to study adsorption phenomena in electrochemical systems. Early work was concerned with strongly IR-absorbing small molecules like CO, CN^-, or SCN^-. While it was of fundamental interest to study their absorption strength, geometry, and potential dependence, they have been also employed as probe molecules to learn about the properties of the various metal electrode surfaces. A more detailed analysis of the electrochemical interphase has been attempted by also studying the adsorption of electrolyte anions like sulfate and even of water. These approaches take advantage of the so-called Stark effect, i.e., the shift of absorption bands in strong electromagnetic fields, which leads to a potential dependent shift of IR bands of species adsorbed at an electrode surface [6].

Mechanistic studies in electrocatalysis are another major application of infrared spectroelectrochemistry. The archetypical subject is the oxidation of methanol, both from the viewpoint of basic studies as well as in the framework of fuel cell catalysis. However, larger molecules like ethanol or even sugars have also been frequently studied [7].

Other major fields of application include conducting polymers, redox-active organic molecules in solution, and inorganic or organometallic complexes [4].

Future Directions

IR spectroelectrochemistry, as mentioned above, is a well-established technique. However, scientists in spectroelectrochemistry generally strive to extend the current limitations. This includes multi-method approaches combining different spectroelectrochemical techniques as well as enhanced temporal and spatial resolution [4]. Combination of infrared spectroelectrochemistry with other physical techniques, e.g., scanning electrochemical microscopy, has also been recently described and will have impact on future research [12].

Cross-References

▶ Electrode Catalysts for Direct Methanol Fuel Cells
▶ Electrogenerated Reactive Species
▶ Formic Acid Oxidation

References

1. Plieth W, Wilson GS, Guiterrez de la Fe C (1998) Spectroelectrochemistry: a survey of in situ spectroscopic techniques. Pure Appl Chem 70:1395–1414
2. Holze R (2009) Surface and interface analysis. In: Springer series in chemical physics, vol 74. Springer, Berlin/Heidelberg
3. Kaim W, Fiedler J (2009) Spectroelectrochemistry: the best of two worlds. J Chem Soc Rev 38:3373–3382
4. Dunsch LJ (2011) Recent advances in in situ multi-spectroelectrochemistry. Solid State Electrochem 15:1631–1646
5. Bewick A, Pons S (1985) Infrared Spectroscopy of the Electrode-Electrolyte Solution Interface. In: Clark RE, Hester RJH (eds) Advances in infrared and raman spectroscopy, vol 12. Wiley Heyden, New York
6. Ashley K, Pons S (1988) Infrared spectroelectrochemistry. Chem Rev 88:673–695
7. Iwasita T, Nart FC (1997) In situ infrared spectroscopy at electrochemical interfaces. Prog Surf Sci 55:271–340
8. Bewick A, Kunimatsu K, Pons BS, Russell JW (1984) Electrochemically modulated infrared spectroscopy (EMIRS) experimental details. J Electroanal Chem 160:47–61
9. Seki H (1988) In situ optical vibrational spectroscopy of simple ions at the electrode-electrolyte interface. Am Chem Soc Symp Series 373:322–337
10. Kunimatsu K, Seki H, Golden WG, Gordon JG II, Philpott MR (1985) Electrode/electrolyte interphase study using polarization modulated FTIR reflection-absorption spectroscopy. Surf Sci 158:596–608

11. Park S, Wasileski SA, Weaver MJ (2001) Electrochemical infrared characterization of carbon-supported platinum nanoparticles: a benchmark structural comparison with single-crystal electrodes and high-nuclearity carbonyl clusters. J Phys Chem B 105:9719–9725
12. Wang L, Kowalik J, Mizaikoff B, Kranz C (2010) Combining scanning electrochemical microscopy with infrared attenuated total reflection spectroscopy for in situ studies of electrochemically induced processes. Anal Chem 82:3139–3145

Insertion Electrodes for Li Batteries

Yuping Wu
Department of Chemistry, Fudan University, Shanghai, China

Synonyms

Intercalation electrode

Introduction

In secondary batteries, during the charge and discharge process, not only redox reactions, i.e., gain and/or loss of electrons, happen, but also insertion process is accompanied. The latter is the main difference from other traditional electrodes for batteries [1]. Some insertion electrode can be positive electrode and some ones can be negative electrode. For example, when they are used for lithium-ion batteries, both electrodes can be insertion electrodes.

In terms of the positive electrode, which is also called as cathode, during the charge process, whether lithium metal or graphite is used as the reference electrode, lithium ions move from the positive electrode, electrons also lose via the outside circuit, and potential of the insertion electrode or the voltage of the battery increases. During the discharge process, the reverse process happens.

In terms of the negative electrode, which is also called as anode, during the charge process, i.e., the process for voltage increase, when lithium metal is usually used as the reference electrode, lithium ions move from the negative electrode, electrons also lose via the outside circuit, and potential of the insertion electrode or the voltage of the battery increases. During the discharge process, the reverse process happens.

When both insertion electrodes are comprised of a battery such as lithium-ion battery, during the charge and discharge process, the above description for the positive electrode is still the same. However, for the negative electrode, it will be different. During the charge process of lithium-ion battery, lithium ions will insert into the negative electrode together with the gain of electrons via the outside circuit. The potential of the negative electrode will decrease. This process is the same process for the discharge process of the negative electrode when lithium metal is used as the reference electrode. During the discharge process, the reverse process will happen.

Some Insertion Materials

There are quite some insertion materials. Here it is classified based on their elements and includes oxides, carbonaceous materials, multiatom anion compounds, inverse spinels, sulfides, nitrides, silicon, and tin alloys [1]. Here the main topic will be related to lithium insertion. For the other kinds of insertion compounds such as sodium, potassium, sulfate, and the like, please refer to other references [2].

Oxides

Oxides include $LiCoO_2$, $LiNiO_2$, $LiMn_2O_4$, MnO_2, V_2O_5, LiV_3O_8, Fe_2O_3, MoO_3, MoO_2, Co_3O_4, CoO, NiO, Cu_2O, CuO, SnO, SnO_2, and the like.

In the case of $LiCoO_2$, $LiNiO_2$, $LiMn_2O_4$, MnO_2, V_2O_5, LiV_3O_8, and MoO_3, lithium ions can reversibly remove/insert from/into them together with the oxidation/reduction of the metal elements. Since their redox potentials are above 2 V (vs. Li/Li^+), they can be used as positive electrode for lithium-ion batteries and hybrid supercapacitors. These compounds especially $LiCoO_2$, $LiNiO_2$, and $LiMn_2O_4$ are very stable for lithium insertion and removal in the organic electrolytes, and their cycling life is very good. The insertion and removal of lithium ions will not lead to marked change of the host structures. In order to improve their performance including reversible capacity and cycling at elevated temperature, they can be doped and coated.

As to Fe_2O_3, MoO_2, Co_3O_4, CoO, NiO, Cu_2O, CuO, SnO, SnO_2, and the like, lithium ions can also insert and remove reversibly. However, their average redox potentials are below 2 V (vs. Li/Li^+), and they can be used as negative electrodes for lithium-ion batteries. The insertion and removal of lithium ions will lead to the change of the host structures. As a result, they should be coated to achieve good cycling performance.

Carbonaceous Materials

Carbonaceous materials include graphitic carbons and disordered carbons. The former is from heat treatment at a temperature above 2,000 °C, and the content of graphite crystal is very large. In the case of the disordered carbons, they are heat-treated at temperature below 1,000 °C, and the graphite crystallites are very small and their content is not much. Formerly accordingly to whether become graphitized easily or difficultly, carbons are divided into soft carbons and hard carbons. This classification is just based on the precursors and could not differentiate the final carbonaceous materials.

In graphitic carbons, the insertion/removal of lithium ions is below 0.25 V (vs. Li/Li^+). To be stable in the organic electrolytes, a solid-electrolyte interface (SEI) film is needed to form prior to the insertion of lithium ions, which is from the decomposition of the organic electrolytes at about 0.8 V (vs. Li/Li^+). This film can prevent the further reaction of the inserted (reduced) lithium with the organic electrolytes. Several lithium insertion (intercalation) compounds are formed, LiC_{24} (stage IV), LiC_{18} (stage III), LiC_{12} (stage II), and LiC_6 (stage I), and their formation potentials are between 0.25 and 0.05 V (vs. Li/Li^+). The final stage is LiC_6 at ambient condition, which means that its theoretic capacity for lithium storage is 372 mAh/g. Some examples are mesocarbon microbeads (MCMB) from Osaka Gas Company, Japan, and CMS from Shanshan Company, China.

In amorphous carbons, there are a large amount of micropores due to the low heat-treatment temperature. These micropores can be sites for lithium insertion, and the stored lithium can form Li_x (x is or greater than 2) molecules. In the graphite crystallites, lithium can be inserted like in graphite. Consequently, their reversible lithium capacity can be above the theoretic value of graphite. It is anticipated that the capacity can be up to 1,000 mAh/g. However, due to the unstable disordered areas, the cycling is not good. During charge and discharge, there is large voltage hysteresis. Since there are quite some micropores, their charge rate can be faster than that of graphitic carbons [3].

Multiatom Anion Compounds

These compounds mainly include $LiMPO_4$, Li_2MSO_4, Li_2MSiO_4, Li_2MBO_3, and Li_2MTiO_4 (M = Fe, Mn, Ni, and Co).

Among $LiMPO_4$, the well known is $LiFePO_4$, which belongs to olive structure. This compound does not have high electronic and ionic conductivities. After carbon coating, doping, and nanostructuring, its reversible capacity, rate capability, and cycling life have been greatly improved, which greatly push forward the development of electric vehicles. Its lithium removal/insertion (redox) potentials are 3.6 and 3.4 V (vs. Li/Li^+), respectively. In the case of $LiMnPO_4$, its electrochemical performance is similar to that of $LiFePO_4$, and its redox potential is 4.2/4.0 V (vs. Li/Li^+), respectively, higher than those for $LiFePO_4$.

Among Li_2MSO_4, only $LiFeSO_4F$ is reported so far since the sulfate is not stable in ambient condition and can absorb water very easily. Its redox potential is about 3.7/3.5 V (vs. Li/Li^+), a little higher than those for $LiFePO_4$.

Li_2MSiO_4 can allow the removal/insertion of two lithium ions, and its theoretic capacity can be higher. For example, the theoretic capacities for Li_2MnSiO_4, Li_2CoSiO_4, and Li_2NiSiO_4 are 333 325 and 326mAh/g, respectively. However, their electronic conductivity is very low, and their crystal structure is not stable when two lithium ions are removed; their available reversible capacity is generally below 200 mAh/g with poor cycling.

When Si in Li_2MSiO_4 is substituted with B, Li_2MBO_3 will be achieved. Their reversible capacity can be above 200 mAh/g with good cycling when carbon is coated. Their redox potentials are a little lower than those for the corresponding phosphates. The well-studied one is $LiFeBO_3$.

When Si is substituted with Ti, Li_2MTiO_4 can be achieved. However, there is very rare report on this kind of insertion compounds.

Inverse Spinels

These inverse spinels can be represented by $LiMn_{2-x}M_xO_4$ (M = Ni, V, Cr, Cu, Co, and Fe). There are two redox plateaus during the charge and discharge curves. One is around 4 V and the other is around 5 V. When M is Ni, V, Cr, Cu, Co, and Fe, the higher plateaus are situated at 4.7, 4.8, 4.8, 4.9, 5.0, and 4.9 V (vs. Li/Li^+). With the increase of x, the part at around 5 V increases.

Sulfides

MoS_2 and TiS_2 are two main sulfides as insertion electrodes since they were formerly regarded as good cathode materials for Li-metal rechargeable batteries. Of course, their reversibility for lithium insertion and removal is very good and their voltage is about 2 V (vs. Li/Li^+). Since their voltage is neither high nor low, the researches on them become very rare.

Nitrides

Nitrides can be represented by $Li_{3-x}M_xN$ (M = Co, Ni, and Cu), which is from the reaction of Li_3N with M. They are layered structure. All the lithium in the Li layer can be reversibly removed and inserted, and half lithium in Li-M layer can be reversibly removed and inserted. If more lithium is removed, the layered structure will be destroyed. Their reversible capacity can be ranged from 500 to 1,000 mAh/g, and charge voltage will be higher than that for graphite, about 1.1 V (vs. Li/Li^+).

Silicon Alloys

Lithium can be inserted into silicon-based alloys, and the utmost component is $Li_{4.4}Si$, which means that the reversible capacity for Si can be 4,200 mAh/g. The insertion potential is below 1 V (vs. Li/Li^+). However, the volume expansion is very large, about 400 %, and their cycling performance is very poor. Carbon coating, alloying, and nanostructuring have been tried to improve the cycling performance.

Tin Alloys

Sn is similar to Si and belongs to IVA group elements. Its electrochemical performance is also similar to Si. Since the atomic weight of Sn is much larger than that of Si, the reversible theoretic capacity is only 994 mAh/g.

Applications

Insertion electrodes can build up lithium-ion batteries with different chemistry. $C//LiCoO_2$ and $C//LiMn_2O_4$ are two well used as power sources for electronics. The chemistry of $C//LiFePO_4$ is mainly used for plug-in electric vehicles or pure electric vehicles [1].

Insertion electrodes can also be used to build up aqueous rechargeable lithium-ion batteries (ARLBs) with excellent rate capability and cycling performance [4, 5].

Hybrid supercapacitors can also be assembled by using insertion compounds as one electrode, and the other electrode can be activated carbon (AC). For example, $AC//LiCoO_2$ and $AC//LiMn_2O_4$ can be assembled using organic and aqueous electrolytes. $MoO_3//AC$ can also be assembled by using aqueous electrolytes since their energy density can be more higher [6].

Future Directions

To achieve energy storage systems with larger energy density and higher power density, insertion electrodes of higher redox potential and multi-electron redox process are one direction. The other is to find insertion electrodes with low redox potential but larger reversible amount of lithium.

Cross-References

▶ Aqueous Rechargeable Lithium Batteries (ARLB)
▶ Hybrid Li-Ion Based Supercapacitor Systems in Organic Media
▶ Lithium-Ion Batteries

References

1. Wu YP, Dai XB, Ma JQ, Cheng YJ (2004) Lithium ion batteries: applications and practice. Chemical Industry Press, Beijing
2. Yang ZG, Zhang JL, Kintner-Meyer M, Lu XC, Choi D, Lemmon JP, Liu J (2011) Electrochemical energy storage for green grid. Chem Rev 111:3577–3613
3. Wu YP, Wan CR, Fang SB, Jiang Y (1999) Mechanism of lithium storage in low temperature carbon. Carbon 37:1901–1908
4. Tang W, Liu LL, Zhu YS, Sun H, Wu YP, Zhu K (2012) An aqueous rechargeable lithium battery of excellent rate capability based on nanocomposite of MoO_3 coated with PPy and $LiMn_2O_4$. Energy Environ Sci 5:6909–6913
5. Tang W, Gao XW, Zhu YS, Yue YB, Shi Y, Wu YP, Zhu K (2012) Coated hybrid of V_2O_5 nanowires with MWCNTs by polypyrrole as anode material for aqueous rechargeable lithium battery with excellent cycling performance. J Mater Chem 22:20143–20145
6. Tang W, Liu LL, Tian S, Li L, Yue YB, Wu YP, Zhu K (2011) Aqueous supercapacitors of high energy density based on MoO_3 nanoplates as anode material. Chem Commun 47:10058–10060

Intercalation Electrode

▶ Insertion Electrodes for Li Batteries

Interconnectors

Teruhisa Horita
Fuel Cell Materials Group, Energy Technology Research Institute, National Institute of Advanced Industrial Science and Technology (AIST), Tsukuba, Ibaraki, Japan

Introduction

Interconnectors are one of the important components in Solid Oxide Fuel Cells (SOFCs). They should separate fuel and air as well as electronic conduction [1–4]. Candidate materials have relatively high electronic conductivity (higher than 10 Scm^{-1}) at operating temperatures (873–1,273 K). The interconnectors should meet the following requirements:

1. Dense and gas tightness both in fuel and air atmospheres
2. High electronic conductivity without oxygen electrochemical leak
3. Chemical stability both in fuel and oxidant atmospheres
4. Thermochemical compatibility with the other cell components

So far, several kinds of candidate materials are investigated. They are classified into two categories: One is metallic interconnectors and the other is oxide ceramic one. In this section, chemical and physical properties of each material are introduced.

Metallic Interconnectors

Candidate Metallic Materials

In recent decades, the operation temperature of SOFCs was reduced to lower than 1,073 K. In such a temperature zone, metallic interconnectors can be applied instead of oxides [4, 5]. The application of metallic materials is expected to reduce the materials cost effectively. In addition, they have high thermal and electrical conductivities in comparison with oxide ceramics, which has an advantage of designing compact SOFC systems.

There are several candidate metallic materials proposed as follows [4, 5]; Ni-Cr alloy, Fe-Cr alloy, and Cr-based alloys. Traditionally, Ni-Cr alloys are high temperature resistant materials, such as Inconnel, Hastelloy, and Haynes. Fe-Cr (ferritic) alloys are often used as high temperature metallic components, such as stainless steel (SUS), E-Brite, ZMG-series, and Crofer-series. Cr-based alloys are specially developed for SOFC interconnects, which contains oxide materials, such as Cr-Fe-Y_2O_3.

The key issue for applying the metallic materials is the formation of oxide scales during operation, basically Cr_2O_3 protective oxides on the surface of metals when addition of Cr above 17 wt% [6–8]. The Cr_2O_3-based oxide scales are inevitable to protect the alloys for the further oxidation. However, a thick oxide scale formation can reduce the conductivity of oxide scale/alloy interfaces and surface morphology of metals. Therefore, a thin and compact Cr_2O_3-based oxide scale formation is favorable in terms of long-term stability. For controlling the oxide scale growth on the metal surface, some minor elements were added into the metals. Addition of aluminum (Al) and silicon (Si) can show better adhesion of oxide scales to the alloys [5–9]. But much amount of these elements will reduce the electrical conductivity by forming insulating layers at the oxide scale/alloy interfaces. Addition of molybdenum (Mo) and tungsten (W) will better match the thermal expansion coefficient to the oxide ceramics [10]. Addition of manganese (Mn) will generate Mn-Cr spinel structures on the top of oxide scales, which improve surface conductivity on Cr_2O_3-based oxide scales [5–9]. Many cases of the growth of oxide scales are followed by a parabolic relationship between thickness (d) and operation time (t): $d^2 = k_p t$ (k_p: growth rate constant). The growth rate constants of oxide scales are controlled to be 10^{-14} cm^2s^{-1} at 1,073 K when addition of several elements into Fe-Cr alloys. Recently, the growth rate of oxide scales was successfully reduced to 10^{-15} cm^2s^{-1} levels, which is estimated to be 2-μm thick after long-term operation of 10,000 h [11]. This value is expected to be low enough to apply the metallic materials to SOFC interconnectors.

Cr Poisoning on the Cathode Performance

Another issue for applying metals to the interconnectors is the vaporization of Cr from the alloys [12, 13]. Even though the Cr vapor pressures are reduced by the formation of Cr_2O_3 oxide scales, a small amount of vapor pressures of Cr was measured, which was lower than 10^{-8} atm levels at 1,073 K (10 ppb levels). Cr vapors can be transported to cathode and deposited at the active sites of oxygen reduction. Because the reduction of Cr valence in CrO_3 vapor occurs ($Cr^{6+} \rightarrow Cr^{3+} + 3e^-$) at the electrochemical active sites. This reaction is represented in the following reaction:

$$2CrO_3(g) + 6e^- \rightarrow Cr_2O_3(s) + 3/2O^{2-} \quad (1)$$

The above reaction was promoted when using (La,Sr)MnO_3-based cathode because the oxygen reduction was only active at the gas/(La,Sr) MnO_3/electrolyte interfaces (triple phase boundaries) [14]. Other cathode, such as (La,Sr) (Co,Fe)O_3, shows relatively high chemical reactivity with Cr vapor to form $SrCrO_4$ on the surface. The difference of reaction products and electrochemical degradation was determined by the reactivity and electrochemical activities.

Coating of Fe-Cr Alloys

Recent trend of application of metallic interconnects is to apply the coating technology [15, 16]. Fe-Cr alloys are often the substrate of coatings because of their reasonable cost. For the coating, the coating materials should be compatible with Fe-Cr alloys. An important point for forming the protective oxide coating is to match the thermal expansion coefficient and chemical compatibility between oxide coating and alloys. So far, several coating materials were evaluated and some candidates were reported. Among them, Co-Mn-spinel is one such candidate due to the similarity of thermal expansion behaviors and the chemical compatibility between oxide scales and alloys [16, 17]. Also, the electronic conductivity of Co-Mn spinel shows relatively high values which expand the reaction zones around the cathode/ electrolyte interfaces.

Oxide Interconnectors

Candidate oxide interconnectors are (La, AE)(Cr, Mg)O_3 or (Sr, La)TiO_3 ceramics (AE: alkaline earth elements). They should be dense and separate fuel and air as well as electronic conduction at operating high temperatures. Among several candidate ceramic materials, Lanthanum chromite–based perovskite oxides ($LaCrO_3$) have been widely recognized as promising interconnect materials [18]. In this section, $LaCrO_3$ is focused on the description of the entry. The composition of $LaCrO_3$ was modified by doping of lower valence alkaline ions, such as Ca^{2+}, Mg^{2+}, and Sr^{2+} at the La^{3+} or Cr^{3+} sites. The substitution of La-site and Cr-site with the other elements can decrease the sintering temperature and increase the electronic conductivity. So far, a number of papers have been published regarding the physical and chemical properties of doped $LaCrO_3$. Densification of (La,AE)CrO_3 is technologically important, and some densification mechanisms have been proposed [19–21].

Electronic Conductivity

Doped $LaCrO_3$ is a p-type conductor, and the electronic conductivity increases with a concentration of low valence cations, such as Sr^{2+}, Ca^{2+}, in La^{3+} sites in the following reaction:

$$LaCrO_3 + xAEO + 0.25xO_2 \rightarrow La_{1-x}AE_xCr^{3+}{}_{1-x}Cr^{4+}{}_xO_3 + 0.5xLa_2O_3 \quad (2)$$

Electronic hole will be formed on Cr^{4+} sites, and the conduction mechanism is a small polaron hopping process via Cr^{4+} sites. The electronic conductivity is about 10–100 Scm^{-1} at 1,273 K in air [22, 23]. The electronic conductivity increases with increasing temperature, suggesting the semiconductor temperature dependence. An increasing of the Ca concentration in $La_{1-x}Ca_xCrO_{3-\delta}$ enhanced the electronic conductivity due to the increase of Cr^{4+} concentration. There are some deviations of the electrical conductivity among the examined alkaline earth elements: Ca-doped $LaCrO_3$ shows the larger electrical conductivity than Sr-doped $LaCrO_3$. This difference was reported to be due to the difference of lattice distortion and phase stability. The activation energy for conductivity was 0.12–0.14 eV and the mobility was 0.066–0.075 $cm^2/V/s$ at 1,173–1,323 K.

The electronic conductivity decreases with a reduction of oxygen partial pressure because of the decrease of Cr^{4+} concentration in a reducing atmosphere as follows:

$$La_{1-x}AE_xCr^{3+}_{1-x}Cr^{4+}_xO_3 \rightarrow La_{1-x}AE_xCr^{3+}O_{3-\frac{x}{2}} + \frac{x}{4}O_2 \quad (3)$$

As predicted above, the electrical conductivity decreases with a reduction of oxygen partial pressures. The electrical conductivity is proportional to $p(O_2)^{1/4}$, which is consistent with the defect chemistry of $La_{1-x}Ca_xCrO_{3-\delta}$ [23]. A doping of B-site has been also considered by several authors [1–5]. Typical dopant cation is Mg^{2+} that is replaced into Cr^{3+} sites. This substitution also increases the concentration of Cr^{4+}, eventually increasing the electrical conduction.

Defect Chemistry and Oxygen Electrochemical Leak

Doped $LaCrO_3$ generates Cr^{4+} at high oxygen partial pressures and oxygen vacancies in reducing atmospheres. The important function of interconnects is electronic conduction without electrochemical oxygen leak. Thus, the defect chemistry–associated Cr^{3+}/Cr^{4+} transition and oxygen vacancies formation should be clarified. Several authors have treated the defect chemistry of doped $LaCrO_3$ (Weber et al. and Mizusaki et al. treated Sr-doped $LaCrO_3$ [22, 24], and Yasuda et al. and Onuma et al. treated Ca-doped $LaCrO_3$, [23, 25], and Oishi et al. treated LaCr(M)O_3 [26]). Using Kröger-Vink notation, the reaction for oxygen vacancy formation can be expressed by:

$$2Cr^{\cdot}_{Cr} + O^{x}_{o} = 2Cr^{x}_{Cr} + V^{\cdot\cdot}_{o} + \frac{1}{2}O_2(g) \quad (4)$$

The equilibrium constant for the above reaction can be written as:

$$K = \frac{[Cr^{x}_{Cr}]^2 [V^{\cdot\cdot}_{O}]}{[Cr^{\cdot}_{Cr}]^2 [O^{x}_{O}]} \cdot (p(O_2))^{\frac{1}{2}} \quad (5)$$

The oxygen vacancies can be formed in the $LaCrO_3$ lattice in reducing atmospheres, which are the diffusion paths of oxide ions. Since the interconnect material is placed in a large oxygen potential gradient, oxygen can permeate through the $LaCrO_3$-based materials via oxygen vacancies ($V^{\cdot\cdot}_{O}$) [27–31]. When oxide ions can migrate from high- to low-oxygen partial pressures, electrons can move in the opposite direction. In accordance with the convention, the oxygen permeation current density passes through the dense LaCrO3 base oxides. The oxygen permeation current density can be calculated from the following equation:

$$J_{O2} - \left(Am^{-2}\right) = -\frac{1}{4FL}\int_{p_{O2}(x=0)}^{p_{O2}(x=L)} \frac{(\sigma_{O^{2-}}\sigma_{e-})}{(\sigma_{O^{2-}} + \sigma_{e-})} d\mu_{O_2} \quad (6)$$

where F is the Faraday constant, μ_{O2} is the chemical potential of oxygen, σ_{O2-} is the oxide ionic conductivity, σ_e is the electronic conductivity, and L is the thickness of $LaCrO_3$. If the electronic conductivity is high enough, the above equation can be simplified as:

$$J_{O2} - \left(Am^{-2}\right) = -\frac{1}{L}\int_{p_{O2}(x=0)}^{p_{O2}(x=L)} \sigma_{O^{2}} \frac{RT}{4F} d\ln P_{O_2} \quad (7)$$

where R is the gas constant and T is the temperature. The above equation indicates that oxygen permeation current density can be calculated from the conductivity of oxide ion through the dense $LaCrO_3$. To evaluate the conductivity of oxide ion, the following equation can be applied:

$$\sigma_{O2} - \left(\Omega^{-1}m^{-1}\right) = \frac{4F[V^{\cdot\cdot}_{O}]D_V}{RTV_m} \quad (8)$$

where R is the molar gas constant ($J\,mol^{-1}\,K^{-1}$), T is temperature (K), $[V^{\cdot\cdot}_{O}]$ is the oxygen vacancy mole fraction, D_v is the oxygen vacancy diffusion coefficient (m^2s^{-1}), and V_m is the molar volume of $LaCrO_3$. The vacancy concentration can be determined from the experimental and calculated oxygen non-stoichiometry data. A precise analysis of the oxygen chemical potential distribution in the $LaCrO_3$-based oxides and oxygen electrochemical permeation (leak) were examined. At 900 °C (1,173 K), the permeation current density is expected be less than 80 mA/cm^2 in any total current densities examined (in the case of $La_{0.7}Ca_{0.3}CrO_{3-\delta}$). However, at high temperatures above 950 °C (1,223 K), the leakage current densities are more than 100 mA/cm^2, which are above 10 % of total current densities [29, 30]. This value is significantly large compared with the operating current densities. For the experimental determination of oxygen permeation through the $LaCrO_3$-based oxide ceramics, electronic blocking electrochemical method and isotope oxygen exchange method ($^{16}O/^{18}O$ exchange) were applied [31, 32]. The measured oxygen permeation current density was about 3–10 mA/cm^2 at 10^{-13} Pa (about 10^{-18} atm) at the temperature of 1,273 K (the thickness is assumed to be 3 mm). The measured current density is considerably small compared with the calculated value. This was due to the surface oxygen reactivity of $LaCrO_3$. The oxygen permeation needs ionization process to oxide ion (O^{2-}) from oxygen molecules. A low surface reactivity can reduce the permeation flux through the $LaCrO_3$ and eventually reduces the permeation current density. The oxygen vacancy diffusion coefficients in $La_{0.7}Ca_{0.3}CrO_3$ were determined by isotope oxygen exchange method: The oxygen vacancy diffusion coefficient (D_v) was measured to be around $10^{-5}\ cm^2s^{-1}$ at 1,273 K [32]. The D_v values with different Ca concentration are almost at the same level ($10^{-5}\ cm^2s^{-1}$) and the activation energy for D_v is around 77–142 kJ/mol [29].

Future Directions and Summary

Interconnectors for SOFCs have been developed in two types of materials: metals and oxides. In the developing SOFC technology, many companies apply metallic interconnects with oxide coatings. Lowering operation temperature needs more stable and Cr evaporation-less metallic interconnectors. In that sense, coating materials and technology will be the key for developing the technology.

Cross-References

► Solid Oxide Fuel Cells, History
► Solid Oxide Fuel Cells, Introduction

References

1. Minh NQ (1993) Ceramic fuel cells. J Am Ceram Soc 76:563–588. doi: 10.1111//j.1151-2916.1993.tb03645.x
2. Minh NQ, Takahashi T (1995) Science and technology of ceramic fuel cells. Elsevier, Amsterdam
3. Singhal SC (2000) Advances in solid oxide fuel cell technology. Solid State Ion 135:305–313
4. Anderson HU, Tietz F (2003) Interconnects. In: Singhal SC, Kendall K (eds) High temperature solid oxide fuel cells. Elsevier Advanced Technology, London, Chapter 7
5. Fergus JW (2005) Metallic interconnects for solid oxide fuel cells. Mater Sci Eng A397:271–283
6. Horita T, Xiong Y, Yamaji K, Sakai N, Yokokawa H (2003) Evaluation of Fe-Cr alloys as interconnects for reduced operation temperature SOFCs. J Electrochem Soc 150:A243–A248
7. Horita T, Yamaji K, Xiong Y, Kishimoto H, Sakai N, Yokokawa H (2004) Oxide scale formation of Fe-Cr alloys and oxygen diffusion in the scale. Solid State Ion 175:157–163
8. Hoirta T, Kishimoto H, Xiong Y, Xiong Y, Sakai N, Brito ME, Yokokawa H (2006) Oxide scale formation and stability of Fe-Cr alloy interconnects under dual atmospheres and current flow conditions for SOFCs. J Electrochem Soc 153:A2007–A2012
9. Horita T, Kishimoto H, Yamaji K, Sakai N, Xiong Y, Brito ME, Yokokawa H (2006) Effects of silicon concentration in SOFC alloy interconnects on the formation of oxide scales in hydrocarbon fuels. J Power Sources 157:681–687
10. Horita T, Kishimoto H, Yamaji K, Xiong Y, Sakai N, Brito ME, Yokokawa H (2008) Evaluation of laves-phase forming Fe-Cr alloy for SOFC interconnects in reducing atmosphere. J Power Sources 176:54–61
11. Yasuda N, Uehara T, Tanaka S, Yamamura K (2011) Development of New alloys for SOFC interconnects with excellent oxidation resistance and reduced Cr-evaporation. ECS Trans 35:2437–2445
12. Stanislowski M, Wessel E, Hipert K, Markus T, Singheiser L (2007) Chromium vaporization from high-temperature alloys I. Chromia-forming steels and influence of outer oxide layers. J Electrochem Soc 154:A295–A306
13. Stanislowski M, Froitzheim J, Niewolak L, Quadakkers WJ, Hilpert K, Markus T, Singheiser L (2007) Reduction of chromium vaporization from SOFC interconnectors by highly effective coatings. J Power Sources 164:578–589
14. Horita T, Xiong Y, Kishimoto H, Yamaji K, Brito ME, Yokokawa H (2010) Chromium poisoning and degradation at (La, Sr)MnO3 and (La, Sr)FeO3 cathodes for solid oxide fuel cells. J Electrochem Soc 157: B614–B620
15. Shaigan N, Qu W, Ivey DG, Chen W (2010) A review of recent progress in coatings, surface modifications and alloy developments for solid oxide fuel cell ferritic stainless steel interconnects. J Power Sources 195:1529–1542
16. Yang Z, Xia GG, Maupin GD, Stevenson JW (2006) Evaluation of perovskite overlay coatings on ferritic stainless steels for SOFC interconnect applications. J Electrochem Soc 153:A852–A1858
17. Horita T, Kishimoto H, Yamaji K, Xiong Y, Brito ME, Yokokawa H, Baba Y, Ogasawara K, Kameda H, Matsuzaki Y, Yamashita S, Yasuda N, Uehara T (2008) Diffusion of oxygen in the scales of Fe-Cr alloy interconnects and oxide coating layer for solid oxide fuel cells. Solid State Ion 179:2216–2221
18. Fergus JW (2004) Lanthanum chromite-based materials for solid oxide fuel cell interconnects. Solid State Ion 171:1–4
19. Sakai N, Kawada T, Yokokawa H, Dokiya M, Iwata T (1990) Thermal expansion of some chromium deficient lanthanum chromites. Solid State Ion 40(41):394–397
20. Chick LA, Liu J, Stevenson JW, Armstrong TR, McCready DE, Maupin GD, Coffery GW, Coyle CA (1997) Phase transitions and transient liquid-phase sintering in calcium-substituted lanthanum chromite. J Am Ceram Soc 80:2109–2120
21. Mori M, Yamamoto T, Ichikawa T, Takeda Y (2002) Dense sintered conditions and sintering mechanisms for alkaline earth metal (Mg, Ca, and Sr)-doped $LaCrO_3$ perovskites under reducing atmosphere. Solid State Ion 148:93–101
22. Weber WJ, Griffin CW, Bates JL (1987) Effects of cation substitution on electrical and thermal transport properties of $YCrO_3$ and $LaCrO_3$. J Am Ceram Soc 70(4):265–270
23. Yasuda I, Hikita T (1993) Electrical conductivity and defect structure of calcium-doped lanthanum chromites. J Electrochem Soc 140:1699–1704
24. Mizusaki J, Yamauchi S, Fueki K, Ishikawa A (1984) Nonstoichiometry of the perovskite-type oxide $La_{1-x}Sr_xCrO_{3-\delta}$. Solid State Ion 12:119–124
25. Onuma S, Yashiro K, Miyoshi S, Kaimai A, Matsumoto H, Nigara Y, Kawada T, Mizusaki J, Kawamura K, Sakai N, Yokokawa H (2004) Oxygen nonstoichiometry of perovskite-type oxide $La_{1-x}Ca_xCrO_{3-\delta}$($x = 0.1, 0.2, 0.3$). Solid State Ion 174:287–293

26. Oishi M, Yashiro K, Hong JO, Nigara Y, Kawada T, Mizusaki J (2007) Oxygen nonstoichiometry of B-site doped $LaCrO_3$. Solid State Ion 178:307–312
27. van Hassel BA, Kawada T, Sakai N, Yokokawa H, Dokiya M (1993) Oxygen permeation modeling of $La_{1-y}Ca_yCrO_{3-\delta}$. Solid State Ion 66:41–47
28. van Hassel BA, Kawada T, Sakai N, Yokokawa H, Dokiya M, Bouwmeester HJM (1993) Oxygen permeation modeling of perovskite. Solid State Ion 66:295–305
29. Yasuda I, Hishinuma M (1996) Electrochemical properties of doped lanthanum chromites as interconnectors for solid oxide fuel cells. J Electrochem Soc 143:1583–1590
30. Yasuda I, Hishinuma M (1994) Precise determination of the chemical diffusion coefficient of calcium-doped lanthanum chromites by means of electrical conductivity relaxation. J Electrochem Soc 141:1268–1273
31. Sakai N, Yamaji K, Horita T, Yokokawa H, Kawada T, Dokiya M, Hiwatashi K, Ueno A, Aizawa M (1999) Determination of the oxygen permeation flux through $La_{0.75}Ca_{0.25}CrO_{3-\delta}$ by an electrochemical method. J Electrochem Soc 146:1341–1345
32. Kawada T, Horita T, Sakai N, Yokokawa H (1995) Dokiya M (1995) experimental determination of oxygen permeation flux through bulk and grain boundary of $La_{0.7}Ca_{0.3}CrO_3$. Solid State Ion 79:201–207

Interfaces Modified with Electroactive Biological Species

Robert Forster
School of Chemical Sciences National Center for Sensor Research, Dublin City University, Dublin, Ireland

Introduction

The shuttling of electrons in living systems is relatively efficient because the electroactive biomolecules exist within highly organised biomolecular superstructures. For example, the liver cytochrome enzyme P450, cyt-P450, exists within a membrane and one or more reductase enzymes bind and deliver electrons from NAD(P)H for the oxidative metabolism of lipophilic molecules including drugs. The two key features are that biomolecules at interfaces play significant roles, and that transporting electrons into, and out of, active sites requires significant tailoring of the local microenvironment. In this article, we discuss interfaces that are modified with electroactive biomolecules and consider how electrochemical communication can be established between intrinsically electroactive biomolecules, such as redox enzymes, nucleic acids and proteins, and a conducting electrode surface. These modified interfaces are centrally important to the broad field of bioelectronics and play crucial roles in the construction of electrochemical biosensors.

Typically, for immobilised biomolecules, such as enzymes, their redox active sites cannot communicate directly with an electrode because the biocatalytic centre is insulated by a protein shell. Therefore, a mediator, capable of shuttling electrons between the active site and the electrode, is typically used. There are two significant approaches to "wiring" biomolecules so as to establish electrical contact. The first approach is to use an electron transfer mediator that can either physically diffuse into, and out of, the active site region transporting electrons or provide a redox pathway along which electrons can "diffuse", e.g., a polymer chain with coordinated metal complexes. The second approach involves directly coupling redox active moieties, e.g., ferrocenes, within the proteins so as to enable electrons to 'hop' through the protein matrix. However, just as the successful functioning of cyt-P450 in natural systems demands that it be bound within a membrane, synthetic systems must provide an optimised microenvironment for the electroactive biomolecule if full functionality is to be achieved [1]. Most biomolecules that are freely mobile in solution are dimensionally unoriented, mechanically fragile, and rather unstable under normal conditions. Assembling these molecules at interfaces can provide an optimised orientation, e.g. with respect to substrate binding, rigidity, optical and/or electronic addressability as well as stability, all of which are critical for practical applications. This synthetic superstructure approach using MET was first applied in 1987, when glucose oxidase was electrically 'wired' by covalently binding mediators

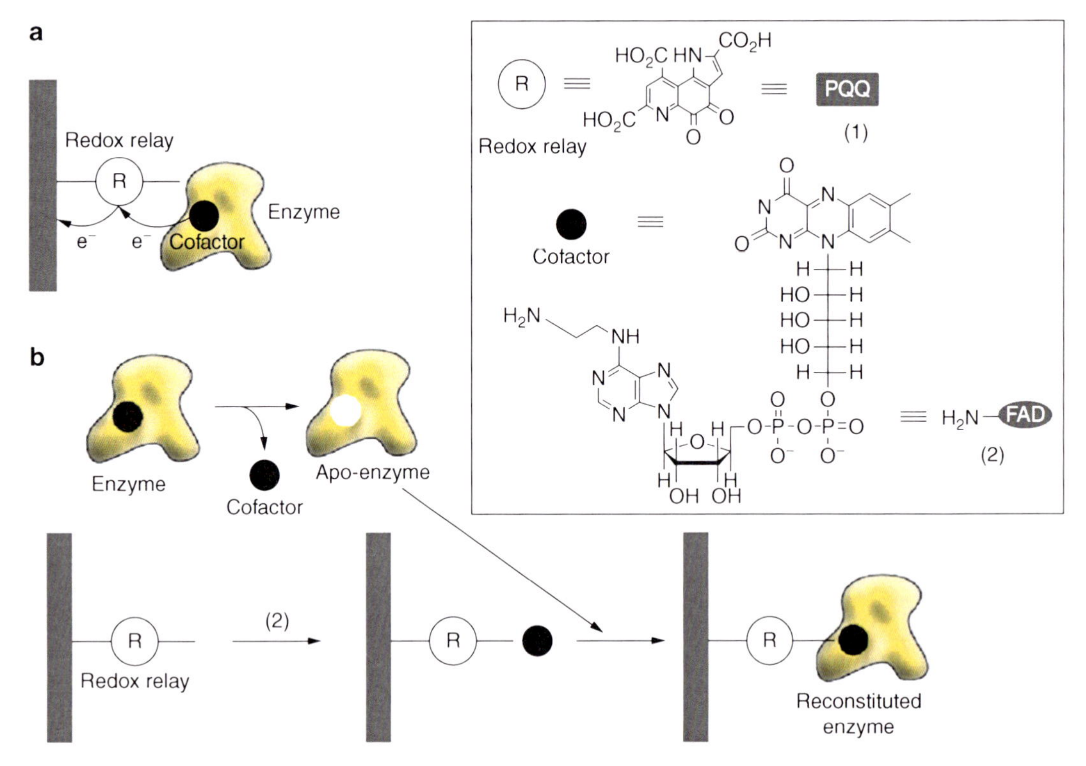

Interfaces Modified with Electroactive Biological Species, Fig. 1 Electrical wiring of redox enzymes. (**a**) Optimal configuration for the electrical contacting of a redox enzyme with the electrode. (**b**) Reconstitution of an apo-enzyme on a relay–cofactor monolayer for the alignment and electrical wiring of a redox enzyme. The structures of the redox relay molecule PQQ and cofactor amino-FAD are shown in the inset (Reproduced with permission from Ref. [1])

between FAD/FADH2 sites and the electrode [2]. The bound centres split the electron tunnelling distance by relaying the holes from the electrode to the FADH2 and the first mediated electron transfer to an interfacial electroactive biomolecule was achieved.

Electrical Wiring of Redox Enzymes in Monolayer Systems

Figure 1a outlines the desired surface architecture of an immobilised redox protein so as to maintain effective electrochemical communication with the electrode. First, the enzyme must be aligned on the electrode surface in a specific orientation where the redox centre is close to the electrode surface, but the binding site remains accessible. Second, the electron transfer mediating units need to be positioned so that they can facilitate electron transfer between the biocatalyst and the electrode.

The nature of the "wire" can vary significantly from nanoparticles and carbon nanotubes to electronically conducting molecular wires. For example, oligophenylacetylenes are rigid rod, conjugated molecular wires capable of sustaining efficient charge transport. These molecular wires have been functionalized with a head group that acts as an inhibitor of the redox protein of interest so that the association event aligns the redox protein on the electrode, and the oligophenylacetylenes facilitates the electron transport between the enzyme and the electrode [3]. For example, this approach was used to successfully wire amine oxidase, which has a quinone active centre and catalyzes the oxidation of amines to aldehydes using Cu^{2+} and the topaquinone redox centre of the enzyme. This enzyme binds diethylaniline and an aniline functionalised molecular wire with a thiol terminus can be used to bind the enzyme to the electrode. Following immobilisation on an electrode, the quinone unit of the protein undergoes quasi-reversible electron transfer indicating that the active site is in electrochemical communication with the electrode surface [3].

Supramolecular assemblies/architectures represent an alternative approach to electrically connecting redox proteins with electrodes. For example, bis-bipyridinium cyclophane has been threaded onto a molecular wire assembled on an electrode by the formation of an intermediary pdonor–acceptor bond with the bis-imine-benzene site of the molecular wire [4]. The wire was then 'stoppered' by an adamantane stopper unit to form a supramolecular assembly. The charged cyclophane could then be moved along the wire by electrochemically changing its oxidation state. Reduction of the cyclophane to the bis-radical cation removed the electron acceptor properties of the threaded ring, and the reduced acceptor was electrostatically attracted by the electrode. This resulted in translocation of the reduced cyclophane to the electrode with a rate constant corresponding to $k = 320\ s^{-1}$. Oxidation of the reduced cyclophane reversed the direction of movement.

Metal Nanoparticles and Carbon Nanotubes as Mediators

Willner and co-workers introduced the concept of using metal nanoparticles as the electrical connector between a reaction centre of an enzyme and an electrode [5]. For example, as shown in Fig. 2, a 1.4 nm gold nanoparticle 'wires' one of the two reaction centres of the enzyme glucose oxidase to a gold macroelectrode. The wired enzyme catalyzes the oxidation of glucose even in the absence of molecular oxygen which is the natural co-reactant of the glucose oxidase.

In this approach, an Au-NP with a *single* N-hydroxysuccinimide group attached is first reacted with N6-(-aminoethyl)-flavin adenine dinucleotide to yield the FAD-functionalized Au-NP. The apo-GOx (GOx without the FAD/FADH2 cofactor) is first reconstituted with the FAD-functionalized Au-NP and then assembled on the macroelectrode by reacting the available thiol moiety. Then, FAD-functionalized Au-NPs are assembled on the electrode and reacted with apo-GOx. Electrons move to the gold electrode as glucose is oxidised. Meanwhile, holes or electron vacancies flow in the reverse direction, from the electrode to glucose, fulfilling the role normally played by molecular oxygen. This approach reduces the effective distance that holes must travel in a single step and, since electron and hole tunnelling depends exponentially on distance, the electrocatalytic rate is dramatically increased, i.e., more than 1,000-fold. The 5,000 s^{-1} turnover rate of the gold-particle wired glucose oxidase is among the highest ever observed for the enzyme. It is about sevenfold higher than that observed for the enzyme using its natural co-reactant, molecular oxygen.

Carbon nanotubes, CNTs, provide an additional nanostructure that can mediate electron transfer to an immobilised electroactive biomolecule [6, 7]. First, shortened, carboxylic-acid-terminated CNTs were produced by thermal treatment with a concentrated H_2SO_4/HNO_3 mixture in an ultrasonic bath followed by fractionation according to CNT length. These CNTs were then covalently linked to a cysteamine-functionalized electrode to obtain an upright CNT forest. The aminoethyl-FAD cofactor was then bound to the free end of the CNTs, and apo-glucose oxidase was reconstituted on the cofactor site. While the CNTs successfully wired the reconstituted glucose oxidase, and glucose could be oxidised irrespective of the length of the CNTs, the overall efficiency decreased as the length of the CNT decreased. This decrease was attributed to the lower conductivity of the longer CNTs especially due to defects.

Metallopolymer Mediators

Three-dimensional wires or networks of mediators have the central advantage that multiple

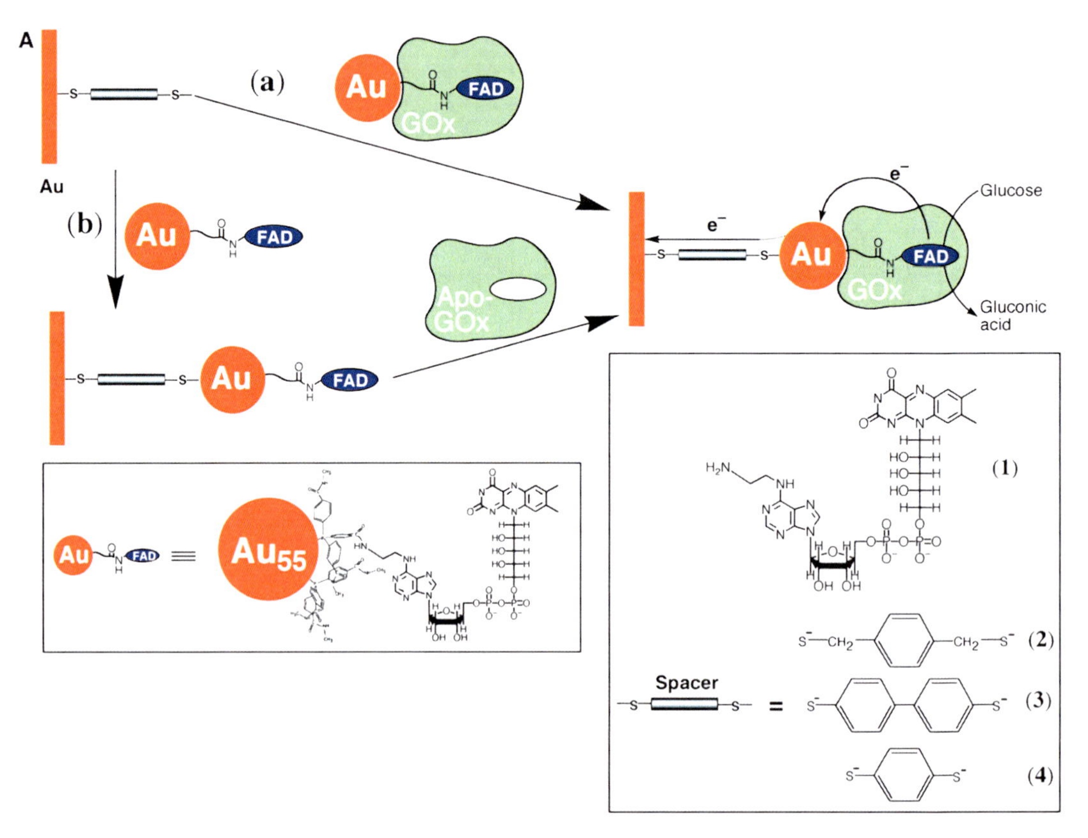

Interfaces Modified with Electroactive Biological Species, Fig. 2 Assembly of Au-NP–reconstituted GOx electrode by (**a**) the adsorption of Au-NP–reconstituted GOx to a dithiol monolayer associated with a Au electrode and (**b**) adsorption of Au-NPs functionalized with FAD on the dithiol-modified Au electrode followed by the reconstitution of apo-GOx on the functional NPs (Reproduced with permission from Ref. [5])

layers of an oxido-reductase enzyme can be electrically connected to an electrode [8–10]. Redox-active polymer films are ideally suited to developing these systems. The properties of many of these electroactive polymers, e.g., their conductivity, charge distribution, and shape, can be changed in a controlled and reproducible way in response to environmental stimuli, e.g., a change in the nature of the contacting solution or an applied voltage, light intensity, or mechanical stress.

Redox hydrogels containing coordinated metal complexes that have a formal potential which enables mediated electron transfer to and from interfacial electroactive biomolecules are highly attractive for biosensor development [11]. For example, redox hydrogels are the only known electron conducting phase in which glucose, gluconolactone, and water-soluble ions dissolve and diffuse. In particular, polycationic metallopolymers form electrostatic adducts with glucose oxidase which is a polyanion at neutral pH. Strands of the polymer can approach the protein-buried FADH2 centres so that electron and hole tunnelling can occur [12]. Because the redox hydrogels envelope the redox enzymes, they electrically connect the enzymes' reaction centres to the electrode irrespective of the spatial orientation of the enzyme and also connect multiple enzyme layers. Hence, the current densities are usually about 10-fold higher, and in some cases 100-fold higher, than they are for enzyme monolayers. This approach leads to very high current densities, of the order of mA cm^{-2},

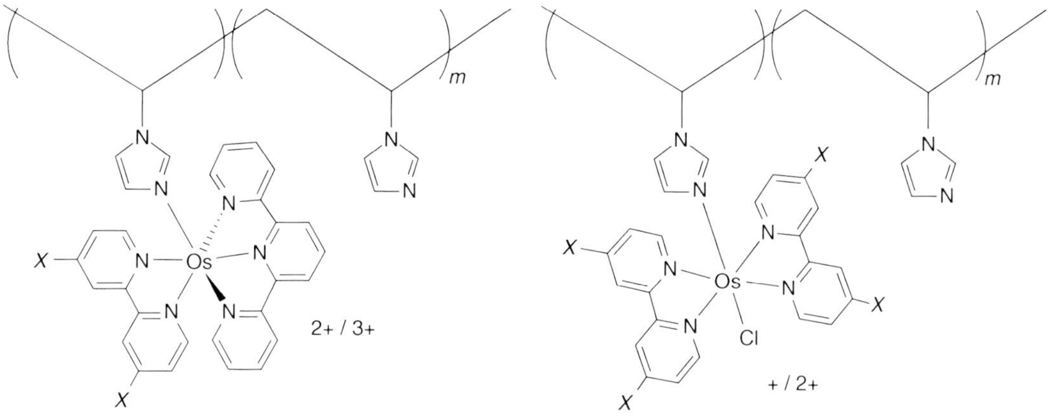

Interfaces Modified with Electroactive Biological Species, Fig. 3 Osmium containing redox hydrogels used for mediated electron transfer to the laccase enzyme of Trametes versicolor. X can be chloride, proton, methyl, or methoxy

reflecting efficient 3-D electrocatalysis. Moreover, the maximum rate of glucose electro-oxidation is observed at a very low potential difference, e.g., 200–300 mV, between the reversible potential of the FAD/FADH2 centre and the anode.

A key issue in developing supramolecular redox systems of this kind is determining the exact mechanism of charge transport through the film and maximising its rate [13–18]. In common with proteins, the redox potential of the coatings is defined by the metal centre or the nature of the conducting polymer, but charge transport through the layer depends strongly on the nature of the polymer layer and its interaction with the contacting sample. In this sense, both polymer-modified electrodes and redox-active proteins are true supramolecular assemblies. First, they both contain linked components each of which largely retains its own identity, e.g., redox centres and a polymer backbone. Second, the assembly is capable of performing a function beyond that possible using the building blocks in isolation, e.g., selective catalysis. Third, the properties of the active sites are modulated by the physicochemical properties of the molecular scaffolding; e.g., the rate of ion transport in response to a change in the oxidation state of the metal centres is modulated by the polymer properties.

Calabrese Barton built on Heller's pioneering work of using osmium containing redox hydrogels to wire enzymes by introducing a novel approach to obtaining enzyme kinetic data for wired systems where multiple physical processes may simultaneously limit the device performance. Oxygen-reducing enzyme electrodes were prepared from laccase derived from Trametes versicolor. As shown in Fig. 3, osmium containing metallopolymers were synthesized with redox potentials ranging from 0 V to nearly the redox potential of the laccase itself, approximately 0.82 V (SHE). Particular attention was paid to driving forces in the 0–400 mV range where the maximum mediated rate constant is expected. The experimental current density generated by the modified electrodes were analyzed using a one-dimensional numerical model to obtain kinetic parameters. Significantly, the bimolecular rate constant for mediation is found to vary with mediator redox potential from 250 s^{-1} M^{-1} when mediator and enzyme are close in redox potential to 9.4×10^4 s^{-1} M^{-1} when a large driving force was present. The value of the bimolecular rate constant for the simultaneously occurring laccase-oxygen reaction is found to be 2.4×10^5 s^{-1} M^{-1}. The relationship between the mediator-enzyme overpotential and the bimolecular rate constant allowed the optimum mediator redox potential for maximum power output of a biofuel cell to be predicted. For the laccase of T. versicolor ($E' = 0.82$ V), the optimum mediator potential is 0.66 V (SHE).

Practical Applications

Electroactive biomolecules at interfaces have transformed healthcare particularly through portable glucose monitoring for diabetics using electrochemical sensors. There is intense interest in developing implantable sensors that monitor key biomarkers with frequencies as high as one per minute with a reasonable time period (of the order of a week) between sensor changes [19]. Systems of this kind provide powerful insights into an individuals longitudinal health status and can also monitor the efficacy of therapy. In this regard, there has been significant success and electrochemical glucose monitoring systems [20] now exist that can remain implanted for up to 5 days in the adipose tissue while measuring the glucose concentration every minute. The sensor head is connected to a transmitter that measures the blood glucose concentration and transmits the readings to a hand held display. The sensing electrode comprises a miniature (0.1 mm^2), disposable, amperometric glucose sensor, with an electro-oxidation catalyst made from a crosslinked adduct of glucose oxidase and a redox hydrogel containing a polymer-bound $Os^{2+/3+}$ complex. The dynamic range is from 20 to 500 mg/dL and because GOx is highly selective for glucose, a one-point calibration is possible.

Providing power for implanted sensors and for other autonomous applications represents a significant challenge that interfacial electroactive biomolecules offer significant potential for solving [21]. Following the pioneering work of Heller and co-workers on developing biofuel cells to generate electricity within grapes [22], biofuel cells have been developed that operate within snails [23] and cockroaches [24]. Fuel cells of this type could power implanted sensors and associated communications.

Future Directions

This short article has attempted to convey a sense of the many significant new insights and opportunities that have been made possible using interfaces modified with electroactive biological species. The successful coupling of nanostructured materials ranging from self-assembled monolayers, polymer films and nanoparticles with naturally existing, structurally organised, biomolecular systems, has led to the development of practical sensors and biofuel cells. These systems include ultrasensitive devices capable of detecting a few thousand target analytes in a complex sample such as blood as well as robust systems that can be implanted. They have been applied to diagnostics, the detection of pathogenic organisms, food safety, environmental measurements and clinical applications. The key issue to be addressed in the future is perhaps the increasing demand for multi-analyte detection with higher sensitivity and selectivity that will allow molecules to be monitored in real-time, in real-world samples, at a minimal cost. These sensors will uniquely enable biomarkers of disease to be validated in statistically meaningful populations of patients, facilitate geographically disperse environmental monitoring and enable high-throughput, multianalyte detection in industry.

References

1. Willner B, Katz E, Willner I (2006) Electrical contacting of redox proteins by nanotechnological means. Curr Opin Biotechnol 17:589–596
2. Degani Y, Heller A (1987) Electrical communication between chemically modified enzymes and metal electrodes. I. Electron transfer from glucose oxidase to metal electrodes via electron relays, bound covalently to the enzyme. J Phys Chem 91(6):1285–1289
3. Hess CR, Juda GA, Dooley DM, Amii RN, Hill MG, Winkler JR, Gray HB (2003) Gold electrodes wired for coupling with the deeply buried active site of Arthrobacter globiformis amine oxidase. J Am Chem Soc 125:7156–7157
4. Katz E, Sheeney-Ichia L, Willner I (2004) Electrical contacting of glucose oxidase in a redox-active rotaxane configuration. Angew Chem Int Ed 43:3292–3300
5. Xiao Y, Patolsky F, Katz E, Hainfeld JF, Willner I (2003) "Plugging into enzymes": nanowiring of redox enzymes by a gold nanoparticle. Science 299(21):1877–81

6. Patolsky F, Weizmann Y, Willner I (2004) Long-range electrical contacting of redox enzymes by SWCNT connectors. Angew Chem Int Ed 43:2113–2117
7. Katz E, Willner I (2004) Biomolecule-functionalized carbon nanotubes: applications in nanobioelectronics. Chem Phys Chem 5:1184–2104
8. Rusling JF, Forster RJ (2003) Electrochemical catalysis with redox polymer and polyion–protein films. J Colloid Interface Sci 262:1–15
9. Gregg BA, Heller A (1990) Cross-linked redox gels containing glucose oxidase for amperometric biosensor applications. Anal Chem 62(3):258–63
10. Forster RJ, Vos JG (1990) Synthesis, characterisation and properties of a series of osmium and ruthenium containing metallopolymers. Macromolecules 23:4372
11. Heller A, Feldman B (2008) Electrochemical glucose sensors and their applications in diabetes management. Chem Rev 108:2482–2505
12. Heller A (2006) Electron-conducting redox hydrogels: design, characteristics and synthesis. Curr Opin Chem Biol 10:664
13. Forster RJ, Kelly AJ, Vos JG, Lyons MEG (1989) The effect of supporting electrolyte and temperature on the rate of charge propagation through thin films of $[Os(bpy)_2PVP_{10}Cl]Cl$ coated on stationary electrodes. J Electroanal Chem 270:365
14. Forster RJ, Vos JG (1994) Ionic interactions and charge transport properties of metallopolymer films on electrodes. Langmuir 10:4330
15. Brennan JL, Keyes TE, Forster RJ (2011) Electrochemical properties of ruthenium metallopolymer: monolayer-protected gold cluster nanocomposites. J Electroanal Chem 662(1):30–35
16. O'Reilly EJ, Dennany L, Griffith D, Moser F, Keyes TE, Forster RJ (2011) Ground and excited state communication within a ruthenium containing benzimidazole metallopolymer. Phys Chem Chem Phys 13(15):7095–7101
17. Venkatanarayanan A, O'Connell B, Devadoss A, Keyes TE, Forster RJ (2011) Potential modulated electrochemiluminescence of ruthenium metallopolymer films. Electrochem Commun 13(5):396–398
18. Devadoss A, Dickinson C, Keyes TE, Forster RJ (2011) Electrochemiluminescent metallopolymer-nanoparticle composites: nanoparticle size effects. Anal Chem 83(6):2383–2387
19. Heller A (1999) Implanted electrochemical glucose sensors for the management of diabetes. Ann Rev Biomed Eng 1:153–175
20. Heller A, Feldman B (2010) Electrochemistry in diabetes management. Acc Chem Res 43(7):963–973
21. Gao F, Viry L, Maugey M, Poulin P, Mano N (2010) Engineering hybrid nanotube wires for high-power biofuel cells. Nat Commun 1(1):2–7
22. Mano N, Mao F, Heller A (2003) Characteristics of a miniature compartment-less glucose-O_2 biofuel cell and its operation in a living plant. J Am Chem Soc 125(21):6588–6594
23. Halámková L, Halámek J, Bocharova V, Szczupak A, Alfonta L, Katz E (2012) Implanted biofuel cell operating in a living snail. J Am Chem Soc 134(11):5040–5043
24. Rasmussen M, Ritzmann RE, Lee I, Pollack AJ, Scherson D (2012) An implantable biofuel cell for a live insect. J Am Chem Soc 134(3):1458–1460

Ion Channels, Natural Nanovalves

Robert Eisenberg
Department of Molecular Biophysics and Physiology, Rush University Medical Center, Chicago, IL, USA

Ion Channels

Ion channels are proteins with holes down their middle that control the flow of ions and electric current across otherwise impermeable biological membranes [1–4]. The flow of Na^+, K^+, Ca^{2+}, and Cl^- ions has been a central issue in biology for more than a century. The flow of current is responsible for the signals of the nervous system that propagate over long distances (meters). The concentration of Ca^{2+} is a "universal" signal that controls many different systems inside cells. The concentration of Ca^{2+} and other messenger ions has a role in life rather like the role of the voltage in different wires of a computer. Ion channels also help much larger solutes (e.g., organic acids and bases, perhaps polypeptides) to cross membranes, but much less is known about these systems.

Ion channels can select and control the movement of different types of ions, because the holes in channel proteins are a few times larger than the (crystal radii of the) ions themselves. Biology uses ion channels as selective valves to control flow and thus concentration of crucial chemical signals. For example, the concentration of Ca^{2+} ions determines whether muscles contract or not. Ion channels have a role in biology similar to the role of transistors in computers and technology [5]. Ion channels control concentrations important to life in the way transistors control voltages that are important to computers.

Ion channels are not always open. Single channel molecules switch suddenly (in less than one microsecond) between definite open and closed states. Single channel molecules open and close stochastically (in a process called spontaneous "gating") according to well-defined probability distributions [2] that are usually well described by Markov models (with a few rate constants that depend dramatically on conditions [6]).

The biological properties of channels are determined by the properties of the ensemble of channels. The gating properties of the ensemble of channels are not noticeably stochastic. The properties of the ensemble of voltage-dependent channels in nerve fibers are well described by partial differential equations [7] invented to allow computation of the propagating voltage signal of nerve fibers [1]. Different equations are needed for different ions (e.g., Na^+ and K^+) because the ions flow through different proteins that have different structures and properties. The currents through these different channel proteins combine according to the same equations used to describe transmission lines in engineering; the equations were first used by Kelvin to describe the spread of telegraph signals in an insulated cable under the Atlantic Ocean. The solutions of the cable equations of Hodgkin and Huxley are the signals of the nervous system. The signals arise from the coupling of macroscopic boundary conditions and conservation laws (of the transmission line equations) to the gating properties of ensembles of channels (described by the Hodgkin Huxley equations) and the selectivity properties of the hole in single channel proteins (see below). The biological function of nerve is understood and computed by a multiscale analysis, with each scale linked in an explicit and calibrated way to its neighboring scales. It is likely that such multiscale analysis will be needed for many biological systems.

Different types of channels gate in response to different stimuli because different channel proteins have different structural modules to sense the stimuli. Some channels have modules that respond to voltage. Others have modules that respond to temperature. Still others have modules that respond to specific chemicals, e.g., capsaicin, the ingredient that makes chili peppers taste hot. New channels with new types of sensitivity are constantly being discovered. The diversity is remarkable [8]. A substantial fraction of all proteins in a human are ion channels.

Ion channels are so important biologically and medically that they are studied in thousands of laboratories every day by molecular biologists and physicians interested in understanding and controlling disease. Many of the most important drugs in clinical medicine work by controlling the gating of channels. A common strategy of the pharmaceutical industry is to identify agents that control channels and then see what those agents do clinically.

Ion channels are proteins and so can be identified and classified by the powerful methods of molecular and structural biology. These methods are used throughout the worlds of biotechnology and pharmaceutical science, as well as widely in medical and clinical science. These methods readily provide information with atomic resolution not available for most noncrystalline physical systems. The location of individual atoms is known in thousands of different proteins. Remarkably, the chemical nature of small groups of atoms (the side chains of amino acids that make up the protein) can be easily changed by the technique of site-directed mutagenesis. Such mutation experiments often discover small groups of side chains that control the biological function of proteins and channels. The genome (and evolution) evidently controls function in this way.

The selectivity of ion channels is often determined by only a handful of amino acids. Thus, the amino acids Glu-Glu-Glu-Glu (EEEE) determine the selectivity of the calcium channel of the heart [9]; the amino acids Asp-Glu-Lys-Ala (DEKA) determine the selectivity of the sodium channel of nerve and muscle cells [10]. The structure of ion channels is notoriously difficult to determine because channel proteins do not easily crystallize. They are normally found in lipid membranes, and so methods suitable for crystallizing soluble proteins are not too helpful. Several structures have been determined, most notably, the KcsA potassium channel [4].

Ion channels are unusual proteins. The current through a single channel is the same whether the channel is open for a very short time (microseconds) or for a very long time (seconds). We conclude that an ion channel does not change structure significantly, once it is open.

Ion channels are unusual proteins because an important part of their function occurs without conformational changes. Current flow through ion channels is driven by electrodiffusion through a hole of one conformation, once the channel opens. Thus, an open ion channel is a physical object that can be analyzed by the techniques of classical physical chemistry. We do not need to understand vaguely defined conformation changes or mysterious allosteric effects. We do not need to understand the opening process to understand the open channel.

Ion channels allow atomic scale structures to control macroscopic function, so it is natural to seek understanding with models that include all atomic detail using the methods of molecular dynamics to compute atomic motion. The problem with this approach is that atoms move a great deal, at the speed of sound, as a first approximation [11], and so computations must resolve 10^{-16} s. Little biology happens faster than 10^{-4} s. Atomic scale computations must extend over 12 orders of magnitude in time if they are to compute biological function.

Current flow through channels depends on bulk concentrations of ions ranging from 10^{-7} M (or even much less) to 0.5 M. Concentrations of ions in and near ion channels (and in and near active sites of enzymes) are much higher, even larger than 10 M because the systems are so small and have large densities of permanent charge from acid and base side chains of proteins. Computations of these concentrations must include nonideal properties of highly concentrated interacting solutions and side chains because these are known to be of great importance in ionic mixtures or solutions greater than 50 mM concentration. Simulations in full atomic detail must also include staggering numbers of water molecules to deal with the trace concentrations of 10^{-11} to 10^{-7} M of signaling ions, e.g., Ca^{2+}.

Molecular dynamics of biological function must be done in nonequilibrium conditions where flows occur, because almost all biology occurs in these conditions. Simulations have difficulty in dealing with the action potentials of nerve and muscle fibers. Molecular dynamics cannot compute the billions of trajectories of ions that cross membranes to make action potentials, lasting milliseconds to nearly a second, flowing centimeters to meters down nerve axons. Simulations at present cannot deal with flows that are controlled by channels on the atomic scale but couple to boundary conditions on the macroscopic scale, millimeters away from the channel. Many biological systems use atomic scale structures this way to control macroscopic function.

Simulations must deal with all these issues of scale at once, because biology uses them all at once. For these reasons, molecular dynamics cannot deal directly with biological function, as of now. A multiscale approach with explicit models and calibrated links between scales seems unavoidable, as in the analysis of the propagating voltage signal of nerve fibers, and in engineering technology [6], in general.

Multiscale reduced models of ion channels are feasible. Reduced models of some open ionic channels have proven surprisingly successful [6]. In several cases, it has been possible to understand, and predict experimental results before they were done. Channels have been built that behave as expected [12]. These reduced models include surprisingly little atomic detail; for example, they treat water as a continuum dielectric and side chains as spheres. It is not clear why a sensitive biological function like selectivity can be explained with such little regard to the atomic details of hydration and solvation, but the evidence is clear. In several important cases, the explanation is successful.

Specifically, the selectivity of the ryanodine receptor of cardiac and skeletal muscle can be understood with a model with less than a dozen parameters that never change value [13]. Detailed properties of the current through the channel were successfully predicted (in quantitative detail, with errors of a few per cent) with this model, often before the experiments were performed. Predictions were successful after drastic mutations, and in many (>100) solutions, of widely varying composition.

The calcium channel of cardiac muscle has a complex pattern of binding of ions that can be understood (over four orders of magnitude of concentration in many types of solutions, containing Na^+, K^+, Rb^+, Cs^+, Ca^{2+}, Ba^{2+}, Mg^{2+}, and so on) with a model containing two or three parameters [14]. Specific mutations change this model of a calcium channel into a sodium channel [15] just as the mutations change the selectivity in experiments. A reduced model can explain these data. The model has just three parameters that never change value, namely, the diameter of the channel, the dielectric coefficient of the solution, and the dielectric coefficient of the protein. Ion diameters in the model never change value. Reduced models of this type have not yet accounted for the selectivity of potassium channels, and it is not clear if they can.

Future Directions

Biologists and physicians will continue to discover and describe the thousands of types of channel proteins that make life possible. Physical scientists will seek models that include enough atomic detail to explain the role of structure, while mathematicians and computer scientists struggle to deal with the motions of those atoms and other problems of scale. Physical chemists will seek simple principles that control the selectivity of ion channels and will try to apply those principles to electrochemical systems of technological interest. The mystery of gating will eventually be resolved. Perhaps, the steady time-independent current of a single open channel will emerge from the analysis of the stability of a coupled ion, channel, bath system, as some type of nonlinear "eigenstate." Biologists will study information from the multiplicity of channels, seeking the pattern(s) that evolution has used to create function, from structure and physics.

Physical scientists will approach main questions: How can multiscale calibrated models of ion channels be built, computed, and tested? How do specific structures control the opening and closing of channels? How do proteins sense voltage? What is opening and closing when channels open? Why do reduced models of selectivity work for some channels, and not others? How can we increase the flow of current through channels so that they perform better? How can we mimic the properties of biological channels in technologically useful systems?

Ion channels are unique objects: They have enormous biological and clinical importance. They use simple physical forces to perform those functions. Ion channels are likely to be one of the first important biological systems that can be understood in full detail, in the tradition of physical science. Understanding ions in channels may be as important to the future of mankind as understanding holes and "electrons" in semiconductors has been to our recent past.

Cross-References

▶ Cell Membranes, Biological
▶ Specific Ion Effects, Evidences

References

1. Hodgkin AL (1992) Chance and design. Cambridge University Press, New York
2. Sakmann B, Neher E (1995) Single channel recording. Plenum, New York
3. Ashcroft FM (1999) Ion channels and disease. Academic, New York
4. MacKinnon R (2004) Nobel lecture. Potassium channels and the atomic basis of selective ion conduction. Biosci Rep 24(2):75–100
5. Eisenberg B (2012) Ions in fluctuating channels: transistors alive. Fluct Noise Lett 11(2):76–96. Earlier version available on http://arxiv.org/as q-bio/0506016v2
6. Eisenberg B (2011). Crowded charges in ion channels. In: Rice SA (ed) Advances in Chemical Physics. John Wiley & Sons, Inc., New York, pp 77–223. Also available at http:\\arix.org as arXiv 1009.1786v1001
7. Hodgkin AL (1958) Ionic movements and electrical activity in giant nerve fibres. Proc R Soc Lond Ser B 148:1–37
8. Abelson JN, Simon MI, Ambudkar SV, Gottesman MM (1998) ABC transporters: biochemical, cellular, and molecular aspects. Academic, New York

9. Sather WA, McCleskey EW (2003) Permeation and selectivity in calcium channels. Annu Rev Physiol 65:133–159
10. Payandeh J, Scheuer T, Zheng N, Catterall WA (2011) The crystal structure of a voltage-gated sodium channel. Nature 475(7356):353–358
11. Berry S, Rice S, Ross J (1963) Physical chemistry. Wiley, New York
12. Vrouenraets M, Wierenga J, Meijberg W, Miedema H (2006) Chemical modification of the bacterial porin OmpF: gain of selectivity by volume reduction. Biophys J 90(4):1202–1211
13. Gillespie D (2008) Energetics of divalent selectivity in a calcium channel: the ryanodine receptor case study. Biophys J 94(4):1169–1184
14. Boda D, Valisko M, Henderson D, Eisenberg B, Gillespie D, Nonner W (2009) Ionic selectivity in L-type calcium channels by electrostatics and hard-core repulsion. J Gen Physiol 133(5):497–509
15. Boda D, Nonner W, Valisko M, Henderson D, Eisenberg B, Gillespie D (2007) Steric selectivity in Na channels arising from protein polarization and mobile side chains. Biophys J 93(6):1960–1980

Ion Extraction

Pierre Bauduin and Luc Girard
Institut de Chimie Separative de Marcoule, UMR 5257 – ICSM Site de Marcoule, CEA/CNRS/UM2/ENSCM, Bagnols sur Ceze, France

Definition

The term "ion extraction" refers to the process of extracting one or several ions from a liquid phase, usually an aqueous phase, to another phase that can be either solid or liquid. The general aim is to separate ions from a native solution or to concentrate ions in order to handle smaller volumes. When different types of ions need to be recovered from their mixture, i.e., performing a selective extraction of each ion, the term "ion separation" is used.

During the last decades, the market for metal resources has been greatly stimulated by the emergence of new technologies and by the world population increase. On the other hand, the limited metal resources as well as fossil fuels will constraint governments to set up restriction measures and will undoubtedly lead to limit export and raise the price. Moreover, the extraction of metal ions from ores leads to toxic wastes whose disposal is now becoming expensive because of increasingly environmental protection regulations. Most of these wastes are classified as hazardous and toxic mainly due to the presence of different metals such as cadmium, chromium, lead, and arsenic whose extraction will be required in the future. As a consequence, there will be a growing need for the development of new efficient and eco-friendly recycling systems based on ion extraction process in order to avoid future mining of underground resources.

The recovery of metals from ores and recycled or residual materials is usually performed by the use of aqueous chemistry. The overall process is covered by the field of hydrometallurgy and is typically divided into three general parts:

1. Leaching, which converts metals into soluble salts in aqueous media
2. Solution concentration and purification, which are performed by the use of ion extraction techniques, and
3. Metal recovery from the purified solution, for example, by electrolysis

An attempt is made here to give a brief description of important ion extraction methods: solvent extraction, ion exchange, ion flotation, cloud point extraction, precipitation, and membrane separation. All these techniques are based on the more or less specific recognition of ions with chemical/physical sites. Depending on the technique, this recognition can originate from

1. electrostatics including ion-pairing properties, surface effects, e.g., charge density, polarizabilities of the species and/or
2. physical effects, i.e., size and shape recognition.

Steric hindrance in a porous structure may strongly influence ion selectivity. Moreover, the bulk physicochemical properties of the ions in water may play an important role in the ion extraction/separation process. To give an example, ion hydration, which is characterized by hydration entropy and enthalpy, is known to be involved in many specific ion effects in solution.

Solvent Extraction

Solvent extraction is the general term referring to the distribution of a solute between two immiscible liquid phases in contact, usually water and oil (organic solvent) also called "diluent" in hydrometallurgy [1]. When coexisting solutes (metal ions) have different distribution ratios (D), defined as the ratio between the solute concentration in the organic phase and its concentration in the aqueous phase, see Eq. 1, separation of these solutes can be obtained. D depends on the thermodynamic properties of the system, like temperature, concentration, pressure, and so on, and can be related to the Gibbs free energy of the extraction process (ΔG).

$$D_M = \frac{\overline{[\mathrm{M}]}_{t,org}}{[\mathrm{M}]_{t,aq}} \qquad \Delta G = -RT \ln D_M \tag{1}$$

with $[\mathrm{M}]_t$ being the sum of the concentrations of the metal species in its aqueous medium or in its complexed organic form (denoted with horizontal bar).

Turbulent stirring of the two liquid phases is usually employed to increase surface contact between oil and water so as to reach rapid distribution of the solute. This emulsification process is then followed by the settling of the two liquid phases which has to be fast and efficient for industrial applications. The extraction process is performed in the presence of hydrophobic extractant molecules that aim to enable the solute transfer from the water phase to the organic phase, see Fig. 1. Hydrophobic extractants are constituted by a polar complexing part and a hydrophobic part. Extractants are designed to achieve selectivity for ion separation.

Owing to their amphiphilic structures, extractant molecules show self-assembly properties in the organic phase. Extractants form spontaneously small reverse micelles having low aggregation numbers less than ten molecules, see Fig. 1, in the organic phase. The core of the aggregate is less than 0.5 nm; therefore, complexing agents are always in first or second coordination spheres of the extracted ions.

Ion Extraction, Fig. 1 Scheme of an ion extraction system in the water/oil interface region. Ions in the aqueous phase are transferred in extractant reverse micelles present in the organic phase (Courtesy Philippe Guilbaud *CEA* Marcoule, France)

Therefore, metal ions are usually completely dehydrated in the extractant reverse micelles.

Different types of ion extractant are available, and they could be classified according to their extraction mechanism as follows [2]:

1. Extraction by cation exchange, e.g., with acidic function like in HDEHP (bis (ethylhexyl) phosphoric acid)
2. Extraction by solvation, e.g., with a neutral function like in TBP (tri-*n*-butyl phosphate), TOPO (trioctyl phosphine oxide)
3. Extraction by formation of ion pairs, e.g., amine salts like TOA (tri-*n*-octylamine)
4. Extraction by complex synergistic effects, e.g., enhanced U(IV) extraction by coupling of an acidic extractant with a neutral one (HDEHP-TOPO).

The chemical behavior of these four classes of extractant molecules in solution gives the mechanism of transfer of the ion into the organic phase: *1* and *3* are based more on the ionizability of the molecule and so roughly on the electrostatic interaction with the metal ion, whereas *2* is based on the competition of the extractant with the first solvation shell of the metal ion which once replaced facilitates the transfer of the hydrophobic ion complex into an organic phase; *4* combines both mechanisms.

Ion Exchange

Ion exchange is, with solvent extraction, the most used process for ion extraction. It is based on the competitive adsorption process between two ions at a charged solid surface. This process is reversible; the ion exchanger can be regenerated or loaded by washing with the appropriate ions [3, 4].

The materials used as ion exchangers can have different chemical nature: functionalized porous or gel polymer known as ion-exchange resins, zeolites, clays (montmorillonite), and soil humus. Ion exchangers are anionic, cationic, or amphoteric respectively when cations, anions, or both have to be separated from the solution. However, the amphoteric exchange can be more efficiently performed in *mixed beds* containing a mixture of anion and cation exchanger, or passing the treated solution through several different ion-exchange materials. Zeolites are all-inorganic microporous materials and are widely used in ion extraction [3]. They are aluminosilicate minerals with a porous structure that can accommodate a wide variety of cations and shows ion selectivity according to the cation and pore sizes. Ion-exchange resins are porous support structures, usually small beads (1–2 mm), made of an insoluble organic polymer substrate (most of them are based on cross-linked polystyrene). The pores on the surface are used to easily trap and release ions. The trapping of ions takes place only with the simultaneous release of other ions.

The four main types of ion-exchange resins differ in their functional groups: strongly acidic (typically sulfonic acid groups, e.g., sodium polystyrene sulfonate or polyAMPS), strongly basic (quaternary amino groups, e.g., trimethylammonium groups, e.g., polyAPTAC), weakly acidic (mostly, carboxylic acid groups) and weakly basic (primary, secondary, and/or ternary amino groups, e.g., polyethylene amine). There are also specialized resin types containing chelating moieties (iminodiacetic acid, thiourea, and many others). Depending on the nature of the functional group, ion exchangers can be non-ion-specific or can have binding specificity for certain ions or classes of ions [4]. The active groups can be introduced after polymerization, or substituted monomers can be added during the polymerization process. Ion-exchange resins are not only designed as bead-shaped materials but are also produced as membranes (see the section on "Membrane Filtration").

Ion Flotation

Ion flotation (or foam fractionation) is an extraction/separation process used to concentrate ionic species present in a dilute aqueous solution [5]. This is a simple and cost-effective technique that can be used for the concentration of valuable materials or removal of toxic materials from very large volumes of very dilute solutions, an ideal proposition for waste-water treatment [6]. Ion flotation is a subcategory of the more general flotation process, also called froth flotation. Froth flotation is a process for separating solid minerals from gangue by taking advantage of differences in their hydrophobicity. Hydrophobicity differences between valuable minerals and waste gangue are increased through the use of surfactants that adsorb on the mineral surface. The selective separation of the minerals makes processing complex ores economically feasible (Fig. 2).

Ion flotation involves the addition of an ionic surfactant (collector) to a solution containing ions of opposite charge (colligend). Gas is then bubbling into the solution. As the bubbles rise, the surfactant molecule, which consists of a hydrophobic tail and a charged hydrophilic head group, adsorbs at the surface of the bubble, and the bubble surface becomes charged. Ions in the solution adsorb then at the bubble surface by an ion-exchange process with the surfactant counter ion:

$$Y^{-}_{\text{interface}} + X^{-}_{\text{aqu.}} \rightleftharpoons Y^{-}_{\text{aqu.}} + X^{-}_{\text{interface}}$$

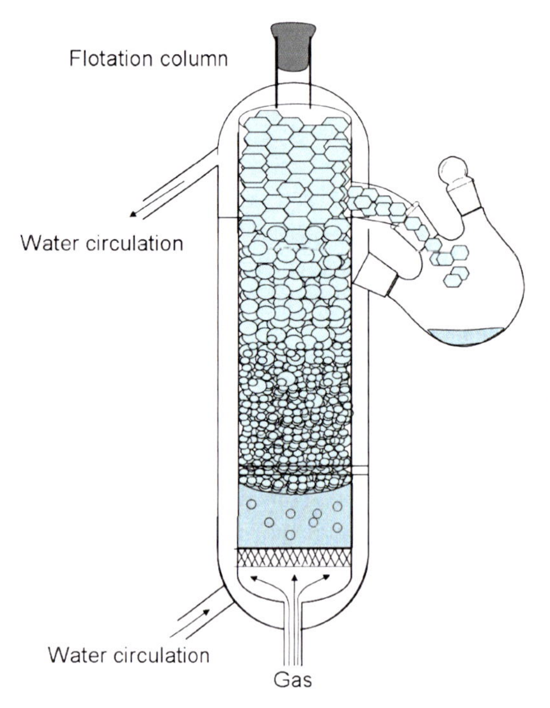

Ion Extraction, Fig. 2 Schematic representation of an ion flotation lab setup (Courtesy Caroline Bauer MedesisPharma, France)

Owing to the foam's large surface area over liquid volume ratio, the liquid that results upon collapse of the foam is manifold enriched in the ion compared to the initial solution. The ion extraction and separation largely depends on the selectivity of the charged surfactant interface for the ion in the presence of competing counterions.

Cloud Point Extraction

Cloud point extraction (CPE) is an analytical tool sometimes presented as a solvent-free alternative to liquid-liquid extraction. CPE is based on the property of many nonionic surfactants, mainly poly-ethoxylated surfactants, in aqueous solutions to form micelles and to undergo liquid-liquid phase separation when heated to a certain temperature called "cloud point" [7]. For analytical purpose, CPE is useful as pre-concentration method that has many advantages, such as low cost, safety, simple procedure, rapid and high capacity to concentrate a wide variety of analytes.

Any species that interacts with the micellar system may be extracted from the initial solution. In order to extract ions, the addition of a suitable hydrophobic ligand or extractant, that is solubilized in the non-ionic surfactant micelles, is needed. An approach based on the use of a surfactant functionalized by a complexing part has been recently proposed [8]. The use of such a surfactant enables to increase the extraction effectiveness and the ion separation factor compared to the system when the surfactant is simply mixed with the ligand.

The process of CPE is divided in three steps: (1) solubilization of the analytes in the micellar aggregates, (2) clouding by increasing temperature, and (3) phase separation. Above the cloud point of the surfactant, the solution separates into two distinct phases: one water-rich phase containing the surfactant at a concentration below, or equal to, its critical micelle concentration and the other one a surfactant-rich phase containing the ion-extractant complex.

Precipitation

Chemical precipitation is widely used for heavy metal removal from inorganic effluent [9]. The main principle is based on adjusting pH to basic conditions (pH >9), in order to change the dissolved metal ion speciation and to convert the metal ions into an insoluble solid phase via a chemical reaction with a precipitant agent such as lime. Typically, the metal precipitates from the solution in its hydroxide form with a general reaction described by the following equation:

$$M^{n+}_{\mathrm{aqu.}} + n\mathrm{OH}^{-}_{\mathrm{aqu.}} \rightleftharpoons M(\mathrm{OH})_n \tag{2}$$

where M^{n+} and OH^- represent the dissolved metal ions and the precipitant, respectively, while $M(OH)_n$ is the insoluble metal hydroxide.

Selective metal ion precipitation from acidic aqueous waste solutions (liquid-solid extraction) was also proposed, e.g., as a simple way to separate actinides by using cationic surfactant as precipitant [10]. This approach has some interesting advantages over solvent extraction, because several steps are omitted like stripping of the extracted species or solvent washing. Moreover, the amount of waste is decreased considerably since no contaminated organic solvent is produced.

Membrane Filtration

Membrane filtration is capable of removing not only suspended solid or organic compounds, but also inorganic compounds such as heavy metal ions. Depending on the nature of the inorganic compound, three different processes can be employed: ultrafiltration (UF), nanofiltration (NF), and reverse osmosis (RO) [11].

UF uses permeable membranes with membrane weight of separating compounds of 1,000–100,000 Da and a pore size of 5–20 nm, allowing the passage of water and low molecular weight solids, while retaining macromolecules which have a size larger than the pore size. The membranes are made of cellulose acetate, polyamide, or silica and alumina.

The ion separation concern several ions such as Co(II), Ni(II), Zn(II), Cr(III), and Cd(II). Depending on the membrane characteristics, it can achieve more than 90 % of removal efficiency with metal concentration from 10 to 112 mg/L. In principle, the separation could be enhanced by using a surfactant such as sodium dodecyl sulfate (SDS) to form micelles or a water-borne polymer such as chitosan to complex ions. However, the membrane fouling problems have hindered this technique from a wider industrial waste-water treatment.

NF involves steric and electrical (Donnan) effects. The interest of the NF membranes lies in its small pore ($\sim$1 nm) and surface charge, which allows charged solutes smaller than the membrane pores to be rejected along with the bigger neutral solutes and salts. The membranes can be organic or inorganic, e.g., polyvinyl alcohol or TiO_2. NF can be assisted by chelating agents such as DTPA, EDTA, HEDTA, e.g., in the actinides(III)/lanthanides(III) separation [12]. Although NF is able to treat inorganic effluents with a metal concentration of 2 g/L in a rather wide pH range (3–8) at pressure of 3–4 bars, it is less intensively investigated than UF and RO for heavy metal removal.

In RO process, water can pass through the membrane, while heavy metal is retained. By applying a higher hydrostatic pressure than the osmotic pressure of the feeding solution, cationic compounds can be removed from water. RO is the most effective membrane separation technique for metal ions' removal from inorganic solution with 97 % of rejection percentage with a metal concentration ranging from 21 to 200 mg/L. RO works (depending on the characteristics of the membrane) in a wide pH range (3–11) at 4.5–15 bars of pressure. The membranes are made of polyamide or sulfonated polysulfone. There are many advantages of the RO: The metal removal efficiency is tuned by the pressure, the high flux rate, and the chemical, thermal, biological stability of the membrane. The main limitations of RO are the fouling of the small pores which might be irreversible and the high energy consumption.

Apart from these techniques, electrodialysis (ED) is also considered as a membrane or an electrochemical technique [13]. It is used to transport ions from one solution through an ion-exchange membrane to another solution under the influence of an applied electric potential. The membranes are made of thin sheets of polymer materials with cationic or anionic properties. ED is performed in an electrodialysis cell consisting of a feed and a concentrate compartment formed by an anion and a cation-exchange membrane placed between two electrodes. For electrodialysis processes, multiple electrodialysis cells are arranged together with alternating anion and cation-exchange membranes to form an ED stack. The anions/cations pass in the diluate stream and migrate toward the anode/cathode through the positively/negatively charged anion-/cation-exchange membranes, but are prevented from further migration toward the anode/cathode by the negatively/positively charged exchange membrane and therefore stay in the concentrate stream. The major applications of ED are the desalination of brackish water or seawater and the production of pure and ultrapure water by electro-deionization.

Cross-References

▶ Ions at Solid-Liquid Interfaces
▶ Membrane Processes, Electrodialysis

References

1. Rydberg J, Musikas C, Choppin GR (1992) Principles and practices of solvent extraction. Marcel Dekker, New York
2. Danesi PR, Chiariza R, Coleman CF (1980) The kinetics of metal solvent extraction. CRC Crit Rev Anal Chem 10(1):1–126
3. Auerbach SM, Carrado KA, Dutta PK (2003) Handbook of zeolite science and technology. Marcel Dekker, New York
4. Korkisch J (1989) Handbook of ion exchange resins: their application to inorganic analytical chemistry. CRC Press, Florida
5. Sebba F (1962) Ion flotation. Elsevier, Amsterdam
6. Zabel T (1984) Flotation in water treatment. In: Ives KJ (ed) The scientific basis of flotation. Springer, The Hague, pp 349–378
7. Almeida Bezerra M, Zezze Arruda M, Costa Ferreira SL (2005) Cloud point extraction as a procedure of separation and pre-concentration for metal determination using spectroanalytical techniques: a review. Appl Spectrosc Rev 40(4):269–299. Silva MF, Cerutti ES, Martinez LD (2006) Microchim Acta 155:349–364
8. Larpent C, Prevost S, Berthon L, Zemb T, Testard F (2007) Nonionic metal-chelating surfactants mediated solvent-free thermo-induced separation of uranyl. New J Chem 31(8):1424–1428
9. Benefield LD, Morgan JM (1999) Chemical precipitation. In: Letterman RD (ed) Water quality and treatment. McGraw-Hill, New York
10. Strnad J, Heckmann K (1992) Separation of actinides with alkylpyridinium salts. J Radioanal Nucl Chem 163:47–57
11. Kurniawan TA, Chan GYS, Lo WH, Babel S (2006) Physico-chemical treatment techniques for wastewater laden with heavy metals. Chem Eng J 118:83–98
12. Borrini J, Bernier G, Pellet-Rostaing S, Favre-Reguillon A, Lemaire M (2010) Separation of lanthanides(III) by inorganic nanofiltration membranes using a water soluble complexing agent. J Membr Sci 348:41–46
13. Davis TA (1990) Electrodialysis. In: Porter MC (ed) Handbook of industrial membrane technology. Noyes, Park Ridge

Ion Mobilities

Roland Neueder
Institute of Physical and Theoretical Chemistry, University of Regensburg, Regensburg, Germany

Introduction

Many properties of electrolyte solutions (e.g., electric conductance, ohmic losses, and power capability of a battery; electrophoresis and polarography; polarization and depolarization of electric double layers; electroplating and electrodeposition) depend directly or indirectly on the ion mobility of its constituents. This entry deals with the ion mobility in aqueous and nonaqueous solutions. For other systems, the reader is referred to monographs of, e.g., J. Maier [1] for solids or of K. Kontturi et al. [2] for membranes or electrodes.

Charge Flow in an External Electric Field

An electrolyte solution which is not in equilibrium is exposed to generalized forces that are responsible for irreversible processes, such as transport or relaxation processes. A gradient of the chemical potential of the considered ions is the source of such a force, producing a particle flow that leads to diffusion and to electric conductance. Neglecting activity coefficients (dilute solutions) the flow of ion i is given by the relation (with the convection term omitted)

$$\vec{J}_i = \underbrace{-D_i \vec{\mathrm{grad}}\, c_i}_{\text{diffusion}} + \underbrace{D_i \frac{c_i z_i F}{RT} \vec{E}}_{\text{migration}} \qquad (1)$$

with D_i, z_i, and c_i as diffusion coefficient, charge number, and concentration of ion i. $\vec{E}$ is the applied electric field strength and F and R are Faraday and gas constant, respectively. The ionic mobility u_i is the velocity v_i of an ion i under unit electric field ($u_i = v_i/E$) and is related to the single ion conductivity λ_i via Faraday constant.

$$\lambda_i = u_i F \tag{2}$$

From the migration part of Eq. 1 and Ohm's law, the following relation between single ion conductivity and diffusion coefficient D_i can be derived:

$$\lambda_i = D_i \frac{|z_i| F^2}{RT} \tag{3}$$

Equation 3 is a very important link between the respective limiting values λ_i^∞ and D_i^∞ that are characteristic ionic transport properties without disturbance by ionic interaction (infinite dilution). Examples in different solvents [3] are given in Table 1.

For practical use, an extensive collection of λ_i^∞-values in aqueous and nonaqueous solutions are given in Ref. [3].

The conductivity can be expressed by three contributions:

$$\lambda_i = \lambda_i^\infty - \lambda_i^{el} - \lambda_i^{rel} \tag{4}$$

with λ_i^∞ from the unperturbed ion movement at infinite dilution and two concentration-dependent terms associated with the ionic atmosphere that lower the ionic mobility. The average effect of the hydrodynamic interactions of solute and solvent is represented by the electrophoretic term λ_i^{el} (central ion and ionic atmosphere are moving in opposite direction), whereas the relaxation term λ_i^{rel} takes into account the relaxation of the deformation of the ionic cloud around the central ion (in front of the moving ion the ionic atmosphere has to build up, whereas behind the moving ion it has to decay).

A series expansion of the respective expressions leads to the following dependence on ionic strength I:

$$\lambda_i = \lambda_i^\infty - S_i\sqrt{I} + E_i\, I \ln I + J_{1i} I - J_{2i} I^{3/2},$$
$$I = \frac{1}{2}\sum z_i^2 c_i \tag{5}$$

Ion Mobilities, Table 1 Limiting ionic conductivities at 25 °C

$\lambda_i^\infty \cdot 10^4/(\mathrm{m^2\Omega^{-1}mol^{-1}})$					
Ion	Water	Ethanol	PC	Sulfolane (30 °C)	Nitromethane
H^+	350.0	–	–	–	–
Li^+	38.8	17.1	7.1	4.3	53.9
K^+	73.5	23.4	11.1	4.1	58.1
Bu_4N^+	19.3	19.8	9.0	2.8	34.1
Cl^-	76.3	23.7	18.7	9.3	62.5

Complete expressions for the coefficients of this equation can be found in Refs. [3, 5] for symmetrical electrolytes and in Ref. [4] for unsymmetrical electrolytes.

Figure 1 shows experimental single ion conductivities of potassium and thiocyanate ions in methanol as an example. The lines are the theoretical curves, calculated with the help of Eq. 5.

For a given solvent, the limiting value of the single ion conductivity, λ_i^∞, is independent of the counter-ion present in the solution and characterizes the solvated ion i. Therefore the limiting equivalence conductance of an electrolyte, Λ^∞, can easily be calculated from the ion conductivities, e.g., for a binary electrolyte

$$\Lambda^\infty = \lambda_+^\infty + \lambda_-^\infty \tag{6}$$

Transference Numbers

The transference number t_i (i = + or −) is that fraction of the total electric current that is carried by the i-ions and is related to single ion conductivities according to

$$t_i = \frac{\lambda_i}{\sum_j \lambda_j} \tag{7}$$

The concentration dependence of the transference number may also be represented with the help of truncated series expansions

$$t_i = \frac{1}{2} + \left(t_i^\infty - \frac{1}{2}\right)\left[1 + \frac{S_2\sqrt{I}}{\Lambda^\infty} + \frac{E_2 I \ln I}{\Lambda^\infty} + B_2 I\right] \tag{8}$$

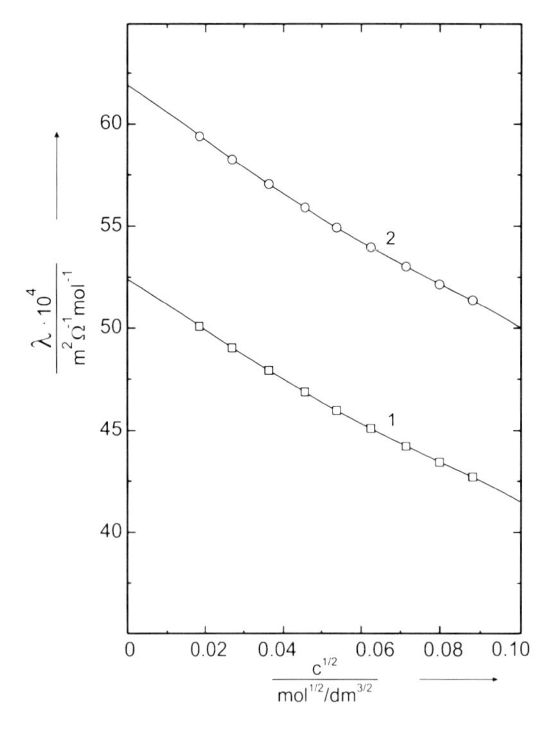

Ion Mobilities, Fig. 1 Single ion conductivities (1: K^+, 2: SCN^-) of the system potassium thiocyanate in methanol; data and parameters from Ref. [5]

Ion Mobilities, Table 2 Anchor values for the Walden product [8]

$\lambda_i^\infty \eta \cdot 10^5/(m^2\ \Omega^{-1}\ mol^{-1}\ Pa \cdot s)$		
0.213 (Bu_4N^+)	0.191 (Am_4N^+)	0.194 (i-$Pent_4N^+$)
0.173 (Hex_4N^+)	0.201 (i-$Pent_3BuN^+$)	0.190 (Ph_4As^+)
0.201 ($Ph4B^-$)	0.199 (i-$Pent4B^-$)	

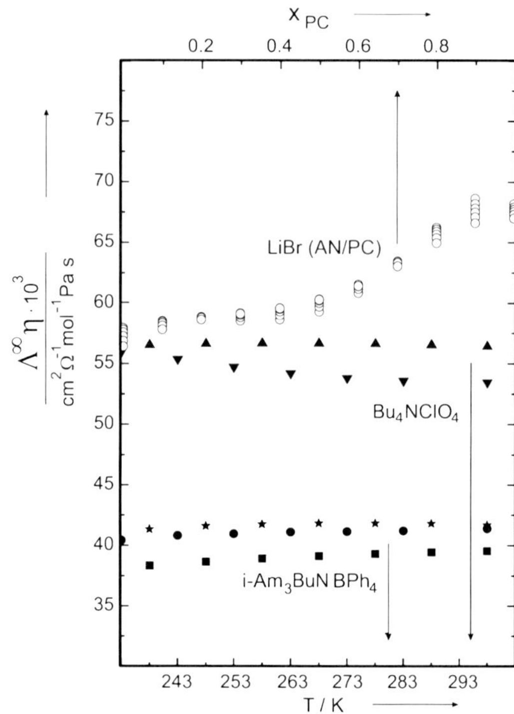

Ion Mobilities, Fig. 2 Walden product ($\Lambda^\infty \eta$) as function of temperature and solvent composition. Butyl triisoamylammonium tetraphenylborate in ethanol (★), 1-propanol (•), and acetonitrile (■). Tetrabutylammonium perchlorate in 1-propanol (▼) and acetonitrile (▲). LiBr (○) in acetonitrile-propylene carbonate mixtures at temperatures from 25 °C to 75 °C

For details on the derivation and the respective coefficients (S_2, E_2, B_2) see Ref. [3].

Some empirical assumptions have been made to avoid the difficult and time-consuming experiments (Hittorf or moving boundary method) to determine transference numbers.

One successful assumption is the splitting of the molar conductivity of a reference electrolyte, such as tetrabutylammonium tetraphenylborate or tetraisoamylammonium tetraisoamylboride, into equal parts $\left(\lambda_+^\infty = \lambda_-^\infty = \frac{1}{2}\Lambda^\infty\right)$ [6].

Using Kohlrausch's additivity law of the independent migration of ions, single ion limiting conductivities of all other ions can be obtained.

Another assumption for the determination of ionic conductivities uses Walden's rule [7], which correlates the conductivity of ions in different solvents with the help of the respective viscosities. Using this assumption, Krumgalz [8] proposed the apparent constancy of the Walden product $\lambda_i^\infty \eta$ for various ions in different solvents and established his so-called anchor values (see Table 2).

The constancy of the Walden product with respect to different temperatures and, in a sufficient way, with respect to different solvents can be seen in Fig. 2 (data from Ref. [9]). For LiBr in acetonitrile-propylene carbonate mixtures, the Walden product is constant up to a mole fraction of 0.5. With increasing amount of propylene carbonate, the radii of the solvated ions changes due to a modified solvation shell leading to an increase of the Walden product (increase of about 10 % for pure propylene carbonate).

Future Directions

Ionic transport properties like ion mobilities and the related conductivities λ_i and diffusion coefficients D_i are measurable quantities that permit the development of helpful concepts of electrolyte solution theories. At dilute and moderate concentrations, the models in use are very satisfying, whereas at high concentrations they still need some improvement. Nevertheless, promising results can be found in the application of the mean spherical approximation (MSA) theory to transport processes.

Cross-References

▶ Conductivity of Electrolytes
▶ Electrolytes, Classification
▶ Electrolytes, History
▶ Ion Properties
▶ Non-Aqueous Electrolyte Solutions
▶ Thermodynamic Properties of Ionic Solutions - MSA and NRTL Models

References

1. Maier J (2005) Physical chemistry of ionic materials. Ions and electrons in solids. Wiley, New York
2. Kontuturi K, Murtomäki L, Manzanares JA (2008) Ionic transport processes in electrochemistry and membrane science. Oxford University Press, New York
3. Barthel JMG, Krienke H, Kunz W (1998) Physical chemistry of electrolyte solutions. Modern aspects, vol 5, Topics in physical chemistry. Springer, Berlin
4. Tsurko EN, Neueder R, Barthel J, Apelblat A (1999) Conductivity of phosphoric acid, sodium, potassium, and ammonium phosphates in dilute aqueous solutions from 278.15 K to 308.15 K. J Solut Chem 28:973–999
5. Barthel J, Neueder R (1992) Electrolyte data collection. In: Eckerman R, Kreysa G (eds) DECHEMA data series, Part 1, XII. Dechema, Frankfurt
6. Coetzee JF, Cunningham GP (1965) Evaluation of single ion conductivities in acetonitrile, nitromethane, and nitrobenzene using tetraisoamylammonium tetraisoamylboride as reference electrolyte. J Am Chem Soc 87:2529–2534
7. Schreiner C, Zugmann S, Hartl R, Gores HJ (2010) Fractional Walden rule for ionic liquids: examples from recent measurements and a critique of the so-called Ideal KCl line for the Walden plot. J Chem Eng Data 55:1784–1788
8. Krumgalz BS (1983) Separation of limiting equivalent conductances into ionic contributions in non-aqueous solutions by indirect methods. J Chem Soc Faraday Trans I 79:571–587
9. Barthel J, Neueder R (1992–2003) Electrolyte data collection. In: Eckerman R, Kreysa G (eds) DECHEMA data series, Part 1a to Part 1h, vol XII. Dechema, Frankfurt

Ion Properties

Yizhak Marcus
Department of Inorganic and Analytical Chemistry, The Hebrew University, Jerusalem, Israel

Introduction

Ions are particles that carry electrical charges, but when in condensed phases, they exist as electrically neutral combinations of positively and negatively charged particles: cations and anions. Although water is the most important solvent, ions do exist also in other environments: in nonaqueous and mixed solvents, and in condensed phases without any solvents as (room temperature) ionic liquids or molten salts.

An *electrolyte* is such a neutral combination that can exist as a chemical substance capable of dissociating (nearly completely) into its constituent ions in a suitable environment. In the following, the subscript $_2$ symbolizes a quantity pertaining to an electrolyte, $_I$ symbolizes one pertaining to an ion, and a generalized ion is designated by $I^{z\pm}$.

Isolated Ions

Isolated ions exist in an ideal gaseous state, devoid of interactions with their surroundings. Their amount of *charge* is a multiple, z_I, of the elementary units $e = 1.60218 \times 10^{-19}$ C. Their *mass* M_I is generally specified per Avogadro's number $N_A = 6.02214 \times 10^{23}$ mol^{-1} (a mole) of ions, in g mol^{-1}. The *shape* of monatomic isolated ions is spherical, but ions that consist of several atoms may have any shape, common ones being planar (NO_3^-), tetrahedral (NH_4^+), octahedral ($Fe(CN)_6^{4-}$), elongated (SCN^-), or more irregular ($CH_3CO_2^-$). The *size* (radius) of an isolated ion is difficult to specify, because its peripheral electrons extend indefinitely.

For many ions, the standard molar *Gibbs energy* and *enthalpy of formation*, $\Delta_f G°(I^{z\pm},g)$ and $\Delta_f H°(I^{z\pm},g)$, from the elements in their standard states, and their standard molar *entropy* and constant pressure *heat capacity*, $S°(I^{z\pm},g)$ and $C_P°(I^{z\pm},g)$, all at $T = 298.15$ K, are reported in the book by Marcus [1]. The ionization potential, $\sum I_p$, is the required investment of energy to produce a cation from a neutral particle. The electron affinity, EA, is the energy released when a neutral particle forms an anion by electron capture. These energies, in electron-volt units (1 eV/particle = 96.483 kJ mol^{-1}), have also been reported [1] for many ions.

Most ions are diamagnetic and their molar *magnetic susceptibilities*, χ_{Im}, range from a few to several tens of -10^{-12} m^3 mol^{-1} [1]. Paramagnetic ions have n unpaired electron, their molar magnetic susceptibilities are $\chi_{Im} = +1.676n(n + 2) \times 10^{-9}$ m^3 mol^{-1} at $T = 298.15$ K. The *polarizability* of an ion $\alpha_I = (3/4\pi N_A)R_{I\infty} = 3.964 \times 10^{-25} R_{I\infty}$ is of the order of 10^{-30} m^3 per ion and the *molar* refractivity $R_{I\infty}$ is obtained experimentally from the *refractive index* n_D of neutral species. It is additive for the constituting ions and individual ionic values are based on the arbitrary value of Na^+ ions: 0.65×10^{-6} m^3 mol^{-1} [1].

Dissolved Ions

A solution of an infinitesimal amount of electrolyte in a finite amount of solvent approximates *infinite dilution*. The ions are then remote from each other and are surrounded by solvent only, and therefore, the individual ionic quantities are additive. They are weighted by their stoichiometric coefficients in the electrolyte: ν_+ cations C^{z+} and ν_- anions A^{z-}, where $\nu_+ z_+ = \nu_- |z_-|$.

Cations in aqueous solution have the water molecules oriented toward them, and for small multivalent cations such as Mg^{2+} and transition metal cations, this results in a coordinate bond in the first *hydration shell*. These water molecules form hydrogen bonds with molecules in a second hydration shell. Other cations, such as Ca^{2+} and the alkali metal ones, do not have a definite number of coordinated water molecules but a distribution with a fractional average number.

Anions in aqueous solutions have the water molecules pointing hydrogen atoms toward them, resulting in hydrogen bonds, multivalent anions, such as CO_3^{2-}, being strongly hydrated. Anions such as HSO_4^- or $H_2PO_4^-$ also donate hydrogen bonds to adjacent water molecules.

Ions with hydrophobic groups in their periphery around a buried charge, such as $(C_6H_5)_4B^-$, are generally only poorly hydrated, but they enhance the water structure.

Sizes of Ions

The *interionic distances* in crystals, of the order of 0.2–1.0 nm, are measured by diffraction of x-rays and neutrons with an accuracy of ±0.0001 nm, being fairly regularly additive in the sizes of the ions. The splitting of the interionic distances into ionic radii r_I depends on the coordination number and geometry [2]. A "selected" set of *ionic radii*, r_I, for those coordination numbers that correspond with the hydration of ions has been provided [1]. Radii for polyatomic ions deviating strongly from a globular shape are ambiguous.

Diffraction methods have also been used in aqueous salt solutions for establishing the distances $d_{I\text{-}O}$ between the centers of ions and those of the oxygen atoms of adjacent water molecules. They have a mean uncertainty of ±0.002 nm, and if the radius of a water molecule, $r_W = 0.138$ nm, is deducted from the $d_{I\text{-}O}$ values, the results correspond quite well with the r_I set for the "bare" ions, as noted by Marcus [3].

The actual volume that should be assigned to an electrolyte in solution is its *partial* molar volume, $V_2 = {}^{\varphi}V_2 + m_2(\partial^{\varphi}V_2/\partial m_2)_T$, where ${}^{\varphi}V_2 = (V - n_1V_1^*)/n_2$ is its *apparent molar volume* in a solution of volume V made up from n_2 mol of electrolyte and n_1 mol of water. In a solution of density ρ at a molality m_2, ${}^{\varphi}V_2 = M_2/\rho + 1{,}000(\rho - \rho_1^*)/\rho\rho_1^* m_2$, M_2 being the molar mass of the solute and ρ_1^* the density of pure water. On extrapolation to infinite dilution ${}^{\varphi}V_2^{\infty} = V_2^{\infty}$, and the contributions, $V^{\infty}(I^{z\pm},aq)$, of the cations and anions are additive. For splitting the measured V_2^{∞} into such contributions, $V^{\infty}(H^+,aq) = -5.4\ cm^3\ mol^{-1}$ at 298.15 K has been accepted as a reasonable estimate. The ionic values reported in [1] have uncertainties of at least $\pm 0.2z_I\ cm^3\ mol^{-1}$. Note that for some cations, and in particular for multivalent ones, the values of $V^{\infty}(I^{z\pm},aq)$ are negative, such cations cause a large electrostriction. The (negative) *molar electrostriction* caused by an ion at infinite dilution, V_{Ielec}^{∞}, is the difference between $V^{\infty}(I^{z\pm},aq)$, and its intrinsic volume, $V_{Iintr} = (4\pi N_A/3)(kr_I)^3$ with the factor $k = 1.23$.

Thermodynamics of Aqueous Ions

Experimental thermodynamic quantities pertain to entire electrolytes or to combinations (sums or differences) of ions that are neutral. Splitting of the electrolyte values for assignment of values to individual ions requires some extra thermodynamic assumption.

The standard partial molar (constant pressure) heat capacity of an electrolyte, C_{P2}^{∞}, is preferably obtained by flow microcalorimetry, accurate to $\pm 3\ J\ K^{-1}\ mol^{-1}$. Absolute standard molar ionic heat capacities, C_{PI}^{∞}, are based on $C_P^{\infty}(H^+,aq) = -71 \pm 14\ J\ K^{-1}\ mol^{-1}$ at 298.15 K. This value is due to equating the C_P^{∞} of aqueous Ph_4P^+ and BPh_4^- ions (the *TPTB assumption*), because of their chemical similarity, similar sizes, and well-buried charges. Since these C_{PI}^{∞} are large, also the uncertainty involved in equating them is large. The values of multi-charged ions are large and negative, and those of polyatomic ions are more positive (or less negative) than those of monatomic ions of the same charge sign [1].

The standard molar entropies of aqueous electrolytes, S_2^{∞}, are preferably obtained from the temperature coefficients of the electromotive forces of galvanic cells. The absolute values for individual ions are based on $S^{\infty}(H^+,aq) = -22.2 \pm 1.4\ J\ K^{-1}\ mol^{-1}$ at 298.15 K, from data for thermocells [1]. The S_I^{∞} values increase with the masses of the ions but are small or negative for multi-charged ions.

The difference between S_I^{∞} in the aqueous solution and S_I° of the ion in the ideal gas phase is its standard molar entropy of hydration $\Delta_{hydr}S_I^{\infty} = S_I^{\infty} - S_I^{\circ}$. These values are related to the effect that ions have on the structure of water [4].

When an ion is transferred, in a thought process, from its isolated state into water at infinite dilution, a large amount of energy is released due to the interaction of the ion with the surrounding water. The net relevant energetic amount is the change in the enthalpy, $\Delta_{hydr}H_I^{\infty}$. The values for complete electrolytes, $\Delta_{hydr}H_2^{\infty}$, are obtained from their heats of solution and lattice energies. In order to split the experimentally available values dealing with entire electrolytes into the ionic contribution, the value $\Delta_{hydr}H^{\infty}(H^+, aq) = -1{,}103 \pm 7\ kJ\ mol^{-1}$ is used, resulting from equating the $\Delta_{hydr}H^{\infty}$ of the Ph_4P^+ and BPh_4^- ions (the TPTB assumption). Contrary to the case of the heat capacities, the latter $\Delta_{hydr}H_I^{\infty}$ are small compared with those of small ions, so that the uncertainty involved in equating them is also small. The values for various ions [1] are accurate to within $\pm 7z_I\ kJ\ mol^{-1}$ and are all negative, as expected (heat is released). They are of similar magnitude for singly charged ions, whether cations or anions, becoming less negative with increasing sizes, but becoming considerably more negative for multi-charged ions, by a factor of the order of z^2.

The "*softness*" of ions is obtained from their hydration enthalpies and the ionization potentials $\sum I_p$ for cations and electron affinities EA for anions. For the former, $\sigma_+ = [\sigma_{M+} - \sigma_{H+}]/\sigma_{H+}$, where $\sigma_{M+} = [\sum I_p - \Delta_{hydr}H_{M+}^{\infty}]/z_+$, and for the latter $\sigma_- = [\sigma_{X-} - \sigma_{OH-}]/\sigma_{H+}$, where

$\sigma_{X-} = [EA - \Delta_{hydr}H_{X-}^{\infty}]/|z_-|$ [1]. On this scale, Hg^{2+} and SCN^- are very soft whereas Li^+ and PO_4^{3-} are very hard. In interactions of ions with ligands and solvents, soft-soft and hard-hard interactions are preferred to those with differing kinds of softness or hardness.

The standard molar ionic Gibbs energy of hydration $\Delta_{hydr}G_I^{\infty}$ has traditionally been estimated from the *Born expression*. An isolated gaseous ion $I^{z\pm}(g)$ is discharged, the neutral particle is transferred into bulk water, where the permittivity is $\varepsilon_r^* = 78.4$ at 298.15 K, and is then charged up to the original value, producing the infinitely dilute aqueous ion, $I^{z\pm}$,aq$^{\infty}$. The net effect of this idealized process is $\Delta_{hydr}G_I^{\infty} = (N_A e^2/4\pi\varepsilon_0)z_I^2 r_I^{-1}(1 - 1/\varepsilon_r^*)$. The problem with this application of the Born equation is its use of the same r_I for the ion in the aqueous solution and the isolated state and ε_r^* of pure water for the immediate surroundings of the ion, where dielectric saturation occurs. To counter these problems, a quantity Δr is added to r_I and/or a spatial region adjacent to the ion with dielectric saturation ($\varepsilon_r^* \approx n_D^2$) and beyond with ε_r^* are used. Then reasonably correct $\Delta_{hydr}G_I^{\infty}$ values are obtained [1], consistent with $\Delta_{hydr}G_I^{\infty} = \Delta_{hydr}H_I^{\infty} - T\Delta_{hydr}S_I^{\infty}$. The value for the hydrogen ion, $\Delta_{hydr}G^{\infty}(H^+,aq) = -1{,}056 \pm 6$ kJ mol^{-1}, is consistent with this requirement.

"*Real*" standard molar Gibbs energies of hydration, $\Delta_{hydr}G_I^{\infty R}$, are obtained from the electromotive force of cells consisting of a downward flowing jet of aqueous solution in in the center of a tube, along the inner surface of which another solution flows, with a small vapor gap between them. The difference $\Delta_{hydr}G_I^{\infty R} - \Delta_{hydr}G_I^{\infty} = z_I F\Delta\chi$, where $F = 96{,}485.3$ C mol^{-1} is Faraday's constant and $\Delta\chi$ is the surface potential of water. The uncertainties connected with the value of $\Delta\chi = 0.10 \pm 0.08$ V [5] make the use of the measurable "real" standard molar ionic Gibbs energies of hydration unattractive for obtaining individual $\Delta_{hydr}G_I^{\infty}$ values.

Thermodynamic quantities relate to the *transfer* of ions from a source solvent, say water, W, to a target solvent or solvent mixture, S. Values of $\Delta_t G_2^{\infty}$ for electrolytes can be determined from the relative solubilities in W and in S of sparingly soluble electrolytes, or electrochemically from the electromotive force of cells. The corresponding $\Delta_t H_2^{\infty}$ can be determined from their relative heats of solution in W and in S, and the $\Delta_t S_2^{\infty}$ can be determined from the temperature coefficients of the $\Delta_t G_2^{\infty}$ values. The standard molar volumes of transfer $\Delta_t V_2^{\infty}$ are obtained from the densities of the solutions in water and in S.

An extra thermodynamic assumption is required in order to describe the transfer properties of individual ions, and consistent values, obeying $\Delta_t G_I^{\infty} = \Delta_t H_I^{\infty} - T\Delta_t S_I^{\infty}$, are obtained when the transfer quantities of Ph_4As^+ and BPh_4^- are assumed to be equal (the *TATB assumption*) as is widely employed [6]. The results for transfer of ions into many pure organic solvents have been summarized by Marcus [1] and those for transfer into mixed aqueous-organic solvents by him with coworkers [7].

The sign of $\Delta_t G_I^{\infty}(I^{\pm}, W \rightarrow S)$ determines whether $I^{\pm}$ prefers to be in an aqueous environment ($\Delta_t G_I^{\infty} > 0$) or has S as its near environment ($\Delta_t G_I^{\infty} < 0$). The transfer of many ions from water into most organic solvents is an unfavorable process. Exceptions include the transfer of Ag^+ and Cu^+ into nitriles and of "soft" cations into solvents with thio donor atoms. Strong electron-pair donor solvents also have $\Delta_t G_I^{\infty}(I^{z+}$, aq $\rightarrow$ S$) < 0$. Bulky ions do prefer the less structured environment of most organic solvents.

The environment of an ion in a solvent mixture, $S_A + S_B$, generally has a composition differing from that of the bulk mixture due to *preferential* solvation of the ion. *Selective solvation* occurs when the preferential solvation in the mixture is practically complete. The fraction of the solvation shell of the ion occupied by each solvent component can be ascertained by one of the two approaches [8]. Both take the excess Gibbs energy of mixing, G_{AB}^E, into account, expressing the mutual interactions of the solvent components. One, the quasi-lattice quasi-chemical (QLQC) approach, involves a model; the other, the inverse Kirkwood-Buff integral method (IKBI), is rigorous but requires very accurate data since it involves derivatives of $\Delta_t G_I^{\infty}(I^{\pm}$, aq $\rightarrow S_A + S_B)$ and of G_{AB}^E.

Solvation Numbers

Hydration numbers, h_I, are the time-average numbers of water molecules residing in the first (and second, if formed) hydration shell of ions. If directional coordinate bonds are formed with the water molecules in the first hydration shell, h_I equals the coordination number.

An ion compresses the water molecules in its hydration shell by its electric field, causing *electrostriction*. The (negative) molar electrostriction caused by an ion at infinite dilution, V_{Ielec}^{∞}, is the difference between $V^{\infty}(I^{z\pm}$, aq) and its intrinsic volume, V_{Iintr}. The hydration number of the ion is $h_{Ielec}^{\infty} = V_{Ielec}^{\infty}/\Delta V_{Welec}$, where $\Delta V_{Welec} = -2.9$ cm^3 mol^{-1} at 298.15 K is the molar compression of electrostricted water. Alternatively, the ion and the water in its first hydration shell may be considered to be incompressible by a moderate external pressure. Then the hydration number is $h_{Icomp}^{\infty} = 1 - (\partial V_I^{\infty}/\partial P)_T/\kappa_W^* V_W^*$. Here $(\partial V_I^{\infty}/\partial P)_T$ is the (negative) partial molar compression of the ion at infinite dilution, and κ_W^* is the isothermal compressibility of water. Individual ionic values of $(\partial V_I^{\infty}/\partial P)_T$ are based on that for chloride, $[\partial V(Cl^-,aq)^{\infty}/\partial P]_T = -16.5 \pm 1.5$ cm^3 GPa^{-1} mol^{-1} at 298.15 K. For ions for which no other value of the hydration number is available, the approximation $h_I^{\infty} = 0.360|z|/(r_I/\text{nm})$ can be used [1].

Hydration numbers diminish as the concentration of the electrolyte increases, mildly at low concentrations but strongly when the hydration shells of oppositely charged ions start to overlap.

Solvation numbers of ions in solvents other than water may also be derived as above from V_{Ielec}^{∞} divided by the solvent ΔV_{Selec} [9].

Transport Properties

The rate of movement of individual ions, whether in the absence or presence of fields, can be determined experimentally, contrary to their thermodynamic quantities. The inherent movement of ions in the absence of a field is their *self-diffusion*. Its rate is commonly obtained from the conductivity or by NMR. The values of the self-diffusion coefficient at infinite dilution, D_I^{∞}, are of the order of 10^{-9} m^2 s^{-1}. The more strongly hydrated an ion is, the lower is its rate of self-diffusion. An exception is the hydrogen ion that diffuses in water by the *Grotthuss mechanism*, the positive charge hopping from one water molecule to the next.

The most characteristic properties of ions are their abilities to move in solution in the direction of an electrical field gradient imposed externally. The conductivity of an electrolyte solution is readily measured accurately with a ~1 kHz alternating potential in a virtually open circuit, in order to avoid electrolysis. The *molar conductance* of a completely dissociated electrolyte is $\Lambda_2 = \Lambda_2^{\infty} - Sc_2^{1/2} + Ec_2 \ln c_2 + J'(R')c_2 - J''(R'')c_2^{3/2}$, where S, E, J', and J'' are explicit expressions, containing contributions from ionic atmosphere relaxation and electrophoretic effects, the latter two depending also on ion-distance parameters R. The infinite dilution Λ_2^{∞} can be split into the limiting molar *ionic conductivities* λ_I^{∞} by using experimentally measured *transport numbers* extrapolated to infinite dilution, t_+^{∞} and $t_-^{\infty} = 1 - t_+^{\infty}$. For a binary electrolyte, $\Lambda_2^{\infty} = \lambda_+^{\infty} + \lambda_-^{\infty}$ and $\lambda_+^{\infty} = t_+^{\infty}\Lambda_2^{\infty}$. Values of the limiting ionic molar conductivities in water at 298.15 K [1] are accurate to ± 0.01 S cm^2 mol^{-1} ($S = \Omega^{-1}$).

The *mobilities*, u_I^{∞}, of ions at unit electric field gradient (1 V m^{-1}) are proportional to the limiting ionic molar conductivities, $u_I^{\infty} = \lambda_I^{\infty}/|z|F$ as are also the self-diffusion coefficients, $D_I^{\infty} = RT\,\lambda_I^{\infty}/z^2F^2$. Ion mobilities increase with increasing temperatures and between 0 and 100 °C, a fivefold increase in Λ_2^{∞} occurs, mainly due to the corresponding decrease of the viscosity of the solvent, η_W^*.

A *Stokes radius* may be assigned to an ion, $r_{ISt} = (F^2/6\pi N_A)|z_I|/\eta_W^* \lambda_I^{\infty}$. Ionic Stokes radii are commensurate with r_I but not directly related to them; for some ions even $r_{ISt} < r_I$, although they ought to be larger, pertaining to the hydrated ions. The Walden products $\eta_I^* \lambda_I^{\infty}$ are approximately constant for tetraalkylammonium cations in various solvents, hence their Stokes radii are not sensitive to the natures of the solvents. It should be noted that Stokes radii, although formally calculated as above, have no real physical significance.

Electrolytes affect the dynamic viscosity of the solution, η, according to the Jones-Dole expression $[(\eta/\eta_W^*) - 1] = Ac_2^{1/2} + Bc_2$. The A coefficients can be calculated from the conductivities, but the B coefficients are empirical. The B coefficients are split into the individual ionic effects, B_I, according to their mobilities: $B_+/B_- \approx u_+/u_-$ and $B(K^+, aq) = B(Cl^-, aq)$ is used. The B_I at 298.15 K [1] are positive for small and multivalent ions but negative for univalent large ions. These algebraic signs have led to the classification of ions into *water structure makers* (*kosmotropes*, $B_I > 0$) and *water structure breakers* (*chaotropes*, $B_I < 0$). The B_I agree with other measures of effects of the ions on the structure of water, e.g., those derived from the relaxation of NMR signals.

Cross-References

- ► Conductivity of Electrolytes
- ► Electrolytes, Thermodynamics
- ► Ion Mobilities
- ► Specific Ion Effects, Evidences

References

1. Marcus Y (1997) Ion properties. Dekker, New York
2. Shannon RD, Prewitt CT (1969) Effective ionic radii in oxides and fluorides. Acta Cryst B 25:925; 1970, 26:1046
3. Marcus Y (1988) Ionic radii in aqueous solutions. Chem Rev 88:1475
4. Marcus Y (2009) The effects of ions on the structure of water: structure breaking and –making. Chem Rev 109:1346
5. Parfenyuk VI (2002) Surface potential at the gas-aqueous solution interface. Colloid J 64:588
6. Marcus Y (1986) Thermodynamics of transfer of single ions from water to non-aqueous and mixed solvents. 4. The selection of the extrathermodynamic assumptions. Pure Appl Chem 58:1721
7. Kalidas C, Hefter GT, Marcus Y (2000) Gibbs energies of transfer of cations from water to aqueous organic solvents. Chem Rev 100:819; Hefter GT, Marcus Y, Waghorne WE (2002) Enthalpies and entropies of transfer of electrolytes and ions from water to mixed aqueous organic solvents. Chem Rev 102:2773; Marcus Y (2007) thereafter Gibbs energies of transfer of anions from water to mixed aqueous organic solvents. Chem Rev 107:3880
8. Marcus Y (2002) Solvent mixtures. Dekker, New York
9. Marcus Y (2005) J Phys Chem B 109:18541

Ionic Liquids

Werner Kunz[1] and Heiner Jakob Gores[2]
[1]Institut für Biophysik, Fachbereich Physik, Johann Wolfgang Goethe-Universität Frankfurt am Main, Frankfurt am Main, Germany
[2]Institute of Physical Chemistry, Münster Electrochemical Energy Technology (MEET), Westfälische Wilhelms-Universität Münster (WWU), Münster, Germany

Definition und Main Properties

According to the now commonly accepted definition, Ionic Liquids (ILs) are salts – completely composed of ions – that are liquid at temperatures below 100 °C [1]. In principle, millions of combinations of cations and anions could be conceived to fulfill this condition [1]. Either the cation or the anion or both must be bulky or unsymmetrical enough to prevent easy crystallization. Large electronically delocalized molecular ions and unfavorable packing leads to low lattice energies, closer to the energies of the liquid state when compared with alkali halogenides such as LiF. A most prominent, historical example is ethylammonium nitrate (EAN) having a melting point of about 12 °C [2], see Fig. 1. EAN is a also a prominent example of a subclass of Ionic Liquids, the so-called room-temperature molten salts (RTMS), now room-temperature ILs, RTILs, and it is also one of the earliest examples of ILs, known since more than a 100 years. In this context, it is interesting to note that all nitrate salts have relatively low melting points compared to halogenides. As a consequence, it is straightforward that the combination of nitrate with a slightly dissymmetric cation such as ethylammonium is sufficient to get a salt with a melting point (m.p.) as low as 11 °C.

Ionic Liquids, Fig. 1 Ethylammonium nitrate

Ionic Liquids, Fig. 2 NaCl, m.p. 801 °C; $NaAlCl_4$, m.p. 185 °C; emim $AlCl_4$, m.p. 9 °C

The cation is unsymmetrical and the anion has a flat structure. Combined together this makes crystallization of the salt difficult, which explains the low melting point. By contrast, sodium chloride can easily be packed, and hence, the melting point is as high as 801 °C. Replacing chloride by tetrachloroaluminate reduces the melting point to 185 °C, replacing finally the cation sodium by ethylmethylimidazolium (emim), resulting in a melting point of about 9 °C for ethylmethylimidazolium tetrachloroaluminate, see Fig. 2. Of course, it is possible to get ILs based on an anion with reduced symmetry as well. The sequence tetraethyl ammonium tetrafluoroborate (m.p. about 375 °C), tetraethylammonium mono(oxalate) difluoroborate (m.p. = 33 °C), and ethylmethylimidazolium mono(oxalate) difluoroborate (m.p. = 22 °C [3]) shows that reducing the symmetry of the anion alone results in a huge drop of the melting point, whereas a small reduction of the melting point is observed when the symmetrical cation is replaced by the less symmetrical one as well (Fig. 3).

Today, Ionic Liquids are of significant interest mainly because of several features:

- The nearly infinite number of possible cation-anion combinations is supposed to make it possible to find a specifically designed solvent for a particular application. Therefore, they are often called "designer solvents" [4].
- Due to their ionic nature, they have very low vapor pressures. As a result, they are not Volatile Organic Compounds (VOCs). This argument is often used to consider them as "green" solvents, because they do not pollute the atmosphere [4, 5].
- The temperature range, over which they are liquid, is often quite extended, much more than for many conventional solvents. The liquid temperature range can be more than 300 °C [6].
- Such a large temperature range is of course only possible, if the Ionic Liquids are stable. And indeed, the temperature stability range as well as the electrochemical stability (both against oxidation and reduction) is very pronounced for many of the commonly used Ionic Liquids. The same is true for their high non-inflammability [1, 4, 5, 7].
- Depending on their chemical composition, Ionic Liquids can show an outstanding solvation potential. They are good solvents for many organic and inorganic compounds. Even cellulose can be dissolved, because the strong electrostatic interactions can cut the very strong hydrogen bonds in this biopolymer. They can be designed to be miscible or immiscible with other solvents, thus providing opportunities for separation processes [8].
- Many of the involved ions are poorly coordinating. Consequently, highly polar but noncoordinating solvents can be conceived, an important feature for chemical reactions [8].

However, there are also some drawbacks:

- Ionic Liquids are often very viscous. This is not unexpected. When the viscosities of classical high temperature molten salts, such as NaCl, is extrapolated down to room temperature, the viscosity of RTILs can be estimated. These viscosities are in the range of the measured values. The viscosities can be many decades higher than the value of water. Many ILs are like honey [9, 10]. Ionic Liquids are usually electrochemically stable, but they are often hygroscopic and chemically not always very stable. For example, at temperatures above 80 °C, choline salts rapidly decompose [11].
- Many ILs are relatively easy to synthesize, but the following purification steps are often cumbersome and expensive [12].
- Toxicities, biodegradation, and bioaccumulation are not yet sufficiently known for most of the Ionic Liquids. As a consequence,

Ionic Liquids, Fig. 3 Tetraethyl ammonium tetrafluoroborate, m.p. about 375 °C; tetraethyl ammonium mono(oxalate) difluoroborate, m.p. 33 °C; ethylmethylimidazolium mono(oxalate)difluoroborate, m.p. 22 °C

it is difficult to estimate: How far they can be considered as green alternatives to classical solvents [13, 14].

Transport Properties

The Walden Rule states that the product of the limiting molar conductivity Λ_m^0 and the pure solvent's viscosity η constant for infinitely diluted electrolyte solutions. It is a nice approximation for dilute solutions but shows rather large deviations for pure salts. The modified Walden rule, see Eq. 1a, introduces an additional empirical parameter, α, to cope with this deviation, see Eq. 1b, 1c.

$$\begin{aligned} &(a)\ \eta\Lambda_m{}^0 = \text{const} \\ &(b)\ \eta^{\alpha}\Lambda_m = \text{const} \\ &(c)\ \log\Lambda_m = \log\ \text{const} + \alpha\log\eta^{-1} \\ &(d)\ \Lambda_m = \kappa V_m \end{aligned} \qquad (1)$$

The molar conductivity Λ_m of the liquid salt is obtained from the specific conductivity κ and the molar volume V_m (Eq. 1d) of the salt that is available from density measurements. The modified Walden rule is applied for comparing ILs. The deviation from the Walden rule represented by α is interpreted in terms of ionicity. A small ionicity is caused by strong ion-ion interaction.

For an example and a critique of the application of the modified Walden rule, see Ref. [15].

By mixing an IL based on iodide as the anion, iodine and another IL, the diffusive transport behavior of triiodide, a crucial parameter for the performance of dye solar cells, can be improved. The diffusion coefficient measured by four independent electrochemical methods does not follow the Stokes-Walden rule but deviates by nearly an order of magnitude. This is caused by a charge transfer mechanism in addition to the diffusive transport [16, 17].

History

Probably the very first Ionic Liquid described in literature was ethanolammonium nitrate with a melting point of about 52–55 °C [18]. A further early example is EAN, which was first described by Paul Walden in 1914 [2]. At that time and for many decades, Ionic Liquids were considered only as a curiosity without any practical use. The situation began to change in the 1950s when electrochemists got interested in organic chloroaluminates [19], for example, $Cl^-/AlCl_3/AlEtCl_2$. However, these salts are very sensitive to hydrolysis. Among many other isolated attempts, also substituted imidazolium salts have been considered. Especially with this type of cations in the salts, the fashion of Ionic Liquids began in the 1990s [20]. Most of the modern studies and also some applications are done today with alkylimidazolium-based Ionic Liquids.

In the last 20 years, an enormous amount of scientific papers and patents is devoted to Ionic Liquids, their characteristics, and potential applications. The motivation of them is not always evident. Of course, the nearly unlimited cation-anion combination possibilities are a challenge, but this should not be the only motivation.

Classification

Since Ionic Liquids represent a vast class of various substances, it is not easy to propose a classification. Here are some possibilities:

- *Protic or aprotic Ionic Liquids* [21]. Protic ionic liquids (PILs), as all other protic solvents can give away protons and, more important, can form hydrogen bonds. As a result, they are not necessarily fully dissociated and can be distillated more easily (though still under extreme conditions) than aprotic Ionic Liquids (AILs). Further, they are considered to be poor ILs in the sense that their electrical conductivity is lower than what is predicted by the Walden plot. EAN is a representative example of a PIL.

  An example of an AIL is 1-*n*-butyl-3-methyl-imidazolium hexafluoroborate (bmim PF_6) with a melting point of −8 °C [22] and a decomposition temperature higher than 180 °C. AILs are more difficult to synthesize, and the anions such as PF_6^- are usually subject to hydrolysis in contact with air, but they show more ionicity than PILs. Therefore, they can be of interest, for example, for electrochemical devices [21].
- *Ionic Liquids with different unsymmetrical cations*. The most important classes are Ionic Liquids having as cation either imidazolium derivatives or quaternary ammonium ions; pyrrolidimium-, piperidimium-, or morpholinium-based derivatives; or quaternary phosphonium ions [23]. As counterions, simple halogenides are possible or more hydrophobic ions such as PF_6^- or bis ((trifluoromethyl) sulfonyl) amide $(N(SO_2CF_3)_2)^-$.
- *Ionic Liquids with simple cations, but special anions*. Although less often considered, the low melting point can also result from particular structures of the anion. An example is the class of alkyl-oligoether carboxylates with counterions as simple as sodium [24, 25]:

$$CH_3O[CH_2CH_2O]_3CH_2COO^-Na^+$$

These ions owe their low melting points to the high flexibility of the ethylene oxide groups and also the strong interactions of the oxygen atoms with the counterions.

It should be stressed here that in general, determination of melting points is a rather difficult task. Our studies with a novel equipment show that published data and our results deviate by more than 10 K [26]. This may be caused by impurities especially water, organic solvents, and chloride. Despite several hints in literature showing the effect of impurities, even now papers are submitted (and even published) that do not specify the purity of the investigated IL. Readers are advised to consult Ref. [27].

Ionic Liquids and Surfactants

Ionic Liquids can be mixed with surfactants and other solvents to give all types of association colloids that are known with water and other surfactants [28]. Microemulsions can be made as well as different types of lyotropic liquid crystals and emulsions. The Ionic Liquids can play the role of the polar (pseudo-) phase, the apolar one, or of the surfactant. Even two of the three components can be Ionic Liquids. Possible advantages are in the wide temperature range over which such systems show a reasonable stability and the low vapor pressure. Further advantages are the reduced viscosity compared to pure Ionic Liquids (in the cases where the ILs are the inner phase) and the possibility to work in water-free systems [29–31].

Potential and Real Applications

Ionic Liquids have been proposed for many chemical reactions, as solvents and catalysts [8, 23], in electrochemistry [32], for electroplating [33], for batteries and fuel cells [7], and in photoelectrochemistry for dye-sensitized solar cells [34]. They can be used to extract natural compounds of high value and to optimize biotransformation processes that require two-phase systems. Also for pharmaceutical applications, they were proposed.

However, today, only about a dozen of industrial or semi-industrial processes involving

ILs are known [35, 36]. Two prominent examples are the BASIL process from BASF [37] and the dissolution of cellulose to make high-value fibers out of it [38].

The BASIL process was the very first industrial application and it started only in 2002. It stands for Biphasic Acid Scavenging utilizing Ionic Liquids. In this process, 1-methylimidazole is used as a scavenger for HCl that is released during the synthesis of dialkoxyphenylphosphines. When HCl is scavenged with a classical trialkylamine, a solid salt is formed that is dispersed in the reaction mixture forming a suspension. Suspensions are a problem for large-scale processes, because the behavior and the separation of the salt particles are difficult to control. In the BASIL process, this problem is overcome: 1-methylimidazole is protonated, and the resulting 1-methyl-imidazolium chloride is an Ionic Liquid that does not mix with the product. As a result, the IL can be easily separated from the pure product and also easily recycled by simple deprotonation processes. This important feature allowed the productivity of this process to be improved by a factor of 8×10^4.

Recently introduced also by the BASF [38], different solutions of cellulose in Ionic Liquids have been launched under the trademark Cellionics. The solutions can contain up to 25 wt% of cellulose, they are long-term stable, and they can provide cellulose fibers with interesting properties.

Future Directions

It is difficult to estimate the future of Ionic Liquids and their relevance for real applications. Certainly, the research efforts will continue in a large number of laboratories all over the world. The topic is very attractive, because so many combinations are possible and each new combination and characterization of it leads to a new publication.

On the other hand, large-scale applications are not so easy to make, because of the aforementioned drawbacks and disadvantages of ILs. In addition, serious purification problems prevent researchers from getting valid data of ILs. Maybe, only few applications will result from the huge number of academic research efforts. Currently, the use of ILs in lithium ion batteries as a substitute for flammable solvents is developed, perhaps leading to a new large-scale application.

Cross-References

- ▶ Electrosynthesis in Ionic Liquid
- ▶ Ionic Liquids for Supercapacitors
- ▶ Ionic Liquids, Biocompatible
- ▶ Ionic Liquids, Structures

References

1. Stark A, Seddon KR (2007) Ionic liquids. In: Seidel A (ed) Kirk-Othmer encyclopedia of chemical technology. Wiley, Hoboken, pp 836–920
2. Walden P (1914) Über die Molekulargrösse und elektrische Leitfähigkeit einiger geschmolzener Salze. Bull Acad Imper Sci St Petersburg, VI série 8(6):405–422
3. Schreiner C, Amereller M, Gores HJ (2009) Chloride-free method to synthesise new ionic liquids with mixed borate anions. Chem Eur J 15:2270–2272
4. Plechkova NV, Seddon KS (2007) Ionic liquids: "designer" solvents for green chemistry. In: Tundo P, Perosa A, Zecchini F (eds) Methods and reagents for green chemistry: an introduction. Wiley, New York, pp 105–130
5. Deetlefs M, Seddon KR (2006) Ionic liquids: fact and fiction. Chim Oggi-Chem Today 24:16–23
6. Zhang S, Sun N, He X, Lu X, Zhang X (2006) Physical properties of ionic liquids: database and evaluation. J Phys Chem Rev 35:1475–1516
7. Armand M, Endres F, MacFarlane DR, Ohno H, Scrosati B (2009) Ionic-liquid materials for the electrochemical challenges of the future. Nat Mater 8:621–629
8. Welton T (1999) Room-temperature ionic liquids. Solvents for synthesis and catalysis. Chem Rev 99:2071–2208
9. Mantz RA, Trulove PC (2008) Viscosity and density of ionic liquids. In: Wasserscheid P, Welton T (eds) Ionic liquids in synthesis, vol 1, 2nd edn. Wiley-VCH, Weinheim, pp 72–88
10. Fletcher SI, Sillars FB, Hudson NE, Hall PJ (2010) Physical properties of selected ionic liquids for use as electrolytes and other industrial applications. J Chem Eng Data 55:778–782

11. Petkovic M, Ferguson JL, Nimal Gunaratne HQ, Ferreira R, Leitão MC, Seddon KS, Rebelo LPN, Pereira CS (2010) Novel biocompatible cholinium-based ionic liquids – toxicity and biodegradability. Green Chem 12:643–649
12. Stark A, Behrend P, Braun O, Müller A, Ranke J, Ondruschka B, Jastorff B (2008) Purity specification methods for ionic liquids. Green Chem 10:1152–1161
13. Petkovic M, Seddon KR, Rebelo LPN, Silva Pereira C (2011) Ionic liquids: a pathway to environmental acceptability. Chem Soc Rev 40:1383–1403
14. Pham TPT, Cho CW, Yun YS (2010) Environmental fate and toxicity of ionic liquids: a review. Water Res 43:516–521
15. Schreiner C, Zugmann S, Hartl R, Gores HJ (2010) Fractional Walden rule for ionic liquids: examples from recent measurements and a critique of the so-called ideal KCl line for the Walden plot. J Chem Eng Data 55:1784–1788
16. Zistler M, Wachter P, Wasserscheid P, Gerhard D, Hinsch A, Sastrawan R, Gores HJ (2006) Comparison of electrochemical methods for triiodide diffusion coefficient measurements and observation of non-Stokesian diffusion behaviour in binary mixtures of two ionic liquids. Electrochim Acta 52:161–169
17. Wachter P, Schreiner C, Zistler M, Gerhard D, Wasserscheid P, Gores HJ (2008) A microelectrode study of triiodide diffusion coefficients in mixtures of room temperature ionic liquids, useful for dye-sensitised solar cells. Microchim Acta 160:125–133
18. Gabriel S, Weiner J (1888) Ueber einige Abkömmlinge des Propylamins. Ber Deutsche Chem Ges 21:2669–2679
19. Hurley FH, Wier TP Jr (1951) Electrodeposition of metals from fused quaternary ammonium salts. J Electrochem Soc 98:203–206
20. Wilkes JS, Zaworotko MJ (1992) Air and water stable 1-ethyl-3-methylimidazolium based ionic liquids. J Chem Soc, Chem Commun 965–967. For the history of ionic liquids, see: Wilkes JS (2002) A short history of ionic liquids – from molten salts to neoteric solvents. Green Chem 4: 73–80
21. Angell CA, Byrne N, Belieres JP (2007) Parallel developments in aprotic and protic ionic liquids: physical chemistry and applications. Acc Chem Res 40:1228–1236
22. Carda-Broch S, Berthod A, Armstrong DW (2003) Solvent properties of the 1-butyl-3-methylimidazolium hexafluorophosphate ionic liquid. Anal Bioanal Chem 375:191–199
23. Hallett JP, Welton T (2011) Room-temperature ionic liquids: solvents for synthesis and catalysis. 2. Chem Rev 111:3508–3576
24. Zech O, Kellermeier M, Thomaier S, Maurer E, Klein R, Schreiner C, Kunz W (2009) Alkali metal oligoether carboxylates – a new class of ionic liquids. Chem Eur J 15:1341–1345
25. Zech O, Hunger J, Sangoro R, Iacob C, Kremer F, Kunz W, Buchner R (2010) Correlation between polarity parameters and dielectric properties of [Na][TOTO] – a sodium ionic liquid. Phys Chem Chem Phys 12:14341–14350
26. Wachter P, Schreiner C, Schweiger HG, Gores HJ (2010) Determination of phase transition points of ionic liquids by combination of thermal analysis and conductivity measurements at very low heating and cooling rates. J Chem Thermodyn 42:900–903
27. Seddon KR, Stark A, Torres MJ (2000) Influence of chloride, water, and organic solvents on the physical properties of ionic liquids. Pure Appl Chem 72:2275–2287
28. Greaves TL, Drummond CJ (2008) Ionic liquids as amphiphile self-assembly media. Chem Soc Rev 37:1709–1726
29. Zech O, Kunz W (2011) Conditions for and characteristics of nonaqueous micellar solutions and microemulsions with ionic liquids. Soft Matter 7:5507–5513
30. Zech O, Harrar A, Kunz W (2011) Nonaqueous microemulsions containing ionic liquids – properties and applications. In: Kokorin A (ed) Ionic liquids: theory, properties, new approaches. Rijeka, Croatia, pp 245–270 (open access)
31. Kunz W, Zemb T, Harrar A (2012) Using ionic liquids to formulate microemulsions: current state of affairs. Curr Opin Coll Interf Sci 17:205–211
32. Ohno H (ed) (2005) Electrochemical aspects of ionic liquids. Wiley, Hoboken
33. Endres F, Abbott AP, McFarlane DR (2008) Electrodeposition from ionic liquids. Wiley-VCH, Weinheim
34. Hinsch A, Behrens S, Berginc M, Boennemann H, Brandt H, Drewitz A, Einsele F, Fassler D, Gerhard D, Gores H, Haag R, Herzig T, Himmler S, Khelashvili G, Koch D, Nazmutdinova G, Opara-Krasovec U, Putyra P, Rau U, Sastrawan R, Schauer T, Schreiner C, Sensfuss S, Siegers C, Skupien K, Wachter P, Walter J, Wasserscheid P, Wuerfel U, Zistler M (2008) Material development for dye solar modules: results from an integrated approach. Prog Photovolt 16:489–501
35. Rogers RD, Seddon KR (eds) (2005) Ionic liquids IIIB: fundamental progress, challenges, and opportunities transformation and processes. Amercian Chemical Society, Washington DC
36. Plechkova NV, Seddon KR (2008) Application of ionic liquids in the chemical industry. Chem Soc Rev 37:123–150
37. Maase M, Massonne K (2005) Biphasic acid scavenging utilizing ionic liquids: the first commercial process with ionic liquids. In: Rogers RD, Seddon KR (eds) Ionic liquids IIIB: fundamental progress, challenges, and opportunities transformation and processes. American Chemical Society, Washington, DC, pp 126–132 (Chap 10)
38. Maase M, Massonne K, Uerdingen E, Vagt U (2006) BASIL™ – BASF's processes based on ionic liquids. Aldrich Chem Files 6(9):3

Ionic Liquids for Supercapacitors

Minato Egashira
College of Bioresource Sciences, Nihon University, Fujisawa, Kanagawa, Japan

Introduction

Different from conventional electrochemical power sources, batteries, electrolyte for supercapacitor is the component which limits the energy density of a cell by limiting its working voltage and specific capacitance. The energy density of a supercapacitor cell U is defined by its specific capacitance C and working voltage V via the following equation:

$$U = (1/2)CV^2$$

The specific capacitance of a capacitor cell depends on the number of ions accumulated on the surface of unit mass of porous electrode. On the other hand, the working voltage of a capacitor cell is largely influenced by the anodic and cathodic decomposition potential limits of electrolyte. For practical energy storage devices, safety issue is also necessary to be considered. In this viewpoint, the selection of nonflammable components is important to improve intrinsic safety (against firing).

Both aqueous and nonaqueous electrolyte systems have been utilized for commercial supercapacitors. Aqueous electrolytes suffer from the working voltage limitation of cells to 1 V. On the other hand, the flammable nature of organic solvents is a limiting factor for the development of nonaqueous electrolytes. Electrolyte having wide electrochemical window and nonflammability is desirable for high energy application.

Ionic liquids, defined as molten salts at ambient temperature, have attracted researchers on wide area, because their "ionic" nature is different from both aqueous and nonaqueous "molecular" media and most ionic liquids consist of cations and anions with organic frameworks by which the properties of ionic liquids are controlled [1]. For the use in electrochemistry, ionic liquids have advantages such as high conductivity, wide electrochemical window, and nonflammability [2]. Some researchers have focused on their electrochemical stability because they have assumed that the electrochemical decomposition of organic solvent electrolytes is induced by the solvent species. In the following sections, the kinds of ionic liquids applicable for supercapacitor electrolyte and their characteristics are to be reviewed briefly. At first, the unique structure of the double layer formed in ionic liquids is mentioned. Their application in double-layer capacitors (EDLC) is mainly described at following section.

Double Layer Formed in Ionic Liquid

It is reasonable to assume that ionic liquids provide electric double layer different from that formed in conventional aqueous or organic solvent electrolytes because the solvent, which plays important role on Gouy-Chapman-Stern (GCS) theory of double-layer formation, is absent. Alternative theoretical model for the interface between electrode and ionic liquid electrolyte has been proposed based on the mean-field approach together with the assumption of the limitation of local ion densities at double-layer region [3]. According to the theory, the differential capacitance shows potential dependence different from the GCS model, that is, the capacitance shows bell-like or camel-like (having two maxima) potential dependence with maximum or local minimum at the potential of zero charge (PZC). While experimental data of potential dependence of double-layer capacitances in various ionic liquids would not correspond to this model, some peculiar behavior, such as the hump and parabolic curve, may be explained by taking into consideration of this model [4]. Recently there have been interesting electrochemical or spectroscopic attempts to confirm the structure of double layer in ionic liquids [5, 6].

Models for the structure of double layer in pores have also been proposed [7].

Ionic Liquids for the Use in Double-Layer Capacitors

Ionic liquid electrolyte when applied in high-energy EDLC is required to have wide electrochemical window, high conductivity, and high thermal stability which assure the safety of EDLC cell. Many attempts have been made to utilize various ionic liquids for EDLC [8–24]. Some cations and anions frequently investigated for double-layer capacitors are illustrated in Fig. 1. The performances of a capacitor cell depend on the properties of ionic liquid, which are influenced by cation and anion species as described as follows. Ionic liquids with 1-alkyl-3-methyl imidazolium cation have relatively low viscosity and high conductivity, while their low cathodic stability is the limiting factor for high energy density [8, 11, 13, 15]. In contrast, the use of the ones having (linear and cyclic) quaternary ammonium cations leads to the improvement of stability toward negative electrode by the sacrifice of the fluidity. The properties of several ionic liquids are listed in Table 1. While the properties of an EDLC cell are influenced by porous structure of electrode as well as electrolyte, the EDLC performances of cells with these ionic liquid electrolytes are mostly as expected by the properties of electrolyte. Ionic liquid electrolytes often provide similar or slightly smaller specific capacitance but wider working voltage up to 4 V compared with conventional organic solvent electrolyte (2.5 V) [11, 14, 15, 19]. Among these ionic liquids, EMIF$(HF)_n$ provides interesting cell properties. This ionic liquid not only has markedly low viscosity and high conductivity but provides unusual potential dependence of capacitance [18]. Examples of power and energy densities of test cells containing ionic liquid electrolyte are summarized in Table 2.

The specific capacitances of conventional porous carbon electrodes in ionic liquid electrolytes are often low, and under high-rate discharge the capacitances exhibit severe degradation, due to the difficulty of immersion of high-viscosity ionic liquid in the pores of electrode. Therefore in many reports concerning EDLC containing ionic liquid electrolyte, the operation temperature as high as 60 °C is applied [15, 20], or ionic liquid is mixed with various amount of organic solvents such as acetonitrile (AN) or propylene carbonate (PC) [11, 12, 14, 16, 19]. The extent of viscosity decrease by the temperature increase is significant for ionic liquids. Therefore the operation of high temperature appears to be adequate for devices containing ionic liquid electrolyte. The mixing of ionic liquid and organic solvent is similar thing to that the dissolution of liquid salt into organic solvent. The considerable advantages of using ionic liquid as electrolytic salt are to increase salt concentration, in order to inhibit starvation effect and to assure the operation at low temperature. A commercialized example for this concept is the attractive contribution by Nisshinbo Co. named as "N's CAP." They have first prepared ionic liquids having *N*, *N*-diethyl-*N*-methyl-*N*-(2-methoxyethyl) ammonium (DEME) cation for this purpose. As indicated in Table 1, such ionic liquids have rather good fluidity with excellent electrochemical window. From their website, four kinds of capacitor modules are lined up with working voltages of 15, 100, 200, and 400 V. The capacitance, internal resistance, and maximum current of the 15 V cell (weight 1.9 kg, volume 1.4 l) are 200 F, 8 mΩ, and 600 A, respectively. They have reported that their test cell maintains ca. 95 % of room-temperature capacitance even at −40 °C [10, 25, 26].

Future Directions

It is expected to be adequate EDLC setting with viscous ionic liquid electrolyte to combine porous electrode in order to assure facile ion transfer [20]. One candidate is carbon nanotube (CNT), which provides "open" mesopores from interfibrous spaces. In such open pores, ions in

Ionic Liquids for Supercapacitors, Fig. 1 Structures of ions in ionic liquids for supercapacitor

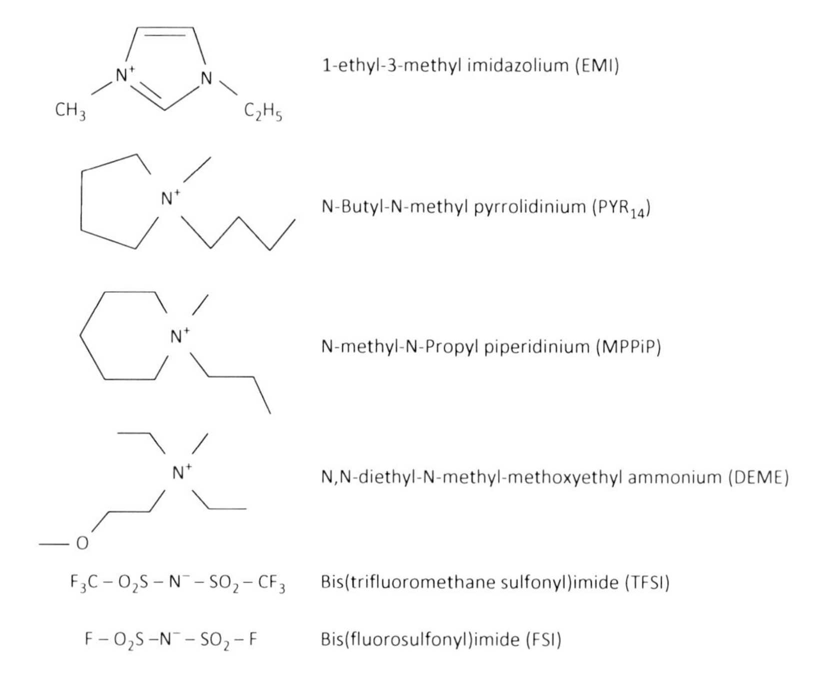

Ionic Liquids for Supercapacitors, Table 1 Properties of typical ionic liquids for supercapacitor

Ionic liquid	Viscosity[a]/mPa s	Conductivity[a]/mS cm^{-1}	References
$EMIBF_4$	41.0	12.3	[10]
EMITFSI	28.0	8.4	[16]
EMIFSI	17.9	15.5	[17]
$EMIF(HF)_{2.3}$	4.9	100	[18]
$PYR_{14}TFSI$	98 (60 °C)	6.0 (60 °C)	[20]
MPPipTFSI	117	1.5	[16]
$DEMEBF_4$	1,200	4.8	[10]
DEMETFSI	120	3.5	[10]

[a]At 25 °C otherwise notified

Ionic Liquids for Supercapacitors, Table 2 Performances of test EDLC cells containing ionic liquids

Electrolyte	Electrode	Power density[a]/kW kg^{-1}	Energy density[a]/Wh kg^{-1}	References
$EMIBF_4$	AC (2,500 m^2 g^{-1})	30.0	54	20
EMITFSI	AC	11.6	48	20
$EMIBF_4$-AN	AC	69.1	39	20
MPPipTFSI	AC	4.1	55	20
MPPipTFSI-AN	AC	28.7	46	20
EMITFSI	Xerogel carbon (600 m^2 g^{-1})	9.5 (60 °C)	31 (60 °C)	21
$PYR_{14}TFSI$	Xerogel carbon	11.4 (60 °C)	30 (60 °C)	21
EMITFSI	Aligned CNT	315	148	24

[a]At 25 °C otherwise notified

ionic liquid are easily accessible to the electrode surface [21–23]. In particular, vertically aligned CNT has highly accessible structure for viscous ionic liquids. The combination of ionic liquid electrolyte with CNT electrodes has exhibited excellent high-rate performances [23]. Recent development on the nanoscale design of carbon material is anomalous, represented by the studies on graphene [24]. Such novel nano-carbons are expected to contribute for the realization of practical high-performance EDLCs containing neat ionic liquid electrolyte.

The number of studies which utilize ionic liquid electrolyte in redox capacitor system is still small, probably due to the difficulty to reproduce the pseudo-capacitive reaction in ionic liquid media. While the principle of pseudo-capacitance of conductive polymer electrodes permits to utilize ionic liquid electrolytes, high viscosity and rather "inactive" ions of ionic liquid may make their pseudo-capacitive reaction slow. The combination of nanostructured conductive polymer electrode and ionic liquid electrolyte is expected to be effective [27]. It is far difficult that ionic liquids are utilized in transition metal-based redox capacitors where proton frequently participates in the reaction mechanisms. Some anions such as thiocyanate have been reported to provide pseudo-capacitance of manganese oxide [28]. The pseudo-capacitance of hydrous ruthenium oxide is based on the adsorption of proton on the electrode surface and thus requires proton in electrolyte. Therefore ionic liquids having proton have been attempted to be utilized with ruthenium oxide electrode [29]. Recent report that 1,3-substituted imidazolium cations such as EMI promote pseudo-capacitive reaction of ruthenium oxide is interesting on the viewpoint of the establishment of the pseudo-capacitive system based on chemical nature of ionic liquids [30].

Cross-References

► Gel Electrolyte for Supercapacitors, Nonaqueous
► Ionic Liquids
► Ionic Liquids, Structures
► Super Capacitors

References

1. Welton T (1999) Room-temperature ionic liquids solvents for synthesis and catalysis. Chem Rev 99:2071–2083
2. Fernicola A, Scrosati B, Ohno H (2006) Potentialities of ionic liquids as new electrolyte media in advanced electrochemical devices. Ionics 12:95–102
3. Kornyshev AA (2007) Double-layer in ionic liquids paradigm change? J Phys Chem B 111:5545–5557
4. Alam M, Lslam M, Okajima T, Ohsaka T (2007) Measurements of differential capacitance at Mercury/room-temperature ionic liquids interfaces. J Phys Chem C 111:18326–18333
5. Nanbu N, Sasaki Y, Kitamura F (2003) In situ FT-IR spectroscopic observation of a room-temperature molten salt|gold electrode interphase. Electrochem Commun 5:383–387
6. Baldelli S (2008) Surface structure at the ionic liquid-electrified metal interface. Acc Chem Res 41:421–431
7. Huang J, Sumpter BG, Meunier V (2008) A universal model for nanoporous carbon supercapacitors applicable to diverse pore regimes carbon materials and electrolytes. Chem Eur J 14:6614–6626
8. McEwen AB, Ngo HL, LeCompte K, Goldman JL (1999) Electrochemical properties of imidazolium salt electrolytes for electrochemical capacitor applications. J Electrochem Soc 146:1687–1695
9. Ue M, Takeda M, Takahashi T, Takehara M (2002) Ionic liquids with low melting points and their application to double-layer capacitor electrolytes. Electrochem Solid State Lett 5:A119–A121
10. Sato T, Matsuda G, Takagi K (2004) Electrochemical properties of novel ionic liquids for electric double layer capacitor applications. Electrochim Acta 49:3603–3611
11. Lewandowski A, Galinski M (2004) Carbon-ionic liquid double-layer capacitors. J Phys Chem Solids 65:281–286
12. Fracowiak E, Lota G, Pernak J (2005) Room-temperature phosphonium ionic liquids for supercapacitor application. Appl Phys Lett 86:164104
13. Ania CO, Pernak J, Stefaniak F, Raymundo-Pinero E, Beguin F (2006) Solvent-free ionic liquids as in situ probes for assessing the effect of ion size on the performance of electrical double layer capacitors. Carbon 44:3126–3130
14. Lewandrowski A, Galinski M (2007) Practical and theoretical limits for electrochemical double-layer capacitors. J Power Sources 173:822–828
15. Lazzari M, Mastragostino M, Soavi F (2007) Capacitance response of carbons in solvent-free ionic liquid electrolytes. Electrochem Commun 9:1567–1572

16. Lewandowski A, Olejniczak A (2007) *N*-Methyl-*N*-propyl piperidinium bis(trifluoromethanesulphonyl) imide as an electrolyte for carbon-based double layer capacitors. J Power Sources 172:487–492
17. Handa N, Sugimoto T, Yamagata M, Kikuta M, Kono M, Ishikawa M (2008) A neat ionic liquid electrolyte based on FSI anion for electric double layer capacitor. J Power Sources 185:1585–1588
18. Senda A, Matsumoto K, Nohira T, Hagiwara R (2010) Effects of the cathodic structures of fluorohydrogenate ionic liquid electrolytes on the electric double layer capacitance. J Power Sources 195:4414–4417
19. Lewandowski A, Olejniczak A, Galinski M, Stepniak I (2010) Performance of carbon-carbon supercapacitors based on organic aqueous and ionic liquid electrolytes. J Power Sources 195:5814–5819
20. Lazzari M, Mastragostino M, Pandolfo AG, Ruiz V, Soavi F (2011) Role of carbon porosity and ion size in the development of ionic liquid based supercapacitors. J Electrochem Soc 158:A22–A25
21. Barisci JN, Wallace GG, MacFarlane DR, Baughman RH (2004) Investigation of ionic liquids as electrolytes for carbon nanotube electrodes. Electrochem Commun 6:22–27
22. Katakabe T, Kaneko T, Watanabe M, Fukushima T, Aida T (2005) Electric double-layer capacitors using "Bucky Gels" consisting of an ionic liquid and carbon nanotubes. J Electrochem Soc 152:A1913–A1916
23. Lu W, Qu L, Henry K, Dai L (2009) High performance electrochemical capacitors from aligned carbon nanotube electrodes and ionic liquid electrolytes. J Power Sources 189:1270–1277
24. Zhu Y, Murali S, Stoller MD, Ganesh KJ, Cai W, Ferreira PJ, Pirkle A, Wallace RM, Cychosz KA, Thommes M, Su D, Stach EA, Ruoff RS (2011) Carbon-based supercapacitors produced by activation of graphite. Science 332:1537–1541
25. http://www.nisshinbo.co.jp/r_d/capacitor/index.html
26. Sato T (2004) Some properties of an EDLC using an ionic liquid. Electrochemistry 72:711–715
27. Wang K, Huang J, Wei Z (2010) Conducting polyaniline nanowire arrays for high performance supercapacitors. J Phys Chem C 114:8062–8067
28. Lee MT, Tsai WT, Deng MJ, Cheng HF, Sun IW, Chang JK (2010) Pseudocapacitance of MnO_2 originates from reversible insersion/desersion of thiocyanate anions studied using in situ X-ray absorption spectroscopy in ionic liquid electrolyte. J Power Sources 195:919–922
29. Rochefort D, Pont AL (2006) Pseudocapacitive behavior of RuO_2 in a proton exchange ionic liquid. Electrochem Commun 8:1539–1543
30. Egashira M, Matsuno Y, Yoshimoto N, Morita M (2010) Pseudo-capacitance of composite electrode of ruthenium oxide with porous carbon in non-aqueous electrolyte containing imidazolium salt. J Power Sources 195:3036–3039

Ionic Liquids, Biocompatible

Cristina Pereira, Rui Ferreira, Helga Garcia and Marija Petkovic
Instituto de Tecnologia Química e Biológica, Universidade Nova de Lisboa, Oeiras, Portugal

Introduction and Description

Biocompatibility should be considered for each *ionic liquid* and in the context of its application. For that reason, several views of their biocompatibility may arise. General understanding of biocompatibility is a "quality of not having toxic or injurious effects on biological systems" (Dorland's medical dictionary). This term is now essentially applied to materials used for medical devices (ISO 10993). Biocompatibility evaluation can however be analyzed accordingly to different areas of research and application, and discussed within regulatory and environmental requirements.

Interpretation of ionic liquids biocompatibility should consider extant eco-toxicological knowledge [1, 2]. In addition, the ionic liquid stability during function; and afterward its environmental persistence (10th Principle of *Green Chemistry*) [3] should be also accounted for.

Ionic liquids are increasingly attracting interest in both academia and industry, especially with the aim to reduce or eliminate hazards associated with traditional *solvents*. Ionic liquids' negligible vapor pressure and nonflammability (with some rare exceptions) have led to their classification as "green" and non toxic solvents. Common generalizations that ionic liquids are either "green" or "toxic" solvents should however be avoided: Both extremes are totally misleading. Some critical aspects of ionic liquids' toxicity and biodegradability were recently reviewed [1, 2]. The understanding of ionic liquids (their core chemistry, syntheses and purification methods) has advanced significantly over the past decade, and is currently set on solid ground, opening doors to the design of biocompatible ionic liquids [4], incorporating (*inter alia*)amino acids [5], carboxylic acids [6, 7], non-nutritive sweeteners [8], or glucose [9].

At an industrial scale, ionic liquids' safe usage needs to account for a broad diversity of aspects, including human health risk and workplace safety. Personnel involved in synthesis, transport, handling and disposal of ionic liquids is under increased risk due to the presence of bulk material and higher possibility of exposure.

Even though ionic liquids' use generally reduces the risk of atmospheric pollution, both aquatic and terrestrial environments are under risk. Ionic liquids might end up in the environment in the case of accidental spill or waste disposal. In this scenario, their physicochemical properties and concentration will influence their bioavailability, and consequently their biodegradability. Their environmental fate is a complex situation which crosses numerous unknown abiotic and biotic factors. Jastorff and coworkers have proposed a multidimensional risk analysis, correlating five distinct indicators, namely release, spatiotemporal range, bioaccumulation, biological activity and uncertainty, which can be used for predicting the environmental impact of ionic liquids [10]. For most ionic liquids, the proposed indicators are yet to be comprehensively addressed.

From an environmental perspective, biocompatibility of ionic liquids is difficult to define, especially due to the complexity and dynamics of physical and chemical processes which are taking place in the contaminated ecosystem. The presence of ionic liquids in terrestrial or aquatic environments will affect organisms of different complexity in distinct ways, consequently affecting the food chain and the ecosystem equilibrium. Even when present at sublethal concentrations, toxicants can have severe consequences on multiple organisms. Filamentous fungi were reported to be the most tolerant group of microorganisms, able to grow in media supplemented with very high concentrations of a broad diversity of ionic liquids [6, 11, 12]. This emphasizes their potential role in the environmental mitigation of ionic liquids. In addition, it might suggest the exploitation of ionic liquid-tolerant fungal catalysts for industrial, pharmacological and ecological usages.

Numerous and diverse, most eco-toxicological studies undertaken until now clearly demonstrate that different organisms exhibit fairly diverse susceptibilities to ionic liquids [1, 2]. The reported toxicity trends are generally highly consistent, suggesting a similar mode of toxicity. Some structure-activity relationships are now apparent. It is generally accepted that aromatic cations, containing imidazolium and pyridinium rings, present high toxicity, especially those carrying long alkyl chains. On the contrary, nitrogen-containing alicyclic cations (e.g., morpholinium, piperidinium, or pyrrolidinium) and quaternary ammonium cations generally display low toxicity [1]. At the level of the anion, preference should be also given to benign ones, such as saccharinate, short chain length alkanoates, amino acids, or acesulfamate [2]. The incorporation of hydrolysable moieties, such as ester or ether bonds, is also generally accepted to have a considerable positive effect on their biodegradability [2, 13]. Recently, it was clearly demonstrated that ionic liquids of low toxicity, namely some cholinium alkanoates, show high biodegradability potential and solvent quality [6, 14].

From a pure biotechnological perspective, biocompatibility of ionic liquids is often evaluated in a confined system. At the outset, evaluation of biocompatibility will consider the integrity (i.e., plasma membrane damage) and the efficiency of the whole-cell biocatalysts (e.g., bacteria, yeast, fungi) during exposure to the ionic liquid [15]. Ionic liquids' application in whole-cell biocatalysis might exploit single phase or biphasic systems, based on water-miscible [16] and water-immiscible ionic liquids [17], respectively. To present, most studies have focused on biphasic systems, where the phase containing the ionic liquid is used to prevent accumulation of toxic products or substrates in the aqueous phase where the cells grow. The concentration of ionic liquid solubilized in the aqueous phase (even if vestigial in the biphasic systems) plays a critical role in evaluating the biocompatibility of the solvent.

Innovative applications of ionic liquids in pharmaceutical industry have been also proposed.

They can be used as replacements for conventional organic solvents and contribute to reduction of the *Environmental Factor*, which expresses the amount of generated waste over the amount of product [18]. Moreover, their use as reservoirs for controlled drug delivery [19, 20] or protein stabilization media was proposed [21]. Although rather exploratory, ionic liquids carrying at the level of the cation or the anion, or both, an active pharmaceutical ingredient were suggested to have improved the compound pharmacokinetics [22]. Biocompatibility analysis of any ionic liquid-based pharmacological drug, needs to account for particular, or often multiple, metabolic pathways and potential targets.

To date, most ionic liquids used in electrochemistry, though not always inherently biocompatible and biodegradable, are beneficial for the processes' environmental sustainability. Improvements regarding security issues arise directly from their negligible vapor pressure, which additionally allows safer operation at higher temperatures without atmosphere contamination or pressure buildup [23]. Ionic liquids might be used in substitution of some hazards compounds normally used in *aqueous electrolytes*, such as strong acids, alkalis, or toxic complexants. In many *electrodeposition* systems, chromic acid and cyanides ($[Ag(CN)_2]^-$) are still used. Abbott et al. proposed a sustainable *electropolishing* process which was based on cholinium chloride and ethylene glycol [24]. This alternative process avoids the treatment of large quantities of acidic and metal-contaminated wastewaters.

Historically, ionic liquids' initial advances in electrochemistry were encouraged by difficulties and safety issues in the aluminum deposition process known as SIGAL (Siemens Galvano-Aluminium). Major concerns were related with the flammability of the aluminum precursors and of the volatile organic solvents used. In the search for low melting, nonvolatile, and nonaqueous electrolytes, pyridinium [25] and imidazolium chloroaluminates (III) were investigated [26]. These ionic liquids are able to dissolve various metal salts. Their biocompatibility is questionable due to their potential toxicity and because they are also corrosive and unstable in air and/or moisture [27]. The same applies to hexafluorophosphate (PF_6) based ionic liquids, which are known to react with water to produce hydrofluoric acid [27]. To thwart the chemical instability of the anions in the presence of water, preserving the ionic liquid large electrochemical window, the use of other hydrophobic anions, such as trifluoromethanesulfonate (OTf) [27] and bis (trifluoromethylsulfonyl)amide (NTf_2) [23], was investigated. Combining 1-ethyl-3-methylimidazolium cation with NTf_2 anion results in a water-immiscible ionic liquid (melting point -15 °C) that shows no decomposition or significant vapor pressure (up to ~300–400 °C) [23]. Both, fluorinated anion and imidazolium-based cation, contribute to the overall toxicity of the ionic liquid. Even though it might increase the overall sustainability of a specific process, due to the ions' toxicity and recalcitrant nature, this ionic liquid cannot be consider biocompatible [1].

At present, in electrochemistry, the most commonly studied ionic liquids belong to the family of the imidazolium-based ones. This is due to their low melting points and high electrochemical stability. The potential application of these ionic liquids in electrodeposition processes [26] and development of some electrochemical devices, such as *biosensors* [28] and *lithium batteries* [29], have been largely considered. Evidently, when accounting for toxicity and environmental persistence, these solvents do not fully address these important principles of green chemistry.

As aforementioned, improvements in the ionic liquids field led to "greener" alternatives such as the use of ammonium, phosphonium, or sulfonium cations. The selection of benign cations, such as cholinium, constitutes, still today, one of the most important advances toward their conscious design. The quaternary ammonium salt, cholinium chloride, classified as a provitamin in Europe and widely used as animal feed supplement, has been extensively investigated for electrochemical applications [30, 31]. Cholinium chloride environmental sustainability is also made known by the synthesis process: a one step solvent-free reaction of hydrogen chloride, ethylene oxide and trimethylamine.

Some eutectic mixtures composed of biocompatible salts, including cholinium chloride, showed very promising results in applied electrochemistry. Successful examples include the electrodeposition of trivalent chromium, instead of its toxic hexavalent homologous (cholinium chloride and $CrCl_3.6H_2O$ mixture) and the electrodeposition of magnesium [31] and zinc alloys [32] (cholinium chloride and ethylene glycol or urea mixtures). These data might, in a near future, favor their industrial implementation.

The dissolution of enzymes, without loss of activity, is possible in several ionic liquids. This opens the avenue for enzyme-catalyzed processes in ionic liquids. In this context, once more, biocompatible cholinium-based ionic liquids, namely, cholinium dihydrogen phosphate, was shown to be an excellent solvent for different proteins maintaining its activity for long time storage at room temperature [33]. This is especially interesting when considering ionic liquids as replacement for *aqueous electrolyte solutions* in biosensor devices [34]. The preparation of *enzyme-modified electrodes* for an accurate detection of biological molecules [28] or the use of ionic liquid-based *biofuel cells* for energy production [23] is also gaining remarkable attention.

Future Directions

Ionic liquids constitute an exceedingly heterogeneous group of compounds, since the versatility of their synthesis allows formulation of millions of possible cation-anion combinations. This is clearly an advantage when compared to conventional molecular organic solvents. Their conscious design and use of structure-activity relationships are essential tools to deliver safer formulations with enhanced technical performance. True greenness should incorporate not only a sustainable synthesis, but also low toxicity and environmental persistence (Green Chemistry principles 2, 3 and 10) [3]. In other words, one should consider not only ionic liquids' impact during synthesis and application, but also their lifetime and degradation products.

Along with the growing number of studies on the toxicity and biodegradability of ionic liquids, major correlations between the chemical structure of the ions and their observed toxicities are becoming apparent. However, there are still major questions to be resolved, such as the ionic liquid modes of toxicity, biodegradation pathways, and behavior concerning biosorption. More significant and extensive toxicological testing will favor the design of biocompatible ionic liquids, and potentially their use in a wider range of technological applications. This is especially true in the context of their registration under legislation such as that required by the European Community regulation on chemicals and their safe use – *REACH* (Registration, Evaluation, Authorisation and Restriction of CHemical substances). REACH registration, which is now mandatory for any chemical produced in the quantity over one tonne per year, aims to increase the awareness of the industry on hazards and risk management. This compiles with the principles of Green Chemistry, that are continuously inspiring advances in ionic liquids. There are no doubts that we are witnessing an increasing interest in biocompatible ionic liquids in a broad range of electrochemistry applications, from biosensor devices to biofuel cells.

Cross-References

▸ Biosensors, Electrochemical
▸ Electrodeposition of Electronic Materials for Applications in Macroelectronic- and Nanotechnology-Based Devices
▸ Ionic Liquids
▸ Ionic Liquids, Structures
▸ Lithium-Ion Batteries
▸ Non-Aqueous Electrolyte Solutions

References

1. Petkovic M, Seddon KR, Rebelo LPN, Silva Pereira C (2011) Ionic liquids: a pathway to environmental acceptability. Chem Soc Rev 40:1383–1403
2. Coleman D, Gathergood N (2010) Biodegradation studies of ionic liquids. Chem Soc Rev 39:600–637

3. Anastas P, Warner J (1998) Green chemistry: theory and practice. Oxford University Press, New York
4. Imperato G, König B, Chiappe C (2007) Ionic green solvents from renewable resources. Eur J Org Chem 7:1049–1058
5. Fukumoto K, Yoshizawa M, Ohno H (2005) Room temperature ionic liquids from 20 natural amino acids. J Am Chem Soc 127:2398–2399
6. Petkovic M, Ferguson JL, Gunaratne HQN, Ferreira R, Leitão MC, Seddon KR, Rebelo LPN, Silva Pereira C (2010) Novel biocompatible cholinium-based ionic liquids-toxicity and biodegradability. Green Chem 12:643–649
7. Klein R, Zech O, Maurer E, Kellermeier M, Kunz W (2011) Oligoether carboxylates: task-specific room-temperature ionic liquids. J Phys Chem B 115:8961–8969
8. Carter EB, Culver SL, Fox PA, Goode RD, Ntai I, Tickell MD, Traylor RK, Hoffman NW, Davis JH (2004) Sweet success: ionic liquids derived from non-nutritive sweeteners. Chem Commun 6:630–631
9. Poletti L, Chiappe C, Lay L, Pieraccini D, Polito L, Russo G (2007) Glucose-derived ionic liquids: exploring low-cost sources for novel chiral solvents. Green Chem 9:337–341
10. Jastorff B, Störmann R, Ranke J, Mölter K, Stock F, Oberheitmann B, Hoffmann W, Hoffmann J, Nuchter M, Ondruschka B, Filser J (2003) How hazardous are ionic liquids? Structure-activity relationships and biological testing as important elements for sustainability evaluation. Green Chem 5:136–142
11. Deive FJ, Rodríguez A, Varela A, Rodrígues C, Leitão MC, Houbraken JAMP, Pereiro AB, Longo MA, Ángeles Sanromán M, Samson RA, Rebelo LPN, Silva Pereira C (2010) Impact of ionic liquids on extreme microbial biotypes from soil. Green Chem 13:687–696
12. Petkovic M, Ferguson J, Bohn A, Trindade JR, Martins I, Carvalho M, Leitão MC, Rodrigues C, Garcia H, Ferreira R, Seddon KR, Rebelo LPN, Silva Pereira C (2009) Exploring fungal activity in the presence of ionic liquids. Green Chem 11:889–894
13. Ranke J, Stolte S, Stoermann R, Arning J, Jastorff B (2007) Design of sustainable chemical products – the example of ionic liquids. Chem Rev 107:2183–2206
14. Garcia H, Ferreira R, Petkovic M, Ferguson JL, Leitao MC, Gunaratne HQN, Seddon KR, Rebelo LPN, Silva Pereira C (2010) Dissolution of cork biopolymers in biocompatible ionic liquids. Green Chem 12:367–369
15. Weuster-Botz D (2007) Process intensification of whole-cell biocatalysis with ionic liquids. Chem Rec 7:334–340
16. Hussain W, Pollard DJ, Lye GJ (2007) The bioreduction of a beta-tetralone to its corresponding alcohol by the yeast *Trichosporon capitatum* MY1890 and bacterium *Rhodococcus erythropolis* MA7213 in a range of ionic liquids. Biocatal Biotransform 25:443–452
17. van Rantwijk F, Sheldon RA (2007) Biocatalysis in ionic liquids. Chem Rev 107:2757–2785
18. Sheldon R (2001) Catalytic reactions in ionic liquids. Chem Commun 23:2399–2407
19. Jaitely V, Karatas A, Florence AT (2008) Water-immiscible room temperature ionic liquids (RTILs) as drug reservoirs for controlled release. Int J Pharm 354:168–173
20. Moniruzzaman M, Tahara Y, Tamura M, Kamiya N, Goto M (2010) Ionic liquid-assisted transdermal delivery of sparingly soluble drugs. Chem Commun 46:1452–1454
21. Byrne N, Wang L-M, Belieres J-P, Angell CA (2007) Reversible folding-unfolding, aggregation protection, and multi-year stabilization, in high concentration protein solutions, using ionic liquids. Chem Commun 26:2714–2716
22. Stoimenovski J, MacFarlane DR, Bica K, Rogers RD (2010) Crystalline*vs.* ionic liquid salt forms of active pharmaceutical ingredients: a position paper. Pharm Res 27:521–526
23. Armand M, Endres F, MacFarlane DR, Ohno H, Scrosati B (2009) Ionic-liquid materials for the electrochemical challenges of the future. Nat Mater 8:621–629
24. Abbott AP, Capper G, McKenzie KJ, Ryder KS (2006) Voltammetric and impedance studies of the electropolishing of type 316 stainless steel in a choline chloride based ionic liquid. Electrochim Acta 51:4420–4425
25. Chum HL, Koch VR, Miller LL, Osteryoung RA (1975) Electrochemical scrutiny of organometallic iron complexes and hexamethylbenzene in a room-temperature molten-salt. J Am Chem Soc 97:3264–3265
26. Wilkes JS, Levisky JA, Wilson RA, Hussey CL (1982) Dialkylimidazolium chloroaluminate melts – a new class of room-temperature ionic liquids for electrochemistry, spectroscopy, and synthesis. Inorg Chem 21:1263–1264
27. Plechkova NV, Seddon KR (2008) Applications of ionic liquids in the chemical industry. Chem Soc Rev 37:123–150
28. Silvester DS (2011) Recent advances in the use of ionic liquids for electrochemical sensing. Analyst 136:4871–4882
29. Lewandowski A, Swiderska-Mocek A (2009) Ionic liquids as electrolytes for Li-ion batteries-an overview of electrochemical studies. J Power Sources 194:601–609
30. Haerens K, Matthijs E, Chmielarz A, Van der Bruggen B (2009) The use of ionic liquids based on choline chloride for metal deposition: a green alternative? J Environ Manage 90:3245–3252
31. Abbott AP, Ryder KS, Koenig U (2008) Electrofinishing of metals using eutectic based ionic liquids. Trans Inst Met Finish 86:196–204

32. Abbott AP, Capper G, McKenzie KJ, Ryder KS (2007) Electrodeposition of zinc-tin alloys from deep eutectic solvents based on choline chloride. J Electroanal Chem 599:288–294
33. Fujita K, MacFarlane DR, Forsyth M, Yoshizawa-Fujita M, Murata K, Nakamura N, Ohno H (2007) Solubility and stability of cytochrome c in hydrated ionic liquids: effect of oxo acid residues and kosmotropicity. Biomacromolecules 8:2080–2086
34. Opallo M, Lesniewski A (2011) A review on electrodes modified with ionic liquids. J Electroanal Chem 656:2–16

Ionic Liquids, Structures

Hermann Weingärtner
Lehrstuhl für Physikalische Chemie II,
Ruhr-Universität Bochum, Bochum, Germany

General Aspects

Inorganic salts, such as NaCl, melt at very high temperatures, which render their routine use as solvents impossible. Ionic Liquids – low-melting organic salts – open a window for the molten state at ambient conditions, providing ionic media with remarkable material and solvent properties [1–4]. The low melting points of Ionic Liquids are essentially founded in the bulky size, low molecular symmetry, and/or conformational flexibility of molecular ions, which favor the liquid state by leading to small lattice energies and large entropy effects [5].

The currently known cation and anion families and their manifold substitution patterns provide a huge number of potential cation-anion combinations, which allow to tune their properties to given applications. As it is impossible to experimentally characterize even a small fraction of these, and as many well-established rules for properties of conventional solvents are difficult to transcribe to Ionic Liquids, a molecular-based understanding of the properties of Ionic Liquids is mandatory. Among others, such an understanding has to include:

- The atomic and electronic structure of the ions
- The molecular interactions between the ions
- The structure of isolated ion pairs and ion clusters
- The structure of the liquid phase
- The structural behavior on mesoscopic length scales.

Despite some important structural information deduced from spectroscopic methods scattering techniques, theoretical and computational methods are necessary to analyse the complex molecular forces between the ions and interpret the experimental data. Increasing computer power will enable the refinement of the systems studied, bring the models closer to real systems [6].

Ion Structures and Interionic Interactions

The properties of simple salts, such as NaCl, are controlled by long-range Coulomb forces between the net charges of the ions, complemented by short-range, dispersive/repulsive van der Waals forces. Charged neighbors screen, however, the long-range Coulomb forces of a central ion with particles further away. The screened Coulomb potential of the ions forms the major basis for understanding the liquid-phase properties of simple salts.

Molecular ions give rise to additional short-range contributions to the interaction potential. Some major consequences for the interionic interactions are: [3]:

- Large distances between the charge centers of bulky ions soften the Coulomb forces.
- The large size and high polarizability of the ions render the dispersive and repulsive van der Waals forces more significant than in simple salts.
- Ions with asymmetric charge distribution have electric dipole moments and give rise to highly directional dipole-dipole interactions, which control the relative dielectric permittivity (dielectric constant) and solvation capability of the Ionic Liquid.
- The electronic structures can result in strong and highly directional specific forces, such as the hydrogen bonding or effects involving the p electrons of aromatic rings.

From this perspective, the major factors allowing to tune the properties of Ionic Liquids are:

- The propensity to form hydrogen bonds between cations and anion
- The hydrophobicity and functionalization of side chains
- The coordinating ability of the anions
- Conformational equilibria of the ions.

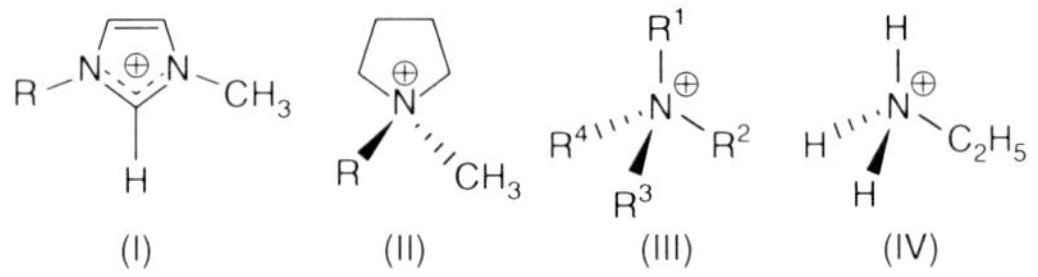

Ionic Liquids, Structures, Fig. 1 Widely used cations of ionic liquids: (**I**) 1-alkyl-3-methylimidazolium, (**II**) *N*-alkyl-*N*-methylpyrrolidinium; (**III**) tetralkylammonium, (**IV**) ethylammonium

Hydrogen Bonding

Many cations have the propensity to form H-bonds with anions. H-bonds are particularly pronounced in Protic Ionic Liquids (PILs), which are formed by reaction of a Brønsted acid with a Brønsted base. Due to the exchangeable proton PILs can form strong H-bonds [4, 7]. A prominent example is ethylammonium nitrate, which is able to form a H-bonded network, which is sometimes said to resemble the three-dimensional network of water. As a result of H-bonds between cations and anions, many PILs are incompletely dissociated. This low ionicity is a major determinant of the properties of PILs.

The majority of Ionic Liquids are classified as aprotic solvents. Cations, such as tetraalkylammonium or *N*,*N*-dialkylpyrrolidinium ions, cannot act as H-bond donators. On the other hand, salts involving 1-alkyl-3-methylimidazolium ions, albeit being classified as *aprotic* Ionic Liquids, may reveal weak H-bonds to some anions, which are founded in a slight acidity of the hydrogens of the aromatic ring [8]. This H-bond characteristics form the key for understanding the low melting points and comparatively low viscosities of 1-alkyl-3-methylimidazolium salts, the discovery of which has formed a major impetus for the rapid development of the Ionic Liquid field in the 1990s. The existence and nature of such H-bonds is, however, subject to controversial debate [9] (Fig. 1).

Hydrophobicity and Functionalization of Side Chains

Most Ionic Liquids involve *n*-alkyl chains, the chain length of which can be used to tune the hydrophobicity of the salts and their miscibility with water. Long alkyl chains, with more than 12 carbon atoms, say, can result in liquid-crystalline phases [10]. For specific applications, the side chains can be functionalized, for example, by introducing polar, H-bonding, fluorinated, or chiral groups.

Coordinating Ability of Anions

Anions range from simple ions, such as Cl^-, $[BF_4]^-$ or $[PF_6]^-$, to complex organic ions, such as bis(trifluoromethanesulfonyl)imide ($[NTf_2]^-$). Because the atomic and electronic structures of anions can be very different, it is difficult to systematize anion effects. To some extent, the strength of cation-anion interaction is, however, correlated with the coordinating ability and Lewis basicity of the anion. In this context, weakly coordinating anions are of particular interest. Such anions can form highly polar yet non-coordinating solvents, which provide ideal conditions for many chemical reactions. Weakly coordinating anions, including $[NTf_2]^-$ as a prominent example, are often fluorinated because the low electron polarizability of the fluorine atoms reduces the van der Waals interactions.

Conformational Equilibria

Many ions exhibit conformational equilibria. The different steric and electronic environments of conformers are sensitive to packing effects and can lead to significantly different structures of the liquid and solid phases and to polymorphism of the solid phase. Typical examples are torsional motions of alkyl groups, among others leading to *trans-trans* and *trans-gauche* conformations of the *n*-butyl chain. Another prominent example is the *trans-cis* isomerism of $[NTf_2]^-$ [11], which

Ionic Liquids, Structures, Fig. 2 *Trans-cis* isomerism of the bis(trifluoromethanesulfonyl)imide anion

adds to the configurational entropy, and has far-reaching consequences on the liquid-phase properties of $[NTf_2]^-$ based Ionic Liquids (Fig. 2).

Ion Pairs and Ion Cluster in the Vapor Phase

Probably the most striking feature of Ionic Liquids is the very low vapor pressure. Although reliable vapor pressure data are available for some time [12], the low pressures prevent a straightforward experimental characterization of the molecular properties of the vapor phase. Thus, the ion speciation in the vapor phase and the properties of these species necessarily result from computational studies. The following features are now well established:

- The ion speciation in the vapor phase is dominated by neutral ion pairs complemented by some neutral clusters of larger size.
- Free ions and charged ion clusters are practically absent [13].
- The computed binding energies in energy-minimized structures of the isolated cation-anion pairs have values up to 400 kJ mol^{-1}, which are an order of magnitude larger than the typical binding energies of pairs of uncharged molecules [14].
- The large binding energies almost exclusively result from electrostatic forces [14].
- The cation-anion configurations in the isolated pair are highly salt-specific, and are difficult to predict by plausible arguments. The rationale is that a given ion has to maximize interactions with many atoms of the counterion.

An illustrative example is 1-butyl-3-methylimidazolium tetrafluoroborate. In the isolated ion pair, the $[BF_4]^-$ ion is located over the aromatic ring of the cation, with three fluorine atoms forming a triangle with short contacts to the hydrogen at carbon C^2 between the two nitrogens and to the hydrogen atoms of the alkyl groups.

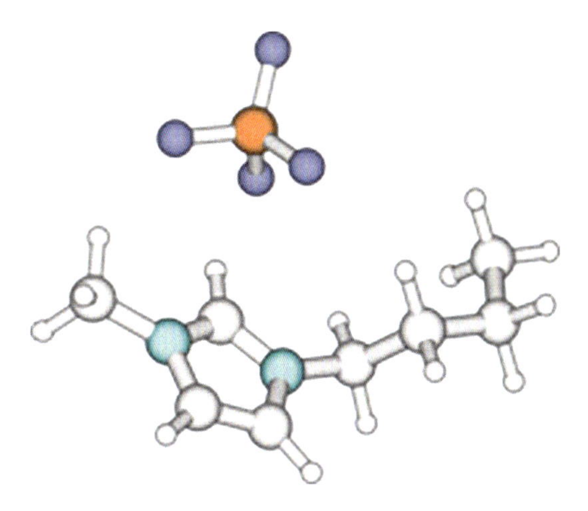

Ionic Liquids, Structures, Fig. 3 Structure of the isolated cation–anion pair of 1-butyl-3-methylimidazolium tetrafluoroborate

In going from ion pairs to larger clusters and eventually to the liquid phase, the number of interaction centers between the ionic constituents largely increases and cooperative effects will come into play. This increased complexity results in a quantitative change of the structure of cation-anion configurations, and may even affect the electronic structure of the ions themselves [15]. As a consequence, isolated cation-anion pairs usually do not signal cation-anion configurations in the liquid phase. Illustrative examples are 1-alkyl-3-methylimidazolium chlorides, where in the liquid phase, the preferred positions of the chloride with regard to the imidazolium ring and the acid C–H bonds are substantially different from those in isolated pairs. The study of ion clusters of different sizes can elucidate the role of the intermolecular interactions for these structural changes [15] (Fig. 3).

The Structure of the Liquid Phase

In contrast to the long-range structural order in crystals, the phrase "liquid structure" usually concerns local structural characteristics. Owing to the complex atomic structures of the ions and resulting broad distance distributions, the liquid phase decisive X-ray and neutron scattering experiments are difficult. Meaningful experimental data have been reported only in a few cases [11]. The available structural information is therefore almost exclusively based on atomistic (molecular dynamics or Monte Carlo) simulations [6].

The local ion configurations are highly salt-specific, while some long-range structural characteristics are generic for Ionic Liquids:

- The most outstanding observation is the existence of comparatively long-range charge-ordered structures. The cation-anion distribution and the like-ion (cation–cation and anion–anion) distributions show pronounced oscillations up to 20 Å or even longer [6, 11]. This ordering extends over a much longer range than observed in molecular liquids and indicates the existence of several distinct solvation shells around a given ion.
- The oscillations of the cation-anion distribution are out-of-phase with those of the like-ion distributions, signaling alternating layers of cations and anions.
- The global structural features of the liquid, glassy and crystalline phases of a Ionic Liquid are often similar, but the local ion configurations of these phases sensitively depend on packing effects and are often not identical.

A controversially debated subject is the existence (or not) of neutral ion pairs, causing the low ionicity of some Ionic Liquids [3]. The liquid structures show that cations and anions interact with many counterions. Neutral cation-anion configurations, the correlated translational motions of which are responsible for the low electrical conductance, only exist on fast (subpicosecond) time scales [8, 9].

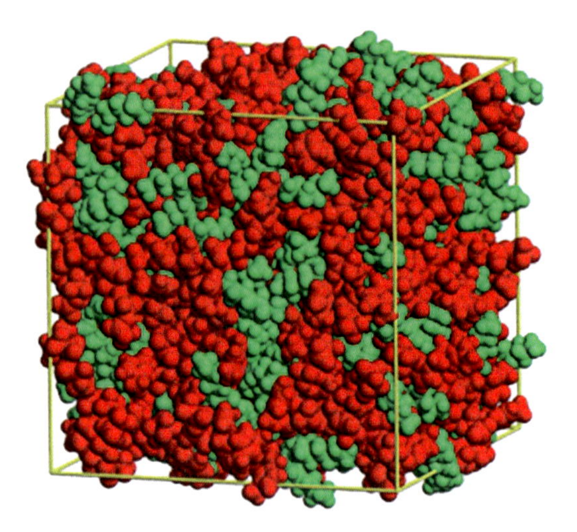

Ionic Liquids, Structures, Fig. 4 Microheterogeneity in 1-butyl-3-methylimidazolium bis(trifluoromethylsulfonyl)imide. Polar regions involving the head groups of the cations and the anions are shown in *red* color. Apolar regions of the alkyl groups are shown in *green* color (Figure provided by courtesy by Prof. Agilio A. H. Padua)

Phenomena at Mesoscopic Length Scales

Recent work has shown that Ionic Liquids possess micro-heterogeneous structures on mesoscopic length scales, which render their morphology much more complex than expected on the basis of the properties of simple molten salts and are not present in simple molecular solvents. The existence of a high degree of order of Ionic Liquids over mesoscopic spatial scales is, for example, witnessed by small-angle X-ray and neutron scattering experiments [16] and molecular dynamics simulations [17]. A straightforward explanation is the nanoscale segregation of apolar moieties in a polar network formed by charged head groups of the cations and by anions [16]. The resulting hydrophilic and hydrophobic patches of Ionic Liquids on mesoscopic scales have intriguing consequences for solvation because they enable a dual solvent behavior: Ionic Liquids can incorporate nonpolar solutes in their nonpolar domains, while hydrophilic domains can solvate polar solutes. Thus, ILs can simultaneously dissolve species of very different nature [18] (Fig. 4).

Future Directions

Despite some important structural information deduced from spectroscopic methods and scattering techniques, theoretical and computational methods are necessary to analyse the complex molecular forces between the ions and interpret the experimental data. Increasing computer power will enable the refinement of the systems studied, bring the models closer to real systems.

Cross-References

- ▶ Ionic Liquids
- ▶ Ionic Liquids, Biocompatible

References

1. Wasserscheid P, Welton T (2008) Ionic liquids in synthesis, 2nd edn. VCH, Weinheim
2. Plechkova NV, Seddon KR (2008) Application of ionic liquids in the chemical industry. Chem Soc Rev 37:123–150
3. Weingärtner H (2008) Understanding ionic liquids at the molecular level: facs, problems and controversies. Angew Chem Int Ed 46:2–19
4. Angell CA (2012) Ionic liquids: past, present and future. Faraday Discuss 154:9–27
5. Krossing I, Slattery JM, Daguenet C, Dyson PJ, Oleinikova A, Weingärtner H (2006) Why are ionic liquids liquid? A simple explanation based on lattice and solvation energies. J Am Chem Soc 128:13427–13434
6. Maginn EJ (2007) Atomistic simulation of the thermodynamic and transport properties of Ionic Liquids. Acc Chem Res 40:1200–1207
7. Greaves TL, Drummond CJ (2008) Protic ionic liquids: properties and applications. Chem Rev 108:206–237
8. Wulf A, Fumino K, Ludwig R (2010) Spectroscopic evidence for an enhanced anion-cation interaction from hydrogen bonding in pure imidazolium Ionic Liquids. Angew Chem Int Ed 49:449–453
9. Zhao W, Leroy F, Heggen B, Zahn S, Kirchner B, Balasubramanian S, Müller-Plathe F (2009) Are there stable ion-pairs in room temperature ionic liquids? Molecular dynamics of 1-butyl-3-methylimidazoliun hexafluorophosphate. J Am Chem Soc 131: 15825–15833
10. Holbrey D, Seddon KR (1999) The phase behavior of 1-alkyl-3-methylimidazolium tetrafluoroborates; ionic liquids and ionic liquid crystals. J Chem Soc Dalton Trans 2133–2140
11. Deetlefs M, Hardacre C, Nieuwenhausen M, Padua AAH, Sheppard O, Soper AK (2006) Liquid structure of the ionic liquid 1,3-dimethylimidazolium bis{(trifluoromethyl)sulfonyl}amide. J Phys Chem B 110:12055–12061
12. Ludwig R, Kragl U (2007) Do we understand the volality of ionic liquids? Angew Chem Int Ed 46:6582–6584
13. Rai N, Maginn EJ (2012) Critical behaviour and vapour-liquid coexistence of 1-alkyl-3-methhylimidazolium (trifluoromethylsulfonyl) amide ionic liquids via Monte Carlo simulations. Faraday Discuss 154:53–69
14. Tsuzuki S, Tokuda H, Hayamizu K, Watanabe M (2005) Magnitude and directionality of interactions in ion pairs: relationship with conductivity. J Phys Chem B 109:16474–16481
15. Wendler K, Dommert F, Zhao YY, Berger R, Holm C, Delle Site L (2012) Ionic Liquids studied across different scales: a computational perspective. Faraday Discuss 154:111–132
16. Russina O, Triolo A (2012) New experimental evidence supporting the mesoscopic segregation model in room temperature ionic liquids. Faraday Discuss 154:97–109
17. Canongia Lopes JN, Padua AAH (2006) Nanostructural organization in Ionic Liquids. J Phys Chem B 110:3330–3335
18. Canongia Lopes JN, Costa Gomes MF, Padua AAH (2006) Nonpolar, polar and associated solutes in Ionic Liquids. J Phys Chem B 110:16816–16818

Ionic Mobility and Diffusivity

Charles W. Monroe
Department of Chemical Engineering, University of Michigan, Ann Arbor, MI, USA

Fundamental Concepts

Theories of mass transport in electrolytes or electrolytic solutions take into account that motion of dissolved species i can be driven by gradients in electric potential Φ (migration), as well as by gradients in molar concentration c_i (diffusion) and by motion of material at the bulk velocity v (convection). The most commonly deployed model for electrolyte transport is the Nernst-Planck theory [1], developed in detail by Levich [2]. Within this theory, one constituent of the solution – typically a neutral species in relative excess – is identified as a 'solvent'. The total molar flux of any remaining 'solute' species i, $\vec{N}_i$, is then expressed relative to a stationary coordinate frame as

Ionic Mobility and Diffusivity, Table 1 Ion transport properties at infinite dilution in water at 25 °C and 1 atm [5]

Cation	$D_i \times 10^9$ m^{2}s^{-1}	$u_i \times 10^{12}$ mol•s•kg^{-1}	Anion	$D_i \times 10^9$ m^{2}s^{-1}	$u_i \times 10^{12}$ mol•s•kg^{-1}
Ag^+	1.648	0.665	Br^-	2.080	0.839
Al^{3+}	0.541	0.218	CN^-	2.077	0.838
Ba^{2+}	0.847	0.342	Cl^-	2.032	0.820
Be^{2+}	0.599	0.242	ClO_2^-	1.385	0.559
Ca^{2+}	0.792	0.319	ClO_3^-	1.720	0.694
Ce^{3+}	0.620	0.250	ClO_4	1.792	0.723
Co^{2+}	0.732	0.295	CrO_4^{2-}	1.132	0.457
Cr^{3+}	0.595	0.240	F^-	1.475	0.595
Cu^{2+}	0.714	0.288	$Fe(CN)_6^{4-}$	0.735	0.296
Fe^{2+}	0.719	0.290	$Fe(CN)_6^{3-}$	0.896	0.361
Fe^{3+}	0.604	0.244	HCO_3^-	1.185	0.478
H^+	9.311	3.756	HSO_3^-	1.545	0.623
K^+	1.957	0.789	HSO_4^-	1.385	0.559
Li^+	1.029	0.415	I^-	2.045	0.825
Mg^{2+}	0.706	0.285	MnO_4^-	1.632	0.658
Mn^{2+}	0.712	0.287	NO_2^-	1.912	0.771
NH_4^+	0.957	0.789	NO_3^-	1.902	0.767
$N_2H_5^+$	1.571	0.634	OH^-	5.273	2.127
Na^+	1.334	0.538	PF_6^-	1.515	0.611
Ni^{2+}	0.661	0.267	PO_4^{3-}	0.824	0.332
Pb^{2+}	0.945	0.381	SO_3^{2-}	0.959	0.387
Zn^{2+}	0.703	0.284	SO_4^{2-}	1.065	0.430

$$\vec{N}_i = -D_i \nabla c_i - F z_i u_i c_i \nabla \Phi + c_i \vec{v}, \quad (1)$$

where z_i is the number of charge equivalents carried by i, F is Faraday's constant, R is the gas constant, and T is the absolute temperature. Two transport properties appear in the Nernst-Planck constitutive law, and form the basis for the definitions of the concepts of the *mobility* and the *diffusivity* of a dissolved ionic species. The ionic mobility u_i (units of mol•s/kg) quantifies how electric fields force molar flux of i by migration; the ionic diffusivity D_i (units of m^2/s) quantifies how concentration gradients drive molar flux of i by diffusion. Both properties are frequently tabulated and have been reported by a number of classic sources [3–5]; exemplary values for some ions in aqueous solutions at ambient temperature and pressure are given in Table 1.

To prevent inconsistencies that arise when attempting to define a diffusivity or mobility of a solvent, $\vec{v}$ in Eq. 1 must be recognized as the solvent velocity. With that in mind, the diffusivity and mobility should be viewed as properties that quantify binary interactions, in the sense that the diffusivity or mobility quantifies a drag force that the solvent exerts in response to a concentration or potential gradient imposed on species i. (In multi-solvent systems, such as the mixed solvents used in Li-ion batteries, a particular solvent or the solvent mixture as a whole can be used to define $\vec{v}$, depending on whether or not the constituent solvents are treated as individual species).

The Nernst-Planck equation is often employed by practitioners because of its similarity to Fick's law and its convenient separation of diffusion and migration terms. It should be borne in mind, however, that the theory is inconsistent with the basic requirements of irreversible thermodynamics [6, 7]. Nernst-Planck theory uses $n + k - 1$ properties to characterize transport in an isothermal, isobaric n-species system containing

k charged species. Onsager's reciprocal relation for isothermal isobaric diffusion [8] suggests instead that a system containing n constituents should afford $n^*(n-1)/2$ independent transport coefficients. (These two equations only yield the same number in the case of a phase containing three types of ion dissolved in a single neutral solvent, which is a coincidence at best).

Another problem with constitutive law 1 is the assumption that a diffusive flux of i can only be driven by its own concentration gradient. This model does not account for all the possible sources of diffusional drag. For instance, in a solution containing an electrolyte with an additional neutral solute, as well as a solvent, flux of the neutral solute could lead to drag forces that induce gradients in the electrolyte concentration. The recognition of this sort of phenomenon suggests that in principle additional terms, associated with solute/solute interactions, should appear in Eq. 1.

The difficulties mentioned can be resolved to some extent by recognizing that Eq. 1 holds only in the limit of extreme dilution – a situation wherein solute/solute interactions are negligible and the diffusion and mobility relate through solvent properties. In this limit, statistical treatments combine with force balances to yield an Einstein-Smoluchowski relation [9, 10] (for electrolytic diffusion this is also sometimes called the Nernst-Einstein equation [11])

$$D_i = u_i RT, \qquad (2)$$

where R is the gas constant and T is the absolute temperature. This relationship between ionic diffusivity and mobility holds in the limit that species-species interactions occur primarily between solutes and solvent. With the relation given by Eq. 2 taken into consideration, the terms "ionic diffusivity" and "ionic mobility" become interchangeable: both properties quantify the drag exerted by solvent molecules in response to an applied thermodynamic driving force. This driving force comprises contributions from the electric field, $-\nabla\Phi$, or concentration-gradient induced osmotic pressure ($-RT\nabla c_i$ for an ideal solute).

Onsager provided a more complete theory of molecular mobility that accounts for solute/solute interactions, making possible analyses of concentrated multicomponent systems. Based on inspection of the second law of thermodynamics in a transient form [7, 12], a thermodynamic force d_i driving the diffusion of species i in a concentrated solution can be identified as

$$d_i = -c_i\left(\nabla\mu_i + \bar{S}_i\nabla T - \frac{M_i}{\rho}\nabla p\right), \qquad (3)$$

where μ_i is the chemical or electrochemical potential of i, $\bar{S}_i$ its partial molar entropy, and M_i its partial molar mass; ρ is the solution density and p the absolute pressure. The first law of thermodynamics and the extensivity of internal energy imply that all the diffusion driving forces within a volume element of an n-component solution balance, according to the Gibbs-Duhem equation $\sum_i d_i = 0$.

An extended Stefan-Maxwell diffusion equation can be constructed by setting up a homogeneous, linear relationship between the vector of diffusion driving forces and the vector of species velocity differences,

$$d_i = \sum_{j\neq i} \frac{RTc_ic_j}{c_T D_{ij}}\left(v_i - v_j\right), \text{ or}$$
$$d_i = \sum_{j\neq i} \frac{c_ic_j}{c_T u_{ij}}\left(v_i - v_j\right), \qquad (4)$$

where v_i is the velocity of species i and $c_T = \sum_i c_i$ is the total molar concentration of the solution. Equation 4 can be viewed as a general equation for drag – the force on species i per unit volume relates to each species velocity difference through a drag coefficient, which quantifies i/j interactions. The flux law on the left of Eq. 4 defines the Stefan-Maxwell coefficients D_{ij} that quantify the diffusivity of species i through j; the equation on the right defines a corresponding relative mobility u_{ij}. These properties are generally functions of local temperature, pressure, and composition in electrolytic solutions [13]. It can be proved that this formulation results in symmetric transport coefficients; a microscopic analysis shows that the Onsager reciprocal relation for this flux law is $D_{ij} = D_{ji}$ (or $u_{ij} = u_{ji}$).

Extended Stefan-Maxwell constitutive laws for diffusion Eq. 4 resolve a number of fundamental problems presented by the Nernst-Planck transport formulation Eq. 1. A thermodynamically proper pair of fluxes and driving forces is used, guaranteeing that all the entropy generated by transport is taken into account. The symmetric formulation of Eq. 4 makes it unnecessary to identify a particular species as a solvent – every species in a solution is a 'solute' on equal footing. Use of velocity differences reflects the physical criterion that the forces driving diffusion of species i relative to species j be invariant with respect to the convective velocity. Finally, all possible binary solute/solute interactions are quantified by distinct transport coefficients; each species i in the solution has a diffusivity or mobility relative to every other species j, D_{ij} or u_{ij}, respectively.

The proper number of transport properties can be shown to appear naturally in the extended Stefan-Maxwell formulation as follows: one of Eq. 4 (say, the one quantifying d_n) depends on the others through the Gibbs-Duhem equation, leaving $n-1$ independent flux laws for species transport remaining; species i is skipped in the sum over velocity differences, showing that in those $n-1$ independent flux laws appear $(n-1)^2$ transport coefficients; the symmetry of microscopic correlations implied by the Onsager reciprocal relation then implies that $D_{ij} = D_{ji}$, reducing the number of independently specifiable transport coefficients by $\frac{1}{2}(n-2)(n-1)$ [14, 15]. Thus there remain $\frac{1}{2}n(n-1)$ distinct transport coefficients in the extended Stefan-Maxwell flux laws for an isothermal, isobaric n-component system.

It is a drawback that transport constitutive laws in the form of Eq. 4 make it difficult to identify the effect of convection. Also, analyses based on analogies to systems described by Fourier's law, Ohm's law, or Fick's law are confounded by the fact that Eq. 4 is explicit in the diffusion driving forces, rather than the fluxes (cf. Eq. 1, a more familiar formulation in which $\vec{N}_i = c_i \vec{v}_i$ stands alone on the left side). The measurement of mobilities usually requires fitting transient experimental data with formulas based on analytical solutions of the transport equations. Such formulas are much easier to derive with flux-explicit transport laws.

Inversion of the extended Stefan-Maxwell equations is challenging because if the natural set of n flux laws is used (as it was by Onsager), the linear dependence that follows from the Gibbs-Duhem equation implies that the transport-coefficient matrix is singular. (Singularity of the n-dimensional transport matrix also follows from the principle of driving-force invariance with respect to reference velocity [8].) A general procedure to invert the Stefan-Maxwell equations was first provided by Helfand [16], and has been implemented by Newman for transport in multicomponent systems containing ionic species [11].

Nevertheless, the most rigorous, thermodynamically complete understanding of ion mobility or diffusivity in concentrated electrolytic solutions is provided by the extended Stefan-Maxwell transport theory, which can be applied to electrolytic solutions [13, 17–22], ionic melts or ionic liquids [23, 24], and ion-exchange membranes [25–28]. The diffusion driving force in any system involving ion transport is taken to be a gradient of electrochemical potential μ_i, typically expressed in terms of a "chemical part" and an "electrical part" as

$$\mu_i = \mu_i^\theta + RT\ln(\gamma_i y_i) + z_i F\Phi. \quad (5)$$

Here μ_i^θ is the electrochemical potential of i in a secondary reference state (typically specified by choosing a reference electrode of a given kind [29]), γ_i is an activity coefficient on a mole-fraction basis, y_i is the mole fraction of i, F is Faraday's constant, and Φ is the electric potential. (Note that y_i is a sensible basis for composition because it leads the electrochemical potentials to satisfy the Gibbs-Duhem equation in the limit of infinite dilution, where γ_i goes to unity for all species.) For thermodynamic consistency it should be borne in mind that the balance of net charge in equilibrated phases requires overall electroneutrality, and that chemical potentials can therefore only be measured for neutral combinations of ions. This difficulty can be resolved in a satisfactory way by defining Φ as a "quasi-electrostatic" potential. A quasi-electrostatic potential referred to ionic species m is defined by letting f_m be 1 and defining all the other ionic activity coefficients through appropriate dissociation equilibria [30].

The Nernst-Planck theory (under the Nernst-Einstein Eq. 4) can be derived from the extended Stefan-Maxwell equation by taking Φ to be a quasi-electrostatic potential referred to one ion m and taking the limit of extreme dilution. Thus it can be seen formally that Nernst-Planck theory neglects solute-solute interactions, and applies strictly only in the limit of infinite dilution. In an n-component electrolytic phase, transport can be quantified using $\frac{1}{2}n(n-1)$ independent species mobilities, which quantify the binary interactions between each pair of species.

Measurement

Ionic mobilities or diffusivities are measured experimentally by combining a number of independent dynamic experiments to isolate the transport properties of interest. Auxiliary experiments must also be performed to establish the thermodynamic properties of the solution. For the simplest case of a binary electrolytic solution (comprising an electrolyte and a neutral solvent) of a binary electrolyte (comprising one anion and one cation), there are three species (solvent 0, cations +, and anions –). This case requires that a single thermodynamic characterization be implemented to quantify the electrolyte activity as a function of composition. Subsequently, three independent dynamic measurements must be implemented to quantify the three independent relative diffusivities D_{+0}, D_{-0}, and D_{+-}.

The local electroneutrality that holds within electrolytic solutions impedes direct measurements of the Stefan-Maxwell coefficients. Instead, the values of D_{ij} are usually extracted from measurements of other properties that can be measured directly. The properties typically measured are the ionic conductivity κ, the Fickian diffusion coefficient of the electrolyte D, and the cation transference number of the electrolyte t_+^0. These relate to the Stefan-Maxwell coefficients through

$$\frac{1}{\kappa} = -\frac{RT}{c_T z_+ z_- F^2}\left[\frac{1}{D_{+-}} - \frac{c_0 z_-}{c_+(z_+ D_{0+} - z_- D_{0-})}\right] \tag{6}$$

$$D = \frac{c_T \chi}{c_0}\left[\frac{D_{0+} D_{0-}(z_+ - z_-)}{z_+ D_{0+} - z_- D_{0-}}\right] \tag{7}$$

$$t_+^0 = \frac{z_+ D_{0+}}{z_+ D_{0+} - z_- D_{0-}}. \tag{8}$$

Here the thermodynamic factor χ relates to the derivative of the electrolyte activity coefficient γ_{+-} with respect to molality m through

$$\chi = 1 + \left(\frac{\partial \ln \gamma_{+-}}{\partial \ln m}\right)_{T,p},$$

and can be obtained by thermodynamic measurements of the concentration overpotential in concentration cells (state-of-the-art examples of these experiments can be found in references [22] and [21]).

Conductivity measures the ionic current in response to an electric field in a system without concentration gradients; it is relatively straightforward to measure using a DC conductivity meter or an AC impedance spectrum extrapolated to the high-frequency limit. Diffusivities measure the characteristic time for diffusion of the neutral electrolyte; they can be established using the restricted diffusion experiment of Harned [31], in which some voltage-dependent characteristic of a cell with an initial concentration gradient is tracked as the concentration profile relaxes to equilibrium. The transference number quantifies the fraction of ionic current carried by co-current motion of cations; it can be measured by the Hittorf [3], moving-boundary [32], or galvanostatic polarization [33] methods.

Cross-References

- ► Electrolytes, Classification
- ► Electrolytes, History
- ► Electrolytes, Thermodynamics
- ► Ion Mobilities
- ► Ion Properties

References

1. Planck M (1890) Ueber die Potentialdifferenz zwischen zwei verdünnten Lösungen binärer Electrolyte. Annalen der Physik und Chemie 40(8):561–576
2. Levich B (1942) The theory of concentration polarization. Acta Physicochimica URSS 17(5–6):257–307
3. MacInnes DA (1961) The principles of electrochemistry, 2nd edn. Dover Books, New York
4. Robinson RA, Stokes RH (1968) Electrolyte solutions, second edition (revised). Butterworths, London
5. Vanysek P (2011) Conductivity ionic diffusion at infinite dilution. In: Haynes WM, Lide DR (eds) CRC handbook of chemistry and physics, 92nd edn. CRC Press/Taylor and Francis Group, Boca Raton/Florida, pp 77–79
6. Kuiken GDC (1994) Thermodynamics of irreversible processes: applications to diffusion and rheology. Wiley, Sussex
7. de Groot SR, Mazur P (1984) Non-equilibrium thermodynamics. Dover, Mineola
8. Onsager L (1945) Theories and problems of liquid diffusion. Ann N Y Acad Sci 46(5):241–265
9. von Smoluchowski M (1906) Zur kinetischen Theorie der Brownschen Molekularbewegung und der Suspensionen. Ann Phys 326(14):756–780
10. Einstein A (1905) Über die von der molukularkinetischen Theorie der Wärme geforderte Bewegung von in ruhenden Flüssigkeiten suspendierten Teilchen. Ann Phys 322(8):549–560
11. Newman J, Thomas-Alyea KE (2004) Electrochemical systems, 3rd edn. Wiley, Hoboken
12. Hirschfelder JO, Curtiss CF, Bird RB (1964) Molecular theory of gases and liquids
13. Newman J, Bennion D, Tobias CW (1965) Mass transfer in concentrated binary electrolytes. Berichte Der Bunsen-Gesellschaft Fur Physikalische Chemie 69(7):608
14. Monroe CW, Newman J (2009) Onsager's shortcut to proper forces and fluxes. Chem Eng Sci 64(22):4804–4809
15. Monroe CW, Newman J (2006) Onsager reciprocal relations for Stefan-Maxwell diffusion. Ind Eng Chem Res 45(15):5361–5367
16. Helfand E (1960) On inversion of the linear laws of irreversible thermodynamics. J Chem Phys 33(2):319–322
17. Ma YP, Doyle M, Fuller TF, Doeff MM, Dejonghe LC, Newman J (1995) The measurement of a complete set of transport properties for a concentrated solid polymer electrolyte solution. J Electrochem Soc 142(6):1859–1868
18. Doeff MM, Edman L, Sloop SE, Kerr J, Jonghe LCD (2000) Transport properties of binary salt polymer electrolytes. J Power Sources 89(2):227–231
19. Doeff MM, Georen P, Qiao J, Kerr J, Jonghe LCD (1999) Transport properties of a high molecular weight poly(propylene oxide)-LiCF3SO3 system. J Electrochem Soc 146(6):2024–2028
20. Ferry A, Doeff MM, DeJonghe LC (1998) Transport property measurements of polymer electrolytes. Electrochim Acta 43(10–11):1387–1393
21. Nyman A, Behm M, Lindbergh G (2008) Electrochemical characterisation and modelling of the mass transport phenomena in LiPF6-EC-EMC electrolyte. Electrochim Acta 53(22):6356–6365
22. Valøen LO, Reimers JN (2005) Transport properties of LiPF6-based Li-ion battery electrolytes. J Electrochem Soc 152(5):A882–A891
23. Wang MH, Newman J (1995) The electrical conductivity of sodium polysulfide melts. J Electrochem Soc 142(3):761–764
24. Thompson SD, Newman J (1989) Differential diffusion coefficients of sodium polysulfide melts. J Electrochem Soc 136(11):3362–3369
25. Heintz A, Wiedemann E, Ziegler J (1997) Ion exchange diffusion in electromembranes and its description using the Maxwell-Stefan formalism. J Membr Sci 137(1–2):121–132
26. Wiedemann E, Heintz A, Lichtenthaler RN (1998) Transport properties of vanadium ions in cation exchange membranes: determination of diffusion coefficients using a dialysis cell. J Membr Sci 141(2):215–221
27. Delacourt C, Newman J (2008) Mathematical modeling of a cation-exchange membrane containing two cations. J Electrochem Soc 155(11):B1210–B1217
28. Okada T, Moller-Holst S, Gorseth O, Kjelstrup S (1998) Transport and equilibrium properties of Nafion (R) membranes with H + and Na + ions. J Electroanal Chem 442(1–2):137–145
29. Guggenheim EA (1967) Thermodynamics: an advanced treatment for chemists and physicists, 5th edn. North-Holland Publishing Company, Amsterdam
30. Smyrl WH, Newman J (1968) Potentials of cells with liquid junctions. J Phys Chem 72(13):4660
31. Harned HS, Nuttall RL (1949) The differential diffusion coefficient of potassium chloride in aqueous solutions. J Am Chem Soc 71(4):1460–1463
32. MacInnes DA, Cowperthwaite IA, Blanchard KC (1926) The moving-boundary method for determining transference numbers. V. A constant current apparatus. J Am Chem Soc 48:1909–1912
33. Hafezi H, Newman J (2000) Verification and analysis of transference number measurements by the galvanostatic polarization method. J Electrochem Soc 147(8):3036–3042

Ions at Biological Interfaces

Pavel Jungwirth
Institute of Organic Chemistry and Biochemistry, Academy of Sciences of the Czech Republic, Prague, Czech Republic

Historical Overview

The effects of salt ions on the behavior of biomolecules in solutions, such as salting out of proteins, has been traditionally ascribed to ion–water interactions in the aqueous bulk [1]. The ion-specific behavior, expressed, e.g., in the famous Hofmeister series [2, 3], has been then rationalized by classifying ions as either kosmotropes ("structure makers") or chaotropes ("structure breakers") according to their ability to structure water molecules around themselves [4]. According to this picture, cosmotropes, but not chaotropes, organize layers of water molecules around themselves, thus effectively removing the solvent from proteins, which leads to salting out. There is, however, mounting experimental evidence that this picture is incomplete at best and that ions (at least monovalent ones) are not able to strongly affect more water molecules than their immediate hydration shells [5–7]. Alternative or additional explanations of salt action are, therefore, being searched for with the prime suspect (indeed, the elephant in the closet) being the interface between the biomolecule and the surrounding salt solution.

Ion–Protein Interactions

As a simplest possible model, the picture of a protein as homogeneous sphere of a low dielectric constant has been repeatedly invoked in the literature [8, 9]. Within this picture the interface between a globular protein and an aqueous solution should resemble that between air or oil and water. However, when comparing the behavior of ions at the protein/water and air(oil)/water interface, there are striking differences. For example, alkali cations such as sodium or potassium and divalent ions like calcium or sulfate are repelled from the water surface, but they exhibit affinities for the protein/water interface [10]. Calculations and experiments also show that interactions of ions with the protein surface are mostly of a local nature. For example, the alkali cations are primarily attracted by negatively charged moieties such as the carboxylic groups in the side chains of glutamate and aspartate and, to a lesser extent, the amide oxygens of the backbone, whereas sulfate exhibits an affinity for the positively charged groups in the side chains of lysine and arginine [11]. The dielectric similarity between the air/water and protein/water interface is thus of little use here because the ion–protein behavior is dominated by local interactions with charged and polar groups at the protein surface in the presence of explicit water molecules rather than by the average dielectric properties of the protein. An exception is the case of large polarizable monovalent ions which exhibit qualitatively comparable affinities for the air/water interface and for hydrophobic parts of the protein surface, primarily due to cavitation and polarization effects. The situation is, however, more complicated and subtle in the latter case, where not only the hydrophobicity of the nonpolar groups but also their proximity to electron-withdrawing atoms (particularly nitrogen and oxygen in the peptide bond) enhances their interaction with soft anions.

Since the local ion–protein interactions are of prime importance for ion segregation at surfaces of hydrated proteins, one can invoke reductionism as a reasonable first step. Thus, if we understand how different ions interact with individual (terminated) amino acids or even with their side-chain groups and with the amide group in water, we can extrapolate to a rough picture of their segregation at a protein surface [12]. For these cases, the empirical rule of matching water affinities, stating that an ion pairs most efficiently with an oppositely charged ion of comparable hydration enthalpy (i.e., surface charge density, within the simple Born model [13]), can be applied, albeit only as a first estimate [14, 15]. Ion-specific interactions with amino acid residues, as well as complex proteins, are thus governed by two main mechanisms that target distinct surface groups on the macromolecule – ion pairing with charged side-chain groups, as well as the backbone amide moiety, and weak interactions with nonpolar groups.

In summary, solubilities and stabilities of proteins in solutions are governed not only by the macromolecular net charge, salt concentration, and valency, but also by the chemical nature of the dissolved ions [1]. For common anions, the Hofmeister ordering in the ability to precipitate egg white proteins is $F^- > CH_3COO^- > Cl^- > NO_3^- > Br^- > I^- > SCN^-$, while the effect of (alkali) cations is usually less pronounced [2, 3]. The Hofmeister ordering is, however, dependent also on the particular counterion, solution pH, and the protein isoelectric point, pI. It is well know that several proteins such as lysozyme follow the reverse Hofmeister series at low and intermediate pHs, when they are positively charged, which has been rationalized in terms of a fine interplay between the above ion-specific interaction [16, 17]. It should be noted that dispersion interaction is also present and ionspecific [18]; however, except for special cases involving (quasi) aromatic residues (such as the "Coulomb-defying" pairing between like-charge guanidinium groups [19, 20]) they are of a secondary importance.

Ion-Membrane and Ion-Nucleic Acid Interactions

Despite the fact that main focus has always been on ion channels, the influence of physiologically most relevant ions (such as Na^+, K^+, Cl^-, Ca^{++}, or Mg^{++}) on model lipid membranes was also studied in considerable detail [21–24]. Additionally, other ions, such as Li^+, Cs^+, NH_4^+, Ba^{++}, La^{+++}, F^-, Br^-, I^-, NO_3^-, and SCN^-, were also investigated [24–26] in order to elucidate the factors influencing the specific ionic effects observed. This specificity has been known from measurements to be more pronounced for anions than for cations; consequently, more experimental data are available for the former ions [26–29]. Computer simulations are, however, typically more focused on cations, since a proper description of the effects of larger anions often requires the use of resource-consuming polarizable force field [24], while cationic interactions with model membranes have been satisfactorily described using nonpolarizable potentials [23, 30–32]. The strongest cationic effects have been observed both in simulations and experiments for multivalent cations (Mg^{++} and Ca^{++}) and monovalent cations with large charge density (Li^+) [26, 33], which interact appreciably with lipid bilayers [25, 34], rigidifying them [35] and stabilizing their gel phase [35–37]. Association of larger monovalent cations (Na^+, K^+, Rb^+, and Cs^+) with neutral (zwitterionic) lipid bilayer is much weaker; it is, therefore, difficult to measure directly their ion-specific effects [26, 29]. Fluorescence measurements using solvent relaxation techniques have shown weak dehydration and hindered mobility at the glycerol level of DOPC membrane upon addition of 150 mM NaCl [23], and molecular dynamics simulations showed that Na^+ (in contrast to K^+ or Cs^+) exhibits affinity to the headgroups of DOPC membrane [23, 38]. The binding site of Na^+ was found to be the phosphodiester and/or the carbonyl groups of phospholipids, depending on the force field employed [21, 23, 38, 39]. The cationic effects are strongly amplified in negatively charged membranes (e.g., phosphatidylserine bilayers) or in those composed from a mixture of zwitterionic and negatively charged phospholipids [40].

Nucleic acids represent another type of biological systems where specific interactions with ions are of a crucial importance. As a highly charged polyelectrolyte, the negative charges on the phosphate groups need to be compensated by counter cations. Among monovalent cations, specific interactions of Na^+ and K^+ have been investigated in considerable detail. Molecular dynamics and quantum chemical calculations [41–47] point to an interesting observation, namely, that while an isolated monovalent phosphate group exhibits only a weak and roughly comparable affinity to these two alkali cations, in DNA the polyelectrolyte effect enhances both ion binding and ion specificity which goes in favor of Na^+ over K^+. These observations are in general supported by measurements, although the experimental evidence is somewhat internally conflicting [48–50]. Divalent cations interact even more strongly with nucleic acids leading to counterion condensation or even charge reversal in the vicinity of aqueous DNA [45].

Future Directions

Within the last decade significant progress has been made in understanding of the molecular origins of the ion-specific Hofmeister effects. The attention has clearly moved from ion properties in homogeneous aqueous solutions to their behavior at the biomolecule/solution interface. Researchers got a lot of millage from the reductionist approach assuming a local character and additivity in ion–biomolecule interactions. Within this picture the key components responsible for the specific ion–biomolecule interaction are pairing of ions from the solution with charged and highly polar groups at the biomolecular surface, together with the ability of large soft ions to segregate at the interface between the solution and hydrophobic surface groups. In this context, the effects of aromatic groups, as well as that of neighboring electron-withdrawing atoms (nitrogen and oxygen in particular), deserve a closer scrutiny.

Future will likely see the third step of the Hegelian triad of *thesis* (i.e., the bulk origin of ion-specific effects), negated by *antithesis* (i.e., the interfacial origin of ion-specific effects), to be finally replaced by *synthesis* which will interpret ion-specific Hofmeister effects in terms of both bulk and interfacial behavior of ions. This synthesis will also likely lead us to the understanding that cationic and anionic effects cannot be always divided into separate Hofmeister series for cations and anions and that the local reductionist picture can in many cases (particularly if polyvalent ions are involved) serve only as a first approximation and more extended effects may come into play. This synthetic view is likely to prove useful also when further investigating the specific spatial and chemical arrangements nature has engineered to manipulate ions in transmembrane ion channels and pumps.

Cross-References

▶ Specific Ion Effects, Evidences
▶ Specific Ion Effects, Theory

References

1. Baldwin RL (1996) How Hofmeister ion interactions affect protein stability. Biophys J 71:2056–2063
2. Hofmeister F (1888) Zur Lehre von der Wirkung der Salze. Arch Exp Pathol Pharmakol Leipzig 24:247
3. Kunz W, Henle J, Ninham BW (2004) 'Zur Lehre von der Wirkung der Salze' (about the science of the effect of salts): Franz Hofmeister's historical papers. Curr Opin Colloid Interface Sci 9:19–37
4. Dill KA, Bromberg S (2002) Molecular driving forces: statistical thermodynamics in chemistry & biology. Taylor & Francis, London
5. Omta AW, Kropman MF, Woutersen S, Bakker HJ (2003) Negligible effect of ions on the hydrogen-bond structure in liquid water. Science 301:347–349
6. Smith JD, Saykally RJ, Geissler PL (2007) The effects of dissolved halide anions on hydrogen bonding in liquid water. J Am Chem Soc 129:13847–13856
7. Mancinelli R, Botti A, Bruni F, Ricci MA, Soper AK (2007) Hydration of sodium, potassium, and chloride ions in solution and the concept of structure maker/breaker. J Phys Chem B 111:13570–13577
8. Bostrom M, Williams DRM, Ninham BW (2003) Special ion effects: why the properties of lysozyme in salt solutions follow a Hofmeister series. Biophys J 85:686–694
9. Kirkwood JG, Shumaker JB (1952) Forces between protein molecules in solution arising from fluctuations in proton charge and configuration. Proc Natl Acad Sci USA 38:863–871
10. Hrobarik T, Vrbka L, Jungwirth P (2006) Selected biologically relevant ions at the air/water interface: a comparative molecular dynamics study. Biophys Chem 124:238–242
11. Vrbka L, Jungwirth P, Bauduin P, Touraud D, Kunz W (2006) Specific ion effects at protein surfaces: a molecular dynamics study of bovine pancreatic trypsin inhibitor and horseradish peroxidase in selected salt solutions. J Phys Chem B 110:7036–7043
12. Vrbka L, Vondrasek J, Jagoda-Cwiklik B, Vacha R, Jungwirth P (2006) Quantification and rationalization of the higher affinity of sodium over potassium to protein surfaces. Proc Natl Acad Sci USA 103:15440–15444
13. Born M (1920) Zeitschrift fur Physik 1:45
14. Collins KD (2006) Ion hydration: implications for cellular function, polyelectrolytes, and protein crystallization. Biophys Chem 119:271–281

15. Collins KD (1997) Charge density-dependent strength of hydration and biological structure. Biophys J 72:65–76
16. Finet S, Skouri-Panet F, Casselyn M, Bonnete F, Tardieu A (2004) The Hofmeister effect as seen by SAXS in protein solutions. Curr Opin Colloid Interface Sci 9:112–116
17. Horinek D, Netz RR (2007) Specific ion adsorption at hydrophobic solid surfaces. Phys Rev Lett 99:226104
18. Bostrom M, Tavares FW, Finet S, Skouri-Panet F, Tardieu A, Ninham BW (2005) Why forces between proteins follow different Hofmeister series for pH above and below pI. Biophys Chem 117:217–224
19. Mason PE, Neilson GW, Enderby JE, Saboungi ML, Dempsey CE, MacKerell AD, Brady JW (2004) The structure of aqueous guanidinium chloride solutions. J Am Chem Soc 126:11462–11470
20. Vondrasek J, Mason PE, Heyda J, Collins KD, Jungwirth P (2009) The molecular origin of like-charge arginine-arginine pairing in water. J Phys Chem B 113:9041–9045
21. Bockmann RA, Hac A, Heimburg T, Grubmuller H (2003) Effect of sodium chloride on a lipid bilayer. Biophys J 85:1647–1655
22. Filippov A, Oradd G, Lindblom G (2009) Effect of NaCl and $CaCl_2$ on the lateral diffusion of zwitterionic and anionic lipids in bilayers. Chem Phys Lipids 159:81–87
23. Vacha R, Siu SWI, Petrov M, Bockmann RA, Barucha-Kraszewska J, Jurkiewicz P, Hof M, Berkowitz ML, Jungwirth P (2009) Effects of alkali cations and halide anions on the DOPC lipid membrane. J Phys Chem A 113:7235–7243
24. Vacha R, Jurkiewicz P, Petrov M, Berkowitz ML, Bockmann RA, Barucha-Kraszewska J, Hof M, Jungwirth P (2010) Mechanism of interaction of monovalent ions with phosphatidylcholine lipid membranes. J Phys Chem B 114:9504–9509
25. McLaughlin A, Grathwohl C, McLaughlin S (1978) Adsorption of divalent-cations to phosphatidylcholine bilayer membranes. Biochim Biophys Acta 513:338–357
26. Garcia-Celma JJ, Hatahet L, Kunz W, Fendler K (2007) Specific anion and cation binding to lipid membranes investigated on a solid supported membrane. Langmuir 23:10074–10080
27. Kunz W, Lo Nostro P, Ninham BW (2004) The present state of affairs with Hofffieister effects. Curr Opin Colloid Interface Sci 9:1–18
28. Tatulian SA (1987) Binding of alkaline-earth metal-cations and some anions to phosphatidylcholine liposomes. Eur J Biochem 170:413–420
29. Clarke RJ, Lupfert C (1999) Influence of anions and cations on the dipole potential of phosphatidylcholine vesicles: a basis for the Hofmeister effect. Biophys J 76:2614–2624
30. Berkowitz ML, Bostick DL, Pandit S (2006) Aqueous solutions next to phospholipid membrane surfaces: insights from simulations. Chem Rev 106:1527–1539
31. Gurtovenko AA, Vattulainen I (2008) Effect of NaCl and KCl on phosphatidylcholine and phosphatidylethanolamine lipid membranes: insight from atomic-scale simulations for understanding salt-induced effects in the plasma membrane. J Phys Chem B 112:1953–1962
32. Porasso RD, Cascales JJL (2009) Study of the effect of Na+ and Ca^{2+} ion concentration on the structure of an asymmetric DPPC/DPPC plus DPPS lipid bilayer by molecular dynamics simulation. Colloids Surf B Biointerfaces 73:42–50
33. Vernier PT, Ziegler MJ, Dimova R (2009) Calcium binding and head group dipole angle in phosphatidylserine-phosphatidylcholine bilayers. Langmuir 25:1020–1027
34. Akutsu H, Seelig J (1981) Interaction of metal-ions with phosphatidylcholine bilayer-membranes. Biochemistry 20:7366–7373
35. Pabst G, Hodzic A, Strancar J, Danner S, Rappolt M, Laggner P (2007) Rigidification of neutral lipid bilayers in the presence of salts. Biophys J 93:2688–2696
36. Chapman D, Peel WE, Kingston B, Lilley TH (1977) Lipid phase-transitions in model biomembranes - effect of ions on phosphatidylcholine bilayers. Biochim Biophys Acta 464:260–275
37. Binder H, Zschornig O (2002) The effect of metal cations on the phase behavior and hydration characteristics of phospholipid membranes. Chem Phys Lipids 115:39–61
38. Cordomi A, Edholm O, Perez JJ (2008) Effect of ions on a dipalmitoyl phosphatidylcholine bilayer. A molecular dynamics simulation study. J Phys Chem B 112:1397–1408
39. Pandit SA, Bostick D, Berkowitz ML (2003) Molecular dynamics simulation of a dipalmitoylphosphatidylcholine bilayer with NaCl. Biophys J 84:3743–3750
40. Eisenberg M, Gresalfi T, Riccio T, McLaughlin S (1979) Adsorption of mono-valent cations to bilayer membranes containing negative phospholipids. Biochemistry 18:5213–5223
41. Savelyev A, Papoian GA (2006) Polyionic charge density plays a key role in differential recognition of mobile ions by biopolymers. J Am Chem Soc 128:14506–14518
42. Savelyev A, Papoian GA (2007) Free energy calculations of counterion partitioning between DNA and chloride solutions. Mendeleev Commun 17:97–99
43. Savelyev A, Papoian GA (2008) Polyionic charge density plays a key role in differential recognition of mobile ions by biopolymers. J Phys Chem B 112:9135–9145
44. Korolev N, Lyubartsev AP, Nordenskiold L (1998) Application of polyelectrolyte theories for analysis of DNA melting in the presence of Na+ and Mg^{2+} ions. Biophys J 75:3041–3056
45. Korolev N, Lyubartsev AP, Rupprecht A, Nordenskiold L (1999) Competitive binding of Mg^{2+}, Ca^{2+}, Na+, and K+ ions to DNA in oriented

DNA fibers: experimental and Monte Carlo simulation results. Biophys J 77:2736–2749
46. Lyubartsev AP, Laaksonen A (1999) Effective potentials for ion-DNA interactions. J Chem Phys 111:11207–11215
47. Jagoda-Cwiklik B, Vacha R, Lund M, Srebro M, Jungwirth P (2007) Ion pairing as a possible clue for discriminating between sodium and potassium in biological and other complex environments. J Phys Chem B 111:14077–14079
48. Denisov VP, Halle B (2000) Sequence-specific binding of counterions to B-DNA. Proc Natl Acad Sci USA 97:629–633
49. Stellwagen E, Dong Q, Stellwagen NC (2005) Monovalent cations affect the free solution mobility of DNA by perturbing the hydrogen-bonded structure of water. Biopolymers 78:62–68
50. Zinchenko AA, Yoshikawa K (2005) Na+ shows a markedly higher potential than K+ in DNA compaction in a crowded environment. Biophys J 88:4118–4123

Ions at Solid-Liquid Interfaces

Johannes Lyklema
Department of Physical Chemistry and Colloid Science, Wageningen University, Wageningen, The Netherlands

Introduction

In this entry the charge distribution at the interface between solids and electrolyte solutions is considered. Experience has shown that, unless special precautions are taken, the solids acquire a surface charge. An equal but opposite charge accumulates in the solution, adjacent to the solid. Thus, an electric double layer is formed. This double layer formation is a spontaneous process. Relevant questions include "What is the driving force?" "What can be measured?" and "What is the structure of the double layer?" We shall emphasize strong electrolytes and aqueous systems because there the basic features are most pronounced and because such systems are relevant for practice.

As a whole, electrolyte solutions are electroneutral. They contain equal amounts of cationic and anionic charges. Thermodynamically they consist of water and one, or more, dissolved electroneutral electrolytes, which can be acids, bases, and/or salts. The distribution of ions in solution is not random; the systems are far from ideal. The higher the electrolyte concentrations the stronger is the non-ideality. The very fact that many electrolytes do dissolve in water, against the electric attraction between ions of opposite sign, shows that non-electrostatic forces are also operative. The most important of these are entropy maximalization, chemical bond formation, and ion hydration. The first one strives for randomization of the ions; the last one contributes through water binding, which has enthalpic and entropic contributions. In fact, the balance between all these forces determines whether an electrolyte is soluble in water at all.

Origin of Double Layer Formation

Although one automatically couples double layers to electrostatics, it is obvious that their formation cannot be of a purely coulombic nature. Accumulation of many charges of the same sign on a solid surface is electrostatically unfavorable. In thermodynamic terms, the electrical part of the *Gibbs energy of formation*, $\Delta G(\mathrm{el}) > 0$. Hence, a non-coulombic contribution, $\Delta G\,(\mathrm{non\text{-}coul}) < 0$, must also be present and exceed the electrical contribution. This basic statement requires at least two specifications.

- The first is that hydration, or solvent structure-based interactions, is also to some extent of an electrostatic nature. For example, they involve interactions of ions with water dipoles and quadrupoles. Complex formation and chemical bond formation also involve electrical contributions. The difference with interionic coulomb forces between free ions is that this second category is typically short range. However, for the sake of argument, we shall heed the usual, but sloppy, distinction between *electrical* and *chemical* forces where the first category stands for the long-range coulombic interactions between free ions and the latter is a collective noun for all the remaining short-range interactions. In formula,

$$\Delta G(\mathrm{dl}) = \Delta G(\mathrm{el}) + \Delta G(\mathrm{chem}) \quad (1)$$

- The second specification is that we shall only discuss systems of solid particles in electrolyte solutions, ignoring boundaries between macroscopic surfaces like electrodes, which can be charged by applying an external force. We shall also exclude the special case of clay surfaces.

Identification and Measurement of Surface Charges

Establishing which ions are responsible for the surface charge is not always a foregone conclusion. From the previous section, it follows that only ions that have a very strong "chemical" affinity for the surface can become *charge-determining*. Experience has shown that such strong affinities mostly require the ions to fit very snugly onto the surface of the solid. Illustrations are H^+ and OH^- ions for oxides and Ag^+ and I^- for silver iodide.

Almost all ions have at least a small non-coulomb affinity for any surface so that the distinction between ions that can become surface ions and other ions is not sharp. However, for practical reasons, and considering that ΔG (chem) refers to short-range interactions between ions and surfaces, we can at least make a distinction between ions that are close to, or even in direct contact with, the surface and ions that are further away. For the latter category, the interactions with the surface are purely coulombic and long range. These ions give rise to *the diffuse part* of the double layer; see below, the section on double layer models. For surface ions and counterions close to the surface, the chemical contribution cannot be ignored. Regarding the non-coulomb contribution, a further distinction is between very strongly adsorbed ions that become surface ions and ions that remain outside the surface but do feel a small chemical affinity besides the large electrostatic interaction. Such ions are identified as *specifically adsorbing*. Quantitatively, the term specific adsorption is usually reserved for ions with binding Gibbs energies of a few kT, whereas the binding energies of surface ions are of the order of 20 kT or more. As a rule of thumb, a chemical binding Gibbs binding energy of about 1 kT corresponds with an electrical energy of a monovalent ion at a potential of 25 mV. A surface ion, adsorbing with a Gibbs energy of 20 kT per ion, can therefore create a surface charge that is so high that the surface potential ψ^o is 0.5 V. Specifically adsorbing ions can only create a weak surface charge because the binding force is not sufficient to overcome large coulomb repulsion. When high surface charges are needed but charge-determining ions are unavailable, ionic surfactants can be used to attain high particle charges, but in our terminology these ions would not be classified as "surface ions," but rather as strongly chemically adsorbing ions. Typical in-between cases are phosphate ions which can chemisorb onto oxides without fitting very well onto the solid surface but attaching with Gibbs energies of several kT per ion.

Let us for the sake of argument assume that the charge-determining process has been identified and proceed to discuss the measurement of the surface charge density σ^o. It is impossible to measure the adsorption of one single ion. Only the adsorption of electroneutral entities is experimentally accessible. Then it is possible to define an operational definition for σ^o. Consider by way of illustration insoluble oxide particles, for which H^+ and OH^- ions are charge determining. Then the surface charge is given by

$$\sigma^o = F(\Gamma_{\mathrm{HNO3}} - \Gamma_{\mathrm{KOH}}) \quad (2)$$

Here the Γ's stand for the indicated surface concentrations. Equation 2 is operational, meaning that it contains measurable quantities. The difference between the adsorption of acid and base can be measured by titration. (Such titrations only give relative values of σ^o; conversion into absolute values requires a reference point, viz., *the point of zero charge*, see next section). Behind Eq. 2 is the tacit non-operational assumption that adsorption of an electroneutral

HNO_3 molecule implies adsorption of a proton on the surface with the accompanying NO_3^- ion remaining in the solution side of the double layer. Titrations yield σ^o (pH) curves. As pH is a measure of the activities of charge-determining ions, such curves are a kind of semi logarithmic adsorption isotherms.

This last type of argument (thermodynamic measurements can perfectly well be used to establish surface charges provided an interpretational step is added) also applies to the adsorption of *indifferent* electrolytes, i.e., electrolytes not containing specifically adsorbing ions. Such electrolytes are expelled, *or negatively adsorbed*. The phenomenon is called the *Donnan effect*. The origin involves the repulsion from the surface of co-ions, i.e., ions with the same charge sign as the surface. The expulsion refers to only one ionic species, but it is nevertheless experimentally observable as expulsion of electroneutral salt. The connection between the two can be illustrated by considering the simple system of an oxide dispersion in a solution containing only HNO_3 and KNO_3. The acid contains the surface ion H^+, whereas the salt is indifferent. Macroscopically simultaneous adsorption of these two electrolytes takes place. Hence, Γ_{HNO3} and Γ_{KNO3} can be measured, where the former is positive and the latter negative. Interpretationally,

$$\sigma^o = \sigma(H^+) = F\Gamma_{HNO_3} \quad (3)$$

$$\sigma(K^+) = F\Gamma_{KNO_3} \quad (4)$$

$$\sigma^o(NO_3^-) = -F\Gamma_{HNO3} - F\Gamma_{KNO_3} \quad (5)$$

Here, the surface charge in Eq. 3 is positive, the co-ion (K^+) adsorption according to Eq. 4 is negative (Donnan exclusion), and the adsorption of the counterion (NO_3^-) in Eq. 5 is positive, but its contribution to the countercharge is negative because of the negative charge of the nitrate ions. The countercharge attributed by the nitrate ions is negative because in Eq. 5 the acid term is larger than the salt term. The algebraic sum of the three ionic components of charge is zero because the double layer as a whole is electroneutral. In practice, the negative adsorption can either be measured directly or obtained from titrations as a function of the electrolyte concentration; see the general reference.

In summary, having identified the surface ions and indifferent ions, it is possible to obtain information about the surface charge and the ionic components of charge. This information can be used to verify double layer models [1].

Points of Zero Charge

Titrations only give relative surface charges. To make them absolute a reference is needed and the most logical one is the *point of zero charge* (p.z.c.). For oxides it is the pH (called pHo) at which Γ_{H+} and Γ_{OH-} are exactly equal. Experimentally pHo cannot be directly measured, but there is an indirect way, based on the fact that surface charges are screened by the countercharge. The higher the electrolyte concentration, the better the screening. In practice that is observed on both sides of the p.z.c. When pH > pHo, the surface is negative and addition of electrolyte at fixed pH makes it more negative. When pH < pHo, the surface is positive and becomes more positive upon addition of electrolyte. Only at pH = pHo is the surface charge independent of the electrolyte concentration because there is nothing to screen. Routinely the p.z.c. is obtained as the common intersection point of a set of titration curves at different salt concentrations.

For oxides the p.z.c. is a measure of the surface acidity/basicity. Silica is an acid oxide; it has a low pHo, implying that its surface is negative over almost the entire pH range. Hematite is a basic oxide; its pHo is high so that the surface tends to be positively charged. Tables of critically evaluated pHos are available in the literature; see the general reference [2].

Double Layer Models

The most general model of double layers is that of Gouy and Stern. In Fig. 1a–c sketches are given

I

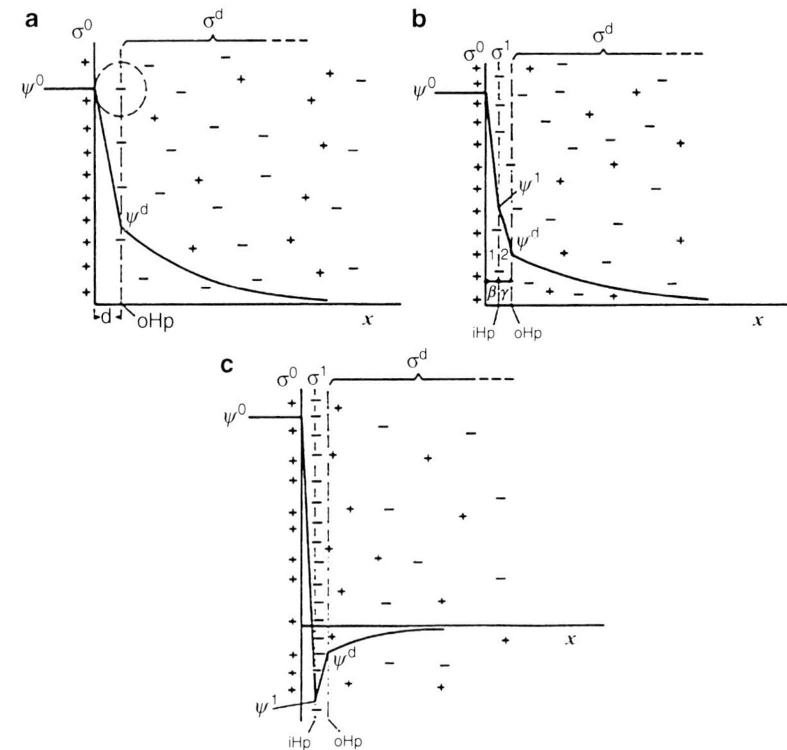

Ions at Solid-Liquid Interfaces, Fig. 1 Examples of Gouy-Stern double layers, considering (**a**) only finite counterion size in the Stern layer, (**b**) ion size and specific adsorption in the Stern layer, and (**c**) ion size and superequivalent adsorption in the Stern layer. All double layers have the same surface potential (ψ^o) but the surface charge (σ^o) increases from (**a–c**) because of better screening. The terms iHp and oHp stand for inner and outer Helmholtz layer, respectively

for three typical cases of the potential distribution [2]. What all three situations have in common is that the counter layer is divided into two parts: a molecular condenser, a few water molecule layers wide, which is the seat of all "chemical" interactions, including ion size effects, and a diffuse outer part in which the ions may be treated as point charges with only coulomb interactions. This latter part can be described by the well-known Poisson-Boltzmann equation. Across the inner part, or Stern layer, the potential decays linearly, but in the diffuse part, the decay is gradual with, for low potentials, a thickness defined by the so-called *Debye length.* The higher the electrolyte concentration, the thinner that part. Situation (a) is the most simple; the inner layer is ion-free and its thickness d is determined by the radius of the hydrated counterion. In case (b) specific adsorption, including loss of hydration water, is assumed to take place in a plane closer to the surface (distance β). Case (c) accounts for overcharging; there is more countercharge in the iHp than needed to compensate the surface charge. Typically this situation is met with very strongly specifically adsorbing ions.

It is noted that the potential ψ^d, where the diffuse part of the double layer starts, cannot be directly obtained from experiment. Although in principle it is possible to evaluate ψ^d from titration data using double layer models, that approach is not recommended because of the many assumptions to be made. However, in practice, ψ^d is a very important parameter because it quantifies the part of the double layer that is relevant for particle interaction. Often therefore this parameter is equated to the *electrokinetic*, or *ζ-potential*, that is the potential obtainable from electrokinetic phenomena like electrophoresis and streaming potentials. The tacit assumption is then that the position of the slip plane is close to that of the oHp [3–5, 7].

Applications

An abundance of applications can be mentioned, including applied electrochemistry, electrode processes, rheology, and phase behavior upon sedimentation, soil structure, and polyelectrolytes [6]. Perhaps the most direct applications are those involving particle interactions, including those encountered in AFM-like techniques [4]. The part that involves only the diffuse layers is well understood and known as DLVO theory. Because of the extension of the diffuse parts of double layers, this is a very important contribution, particularly when the particles are not too close. However, DLVO theory is based on Gouy-Chapman theory, which applies only to weak overlap, low electrolyte concentrations, and low surface charges, so in the general situations, the nondiffuse parts and their adjustments upon interaction (*regulation*) have to be also considered [7]. Current interest also concerns lyotropic or Hofmeister series [8], i.e., sequences of ions arranged by their effectiveness in binding to surfaces or polyelectrolytes, including proteins. The common denominator with the origin of double layers around particles is that in both cases the system as a whole remains electroneutral. Hofmeister series are therefore always series of ion pairs [9, 10].

Future Directions

Two topical issues may be mentioned. The first is the definition of the potentials that are measured by different techniques, say by AFM, electrokinetically and externally imposed, and their relationships [11]. The second is of a more theoretical nature and concerns modeling of the nondiffuse part of the double layer. The classical approach is through Stern theory [2], which in most cases is adequate, although it requires two additional parameters. A more recent development is in terms of *ion correlations*, essentially an advanced statistical theory whereby all coulombic ion-ion and ion-surface interaction pairs are counted and statistically summed [2]. This is a step forward over the smeared-out models of Gouy and Stern. The issue here is that cases must be found where deviations from Gouy theory cannot be interpreted on the basis of the Stern model [12]. So far, only one convincing case has been reported [13].

Cross-References

- ▶ Activity Coefficients
- ▶ Dielectric Properties
- ▶ DLVO Theory
- ▶ Electrical Double-Layer Capacitors (EDLC)
- ▶ Ion Extraction
- ▶ Ionic Liquids
- ▶ Ions at Biological Interfaces
- ▶ Ions in Clays
- ▶ Polyelectrolytes, Films-Specific Ion Effects in Thin Films
- ▶ Specific Ion Effects, Evidences

References

1. Lyklema J (1990) Interfacial electrochemistry. In: Fundamentals of interface and colloid science, vol I. Academic/Elsevier, Chap 5
2. Lyklema J (1995) Electric double layers, origin, thermodynamics, models, applications. In: Fundamentals of interface and colloid science, vol II. Academic/Elsevier, Chap 3
3. Lyklema J (1995) Electrokinetics. In: Fundamentals of interface and colloid science, vol II. Academic/Elsevier, Chap 4
4. Lyklema J (2005) Pair interactions. In: Fundamentals of interface and colloid science, vol IV. Academic/Elsevier, Chap 3
5. Lyklema J (2005) Dynamic aspects. In: Minor M, Van Leeuwen H (eds) Dynamics and Kinetics Fundamentals of interface and colloid science, vol IV. Academic Press/Elsevier, Chap 4
6. Lyklema J (2005) Polyelectrolytes. In: Fundamentals of interface and colloid science, vol V, Academic/Elsevier, Chap 2
7. Lyklema J, Duval JFL (2005) Hetero-interaction between Gouy-Stern double layers: charge and potential regulation. Adv Colloid Interface Sci 114–115:27–45
8. cross ref Kunz W. Specific ion effects – experimental evidences. This book
9. Lyklema J (2009) Simple Hofmeister series. Chem Phys Lett 467:217–222
10. Lyklema J (2003) Lyotropic sequences revisited. Adv Colloid Interface Sci 100–102:1–12
11. Lyklema J (2011) Surface charges and electrokinetic charges: distinctions and juxtapositionings. Colloid Surf Sci A Physicochem Eng Aspects 376:2–8

12. Lyklema J (2009) Quest for ion-ion correlations in electric double layers and overcharging phenomena. Adv Colloid Interface Sci 147–148:205–213
13. Wennersson E, Kjellander R, Lyklema J (2010) Charge inversion and Ion-ion correlations effects at the mercury-aqueous electrolyte interface; toward the solution of a long-standing issue. J Phys Chem 114:1849–1866

Ions in Clays

Pierre Turq[1], Benjamin Rotenberg[1], Virginie Marry[1] and Jean François Dufreche[2]
[1]Laboratoire Physicochimie des Electrolytes, Colloïdes et Sciences Analytiques, CNRS, ESPCI, Université Pierre et Marie Curie, Paris, France
[2]Institut de Chimie Séparative de Marcoule and Université Montpellier, Marcoule, France

Introduction and Description

Clay minerals are layered aluminosilicates (mixed silicon and aluminum oxides), which are ubiquitous in soils and the underground. Due to their behavior with respect to ions and water, they play an important role in many environmental and industrial processes, which exploit their mechanical (swelling), catalytic, or retention (e.g., cation exchange) properties. As an example, their ability to retain ions explains their consideration as part of natural (argillite rocks) and engineered (bentonite buffers) barriers for the geological disposal of toxic and radioactive waste. Clays also play a crucial role in the context of natural gas reservoirs or possible future carbon dioxide repositories, as they are a major component of cap rocks above these reservoirs. The properties of clay minerals are intimately related to their chemical composition and structure, which results in most cases in a permanent negative charge compensated by counterions. We discuss here the origin of this charge, the implications of the presence of counterions, as well as complementary strategies to model these complex materials on various scales.

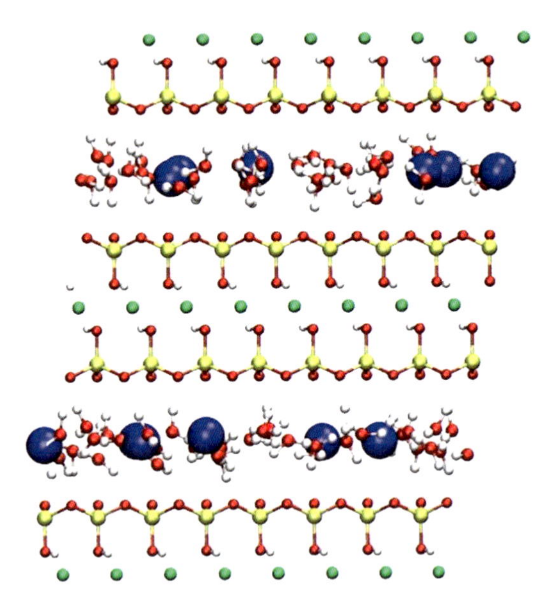

Ions in Clays, Fig. 1 Structure of montmorillonite clays. An octahedral (O) aluminum oxide layer is sandwiched between two tetrahedral (T) silicon oxide layers. Substitutions of Al(III) by Mg(II) in the T layer result in a permanent negative charge compensated by counterions (here sodium Na^+), located in the interlayer pores and can be hydrated by water molecules. Al and Mg atoms are in *green*, Si in *yellow*, Na in *blue*, oxygen in *red*, and hydrogen in *white*

The basic structure of clay minerals consists of parallel sheets of mixed silicon and aluminum and oxides of different chemical compositions, as illustrated in Fig. 1. Some are neutral and do not carry any permanent charge: This is, e.g., the case of kaolinite (in which each sheet consists of one tetrahedral silicon oxide layer, T, and one octahedral aluminum oxide layer, O) used to make china, or talc (with one O layer sandwiched between two T layers) which has many industrial applications and is found in almost every household. Most others have in their structure chemical variations of composition, leading to charge defect or excess. As an example, the *substitution* of some tetravalent silicon ions by lower valence elements such as aluminum (III) leads to a charge defect, giving to the clay sheet a global negative charge, which has to be compensated by positive *counterions*. Alternatively, aluminum (III) can be replaced by lower valence ions such as bivalent magnesium or iron (II), also resulting in a negative charge compensated by counterions. The number of substitutions and their location in the T or O layers defines the various families (smectites, illites, etc.) and types (montmorillonite, beidellite, laponite, etc.) of clay minerals.

Given the size of clay particles (10–1,000 nm), they are found in solution as colloidal dispersions or gels. At low water content, they can be obtained as dry powders, and can form solid porous materials upon compaction. In all these regimes, their properties crucially depend on the charge density and on the nature of the counterions. Most counterions are mono- or divalent, usually alkaline (most commonly sodium Na^+ or potassium K^+) or alkaline earth cations (most commonly calcium Ca^{2+}). They are not incorporated in the clay layers. Rather, they are located near the surface, either between different layers, in the so-called *interlayer* porosity, or on the *external surfaces* of clay stacks (typically ~10 layers). Such stacks are called particles, and their assembly to form a porous material then leaves voids called *interparticle* porosity, with sizes between a few nanometers to tens of nanometers, which are usually saturated by an electrolyte solution.

In the presence of water, the counterions may hydrate, depending on their nature. The solvation of interlayer counterions results in an increase of the interlayer distance, known as *swelling*. Such a water uptake is observed with Na^+-montmorillonite, but not with its K^+ exchange counterpart. At low relative humidity, it occurs stepwise, with the formation of two to three discrete water layers (crystalline swelling), followed by a continuous increase in the interlayer distance (osmotic swelling), from a dry powder to an aqueous suspension. The understanding of clay swelling, its thermodynamics and its ion-specificity, has strongly benefitted from molecular simulations [1, 2]. Cations in hydrated interlayers are also mobile and diffusion through interlayers is in fact a major transport pathway for cations through compacted clay rocks, because larger pores are then usually not connected to each other [3]. The diffusion of these ions is difficult to probe experimentally (while that of interlayer water can be investigated by Quasi-Elastic Neutron Scattering [4]), even though some information can be gained using dielectric spectroscopy [5, 6]. Thus, molecular simulation has been essential to determine the interlayer diffusion coefficient, its dependence on the nature of the counterion, the density and location (in the T or O layer) of substitution, the water content, or the temperature [7–11].

Interlayer ions can also be replaced by other cations. This *ion exchange* is one of the major contributors to the retention of cations by clays. It was believed until recently that the driving force for this exchange, which is thermodynamically favorable in the case of the replacement of smaller alkaline ions by larger ones (e.g., Na^+ cesium Cs^+), was some specific interactions between the cations and the surface. Using microcalorimetry experiments combined with molecular simulation, it was however possible to demonstrate that the interactions with the surface play a minor role (in fact, it is not in favor of the exchange) and that the thermodynamics of this exchange is dominated by the larger hydration free energy of the smaller ion released in the aqueous phase [12]. The other main mechanism for the retention of cations is the *sorption on edge sites*, which are found on the lateral surfaces of clay particles. These sites, typically silanol SiOH or aluminol AlOH groups, arise from the broken bonds due to the finite lateral extent of the particles. Sorption on clay edges then depends on *p*H, which controls the protonation state of these sites. Since titration experiments only provide a global measure of the acidity, it does not allow inferring the acidity constant pK_a of each site. Recently, such a determination has been possible using ab-initio simulations [13].

The charged external surface, together with the external solution containing the compensating counterions and other electrolyte ions, plays a key role in the properties of clay particles. The interplay between short-range specific interactions at the surface (which depends on the nature of the ions and the location of the substitution in the mineral layers), on the one hand, and long-range electrostatic interactions, on the other hand, governs the stability and rheology of colloidal clay suspensions [14, 15]. From the modeling point of view, molecular simulation is necessary to capture the former, specific interactions, and has been successfully applied to describe ionic solutions in the first few nanometers from the surface [16–21]. In order to describe larger pores or clay dispersion, with distances larger than ten nanometers, one then needs to resort to simpler models, whereby the solvent is treated as a continuum. In the Poisson-Boltzmann theory, ions are further treated as point particles and the correlation between ions is neglected: They interact only via the mean-field electrostatic potential. Such drastic assumptions have the advantage of being tractable analytically or numerically, thus allowing for the determination of a complete phase diagram with limited computational cost.

Turning now to dynamical properties, the external surface charge also gives rise to coupled *electrokinetic phenomena*. When an electric field is applied to the suspension, it acts not only on the clay particles, but also on the diffuse layer in their vicinity, which is electrically charged, thereby generating fluid flow. Thus, the electrophoretic mobility of clay particles results from the coupling between electric and hydrodynamic phenomena. Similarly, in a porous clay rock, applying an external electric field will drive the fluid (electro-osmosis), while applying a pressure gradient will induce the transport of ions, hence an electric current (streaming current). Once again, these complex phenomena depend on the nature of the counterion, the charge density of the clay surface, and their localization inside the mineral, features which require a molecular description. Nevertheless, continuous models combining the Poisson-Boltzmann theory for the ionic distribution with the Navier-Stokes equation for fluid flow (Helmholtz-Smoluchowski theory) are much simpler to use, in particular when one want to upscale the transport equations from the pore scale to the macroscopic sample scale [22].

The assumptions underlying the continuous Poisson-Boltzmann and Helmholtz-Smoluchowski theories have further been tested against reference molecular simulations. In the case of monovalent ions, the counterion density profiles are reasonably well described by the Poisson-Boltzmann theory, provided that the surface charge density is not too large. Its major limitation is the lack of oscillations in the first nanometer from the surface, which is a consequence of the layering of the solvent in this region. However, the electro-osmotic flow profiles are very sensitive to the details of the ionic distribution. In addition, for the dynamics, it is important to know the hydrodynamic boundary conditions at the solid/liquid interface. As illustrated in Fig. 2, which reports the electro-osmotic flow profile between two Na-montmorillonite surfaces separated by a 4.5-nm pore, the continuous theories are insufficient to reproduce the reference molecular simulation results, even using slip boundary conditions. However, a reasonably accurate description of the flow can be obtained by solving the Stokes equation using slip boundary conditions and the ionic density profiles from molecular simulations are introduced [23].

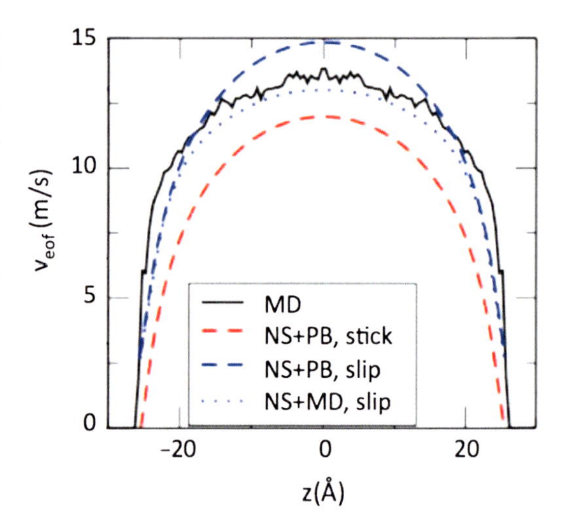

Ions in Clays, Fig. 2 Electro-osmotic flow profile between two Na-montmorillonite surfaces separated by a 4.5-nm pore containing water. Reference molecular dynamics (MD) simulations allow to test the validity of continuous models based on the Navier-Stokes (NS) and Poisson-Boltzmann (PB) equations. Such equations must be solved for given boundary conditions (stick or slip) at the solid/liquid interface

Future Directions

In the future, new spectroscopic techniques such as time-resolved X-ray scattering should provide a first direct look into the microscopic scale dynamics of cations in clays. From the modeling point of view, the understanding of multivalent ions and their description in continuous theories remains a challenge. Adopting a multiscale strategy, whereby each level of description is calibrated on a more fundamental one, from ab-initio and molecular simulations to macroscopic models via mesoscopic descriptions, will be essential to achieve this goal.

Cross-References

- ▶ Charged Colloids
- ▶ Ions at Solid-Liquid Interfaces

References

1. Boek ES, Coveney PV, Skipper NT (1995) - Monte-Carlo molecular modelling studies of hydrated Li-, Na- and K-smectites: understanding the role of potassium as a clay swelling inhibitor. J Am Chem Soc 117:12608–12617
2. Hensen EJM, Tambach TJ, Bliek A, Smit B (2001) Adsorption isotherms of water in Li-, Na- and K-montmorillonite by molecular simulation. J Chem Phys 115:3322–3329
3. Glaus MA, Baeyens B, Bradbury MH, Jakob A, Van Loon LR, Yaroshchuk A (2007) Diffusion of ^{22}Na and ^{85}Sr in montmorillonite: evidence of interlayer diffusion being the dominant pathway at high compaction. Environ Sci Technol 41:478–485
4. Malikova N, Dubois E, Marry V, Rotenberg B, Turq P (2010) Dynamics in clays – combining neutron scattering and microscopic simulation. Z Phys Chem 244:153–181
5. Rotenberg B, Cadène A, Dufrêche JF, Durand-Vidal S, Badot JC, Turq P (2005) An analytical model for probing ion dynamics in clays with broadband dielectric spectroscopy. J Phys Chem B 109:15548–15557
6. Cadène A, Rotenberg B, Durand-Vidal S, Badot JC, Turq P (2006) Dielectric Spectroscopy as a probe for dynamic properties of compacted smectites. Phys Chem Earth 31(10–14):505–510
7. Sutton R, Sposito G (2001) Molecular simulation of interlayer structure and dynamics in 12.4 ÅCs-smectite hydrates. J Coll Interface Sci 237:174–184
8. Marry V, Turq P, Cartailler P, Levesque D (2002) Microscopic simulation for structure and dynamics of water and counterions in a monohydrated montmorillonite. J Chem Phys 117:3454–3463
9. Marry V, Turq P (2003) Microscopic simulations of interlayer structure and dynamics in bihydrated heteroionic montmorillonite. J Phys Chem B 107:1832–1839
10. Malikova N, Marry V, Dufrêche JF, Turq P (2004) Na/Cs-montmorillonite: temperature activation of diffusion by simulation. Curr Opin Coll Interface Sci 9:124–127
11. Michot LJ, Ferrage E, Jimenez-Ruiz M, Boehm M, Delville A (2012) Anisotropic features of water and ion dynamics in synthetic Na- and Ca-smectites with tetrahedral layer charge. A combined quasi-elastic neutron-scattering and molecular dynamics simulations study. J Phys Chem C 116:16619–16633
12. Rotenberg B, Morel JP, Marry V, Turq P, Morel-Desrosiers N (2009) On the driving force of cation exchange in clays: insights from combined microcalorimetry experiments and molecular simulation. Geochim Cosmochim Acta 73:4034–4044
13. Tazi S, Rotenberg B, Salanne M, Sprik M, Sulpizi M (2012) Absolute acidity of clay edge sites from ab-initio simulations. Geochim Cosmochim Acta 94:1–11
14. Paineau E, Bihannic I, Baravian C, Philippe AM, Davidson P, Funari SS, Rochas C, Michot LJ (2001) Aqueous suspensions of natural swelling clay minerals. 1. Structure and electrostatic interactions. Langmuir 27:5562–5573
15. Paineau E, Michot LJ, Bihannic I, Baravian C (2001) Aqueous suspensions of natural swelling clay minerals. 2. Rheological characterization. Langmuir 27:7806–7819
16. Greathouse JA, Cygan RT (2005) Molecular dynamics simulation of Uranyl(VI) adsorption equilibria onto an external montmorillonite surface. Phys Chem Chem Phys 7:3580–3586
17. Greathouse JA, Cygan RT (2006) Water structure and aqueous Uranyl(VI) adsorption equilibria onto external surfaces of beidellite, montmorillonite and pyrophillite: results from molecular simulations. Environ Sci Technol 40:3865–3871
18. Marry V, Rotenberg B, Turq P (2008) Structure and dynamics of water at a clay surface from molecular dynamics simulation. Phys Chem Chem Phys 10:4802–4813
19. Tournassat C, Chapron Y, Leroy P, Bizi M, Boulahya F (2009) Comparison of molecular dynamics simulations with triple layer and modified Gouy-Chapman models in a 0.1 M NaCl-montmorillonite system. J Coll Interface Sci 339:533–541
20. Rotenberg B, Marry V, Malikova N, Turq P (2010) Molecular simulation of aqueous solutions at clay surfaces. J Phys Condense Matter 22:284114
21. Bourg IC, Sposito G (2011) Molecular dynamics simulations of the electrical double layer on smectite

surfaces contacting concentrated mixed electrolyte ($NaCl$-$CaCl_2$) solutions. J Coll Interface Sci 360:701–715
22. Moyne C, Murad MA (2006) A two-scale model for coupled electro-chemo-mechanical phenomena and Onsager's reciprocity relations in expansive clays: I homogenization analysis. Transp Porous Media 62:333–380
23. Botan A, Marry V, Rotenberg B, Turq P, Noetinger B (2013) How electrostatics influences hydrodynamic boundary conditions: Poiseuille and electro-osmostic flows in clay nanopores. J Phys Chem C 117:978–985

iR-Drop Elimination

Manuel Lohrengel
University of Düsseldorf, Düsseldorf, Germany

Introduction

Potentiostatic setups control the potential drop between electrode and electrolyte. This requires a probe to measure the potential of the electrolyte, the so-called reference electrode. The probe tip, the sensing point, is positioned somewhere in the electrolyte. This is illustrated in Fig. 1, showing the equivalent circuit of a common electrochemical cell.

The interfaces of working (WE) and counter electrode (CE) are symbolized by their impedances Z_{WE} and Z_{CE}, respectively, often simplified by resistor and capacity in parallel. A cell current I induces potential drops within the electrolyte, the potential drop U_Ω across R_Ω between WE and reference electrode (RE) and a second one across R_{El} between RE and CE. Electrochemist wants to control the potential drop across Z_{WE}, but in fact the drop across $Z_{WE} + R_\Omega$ is controlled by the potentiostat, with a deviation:

$$U_\Omega = I \cdot R_\Omega \tag{1}$$

R_Ω will be in the range from 0.5 to 100 Ω in common cells. This means, the error according to Eq. 1 will be negligible for currents below some 100 μA but will become significant for larger currents.

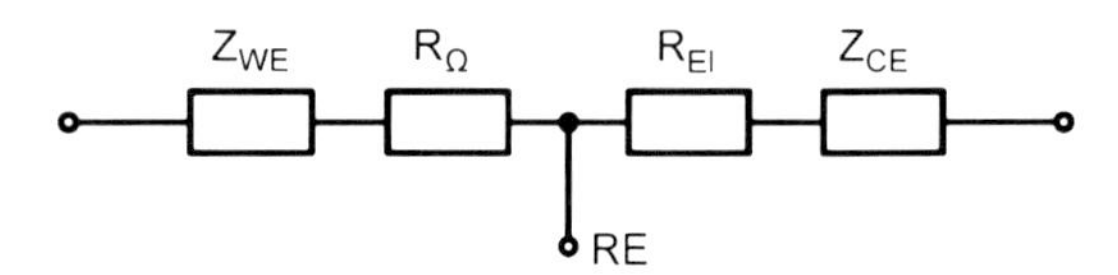

iR-Drop Elimination, Fig. 1 Equivalent circuit of an electrochemical cell

Moreover, R_Ω will dominate the settling time of the potentiostat in pulse experiments. The capacitive part C_{WE} of Z_{WE} is charged via R_Ω. This means a time constant τ:

$$\tau = R_\Omega \cdot C_{WE} \tag{2}$$

which cannot be shortened by the main amplifier of the potentiostat [1].

Concepts

The effects and compensation of the potential drop U_Ω across R_Ω, usually called iR-drop, are permanently discussed in literature. Comprehensive information is found in [1–8].

According to [9], three strategies can be followed to reduce this error:

- Optimization of cell design to minimize R_Ω
- Online techniques, typically an electronic compensation of U_Ω
- Post-factum techniques, which means manipulation of the recorded data after the experiment.

Optimization of Cell Design

A very simple concept to reduce effects of iR-drop is using electrolytes of high conductivity.

Moreover, the reference electrode tip, usually the orifice of the Haber-Luggin capillary, should be positioned close to the working electrode. Insulating objects in the electrolyte, however, mean distortions of the current distribution and, thus, local potential changes. Therefore, the capillary should not come too close to the electrode (screening effect, [10, 11]). A more distant position, however, means a larger electrolyte resistance R_Ω between capillary mouth and electrode. A distance equal to the diameter of the capillary is a good compromise [1, 12].

Concepts to minimize effects of R_Ω by capillary or cell modification were presented in [13, 14]. Thin wires (diameters around 10 μm) can be positioned very close to the working electrode with negligible shielding effect. They are either charged with stable redox systems (e.g., palladium/hydrogen [15] or gold/gold oxide [16]) or connected via capacitors in parallel to conventional reference electrode systems [17] to reduce U_Ω or τ, respectively Eqs. 1 and 2.

Tiny cells [18] or ultramicroelectrodes [19–29], e.g., in scanning electrochemical microscopes (SECM, [28]), are used in other concepts. Ultramicroelectrodes are mainly used in systems where diffusion from or to the electrode is dominant. R_Ω reduces rapidly with decreasing electrode radius r and depends finally only on electrolyte conductivity, independent of radius or shape in steady state diffusion regime [22, 30, 31].

It is necessary to position the reference electrode behind the counter electrode in few cases. This seems to be disadvantageous but is acceptable in special microcells where the electrolyte resistance is extremely small, due to the reduced distance between counter and working electrode (e.g., some μm, [32]).

Online Techniques

The most common online technique for compensation of iR-drop is positive feedback [33–47]. The corresponding circuits depend on the concept of the potentiostat and are presented for the adder potentiostat (Fig. 2) and for the hybrid type (separate adder, Fig. 3). The simple idea behind is to add the potential drop U_Ω across R_Ω to the desired voltage so that they compensate each other. The current is measured by a current-to-voltage converter (CV in Figs. 2 and 3). The output of CV is

$$U_{CV} = -I \cdot R_{CV} \tag{3}$$

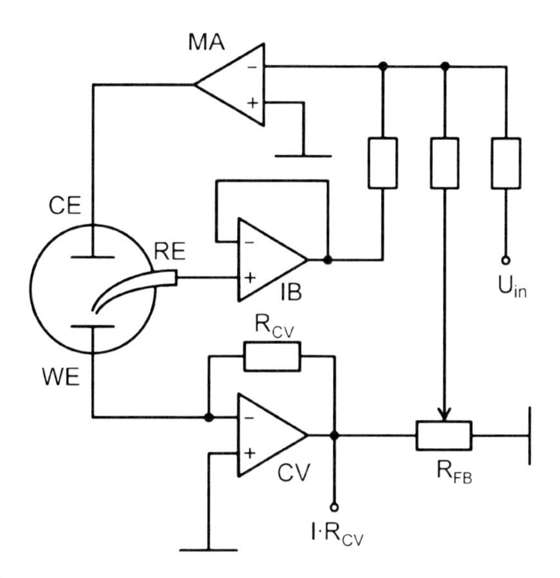

iR-Drop Elimination, Fig. 2 Potentiostat (adder concept) with iR compensation by positive feedback with main amplifier *MA*, current-to-voltage converter *CV*, and impedance buffer *IB*

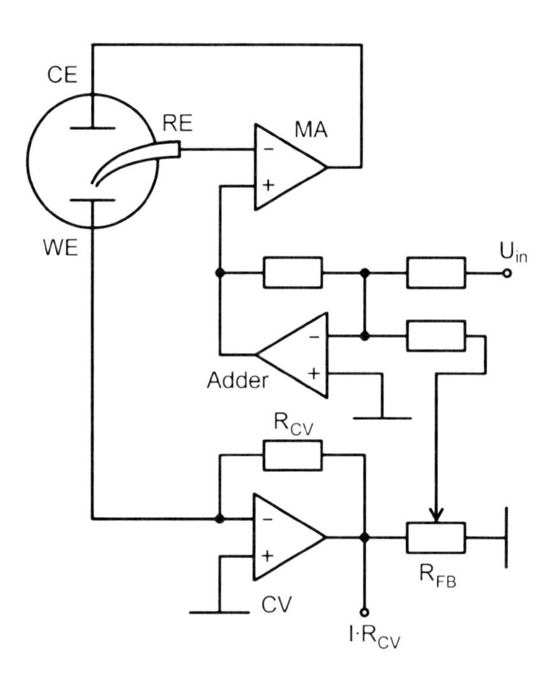

iR-Drop Elimination, Fig. 3 Potentiostat with iR compensation by positive feedback (concept with separate adder) with main amplifier *MA*, current-to-voltage converter *CV*, and adder [31] and feeds the potentiometer R_{FB}. If this fraction is just

$$\frac{R_\Omega}{R_{CV}} \cdot U_{CV} = -I \cdot R_\Omega = -U_\Omega \tag{4}$$

then it means a perfect compensation (the sign of U_Ω is again inverted in the adder, Fig. 3, or in MA, Fig. 2).

A completely different concept is based on the idea to connect the working electrode not to ground but to a potential $-U_\Omega$ [48, 49]. This means, the potential drop in the electrolyte U_Ω and this potential offset $-U_\Omega$ compensate each other completely. This requires a special operational amplifier circuit, a so-called negative resistance (Fig. 4) with a transfer function:

$$I \cdot R_{NR} = -U_\Omega \tag{5}$$

R_{NR} has to be adjusted to be equal to R_Ω.

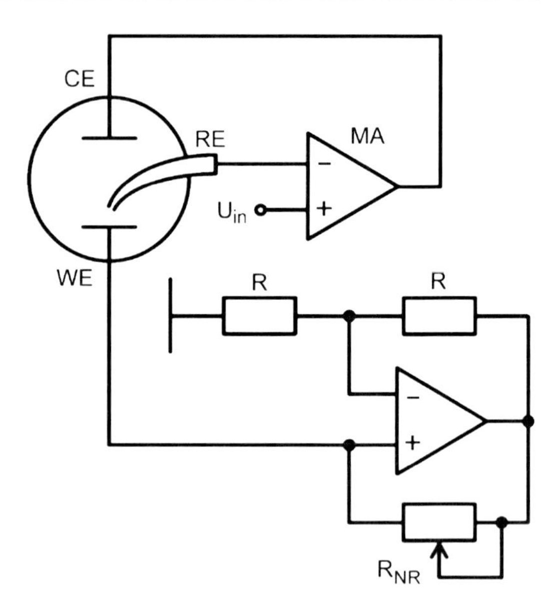

iR-Drop Elimination, Fig. 4 Potentiostat with iR compensation by a negative resistance

Perfect compensation in a wide frequency range is hardly realized, as positive feedback, together with internal phase shifts, transforms the potentiostatic circuit into an oscillator and potential control is lost. Frequency and amplitude of these oscillation is usually large (typically > 10 kHz, several volts at the counter electrode output), but in some cases damped oscillations of only some mV could be monitored at the reference electrode. This should be controlled with an oscilloscope. This means, R_Ω can be compensated only in part, typically 50–90 % and enable sweep rates up to 1,000 V/s with conventional macroelectrodes [9, 50]. A combination of positive feedback and ultramicroelectrodes allow sweep rates up to 1 MV/s [21, 51, 52].

So far, R_Ω is assumed to be as simple as in Fig. 1. This is only true if the current distribution in the cell is absolutely uniform. The shapes of the vessel and the electrodes will prevent this in most real cells. Therefore, R_Ω will become position dependent and cannot be represented by a single value [9].

Several methods were proposed to determine R_Ω. A common way is to compensate by turning potentiometer R_{FB} until the potentiostat oscillates and then to turn back until the system becomes stable again [53]. More precise values can be taken from techniques such as potentiostatic pulses [47], galvanostatic pulses (resolution some μs [1, 54]), or impedance spectroscopy at frequencies > 10 kHz [55]. Prerequisite, however, are layer-free electrodes in the double layer region, e.g., noble metals. This is illustrated in Fig. 5. If the experiment corresponds to a frequency f, the parallel capacities of interface and layer, C_{WE} and C_L, must short-circuit R_{WE} and R_L, which means

$$\frac{C_{WE} + C_L}{2\pi f \cdot C_{WE} \cdot C_L} \ll R_\Omega \tag{6}$$

This is not guaranteed in many systems, e.g., passive electrodes.

In most cases, R_Ω is assumed to be constant, but variations of geometry, such as deposition or dissolution, or concentration in the electrolyte will change R_Ω. These cases require continuous measurements of R_Ω. Some concepts are simplified impedance measurements [56–58]. A small sinusoidal signal of higher frequency (typically >10 kHz) is superimposed and the system response is interpreted.

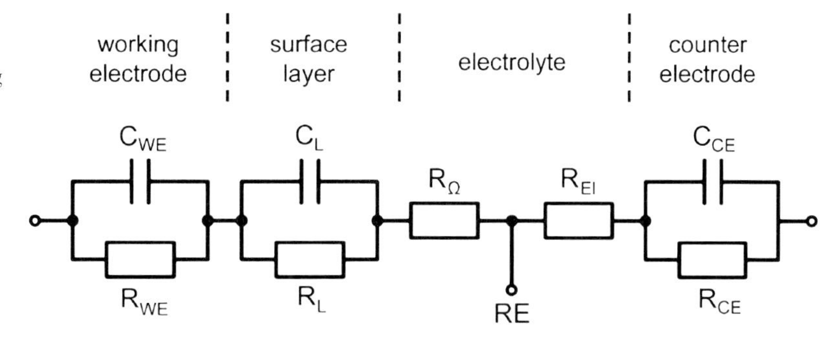

iR-Drop Elimination, Fig. 5 Equivalent circuit of a layer-covered working electrode

Other concepts interrupt potential control from time to time, which means no current for a short period [59–62]. The capacitances C_{WE} (and C_L, if present) store the actual potential without drop across R_Ω. This potential must be recorded directly after break, as these capacitances are discharged via R_Ω, R_{WE}, or R_L. Moreover, this relaxation can be used to determine R_Ω, as discharge is dominated by R_Ω, as R_Ω is usually much smaller than R_{WE} or R_L.

Post-Factum Techniques

Post-factum techniques mean a subsequent manipulation of the experimental data by mathematical methods [9]. Experimental data, for example from cyclic voltammetry, consist usually of potential-current couples (U_{exp} and I_{exp}). Therefore, calculations of "true potentials" U according to

$$U = U_{exp} - U_\Omega = U_{exp} - I_{exp} \cdot R_\Omega \qquad (7)$$

become possible. This means, however, a current-dependent shift of the potential data. As a result, the potential data are no more equidistant in time. This means, the constant sweep, which is prerequisite of cyclic voltammetry, is formally transformed into a current- and time-dependent sweep which would yield different results in a real experiment.

Other, especially mathematic, concepts use iterations techniques [63], deconvolution of amplifier delays [21, 52], or assume special properties of the investigated system such as structure of the interface, kinetics of the contributing reactions, or linear response [9].

Future Directions

An elimination of iR-drop is still challenging. This starts with electrode (microelectrodes) and cell design (position of electrodes, electrolyte conductivity). Electronic positive feedback and post-factum deconvolution of amplifier limits are necessary and allow in special systems sweep rates up to 10^6 V/s. A universal setup for random electrodes (micro and macro, technical electrodes) and random systems (layer formation and removal, gas reactions, porous systems, electrolytes of low conductivity, extreme currents, etc.) with complete elimination of iR-drop is still missing.

Cross-References

- ▶ Electrochemical Impedance Spectroscopy (EIS) Applications to Sensors and Diagnostics
- ▶ Potentiostat
- ▶ Reference Electrodes

References

1. Vetter KJ (1967) Electrochemical kinetics theoretical and experimental aspects. Academic, New York
2. Smith DE (1971) Recent developments in alternating current polarography. Crit Rev Anal Chem 2:247–343
3. Harrar JE, Pomernacki CL (1973) Linear and nonlinear system characteristics of controlled- potential electrolysis cells. Anal Chem 45:57–79
4. Macdonald DD (1977) Transient techniques in electrochemistry. Plenum, New York
5. Garreau D, Saveant JM (1978) Resistance compensation and faradaic instability in diffusion controlled processes. J Electroanal Chem 86:63–73

6. Bard AJ, Faulkner LR (1980) Electrochemical methods: fundamentals and applications. Wiley, New York
7. Roe DK (1984) Overcoming solution resistance with stability and grace in potentiostatic circuits. In: Kissinger PT (ed) Laboratory techniques in electroanalytical chemistry. Marcel Dekker, New York
8. Andrieux CP, Hapiot P, Saveant JM (1990) Fast kinetics by means of direct and indirect electrochemical techniques. Chem Rev 90:723–738. doi:10.1021/cr00103a003
9. Britz D (1978) iR compensation in electrochemical cells. J Electroanal Chem 88:309–352
10. Bewick A (1968) Analysis of the use of "IR" compensators in potentiostatic investigations. Electrochim Acta 13:825–830
11. Barnartt S (1961) Magnitude of IR-drop corrections in electrode polarization measurements made with a Luggin-Haber capillary. J Electrochem Soc 108:102–104
12. Piontelli R, Bianchi C, Bertucci U, Guerci C, Rivolta B (1954) Methods of measurement of polarization voltage II. Z Elektrochem 58:54–64
13. Cahan BD, Nagy Z, Genshaw MA (1972) Cell design for potentiostatic measuring system. J Electrochem Soc 119:64–69. doi:10.1149/1.2404134
14. Piontelli R (1955) Basis and examples of applications of new methods for measurement of overvoltages. Z Elektrochem 59:778–784
15. Ives DJG, Janz GJ (1961) Reference electrodes. Academic, New York
16. Hassel AW, Fushimi K, Seo M (1999) An agar-based silver|silver chloride reference electrode for use in micro-electrochemistry. Electrochem Comm 1:180–183. doi:10.1016/S1388-2481(99)00035-1
17. Kluger K, Lohrengel MM (1991) Mobility of ionic space charges in thin insulating films. Ber Bunsenges Physik Chem 95:1458–1461
18. Beck F, Guthke H (1969) Entwicklung neuer Zellen für elektro-organische Synthesen. Chemie Ing Techn 41:943–950. doi:10.1002/cite.330411702
19. Montenegro MI, Queiros MA, Daschbach JL (1991) Microelectrodes: theory and applications. Kluwer, Dordrecht. ISBN 0-7923-1229-5
20. Heinze J (1981) Diffusion processes at finite (micro) disk electrodes solved by digital simulation. Angew Chem 124:73–86
21. Wipf DO, Wightman RM (1988) Submicrosecond measurements with cyclic voltammetry. Anal Chem 60:2460–2464. doi:10.1021/ac00173a005
22. Bond AM, Luscombe D, Oldham KB, Zoski CG (1988) A comparison of the chronoamperometric response at inlaid and recessed disc microelectrodes. J Electroanal Chem 249:1–14
23. Wipf DO, Michael AC, Wightman RM (1989) Microdisk electrodes: Part II. Fast-scan cyclic voltammetry with very small electrodes. J Electroanal Chem 269:15–25
24. Zoski CG (1990) A survey of steady-state microelectrodes and experimental approaches to a voltammetric steady state. J Electroanal Chem 296:317–333
25. Nomura S, Nozaki K, Okazaki S (1991) Fabrication and evaluation of a shielded ultramicroelectrode for submicrosecond electroanalytical chemistry. Anal Chem 63:2665–2668. doi:10.1021/ac00022a022
26. Forster RJ (1994) Microelectrodes: new dimensions in electrochemistry. Chem Soc Rev 23:289–297. doi:10.1039/CS9942300289
27. Tschuncky P, Heinze J (1995) A method for the construction of ultramicroelectrodes. Anal Chem 67:4020–4023. doi:10.1021/ac00117a032
28. Cornut R, Lefrou C (2008) New analytical approximation of feedback approach curves with a microdisk SECM tip and irreversible kinetic reaction at the substrate. J Electroanal Chem 621:178–184. doi:10.1016/j.jelechem.2007.09.021
29. Robinson DL, Hermans A, Seipel AT, Wightman RM (2008) Monitoring rapid chemical communication in the brain. Chem Rev 108:2554–2584. doi:10.1021/cr068081q
30. Oldham KB (1987) All steady-state microelectrodes have the same "iR drop". J Electroanal Chem 237:303–307
31. Bruckenstein S (1987) Ohmic potential drop at electrodes exhibiting steady-state diffusional currents. Anal Chem 59:2098–2101. doi:10.1021/ac00144a020
32. Lohrengel MM, Moehring A (2002) Electrochemical microcells and surface analysis. In: Schultze JW, Osaka T, Datta M (eds) Electrochemical microsystem technologies, vol 2, New trends in electrochemical technologies series. Taylor & Francis, London
33. Booman GL, Holbrook WB (1965) Optimum stabilization networks for potentiostats with application to a polarograph using transistor operational amplifiers. Anal Chem 37:795–802. doi:10.1021/ac60226a006
34. Hayes JW, Reilley CN (1965) Operational-amplifier, alternating-current polarograph with admittance recording. Anal Chem 37:1322–1325. doi:10.1021/ac60230a009
35. Gerischer H, Staubach KE (1957) Der elektronische Potentiostat und seine Anwendung zur Untersuchung schneller Elektrodenreaktionen. Z Elektrochem 61:789–794. doi:10.1002/bbpc.19570610705
36. Lauer G, Osteryoung RA (1966) Effect of uncompensated resistance on electrode kinetic and adsorption studies by chronocoulometry. Anal Chem 38:1106–1112. doi:10.1021/ac60241a002
37. Pilla AA, Roe RB, Herrmann CC (1969) High speed non-faradaic resistance compensation in potentiostatic techniques. J Electrochem Soc 116:1105–1112. doi:10.1149/1.2412225
38. Pilla AA (1971) Influence of the faradaic process on nonfaradaic resistance compensation in potentiostatic techniques. J Electrochem Soc 118:702–707. doi:10.1149/1.2408148
39. Wells E (1971) Question of instrumental artifact in linear sweep voltammetry with positive feedback

ohmic drop compensation. Anal Chem 43:87–92. doi:10.1021/ac60296a010

40. Amatore C, Lefrou C, Pflüger F (1989) On-line compensation of ohmic drop in submicrosecond time resolved voltammetry at ultramicroelectrodes. J Electroanal Chem 270:43–59. doi:10.1016/0022-0728(89)85027-2
41. Brown ER, McCord TG, Smith DE, DeFord DD (1966) Some investigations on instrumental compensation of nonfaradaic effects in voltammetric techniques. Anal Chem 38:1119–1129. doi:10.1021/ac60241a004
42. Brown ER, Smith DE, Booman GL (1968) Operational amplifier potentiostats employing positive feedback for IR compensation I Theoretical analysis of stability and bandpass characteristics. Anal Chem 40:1411–1423. doi:10.1021/ac60266a024
43. Brown ER, Hung HL, McCord TG, Smith DE, Booman GL (1968) Operational amplifier potentiostats employing positive feedback for IR compensation II Application to ac polarography. Anal Chem 40:1424–1432. doi:10.1021/ac60266a025
44. Sarma NS, Sankar L, Krishnan A, Rajagopalan SR (1973) IR compensation in potentiostat. J Electroanal Chem 41:503–504. doi:10.1016/S0022-0728(73)80427-9
45. Britz D (1980) 100 % ir compensation by damped positive feedback. Electrochim Acta 25:1449–1452. doi:10.1016/0013-4686(80)87160-X
46. Meyer JJ, Poupard D, Dubois JE (1982) Potentiostat with a positive feedback IR compensation and a high sensitivity current follower indicator circuit for direct determination of high second-order rate constants. Anal Chem 54:207–212. doi:10.1021/ac00239a014
47. He P, Faulkner LR (1986) Intelligent, automatic compensation of solution resistance. Anal Chem 58:517–523. doi:10.1021/ac00294a004
48. Gabrielli C, Keddam M (1974) Progres recents dans la mesure des impedances electrochimiques en regime sinusoidal. Electrochim Acta 19:355–362. doi:10.1016/0013-4686(74)87009-X
49. Lamy C, Herrmann CC (1975) A new method for ohmic-drop compensation in potentiostatic circuits: stability and bandpass analysis, including the effect of faradaic impedance. J Electroanal Chem 59:113–135
50. Schultze JW, Lohrengel MM (1978) Ageing effects in monomolecular oxide layers on gold. Ber Bunsenges Physik Chem 80:552–556
51. Wightman RM, Wipf DO (1990) High-speed cyclic voltammetry. Acc Chem Res 23:64–70. doi:10.1021/ar00171a002
52. Amatore C, Lefrou C (1992) New concept for a potentiostat for on-line ohmic drop compensation in cyclic voltammetry above 300 kV s − 1. J Electroanal Chem 324:33–58. doi:10.1016/0022-0728(92)80034-2
53. Whitson PE, VandenBorn HW, Evans DH (1973) Acquisition and analysis of cyclic voltammetric data. Anal Chem 45:1298–1306. doi:10.1021/ac60330a016
54. Yarnitzky C, Klein N (1975) Dynamic compensation of the overall and uncompensated cell resistance in a two- or three-electrode system. Transient techniques. Anal Chem 47:880–884. doi:10.1021/ac60356a030
55. Macdonald JR (1987) Impedance spectroscopy. Wiley, New York
56. Devay J, Lengyel B, Meszaros L (1973) Method and apparatus for the automatic compensation of the ohmic potential drop. Zash Met 9:276–281
57. Yarnitzky C, Friedman Y (1975) Dynamic compensation of the overall and uncompensated cell resistance in a two- or three-electrode system. Steady state techniques. Anal Chem 47:876–880. doi:10.1021/ac60356a050
58. Bezman R (1972) Sampled-data approach to the reduction of uncompensated resistance effects in potentiostatic experiments. Anal Chem 44:1781–1785. doi:10.1021/ac60319a002
59. McIntyre JDE, Peck WF (1970) An interrupter technique for measuring the uncompensated resistance of electrode reactions under potentiostatic control. J Electrochem Soc 117:747–751. doi:10.1149/1.2407622
60. Williams LFG, Taylor RJ (1980) iR correction Part I A computerised interrupt method. J Electroanal Chem 108:293–303. doi:10.1016/S0022-0728(80)80338-X
61. Britz D, Brocke WA (1975) Elimination of iR-drop in electrochemical cells by the use of a current-interruption potentiostat. J Electroanal Chem 58:301–311. doi:10.1016/S0022-0728(75)80088-X
62. Amatore C, Maisonhaute E, Simonneau G (2000) Ohmic drop compensation in cyclic voltammetry at scan rates in the megavolt per second range: access to nanometric diffusion layers via transient electrochemistry. J Electroanal Chem 486:141–155. doi:10.1016/S0022-0728(00)00131-5
63. Wipf DO (1996) Ohmic drop compensation in voltammetry: iterative correction of the applied potential. Anal Chem 68:1871–1876. doi:10.1021/ac951209b

K

Kolbe and Related Reactions

István Markó and François Chellé
Universite Catholique de Louvain,
Louvain-la-Neuve, Belgium

Introduction

In 1834, while studying the conductivity of acetates, M. Faraday observed that an inflammable gas was produced at the anode [1]. However, being more interested in physics than in chemistry, he reported this phenomenon but did not identify the gas. Fifteen years later, in 1849, W.H. Kolbe reinvestigated this transformation [2]. He characterized ethane as the product formed at the anode and recognized the nature and utility of this electrochemical process. The "Kolbe" reaction, the electrochemical oxidative decarboxylation-dimerization of carboxylic acids, is a powerful method for the generation of C-C bonds under particularly mild conditions (Scheme 1).

The importance of this process was rapidly appreciated. For example, in 1855, A. Wurtz obtained mixed dimers by electrolyzing two different acids [3], while A.C. Brown and J. Walker prepared α,ω-diesters from readily available fatty acid derivatives [4].

Studies on the scope and the mechanism of the Kolbe reaction reached their pinnacle in the 1960s and 1970s, and seminal contributions were made by a number of important electrochemists, such as Weedon [5], Conway [6], Eberson [7], and Brennan and Brettle [8], to cite only but a few. Following the advent of organometallic chemistry and the excitement of performing enantioselective reactions by metal-catalyzed processes, interest in electro-organic chemistry dwindled down. Only a few dedicated laboratories pursued their work in this domain.

Today, environmental concerns have impelled regained interest in this area of chemistry, and the Kolbe and related reactions are gradually being rejuvenated. It is the purpose of this brief entry to summarize some of the most important achievements in this electrochemical process. Of course, in view of its brevity, only key features of this transformation can be presented. For the interested reader, a few pertinent reviews are mentioned in the literature section [9–12].

The Kolbe Dimerization

During the electrolysis of a carboxylic acid **3**, an electron is removed (at the anode) from the carboxylate anion **1**, leading to the corresponding carboxyl radical **5**. This unstable species rapidly loses CO_2, generating the C-centered radical **6**. Coupling between two radicals affords the dimer **2** (Scheme 2) [13].

Under typical Kolbe conditions, excellent yields of products can usually be obtained provided a few simple rules are followed. Indeed, the main experimental parameters affecting the yield are the current density, the pH of the solution, the

G. Kreysa et al. (eds.), *Encyclopedia of Applied Electrochemistry*, DOI 10.1007/978-1-4419-6996-5,

presence and nature of the ionic additives, the solvent, and the anode material. Thus, high current density and elevated carboxylate concentrations favor the formation of the desired dimers. This results from a high radical concentration at the surface of the electrode that increases the rate of the dimerization. At high current density (between 0.25 and 1.0 A/cm^2), the rate of radical combination is nearly proportional to the concentration of the radical species generated on the electrode. Furthermore, when a critical potential, located between 2.0 V and 2.8 V (versus NHE, normal hydrogen electrode), is reached, discharge of the carboxylate proceeds almost exclusively and side reactions are usually suppressed. The influence of the anode material has been studied in detail. Platinum (overwhelmingly employed), iridium, and vitreous carbon are suitable for dimer formation; however, other materials such as palladium, gold, and lead dioxide can also be employed though only in methanol or glacial acetic acid [26–29].

A weakly acidic medium favors the Kolbe product. Therefore, the carboxylic acid is partially neutralized to the extent of 2–5 %. The concentration of the carboxylate anion remains constant during the whole electrolysis process since the base is regenerated at the cathode at the same rate as the carboxylate is consumed at the anode. While water has been used before, methanol or aqueous methanol is now the solvent of choice. Obviously, the selection of the best solvent rests mostly on experimental investigations. Temperature is usually not a critical parameter though an increase in temperature results in higher yields of Kolbe dimer. However, beyond 50 °C, side reactions can occur. Finally, ionic additives have a negative impact on the production of the Kolbe dimer. This results from a competitive adsorption of the additive on the electrode surface. The amount of carboxylate is thus lowered and the radical dimerization significantly decreased. In some cases, it can even be completely suppressed. In summary, high current density, a platinum anode, a weakly acidic medium (water or methanol), no additives, and a temperature below 50 °C are conditions that favor the formation of Kolbe dimers. Some selected examples are collected in Scheme 3 [14–16].

As can be seen, the reaction affords excellent yields of the desired products and tolerates a large number of functionalities, including esters, amides, ethers, alkenes, free alcohols, nitriles, and ketals. The electrolysis of two different carboxylic acids, as originally described by Wurtz, generates three products: the two symmetrical and the nonsymmetrical dimers (Scheme 4).

By adjusting the relative ratios of the co-acids, the mixed adduct, accompanied by only one symmetrical dimer, can be obtained in good yields. A ratio of 1:5 in the sacrificial co-acid is usually enough to minimize or suppress the undesired dimerization. Using this procedure, a wide variety of otherwise difficult to prepare adducts can be smoothly assembled, as illustrated by the conversion of **18** into **20** (Scheme 5) [17].

It is noteworthy that a CF_3 group can be attached to another radical species, leading to fluorinated derivatives such as **22** [18]. The electrolysis of **23** is remarkable in many ways, not the least of them being the capture of a radical beta to a chlorine atom. Despite the inevitable competing elimination of chloride radical, coupling to **25** proceeds in 70 % yields[19].

$$2\ RCOO^{\ominus}\ (\mathbf{1}) \xrightarrow{\text{Electrolysis}} R{-}R\ (\mathbf{2}) + 2\ CO_2\ (\mathbf{3}) + 2\ e^{\ominus}$$

Kolbe and Related Reactions, Scheme 1 The Kolbe reaction

$$RCOOH\ (\mathbf{4}) \underset{}{\overset{OH^{\ominus}}{\rightleftharpoons}} RCOO^{\ominus}\ (\mathbf{1}) \xrightarrow{-e^{\ominus}} RCOO^{\odot}\ (\mathbf{5}) \xrightarrow{-CO_2} R^{\odot}\ (\mathbf{6}) \xrightarrow{2x} R{-}R\ (\mathbf{2})$$

Kolbe and Related Reactions, Scheme 2 Mechanism of the Kolbe reaction

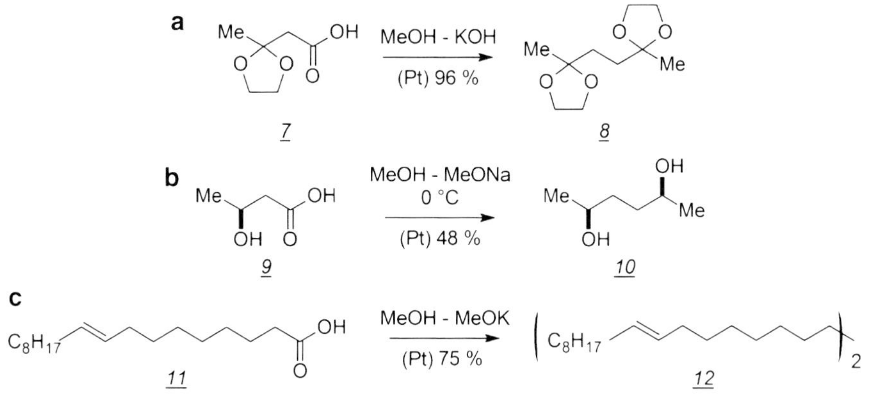

Kolbe and Related Reactions, Scheme 3 Synthetic applications

Kolbe and Related Reactions, Scheme 4 Mixed Kolbe reactions

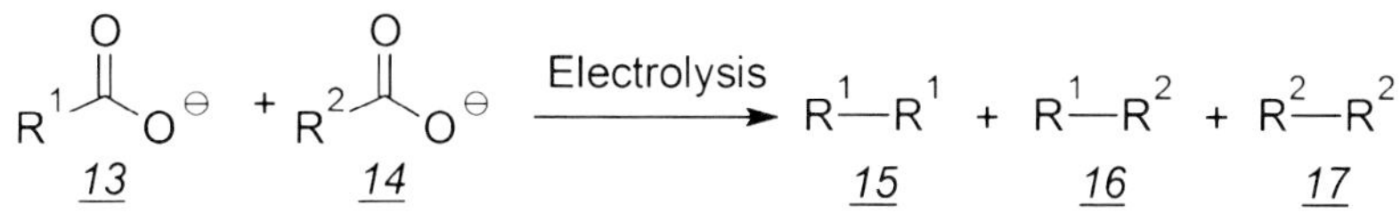

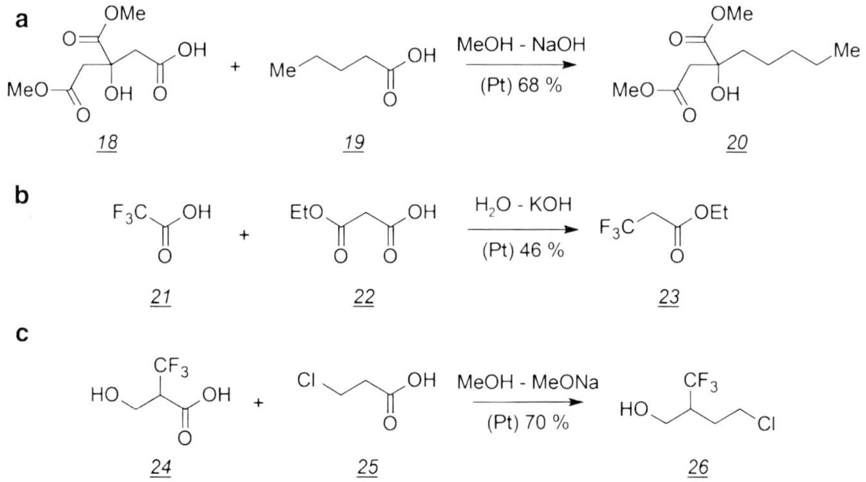

Kolbe and Related Reactions, Scheme 5 Examples of mixed Kolbe reactions

The C-centered radical, generated during Kolbe electrolysis, can also react with various olefins, affording addition products **27**. These radical intermediates can either dimerize, leading to adduct **29**, or recombine with radical **6**, leading to product **28**. Both reaction pathways are illustrated in Scheme 6 [20].

While intermolecular capture of **6** occurs with moderate yields, the intramolecular cyclization of radicals generated under Kolbe electrolysis provides mono-, bi-, or tricyclic structures in good to excellent yields (Scheme 7).

Schäfer demonstrated that ring closure of oxygen and nitrogen-containing precursors provided the corresponding tetrahydrofurans and pyrrolidines in moderate to good yields [21, 22]. The triquinane derivative **39** was also prepared in a similar way [23]. Markó et al. revealed that

a

R $^{\odot}$

R R R ^{1} 29

O

$-e^{\ominus}$

R $^{\odot}$ + R ^{1} R R ^{1}

R O $^{\ominus}$ $-CO_2$

1 6 27 28

R R ^{1} 30

2 x R ^{1}

R

b

MeO OMe

MeO OH + Ph 38 % O O

O O Ph Ph

31 32 33

c

CF_3

F_3C OH + O F_3C O + F_3C O

O NH_2 37 % NH_2 9 % NH_2

21 34 35 36

Kolbe and Related Reactions, Scheme 6 Additions on alkenes

a

O R

OH R-CO_2H X = O, NCHO, NAc

(Pt) 45 - 72 % R = CH_3, C_5H_{11}, $(CH_2)_4CO_2Me$, ...

X X

37 38

b

O

O H Me

CH_3CO_2H

(Pt) 45 %

OH

O 39 40

c

OH

O CH_3CO_2H Me H Me H

MeOH

O O (Pt) 65 % O + O

O O O O H O H O

41 42 9:1 43

d

O

O OH Me O O

O Me R-CO_2H Me R = Me, Et, *i*-Pr

O (Pt) 70-90 % ( )n O Me n = 1,2

n Me Me

O R O

44 45

Kolbe and Related Reactions, Scheme 7 Intramolecular Kolbe cyclisations adjusting the electronic characters of the radical and the alkene led to substantially improved transformations [24]. Noteworthy among them is the assembly of five- and six-membered carbocycles in excellent overall yields [25]. It is important to mention that in all cases, a co-acid is employed and the final radical recombination proceeds efficiently. This feature is unique to radical reactions performed under Kolbe electrolysis conditions. Examples of such processes under "normal" radical conditions are quite rare.

The Hofer-Moest Reaction

As mentioned previously, the structure of the carboxylic acid substrate has a profound effect on the course of the electrochemical oxidation. Indeed, if an electron-donating substituent is located alpha to the carboxylic acid function, a second electron can be removed rapidly from the initially generated radical, affording a carbocationic species **45**. Subsequent capture by a nucleophile – i.e., the solvent – leads to adduct **46**. Rearrangement and elimination products, typically observed when cationic intermediates are involved, are also isolated (Scheme 8).

Like the Kolbe dimerization, the success of the Hofer-Moest reaction depends upon the judicious control of a number of experimental variables. These include among others: (1) the use of low current density, (2) carbon electrodes, and (3) the presence of additives. The low current density ensures that the concentration of radical species on the surface of the electrode is kept low so as to favor the abstraction of the second electron. Carbon electrodes proved particularly adapted to such two-electron oxidation processes. The addition of additives leads to competitive adsorption of substrate and additives on the surface of the electrodes, resulting in a decrease of the local concentration of the radical species.

A vast literature is available on the Hofer-Moest reaction. The generation of oxonium, iminium, and thionium cations, by two-electron oxidative decarboxylation, has been reported (Scheme 9) [26–29].

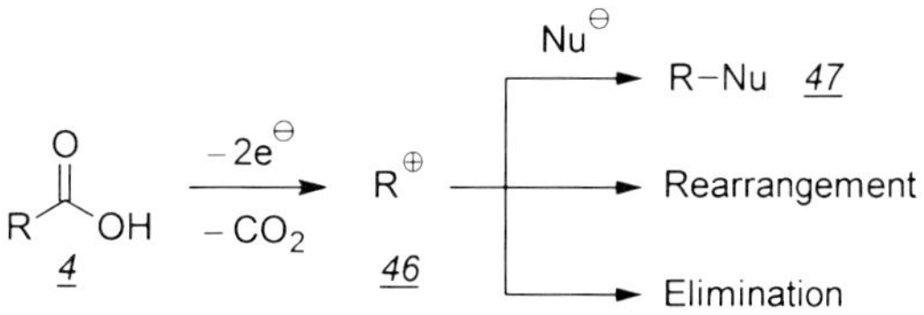

Kolbe and Related Reactions, Scheme 8 The Hofer-Moest reaction

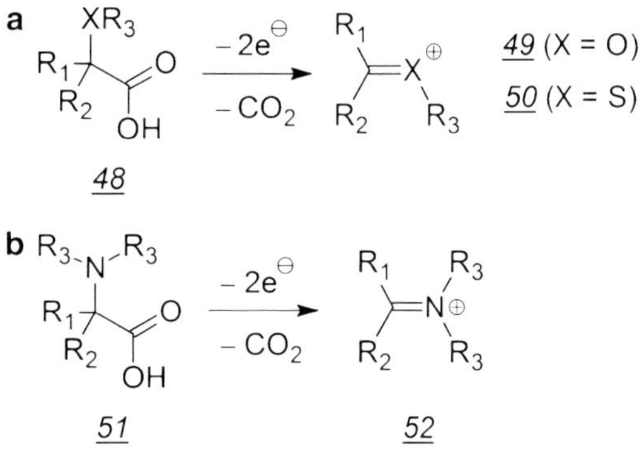

Kolbe and Related Reactions, Scheme 9 Onium salts preparation

The subsequent capture of these cations provides a wide range of functionalities, often difficult to install by alternative, non-electrochemical techniques. A selection of representative examples is displayed in Scheme 10.

The oxidation of the sugar-containing carboxylic acid **52**, in the presence of acetate, leads to the unsymmetrical bis-acetal **53** [30]. In a similar manner, O,N-acetals can be efficiently constructed from alpha-amino-acids [24]. This process has been employed as a key step in the synthesis of some beta-lactam antibiotics [31]. The oxidative decarboxylation of alpha-alkoxy acids **58** in methanol provides a simple and efficient route to the formation of MOM ethers and obviates the use of the highly toxic MOMCl [32]. Furthermore, Markó et al. demonstrated that Hofer-Moest reaction of dialkoxy carboxylic acids **60** offers an easy and general route to variously functionalized orthoesters [33]. These are usually difficult to prepare by alternative methodologies and some of them can only be assembled using this electrochemical process.

Kolbe and Related Reactions, Scheme 10
Applications of the Hofer-Moest reaction

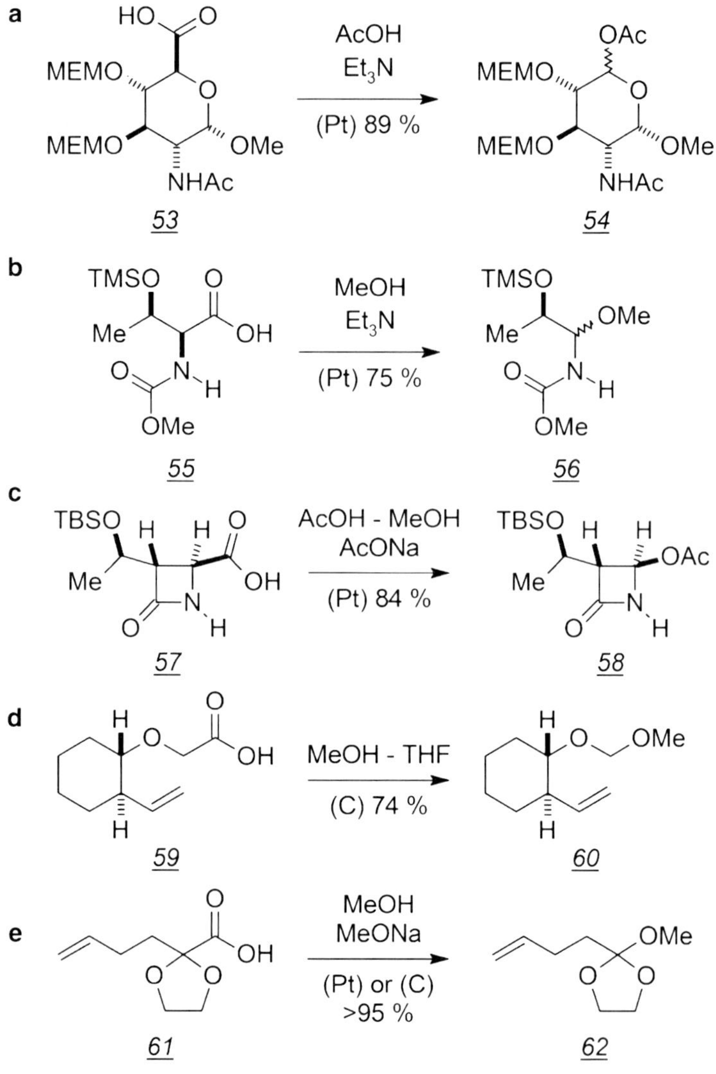

a: 53 → 54 (AcOH, Et_3N, (Pt) 89 %)

b: 55 → 56 (MeOH, Et_3N, (Pt) 75 %)

c: 57 → 58 (AcOH - MeOH, AcONa, (Pt) 84 %)

d: 59 → 60 (MeOH - THF, (C) 74 %)

e: 61 → 62 (MeOH, MeONa, (Pt) or (C) >95 %)

Miscellaneous

The two-electron oxidation process can be diverted to provide an interesting route to variously substituted alkenes. The presence of a function (COOH, $SiMe_3$, SR, NR_2) beta to the carboxylic acid is useful in controlling the regiochemistry of the olefination reaction. However, it is not a prerequisite (Scheme 11).

Strained olefins can be easily prepared by electrochemically induced oxidative decarboxylation, followed by generation of the corresponding carbocation and subsequent elimination of the trimethylsilyl group [34]. In the case of **66**, a different mechanism probably takes place, as the sulfur is far easier to oxidize than the carboxylate anion [35].

The occurrence of carbocationic intermediates in the Hofer-Moest variant of the Kolbe reaction can lead to numerous rearrangement processes. For example, electrolysis of carboxylic acid **68** in methanol results in the oxidative ring opening of the norbornane skeleton[30]. In a similar manner, Groß-type fragmentation of the decaline derivative **70** affords the 10-membered-ring ketone **71** in good yield (Scheme 12) [31].

Kolbe and Related Reactions, Scheme 11 Oxidative elimination

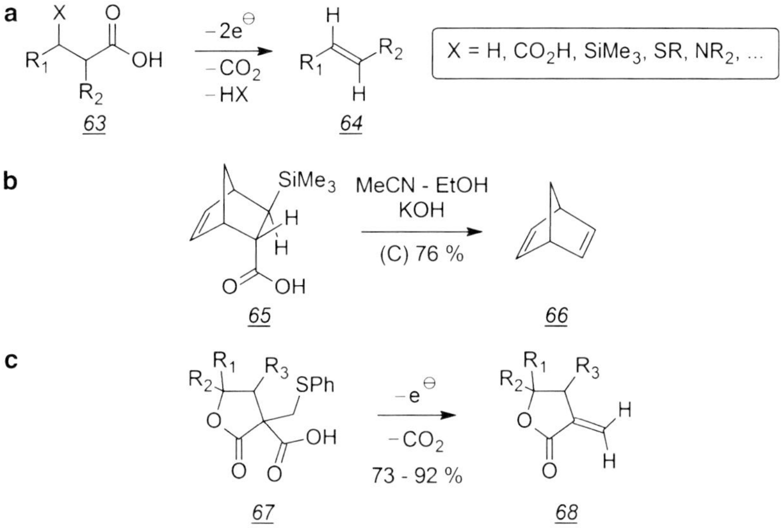

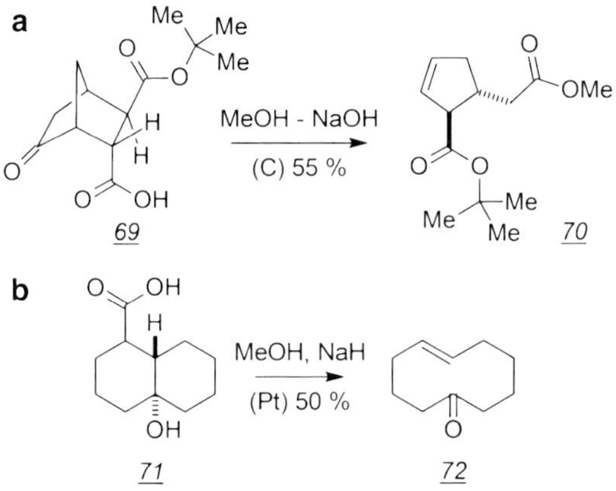

X = H, CO_2H, $SiMe_3$, SR, NR_2, ...

Kolbe and Related Reactions, Scheme 12 Oxidative fragmentation

The combination of Kolbe electrolysis and Hofer-Moest cationic decarboxylation can afford a remarkably facile route to the preparation of otherwise difficult to access products. For example, electro-oxidation of ammonium salt **72** in methanol afforded initially the C-centered radical that underwent a *5-exo-trig* cyclization. The intermediate radical species – a capto-dative radical – was further oxidized to the corresponding, highly reactive, oxonium cation. Trapping with MeOH provided ketal **73** in 72 % yield. Saponification led to the acid, which was submitted again to a Hofer-Moest oxidative decarboxylation, resulting in the quantitative formation of the orthoester **74**, bearing a ketal function [24]. Such compounds are rather unique and their preparation demonstrates the synthetic power of organic electrochemistry (Scheme 13).

Conclusions and Perspectives

The Kolbe and Hofer-Moest reactions are particularly useful transformations that generate, from readily available carboxylic acids, a wide variety of structures bearing varied functionalities and substitution patterns. From the initial preparation of symmetrical dimers, by a C-C radical coupling process (a rather unusual event by itself), the Kolbe reaction has enabled the synthesis of mixed adducts by using two different carboxylic acids as substrates. The intramolecular version provided an efficient entry into 5- and 6-membered carbo- or heterocycles.

The Hofer-Moest variant has opened new vistas and led to the preparation of a range of products originating from carbocationic pathways. These include cation capture adducts, rearrangement compounds, and elimination products. Finally, combining the Kolbe and Hofer-Moest reactions afforded an original entry to yet another set of uniquely functionalized structures.

Organic electrochemistry is on the rise again and the increased requirements for ecologically

Kolbe and Related Reactions, Scheme 13 Combined Kolbe and Hofer-Moest reactions

MeOH - NaOMe (Pt) 72 %

1. KOH, MeOH 2. MeOH, (Pt) ~ 95 %

73 74 75

benign, yet economically viable, processes shall restore its due place. Electrochemistry has a bright and exciting future.

Future Directions

Organic electrochemistry is again at a crossroad. The increased awareness of chemists towards environmental issues and, hence, the desire to perform ecologically benign transformation, should result in an even greater development of electrochemical processes. Controlling the selectivity and the innocuousness of these reactions will be a major challenge. At the same time, the generation of highly reactive or unstable intermediates - though synthetic useful ones – will constitute another rich area of investigation, as will be the conversion of CO2 into synthetically important building blocs – other than methanol and methane – and the advent of asymmetric organic electrochemistry.

Cross-References

- ► Electrocatalysis - Basic Concepts, Theoretical Treatments in Electrocatalysis via DFT-Based Simulations
- ► Electrochemical Functional Transformation
- ► Electrodeposition of Electronic Materials for Applications in Macroelectronic- and Nanotechnology-Based Devices
- ► Elements of Electrocatalysts for Oxygen Reduction Reaction

References

1. Faraday M (1834) Siebente Reihe von Experimental-Unter-suchungen über Elektricität. Pogg Ann 33:433
2. Kolbe H (1849) Untersuchungen über die Elektrolyse organischer Verbindungen. Liebigs Ann Chem 69:257–294
3. Würtz A (1855) Sur une nouvelle classe de radicaux organiques. Ann Chim Phys 44:275–313
4. Brown AC, Walker J (1891) Elektrolytische Synthese zweibasischer Säuren. Liebigs Ann Chem 261:107–128
5. Weedon BCL (1960) The kolbe electrolytic synthesis In: Raphael, AR, Taylor, EC, Wynberg, H (eds) Advances in organic chemistry: methods and results. Interscience Publishers Inc, New York, 1, p. 1
6. Vijh AK, Conway BE (1967) Electrode kinetic aspects of the kolbe reaction. Chem Rev 67:623–664
7. Eberson L (1967) Mechanism of the kolbe electrosynthesis. Electrochim Acta 12:1473–1478
8. Brennan MPJ, Brettle R (1973) Anodic oxidation. Part XI. Carbon anodes in electrosyntheses based on carboxylate ions. J Chem Soc, Perkin Trans I: 257–261
9. Schäfer HJ (1990) Electrochemistry IV topics. Current Chem 152:91–151
10. Utley J (1997) Trends in organic electrosynthesis. Chem Soc Rev 26:157–167
11. Sperry JB, Wright DL (2006) The application of cathodic reductions and anodic oxidations in the synthesis of complex molecules. Chem Soc Rev 35:605–621
12. Schäfer HJ (2012) Electrochemical conversion of fatty acids. Eur J Lipid Sci Technol 114:2–9
13. Andrieux CP, Gonzalez F, Savéant J-M (2001) Homolytic and heterolytic radical cleavage in the kolbe reaction; electrochemical oxidation of arylmethyl carboxylate ions. J Electroanal Chem 498:171–180
14. Utley JHP, Yates GB (1978) Electro-organic reactions. Part 11. Mechanism of the kolbe reaction; the stereochemistry of reaction of anodically generated cyclohex-2-enyl radicals and cations. J Chem Soc, Perkin Trans II: 395–400
15. Burke MJ, Feaster JE, Harlow RL (1991) New chiral phospholanes; synthesis, characterization, and use in asymmetric hydrogenation reactions. Tetrahedron Asymm 2:569–592
16. Peterson J (1905) Reduction of oleic acid to stearic acid by electrolysis. Z Elektrochem 11:549–553
17. Schäfer HJ, Kratschmer S, Weiper A, Klocke E, Plate M, Maletz R (1998) Conversion of biomass derived products by anodic activation. In: Torii S (ed) Novel trends in electroorganic synthesis. Springer, Tokyo, pp 187–190

18. Renaud RN, Champagne PJ (1975) Electrochemical oxidation of trifluoroacetic acid in an organic substrate. III. In the presence of substituted malonic acid half esters and unsaturated carboxylic acid esters. Can J Chem 53:529–534
19. Kubota T, Aoyagi R, Sando H, Kawasumi M, Tanaka T (1987) Preparation of Trifluoromethylated Compounds by Anodic Oxidation of 3-Hydroxy-Trifluoromethylpropionic acid. Chem Lett: 1435–1438
20. Schäfer HJ, Pistorius R (1972) Single-step synthesis of 1, n-dicarboxylic diesters by kolbe electrolysis of oxalic and malonic half esters in the presence of olefins. Angew Chem Int Ed Engl 11:841–842
21. Becking L, Schäfer HJ (1988) Pyrrolidines by intramolecular addition of kolbe radicals generated from β-allylaminoalkanoates. Tetrahedron Lett 29:2797–2800
22. Becking L, Schäfer HJ (1988) Synthesis of a prostaglandin precursor by mixed kolbe electrolysis of 3-(Cyclopent-2-enyloxy)propionate. Tetrahedron Lett 29:2801–2802
23. Matzeit A, Schäfer HJ, Amatore C (1995) Radical tandem cyclizations by anodic decarboxylation of carboxylic acids. Synthesis: 1432–1444
24. Lebreux F, Markó IE unpublished results
25. Lebreux F, Buzzo F, Markó IE (2008) Synthesis of five- and six-membered-ring compounds by environmentally friendly radical cyclizations using kolbe electrolysis. Synlett: 2815–2820
26. Sato N, Sekine T, Sugino K (1968) Anodic processes of acetate ion in methanol and glacial acetic acid at various anode materials. J Electrochem Soc 115:242–246
27. Torii S, Tanaka H (1991) Carboxylic acids. In: Lund H, Hammerich O (eds) Organic electrochemistry. Marcel Dekker, New York, pp 499–543
28. Yoshikawa M, Kamigauchi T, Ikeda Y, Kitagawa I (1981) Chemical transformation of uronic acids leading to aminocyclitols. IV. Synthesis of hexaacetyl-streptamine from N-Acetyl-D-glucosamine by means of electrolytic decarboxylation. Chem Pharm Bull 29:2582–2586
29. Renaud P, Seebach D (1986) Preparation of chiral building blocks from amino acids and peptides via electrolytic decarboxylation and $TiCl_4$-induced aminoalkylation. Angew Chem Int Ed Engl 25:843–844
30. Mori M, Kagechika K, Tohjima K, Shibasaki M (1988) New synthesis of 4-acetoxy-2-azetidinones by use of electrochemical oxidation. Tetrahedron Lett 29:1409–1412
31. Bastug G, Eviolitte C, Markó IE (2012) Functionalized orthoesters as powerful building blocks for the efficient preparation of heteroaromatic bicycles. Organic Lett 14:3502–3507
32. Hermeling D, Schäfer HJ (1984) 3-trimethylsilylacrylic acid as an acetylene equivalent in diels-alder reactions; olefins via anodic decarboxylation-desilylation. Angew Chem Int Ed Engl 23:233–235
33. Torii S, Okamoto T, Oida T (1978) An improved synthesis of a-methylene g-Lactones by electrolysis of a-carboxy-a-phenylthiomethyl-g-butyrolactones. J Org Chem 43:2294–2296
34. Michaelis R, Müller U, Schäfer HJ (1987) Anodic grob Fragmentation of bicycloalkylcarboxyl acids to specifically disubstituted cycloalkenes. Angew Chem Int Ed Engl 26:1026–1027
35. Wharton PS, Hiegel GA, Coombs RV (1963) trans-5-cyclodecenone. J Org Chem 28:3217–3219

Kröger-Vinks Notation of Point Defects

Ulrich Guth
Kurt-Schwabe-Institut für Mess- und Sensortechnik e.V. Meinsberg, Waldheim, Germany
FB Chemie und Lebensmittelchemie, Technische Universität Dresden, Dresden, Germany

In order to express the defects in solids different notations were proposed. SCHOTTKY [1] has introduced the first notation which is an absolute one. KRÖGER and VINK [2–4] developed a different notation which becomes generally accepted and denote structural elements on real sites compared with the normal occupied sites (Table 1).

Kröger-Vinks Notation of Point Defects, Table 1 Structural elements for description of a defect crystal AB (Kröger-Vink notation)

Structural element	Notation
Regular occupied site	$A_A^x B_B^x$
Substitution defects	
Lattice site occupied by other ion or atom	$A_B^x, A_B^\bullet, B_A^x, B_A^{\prime}$
Lattice site occupied by foreign ion or atom	$F_A^x, F_A^\bullet, F_B^x, F_B^{\prime}$
Empty lattice sites	$V_A^x, V_A^{\prime}, V_B^x, V_B^\bullet$
Interstitial sites	
Interstitial occupied by own ion or atom	$A_i^x, A_i^\bullet, B_i^x, B_i^{\prime}$
Interstitial occupied by foreign ion or atom	$F_i^x, F_i^\bullet, F_i^{\prime}$
Interstitial vacancy (empty interstitial)	V_i^x
Electronic defects	
Excess electrons	$e^{\prime}$
Defect electrons (holes)	$h^\bullet$

Kröger-Vinks Notation of Point Defects, Table 2 Relation between the absolute and relative notation in defect crystal AB

Structural element –vacancy		Regular structural element =		Building element		
Absolute notation				Relative notation		Kröger-Vink notation
Sodium deficiency in sodium chloride						
Na^+ Cl^- Na^+		Na^+ Cl^- Na^+		0 0 0		V'_{Na}
Cl^- Cl^-	-	Cl^- Na^+ Cl^-	=	0 o 0	=	
Na^+ Cl Na'		Na^+ Cl^- Na^+		0 0 0		
Vacancy on Na^+ site		-Na^+ on Na^+ site		Na^+ vacancy		V'_{Na}
Oxygen *deficiency in zirconia*						
Zr^{4+} O^{--} Zr^{4+}		Zr^{4+} O^{--} Zr^{4+}		0 0 0		$V_O^{\cdot\cdot}$
O^{--} Zr^{4+}	-	O^{--} Zr^{4+} O^{--}	=	0 0 O^{++}	=	
Zr^{4+} O^{--} Zr^{4+}		Zr^{4+} O^{--} Zr^{4+}		0 0 0		
Vacancy on O^{--} site		O^{--} on O^{--} site		O^{--} vacancy		$V_O^{\cdot\cdot}$
Barium substitution in sodium chloride						
Na^+ Na^+ Na^+		Na^+ Na^+ Na^+		0 0 0		$Ba_{Na}^{\cdot}$
Na^+ $Ba^{++}Na^+$	-	Na^+ Na^+ Na^+	=	0 Ba^+ 0	=	
Na^+ Na^+ Na^+		Na^+ Na^+ Na^+		0 0 0		
Barium ion on sodium ion site		Sodium ion on sodium ion site		Ba^+ ion		$Ba_{Na}^{\cdot}$

The resulting charge as compared with the charge of a regular occupied site is written as an exponent. x means the charge is not changed regarding to the regular charge; it is often omitted. If a negative charge is missing then a positive (the opposite) charge is active and denoted as a point $^{\bullet}$. On the other hand, a negative charge is marked as a prime′.

The relation between the real structural element, the regular structure element, und the building element according Kröger-Vink becomes clear in the Table 2.

The equations written in Kröger-Vink symbols with building elements can be used in mass action law. As an example the incorporation of oxide ions on interstitials (Anti-Frenkel disorder) can be expressed as follows:

$$O_O^x + V_i \rightleftharpoons O_i'' + V_O^{\cdot\cdot}$$

The concentration of oxide ions and vacancies on interstitials does not change markedly due to the defect formation. The mole fraction activity is zero.

$$\text{Null} \rightleftharpoons O_i'' + V_O^{\cdot\cdot}$$

Therefore the mass law action is (the square brackets symbolize the concentration) $K_F = [O_i''] [V_O^{\cdot\cdot}]$ and with $[O_i''] = [V_O^{\cdot\cdot}] = \sqrt{K_F}$ the electrical conductivity that can be measured is proportional to the concentration of oxide ion vacancies

$$\sigma \propto [V_O^{\cdot\cdot}] \propto c_{V^{\cdot\cdot}}$$

$$\sigma = \text{const}\, c_{V^{\cdot\cdot}}$$

Cross-References

- ▶ Defect Chemistry in Solid State Ionic Materials
- ▶ Defects in Solids

References

1. Schottky W (1958) Halbleiterprobleme Vol. IV, p. 235. Vieweg, Braunschweig
2. Hauffe K (1966) Reaktionen an und in festen Stoffen. Springer, Berlin/Göttingen/Heidelberg
3. Kröger FA, Vink HJ (1956) In: Seitz F, Turnbull D (eds) Solid state physics: advances in research and applications, vol 3. Academic, New York, p 307
4. Kröger FA (1974) The chemistry of imperfect crystals. North-Holland, Amsterdam

L

Lead Acid Batteries

Yoshiaki Yamaguchi
Technical Development Division, Global Technical Headquarters, GS Yuasa International Ltd., Kyoto, Japan

Introduction

Lead acid battery was invented in 1859 by Gaston Plante, and has been widely used throughout the world for more than 150 years [1]. At present, all automobiles are equipped with one or more lead-acid battery. As for industrial application, lead-acid batteries have served as a backup for telecommunication system, office, and medical emergency power supply equipment [2]. Those are also used as traction battery for the electric forklifts [3]. In addition, those are used for electric moped in China and Asian area in recent years. In such ways, lead-acid batteries have become an inseparable device for our life.

Here is brief explanation of lead-acid battery principle and its structure, features of those for each usage, and recent market and development trend.

Principle and Features of Lead-Acid Battery

The reaction principle of lead-acid battery remains unchanged for over 150 years from the invention.

As shown in reaction formula for the discharging of battery, at the negative electrode, metallic lead reacts with the sulfate ions in water solution to produce lead sulfate and release electrons (Formula 1). At the positive electrode, lead dioxide reacts also with the sulfate ions in solution to produce lead sulfate and water (Formula 2). For the charging of the battery, the inverse reactions occur at the negative and positive electrodes.

Lead-acid batteries actuate each kind of load by utilizing these electron transfers initiated by negative and positive reactions.

$$Pb + H_2SO_4 \rightarrow PbSO_4 + 2H^+ + 2e \quad (1)$$

$$PbO_2 + H_2SO_4 + 2H^+ + 2e \rightarrow PbSO_4 + 2H_2O \quad (2)$$

However the battery structure has changed substantially from initial ones. In the early days of lead acid batteries, the corrosion layers formed on the surface of lead sheet were used as active materials. But at present, the pasted type electrodes, which are made from lead-oxide paste and lead-alloy grid, are used generally. Then such pasted type electrodes are charged in sulfuric acid to make positive and negative plates and have much larger effective surface area which leads to larger capacity compared to the batteries of the early days.

Furthermore, separator materials for the lead-acid battery have changed from wood to paper,

G. Kreysa et al. (eds.), *Encyclopedia of Applied Electrochemistry*, DOI 10.1007/978-1-4419-6996-5,

synthetic resin, and fine glass fiber mat called absorptive glass mat (AGM), and such changes have contributed to improve internal resistance and durability of the lead-acid battery. AGM separators also add the functions of non-fluidization of electrolyte and make oxygen gas transmit from the positive electrode to the negative electrode through the micro pores of the separator during over charging state, and after that the oxygen gas is finally reduced to water at the negative electrode (Formula's 3 and 4).

$$2Pb + O_2 + 2H_2SO_4 \rightarrow 2PbSO_4 + 2H_2O \quad (3)$$

$$PbSO_4 + 2H^+ + 2e \rightarrow Pb + H_2SO_4 \quad (4)$$

The battery container material has shifted to ABS (acrylonitrile–butadiene–styrene copolymer) and PP (polypropylene) resin from the wood or ebonite, to attain smaller and lighter battery design [1].

The biggest feature of lead-acid battery is the fact that it is mostly made of the lead and lead alloy. The positive and negative active materials, grid, weld parts (strap), and terminal are made of the lead and lead alloy. It indicates that the battery maker can make most of the materials and parts for lead-acid battery in-house, just by purchasing lead and lead alloy ingots, without having to purchase the materials from many suppliers like the other battery systems. Therefore, it enables to produce lead-acid batteries in all over the world without creating particular supply chain or purchasing main materials from foreign countries.

Furthermore, lead-acid batteries are made of very simple materials such as lead (alloy), resin, glass, water, and sulfuric acid. And because melting point of the lead is 327.46 °C which is lower than other metals, the used lead-acid batteries can easily be recycled to retrieve lead material.

Meanwhile, the lead-acid-batteries have disadvantages which are the heaviness of battery and the pollution due to toxicity of, since the main materials of lead acid battery are lead and lead alloy after all. Specific energy density of lead acid-batteries is 30–40 Wh/kg, which is only about one-third of Lithium ion batteries. Therefore the lead-acid batteries are not used for personal computer, mobile phone, and electric vehicle at present in most cases, and also considered not to be used for such mobile devices for the future.

During production of the lead-acid battery, a plant may cause the environmental problems by lead around it because main materials of the lead-acid battery are lead and lead alloy. Especially in the small-scale lead-acid battery plant, the facilities and the management for the environmental measures are not enough in most cases. The air and water containing lead dust and ion would be exhausted directly from the plant without treatment, and the lead will give harm to health of the people living near by the plant. The size of the plants in the USA, Europe, and Japan are relatively large, and they exhaust air and water after having removed lead by using dust collector and water treatment equipment. Therefore, it is possible to prevent environmental pollution by lead. The most important thing is to manage the emission of lead definitely by the specialized equipment and regulation at process of production of the lead battery.

Industrial Batteries

Valve-regulated lead-acid (VRLA) batteries, using AGM separator as shown in Fig. 1, are used as a backup for telecommunication, office, and medical emergency power supply [2]. These VRLA batteries are usually connected with commercial power supply and kept in full-charged state with float charging at constant voltage. In emergency situations like power outage, they supply electricity by discharging to the loads like telecommunication system, computer, and emergency light, and they enable the usage of such loads continuously for some time. The expected life of these batteries is relatively

Lead Acid Batteries, Fig. 1 Valve-regulated lead-acid (VRLA) batteries using AGM separator

long some of those are used for more than 10 years. Due to such usage in these fields, the requirement of reliability and durability for the battery is very high, and there are high demands for large capacity battery which has more than 500 Ah/cell.

Following the Great East Japan Earthquake on March 11, 2011, all nuclear power plants throughout Japan were shut down. Thereby the electricity might be short at the time of demand peak in summer in Japan. If such power shortage trend continues, the cycle use battery may be needed, which can store electricity during the night time and discharge electricity during the day time of peak power demand. In addition, it is considered that the photovoltaic power generation will be used widely in near future. The surplus electricity generated by photovoltaic system is stored to the battery, and this electricity is used when there is the demand. For such application, it is thought that the lead-acid battery will compete with the lithium-ion battery, and lead-acid battery might be accepted from market because of its low cost and easiness to produce batteries with large capacity.

Automotive Battery

Today almost all automobiles are using lead-acid batteries which are mainly flooded paste type as shown in Fig. 2. The grids of their electrode consist of Pb-Sb alloy or Pb-Ca alloy. The Pb-Sb alloy grid is normally manufactured by gravity casting, and the Pb-Ca alloy grid is also manufactured by gravity casting and can also be manufactured by expanding or stamping technology. The grids of latter two types are made from lead alloy sheet rolls.

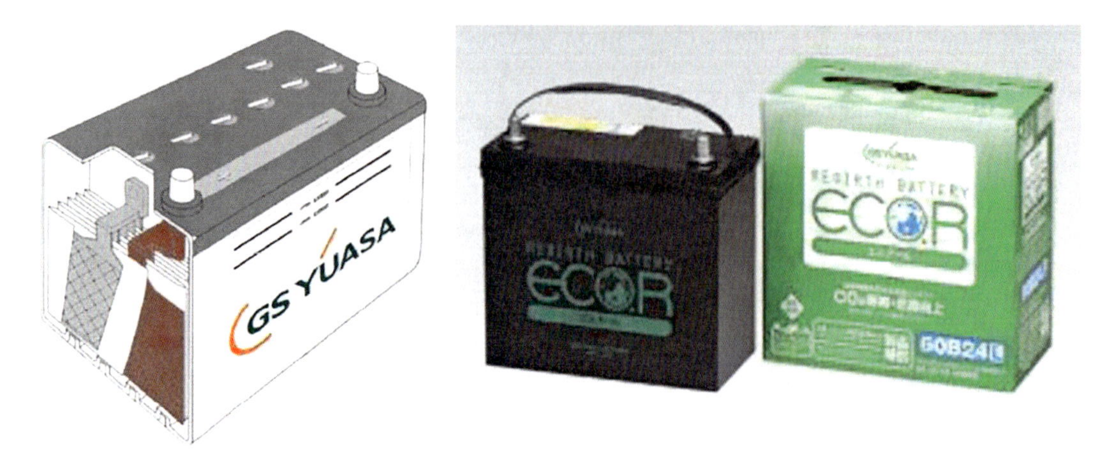

Lead Acid Batteries, Fig. 2 Flooded paste-type automotive battery

The separators of automotive batteries are paper-based type with glass fiber sheet attached, the synthetic resin fiber type, or the polyethylene type. The structure of recent mainstream for automotive battery is using the paste-type electrodes which consist of expanding or stamping grid, and the electrodes are enveloped with polyethylene separators. In Europe, the VRLA batteries using AGM separator is also used recently for the automotive as well.

Generally, the lead-acid batteries for automobiles have been used at fully charged condition by charging at 14–15 V constantly. In other words, part of energy from the fuel of automobiles has been idly consumed for the charging of battery. However by a rise of the recent environmental problem awareness, in the newly developed automobiles, the automobiles using the "charge control system" or "idling stop system" are increasing recently. The former is a method to control the unnecessary charge to the battery, and latter is a method to stop the engine when an automobile is stopped and then to restart the engine automatically when it runs again. These systems are the effective means to reduce fuel consumption, but the depth of discharge of the lead-acid battery is increased compared to conventional usage without such systems. It is considered that such trend will spread more, and many automobiles will use such systems to improve fuel efficiency in the recently [4–6].

Future Directions for Lead-Acid Battery

As mentioned above, conventional lead-acid batteries were used at fully charged state and were only discharged at the time when need. However it is considered that those will be used more like cyclic use for industrial and automotive in the future.

In the past, the lead-acid battery was called the secondary battery, but in the same time when cyclic use was applied by repeating discharge and charge, it had the problem that the life became short. It was caused by decline of the bonding strength of positive active material and by occurrence of the sulfation which are due to large sized lead sulfate crystals on negative active material that cannot be reduced to metallic lead any more [4]. To resolve these problems, improvement to durability of the active materials will be researched for the future lead-acid battery [7].

Conclusion

The lead-acid battery is used worldwide and indispensable for our life at present, and it is also considered that its needs will continue in the future because production of the lead-acid battery is relatively simple and the cost is low. Moreover, the application field for cyclic use is spread to solve energy and environmental problems, therefore development of the new lead-acid battery is also required for applying to such fields. During the history of the lead-acid battery for more than 150 years, many researches and developments have been done for materials like separator and container. However, there are still many unexplained phenomena about the active materials and its reaction, which is a core of the lead-acid battery. I hope that these problems of the active materials will be solved by research and development, and the new lead-acid battery which can be used for cyclic use with the high charge acceptance at low cost will come true in the future [7].

Cross-References

- ▸ Electrolytes for Rechargeable Batteries
- ▸ Electrolytes, Classification
- ▸ Electrolytes, History

References

1. Tatsuo Nagayasu, Toshiki Yoshioka (2007) History and future on technology development from environmental aspect for lead-acid battery. GS Yuasa Tech Rpt 4(1):9–13
2. Masayuki Maeda, Takuji Nakamura, Kazuya Akamatsu, Shuichi Manya, Tatsuo Nagayasu (2006) Front terminal type of valve regulated lead-acid battery "PWL12V125FS" for telecommunication applications. GS Yuasa Tech Rpt 3(2):29–33
3. Yoshitsugu Ishikura, Tomoyuki Enomoto, Tatsuo Nagayasu (2008) Development of lead-acid traction battery with new connecting structure conformable to DIN standard. GS Yuasa Tech Rpt 5(1):16–20
4. Shigeharu Osumi, Masaaki Shiomi (2008) Recent technological developments of lead-acid batteries. GS Yuasa Tech Rpt 5(1):8–15
5. Taisuke Takeuchi, Ken Sawai, Takuji Matsumura, Tomohiro Imamura, Shinji Ishimoto, Shigeharu Osumi (2007) Improvement in fuel efficiency of vehicle equipped with new automotive lead-acid battery of higher charge acceptance. GS Yuasa Tech Rpt 4(1):22–27
6. Hidetoshi Wada, Masaaki Hosokawa, Takao Ohmae (2012) Technical transition of lead-acid battery for idling stop vehicles. GS Yuasa Tech Rpt 9(2):16–23
7. Yuki Arai, Kohei Fujita, Yuichi Okada, Tetsuo Takama, Shigeharu Osumi (2011) Influence of metal ions in electrolyte on regenerative charge acceptance of lead-acid battery. GS Yuasa Tech Rpt 8(2):22–28

Lithium Primary Cells, Liquid Cathodes

Thomas B. Reddy
Department of Materials Science and Engineering, Rutgers, The State University of New Jersey, Piscataway, NJ, USA

Introduction

Lithium has been recognized to be attractive as an anode material for nearly a century. There is an apocryphal story that a vial of lithium metal was found on the desk of Thomas A. Edison at the time of his death in 1931. Lithium metal possesses the following characteristics which are desirable for battery use:

1. High negative electrode potential which provides high-voltage cells.
2. High specific energy and energy density which allows the construction of primary cells that are two to five times better than conventional alkaline cells from an energy standpoint.
3. Wide range of operating temperatures. Operation from +70 to −40 °C is typical and operation from +150 to −80 °C is possible with some systems.
4. High-power output. Lithium systems are capable of high-power output using spiral-wound electrodes or other high-surface-area designs.
5. Flat discharge curve. Lithium primary cells with soluble cathode reactants typically show a flat discharge curve which provides constant power output.
6. Long storage time. Because the Li metal in this class of cells is passivated by the soluble cathode reactant in storage, long shelf lives of 10–20 years are achievable.

Lithium Primary Cells with Soluble Cathode Reactants

Development of Lithium/Sulfur Dioxide Cells

Experiments to develop this type of cell began in the 1960s and led to the Lithium-sulfur dioxide

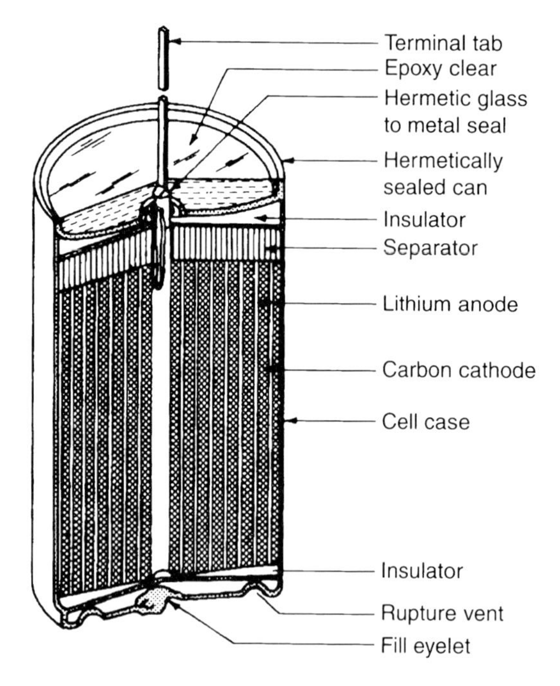

Lithium Primary Cells, Liquid Cathodes, Fig. 1 Lithium/sulfur dioxide cell

cell at the Central Research Laboratory of the American Cyanamid Company (now Cytec Industries) in Stamford, Connecticut. This research effort was intended to develop a rechargeable Li battery but resulted instead in the first commercially viable Li primary cell. As presently constructed, the Li/SO_2 cell uses a Teflon-bonded porous carbon cathode on an expanded Al metal grid, a lithium-foil anode with a copper-stripe current collector embossed in the surface, and an electrolyte of condensed sulfur dioxide (ca 70 %), acetonitrile (ca 23 %), and the balance lithium bromide as the electrolyte salt. This electrolyte exhibits a vapor pressure of $3-4 \times 10^5$ Pa at room temperature, necessitating the use of an hermetically sealed container, typically nickel-plated cold-rolled steel, which is welded shut. Figure 1 shows a cutaway view of such a cell. The electrodes are wound in a jelly-roll configuration using a separator of microporous polypropylene, typically 0.001 in. thick. Li/SO_2 and the other types of Li cells with soluble reactants are unique in that the cathode active material, in this case sulfur dioxide, is in direct contact with the Li metal anode without significant self-discharge.

In the case of the Li/SO_2 cell, the discharge reaction when connected to an external load has been shown to be: $2\,Li + 2\,SO_2 \rightarrow Li_2\,S_2O_4 \downarrow$. The product is insoluble lithium dithionite ($Li_2S_2O_4$) which precipitates in the pores of the carbon cathode. The ability of the cathode to accommodate this precipitate normally limits cell performance. The same reaction occurs on open circuit, resulting in the formation of the passive film of lithium dithionite on the Li surface which provides a long shelf life for this system. When the cell is activated with an external load, the passive film breaks down, allowing high-rate discharge since the conductivity of the electrolyte is ca 0.05 Scm^{-1} at 25 °C [1]. Since the electrolyte is pressurized, the cell contains a double-convolution vent in the bottom of the case which is designed to activate at ca 350 psi (2.41 MPa) during an abusive condition, such as short circuit or overheating. The vent activates at a temperature of 90 °C, well above the upper limit of normal operation, 130 °F or 54.4 °C. A critical component of this cell is the glass-to-metal (GTM) seal which must employ a corrosion-resistant glass to prevent lithiation of the glass from the 3-V potential across the terminals. To avoid safety problems which occurred when Li/SO_2 technology was in its infancy, a balanced cell design is employed with the coulombic ratio of Li/SO_2 in the range of 0.9–1.05. This design feature ensures that any excess Li metal left in the cell after discharge is kept in the passive state by excess SO_2 in the electrolyte, thus avoiding chemical reactions between the acetonitrile (CH_3CN) in the electrolyte and the residual Li metal. Under most conditions, cell capacity is limited by the ability of the cathode to accommodate the precipitated reaction product on discharge.

Performance

The nominal voltage of a Li/SO_2 cell is 3.0 V, but the actual open-circuit voltage (OCV) is 2.95 V. Depending on discharge rate and temperature, the Li/SO_2 cell will operate in the range of 2.7–2.9 V. A cutoff voltage of 2.0 V is normally employed. Figure 2 [1] shows the discharge performance of a high-rate Li/SO_2 D-cell at four currents from 1.5 mA to 3.0 A. Note that the initial voltage on

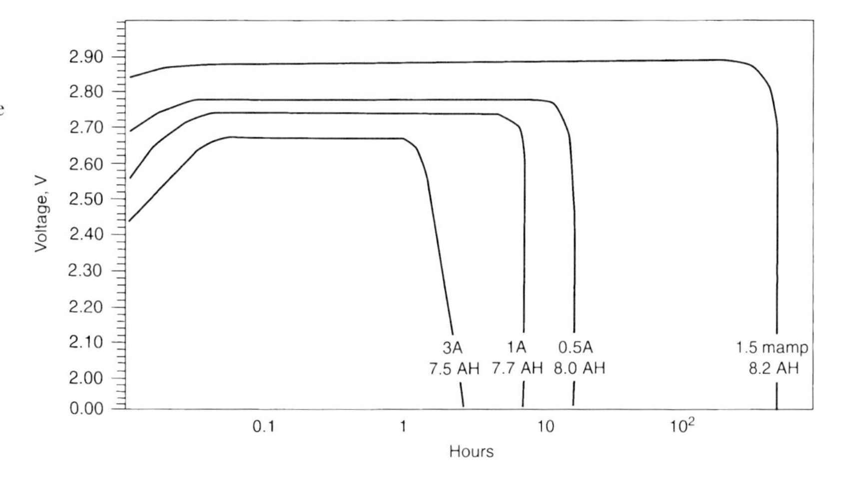

Lithium Primary Cells, Liquid Cathodes, Fig. 2 Discharge characteristics of high-rate Li/SO_2 D-size battery at four rates at 23 °C

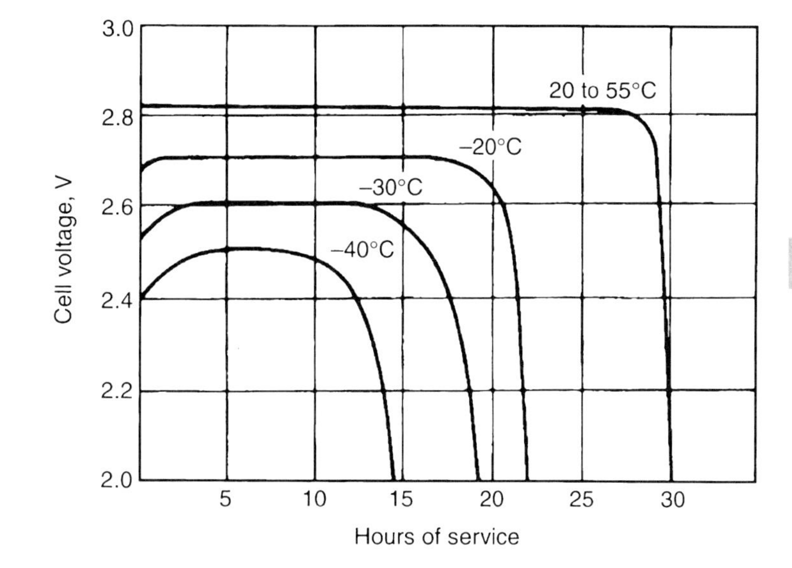

Lithium Primary Cells, Liquid Cathodes, Fig. 3 Typical discharge characteristics of Li/SO_2 battery at various temperatures, C/30 discharge rate

activation is somewhat depressed due to the voltage delay caused by the breakdown of the passive film on the Li metal surface. Note also the flat discharge curve after depassivation. Figure 3 exhibits the discharge characteristics of a Li/SO_2 cell at four temperatures on a C/30 discharge rate and demonstrates the ability of this system to operate over a wide temperature range. Figure 4 [1] shows the effect of rate and temperature on Li/SO_2 cell performance. At higher rates and lower temperature, cell capacity is limited by the ability of the porous carbon cathode to accommodate the $Li_2S_2O_4$ precipitate which plugs the outer pores and limits cell capacity.

Li/SO_2 cells and batteries are capable of very-high-rate discharge on pulse loads. A squat D-cell using a high-rate design delivered pulses as high as 37.5 A [2]. Continuous discharge at rates higher than 2C is not recommended since this may lead to overheating and vent activation.

A design optimization study [3] has resulted in the production of Li/SO_2 D-cells with a capacity of 9.1 Ah at room temperature on 250 mA discharge and 8.8 Ah on a 2 A drain. The latter capacity compares to a 7.75 Ah capacity for a standard D-cell. This result was obtained by varying the aspect ratio of both anode and cathode along with the use of three types of carbon in the cathode. A central cathode tab was also employed. These cells were found to produce less heat than standard cells when discharged from 2.0 V, the normal cutoff, to 0.0 V.

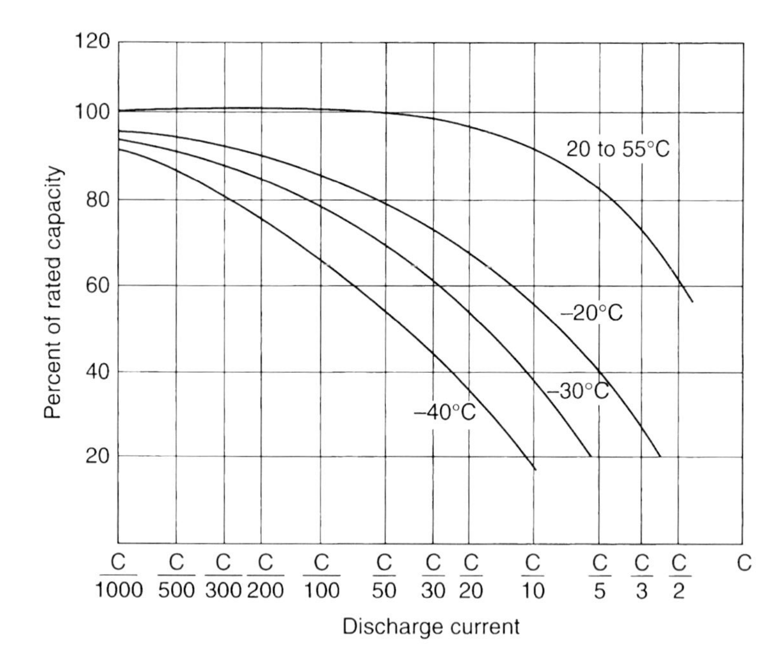

Lithium Primary Cells, Liquid Cathodes, Fig. 4 Performance of Li/SO_2 batteries as a function of discharge temperature and load

Shelf Life and Voltage Delay

The Li/SO_2 cell provides excellent storage characteristics and is noted for its low self-discharge at temperatures up to +70 °C. This is due to the use of an hermetically sealed design and the passive film on the Li surface which inhibits self-discharge reactions.

Storage data [3] on 2-year-old BA-5590 batteries consisting of ten Li/SO_2 D-cells discharged in series at 2 A at +21 and −30 °C showed a capacity loss of 6.5 % at the higher temperature but no loss at the lower value. Storage data [4] for BA-5598 batteries, consisting of five squat D-cells in series, for 14 years showed an 8 % capacity loss when discharged on 2 A at room temperature but virtually no loss at cold temperature. A reduced discharge voltage was observed under both conditions, however.

Cell and Battery Types and Applications

Li/SO_2 cells are produced in a wide range of sizes, both standard (ANSI/IEC) and custom sizes, from ½ AA with a 0.45 Ah capacity to a DD with a 16.5 Ah capacity. Some of these cells are assembled into standard US military battery types as seen in Table 1. Others have been developed for specific military or industrial applications. The Li/SO_2 system provides a high-energy output, operates over a wide range of discharge rates and temperatures, and possesses excellent storage properties. These characteristics make it an ideal system for use in military batteries, some of which are shown in Table 1. These batteries operate a variety of military electronics equipment, including radio transceivers, chemical agent detectors, digital message devices and electronic counter measures. Other military devices using Li/sulfur dioxide cells and batteries include night vision devices, sonobuoys and munitions.

Nonmilitary applications include industrial controllers and weather balloons. Commercial applications have been limited by the hazardous nature of the contents and restrictions on shipment by common carrier.

Lithium/Oxyhalide Cells

The oxyhalides, thionyl chloride ($SOCl_2$) and sulfuryl chloride (SO_2Cl_2), have been employed in lithium primary cells by themselves and with macro additives to improve performance. The basic technology was developed after the Li/sulfur dioxide system by the Eveready Battery Company (now Energizer), the

Lithium Primary Cells, Liquid Cathodes, Table 1 US military lithium/sulfur dioxide non-rechargeable batteries (MIL-PRF-49,471[a])

Type designation	Open-circuit voltage (series/parallel) (V)	Nominal voltage (series/parallel) (V)	Nominal energy[b] (Wh)	Weight (g)	Typical applications
BA-5093/U	27	23.4	77.2	635	Respirators
BA-5557A/U	30/15	16/13	54	410	Digital message devices
BA-5588A/U	15	13	35	290	PRC-68 and PRC-126 radios; respirators
BA-5590A/U[c]	30/15	26/13	185	1,021	SINCGARS radios; chemical agent detectors; satellite radios; jammers; loudspeakers; range finders; counter measures
BA-5590B/U[c]	30/15	26/13	185	1,021	Same as BA-5590A/U
BA-5598A/U	15	13	87	650	PRC-77 radios; direction finders; sensors
BA-5599A/U	9	7.8	50	450	Test sets; sensors

Notes: [a]MIL-PRF-49471 will be replaced by MIL-PRF-32271 in DOD procurements
[b]Nominal energy rating for temperature range of 25 ± 10 °C (77 ± 18 °F)
[c]The BA-5590A/U has a built-in state of charge indicator

GTE Research Labs, and the US Army CECOM at Fort Monmouth, NJ.

The oxyhalide cells are similar to the Li/SO_2 chemistry in their mode of operation but differ in that both compounds are liquid at room temperature and do not require the use of a cosolvent as required with Li/SO_2 cells. The similarities to Li/SO_2 include the utilization of a liquid phase cathode reactant which produces a passive film on the Li anode surface and the formation of a precipitate in the pores of a passive carbon cathode on discharge.

Lithium/Thionyl Chloride Cells

The $Li/SOCl_2$ primary cell has a high open-circuit voltage (OCV) of 3.65 V and a very high specific energy, up to 590 Wh/kg, and energy density, achieving 1,100 Wh/l in low-rate designs.

The basic cell reaction is

$$2SOCl_2 + 4Li \rightarrow 4LiCl \downarrow + SO_2 + S.$$

These cells are produced in cylindrical bobbin and spiral-wound designs as well as disc types. Large prismatic designs have also been produced in the past. These designs employ porous carbon cathodes with a nickel or stainless-steel current collector, a lithium anode, and a highly porous separator, typically a glass-mat type, since most conventional polymeric separators are attacked by thionyl chloride which is highly reactive. Relatively few salts are soluble in $SOCl_2$, and lithium tetrachloroaluminate ($LiAlCl_4$) is normally employed as the electrolyte salt.

The LiCl produced on discharge precipitates in the pores of the carbon cathode current collector. The SO_2 and elemental sulfur are initially soluble in the $SOCl_2$ electrolyte, but the sulfur may eventually precipitate in the carbon cathode. Considerable controversy exists in the technical literature as to whether anode-limited or cathode-limited high-rate designs are safer. The anode passivation problem, particularly after high-temperature storage, is more severe with $Li/SOCl_2$. As a result of the anode passivation, an initial voltage depression occurs with this technology, particularly at high rates and low temperatures. Anode coating techniques have been developed to ameliorate this problem. Another advantage of the $Li/SOCl_2$ cell is its wide operating temperature range since thionyl chloride is a liquid from below −110 °C to its boiling point of 78.8 °C.

Bobbin Cell Design

Figure 5 shows the design of a low-rate hermetically sealed, bobbin D-cell showing the cell

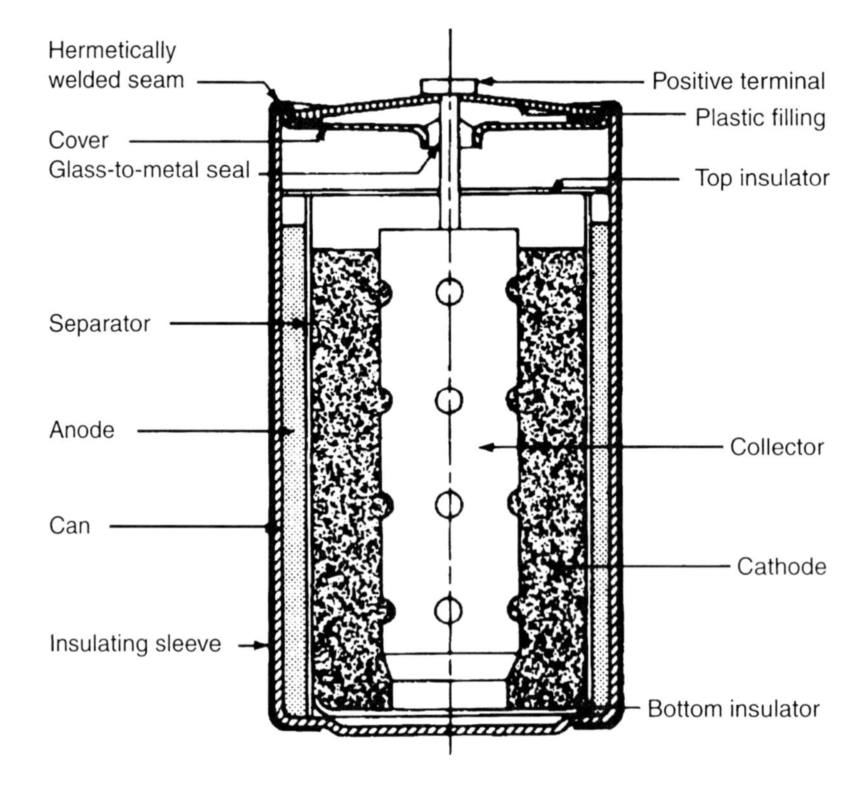

Lithium Primary Cells, Liquid Cathodes, Fig. 5 Cross section of bobbin-type Li/$SOCl_2$ D-cell [4]

components, including an annular current collector inserted in the carbon cathode. The Li metal anode is swaged to the inner surface of the case. Smaller sizes employ a solid cylindrical cathode with a pin-type current collector inserted in it. These bobbin cells use a nickel-plated steel can and are only capable of low-rate discharge. They polarize severely if subjected to higher currents and thus are safe to operate. Figure 6a shows the discharge characteristics of a bobbin-type D-cell at 25 °C on different resistive loads. Figure 6b shows the D-cell capacity as a function of rate and temperature.

Spiral-Wound Cells

Cells of this type are capable of medium to moderately high-power output and have been developed for military and selected industrial applications. Figure 7 shows the design of such a cell in a stainless-steel case which is welded shut. Both cathode and anode employ grid-type current collectors.

This design incorporates several safety features including a corrosion-resistant glass-to-metal seal for the positive terminal, a safety vent in the cell cover (top shell), and a fuse to prevent external short circuits and over current use. Figure 8 displays the discharge characteristics of such a D-cell at several rates up to 2.0 Ω (1.6 A). Note the flat discharge curves event at the higher rates. These cells are available in sizes from 1/3 C to D.

Other Designs

Lithium/thionyl cells have been designed in several other form factors, such as large disk types with nominal capacities up to 2,400 Ah currently available. These cells provide a specific energy and an energy density of 523 Wh/kg and 1,043 Wh/l at 8 A and 434 Wh/kg and 871 Wh/l at 50 A. These products have been developed for underwater applications by the US Navy and incorporate several unique design features [1].

Additionally, very large prismatic Li/$SOCl_2$ cells have been developed with capacities up to 16,500 Ah for use in nine-cell batteries for Missile Extended System Power (MESP) as backup power in ICBM silos. A typical 10,000 Ah cell weighed 71 kg and provided a specific energy of 480 Wh/kg

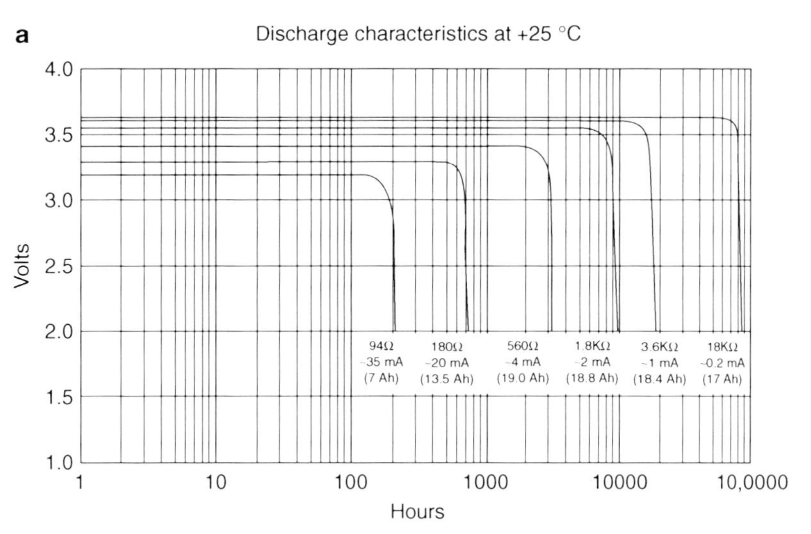

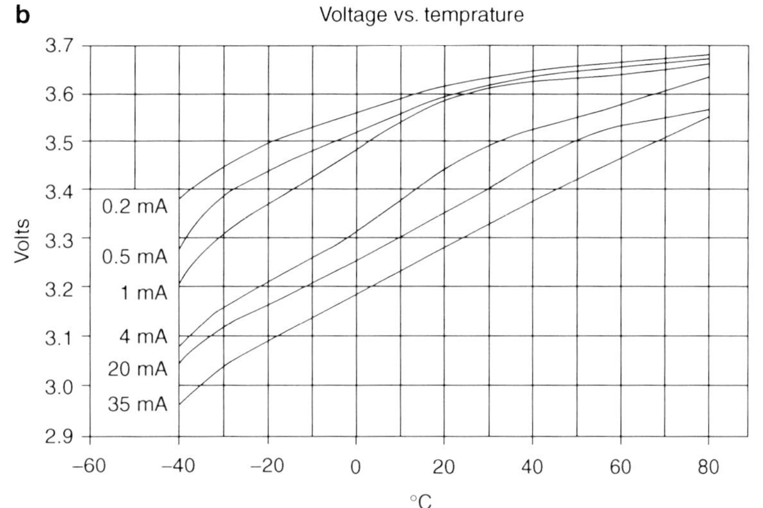

Lithium Primary Cells, Liquid Cathodes, Fig. 6 (**a**) Discharge characteristics of high-capacity Li/$SOCl_2$ cylindrical D-size bobbin cell at +25 °C. (**b**) Operating (plateau) voltage of the same battery as a function of temperature at various drain rates [4]

and an energy density of 950 Wh/l at low rate (200–300 h). These batteries have been decommissioned but still remain the largest lithium primary cells ever constructed.

Applications for Lithium/Thionyl Chloride Cells

Low-rate cylindrical cells are employed for CMOS memories, radio-frequency ID tags, utility meters, toll collection systems, and wireless security systems, among others. High-rate cylindrical cells, disk-type cells and very large prismatic batteries have been used for military applications. Use in consumer products has been limited by safety and cost considerations. Intermediate and low-rate cells are employed in oil exploration measure-while-drilling (MWD) equipment since they are capable of operating at high temperature.

Lithium/Sulfuryl Chloride Cells

In most characteristics, the Li/SO_2Cl_2 cell is similar to Li/$SOCl_2$. This cell employs a Li metal anode, a porous carbon cathode, a nonwoven glass separator, and an electrolyte of $LiAlCl_4$ in SO_2Cl_2. The cell reaction is $2Li + SO_2Cl_2 \rightarrow 2LiCl + SO_2$. The open-circuit voltage is 3.95 V, the highest for any lithium primary system.

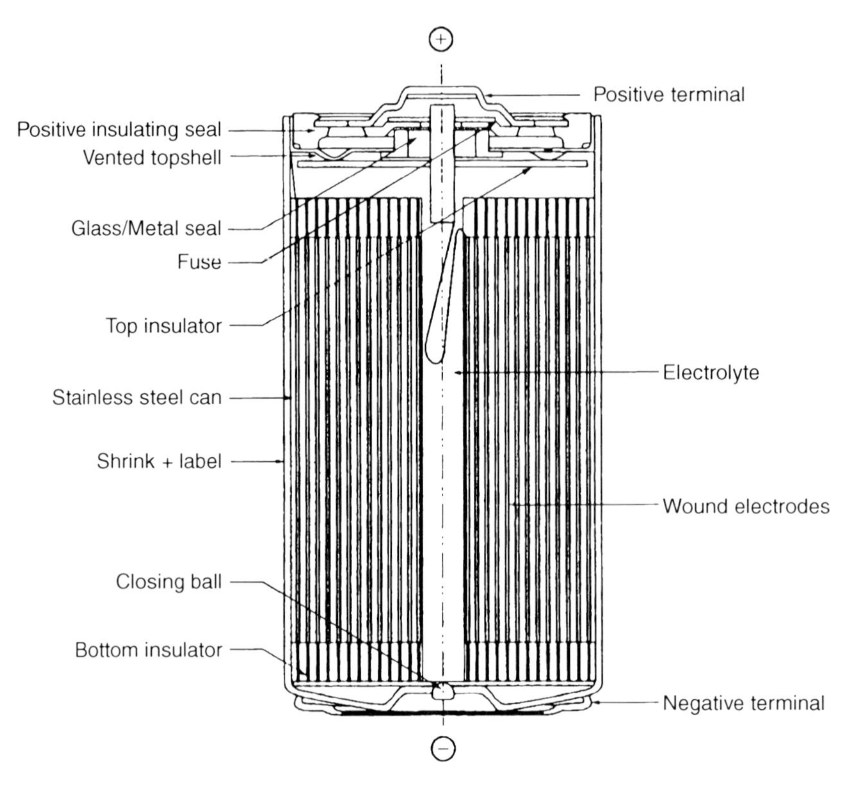

Lithium Primary Cells, Liquid Cathodes, Fig. 7 Cutaway view of lithium/thionyl chloride spirally wound electrode D-cell battery [5]

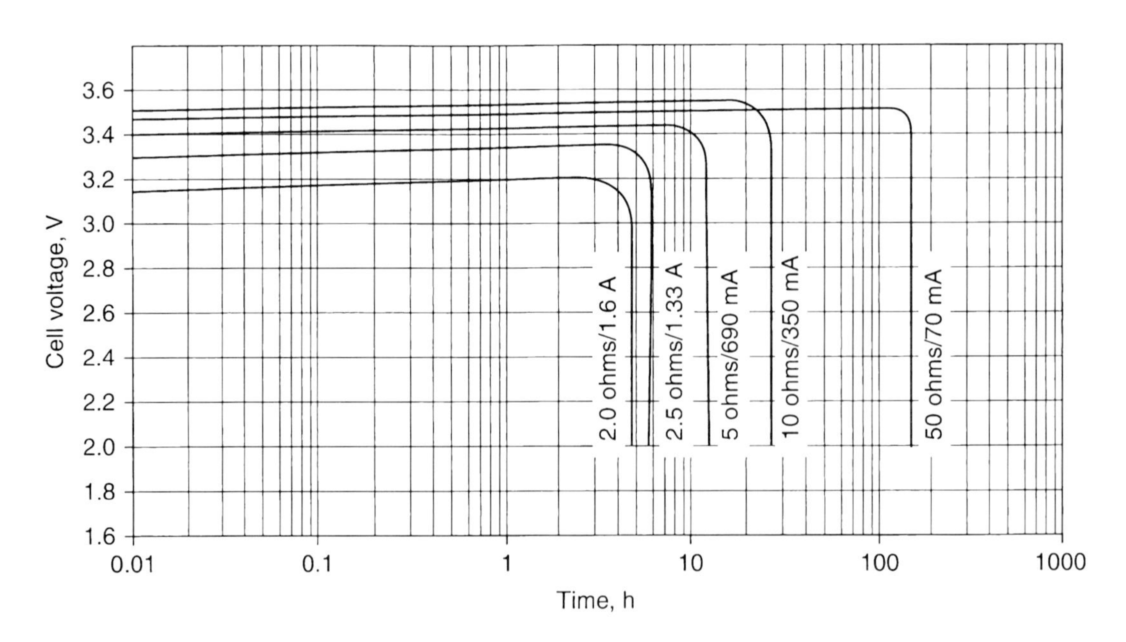

Lithium Primary Cells, Liquid Cathodes, Fig. 8 Discharge characteristics of spirally wound Li/$SOCl_2$ D-size battery, medium discharge rate at 20 °C [5]

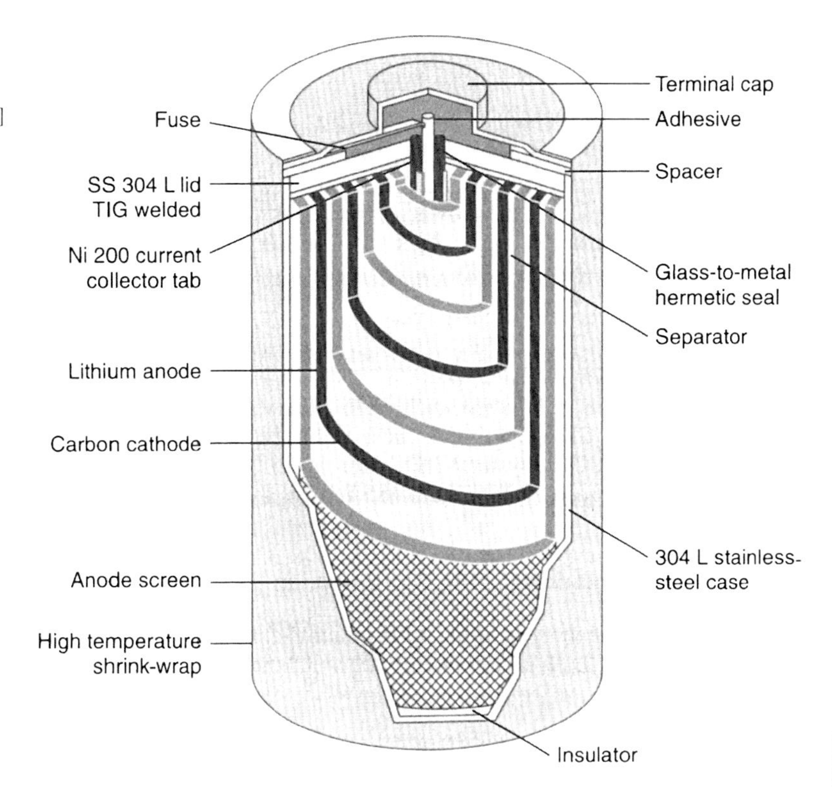

Lithium Primary Cells, Liquid Cathodes, Fig. 9 Cross section of lithium/oxychloride cell [6]

Sulfuryl chloride exists in equilibrium with chlorine and sulfur dioxide, and the presence of chlorine results in variations in voltage with temperature and during storage. For this reason, Li/SO_2Cl_2 cells have never been commercialized. This problem is reduced by the use of a chlorine additive as discussed below.

Lithium Oxyhalide Cells with Additives

Lithium/Thionyl Chloride with Bromine Monochloride (BCX)

The addition of BrCl to the thionyl chloride cell enhances cell performance to increase the OCV to 3.9 V, operate over a wide temperature range, and provide an energy density up to 1,000 Wh/l. The structure of a typical BCX cell is shown in Fig. 9. Two layers of nonwoven glass separator are employed to provide added safety since this cell does not incorporate a safety vent. The glass-to-metal seal is designed to crack if an overpressure or over-temperature condition occurs. Cells are available in sizes from AA (2.0 Ah) to DD (30.0 Ah). Typical discharge characteristics for a BCX D-cell on four loads from 3 to 75 Ω are depicted in Fig. 10. The initial part of the curve is associated with the reduction of BrCl. Discharge voltages on the lower plateau range from 3.2 to 3.5 V. Capacity loss is rated at 3 %/year at 25 °C.

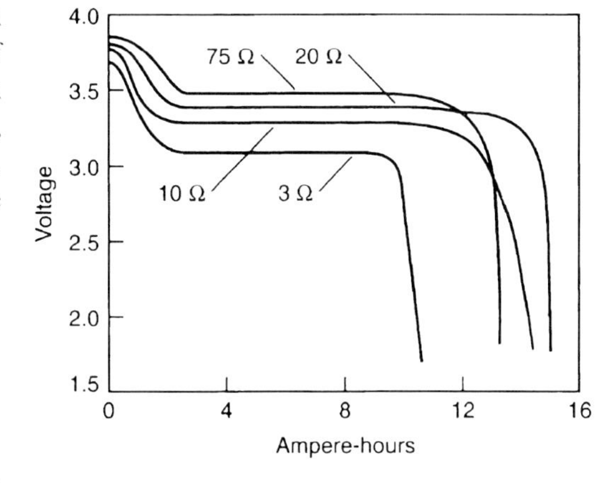

Lithium Primary Cells, Liquid Cathodes, Fig. 10 Performance characteristics of $Li/SOCl_2$ with BrCl additive. D-size batteries. Discharge characteristics at 20 °C [6]

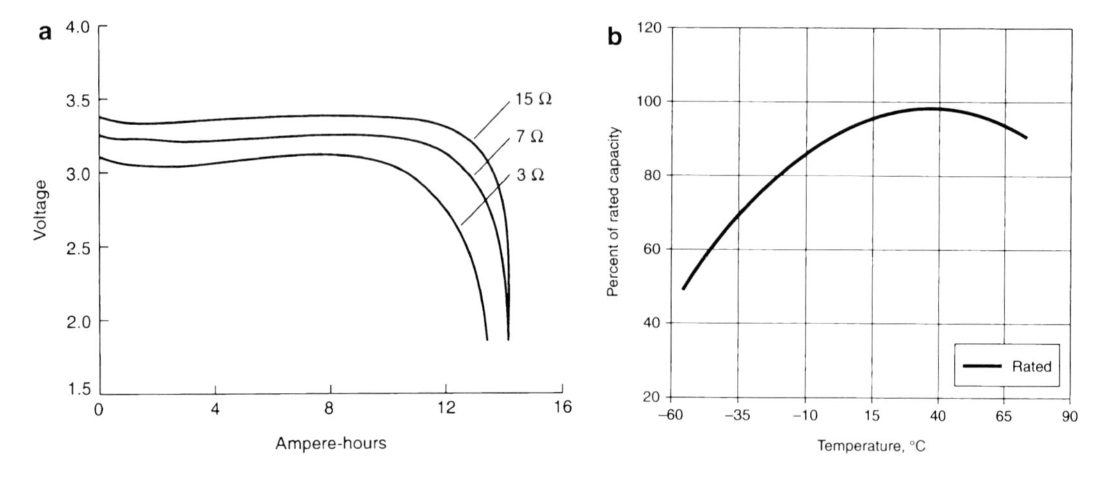

Lithium Primary Cells, Liquid Cathodes, Fig. 11 Performance characteristics of Li/SO_2Cl_2 with Cl_2 additive in D-size batteries. (**a**) Discharge at 20 °C. (**b**) Capacity vs. discharge temperature; 100 % capacity delivered at 20 °C [6]

Lithium/Sulfuryl Chloride with Chlorine (CSC)
This chemistry provides a very high OCV of 3.9 V and an energy density up to 1,050 Wh/l. The use of the chlorine additive serves to reduce the voltage delay on activation and provides a wide operating temperature range. The cell design is similar to that shown in Fig. 9. Typical discharge curves for a CSC D-cell are shown in Fig. 11a on resistive loads of 3, 7, and 15 Ω at 20 °C. No voltage delay is observed and the discharge curves are all flat, indicating good performance under these conditions. Rated capacity vs. temperature is shown in Fig. 11b. Capacity loss is rated at 3 %/year at 25 °C. CSC-type cells are available in C, D, and DD sizes with rated capacities of 7.0, 15.0, and 30.0 Ah, respectively. More detailed information on CSC-type cells can be found in Reference [7].

Both BCX and CSC Li/oxyhalide cells are employed in specialized applications such as oceanographic, space, memory backup, and industrial and telecommunications devices.

Safety Considerations

Lithium primary cells and batteries with liquid cathode reactants are all very-high-energy devices, contain toxic and flammable components, and must be treated accordingly. A detailed discussion of safety-related considerations can be found in Section 14.4 of reference [1]. Manufacturer's performance data sheets and materials safety data sheets (MSDSs) must be consulted before use of these products is considered.

Future Direction

Lithium primary batteries with liquid cathodes are a relatively mature technology. Incremental improvements in capacity and performance may occur through design modifications and the use of new materials such as improved carbons in the passive cathode. The U.S. Army is adopting Lithium/Manganese Dioxide replacements for some of the Lithium/Sulfur Dioxide Batteries listed in Table 1 in certain applications. These replacements provide higher capacity and energy at room temperature but not at lower temperature. See the chapter on Lithium Primary Cells: Solid Cathodes in this work.

Cross-References

► Lithium Solid Cathode Batteries
► Primary Battery Design

References

1. Reddy TB (2011) Lithium Primary Batteries. In: Reddy TB (ed) Linden's handbook of batteries, 4th edn. McGraw-Hill, New York, Chapter 14
2. Mathews M (2000) Proceedings of the 39th power sources conference, Cherry Hill, NJ, pp 77–80
3. Sink M (1998) Proceedings of the 38th power sources conference, Cherry Hill, NJ, pp 187–190
4. Tadiran Batteries, Port Washington, NY 11050 USA
5. SAFT Batteries, Cockeysville, MD, USA
6. Electrochem Solutions Div., Greatbatch, Inc., Raynham, MA, USA
7. Spillman DM, Takeuchi ES (1998) Proceedings of the 38th power sources conference, Cherry Hill, NJ, pp 199–202

Lithium Solid Cathode Batteries

Andrew Webber
Technology, Energizer Battery Manufacturing Inc., Westlake, OH, USA

Introduction

Several types of primary batteries have been developed that use lithium-metal anodes and solid cathodes. This entry reviews the more common commercial systems, namely Li-FeS_2, Li-MnO_2, and Li-CF_x. Readers are referred to the relevant sections for information on Li-V_2O_5 and Li-$Ag_2V_4O_{11}$ cells that are used for reserve and medical battery applications, respectively. There has been a wide range of cathodes developed in the laboratory and also marketed for specialty applications [1], but most have never been produced commercially. (Li-CuO cells were made for some military applications [2], but production was discontinued in the mid-1990s). Before going into details on the aforementioned three types mostly used in consumer applications, we will cover the main characteristics that they have in common.

General Attributes

The general characteristics of cells using lithium-metal anodes and solid cathodes are as follows:

- Extremely high energy density and especially specific energy, leading to long run times in devices
- Voltage ranges from 1.5 to 3 V, dictated by the cathode chemistry
- Flat discharge profile
- Good performance at low and high temperatures
- Very long shelf life
- In some cases, good power density
- Extremely well-sealed; cells do not leak or vent under normal use or routine misuse, even after extensive aging
- No heavy metals such as lead, cadmium, or mercury
- Non-rechargeable
- Higher cost than many aqueous systems; both the materials and the manufacturing process are more expensive
- Potential safety concerns with lithium-metal anodes and a flammable electrolyte. However, they have an excellent safety record in the field, and the safety of the solid cathode cells is generally regarded as being much better than that of lithium cells using liquid cathodes (thionyl chloride or sulfur dioxide) used primarily in industrial/military applications.

These and other characteristics have made them very well suited for industrial and military applications, as well as for consumer goods. The most common cathode chemistries used in consumer goods are 1.5 V Li-FeS_2 batteries, that are a direct replacement for alkaline AA and AAA batteries, and 3 V Li-MnO_2 cells that come in various shapes and sizes [1]. The 3 V Li-CF_x batteries are currently used mainly for memory back-up applications.

Lithium is the most electronegative anode used in batteries (standard potential of −3.05 V) and has a very low atomic weight (6.941 g/mol). The former gives the cells high voltage while the latter results in anodes with very high specific capacity (Ah/kg). This combination of high cell voltage and high specific capacity yields cells with extremely high specific energies (Wh/kg). Lithium metal has a low density (0.534 g/mL), but the larger active volume this causes is more than offset by the fact that lithium-metal cells use lithium-metal foil (or billet) as an anode so that the anode has 100 % packing efficiency. This, combined with the high specific capacity and

Lithium Solid Cathode Batteries, Table 1 Characteristics of common lithium-metal cells using solid cathodes[a]

Cathode	Nominal voltage (V)	F/Mol	Currently available battery types	Typical CCV (V)	Energy density (Wh/L)		Specific energy (Wh/kg)		Drain rate and cell size for "actual"
					Theoretical	Actual	Theoretical	Actual	
FeS_2	1.5	4	AA, AAA, CRV3	1.45	2,022	569	1,052	309	200 mW AA
MnO_2	3	1	2/3A, AA, 1/2AA, CR2, 2CR5, CRV3, 9 V, various miniature cells. Larger cells for military and industrial applications	2.8	2,380	499	799	228	200 mW 2/3A
CF_x	3	1	2/3A, AA, various miniature cells. Larger cells for military and industrial applications	2.75	2,979	515	1,942	290	2.7 mA 2/3A

[a]The theoretical energy density and specific energies are calculated using the typical closed circuit voltage (CCV) values rather than the OCV or nominal voltage often used. "Actual" data are representative examples only for some cylindrical batteries; there can be considerable variability in energy density with discharge rate and among different cell sizes and manufacturers. Note that the Li/CF_x drain rate is much lower than that for the other cell systems shown

high cell voltage, ensures that these cells also deliver very high energy density (Wh/L), on a volume basis. Table 1 shows some key characteristics of the three systems.

Lithium-metal batteries exhibit a very flat discharge profile that ensures reliable device operation throughout the battery discharge, although the shape of the discharge curve can make it more difficult for some devices to reliably judge the remaining run time.

Lithium batteries using solid cathodes typically use a thin lithium-metal foil or disk as the anode; a transition metal oxide, metal sulfide, or a fluoride as the cathode; and an organic electrolyte. The cathode material is either coated onto a foil substrate (often aluminum) or embedded into an expanded metal or perforated metal substrate. The anode and cathode are separated by a microporous plastic membrane separator, usually polyethylene and/or polypropylene.

Since lithium can react violently with water, the cells are completely sealed and are made in dry rooms/dry boxes where humidity is strictly controlled to protect the lithium; this is necessary to ensure a quality product and maintain a safe production facility. Lithium-metal cells with solid cathodes all use aprotic, nonaqueous electrolytes, and this is one reason why they can perform extremely well at low temperatures (typically −20 °C, and as low as −40 °C). These electrolytes are typically a blend of ethers (commonly 1,2-dimethoxyethane or 1,3-dioxolane) and carbonates (propylene, ethylene, and butylene carbonate) or other high-boiling point solvents (sulfolane and γ-butyrolactone) with a lithium salt, such as LiI, $LiClO_4$, $LiCF_3SO_3$, $Li(CF_3SO_2)_2N$, $LiBF_4$ and in one case $LiAsF_6$.

Generally, the electrolytes are relatively innocuous and non-corrosive and this is one reason why they are supplanting liquid cathode cells from many applications. However, they cannot match the conductivity of aqueous systems. In addition, lithium-metal anodes, while compact, have a relatively low surface area. Partly for these reasons, the systems have poor-to-moderate *inherent* rate capability and bobbin construction or miniature cells of this type are usually only capable of delivering μA currents. However,

Lithium Solid Cathode Batteries, Fig. 1 Cross-sectional view of an AA Li-FeS_2 cell showing the jellyroll construction (Courtesy of Energizer Battery Manufacturing Inc.)

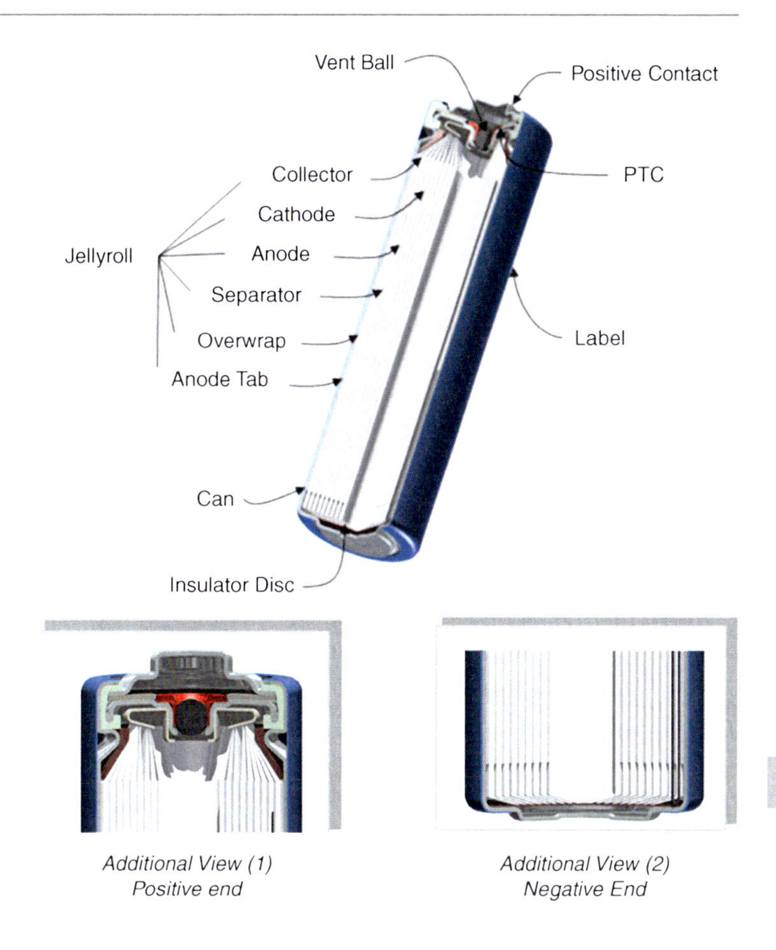

these limitations can be overcome using high interfacial area designs (Fig. 1) and some of these cells can greatly outperform aqueous batteries on high drain applications.

The electrolyte is thermodynamically unstable to lithium, but the cells actually exhibit extremely long shelf life (15–40 years). This is because the lithium reacts with the electrolyte instantly to form a very thin, ionically conducting, protective layer on the surface, called the solid-electrolyte interphase or SEI layer. Moreover, the cells do not normally generate gases as can be the case for some aqueous-based battery systems. Typically, lithium-metal cells with solid cathodes use a plastic gasket, often polypropylene, that is compressed between metal components to form an air-tight, leak-resistant seal. Partly for these reasons, the cells also perform well at high temperatures, up to 60 °C. Some specialty cells can operate at even higher temperatures, although usually with significant trade-offs, such as poor performance at high rate or low temperature.

During discharge, the lithium-metal anode is oxidized to form lithium ions in the electrolyte. These then migrate through the separator to the cathode whereupon they react with the cathode to form the final products. Thus, at the end of discharge, the lithium-metal anode has almost disappeared while the cathode has expanded and for the most part filled up the space vacated by the discharged anode.

Specific Systems

The three main commercial systems are Li-FeS_2, Li-MnO_2, and Li-CF_x. The Li-MnO_2 market is still the largest of these systems, although the Li-FeS_2 has been the fastest growing. The Li-CF_x system has received renewed interest because new forms of the cathode have recently been developed to overcome some of the shortcomings of conventional Li-CF_x cells.

Li-FeS_2 Batteries

Li-FeS_2 cells are the most recent of the three lithium primary systems used in consumer devices. They were introduced by Energizer, first as button cells in mid-1970s and later as AA cells in 1989 and AAA cells in 2004 [3]. Currently, Li-FeS_2 cells are manufactured in AA, AAA, and CRV3 sizes and all employ a jellyroll construction (Fig. 1). Li-FeS_2 batteries operate at a relatively low voltage (~1.5 V), but more than compensate for this by having a cathode that provides 4 F/mol during discharge. This uniquely high capacity results from discharging both the iron and sulfur species. The discharge reaction at very low drain rates and/or high temperatures proceeds by a two-step reduction [4–7]. In the first step, an intermediate Li_2FeS_2 is formed and then this further reduces to the final products, Li_2S and elemental iron, as shown below.

Anode Reactions:

$$Li \rightarrow Li^+ + e^-$$

Cathode Reactions Slow Discharge:

$$FeS_2 + 2\ Li^+ + 2\ e^- \rightarrow Li_2FeS_2$$

$$Li_2FeS_2 + 2\ Li^+ + 2\ e^- \rightarrow 2\ Li_2S + Fe$$

In the first step, persulfide is reduced to sulfide, and in the second step, iron (II) is reduced to metallic iron. Each of the steps represents a two-electron reduction. However, the second step is kinetically faster than the first and under most applications, the reduction appears to be a single four-electron reduction, as summarized below. The reaction still follows the sequence shown above for slow discharge; it is just that the intermediate Li_2FeS_2 discharges as quickly as it is formed.

Cathode Reaction Normal Discharge:

$$FeS_2 + 4\ Li^+ + 4\ e^- \rightarrow 2\ Li_2S + Fe$$

This leads to different voltage profiles depending on the drain rate and temperature, as shown in Fig. 2, although it must be emphasized that under the majority of standard tests, the discharge follows the single-step route and exhibits a single, flat discharge plateau. At room temperature, the two-step discharge is seen if the current drain is 10 mA or below for the AA cell size. Note that the final product incorporates highly conductive, metallic iron into the cathode that helps the cell maintain its high rate capability throughout the entire discharge process. During discharge, the cathode expansion more than compensates for the loss in anode volume. Consequently, the cell electrode stack expands and this must be factored into the cell design to ensure safe and reliable operations [8].

Li-FeS_2 cells are unique in that they are the only lithium batteries suitable for the ubiquitous world of 1.5 V devices and as such are direct, "drop-in" replacements for everyday alkaline, carbon/zinc, and rechargeable nickel metal hydride (NiMH) batteries. The four-electron discharge provides more energy on medium to high discharge rates than any other system in an AA-size or similar cell [8–10]. Other systems such as liquid cathode cells and Li-CF_x cells may provide greater energy, but only at far lower drain rates.

Commercial Li-FeS_2 cells use natural pyrite as the cathode. They are also manufactured in very high volume, and partly for these reasons, Li-FeS_2 cells retail at much lower costs than other lithium cells. This is a critical feature as they must compete against low cost alkaline batteries. The jellyroll construction, especially when combined with an all-ether based electrolyte [11], results in an extremely high power capability (Fig. 3) [3, 8–10]. This, along with the higher operating voltage than alkaline cells, makes Li-FeS_2 far more powerful than alkaline cells (Fig. 4). Consequently, they are especially well suited to high rate applications and many high tech devices that make full use of the higher operating voltage of the system. In particular, they are the primary battery of choice for digital still cameras where, despite their higher unit price, Li-FeS_2 cells provide far better performance (up to eight times for some cells) at

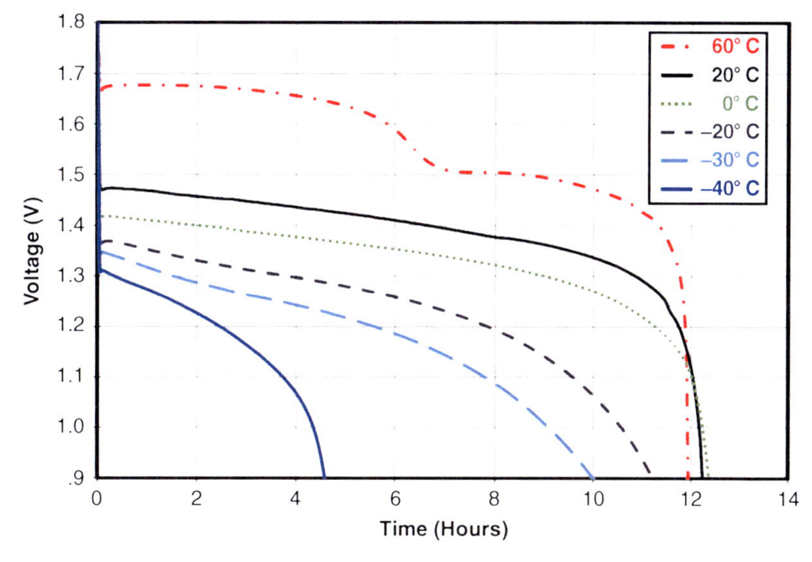

Lithium Solid Cathode Batteries, Fig. 2 Effect of temperature on AA Li-FeS_2 discharge curve, 250 mA constant current (courtesy of Energizer Battery Manufacturing Inc.)

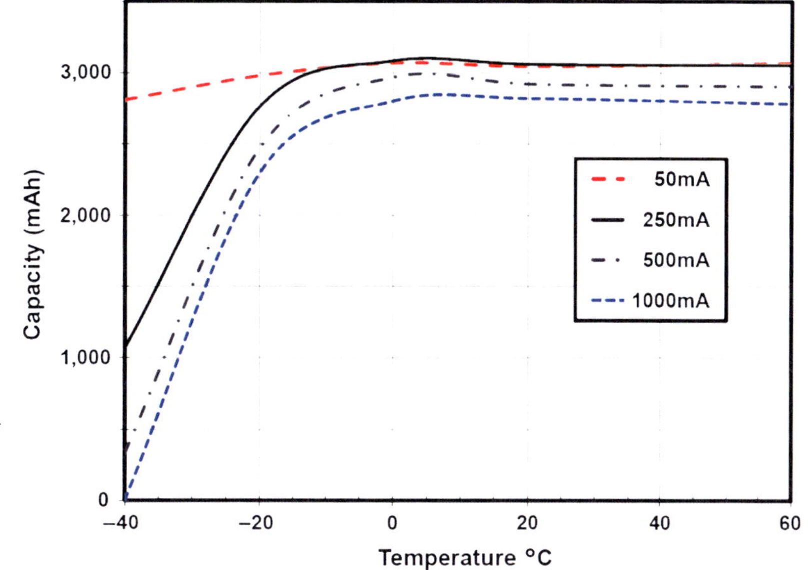

Lithium Solid Cathode Batteries, Fig. 3 Effect of temperature and drain rate on AA Li-FeS_2 capacity to a 0.9 V cut (courtesy of Energizer Battery Manufacturing Inc.)

a lower cost/picture [12]. Note that the low density of lithium metal results in AA cells that weigh about 37 % less than comparable alkaline cells [3, 8, 10]. Their low weight has made Li-FeS_2 cells very popular for night vision goggles, binoculars, and other multicell applications.

At low drain rates, they actually have similar capacity to alkaline cells, but still provide significantly more energy (Fig. 4). At high drain rates AAA Li-FeS_2 cells provide more energy than even the much larger and heavier alkaline AA cells. AA Li-FeS_2 cells have also been used to replace C- and D-size alkaline batteries through device redesign or by using upsizers in several devices, such as trail cameras [10].

These advantages over alkaline become even more pronounced at low temperature (Fig. 5). Li-FeS_2 cells perform extremely well even at high rates down to −20 °C and at low and moderate drain rates to −40 °C [3, 10]. Li-FeS_2 cells that use an all-ether based electrolyte outperform other lithium solid cathode cells at low

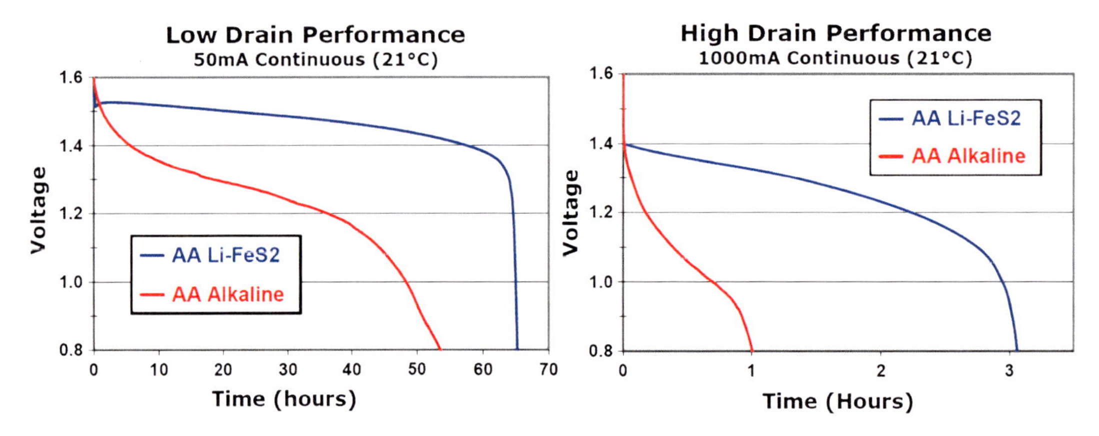

Lithium Solid Cathode Batteries, Fig. 4 Comparison of 1.5 V primary cell systems – AA alkaline Zn-MnO_2 versus Li-FeS_2 (courtesy of Energizer Battery Manufacturing Inc.)

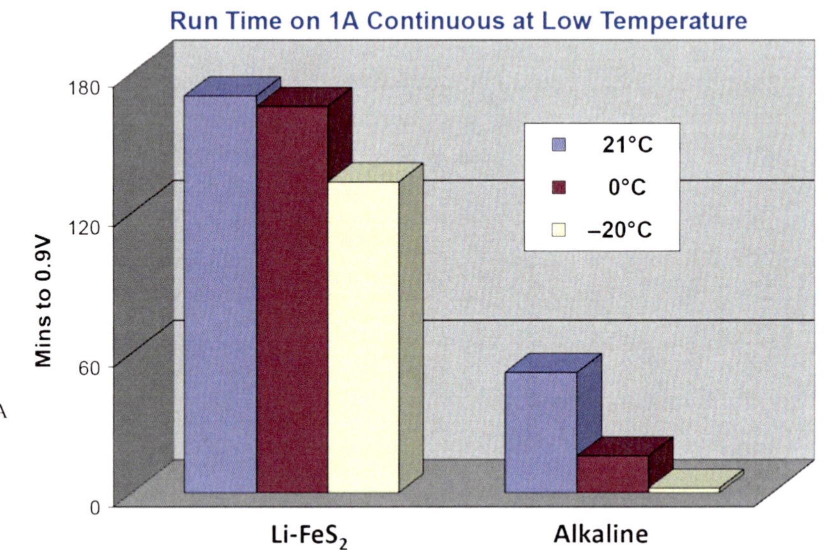

Lithium Solid Cathode Batteries, Fig. 5 Comparison of AA Li-FeS_2 cells and alkaline cells at low temperature (courtesy of Energizer Battery Manufacturing Inc.)

temperatures, in part because this electrolyte has been shown to *increase* in conductivity as the temperature decreases, in contrast to conventional electrolytes [3]. Advances in electrolytes, materials, formulation, and processing have led to a steady increase in high rate capacity for these cells over the last decade (Fig. 6).

A combination of 15-year shelf data, extended high-temperature storage at 60 °C, and microcalorimetry studies show that the Li-FeS_2 cells have an ambient shelf life of over 30 years [10, 13]. For example, there is almost no discernible loss in capacity or voltage after over 15-year ambient storage (Fig. 7). There is no evidence of voltage dips or passivation either before or after this extended storage, even on a high rate 1A test. They can withstand storage and operation up to 60 °C; service loss after 1 year at 60 °C is typically only a few percent (Table 2).

Li-FeS_2 cells have had an outstanding safety record in the field as a result of numerous redundant safety features and sophisticated production processes that incorporate 100 % inspection at multiple stages of production [8]. Their excellent safety performance, even under abuse conditions, was one of the reasons why Energizer's AA Li-FeS_2 cells were chosen by NASA to power impact sensors in the wings of the space shuttle

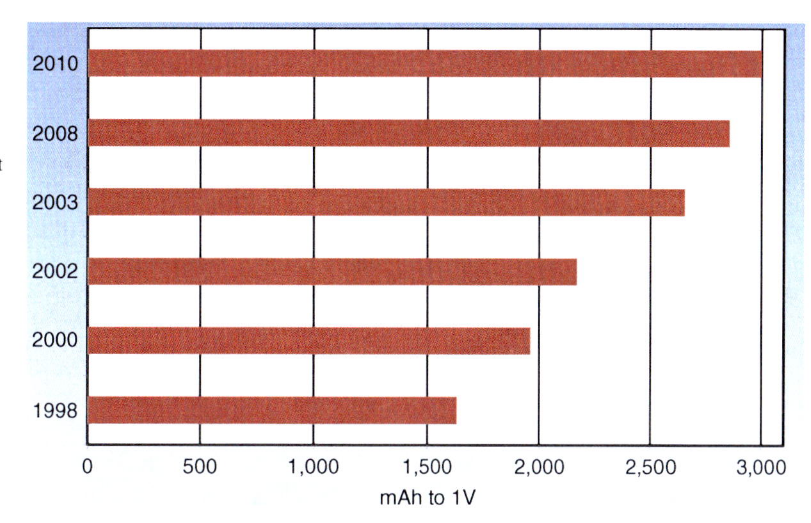

Lithium Solid Cathode Batteries, Fig. 6 Improvements in AA Li-FeS_2 performance since 1998. Capacity on a 1,000 mA continuous test (courtesy of Energizer Battery Manufacturing Inc.)

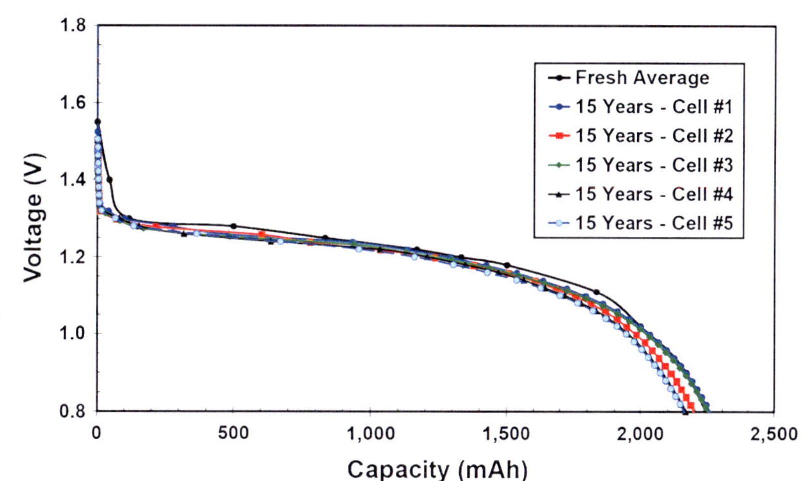

Lithium Solid Cathode Batteries, Fig. 7 High rate discharge test of AA Li-FeS_2 cells fresh and after 15-year ambient storage. Note that current versions of this cell deliver over 3,000 mAh (courtesy of Energizer Battery Manufacturing Inc.)

(about 100 were used for each flight) [14–16]. Significant improvements in capacity, low temperature performance, and reliability have been demonstrated in prototype cells [3, 8–10], while at the same time passing all standard safety and abuse tests.

Li-MnO_2 Batteries

Li-MnO_2 batteries also provide excellent energy density, long shelf life, and decent rate performance. Li-MnO_2 cells have a high operating voltage, 2.7–2.9 V, but the discharge reaction is only 1 F/mol, as shown below. The cathode active material used in making the cell is heat-treated electrolytic manganese dioxide (EMD); the heat treatment and the original EMD used dictate battery performance [17, 18]. The cathode discharge reaction involves the reduction of Mn(IV) to Mn(III).

Anode Reaction:

$$Li \rightarrow Li^+ + e^-$$

Cathode Reaction:

$$MnO_2 + Li^+ + e^- \rightarrow LiMnO_2$$

The high voltage is an asset when high pack voltages are needed, although recent advances in inexpensive, efficient DC/DC converters have

Lithium Solid Cathode Batteries, Table 2 Service of AA Li-FeS_2 cells after 1-year 60 °C storage as a percentage of fresh performance (data provided by Energizer Battery Manufacturing Inc.)

Discharge test	Service maintenance after 1 year at 60 °C
Digital still camera test	93 %
1A constant current	95 %
1 W constant power	95 %
250 mA 1 h/day	95 %

rendered this somewhat moot. A disadvantage for the high voltage is that they cannot be used in billions of everyday devices designed around the 1.5 V platform. Nevertheless, the Li-MnO_2 market is largest of the lithium primary systems and cells are available in a wide range of types, such as coin cells (CR2016, CR2032), cylindrical (CR123 or 2/3A, CR223, 2CR5, CR2, and CRV3), and a prismatic 9 V version. Consumer cells often have the "CR" designation.

The cylindrical cells use a jellyroll construction similar to that already described for Li-FeS_2 cells (Fig. 1). However, the cathode is typically thicker and embedded into a stainless steel Exmet screen rather than coated onto foil. Specialty versions of cylindrical cells using a bobbin construction are also manufactured; these cells have higher capacity, but can only discharge at very low drain rates. Miniature cells use much of the same materials, but the electrodes are punched out and stacked together with a disk of separator material (Fig. 8). The 9 V battery uses a system of three unit cells wired in series using folded electrodes and a common case (Fig. 9).

Li-MnO_2 cells were one of the earliest lithium cells developed using solid cathodes. Originally, they were aimed at powering wrist watches (coin cells) and 35-mm film cameras (cylindrical). Coin cells have seen considerable growth in recent years; in addition to their traditional usage in watches and calculators, they are becoming common in toys (sound and light effects), small lights, and remote controls. At the same time, cylindrical cell usage for Li-MnO_2 in the consumer area has been adversely affected by the shift from film-based to digital cameras. Cylindrical cells are used in some industrial applications and also for clocks and some high-end lights, where the voltage is high enough to power an LED light using a single cell, although again, this can now be done with lower voltage cells using a DC/DC converter. There are extensive programs in place to develop and deploy Li-MnO_2 battery packs for the military, mainly as a replacement for more hazardous Li-SO_2 packs.

The expanding use of Li-MnO_2 coin cells in consumer devices has raised concerns about the hazard of swallowing these cells, mostly by small children. In rare cases, the battery may become lodged in the esophagus whereupon the high voltage of these batteries can generate caustic materials in the saliva, leading to serious injuries and even death [19]. This has led to industry and government initiatives to warn the public and medical professionals about the dangers of swallowing batteries and the need to address any cases of battery ingestion promptly [20–22]. Moreover, steps are being taken to restrict access to battery compartments of toys and many other devices by small children [22, 23].

The 9 V prismatic battery design leverages the high voltage of the cells to generate 9 V output using only three cells instead of the six alkaline cells in a conventional 9 V battery. The Li-MnO_2 9 V batteries are targeted mainly at smoke alarms and industrial/medical/security applications. Much larger D-size Li-MnO_2 cells have also been developed, mainly for military use.

Li-MnO_2 cells have decent rate capability, in part because of the conductive nature of the cathode material. Performance at high rate and/or low temperature is usually much better than Li-CF_x cells, although neither can match the high rate or low temperature performance of Li-FeS_2 cells. Discharge curves for Li-MnO_2 cells are very flat and do not normally exhibit any voltage dips when first placed on load (Fig. 10).

Most commercial Li-MnO_2 cells also operate over a wide temperature range, typically −20 °C to 60 °C (Fig. 11). They can also run at lower temperatures, down to −40 °C, but usually only at low discharge rates. In some cases, cells can be used up to 70 °C with caution. While low

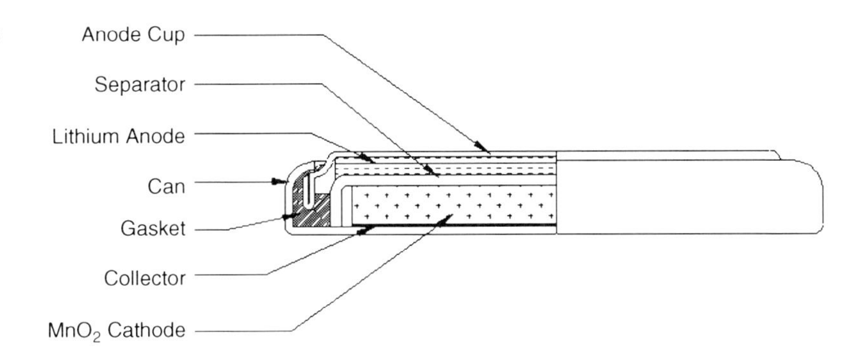

Lithium Solid Cathode Batteries, Fig. 8 Construction of a Li-MnO_2 coin cell (courtesy of Energizer Battery Manufacturing Inc.)

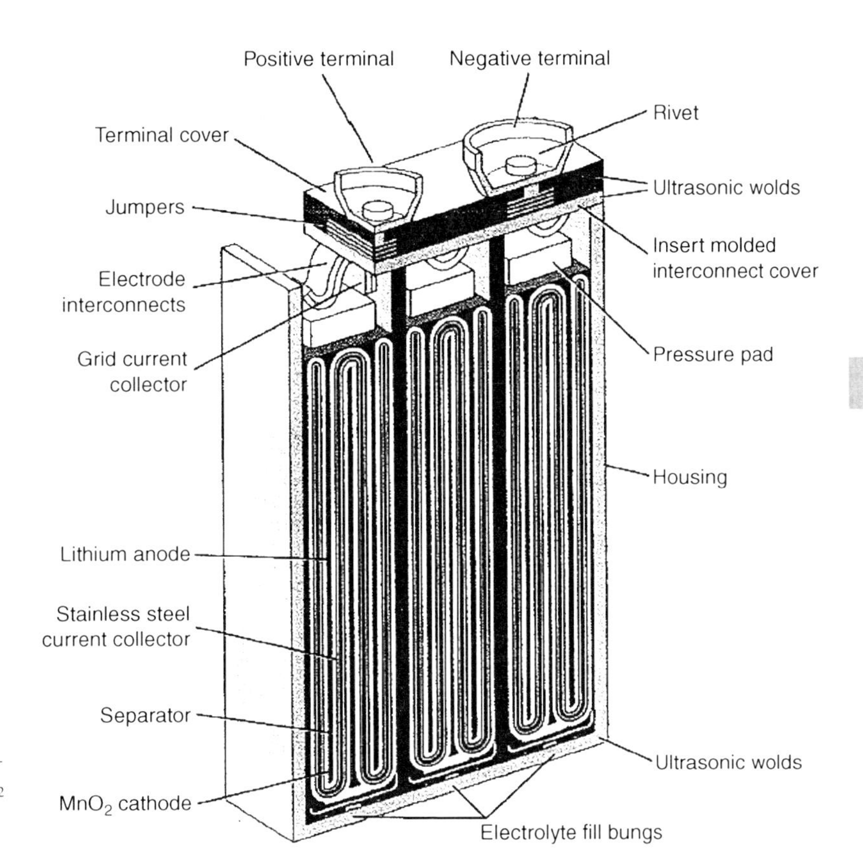

Lithium Solid Cathode Batteries, Fig. 9 Cross-section of a 9 V Li-MnO_2 battery (courtesy of Ultralife Corp.)

temperature performance is often not quite as good as some other lithium systems, it is more than adequate for almost all consumer and some military/industrial applications. Moreover, specialized versions for military and industrial applications, mainly large cylindrical cells (D-size), have been developed with improved low temperature service [24].

Shelf life is also excellent with Li-MnO_2 cells; service maintenance has been reported to be as high as 97 % after 5-year ambient storage. Unlike Li-FeS_2 and Li-CF_x cells, the Li-MnO_2 electrode stack does not show a net expansion during discharge. This makes the MnO_2 system well suited for coin cells and pouch cells, where it can be difficult to constrain expansion. The cost of Li-MnO_2 cylindrical cells lies between that of Li-FeS_2 and Li-CF_x cells.

$$\text{Cost: Li} - FeS_2 < \text{Li} - MnO_2 < \text{Li} - CF_x$$

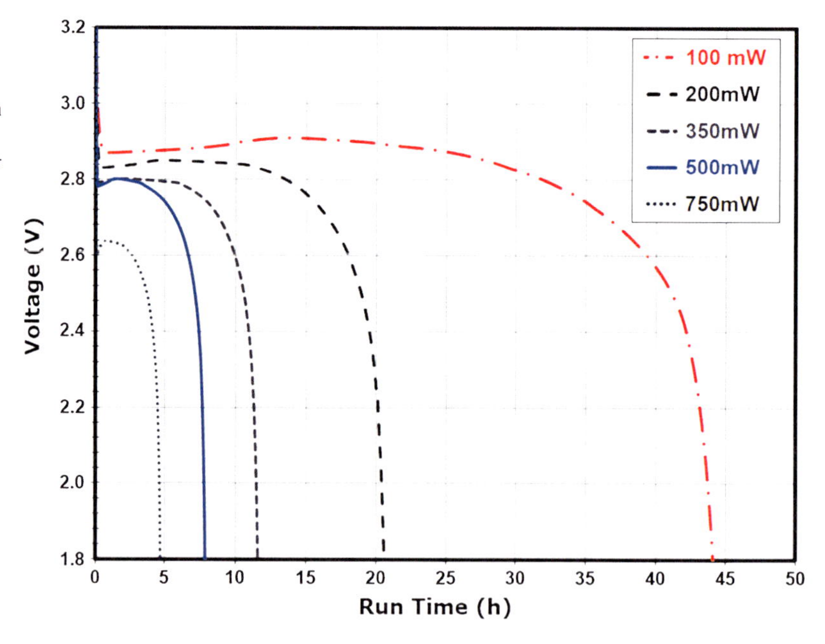

Lithium Solid Cathode Batteries, Fig. 10 Discharge curves of CR123 Li-MnO_2 cells on various constant power tests (courtesy of Energizer Battery Manufacturing Inc.)

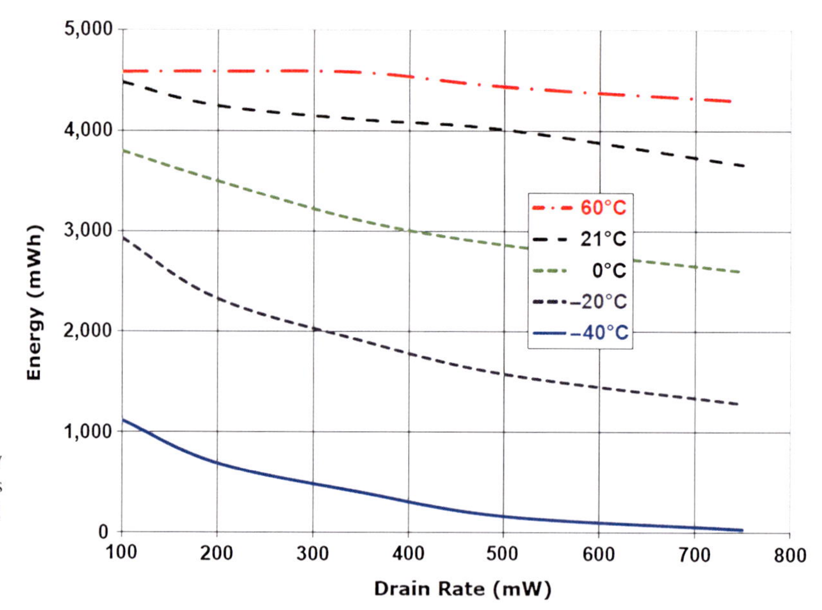

Lithium Solid Cathode Batteries, Fig. 11 Energy of CR123 Li-MnO_2 cells as a function of drain rate and temperature (courtesy of Energizer Battery Manufacturing Inc)

Li-CF_x Batteries

Li-CF_x batteries are another 3 V lithium chemistry that provides excellent energy density and long shelf life, but historically poor rate and low temperature performance. CF_x cells have a high operating voltage, 2.6–2.9 V, but the discharge reaction is normally only one F/mol, as shown below. CF_x is a synthetic material made by a high-temperature reaction of fluorine gas with graphite. CF_x has a covalent structure with C-F bonding, sp2 and/or sp3, depending on the value of x and the reaction conditions used for the synthesis [25]. The discharge reaction shown is a simplification; the reaction actually involves incorporation and subsequent release of solvent molecules from the cathode and LiF formation, dissolution, and recrystallization [26, 27]. The role of the solvent is one reason why the CF_x

cathode expands considerably during discharge and Li-CF_x cells must be designed to handle this expansion.

Anode Reaction:

$$Li \rightarrow Li^+ + e^-$$

Cathode Reaction:

$$CF_x + x\ Li^+ + x\ e^- \rightarrow x\ LiF + C$$

Note that care must be taken when synthesizing CF_x as deflagrations are possible due to a delicate interplay of temperature, gas formation, and pressure [25]. This synthesis makes CF_x an expensive material, and this has been a serious drawback in its use as a primary battery material. Li-CF_x cells are mainly used for specialty applications where their performance benefits can outweigh their costs.

Recent advances have split this field into "conventional" CF_x cells and cells using subfluorinated material in an attempt to address some of the performance issues of conventional CF_x cells. For clarity, we will specify the conventional material ($x = 1$) as CF_x and the subfluorinated materials as $SFCF_x$ [28].

Advantages and Disadvantages of Conventional Li-CF_x Cells

The relatively low molecular weight of CF_x gives it a high specific capacity for a 1 F/mol reduction. This, combined with its high electrochemical potential, gives the Li-CF_x system very high theoretical specific energy (Wh/kg, see Table 1) that results in cells with long run times when discharged at low rate. The low density of CF_x makes the system somewhat less attractive on a volume basis, especially when one factors in the additional non-active carbon needed in the cathode to overcome the poor conductivity of CF_x. The system's main advantage is its very high specific energy that is an especially important benefit for large batteries where low weight can be a critical feature. It is less important for smaller cells sizes used in most consumer devices. As with Li-MnO_2 cells, the high cell voltage is an asset when high pack voltages are need, although recent advances in inexpensive DC/DC converters have rendered this somewhat moot. A significant disadvantage for the high voltage is that they cannot be used in billions of everyday devices designed around the 1.5 V platform, but Li-CF_x cells are currently used mainly in industrial applications and for memory back-up in electronic devices where the voltage platform is less of an issue.

Li-CF_x cells are available in a wide range of types under the BR designation: coin cells (e.g., BR2032), cylindrical (BR-A, BR-2/3A, BR-1/2AA and even a C-size version). Larger cell sizes (D-size) have been developed for military/industrial applications. Li-CF_x cell construction is basically similar to that used for Li-MnO_2 cells.

Apart from high energy (long run time), the main advantages for Li/CF_x cells are low weight, excellent shelf characteristics, and the ability to operate at relatively high temperatures [29]. These characteristics have made them very successful for memory back-up operations for computers and other electronics, both in the consumer and the industrial/military field. Some versions using a nonvolatile electrolyte ($LiBF_4$ in γ–butyrolactone) are available that are rated to 85 °C and even 100 °C. However, those cells designed for high temperature pay a significant additional penalty in discharge rate capability and low temperature performance due to use of such a viscous electrolyte.

Under low drain rates, Li-CF_x batteries provide a flat discharge profile (Fig. 12) and the energy delivered under those conditions exceeds that for other cells systems. However, the conductivity of CF_x used to make these cells rapidly decreases as x approaches or exceeds one as the carbon-fluorine matrix loses its aromaticity and the carbons shift from sp2 to sp3 configurations. Conventional Li-CF_x cells use $CF_{x\sim1}$ with very low conductivity and consequently, they have poor rate capability, far worse than Li-MnO_2 or Li-FeS_2 cells. Studies have shown that the high impedance of Li-CF_x cells is linked to the cathode [26]. Moreover, the rate limitation becomes aggravated at low temperatures, such that Li-CF_x cells are typically not recommended for temperatures below −20 °C and even at −20 °C,

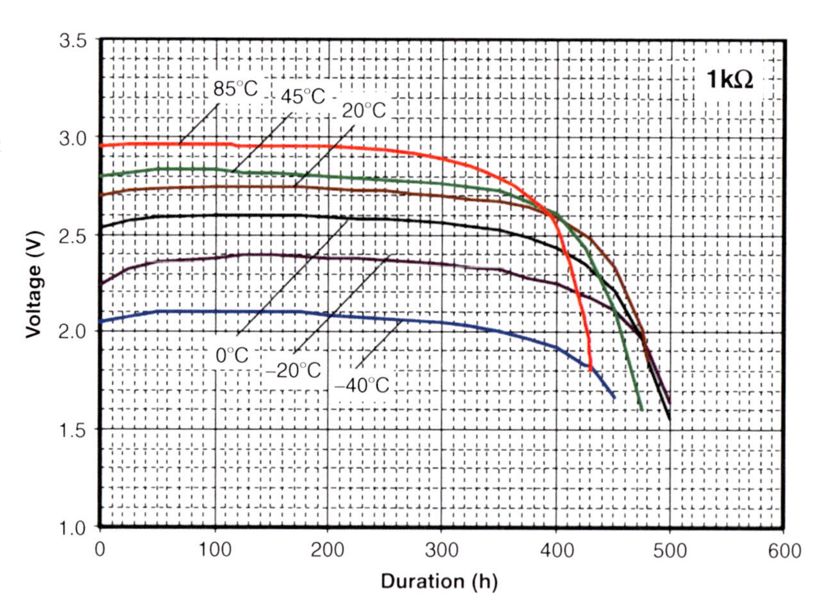

Lithium Solid Cathode Batteries, Fig. 12 Discharge profile of a BR-2/3A battery on a 1,000 Ω constant load test at various temperatures (Courtesy of Panasonic Corporation). This test corresponds to a 2–3 mA drain rate

operating voltage is suppressed. In addition, one may see signs of a front-end voltage dip at low temperature and/or higher discharge rates, although this is not usually observed in most of the applications for which this system is currently used.

Larger batteries being developed for the military use special formulations and electrolytes [28, 30] intended to alleviate these characteristics, with varying degrees of success. In particular, a concept originally proposed by Energizer [31, 32] of mixing MnO_2 with CF_x is being extensively evaluated by many companies [33–35], both for performance and cost reasons. Another consequence of the high resistivity of the CF_x cathode is that the cells can become hot during discharge at higher rates. This is especially a concern for larger cells and is another reason why the hybrid approach is being investigated. The self-heating is not normally an issue for the smaller commercial cell sizes.

Li-$SFCF_x$ Cells

As mentioned above, there has recently been a resurgence of interest in this system mainly because of improvements in rate capability that have been achieved by using subfluorinated materials, along with modifications to the synthesis methods [29, 36, 37]. In the subfluorinated materials, $SFCF_x$, the F:C ratio is less than one, an example being $CF_{0.65}$. (Note that these and conventional CF_x materials actually have a polymeric structure not reflected by the formulae). By subfluorinating the material, it retains more of the sp2 character; therefore, $SFCF_x$ has significantly higher conductivity than CF_x. In addition, the reaction kinetics are also enhanced. While a downside is that the total cathode capacity decreases as the fluorine content drops, this approach can lead to an overall increase in energy density in practical applications. Essentially, this approach trades a small amount of input capacity against a substantial gain in cathode efficiency and utilization [29].

The benefits of this approach on some tests are very marked. Proponents of this technology have claimed a major boost in power density (Fig. 13) and extended the system operating temperature down to −60 °C [38]. Although the cathode material is still very expensive, research efforts appear to have helped to lower the cost of the synthesis, boosting the performance/cost ratio. It will be interesting to see whether this technology will succeed in the price-sensitive, consumer marketplace. Currently, it is being targeted at coin cells for 3D television glasses that have relatively high power demands for a coin cell.

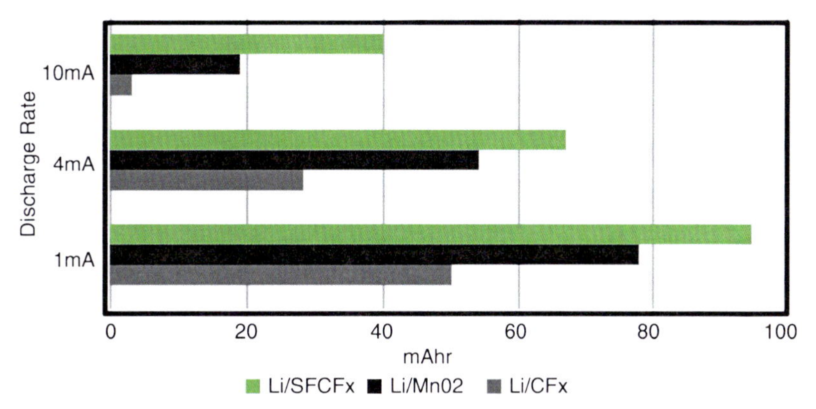

Lithium Solid Cathode Batteries, Fig. 13 Capacity to 2.0 V for 2016 coin cells at ambient temperature showing the higher rate capability of the Li-$SFCF_x$ cells over the other coin cells (Courtesy of Contour Energy Systems)

Cross-References

- ▶ Aerospace Applications for Primary Batteries
- ▶ Alkaline Primary Cells
- ▶ Conductivity of Electrolytes
- ▶ Lithium Primary Cells, Liquid Cathodes
- ▶ Ni-Metal Hydride Batteries
- ▶ Primary Batteries for Medical Applications
- ▶ Primary Batteries for Military Applications
- ▶ Primary Batteries, Comparative Performance Characteristics
- ▶ Primary Batteries, Selection and Application
- ▶ Primary Battery Design
- ▶ Role of Separators in Batteries

References

1. Reddy TB (2011) Linden's handbook of batteries, 4th edn. McGraw Hill, New York
2. Bates R, Jumel Y (1983) Lithium-cupric oxide cells. In: Gabano J-P (ed) Lithium batteries. Academic, London, pp 73–95
3. Webber A (2004) Improvements in Energizer's L91 Li-FeS_2 AA cells. In: Proceedings 41st Power Sources Conference, Philadelphia, 14–17 June 2004, pp 25–28
4. Fong R, Dahn JR, Jones CHW (1989) Electrochemistry of pyrite-based cathodes for ambient temperature lithium batteries. J Electrochem Soc 136:3206–3210
5. Jones CHW, Kovacs PE, Sharma RD, McMillan RS (1991) An ^{57}Fe Mössbauer study of the intermediates formed in the reduction of FeS_2 in the Li/FeS_2 battery system. J Phys Chem 95:774–779
6. Shao-Horn Y, Horn QC (2001) Chemical, structural and electrochemical comparison of natural and synthetic FeS_2 pyrite in lithium cells. Electrochim Acta 46:2613–2621
7. Shao-Horn Y, Osmialowski S, Horn QC (2002) Reinvestigation of lithium reaction mechanisms in FeS_2 pyrite at ambient temperature. J Electrochem Soc 149:A1547–A1555
8. Marple JW, Feddrix FH (2008) Energizer's lithium iron disulfide commercial batteries: continuous service improvement enabled by a dedicated focus on safety and reliability. In: Proceedings 43rd Power Sources Conference, Philadelphia, 7–10 July, pp 557–560
9. Feddrix FH (2009) Advances in lightweight, AA Li-FeS_2 primary cells: improvements in energy density and reliability. Tactical Power Sources Summit, Washington, DC, 26–28 January
10. Webber A, Marple JW, Zheng G, Huang W, Feddrix FH (2010) Energizer's lithium iron disulfide commercial batteries: high specific energy and reliability under extreme conditions. In: Proceedings 44th Power Sources Conference, Las Vegas, 14–17 June, pp 417–419
11. Webber A (1996) Nonaqueous cell having a lithium iodide-ether electrolyte. US Patent 5514491
12. Drier T (2006) The new power generation. PC Magazine, May 3 http://www.pcmag.com/article2/0,2817,1953699,00.asp. Last accessed 22 Nov 2011
13. Webber A, Kaplin DA (2004) Extended shelf life of Energizer L91 Li-FeS_2 AA cells. In: Proceedings 41st Power Sources Conference, Philadelphia, 14–17 June, pp 53–56
14. Jeevarajan JA, Baldwin L, Bragg B (2002) Safety and abuse testing of Energizer LiFeS_2 AA cells. NASA battery workshop, 19–21 November http://ntrs.nasa.gov/archive/nasa/casi.ntrs.nasa.gov/20030112597_2003133523.pdf. Last accessed 22 Nov 2011
15. Jeevarajan JA (2005) Comparison of safety of two primary lithium batteries for the orbiter wing leading edge impact sensors. First IAASS Conf, pp 375–380
16. Jeevarajan, JA (2006) Comparison of safety of two primary lithium batteries for the orbiter wing leading edge impact sensors. In: Abstracts 208th Meeting of the Electrochemical Society 1979. Published as ECS Trans, vol 1, pp 7–12

17. Sarciaux S, Le Gal La Salle A, Verbaere A, Piffard Y, Guyomard D (1999) γ-MnO_2 for Li batteries part I. γ-MnO_2: relationships between synthesis conditions, material characteristics and performances in lithium batteries. J Power Sources 81–82:656–660
18. Dose WM, Donne SW (2011) Heat treated electrolytic manganese dioxide for Li/MnO_2 batteries: effect of precursor properties. J Electrochem Soc 158: A1036–A1041
19. Litovitz T, Whitaker N, Clark L (2010) Preventing battery ingestions: an analysis of 8648 cases. Pediatr 125:1178–1183
20. National Capital Poison Center (2011) Swallowed a button battery? Battery in the nose or ear? http://www.poison.org/battery/. Last accessed 7 Dec 2011
21. U.S. Consumer Product Safety Commission (2011) CPSC warns: as button battery use increases, so do battery-related injuries and deaths toddlers and seniors most often injured in battery-swallowing incidents. Release #11-181, 23 March http://www.cpsc.gov/cpscpub/prerel/prhtml11/11181.html. Last accessed 7 Dec 2011
22. National Electrical Manufacturers Association (2011) Risk of serious injury from battery ingestion and non-secure battery compartments. http://ftpcontent.worldnow.com/wthr/PDF/nema.pdf. Last accessed 22 Nov 2011
23. The Consumer Electronics Association (CEA) is working with Underwriters Laboratory to devise improved standards in this area, probably through a revision of UL standard 60065
24. Raman NS, Davenport AJ Sink MS (2010) Development of high energy lithium manganese dioxide primary cell for military applications. In: Proceedings 44th Power Sources Conference, Las Vegas, 14–17 June, pp 111–113
25. Meshri DT, Hage DB (1999) Fluorine reactions with structured carbon. In: Houben-Weyl (ed) methods in organic chemistry, vol E10. organo-fluorine compounds, 4th edn. Thieme Medical
26. Zhang SS, Foster D, Wolfenstine J, Read J (2009) Electrochemical characteristic and discharge mechanism of a primary Li/CF_x cell. J Power Sources 187:233–237. doi:10.1016/j.jpowsour.2008.10.076
27. Read J, Collins E, Piekarski B, Zhang S (2011) LiF formation and cathode swelling in the Li/CF_x battery. J Electrochem Soc 158:A504–A510
28. Whitacre JF, West WC, Smart MC, Yazami R, Surya Prakash GK, Hamwi A, Ratnakumar BV (2007) Enhanced low-temperature performance of Li-CF_x batteries. Electrochem Solid-State Lett 10: A166–A170
29. De-Leon S (2011) Li/CF_x batteries the renaissance Li-CF_x - historic point. http://www.sdle.co.il/AllSites/810/Assets/li-cfx%20-%20the%20renaissance.pdf. Last accessed 3 Jan 2011
30. Sun D, Ramanathan T, Destephen M, Higgins R (2010) Development of low temperature electrolyte for Li/CF_x batteries. In: Proceedings 44th Power Sources Conference, Las Vegas, 14–17 June, pp 123–125
31. Leger VZ (1982) Nonaqueous cell having a MnO_2/poly-carbon fluoride cathode. US Patent 4327166
32. Marple JW (1987) Performance characteristics of Li/MnO_2-CF_x hybrid cathode jellyroll cells. J Power Sources 19:325–335
33. Raman NS, Davenport AJ, Sink MS (2010) Development of high energy hybrid solid cathode primary system for military applications. In: Proceedings 44th Power Sources Conference, Las Vegas, 14–17 June, pp 107–109
34. Wang X, Zhang X, Schoeffel S, Modeen D (2010) Safe Li-CF_x/MnO_2 hybrid chemistry primary battery for wearable power. In: Proceedings 44th Power Sources Conference, Las Vegas, 14–17 June, pp 115–118
35. Zhang D, Ndzebet E, Yang M (2010) Advanced battery chemistry for portable power. In: Proceedings 44th Power Sources Conference, Las Vegas, 14–17 June, pp 119–121
36. Lam P, Yazami R (2006) Physical characteristics and rate performance of $(CF_x)_n$ ($0.33 < x < 0.66$) in lithium batteries. J Power Sources 153:354–359
37. Whitacre J, Yazami R, Hamwi A, Smart MC, Bennett W, Surya Prakash GK, Miller T, Bugga R (2006) Low operational temperature Li-CF_x batteries using cathodes containing sub-fluorinated graphitic materials. J Power Sources 160:577–584
38. http://www.contourenergy.com/technology/fluoronetic.html. Last accessed 22 Nov 2011

Lithium-Air Battery

Tatsumi Ishihara
Department of Applied Chemistry, Faculty of Engineering, International Institute for Carbon Neutral Energy Research (WPI-I2CNER), Kyushu University, Nishi ku, Fukuoka, Japan

Introduction

There are strong demands for electric source with large capacity from electric vehicle and leveling electric power from renewable energy source like photo voltaic cell or wind mill power generator. In general battery, active materials which shows redox during charge and discharge are requested for both electrodes. Metal-air batteries have been attracting considerable attentions because of their

Lithium-Air Battery, Table 1 Theoretical electromotive force and theoretical energy density of various metal-air batteries

Type of metal-air battery	Theoretical electromotive/V	Theoretical energy density/ Wh/kg[a]	Theoretical capacity/ mAh/g
Li/O_2	2.91	11,140	3,828
Na/O_2	1.94	2,260	1,165
Ca/O_2	3.12	4,180	1,340
Mg/O_2	2.93	6,462	2,200
Zn/O_2	1.65	1,350	818
Al/O_2	2.70	4,021	1,490
Fe/O_2	1.32	1,584	1,200

[a]Energy density without oxygen

extremely large specific capacity. The reason for such a large specific capacity is that these cells consist of metal as an anode and an air electrode for activation of oxygen in air; oxygen is used as active materials and not necessary to be stored in battery. Hence, these metal-air batteries work as a half cell and have a simple structure. The capacity and theoretical open circuit potential for some metal-air batteries are compared in Table 1. Among the various metal-air battery systems, the lithium-air battery is the most attractive one because it has the highest energy density per unit weight. The cell discharge reaction occurs between Li and oxygen to yield Li_2O or Li_2O_2 theoretically with a discharge voltage of ca. 3.0 V and a specific energy density of 5,200 Wh/kg-Li. The theoretical specific energy (excluding oxygen) is 11.140 kWh/kg-Li, which is much higher than that of other advanced batteries and even higher than that of methanol direct fuel cells. For large part of metal-air battery, alkaline aqueous solution like 5 M KOH is used for electrolyte; however, in case of Li-air battery, organic electrolyte like carbonate or ether is generally used.

Abraham and Jiang first reported a Li-air battery using a nonaqueous electrolyte at 1996 [1]. They suggested that lithium peroxide is a discharge product based on $2(Li^+ + e^-) + O_2 \rightarrow Li_2O_2$, which resulted in a theoretical voltage of 2.96 V. However, because of low oxygen solubility in a nonaqueous electrolyte, the reported power density of an Li-air battery using a nonaqueous electrolyte is much lower than the theoretical value [2, 3]. In 2006, Bruce et al. reported that a rechargeable $Li–O_2$ battery using a carbonate-based electrolyte containing Super S carbon with electrolytic MnO_2 as the catalyst had an initial discharge capacity of 1,000 mAh/g (based on the weight of carbon) at a current rate of 70 mA/g and a capacity retention rate of 60 % after 50 cycles [4]. After report by Bruce et al., Li-air battery was studied extensively. Since the discharge products of Li oxide is low electrical conductivity, two types of Li-air battery is now studied as shown in Fig. 1. One is Li-air battery with dry air electrode (see Fig. 1a) and the other uses nonaqueous and aqueous electrolyte which is separated by solid-state Li ion conductor (see Fig. 1b). In the following part, Li-air battery development will be introduced based on two types.

Li-Air Battery Using Nonaqueous Electrolyte

This type of Li-air battery is close to the principle of metal-air battery. Ideal electrode reaction is as follows:

$$\text{(Li electrode) } Li = Li^+ + e$$
$$\text{(air electrode) } 2Li^+ + O_2 + 2e = Li_2O_2 (\text{or } Li_2O)$$

Dobley et al. and Kuboki et al. employed liquid organic solvents as the electrolyte [5, 6]. When employing an organic or ionic liquid-based electrolyte solution, the cell reaction produces insoluble Li_2O or Li_2O_2, which precipitates in the pores of the porous carbon-based air electrode to block the further intake of oxygen, abruptly terminating the discharge reaction. Recently, a lithium-air rechargeable battery employing MnO_2 as an air electrode was reported [7, 8]. Recently, Mizuno et al. also reported that a rechargeable $Li–O_2$ battery using propylene carbonate (PC) electrolyte containing Ketjenblack (EC600JD) carbon with MnO_2 as the catalyst had an initial discharge capacity of 820 mAh/g-cat. and a capacity retention rate of

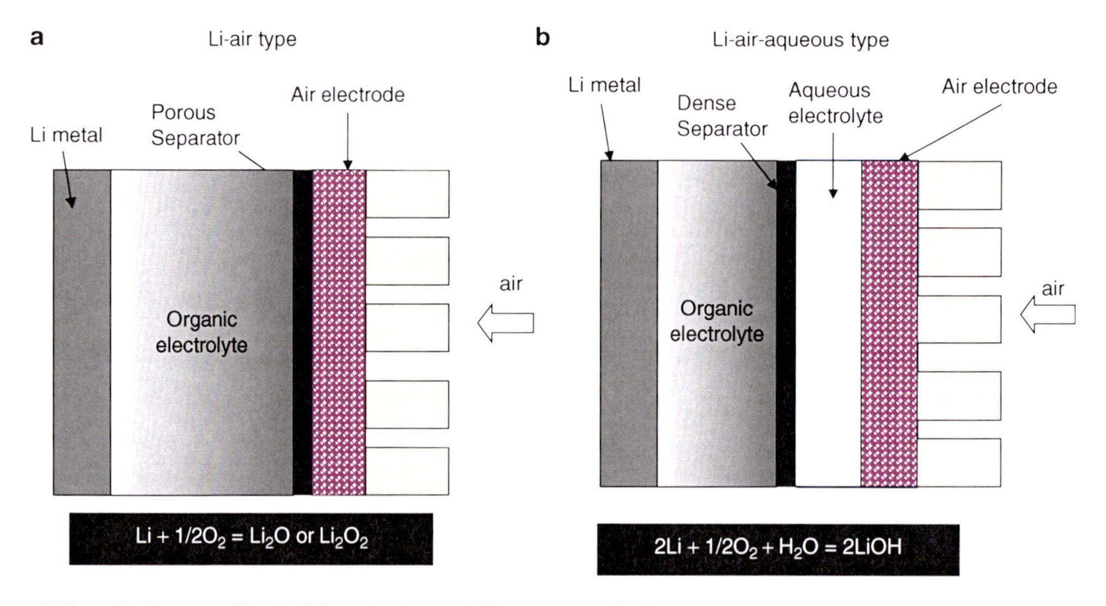

Lithium-Air Battery, Fig. 1 Schematic image of Li-air (**a**) and Li-air-aqueous battery (**b**)

60 % after 100 cycles, which is one of the longest life cycle reported for a rechargeable Li–O_2 to date [9]. However, decrease in the charge potential is required because of low energy efficiency for charge and discharge (62 % of the reported cell) and also leads to the decomposition of electrolyte resulting in the formation of Li_2CO_3 [10]. In principle, discharge products should be Li oxide; however, because of oxidative decomposition of electrolyte, mainly during charge at potential higher than 4 V, there are many cases of Li_2CO_3 formed [11, 12]. For this purpose, further improvement in the catalytic activity of an air electrode is requested from larger discharge capacity and the decrease in charge potential. MnO_2 can be used as an alternative low-cost electrocatalyst for oxygen reduction/evolution reactions [13]. However, MnO_2 prepared by the traditional ceramic route has limited electrocatalytic activity because of its large particle size and low specific surface area [13]. In addition, decomposition of electrolytes containing propylene carbonate (PC) leads to the formation of Li_2CO_3; this is the main reason for the high charge potential [12]. The authors reported that mesoporous α-MnO_2 modified with a Pd cathode air electrode for a rechargeable Li-air battery, which showed higher reversible capacity as well as higher current density than the standard Li-air battery using carbon air electrode [13, 14]. Figure 2 shows charge and discharge curves of mesoporous β-MnO_2 for air electrode. For this cell, ethylenecarbonate and diethylenecarbonate mixture at 3:7 volumetric ratio was used. Charge and discharge was performed at 2.7 and 3.6 V, respectively. A mixture of Pd and mesoporous β-MnO_2 shows high oxidation activity and Li^+ undergoes reduction to form Li_2O_2 or Li_2O, even in a dry atmosphere by Raman spectroscopy. We also pointed out negligible decomposition of organic electrolyte by decreasing the charge potentials, and so, as shown in Fig. 3 [14], stable cycle charge and discharge capacity was achieved.

At present, many efforts are paid for development of new nonaqueous electrolyte which is stable during charge and discharge. During charge, contribution of oxygen molecule radical is suspected [14], and it is considered that propylene carbonate is easily oxidized with oxygen molecule radical resulting in decreased cycle stability. It is reported that dimethyl ether (DME) is stable for Li metal and no CO_2 formation was detected by mass spectrometer [15]. However, high volatility of DME is another issue for electrolyte of Li-air battery which is open structure.

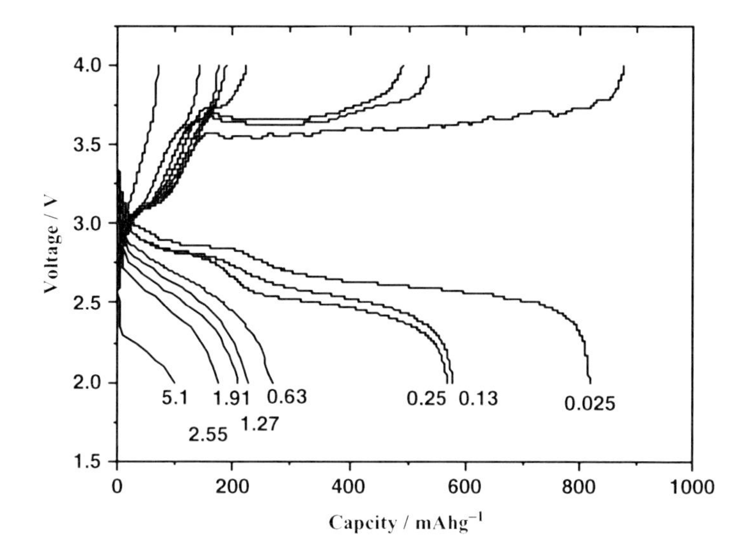

Lithium-Air Battery, Fig. 2 Charge and discharge curves of mesoporous β-MnO_2 for air electrode at different current density. Number in figure is the current density (mA/cm^2) (Ref. [14])

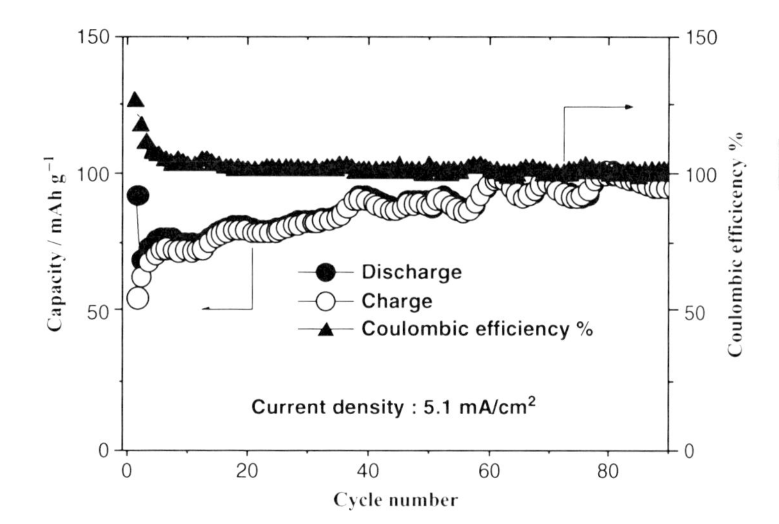

Lithium-Air Battery, Fig. 3 Cycle property of charge and discharge capacity of Li-air battery using mesoporous β-MnO_2 for air electrode (Ref. [14])

Recently, tetraethylene glycol dimethyl ether (TEGDME) is reported as a stable electrolyte for Li-air battery, and stable cycle performance was reported over 50 cycles, although discharge was cut off at rather small capacity [16, 17]. Apart from organic ether, dimethyl sulfone, which is highly hydrophilic nature, is also proposed for Li-air battery, and superior cycle stability is reported over 100 cycles, although discharge capacity is suppressed to a small value [18]. Ionic liquid is also examined for stable electrolyte of Li-air battery. Consequently, by using a suitable electrolyte with high stability against oxidation and reduction, low volatility, and hydrophobic property, Li-air rechargeable battery using nonaqueous electrolyte is highly promising as a new battery with huge volumetric power density.

Li-Air Battery with Nonaqueous and Aqueous Electrolyte

Since discharge products of Li oxide is insulator, electrical conductivity of air electrode at the final stage of discharge became much larger resulting in the limited discharge capacity. Therefore, dissolving of discharge Li products into aqueous

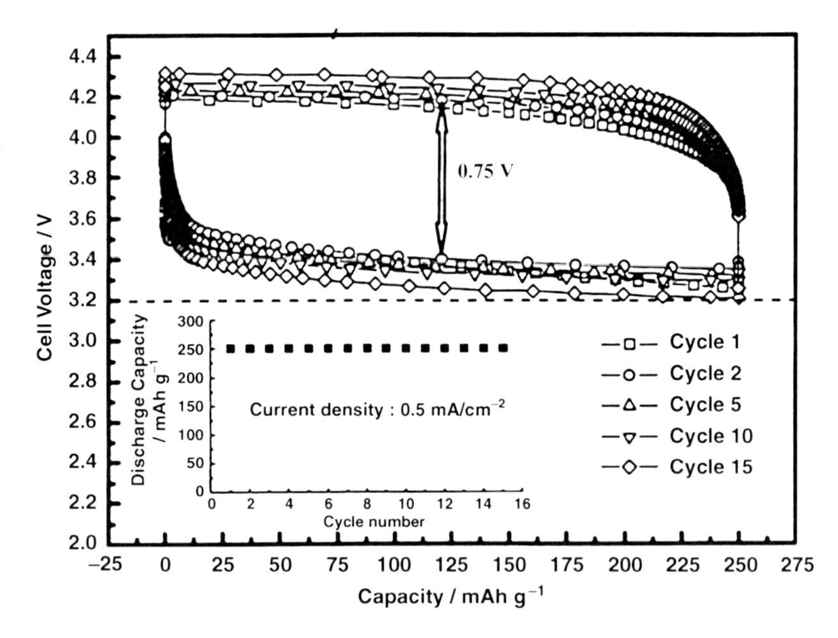

Lithium-Air Battery, Fig. 4 Discharge and charge curves for Li-air-aqueous battery. Acetic acid was used for aqueous electrolyte (Ref. [19])

solution was also studied, and this type of Li-air battery with aqueous and nonaqueous electrolyte, which is denoted as Li-air aqueous battery in the following part, is also called Li-air battery. As shown in Fig. 1b, in order to present reaction of Li with water, nonaqueous electrolyte, typically carbonate-based electrolyte, is requested for Li anode side, and acidic aqueous electrolyte is requested for air electrode side. The inorganic Li ion conducting dense electrolyte is requested for separating two electrolytes. The electrode reaction for this battery is shown as follows:

Anode; $Li = Li^+ + e$

Cathode; $1/2O_2 + H_2O + 2e = 2OH^-$ and then

$Li^+ + OH^- = LiOH$ (dissolved in aqueous solution)

Li-air aqueous battery has also been studied extensively and partially put into commercial. Polyplus Co., Ltd. successfully develops Li electrode covered with Li ion conducting ceramics and close to commercial area [19]. Figure 4 shows the discharge and charge curves for Li-aqueous battery reported by Imanishi et al. [20]. They used Li phosphate glass for Li ion conducting separator, and the observed charge and discharge potential is close to the theoretical value which is also shown in Table 1. The stable capacity during charge and discharge was also reported as shown in Fig. 4.

The main issue for Li-air aqueous battery is stability of Li ion conducting solid over wide potential and pH. In general, Li ion conducting glass (typically phosphate glass) or oxide (typically Ti-based mixed oxide) is not stable in basic or acidic solution, respectively. Since discharge product of Li hydroxide is dissolved in aqueous electrolyte and pH is changed to basic with discharge, conductivity of aqueous electrolyte and also charge carrier will change during discharge. In addition, solid separator may also dissolve in electrolyte. Therefore, discharge capacity of Li-air aqueous electrode may be determined by chemical stability of solid separator. In addition, solubility of Li(OH) in aqueous solution also determined the discharge capacity. Therefore, fairly large amount of aqueous electrolyte is requested for achieving the high discharge capacity. In addition, keeping aqueous electrolyte in battery is also another issue because the battery is open structure to air.

In any way, much higher energy density than that of the current Li ion battery is also demonstrated on Li-air aqueous battery, but cycle stability for charge and discharge is also an issue for the commercial area. For this, development of Li ion conducting ceramic separator with high chemical stability is the key issue.

Future Directions

At present, Li ion rechargeable battery is mainly used, development of new and safety battery is strongly required.

Conclusion

In this entry, principle, advantage, and the state of the art of Li-air battery are explained. The largest advantage of Li-air battery is huge capacity which is the largest among the batteries principally. However, because of open structure, there are many issues to request to solve for practical use. The most important issue to overcome is the cycle stability for charge and discharge as a rechargeable battery. This could be achieved by decreasing charge potential which is also effective for improving energy efficiency. Although Li-air battery has many issues, Li-air battery is highly interesting as the next rechargeable battery from capacity and simple structure.

Cross-References

▶ Metal-Air Batteries

References

1. Abraham KM, Jiang Z (1996) A polymer electrolyte-Based rechargeable lithium/oxygen Battery. J Electrochem Soc 143:1–5
2. Read J (2002) Characterization of the lithium/oxygen organic electrolyte battery. J Electrochem Soc 149: A1190–A1195
3. Read J, Mutolo K, Ervin M, Behl W, Wolfenstine J, Driedger A, Foster D (2003) Oxygen transport properties of organic electrolytes and performance of lithium/oxygen battery. J Electrochem Soc 150: A1351–A1356
4. Ogasawara T, Debart A, Holfazel M, Novak P, Bruce PG (2006) Rechargeable Li2O2 electrode for lithium batteries. J Am Chem Soc 128:1390–1393
5. Dobley A, Rodriguez R, Abraham KM (2004) High capacity cathodes for lithium-air batteries. In: 206th Meeting of the Electrochemical Society, Honolulu, USA, Abstract #496, 4–8 Oct 2004
6. Kuboki T, Okuyama TT, Ohsaki T, Takami N (2005) Lithium-air batteries using hydrophobic room temperature ionic liquid electrolyte. J Power Sources 146:766–769
7. Debert A, Bao J, Armstrong G, Bruce PG (2007) An O2 cathode for rechargeable lithium batteries: The effect of a catalyst. J Power Sources 174:1177–1182
8. Debart A, Paterson AJ, Bao J, Bruce PG (2008) α-MnO2 Nanowires: A Catalyst for the O2 Electrode in Rechargeable Lithium Batteries. Angew Chem Int Ed 47:4521–4524
9. Mizuno F, Nakanishi S, Kotani Y, Yokoishi S, Iba H (2010) Rechargeable li-air batteries with carbonate-based liquid electrolytes (E). Electrochem 78:403–405
10. Cheng F, Shen J, Peng B, Pan Y, Tao Z, Chen J (2011) Rapid room-temperature synthesis of nanocrystalline spinels as oxygen reduction and evolution electrocatalysts. Nat Chem 3:79–84
11. Thapa AK, Saimen K, Ishihara T (2010) Pd/MnO2 air Electrode catalyst for rechargeable lithium/air battery. Electrochem Solid-State Lett 13:A165–A167
12. Freunberger SA, Chen Y, Peng Z, Griffin JM, Hardwick LJ, Bard F, Novak P, Bruce PG, (2011) Reactions in the rechargeable lithium–o2 battery with alkyl carbonate electrolytes. J Am Chem Soc 133:8040–8047
13. Thapa AK, Ishihara T (2011) Mesoporous α-MnO2/Pd catalyst air electrode for rechargeable lithium–air battery. J Power Sources 196:7016–7020
14. Thapa AK, Hidaka Y, Hagiwara H, Ida S, Ishihara T (2011) Mesoporous β-MnO2 Air Electrode Modified with Pd for Rechargeability in Lithium-Air Battery. J Electrochem Soc 158:1483–1489
15. McCloskey BD, Bethune DS, Shelby RM, Girishkumar G, Luntz AC (2011) Figure 1 of 7 Solvents' Critical Role in Nonaqueous Lithium–Oxygen Battery Electrochemistry. J Phy Chem Lett 2:1161–1166
16. Laoire CÓ, Mukerjee S, Plichta EJ, Hendrickson MA, Abrahama KM (2011) Rechargeable Lithium/TEGDME-LiPF6/O2 Battery. J Electrochem Soc 158:A302–A308
17. Jung HG, Hassoun J, Park JB, Sun YK, Scrosati B (2012) An improved high-performance lithium–air battery. Nat Chem 4:579–585
18. Peng Z, Freunberger SA, Chen Y, Bruce PG (2012) A Reversible and Higher-Rate Li-O2 Battery. Science 337:563–566
19. Visco SJ, Nimon E, Jonghe LD (2010) Next generation Li-Air and Li-S batteries based on ceramic protected li electrodes The 15th international meeting on lithium batteries, Abstract #831
20. Zhang T, Imanishi N, Shimonishi Y, Hirano A, Takeda Y, Yamamoto O, Sammes N (2010) A novel high energy density rechargeable lithium/air battery. Chem Commun 46:1661–1664

Lithium-Ion Batteries

Akira Yoshino
Yoshino Laboratory, Asahi Kasei Corporation, Fuji, Shizuoka, Japan

Introduction

Defined narrowly, the lithium-ion battery (LIB) is "a secondary battery using nonaqueous electrolyte with carbonaceous material as a negative electrode and a metal oxide compound containing lithium (usually $LiCoO_2$) as a positive electrode [1]." The LIB's electrochemical system is shown in Fig. 1.

During charging, Li ions separate from the $LiCoO_2$ of the positive electrode and intercalate in the carbonaceous material of the negative electrode, while the electrochemical reaction causes electrons to flow from the positive electrode to the negative electrode. The opposite reaction occurs during discharging. The principle of the LIB is fundamentally different from that of other rechargeable batteries, as no chemical transformation occurs and only ions and electrons shuttle back and forth.

Features of the LIB

The major features of the LIB are high electromotive force (4.2 V), small size and light weight, possibility of high-current discharge, low self-discharge rate, charging and discharging efficiency near 100 %, and an absence of hazardous substances. These features are made possible by the use of a nonaqueous electrolyte solution. As shown in Fig. 2, the LIB has much higher specific energy (Watt-hour capacity per unit of mass) and energy density (Watt-hour capacity per unit of volume) than nickel-cadmium or nickel-metal hydride secondary batteries, which use an aqueous electrolyte solution.

Components of the LIB

The main materials which are used for each of the basic components of the LIB are shown in Table 1.

Positive Electrode

In addition to $LiCoO_2$ [2], lithium-containing metal oxide compounds which can be used for LIB positive electrodes include $LiNiO_2$ and $LiMn_2O_4$ [3]. Recently, ternary systems such as $Li(Ni_{1/3}Mn_{1/3}Co_{1/3})O_2$ have also been developed as high-capacity positive electrode materials [4]. In each case, the positive electrode material is the key to obtaining electromotive force in the 4 V ranges, which is a basic characteristic of most LIBs today. These compounds are manufactured by calcining oxides of Co, Ni, Mn, or a combination thereof together with lithium salts such as lithium carbonate at 750–900 °C. In addition, phosphates such as $LiFePO_4$ [5] are also finding some use as LIB positive electrode material due to the outstanding stability they afford, although electromotive force is somewhat lower at around 3.5 V.

Negative Electrode

The most prominently used carbonaceous material for LIB negative electrodes is highly crystalline graphite, either synthetic graphite or modified natural graphite. Synthetic graphite is manufactured by heat-treating petroleum pitch or coal pitch. Based on the properties of the pitch material and the heating conditions, it is possible to control the microstructure of the graphite obtained. A variety of synthetic graphite materials with different characteristics are thus available. Natural graphite has the advantage of being more inexpensive than synthetic graphite, but it is unsuitable for use as negative electrode material without modification. By forming a layer of synthetic graphite on the surface of natural graphite to obtain a "core/shell" structure, it is possible to produce modified natural graphite as negative electrode material at lower cost than synthetic graphite.

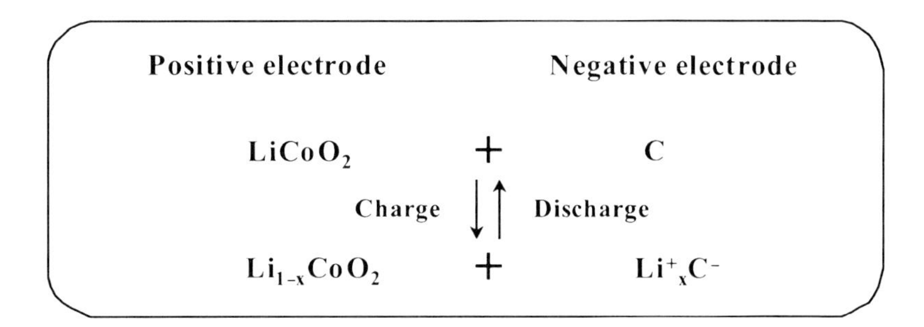

Lithium-Ion Batteries, Fig. 1 Electrochemical system of the LIB

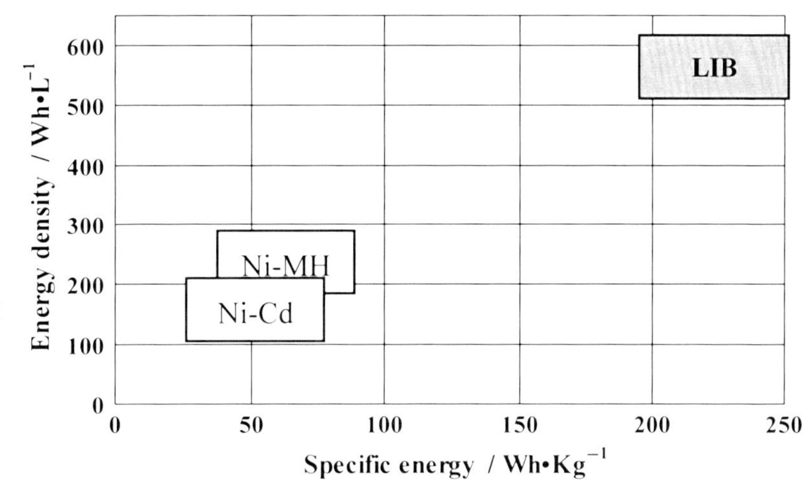

Lithium-Ion Batteries, Fig. 2 Specific energy and energy density of secondary batteries. *Ni-Cd* nickel-cadmium battery, *Ni-MH* nickel-metal hydride battery, *LIB* lithium ion battery

Lithium-Ion Batteries, Table 1 Main materials used for LIB components

Positive electrode	$LiCoO_2$, $LiMn_2O_4$
Negative electrode	Graphite, hard carbon
Electrolyte solvents	Ethylene carbonate
	Ethyl methyl carbonate
Electrolyte salts	$LiPF_6$, $LiBF_4$
Separators	Polyethylene
Binder resins	Polyvinylidene fluoride, SB latex

Another carbonaceous material which finds some use as negative electrode material is hard carbon. Because hard carbon contains heteroelements such as oxygen and nitrogen in the carbon skeleton, it does not easily form graphite even when heat treated at high temperature. In contrast to the mostly flat discharge profile of graphite negative electrode, the discharge profile of hard carbon negative electrode is characterized by a steady slope, which makes hard carbon more suited to fast charging. Alloys of silicon, tin, etc. are also being developed as negative electrode materials with higher capacity than carbon.

Binder Resin

LIB electrodes are structured with layer particles of active electrode material on metal foil current collectors, as shown in Fig. 3. A binder resin provides the mechanical strength which binds the particles to each other and to the current collectors. Electrical contact between particles and between the particles and the current collectors must be maintained, so the binder resin must be evenly dispersed throughout the electrodes. Uniform gaps in the binder resin are also needed so that electrolyte solution can penetrate the

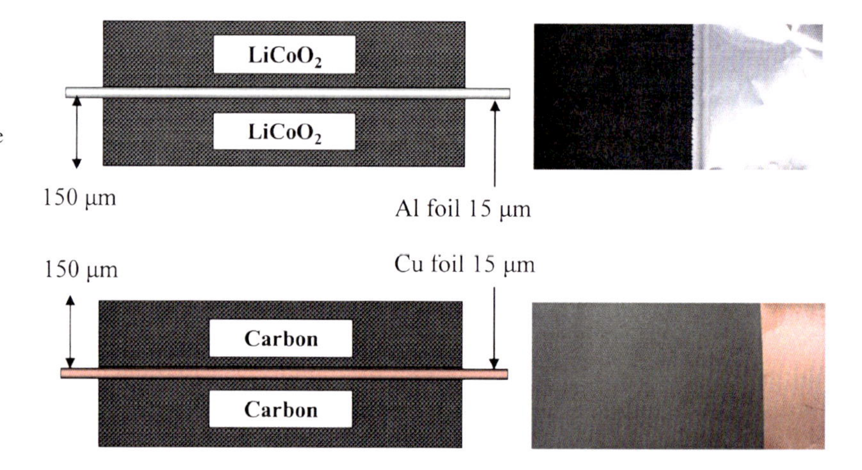

Lithium-Ion Batteries, Fig. 3 Structure of LIB electrodes. *Upper figure*: positive electrode, *lower figure*: negative electrode

electrode layers to enable ion transport. Electrodes are formed by coating slurry of binder resin and active electrode material on the metal foil current collectors, which are then dried.

Electrolyte Solution

A mixture of cyclic carbonate esters (such as ethylene carbonate and propylene carbonate) and linear carbonate esters (such as dimethyl carbonate and diethyl carbonate) is used as electrolyte solvent. Compounds such as $LiPF_6$ and $LiBF_4$ are used as electrolyte salts. LIB electrolyte solutions need to have both high dielectric constant and low viscosity in order for Li ions to transport freely. A high dielectric constant enables the dissolved electrolyte salt to easily dissociate. Low viscosity enables dissociated ions to move easily through the solution. Cyclic carbonate esters have high dielectric constant, while linear carbonate esters have low viscosity. Suitable electrolyte solutions are obtained by using a mixture of the two types of solvent.

Separators

LIB separators must provide electrical insulation between the positive and negative electrodes while permitting ion transport between them. They are microporous polyolefin films 10–30 μm thick with pores of 0.01–0.1 μm diameter. Most LIB separators are made of high-density polyethylene, although polypropylene is also used to a certain extent.

Future Directions

Much attention is now focused on whether or not solid electrolytes can be commercialized. Inorganic solid electrolytes have achieved ionic conductivity equivalent to current liquid electrolyte solutions. If such solid electrolytes can be commercialized, they are expected to revolutionize LIB electrode structure and battery characteristics.

Emerging Trends

While the LIB has become widespread in mobile phones, laptop computers, and other portable electronic devices, it is now increasingly being used in larger applications such as electric vehicles and stationary power storage systems. Most technological development is now focused on meeting the performance requirements of such new applications.

References

1. Yoshino A, Sanechika K, Nakajima T (1985) U.S. Patent 4,668,595
2. Mizushima K, Jones PC, Wiseman PJ, Goodenough JB (1980) LixCoO2 (0<x<-1): a new cathode material for batteries of high energy density. Mater Res Bull 15:783
3. Thackeray MM, David WIF, Bruce PG, Goodenough JB (1983) Lithium insertion into manganese spinels. Mater Res Bull 18:461

4. Ohzuku T, Makimura Y (2001) Layered lithium insertion material of LiCo1/3Ni1/ 3Mn1/3O2 for lithium-ion batteries. Chem Lett 7:642
5. Padhi AK, Nanjyundaswamy KS, Goodenough JB (1997) Phospho-olivines as positive-electrode materials for rechargeable lithium batteries. J Electrochem Soc 144:1188

Lithium-Sulfur Batteries

Hikari Sakaebe
Research Institute for Ubiquitous Energy Devices, National Institute of Advanced Industrial Science and Technology (AIST), Ikeda, Osaka, Japan

Introduction

Li-S battery system is advantageous due to theoretically higher energy density (2,600 Wh kg^{-1} assuming the redox reaction of $2Li + S \rightleftarrows Li_2S$), and thus research and development for sulfur electrode had started in 1960–1970s [1–3].

The difficulty in Li-S battery operated at ambient temperature lies in two points: conductivity and the solubility of partly lithiated polysulfides, Li_xS_8. S powder is usually a yellow powder and it has poor conductivity. So the conduction pathway should be ensured. At present, preparation of composite of S and carbons has been attracting many researcher for improvement. Present status will be outlined later in the next section.

It was reported that polysulfide, Li_xS_8, formed during the way of the discharge and charge of the battery and undergoes electrochemical reaction in the liquid phase in the organic solvent [4, 5]. This causes a worse efficiency and a poor cycleability. As shown in Fig. 1, polysulfides are thought to undergo the cycling between S electrode and Li-metal electrode. This is usually called as "redox-shuttle" phenomena. Mainly from the latter reason, R&D of Li-S batteries did not progress so much.

In 1990s, Li-ion batteries have been put into market. Rocking-chair system was applied to this battery with carbonaceous negative electrode, and positive electrode material without Li became difficult to be used. Research activity of Li-S battery was also toned down then; however, the situation turned around 2006.

Because of the consciousness for environment and the economic situation for fuels, electric vehicles (EV) have become reconsidered. Higher energy density over 500 Wh kg^{-1} is required for widespread of EV. Energy density of Li-ion battery system with rocking-chair mechanism is limited up to c.a. 250 Wh kg^{-1}. Theoretically Li-air, Li-S, and Zn-air can only be the candidates for this kind of high-energy battery overcoming the limitation of Li-ion system. So, Li-S system has been revived.

Strategy of recent R&D of Li-S is classified as follows:

1. Composite of metal and sulfur
2. Composite of carbon and S or Li_2S
3. Investigation of electrolyte
4. Li electrode protection
5. Organic compound containing S

Introduction and recent progress of 1–4 categories will be explained.

Composite of Metal and Sulfur

One useful technique to utilize sulfur with poor conductivity can be the formation of metal composite. Transition metal sulfide such as CuS, NiS, NiS_2, FeS_2, TiS_2, MoS_2, and VS_2 have been applied to the nonaqueous electrolyte system in 1970s. TiS_2 and MoS_2 undergo topochemical reaction with Li. They were applied to the cathode materials of Li-metal battery in 1980s. Other sulfide shows different discharge mechanisms in the deep discharge, so-called conversion reaction. These materials were tested in organic electrolyte system and molten salt system.

Figure 2 shows the charge–discharge profiles of NiS and NiS_2 [6]. These sulfides have plateau at 1.5–2.0 V. Average potential was lower than S. In the case of conversion system in organic electrolyte, reversibility is a serious problem. NiS_2 cycling properties were reported with quite

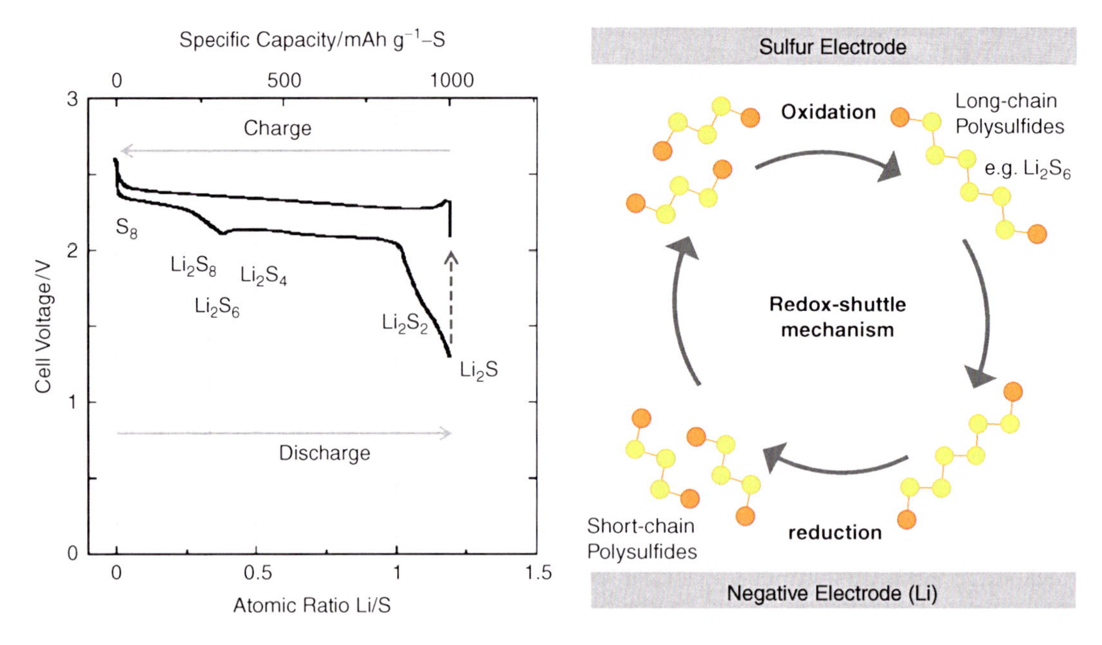

Lithium-Sulfur Batteries, Fig. 1 Typical discharge-charge curves of sulfur in organic electrolyte and the schematic illustration of "redox-shuttle" mechanism

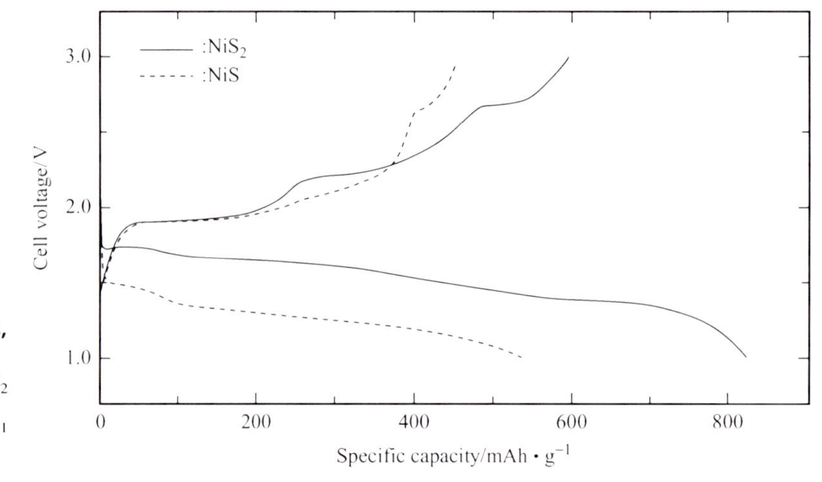

Lithium-Sulfur Batteries, Fig. 2 Initial discharge and charge profiles for NiS_2 and NiS samples cells at 175 (0.2C) and 177 mA·g^{-1} (0.3C), respectively

high initial capacity c.a. 800 mAh g^{-1}; however, capacity degraded quickly within 10 cycles. These metal sulfides also were applied to the solid electrolyte system [7]. Quite recently, it was found that formation of amorphous metal sulfide was effective to increase sulfur content for TiS_x ($x = 3, 4$) and that amorphous TiS_4 showed high capacity (over 570 mAh g^{-1}) above 2 V [8]. Cycling properties is not enough at this moment, however, dissolution of polysulfides from electrode was not significant even after cycles. Making use of such composites seems a promising technique to construct the metal sulfide batteries.

Composite of Carbon and S or Li_2S

Metal sulfide can be a one solution for the conductivity improvement of sulfur; however, energy

Lithium-Sulfur Batteries, Table 1 Selected reports for S-C composites

First author	Conductive support	Initial capacity/ mAh g^{-1} -S	S content in electrode/%	Initial capacity/ mAh g^{-1} -electrode	Capacity after cycles/ mAh g^{-1}-electrode (cycles)	References
J. Wang	PAN $[C_3NH_3]_n$		53.41	800	600(59)	[9]
J. L.Wang	Active carbon		21	800	400	[10]
J. Wang	Mesoporous carbon	1,300	16	208	80	[11]
X. Ji	Mesoporous carbon(CMK-3)	1,320	50	660	550	[12]
Feng Wu	Polyaniline, MWCNT	1,334	49	654	456(80)	[13]
L. Ji	Graphene oxide	1,400	46.2	647	439(50)	[14]
X. Liang	Polypyrrole, MWCNT	1,300	50	650	400(100)	[15]

density decreases by making a composite with metal. And instead of metal, carbon can be an effective substrate because of good conductivity and light weight. Quite recently a number of papers have published concerning S-C composite [9–15]. Controlling the pore size and distribution, reversibility of the S-C electrode could be improved dramatically. In addition to it, this way of composite possibly prevents the dissolution of polysulfide into the electrolyte solution. Selected reports that clearly described the sulfur content in the electrode were listed in Table 1. Cycleability was greatly improved by this technique. Sulfur content in the electrode cannot be so high in these cases, and achievement of higher energy battery requires further improvement in the balance of cycleability and electrode composition.

Not only for S, but also Li_2S was applied to be a component of such a kind of composite [16]. In this case, positive electrode contains sufficient amount of Li, carbon or alloy negative electrode can be used.

Investigation of Electrolyte

Polysulfide, Li_xS_8 (x = 2–8) is known to be generally soluble into organic electrolyte, and electrolyte composition has been intensively investigated in order to prevent the dissolution. TEGDME (tetraethylene glycol dimethyl ether, tetraglyme), ionic liquids, and solid electrolyte were proposed as candidate electrolyte. As a liquid electrolyte, ionic liquid may be a solution. 1-Ethyl-3-methylimidazolium bis(trifluoromethylsulfonyl)imide (EMI [TFSI]),1-butyl-3-methylimidazolium bis (trifluoromethylsulfonyl)imide (BMI[TFSI]), 1,2-dimethyl-3-propylimidazolium bis(trifluoromethylsulfonyl)imide (DMPI[TFSI]), and so on were applied to Li-S system [11, 17]. Li-glyme complex, similar system as ionic liquid, is also a candidate because it prevents the dissolution of polysulfides [18]. On the other hand, redox process Li_2S_6–Li_2S_3 is easier in solution phase. And DOL (1,3-dioxolan), DEE (1,2-diethoxyethane), and other ether systems are frequently used recently to obtain a full capacity of sulfur despite of the solubility. This is because the S-C composite described in the previous section predominates the electrolyte study. Carbonate solvent is common for Li-ion battery; however, this system is easily decomposed in Li-S system.

Solid electrolyte is also a good candidate. Sulfur-containing electrolyte such as Li_2S-P_2S_5 or Li_2S system has a good compatibility with S electrode. Big problem of the system is a reactivity with Li metal. Alternative anode like graphite or alloys should be used.

Li Electrode Protection

Li metal originally suffers from the dendrite growth problem that causes shorter life and safety problem. This point is not discussed in this article as so many reports and reviews can be found elsewhere. Li electrode in Li-S battery has inherent problem. As stated in redox-shuttle mechanism of polysulfides in Fig. 1, reduction of polysulfides at the Li electrode surface is a problem. If polysulfides are deposited as Li_2S, capacity is lost and Li electrode efficiency becomes worse. It is very important to prevent Li_2S deposition on the Li electrode to maintain cycles. Solid electrolyte is a one solution; however, brittle solid electrolyte cannot tolerate the volume change of Li electrode. Practical way to improve is thought to be a use of electrolyte additives. $LiNO_3$ is well known to depress the redox-shuttle phenomena [19]. The effect of this additive was concluded in the paper as a formation of protective layer on Li surface. $LiNO_3$ possibly oxidized the polysulfides and at the same time, formed the film by decomposition of itself. As a result, polysulfides were prevented from contacting with Li electrode.

Future Directions

In the past, several reports for the practical-sized Li-S battery were found. For example, high energy cell over 300 Wh kg^{-1} was manufactured by Sion Power, and it was proved not to reach the thermal runaway [20]. But at this stage, there has not been a commercially available Li-S cells. It still will need time to put this system into the market. One of the main drawbacks is the lower density of the material and electrolyte in addition to the problem already stated in this article. Volumetric energy density cannot be so advantageous as gravimetric energy density because of the lower density.

Anyway, this system is very attractive for material researcher and still a great number of paper are published concerning Si-C composite. Future technical breakthrough can be made by such an intensive research activities.

References

1. Jasinski R, Burrows B (1969) Cathodic discharge of nickel sulfide in a propylene carbonate-LiClO electrolyte. J Electrochem Soc 116:422–424
2. Margaret VM, Sawyer DT (1970) Electrochemical reduction of elemental sulfur in aprotic solvents. Formation of a stable Ss- species. Inorg Chem 9:211–215
3. Robert PM, Sawyer DT (1972) Electrochemical reduction of sulfur dioxide in dimethylformamide. Inorg Chem 11:2644–2647
4. Dominko R, Demir-Cakan R, Morcrette M, Tarascon JM (2011) Analytical detection of soluble polysulphides in a modified Swagelok cell. Electrochem Commun 13:117–120
5. Bruce PG, Freunberger SA et al (2012) Li-O-2 and Li-S batteries with high energy storage. Nat Mater 11:19
6. Takeuchi T, Sakaebe H, Kageyama H, Sakai T, Tatsumi K (2008) Preparation of NiS2 using spark-plasma-sintering process and its electrochemical properties. J Electrochem Soc 155:A679–A684
7. Hayashi A, Ohtsubo R, Nagao M, Tatsumisago M (2010) Characterization of Li2S-P2S5-Cu composite electrode for all-solid-state lithium secondary batteries. J Mater Sci 45:377–381
8. Sakuda A, Taguchi N, Takeuchi T, Kobayashi H, Sakaebe H, Tatsumi K, Ogumi Z (2013) Amorphous TiS_4 positive electrode for lithium-sulfur secondary batteries. Electrochem Commun 31:71–75
9. Wang J, Yang J, Xie J, Xu N (2002) A novel conductive polymer–sulfur composite cathode material for rechargeable lithium batteries. Adv Mater 14:963–965
10. Wang JL, Yang J, Xie JY, Xu NX, Li Y (2002) Sulfur–carbon nano-composite as cathode for rechargeable lithium battery based on gel electrolyte. Electrochem Commun 4:499–502
11. Wang J, Chew SY, Zhao ZW, Ashraf S, Wexler D, Chen J, Ng SH, Chou SL, Liu HK (2008) Sulfur–mesoporous carbon composites in conjunction with a novel ionic liquid electrolyte for lithium rechargeable batteries. Carbon 46:229–235
12. Xiulei Ji, Kyu Tae Lee, Nazar LF (2009) A highly ordered nanostructured carbon–sulphur cathode for lithium–sulphur batteries. Nat Mater 8:500–506
13. Feng Wu, Junzheng Chen, Renjie Chen, Shengxian Wu, Li Li, Shi Chen, Teng Zhao (2011) Sulfur/polythiophene with a core/shell structure: synthesis and electrochemical properties of the cathode for rechargeable lithium batteries. J Phys Chem C 115:6057–6063
14. Liwen Ji, Rao M, Haimei Zheng, Liang Zhang, Yuanchang Li, Wenhui Duan, Jinghua Guo, Cairns EJ, Yuegang Zhang (2011) Graphene oxide as a sulfur immobilizer in high performance lithium/sulfur cells. J Am Chem Soc 133:18522–18525
15. Liang X, Wen Z, Liu Y, Zhang H, Jin J, Wu M, Wu X (2012) A composite of sulfur and polypyrrole–multi

walled carbon combinatorial nanotube as cathode for Li/S battery. J Power Sources 206:409–413
16. Takeuchi T, Sakaebe H, Kageyama H, Senoh H, Sakai T, Tatsumi K (2010) Preparation of electrochemically active lithium sulfide–carbon composites using spark-plasma-sintering process. J Power Sources 195:2928–2934
17. Kim S, Jung Y, Park S-J (2007) Effect of imidazolium cation on cycle life characteristics of secondary lithium–sulfur cells using liquid electrolytes. Electrochim Acta 52:2116–2122
18. Tachikawa N, Yamauchi K, Takashima E, Park J-W, Dokko K, Watanabe M (2011) Reversibility of electrochemical reactions of sulfur supported on inverse opal carbon in glyme–Li salt molten complex electrolyte. Chem Commun 47:8157–8159
19. Liang X, Wen Z, Liu Y, Wu M, Jin J, Zhang H, Wu X (2011) Improved cycling performances of lithium sulfur batteries with LiNO3 – modified electrolyte. J Power Sources 196:9839–9843
20. Mikhaylik YV, Kovalev I, Schock R, Kumaresan K, Jason Xu, Affinito J (2010) High energy rechargeable Li-S cells for EV application: status, remaining problems and solutions. ECS Trans 25:23–34

M

Macroscopic Modeling of Porous Electrodes

Adam Z. Weber
Lawrence Berkeley National Laboratory, Berkeley, CA, USA

Introduction

It is well known that for optimal performance of electrochemical energy storage and conversion devices, it is necessary to have a nonplanar electrode to increase reaction area. One requires a porous electrode with multiple phases that can transport the reactant and products in the electrode while also undergoing reaction [1]; an analogy in heterogeneous catalysis is reaction through a catalyst particle [2]. For traditional devices, porous electrodes are often comprised of an electrolyte (which can be solid or liquid) that carries the ions or ionic current and a solid phase that carries the electrons or electronic current. In addition, there may be other phases such as a gas phase (e.g., fuel cells). Schematically one can consider the porous electrode as a transmission-line model as shown in Fig. 1.

The key to modeling a porous electrode is determining the expressions for the various resistances. As discussed below, for charge-transfer resistance a kinetic expression is used, for the electronic- and ionic-phase resistances transport equations are used, and the capacitance is often taken from experiment and is due to charging/discharging of the double layer.

Conservation Equations

A general conservation equation can be written as

$$\frac{\partial \psi}{\partial t} + \nabla \cdot N_{\psi} = S_{\psi} \quad (1)$$

where the variable ψ corresponds to any variable or property that is conserved within the control domain. The first term corresponds to the rate of accumulation of ψ and the second term to the contribution from the flux (N) of ψ, while the third term, S_{ψ}, corresponds to the source term for ψ. Depending on the type of physical quantity and subdomain in which the system is solved, the flux and source terms can be split into multiple contributing components. Transport through porous structures and all of the domains in this chapter is from a macrohomogeneous perspective using volume-averaged conservation equations.

Material

The conservation of material can be written as in Eq. 1 except that the physical quantity ψ could be p – partial pressure, c – concentration, x – mole fraction, or ρ – density. However, for the case of a mixture in a multiphase system, it is necessary to write material balances for each component in

G. Kreysa et al. (eds.), *Encyclopedia of Applied Electrochemistry*, DOI 10.1007/978-1-4419-6996-5,

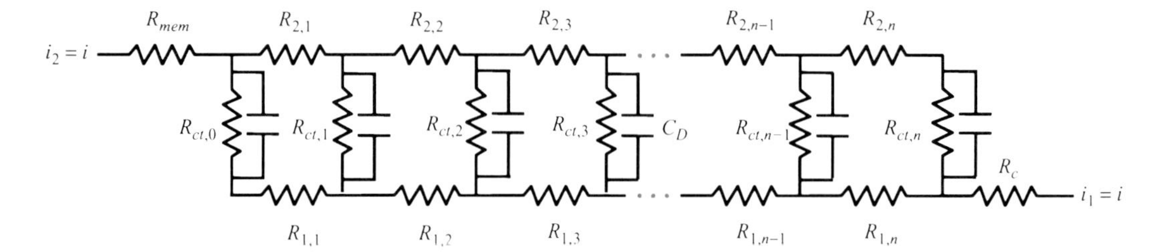

Macroscopic Modeling of Porous Electrodes, Fig. 1 Simple equivalent-circuit representation of a porous electrode. The total current density, i, flows through the separator or membrane to the electrolyte phase (2) and then into the solid or electronic phase (1) including a contact resistance. In between, the current is apportioned based on the resistances in each phase and the charge-transfer resistances, which also contain a capacitance

each phase k, which in summation still govern the overall conservation of material,

$$\frac{\partial \varepsilon_k c_{i,k}}{\partial t} = -\nabla \cdot \mathbf{N}_{i,k} - \sum_h a_{1,k} s_{i,k,h} \frac{i_{h,1-k}}{n_h F} + \sum_l s_{i,k,l} \sum_{p \neq k} a_{k,p} r_{l,k-p} + \sum_g s_{i,k,g} \varepsilon_k R_{g,k} \quad (2)$$

In the above expression, ε_k is the volume fraction of phase k, $c_{i,k}$ is the concentration of species i in phase k, and $s_{i,k,l}$ is the stoichiometric coefficient of species i in phase k participating in heterogeneous reaction l, $a_{k,p}$ is the specific surface area (surface area per unit total volume) of the interface between phases k and p, $i_{h,l-k}$ is the normal interfacial current transferred per unit interfacial area across the interface between the electronically conducting phase and phase k due to electron-transfer reaction h, and is positive in the anodic direction by convention.

The term on the left side of the equation is the accumulation term, which accounts for the change in the total amount of species i held in phase k within a differential control volume over time. The first term on the right side of the equation keeps track of the material that enters or leaves the control volume by mass transport as discussed in later sections. The remaining three terms account for material that is gained or lost (i.e., source terms, S_ψ, in Eq. 1). The first summation includes all electron-transfer reactions that occur at the interface between phase k and the electronically conducting phase 1; the second summation accounts for all other interfacial processes that do not include electron transfer like evaporation or condensation; and the final term accounts for homogeneous chemical reactions in phase k. It should be noted that in terms of an equation count, for n species there are only $n - 1$ conservation equations needed since one can be replaced by the sum of the other ones, or, similarly, by the fact that the sum of the mole fractions equals 1,

$$\sum_i x_i = \sum_i \frac{c_{i,k}}{c_{T_k}} = 1 \quad (3)$$

where c_{T_k} is the total molar concentration of species in phase k.

Charge

The conservation equation for charged species is an extension of the conservation of mass. Taking Eq. 2 and multiplying by $z_i F$ and summing over all species and phases while noting that all reactions are charge balanced yields

$$\frac{\partial}{\partial t} F \sum_k \sum_i z_i c_{i,k} = -\nabla \cdot F \sum_k \sum_i z_i \mathbf{N}_{i,k} \quad (4)$$

where the charge and current densities can be defined by

$$\rho_e = F \sum_k \sum_i z_i c_{i,k} \quad (5)$$

and

$$\mathbf{i}_k = F\sum_i z_i\mathbf{N}_{i,k} \quad (6)$$

respectively. Because a large electrical force is required to separate charge over an appreciable distance, a volume element in the electrode will, to a good approximation, be electrically neutral; thus one can assume electroneutrality for each phase

$$\sum_i z_i c_{i,k} = 0 \quad (7)$$

While this relationship applies for almost all macroscopic models, there are cases where electroneutrality does not strictly hold (as shown in Fig. 1), including for some transients and impedance measurements, where there is charging and discharging of the double layer, as well as simulations at length scales within the double layer (typically on the order of nanometers). For these cases, the correct governing equation for charge conservation results in Poisson's equation,

$$\nabla^2\Phi = \frac{\rho_e}{\varepsilon_0} \quad (8)$$

where ε_0 is the permittivity of the medium. It should be noted that Eq. 8 or Eq. 7 are often used to determine the potential in the system.

To charge the double layer, one can derive various expressions for the double-layer capacitance depending on the adsorption type, ionic charges, etc. [1], where the double-layer capacitance is defined as

$$C_D = \left(\frac{\partial q}{\partial\Phi}\right)_{\mu_i,T} \quad (9)$$

where q is the charge in the double layer and the differential is at constant composition and temperature. To charge the double layer, one can write an equation of the form

$$i = C_d\frac{\partial\Phi}{\partial t} \quad (10)$$

where the charging current will decay with time as the double layer becomes charged.

Energy

The overall conservation of thermal energy can be written as [3]

$$\sum_k \rho_k\hat{C}_{p_k}\frac{\partial T}{\partial t} = -\sum_k \rho_k\hat{C}_{p_k}\mathbf{v}_k\cdot\nabla T + \nabla\cdot\left(k_T^{\mathrm{eff}}\nabla T\right) + \sum_k\frac{\mathbf{i}_k\cdot\mathbf{i}_k}{\kappa_k^{\mathrm{eff}}} + \sum_h i_h(\eta_h + \Pi_h) - \sum_h \Delta H_h r_h \quad (11)$$

where Π is the Peltier coefficient, η is the reaction overpotential as discussed below, C_p is the heat capacity, H is the enthalpy of non-electrochemical reactions (e.g., evaporation/condensation), and k_T is the thermal conductivity. In Eq. 11, the first term on the right side is energy transport due to convection, the second is energy transport due to conduction, the third is the ohmic heating, the fourth is the reaction heats, and the last represents reactions in the bulk which include such things as phase-change latent heats. For many electrochemical systems, in terms of magnitude, the major heat generation sources are the reactions and the main mode of heat transport is through conduction.

Momentum

The momentum or volume conservation equation is highly coupled to the mass or continuity conservation equation (2). Newton's second law governs the conservation of momentum and can be written in terms of the Navier–Stokes equation [4]

$$\frac{\partial(\rho_k\mathbf{v}_k)}{\partial t} + \mathbf{v}_k\cdot\nabla(\rho_k\mathbf{v}_k) = -\nabla p_k + \mu_k\nabla^2\mathbf{v}_k \quad (12)$$

where μ_k and $\mathbf{v}_k$ are the viscosity and mass-averaged velocity of phase k, respectively, where the latter is related to the fluxes in the system,

$$\mathbf{v}_k = \frac{\sum_{i \neq s} M_i \mathbf{N}_{i,k}}{\rho_k} \quad (13)$$

The transient term in the momentum conservation equation represents the accumulation of momentum with time and the second term describes convection of the momentum flux. The two terms on the right side represent the divergence of the stress tensor.

For porous materials where there is flow (inherent in many porous electrodes), the Navier–Stokes equations are not used and instead one uses the more empirical Darcy's law for the momentum transport equation,

$$\mathbf{v}_k = -\frac{k_k}{\mu_k} \nabla p_k \quad (14)$$

where k_k is effective permeability of phase k. Since Darcy's law is first-order and Navier–Stokes is second order, one loses a boundary condition which is the no-slip condition at the interior walls of the porous medium. There are extensions to Darcy's law that try and address this, but these approaches are beyond the scope of this chapter and the reader is referred to texts on this subject (e.g., [5]).

Material Transport Equations

In the above conservations equations, expressions for the fluxes are required. This has already been accounted for in the energy and momentum balances (Eqs. 11 and 12, respectively) to provide second-order equations; the reason being that they are highly coupled and remain general for many systems. However, understanding the fluxes and transport expressions for the material species including ions (Eq. 2) is critical in determining the resistances in the ionic and electronic phases and the overall response of the porous electrode; thus, they are discussed in more detail.

For transport of charged species, one can use either a dilute-solution or concentrated-solution approach. In general, the concentrated-solution approach is more rigorous but requires more knowledge of all of the various interactions. For the dilute-solution approach, one can use the Nernst-Planck equation,

$$\mathbf{N}_{i,k} = -z_i u_i F c_{i,k} \nabla \Phi_k - D_i \nabla c_{i,k} + c_{i,k} \mathbf{v}_k \quad (15)$$

where u_i is the mobility of species i. In the equation, the terms on the right side correspond to migration, diffusion, and convection, respectively. For uncharged species, this first term on the right side can be neglected and the equation of convective-diffusion is obtained.

Multiplying Eq. 15 by $z_i F$ and summing over the species i in phase k,

$$F \sum_i z_i \mathbf{N}_{i,k} = -F^2 \sum_i z_i^2 u_i c_{i,k} \nabla \Phi_k - F \times \sum_i z_i D_i \nabla c_{i,k} + F \times \sum_i z_i c_{i,k} \mathbf{v}_k \quad (16)$$

and noting that the last term is zero due to electroneutrality (convection of a neutral solution cannot move charge) and using the definition of current density (Eq. 6), one gets

$$\mathbf{i}_k = -\kappa_k \nabla \Phi_k - F \sum_i z_i D_i \nabla c_{i,k} \quad (17)$$

where κ_k is the conductivity of the solution of phase k

$$\kappa_k = F^2 \sum_i z_i^2 c_{i,k} u_i \quad (18)$$

When there are no concentration variations in the solution, Eq. 18 reduces to Ohm's law,

$$\mathbf{i}_k = -\kappa_k \nabla \Phi_k \quad (19)$$

Ohm's law is typically used for modeling electron conduction, typically with an effective conductivity that accounts for the porous nature of the electrode,

$$\mathbf{i}_1 = -\sigma_0 \varepsilon_1^n \nabla \Phi_1 \quad (20)$$

where ε_1 and σ_0 are the volume fraction and electrical conductivity of the electronically conducting phase, respectively, and n is a power depending on the tortuosity of the phase (a Bruggeman factor of $n = 1.5$ is often assumed [6]).

This dilute-solution approach does not account for interaction between the solute molecules which could be dominant for many concentrated ionic solutions. Also, this approach will either use too many or too few transport coefficients depending on if the Nernst-Einstein relationship is used to relate mobility and diffusivity,

$$D_i = RTu_i \tag{21}$$

which only rigorously applies at infinite dilution. Thus, the concentrated-solution-theory approach is recommended, however the Nernst-Planck equation can be used in cases where most of the transport properties are unknown or where the complex interactions and phenomena being investigated necessitate simpler equations.

The concentrated-solution approach starts with the original equation of multicomponent transport [7]

$$\mathbf{d}_i = c_i \nabla \mu_i = \sum_{j \neq i} K_{i,j} (\mathbf{v}_j - \mathbf{v}_i) \tag{22}$$

where $\mathbf{d}_i$ is the driving force per unit volume acting on species i and can be replaced by a electrochemical-potential (μ) gradient of species i, and $K_{i,j}$ are the frictional interaction parameters between species i and j. The above equation can be analyzed in terms of finding expressions for $K_{i,j}$'s, introducing the concentration scale including reference velocities and potential (reference ion) definition, or by inverting the equations and correlating the inverse friction factors to experimentally determined properties. Which route to take depends on the phenomena being studied [1] and is beyond the scope of this chapter. Replacing the $K_{i,j}$'s with a thermodynamic diffusion coefficient, $D_{i,j}$, and a concentration scale results in

$$\mathbf{N}_{i,k} = -\frac{D_{i,0}}{RT} c_{i,k} \nabla \mu_{i,k} + c_{i,k} \mathbf{v}_0 \tag{23}$$

where 0 denotes the solvent. This equation is very similar to the Nernst-Planck Equation 14, except that the driving force is the thermodynamic electrochemical potential, which contains both the migration and diffusive terms. In addition, since the electrochemical potential is used, this is undefined for a single ion species. Thus, one needs to define a reference ion as discussed in [1].

Charge-Transfer Expression

To model the charge-transfer reactions within the porous electrode (see Fig. 1), kinetic expressions are used. A typical electrochemical reaction can be expressed as

$$\sum_k \sum_i s_{i,k,h} M_i^{z_i} \rightarrow n_h \mathrm{e}^- \tag{24}$$

where $s_{i,k,h}$ is the stoichiometric coefficient of species i residing in phase k and participating in electron-transfer reaction h, n_h is the number of electrons transferred in reaction h, and $M_i^{z_i}$ represents the chemical formula of i having valence z_i.

The ion-transfer rate is equal to the electrochemical reaction rate at the electrodes (which is the source term, or transfer current density in Eq. 2). According to Faraday's law, the flux or species i in phase k and rate of reaction h is related to the current as

$$N_{i,k} = \sum_h r_{h,i,k} = \sum_h s_{i,k,h} \frac{i_h}{n_h F} \tag{25}$$

where i_h refers to the transfer current density, i.e. current (i) per unit geometric area of the electrode. The rate of a chemical reaction is related to its concentration and temperature through an Arrhenius relationship,

$$r_h = k \exp\left(\frac{-E_a}{RT}\right) \prod_i \left(\frac{a_i}{a_i^{\mathrm{ref}}}\right)^{m_i} \tag{26}$$

where k is the rate constant, m_i is the order of reaction for species i, and a_i is the activity of reactant i, which as discussed above requires that an appropriate reference ion is chosen as the activity of a single ion is undefined [1]. For modeling purposes, especially with the multi-electron transport of species, it is often easiest to use a semi-empirical equation to describe the reaction rate, namely, the Butler-Volmer equation [1, 8],

$$i_h = i_{0_h}\left[\exp\left(\frac{\alpha_a F}{RT}\left(\Phi_k - \Phi_p - U_h^{\text{ref}}\right)\right)\prod_i^a\left(\frac{a_i}{a_i^{\text{ref}}}\right)^{s_{i,k,h}} - \exp\left(\frac{-\alpha_c F}{RT}\left(\Phi_k - \Phi_p - U_h^{\text{ref}}\right)\right)\prod_i^c\left(\frac{a_i}{a_i^{\text{ref}}}\right)^{-s_{i,k,h}}\right] \tag{27}$$

where i_h is the transfer current between phases k and p due to electron-transfer reaction h, the products are over the anodic and cathodic reaction species, respectively, α_a and α_c are the anodic and cathodic transfer coefficients, respectively, and i_{0_h} and U_h^{ref} are the exchange current density per unit catalyst area and the potential of reaction h evaluated at the reference conditions and the operating temperature, respectively.

In the above expression, the composition-dependent part of the exchange current density is explicitly written, with the multiplication over those species in participating in the anodic or cathodic direction. The reference potential is determined by thermodynamics as described elsewhere [1], and can commonly be determined using a Nernst equation,

$$U = U^0 - \frac{RT}{z_i F}\ln\left(\prod a_i^{s_i}\right) \tag{28}$$

where s_i is the stoichiometry of species i; and the activity of the species is often approximated by its local concentration of the species. If the reference conditions are the same as the standard conditions, then U^{ref} has the same numerical value as U^0.

The term in parentheses in Eq. 27 can be written in terms of an electrode overpotential

$$\eta_h = \Phi_k - \Phi_p - U_h^{\text{ref}} \tag{29}$$

If the reference electrode is exposed to the conditions at the reaction site, then a surface or kinetic overpotential can be defined

$$\eta_{s_h} = \Phi_k - \Phi_p - U_h \tag{30}$$

where U_h is the reversible potential of reaction h. The surface overpotential is the overpotential that directly influences the reaction rate across the interface. Comparing Eqs. 29 and 30, one can see that the electrode overpotential contains both a concentration and a surface overpotential for the reaction.

For very high exchange current densities (i.e., rapid reactions), a linearized form of Eq. 27 can often be used. For very slow reaction kinetics, either the anodic or cathodic term dominates the kinetics, and so the other term is often ignored, yielded what is known as a Tafel equation for the kinetics. Often, more complicated expressions than that of Eq. 27 are used. For example, if the elementary reaction steps are known, one can write down the individual steps and derive the concentration dependence of the exchange current density and the kinetic equation. Other examples include accounting for surface species adsorption or additional internal or external mass transfer to the reaction site [9]. All of these additional issues are beyond the scope of this chapter, and often an empirically based Butler-Volmer equation is used for modeling the charge transfer in porous electrodes.

Boundary Conditions and Implementation

To model a porous electrode involves the material, energy, momentum, and charge-balance expressions where the species fluxes are accounted for by the transport equations. In addition, the equations require boundary conditions in order to solve, which can vary depending on the electrode. As a simple example, let's take a dilute solution of two cations (A, B) and an anion (M), and assume Darcy's law and electroneutrality hold. The modeling variables and equations including possible boundary conditions are shown in Table 1. It should be noted that for concentrated solutions the concentrated-solution theory equations would be needed as well as an additional equation for the solvent.

Macroscopic Modeling of Porous Electrodes, Table 1 List of equations and boundary conditions for a porous electrode with an electronic conducting phase and an ionic conducting phase comprised of cations A and B and anion M

Variable	Phenomena	Equation	Boundary condition
Primary			
c_A	Nernst-Planck	15	Concentration
c_B	Nernst-Planck	15	Concentration
c_M	Sum of mole fraction	3	–
Φ_1	Ohm's law	20	Reference potential
Φ_2	Electroneutrality	7	–
p	Darcy's law	14	Pressure
T	Energy	11	Temperature
Secondary			
$\mathbf{v}$	Definition	13	–
$\mathbf{N}_A$	Mass-balance	2	Flux or concentration at other boundary
$\mathbf{N}_B$	Mass-balance	2	Flux or concentration at other boundary
$\mathbf{N}_M$	Mass-balance	2	Flux
$\mathbf{i}_1$	Charge balance	4	Current or potential
$\mathbf{i}_2$	Definition	6	–

In terms of the charge transfer, the charge conservation equation (4) (assuming electroneutrality, i.e., no double-layer effects) would be

$$\nabla \cdot \mathbf{i}_2 = -\nabla \cdot \mathbf{i}_1 = a_{1,2} i_{h,1-2} \tag{31}$$

where $a_{1,2}$ is the reaction area per unit volume and the transfer current density is given by a kinetic expression (Eq. 30). This transfer current density is also used in the material-balance equations 2.

From Fig. 1 and the above equations, one can see that the reaction distribution will depend on the various transport and reaction phenomena. For limiting cases, if one of the reactants is limiting in terms of transport or concentration, then the reaction or transfer current-density profile will be exponential towards the place with the incoming limiting reactant. If both are equally limiting, one obtains a parabolic reaction profile. If the reaction itself is limiting, then one obtains a uniform reaction profile.

Future Directions

The above discussion outlines the modeling methodology and equations for understanding porous electrodes from a macrohomogeneous view. As noted throughout, while the general equations hold, there are many nuances that can go into the expressions. For example, one can consider agglomerate models, coupling of the equations at the macro and mesoscales with phenomena and models at the nano and particle scales, etc. It is clear that reaction within a porous electrode is complex and nonlinear, making mathematical simulation of it very useful.

Cross-References

▶ Electrochemical Impedance Spectroscopy (EIS) Applications to Sensors and Diagnostics
▶ Thermal Effects in Electrochemical Systems

References

1. Newman J, Thomas-Alyea KE (2004) Electrochemical systems, 3rd edn. Wiley, New York
2. Fogler HS (1992) Elements of chemical reaction engineering, 2nd edn. Prentice-Hall, Upper Saddle River
3. Weber AZ, Newman J (2004) Modeling transport in polymer-electrolyte fuel cells. Chem Rev 104:4679–4726
4. Bird RB, Stewart WE, Lightfoot EN (2002) Transport phenomena, 2nd edn. Wiley, New York
5. Dullien FAL (1992) Porous media: fluid transport and pore structure, 2nd edn. Academic, New York

6. Bruggeman DAG (1935) Calculation of various physics constants in heterogenous substances I Dielectricity constants and conductivity of mixed bodies from isotropic substances. Annalen Der Physik 24:636–664
7. Pintauro PN, Bennion DN (1984) Mass-transport of electrolytes in membranes. 1. Development of mathematical transport model. Ind Eng Chem Fundam 23:230–234
8. Vetter KJ (1967) Electrochemical kinetics. Academic, New York
9. Weber AZ, Balasubramanian S, Das PK (2012) Proton exchange membrane fuel cells. In: Sundmacher K (ed) Advances in chemical engineering: fuel cell engineering: model-based approaches for analysis, control and optimization, vol 41. Elsevier, Amsterdam, pp 65–144

Magnesium Smelter Technology

Geir Martin Haarberg
Department of Materials Science and Engineering, Norwegian University of Science and Technology (NTNU), Trondheim, Norway

Introduction

The annual production of primary magnesium metal was about 800,000 metric tons in 2010, while the production capacity was about 1.3 million tons [1]. Today, magnesium is mainly produced by the Pidgeon process, which involves the reduction of MgO by silicon in the form of ferrosilicon. Electrolysis was the dominant production route in the 1990s. The thermal process is presently more economic, but electrowinning may be more sustainable and could again become more important in the near future. However, electrolysis is still important for producing magnesium for reducing $TiCl_4$ within the Kroll process for titanium production. Information about innovations and performance data related to magnesium electrolysis has traditionally been rather secretive. Therefore, the literature is scarce. There are some useful review articles [2–4].

Small quantities of electrolytic magnesium were produced by Davy and Faraday in the first half of the 1800s. The modern industrial electrolysis cells were developed by IG Farben from the 1920s.

MgO does not have appreciable solubility in molten salts, so the selected raw material added to the electrolyte is $MgCl_2$. Current technologies all use anhydrous $MgCl_2$ either in the form of liquid or solid pellets as the feedstock. The molten electrolyte is a mixture of chlorides, mainly NaCl and $CaCl_2$ at ~750 °C. Liquid magnesium droplets are formed in a vertical electrode design. Both monopolar and bipolar cell technologies are available. The current can exceed 400 kA, and the best cells have an energy consumption of ~13 kWh/kg Mg.

Raw Materials and Electrolyte

The main ores rich in magnesium are magnesite ($MgCO_3$), dolomite, ($MgCO_3 \cdot CaCO_3$), and carnallite ($MgCl_2 \cdot KCl \cdot 6H_2O$). Also natural brines and seawater represent as important sources of magnesium. Anhydrous magnesium chloride is produced by dehydration of magnesium chloride solutions or carbochlorination of magnesium oxide. Anhydrous $MgCl_2$ is very hygroscopic, so dissolved oxygen and hydrogen containing species will be present in the molten electrolyte [5].

In melts containing $MgCl_2$, hydroxides either decompose or react causing the formation of MgO, HCl, MgOHCl, and H_2O. The following equilibria will be established between these species:

$$H_2O + MgCl_2 = MgO + 2HCl \qquad (1)$$

$$H_2O + MgCl_2 = MgOHCl + HCl \qquad (2)$$

$$MgOHCl = MgO + HCl \qquad (3)$$

The solubilities of HCl and H_2O depend on the electrolyte composition. In melts containing $MgCl_2$, the solubility of H_2O is linked to hydrolysis reactions.

In melts containing $MgCl_2$, the oxide solubility is very low due to the high stability of MgO. Oxide solubilities in $MgCl_2$ containing melts have been reported [6]. The solubility increases by increasing activity of $MgCl_2$. MgOHCl, which is formed in melts containing $MgCl_2$, is a relatively stable dissolved species. HCl and H_2O may be formed due to chemical reactions but are generally unstable in the absence of atmosphere containing these species. MgO is present as a dissolved complex [6].

$$MgO + MgCl_2 = Mg_2OCl_2 \quad (4)$$

An important aspect of the process is the fact that dissolved MgOHCl is electroactive and may be reduced at the cathode according to the following reaction [7]:

$$2MgOHCl + 2e = H_2 + 2MgO + 2Cl^- \quad (5)$$

Hence, the formation and possible precipitation of MgO will take place at the cathode.

The selected electrolyte composition is based on the effects of the various components on the physicochemical properties of the molten electrolyte. However, in some cases, the $MgCl_2$ feedstock contains small amounts of another metal chloride such as KCl. Electrical conductivity, density, vapor pressure, interfacial tension, and metal solubility are important properties. A typical composition can be 45 % NaCl, 35 % $CaCl_2$, 10 % KCl, and 10 % $MgCl_2$. The content of $MgCl_2$ must not be too low to avoid codeposition of sodium, although some sodium will always be deposited at a low activity.

Magnesium is soluble in molten chlorides containing $MgCl_2$, which is important for the current efficiency. It is known that dissolved magnesium is present as a subvalent ion Mg_2^{2+} [8].

Electrode Processes and Current Efficiency

The main cell reaction is the electrodecomposition of dissolved $MgCl_2$.

$$MgCl_2(diss) = Mg(l) + Cl_2(g) \quad (6)$$

The cathode process involves the nucleation of Mg on the steel cathode. The electron transfer is very fast, and the small contribution to the cathodic overvoltage is due to the diffusion of Mg (II) species through the boundary layer near the cathode [9].

The anode process involves the formation of adsorbed chlorine and the subsequent desorption, which will be accompanied by a considerable overvoltage of about 0.2 V at the applied current density. The presence of dissolved oxygen containing species gives rise to a slow but significant consumption of the graphite anodes due to the formation of CO_2 [10].

The main reason for the loss in current efficiency is the recombination reaction between the products, most likely in the form of dissolved species.

$$Mg(diss) + Cl_2(diss) = MgCl_2(diss) \quad (7)$$

Under normal conditions, the rate of this back reaction is probably controlled by diffusion of dissolved Mg. The formation of Mg droplets is very important for the success of the process. Very small droplets may have a long residence time in the electrolyte and are likely to cause a more significant loss in the current efficiency. Such conditions may occur when the wetting properties at the cathode deteriorate which may take place at high contents of moisture. The presence of certain dissolved impurity elements such as silicon and boron may also cause inferior wetting of Mg [3].

Good wetting of Mg to the steel cathode substrate may occur at low contents of dissolved hydrogen and oxygen containing species. However, the formation of very large Mg droplets or a film of liquid Mg on the cathode may have an adverse effect on the operation.

Cell Technologies and Operating Conditions

Both products, liquid Mg and gaseous Cl_2, are lighter than the electrolyte. Therefore, it is challenging to avoid recombination and loss in current efficiency. Older electrode designs were equipped with diaphragms or separators near the cathode [4]. Modern designs have utilized the so-called gas lift to separate the liquid magnesium droplets from the chlorine bubbles [11]. The upward circulation of the electrolyte is set up by the massive evolution of chlorine gas. Magnesium droplets are formed on the cathodes of steel. These droplets are detached from the cathode and rise up toward the surface. Due to the lower buoyancy, the majority of the Mg droplets will not reach the surface but enter the metal collecting compartment. One of the successful monopolar designs for achieving a good separation of the products is shown in Fig. 1. Modern cells operate at ~400 kA and a current efficiency up to ~95 % and an energy consumption of ~15 kWh/kg Mg. The cathodic current density is ~0.8 A/cm^2. The temperature is ~700 °C. The interelectrode distance is fixed, so there is little room for adjusting the heat production. The cells are commonly equipped with separate heating options through auxiliary AC electrodes placed in the metal collection compartment.

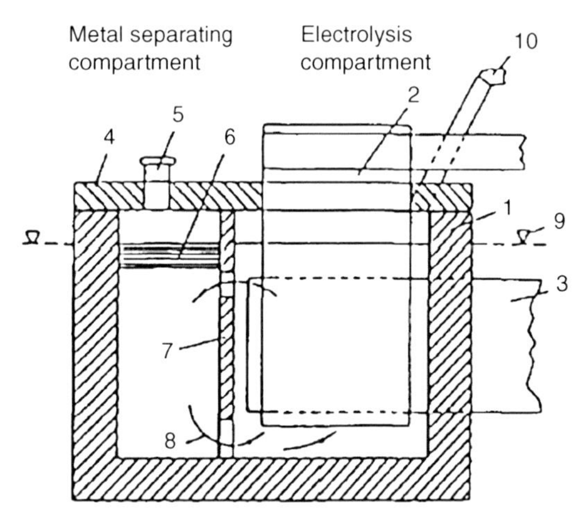

Magnesium Smelter Technology, Fig. 1 Monopolar magnesium electrolysis cell design [11]. *1* refractories, *2* graphite anode, *3* steel cathode, *4* refractory cover, *5* metal outlet, *6* metal, *7* partition wall, *8* electrolyte flow, *9* electrolyte level, *10* chlorine outlet

Modern bipolar technologies were developed in the 1980s [12, 13]. Figure 2 shows a bipolar magnesium cell from a patent [12] where the electrodes are made from graphite which are coated by steel on the cathode side. Bipolar cells have shorter interelectrode distances, but the current efficiency is much lower due to bypass current, so the energy consumption is a little higher than that of the monopolar cells.

The strong circulation of the electrolyte near the cathode increases the rate of the diffusion of Mg (II) species through the boundary layer. This situation helps to achieve a low cathodic overpotential and a decrease in the reduced Na codeposition.

The cell life can be about 4–5 years, and the service life of the graphite anodes can match the lifetime of the cell, while the steel cathodes can be used for an additional period [11]. The presence of dissolved oxygen containing species gives rise to the formation of CO_2 at normal operation anode potentials.

The formation of sludge containing Mg droplets, precipitated MgO, and electrolyte may take place at high contents of oxide and moisture.

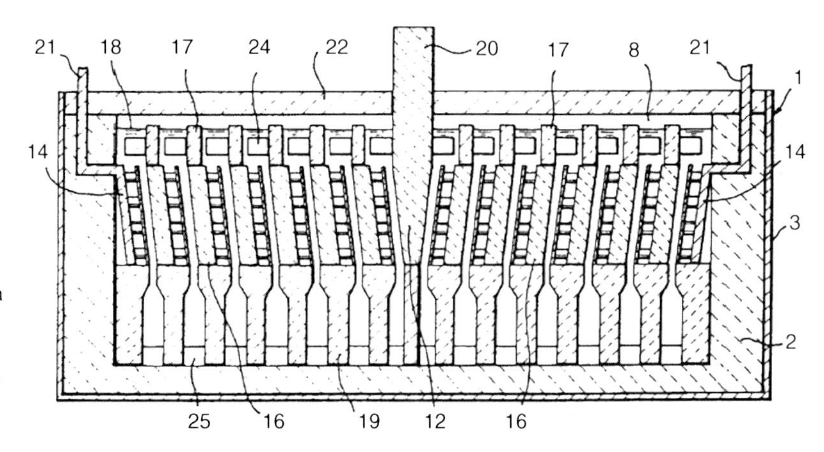

Magnesium Smelter Technology, Fig. 2 Bipolar magnesium electrolysis cell [12]. *12* anode, *14* cathode, *20* terminal anode connection, *21* terminal cathode connection

Future Directions

Electrolysis must be developed and optimized to be able to compete with thermal routes such as the Pidgeon process. Better control of the levels of dissolved oxygen and hydrogen containing species is of crucial importance for achieving a high current efficiency. The cell life and especially the service life of the anodes must be increased.

Cross-References

► Electrolytes, Classification
► High-Temperature Molten Salts

References

1. US Geological Survey. http://minerals.usgs.gov/minerals/pubs/commodity/magnesium/myb1-2010-mgmet.pdf
2. Strelets KL (1977) Electrolytic production of magnesium. Keterpress Enterprises, Jerusalem
3. Kipouros GJ, Sadoway DR (1987). In: Mamantov G (ed) Advances in molten salt chemistry, vol 6. Elsevier, Amsterdam
4. Høy-Petersen N (1990) From past to future. Light Metal Age 48:14–16
5. Haarberg GM, Tunold R, Osen KS (2001) Voltammetric characterization of dissolved oxygen and hydrogen containing species in chloride melts. In: Rosenkilde C (ed) Jondal 2000, Proceedings, international symposium, p 147
6. Boghosian S, Godø A, Mediaas H, Ravlo W, Østvold T (1991) Acta Chem Scand 45:145
7. Vilnyanski YE, Savinkova EI (1957) J Appl Chem USSR 28:827
8. van Norman JD, Egan JJ (1963) J Phys Chem 67:2460
9. Martinez AM, Børresen B, Haarberg GM, Castrillejo Y, Tunold R (2004) Electrodeposition of magnesium from $CaCl_2$-NaCl-KCl-$MgCl_2$ melts. J Electrochem Soc 151:C508–C513
10. Mohamedi M, Børresen B, Haarberg GM, Tunold R (1999) Anodic behaviour of carbon electrodes in CaO-$CaCl_2$ melts at 1123 K. J Electrochem Soc 146:1472
11. Wallevik O, Amundsen K, Faucher A, Mellerud T (2000) Magnesium electrolysis – a monopolar viewpoint. In: Kaplan HI, Hryn J, Clow B (eds) Magnesium technology 2000. The Minerals, Metals & Materials Society, Warrendale, pp 13–16
12. Ishizuka H (1985) Method for electrolytically obtaining magnesium metal. US Patent 4,495,037
13. Sivilotti OG (1985) Metal production by electrolysis of a molten electrolyte. US Patent 4,514,269

Magnetic Resonance Spectroscopy

Mario Krička and Rudolf Holze
AG Elektrochemie, Institut für Chemie, Technische Universität Chemnitz, Chemnitz, Germany

Magnetic resonance spectroscopies are methods capable of detecting transitions of spin orientations of electrons or atomic nuclei between states separated energetically under the influence of an external magnetic field. Transitions involving the spin of the nucleus of an atom with a nonzero magnetic moment are studied with nuclear magnetic resonance spectroscopy (NMR), whereas transitions involving the spin of unpaired electrons in paramagnetic samples are investigated with electron spin resonance spectroscopy (ESR) (Frequently this method has been called electron paramagnetic resonance spectroscopy (EPR) because the presence of one or several unpaired electrons being a precondition for this spectroscopy is also closely related to the phenomenon of paramagnetism.) [1]. Both methods are widely employed in analytical chemistry. NMR preferably of protons and ^{13}C-atoms and further selected atoms is presumably the most important method in analytical organic chemistry. ESR is used less frequently in studies of free radicals (chemical species with unpaired electrons) observed typically in organic reactions and in investigations of transition metal ions and paramagnetic substances.

Because in many electrochemical reactions, particularly in electroorganic ones, radicals are formed as reactive intermediates, ESR has been applied frequently to studies of the mechanism and the kinetics of these reactions [1–3]. Although possible, NMR spectroscopy has been used infrequently and only in very recent experiments mainly because of the considerably larger experimental effort [4]. With NMR spectroscopy information about surface structure, surface diffusion and electron spillover from the metal electrode onto an adsorbate can be obtained. So far an application as broad as that of ESR has not materialized. Both methods have been employed infrequently to monitor the concentration of species involved in electrochemical reactions in order to establish reaction kinetic parameters, activation energies, etc. These applications do not fall within the scope of this entry, they are not considered here.

From the fundamentals of magnetic resonance spectroscopy, basic building blocks and functional elements of the experimental can be derived easily: The sample has to be brought into a magnetic field of appropriate strength. The energy corresponding to the difference between the two states of the nuclear or electron spin is supplied as electromagnetic radiation of suitable frequency, generally in the radio-frequency range or, more precisely, in the microwave range for ESR and in the UHF-range for NMR (UHF: ultra high frequency) [2]. Because of the modes of propagation of radiation in the microwave range, waveguides have to be used instead of cables. The sample is inserted into a cavity at the end of a waveguide in between the poles of the magnet. Since the actual resonance conditions are shifted from the reference values stated above, either the strength of the magnetic field (i.e., the magnetic flux) or the radiation frequency has to be tuned in order to detect the actual resonance. A modulation of the frequency of the radiation supplied at constant intensity (cw: continuous wave) with the necessary precision and stability is somewhat difficult to obtain for reasons related to high-frequency electronics. A direct modulation of the magnetic field by simply changing the current flowing across the coils of the electromagnets employed frequently in NMR and practically exclusively in ESR is cumbersome because of the need to control pretty large electric currents. In the case of very strong magnets using superconducting coils this is impossible anyway. A more effective approach is the addition of small coils attached to the poles of the main magnet. The additional magnetic flux provided by these modulation coils can be controlled easily. Consequently the resonance conditions are probed by changing the magnetic field slowly. Absorption of electromagnetic radiation can be detected by various means. The actual absorption in particular with ESR is rather small because of the small difference in occupation of both states. An increase of sensitivity can be obtained by applying sophisticated amplification and detection methods (phase sensitive detection with lock-in amplifiers). In the case of ESR, this results in spectra being equivalent to the first derivative of the actual absorption spectrum. More recently the Fourier transform (FT) technique as applied to vibrational spectroscopy has been adapted for ESR and NMR. The electromagnetic radiation is supplied as a pulse. Detection and data manipulation is more complex, the advantage is a greatly enhanced sensitivity. This is described in more detail in textbooks of spectroscopy.

Basically an NMR or ESR spectrometer is composed of a magnet, a radio-frequency generator, an additional generator driving the field modulation coils, the necessary detector and data manipulation and storage electronics. The sample is inserted into the magnet between its poles. Some preparation of the sample is generally necessary; materials containing ESR- or NMR-active substances have to be avoided as sample holder. Various types of cuvettes for solid, liquid, and gaseous samples are in use. In NMR spectroscopy almost exclusively quartz tubes of various diameters free of paramagnetic impurities are used. In ESR spectroscopy, different shapes (tubes, flat cells) of cuvettes are used depending upon the type of cavity (cylindrical ones for the former, rectangular ones for the latter cuvettes).

For electrochemical applications the experimental arrangement is rather simple. Because of the broad application of ESR, this method is treated first; some additional information on NMR in electrochemistry can be found at the end of this entry. In ESR experiments the spectrum can be recorded with the species under investigation created just inside the spectrometer (intramuros generation, subsequently treated as in situ method) or outside the spectrometer (extra muros). In the latter case the sample has to be transferred by means of a flow apparatus or by removal of a small sample from the electrochemical cell, which is put into a standard ESR cuvette.

The latter procedure is similar to an ex situ experiment carrying some inherent sources of error because of, e.g., limited lifetime or subsequent chemical reactions of the species initially created by the electrochemical reaction. Since with respect to the ESR spectrometer no particular design of the cuvette is necessary, the latter procedure will not be discussed in detail.

Generation of species and their detection inside the cavity of the ESR spectrometer need some consideration of experimental requirements of the spectrometer and an appropriate electrochemical cell design. Since most electrochemical reactions proceed at metal electrodes, a very fundamental problem is encountered in any attempt to obtain ESR spectra of radicals still adsorbed, i.e., interacting strongly, on the electrode surface. This interaction between the unpaired electron of the radical and the electrons in the metallic conductor will quench the free spin; no ESR spectrum will be observed. This is different with semiconductor or insulator electrode surfaces. The quenching can be suppressed by coating the electrode with a layer of nonmetallic material (chemically modified electrode), but obviously this may change the interesting properties of the electrode considerably. Nevertheless ESR-active species can be detected as soon as they leave the electrode surface and stay in the electrolyte solution at a sufficiently large concentration for a time long enough in order to allow measuring a spectrum. The first requirements seem somewhat odd at first glance, because ESR is a rather sensitive spectroscopy. Paramagnetic species at a concentration as low as 10^{-10} mol dm^{-3} can be detected easily. Unfortunately the high reactivity of organic radicals tends to keep the stationary concentration low; in addition the components of the electrochemical cell (electrolyte, solvent, electrodes) reduce the sensitivity of the spectrometer considerably in particular by increasing dielectric losses. Detection of radicals being present in only very small concentrations or of short lifetime can be facilitated by using "spin traps." These are mostly organic compounds which form stable radicals by reaction with the radicals formed during the investigated process. In many cases these spin traps contain a nitroso group. The observed ESR spectra are more or less complicated depending on the type of spin trap and the trapped radical. A fairly simple spectrum is obtained by using *t*-nitroso-butane (NtB) as a spin trap (for further details see below). The adduct formed with hydrogen radicals causes the spectrum depicted below (Fig. 1).

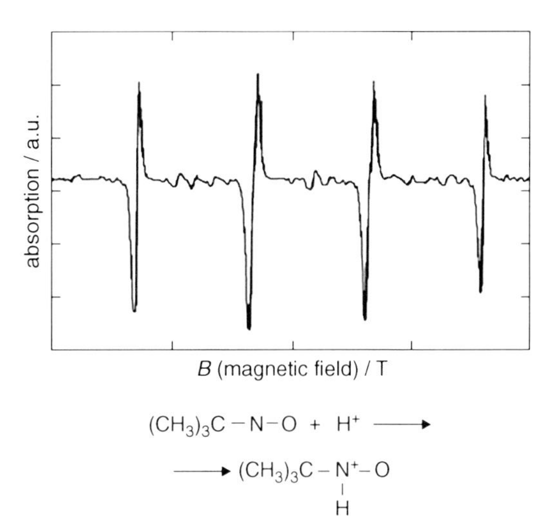

$$(CH_3)_3C-N-O + H^+ \longrightarrow (CH_3)_3C-N^{+}(H)-O$$

Magnetic Resonance Spectroscopy, Fig. 1 ESR spectrum of the hydrogen adduct of NtB generated electrochemically at $E_{SCE} = 0.2$ V in an aqueous solution of 0.5 M $LiClO_4$ [5]

A first cell design with two electrodes was reported by Maki and Geske [6] (Fig. 2).

A platinum wire used as a working electrode was mounted in the center of a quartz tube serving as a cell vessel in the middle of the ESR cavity at the position of highest sensitivity. A platinum wire as the counter electrode was mounted in the tube at a position outside the cavity of the spectrometer. Any species created at this counter electrode will not be detected; in addition the distance between both electrodes reduces the risk of unwanted electrochemical reactions at the working electrode of species created at the counter electrode. The small actual surface area of the working electrode limited the rate of formation of species to be studied, very precise positioning was required. Because of the simple design, nevertheless this cell design has been used continuously despite the obvious drawbacks and limitations [7]. Using a vanadium wire as a working electrode, VO^{2+}

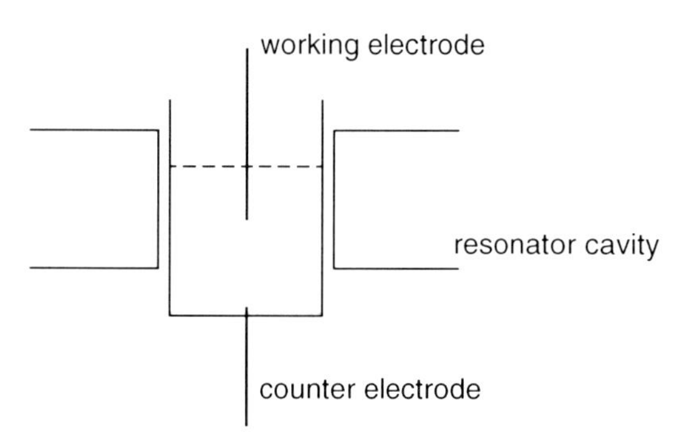

Magnetic Resonance Spectroscopy, Fig. 2 Cell design for ECESR spectroscopy (According to Maki and Geske [6])

ions could be identified as electrooxidation products [8]. Numerous cell designs have been reported; for a critical overview, see [9].

A considerable improvement in terms of electrode potential distribution inside the ECESR cell, available electrolyte solution volume, and ease of manufacturing was provided with the cell design of Allendoerfer et al. [10], subsequently improved by Heinzel et al. [5, 11]. A metal wire coil of the working electrode material is inserted into a quartz glass tube. Because of the limited penetration depth of the microwaves, only the solution volume enclosed by the wire surface and the inner glass wall is probed. The counter and the reference electrode inserted centrally inside the working electrode coil can be made without major constraints caused by cell or working electrode design. Because of the size of the cell, a less common large cylindrical microwave resonator cavity is required. The cell components and their placement inside the resonator between the magnet poles are shown in detail below (Fig. 3).

The superior electrochemical performance of this design is demonstrated with a cyclic voltammogram as shown below (Fig. 4).

The cell has been employed in studies of electrooxidation of organic fuels [5–13] and of nitrogen-containing monomers for the generation of intrinsically conducting polymers [14–16]. In numerous studies films of intrinsically conducting polymers deposited onto the working electrode have been investigated [17–19].

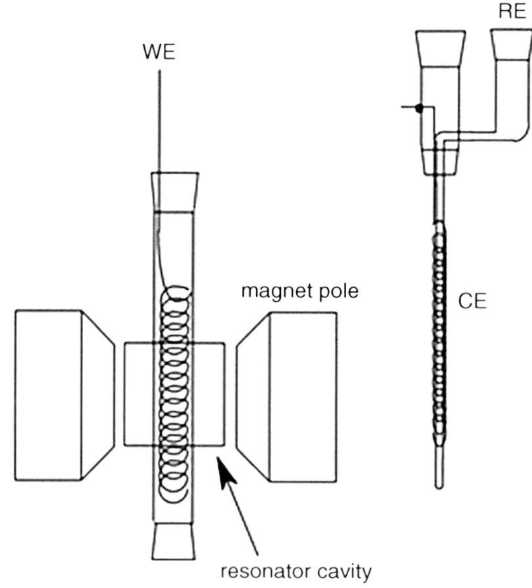

Magnetic Resonance Spectroscopy, Fig. 3 The components of an ECESR cell (According to Heinzel et al. [5])

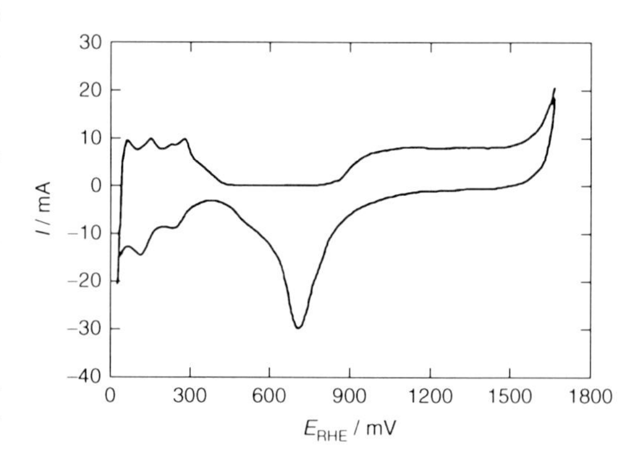

Magnetic Resonance Spectroscopy, Fig. 4 Cyclic voltammogram of a platinum wire-working electrode inside an ECESR cell according to Heinzel et al.; electrolyte solution 0.5 **M** H_2SO_4, $dE/dt = 50$ mV s^{-1}; nitrogen purged [12]

Because of the high degree of delocalization, the radical cations created by oxidation (*p*-doping) of the film show only a single line. The amplitude of the signal corresponds to the concentration of free spins and thus presumably to the number of mobile charge carrier. In a plot of ESR spectra obtained as a function of electrode potential this can be illustrated below (Fig. 5).

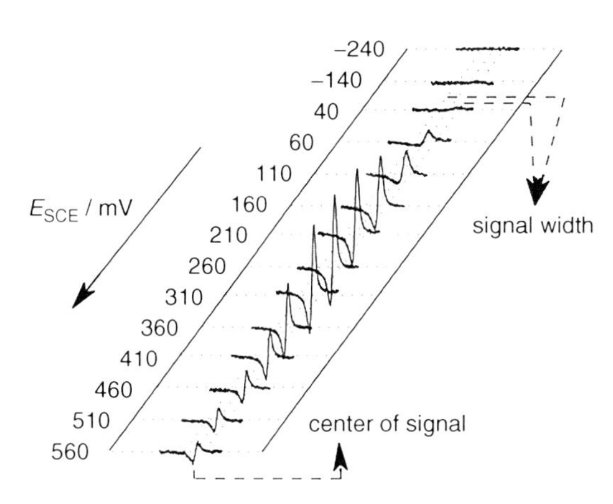

Magnetic Resonance Spectroscopy, Fig. 5 ESR spectra of a film of polyaniline in a solution of 0.1 **M** $LiClO_4$ in acetonitrile

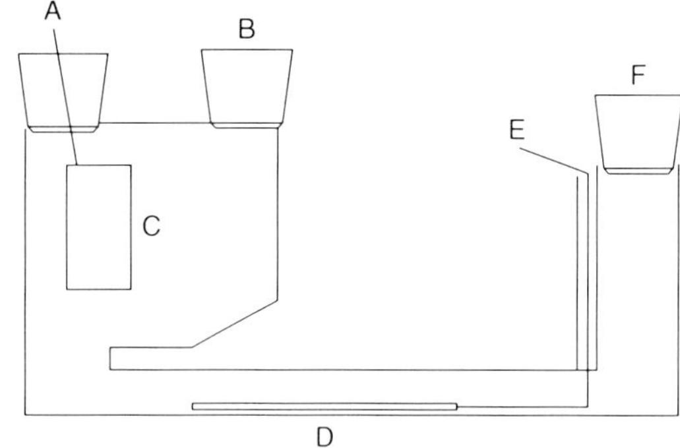

Magnetic Resonance Spectroscopy, Fig. 6 Schematic drawing of an electrochemical cell for in situ NMR spectroscopy [21]. *A* counter electrode connection, *B* reference electrode joint, *C* counter electrode, *D* platinum black working electrode, *E* working electrode connection, *F* purge connection

The striking similarity, which in some cases even extends to small asymmetries in line shape, has caused early erroneous assignments of the line as being caused by mobile electrons as observed with ESR in case of metals by Dyson (Dysonian line, [20]).

Cell designs for in situ NMR spectroscopy with electrochemical cells are scant. Because of the low sensitivity designs with working electrodes having large active surface areas (powder electrodes, metal coated inert supports like silica or alumina) have been described [21, 22]. This is a result of the fact, that in a typical NMR experiment, about 10^{19} spins are required. Assuming 10^{15} surface atoms on 1 cm^2 metal surface area at complete coverage of all adsorption sites, 1 m^2 working electrode surface area is required. A typical design is shown in a schematic drawing below (Fig. 6).

If continuous electrode potential control during acquisition of the NMR spectrum is desired, particular attention has to be paid to proper shielding and decoupling of the electrical wiring between potentiostat and NMR probe head.

Results reported so far pertain in particular to ^{13}C-NMR spectroscopy of adsorbed CO- and CN^- ions on platinum [21, 23, 24]. Besides, information about surface diffusion and electronic adsorbate-surface interaction data on the effect of the strong electric field at the electrochemical interface (typically 10^7 V cm^{-1}) have been reported. With both adsorbates the ^{13}C resonance becomes more shielded at more positive electrode potentials as expected when assuming an adsorbate attachment via the carbon atom. These results were supported with data from ^{195}Pt-NMR spectroscopy [25]. Differences in the electrooxidation of methanol and CO on carbon-supported platinum have been associated with different linewidths of the ^{13}C signal [26].

The use of NMR spectroscopy with various species like H, F, Na, Al, or Li in studies of solid electrolytes has been reviewed elsewhere [27].

Future Directions

The rather expensive instrumentation and the nontrivial use of spectroelectrochemical cells will presumably limit the application of both spectroscopies. Instrumental developments enabling detection of small amounts of reactions intermediates with ECESR may broaden the range of studied systems. Despite the omnipresence of NMR in organic chemistry, widespread use in electrochemistry is hardly conceivable.

Cross-References

▶ Organic Electrochemistry, Industrial Aspects

References

1. Compton RG, Waller AM (1988) In: Gale RJ (ed) Spectroelectrochemistry. Plenum, New York, p 349
2. McKinney TM (1996) In: Bard AJ (ed) Electroanalytical chemistry vol. 19. Marcel Dekker, New York, p 109
3. Holze R (2003) Recent Res Dev Electrochem 6:111
4. Wieckowski A (2003) Extended abstracts of the 203rd meeting of the Electrochemical Society, Paris, 27 April–02 May, Ext Abstr 1249
5. Heinzel A (1985) PhD-dissertation, Universität Oldenburg
6. Maki AH, Geske DH (1959) Anal Chem 31: 1450; (1959) J Chem Phys 30: 1356; (1960) 33: 825; Geske DH, Maki AH (1960) J Am Chem Soc 82: 2671; (1961) 83: 1852
7. Kaim W, Ernst S, Kasack V (1990) J Am Chem Soc 112:173
8. Ameer MAM, Ghoneam AA (1995) J Electrochem Soc 142:4082
9. Holze R (2007) Surface and interface analysis: an electrochemists toolbox. Springer-Verlag, Heidelberg
10. Allendoerfer RD, Martincheck GA, Bruckenstein S (1975) Anal Chem 47: 890; Allendoerfer RD (1975) J Am Chem Soc 97: 218
11. Heinzel A, Holze R, Hamann CH, Blum JK (1989) Z Phys Chem NF 160:11
12. Holze R (1989) Habilitation thesis, Universität Oldenburg
13. Heinzel A, Holze R, Hamann CH, Blum JK (1989) Electrochim Acta 34:657
14. Hamann CH, Holze R, Köleli F (1990) DECHEMA-Monographie 121 297
15. Holze R, Köleli F, Hamann CH (1989) DECHEMA-Monographie 117 315
16. Holze R, Hamann CH (1991) Tetrahedron 47:737
17. Lippe J (1991) PhD-dissertation, Universität Oldenburg
18. Lippe J, Holze R (1991) Synth Met 41–43:2927
19. Holze R, Lippe J (1992) Bull Electrochem 8:516
20. Dyson FJ (1955) Phys Rev 98:349
21. Wu J, Day JB, Franaszczuk K, Montez B, Oldfield E, Wieckowski A, Vuissoz P-A, Ansermet J-P (1997) J Chem Soc (trans: Faraday) 93 1017
22. Day JB, Wu J, Oldfield E, Wieckowski A (1998) In: Lorenz WJ, Plieth W (eds) Electrochemical nanotechnology. Wiley-VCH, Weinheim, p 291
23. Tong YY, Kim HS, Babu PK, Wieckowski A (2002) J Am Chem Soc 124:468
24. Tong YY, van der Klink JJ (2003) In: Wieckowski A, Savinova ER, Vayenas CG (eds) Catalysis and electrocatalysis at nanoparticle surfaces. CRC Press, Boca Raton, p 455
25. Babu PK, Tong YY, Kim HS, Wieckowski A (2001) J Electroanal Chem 524–525:157
26. McGrath P, Fojas AM, Rush B, Reimer JA, Cairns EJ (2006) Electrochemical society spring meeting 209th, Denver, May 7–11, Ext Abstr #1097
27. Wagner JB Jr (1991) In: Varma R, Selman JR (eds) Techniques for characterization of electrodes and electrochemical processes. Wiley, New York, p 3

General References and Further Reading: On ESR-Spectroscopy

Atherton NM (1993) Principles of electron spin resonance. Ellis Horwood PTR, Prentice Hall/Chichester

Kirmse R, Stach J (1985) ESR-Spektroskopie. Akademie-Verlag, Berlin

Poole CP Jr (1996) Electron spin resonance. Dover, Mineola

Poole CP, Farach HA (1999) Handbook of electron spin resonance (Vol. 1). AIP, New York

Poole CP, Farach HA (1999) Handbook of electron spin resonance (Vol. 2). Springer, New York

Scheffler K, Stegmann HB (1970) Elektronenspinresonanz. Springer, Berlin

Weil JA, Bolton JR (2007) Electron paramagnetic resonance. Wiley, Hoboken

Wertz JE, Bolton JR (1986) Electron spin resonance. Chapman & Hall, New York

General References and Further Reading: On NMR-Spectroscopy

Akitt JW (1983) NMR and chemistry. Chapman & Hall, New York

Becker ED (1980) High resolution NMR. Academic, New York

Bovey FA (1988) Nuclear magnetic resonance spectroscopy. Academic, New York

Harris RK (1986) Nuclear magnetic resonance spectroscopy. Longman Scientific& Technical, Essex

Manganese Oxides

Naoaki Yabuuchi, Masataka Tomita and Komaba Shinichi

Department of Applied Chemistry, Tokyo University of Science, Shinjuku, Tokyo, Japan

Introduction

Electrochemical power sources, rechargeable batteries and supercapacitors (or electrical double-layer capacitor, EDLC), are attractive in the wide range of applications, such as mobile electrical devices, hybrid electric vehicle (HEV), and smart grid. In the past century, considerable research efforts have been done to study the manganese oxides as the electrode materials for the power sources. Crystallization processes of the manganese oxides are highly influenced by many factors, e.g., intergrowth of different structural units, cations' incorporation to the vacant sites, and structural water, leading to the complexity of the crystal structures. Typically, the manganese oxides constituted from mainly tetravalent manganese ions are simply denoted as "manganese dioxides (MnO_2)." The manganese dioxides are widely utilized as a positive electrode with a zinc negative electrode, i.e., carbon-zinc cells (dry cells) and alkaline cells. Electrolytic manganese dioxide (γ-type MnO_2) is typically used in the alkaline cell, which can be described as an intergrowth structure between pyrolusite (β-type MnO_2) and ramsdellite. In the alkaline cell, manganese oxides are electrochemically reduced with proton insertion into the bulk of particles for the charge compensation. This process can be described as follows:

$$Mn^{4+}O_2 + xH^+ + xe^- \leftrightarrow H_xMn_{(1-x)}{}^{4+}Mn_x{}^{3+}O_2 \quad (1)$$

In theory, one mole of proton can be inserted into MnO_2, which is accompanied by the manganese reduction from tetravalent to trivalent state. MnO_2 can also accommodate lithium ions, which are much larger than proton, to the vacant sites (Eq. 2):

$$Mn^{4+}O_2 + xLi^+ + xe^- \leftrightarrow Li_xMn_{(1-x)}{}^{4+}Mn_x{}^{3+}O_2 \quad (2)$$

Heat-treated MnO_2 is often used as the positive electrode materials for a primary lithium cell. Spinel-type $LiMn_2O_4$ is applied as the positive electrode material in the rechargeable and large-scale lithium-ion cells.

Manganese Dioxides for Supercapacitor

Since 1995, amorphous hydrated ruthenium dioxide, $RuO_2{\bullet}nH_2O$, has been extensively studied as the electrode materials for the redox capacitor [1–3]. $RuO_2{\bullet}nH_2O$ composite electrode (~100 μm thickness) exhibits extremely high specific capacitance (typically >700 F g^{-1}) in acidic media (e.g., H_2SO_4 aqueous solution). Although the ruthenium dioxide is one of the attractive electrode materials, the manganese dioxides are also worth studying as the electrode materials for the redox capacitor, i.e., material abundance in the earth's crust and its environmental benign. The use of manganese dioxides for the supercapacitor has been first reported by Goodenough's research group in 1999 [4]. Amorphous hydrated manganese dioxides, $MnO_2{\bullet}nH_2O$, were synthesized by simple precipitation reaction between $KMnO_4$ and Mn $(CH_3COO)_2$. $MnO_2{\bullet}nH_2O$ has large surface area (200–300 m^2 g^{-1} by BET measurement), and a capacitor-like electrochemical response in KCl aqueous electrolyte was reported. In addition to the precipitation technique from $KMnO_4$, many solution-based techniques can be used to prepare the hydrated manganese dioxides with high surface area, such as sol–gel and electrodeposition methods [5–8]. Mn_3O_4, which has originally low surface area, is also utilized after structural conversion into nanostructured birnessite-type manganese dioxide by the electrochemical stimulation process [9, 10].

Typical particle morphology of amorphous hydrated manganese dioxides, $MnO_2{\bullet}nH_2O$, prepared by the simple precipitation, and its electrochemical behavior in aqueous electrolyte solution are shown in Fig. 1.

As shown in Fig. 1b, a specific capacitance of $MnO_2{\bullet}nH_2O$ reaches 200 F g^{-1}, which corresponds to 60–100 μF cm^{-2} as normalized capacitance based on the surface area of manganese dioxides. This capacitance is approximately three times larger value than that of expected value based on a classical electrical double-layer charging (typically ~20 μF cm^{-2} for a flat electrode). It is noted that surface area of the activated carbon is ranged from 1,000 to 3,000 m^2 g^{-1}, [11] from which the area-normalized capacitance is calculated to be 10–15 μF cm^{-2}. The observed area-normalized capacitance for of $MnO_2{\bullet}nH_2O$ is anomalously large compared to the conventional electrical double-layer charging. Rate dependency of the

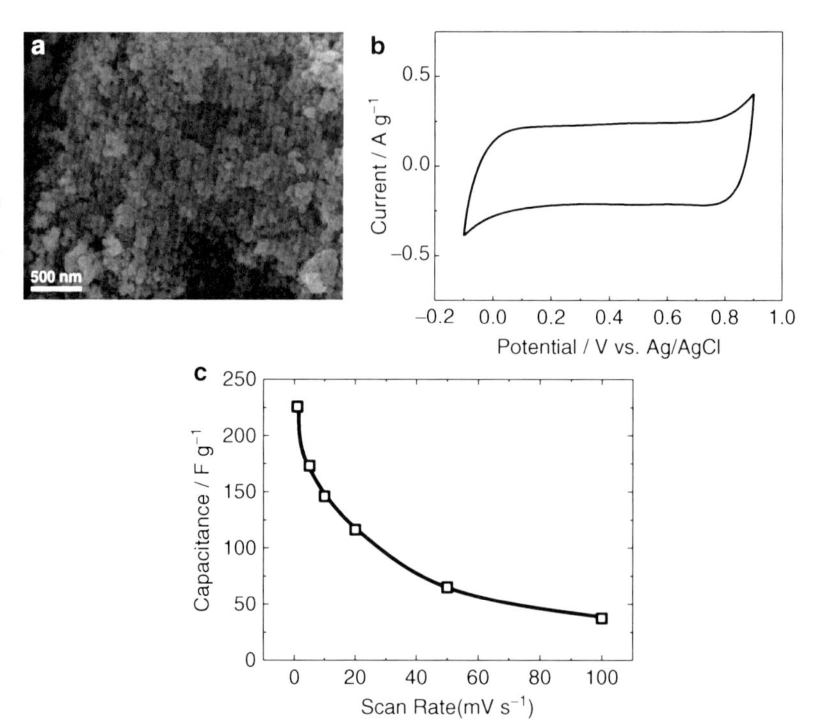

Manganese Oxides, Fig. 1 (**a**) SEM image of the amorphous hydrated manganese dioxides, $MnO_2 \cdot nH_2O$. Electrochemical behavior of $MnO_2 \cdot nH_2O$ electrode in a 1.0 mol dm^{-3} Na_2SO_4 aqueous solution, (**b**) cyclic voltammogram at a rate of 1.0 mV s^{-1}, and (**c**) rate capability at different rates of 1.0–100 mV s^{-1}

specific capacitance for the amorphous $MnO_2 \cdot nH_2O$ is shown in Fig. 1c. The specific capacitance of 225 F g^{-1} observed at 1.0 mV s^{-1} declines as a function of the scan rate. The specific capacitance decreases to 37 F g^{-1} at 100 mV s^{-1}. The area-normalized capacitance is calculated to be 18.5 μF cm^{-2}, which is good agreement with the conventional specific capacitance as described above. From these observations, it is estimated that Faradaic reaction (or charge-transfer reaction) partly contributes to the capacitor-like response of manganese dioxides, i.e., "pseudocapacitance," especially at lower rate (<50 mV s^{-1}). The utilization of both the electrical double-layer charging as the non-Faradaic process and reversible redox of manganese oxides as the Faradaic process is the basic concept for the "redox capacitors" in comparison to the conventional supercapacitors (based on the activated carbons) and batteries. There are merits for the redox capacitors with the utilization of pseudocapacitance, that is, the higher capacity than conventional supercapacitors and the higher power than batteries. When we assume that one electron transfer per manganese ion with capacitor-like behavior is achieved as shown in the Eqs. 1 and 2, the specific capacitance would reach more than 1,000 F g^{-1}. From these results, for the $MnO_2 \cdot nH_2O$, the Faradaic reaction occurs only at the surface (Eq. 3) or is limited to near the surface with the slow kinetics on the transfer of cations, such as sodium, potassium, and possibly proton:

$$\left(Mn^{4+}O_2\right)_{surface} + xC^+ + xe^- \leftrightarrow C^+{}_x\left(Mn_{1-x}{}^{4+}Mn_x{}^{3+}O_2\right)_{surface} \quad (3)$$

where C^+ denotes alkaline cations (and proton), Li^+, Na^+, K^+, etc., which depend on the electrolyte solution utilized.

Analysis on the manganese dioxide thin film by X-ray photoelectron spectroscopy (XPS) sheds light on the reaction mechanisms of the manganese dioxides in aqueous solution [12]. It was revealed that Mn 3 s core level shifted, depending on the applied potential. Furthermore, formation of the hydroxide was found from O 1 s core level when the manganese dioxide thin film was discharged to 0 V versus Ag/AgCl, suggesting that proton was partly involved in the charge compensation. Na ions were also found by XPS, indicating insertion of Na ions near the surface of manganese dioxides. These observations prove that the electrochemical performance of manganese oxides is affected by not only its crystal structure but also the morphology and surface structures of the manganese dioxides.

Toward Improvement in Electrode Performance

As the electrode materials for the supercapacitors, electrical conductivity is one of the important physical properties to achieve higher power density. Ruthenium oxide shows excellent electrical conductivity, e.g., $\sim 2 \times 10^4$ S cm^{-1} for the polycrystalline RuO_2 [13] and ~ 2 S cm^{-1} as the hydrous form [14]. The electrical conductivity of the $RuO_2 \cdot nH_2O$ is comparable to the carbon-based electrode materials. In this regard, the major drawback of manganese dioxides as the electrode materials for the redox capacitors seems to be poor electrical conductivity (5×10^{-6} S cm^{-1} for cryptomelane [15]). The poor electrical conductivity inevitably results in the restricted solid-state diffusion rate for the electrons and/or ions, and thus both power and energy density are reduced as the electrode materials. In fact, manganese dioxides prepared as the thin film electrode (<100 nm) demonstrate clear improvement of the specific capacitance of the manganese electrode, which becomes comparable value (~ 700 F g^{-1}) with that of ruthenium dioxides [6, 16]. This is also consistent with the XPS study of the manganese oxides. The improvement of the energy density has been achieved even as composite electrode by embedding manganese dioxides into the mesoporous carbon [17] and ultraporous carbon [18] frameworks, which allow the three-dimensional path for electron conduction with maximized interfacial area between electrolyte solution and manganese oxides in the composite structure. The specific capacitance of 600 F g^{-1} can be achieved based on the embedded manganese dioxides in the carbon form [17].

Dissolution of the manganese ions into the electrolyte is also a major drawback as the electrode materials. As the electrolyte solution, concentrated acidic or alkaline media, e.g., H_2SO_4 and KOH solutions, are often utilized for the carbon and ruthenium dioxide supercapacitor because of the superior ionic conductivity. Manganese dioxides are, however, soluble in the acidic media unlike ruthenium dioxide. Alternatively, 1–2 mol dm^{-3} Na_2SO_4 and KCl aqueous solutions are often used as the electrolyte for the manganese dioxide-based supercapacitor. It should be noted that such neutral electrolyte solutions with environmental-friendly and safety salts offer great advantage compared with the concentrated acid/alkaline electrolytes. However, the loss of manganese through partial dissolution of the manganese dioxides is typically observed even in such mild electrolyte solution because of the disproportionation of trivalent manganese during electrochemical cycling [6, 7]. Recently we have reported that the electrolyte additive is highly beneficial to reduce the dissolution of manganese ions [9, 10]. In general, manganese carbonate and phosphate species are insoluble in water. Therefore, the carbonate or phosphate additive into an electrolyte can suppress the dissolution of manganese oxides similar to parkerizing (also called phosphating), which is a method of protecting a steel surface from corrosion and increasing its resistance to wear through the application of an electrochemical phosphate conversion coating. Indeed, electrolyte additives, such as $NaHCO_3$ or Na_2HPO_4, significantly improve the cycleability of the manganese dioxides as electrode materials for supercapacitors by suppressing the loss of manganese ions during the electrochemical cycling [9, 10].

Asymmetric Supercapacitors with Activated Carbon Electrodes

A potential window of the manganese dioxides in mild aqueous electrolyte solution is typically limited in the range of − 0.1 to 0.9 V versus Ag/AgCl because of the decomposition of water. This fact indicates that the energy density of the manganese oxides as a conventional

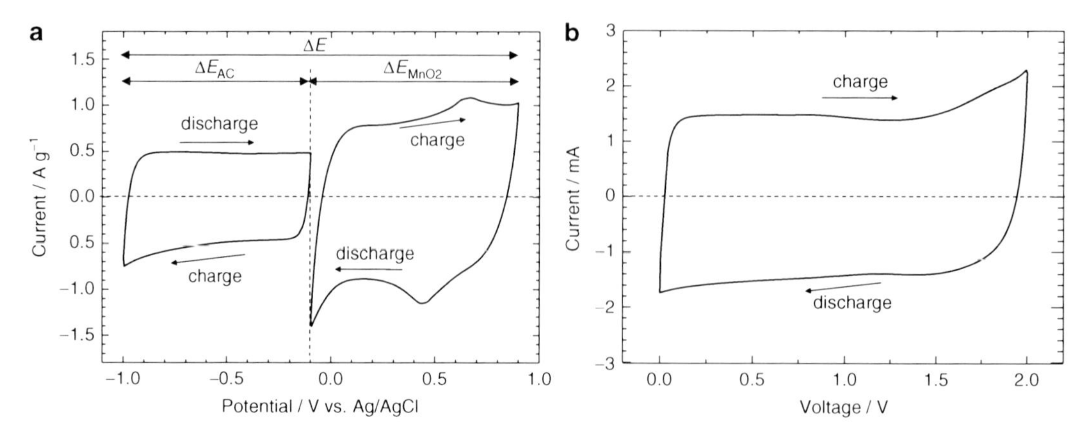

Manganese Oxides, Fig.2 Cyclic voltammograms of (**a**) birnessite electrodes ($-0.1 \sim 0.9$ V) prepared by the electrochemical stimulation method and activated carbon ($-1.0 \sim -0.1$ V) and (**b**) the asymmetric supercapacitor of the birnessite/1 mol dm^{-3} Na_2SO_4 aq./activated carbon cell

symmetrical capacitor with the maximum terminal voltage of is reduced to approximately 1 V. The energy (E) accumulated in the capacitor is represented by the following formula:

$$E = \frac{1}{2}CV^2 \tag{4}$$

where C and V denote capacitance and voltage of the capacitor, respectively. Since the energy density of the capacitor is proportional to the square of terminal voltage, this results in the lowering energy density for the symmetrical supercapacitor consisting of manganese dioxide electrodes because of the limited potential window. The concept of asymmetric (or hybrid) supercapacitor was, therefore, introduced to improve the energy/power density as the supercapacitor [19–21]. For the asymmetric capacitor, the manganese dioxides electrode is used as the positive electrode with the different negative electrodes, which can be cycled below 0 V versus Ag/AgCl without the hydrogen gas generation in water-based electrolyte. Activated carbon electrodes can be used as the negative electrodes, because the electrodes show higher overpotential for hydrogen evolution (~ -1.0 V vs. Ag/AgCl in neutral solution).

Cyclic voltammograms of an asymmetric supercapacitor, consisting of the manganese dioxides as positive electrode and the activated carbon as negative electrode, and their half cells are shown in Fig. 2. The asymmetric supercapacitor can be charged to approximately 2 V (Fig. 2b), and its energy density reaches ca. 21–29 Wh kg^{-1}, [19–21] which is nearly ten times larger energy density than that of the symmetrical supercapacitor based on the manganese dioxides. The high oxygen and hydrogen overpotentials for the manganese dioxide and activated carbon electrodes, respectively, allow it to design 2.0 V supercapacitor with environmentally benign electrolyte. This inevitably contributes to the increase in energy density of the asymmetric supercapacitor as expected from Eq. 4. Though the specific capacitance of the negative and positive electrodes is lower than $RuO_2 \cdot nH_2O$ electrode, available energy density is comparable with the ruthenium symmetrical supercapacitor (27 Wh kg^{-1}) [1] because of its higher terminal voltage as the aqueous-based electrolyte systems.

Summary

Recent progress and development in the manganese (di)oxides have been briefly reviewed as the electrode materials for supercapacitors. Manganese oxides show a variety of the crystal structures, surface structures, and particle morphology, depending on the synthesis condition of the materials, which results in the distinctive electrochemical properties as the electrode materials. Asymmetric supercapacitors consisting of the manganese dioxide and activated carbon electrodes have to be potentially interesting devices, i.e., simple, low cost, and environmental friendliness. Further improvement of the electrode performance of manganese oxides is the great challenge to further enhance the energy and power density of the supercapacitor in the future.

Cross-References

- ▶ Activated Carbons
- ▶ Conductivity of Electrolytes
- ▶ Electrical Double-Layer Capacitors (EDLC)
- ▶ Electrolytes for Electrochemical Double Layer Capacitors
- ▶ Nickel Oxide Electrodes
- ▶ Redox Capacitor
- ▶ Ruthenium Oxides as Supercapacitor Electrodes
- ▶ Super Capacitors

References

1. Zheng JP, Jow TR (1995) A new charge storage mechanism for electrochemical capacitors. J Electrochem Soc 142:L6–L8
2. Zheng JP, Cygan PJ, Jow TR (1995) Hydrous ruthenium oxide as an electrode material for electrochemical capacitors. J Electrochem Soc 142:2699–2703
3. Sugimoto W, Iwata H, Yasunaga Y, Murakami Y, Takasu Y (2003) Preparation of ruthenic acid nanosheets and utilization of its interlayer surface for electrochemical energy storage. Angew Chem Int Ed 42:4092–4096
4. Lee HY, Goodenough JB (1999) Supercapacitor behavior with KCl electrolyte. J Solid State Chem 144:220–223
5. Pang SC, Anderson MA (2000) Novel electrode materials for electrochemical capacitors: Part II. Material characterization of sol-gel-derived and electrodeposited manganese dioxide thin films. J Mater Res 15:2096–2106
6. Pang SC, Anderson MA, Chapman TW (2000) Novel electrode materials for thin-film ultracapacitors: Comparison of electrochemical properties of sol-gel-derived and electrodeposited manganese dioxide. J Electrochem Soc 147:444–450
7. Reddy RN, Reddy RG (2003) Sol-gel MnO2 as an electrode material for electrochemical capacitors. J Power Sources 124:330–337
8. Subramanian V, Zhu HW, Wei BQ (2008) Alcohol-assisted room temperature synthesis of different nanostructured manganese oxides and their pseudocapacitance properties in neutral electrolyte. Chem Phys Lett 453:242–249
9. Komaba S, Ogata A, Tsuchikawa T (2008) Enhanced supercapacitive behaviors of birnessite. Electrochem Commun 10:1435–1437
10. Komaba S, Tsuchikawa T, Ogata A, Yabuuchi N, Nakagawa D, Tomita M (2011) Nano-structured birnessite prepared by electrochemical activation of manganese(III)-based oxides for aqueous supercapacitors. Electrochim Acta 59:455
11. Xing W, Qiao SZ, Ding RG, Li F, Lu GQ, Yan ZF, Cheng HM (2006) Superior electric double layer capacitors using ordered mesoporous carbons. Carbon 44:216–224
12. Toupin M, Brousse T, Belanger D (2004) Charge storage mechanism of MnO2 electrode used in aqueous electrochemical capacitor. Chem Mater 16:3184–3190
13. Abe O, Taketa Y (1991) Electrical-conduction in thick-film resistors. J Phys D: Appl Phys 24:1163–1171
14. Fletcher JM, Gardner WE, Greenfie BF, Holdoway MJ, Rand MH (1968) Magnetic and other studies of ruthenium dioxide and its hydrate. J Chem Soc Inorg Phys Theor 653–657
15. Sharma PK, Moore GJ, Zhang F, Zavalij P, Whittingham MS (1999) Electrical properties of the layered manganese dioxides MxMn1-yCoyO2, M = Na, K. Electrochem Solid State Lett 2:494–496
16. Broughton JN, Brett MJ (2004) Investigation of thin sputtered Mn films for electrochemical capacitors. Electrochim Acta 49:4439–4446
17. Dong XP, Shen WH, Gu JL, Xiong LM, Zhu YF, Li Z, Shi JL (2006) MnO2-embedded-in-mesoporous-carbon-wall structure for use as electrochemical capacitors. J Phys Chem B 110:6015–6019
18. Fischer AE, Pettigrew KA, Rolison DR, Stroud RM, Long JW (2007) Incorporation of homogeneous, nanoscale MnO2 within ultraporous carbon structures via self-limiting electroless deposition: Implications for electrochemical capacitors. Nano Lett 7:281–286
19. Hong MS, Lee SH, Kim SW (2002) Use of KCl aqueous electrolyte for 2 V manganese oxide/activated carbon hybrid capacitor. Electrochem Solid State Lett 5:A227–A230
20. Brousse T, Toupin M, Belanger D (2004) A hybrid activated carbon-manganese dioxide capacitor using a mild aqueous electrolyte. J Electrochem Soc 151: A614–A622
21. Khomenko V, Raymundo-Pinero E, Beguin F (2006) Optimisation of an asymmetric manganese oxide/activated carbon capacitor working at 2 V in aqueous medium. J Power Sources 153:183–190

Membrane Processes, Electrodialysis

Vicente Montiel, Vicente García-García,
Eduardo Expósito, Juan Manuel Ortiz and
Antonio Aldaz
Instituto Universitario de Electroquímica,
University of Alicante, Alicante, Spain

Introduction

Electrodialysis is an electrochemical technology of separation dating back to 1890, when Maigrot and Sabates patented a three compartment electrochemical cell [1]. This separation process is characterized by the selective transport of ions across a set of ion exchange membranes due to the existence of an electric field inside the membranes. The operating principle of "conventional electrodialysis" (Fig. 1) is based on an alternative placement of cation and anion exchange membranes inside a stack ("electrodialyzer"). Thus, establishing a difference of potential between two electrodes (cathode and anode) housed at the two ends of the stack inside electrode plates, cations migrate towards the cathode, and anions migrate towards the anode through the cation and anion exchange membranes, respectively. Figure 1 shows that in conventional electrodialysis, three electrolytic streams flow across the set of anion and cation exchange membranes. Note that while the concentration of ions decreases in one of these stream (diluate), it increases in the adjacent one (concentrate). The third electrolytic stream (electrolyte) is in contact with the electrodes where water oxidation (oxygen evolution) and water reduction (hydrogen evolution) processes occur.

The assembly formed by a cation exchange membrane, an anion exchange membrane, spacers and gaskets, is called a "cell pair" or "cell." An electrodialyzer can consist of up to 300 cells. The difference of potential applied is usually set between 0.75 and 1.5 V/cell (depending on the membranes used and the concentration of the solutions). It is obvious, then, that an electrodialyzer works with high values of the difference of potential between the anode and cathode (Fig. 2).

As you may easily guess, the properties of ion exchange membranes have a strong influence on the efficiency of an electrodialysis process. Among the most important properties of membranes are: high permselectivity, low electrical resistance, and high mechanical, chemical and thermal stabilities. These properties are determined by different parameters, such as density of the polymer matrix, hydrophilic or hydrophobic character, type and concentration of fixed charges in the polymer matrix and the morphology of the membrane.

Dating back to the 1980s [1], a new type of ion exchange membrane, called a bipolar membrane, was developed. This type of membrane is characterized by being composed of three layers: (i) a cation exchange membrane, (ii) an anion exchange membrane, and (iii) a hydrophilic layer located between the two previous layers (Fig. 3). When an electric field is applied through the bipolar membrane, all of the charged species initially present in the hydrophilic layer migrate through the adjacent ion exchange layers. Once the charged species have migrated, the electric field causes the dissociation of water molecules into H^+(in fact "hydronium" H_3O^+) and OH^-; they migrate through the ion exchange layers to the solutions in contact with the bipolar membrane (water splitting). The overall effect is that the bipolar membrane generates H^+ and OH^- ions, as do two electrodes, an anode and a cathode. The two main differences are: first, a bipolar membrane does not generate gases (O_2 and H_2), and second, the potential difference is significantly lower than that required when two electrodes are used.

The use of bipolar membranes with the conventional cation and/or anion exchange membranes allows very interesting applications to develop, different from the dilution or concentration of solutions characteristic of "conventional electrodialysis."

Applications

Conventional Electrodialysis

Grouped in this section are the electrodialysis processes involving both cation and anion

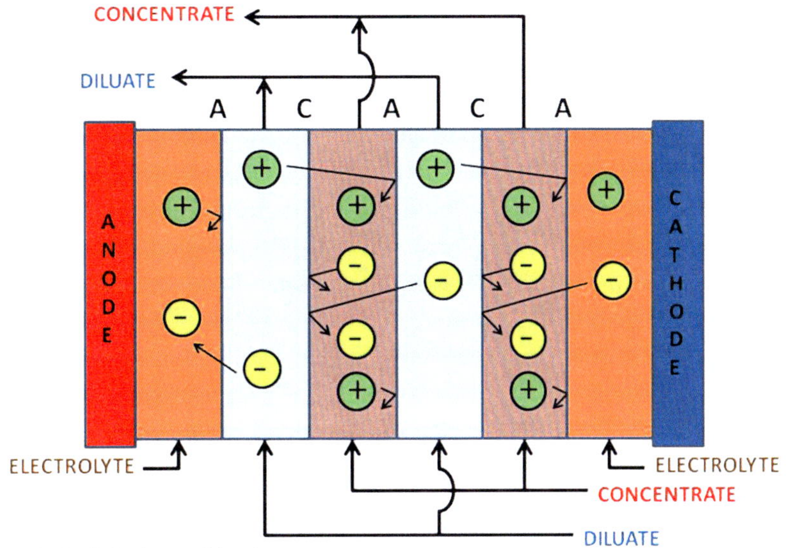

Membrane Processes, Electrodialysis, Fig. 1 Schematic of the conventional electrodialysis process

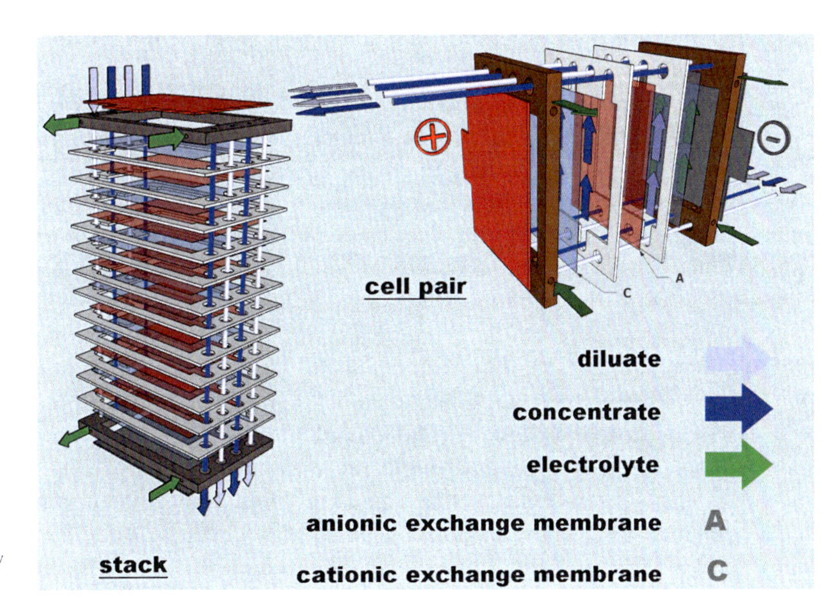

Membrane Processes, Electrodialysis, Fig. 2 An exploded view of an electrodialyzer

exchange membranes, a diluate solution and a concentrate solution. Among the most important processes can be cited:

Production of Table Salt and Drinking Water

From sea water (30–40 g/L NaCl) and applying conventional electrodialysis technology, there can be obtained both a final diluate with a saline content suitable for further purification and a final concentrate with a NaCl concentration close to 200 g/L (from which table salt can be obtained). The peculiarity of this electrodialysis process is the use of monoselective ion exchange membranes. Monoselective ion exchange membranes differ from normal ion exchange membranes in that they do not allow

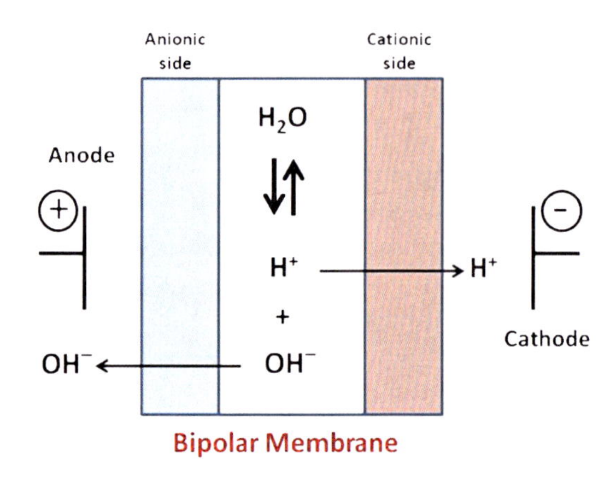

Membrane Processes, Electrodialysis, Fig. 3 Schematic operation of a bipolar membrane

multiple charged ions to pass through. This will prevent the passage of multicharged ions like Ca^{2+} and SO_4^{2-} (present in sea water) from diluate to concentrate, avoiding the possible precipitation of $CaSO_4$ in the membranes. This allows cleaning tasks to be carried out without having to disassemble the electrodialyzer and significantly extends the life of the membranes. This process has been mainly developed in Japan, which has no natural sources of crystallized NaCl, with a production capacity of 30,000 t of table salt per year and a power consumption of 150 kWh/t (table salt) [2].

Desalination of Brackish Water

Most of the electrodialysis plants that desalinate brackish water (containing less than 5,000 mg/L of total dissolved solids) operate in polarity reversal mode. This is useful because impurities can build up on the surface of the membranes. Therefore, reversing the polarity between the electrodes lifts the impurities off the membranes, minimizing the fouling of the membranes and increasing their usable lifetime. The production capacity of these plants is usually average, not surpassing the 20,000 m^3/day of desalinated brackish water. These electrodialysis plants are very suitable for installation in remote locations with difficulties in drinking water supply. This application allows: (i) the obtention of drinking water (total dissolved solids content below 500 mg/L), (ii) the reduction of nitrate and/or fluoride concentrations in contaminated groundwater to concentrations that are not harmful to human health, (iii) the reduction of the lime present in water in order to use it as boiler feed water, and (iv) the possibility for it to be used as a pretreatment step prior to an ion exchange step, thus reducing the cost of regeneration of the resins.

Desalination of Organic Compounds and Food Products

Food manufacturing provides several examples of the application of conventional electrodialysis, because this technique allows the separation of non-electrolytes from electrolytes. The main application in this field is the demineralization of whey obtained as a byproduct in the manufacture of cheese [3]. The high salt content in cheese serum makes it unworkable both for human consumption and for livestock feed. To desalinate the cheese whey, it has been necessary to adapt the electrodialysis systems, incorporating demanding washing procedures and mechanical cleaning of the membranes. In spite of this, the treatment is feasible from a technical and economic point of view. Included among the substances in this whey is lactic acid, present in the sodium lactate. Using a pretreatment in which fats and proteins are removed and lactose is fermented to sodium lactate, followed by acidification to pH 2–3, the inorganic salts generated can be then removed by conventional electrodialysis, achieving a lactic acid solution of the desired concentration [4]. On the other hand, in the synthesis of an amino acid, a mother liquor with a high salt content can be produced. The disposal of this mother liquor not only causes an environmental problem due its high salinity and COD (chemical oxygen demand) content but also an economic loss due to the high price of the unrecovered amino acid. To avoid this problem, an electrodialysis process can be applied,

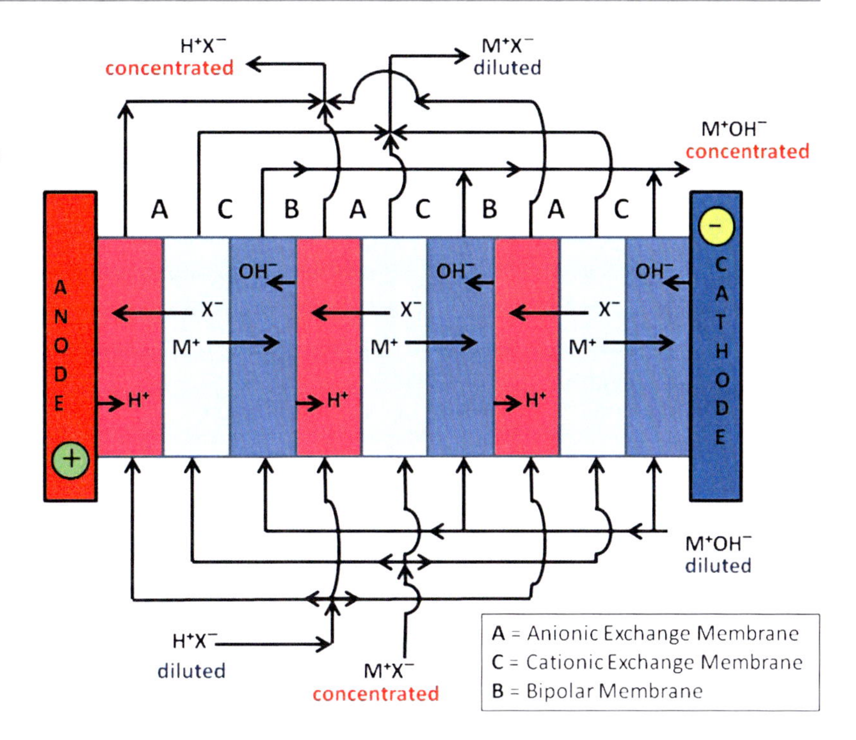

Membrane Processes, Electrodialysis, Fig. 4 Schematic electrodialysis process using bipolar membranes

allowing the recovery of most of the amino acid in the form of a low salinity stream. This low salinity stream can be incorporated into the main process, and in this way the amino acid can be recovered [5]. It is also possible to separate a mixture of amino acids using an electrodialysis process [6].

Electrodialysis Using Bipolar Membranes

Bipolar membranes were first commercialized in 1977 [7]. Since then, a wide number of possible applications for electrodialysis using bipolar membranes have been developed. Figure 4 shows the typical structure of an electrodialyzer employing bipolar membranes. Choosing the arrangement of the electrodialyzer properly, the process converts the neutral salt MX into the conjugate acid of the anion (HX) and the conjugate base of the cation (MOH). The main requirements for a bipolar membrane are: (i) a low electrical resistivity at high current densities, (ii) low transport number of the co-ions, (iii) high ion selectivity, and (iv) high chemical and thermal stabilities in both strong acidic and basic media [8].

The main applications include:

Preparation of Acids and Bases

This is the main application. However, sometimes this process may not be feasible because highly concentrated solutions of acids and bases cannot be obtained for various reasons [7]. These include: (i) low permselectivity of bipolar membranes at high acid and base concentrations, causing salt contamination of the desired products and low values of current efficiency, (ii) the short life of bipolar membranes if the temperature exceeds 40 °C, (iii) the presence of divalent cations, e.g., Ca^{2+} and Mg^{2+}, should there be less than 1 mg/L to prevent the precipitation of their hydroxides on the anion side of the bipolar membrane, and (iv) the presence of aromatic compounds that may swell the bipolar membrane ("swelling"). Examples of this application may include the following:

- NaOH and H_2SO_4 recovery from Na_2SO_4 generated as wastewater from different chemical processes, such as the neutralization of a waste acid, regeneration of ion

exchange resins, flue gas desulfurization and manufacture of rayon [9].
- Recovery of HF and HNO_3 from an effluent coming from a steel pickling bath containing KF and KNO_3. The process of electrodialysis with bipolar membranes also allows the generation of KOH that can be used to neutralize the acid mixture from the pickling step [10].
- Recovery of organic acids and amino acids. One of the applications of electrodialysis with bipolar membranes with a bright future is the obtention of organic acids present in anionic form in fermentation broths. Among them we can cite as representative examples lactic acid and itaconic acid. This process also generates the base (NaOH, NH_4OH, etc.), which must be added to the fermentation broth to transform the acid generated in its anionic form. This is a necessary step to prevent inhibition of the fermentation process. Other notable examples are the recovery of gluconic and ascorbic acids and the amino acid glycine from the corresponding sodium salt.

Use in Non-aqueous Solvents

The electrodialysis technology using bipolar membrane electrodialysis can also be applied in non-aqueous solvents, which necessarily must be water-free. Examples include the obtention of sodium methoxide from a solution of sodium acetate in methanol [11], where acetic acid is generated as a byproduct.

Electrodeionization

The electrodeionization process is very similar to conventional electrodialysis. The main difference is that the diluate compartment is filled with a mixed ion exchange resin bed. Ions present in the diluate diffuse into the resin and are exchanged with H^+ or OH^-. When a difference of potential is applied, the ions exchanged migrate from the resin bed to the adjacent concentrate compartments. The resin allows the dissolution of diluate to keep acceptable values of electrical conductivity although the salt content is very low, because water dissociates into H^+ or OH^- at the point of contact of the diluate with the cationic and anionic resin beds. In this way, you can get completely deionized water, integrating an electrodeionization unit as the last stage of the production process.

Future Directions

1. Direct renewable energy coupling as electricity suppliers for the operation of electrodialysis systems: On this matter, photovoltaic solar energy can be the most promising because it provides a direct current that can be applied directly to the electrodialyzer, avoiding the use of batteries as an intermediate step. Obviously, to obtain a certain amount of desalinated water, the electrodialyzer size depends on whether or not the system has batteries for energy storage. Direct coupling between photovoltaic solar panels and an electrodialysis system is particularly useful for remote locations with access to brackish water wells and connection difficulties with the conventional electricity grid. Not using batteries solves future problems of potentially toxic waste management. The use of low power wind generators that provide a direct current can also be a suitable method to turn electrodialysis into a green technology. Obviously, both types of power generation can be combined to try to overcome problems arising from lack of solar light or wind.
2. Development of new operating procedures to avoid, minimize or reverse processes of "fouling" of the ion exchange membranes: These processes hamper the maintenance works of the electrodialyzer, decrease the life of the membranes, and therefore increase the costs of operation. In this regard, initiatives such as the use of a pulsed electric field can be promising.
3. Application of conventional electrodialysis and electrodialysis with bipolar membranes in the development of new processes: desalination of wastewater from organic synthesis processes to allow recovery and reuse in the system; obtention of high value products that can increase their pharmacological activity by substitution of the anions or cations present in natural sources.

Cross-References

► Conductivity of Electrolytes
► Electrochemical Reactor Design and Configurations
► Electrogenerated Acid
► Electrogenerated Base
► Green Electrochemistry
► Ion Mobilities
► Membrane Technology

References

1. Bazinet L (2004) Electrodialytic phenomena and their applications in the dairy industry: a review. Crit Rev Food Sci Nutr 44:525–544. doi:10.1080/10408690490489279
2. Strathmann H (2010) Electrodialysis, a mature technology with a multitude of new applications. Desalination 264:268–288. doi:10.1016/j.desal.2010.04.069
3. Davis TA, Genders D, Pletcher D (1997) A first course in ion permeable membrane. The Electrochemical Consultancy, Romsey
4. Thang VH, Koschuh V, Kulbe KD et al (2005) Detailed investigation of an electrodialytic process during the separation of lactic acid from a complex mixture. J Membr Sci 249:173–182. doi:10.1016/j.memsci.2004.08.033
5. Montiel V, García-García V, González-García J et al (1998) Recovery by means of electrodialysis of an aromatic amino acid from a solution with a high concentration of sulphates and phosphates. J Membr Sci 140:243–250. doi:10.1016/S0376-7388(97)00275-5
6. Kikuchi K, Gotoh T, Takahashi H et al (1995) Separation of amino acids by electrodialysis with ion-exchange membranes. J Chem Eng Japan 28:103–109. doi:10.1252/jcej.28.103
7. Strathmann H (2004) Ion-exchange membrane separation process, vol 9, Membrane science and technology series. Elsevier, Amsterdam/Boston
8. Gineste JL, Pourcelly G, Lorrain Y et al (1996) Analysis of factors limiting the use of bipolar membranes: a simplified model to determine trends. J Membr Sci 112:199–208. doi:10.1016/0376-7388(95)00284-7
9. Tanaka Y (2007) Ion-exchange membrane fundamentals and applications, vol 12, Membrane science and technology series. Elsevier, Amsterdam/Oxford
10. Pourcelly G, Gavach C (2000) Electrodialysis water splitting – application of electrodialysis with bipolar membranes. In: Kemperman AJB (ed) Handbook of bipolar membrane technology. Twenty University Press, Enschede
11. Sridhar S (1996) Electrodialysis in a non-aqueous medium: production of sodium methoxide. J Membr Sci 13:73–79. doi:10.1016/0376-7388(95)00217-0

Membrane Technology

Helmut Ullmann[1], Vladimir Vashook[1,2] and Ulrich Guth[1,2]
[1]FB Chemie und Lebensmittelchemie, Technische Universität Dresden, Dresden, Germany
[2]Kurt-Schwabe-Institut für Mess- und Sensortechnik e.V. Meinsberg, Waldheim, Germany

Definition

Membranes could be considered as layer structures which can separate two fluids (gas or liquid) and which have different permeabilities for these fluids or their components. Depending on structure, the membranes could be classified in dense and porous ones. According to IUPAC, the porous membranes can be distinguished into micropores (pore diameter is <2 nm), mesopores ($2 \text{ nm} < d < 50 \text{ nm}$), and macropores ($d > 50$ nm), respectively. The dense membranes have no open porosity and are impenetrable for particles in gas molecules.

Dense membranes can be divided into ceramic membranes, metal membranes, and liquid-immobilized membranes. These include materials which allow preferential passage of hydrogen or oxygen, in the form of either ions or atoms. Liquid-immobilized membranes consist of a porous support in which a semipermeable liquid is immobilized which fills the pores completely. Interesting examples are molten salts immobilized in porous steel or ceramic supports, semipermeable for oxygen or ammonia [1].

Dense Ceramic Membranes

The first ceramic oxygen membranes were discovered by *Nernst* [2] in 1899 in the form of mixtures of zirconia and rare-earth metal oxides. Basically, oxygen transport in oxide ceramics can be realized in three variants (Fig. 1). Materials

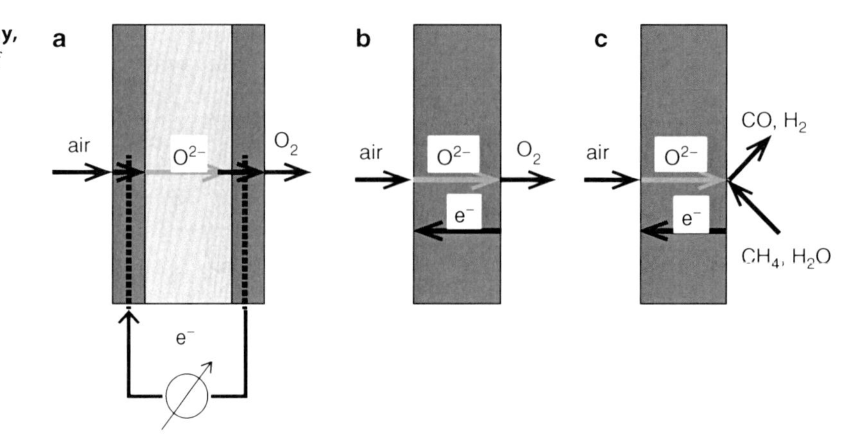

Membrane Technology, Fig. 1 Modifications of process applications of membranes with high oxide-ion transport (**a**) oxygen pump with an electrical current (**b**) transport membrane (**c**) heterogeneous catalysis (**b**) and (**c**) without an electrical field

with pure oxide-ion conductivity are used as solid electrolytes to build electrochemical cells. Equipped with electrodes and an external current circuit, they generate electric current by electrochemical combustion of methane or hydrogen in solid oxide fuel cells [3]. By utilizing an external electric voltage, it becomes also possible to pump oxygen out of a gas mixture into another volume (Fig. 1a). By this way, pure oxygen (99.99 %) can be provided for medical supply. First experiments of air separation to separate oxygen from air were performed by means of zirconia electrolytes. However, oxidation of methane was not successful by such an arrangement because oxygen flux through this material is too low [4]. This situation had been changed after the discovery by *Teraoka* [5, 6] of the mixed ionic-electronic conductors (abbreviated MIEC or mixed conductors) of the perovskite-type structure, which show oxygen fluxes higher by more than one order of magnitude as compared to the former materials.

Materials with these properties opened the gates toward development of ceramic oxygen membranes that could be of practical interest. The external current circuit becomes obsolete since the electronic current flows within the ceramic membrane (Fig. 1b). A so-called permeation flux of oxygen flows through the gas-tight ceramic membrane under the influence of an oxygen partial pressure gradient, as a combined current of ions and electrons. On the basis of this principle, oxygen membranes are operated for the production of pure oxygen from air, accumulating the oxygen at the product side in a vacuum. Modifying the principle as a membrane reactor, the oxygen at the product gas side may be used for processes of partial oxidation, e.g., of natural gas to syngas. This process had been demonstrated in many variations at laboratory scale [7].

Syngas can be produced by steam reforming of methane at pressures of 15–30 atm. and temperatures of 850–900 °C, respectively. The reforming reaction $CH_4 + H_2O \rightarrow 3H_2 + CO$ is endothermic. Syngas with a high H_2/CO ratio is obtained. The partial catalytic oxidation reaction at temperatures of >700 °C under conditions of controlled oxygen supply is exothermic. A stoichiometric syngas, to be used readily for important chemical syntheses such as *Fischer-Tropsch* or methanol synthesis processes, is formed. The partial oxidation of methane by means of ceramic oxygen-permeable membranes enables combination of two process steps within one reactor, *videlicet*, oxygen separation from air with production of syngas. The combined reaction is exothermic, and the H_2/CO ratio can be controlled by appropriate adding of H_2O or CO_2 to methane, to obtain mixtures with values of that ratio as required by specific final products. By additional feeding of water vapor to the syngas, the shift reaction can be driven toward $CO_2 + H_2$ for H_2 production.

Oxygen Flux Through Mixed Conducting Ceramic Membranes

Ceramic materials for oxygen-permeable membranes should exhibit primarily a high oxygen-permeation flux. Within the membrane, the oxygen is transported as a flux of both oxide ions and electrons (Fig. 2), driven by the different oxygen partial pressures at the two sides of the membrane (*Wagner's theory of permeation* by bipolar diffusion). The oxygen flux can be calculated by the *Wagner's* equation [8]

$$J_{O_2} = \frac{RT}{16 \cdot F^2 \cdot L} \cdot \int_{\ln pO''_2}^{\ln pO'_2} \frac{\sigma_e \cdot \sigma_i}{\sigma_e + \sigma_i} \cdot d \ln pO_2 \quad (1)$$

where F is the *Faraday* constant, R *is the* gas constant, σ_e and σ_i *are the* electronic and ionic conductivities, respectively, and pO'_2 and pO''_2 are the oxygen partial pressures in the surface layers of the membrane.

The flux of oxygen through a ceramic membrane is controlled by both diffusion of the oxide ion within the polycrystalline solid and the rate of surface exchange as the transformation between oxide ions and molecular oxygen. These two processes exhibit different rates and activation energies, which means that one of these processes controls the oxygen flux in different temperature ranges. The activation energy of surface exchange (coefficient k_s) is often found to be larger than that of the solid-state diffusion of oxide ions (coefficient D_O^{chem}).

Thus, at lower temperature, the oxygen flux is limited by the rate of surface exchange, at higher temperatures, the oxygen flux is limited by that of solid-state diffusion (Fig. 3). So far, only some experimental results on these coefficients are available (Table 1). The transition between these two limiting processes must be determined experimentally for each material composition, in dependence on both temperature and membrane thickness. The surface exchange rate increases with surface area and number of catalytic active sites at the surface layer. The surface exchange kinetics is frequently characterized by the ratio $h = k_s/D_O^{chem}$. In materials with high oxide-ion

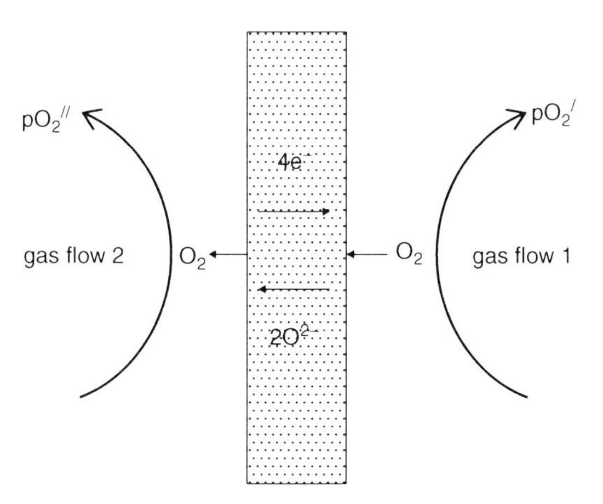

Membrane Technology, Fig. 2 Oxygen-permeation flux through a ceramic membrane (schematically). Gas flow 1 is of high oxygen pressure (air), gas flow 2 is vacuum or methane

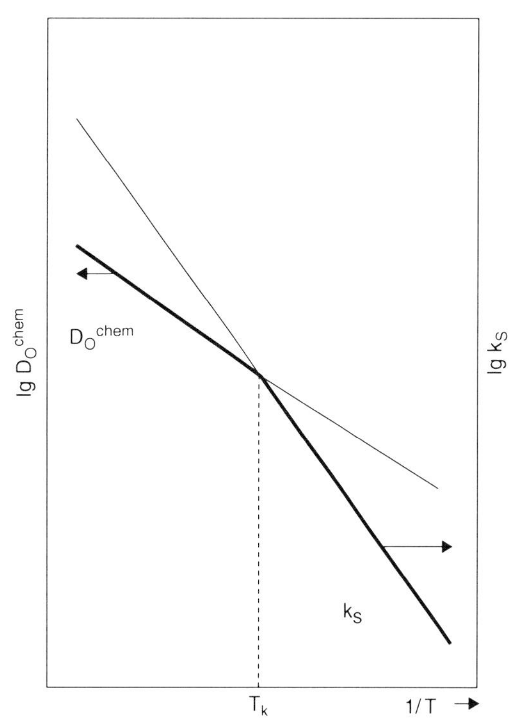

Membrane Technology, Fig. 3 Temperature dependencies of diffusion coefficients D_O^{chem}, surface exchange coefficients k_s, and the critical temperature T_k of transition. In the range of the solid lines, the corresponding coefficient limits the oxygen flux

Membrane Technology, Table 1 Coefficients k_s and D_O^{chem} of oxygen at 800 °C in air and activation energies E_a

Composition	$k_s(cm^*s^{-1})/E_a(eV)$	$D_O^{chem}(cm2^*s^{-1})/Ea(eV)$	Ref.
La_2NiO_4	–/1.61	–/0.85	[9]
$La_{1.9}Sr_{0.1}NiO_4$	–/1.29	–/0.57	[9]
$La_{1.9}Sr_{0.1}Ga_{0.8}Mg_{0.2}O_{3-x}$	10^{-7}/1.6	3^*10^{-7}/0.74	[10, 11]
$Ce_{0.7}Gd_{0.3}O_{2-x}$	5^*10^{-7}/2.4	8^*10^{-6}/(0.9–1.2)	[12]
$La_{0.8}Sr_{0.2}MnO_3$ (900 °C, 1 atm.)	5^*10^{-8}/–	10^{-13}/2.8	[13]
$La_{0.8}Sr_{0.2}CoO_3$ (900 °C, 1 atm.)	2^*10^{-6}/–	10_{-8}/2.2	[13]
$La_{0.5}Sr_{0.5}CoO_3$ (900 °C, 1 atm.)	2^*10^{-5}/–	10^{-6}/1.4	[13]
$La_{0.6}Sr_{0.4}Co_{0.6}Fe_{0.4}O_3$	10^{-5}/–	10^{-6}/(0.9–1.2)	[14]

diffusion and high catalytic activity, for example, in lanthanum-strontium-cobaltites, the oxygen flux is limited by solid-state diffusion at higher temperature. In contrast to that behavior, surface exchange remains the limiting process on solid electrolyte materials, such as cubic stabilized zirconia or doped ceria, up to high temperatures.

Some Materials for Oxygen-Transport Membranes

These materials may be conveniently categorized into oxides with pure ionic conductivity and oxides with *mixed ionic and electronic conductivity*.

Best known examples of oxides with pure ionic conductivity are yttria-stabilized zirconia (YSZ), gadolinia-doped ceria (CGO) with fluorite structure, and strontia-doped lanthanum gallates (LSG) with perovskite structure. These oxides exhibit high thermodynamic stability because of the absence of transition-metal ions. These ionic conductors differ in their high oxide-ionic conductivity as follows: LSG > CGO > YSZ. They show low electronic conductivity and low catalytic activity. For their utilization as oxygen membrane materials, an electronic part of conductivity is required.

Electronic conductivity of these compositions could be established by doping with transition-metal oxides or by forming of ceramic-metal composites (cermets), e.g., by adding silver or platinum metals. These cermets provide for high oxygen-diffusion rate within the oxide grains, and high electronic conductivity within the metallic part of the composite. Optimum conductivity in YSZ is obtained at about 40 vol.% of Ag [15], for example.

ABO_3 perovskite-type oxides with transition-metal ions at the B-site have high ionic and electronic transport in the form of p or n semi-conductivity (*mixed ionic and electronic conductivity*), caused by different oxidation states of the transition-metal cation. For dense ceramic membranes, perovskite-type oxides with the following cations are preferred: A = Ln (lanthanide ion), Ca, Sr, Ba; B = Cr, Mn, Fe, Co, Ni, Cu.

For high oxide-ionic transport, the following rules for ionic charge and *radii* of the cations can be given. Firstly, B-cations should be possibly large (Co > Ni > Fe > Mn > Cr) and in a possibly low oxidation state (2+ better than 3+ better than 4+), for a weak bonding of the oxide ion. Secondly, the smaller the A-ions (Ca < Sr < Ba; Ce < Pr < Gd < La), the more space is available for oxygen mobility, but the higher is the electrostatic attraction of the oxide ions. The concentration of the mobile ionic species (oxygen vacancies) can be increased also by an A/B-sub-stoichiometry (of A < 1) [16, 17].

Cross-References

- ▶ Membrane Processes, Electrodialysis
- ▶ Solid Electrolytes
- ▶ Solid Electrolytes Cells, Electrochemical Cells with Solid Electrolytes in Equilibrium

References

1. Bouwmeester HJM, Burggraaf AJ (1996) Dense Ceramic Membranes. In: Gellings PJ, Bouwmeester JM (eds) The CRC handbook of solid state electrochemistry. CRC Press, Boka Raton, pp 481–553
2. Nernst W (1899) On hydrogen generation (in German) Z. Elektrochemie 6:37–41
3. Steele BCH (2000) Materials for IT-SOFC stacks. 35 years R&D: the inevitability of gradualness? Solid State Ion 134:3–20
4. Alqahtany H, Eng D, Stoukides M (1993) Synthesis gas production from methane over an iron electrode in a solid electrolyte cell. J Electrochem Soc 140:1677–1681
5. Teraoka Y, Zhang H M, Furukawa S, Yamazoe M (1985) Oxygen permeation through perovskite-type oxides. Chem Lett 14:1743–1746
6. Teraoka Y, Nobunaga T, Yamazoe N (1988) Effect of cation substitution on the oxygen semipermeability of perovskite-type oxides. Chem Lett 17:503–506
7. Dyer PN, Richards RE, Russek SL (2000) Ion transport membrane technology for oxygen separation and syngas production. Solid State Ion 134:21–33
8. Wagner C (1975) Equations for transport in solid oxides and sulfides of transition metals. Prog Solid State Chem 10:3–16
9. Skinner SJ, Kilner JA (2000) Oxygen diffusion and surface exchange in $La_{2-x}Sr_xNiO_{4+\delta}$. Solid State Ion 135:709–712
10. Huang P, Petric A (1996) Superior oxygen ion conductivity of lanthanum gallate doped with strontiom and magnesium. J Electrochem Soc 143:1644
11. Ishihara T, Honda M, Nishiguchi H, Takita Y (1997) In: Proceedings of the 5th international symposium. SOFC, Aachen, 2–5 June 1997, p 301
12. Ruiz-Trejo E, Sirman JD, Baikov Ju M, Kilner JA (1998) Oxygen ion diffusivity, surface exchange and ionic conductivity in single crystal Gadolinia doped Ceria. Solid State Ion 113–115:565–569
13. De Souza RA, Kilner JA (1998) Oxygen transport in $La_{1-x}Sr_xMn_{1-y}Co_yO_{3\pm\delta}$ perovskites: part I. Oxygen tracer diffusion. Solid State Ion 106:175–187
14. ten Elshoff JE, Langhorst MHR, Bouwmeester HJM (1997) Chemical diffusion and oxygen exchange of $La_{0.6}Sr_{0.4}Co_{0.6}Fe_{0.4}O_{3-\delta}$. Solid State Ion 99:15–22
15. Minh NQ (1993) Ceramic fuel cells. J Am Ceram Soc 76:563–588
16. Ullmann H, Trofimenko N, Tietz F, Stöver D, Ahmad-Khanlou A (2000) Correlation between thermal expansion and oxide ion transport in mixed conducting perovskite-type oxides for SOFC cathodes. Solid State Ion 138:79–90
17. Nakamura T, Petzow G, Gauckler LJ (1979) Stability of the perovskite phase $LaBO_3$ (B = V, Cr, Mn, Fe, Co, Ni) in reducing atmosphere. I. Experimental results. Mater Res Bull 14:649–659

Mercury Drop Electrodes

Ying-Hui Lee and Chi-Chang Hu
Chemical Engineering Department, National Tsing Hua University, Hsinchu, Taiwan

Introduction and Description

Mercury has been widely used in electrochemistry, especially as a working electrode due to its unique electrochemical properties. The employment of a mercury electrode as the working electrode is called polarography, which was invented by Professor Jaroslav Heyrovský in 1922 [1]. Prof. Heyrovský was awarded Nobel Prize in 1959 for the significance polarography has established in electrochemical analysis. In this entry, we will introduce different types of mercury drop electrodes, the associated techniques for utilizing mercury electrodes, and the applications of mercury electrodes for the determination of various species.

Mercury electrodes are beneficial for the measurement of both inorganic and organic materials. The species to be determined on the mercury electrode should demonstrate electroactivity within the available potential range, or catalytic activity, or the ability to be adsorbed on mercury. The electroactive species must be soluble in a conductive solvent free of other interfering materials. Mercury electrodes enable negative potentials down to −2.5 V versus SCE since they have high overpotential for hydrogen evolution, but the positive potential is limited to ca. +0.4 V due to the dissolution of mercury beyond this potential. The utilization of nonaqueous electrolyte could further extend the negative potential down to −3.0 V. Numerous reduction reactions can occur in the wide negative potential range, which is useful for the determination of many reducible species. However, the highest positive potential of +0.4 V limits the mercury electrodes to measure less easily oxidized species such as phenols and anilines [2].

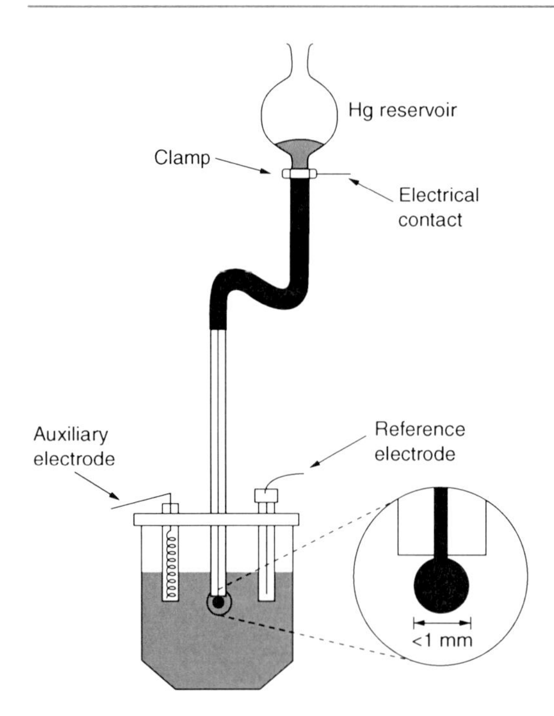

Mercury Drop Electrodes, Fig. 1 Classical design of the dropping mercury electrode (Wang [3]. Reprinted by permission of John Wiley & Sons)

There are several types of mercury electrodes. The dropping mercury electrode (DME), the hanging mercury drop electrode (HMDE), and the mercury film electrode (MFE) are some of the most frequently used electrodes.

The most representative mercury electrode in polarography is DME. The classical design of DME consists of a long glass capillary tubing (12–20 cm) connected to a flexible tube which is attached to an elevated mercury reservoir, as shown in Fig. 1 [3]. The level of mercury reservoir is sufficiently high with respect to the end of capillary, leading to the formation of mercury drop at the tip of the capillary by gravity. With the adjustment of the mercury reservoir height and the assistance of drop knocking device, a steady mercury drop flow with typical drop lifetime ranging from 2 to 6 s is obtained. The constant production of fresh mercury drops (i.e., continuous renewal of mercury surface) eliminates the problem of electrode passivation which is commonly found in solid electrodes during electroanalysis. Proper maintenance of the capillary, such as preventing the electrolyte infusion and the formation of air bubbles, can ensure the successful operation of the DME. The DME exhibits the advantages of simplicity, reliability, and renewable surface, which is suitable for many cases of electroanalysis. Nevertheless, the DME is less frequently used nowadays and is being replaced by the HMDE due to the problems such as high consumption of mercury and higher charging current (i.e., higher background current).

In contrast to the DME which steadily releases mercury drop during the analysis (i.e., continuously changing electrode surface area), the HMDE produces a static mercury drop with controlled geometry (i.e., constant electrode surface area). The constant surface area is desired to minimize the charging current to improve the detection limit. The design of the HMDE demonstrates low consumption of mercury, high reproducibility, and the possibility of adsorptive accumulation of analytes on mercury surface. However, drawbacks also exist for the HMDE. Sophisticated electronics and mechanics are required for precise drop production and disposal. Unlike solid electrodes, the HMDE is mechanically unstable, not ideal for on-site analysis and flow-through applications. Moreover, permanent modification by chemical reagents on the HMDE is not suitable to improve the sensitivity and selectivity [4].

Considering the limitations faced by mercury drop electrodes, the MFE is developed as an alternative. A very thin layer of mercury (10–100 μm) is deposited on conductive and inert substrates such as glassy carbon and noble metal supports. The MFE provides mechanical stability, a larger surface-to-volume ratio, the ability of chemical modification on the surface, and the possibility of different cell configurations which are useful in hydrodynamic techniques. In addition, the fabrication of the MFE further minimizes the consumption of mercury; only small amount of mercury is required. Nevertheless, due to the fact that the mercury film deposited on the substrates is actually composed of many droplets, the non-pure mercury surface renders the MFE a lower overpotential for hydrogen evolution and higher background current, which results in the limited potential range (i.e., lower selectivity) and lower sensitivity of the MFE.

Despite the environmental issue concerning the use of mercury, small amount of mercury at the room temperature is actually innocuous. Only when mercury is heated above 100 °C to produce vapor or organomercury compounds, the human health will be affected. Mercury electrodes can be safe and promising tools for electroanalysis when operated with care. Different electrochemical techniques associated with the use of mercury electrodes for the improvement of measurement sensitivity are introduced in the following section.

Techniques for Utilizing Mercury Drop Electrodes

DC Polarography (DCP)

For the conventional DC polarography, the current response changes with the applied linear potential ramp until it reaches the limiting current (diffusion-controlled). By measuring the half-wave potential (the potential where the current is one-half of its limiting value), species responsible for the polarographic wave can be determined, whereas the limiting current obtained can be related to the concentration of the analyte. The limiting current for the mercury drop is expressed as the following:

$$i_l = 708n\mathrm{D}^{1/2}\mathrm{m}^{2/3}\mathrm{t}^{1/6}\mathrm{C} \quad (1)$$

where n, D, m, t, C are number of electrons, the diffusion coefficient, mercury mass flow rate, drop lifetime, and the concentration. However, the real current response contains background current which is mainly ascribed to the charging current.

$$i_{\mathrm{total}} = i_l(\mathrm{t}) + i_c(\mathrm{t}) = k\,\mathrm{t}^{1/6} + k'\mathrm{t}^{-1/3} \quad (2)$$

The charging current is contributed by the double-layer charge at the interface between the electrode and the electrolyte. Although it can be minimized by increasing the lifetime of mercury drop, the significance of the charging current for the electroanalysis depends on its value relative to the analytical current (i_l). When the concentration of the analyte is low, the charging current becomes comparable to the analytical current, which limits the detection limit of DC polarography to 1×10^{-5}–5×10^{-6} M. Therefore, pulse voltammetric techniques are developed aimed at offering increased ratio between the analytical current and the charging current to achieve lower detection limit.

For pulse polarography, a sequence of potential steps is applied to the mercury electrode, each with the duration of ca. 50 ms. The pulse is synchronized with the mercury drop growth by controlling the drop time through the knocking device. During each potential step, the charging current decays rapidly to a negligible value whereas the analytical current only decays slightly. Consequently, clear discrimination between the analytical current and the charging current can be achieved by measuring the current at the end of pulse life. The difference between various pulse techniques lies in the excitation waveform and the current sampling regime, which will be introduced in the following paragraphs.

Normal-Pulse Polarography (NPP)

Normal-pulse polarography provides a series of pulses with increasing amplitude, which increases linearly with each mercury drop. Between each pulse, the electrode is maintained at the base potential at which no electrochemical reaction of the analyte occurs. To minimize the value of the charging current, the current response is measured about 40 ms after the pulse is applied. Moreover, larger analytical current can be obtained since thinner diffusion layer contributed by short pulse duration results in greater flux of the analyte. Therefore, the limiting current measured in NPP is 5–10 times larger than that in DCP, as expressed in the following equation:

$$\frac{i_{l,\mathrm{NPP}}}{i_{l,\mathrm{DCP}}} = \left(\frac{3t_d}{7t_m}\right)^{1/2} \quad (3)$$

where t_d and t_m represent the drop lifetime in DCP and the sampling time after the pulse in NPP, respectively. The increase of analytical signal improves the sensitivity of mercury electrodes.

Differential-Pulse Polarography (DPP)

Differential-pulse polarography provides fixed magnitude pulses which is superimposed on a linear potential ramp to mercury electrodes at the time just before the end of the mercury drop. The current is sampled twice, one before the pulse, and the other near the end of the pulse. The current difference ($i(t_2)-i(t_1)$) is plotted against the applied potential to obtain differential-pulse polarograms. The height of the current peak in the polarograms is directly proportional to the concentration of the analyte. The differential-pulse operation leads to the effective correction of the background current since the differential charging current is almost negligible, as shown in Eq. 4:

$$\Delta i_c \cong -0.00567 C_i \Delta E m^{2/3} t^{-1/3} \tag{4}$$

where C_i and ΔE are the integral capacitance and pulse amplitude. The background current contribution is one order of magnitude smaller than that in NPV, allowing the detection limit of 10^{-7}–10^{-8} M for the measurement of trace levels of inorganic and organic species. Furthermore, DPP improves the resolution of current response (peaks separated by 50 mV can be distinguished) and provides more information about the chemical form in which the analyte appears.

Square-Wave Polarography (SWP)

Square-wave polarography provides large-amplitude symmetric square waves which are superimposed on the base staircase potential to the mercury electrode. The current is sampled twice, one at the end of forward pulse, and the other at the end of the reverse pulse. The current difference ($i(t_2)$ – $i(t_1)$) is plotted against the applied base staircase potential to obtain polarograms. The height of the current peak in the polarograms is proportional to the concentration of the analyte. The larger net current obtained from the difference between forward and reverse current leads to the higher sensitivity of SWP (detection limit of 10^{-8} M) compared to that of DPP. Besides the sensitivity, the speed of SWP is the major advantage. The polarograms can be recorded within a few seconds while it takes about 2–3 min for DPP.

Stripping Polarography

Stripping analysis involves two steps, preconcentration (the deposition step) and stripping (the measurement step). The ions of interest are first collected into the mercury electrode for preconcentration; during the stripping, the collected ions will be oxidized or reduced back to the solution for the measurement. With the combination of effective preconcentration and stripping, extremely high signal-to-background ratio is generated, which results in the detection limit down to 10^{-10} M. Different stripping analyses have been developed, including anodic stripping voltammetry (ASV), potentiometric stripping analysis (PSA), cathodic stripping voltammetry (CSV), and adsorptive cathodic stripping polarography (AdCSV). ASV is the most widely used form of stripping. Different sensitivity can be obtained by adjusting the deposition time of metals into the mercury electrodes during preconcentration, e.g., from the deposition time of less than 0.5 min for the detection limit of 10^{-7} M to 20 min for 10^{-10} M. The sensitivity toward numerous trace elements can be further enhanced by AdCSV. With the short adsorption time of 1–5 min for preconcentration and the effective reduction step for the measurement, extremely low detection limit of 10^{-10}–10^{-11} M can be achieved; even 10^{-12} M is possible by coupling the adsorption process with catalytic reactions. Different polarographic/voltammetric techniques associated with the mercury electrodes for various detection limits are summarized in Table 1 [5].

Applications of Mercury Drop Electrodes

Mercury drop electrodes exhibit great sensitivity and selectivity for the determination of various inorganic and organic species, which are

Mercury Drop Electrodes, Table 1 Basic parameters of polarographic and voltammetric techniques (Barek et al. [5]. Reprinted by permission of Taylors & Francis, http://www.tandf.co.uk/journals)

Technique	*Applied potential program*	*Current response*	*Working electrode*	*LOD*
TAST		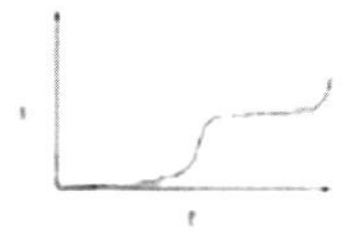	DME	~ 10^{-6} M
NPP (NPV)	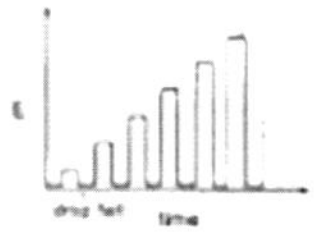	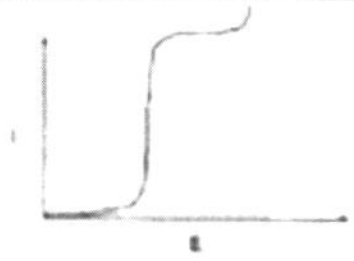	DME (HMDE)	~ 10^{-7} M (~10^{-7} M)
SCV	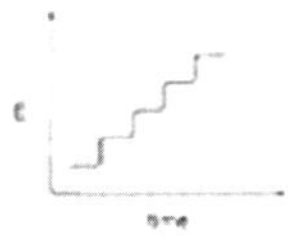	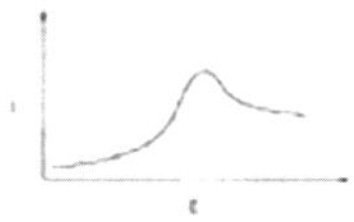	HMDE	~ 10^{-7} M
DPP (DPV)	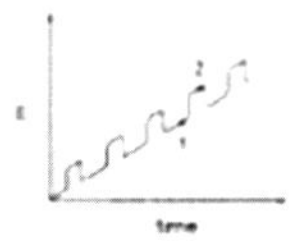	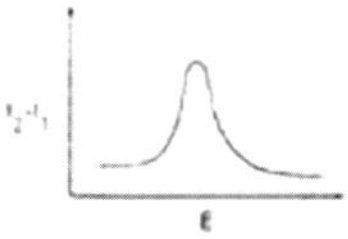	DME (HMDE)	~ 10^{-7} M ~ 10^{-8} M
SWP (SWV)		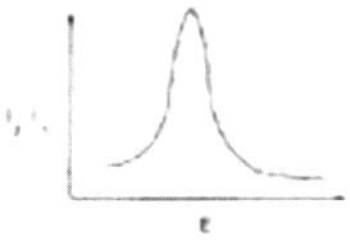	DME (HMDE)	~ 10^{-8} M ~ 10^{-8} M
ACP (ACV)	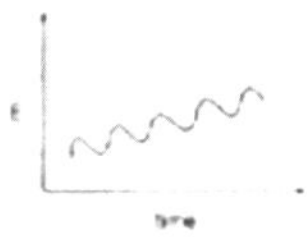	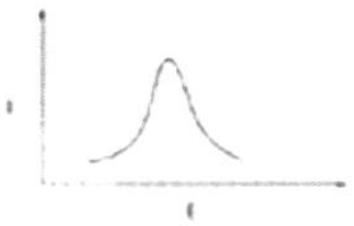	DME (HMDE)	~ 10^{-7} M (~10^{-8} M)
ASV [a,c,d] (CSV) [a,b,c,d]	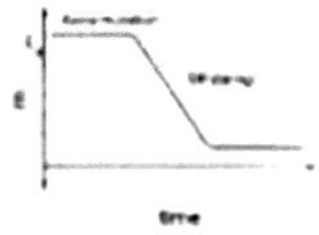	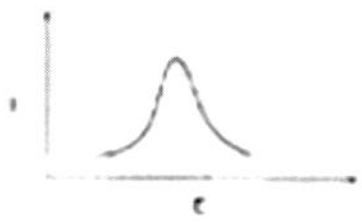	HMDE, MFE	~10^{-10} M (~10^{-9} M)
AdSV [b,d]	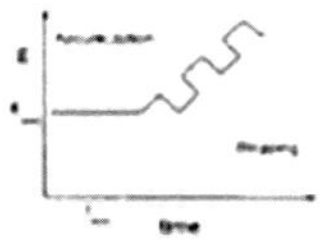	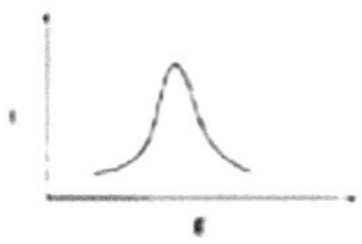	HMDE, MFE	~10^{-11} M ~10^{-12} M
PSA	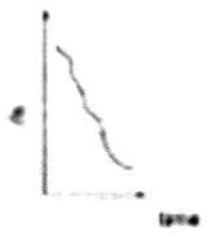	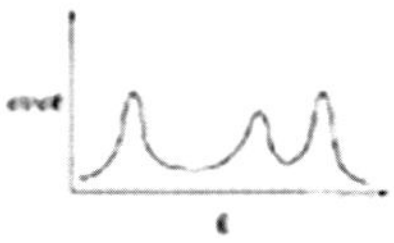	MFE	~ 10^{-12} M

Mercury Drop Electrodes, Table 2 Selected determination of heavy metal ions at mercury drop electrodes

Species	Matrix	Electrode	Technique	Detection limit	Ref.
Cr^{6+}	Natural samples	HMDE	DP-AdCSP	0.13 nM	[13]
Ni^{2+}	Sea water	HMDE	SW-AdCSP	5.00 nM	[14]
As^{3+}	Natural water	HMDE	SW-CSP	0.06 nM	[12]
Se^{4+}	Water samples	HMDE	DP-CSP	4.43 nM	[15]
Cu^{2+}	Waste water	HMDE	DP-ASP	0.16 μM	[16]
Pb^{2+}				0.09 μM	
Cd^{2+}				0.06 μM	
Co^{2+}	Sea water	HMDE	DP-AdCSP	3.00 pM	[17]
Se^{4+}	Human hair	MFE on GCE	DP-CSP	1.40 nM	[18]
Zn^{2+}	Water samples	MFE on GCE	SW-ASP	0.26 μM	[19]
Cu^{2+}				0.28 μM	
Pb^{2+}				9.65 nM	
Cd^{2+}				8.80 nM	
Tl^{+}	Water, soil, sediments	MFE on GCE	DP-ASP	0.25 pM	[20]

DP differential pulse, SW square wave, AdCSP adsorptive cathodic stripping polarography, CSP cathodic stripping polarography, ASP anodic stripping polarography

extremely useful from the environmental perspective. The following introduces examples utilizing mercury electrodes for the monitoring of pollutants in the environment.

Inorganic Species

About 20 heavy metals, including Pb, Sn, As, Cu, Zn, Cd, Bi, Sb, Tl, Mn, etc., can be measured by mercury electrodes. With the choice of suitable polarographic techniques, sensitivity required for certain metals of interest can be achieved. Stripping strategies are especially useful for the determination of trace metals in the pollutants. Take the detection of toxic As^{3+} as an example [6, 7]. Compared to metals such as Pb, Cu, Cr, and Zn, As is slightly soluble in mercury, which limits the measurement employing DC polarography due to the insufficient sensitivity for trace analysis. The applications of differential-pulse polarography in several electrolytes have been reported to achieve the determination of As^{3+} with low detection limits and wide linear concentration range on mercury drop electrodes [8–10]. The most suitable techniques for the detection of trace As^{3+} in real samples would be stripping analysis. Due to the low solubility of As^{3+} in mercury, the use of anodic stripping polarography is less common. Cathodic stripping polarography is perhaps the most popular method to measure As^{3+}. In the presence of copper or selenium [11, 12], the sensitivity for the detection of As^{3+} can be further improved through the following process:

$$\text{Preconcentration}: \; 2\,As^{3+} + 3\,CuHg + 6\,e^- \rightarrow Cu_3As_2 + 3\,Hg$$

$$\text{Stripping}: \; Cu_3As_2 + 12\,H^+ + 12\,e^- + 3\,Hg \rightarrow AsH_3 + H_2 + 3\,MHg$$

The determinations of As^{3+} along with other common heavy metal ions using mercury drop electrodes are summarized in Table 2.

Organic Species

Nowadays, many organic materials used in pharmaceuticals, dyes, or pesticides have become hazardous wide spread pollutants in the effluent. These organic species which exhibit electrochemical activities can be determined by mercury electrodes, especially trace amounts of organic compounds in pharmaceutical and biomedical applications. High sensitivity is easily achieved by employing adsorptive accumulation as the preconcentration step in the stripping technique. The determinations of common organic compounds are summarized in Table 3.

Mercury Drop Electrodes, Table 3 Selected determination of organic compounds at mercury drop electrodes

Species	Matrix	Electrode	Technique	Detection limit	Ref.
Diacetyl	Beer	HMDE	SW-AdCSP	10.0 $\mu g\ l^{-1}$	[21]
Fluvoxamine	Pharmaceutical products	HDME	SW-AdCSP	1.50 $\mu g\ l^{-1}$	[22]
Cefaclor	Human urine	HDME	CSP	2.90 $\mu g\ l^{-1}$	[23]
RR120 dye	Human serum albumin	HMDE	DP-AdCSP	0.06 $\mu g\ l^{-1}$	[24]
Cyanide	Waste water	HMDE	DPP	1.70 $\mu g\ l^{-1}$	[25]
C.I. Reactive Orange 13	Water samples	HMDE	DP-AdCSP	0.76 $\mu g\ l^{-1}$	[26]
Methyl-parathion (pesticide)	Spiked water	HMDE	SW-AdCSP	2.00 $\mu g\ l^{-1}$	[27]
Atrazine (herbicide)	River water	HMDE	SWV	2.00 $\mu g\ l^{-1}$	[28]
Triflumizole (fungicide)	Soil, natural water	DME	DPP	0.25 $ng\ l^{-1}$	[29]

DP: differential pulse, SW-: square wave, AdCSP: adsorptive cathodic stripping polarography, CSP: cathodic stripping polarography, DPP: differential-pulse polarography, SWV: square-wave polarography

The utilization of mercury electrodes plays an important role in electrochemical analysis, especially for the environmental monitoring of various hazardous species. The polarography employing mercury electrodes exhibit the advantages of low cost, high sensitivity (as low as pM levels using AdCSP), and high speed (routine analysis with less than 3 min), despite the limited selectivity due to the oxidation of mercury at potentials more positive than +0.4 V. Sufficient cognition about mercury and proper operation with the device are required to further extend the applications of mercury electrodes in the near future.

Cross-References

▶ Polarography

References

1. Heyrovsky M (2011) Polarography-past, present, and future. J Solid State Electr 15:1799
2. Zuman P (2000) Role of mercury electrodes in contemporary analytical chemistry. Electroanalalysis 12:1187
3. Wang J (2006) Analytical electrochemistry. Wiley, New Jersey, p 250
4. Vyskocil V, Barek J (2009) Mercury electrodes-possibilities and limitations in environmental electroanalysis. Crit Rev Anal Chem 39:173
5. Barek J, Fogg AG, Muck A, Zima J (2001) Polarography and voltammetry at mercury electrodes. Crit Rev Anal Chem 31:291
6. Cavicchioli A, La-Scalea MA, Gutz IGR (2004) Analysis and speciation of traces of arsenic in environmental food and industrial samples by voltammetry: a review. Electroanalysis 16:697
7. Hung DQ, Nekrassova O, Compton RG (2004) Analytical methods for inorganic arsenic in water: a review. Talanta 64:269
8. Dubey RK, Puri BK, Hussain MF (1997) Determination of arsenic in various environmental and oil samples by differential pulse polarography after adsorption of its morpholine-4-carbodithioate on to microcrystalline naphthalene or morpholine-4-dithiocarbamate-CTMAB-naphthalene adsorbent. Anal Lett 30:163
9. Higham AM, Tomkins RPT (1993) Determination of trace quantities of selenium and arsenic in canned tuna fish by using electroanalytical techniques. Food Chem 48:85
10. Greschonig H, Irgolic KJ (1992) Electrochemical methods for the determination of total arsenic and arsenic compounds. Appl Organomet Chem 6:565
11. Henze G, Wagner W, Sander S (1997) Speciation of arsenic(V) and arsenic(III) by cathodic stripping voltammetry in fresh water samples. Fresen J Anal Chem 358:741
12. Li H, Smart RB (1996) Determination of subnanomolar concentration of arsenic(III) in natural waters by square wave cathodic stripping voltammetry. Anal Chim Acta 325:25
13. Grabarczyk M (2008) A catalytic adsorptive stripping voltammetric procedure for trace determination of Cr(VI) in natural samples containing high concentrations of humic substances. Anal Bioanal Chem 390:979
14. Whitworth DJ, Achterberg EP, Nimmo M, Worsfold PJ (1998) Validation and in situ application of an automated dissolved nickel monitor for estuarine studies. Anal Chim Acta 377:217
15. de Carvalho LM, Schwedt G, Henze G, Sander S (1999) Redoxspeciation of selenium in water samples by cathodic stripping voltammetry using an automated flow system. Analyst 124:1803

16. dos Santos ACV, Masini JC (2006) Development of a sequential injection anodic stripping voltammetry (SI-ASV) method for determination of Cd(II), Pb(II) and Cu(II) in wastewater samples from coatings industry. Anal Bioanal Chem 385:1538
17. Vega M, van den Berg CMG (1997) Determination of cobalt in seawater by catalytic adsorptive cathodic stripping voltammetry. Anal Chem 69:874
18. Wang Y, Liu ZQ, Yao GJ, Zhu PH, Hu XY, Yang C, Xu Q (2009) An electrochemical assay for the determination of Se (IV) in a sequential injection lab-on-valve system. Anal Chim Acta 649:75
19. Suteerapataranon S, Jakmunee J, Vaneesorn Y, Grudpan K (2002) Exploiting flow injection and sequential injection anodic stripping voltammetric systems for simultaneous determination of some metals. Talanta 58:1235
20. Lukaszewski Z, Jakubowska M, Zembrzuski W, Karbowska B, Pasieczna A (2010) Flow-Injection Differential-pulse anodic stripping voltammetry as a tool for thallium monitoring in the environment. Electroanalysis 22:1963
21. Rodrigues PG, Rodrigues JA, Barros AA, Lapa RAS, Lima JLFC, Cruz JMM, Ferreira AA (2002) Automatic flow system with voltammetric detection for diacetyl monitoring during brewing process. J Agr Food Chem 50:3647
22. Nouws HPA, Delerue-Matos C, Barros AA, Rodrigues JA, Santos-Silva A (2005) Electroanalytical study of fluvoxamine. Anal Bioanal Chem 382:1662
23. Rodrigues LNC, Zanoni MVB, Fogg AG (1999) Indirect polarographic and cathodic stripping voltammetric determination of cefaclor as an alkaline degradation product. J Pharm Biomed 21:497
24. Guaratini CCI, Zanoni MVB, Fogg AG (2002) Cathodic stripping voltammetric detection and determination at a hanging mercury-drop electrode of dye contaminants in purified biomaterials: study of the human serum albumin and reactive dye 120 system. Microchem J 71:65
25. LaFuente JMG, Martinez EF, Perez JAV, Fernandez SF, Ordieres AJM, Uria JES, Sanchez MLF, Sanz-Medel A (2000) Differential-pulse voltammetric determination of low mu gl(-1) cyanide levels using EDTA, Cu(II) and a hanging mercury drop electrode. Anal Chim Acta 410:135
26. Zima J, Barek J, Moreira JC, Mejstrik V, Fogg AG (2001) Electrochemical determination of trace amounts of environmentally important dyes. Fresenius J Anal Chem 369:567
27. dos Santos LBO, Masini JC (2008) Square wave adsorptive cathodic stripping voltarnmetry automated by sequential injection analysis - Potentialities and limitations exemplified by the determination of methyl parathion in water samples. Anal Chim Acta 606:209
28. dos Santos LBO, Abate G, Masini JC (2004) Determination of atrazine using square wave voltammetry with the Hanging Mercury Drop Electrode (HMDE). Talanta 62:667
29. Inam R, Guleman EZ, Sarigul T (2006) Determination of triflumizole by differential pulse polarography in formulation, soil and natural water samples. Anal Chim Acta 579:117

Metal Ion Removal by Cathodic Reduction

Luís Augusto M. Ruotolo
Department of Chemical Engineering, Federal University of São Carlos, São Carlos, SP, Brazil

Introduction

The removal of metal ions occurs through their electrodeposition on the solid–liquid interface on the cathode surface area. Two distinct situations must be recognized considering metal removal by cathodic reduction: (1) electrowinning, in which concentrated solutions (usually greater than 30 g dm^{-3}) are used for metal production, and (2) dilute solutions, typically between 1 and 1,000 mg dm^{-3} [1, 2]. Electrowinning of base metals such as copper, zinc, nickel, and cadmium is produced by electrodeposition using plate cathodes from aqueous solutions obtained by alkali or acid leaching of their ores. On the other hand, metal removal from dilute solutions can be motivated mainly by three different reasons: (1) it could be necessary to achieve the concentration limits imposed by environmental legislation concerning waste discharge in rivers or sewage, (2) the process conditions per se impose the need for metal recycling, or (3) it could be economically interesting to recover valuable metals [3, 4].

This essay is focused on the application of metal removal from dilute solutions in which environmental concern is still the main motivation for metal recovery.

Nowadays there has been great concern about the risks and dangers to humans and the environment that can be caused by industrial effluents and wastes generated by activities of modern life. Among them, pollution associated with inadequate disposal of aqueous effluents containing heavy metals is still considered one of the major challenges in the field of effluent treatment and environmental protection.

A large number of anthropogenic sources of metal ion contamination can be recognized nowadays, but industrial processes are responsible for the production of large volumes of wastewater. Notable among industrial effluents containing metal ions are those of the ore leaching process, electroplating, scrap processing, photographic processing, production and recycling of batteries, and organic and inorganic chemical processes [5]. Hence, from the environment and human health point of view, it should be considered that effluents contaminated with metal ions, even when they are present in very low concentrations, still constitute very hazardous effluents. Water and soil heavy metal pollution is considered nowadays one of the most serious environmental problems since the metals can accumulate in the food chain and cause serious hazards to animals and humans [6].

Considering the aforementioned reasons, the necessity of metal ion removal from liquid effluents becomes evident. The conventional technique applied for effluent treatment containing metal ions consists of increasing the pH by adding calcium oxide or another chemical. The metal ion is precipitated as an insoluble hydroxide and removed by flotation or sedimentation, concentrated, and disposed of in special class landfills, which represents an important cost for the treatment process. The main drawback of this technique is the high and irreversible consumption of chemicals and the fact that the problem is only transferred from the liquid to the solid phase [7].

In order to overcome the disadvantages associated with the conventional process, the development of new technologies is a very active research field in the area of effluent treatment. Ion exchange and membrane processes have been intensively investigated nowadays; however, these processes only concentrate the metals ions, which must be removed by chemical precipitation, and so the problem persists. The electrochemical process constitutes an interesting alternative, since the metal can be removed from the liquid phase by applying an electric current. Since the electron is the only reagent used in most cases, this process has been considered a "clean technology." Furthermore, in most cases, the metal ion can be recovered in its solid form and, consequently, recycled. Other advantages of the electrochemical process include low operational cost, reduced workmanship, partial or total elimination of areas for solid waste disposal, and simplicity in automating the process [8].

Cathodic Reduction

In an electrochemical reaction, ion in solution + electron from electrode → reduced metal; the metal is recovered, in most cases, in its solid form, and the water can also occasionally be reused. From an economic point of view, besides the aforementioned advantages, it also must be recognized that the recovery of precious metals such as gold and silver would be very attractive, since the capital and operational costs involved in the electrochemical process could be easily overcome by the price for the recovery metal. On the other hand, for metals like zinc, copper, or cadmium, for example, the electrodeposition process must be optimized in order to make the electrochemical technology competitive with the precipitation process, which is by far the most inexpensive, despite all mentioned disadvantages. In this sense, the electrochemical technology needs to be well understood under the kinetics and mass transfer aspects in order to allow for adequate reactor design and to maximize the reaction rate and current efficiency with consequent minimization of energy consumption [8]. Regarding the environmental aspects, electrochemical reduction for metal removal has been recognized as "environmentally compatible," since in most cases the use of chemicals is not necessary, and the electron can be considered a clean reagent [9].

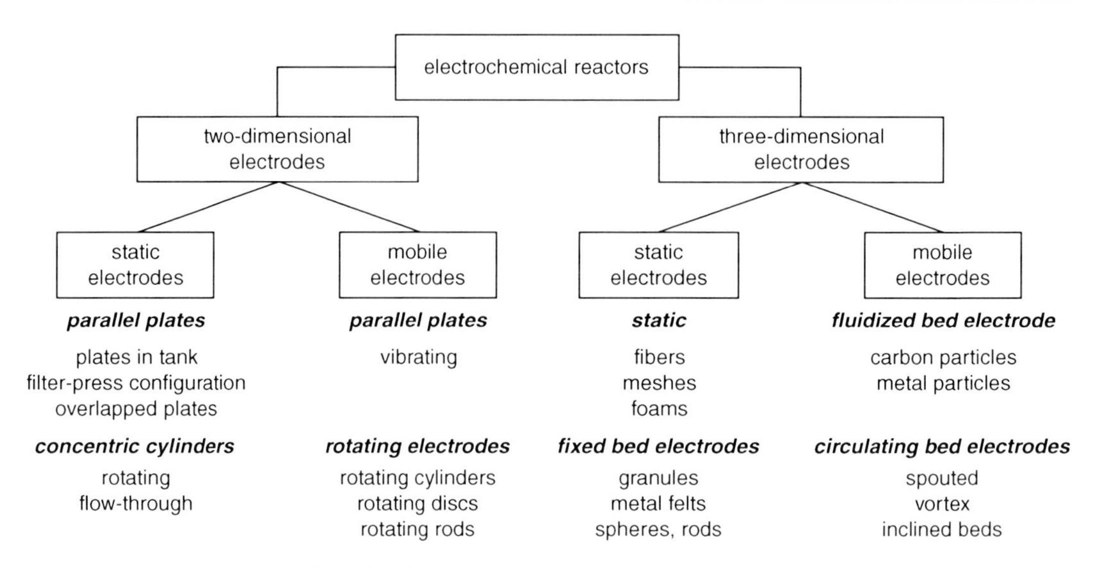

Metal Ion Removal by Cathodic Reduction, Fig. 1 Electrochemical reactor classification considering its geometry and fluid dynamics

In most industrial electrochemical processes (such as in chlor-alkali and metal electrowinning industries), the use of plate (two-dimensional) electrodes is quite adequate since the electroactive species are present at very high concentrations. Although the main advantages of two-dimensional electrodes are addressed as to their construction and operational simplicity, surface area and mass transfer limitations would be impeditive for their use in wastewater treatment since the typical metal concentrations are very low [2]. In order to overcome such disadvantages of two-dimensional electrodes, the development of three-dimensional electrodes was proposed.

Three-Dimensional Electrodes

The conception of the three-dimensional electrode occurred almost simultaneously in France and England around 1966 as a necessity for precious metal recovery from dilute solutions occurring in industrial processes and for the removal of heavy metals from wastewaters. In this class of electrodes, the reaction still occurs at the liquid–solid interface, but the surface area is provided by the electrode volume. Porous or three-dimensional electrodes have been recognized for improving the electrodeposition of metal ions from dilute solutions due to the high mass transfer coefficients provided by turbulent flow and high specific surface area [10].

There is vast literature detailing the use of different electrochemical reactors used for metal removal [2, 6]. The main types of electrochemical reactors can be classified first according to the kind of electrode as two or three-dimensional. The second classification considers the movement of the electroactive material with relation to a fixed referential. Thus the electrodes can be classified as static or mobile. Figure 1 shows the major electrode classifications according to geometry and fluid dynamics [1].

Porous or particulate fixed bed electrodes have been recognized as the more efficient three-dimensional cathodes used for metal electrodeposition from dilute solutions, mainly due to their uniform effective conductivity of the solid phase. However, electrode clogging due to metal electrodeposition restricts their use as only for very dilute solutions in which the long operational time would justify their use [11, 12]. In order to overcome this limitation and make the electrochemical process continuous, mobile electrodes, such as fluidized, spouted, and inclined bed electrodes, were proposed [13–16]. In these electrodes, the conductivity of the

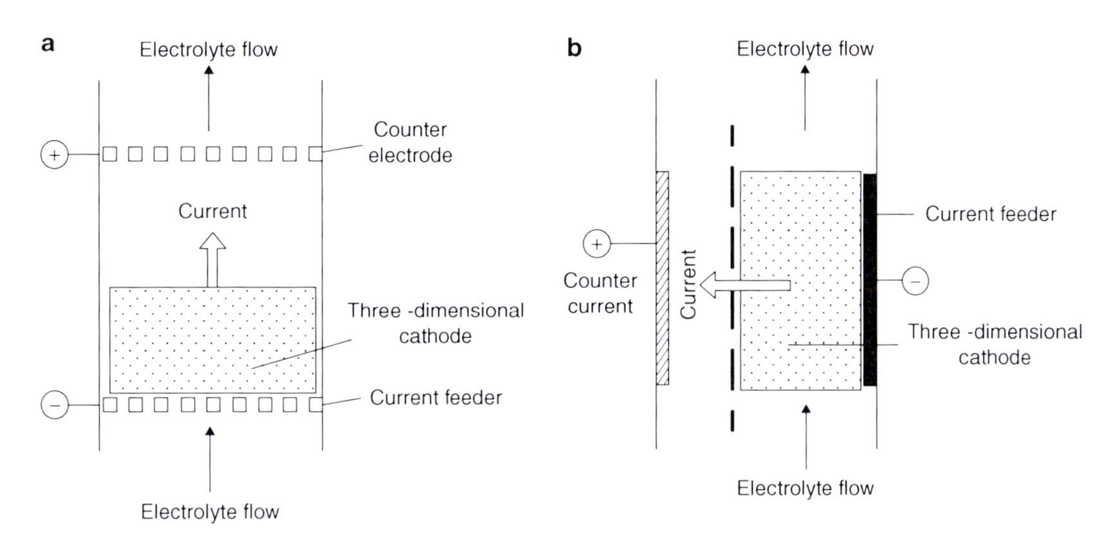

Metal Ion Removal by Cathodic Reduction, Fig. 2 Electrode configurations according to the current and electrolyte flow: (**a**) flow-through and (**b**) flow-by

particulate cathode depends on the frequency of particle collisions or contact. Thus, the potential difference between the solid and liquid phase results in very nonuniform electrode potential inside the porous bed. This leads to different electrodeposition rates that promote metal electrodeposition on the current feeder, hydrogen evolution close to the counter electrode, and, eventually, zones of metal dissolution [17, 18].

Figure 2 shows a schematic view of a typical simple three-dimensional electrode with its main components: porous cathode, current feeder, counter electrode, and membrane or separator. Regarding the constructive aspects, the three-dimensional electrode can be classified as flow-through or flow-by, depending on the relative direction of current and electrolyte flow, as illustrated in Fig. 2. The flow-though configuration is convenient in many cases only for laboratory studies, since electrode thickness is limited by the irregular potential and current distribution which make scale-up difficult. Moreover, the flow-by configuration can also proportionate long residence times and high conversions per electrolyte pass [2].

Three-dimensional electrodes show more problems associated with potential and current distribution than two-dimensional electrodes. The anisotropy of porous or particulate

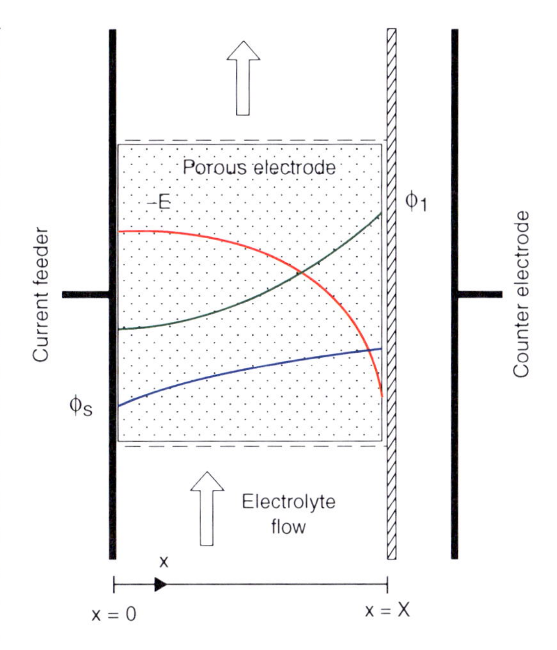

Metal Ion Removal by Cathodic Reduction, Fig. 3 Schematic representation of electrode potential inside a three-dimensional electrode

electrodes concerned with the electrode/electrolyte conductivities, electrolyte flow, and electroactive species concentration has been pointed out as the main reason for solid (ϕ_s) and liquid (ϕ_l) phase potential variations, as shown in Fig. 3 [10, 19]. This potential profile results in zones with different electrochemical activities inside the three-dimensional electrode, and consequently the electrodeposition rate will not be uniform [20].

There is extensive literature about the measurement and modeling of electrode potential profiles in three-dimensional electrodes [18, 21, 22]. Although many models are quite complete (and many times complex), their effective quantitative use is very restricted, since many parameters are unknown and very difficult to measure, mainly due to the complexity of actual industrial effluents. In many cases, optimization has been achieved using statistical approaches, such as surface response or multi-response methodologies [23].

The optimization of metal electrodeposition in terms of reaction rate of metal removal is obtained under conditions of limiting current (i_L). Hence, according to Eq. 1, the best values of i_L are obtained maximizing the mass transfer coefficient (k_m) and the electrode area (A) [16, 24]. These are the main reasons why three-dimensional electrodes are recommended for cathodic reduction of metals from dilute solutions, since electrode area is enhanced due to the use of the electrode volume; k_m is improved by the hydrodynamic turbulence provided by the three-dimensional matrix [25]. A number of cell designs were developed in order to obtain high k_m, many of them mentioned in Fig. 1.

$$i_L = z\ F\ k_m A\ c \tag{1}$$

As metal concentration drops along electrolysis time, the i_L decreases, and one strategy would be to operate the process under limiting current conditions [26]. Nevertheless, the use of i_L also has important consequences regarding the deposit morphology, which will be rough and powdery [1]. Hence, there should be a compromise between the applied current and deposit morphology. Moreover, it is difficult to maintain the uniformity of i_L inside the three-dimensional electrode due to the aforementioned potential and current distribution [12].

Future Directions

The future of cathodic metal removal seems to be fruitful, since it is a mature technology, and there is a wide variety of cell designs commercially available nowadays. A promising field of application is recovery of precious metals from, e.g., spent catalysts and printed circuit boards, in which cell design and cell potential are usually not critical due to the value of the metal. Selective metal electrodeposition from a mixture of different ions is still a challenge, especially when the electrodeposition overpotentials are very close.

Reactor design still has been studied in order to achieve high k_m and make the recovery process continuous, e.g., new designs for spouted and inclined bed electrodes. On the other hand, process optimization in regard to current efficiency, space-time yield, and energy consumption strongly depends on the characteristics of the effluent to be treated. Conductivity, pH, and the presence of solids and complexing agents will determine the necessity of physical or chemical pretreatments in order to facilitate or to even make the electrodeposition possible. In other words, the cathodic removal of metal ions must be considered case by case, considering the electrochemical characteristics of the electroactive species present in the wastewater in order to make the electrodeposition process feasible in terms of kinetics and energy consumption.

Cross-References

▶ Electrochemical Cells, Current and Potential Distributions
▶ Three-Dimensional Electrode

References

1. Pletcher D, Walsh FC (1990) Industrial electrochemistry. Chapman and Hall, London
2. Ferreira BK (2008) Three-dimensional electrodes for the removal of metals from dilute solutions: a review. Miner Process Extr Metall Rev 29:330–371. doi:10.1080/08827500802045586

3. Goodridge F, Scott K (1994) Electrochemical process engineering. Plenum, New York
4. Heitz E, Kreysa G (1986) Principles of electrochemical engineering. VCH, Weinheim
5. Rajeshwar K, Ibanez JG (1996) Environmental electrochemistry: fundamentals and applications in pollution abatement. Academic Press, San Diego
6. Jüttner K, Galla U, Schmieder H (2000) Electrochemical approaches to environmental problems in the process industry. Electrochim Acta 45:2575–2594. doi:10.1016/S0013-4686(00)00339-X
7. Kurniawan TA, Chan GYS, Lo W-H, Babel S (2006) Physico-chemical treatment techniques for wastewater laden with heavy metals. Chem Eng J 118:83–98. doi:10.1016/j.cej.2006.01.015
8. Janssen LJJ, Koene L (2002) The role of electrochemistry and electrochemical technology in environmental protection. Chem Eng J 85:137–146. doi:10.1016/S1385-8947(01)00218-2
9. Rajeshwar K, Ibanez JG, Swain GM (1994) Electrochemistry and the environment. J Appl Electrochem 24:1077–1091
10. Newman JS (1991) Electrochemical systems. Prentice Hall, Englewood Cliffs
11. Olive H, Lacoste G (1980) Application of volumetric electrodes to the recuperation of metals in industrial effluents – II. Design of an axial field flow through porous electrodes. Electrochim Acta 25:1303–1308. doi:10.1016/0013-4686(80)87138-6
12. Doherty T, Sunderland JG, Roberts EPL, Pickett DJ (1996) An improved model of potential and current distribution within a flow-through porous electrode. Electrochim Acta 41:519–526. doi:10.1016/0013-4686(96)81774-9
13. Germain S, Goodridge F (1976) Copper deposition in a fluidized bed cell. Electrochim Acta 21:545–550. doi:10.1016/0013-4686(76)85148-1
14. Scott K (1988) A consideration of circulating bed electrodes for the recovery of metal from dilute solutions. J Appl Electrochem 18:504–510. doi:10.1007/BF01022243
15. Fleischmann M, Kelsall GH (1984) A parametric study of copper deposition in a fluidized bed electrode. J Appl Electrochem 14:269–275. doi:10.1007/BF01269926
16. Stankovic VD, Stankovic S (1991) An investigation of the spouted bed electrode cell for the electrowinning of metal from dilute solutions. J Appl Electrochem 21:124–129. doi:10.1007/BF01464292
17. Beenackers AACM, Van Swaaij WPM, Welmers A (1977) Mechanism of charge transfer in the discontinuous metal phase of a fluidized bed electrode. Electrochim Acta 22:1277–1281. doi:10.1016/0013-4686(77)87010-2
18. Hutin D, Coeuret F (1977) Experimental study of copper deposition in a fluidized bed electrode. J Appl Electrochem 7:463–471. doi:10.1007/BF00616757
19. Volkman Y (1979) Optimization of the effectiveness of a three-dimensional electrode with respect to its ohmic variables. Electrochim Acta 24:1145–1149. doi:10.1016/0013-4686(79)87062-0
20. Lanza MRV, Bertazzoli R (2000) Removal of Zn(II) from chloride medium using a porous electrode: current penetration within the cathode. J Appl Electrochem 30:61–70. doi:10.1023/A:1003836418682
21. Sun YP, Scott K (1995) An efficient method for solving the model equations of a two dimensional packed bed electrode. J Appl Electrochem 25:755–763. doi:10.1007/BF00648630
22. Ruotolo LAM, Gubulin JC (2011) A mathematical model to predict the electrode potential profile inside a polyaniline-modified reticulated vitreous carbon electrode operating in the potentiostatic reduction of Cr(VI). Chem Eng J 171:1170–1177. doi:10.1016/j.cej.2011.05.017
23. Ruotolo LAM, Gubulin JC (2005) A factorial-design study of the variables affecting the electrochemical reduction of Cr(VI) at polyaniline-modified electrodes. Chem Eng J 110:113–121. doi:10.1016/j.cej.2005.03.019
24. Ralph TR, Hitchman ML, Millington JP, Walsh FC (1996) Mass transport in an electrochemical laboratory filterpress reactor and its enhancement by turbulence promoters. Electrochim Acta 41:591–603. doi:10.1016/0013-4686(95)00346-0
25. Ponde-de-León C, Low CTJ, Kear G, Walsh FC (2007) Strategies for the determination of the convective-diffusion limiting current from steady state linear sweep voltammetry. J Appl Electrochem 37:1261–1270. doi:10.1007/s.10800-007-9392-3
26. Britto-Costa PH, Ruotolo LAM (2011) Electrochemical removal of copper ions from aqueous solution using a modulated current method. Sep Sci Technol 46:1205–1211. doi:0.1080/01496395.2010.546384

Metal-Air Batteries

Richard L. Middaugh
Independent Consultant, formerly with Energizer Battery Co, Rocky River, OH, USA

Introduction and Description

The only commercially important primary metal-air batteries today are zinc-air batteries. They dominate the hearing aid battery market, having largely replaced silver oxide (Zn/Ag_2O) and mercuric oxide (Zn/HgO) batteries. They provide nearly twice the energy of silver oxide batteries in the same size, without the cost of precious metal cathode material, and do not have the environmental impact of mercuric oxide batteries (now banned in most jurisdictions). Like these metal oxide batteries, they have a flat voltage discharge curve well suited to digital electronics used in hearing aids. The same attributes have led to the use of zinc-air 9-V batteries in some medical telemetry devices, although that market remains a small and diminishing one.

In recent years, zinc-air batteries have been used also in some military applications, because of their high energy, safety, and low cost (e.g., from the Electric Fuel division of Arotech) [1]. Prior to the digital electronics boom, zinc-air batteries were the batteries of choice for buoys and remote railroad signaling devices, because of their low cost and low maintenance requirements.

The main advantages of metal-air batteries are their "free" cathode (positive electrode material), O_2 from the air, and their high specific energies (Wh/kg) and energy densities (Wh/L). The origin of the high energy can be seen in the 2-electron per formula atom cathode reaction:

$$O_2 + 2H_2O + 4e^- \rightarrow 4OH^-,$$

compared to other cathode reactions with lower e^-/formula atom ratios:

$$MnO_2 + H_2O + e^- \rightarrow MnOOH + OH^- \quad (1e^-/3 \text{ formula atoms})$$

$$Ag_2O + H_2O + 2e^- \rightarrow 2Ag + 2OH^- \quad (2e^-/3 \text{ formula atoms})$$

$$FeS_2 + 4Li^+ + 4e^- \rightarrow Fe + 2Li_2S \quad (4e^-/3 \text{ formula atoms})$$

The cell electrode reactions and overall reaction can be written in general form as:

$$M + 2\,OH^- \rightarrow MO + H_2O + 2\,e^- \text{(anode)}$$

$$\tfrac{1}{2}\,O_2 + H_2O + 2\,e^- \rightarrow 2\,OH^- \text{ (cathode)}$$

$$M + \tfrac{1}{2}\,O_2 \rightarrow MO \text{ (overall)}$$

Metal-Air Batteries, Table 1 Specific energies of metal-air batteries

Type	Cell reaction	Nominal voltage	Wh/kg
Silver-zinc	$Zn + Ag_2O$ –> $ZnO + 2Ag$	1.55	280
Alkaline Zn/MnO_2	$Zn + H_2O + 2MnO_2$ –> $ZnO + 2MnOOH$	1.5	314
Li/MnO_2	$Li + MnO_2$ –> $LiMnO_2$	3	856
Zn-air	$2\,Zn + O_2$ –> $2\,ZnO$	1.4	922
Li/FeS_2	$4Li + FeS_2$ –> $2\,Li_2S + Fe$	1.5	1,088
Iron-air	$3Fe + 2O_2$ –> Fe_3O_4	1.3	1,204
Magnesium-air	$2\,Mg + O_2$ –> $2MgO$	1.5	1,995
Aluminum-air	$4Al + 3O_2$ –> $2Al_2O_3$	1.5	2,366

The thermodynamic voltage window for aqueous electrolytes is only 1.23 V, so the metal electrode must be stabilized kinetically rather than thermodynamically in the aqueous electrolyte if a cell voltage greater than 1.23 V is to be achieved. Zinc is unique among metals in its kinetic stability in alkaline electrolytes while also being very active (having very low overvoltage for anodic dissolution). Metals such as Fe, Al, and Mg can be made reasonably stable in aqueous electrolytes. However, when activated by anodic discharge, they undergo direct reaction with H_2O to form H_2, as well. This limits their applicability largely to reserve batteries, despite their attractive specific energies [1]. To date, Zn is the only metal with practical kinetic stability and high overvoltage (low rate) for H_2 evolution during anodic discharge.

Although the manufacturer does not need to package the cathode material in the battery, the battery must be designed to hold the reaction products (MO). Table 1 shows comparative theoretical specific energies of the zinc-air system and some competing primary battery systems (using average low-rate discharge voltages). Aluminum- and magnesium-air batteries have operating voltages far below the calculated thermodynamic voltages. Addition of inactive electrolyte, conductors, binders, separators, collectors, containers, and sealing materials lowers the overall deliverable battery energies to 30–45 % of theoretical energies. The "nominal voltage" and specific energy (Wh/kg) are somewhat arbitrary, since the actual voltage delivered on discharge depends on the load (current) and duty cycle (continuous vs. intermittent).

The disadvantages of metal-air systems limit the applications chosen for them. The main two disadvantages come from being open to the air to use O_2 as a "free" cathode. One is H_2O evaporation to or absorption from the ambient atmosphere, depending on the ambient relative humidity. The other is the risk of absorption of other ambient air components that may degrade the cell performance. Chief among the latter is CO_2, which can neutralize OH^- ions present in alkaline electrolytes, lowering the electrolyte pH:

$$CO_2 + OH^- \rightarrow HCO_3^-$$

This lowers the rate-limiting concentration of OH^- ions and can affect the anode reaction to produce a heavier and more voluminous metal bicarbonate or carbonate instead of the oxide. The presence of HCO_3^- ions can also result in the formation of anode surface carbonate films that passivate or inhibit the anode discharge reaction, severely limiting the anode discharge reaction's rate-carrying ability.

Engineering design approaches can mitigate the system disadvantages, depending on the intended application [2]. To delay the effects of openness to the ambient atmosphere, typical battery product packaging seals off the air access until the user removes the seal to activate the battery. The most common second design feature is to limit the size of the air access holes so that only enough O_2 to support the power requirements of the application can diffuse into the cathode structure. For typical hearing aid power demands, this hole size restricts H_2O and CO_2 transport to levels that can preserve the electrolyte volume and functionality for up to 1 month of service life in all but the most extreme ambient relative humidities.

The strong bond of the O_2 molecule (498 kJ mol^{-1}) is associated with a high activation energy or overvoltage for reduction of O_2. Expensive noble metal catalysts such as Pt can lower the activation energy or overvoltage significantly to give useful electrode voltages at practical discharge rates. To date, lower-cost catalysts approaching Pt effectiveness have been found only for alkaline electrolytes. For this reason, and because the oxygen reduction reaction is inherently faster in alkaline electrolytes, the following discussion focuses on alkaline electrolytes.

Air cathodes pose a demanding design engineering requirement: formation and maintenance of a high-area three-phase interface for the O_2 reduction reaction. Gaseous O_2, liquid H_2O, and solid catalyst/conductor must be collocated simultaneously for the electron-transfer reaction to take place:

$$O_2 + 2H_2O + 4e^- \rightarrow 4\ OH^-$$

Typical electrodes contain finely dispersed catalyst(s) on a high-area oxidation-resistant form of carbon conductor mixed with PTFE or other "wet-proofing" material and a metal foil or screen that serves as a material carrier and electronic charge collector. The carbon/PTFE mixture containing a removable solvent is applied to the carrier and dried to form a porous structure that allows O_2 access to the liquid/solid interface. Both liquid and gas must be able to penetrate the solid electrode structure for the reaction to proceed at practical rates.

The large zinc-air batteries used in buoys, railroad signals, and other remote devices are built in "tanks" with a large excess of KOH electrolyte, so that evaporation or absorption of H_2O will not starve or flood the cell and so that CO_2 absorption will not lower the OH^- concentration below useful levels for an extended time in service. The relatively low cost of zinc anodes and the "free" O_2 cathode material have been significant advantages for these applications in remote locations.

Additional zinc-air battery designs have been patented over the last century with various ways of increasing and decreasing the air access to match intermittent periods of use in an application and to extend calendar life in service for metal-air battery designs [3].

The zinc anode half-reaction (in alkaline electrolyte) proceeds through the zincate ion solubility equilibrium:

$$Zn + 4OH^- \rightarrow [Zn(OH)_4]^{2-} + 2e^-$$

$$[Zn(OH)_4]^2 \leftrightarrow ZnO + H_2O + 2OH^-$$

To give the overall reaction,

$$Zn + 2OH^- \rightarrow ZnO + H_2O + 2e^-$$

The relatively high solubility of ZnO in KOH electrolyte and its equilibrium allow the solid product ZnO to deposit in parts of the cell other than in the anode compartment. Solid ZnO, a semiconductor, can deposit in the separator pores and create a high-resistance internal short circuit, or in the pores of the cathode structure, partially blocking electrolyte access to the active surface of the cathode. Separator design and addition of precipitation agents such as CaO or $CaCO_3$ have been used to prevent these problems, especially in the older buoy battery "tank" designs and in rechargeable cell designs.

There has been considerable recent research and development of lithium-air batteries, but these efforts have been limited to rechargeable battery designs using as yet more expensive materials and designs than are considered practical for primary batteries.

Future Directions

For metal-air batteries to achieve broader commercial success, progress is needed on three issues:

1. For metals other than zinc to be used, electrolytes or electrode coatings must be found that allow the metal to be kinetically inert to electrolyte (water) reduction both at rest and during discharge, without also imposing significant voltage polarization that detracts from effective discharge performance.
2. Solvent (water) transpiration (in or out) and "impurity" entry (such as CO_2) must be prevented while also permitting adequate transport of O_2 to the air electrode for a wide range of discharge rates, in order to extend practical intermittent applications lasting more than a month or so.
3. Improved oxygen reduction catalysts with lower costs and better long-term stability in the cell environment are needed to minimize voltage losses on discharge.

Cross-References

▸ Lithium-Air Battery

References

1. Atwater T, Dobley A (2011) Metal air batteries. In: Reddy T, Linden D (eds) Linden's handbook of batteries, 4th edn. McGraw Hill, New York
2. Passaniti J, Carpenter D, McKenzie R (2011) Button cell batteries. In: Reddy T, Linden D (eds) Linden's handbook of batteries, 4th edn. McGraw Hill, New York
3. Several concepts can be found in J C Derksen. US patent 2,468,430

Micro- and Nanoelectrodes

Robert Forster
School of Chemical Sciences National Center for Sensor Research, Dublin City University, Dublin, Ireland

Introduction

Microelectrodes [1] are miniaturized working electrodes where the critical electrode dimension is on the micron scale yet remains much greater than the thickness of the electrical double layer, which is typically 10–100 Å. Nanoelectrodes are even smaller, with at least one dimension below 100 nm and their size can rival the thickness of the electrical double layer and even approach that of a single molecule [2]. Shrinking the size of

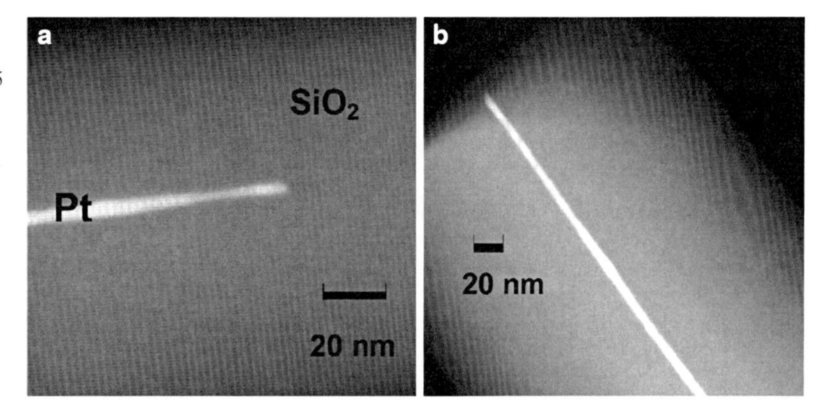

Micro- and Nanoelectrodes, Fig. 1 TEM image of a 1.5 nm radius Pt nanowire (**a**) and a 3nm radius Pt nanoelectrode (**b**) sealed in SiO2

the working electrode to the micro- and even nano-dimension dramatically changes the response observed and offers important advantages. For example, these small electrodes respond more quickly to changes in the applied potential, exhibit fast mass transport, and can provide lower limits of detection. The mass-transport rate increases as the electrode size decreases and radial diffusion dominates, which allows steady-state voltammetric responses to be observed without the need for convective mass transport. While mass transport continues to become more efficient at the nanoscale, these ultrasmall electrodes often bring additional complications such as migration of the analyte in the electric field as the depletion layer thickness approaches the Debye length. Other advantages include the ability to make measurements in highly resistive solutions because of the significantly smaller currents observed. Moreover, they facilitate measurements at low temperatures and they enable "redox mapping" to be performed with high spatial resolution [3, 4]. These properties have led to their widespread application in areas such as neurophysiology, lithography, chemical analysis, and sensor development.

Figures 1(a) and 1(b) show transmission electron microscopy (TEM) images of a 1.5 nm radius Pt nanowire and a 3 nm radius Pt nanoelectrode sealed in SiO2, respectively [5].

Disks are the most common miniaturized electrode geometry largely because of the relative ease with which a microwire or fiber can be sealed in an insulting barrel. Other geometries include cylinders, arrays, bands, and rings, and less commonly spheres, hemispheres, and more unusual assemblies. The most popular materials include platinum, carbon fibers, and gold, although mercury, iridium, nickel, silver, and superconducting ceramics have also been used. Nanoelectrodes arose from the development of nanoband electrodes that can have bandwidths as small as 5 nm [6]. Penner et al. [7] introduced glass-coated hemispherical nanoelectrodes that approached true molecular dimensions (~1 nm). Although their fabrication and characterization remain challenging, these ultrasmall probes are now an integral part of the field.

Steady-State Voltammetry at Micro- and Nanoelectrodes

For slow scan rates, i.e., a few volts per second or slower, the voltammetric response of a miniaturized electrode is a sigmoid-shaped steady-state voltammogram [8]. Peak-shaped responses are observed only when the scan rate is sufficiently high so that the depletion layer thickness is smaller than the critical dimension of the electrode. For example, a 50-nm-radius nanoelectrode requires the sweep rate to be of the order of 10,000 Vs^{-1} to show the classical peak-shaped responses typically observed for electrochemically reversible systems at macroelectrodes at mVs^{-1} scan rates. The theory describing mass transport to micro- and

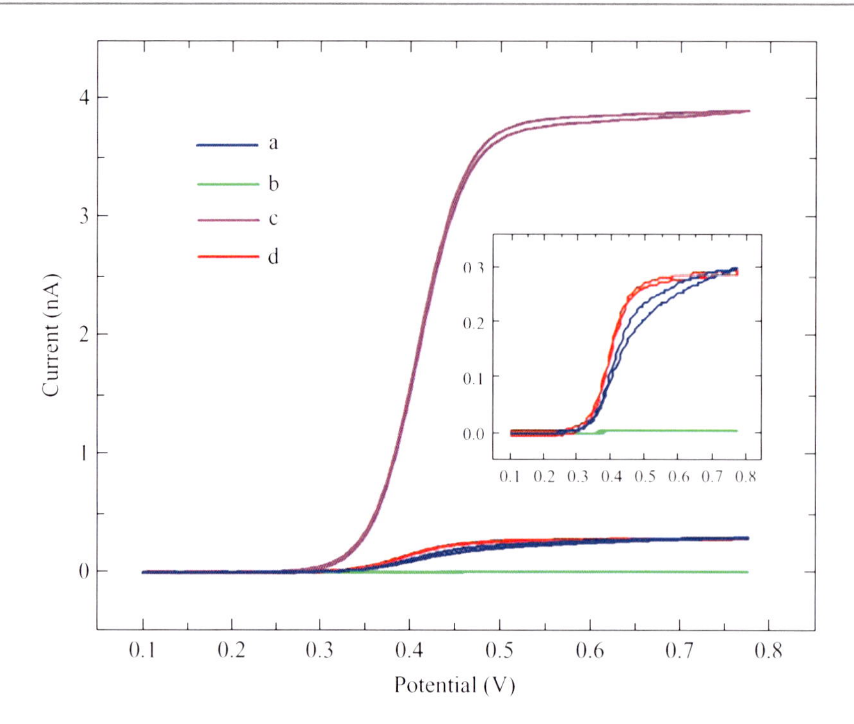

Micro- and Nanoelectrodes, Fig. 2 The cyclic voltammograms in 5-mM ferrocene corresponding to different stages in the fabrication of a 60-nm-radius Au nanoelectrode. a Pt nanoelectrode (radius) ~60 nm, b after etching, c after electrodeposition of Au, and d after polishing the excess Au. Scan rate: 50 mV/s. The *inset* shows a magnified portion of curves a, b, and d (Reproduced with permission from Ref. [10])

nanoelectrodes as well as to arrays is now at an advanced state of development [9].

Figure 2 shows a steady-state cyclic voltammogram of a 5-mM solution of ferrocene in acetonitrile at a 60-nm-radius Pt nanoelectrode [10]. The solution contains a high concentration (0.2 M) of tetra-n-butyl ammonium hexafluorophosphate, $TBAPF_6$, and the response for this electrochemically reversible couple is under radial diffusion control.

Figure 2 shows that a steady-state limiting current, i_{ss}, is observed. The magnitude of i_{ss} depends on the size and shape of the electrode. For a disk-shaped electrode, i_{ss} is described by Eq. 1:

$$i_{ss} = 4nFDCr \qquad (1)$$

where n is the number of electrons transferred, F is Faraday's constant, D is the diffusion coefficient ($cm^2\ s^{-1}$), C is the analyte concentration ($mol\ cm^{-3}$), and r is the electrode radius (cm).

Thus, the magnitude of i_{ss} can provide an insight into the electrode size and shape when a redox-active probe of known concentration and diffusion coefficient is used. Alternatively, once the electrode geometry is known, the limiting current can be used to determine the concentration of redox species. Moreover, the shape of the voltammogram can be analyzed to provide useful information about the rate of heterogeneous electron transfer across the electrode/solution interface.

Properties of Microelectrodes

Reduced Ohmic Effects

When Faradaic and charging currents flow through a solution, they generate a potential that acts to diminish the applied potential by an amount iR, where i is the total current and R is the cell resistance. This is an undesirable process that leads to distorted voltammetric responses.

The solution resistance for a disk-shaped ultramicroelectrode is inversely proportional to the electrode radius:

$$R = \frac{1}{4\kappa r} \tag{2}$$

where κ is the conductivity of the solution and r is the radius of the microdisk. Equation 2 shows that R *increases* as the electrode radius *decreases*. However, the currents observed at microelectrodes are typically six orders of magnitude smaller than those observed at macroelectrodes. These small currents often completely eliminate ohmic drop effects even when working in organic solvents. For example, the steady-state current observed at a 5-μm-radius microdisk is approximately 2 nA for a 1.0-mM solution of ferrocene. Taking a reasonable value of 0.01 Ω^{-1} cm^{-1} as the specific conductivity, then Eq. 2 indicates that the resistance will be of the order of 50,000 Ω. This analysis suggests that the iR drop in this organic solvent is a negligible 0.09 mV. In contrast, for a conventional macroelectrode the iR drop would be of the order of 5–10 mV. Under these circumstances, distorted current responses and shifted peak potentials would be observed in cyclic voltammetry if conventional macroelectrodes were used.

Interfacial Capacitance

When the potential that is applied to an electrode is changed, e.g., in cyclic voltammetry or chronoamperometry, this causes the charge on the metal side of the interface to change triggering reorganization of the ions and solvent dipoles in the double layer on the solution side of the interface. This process causes electrons to flow into or out of the surface giving rise to a charging or capacitive response. The double layer capacitance for a disk-shaped ultramicroelectrode is proportional to the area of the electrode surface and is given by

$$C = \pi r^2 C_O \tag{3}$$

where C_O is the specific double layer capacitance of the electrode. Thus, shrinking the size of the electrode causes the interfacial capacitance to decrease with decreasing r^2. These low capacitive currents are particularly important for analytical applications where the ability to discriminate a Faradaic signal above background charging often dictates the limit of detection that can be achieved.

Electrode Response Times

Beyond chemical analysis, a major application of nano- and microelectrodes is to measure the rates of heterogeneous electron transfer and homogeneous chemical reactions [11]. Therefore, it is important to consider the effect of shrinking the electrode size on its response time. Every electrochemical measurement has a lower timescale limit that is imposed by the RC cell time constant, i.e., the product of the solution resistance, R, and the double layer capacitance, C, of the working electrode. Meaningful electrochemical data can only be extracted at timescales that are typically five to ten times longer than the RC time constant [12]. Therefore, an important objective when seeking to make high-speed transient measurements is to minimize the cell time constant.

The existence of the double layer capacitance at the working electrode complicates electrochemical measurements at short timescales. Specifically, the double layer capacitance must be charged through the solution resistance in order to change the potential across the Faradaic impedance, and this process cannot be achieved instantaneously.

The time constant for this charging process is given by Eq. 4:

$$RC = \frac{\pi r C_o}{4\kappa} \tag{4}$$

and is typically hundreds of microseconds for a conventional millimeter-sized electrode placing a lower limit on the useful timescale of the order of several milliseconds. In contrast, microelectrodes exhibit cell time constants on the order of microseconds, while nanoelectrodes can achieve cell time constants of the order of a few tens of nanoseconds. However, stray capacitance, e.g., from imperfect seals between the conductor and

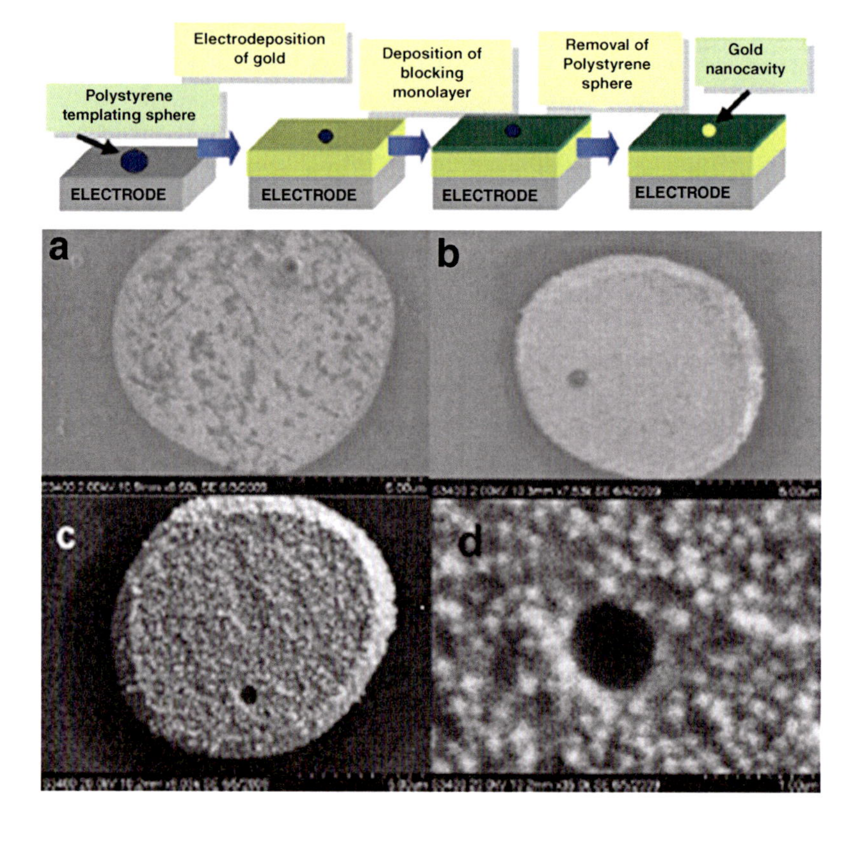

Micro- and Nanoelectrodes, Fig. 3 Scheme showing the formation of a nanocavity electrode and SEM images of (**a**) a single-polystyrene sphere on a 5-mm-radius gold electrode, (**b**) gold which has been electrodeposited around the sphere to a thickness of 600 nm, (**c**) electrode surface after sphere removal, showing a circular pore where the sphere was present, and (**d**) higher magnification image of the pore (Reproduced with permission from Ref. [13])

insulator, the leads, or the electrical connections, often causes the experimental response time to significantly exceed that predicted by theory.

Applications

Toward Single-Molecule Electrochemistry

A key objective in contemporary science is to develop techniques capable of detecting small numbers of molecules or even single molecules.

Cavity Microelectrodes

Nanocavity electrodes offer particular advantages for molecular detection since the cavity can be easily functionalized, e.g., packed with an enzyme, to create ultralow volume sensors that are mechanically stable under real world conditions. We recently described the use of nanosphere lithography, combined with self-assembly of a blocking surface active agent to create recessed nanocavity electrodes whose radius can be reproducibly controlled from 100 nm up to several microns [13]. The volume of a 100-nm-radius cavity is of the order of 3 fL, i.e., for mM concentrations of analyte, the cavity will contain only a couple of 1,000 analyte molecules.

As illustrated in Fig. 3, a single-polystyrene sphere was first deposited on a microelectrode and then gold was electrodeposited around the sphere. A blocking layer was then adsorbed at the upper planar gold surface before sphere removal. This blocking layer confines the electrochemical reactions to the inside of the nanocavity. After sphere removal, a spherical cap recess is obtained, the dimensions of which are controlled by the templating sphere size and the thickness of gold electrodeposited.

The creation of a spherical cavity that is accessible for redox reactions through a small pore ought to exhibit interesting and useful properties. For example, the flux into the cavity will be controlled by the pore diameter, whereas under semi-infinite linear diffusion conditions the current measured depends on the electrode area. In this sense, the radial diffusion field observed for a planar microdisk in contact with a solution containing a redox probe is inverted.

When the sphere is removed, but the alkanethiol remains assembled on the upper gold surface, a sigmoidal voltammogram is observed for a solution phase redox probe, ferrocyanide, whose current magnitude is consistent with the small electrochemically active area of the nanocavity electrode. The interior surface of these nanocavities can be selectively modified with a surface active, phosphorescent metal complex such as $[Ru(bpy)_2Qbpy]^{2+}$ where bpy is 2,2′-bipyridyl and Qbpy is 2,2′:4,4″-4′4″-quarterpyridyl. Significantly, an intense, localized, emission is observed due to an enhanced emission of the ruthenium complexes adsorbed on the nanocavity walls. Luminescence lifetime imaging shows that the average lifetime of the dye adsorbed on the cavity walls is approximately 20 ns, while the lifetime of the dye across the remainder of the electrode surface is 382 ns. This reduction in emission lifetime with simultaneous increase in emission intensity is consistent with a plasmonic increase in radiation rate.

Protein film voltammetry, e.g., using electrocatalytic hydrogenases, opens up the possibility of dramatically enhancing the current associated with the binding of small numbers of biomolecules. Heering and Lemay [14] recently reported on the immobilization of a small number of redox enzymes at nanoelectrodes. As the electrode potential is modulated so as to control the chemistry of the active site, and the catalytic activity of the enzyme can be monitored by detecting the electrical current flow. As illustrated in Fig. 4, using lithographically fabricated Au nanoelectrodes with dimensions of the order of 10000 nm2 as a platform, a distinct catalytic response was observed from less than 50 molecules of highly active [NiFe]-hydrogenase from Allochromatium vinosum. Using lithographically fabricated Au nanoelectrodes with dimensions of the order of $70 \times 70\ nm^2$ as a platform, a distinct catalytic response was observed from less than 50 molecules of highly active [NiFe]-hydrogenase from Allochromatium vinosum. These results strongly suggest the feasibility of using bioelectrochemistry as a new tool for studying redox enzymes at the single-molecule level. However, single-molecule detection places significant demands on the current measuring capability of the instrument since the current expected for a single molecule will be in the femtoamp range.

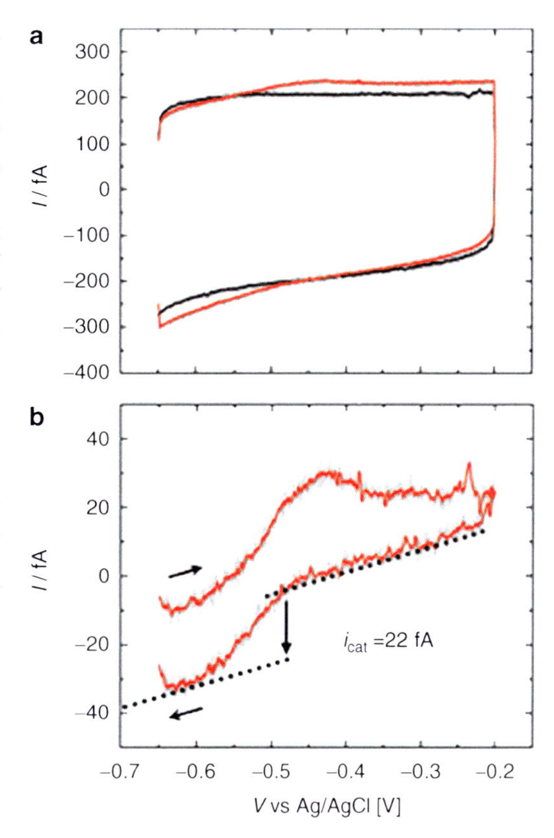

Micro- and Nanoelectrodes, Fig. 4 PFV of Av H2ase immobilized on a $100 \times 100\ nm^2$ (design value) fabricated Au electrode, pretreated 30 min with 30 g L^{-1} polymyxin. The electrode was insulated with a 10 nm, SiO_2/60 nm, Si_3N_4/20 nm, and SiO_2 layer. (**a**) Curves recorded with only buffer (*black*) and after the introduction of 250 mL of 500-nM H2ase in buffer (*red*). (**b**) Difference between the red and the black curves from (**a**) the buffer contained 50-mM MES buffer and 100-mM NaCl, pH = 5.7. Scan speed: 1.5 mVs^{-1}. The *light gray* lines represent the raw data, while the *red lines* are the same data smoothed over a window of 2 s (Reproduced with permission from Ref. [14])

Single-Nanoparticle Electrodes

The ability to fabricate nanoelectrodes whose radius is *smaller* than that of a single nanoparticle opens up the possibility of probing the electrocatalytic and other properties of single

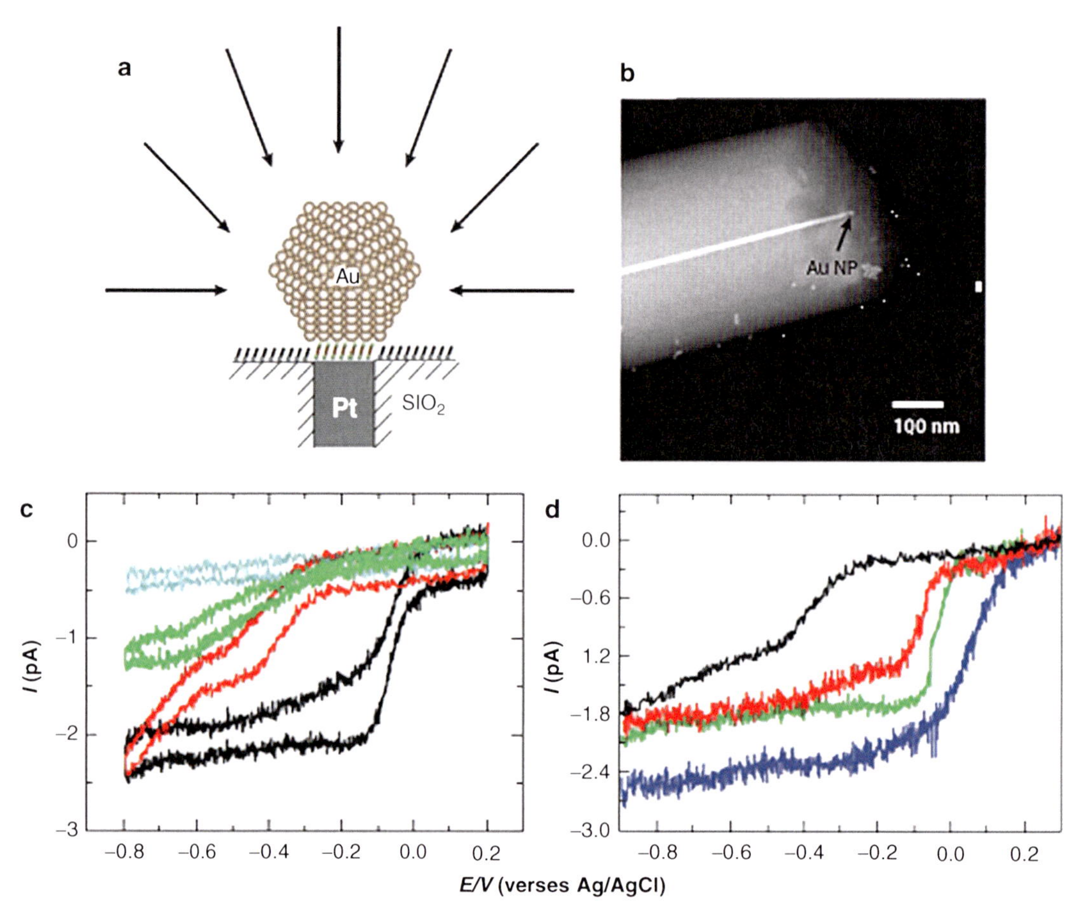

Micro- and Nanoelectrodes, Fig. 5 (**a**) Schematic of a single-nanoparticle Au electrode and the radial-type diffusion profile. (**b**) Transmission electron microscope image of a single-Au nanoparticle (*Au NP*) immobilized on a Pt nanoelectrode. (**c**) Voltammetric responses in an O_2-saturated 0.10 M KOH solution of an 18-nm Au single-nanoparticle electrode (*SNPE*) (*black*), a 3-aminopropyltrimethoxysilane (*APTMS*)-modified Pt nanoelectrode (*green*), the bare Pt nanoelectrode (*red*), and an 18-nm Au SNPE after the nitrogen is bubbled (*cyan*). (**d**) Voltammetric responses in an O_2-saturated 0.10-M KOH solution of a bare 7-nm-diameter Pt nanoelectrode (*black*), a 14-nm Au SNPE (*red*), an 18-nm Au SNPE (*green*), and a 24-nm Au SNPE (*blue*). The scan rate was 10 mVs^{-1} (Reprinted with permission from Ref. [15])

immobilized nanoparticles. In this way, the problems caused by averaging over many different nanoparticle sizes, shapes, and perhaps surface chemistry can be avoided.

Figure 5 illustrates the approach used by Cox and Zhang to create a single-nanoparticle electrode (SNPE) [15]. The nanoelectrode surface was first modified by a silane or thiol linker and then immersed in a solution of Au nanoparticles. The radius of the disk electrode is smaller than that of the nanoparticle, facilitating the attachment of a single nanoparticle. Figure 5b shows a TEM image of a single-Au nanoparticle immobilized on a Pt-nanoelectrode surface. Significantly, this approach reveals that the catalytic activity of Au nanoparticles for the oxygen reduction reaction (ORR) depends on their size. Figure 5c shows the current–voltage response of Pt nanoelectrodes and Au SNPEs of various sizes in a KOH solution. Strikingly, in contrast to the two-step process for oxygen reduction observed at the Pt electrode, the Au SNPEs exhibit a one-step process for oxygen reduction with an increased limiting current. Moreover, the half-wave potential for oxygen reduction is also shifted to −0.07 V, which suggests that, in this case, the Au SNPE demonstrates good electrocatalytic activity for oxygen reduction.

Bioanalysis

The advantages of microelectrodes are exploited in many different areas of electroanalysis, with the environmental, food quality assurance, and biomedical applications being the most active. Portability, simplified instrumentation, and label-free detection are some of the key advantages of electrochemistry over spectroscopy. In particular, microelectrodes open up the possibility of probing redox processes in small sample volumes or physically small spaces. For example, the redox properties of sample volumes as small as a few picoliters have been interrogated. Experiments of this kind are possible because, as described by Eq. 5, for solution phase reactants, the depletion layer thickness, d, depends on the experimental timescale:

$$\delta(Dt)^{1/2} \quad (5)$$

For a 50-μm-radius microelectrode and a 1-ms electrolysis time, the volume that is depleted of reactant will be less than 10 p ℓ!

Cell secretion is conventionally assayed using techniques that average the response across a large cell population with low temporal resolution. Measuring fast cell secretion, the heterogeneity of response between individual cells, probing the heterogeneous functions of cell membrane domains, and detecting trace amount of released molecules (zepto-moles from single-vesicle release) are difficult if not impossible using these conventional approaches. Micro- and nanoelectrodes can provide profound spatiotemporal insights into these bioprocesses that underpin fundamental mechanistic insights, disease diagnosis, and drug discovery. Wightman and co-workers initiated the use of carbon fiber microelectrodes to detect neuronal secretion and probed the frequency, intensity, and characteristics of exocytotic events [16].

Since this seminal work, there has been a continuous effort to apply smaller and smaller electrodes in order to record the responses of cellular structures at higher spatial and temporal resolution. Higher resolution is needed to precisely define where exocytosis occurs, to minimize overlap between discrete release events, and to avoid diffusional broadening. For example, Wu and colleagues [17, 18] developed exceptionally low noise carbon fiber nanoelectrodes (tip diameter ~100 nm) and used them to monitor exocytosis from PC12 cells with single-vesicle spatial resolution. Significantly, these investigations reveal that the majority of the cell surface is not active for exocytosis and even when the nanoelectrode was located directly above the fusion site, the release frequency was low. Moreover, these ultrasmall probes reveal that multiple fusion events can occur at the same site and sequential release at those hot spots played the major role in the dopamine release from PC12 cells. These insights are only possible because of the high spatial and temporal resolution of the nanoelectrodes and their high sensitivity due to their high signal-to-noise ratios. Another key issue in bioelectrochemistry is sensitivity [19]. In the case of localized release, e.g., neurotransmitter release through exocytosis, the local concentration may be high, but the total number of molecules released is very small. In this regard, the high mass rather than concentration sensitivity of electrochemical techniques and the ability to routinely measure small currents, ≤ 1 pA, are important.

A significant challenge is to use amperometry or voltammetry to detect redox-inactive molecules. Enzyme-modified electrodes can play a significant role in this area. For example, a combination of horseradish peroxidase and L-glutamate oxidase [20] immobilized within a redox hydrogel can be used to monitor L-glutamate secretion from hippocampal neurons [21].

Future Directions

This short article has attempted to convey a sense of the many significant new insights and opportunities that have been made possible by micro- and nanoelectrodes. This revolution in electrochemistry has greatly extended the range of solvents, temperatures, length scales, and timescales under which it is now possible to obtain direct information about redox processes. In particular, today ultrasmall electrodes allow experiments, e.g., voltammetry in oil or concrete, or measuring the release of a few hundred molecules in a single exocytotic event, to be performed that would simply have been impossible a few years ago. This advance has not only revolutionized the field internally, it has broadened the impact of electrochemistry into new dimensions of space and time. The ability to fabricate nanoelectrodes and arrays opens up exciting new prospects in biology including elucidating structure–function relationships of subcellular structures and investigations into the transport properties of single ion-channel proteins on cells.

Cross-References

► Electrode

References

1. Forster RJ, Keyes TE (2007) Microelectrodes: fundamentals and applications. In: Zoski C (ed) The handbook of Electrochemistry. Elsevier, Oxford, UK
2. Cox JT, Zhang B (2012) Nanoelectrodes: recent advances and new directions. Annu Rev Anal Chem 5:253–272
3. Mirkin MV, Nogala W, Velmurugan J, Wang Y (2011) Scanning electrochemical microscopy in the 21st century. Update 1: five years after. Phys Chem Chem Phys 13(48):21196–21212
4. Takahashi Y, Shevchuk AI, Novak P, Zhang Y, Ebejer N, Macpherson JV, Unwin PR, Pollard AJ, Roy D, Clifford CA, Shiku H, Matsue T, Klenerman D, Korchev YE (2011) Multifunctional nanoprobes for nanoscale chemical imaging and localized chemical delivery at surfaces and interfaces. Angewandte Chemie Int Ed 50(41):9638–9642
5. Li Y, Bergman D, Zhang B (2009) Preparation and electrochemical response of 1–3 nm Pt disk electrodes. Anal Chem 81:5496–5502
6. Morris RB, Franta DJ, White HS (1987) Electrochemistry at Pt band electrodes of width approaching molecular dimensions—breakdown of transport equations at very small electrodes. J Phys Chem 91:3559–3564
7. Penner RM, Heben MJ, Longin TL, Lewis NS (1990) Fabrication and use of nanometer-sized electrodes in electrochemistry. Science 250:1118–1121
8. Forster RJ (2003) Microelectrodes- pushing the boundaries of electrochemistry. In: Unwin PA, Bard AJ (eds) Encyclopedia of electrochemistry. Wiley, New York, pp 160–195
9. Henstridge MC, Compton RG (2012) Mass transport to micro- and nanoelectrodes and their arrays: a review. Chem Rec 12:63–71
10. Jena BK, Percival SJ, Zhang B (2010) Au disk nanoelectrode by electrochemical deposition in a nanopore. Anal Chem 82:6737–6743
11. Forster RJ (2000) Ultrafast electrochemical techniques. In: Meyers R (ed) Encyclopedia of analytical chemistry. Wiley, New York, pp 10142–10171
12. Bard AJ, Faulkner LR (1980) Electrochemical methods: fundamentals and applications. Wiley, New York
13. Mallon CT, Zuliani C, Keyes TE, Forster RJ (2010) Single nanocavity electrodes: fabrication, electrochemical and photonic properties. Chem Commun 46:7109–7111
14. Freek JM, Hoeben F, Meijer S, Dekker C, Albrachtm SPJ, Heering HA, Lemay SG (2008) Toward single-enzyme molecule electrochemistry: [NiFe]-hydrogenase protein film voltammetry at nanoelectrodes. ACS Nano 2(12):2497–2504
15. Li Y, Cox JT, Zhang B (2010) Electrochemical responses and electrocatalysis at single Au nanoparticles. J Am Chem Soc 132:3047–3054
16. Cahill P, Walker Q, Finnegan J, Mickelson G, Travis E, Wightman R (1996) Microelectrodes for the measurement of catecholamines in biological systems. Anal Chem 68:3180–3186
17. Wu WZ, Huang WH, Wang W, Wang ZL, Cheng JK, Xu T, Zhang RY, Chen Y, Liu J (2005) Monitoring dopamine release from single living vesicles with nanoelectrodes. J Am Chem Soc 127:8914–8915
18. Huang Y, Cai D, Chen P (2011) Micro- and nanotechnologies for study of cell secretion. Anal Chem 83:4393–4406
19. Cannon DM, Winograd N, Ewing AG (2000) Quantitative chemical analysis of single cells. Ann Rev Biophys Biomol Struct 29:239–263
20. Mikeladze E, Schulte A, Mosbach M, Blochl A, Csoregi E, Solomonia R, Schumann W (2002) Redox hydrogel-hased bienzyme microelectrodes for amperometric monitoring of L-glutamate. fElectroanalysis 14:393–399
21. Kurita R, Hayashi K, Horiuchi T, Niwa O, Maeyama K, Tanizawa K (2002) Differential measurement with a microfluidic device for the highly selective continuous measurement of histamine released from rat basophilic leukemia cells (RBL-2H3). Lab Chip 2:34–38

Micro-/Nanofabrication for Chemical Sensors

Peter J. Hesketh
School of Mechanical Engineering, Georgia Institute of Technology, Atlanta, GA, USA

Nomenclature

AFM	Atomic force microscope
ALD	Atomic layer deposition
CVD	Chemical vapor deposition
EFAB	Electroplated fabrication
FET	Field effect transistor
GLAD	Glancing angle deposition
IDA	Interdigitated electrode array
LPCVD	Low pressure chemical vapor deposition
MEMS	Microelectromechanical systems
NEMS	Nanoelectromechanical systems
PECVD	Plasma-enhanced chemical vapor deposition
RIE	Reactive ion etching
SEM	Scanning electron microscope
UV	Ultraviolet
VLSI	Very large-scale integrated circuit

Introduction and Overview

The methods for miniaturization of chemical and biosensors are based on an extension of VLSI fabrication techniques, however with a broader range of materials [1–6]. The range of materials is beyond what is normal for IC electronic devices because additional functionality is needed. These materials include electrochemically active metals with catalytic properties, conductive oxides, and high-temperature materials. Examples of metal oxides include SnO_2, WO_3, and TiO_2, and other catalytic metals include Pt, Ru, Ir, Pd, and Ag needed for electrochemical sensors [7, 8]. As the dimensions of semiconductor devices continue to move to smaller gate lengths, nanoscale fabrication techniques are now developed. Hence, structures for sensors and micro-/nanoelectromechanical systems (M/NEMS) can take advantage of the developed nanoscale fabrication methods. In addition, new scaling laws where unexpected physical effects can be investigated provide opportunities for enhance sensing [9]. For example, plasmonic effects predominate at a nanometer-size noble metal structures providing new sensors and electro-optical systems [10]. In fact, the scaling of the sensor element is fast approaching the dimensions of higher molecular weight chemical compounds, opening up more possibility for single-molecule detection [11].

In this entry we describe the basic processes and lithography which is the foundation of miniaturization, followed by a description of a range of methods for depositing thin films. Throughout the entry examples will be provided to illustrate the versatility of the micro-/nanoscale fabrication processes for the manufacture of chemical sensors.

Patterning

Lithography is the method used to transfer or print a pattern into a photoresist layer coated on the surface of a substrate. It involves the use of a light source of short wavelength, typically ultraviolet (UV), an exposure system to align the mask to the wafer, and a resist to transfer the pattern; see Fig. 1. The photoresist consists of three components: a resin, a light-sensitive photoactive compound, and a solvent. The wavelength of the light source should match the adsorption for the photoactive compound in the resist for good sensitivity. With a positive resist, the exposed regions are more soluble in the developer, whereas for a negative resist, they are less soluble in the developer. The resist is spin coated on to the wafer/substrate to a uniform thickness. The exposure dose from the UV light source is calculated based upon the sensitivity, and the development is carried out after a soft baking step. Details are provided by the manufacturer (see, e.g., Shipley web site or Futurex web site). The dose and sensitivity of a positive resist are shown in Fig. 2. A resist needs a minimum dose for

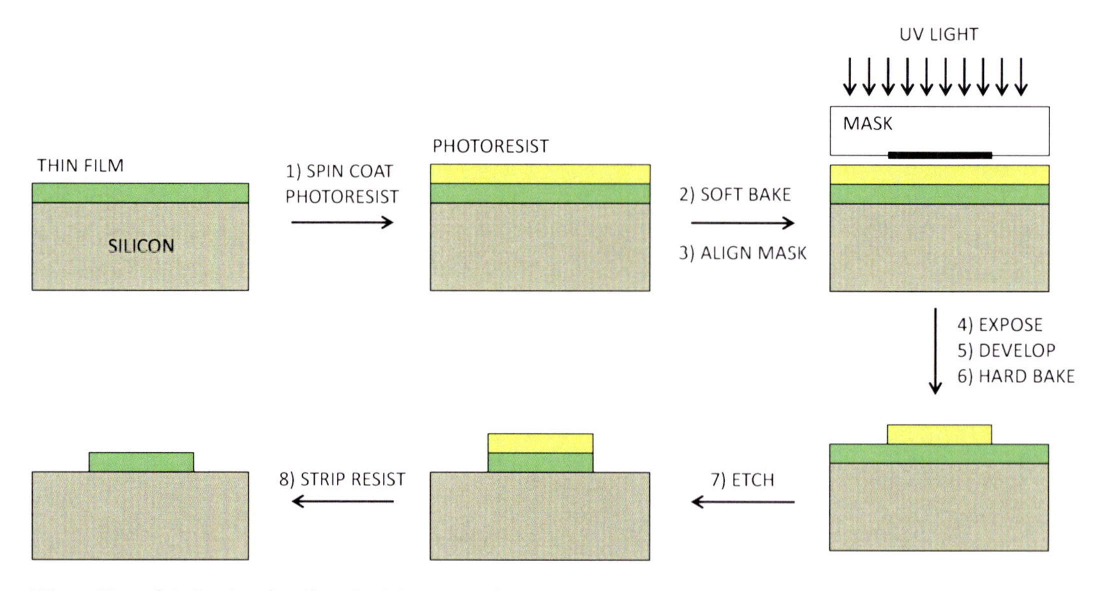

Micro-/Nanofabrication for Chemical Sensors, Fig. 1 Schematic diagram showing the photolithography process

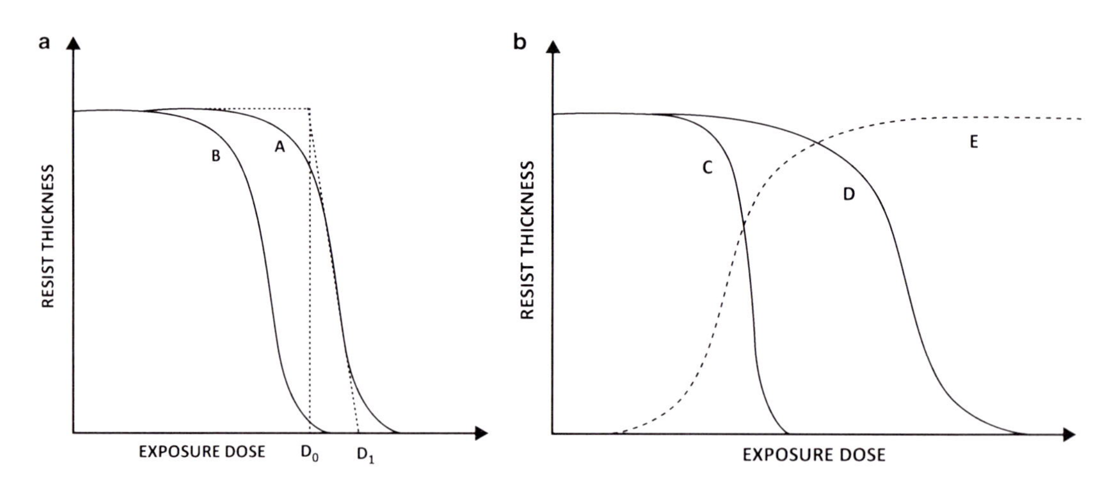

Micro-/Nanofabrication for Chemical Sensors, Fig. 2 Photoresist dose thickness curves showing (**a**) positive resist B is of higher sensitivity than resist A, (**b**) positive resist C is of higher contrast than resist D, and resist E is a negative resist

dissolution to begin to take place, D_0, and the resist is completely developed when dose reaches D_1, indicated on the figure. A resist with higher sensitivity requires lower dose and hence exposure times. A resist with higher contrast has a more rapid change in thickness with dose, as shown in Fig. 2b. The contrast γ is defined as:

$$\gamma = \frac{1}{\log_{10}D_1 - \log_{10}D_0} \quad (1)$$

The resolution of the resist patterning is a function of the resist contrast and the optical diffraction effects produced by the mask aligner. For a description of Fraunhofer and Fresnel diffraction effects which limit resolution as a function of the optical modulation transfer function of the mask aligner, see Campbell [12] and Mack [13]. In general, a shorter wavelength is key to achieve smaller feature sizes, for example, deep UV is able to generate features in submicron dimensions. The selection of positive or negative resist for a given fabrication step often depends upon the required sidewall profile in the resist and the resist's ability to stand up to the subsequent processing step as it is a temporary layer which will be removed in a chemical stripper once the pattern is transferred into the film or substrate material by etching. Electron beam lithography and ion beam lithography offer access to nanoscale feature definition and can also be used for etching and CVD [2, 14].

For the fabrication of M/NEMS and chemical sensors, often thicker layers of resist are needed to build up channels or to define fluid reservoirs. In these cases SU-8 [15] or polyimide resist layers can be used [1]. Thicker layers are frequently used for molding or for microfluidic devices [16]; for stereolithography [11], the desired pattern is defined in resin by scanning a laser through a thin layer of the material in a sequential fashion.

Doping of Silicon

Diffusion and ion implantation are two methods of forming regions of n-type or p-type silicon [5], in particular for the fabrication of field effect transistors (FETs), conductive regions for heating elements, and p/n junction diodes for sensors. Ion implantation produces a more uniform doping than a diffusion process. In diffusion doping the concentration of dopant is set by the surface concentration and the diffusion temperature. This is normally carried out either by a gas-phase delivery of dopant or a solid source or a spin-on dopant to define the concentration at the silicon surface [3, 12]. The doping is carried out in a two-step process: The first step is to establish a high surface concentration in the wafer, near the solid-solubility limit in silicon. When the concentration at the surface of the wafer is kept constant during the "pre-deposition" process, the resultant profile is an *erf* function of depth. The second step is a "drive-in" where the dopant is redistributed by diffusion into the silicon at a high temperature. The dopant "drive-in" is carried out without any additional dopant present at the surface and results in a Gaussian distribution as a function of depth. The doping profile produced is a function of the temperature and source concentration. Phosphorous for n-type and boron for p-type are the frequently used dopants.

Ion implantation provides more precise definition and more uniform dopant profiles. It is carried out by accelerating an ionized beam of dopant atoms to a sufficiently high kinetic energy in a vacuum chamber so that when they collide with the silicon wafer, they are implanted to a controlled depth below the surface. The depth and distribution are defined by the range and straggle. These are available in handbooks [17] to predict the dopant distribution and are carried out usually at a commercial vendor. The ion implantation process results in significant surface damage which needs to be removed by annealing the silicon at high temperature, typically 1,000 °C for 1 h.

Doping can also be used to modify etching processes for the purpose of etch stop in anisotropic etching for cantilever or diaphragm formation [1, 6], and for electrochemical etching or for formation of porous silicon by anodization [18].

Film Growth/Deposition

One of the most useful features of silicon is that it forms silicon dioxide, insulating layers quite readily by heating in an oxygen or steam ambient. The film thickness is highly predictable and can be calculated with the well-known Deal and Grove model [19]. This describes the growth rate as a function of wafer crystallographic orientation and processing conditions. The oxide, in addition to being insulation, is very useful as a masking film during etching and as a sacrificial layer for release of cantilevers and membranes. Finally, it provides a silica surface useful for chemical surface modification.

Physical vapor deposition involves a nucleation and growth process for both a sputtered and an evaporated film. In the evaporation process, the deposition rate is based upon the vapor pressure of the metal [20]. A vapor is formed which condenses onto the cooler substrate or wafer. Atoms

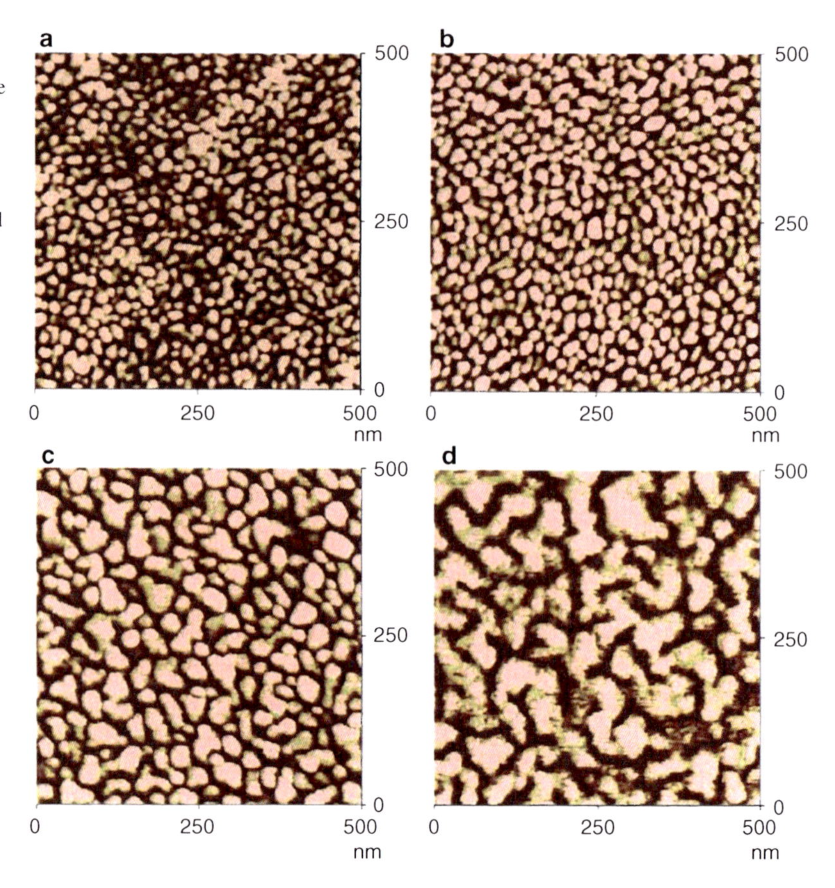

Micro-/Nanofabrication for Chemical Sensors, Fig. 3 AFM tapping mode images of evaporated gold film deposited on silicon dioxide for average film thicknesses of (**a**) 2.5 nm, (**b**) 3.5 nm, (**c**) 7.5 nm, and (**d**) 10 nm [21]

condense onto the surface and move to form clusters which grow in diameter as they form islands by addition of more atoms. The initial size of the islands is a function of the substrate material-adatom interaction and substrate temperature. Each island has a certain capture radius for which adatoms reach the island. Figure 3 shows AFM images of ultrathin gold films formed by evaporation onto silicon oxide-coated silicon wafers, at close to room temperature. As the thickness increases, the islands grow in diameter, becoming an interconnected network and finally a porous film. Eventually the open areas fill in, and a continuous thin film is formed. The pressure inside the vacuum system is normally set in the 10^{-6}–10^{-8} Torr so that the mean free path between gas particle collisions based upon the kinetic theory of gases [2] is of similar or larger dimensions than the dimensions of the chamber. Hence, the evaporating atoms have a low probability of colliding with any gas molecules before they reach the wafer or substrate surface. This has the important advantage that high-purity films can be deposited by this fairly simple process. The orientation of the deposited film is a line of sight projection from the source. Several novel structures can be formed using this property, in particular shadowing to make electrodes on the side of grooves; for example, see Fig. 4. Two evaporation steps with a lift-off resist layer are used to define platinum electrodes on the sides of grooves for biosensing [22]. Angled deposition has also been used for nanofabrication, in particular to create high-surface area interfaces known as glancing angle deposition (GLAD) [23] with application for improved energy efficiency and for surface-enhanced Raman biosensing.

For the deposition of alloys and for improved step coverage, sputtering is preferred. However, sputtering uses an inert plasma for deposition so that some impurities from the plasma gas are included in the film during growth. The deposition

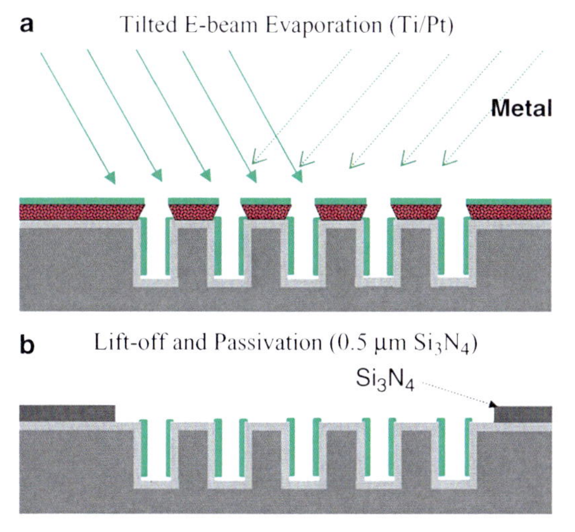

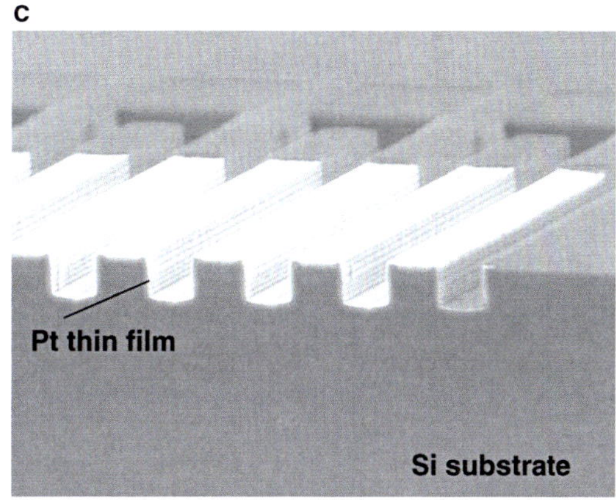

Micro-/Nanofabrication for Chemical Sensors, Fig. 4 Schematic diagram indicating the method for fabrication of a comb-type IDA electrode (**a**) Two angled evaporations of Ti adhesion layer and Pt layer to coat sides of each groove, (**b**) resulting structure after the lift-off of photoresist layer, and (**c**) SEM image of completed comb electrode array [22]

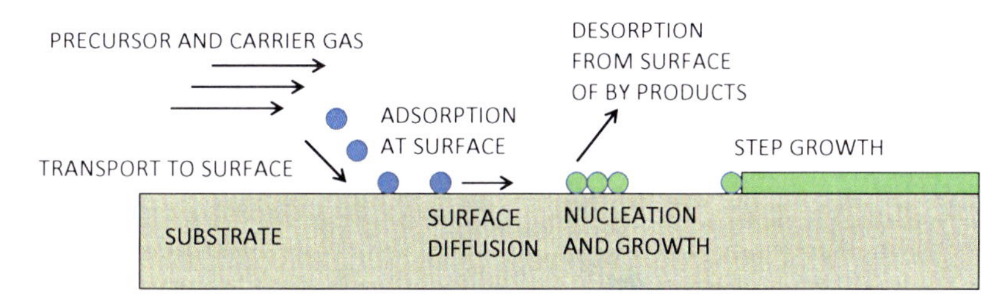

Micro-/Nanofabrication for Chemical Sensors, Fig. 5 Schematic diagram showing the processes involved in chemical vapor deposition, transport to the surface of reactants, adsorption, reaction/nucleation and growth, desorption of by-products, and transport away from surface

rate is a function of the ion bombardment rate of the target from the plasma and energy of impact, which is controlled by the pressure, energy (i.e., power input to the plasma), and bias voltage between the plasma and the target during sputtering. One important advantage of sputtering is that film stress and microstructure can be adjusted based upon the deposition conditions substrate temperature, and ion bombardment [3].

In chemical vapor deposition (CVD), a gas-phase precursor is necessary. In a typical hot-wall reaction chamber, the diluted gas mixture is introduced over the substrate and is adsorbed on the substrate, a chemical reaction occurs at the surface, and the by-products are desorbed; see Fig. 5. The process conditions, temperature, pressure, gas composition, are adjusted so that the surface reaction is promoted and any gas-phase reactions are not likely to occur, which, should they occur, would result in particulate formation. Highly uniform films with good step coverage are produced because at high temperatures diffusion to the surface is rapid and at low pressure the mean free path is larger compared to the dimension of any surface features. High-quality uniform composition films with good control of stoichiometry are deposited. Many different films can be deposited by CVD. For example, in the growth of silicon nitride, the proportion of silicon in the film can be controlled based upon the gas mixture composition. A silicon-rich film, which has lower stress than stoichiometric Si_3N_4, can be deposited

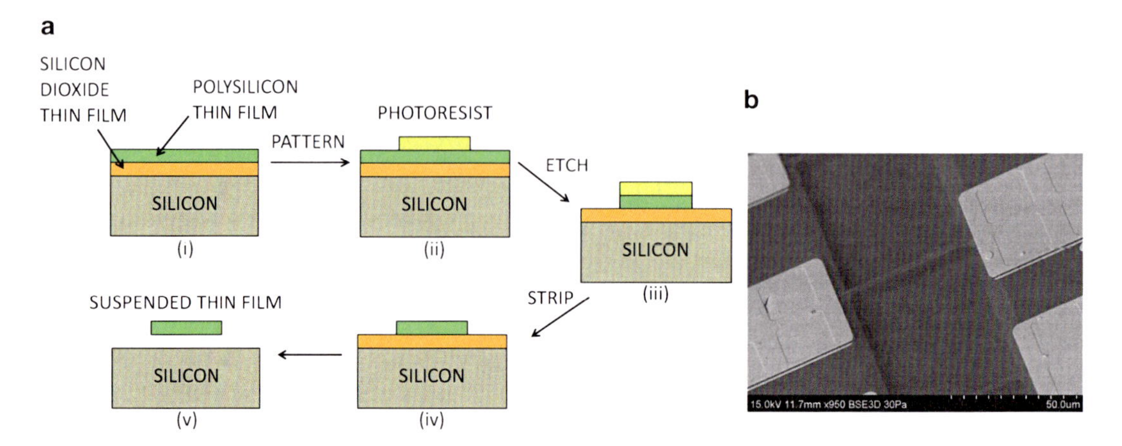

Micro-/Nanofabrication for Chemical Sensors, Fig. 6 (**a**) Schematic diagram of surface micromachining process, which involves (i) film deposition, (ii) patterning, (iii) etching of structural layer, (iv) stripping resist, and (iv) etching of sacrificial layer. (**b**) SEM micrograph of release polysilicon beam for thermal conductivity gas sensor, designed by J. R. Stetter at KWJ Engineering and fabricated in the Nanotechnology Research Center at Georgia Institute of Technology

by adjusting the dichlorosilane to ammonia ratio, typically the order of 6:1 for stress-free films [4].

For the growth of polysilicon from silane, control of the gas concentration and temperature is important to enable reaction rate-limited growth. When diffusion-limited growth occurs, the delivery of reactants across the boundary layer will result in nonuniform thickness which is therefore a function of position on the wafer. Reaction rate-limited growth produces uniform film thickness and is indicated by the larger effective activation energy for the temperature dependence of the growth rate. By plotting the log {growth rate} versus inverse temperature, the resultant slope is proportional to the effective activation energy. For example, with silane in the temperature range 835–925°C, an activation energy of 1.6 eV was observed, which is similar in magnitude to the bond-bond energy of Si-Si (2.2 eV). This confirms the growth is reaction rate limited at the surface of the substrate. Polysilicon is an important material for VLSI and for MEMS. It is used as electrodes in electrostatic motors, for heater elements in miniature thermal sensors, for microresonators, and for sensor films. Through careful selection of the growth conditions, films in tension or compression can be deposited, which is useful to make freely suspended structures, such as membranes and cantilevers. A microthermal gas sensor built of polysilicon and suspended over silica is shown in Fig. 6, designed by Dr. J. R. Stetter at KWJ Engineering. Table 1 lists some typical CVD processes commonly used in microfabrication for electrochemical sensors. For gas sensors, metal oxides have unique sensitivity to various gases such as SnO_2, WO_3, and TiO_2 and some conductive metal oxides, including IrO_2, RuO_2, and indium tin oxide [24]:

As an alternative to using the high temperatures necessary in LPCVD, plasma-assisted growth can be used to lower the temperature (200–400°C), and the plasma is necessary to ionize the reactants. However, this is at the expense of film quality, due to introducing defects and hydrogen gas impurities in the film. During plasma-enhanced chemical vapor deposition (PECVD), the silane concentration is lower, typically 5 %. A wide range of materials can be deposited. Silicon oxide, silicon nitride, and silicon oxynitride films can also be densified after growth to improve their properties, which include refractive index, density, and resistance to etchants such as dilute hydrofluoric acid solutions. Other materials have been grown for sensors, including amorphous SiC for high-temperature devices and diamond-like carbon for electrodes and field emission sensors.

Micro-/Nanofabrication for Chemical Sensors, Table 1 List of materials commonly used in sensors deposited by CVD

Process	Material	Reactants	Temperature range (°C)
LPCVD	Polysilicon	SiH_4/H_2	650–850
LPCVD	SiO_2	$TEOS/O_2$	650–850
LPCVD	SiO_2	$SiCl_2H_4/H_2$, N_2O	900
LPCVD	Si_3N_4	$SiCl_2H_2/H_2$, NH_3	700–900
LPCVD	W	WF_6/H_2	400–500
PECVD	SiO_2	$SiH_4/H_2/O_2$	200–400
PECVD	Si_3N_4	$SiH_4/H_2/NH_3$	200–400

TEOS tetraethoxysilane

In addition, piezoelectric layers such as AlN and ZnO can also be deposited. These and other piezoelectric transducer layers and pyroelectric materials are utilized in thin film sensors, actuators, and electro-optical devices [1].

Atomic layer deposition (ALD) is gaining increasing popularity for the fabrication of nanostructures and ultrathin films. ALD is a layer by layer growth process in which the reaction precursors are introduced one at a time so that only a single layer adsorbs onto the substrate and saturates the available sites in a self-limiting reaction. The excess is removed from the gas phase, and a second reactant is introduced, which reacts at the adsorbed molecules on the wafer surface saturating the surface, again so that excess is removed, and the cycle repeated. Key advantages of ALD are the precise control of the layer thickness through the number of growth cycles, a very smooth surface, and the uniform pinhole-free film which is formed. Control of stoichiometry and doping of films can easily be achieved through modulation of the reactant composition during the growth cycles. In addition, the growth process is able to cover non-flat and reentrant surface topographies with a uniform film thickness. The process is typically carried out at lower temperatures than CVD or PECVD, typically in the range 100–250 °C, so that deposition onto plastics is feasible. The ALD process was originally developed for growth of ZnS for flat panel displays [25] but is now finding widespread use in MEMS devices [26]. It is useful for passivating films and for dielectric layers or to modify the surface chemistry for subsequent covalent linking to a self-assembled monolayer.

Etching

Etching is the process of removal of material from the substrate or a film in a controlled manner. Chemical etching can be carried out in a liquid or gas phase and also with a plasma. Physical etching on the other hand utilizes ions to carry out a sputtering process. A range of different ceramics, metals, polymers, and semiconductors can be etched, and excellent reference books on this topic include Vossen and Kern [27], *Handbook of Metal Etchants* [28], and also for MEMS materials, a review paper [29].

The etching process involves transport of reactants to the surface, adsorption, a reaction, desorption of products, and transport away from the surface. The etch rate is a function of the concentration of reactants and the temperature of the substrate. Selective etching is necessary for good pattern definition, with respect to the substrate and the masking film; see Fig. 7 which shows the profile after isotropic and anisotropic etching has taken place. Anisotropic etching has the advantage of reduced undercut compared to isotropic etching. Some commonly used etchant chemicals are listed in Table 2.

Reactive ion etching (RIE) is a plasma-assisted etch method which has a lot of versatility, due to the influence of changes in the etch conditions on anisotropy selectivity and etch rate. A combination of physical sputtering, chemical etching, and ion-assisted etching provides ability to tailor the etching process for specific film combinations. For example, the Bosch process [30] is frequently used for high-aspect ratio etching in silicon. It is based on switching the etch conditions between a passivating film and etching conditions. Figure 8 shows the sequence for this process and some examples of structures that have been fabricated with this recipe, in particular, high-aspect ratio channels for a miniature gas chromatography system [31] and microgyroscopes [32]. Control of

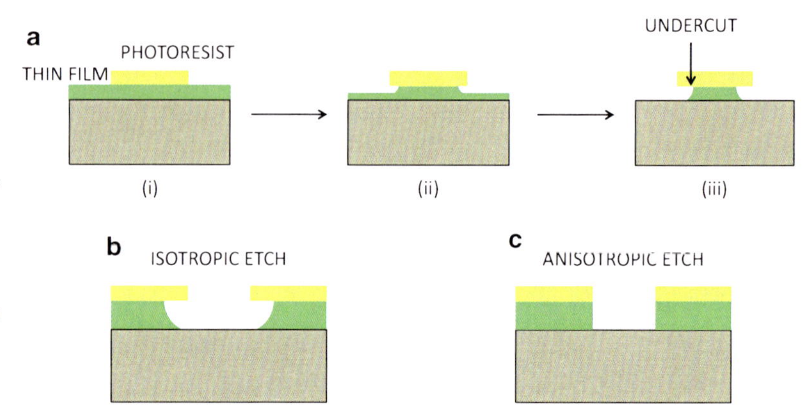

Micro-/Nanofabrication for Chemical Sensors, Fig. 7 (**a**) Schematic diagram of etching process showing (i) photoresist patterning (ii) partway through the etching process and after etch has been completed, with the undercut labeled (iii). (**b**) Cross-sectional diagram of an isotropic etch profile and (**c**) an anisotropic etch profile

Micro-/Nanofabrication for Chemical Sensors, Table 2 Commonly used etchants in chemical sensor fabrication

Process	Chemical	Substrate	Properties	Mask (partial list)
Liquid	HF/HNO_3	Si	Isotropic	SiC
	HF/NH_4F	SiO_2	Isotropic	PR
	KOH	Si	Anisotropic	Si_3N_4
Gas	XF_2	Si	Isotropic	Many options
BOSCH process	SF_6/C_2H_6	Si	Anisotropic	SiO_2 or PR
Plasma	O_2	PR	Isotropic	metal
RIE	CH_4/CHF_4	SiO_2	Isotropic	PR
RIE	SF_6/O_2	Si	Isotropic	PR

PR photoresist

the surface profile can be obtained by modification of the etch chemistry; see, for example, [33].

In a surface micromachining process, a structural layer forms the device and a sacrificial layer is removed, so that overhanging, suspended, or movable elements can be defined; see Fig. 6a. High selectivity is required between these films in order to release the element by dissolving the sacrificial film. Figure 6 shows a 1-μm-wide polysilicon beam released by etching to remove silicon dioxide under the beam. Etch rate of the silicon dioxide is much greater than the silicon nitride layer coating the beam. The depth under beam is approximately eight microns, creating an aspect ratio of 1:8. This forms a bridge suitable for highly sensitive measurement of gas composition [34]. The structure was designed by Dr. J. R. Stetter at KWJ Engineering.

One of the issues which often arise with flexible structures is the contact between the released structures and the substrate, which can provide a large area for stiction to take place. Use of supercritical drying or sublimation has been successful to overcome such problems [1].

Surface micromachining is not limited to the polysilicon/silicon dioxide combination of materials; other combinations are listed in Table 3. With polysilicon and silica, up to five layers have been demonstrated for complex mechanisms in the SUMMiT at Sandia National Laboratories [40]. An example of a very successful metal surface micromachining process is that used at Texas Instruments for their digital mirror array light modulators which are a used in video projection display systems [41].

Laser-based etching has been used to define features by ablation process. It is particularly useful for polymers, dielectrics, metals, and semiconductors. Examples of laser-assisted chemical etching and laser ablation include patterning of polyimide and thin metal films [42].

Electroplating

Electroless plating and electroplating are widely used and are versatile processes for deposition of metal films over a wide range of thicknesses at temperatures close to room temperature. The fundamentals of electroplating are described in the

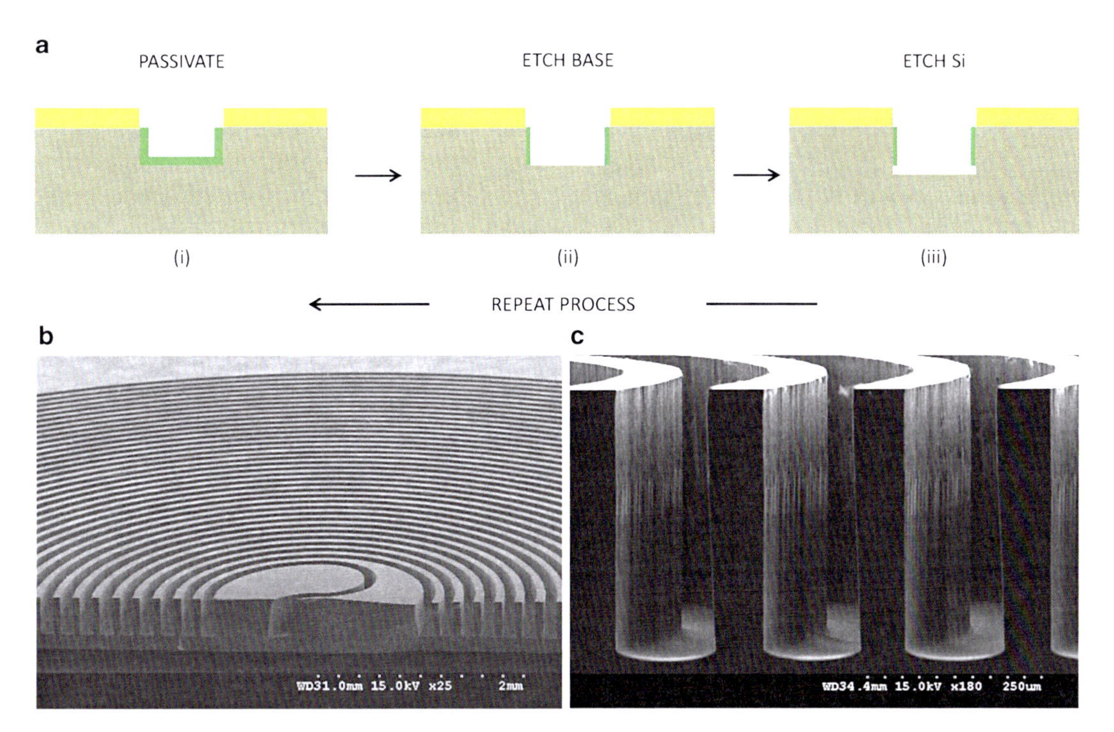

Micro-/Nanofabrication for Chemical Sensors, Fig. 8 (**a**) Schematic diagram of Bosch process which consists of (i) passivation of exposed surface, (ii) etching of passivation from base of the groove, and (iii) etching of silicon substrate while walls remain passivated. This process is repeated to achieve highly anisotropic etching of trenches. (**b**) Example of micromachined silicon channels for a miniature gas chromatography system; here the total length of the spiral column is 1 m. (**c**) SEM micrograph of channels, depth 260 mm, formed in silicon [31]

Micro-/Nanofabrication for Chemical Sensors, Table 3 Surface micromachining material combinations

Structural layer	Sacrificial layer	Process	Reference
Gold	PR	Microrelay	[35]
Si_3N_4	Polysilicon	Microphone	[36]
Si_3N_4	Al	Microphone	[37]
Parylene	PR	Microfluidic devices	[38]
SiO_2/Si_3N_4	Polycarbonate	Fuel cells	[39]

PR photoresist

text by Schlesinger [43], and many useful processes are described in the Handbook on Electroplating [44]. Typically, a conductive seed layer is deposited first, followed by a photoresist mask to define the exposed regions where plating is required. The substrate to be plated needs electrical connections to form the cathode of an electrolytic cell. The potential is applied to a metal anode that is frequently selected to replace the ions which are plated onto the wafer. The reduction reaction occurs at the cathode as the metal deposits from solution in the areas exposed to the electrolyte. At the anode, dissolution takes place, which can replenish the metal concentration in the solution. The most important parameters are the current density, temperature, and bath composition during plating. This sets the rate of deposition and the film quality which is also a function of solution pH, temperature, and additives which modify the surface morphology (brighteners) and reduce stress. For example, saccharine is added when plating Ni to reduce stress. Pulsed plating can also be used to modify the grain size and morphology. The pulses allow time for the ion concentration to recover at the surface and hence minimize the effect of surface geometry on the plating rate. In addition, the pulses can include a period at reverse polarization, which results in some controlled dissolution of the film. Reverse pulse

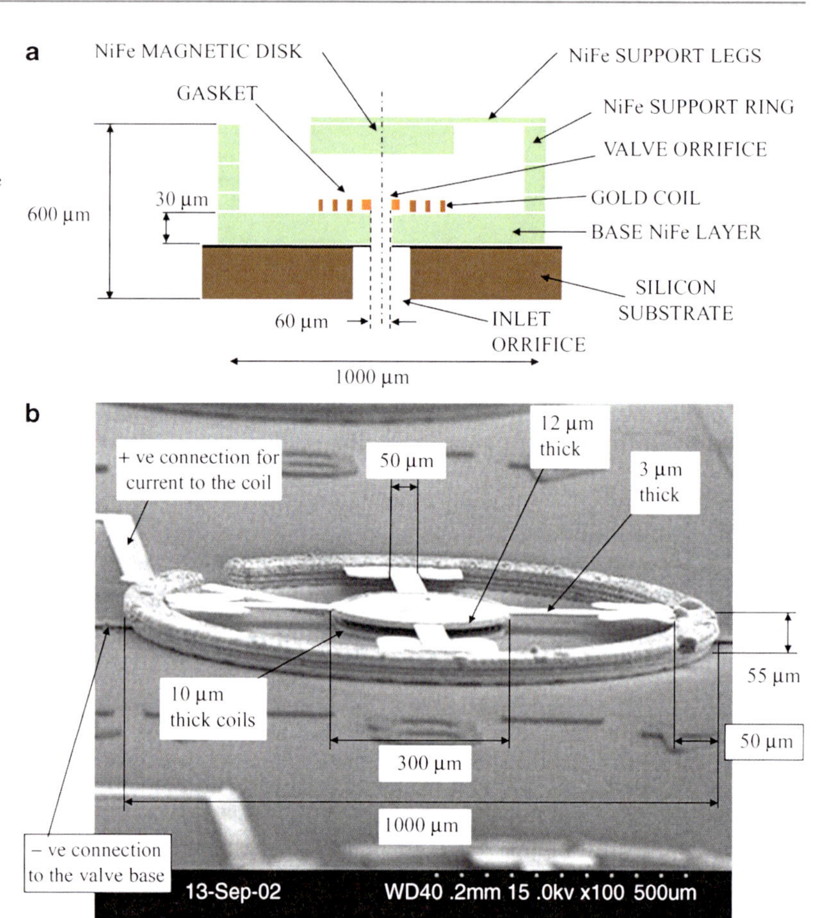

Micro-/Nanofabrication for Chemical Sensors, Fig. 9 (**a**) Schematic diagram of microvalve indicating composition of the layers that comprise the electromechanical valve [46]. (**b**) SEM micrograph of a microvalve [47]

plating is then used to provide control of grain size and stress in the film [43]. Three-dimensional structures can be fabricated, such as the NiFe layers in the MEMS magnetic devices [45]. Complex geometries can be built by repeating the photoresist, seed layer, and plating steps as an additive microfabrication process capable of building high-aspect ratio structures such as the magnetically actuated microvalve shown in Fig. 9 [47]. The commercial process known as EFAB has also been developed based on electroplating [48].

In an electroless plating process, no electrical connection is required to the wafer. A seed layer is however required or a conductive substrate. Nickel is frequently used in electronic devices to provide adhesion layer and diffusion barrier. The nickel is reduced at the conductive substrate and a co-reactant is present in the bath to provide the oxidation reaction, such as sodium hydrophosphite. Hence, the anodic and cathode reactions both occur on the substrate as the film is deposited. The rate is determined by the difference in the electrochemical potential, temperature, and solution concentration. An equilibrium potential is established on the substrate during the film deposition.

Concluding Remarks

There are lots of reasons to be optimistic about the future of nanofabricated chemical sensors. There are many fascinating opportunities to take advantage of, for example, new physical effects at the nanoscale and, in addition, the advantages brought about through scaling dimensions: these advantages include lower power, higher sensitivity, and lower cost. As advances in nanofabrication continue to bring new fabrication methods forward, they will have continued impact as they are taken up for sensor fabrication. One of the key advantages that can be gained through fabrication of arrays of miniature sensors is lower power consumption, saving energy and allowing battery operation for extended periods of time. For example, the sensor shown in Fig. 6 uses nanowatt power and can undergo over 36 billion measurements without changing the battery. It is economical to use nanofabrication methods when the sensors are smaller; however, for the greatest impact, methods such as soft lithography and nanostamping provide access to nanoscale patterning of surfaces but at a fraction of the cost.

Integration of sensors into arrays provides redundancy and multifunctional sensing, combined with self-calibration and built-in regeneration of sensing functions. Different geometry, operating temperatures, and coatings make feasible operation of sensors under different orthogonal operating conditions. For example, cantilever sensor arrays with nanoporous coatings can be combined with electrochemical sensors to emulate olfaction sensing observed in nature [49].

This entry is just an introduction; there have been numerous other developments of advanced micro-/nanofabrication techniques including ion-assisted CVD electron beam etching, for example, to produce nanopores [50] and custom AFM/STM probes for scanning electrochemical imaging [51] and materials for sensing at high temperatures and in corrosive environments.

Cross-References

▶ Sensors

References

1. Madou MJ (2011) Fundamentals of microfabrication and nanotechnology, 3rd edn. CRC Press, Boca Raton, FL
2. Brodie I, Muray JJ (1992) The physics of micro/nanofabrication. Plenum, New York
3. Wolf S, Tauber RN (1999) Silicon processing for the VLSI Era. Process technology, vol 1. Lattice Press, Sunset Beach, CA
4. Campbell SA (2007) Fabrication engineering at the micro and nanoscale. Oxford University Press, New York
5. Ghandhi SK (1994) VLSI fabrication principles. Wiley, New York
6. Korvink JG, Paul O (2005) MEMS: a practical guide to design, analysis, and applications. William Andrew/Springer, New York
7. Janata A (2009) Principles of chemical sensors. Springer, New York
8. Comini E, Faglia G, Sberveglieri G (2009) Solid state gas sensing. Springer, New York
9. van Zee RD, Pomrenke GS (2009) Nanotechnology – enabled sensing. www.nano.gov/NNI-Nanosensors-stdres.pd
10. Maier SA, Atwater HA (2005) Plasmonics: localization and guiding of electromagnetic energy in metal/dielectric structures. J Appl Phys 98:011101
11. Hesketh PJ (2008) BioNanoFluidic MEMS. Springer, New York
12. Campbell SA (2001) The science and engineering of microelectronic fabrication, 2nd ed. Oxford University Press, New York
13. Mack C (2008) Fundamental principles of optical lithography: the science of microfabrication. Wiley, New York
14. Cui Z (2010) Micro-nanofabrication: technologies and applications. Springer, New York
15. http://mems.gatech.edu/msmawebsite/members/processes/processes_files/SU8/SU-8.htm
16. Hardt S, Schonfeld F (2007) Microfliudic technologies for miniaturized analysis systems. Springer, New York
17. Chen WK(2007) The VLSI handbook. CRC Press, Boca Raton, FL
18. Zhang XG (2001) Electrochemistry of silicon and tts oxide. Kluwer, New York
19. Deal BE, Grove AS (1965) General relationship for the thermal oxidation of silicon. J Appl Phys 36(12):3770–3778
20. Mahan JE (2000) Physical vapor deposition of thin films. Wiley, New York
21. Ming Y (1997) Characterization of the growth of ultrathin Pt films by evaporation, in electrical engineering and computer science. University of Illinois, Chicago
22. Kim SK, Hesketh PJ, Li C, Thomas JH, Halsall HB, Heineman WR (2004) Fabrication of comb interdigitated electrode arrays (IDA) for a microbead-based electrochemical immunoassay. Biosens Bioelectron 20:887–894
23. Hawkeye MM, Brett MJ (2007) Glancing angle deposition: fabrication, properties, and applications of micro- and nanostructured thin films. J Vac Sci Technol A 25:1317–1325
24. Ren R, Pearton SJ (2011) Semiconductor device-based sensors for gas, chemical, and biomedical applications. Springer
25. Suntola T (1989) Atomic layer epitaxy. Mater Sci Rep 4(5):261–312

26. Mayer TM, Elam JW, George SM, Kotula PG (2003) Atomic layer deposition of wear-resistant coatings for micromechanical devices. Appl Phys Lett 82:2883–2885
27. Vossen JL, Kern W (1979) Thin film processes. Academic, New York
28. Walker P, Tan WH (1991) CRC handbook of metal etchants. CRC, Boca Raton, FL
29. Williams KR, Muller RS (1996) Etch rates for micromachining processing. J MEMS 5(4):256
30. Marty F, Rousseau L, Saadany B, Mercier B, Francais O, Bourouina T (2005) Advanced etching of silicon based on deep reactive ion etching for silicon high aspect ratio microstructures and three-dimensional micro- and nanostructures. Microelectr J 36(7):673–677
31. Noh HS, Hesketh PJ, Frye-Mason G (2002) Parylene gas chromatographic column for rapid thermal cycling. J Microelectromech Syst 11(6):718–725
32. Zaman MF, Sharma A, Ayazi F (2009) The resonating star gyroscope: a novel multiple-shell silicon gyroscope with sub-5 deg/hr Allan deviation bias instability. IEEE Sens 9(6):616–124
33. Roxhed N, Griss P, Stemme G (2007) A method for tapered deep reactive ion etching using a modified Bosch process. J Micromech Microeng 17:1087–1092
34. Aguilar RJ, Peng Z, Hesketh PJ, Stetter JR (2009) An ultra-low power microbridge gas sensor. ECS Trans 33(8):245–253
35. Nordquist CD, Wanke MC, Rowen AM, Arrington CL, Grine AD, Fuller CT (2011) Properties of surface metal micromachined rectangular waveguide operating near 3 THz. IEEE J Sel Top Quantum Electron 17(1):130–137
36. Hall NA, Murat Okandan F, Levent D (2006) Surface and bulk-silicon-micromachined optical displacement sensor fabricated with the SwIFT-Lite™ process. J Microelectromech Syst 15(4):770–776
37. Scheeper PR, van der Donk AGH, Olthuis W, Bergveld P (1992) Fabrication of silicon condenser microphones using single wafer technology. J Microelectromech Syst 1:147–154
38. Chen P-JC-YS, Yu-Chong T (2006) Design, fabrication and characterization of monolithic embedded parylene microchannels in silicon substrate. Lab Chip 6:803–810
39. Jayachandran Joseph P, Kelleher HA, Sue Ann Bidstrup A, Kohl PA (2005) Improved fabrication of micro air-channels by incorporation of a structural barrier. J Micromech Microeng 15:35–42
40. http://www.mems.sandia.gov/tech-info/summit-v.html
41. Winterton G (2010) The development and fabrication of MEMS in the semiconductor world. ECS Trans 33:285–295
42. Bauerle D (2011) Laser processing and chemistry, 4th edn. Springer, New York
43. Schlesinger M, Paunovic M (2010) Modern electroplating. Wiley/ECS, Pennington, NJ
44. Paunovic M, Schlesinger M (1998) Fundamentals of electrochemical deposition. Wiley
45. Arnold DP, Zana I, Cros F, Allen MG (2004) Vertically laminated magnetic cores by electroplating Ni-Fe into micromachined Si. IEEE Trans Magn 40(4):3060–3062
46. Sutanto JB (2004) CMOS compatible microvalve for fuel cell power generator, in School of Mechanical Engineering. Georgia Institute of Technology, Atlanta
47. Bintoro JS, Hesketh PJ (2005) An electromagnetic actuated on/off microvalve fabricated on top of a single wafer. J Micromech Microeng 15:1157–1173
48. Cohen A, Zhang G, Tseng FG, Frodis U, Mansfeld F, Will P (1999) EFAB: rapid, low-cost desktop micromachining of high aspect ratio true 3-D MEMS. In: Micro electro mechanical systems (MEMS '99). Twelfth IEEE international conference on 1999. IEEE, Orlando
49. Hunter GW, Stetter JR, Hesketh PJ, Chung-Chiun L (2010) Smart sensor systems. Interface 19(4):29–34
50. Ahmadi AG, Zhengchun P, Hesketh PJ, Sankar N (2010) A wafer-scale process for fabricating arrays of nanopore devices for biomolecule analysis. J Microlithogr Microfabr Microsyst 9(3):033011
51. Shin H, Hesketh PJ, Mizaikoff B, Kranz C (2007) Batch fabrication of atomic force microscopy probes with recessed integrate ring microelectrode at a wafer level. Anal Chem 79(13):4769–4777

Microbial Corrosion

▶ Microbiologically Influenced Corrosion

Microbial Electrosynthesis

Dirk Holtmann, Achim Hannappel and Jens Schrader
DECHEMA Research Institute of Biochemical Engineering, Frankfurt am Main, Germany

Introduction

The combination of the advantages of biological components (e.g., reaction specificities or self-replication) and electrochemical processes to bioelectrochemical systems offers the opportunity to develop efficient and sustainable processes. In bioelectrochemical systems (BES), at least one electrode reaction is catalyzed by living microorganisms or isolated compounds, e.g., enzymes. Besides electro-enzymatic processes [1–3], intact microorganisms can be used to produce energy and chemicals. The application of a microbial electrocatalyst in microbial fuel cells (MFC) and microbial electrolysis (MEC) has been investigated for several years. In these applications microorganisms are used to oxidize organic or inorganic substances at an anode to generate electrical power or H_2. The discovery that electrical current can also stimulate microbial metabolism has led to a range of applications in bioremediation [4, 5] and in the production of fuels and chemicals. Notably, the microbial production of chemicals, called microbial electrosynthesis (MEC), provides a highly attractive, novel perspective for future generation of valuable products from electricity or even wastewater [6]. MES can be defined as the microbial conversion of carbon dioxide to organic molecules using electricity [7]. In a wider definition MES is an electricity-driven or electricity-influenced microbial product synthesis [8, 9].

The abiotic electrochemical reduction of carbon dioxide to organic compounds has been studied for a long time [10–12]. However, in practice these processes have not proven applicable, in large part due to the following drawbacks [13]:

- Poor long-term stability of the cathodes
- Cathode expense
- Slow carbon dioxide reduction
- Nonspecific product formation
- Competition with hydrogen production

Incorporating enzyme catalysts on electrodes may promote more specific product formation from electrochemical reduction of carbon dioxide and lower the energy required for reduction [10], but experiments on enzymatic reduction have typically lasted only a matter of hours, reflecting the fact that enzymes adsorbed to electrodes do not show long-term stability. Different microorganisms can reduce carbon dioxide with hydrogen as the electron donor. Processes based on the combination of these microbial reactions with an electrochemical production of hydrogen are difficult to realize due to expensive catalysts and high energy demands [14].

Extracellular Electron Transfer Mechanisms

Extracellular electron transfer (EET) refers to the transport of electrons in and out of a microorganism. In general, there are two possible electron transfer mechanisms between microorganisms and electrode surfaces, namely, direct electron transfer (DET) or mediated electron transfer (MET). DET, in which electrons move between the cell and the electrode via direct contact, relies on the existence of a biofilm or at least a single cell layer on the electrode surface [9]. Advantages of DET are the possibility for direct catalysis and the retention of the biocatalyst in the reaction system. Disadvantages resulting from the biofilm are internal and external diffusion limitations. In MET the presence of soluble shuttle molecules or capacitive particles is necessary. These components should be able to transfer the electrons between the electrode and the microorganism, making any inhibition resulting from diffusion limitations less likely to occur. MET with artificial mediators is often discussed to be problematic because of the cost for the mediator, chemical instability, mediator toxicity, as well as interferences in the downstream process of the products [15]. On the other hand, MET can have different advantages such as avoiding long start-up times to establish a biofilm with corresponding diffusion limitations or an easier integration of MES in existing biotechnological reactor concepts. Certainly the electron transfer rate and the subsequent product formation rate will differ between DET and MET. For an industrial application of MES, both electron transfer principles should be evaluated from an economical and ecological point of view.

To improve MES in the future, more detailed information about electron transfer mechanisms from electrodes to microorganisms is needed. The opposite electron transfer direction – from microorganisms to extracellular substrates, such as minerals or electrodes – has been investigated in detail over last years. It can be assumed that similar principles are involved in MES. Particularly the electron transfer mechanisms of

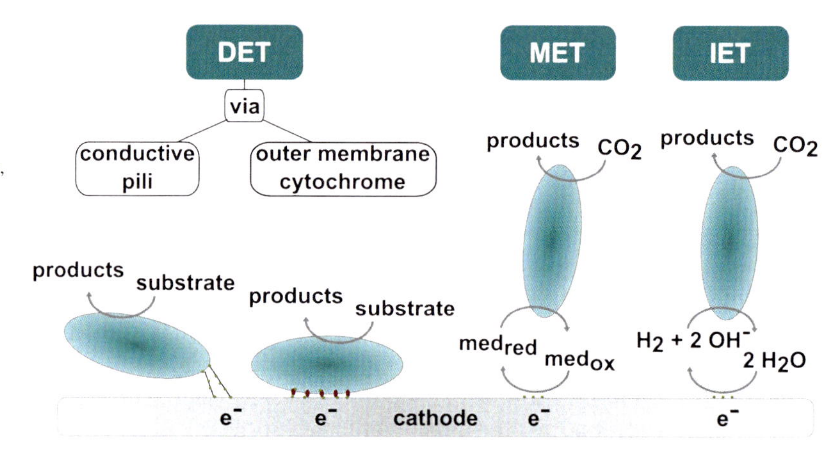

Microbial Electrosynthesis, Fig. 1 Electron transfer mechanisms (*DET* direct electron transfer, *MET* mediated electron transfer, *IET* indirect electron transfer)

Shewanella sp. [16–18] and *Geobacter* sp. [19–21] as model organisms have been investigated leading to the following electron transfer mechanisms described in literature:

- DET – electrically conductive pili form a physical connection between the microorganisms and the substrates [19, 22, 23].
- DET – outer membrane cytochromes [20, 24, 25].
- MET – natural extracellular flavins [17, 18, 26, 27] and artificial compounds [28].

Furthermore, as a variant of MES, it is possible to indirectly transfer electrons from electrodes to microorganisms via the production of H_2 or other carriers. The IET (indirect electron transfer) can be regarded as two independent steps: the electrochemical generation of a suitable electron carrier and the consumption of this natural or artificial carrier by the microorganism. Figure 1 summarizes the different electron transfer principles.

Proteobacteria like *Geobacter sulfurreducens* and *Shewanella oneidensis* are capable of transferring surplus electrons from catabolic metabolism onto solid extracellular electron acceptors like minerals or electrodes. In the following their electron transfer mechanisms are briefly described. Depending on the environmental conditions, multiple or alternative pathways might be dominating. Because of their electrogenic properties, both bacteria are intensively investigated for the reverse electron flow needed in MES.

Starting from the menaquinol pool of the inner membrane, *Geobacter sulfurreducens* transfers electrons through the cytochrome bc-complex to the periplasmic multiheme proteins MacA and PpcA. Via a series of outer membrane multiheme proteins (e.g., OmcB, OmcE, OmcS), electrons are shuttled to conductive pili. Especially, the Lovley group has intensively investigated the fundamental contribution of the mentioned multiheme proteins and their occurrence under different environmental conditions (for excellent review articles, see [21, 29]). *Shewanella oneidensis* mainly uses the Mtr pathway (metal-reducing pathway) as connection to extracellular electron acceptors. The Mtr pathway consists of the conductive multiheme c-type cytochromes MtrA and MtrC and the outer membrane β-barrel protein MtrB. Electrons from intracellular oxidative metabolism are channeled into the extracellular route via the soluble electron carrier menaquinol and the quinol oxidase CymA. Using these components, Ross et al. demonstrated the reversibility of the electron flow for the uptake of electrons from electrode material necessary for MES [16]. Jensen and coworkers successfully introduced electrode communication to genetically modified *E. coli* by introducing the Mtr components MtrA/B/C [30]. Although the Mtr pathway is normally coupled to extracellular riboflavins as soluble electron shuttles, direct interaction with electrodes is also observed in *Shewanella* biofilms (Fig. 2).

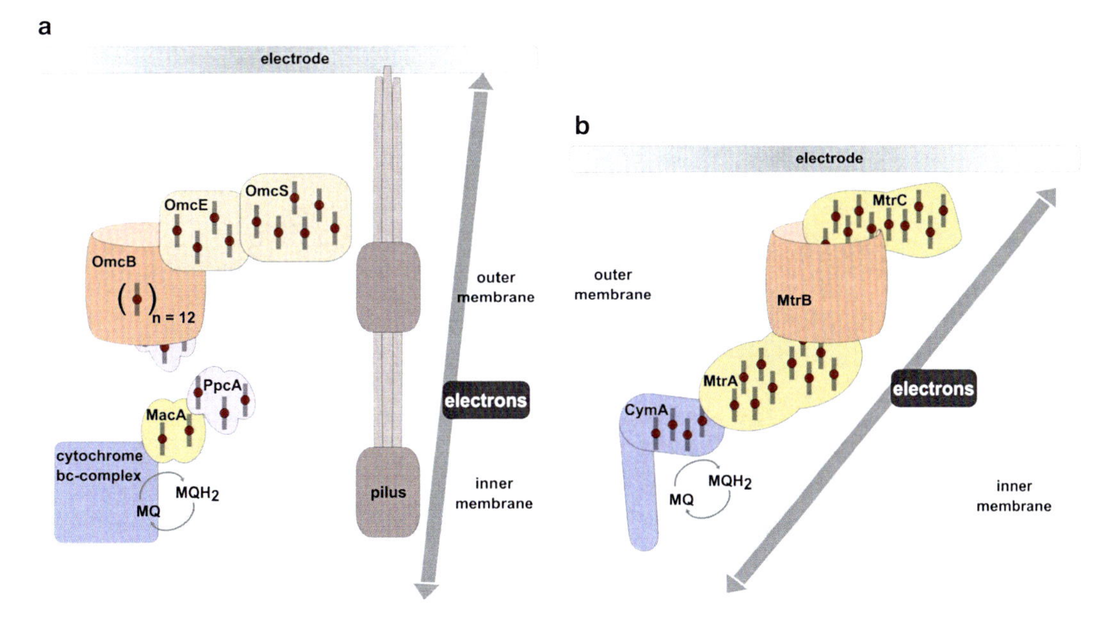

Microbial Electrosynthesis, Fig. 2 (**a**) Electron transfer pathway via multiheme proteins and conductive pili in *Geobacter sulfurreducens* and (**b**) electron transfer pathway via multiheme proteins in *Shewanella oneidensis* (for further details, see text)

Microorganisms and Reaction Systems

Nevin et al. used biofilms of the acetogenic microorganism *Sporomusa ovata* in an H-cell to reduce carbon dioxide and produce acetate [13]. As an example this process will be described in more detail. The anode and cathode chambers were separated with a cation-exchange membrane. To cultivate *S. ovata* an inoculum of the microorganism was grown with hydrogen as the electron donor. The hydrogen-grown culture was introduced into the cathode chamber. A potentiostat provides the energy to extract electrons from water at the anode and poise the cathode at -400 mV (versus standard hydrogen electrode). In a bicarbonate-based medium, carbon dioxide was the sole carbon source. The culture was initially bubbled with a hydrogen-containing gas mixture as an additional electron donor to accelerate the growth of a biofilm on the cathode surface. Once acetate concentration reached 10 mM, 50 % of the medium was replaced by fresh medium. This process was repeated three times. Afterwards the gas phase was changed to a nitrogen- and carbon dioxide-containing gas. Reaction systems with *S. ovata* steadily consumed current while acetate and small amounts of 2-oxobutyrate were produced.

Shorea ovata was also cultivated on different electrode materials to optimize the electron transfer between electrode and biofilm, therefore improving the acetate formation rate [31]. Functionalization of carbon cloth with chitosan, cyanuric chloride, and 3-aminopropyltriethoxysilane resulting in a positively charged electrode enhanced the acetate formation. An increase in productivity over untreated carbon cloth cathodes was also achieved with polyaniline cathodes. However, not all strategies to provide positively charged surfaces were successful, as treatment of carbon cloth with melamine or ammonia gas did not stimulate microbial acetate electrosynthesis. Treating the basis material with metal, in particular gold, palladium, or nickel nanoparticles, also promoted microbial electrosynthesis. *S. ovata*-containing cathodes can also be combined with sulfide oxidation as an alternative electron source [32].

Microbial Electrosynthesis, Table 1 Examples of microorganism using CO_2 in microbial electrosynthesis (*ECE* energy capture efficiency)

Organism	Reaction	Reactor, electrochemical parameters, and productivity	Literature
Sporomusa ovata	$2\ CO_2 + 2\ H_2O \rightarrow CH_3COOH + 2\ O_2$, oxobutyrate is produced as side product	H-cell, – 400 mV versus NHE, unpolished graphite, 86 % ECE, acetate 153 μmol/day, 2-oxobutyrate 6.7 μmol/day (after 6 days)	[13]
Sporomusa silvacetica	$2\ CO_2 + 2\ H_2O \rightarrow CH_3COOH + 2\ O_2$	H-cell, – 400 mV versus NHE, graphite cathode, 48 % ECE, acetate 5.4 μmol/day (after 8 days)	[34]
Sporomusa sphaeroides	$2\ CO_2 + 2\ H_2O \rightarrow CH_3COOH + 2\ O_2$	H-cell, – 400 mV versus NHE, graphite cathode, 84 % ECE, acetate 8.0 μmol/day (after 6 days)	[34]
Clostridium ljungdahlii	$2\ CO_2 + 2\ H_2O \rightarrow CH_3COOH + 2\ O_2$	H-cell, – 400 mV versus NHE, graphite cathode, 82 % ECE, acetate 15.7 μmol/day (after 7 days)	[34]
Clostridium aceticum	$2\ CO_2 + 2\ H_2O \rightarrow CH_3COOH + 2\ O_2$	H-cell, – 400 mV versus NHE, graphite cathode, 53 % ECE, acetate 6.7 μmol/day, 2-oxobutyrate 3.8 μmol/day (after 13 days)	[34]
Moorella thermoacetica	$2\ CO_2 + 2\ H_2O \rightarrow CH_3COOH + 2\ O_2$	H-cell, – 400 mV versus NHE, graphite cathode, 85 % ECE, acetate 11.2 μmol/day (after 8 days)	[34]
Methanobacterium palustre	$CO_2 + 8H^+ + 8e^- \rightarrow CH_4 + 2H_2O$	Single chamber with graphite fiber brush anode, – 500 mV to –800 mV versus NHE, up to 96 % ECE, methane 420 μmol/day (after 40 h)	[33]
Methanogenic mixed culture	$CO_2 + 8H^+ + 8e^- \rightarrow CH_4 + 2H_2O$	H-cell, carbon paper cathode (8 cm^2), methane up to 1,500 μmol/day	[36]
Ralstonia eutropha	Electrochemical reaction: $CO_2 + H^+ + e^- \rightarrow HCOO^-$	Platinum mesh anode separated by ceramic membrane from the indium foil cathode. 1.2 mM isobutanol and 0.6 mM 3-methyl-1-butanol after 100 h	[7]
	Microbial reaction (nonstoichiometric): $HCOO^- \rightarrow C_5H_{12}O + C_4H_{10}O$		

Electrochemical oxidation of sulfide on the anode yields two electrons. The oxidation product, elemental sulfur, was further oxidized to sulfate by *Desulfobulbus propionicus*, generating six additional electrons. In a further system the biocathode was dominated by a single archaeon, *Methanobacterium palustre*. When a current was generated by a biofilm on the anode growing on acetate, methane was produced at the cathode at an overall energy efficiency of 80 % [33]. Furthermore, the Lovley group showed that different acetogenic microorganisms, e.g., further *Sporomusa* species or *Clostridia*, can be used in MES [34]. In a mediated electron transfer process, Neutral Red was electrochemically used as sole source of reducing power [28, 35]. A mutant of *Actinobacillus succinogenes* oxidizes the artificial mediator for the production of succinate. Reduced Neutral Red can also replace hydrogen as the sole electron donor source for microbial growth and production of methane from CO_2. Li et al. used an indirect approach to produce isobutanol and 3-methyl-1-butanol from CO_2 [7]. In a first step CO_2 was electrochemically reduced to formic acid. The alcohols were produced in a second step with a genetically engineered *Ralstonia eutropha* using formic acid as the sole carbon source, which means electricity was the sole energy and CO_2 the sole C source of the overall process. Table 1 summarizes microorganisms, reactions, reactor concepts, and electrochemical parameters as well as the measured productivities investigated in the context of MES so far.

Electrochemical measurements and determination of hydrogen concentration show that an electron transfer via hydrogen is unlikely [33]. The results of the measurements support a mechanism of methane production directly from the current. It was shown that the rate of extracellular electron transfer was strongly dependent on the cathode potential [36]. Methane was produced at potentials more negative than −650 mV versus NHE, both via abiotically produced hydrogen gas and via direct extracellular electron transfer. In a methanogenic culture a 20-fold increase was observed when the potential was decreased from −650 mV to −900 mV. In the same interval, the enhancement of abiotic electron transfer processes was much greater. The authors mentioned that the relative contribution of extracellular electron transfer displayed a maximum at −750 mV. To make sure that a direct electron transfer takes place, the hydrogen production can be measured. The measured hydrogen partial pressures in the headspace of the H-cell with biofilms of *S. ovata* were less than 10 ppm [13]. This concentration is more than two orders of magnitude below the minimum threshold for acetate production from hydrogen by acetogens [37].

Conclusion and Future Directions

The production of biofuels and biochemicals can be regarded as highly electron intensive [9]. To divert fermentative and respiratory pathways to the product of interest, additional electrons "as reducing power" are often required. Due to the thermodynamic properties of carbon dioxide, the energy demand by using CO_2 as carbon source is especially high. Meanwhile, the past decade has seen the breakthrough of sustainable electricity sources such as solar and wind energy. MES is at the nexus of both, as it uses electrical energy as source of reducing power for product formation from CO_2 [9]. MES can be driven by any source of electricity, but when solar power is utilized, the overall reaction powered by light is the same as photosynthesis [38]. However, microbial electrosynthesis is approximately 100-fold more efficient than plants in converting solar energy into organic compounds and can directly produce the desired products, whereas producing fuels from plant or algal biomass requires additional energy inputs and only a fraction of the energy in the biomass is recovered as fuel. Furthermore, MES does not require arable land and the large quantities of water required for growing and processing biomass to fuels and can avoid the negative environmental impact of large-scale biomass production. Thus, MES may transform the bioenergy field because it offers the possibility of converting renewable but intermittent sources of electricity into fuels or other desirable organic compounds that are energy dense and can readily be stored, distributed, and utilized within the existing infrastructure. As shown in Table 1, the productivities of MES are quite low and still far away to enable economic production of desired products. Therefore different investigations and optimizations are necessary:

- Deeper understanding of EET – only the knowledge of the mechanism of the extracellular electron transfer will enable scientists to design adequate electron transfer relays.
- Screening of further microorganisms – currently, there are only a few methanogenic and acetogenic organisms described to be able to be deployed in MES. It is obvious that an intensive screening of microorganisms will help to identify promising candidates for MES. To identify these candidates, suitable screening systems are needed.
- Strain development – to develop MES on a technical scale, the biochemical pathways of naturally electrode-active bacteria should be engineered for the formation of the desired product. For most of these "new" production strains, an adequate set of genetic tools is still missing and therefore needed to be developed. The analyses of the metabolic fluxes as well as of the proteome and metabolome are key technologies to optimize MES. By using the concept of synthetic biology, the electron transfer modules can also be transferred to platform organisms such as *E. coli*.
- Biofilm engineering – different electrochemical and hydrodynamic parameters may alter the microenvironment of biofilms and subsequently affect biofilm integrity and physiological and biochemical activities. An understanding of the fundamental processes which underlie these interactions between the

biofilm and its environment will help to control biofilm formation and its long-term stability. Target-oriented biofilm formation can help to overcome diffusion limitation in biofilms. The biofilm formation can be optimized by surface modifications or applying a suitable potential.
- Electrode materials and scalable reactors – in different bioelectrochemical systems expensive electrode materials such as carbon nanotubes or precious metal electrodes are used. These materials are unconsolidated for large-scale MES. In terms of maximizing productivity and minimizing costs, cheap and reusable three-dimensional electrodes are needed. In a technical electrochemical reactor, the use of an expensive separator such as a membrane should be avoided. During the lab stage, the scalability of the reactor concept should receive attention as important parameter.
- Economic evaluation – one main challenge is to bring MES out of the laboratory to technical application. At the early stage of the development, a rough calculation should at least include the assessment of needed productivities at the given fixed and variable costs (e.g., for reactor, electrodes, membranes, reaction medium, pretreatment of the gas). The volumetric productivities (space-time yields), final product concentrations, and total process times determine the overall process performance. These parameters should be used to define operational windows for the production of bulk chemicals. Furthermore, this theoretical approach allows the identification of limiting process parameters.

MES can be regarded as a promising and bright but challenging technique which is still in its infancy and whose economical feasibility has not been proven so far. In a first step, MES should be used to produce "simple" molecules such as methane, whereas a second generation of MES may allow for the exploitation of the strength of microorganisms, i.e., to produce multicarbon products, such as biopolymers or energy-dense hydrocarbons from CO_2.

Cross-References

- ▶ Biofilms, Electroactive
- ▶ Bioelectrochemical Hydrogen Production
- ▶ Cofactor Substitution, Mediated Electron Transfer to Enzymes
- ▶ Electrocatalysts for Carbon Dioxide Reduction

References

1. Cekic SZ, Holtmann D, Güven G, Mangold K-M, Schwaneberg U, Schrader J (2010) Mediated electron transfer with P450cin. Electroch Commun 12(11):1547–1550
2. Krieg T, Hüttmann S, Mangold K-M, Schrader J, Holtmann D (2011) Gas diffusion electrode as novel reaction system for an electro-enzymatic process with chloroperoxidase. Green Chem 13:2686–2689
3. Lütz S, Vuorilehto K, Liese A (2007) Process development for the electroenzymatic synthesis of (R)-methylphenylsulfoxide by use of a 3-dimensional electrode. Biotechnol Bioeng 98(3):525–534
4. Tiehm A, Lohner ST, Augenstein T (2009) Effects of direct electric current and electrode reactions on vinyl chloride degrading microorganisms. Electrochim Acta 54(12):3453–3459
5. Lohner ST, Becker D, Mangold K-M, Tiehm A (2011) Sequential reductive and oxidative biodegradation of chloroethenes stimulated in a coupled bioelectro-process. Environ Sci Technol 45(15):6491–6497
6. Rabaey K, Rozendal RA (2010) Microbial electrosynthesis – revisiting the electrical route for microbial production. Nat Rev Micro 8(10):706–716
7. Li H, Opgenorth PH, Wernick DG, Rogers S, Wu T-Y, Higashide W, Malati P, Huo Y-X, Cho KM, Liao JC (2012) Integrated electromicrobial conversion of CO_2 to higher alcohols. Science 335(6076):1596
8. Pandit A, Mahadevan R (2011) In silico characterization of microbial electrosynthesis for metabolic engineering of biochemicals. Microb Cell Fact 10(1):76
9. Rabaey K, Girguis P, Nielsen LK (2011) Metabolic and practical considerations on microbial electrosynthesis. Curr Opin Biotechnol 22(3):371–377
10. Cole EB, Bocarsly AB (2010) Photochemical, electrochemical, and photoelectrochemical reduction of carbon dioxide, in carbon dioxide as chemical feedstock. Wiley-VCH Verlag GmbH & Co. KGaA, Weinheim, p 291–316
11. Gattrell M, Gupta N, Co A (2006) A review of the aqueous electrochemical reduction of CO_2 to hydrocarbons at copper. J Electroanal Chem 594(1):1–19
12. Oloman C, Li H (2008) Electrochemical processing of carbon dioxide. ChemSusChem 1(5):385–391
13. Nevin KP, Woodard TL, Franks AE, Summers ZM, Lovley DR (2010) Microbial electrosynthesis: feeding microbes electricity to convert carbon dioxide and

water to multicarbon extracellular organic compounds. mBio 1(2):e00103-10
14. Aulenta F, Reale P, Catervi A, Panero S, Majone M (2008) Kinetics of trichloroethene dechlorination and methane formation by a mixed anaerobic culture in a bio-electrochemical system. Electrochim Acta 53(16):5300–5305
15. Lovley DR, Nevin KP (2013) Electrobiocommodities: powering microbial production of fuels and commodity chemicals from carbon dioxide with electricity. Curr Opin Biotechnol 24:1–6
16. Ross DE, Flynn JM, Baron DB, Gralnick JA, Bond DR (2011) Towards electrosynthesis in Shewanella: energetics of reversing the Mtr pathway for reductive metabolism. PLoS One 6(2):e16649
17. Liu H, Matsuda S, Hashimoto K, Nakanishi S (2012) Flavins secreted by bacterial cells of Shewanella catalyze cathodic oxygen reduction. ChemSusChem 5(6):1054–1058
18. Kotloski NJ, Gralnick JA (2013) Flavin electron shuttles dominate extracellular electron transfer by *Shewanella oneidensis*. mBio 4(1):e00553-12
19. Cologgi DL, Lampa-Pastirk S, Speers AM, Kelly SD, Reguera G (2011) Extracellular reduction of uranium via Geobacter conductive pili as a protective cellular mechanism. Proc Natl Acad Sci 108(37): 15248–15252
20. Bond DR, Strycharz-Glaven SM, Tender LM, Torres CI (2012) On electron transport through Geobacter biofilms. ChemSusChem 5(6):1099–1105
21. Lovley DR (2008) The microbe electric: conversion of organic matter to electricity. Curr Opin Biotechnol 19(6):564–571
22. Malvankar NS, Vargas M, Nevin KP, Franks AE, Leang C, Kim B-C, Inoue K, Mester T, Covalla SF, Johnson JP, Rotello VM, Tuominen MT, Lovley DR (2011) Tunable metallic-like conductivity in microbial nanowire networks. Nat Nano 6(9):573–579
23. El-Naggar MY, Wanger G, Leung KM, Yuzvinsky TD, Southam G, Yang J, Lau WM, Nealson KH, Gorby YA (2010) Electrical transport along bacterial nanowires from *Shewanella oneidensis* MR-1. Proc Natl Acad Sci 107(42):18127–18131
24. Mehta T, Coppi MV, Childers SE, Lovley DR (2005) Outer membrane c-type cytochromes required for Fe(III) and Mn(IV) oxide reduction in *Geobacter sulfurreducens*. Appl Environ Microbiol 71(12):8634–8641
25. Liu Y, Bond DR (2012) Long-distance electron transfer by *G. Sulfurreducens* biofilms results in accumulation of reduced c-type cytochromes. ChemSusChem 5(6):1047–1053
26. von Canstein H, Ogawa J, Shimizu S, Lloyd JR (2008) Secretion of flavins by Shewanella species and their role in extracellular electron transfer. Appl Environ Microbiol 74(3):615–623
27. Marsili E, Baron DB, Shikhare ID, Coursolle D, Gralnick JA, Bond DR (2008) Shewanella secretes flavins that mediate extracellular electron transfer. Proc Natl Acad Sci 105(10):3968–3973
28. Park DH, Laivenieks M, Guettler MV, Jain MK, Zeikus JG (1999) Microbial utilization of electrically reduced neutral red as the sole electron donor for growth and metabolite production. Appl Environ Microbiol 65(7):2912–2917
29. Lovley DR (2006) Bug juice: harvesting electricity with microorganisms. Nat Rev Micro 4(7):497–508
30. Jensen HM, Albers AE, Malley KR, Londer YY, Cohen BE, Helms BA, Weigele P, Groves JT, Ajo-Franklin CM (2010) Engineering of a synthetic electron conduit in living cells. Proc Natl Acad Sci 107(45):19213–19218
31. Zhang T, Nie H, Bain TS, Lu H, Cui M, Snoeyenbos-West OL, Franks AE, Nevin KP, Russell TP, Lovley DR (2013) Improved cathode materials for microbial electrosynthesis. Energ Environ Sci 6(1):217–224
32. Gong Y, Ebrahim A, Feist AM, Embree M, Zhang T, Lovley D, Zengler K (2012) Sulfide-driven microbial electrosynthesis. Environ Sci Technol 47(1):568–573
33. Cheng S, Xing D, Call DF, Logan BE (2009) Direct biological conversion of electrical current into methane by electromethanogenesis. Environ Sci Technol 43(10):3953–3958
34. Nevin KP, Hensley SA, Franks AE, Summers ZM, Ou J, Woodard TL, Snoeyenbos-West OL, Lovley DR (2011) Electrosynthesis of organic compounds from carbon dioxide is catalyzed by a diversity of acetogenic microorganisms. Appl Environ Microbiol 77(9):2882–2886
35. Park DH, Zeikus JG (1999) Utilization of electrically reduced neutral Red by *Actinobacillus succinogenes*: physiological function of neutral red in membrane-driven fumarate reduction and energy conservation. J Bacteriol 181(8):2403–2410
36. Villano M, Aulenta F, Ciucci C, Ferri T, Giuliano A, Majone M (2010) Bioelectrochemical reduction of CO_2 to CH_4 via direct and indirect extracellular electron transfer by a hydrogenophilic methanogenic culture. Bioresour Technol 101(9):3085–3090
37. Cord-Ruwisch R, Seitz H-J, Conrad R (1988) The capacity of hydrogenotrophic anaerobic bacteria to compete for traces of hydrogen depends on the redox potential of the terminal electron acceptor. Arch Microbiol 149(4):350–357
38. Lovley DR, Nevin KP (2011) A shift in the current: new applications and concepts for microbe-electrode electron exchange. Curr Opin Biotechnol 22(3):441–448

Microbial Inhibition of Corrosion

▶ Microbiologically Influenced Corrosion Inhibition

Microbiologically Induced Corrosion

► Microbiologically Influenced Corrosion

Microbiologically Influenced Corrosion

Andrzej Kuklinski and Wolfgang Sand
Fakultät Chemie, Biofilm Centre/Aquatische Biotechnologie, Universität Duisburg-Essen, Essen, Germany

Synonyms

Biocorrosion; Biodeterioration; Microbial corrosion; Microbiologically induced corrosion

Introduction

Microbiologically influenced corrosion (MIC) describes all cases of corrosion caused or influenced by microorganisms present in a certain system. Therefore, it is an interdisciplinary subject embracing the fields of material science, chemistry, microbiology, and biochemistry. This is reflected by the various terms found in literature for MIC such as biocorrosion, biodeterioration, and biodegradation. Besides these, terms like biomineralization or bioleaching are used. Although these terms describe the same phenomenon, they possess a slightly different focus: Biodeterioration is used to indicate a negative biological modification of materials, and biodegradation describes a microbial degradation with the (organic) material serving as a substrate (nutrient) for the microorganisms. Biomineralization indicates that organic materials can be fully degraded to inorganic products like CO_2 and H_2O. Finally, bioleaching is used to describe the dissolution of metal sulphides by microorganisms and the possibility to win precious metals out of low-grade ores. When metallic or constructional materials are damaged, the terms biocorrosion or MIC seem most appropriate, especially considering the definition of corrosion according to DIN EN ISO 8044. Here, corrosion of metals is defined as a reaction of the material with its environment considerably impairing the function of the material itself, its environment or the technical system of which these form a part. Often, the involved processes are of electrochemical nature. These changes will result in corrosion damage, which in turn is defined as an impairment of the function of constructional elements. In this definition, microbial corrosion is limited to their influence on metallic constructional materials. Thus, this review focuses on constructional materials, but also non-metallic organic and inorganic materials are briefly described, as they are subject to both biodeterioration and biodegradation. However, elaborate descriptions of the mechanisms of the latter would exceed the scope of this essay and thus other more extensive reviews are recommended (e.g., [3,6]).

Microorganisms

The term "microorganism" covers several different life forms, from prokaryotes like bacteria, archaea, and cyanobacteria to eukarya such as algae, lichens, yeast and fungi and further to protozoa. Key characteristic for this classification is their size which generally ranges between 0.5 and 10 μm. Thus, single microorganisms are usually not visible by the naked eye and microscopic techniques are necessary in order to visualize and identify them. This is one of the most important facts why MIC has often been ignored or not taken seriously.

It is important to note that all microorganisms involved may occur jointly. Due to their differing physiology, various mechanisms may contribute to biodeterioration. Additionally, some countermeasures (e.g., antibiotics) will be ineffective for such complex microbial consortia, especially when eukaryotes and prokaryotes occur together.

Bacteria, Archaea, and Cyanobacteria

These two distinct groups of microorganisms are morphologically characterized by only a few different shapes: rods, cocci, spirillae, and pleomorphic forms (variable), also fungus-like structures resembling mycelia may occur.

The low variability in cell shapes certainly is more than compensated by a plethora of various metabolic pathways. Virtually all naturally occurring and man-made organic materials are degradable by microorganisms. But also metallic or mineral materials may be affected. They may either serve as energy source, as reactant for metabolic intermediates or products, or just provide the surface (substratum) for attachment.

Another important factor is the high potential in reactivity. Single cells usually will be harmless for construction materials, but (cyano)bacteria and/or archaea possess an enormous ability to exponentially multiply. Under optimal conditions, doubling times of less than 1 h occur frequently, with 20 min for *Escherichia cola* (*E. coli*) being the lowest known value. Thus, also low initial cell numbers may rapidly reach endangering numbers. Additionally, cells possess a high specific surface area. A volume of 1 mL may easily contain 10^{12} cells, and given a typical size of 1×1 µm per cell, a highly reactive surface area of 6 m^2 cm^{-3} results. This enormous surface area is used to take up and metabolize various nutrients and explains why small cells can become so dangerous for technical environments.

Algae

Algae are eukaryotic primary producers living by photosynthesis. According to their different optimal light wavelengths and hence pigmentation, they are divided into green, red, and brown species. The former blue-green algae have been reclassified as bacteria, as they have a prokaryotic cellular organization. Similar to bacteria, algae occur ubiquitously and some species such as the brown algae may become as large as 100 m. Besides purely photosynthetic species, also mixotrophs and saprophytes exist. The former are able to use both light and organic compounds as nutrients, and the latter are able to grow on decomposing organic matter. Compared to bacteria, algae are significantly larger with a typical size of more than 10 µm. They may possess chlorophylls a, b, c, and d as well as phycobilins, carotenes, and xanthophylls. Many algae possess flagella and thus are highly motile.

Lichens

Lichens are plants resulting from the symbiosis of fungi with algae or cyanobacteria. Usually Ascomycota (sac fungi) and only rarely Basidiomycota (higher filamentous fungi) develop. Their symbiotic origin causes that unique metabolic products such as lichenic acids, aliphatic acids, deprides, quinones, and dibenzofuranone derivatives excreted. Most of these products may be released into the environment. Lichens are known to be primary colonizers of many substrata, also in extreme habitats. Likewise to some algae, several species are able to grow endolithically, i.e., well protected inside of rocks. Lichens are able to withstand temperatures of up to 70 °C and several months of desiccation.

Fungi

Another major group of microorganisms with importance for biodeterioration are the eukaryotic fungi. In general, they are saprophytes living on the decomposition of organic matter and thus, many are parasitic and pathogenic. Several fungi are a constant danger for living organisms due to the production of toxins. Vegetative forms are cells, hyphae, and mycelia. Hyphae are long, branching filamentous structures and serve various specific functions such as absorption within host cells, nutrient exchange, water uptake, or formation of trapping structures, e.g., in nematode-trapping fungi. Usually, hyphae are at least 5 µm in diameter. Mycelia are the vegetative part of fungi, consisting of masses of hyphae. Fungi may multiply sexually or asexually, indicating the presence of mitosis and meiosis like in other eukaryotes. They possess a strong secondary metabolism which seems to become important at the end of the growth when substrates are scarce and the cells need to dispose metabolic intermediates. In the context of biodeterioration, several compounds produced by fungi are of importance. Fungi produce and excrete organic acids, alcohols, solvents, EPS, and CO_2 and exhibit outstanding degradative potential for organic compounds. Actually, the first known countermeasures against biodeterioration were directed against fungi: the decay of wood was reduced by charring it in the fire. This way, the material was sterilized and lignin- and cellulose-degrading fungi were killed and a recolonization became impossible.

Biofilms

In virtually all natural and artificial environments, surfaces will be colonized rapidly by microorganisms eventually leading to the formation of biofilms. These are complex microbial communities embedded into a matrix of extracellular polymeric substances (EPS) excreted by the microorganisms itself. They are assumed to be a universal bacterial survival strategy and probably the vast majority of microbial life on earth exists in the form of biofilms.

The formation of biofilms can be described as a 5-step process eventually leading to a highly stable, continuous biofilm (Table 1).

Microbiologically Influenced Corrosion, Table 1 The five steps of biofilm formation

Step	Description	Processes involved
I	Reversible attachment	Formation of a conditioning film on the surface mediating reversible attachment of cells from planktonic phase; can be active by secretion of biosurfactants by the cells or passive by spontaneous adsorption and accumulation of organic substances from the aqueous phase
II	Irreversible attachment	Cells excrete EPS and microcolonies are formed within the EPS. EPS stabilize the microcolonies and promote the irreversible attachment of cells
III	Beginning maturation	Microcolonies grow further and planktonic cells from the surrounding environment are incorporated
IV	Mature biofilm	Continuous biofilm is formed providing, different habitats for different microorganisms. Further planktonic microorganisms may be incorporated
V	Dispersal of planktonic cells	Release of planktonic cells from the biofilm, either by active processes or by passive external factors (e.g., high shear forces)

Biofilms show a highly complex structure depending both on the embedded microbes and external factors. The former influence its structure by excretion of EPS of different composition, exoenzymes, and various other compounds. They also may promote the formation of aerobic and anaerobic zones within the biofilm. Physical (external) factors include the physicochemical environment such as the substratum, nutrients, surrounding medium, and solute transport and diffusion. Hydrodynamic shear forces influence not only the general structure but also transport and reaction rates within the biofilm. When grown fully submerged in steady-flow environments, mushroom-like structures separated by water-filled channels facilitating the rapid distribution of nutrients and metabolites through the whole biofilm are typically found. Biofilms floating at the air-liquid interface, however, usually possess much more complex structures due to the absence of disturbing shear forces. The temperature of the surrounding medium determines the rate of chemical and biochemical processes inside the biofilm as well as transport and diffusion processes.

The microbial community within the biofilm is also strongly dependent on environmental factors. The diffusion rates limited by EPS and density of the biofilm lead to the formation of concentration gradients, thus providing areas with different conditions for microorganisms and enhancing the microbial heterogeneity. Additionally, oxygen gradients throughout the biofilm allow growth of both aerobic (at the upper zones) and anaerobic (at lower zones near to the substratum) organisms. Due to the long retention times, synergistic microbial communities may evolve.

Particularly with regard to biocorrosion and biodeterioration, the influence of biofilms on the embedded microorganisms should be emphasized. Microorganisms living in biofilms may exhibit different morphological, growth, biochemical, and physiological features. Also, changes in genome and proteome (the sum of all genes or, respectively, all expressed proteins) of the cells are frequently observed. Generally, sessile cells show a much higher activity in processes attributed to reproduction and metabolism. Other characteristics typical for biofilm cells are increased EPS production, higher rates in genetic exchange, and thus, better adaption to varying environmental conditions. Additionally, altered biodegradative capabilities and increased production of secondary metabolites are commonly found in sessile microorganisms.

Microbial Deterioration Mechanisms

Despite the staggering diversity of microbial life on earth, biological mechanisms influencing or causing biodeterioration may be summarized in a few main categories. It should be pointed out that one physiological group of organisms may influence biodeterioration by more than one mechanism or even a group of mechanisms. The complexity of microbial biodeterioration is aggravated by the fact that in nature no pure cultures exist but complex consortia of diverse physiological groups. These mixed consortia called biocoenoses may and probably will produce favorable conditions for detrimental organisms even in environments which are normally unfavorable for their growth. Generally, the basic mechanism of biodeterioration will be a chemical attack caused by production and excretion of metabolic compounds by microorganisms.

Excretion of Acids

Several specialized bacterial species are able to produce strong mineral acids. For example, under aerobic conditions, *Acidithiobacilli* oxidize reduced inorganic sulphur compounds and sulphur to sulphuric acid (H_2SO_4). The energy derived from this oxidation is coupled to growth via special enzymes. Beside sulphur and its compounds, these bacteria need only carbon dioxide for the production of cell mass. *Thiobacilli* are able to withstand pH from alkaline to below one values and thus may create strongly acidic environments. Species tolerating neutral or even alkaline pH will be able to grow on alkaline materials like concrete, decreasing the pH, due to their metabolic activity and thus creating favorable environments for species preferring more acidic pH, which will further acidify the environment.

Another important group are nitrifying bacteria deriving their energy from the oxidation of ammonia and nitrite and excreting nitric acid (HNO_3). Two groups of nitrifiers exist:

1. ammonia oxidizers converting ammonia to nitrite and
2. nitrite oxidizers oxidizing nitrite further to nitrate

Like sulphuric acid, nitric acid may react with alkaline materials forming highly soluble salts and thus dissolving the material. Comparable to Thiobacilli, nitrifiers only need CO_2 as a carbon source for their growth.

The third important acid is produced by all life forms: CO_2. It is excreted as the end product of metabolism and reacts with water to carbonic acid, which may dissolve to carbonates. In case of concrete, carbonic acid may dissolve the binding material, lime, and cause serious corrosion problems.

Finally, organic acids such as oxalic, citric, malic, lactic or acetic acid, amino acids, uronic acids, and many more can be produced during microbial metabolism. Usually, these acids are only excreted during an unbalanced metabolic state and may be taken up by the cells during later growth stages or by other microorganisms. However, if large amounts of organic materials occur in the environments, these acids may be produced. Virtually all bacteria, cyanobacteria, algae, lichens, and fungi are able to excrete organic acids.

Chelatization

Organic acids not only attack materials directly but also may act as chelating cations. Because of the high stability of chelate complexes, metals may be dissolved from a crystal lattice resulting in a weakening of the structure. The chelatization of metal ions sometimes is deliberate. Some pathogenic organisms like the fungus *Candida albicans* are able to replenish growth-limiting ferric iron at the expense of the host using chelating compounds such as siderophores [1].

Organic Solvents

Under anaerobic conditions, many microorganisms are able to metabolize organic substances. If suitable electron acceptors such as nitrate, ferrous ion, manganese (IV), or sulphate are not available, fermentation results. Hydrogen is transferred from one organic compound to another one enabling cell growth by substrate chain phosphorylation. The product of fermentation is, thus, another organic compound and sometimes also carbon dioxide. The organic products are often organic acids, as mentioned before, or organic solvents such as ethanol, propanol, or butanol. The latter may react with natural or synthetic materials causing swelling and partial or total dissolution and may eventually lead to a deterioration.

Other Metabolic Compounds

Hydrogen sulphide (H_2S) is another important compound in MIC. H_2S is produced by sulphate-reducing prokaryotes under anaerobic conditions from sulphate, sulphite, sometimes sulphur, and also thiosulphate. As mentioned before, SRP use incompletely reduced sulphur compounds as electron acceptors for their anaerobic respiration.

Hydrogen sulphide influences biodeterioration in various ways. It may be re-oxidized to sulphuric acid under aerobic conditions or in the presence of compounds such as nitrate (act as electron acceptor instead of O_2). Hydrogen sulphide may also react with metal ions and precipitate in form of metal sulphides such as iron sulphide (FeS). Further, H_2S may be oxidized under anaerobic conditions in the light by photosynthetic bacteria, mostly to sulphur, sometimes to sulphate.

Finally, H_2S itself may react as an acid and damage acid-reactive materials. In case of metals, H_2S may cause MIC.

Besides anaerobiosis, H_2S may originate from anaerobic degradation of sulphur-containing amino acids.

Other important compounds for MIC are ammonia and nitrogen oxides. The former may result from microbial degradation of amino acids or urea. Furthermore, ammonia (or more precisely ammonia salts) is a major part of airborne gases or dust particles. By dry or wet deposition, these ammonia salts reach the surface of materials, where they become enriched and, thus, biologically available. The degradation of ammonia by nitrifiers has already been described previously. However, ammonia may also react directly with various materials like copper.

The nitrogen oxides N_2O, NO, and NO_2 result from biological oxidation of ammonia or reduction of nitrite/nitrate to N_2 (denitrification) as well as from industrial processes such as the burning of fossil fuels. Nitrogen oxides may react with various materials. NO_2 is a water-soluble gas dissociating into nitric and nitrous acid, which in turn may attack susceptible materials. NO is less reactive, but may be light-oxidized to NO_2. N_2O, in contrast, is not known to interfere with materials (but as "laughing gas" has well-known properties).

Physical Presence

The pure presence of biofilms alone may already cause detrimental effects on constructional materials. They may cause clogging of porous systems or form a slimy layer on the surface. The high water content retained in the EPS matrix of biofilms may increase the water content of porous materials resulting in an increased risk of a freeze-thaw attack.

Furthermore, biofilms may act insulating or reducing the heat transfer, e.g., in heat exchangers. Another effect is increased drag by biofilm formation on ship hulls, significantly reducing their cruising speed while increasing fuel consumption. Finally, biofilm development may cause all kinds of trouble in technical systems. A common biofilm-related problem can be observed in paper machines, where the inclusion of exopolymers in the pulp may cause the paper web to break in the paper machine. Other problems include resin coatings of cans. If resins used for coating of cans for corrosion protection contain small amounts of microorganisms or their products, incomplete coatings may result and, thus, the risk of corrosion is increased.

Salt Stress. The previously described microbial reactions result in the production and accumulation of salts (except in aquatic environments). Due to their hydrophilic nature, salts are usually hydrated, increasing the water content of porous materials (comparable to biofilms, as described before). This may increase the susceptibility to a physical attack by freezing/thawing, because of the volume change in water/ice crystals.

Furthermore, salt crystals may develop on the surface upon desiccation, which may, e.g., destroy wall paintings on natural stone (by lifting the pigments on top of the stone by the crystals). Finally, salts may form large crystals causing a swelling attack; this effect is often seen on concrete and bricks, where gypsum crystals form ettringite (due to an increase in crystal water).

Exoenzymes and Emulsifying Agents

Biofilms and microorganisms not only produce lipopolysaccharidic exopolymers but also lipoproteins, proteins, and exonucleic acids. Proteins include exoenzymes used for the degradation of high-molecular-weight compounds such as cellulose, which is broken down into soluble, small molecules such as cellobiose and glucose. These may then be taken up by the cells for their metabolism. Similar enzymes exist for other groups of compounds such as esters, amines, and waxes and other poorly soluble and, thus, hardly degradable compounds.

In some cases, microorganisms excrete emulsifying agents to increase the solubility of hydrophobic substances. In case of highly hydrophobic and, thus, insoluble elemental sulphur, an emulsifying agent excreted by bacteria (a phospholipid) causes an increase in dispersibility from 5 to 20,000 μg/L. The same holds true for other hydrophobic materials including hydrocarbon droplets. The increased solubility enables microorganisms to finally biodegrade these compounds.

The mechanisms described above represent the main categories of mechanisms causing biological deterioration of materials. Usually, several of these mechanisms are usually jointly active and may also influence other types of physical or chemical attack. Thus, the microbial share of the total attack can hardly be determined. Much too often, studies on deterioration mechanisms lack proper microbiological analysis.

If corrosion occurs in an environment suitable for life, a microbial contribution needs to be taken into account and, hence, a microbiologist should be consulted.

Microbial Corrosion of Metallic Materials

In many industrial applications, MIC of different metallic constructional materials is a serious problem. It does not represent a novel form of corrosion but is based on the modification of the electrochemical processes by microorganisms. Thus, it is necessary to briefly recall the basic electrochemical mechanisms of corrosion.

In general, metals corrode in the presence of water. This is not limited to corrosion in water-containing solutions, but also includes atmospheric corrosion due to the formation of a humidity film on at the metal surface as well as corrosion in soils due to the humidity of the soil. A detailed discussion of electrochemically influenced corrosion is outside the scope of this essay; hence, only the most important basics needed to understand microbiologically influenced corrosion will be given here.

Basically, electrochemically influenced corrosion is based on the formation of anodic and cathodic sites. At anodic zones, the metal is oxidized and released as metal ions:

$$\mathrm{Me} \rightarrow \mathrm{Me}^{n+} + n\mathrm{e}^-$$

At cathodic zones, compounds such as oxygen are reduced. Cathodic reactions can be summarized as follows:

$$\mathrm{O_2} + 2\mathrm{H_2O} + 4\mathrm{e}^- \rightarrow 4\,\mathrm{OH}^- \times (\text{acidic, neutral or alkaline conditions})$$

$$\mathrm{O_2} + 4\mathrm{H}^+ + 4\mathrm{e}^- \rightarrow 2\mathrm{H_2O}\ (\text{acidic conditions})$$

$$2H^+ + 2e^- \rightarrow H_2 \text{ (acidic solutions)}$$

$$2H_2O + 2e^- \rightarrow H_2 + 2OH^- \times \text{(neutral and alkaline solutions)}$$

The type of cathodic reaction depends on the pH of the solution, the presence or absence of oxygen, and the nature of other oxidizing compounds such as present CO_2. In the pH region of 4 till 10, the rate of oxygen diffusion to a surface controls the corrosion rate of iron. Thus, in this very pH range, only very low corrosion rates would be observed in the absence of oxygen. In this case, the cathodic reaction rate also controls the rate of anodic reactions, as both must be balanced in order to preserve electroneutrality.

Two main types of corrosion are known: "uniform corrosion," where anodic and cathodic sites are virtually inseparable, as well as "localized corrosion," where macroscopic anodic and cathodic sites are physically separable and, thus, observable.

In aqueous environments, microorganisms may influence the electrochemical environment and, consequently, corrosion rates and/or the susceptibility of metals to localized "pitting" corrosion in several ways. However, all known cases of microbiologically influenced corrosion of metals can be attributed to known corrosion mechanisms, which are briefly summarized below.

Formation of Concentration Cells at the Metal Surface

Concentration cells may develop on a surface when a biofilm is distributed heterogeneously over the surface. This leads to areas with higher and lower concentrations of corrosive agents such as oxygen. This effect is, for example, associated with tubercles formed by iron-oxidizing bacteria such as *Gallionella*. Other bacterial groups may accumulate heavy metals in their extracellular polymeric substances, resulting in the formation of ion concentration cells.

Modification of Corrosion Inhibitors

Widely used corrosion inhibitors such as nitrite (for iron and/or mild steel protection) may be destroyed by bacterial conversion to nitrate (e.g., by nitrification). The latter, on the other hand, is widely used as a corrosion inhibitor for aluminum and its alloys. It may, in turn, be transformed to nitrite, ammonia, and N_2.

Production of Corrosive Metabolites

Several bacterial groups produce potentially detrimental metabolites such as inorganic acids (e.g., Acidithiobacillus thiooxidans), organic acids (most bacteria, algae, and fungi), sulphide (sulphate-reducing bacteria), and ammonia, as discussed above.

Destruction of Protective Layers

Various microorganisms are able to degrade organic coatings, which may lead to exposure and subsequent corrosion of the underlying metal surface. Additionally, some microorganisms may cause a breakdown of protective oxide layers on alloyed steel (e.g., by deposition of MnO_2 due to metabolic activity of Mn-oxidizing bacteria).

Stimulation of Electrochemical Reactions

Microbial production of hydrogen sulphide may lead to the evolution of cathodic hydrogen and hydrogen embrittlement.

Hydrogen Embrittlement

Microorganisms may act as a source of hydrogen and/or produce hydrogen sulphide (sulphate-reducing prokaryotes) and, thus, influence the hydrogen embrittlement of metals.

The nature of the metal as well as its environment will strongly influence the corrosion mechanism. Thus, the mechanisms of MIC for different metals are briefly discussed in the following sections.

Iron and Mild Steel

Corrosion Under Anaerobic Conditions

As stated before, very low corrosion rates are expected for iron and mild steel in quasi-neutral pH and the absence of oxygen. Yet, there is a large number of case histories demonstrating severe corrosion damages to buried pipes and marine structures. Often, the corrosion rates are

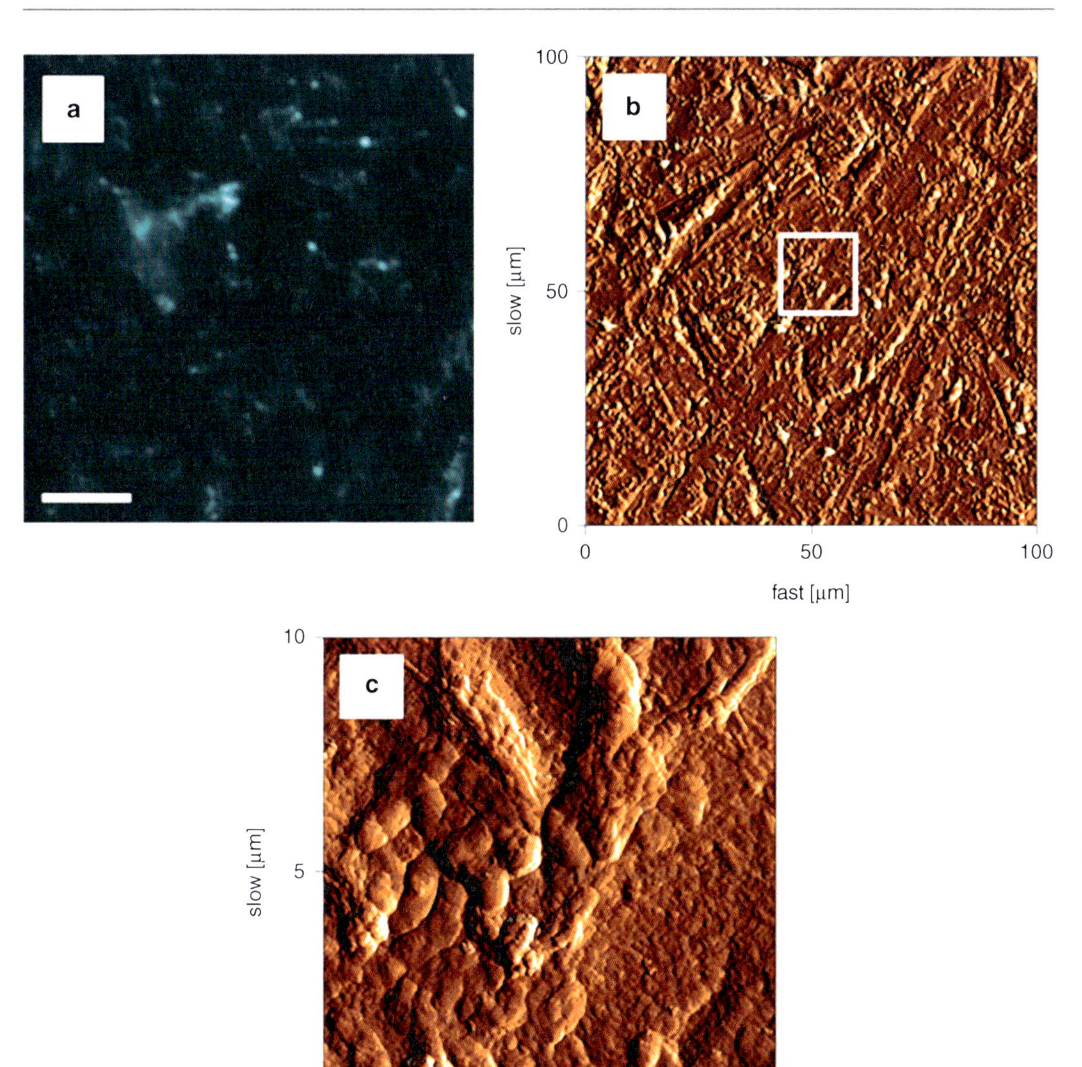

Microbiologically Influenced Corrosion, Fig. 1 Visualization of a sulphate-reducing bacterium *Desulfovibrio vulgaris* on mild steel. (**a**) fluorescence image of DNA-stained cells; (**b**) topographic AFM (atomic force microscope) image of the same area; (**c**) further magnification of white squared area from **b**. Mild steel coupons were incubated under static conditions for 3 days under anaerobic conditions, 37 °C, medium Postgate **c**

magnitudes higher than expected under these conditions. It is acknowledged that sulphate-reducing prokaryotes (SRP) are largely responsible for this phenomenon. Figure 1 shows an illustration of cells of the sulphate-reducing bacterium *Desulfovibrio vulgaris* on the surface of mild steel, visualized by two different combined techniques.

Despite the high number of literature available, the exact mechanism is still subject to discussion. The first description of corrosion by sulphate-reducing bacteria was published already in 1934 by von Wolzogen-Kühr and van der Vlugt. They investigated pitting corrosion on buried cast iron pipes and proposed that SRB are responsible by removing adsorbed hydrogen from the metal surface via their enzyme hydrogenase and using it for the reduction of sulphate according to the following equations:

$$4Fe \rightarrow 4Fe^{2+} + 8e^{-} \text{(anodic reaction)}$$

$$8H_2O \rightarrow 8H^{+} + 8OH^{-} \text{(dissociation of water)}$$

$$8H^{+} + 8e^{-} \rightarrow 8H_{ad} \text{(cathodic reaction)}$$

$$SO_4{}^{2-} + 8\ H_{ad} \rightarrow S^{2-} + 4H_2O \times \text{(cathodic depolarization by SRP)}$$

$$Fe^{3+} + S^{3-} \rightarrow FeS \text{(corrosion products)}$$

$$3Fe^{3+} + 6OH^{-} \rightarrow 3Fe(OH)_2 \times \text{(corrosion products)}$$

$$4Fe + SO_4{}^{2-} + 4H_2O \rightarrow 3Fe(OH)_2 + FeS + 2OH^{-} \text{(overall reaction)}$$

This reaction of the bacteria with the material was referred to as "depolarization." This term underlined that there was an undefined change in the electrochemical response of the system to the presence of the bacteria. The microbial cathodic depolarization represents an alternative path to the two classical mechanisms of hydrogen evolution: chemical desorption or electrochemical desorption. In both hydrogen ions first react with electrons to form adsorbed hydrogen (H_{ad}). In the former mechanism, two atoms of adsorbed hydrogen (H_{ad}) then react with each other to form molecular hydrogen (H_2). In the latter mechanism, H_{ad} reacts with a hydrogen ion (H^+) and one electron to again form H_2. In both cases, the desorption steps are rate-determining as an increased activation energy is required for these reactions compared to the discharge of hydrogen ions.

The classical depolarization theory is highly disputed due to the fact that only hydrogenase-active sulphate reducers would be able to increase corrosion rates under anaerobic conditions. However, several studies demonstrated that also the presence of sulphate-reducing bacteria possessing no hydrogenase significantly increased corrosion rates. Thus, the role of hydrogenases remains questionable. Additionally, other important factors are not taken into account in this classical theory such as the stimulation of the anodic and cathodic reaction by metabolically formed sulphide species. Also, the effect of elemental sulphur, fluctuations in environmental conditions, and the production of other corrosive metabolites are not considered. As a result, several alternative mechanisms to the depolarization theory have been proposed. As a detailed discussion of these mechanisms would go far beyond the scope of this essay, they are briefly summarized in Table 2.

Microbiologically Influenced Corrosion, Table 2 Mechanisms alternative to the classical depolarization theory [6]

Name referred to in literature and references	Main mechanisms	Role of hydrogenase
Depolarization by iron sulphide	Formation of an iron/iron sulphide galvanic cell, iron sulphide acting as the site for cathodic reduction of molecular hydrogen	Secondary through the regeneration of ferrous sulphide
Depolarization by hydrogen sulphide	Cathodic reduction of microbially produced hydrogen sulphide	Secondary through the production of hydrogen sulphide
Elemental sulphur	Formation of a concentration cell with elemental sulphur acting as the reactant	Secondary through the production of elemental sulphur
Iverson's mechanisms	Production of a volatile and corrosive iron phosphate metabolite	Not defined
Local acidification of anodes	Localized acidification of anodes due to the formation of iron sulphide corrosion products	None

Recently, another mechanism for MIC by sulphate-reducing bacteria was postulated by Dinh and coworkers. It was demonstrated that two marine *Desulfobacterium*-like strains possess the ability to directly extract electrons from metallic iron without the need of free hydrogen as a mediator. Presumably, electrons are scavenged directly by an electron uptake system transferring them to the sulphate reduction enzymes [2]. Possibly, the electron transport is mediated by extracellular polymeric substances. This model with a direct extraction of electrons from metallic iron may overcome the problems encountered with the involvement of hydrogenases and explain MIC of metals by SRP.

Corrosion Under Aerobic Conditions

Under aerobic conditions, the formation of inorganic acids such as sulphuric acid by bacteria of the genus *Acidithiobacillus* may cause corrosion of iron. The sulphuric acid is produced by the oxidation of various inorganic sulphur compounds such as elemental sulphur, thiosulphate ($S_2O_3^{2-}$), trithionate ($S_3O_6^{2-}$), tetrathionate ($S_4O_6^{2-}$), and thiocyanate (SCN). Some *Acidithiobacilli* are able to tolerate concentrations of sulphuric acid up to 12 % as well as a pH below 1; these are conditions, under which iron and mild steel become heavily corroded.

Furthermore, the heterogeneous formation of patchy biofilms on the surface may lead to the formation of differential aeration cells. Under respiring colonies or thick parts of biofilms, anodic anaerobic zones will form, whereas at cell-free areas cathodic oxygen-rich areas will be generated. Often, iron-oxidizing bacteria such as *Gallionella*, *Crenothrix*, and *Leptothrix* are associated with this form of corrosion, especially on internal surfaces of water pipes. Corrosion by these bacteria is indicated by the formation of so-called tubercles. They are formed due to the oxidation of ferrous into ferric compounds by microorganisms and the resulting heavy deposition of ferric oxide or hydroxide deposits around microbial colonies.

Finally, degradation of corrosion inhibitors such as aliphatic amines, nitrite, and phosphate-based corrosion inhibitors may result in a damage of iron and mild steel. This mechanism is frequently found in steel pipes in cooling water systems. It results not only in a higher demand for corrosion inhibitors but also in increased bacterial cell numbers due to the constant feed of nutrients and, thus, severe corrosion (if the system is not controlled properly).

Interactions Between Anaerobic and Aerobic Populations

So far, microbial corrosion mechanisms of iron and mild steel have been divided strictly into anaerobic and aerobic environments. However, in biofilms consortia of both aerobic and anaerobic microorganisms may coexist due to the formation of oxygen-free as well as oxygen-rich zones within the biofilm. Additionally, zones with varying oxygen content such as tidal areas will allow to proliferate both aerobic and anaerobic growth. Especially the synergistic interaction of anaerobic sulphate-reducing bacteria with aerobic sulphuric acid-forming *Acidithiobacilli* may lead to severe corrosion damages. This synergistic effect was first described for the corrosion of concrete sewer pipes [4] and later also for harbor steel pilings in tidal areas.

Stainless Steel and Titanium

In many industrial applications ranging from cooling systems in chemical process industries to fresh-water storage and circulation systems in nuclear power plants, MIC of stainless steels has been reported. Unfortunately, most reports do not focus on the role of microorganisms and, consequently, the involved species remain unidentified. Nevertheless, three common features were found in almost all cases. First, MIC is usually reported for alloys containing relatively low amounts of molybdenum. Second, pitting or crevice corrosion generally occurs at or around heat-affected zones of welds and, finally, localized deposits are found in close proximity to the corrosion sites. Not only changes in surface roughness but also in chemical composition of the material at heat-affected zones may explain their increased susceptibility to MIC, as evidenced by reports stating either an increased sensitivity of only the delta-ferrite phases in duplex welds or both, the austenite and ferrite phases, to microbial corrosion. To reduce the risk of MIC, proper surface treatment of welds such as solution annealing and pickling as well as polishing and grinding is a requirement.

Aluminum and Aluminum Alloys

Generally, MIC on aluminum and its alloys is evident in the form of pitting corrosion. It has mainly been reported for fuel tanks of aircrafts and heat exchanger systems with different salinities.

Despite the fact that cases of MIC on aluminum alloys are by far not as well documented as their counterparts on iron and mild steel, possible mechanisms and responsible microorganisms have been identified for MIC of aircraft fuel tanks. Among others, the most predominant species were belonging to the genera *Pseudomonas*, *Aerobacter*, *Desulfovibrio*, and the fungus *Cladosporium*. The corrosion was mainly found in the water phases at the fuel-water interphase. The same mechanisms as already described for MIC of iron and mild steel were identified, comprising the production of corrosive metabolites such as organic acids and hydrogen sulphide, the creation of differential aeration cells, the degradation of the corrosion inhibitor nitrate to nitrite, and finally the removal of metallic atoms from the alloy's basic structure.

Copper and Copper Alloys

MIC of copper and its alloys is not very well documented, probably due to the general belief that copper is toxic to microorganisms. However, many bacterial species are able to withstand high concentrations of copper or to protect themselves against toxic levels by entering the so-called VBNC (viable-but-not-culturable) state. In the latter case, it is not possible to detect these organisms by classical culture methods, but the cells are able to rapidly regrow when copper levels fall below a level, where no toxic effect is observed anymore.

Nevertheless, some case histories have been reported, especially for piping systems and heat exchangers. In the latter, biofilms are known to cause heat transfer problems. Remarkably, corrosion rates increase, when cells inside the biofilms die, possibly due to the production of ammonia and carbon dioxide, which may result in pitting and stress corrosion cracking of copper and its alloys.

Also sulphate-reducing bacteria may cause increased pitting of copper and copper alloys by the production of hydrogen sulphide. The latter forms thick, nonadherent layers of chalcocite (Cu_2S) or covellite (CuS_{1-x}). When the copper sulphide film is partially removed, regions with intact CuS films act as cathodes eventually leading to pitting corrosion of the copper or copper alloy.

Another major field of application for copper (and, thus, for its corrosion) are drinking water systems. Corrosion of copper pipes in potable water installations has been reported in several regions throughout the world. The release of copper ions and, thus, levels exceeding values permitted by drinking water regulations are of great concern. Generally, biofilms in drinking water systems are observed under deposits of copper (II) oxide and basic copper salts. Although a clear correlation between pitting corrosion of copper and microbial activity exists, the exact mechanism remains unknown. It is proposed that in a biofilm copper(I) oxide may be oxidized to copper(II) oxide. In a comprehensive, multidisciplinary approach, Vargas and coworkers [7] were able to refine the role of biofilms in copper corrosion in potable water systems. They identified malachite, a copper hydroxide carbonate with passivating properties, as the main corrosion by-product formed during simulation of MIC of copper in drinking water. Within the malachite scales, pits were observable. The morphology of malachite formed around the biofilms strongly differed from abiotically formed malachite scales. This structure alteration is assumed to be induced by microbial EPS. Additionally, soluble copper was accumulated by mineralization and sorption of copper within the EPS matrix of the biofilm. The sorption capacity of biofilms was found to be up to 18 mg Cu per g of dry biomass. Under turbulent flow, copper was found to be released from the biofilm into the bulk phase, increasing the risk of exceeding limit values of copper for drinking water. Furthermore, biofilms strongly increased the rate of copper oxidation. Usually, the copper surface is passivated by accumulation of corrosion by-products, which controls the release of ions into the bulk phase. Microbial activity leads to local acidification and, thus, dissolution of the passivating scales such as tenorite and malachite. Finally, dissolved oxygen measurements under biotic and abiotic conditions proved that biofilms induce metallic copper oxidation [7]. Additionally, biofilms may both influence bicarbonate levels of the water as well as a shift of chloride/sulphate ratios towards higher chloride levels, increasing both risk and rate of pitting corrosion of copper and its alloys.

Identification of MIC

Identification of MIC as source for an unpredicted/unexpected corrosion damage is often problematic. In most cases, damage symptoms are easily explainable by classical and for maintenance personnel more familiar electrochemical or physical mechanisms. Moreover, most engineers and technicians who are assessing corrosion damages were never introduced to the possibility that microbes may act as causative agent. Additionally, microorganisms are often not taken seriously due to their (small) size. How should these small organisms be able to damage large constructions? Hence, sediments, deposits, or slimy layers tend to be either ignored or neglected as interfering factors, when it comes to the assessment of a damage. The presence of microorganisms, although ubiquitous in nature, seems to be neglected in technical systems.

Keeping in mind the broad variety of microbial diversity described in the beginning of this article, microbial life thrives as long as traces of water are present. Thus, virtually every (technical) system will sooner or later be colonized by microorganisms. This is also true for systems with temperatures ranging from $-10\ ^{\circ}C$ up to $114\ ^{\circ}C$, under highly alkaline or acidic conditions (pH 0–14), extreme pressures of up to 1,000 bar like they occur around deep sea vents, in the presence or absence of oxygen, toxic, mutagenic, or carcinogenic substances, and even under strong ionizing radiation as they exist in storage basins of nuclear power plants. Thus, under all these conditions microbes may occur and be involved in corrosion damages.

When assessing a corrosion damage, a few special markers may give a strong hint on the influence of microbes and, thus, MIC:

1. Slimy layers on the material surface; this can be tested quite easily with the finger. The removal of some slime and burning it off with a fire lighter might cause the smell of burned hair, indicating the presence of proteins.
2. The smell of mud or rotten eggs (H_2S).
3. (Dis)coloration of materials.
4. High water content and a soft nature of deposits.
5. Crevice or pitting corrosion, which often results from microbial activity.

Summarizing, it can be concluded that almost all materials are vulnerable to MIC. Metals such as mild and stainless steel, aluminum, copper, and their alloys; mineral materials such as concrete, brick, stone, ceramics, and glass; natural organic materials such as wood, paper, leather, textiles, starch, and polymers; synthetic organic materials such as plastics, paints, coatings, glues, and polymers; and even hydrocarbons such as oils, waxes, tar, lubricants, fats, and greases may be affected by microorganisms. Only some high molecular, organic polymers seem to be not biodegradable, because microorganisms lack the necessary enzymes to biodegrade them. Additionally, some nondegradable chemically synthesized compounds called xenobiotics exist (often pesticides, antibiotics, etc.).

It cannot be emphasized enough that MIC may play an important role in many cases of (bio) corrosion and biodeterioration. To allow a proper identification of MIC, it is crucial to check the aforementioned points and to consult experts. The corrosion site should be left untouched and under the same conditions at which the corrosion occurred and a cleaning of corroded areas must be avoided. Additionally, good documentation including photographs, videos, sample data, materials, and history is necessary. All this is needed to allow a proper diagnosis of causative microorganisms in order to establish the necessary countermeasures; else, there is a good chance that there will be another failure at the same or neighboring sites in the near future.

Countermeasures Against MIC

The ideal approach to prevent or minimize microbiologically influenced corrosion depends on many factors. These encompass the environment, where MIC occurred (e.g., soil, cooling water, seawater), the type of material damaged, as well as the type of microorganism involved. In practice, several different approaches are usually combined to increase their efficiency. Below, some of the most effective approaches are described briefly. Again, it should be stressed how important a proper identification of MIC and involved microbes is for the choice of effective countermeasures.

Modifying the Material

If technically possible and economically feasible, one simple method to prevent MIC is to choose materials which are not or less susceptible to a microbial attack, e.g., the replacement of mild by stainless steel.

Modifying the Environment

Modifications of the environment such as the avoidance of anaerobic zones or pH control to prevent acid accumulation may help to prevent both classical and microbial corrosion. A simple example is the use of sand, gravel, or chalk as nonaggressive backfill around buried steel pipes, providing improved drainage and thus preventing the formation of anaerobic zones in the surrounding soil. Also, stagnant conditions should be avoided in water systems, as they may lead to an accumulation of aggressive substances or nutrients; increasing the flow velocity will similarly help. The latter methods will also help to reduce the growth of microorganisms and a formation of detrimental biofilms.

Reducing the number of microbes present in a certain system will usually be a good idea. As bacterial life depends on several factors, these provide a good starting point for microbiological methods. Variation of pH, oxygen concentration, temperature, or light conditions may be used to reduce microbial growth. However, care must be taken not to induce changes in the microbial population, which may result in a growth of other detrimental organisms.

Furthermore, some of these countermeasures may increase an electrochemical or a physical corrosion of materials. An increase in flow velocity, for example, may increase erosion of constructional materials, whereas altering the pH may cause corrosion of copper or aluminum alloys. Thus, benefit and potential cost should be evaluated carefully before choosing a certain countermeasure.

Coatings

Organic coatings are widely used for the protection of metals in various technical environments. Yet, one has to keep in mind that all paints themselves are more or less biodegradable and/or may contain biodegradable additions. Additionally, surface preparation is crucial to prevent needles, pores, or voids, where microorganisms could grow, causing highly localized corrosion.

Cathodic Protection

Cathodic protection is a classical method, which is used frequently in case of buried steel pipes, reinforced concrete structures, and marine installations. To prevent anaerobic corrosion of iron by SRP, the potential of the structure is depressed to at least −1,000 mV versus $Cu/CuSO_4$ compared to −850 mV versus $Cu/CuSO_4$, which is usually recommended in the absence of SRP. Cathodic protection may be used in combination with organic coatings. On the one hand, coatings reduce the necessary protection current by several orders of magnitude as compared to uncoated metals, and on the other hand, cathodic protection will efficiently prevent corrosion damage due to coating defects.

Biocides

Biocides are commonly used for many industrial applications, preferably in closed-circuit systems such as cooling water systems. Biocides may be divided into two main categories: (a) oxidizing agents such as chlorine, ozone, and hypochlorite and (b) non-oxidizing agents such as isothiazolines, aldehydes, chlorophenols, and quaternary ammonium salts.

However, microorganisms may adapt to biocides in different ways including the production of biocide-degrading enzymes or changes in the respective target structures inside the cell or the cell wall. To prevent this adaptation, several different biocides are generally used alternatively. One major problem with biocides is, although they are highly effective against planktonic cells (i.e., free microbes in the water phase), that they are much less effective against microorganisms in biofilms. Often, the dosage needs to be increased up to 100-fold to kill the majority of biofilm organisms. Also, non-oxidizing biocides do not remove existing biofilms, but only kill the microbes. Hence, the remaining microorganisms will find a nutrient-rich environment (dead cells), allowing rapid regrowth. Consequently, a repeated frequent treatment is necessary. Besides, the use of biocides is restricted more and more by legislation. Several highly effective biocides will no longer be available for treatment of technical systems in the near future, e.g., due to the European Biocidal Products Directive (BPD).

Physical Methods

Physical methods against MIC include measures such as filtration of water, mechanical biofilm removal (e.g., pigging of pipes), or the use of ultraviolet (UV) radiation. These methods have proven to be useful for specific applications. For example, UV light sterilization has been used in drinking and cooling water systems, replacing the otherwise commonly used chlorination. However, these methods are generally less efficient; for instance, there is no persistence of disinfecting agents in the water system when comparing UV treatment to chlorine addition. Moreover, these methods are costly and an addition to already existing technical systems may be challenging (due to some material's susceptibility to such compounds).

Simulation

Depending on the environment and the type of MIC damage, a simulation of such environments in controlled chambers may help to select suitable countermeasures and materials. Based on the conditions on site as well as the biological consortium in a respective system or damage site, the corrosive effect can be accelerated to test materials for their suitability/resistance under given conditions. Moreover, simulations may be used to develop novel resistant materials. One particularly successful example, where simulation is used, is the simulation of biogenic sulphuric acid attack on mineral materials such as concrete or mortar as well as synthetic metal coatings. Also biogenic nitric acid attack on cement-bound concrete and natural stone has been simulated successfully. However, it must be noted that the design of a proper simulation needs profound knowledge of all processes involved, be they of physical, chemical, or biological nature. If these are well known, conditions optimized for the microorganisms may be created in order to reduce the time span needed for the material's failure to become detectable.

In actual simulations of the biogenic sulphuric acid corrosion on sewage pipelines, which have been described extensively (e.g., [5]), a proper design helped to reduce the time span, until the material failure became obvious, from 8 to 1 year. During these tests it also became obvious that biogenic sulphuric acid attack was stronger/more effective than chemical testing using the same acid. This clearly demonstrates that, even if the damage mechanism (i.e., excretion of sulphuric acid in this particular example) is known, the actual microbial interaction with the material is of utmost importance and cannot be incorporated into purely chemical/physical test methods. Thus, biological testing cannot be replaced by chemical and/or physical simulations using sterile conditions.

Future Directions

This essay is a brief overview mainly based on the more elaborate articles of Sand [3] and Thierry and Sand [6]. We strongly recommend these articles for further reading.

Cross-References

- ▶ Biocorrosion
- ▶ Microbiologically Influenced Corrosion
- ▶ Microbiologically Influenced Corrosion Inhibition

References

1. Baillie GS, Douglas LJ (1998) Iron-limited biofilms of *candida albicans* and their susceptibility to Amphotericin B. Antimicrob Agents Chemother 42(8):2146–2149
2. Dinh HT, Kuever J, Mußmann M, Hassel AW, Stratmann M, Widdel F (2004) Iron corrosion by novel anaerobic microorganisms. Nature 427:829–832
3. Sand W (2001) Microbial corrosion and its inhibition. In: Rehm H-J, Reed G (eds) Biotechnology, 2nd edn. Wiley, Weinheim, pp 265–315
4. Sand W, Bock E (1984) Concrete corrosion in the Hamburg Sewer system. Environ Technol Lett 5(12):517–528
5. Sand W, Bock E, White DC (1987) Biotest system for rapid evaluation of concrete resistance to sulfur-oxidizing bacteria. Mater Perform 26(33):14–17
6. Thierry D, Sand W (2011) Microbially influenced corrosion. In: Marcus P (ed) Corrosion mechanisms in theory and practice, 3rd edn. CRC Press, Boca Raton, pp 737–776
7. Vargas I, Pizarro G, Felis M, Corthorn J, Acevedo S (2012) Multi-technique approach to understand the effects of bacterial communities involved in copper plumbing corrosion. Proceedings EUROCORR 2012. Istanbul, Turkey, pp 9–13

Microbiologically Influenced Corrosion Inhibition

Andrzej Kuklinski and Wolfgang Sand
Fakultät Chemie, Biofilm Centre/Aquatische Biotechnologie, Universität Duisburg-Essen, Essen, Germany

Synonyms

Microbial inhibition of corrosion

Introduction

Microbiologically influenced corrosion inhibition (MICI) represents the counterpart to microbiologically influenced corrosion or biocorrosion (MIC). Whereas in the latter case materials are damaged due to detrimental effects exerted by microorganisms or their metabolic products, the former describes the phenomenon of (bio)corrosion mitigation by direct or indirect action of microorganisms. Although it could be deduced from its name, MICI is not restricted to the protection of materials against microbial attack but also to protect against environmental influences. Primarily, the term is used in conjunction with classical electrochemical corrosion of metals and alloys such as iron, steel, copper, and aluminum. In the broadest sense, it could be extended to the protection of materials such as concrete or plastics against biodeterioration. However, this would go far beyond the scope of this entry.

The mechanisms involved in corrosion mitigation by microorganisms (or their biofilms) are substantially more complex than the ones involved in corrosion protection by "classical" inhibiting agents such as nitrite. The latter are usually highly specific for defined materials and environments and may be described by single mechanisms (e.g., the slowdown of anodic dissolution processes in case of nitrite). Furthermore, these agents are often environmentally detrimental compounds. In contrast, the mechanisms involved in MICI are as diverse as the involved microorganisms, environments, and materials. To understand the complexity of processes, an understanding of the complexity of biofilms is crucial. Biofilms in nature are highly diverse consortia of different microbial species, embedded in a matrix of extracellular polymeric substances (EPS). The EPS are composed of polysaccharides, lipids, proteins, uronic acids, and nucleic acids. Inside the biofilm, oxic and anoxic zones and several synergistic effects, i.e., effects, where all involved microorganisms profit, might occur. For example, a full sulfur cycle with aerobic and anaerobic microorganisms/metabolic types could be established within one single biofilm. In case of metal corrosion, all these biofilm characteristics influence biotic and abiotic processes within the biofilm itself or at the biofilm-material interface. Consequently, the same must be valid for the (bio)corrosion-counteracting effect of MICI. Additionally, several different species and genera from various environmental sources seem to have protective properties for metals. In many cases, it is not possible at present to deduce single

Microbiologically Influenced Corrosion Inhibition, Table 1 Mechanisms involved in MICI

Group	Mechanisms	References
Neutralization of corrosive agents	Oxygen consumption	[19, 25]
	Competition for electron donors	[14]
	Secretion of corrosion inhibitors	[10, 25]
Formation of protective films	Diffusion barriers	[10, 19, 22]
	Passivating (oxide) films	[23, 25]
	Induction of surface composition changes	[2]
	Prevention of microbial adhesion and biofilm formation (also by excretion of antimicrobials)	[10, 19, 25]

electrochemical mechanisms involved in MICI to certain species of microorganisms.

A more reasonable approach to MICI is to divide the involved mechanisms into groups of common characteristics. Zuo [25] proposed a division into three major groups: (a) removal of corrosive agents, (b) inhibition of cell adhesion or biofilm growth of detrimental microbes, and (c) formation of protective layers. Later, Videla and Herrera [22] suggested to classify the mechanisms according to only two major groups: (a) neutralization of detrimental effects of corrosive agents and (b) formation or stabilization of protective films on the material, with the latter comprising formation of passive layers protecting the material from dissolution as well as protective (bio)films inhibiting microbial adhesion. Table 1 summarizes some of the most frequently mentioned mechanisms involved in MICI.

A sharp distinction between the two groups seems not feasible, given that some protective mechanisms could be assigned to both of them. Generally, both groups might – to a certain extent – be distinguished by the fact, whether intimate contact between the microorganisms (biofilms) and the surface is necessary or not. Active processes such as oxygen consumption (first group) inevitably need living (respiring) microorganisms in close proximity to the metal surface. This is not an indispensable prerequisite for the formation of protective films (second group). Metabolic substances excreted by microorganisms may already be sufficient to lead to a significant decrease in corrosion rates [19]. However, an increased protective effect will be achieved, when the cells are directly on the surface. As stated before, MICI effects in natural environments are unlikely to be caused by single mechanisms. For instance, living biofilms might decrease the oxygen content by active respiration, act as a diffusion barrier further reducing the oxygen content, and at the same time produce corrosion inhibiting substances.

Protective Mechanisms

Protective effects of microorganisms have been described in several case studies in recent years. This entry will provide a brief overview about the most important mechanisms described in Table 1 and the involved microorganisms. For a more comprehensive overview, several specialized reviews are recommended such as Little and Ray [10], Zuo [25], and Videla and Herrera [22].

Oxygen Consumption

Both living biofilms and their products alone may act as diffusion barriers against corrosive agents (Fig. 1). It was demonstrated that biofilms of *Pseudomonas* sp. S9 and *Serratia marcescens* were able to strongly reduce the corrosion of mild steel under aerobic conditions. Measurements showed that the concentration of oxygen in the bulk solution was significantly higher than within the biofilm. This leads to a decrease in the cathodic reaction rate due to the removal of the cathodic reactant oxygen [21]. Another study by Jayaraman et al. [8] demonstrated that already a thin layer of microorganisms on the surface is sufficient for corrosion protection. They tested seven different bacterial genera and all of them exhibited protective effects. However, the extent of the protective effect varied greatly between the different species. In both studies, inactivating the biofilm cells (e.g., by glutardialdehyde)

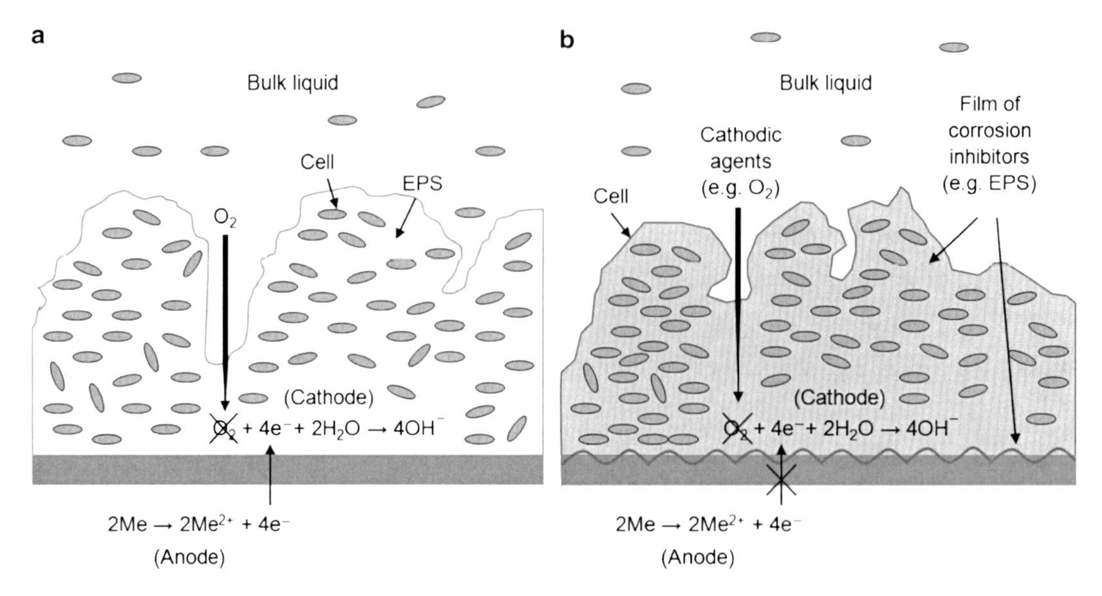

Microbiologically Influenced Corrosion Inhibition, Fig. 1 Illustration of proposed mechanisms of microbiologically influenced corrosion inhibition (modified from [25]). (**a**) Decrease in oxygen concentration near the metal surface by bacterial respiration and diffusion limitation. (**b**) Secretion and accumulation of corrosion inhibitors on the metal surface (e.g., EPS)

diminished the protective effect, underlining the need of living, respiring microorganisms for this type of protection.

Competition for Electron Donors

Microbial dissimilatory Fe(III) reduction may successfully compete with various detrimental bacteria for important substrates such as hydrogen and acetate in anaerobic eco- or technical systems. By removing these substrates, Fe(III) reducers may reduce growth of known MIC-causing microorganisms like sulfate reducers, methanogens, denitrifiers, and fermentative species [14]. The protective effect was demonstrated under both aerobic and anaerobic conditions.

Under aerobic conditions, the reducing power from the substrate degradation is fed into the respiratory chain. The subsequent decrease in oxygen leads to a drop in the cathodic partial reactions and thus decreases the rate of metal dissolution. This effect can be observed electrochemically by a drop in E_{oc} values, which eventually stabilizes due to a decreasing electron flow. In terms of mass loss, up to 96 % less mass loss was observed for mild steel compared to sterile controls. To further elucidate the processes involved in metal protection by iron-respiring bacteria under aerobic conditions, Dubiel and co-workers [5] used mutants *Shewanella oneidensis* MR-1 unable to develop biofilms and/or to reduce iron. It was found that the protective effect is based on a reduction of Fe(III) (ferric) ions to Fe(II) (ferrous) ions in combination with increased oxygen consumption as a direct result of microbial respiration, supporting the theory of Pothekina et al. [14].

Under anaerobic conditions, it is assumed that the protective effect is linked to the ability of microorganisms to transfer electrons and protons to and from the metal surface. This may lead to the formation of a passivating layer of adsorbed atomic hydrogen (H^0) on the surface. In contrast to detrimental organisms such as SRB, these bacteria act as anode and the metal as cathode. It was found that SRB develop very slowly under such conditions [14]. However, the exact mechanism of electron transfer between cells and metal surfaces is still not fully understood [1].

From a practical perspective, organotrophic Fe(III)-reducing bacteria from four different genera (*Pseudomonas*, *Micrococcus*, *Arthrobacter*, and *Vibrio*) were able to reduce MIC by (mainly) sulfate-reducing bacteria in non-sterile industrial systems. It was found that these organisms were not only able to remove corrosion products but also to protect the metals against further corrosion. It is assumed that in this case the removal of corrosion products may destroy ecological niches, which the sulfate-reducing bacteria need to proliferate [14].

Secretion of Corrosion Inhibitors

Some bacteria are able to produce corrosion inhibitors such as siderophores, polyaspartate, or gamma-polyglutamate. Siderophores are iron chelators formed by microorganisms at neutral pH. Siderophores have been demonstrated to cause ennoblement of stainless steel in marine conditions. Thereby, the ennoblement was strongly dependent on metabolic activity of the microorganisms. It is assumed that the excretion of siderophores leads to a strengthening of the passivating film on the steel surface, caused by the intrinsic property of the alloy to profit from the presence of the inhibitor [10].

Several cases of corrosion inhibition by polyaspartate and gamma-polyglutamate have been compiled by Zuo [25] in a review: Genetically engineered biofilms of gamma-polyglutamate-excreting *Bacillus licheniformis* were able to reduce corrosion of aluminum 2024 by up to 90 % in continuous reactor experiments. However, living biofilms were not necessary as the pure substance alone was able to reduce corrosion as well. In contrast, 20-amino-polyaspartate secreted by genetically engineered *Bacillus subtilis* was only able to slightly reduce pitting of aluminum compared to non-polyaspartate-excreting *B. subtilis*. However, the latter already caused a significant reduction of corrosion compared to sterile controls. The inhibitory effect of polyglutamate and aspartate is probably caused by carboxylic aluminum-chelating groups. Although both substances are able to chelate aluminum, they still show a different protective effect. The lower efficacy of polyaspartate may be related to its structural nature and/or the concentration of the respective anionic peptide in solution.

Diffusion Barriers

Extending beyond the aforementioned active removal of oxygen, biofilms may protect materials against a variety of other corrosive agents such as acidic compounds or chloride ions. In contrast to oxygen consumption, living cells are not always necessary; their metabolic products such as extracellular polymeric substances (EPS) alone may already be sufficient to exert protective effects in some cases.

Biofilms of – usually detrimental – sulfate-reducing bacteria were found to be able to protect steel against chloride attack in near-neutral solutions containing 0.1 M (approximately 0.6 %) NaCl [24]. This (unexpected) protective effect is probably related to a separation of the aggressive chloride ions from the metal surface. In this case, the biofilms probably act as a passive diffusion barrier.

In case of acidic compounds, living biofilms or pure, cell-free EPS may counteract their corrosive effect. Biofilms may especially help in highly aggressive sour acidic environments as found in oil fields or offshore platforms, where, additionally, hydrogen embrittlement is a serious problem. The high amounts of hydrogen available directly at the metal surface will result in cracks caused by sulfide poisoning of the hydrogen recombination reaction at the cathodic surface. These effects may be drastically reduced in the presence of an organic film such as bacterial EPS at the metal surface [22].

Passivating Films

Rhodococcus sp. strain C125 and *Pseudomonas putida* Mt2 are able to form vivianite ($Fe_3[PO_4]_2$) films on non-alloyed steel under aerobic conditions, if sufficient amounts of phosphate are present in the bulk solution. This vivianite layer is passivating the steel surface and, thus, able to stop the corrosion process entirely. Additionally, these bacteria are able to re-passivate existing damaged vivianite layers [23].

Induction of Surface Composition Changes

Bacterial biofilms are able to protect materials by inducing changes in the chemical composition of the steel surfaces by excreting so-called biosurfactants. Biosurfactants are surface-active biomolecules, which are able to reduce surface and interfacial tensions to facilitate microbial attachment. They are produced by a large variety of microorganisms in an even larger number of different environments. Consequently, chemical structure and surface properties depend on several different factors such as the type of excreting microorganism, available nutrients, and temperature. Meylheuc and co-workers [11] assessed two biosurfactants produced by *Lactobacillus helveticus* 1181 and *Pseudomonas fluorescens* 495. They found that the surfactants influenced the surface of stainless steel in different ways. Generally, both surfactants produced a more or less homogenous organic layer on the steel surface. More important, however, was the finding that both different surfactants induced a segregation of chromium towards the surface, increasing the resistance of the stainless steel. Additionally, the formation of biofilms by *Listeria monocytogenes* was significantly reduced, probably by modifying the acid–base characteristics of the solid surface. This combined mode of action of biosurfactants clearly demonstrates that similarly to MIC, protective effects of microorganisms are usually not caused by a single mechanism but rather a combination of different mechanisms.

Prevention of the Formation of Detrimental Biofilms

For protection against MIC, microorganisms or microbial products able to reduce the formation of detrimental biofilms are of special interest. This includes growth inhibition of detrimental microorganisms as well as the prevention or reduction of initial adhesion of these microbes (and thus a reduction of biofilm formation).

Growth inhibition is often linked to the production of antimicrobial substances such as Gramicidin-S either natively or after genetic engineering. For example, *Bacillus brevis* biofilms were able to significantly reduce the corrosion rates of mild steel caused by a synergistic couple of SRB (*Desulfosporosinus orientis*) and iron and manganese oxidizers (*Leptothrix discophora*). In contrast, non-antimicrobial-producing biofilms did not show any protective effect [12]. Genetically altered *Bacillus subtilis* biofilms secreting indolicidin, bactenecin, and probactenecin were able to inhibit growth of corrosion-inducing SRB *Desulfovibrio vulgaris* and *D. gigas*. Thus, a significant reduction of corrosion rates in continuous cultures resulted [8]. This study was the first report of an in situ application of genetically modified biofilms for corrosion inhibition. In contrast to artificially dosing antimicrobials into the system, these biofilms produce antimicrobial substances, directly where they are needed, circumventing the various protection mechanisms of biofilm organisms against biocides [25]. However, living and actively respiring biofilms are necessary. The maintenance of a fully functional antimicrobial-excreting biofilm within technical systems may be a challenging task.

The second mechanism – prevention of initial adhesion of detrimental organisms – is of particular interest. A strong advantage of this approach is that detrimental organisms and/or biofilms are not fought after they already formed, which has been proven to be difficult due to their strong resistance against mechanical and chemical treatments. Living biofilms of *Staphylococcus sciuri* have been found to reduce the number of attaching *Listeria monocytogenes* cells on stainless steel by up to 3 log units compared to sterile controls [9]. The population numbers of *Listeria* in the planktonic phase were not affected, suggesting that an antimicrobial effect of *Staphylococcus* can be excluded.

Living biofilms are not a strict prerequisite for adhesion prevention. Excreted compounds such as EPS or biosurfactants alone are often inhibiting. Purified biosurfactants from *Lactobacillus casei*, *L. fermentum*, and *L. acidophilus* were able to reduce the number of pathogenic *Enterococcus faecalis* to biosurfactant-coated glass by up to 77 % compared to uncoated controls [20]. Based on the method used for extracting and purifying biosurfactants, one can assume that bacterial EPS were indeed these surfactants. This is supported by the chemical analysis of these surfactants, which revealed that they consisted mainly of proteins and polysaccharides, two of the main components of EPS. An application of EPS seems to be one of the most promising MICI-based approaches for corrosion inhibition. The artificial induction of living biofilms into technical systems is generally undesirable, as this means also an introduction of a large amount of organic mass into these systems. Additionally, as mentioned before, the maintenance of a living biofilm in a beneficial state is challenging. Several authors studied the effect of EPS on bacterial attachment. For example, Gubner and Beech [7] treated AISI 304 and 316 stainless steels with three different types of EPS: loosely bound EPS released into the medium ("colloidal"), EPS tightly bound to the cell surface ("capsular"), and biofilm exopolymers produced by continuous cultures of *Pseudomonas* NCIMB 2021. The authors demonstrated that all EPS were able to reduce attachment of the same organism to treated surfaces considerably, although to a varying degree. It was found that neither a change in surface hydrophobicity nor roughness was influencing the attachment. Consequently, the chemistry of the EPS extracts was the key parameter for attachment reduction. A similar effect was observed by de Paiva [3]: Stainless steel coated with EPS extracts from *D. alaskensis* significantly reduced attachment of *D. indonesiensis*. The attachment of *D. alaskensis*, however, was in turn facilitated by EPS of *D. indonesiensis*. Attachment reduction or increase as EPS effect may be related to the so-called footprint effect, which describes the finding that microorganisms leave polymeric substances on surfaces after their desorption. It is assumed that these remaining footprints are used to actively label surfaces. For example, *Acinetobacter calcoaceticus* uses footprints to label hydrocarbon droplets as "depleted". The exact nature of footprints remains unknown until now, but they most probably consist of cell-free EPS, exoenzymes, and surface-active compounds such as biosurfactants [13, 16].

Recent Studies

In light of the threat for a ban of several effective biocides due to the European Biocidal Product Directive (BPD), bio-derived products to mitigate MIC have gained interest in recent years. Approaches have been made to use cell-free EPS extracts or derived substances not only to reduce attachment of detrimental organisms but also to generally reduce corrosion by organisms such as SRB or Mn oxidizers. One of the most extensive studies about application of EPS was performed by Stadler et al. [18]. The authors investigated EPS extracts from 14 different microorganisms, including 3 different SRB of the genus *Desulfovibrio*, 5 different *Pseudomonadaceae*, 2 different *Lactobacilli*, *Rhodococcus opacus*, *Citrobacter freundii*, *Enterobacter aerogenes*, and *Arthrobacter* sp., which had been described as protective against MIC or general bacterial attachment in literature before. The authors identified some EPS as promising candidates for further research, including those from *Desulfovibrio alaskensis* and *D. vulgaris*. Based on their findings, Stadler and co-workers [19] investigated the influence of EPS extracts from the latter on the same organism. It was found that films of the EPS extracts on steel strongly decreased attachment of *D. vulgaris* in short time scale as compared to uncoated references. However, the crude EPS films showed a strong heterogeneity, which strongly influenced electrochemical assessment of a possibly protective effect. Dong et al. [4] isolated EPS from thermophilic SRB and assessed their influence on carbon steel corrosion. It was found that the protective effect strongly depended on the concentration of EPS added to the bulk solution (containing 3 % NaCl). Adsorbed EPS layers were able to protect the underlying steel by hindering oxygen reduction. However, it was also found that excessive amounts of EPS might be detrimental by stimulating the anodic dissolution due to the chelation of Fe(II) ions. Finkenstadt and co-workers [6] used two different strains of *Lactobacillus mesenteroides* to produce glucans from sucrose. After dip-coating SAE 1010 steel coupons in EPS solutions, they found the adherence of EPS films to be facilitated by interactions between metal ions and functional chemical groups within the EPS. It was demonstrated that corrosion inhibition was strain specific with dextran-producing bacteria and suggested that the adhesion of EPS to the surface was crucial for its protective effect. The effect was probably caused by complexation of Fe (II) and Fe(III) ions and a subsequent reduction of the amount of electron acceptors available at the interface. Generally, the application of crude EPS extracts seems not feasible due to their complex and hardly predictable chemical composition. Based on these mechanisms, protective effects of EPS and derived pure substances are currently investigated as promising alternatives.

Commercial Applications

Until now there is no large-scale application of MICI-based products known. However, Roux and co-workers [15] used sucrose-derived (1–3, 1–6)-a-D-glucan exopolysaccharides (EPS180) for protecting reinforcing steel rebars in cement under marine conditions. It was used as an admixture applied to the cement without need for coating or painting the rebars themselves. Electrochemical tests revealed the protective effect of EPS180 to be attributed to a modified rebar-cement interface rather than just clogging of the cement pores. Very recently, Scheerder et al. [17] used C6-oxidized-acetylated EPS180 as a zinc phosphate-replacing additive in styrene-acrylic polymer-based anticorrosive paint. This paint was successfully tested under maritime conditions over a period of 1 year and is now commercially available via TNO (Netherlands).

Future Directions

Microbiologically influenced corrosion inhibition seems to be a promising approach against MIC. However, there is still a (considerable) demand for extensive research. Most of the studies available are on an empirical basis, lacking a systematic background. However, current research, especially focusing on applying the principles of MICI by EPS already yielded promising results, such as the aforementioned first commercial product by TNO.

Cross-References

▸ Biocorrosion
▸ Microbiologically Influenced Corrosion

References

1. Bergel A (2007) Recent advances in between biofilms and metals. Adv Mater Res 20–21:329–334
2. Dagbert C, Meylheuc T, Bellon-Fontaine M-N (2006) Corrosion behaviour of AISI 304 stainless steel in presence of a biosurfactant produced by *Pseudomonas fluorescens*. Electrochim Acta 51:5221–5227
3. de Paiva M (2004) Bactérias redutoras de sulfato (BRS): estudo de substâncias poliméricas extracellulares (SPE) e enzimas nos processos de adesão a substratos metálicos e de biocorrosão. PhD thesis, Instituto Oswaldo Cruz. Rio de Janeiro, Brazil
4. Dong ZH, Liu T, Liu HF (2011) Influence of EPS isolated from thermophilic sulphate-reducing bacteria on carbon steel corrosion. Biofouling 27(5):487–495
5. Dubiel M, Hsu CH, Chien CC, Mansfeld F, Newman DK (2002) Microbial iron respiration can protect steel from corrosion. Appl Environ Microbiol 68(3):1440–1445
6. Finkenstadt VL, Côté GL, Willet JL (2011) Corrosion protection of low-carbon steel using exopolysaccharide coatings from *Leuconostoc mesenteroides*. Biotechnol Lett 33:1091–1100
7. Gubner R, Beech IB (2000) The effect of extracellular polymeric substances on the attachment of *Pseudomonas* NCIMB 2021 to AISI 304 and 316 stainless steel. Biofouling 15(1–3):25–36
8. Jayaraman A, Cheng ET, Earthman JC, Wood TK (1997) Axenic aerobic biofilms inhibit corrosion of SAE 1018 steel through oxygen depletion. Appl Microbiol Biotechnol 48:11–17
9. Leriche V, Carpentier B (2000) Limitation of adhesion and growth of *Listeria monocytogenes* on stainless steel surfaces by *Staphylococcus sciuri* biofilms. J Appl Microbiol 88:594–605
10. Little B, Ray R (2002) A perspective on corrosion inhibition by biofilms. Corrosion 58:424–428
11. Meylheuc T, Methivier C, Renault M, Herry J-M, Pradier C-M, Bellon-Fontaine M-N (2006) Adsorption on stainless steel surfaces of biosurfactants produced by gram-negative and gram-positive bacteria:

consequence on the bioadhesive behavior of *Listeria monocytogenes*. Colloids Surf B Biointerfaces 52:128–137
12. Morikawa M (2006) Beneficial biofilm formation by industrial bacteria *Bacillus subtilis* and related species. J Biosci Bioeng 101(1):1–8
13. Neu TR (1992) Microbial "footprints" and the general ability of microorganisms to label surfaces. Can J Microbiol 38:1005–1008
14. Potekhina JS, Sherisheva NG, Povetkina LP, Pospelov AP, Rakitina TA, Warnecke F, Gottschalk G (1999) Role of microorganisms in corrosion inhibition of metals in aquatic habitats. Appl Microbiol Biotechnol 52(5):639–646
15. Roux S, Bur N, Ferrari G, Tribollet B, Feugeas F (2010) Influence of a biopolymer admixture on corrosion behavior of steel rebars in concrete. Mater Corros 61(12):1026–1033
16. Sand W, Gehrke T (2003) Microbially influenced corrosion of steel in aqueous environments. Rev Environ Sci Biotechnol 2:169–176
17. Scheerder J, Breur R, Slaghek T, Holtman W, Vennik M, Ferrari G (2012) Exopolysaccharides (EPS) as anti-corrosive additives for coatings. Prog Org Coat 75:224–230
18. Stadler R, Fürbeth W, Harneit K, Grooters M, Wöllbrink M, Sand W (2008) First evaluation of the applicability of microbial extracellular polymeric substances for corrosion protection of metal substrates. Electrochim Acta 54(1):91–99
19. Stadler R, Wei L, Fürbeth W, Grooters M, Kuklinski A (2010) Influence of bacterial exopolymers on cell adhesion of *Desulfovibrio vulgaris* on high alloyed steel: Corrosion inhibition by extracellular polymeric substances. Mater Corr 61(12):1008–1016
20. Velraeds MM, van der Mei HC, Reid G, Busscher HJ (1996) Inhibition of initial adhesion of uropathogenic *Enterococcus faecalis* by biosurfactants from *Lactobacillus* isolates. Appl Environ Microbiol 62(6):1958–1963
21. Videla HA, Herrera LK (2004) Microbiologically influenced corrosion: looking to the future. Int Microbiol 8:169–180
22. Videla HA, Herrera LK (2009) Understanding microbial inhibition of corrosion. A comprehensive overview. Int Biodeter Biodegr 36:896–900
23. Volkland H-P, Harms H, Müller B, Repphun G, Wanner O, Zehnder AJB (2000) Bacterial phosphating of mild (Unalloyed) steel. Appl Environ Microbiol 66(10):4389–4395
24. Werner SE, Johnson CA, Laycock NJ, Wilson PT, Webster BJ (1998) Pitting of Type 304 stainless steel in the presence of a biofilm containing sulphate reducing bacteria. Corros Sci 40(2–3):465–480
25. Zuo R (2007) Biofilms: strategies for metal corrosion inhibition employing microorganisms. Appl Microbiol Biotechnol 76:1245–1253

MIEC Materials

Hitoshi Takamura
Department of Materials Science, Graduate School of Engineering, Tohoku University, Sendai, Japan

Mixed Ionic and Electronic Conductors (MIECs)

Mixed ionic and electronic conductors (MIECs) are materials in which both the ionic and electronic species carry electricity. In principle, most ionic crystals, like oxides and halides, can be regarded as MIECs since they have both ionic and electronic conductivity, albeit quite small amounts, at finite temperatures; however, for practical applications, MIECs are required to have ionic and electronic conductivities higher than approximately 10^{-4} to 10^{-3}S/cm. Because of the simultaneous conduction of ions and electrons, MIECs can be applied to a variety of solid-state ionics and electronics devices. Figure 1 depicts the many categories of solid-state ionics, solid-state electronics, and MIECs as functions of their ionic and electronic conductivities [1]; insulator and dielectric materials are those with negligible electrical conductivities; high ionic conductivity is required for the electrolytes of batteries and fuel cells while materials with moderate ionic conductivity can be used for potentiometric sensors. The application of solid-state electronics to diodes and transistors, as well as to insulators and dielectrics and superconductors, is common these days. The advantage of MIECs is they can facilitate mass transport, and this ability is driven by their electrochemical potential gradient. As such, there are high expectations that they will be used more extensively in electrodes and separation membranes in the not-so-distant future.

There are several methods available for the production of MIECs. Starting with pure ionic conductors, which are basically electronically insulative, electronic carriers (electrons and/or holes) need to be incorporated into the ionic

MIEC Materials, Fig. 1 Typical applications of MIECs [1]

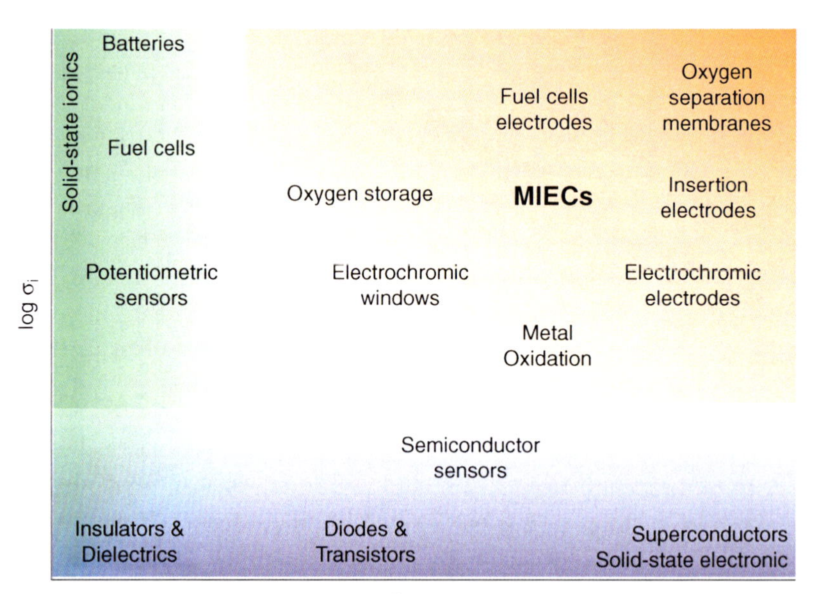

conductors. Even though the generation of electron-hole pairs by thermal and photon activation is possible, the magnitude of induced electronic conductivity is usually limited; therefore, in practice, doping and stoichiometry control are utilized to enhance electronic conductivity just as it is done in semiconductors. The engineering approach, i.e., the physical mixing of two phases, is another effective method for the preparation of MIECs. By mixing an ionic conductor phase and an electronic conductor phase, their interface behaves as an MIEC. These types of MIECs are sometimes referred to as dual-phase MIECs to distinguish them from the doped materials mentioned above. In the engineering approach, a fine microstructure, at a submicron- or nanoscale level, needs to be developed to attain high mixed conductivity since it is important to maximize the number of interfaces between the ionic and electronic phases. It should be noted that the percolation threshold limits the mixing ratio and the resultant mixed conductivity.

To date, a number of MIECs have been reported. Silver chalcogenides, such as Ag_2S and Ag_2Se, are typical MIECs with a strong research history dating back well over 70 years now [2–5]. In this material, silver cations (Ag^+) and electronic carriers (electron, holes) can simultaneously migrate. Their mixed conductivity can be modified by controlling the stoichiometry of silver and the chalcogen. Oxides with a perovskite-type structure are also widely used as MIECs. Among them, (La, Sr)(Co, Fe)$O_{3-\delta}$, which is referred to as LSCF, is well known to exhibit high mixed oxide-ion and electronic conductivity at elevated temperatures above approximately 700 °C [6]. In this material, the Sr on the La sites works as an acceptor to introduce not only electronic holes and but also the oxygen vacancies required for oxide-ion migration. In other words, the ratio of oxide-ion and electronic conductivities (i.e., the transference numbers) varies as function of temperatures and oxygen partial pressure, pO_2. In the case of LSCF, the higher the oxygen partial pressure, the higher the electronic conductivity. Among of the various mixed conductive perovskite-type oxides, $Ba_{0.5}Sr_{0.5}Co_{0.8}Fe_{0.2}O_{3-\delta}$, has the best oxygen transport property [7]. These oxide-based MIECs can be used in the cathode electrodes of solid oxide fuel cells (SOFCs). For the cathode side of SOFCs, oxygen gas has to be incorporated into an oxide-ion electrolyte. In the case of electronic conductors such as (La, Sr)MnO_3, the electrode reaction takes place only in the vicinity of the triple phase boundary (TPB) comprising of

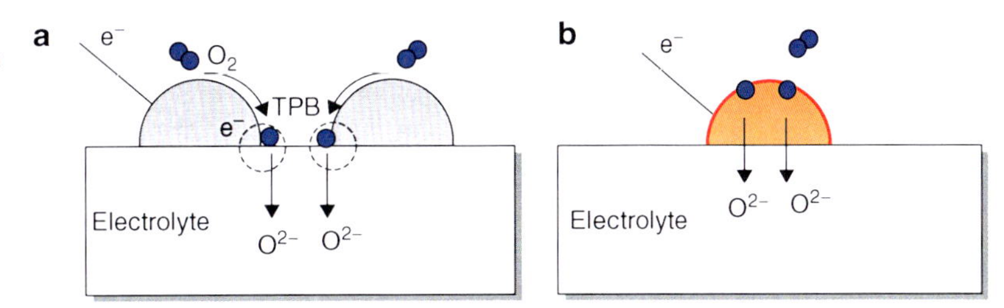

MIEC Materials, Fig. 2 Schematic diagram of SOFC cathode; (**a**) electronic conductors and (**b**) MIECs

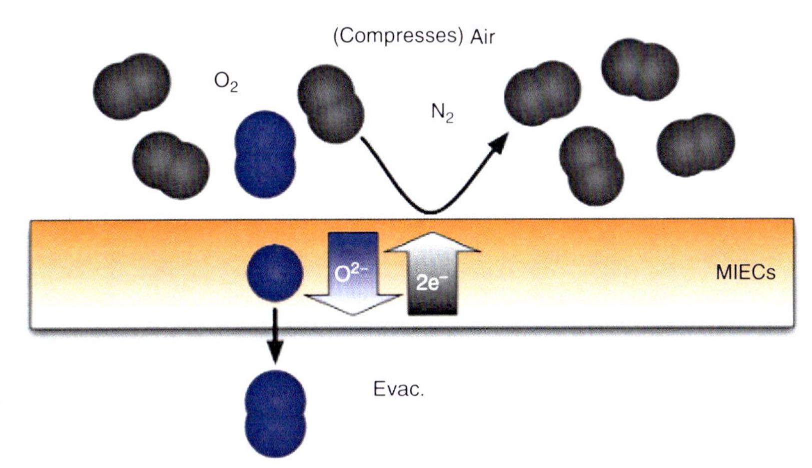

MIEC Materials, Fig. 3 Schematic diagram of oxygen permeable membranes using MIECs

the gas phase, the electrode, and the electrolyte (see Fig. 2a). The perovskite-type MIECs make it possible for the whole surface of the MIECs to contribute to the electrode reactions, including the adsorption and dissociation of O_2 molecules, and the charge transfer processes (Fig. 2b) resulting in lower cathodic polarization.

In addition to this, the oxide-based MIECs can be applied to oxygen separation from air. As shown in Fig. 3, because of the simultaneous conduction of oxide-ion and electrons, oxygen can be transported from the high to low pO_2 side just by using the pO_2 gradient. This type of oxygen separation membrane can be used not only for pure oxygen production but also as a membrane reactor, for example, producing hydrogen from hydrocarbons [8].

Future Directions

In the prospective future, it can be expected that MIECs will be further integrated into nanoscale devices. One example is an atomic-scale switch utilizing silver chalcogenides [9]. The use of Ag_2S will make it possible to precisely control the amount of metallic silver between two electrodes at an atomistic scale in effect functioning as a quantized conductance atomic switch (QCAS). Another MIEC, TiO_2, has been utilized to develop a memristor (memory-resistor), in which the migration of oxide-ions and electrons plays an important role [10–12]. In addition, nanotechnology can be utilized to control mixed conductivity itself. In the case of ionic crystals, the concentration of mobile species in the vicinity of the grain boundaries and interfaces can be controlled by the space-charge effects. The enhancement of ionic or electronic conductivity by several orders of magnitude has been reported for nanoscaled fluorides and oxides [13–16]. There is no doubt that MIECs, which are capable of fast mass (ion) transport, will play an important role in the development of various energy conversion and storage devices.

Cross-References

► Mixed Conductors, Determination of Electronic and Ionic Conductivity (Transport Numbers)
► Solid Electrolytes

References

1. Tuller HL (1997) Semiconduction and mixed ionic-electronic conduction in nonstoichiometric oxides: impact and control. Solid State Ionics 94:63–74
2. Wagner C, Elektrochem Z (1934) Über die elecktromotorische kraft der kette: Ag|AgJ|Ag_2S|Pt(+S). Angew Phys Chem 40:364–365
3. Miyatani S (1960) Electrical Properties of Pseudo-binary Systems of Ag_2VI's; $Ag_2Te_xSe_{1-x}$, $Ag_2Te_xS_{1-x}$, and $Ag_2Se_xS_{1-x}$. J Phys Soc Jpn 15:1586–1595
4. Yokota I (1961) On the theory of mixed conduction with special reference to conduction in silver sulfide group semiconductors. J Phys Soc Jpn 16:2213–2223
5. Nilges T, Lange S, Bawohl M, Deckwart JM, Janssen M, Wiemhöfer H-D, Decourt R, Chevalier B, Vannahme J, Eckert H, Weihrich R (2009) Reversible switching between p- and n-type conduction in the semiconductor $Ag_{10}Te_4Br_3$. Nat Mater 8:101–108
6. Teraoka Y, Zhang HM, Furukawa S, Yamazoe N (1985) Oxygen permeation through perovskite type oxides. Chem Lett 1743–1746
7. Shao ZP, Yang WS, Cong Y, Dong H, Tong JH, Xiong GX (2000) Investigation of the permeation behavior and stability of a $Ba_{0.5}Sr0.5Co_{0.8}Fe_{0.2}O_{3-d}$ oxygen membrane. J Membr Sci 172:177–188
8. Takamura H, Ogawa M, Suehiro K, Takahashi H, Okada M (2008) Fabrication and characteristics of planar-type methane reformer using ceria-based oxygen permeable membrane. Solid State Ionics 179:1354–1359
9. Terabe K, Hasegawa T, Nakayama T, Aono M (2005) Quantized conductance atomic switch. Nature 433:47–50
10. Chua LO (1971) Memristor - The Missing Circuit Element. IEEE Trans Circ Theory CT-18:507–519
11. Strukov DB, Snider GS, Stewart DR, Williams RS (2008) The missing memristor found. Nature 453:80–83
12. Yang JJ, Pickett MD, Li X, Ohlberg DAA, Stewart DR, Williams RS (2008) Memristive switching mechanism for metal/oxide/metal nanodevices. Nat Nanotech 3:429–433
13. Chiang Y-M, Lavik E, Kosacki I, Tuller H, Ying J (1996) Defect and transport properties of nanocrystalline CeO_{2-x}. Appl Phys Lett 69:185–187
14. Tschöpe A, Sommer E, Birringer R (2001) Grain size-dependent electrical conductivity of polycrystalline cerium oxide: I. Experiments. Solid State Ionics 139:255–265
15. Brinkman KS, Takamura H, Tuller HL, Iijima T (2010) The oxygen permeation properties of nanocrystalline CeO_2 thin films. J Electrochem Soc 157: B1852–B1857
16. Sata N, Eberman K, Eberl K, Maier J (2000) Mesoscopic fast ion conduction in nanometre-scale planar heterostructures. Nature 408:946–949

Mixed Conductors, Determination of Electronic and Ionic Conductivity (Transport Numbers)

Ulrich Guth
Kurt-Schwabe-Institut für Mess- und Sensortechnik e.V. Meinsberg, Waldheim, Germany
FB Chemie und Lebensmittelchemie, Technische Universität Dresden, Dresden, Germany

Theoretical Background

Solids which have both ionic and electronic conductivity are called mixed conductors. Most of the known solids show such behavior (► Solid Electrolytes, ► Membrane Technology), depending on temperature and partial pressure of that gas which is in equilibrium with ions in the conductor. For the development of electrolytes, electrode materials, and ceramic membranes, it is important to know which charge carriers are responsible for the current transport and to which extent. The required range of the transference number depends on the desired application of the solids: For solid electrolytes, a very low electronic conductivity is desirable ($t_i \geq 0.99$) whereas for electrode materials, a high electronic conductivity combined with a low ionic conductivity is necessary ($t_i \approx 0.10$). For membranes, those solids are suited which exhibit nearly the same conductivity for electronic and ionic carriers ($t_i \approx 0.5$).

The total conductivity σ_t is the sum of the conductivities of all species which are able to carry a charge:

$$\sigma_t = \sigma_c + \sigma_a + \sigma_e + \sigma_h \quad (1)$$

Mostly only one ion is mobile so that the conductivity of the counter ion σ_c can be neglected. On the other hand, regarding the electronic carriers, either the defect electrons or the holes, predominate under the condition of applications.

$$\sigma_t = \sigma_{ion} + \sigma_e + \sigma_h \quad (2)$$

$$\sigma_t = \sigma_{ion} + \sigma_{el} \quad (3)$$

The transport number (transference number), i.e., the fraction of current or conductivity which is caused of a certain carrier, is:

$$t_{ion} = \frac{I_{ion}}{I_{ion} + I_{el}} = \frac{I_{ion}}{I_t} = \frac{\sigma_{ion}}{\sigma_{ion} + \sigma_{el}} = \frac{\sigma_{ion}}{\sigma_t} \quad (4)$$

Corresponding relation yields for t_{el}. Accordingly, the sum of both transport (transference) numbers amounts to one.

$$t_{ion} + t_{el} = 1 \quad (5)$$

That is different to the transport number in aqueous electrochemistry where electronic conductivity can be neglected.

Experimental Methods

There are different methods to measure the partial conductivities and the transport (transference) numbers [1–4]:

- Determination of the equilibrium voltage (emf) on a gas cell
- Measurement of the ionic or electronic current
- Polarization method according to Hebb-Wagner (steady state)
- Permeation measurements
- Gravimetric measurement according to Tubandt

Emf Method of Transport Number Determination

This method is based on the Wagner formula. For an oxide ion conductor is yielded

$$-E = U_{eq} = \frac{1}{4F}\int_{\mu'_{O_2}}^{\mu''_{O_2}} t_{ion} d\mu_{O_2} \quad (6)$$

where F denotes Faraday's constant and t_{ion} the transport number of oxide ions which must be regarded as a function of the chemical potential of oxygen μ_{O2} in the mixed conductor. Provided that the chemical potentials (the partial pressures) on both electrodes are in the same range, an average transport number $\bar{t}_{ion}$ is determined and can be extracted from the integral as a constant. With the partial pressure dependence of the chemical potential

$$\mu_{O_2} = \mu^O_{O_2} + RT \ln p_{O_2} \quad (7)$$

results

$$-E = U_{eq} = \frac{RT}{4F}\bar{t}_{ion} \ln\frac{p''_{O_2}}{p'_{O_2}} \quad (8)$$

For this method, a galvanic cell, for simplicity, a gas concentration cell with air and oxygen, is used (Fig. 1).

The disk-shape mixed conductor, which can be heated up to 1,200 °C, is pressed gas-tightly between the ends of two alumina tubes establishing two separate compartments for purging the electrodes with different gases. If the ionic charge carriers in the solid are accompanied by significant amounts of electronic carriers, the cell is partially short-circuited so that the value of $U_{eq} = -E$ (emf) becomes smaller. For the application of this method, it must be ensured that the electrode reactions are reversible and sufficiently fast. To check these conditions, it is useful to short-circuit the cell circuit externally, reopen it, and observe the emf and the time needed to get time-independent values during the high ohmic measurement. A fast response and the same final value indicate reversibility. Furthermore, synthetic air has to be applied in order to avoid reactions with traces of combustibles in ambient air, which can influence the cell emf due to the formation of mixed potentials. Especially for the determination of small ionic transference numbers, it is necessary to compensate the thermo emf. That is easily possible by exchanging the gas flows to the electrodes [4, 5].

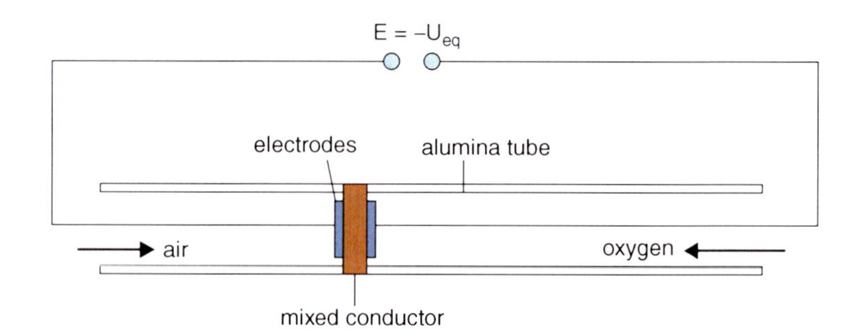

Mixed Conductors, Determination of Electronic and Ionic Conductivity (Transport Numbers), Fig. 1 Experimental setup for emf measurement on a mixed conductor [4, 5]

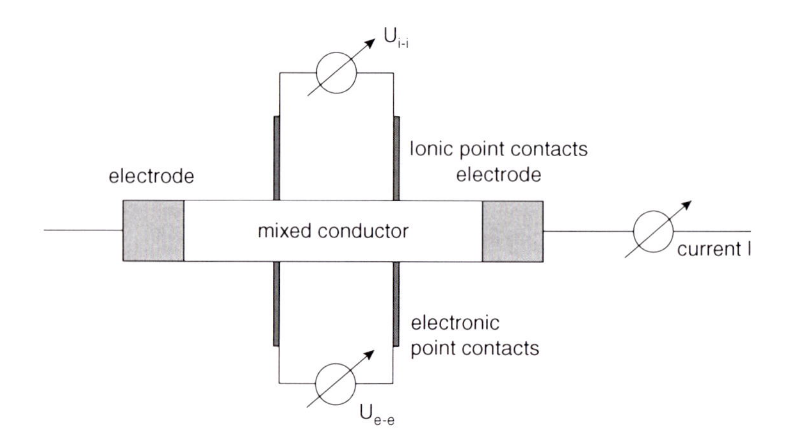

Mixed Conductors, Determination of Electronic and Ionic Conductivity (Transport Numbers), Fig. 2 Measurements of partial conductivities on mixed conductors

Measurement of the Ionic and Electronic Current

In Fig. 2, the basic principle of measurement of partial currents is shown. The current I consists of both electronic and ionic partial currents:

$$I = I_{ion} + I_{el} \tag{9}$$

The ionic contacts (by means of ionic conductors, above, $U_{i-i} = I_{ion}R$) enable the selective measurement of only the ionic current whereas the electronic point contacts (by means of electronic conductors, below, $U_{e-e} = I_{el}R$) respond only to the flow of electronic carriers. The basic principle seems to be simple, but difficulties often come with its experimental implementation. For both kinds of contacts, an interaction with other components like gases has to be avoided. Otherwise, the contact lines between the mixed conductor and ionic point contacts or between the mixed conductor and the electronic point contacts can act as electrodes in terms of electrochemistry and can generate an emf (voltage), adulterating the measurement results. Therefore, the vicinity of contacts has to be protected to prevent the influence of ambient gas (air).

Polarization Method According to Hebb-Wagner (Steady State)

The ionic conductivity can be measured using zirconia microelectrodes based on a Hebb-Wagner analysis of current voltage curves. The experimental setup is shown schematically in Fig. 3. The point contact and the large area of counterelectrodes should be made of yttria-stabilized zirconia with a reversible metal paste electrode, e.g., of silver or platinum. The point contact must be encapsulated with a glass seal to ensure a gas-tight contact. Under steady-state conditions, only an oxide ion current flows through the interface mixed conductor/YSZ.

The chemical potential at the point contact depends linearly on the voltage [1].

The slope of the curve corresponds to $\frac{dU}{dI} = \sigma_{ion} 2\pi a$. For the ionic conductivity σ_{ion}, the following expression results:

$$\sigma_{ion} = \frac{dU/dI}{2\pi a}, \qquad (10)$$

where a is the diameter of the contact.

On the other hand, the electronic conductivities can be determined by means of an oxygen-blocking electrode like this [6]:

$$O_2, Pt/YSZ/Pt(glass)$$

If a polarization voltage is applied, so that the right side of the cell is the cathode, oxygen cannot penetrate through the glass and cannot be reduced to O^{2-} at the interface. Therefore, the electrode reaction with oxygen is blocked. In this case, the total current flow is purely based on electronic charge carriers:

$$I_{tot} = I_e + I_h \qquad (11)$$

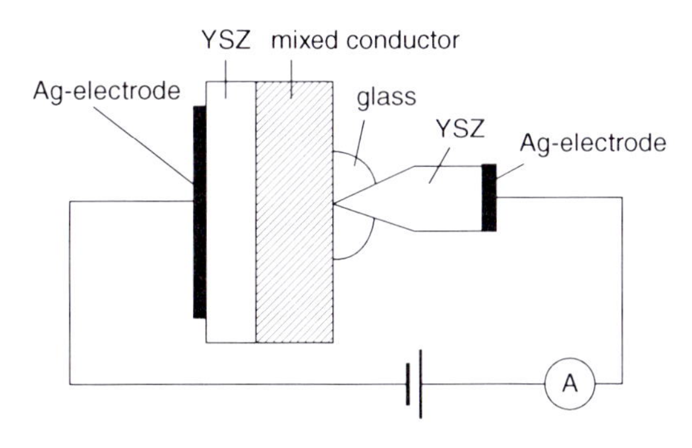

Mixed Conductors, Determination of Electronic and Ionic Conductivity (Transport Numbers), Fig. 3 Polarization setup for ionic current [1]

Permeation Measurements

The permeation flux of oxygen and the flux of charge carriers through a solid (mixed conductor) are connected by ambi-polar (or counter)-diffusion of electronic holes (h°) and oxide ion vacancies ($V_O^{\circ\circ}$). Oxygen can only move through the mixed conductor by means of holes which are formed in the solid [7].

$$\frac{1}{2} O_2(g) + V_O^{\circ\circ}(mixed\ cond.) \rightleftharpoons O_O^x(mixed\ cond.) + 2h^\circ(mixed\ cond.) \qquad (12)$$

Thus, the electronic conductivity is calculated using the permeation flux Π_{O_2} which can be measured by means of a potentiometric determination of oxygen concentration with a solid electrolyte cell. With the temperature and partial pressure dependence (▸ Defects in Solids), the following expression is yielded:

$$\Pi_{O_2} = \frac{RT}{4F^2}\sigma_{h^\circ} = \frac{RT}{4F^2}\sigma_{h^\circ(p_{O_2}=1)} p_{O_2}^{1/4} \exp[E_{a,h}/RT] \qquad (13)$$

With the total conductivity σ_{tot}, the transport number for holes is obtained:

$$t_{h^\circ} = \sigma_{h^\circ}/\sigma_{tot} \qquad (14)$$

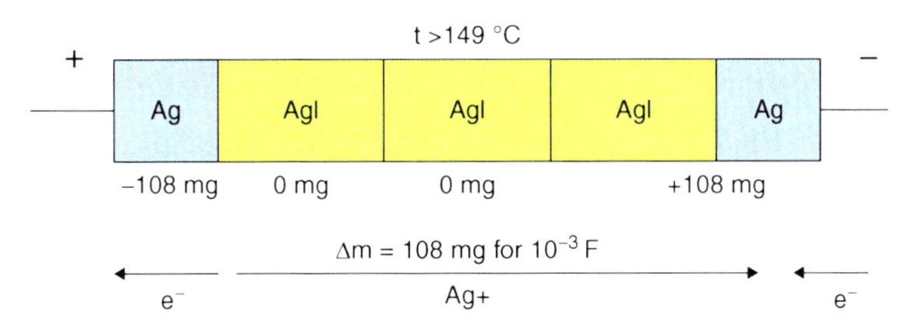

Mixed Conductors, Determination of Electronic and Ionic Conductivity (Transport Numbers), Fig. 4 Tubandt's method for the determination of transport numbers for solid electrolytes and mixed conductors

Determination of the Transport Number with Tubandt's Method

For this method, which is an adaptation of the Hittorf's arrangement for solids, three disk-shaped samples (here AgI) are pressed together which are contacted on both sides with silver electrodes (Fig. 4). After the charge transport of 10^{-3} F (96.485 As/mol) through the cell arrangement, a change in mass can be measured. On the right hand side, the mass of the silver electrode together with the AgI pellet adjacent to silver increases, whereas on the left hand side, the equal amount of mass disappears. Taking into account the molar mass of silver 108 g/mol, the experimental result proves that in AgI, the overall charge is transported by silver ions [3].

Cross-References

- ▶ MIEC Materials
- ▶ Solid Electrolytes

References

1. Shuk P, Guth U (1995) Mixed conductive electrode materials for sensors and SOFC. Ionics 1:106–111
2. Heyne L (1977) Electrochemistry of mixed ionic-electronic conductors. In: Geller S (ed) Solid electrolytes. Springer, Berlin/Heidelberg/New York, pp 202–217
3. Rickert H (1982) Electrochemistry of solids. Springer, Berlin/Heidelberg/New York, pp 96–110
4. Shuk P, Möbius HH (1985) Transport numbers and electrical conductivity of Bi_2O_3 modifications (in German). Z physik Chem (Leipzig) 266:9–16
5. Shuk P, Jacobs S, Möbius HH (1985) Mixed phases of Bi_2O_3 with terbium and praseodymium oxides (in German). Z Anorg Allg Chem 524:144–156
6. Hartung R (1973) Determination of electronic conductivity of solid electrolytes $Zr_{0.82}Y_{0.1}$ $Mg_{0.08}O_{1.87}$ using the Hebb-Wagner's polarization method (in German). Z physik Chem (Leipzig) 254:393–410
7. Hartung R, Möbius HH (1970) Determination of oxidation semi conductivity of zirconia electrolytes by means of permeation data. Z Physik Chem 243:133–138

Modeling and Simulation of Biosensors

Romas Baronas[1] and Juozas Kulys[2]
[1]Faculty of Mathematics and Informatics, Vilnius University, Vilnius, Lithuania
[2]Department of Chemistry and Bioengineering, Vilnius Gediminas Technical University, Vilnius, Lithuania

Introduction

Biosensors are analytical devices in which specific recognition of the chemical substances is performed by biological material. The biological material that serves as recognition element is used in combination with a transducer. The transducer transforms concentration of substrate or product to electrical signal that is amplified and further processed. The biosensors may utilize enzymes, antibodies, nucleic acids, organelles, plant and animal tissue, whole organism, or organs [1–3]. Biosensors containing biological catalysts (enzymes) are called catalytical biosensors. Biosensors of this type are the most abundant, and they found the largest application in medicine, ecology, and environmental monitoring [4, 5].

The action of catalytical biosensors is associated with substrate diffusion into biocatalytical membrane and its conversion to a product. The simulation of biosensor action includes solving the diffusion equations for substrate and product with a term containing a rate of biocatalytical transformation of substrate. The complications of modeling arise due to solving partially differential equations with nonlinear biocatalytical term and with complex boundary and initial conditions [6, 7].

Scheme of Biosensor Action

The biosensor produces a signal when the analyte under determination diffuses from the bulk solution into the biocatalytical membrane. The biocatalyst catalyzes the substrate (S) conversion to the product (P), which is determined by the transducer [8].

The concentration change of S and P is associated with the diffusion and the enzymatic reaction [9]. Following Fick, the compound's concentration change in the biocatalytical membrane can be described by partial differential equations (PDE) of the reaction–diffusion type:

$$\frac{\partial S}{\partial t} = D_e \frac{\partial^2 S}{\partial x^2} - V(S), \tag{1}$$

$$\frac{\partial P}{\partial t} = D_e \frac{\partial^2 P}{\partial x^2} + V(S),\ V(S) = \frac{V_{max} S}{K_M + S}, \quad x \in (0, d),\ t > 0. \tag{2}$$

where x, t stand for space and time, respectively, $S(x, t)$ is the concentration of the substrate S, $P(x, t)$ is the concentration of the reaction product P, $V(S)$ is the reaction rate, d is the thickness of the enzyme membrane, V_{max} is the maximal enzymatic rate, K_M is the Michaelis–Menten constant, and D_e is the diffusion coefficient of compounds in the enzyme membrane that typically is used the same for the substrate, the product, and the mediator, if present [6, 7].

The solution of Eqs. 1 and 2 at corresponding initial and boundary conditions produces the concentration change of S and P in time and membrane thickness.

For the transducer, i.e., amperometric electrode, which is monitoring faradaic current arising when electrons are transferred between substrates, product or enzyme active center, and the electrode, the response R of a biosensor can be written as

$$R(t) = C_1 \frac{\partial P}{\partial x} | x = 0. \tag{3}$$

For the transducers, i.e., ion-selective electrodes and optical fiber, that do not perturb the concentration of the determining compound at the surface, the biosensor response R can be written as

$$R(t) = C_2 P(0, t), \tag{4}$$

or

$$R(t) = C_3 \log P(0, t), \tag{5}$$

where C_1, C_2, and C_3 are the appropriate constants.

The logarithmic dependence (Eq. 5) is characteristic for ion-selective electrodes, whereas for optical and other transducers, linear dependence (Eq. 4) between the response and the concentration is usually realized.

Simple analytical solution of Eqs. 1 and 2 is impossible even for the simplest initial and boundary conditions due to hyperbolic function of the enzymatic rate dependence on substrate concentration [6, 7]. Therefore, the analytical simulation of biosensor action was performed in the simplest cases for which analytical solutions still exist. This approach was used widely, especially at beginning of development of biosensors, to recognize principles of biosensor action. The approximal analytical solution gives information about critical cases. They are useful also to test correctness of digital calculations found at initial and boundary limiting conditions.

When the concentration S_0 to be measured is very small in comparison with K_M,

$$\forall x, t : x \in [0, d], t > 0 : 0 \leq S(x, t) \leq S_0 \ll K_M,$$

the nonlinear function $V(S)$ simplifies to that of the first order $V(S) \approx (V_{max}/K_M)S$.

Practically, the enzyme reaction can be considered first order when the concentration of the detected species is below one-fifth of K_M, i.e., $S_0 < 0.2K_M$, [9]. This case is rather typical for biosensors with high enzyme loading factor.

The nonlinear reaction–diffusion system (1, 2) reduces to a linear one,

$$\frac{\partial S}{\partial t} = D_e \frac{\partial^2 S}{\partial x^2} - kS, \tag{6}$$

$$\frac{\partial P}{\partial t} = D_e \frac{\partial^2 P}{\partial x^2} + kS,\ \ x \in (0, d), t > 0. \tag{7}$$

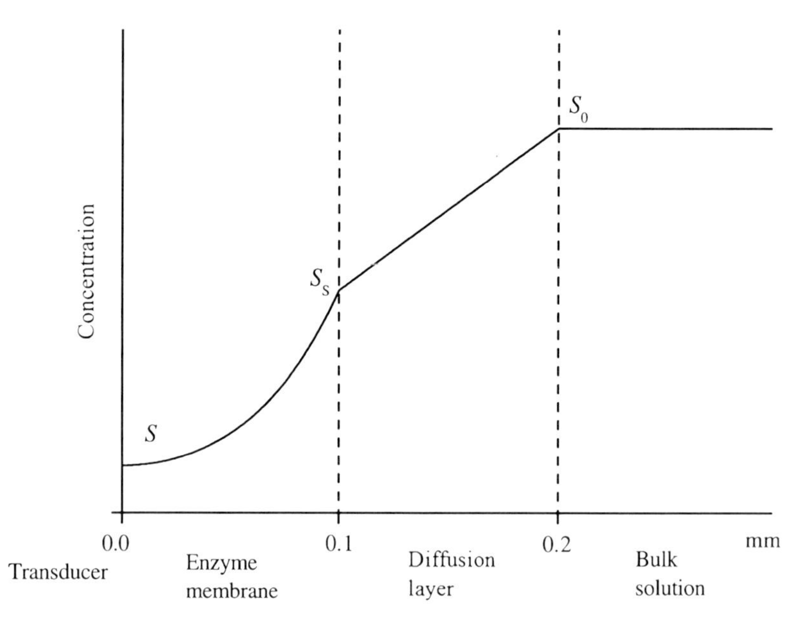

Modeling and Simulation of Biosensors, Fig. 1 The substrate concentration profile in a biosensor at steady-state conditions

where k is the first-order reaction constant (linear enzyme kinetic coefficient), $k = V_{max}/K_M$.

Analytical solutions typically are made at steady state and external and internal diffusion limiting conditions [7].

The steady-state (stationary) (SS) conditions mean that

$$\frac{\partial S}{\partial t} = 0, \ \frac{\partial P}{\partial t} = 0. \tag{8}$$

The external diffusion limitation (EDL) indicates that substrate transport through diffusion (stagnant) layer is a rate-limiting process [10]. At internal diffusion limitation (IDL), the substrate diffusion through the external diffusion layer is fast, and process is limited by the diffusion inside an enzyme membrane. The disadvantage of these approximate solutions is an error at the boundaries between the different approximate treatments. It is helpful to illustrate this approach by reference to a trivial problem of the substrate conversion in the biocatalytical membrane of the biosensor and at the concentration of the substrate less than the Michaelis–Menten constant (K_M). The calculated profile of substrate concentration at steady-state or stationary conditions is shown in Fig. 1.

Abrupt concentration changes of the substrate appear at the boundary of biocatalytical membrane/stagnant layer and at the boundary stagnant layer/bulk solution. This comes from approximate solutions at the boundaries during different approximate treatments. The change of steady-state concentration of the substrate in the membrane can be calculated as

$$\frac{S}{S_s} = \frac{\cosh(\alpha x)}{\cosh(\alpha d)}, \ \alpha^2 = \frac{V_{max}}{K_M D_e}, \tag{9}$$

where S_s is the substrate concentration at the boundary between the enzyme membrane and the stagnant solution.

On the other hand, at the steady state, a substrate flux through the boundary of stagnant layer/bulk solution is equal to the flux through the boundary of biocatalytical membrane/stagnant layer:

$$D_0 \frac{S_0 - S_s}{\delta} = D_e \frac{\partial S}{\partial x} | \ x = d = D_e \alpha \tanh(\alpha d) S_s. \tag{10}$$

A combination of the solutions Eqs. 9 and 10 produces the concentration profile of the substrate in the biocatalytical membrane and the stagnant layer (Fig. 1). It is possible to notice that the greatest error of calculations is at $x = d$ and $x = d + \delta$. However, at the limiting (the internal or the external diffusion limitation) cases, two expressions produce very good approximations to the full equation. Therefore, the modeling of the biosensors at two limiting cases was used to solve different biosensor problems.

Analytical Simulation of Biosensors at Steady State

The steady-state conditions arise when mass transport of substrates in a biocatalytical layer is equal to the rate of the biocatalytical reaction. The action of biosensors utilizing Michaelis–Menten kinetics and the consecutive and parallel substrate conversion catalyzed by single, dual, and multienzyme systems were modeled at SS conditions and at IDL [10]. The main task of these investigations was an evaluation of the biosensor response on the enzymatic and mass transport parameters. High-sensitive biosensors utilizing the cyclic and synergistic substrate conversion were discovered by analyzing complex schemes of the enzyme action, and many of them were experimentally realized [11].

Peculiarities of the modeling of biosensors based on chemically modified electrodes (CME) arise due to the mediator location on the electrode that produces special boundary conditions; it was assumed that a concentration of the mediator on the electrode is constant due to desorption (dissolution) of the mediator [12].

The modeling of biosensor action at external diffusion limitation (EDL) is much easier due to a linear gradient of the substrate concentration in the stagnant layer. For biosensors acting at EDL and at steady-state conditions, the flux of the substrate through the stagnant layer equals to the rate of the biocatalytical reaction on the surface of the transducer. The solution for the mass transport and the biocatalytical processes produces an expression that can be solved algebraically for a simple (Michaelis–Menten) kinetics. The modeling of the substrate conversion by an adsorbed polyenzyme system revealed the dependence of the biosensor response on biocatalytical parameters, limiting the step of process and role of the mass transport.

The modeling of the biosensors utilizing non-Michaelis–Menten kinetics of the enzymes action revealed a multistate possibility at the steady-state conditions [13]. Multi-steady state may have many interesting consequences for the stability of the biosensor response. It can generate oscillations of the concentration and the biosensor response if the negligible perturbation of the enzyme activity or the mass transport occurs.

The conditions of EDL were applied for a simulation of biosensors utilizing microbial cells [14]. The peculiarity of the modeling of the microbial biosensors is a slow substrate and product transport through the microbial cell wall. The modeling was performed by applying formal mathematics as for the biosensor acting at EDL. In the case of the biochemical oxygen demand (BOD) biosensor, the response was simulated assuming that the rate of oxygen consumption by the cells in the layer near the transducer (oxygen electrode) is determined by the substrate concentration [15].

Analytical Simulation of Biosensors at Nonstationary State

The biosensor response at a transition state was modeled solving PDE of the substrate diffusion and with biocatalytical conversion with the corresponding initial and boundary conditions [6, 7]. The analytical solutions, however, exist at very limited cases. The Laplace transformation that is typically used for solving the diffusion equations is no longer applicable for the solution of such problems. Therefore, for the modeling of the diffusion and enzymatic reactions, other methods of PDE solving are used.

Carr [16] used the Furje method to solve Eqs. 1 and 2 at $S_0 << K_M$ and $S_0 >> K_M$ for a potentiometric biosensor. Simulations exposed that the dynamics of the biosensor response weekly depends on the enzyme activity. If the diffusion module αd is larger than 1, the biosensor achieves the difference of 0.1 or 1.0 mV to the steady-state response during a time ($\tau_{0.1}$ or $\tau_{1.0}$):

$$\tau_{0.1} = \frac{2.35d^2}{D_e}, \quad \tau_{0.1} = \frac{1.42d^2}{D_e}.$$

The dynamics of amperometric biosensors was modeled assuming that there exist no EDL and the conversion of the compound in the biocatalytic membrane follows the first-order reaction kinetics [17]. The dynamics of the biosensor response depends on diffusion modules. At IDL the dynamics of the biosensors with consecutive and parallel substrate conversion was similar to the diffusion of the substrate through the biocatalytic membrane, whereas the response of the biosensors with cyclic substrate conversion was 3.5–5.4 times slower.

Numerical Simulation of Biosensors

The governing Eqs. 1 and 2 for biosensors involving Michaelis–Menten kinetics can be solved analytically at limited concentrations at which the reaction term $V(S)$ approaches a linear function. Often, governing equations describing the action of biosensors utilizing complex biocatalytical schemes cannot be linearized for practical values of the model parameters [7].

Mell and Maloy [18] postulated a numerical approach to simulate the steady-state amperometric measurements for an enzyme electrode [18]. Since then, the reaction–diffusion problems describing biochemical processes are often solved numerically [6, 7, 19]. The analytical solutions are often applied to validation of the corresponding numerical solutions.

The finite difference technique is a widely used numerical method for solving the reaction–diffusion problems [7]. When applying this approach, the model equations are transformed into a form such that the differentiation can be performed by numerical calculations. There are several difference schemes that can be used for solving PDEs [7]. Each of these schemes has its own advantages and limitations [20]. Due to a possible singularity of the spatial gradient at $t = 0$ and boundary conditions, the finite difference approximation requires some numerical care. The Crank–Nicolson and backward implicit (with and without linearization) are preferable for simulation of biosensors utilizing Michaelis–Menten kinetics [20].

Simulation of the biosensor action usually involves calculation of the concentrations of the compounds as well as the response for the time interval from the beginning of the action up to the moment called the biosensor response time. The moment of the measurement depends on the type of the device. Devices operating in the stationary mode usually use the time when the absolute response, e.g., anodic or cathodic current, slope value falls below a given small value. Since the biosensor responses usually vary even in orders of magnitude, the response is normalized. In other words, the time T needed to achieve a given dimensionless decay rate ε is accepted as the response time:

$$T = \min_{R(t)>0} \left\{ t : \frac{t}{R(t)} \left| \frac{dR(t)}{dt} \right| < \varepsilon \right\}.$$

Mathematical modeling of biosensors has been successfully used to investigate the kinetic peculiarities of the biosensor action. The numerical simulation became a powerful framework for numerical investigation of the impact of model parameters on the biosensor action and to optimize the biosensor configuration [7]. Recently, the computational modeling of the laccase-based biosensor qualitatively explained and confirmed the experimentally observed synergistic effect of the mediator on the biosensor response [21]. The numerical simulation of an amperometric biosensor based on an enzyme-loaded carbon nanotube (CNT) layer deposited on a perforated membrane highlighted the dependence of the steady-state biosensor current on the anisotropic properties of CNT. It was also shown that the sensitivity of the biosensors based on the CNT electrode can be notably increased by selecting an appropriate geometry of the biosensor [22].

Future Directions

The modeling of biosensors by analytical solution of differential equations is limited by the concentration interval of reactive components, not applicable to the biosensors with complex biocatalytical schemes. The solution for nonstationary state is very complex, and practically it is impossible to get a solution for the complex initial and boundary conditions. The digital modeling of biosensors could cover all concentrations interval, transition state, multiplex geometry, and complex kinetics of biocatalyzer action. For the efficient simulation of practical biosensors, multi-scale numerical approaches and adaptive numerical algorithms have to be developed. Specialized simulators with powerful graphical presentations are under development. Computational models and tools conjugating practical biosensors with methods of computational intelligence such as artificial neural networks are also under development for increasing the sensitivity, selectivity, and stability of analytical systems.

Cross-References

- ► Biosensors, Electrochemical
- ► DNA/Electrode Interface, Detection of Damage to DNA Using DNA-Modified Electrodes
- ► Electrochemical Sensors for Environmental Analysis

References

1. Turner APF, Karube I, Wilson GS (1987) Biosensors: fundamentals and applications. Oxford University Press, Oxford
2. Scheller F, Schubert F (1992) Biosensors. Elsevier, Amsterdam
3. Buerk DG (1995) Biosensors: theory and applications. CRC Press, Lancaster
4. Spichiger-Keller UE (1998) Chemical sensors and biosensors for medical and biological applications. Wiley-VCH, New York
5. Wollenberger U, Lisdat F, Scheller FW (1997) Frontiers in biosensorics 2. Practical applications. Birkhauser, Basel
6. Schulmeister T (1990) Mathematical modeling of the dynamic behavior of amperometric enzyme electrodes. Sel Electrode Rev 12:203–206
7. Baronas R, Ivanauskas F, Kulys J (2010) Mathematical modeling of biosensors: an introduction for chemists and mathematicians. Springer, Dordrecht
8. Gufreund H (1995) Kinetics for the life sciences. Cambridge University Press, Cambridge
9. Guilbault GG (1970) Enzymatic methods of analysis. Pergamon, Oxford
10. Kulys J (1981) Analytical systems based on immobilized enzymes. Mokslas, Vilnius (in Russian)
11. Kulys J, Tetianec L (2005) Synergistic substrates determination with biosensors. Biosens Bioelectron 21:152–158
12. Kulys J (1991) Biosensors based on modified electrodes. In: Turner APF (ed) Advances in biosensors, vol 1. JAI Press, London/Greenwich, pp 107–124
13. Kulys F, Baronas R (2006) Modeling of amperometric biosensors in the case of substrate inhibition. Sensors 6:1513–1522
14. Kulys JJ (1981) Development of new analytical systems based on biocatalysers. Enzyme Microb Technol 3:344
15. Kulys J, Kadziauskiene K (1980) Yeast BOD sensor. Biotechnol Bioeng 22:221–226
16. Carr PW (1977) Fourier analysis of the transient response of potentiometric enzyme electrodes. Anal Chem 49:799–802
17. Kulys JJ, Sorochinskii VV, Vidziunaite RA (1986) Transient response of bienzyme electrodes. Biosensors 2:135–146
18. Mell LD, Maloy T (1975) A model for the amperometric enzyme electrode obtained through digital simulation and applied to the immobilized glucose oxidase system. Anal Chem 47:299–307
19. Kernevez JP (1980) Enzyme mathematics. Studies in mathematics and its applications. Elsevier, Amsterdam
20. Britz D, Baronas R, Gaidamauskaitė E, Ivanauskas F (2009) Further comparisons of finite difference schemes for computational modelling of biosensors. Nonlinear Anal Model Control 14:419–433
21. Kulys J, Vidziunaite R (2009) Laccase based synergistic electrocatalytical system. Electroanalysis 21:2228–2233
22. Baronas R, Kulys J, Petrauskas K, Razumienė J (2011) Modelling carbon nanotube based biosensor. J Math Chem 49:995–1010

Molten Carbonate Fuel Cells, Introduction - Cell Configuration, Features, and Cell Stack Size

Takao Watanabe
Central Research Institute of Electric Power Industry, Yokosuka, Kanagawa, Japan

Introduction

Molten carbonate fuel cell (MCFC) is capable of operation on a various fuel gases including natural gas, renewable biogas, and gasified coal gas because of its high-temperature operation. Internal reforming of hydrocarbon fuel is also possible, resulting in improving fuel utilization and providing higher power generation efficiency. Many MCFC plants are being installed as the stationary cogeneration power supply using various fuels in various countries in the world. The world's largest 2.8 MW fuel cell power plants are composed of MCFCs. The power generation efficiency of the systems including smaller 300 kW units attains 47 % (LHV, net, same as above unless otherwise noted).

Cell Configuration and Principle

Figure 1 shows the principle of the MCFC operation. The electrolyte is a molten carbonate under the operating temperature ranged about 600–650 °C. Hydrogen (H_2) supplied to the anode reacts with carbonate ion (CO_3^{2-}) in the electrolyte. The electron is discharged from the anode and steam (H_2O) and CO_2 are generated at the same time. Reacting with the electron, oxygen (O_2) and CO_2 supplied to the cathode generate CO_3^{2-}. These reactions are expressed as follows:

$$\text{Anode}: H_2 + CO_3^{2-} = H_2O + CO_2 + 2e^- \quad (1)$$

$$\text{Cathode}: CO_2 + 1/2O_2 + 2e^- = CO_3^{2-} \quad (2)$$

$$\text{Total}: H_2 + 1/2O_2 = H_2O \quad (3)$$

Features

The MCFC has not only the features of the general fuel cells such as high efficiency, capacity flexibility, and the superior environmental sustainability but also additional features as follows:

- *Higher Efficiency*: It can form the power generation systems combined with the gas turbine and the steam turbine by using the high-temperature exhaust heat from the MCFC. It can attain much higher power generation efficiency.
- *Fuel Flexibility*: Because the operating temperature is high, electrode reactions take place in fast speed. Nickel (Ni) is used for both electrodes instead of expensive platinum (Pt) catalyst. It means no carbon monoxide (CO) poisoning of Pt. Ni acts as a catalyst and CO in the fuel gas (if there is any) supplies H_2 by the shift reaction ($CO + H_2O = H_2 + CO_2$) relating to the reaction of (1). Therefore, it becomes possible to use the gasification gas of coal, biomass, and waste which contain certain percentage of CO.
- *Internal Reforming*: When methane-based fuel is in use, internal reforming can be applied using steam and heat generated inside the fuel cell. The internal reforming can be expected to improve the generation efficiency by increasing fuel utilization and lowering energy consumption for cooling the MCFC to maintain the operating temperature. The internal reforming can also simplify the system configuration by eliminating the external reformer.
- *CO_2 Concentration*: CO_2 in the cathode inlet gas can be separated with reaction (3) and CO_2 in the anode inlet gas can be concentrated with reaction (1). The MCFC can be utilized as an element for CCS (Carbon Capture and Storage) system.
- *Easier Manufacturing of Large Cells*: The MCFC generally adopts a planar multilayered structure, using metal separators and sheets of active components manufactured by tape-casting method. Therefore, it is rather easy to make it big and to be adopted for large-capacity power plants from the production viewpoint.

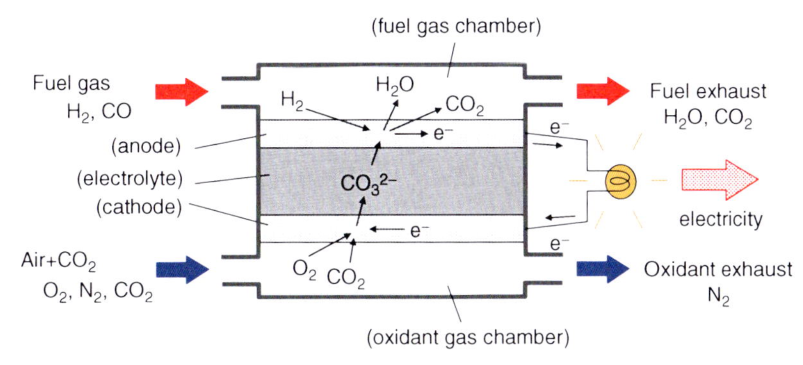

Molten Carbonate Fuel Cells, Introduction - Cell Configuration, Features, and Cell Stack Size, Fig. 1 Principle of the MCFC

Materials

The actual cell consists of the matrix sheet, the electrode (anode and cathode) sheets, and separator.

Matrix

The matrix has a porous ceramic structure consisting of $LiAlO_2$. It is usually impregnated with the electrolyte liquid maintained by capillary force. Mixed carbonate of Li_2CO_3 and K_2CO_3 or of Li_2CO_3 and Na_2CO_3 is used as the electrolyte.

Electrodes

Anode: The anode is required to have high corrosion resistance for melted carbonate under the fuel gas atmosphere. It also should be stable under steam and CO_2 generated at the anode. Therefore, the porous sheet sintered from fine nickel (Ni) particle is used. Chrome (Cr), aluminum (Al_2O_3), etc. are usually added for the improvement of creep resistivity at high temperature.

Cathode: The cathode is operated under a severe condition of oxidative atmosphere, and thus, the metallic oxide is used. Typically, the porous media made of oxidized nickel particle are used.

Separator

Metallic separator is inserted between each cell composed of active components (matrix sandwiched by anode and cathode) when forming a stack. Separator has functions of reactant gas supply, gas sealing, and electrical connection of each cell. The nickel clad stainless steel is used as a center plate of the separator which is exposed to both reducing and oxidizing condition on each side in consideration of corrosion resistance with the carbonate. Perforated fold metal sheets are placed on both sides of the center plate with same materials to form fuel and oxidant gas flow channels. The perforate metal current collector sheets are inserted between electrode and gas flow channels if necessary. Reactant gases are sealed at the outer periphery of the separator touching the electrolyte matrix directly. The wet seal covered with the liquid film of the impregnated electrolyte is formed and prevents gas leakage to the outside.

Cell Stack Size and Manufacturing

Cell Size and Stack Capacity

Each cell and separator has a rectangular shape with about 1 m^2 electrode area which is the biggest comparing to other types of fuel cells. The thickness of the single cell is several mm with the two electrodes and electrolyte matrix (active components). The largest MCFC stack can generate 300 kW level with several hundred cells.

Manufacturing of the Sheet Components and Stack Fabrication

The electrodes and the matrix are made by the tape-casting method, and the formed sheets are cut for required sizes. The tape-casting method is a useful process for simple, easy, quantitative, and cost-effective production. In the process, the slurry to the mixture of the fine particle of the raw material with the solvent is thinly spread by a doctor blade on a substrate sheet and dried. The spread sheet is about 0.5 mm in thickness and about 1 m in width. The electrodes are fired after cutting and laminated with matrix and carbonate powder. After stacking up with separators, the stack is subjected to heat treatment and finally starting power generation.

System Configurations and Applications

The MCFC power generation system is basically composed of the fuel processing system, the air supply system, the MCFC, the inverter, etc. so that fuel and air can be supplied to the stack in appropriate temperature conditions. Because the MCFC can utilize various kinds of fuels, the fuel processing system is different from the kind of fuel. It will contain a steam reformer for natural gas and a gasifier/gas cleanup unit for coal, biomass, or wastes. In the case of using natural gas, internal reforming system can also be formed. Reforming catalyst is placed within MCFC's fuel channel instead of external reformer. Supplied methane reacts with generated steam from anode reaction and produces hydrogen using the heat of reaction within the MCFC. This system can eliminate an external reformer and simplify the system. It is also possible to configure a hybrid power generation system including a gas expander and a steam turbine at MCFC downstream as a bottoming cycle.

The MCFC power plants are generally expected to be stationary applications with high efficiency. Clean natural gas, biogas from waste water (digester gas), and gasified biomass or coal are expected as a fuel. The plant capacity is sub-MW to several MW class as a distributed generators and is several hundred MW as a centralized station with coal gasifier in future.

The natural gas-fueled pressurized external reforming system or ambient pressure internal reforming system has already realized about 47 % efficiency. As for the future large-scale plants, simulation results show that the natural gas-fueled pressurized external reforming system with 1,000 MW capacity would be about 65 %, the pressurized internal reforming system with 700 MW capacity would be about 69 %, and the integrated coal gasification system (IGMCFC) with 600 MW capacity would be about 57 % efficiency [1].

Current Status

The US DOE has promoted the development of MCFC in the United States. In the early 1970s, the final goal was set as a centralized power plants using coal. However, development focus has moved to the MW class small- or middle-sized distributed generators as a cogeneration plant since 1990s. Now the main developer of the MCFC plants is FuelCell Energy (FCE) in the United States. Most of the MCFC plants actually installed so far are internal reforming ambient pressure system manufactured by FCE. The company has developed several lineup with capacity of 300 kW, 1.4 MW, and 2.8 MW so far. Those systems provide 47 % efficiency (LHV, net), and 2.8 MW system is the largest fuel cell plant right now. The systems have been installed in hotel, hospital, college, data center, business building, manufacture, waste water treatment, etc. Those are mainly used as a grid-independent on-site cogeneration. The systems are also installed in fuel cell power park with 10 MW level. Those are used for grid-connected independent power producer (IPP) with application of RPS or FIT system. Main installation is in the USA and Korea and total capacity installed is over 100 MW.

Future Perspective

The market introduction of the MCFC is pushed forward as a cogeneration system of the small or medium size mainly applying the internal reforming system developed in the USA now. However, it concentrates in the country and the state where the promotion plan of the subsidies and grant is well prepared. Therefore, the MCFC market can be understood to be never independent economically at present. As a future direction, fundamental developments such as high efficiency, long life, and lower cost should be continued.

In addition, the power generating systems that have a new function to make the best use of a MCFC's inherent features have been proposed such as DFC/T (an internal reforming hybrid system), DFC-ERG (a pipeline energy collection system), and DFC-H_2 (a tri-generation system) by FCE. Also CRIEPI (Central Research Institute of Electric Power Industry) proposes the CO_2 recovery-type hybrid system using oxygen in place of air and showed the extremely high efficiency of 77 % in 300 MW class [2]. Some of those systems have been demonstrated. It is possible to expand the application field much more by above demonstrations.

Cross-References

▶ Fuel Cells, Principles and Thermodynamics
▶ Solid Oxide Fuel Cells, Introduction

References

1. Watanabe T (2001) Development of molten carbonate fuel cells in Japan – application of Li/Na electrolyte. Fuel Cells 1:1–7
2. Koda E et al (2006) MCFC-GT hybrid system aiming at 70% thermal efficiency. ASME Turbo Expo 2006

Molten Carbonate Fuel Cells, Overview

Shigenori Mitsushima
Yokohama National University, Yokohama, Kanagawa Prefecture, Japan

Introduction

The molten carbonate fuel cell (MCFC) plant is a high-temperature fuel cell power generation system that uses hydrocarbon fuels, including natural gas, methanol, diesel, biogas, coal mine methane, and propane. The MCFC uses molten carbonate as the electrolyte which is immobilized in a porous ceramic matrix and operates at 550–650 °C. Figure 1 shows a schematic diagram of an internal reforming MCFC plant. The hydrocarbon fuel reforms into a hydrogen-containing fuel in the fuel cell stack, and the hydrogen is then electrochemically oxidized in the fuel cells. The spent fuel is oxidized by air, and the produced CO_2 is the source for the cathode gas. The actual electric power generation capacity has been the highest for large scale residential fuel cells systems since 2003. The commercialization of MCFC plants started in 2000. The commercialized cell size is ca. 1 m^2 with ca. 1 kW output, and the unit stack power is ca. 300 kW. The electrochemical reactions are as follows:

$$\text{Anode:}\quad H_2 + CO_3^{2-} = H_2O + CO_2 + 2e^- \qquad (1)$$

$$\text{Cathode:}\quad CO_2 + 1/2\ O_2 + 2e^- = CO_3^{2-} \qquad (2)$$

$$\text{Total:}\quad H_2 + 1/2\ O_2 = H_2O \qquad (3)$$

The open circuit voltage is around 1.05 V which is predicted by the Nernst equation. The operating cell voltage is around 0.8 V. Hydrogen is made by steam reforming of the hydrocarbons in the anode chamber over an internal reforming catalyst (Fig. 2). The steam reaction and shift reaction, which simultaneously occur, are as follows:

$$\text{Steam reforming:}\quad CH_4 + H_2O = 3H_2 + CO \qquad (4)$$

$$\text{Shift reaction:}\quad H_2O + CO = H_2 + CO_2 \qquad (5)$$

Equation 4 is an endothermic reaction, and the heat is supplied from the energy loss of the fuel cell operation. Equation 5 is a very fast chemical reaction and is in equilibrium in the anode chamber, Therefore, CO is a fuel for the MCFCs [1–3].

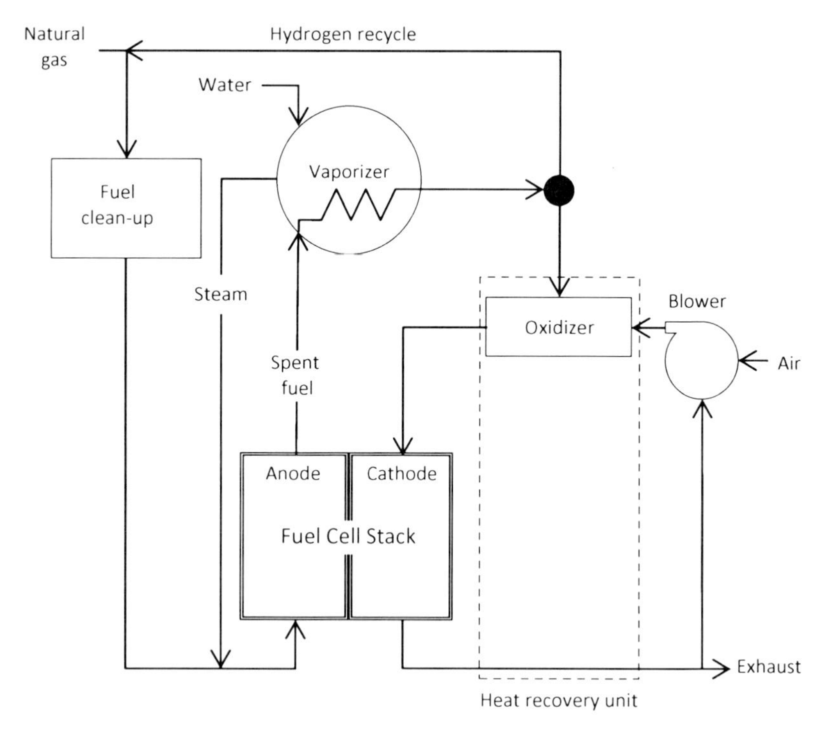

Molten Carbonate Fuel Cells, Overview, Fig. 1 Schematic diagram of internal reforming MCFC plants

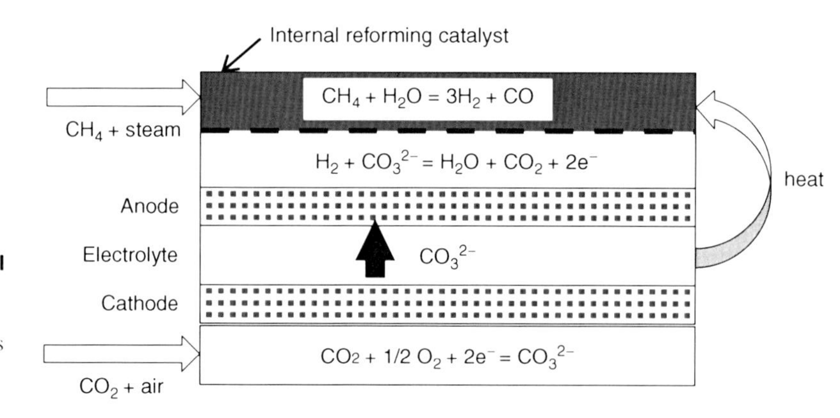

Molten Carbonate Fuel Cells, Overview, Fig. 2 Chemical and electrochemical reactions for internal reforming MCFCs

Figure 3 shows a typical cell construction of the MCFCs. The electrolyte plate is molten carbonate supported by a porous α- or γ-lithium aluminum oxide. The anode and cathode are located on either side of the electrolyte plate, and they are held in a flow field which contains the current collectors, corrugated plates, and bipolar plates [1–4]. The material technology of these components is as follows.

Electrodes

The conventional anode materials are oxide dispersed Ni-based alloys such as Ni–Al, Ni–Cr, and Ni–Al–Cr. The porosity and thickness are around 50 % and 1 mm, respectively. The anode is in a reducing atmosphere at high temperature with compression by stacking of the cells; therefore, the anti-creep property is an important factor for the anode. The aluminum, chromium,

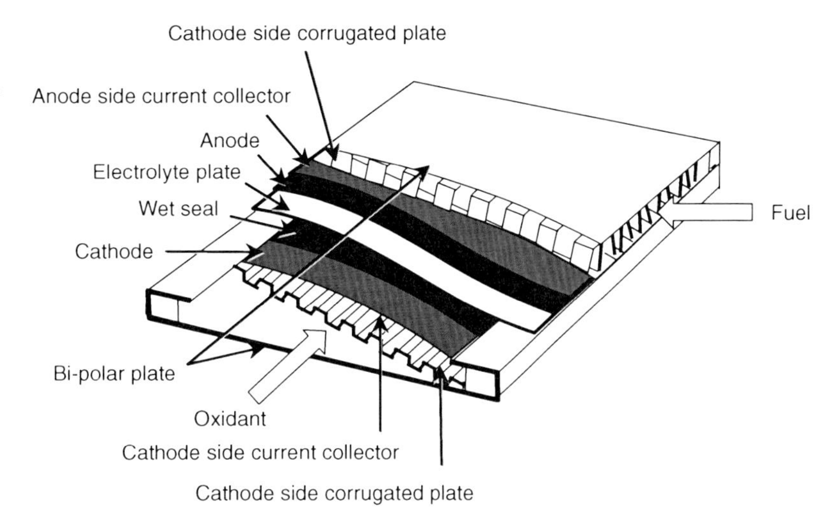

Molten Carbonate Fuel Cells, Overview, Fig. 3 Basic construction of MCFCs

and their oxides are added to improve the anti-creep property [2, 4, 5].

The conventional cathode material is lithiated nickel oxide that is formed by in situ oxidation during cell conditioning. The porosity and thickness are also around 50 % and 1 mm, respectively. The cathode polarization and dissolution/deposition, which are very important factors of the performance and lifetime, are strongly influenced by the cathode pore structure, electrolyte composition, and operating conditions. The nickel oxide slowly dissolves into the electrolyte as Ni^{2+} with an acidic dissolution mechanism as follows:

$$NiO = Ni^{2+} + O^{2-} \quad (6)$$

The Ni^{2+} is reduced to metal nickel in the electrolyte by hydrogen from the anode side and forms Ni shorting. To reduce the Ni shorting, the solubility of Ni^{2+} must be decreased. The $[O^{2-}]$ in the electrolyte determines solubility of Ni^{2+}, and $[O^{2-}]$ is defined by the composition of the electrolyte, temperature, and carbon dioxide partial pressure. More details will be given later. Various alternative materials, such as $LiCoO_2$, $LiCoO_2$-coated NiO, Ni–Fe–MgO, and $LiFeO_2$-based materials have been evaluated for their solubility and electronic conductivity, but further improvement is required to replace the NiO [2–6].

The sintered porous metal sheet electrodes have been installed in the stack as a conventional technology, but to reduce the manufacturing costs, the green sheets of the electrode are installed in the stack as a new technology.

Electrolyte

The electrolyte matrix is porous α- or γ-lithium aluminum oxide with ca. 50 % porosity and 1 mm thickness. To immobilize the molten carbonate by capillary force, the pore diameter is less than 0.5 μm. To fabricate such fine pores, the matrix is made from an ultrafine powder. The stable condition of the α-lithium aluminum oxide is at a lower temperature than that of the γ-lithium aluminum oxide, and the boundary is around 650 °C. Recently, the operating temperature is generally below 650 °C based on the durability; therefore, the α phase is selected. The change in the phase leads to a pore structure change, so fine control of the operating conditions is required. To maintain the fine pores, a uniform particle size is important to decrease the Ostwald ripening, coarsening process [2, 4, 5].

The major compositions of the molten carbonate are Li/K and Li/Na eutectic melts. The Li/K carbonate has a higher oxygen solubility than the Li/Na carbonate; therefore, the cathode polarization is also lower. The Li/Na carbonate has a higher ionic conductivity than the Li/K carbonate; therefore, the ionic resistance polarization is also lower. These properties are criteria for optimizing the cell performance. The properties of the carbonate also influence the durability. The molten carbonate has an equilibrium with the carbon dioxide partial pressure:

$$CO_3^{2-} = CO_2 + O^{2-} \quad (7)$$

$[O^{2-}]$ decreases if the carbon dioxide partial pressure increases; therefore, the carbonate melt changes to acidic. The solubility of NiO then increases according to the equilibrium of Eqs. 6 and 7. Therefore, the durability of the pressurized system is more severe than the ambient pressure system. The cation species also affects the basicity. A higher charge density, which is smaller ion diameter and higher valence, produces more basic property. The Li/Na eutectic carbonate is more basic than the Li/K eutectic; therefore, the solubility in the Li/Na eutectic carbonate is lower than that in the Li/K eutectic carbonate. An alkaline earth metal or rear earth metal additive to the electrolyte decreases the NiO solubility by the control of basicity [2–6].

The electrolyte management is essential for the performance and durability. The function of the electrolyte plate is ionic conduction and gas separation; therefore, the pores of the matrix are fully filled by the electrolyte with a strong capillary force. The gas diffusion electrode requires gas diffusion and an ionic conduction path; therefore, the pore of the porous electrode is partially filled by the electrolyte with a medium capillary force. The amount of electrolyte and relative pore diameter of the electrolyte matrix, anode, and cathode must be then maintained during whole of the lifetime [2, 4, 5, 7–9].

Flow Field

The flow field is constructed of an austenitic stainless steel-based material due to its anticorrosion property and cost. On the anode side, carburization enhances the corrosion rate. In the anodic atmosphere, metallic nickel is stable, and so nickel coating improves the durability of the stainless steel. On the cathode side, chromium easily dissolves from the stainless steel. Therefore, the outer layer and inner layer are porous $LiFeO_2$ and dens chromium-rich oxide layer as the corrosion protection layer, respectively. The corrosion of the current collectors increases the electronic resistance and decreases the amount of electrolyte. Therefore, the material is selected with a lower electronic resistance of the corrosion layer and lower consumption of electrolyte. Based on these requirements, SUS316L is better than SUS310 even it has higher Ni and Cr contents [2, 4, 5].

A characteristic of the corrosion by molten carbonate is complex of high temperature corrosion, which is controlled by diffusion in solid state, and dissolution of corrosion products into molten carbonate. This means that the corrosion rate asymptotes with the dissolution rate; therefore, a decrease in the dissolution rate is essential for the MCFC material [10].

Internal Reforming

The steam reforming is an endothermic reaction; therefore, the reaction is used for hydrogen production and control of the temperature distribution due to the cooling ability by the endothermic reaction. The catalyst for internal reforming is generally a nickel-based material supported on an oxide such as MgO, calcium aluminate, or α-alumina. Molten carbonate poisons the reforming catalyst; therefore, placement and improvement of the catalysts are important for the internal reforming system [2, 5].

Cell Performance and Deterioration

The electrical power generation efficiencies of commercialized residential fuel cell systems using a natural gas-based fuel are 37–40 %, 42 %, 47 %, and 45–50 % in LHV for the PEFCs, PAFCs, MCFCs, and SOFCs in 2011, respectively. Generally, a high-temperature fuel

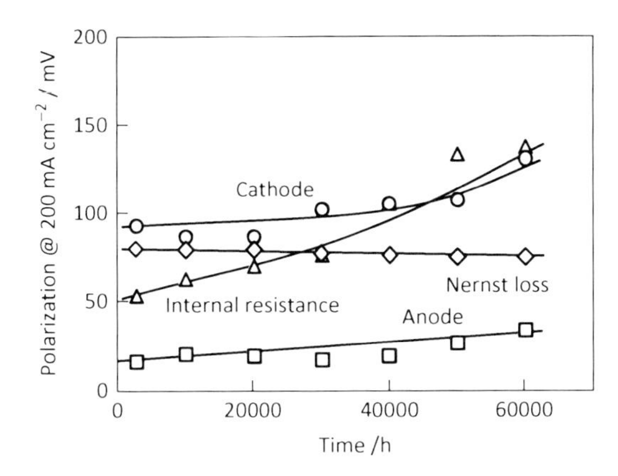

Molten Carbonate Fuel Cells, Overview, Fig. 4 Polarization as a function of time for a small-size single cell at 873 K and 200 mA cm^{-2}

Molten Carbonate Fuel Cells, Overview, Table 1 small size single cell experiment

Electrolyte	Li_2CO_3/Na_2CO_3 = 64/40 %
Pressure	0.297 MPa
Temperature	873 K
Current density	200 mA cm^{-2}
Fuel	30 % H_2O added H_2/CO_2 = 80/20 %
	H_2 utilization = 60 %
Oxidant	10 % H_2O added Air/CO_2 = 90/10 %
	O_2/CO_2

cell has high electric power generation efficiency. Although the energy efficiency is high for high-temperature fuel cells, the cathode reaction and internal resistance polarizations are the major factors affecting the energy loss.

Figure 4 shows the polarization at 200 mA cm^{-2} as a function of time for a small size single cell with frequent electrolyte additions. The conditions of the experiment are summarized in Table 1. In Fig. 4, the cathode, anode, internal resistance, and Nernst loss are polarizations of the cathode reaction, anode reaction, and internal resistance which is mainly the ionic resistance of the electrolyte plate and electronic resistance of the corrosion layer on the current collector and the Nernst loss, which is the difference between open circuit voltage of the inlet gas and effective the open circuit voltage during operation. During the initial period, the cathode polarization is the highest polarization. The cathode polarization and internal resistance increase with aging of the cell components [11].

The electrolyte losses by corrosion of the metal components and decrease in capillary force with particle coarsening of the electrolyte matrix; therefore, the internal resistance and the cathode polarization increase. The cathode polarization increases with not only electrolyte loss but also dissolution of the cathode itself. Improvement of the electrolyte management and cathode stability are key factors for improving the reliability and durability of the MCFC systems [4, 5, 7–11].

Future Directions

The molten carbonate fuel cell systems have one of the highest energy conversion efficiency systems for distributed generation, and reliability is improving. Furthermore, they are free from rare elements such as noble metals and rare earth metals. Therefore, further effort and support to improve reliability and market entry will be required for real commercialization.

Cross-References

▶ Fuel Cells, Principles and Thermodynamics
▶ High-Temperature Molten Salts
▶ Molten Carbonate Fuel Cells, Overview

References

1. Farooque M, Ghezael-Ayagh H (2003) Handbook of fuel cells – fundamentals technology and applications. In: Vielstich W, Lamm A, Gasteiger H (eds), vol 4. Wiley, p 942
2. Yuh C, Hilmi A, Farooque M, Leo T, Xu G (2009) Direct fuel cell materials experience. ECS Trans 17(1):637
3. Mugikura Y (2003) Handbook of fuel cells – fundamentals technology and applications. In: Vielstich W, Lamm A, Gasteiger H (eds), vol 4. Wiley, p 907
4. Fujita Y (2003) Handbook of fuel cells – fundamentals technology and applications. In: Vielstich W, Lamm A, Gasteiger H (eds), vol 4. Wiley, p 969

5. Hoffmann J, Yuh CY, Jopek AG (2003) Handbook of fuel cells – fundamentals technology and applications. In: Vielstich W, Lamm A, Gasteiger H (eds), vol 4. Wiley, p 921
6. Mitsushima S (2009) Handbook of fuel cells – fundamentals technology and applications. In: Vielstich W, Yokokawa H, Gasteiger H (eds), vol 6. Wiley, p 960
7. Mitsushima S, Kuroe S, Matsuda S, Kamo T (1998) Polarization model for molten carbonate fuel cell cathodes. Denkikagaku Electrochemistry 66:817
8. Mitsushima S, Yamaga K, Okada H, Kamo T (1997) Lifetime modeling for molten carbonate fuel cells. Denkikagaku Electrochemistry 65:395
9. Mitsushima S, Takahashi K, Fujimura H, Kamo T (1998) Lifetime analysis method for molten carbonate fuel cell stacks. Denkikagaku Electrochemistry 66:1005
10. Mitsushima S, Nishimura Y, Kamiya N, Ota K (2004) Corrosion model for iron in the presence of molten carbonate. J Electrochem Soc 151:A825
11. Morita H, Kawase M, Mugikura Y, Asano K (2010) Degradation mechanism of molten carbonate fuel cell based on long-term performance: Long-term operation by using bench-scale cell and post-test analysis of the cell. J Power Sources 195:6988

Molten Steel, Measurement of Dissolved Oxygen

Ulrich Guth
Kurt-Schwabe-Institut für Mess- und Sensortechnik e.V. Meinsberg, Waldheim, Germany
FB Chemie und Lebensmittelchemie, Technische Universität Dresden, Dresden, Germany

Theoretical Background

The monitoring of oxygen activity in metal melts is important to shorten the melting processes and to ensure the product quality. Basic principle for the activity measurement is a galvanic cell using solid electrolytes. In molten iron the melt with dissolved oxygen is measuring electrode in cell like this [1]:

$$Fe, O_2(p_{\mathrm{melt}})|MgO - ZrO_2|Me, O_2(p_{\mathrm{ref}}) \quad (1)$$

The oxygen activity is obtained by means of a modified Nernst's equation:

$$U_{\mathrm{eq}} = \bar{t}_{\mathrm{ion}} \frac{RT}{4F} \ln \frac{p_{\mathrm{melt}}}{p_{\mathrm{ref}}} \quad (2)$$

where $\bar{t}_{\mathrm{ion}}$ is the average ionic transference number of ions in solid electrolyte and p the oxygen partial pressures. With increasing temperature and decreasing oxygen partial pressure, the electronic conductivity and therefore electronic transference number rises and the Nernst's voltage is diminished ($\bar{t}_{\mathrm{ion}} + \bar{t}_e = 1$).

The oxygen activity is given by

$$a_O = \mathrm{const} \cdot p_{\mathrm{melt}}^{1/2} \quad (3)$$

For measurements of oxygen dissolved in pure iron, the following equation can be derived from thermodynamic data [2]:

$$\begin{aligned} U_O(O \text{ diss. in } Fe, l)/mV &= -1404 + [0.355 \\ &\quad +0.0992\ lg(c/c_{sa})]T/K \\ &= -818 - [0.105 \\ &\quad -0.0992\ lg(c/w\%O)]T/K \end{aligned}$$

Other relationships with empirical coefficients have been developed which have been taken into account the interaction of oxygen with alloy components [3].

Due to the low oxygen concentration in steel melts, air reference electrodes cannot be used practically. Under this condition of high oxygen gradient, oxygen permeates through the solid electrolytes and increases its activity in the vicinity of the measuring electrode and leads to an incorrect measurement. According to the *Wagner's theory* of permeation, the oxygen flux is given by [4]

$$j_{O_2} = -\frac{RT}{4^2 \cdot F^2 \cdot L} \cdot \int_{\ln pO_2^{/}}^{\ln pO_2^{//}} \frac{\sigma_{el} \cdot \sigma_{\mathrm{ion}}}{\sigma_{el} + \sigma_{\mathrm{ion}}} \cdot d \ln pO_2$$

where σ_{el}, σ_{ion}, and L are the electronic conductivity, the ionic conductivity, and the thickness of the electrolyte, respectively. Due to the given conductivities of solid electrolytes, the permeation flux only depends on the gradient in oxygen concentration on both sides of electrolyte. To avoid the oxygen flux solid metal, metal oxide systems are preferred as references [3]. Provided that the two solid phases are immiscible and pure and the temperature is high enough so that the chemical equilibrium can be established, for *Ni*, *NiO* the chemical reaction can be expressed as

$$NiO \leftrightharpoons Ni + 1/2O_2 \tag{4}$$

The cell reaction can occur without molecular oxygen (▶ *Kröger-Vinks Notation of Point Defects* is used) between oxide ions in *NiO* and the oxide ion vacancies in solid electrolytes according to

$$NiO\,(s) + V_O^{\circ\circ}(\text{zirconia}) + 2e'\,(\text{metal}) \leftrightharpoons O_O^x \times (\text{zirconia}) + Ni(s)$$

With the mass action law results:

$$K_{p,x} = \frac{x_{Ni}}{x_{NiO}} p_{O_2}^{1/2}.$$

For pure substances is the mole fraction $x_{Ni} = 1$ and $x_{NiO} = 1$ so that results

$$K_p(T) = p_{O_2}^{1/2} \tag{5}$$

This value can be calculated by means of standard Gibbs free energy of reaction: $\Delta_r G^\circ$

$$\ln p_{O_2} = -\frac{\Delta_r G^\circ}{RT} \tag{6}$$

That means for a certain temperature, the equilibrium partial pressure of oxygen of an oxide metal mixture is constant. This allows to use such systems as reference systems.

With an air reference for computing the thermodynamic voltage, the *Ni*, *NiO* is obtained:

$$U_O(Ni, NiO)/mV = -1,202 + 0.431T/K \quad \text{for } 650 - 1,100\ ^\circ\text{C}$$

The electrode potential for *Cu*, Cu_2O is given by

$$U_O(Cu,\ Cu_2O)/mV = -871.9 + 0.3775\ \text{T/K} \quad \text{for } 450 - 750\ ^\circ\text{C}$$

Steel Probes

Sensors for measurements of oxygen in molten steel are the most important application in liquid metal sensing. As an electrolyte *MgO* partially stabilized, zirconia is used because of its high thermal shock resistance due to phase transformations.

The sensor consists of a small ceramic tube made of partially stabilized zirconia filled with a mixture of chromium and chromium oxide. The measuring electrode is the metallic melt. Between this electrode and a molybdenum wire in close contact to the Cr, Cr_2O_3 mixture the voltage is measured. A thermocouple (not to seen in Fig. 1) is used to measure the

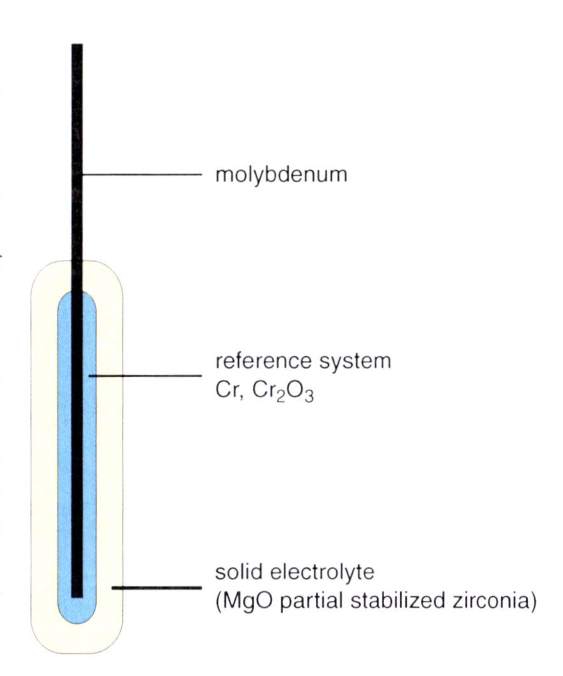

Molten Steel, Measurement of Dissolved Oxygen, Fig. 1 Cross section of a steel probe for dissolved oxygen (schematically)

temperature simultaneously. The sensor is located on the top of a board tube (more than 1 m in length) and once only usable. The lifetime is of course very short but sufficiently long to provide data in few seconds range. The calibration curves are calculated using the Nernst's equation and taking into account the electronic conductivity of the solid electrolyte [2]. The main aim for applying such sensors is the monitoring of the deoxidation process in row iron and the production of ferrous alloys. Oxygen in steel makes it brittle and has to be removed by adding aluminum. Figure 1 shows an alternative setup of such sensors fabricated in a dipping technology. Sensors for dissolved oxygen are produced in a large scale and applied worldwide in steel production.

This principle is also used for determination of oxygen in other metal melts like sodium (cooling circuit in fast breeder reactors), lead, silver, copper, and aluminum [1, 3]. In contrast to discontinuous measurement in molten steel those measurements are carried out continuously.

Cross-References

▸ Sensors
▸ Solid Electrolytes

References

1. Ullmann H (1993) Ceramic gas sensors (in German). Akademie, Berlin
2. Möbius HH (1991) Solid state electrochemical potentiometric sensors for gas analysis. In: Göpel W, Hesse J, Zemel JN (eds) Sensors a comprehensive survey. VCH, New York/Weinheim, pp 1119–1121
3. Fischer WA, Jahnke D (1975) Metallurgical electrochemistry (in German). Stahleisen/Springer, Düsseldorf/Berlin/Heidelberg/New York
4. Gelling PJ, Bouwmeester HJM (1997) The CRC handbook of solid state electrochemistry. CRC Press, Boca Raton/New York/London/Tokyo, Chapter 14

Multiscale Modeling

Alejandro A. Franco
Laboratoire de Réactivité et de Chimie des Solides (LRCS) - UMR 7314, Université de Picardie Jules Verne, CNRS and Réseau sur le Stockage Electrochimique de l'Energie (RS2E), Amiens, France

Introduction

Development of a deep understanding of fuel cells', batteries', and supercapacitors' operation principles will help on achieving significant advances on their controlled design and optimization in both applied science and industry communities.

Porous electrodes are the pivotal components of modern electrochemical power generators. Porous electrodes are inherently multiscale systems as they are made of multiple coexisting materials, each of them ensuring a specific function in their operation (Fig. 1). Such electrodes' structural complexity has been historically driven by the needs of reducing the device cost and of enhancing its efficiency, stability, and safety. The use of nano-engineered materials and chemical additives (in the case of batteries) allowed a significant progress toward these goals. For instance, in the case of polymer electrolyte membrane fuel cells (PEMFCs), the electrodes are constituted by metallic nanoparticles of few nanometers size having the role of electrocatalyst and are supported on carbon particles of few microns size having the role of electronic conductor. The resulting complex structure is in turn embedded within a proton-conducting ionomer, arising into a composite electrode of few micrometers thick (Fig. 1a). A similar viewpoint can be provided for lithium-air batteries (LABs) [43] or lithium-ion batteries (LIBs) where intercalation or conversion materials are frequently coated by carbon to enhance the electronic

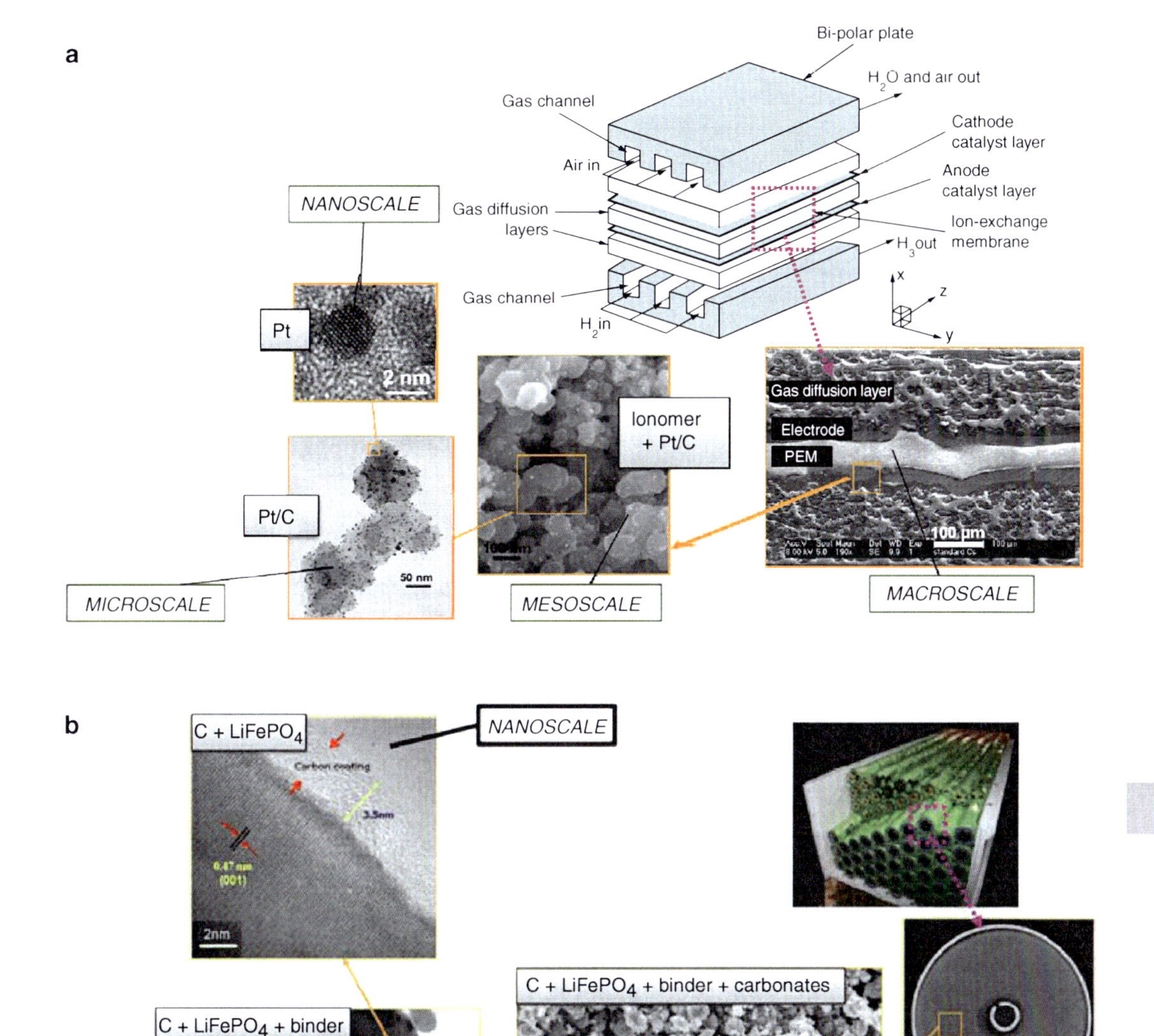

Multiscale Modeling, Fig. 1 Electrode multiscale structure for (**a**) PEM fuel cells [1] and (**b**) lithium-ion batteries [2]

conductivity and surrounded by aprotic electrolytes (Fig. 1b).

During the operation of the electrochemical cell, different physicochemical mechanisms occur at multiple spatial scales, such as the elementary oxidation and reduction reactions, the electronic and ionic charge transport, the uncharged species transport, the thermal management, and the mechanical stresses. The competition between these mechanisms involves multiple (time-evolving) phases:

- Solid (e.g., active materials) and liquid phases (e.g., carbonate electrolyte) as in the case of intercalation and conversion lithium-ion batteries and supercapacitors [3, 4]
- Gas (e.g., reactants) and solid phases (e.g., oxide electrolyte) as in the case of high-temperature fuel cells such as solid oxide fuel cells (SOFCs) [5]
- Solid (e.g., carbon supports), liquid (e.g., water or dimethylsulfoxide electrolytes), and gas phases (e.g., O_2) in devices like low-temperature fuel cells (PEMFCs) [1] and metal-air batteries [6].

Because of these structural and functional complexities, the rationalization of the devices design passes inevitably through the need of developing an overall understanding of the interplaying between the mechanisms and the scales in the components (electrodes, separator, etc.). Thanks to a growing progress in scientific computational techniques, multiscale modeling approaches and numerical simulation are emerging as powerful tools to bridge the gaps between the chemical and structural properties of the materials, the components, and the performance and durability of electrochemical devices for energy conversion and storage (Fig. 2).

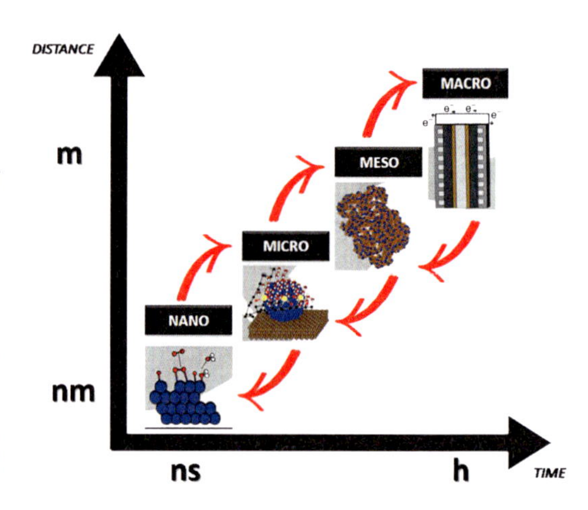

Multiscale Modeling, Fig. 2 Typical relevant scales for the simulation of an electrochemical power generator: elementary reactions and transport processes at the atomistic/molecular level (nanoscale), electrochemical interfaces (e.g., polymer + water/catalyst) at the microscale, transport processes (e.g., in the electrode pores) at the mesoscale, and mechanical/thermal stresses at the device level (macroscale)

Definitions and General Concepts

The current scientific method used in the modeling discipline can be schematized as the *process* in Fig. 3. In this process, the first and second steps consist respectively of defining the physical problem (i.e., the system which will be modeled: the electrode alone? the complete cell? an active particle? etc.), and on indentifying the observables one would intend to simulate with the model (e.g., electrode potential? cell potential? active area evolution?). Then, the third and fourth steps consist respectively of defining the structural model which will be used (i.e., the geometrical assumptions: e.g., 1D, 2D, or fully 3D representation of the electrode?) and the physics to be treated (e.g., detailed electrochemistry? ion transport? both coupled?). These two steps are crucial as they strongly determine the choice of the simulation approach in step five (e.g., quantum mechanics? molecular dynamics? kinetic Monte Carlo? coarse-grain molecular dynamics, continuum fluid dynamics? or combination of several of such approaches?). Then it arises the choice of the mathematical formulations for performant calculations and of the appropriate numerical algorithms, software, and hardware (e.g., parallel computing or not?) to proceed with the simulation of the observables. These observables are then compared with the available experimental data, preferentially obtained with model experiments, i.e., experiments designed to be representative of the model (e.g., geometrical) assumptions. This comparison will allow the model validation or its improvement in terms of its structural/geometrical assumptions or physics accounted for. After several iterations between theory and experiment, one could expect achieving on "producing" a model with predictive capabilities of the electrochemical device operation. The key step in this modeling process is the choice of the modeling

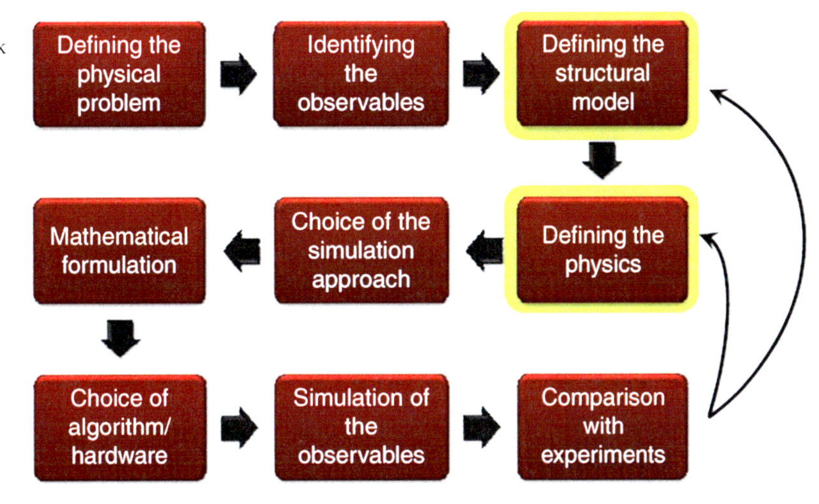

Multiscale Modeling, Fig. 3 The modeling work process

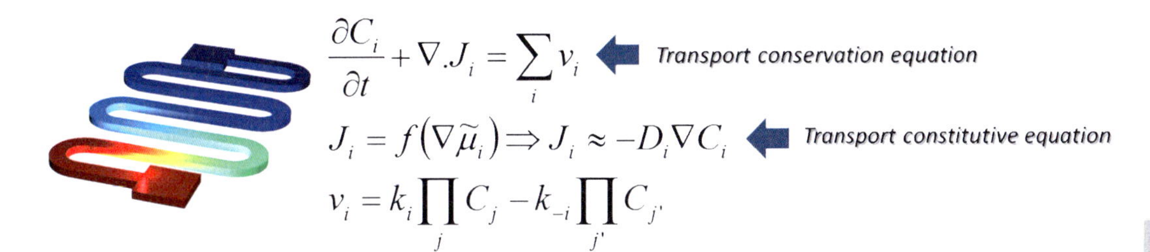

Multiscale Modeling, Fig. 4 Tubular chemical reactor with transport and volume chemical reactions

simulation approach among three categories: *multiphysics*, *multiscale*, and *multiparadigm* modeling approaches [2].

Multiphysics Modeling Approaches

Multiphysics models usually refer in published literature to the models which mathematically describe the interplaying of mechanisms belonging to different physical domains. Multiphysics models include models describing these multiple mechanisms within a single and unique spatial scale (e.g., a model describing the impact of heating on the mechanical stress of a material [7] or a model describing a chemically reactive flow along a tubular reactor (Fig. 4)). Multiphysics models can be, by construction, "multiscale" on time, as it can be built on the basis of mathematical descriptions of multiple mechanisms with different characteristic times (e.g., when the heat dissipation time constant is different to the material deformation time constant).

Multiscale Modeling Approaches

Multiscale models typically refer to multiphysics models accounting for mathematical descriptions of mechanisms taking place at different spatial scales [8]. Multiscale models aim, by construction, to considerably reduce empirical assumptions than can be done in simple multiphysics models. This is because they explicitly describe mechanisms in scales neglected in the simple multiphysics model. Actually, multiscale models have a hierarchical structure: that means that solution variables defined in a lower hierarchy domain have finer spatial resolution than those solved in a higher hierarchy domain. Consequently, physical and chemical quantities of smaller length-scale physics are evaluated with a finer spatial resolution to resolve the impact of the corresponding small-scale geometry. Larger-scale quantities are in turn calculated with coarser spatial resolution, homogenizing the (possibly complex) smaller-scale geometric features.

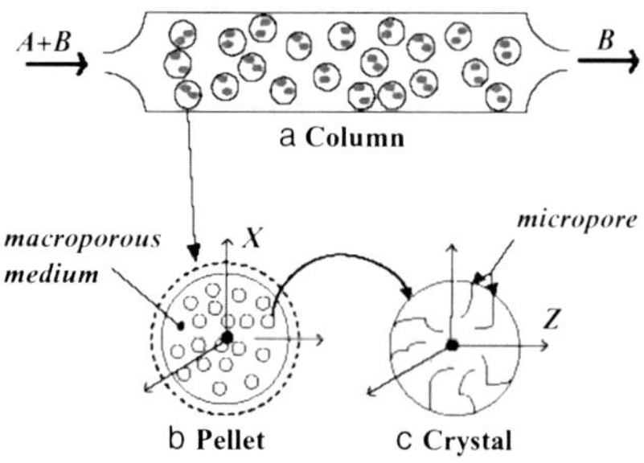

A+B
B
a Column
macroporous medium
X
micropore
Z
b Pellet
c Crystal

$$\frac{\partial C_i^{Column}}{\partial t} + \nabla_y . J_i^{Column} = S_i^{Column/Macrop} = \gamma\, J_i^{Macrop}$$

$$\frac{\partial C_i^{Macropor}}{\partial t} + \nabla_x . J_i^{Macropor} = S_i^{Macrop/Microp} = \gamma'\, J_i^{Microp}$$

$$\frac{\partial C_i^{Microp}}{\partial t} + \nabla_z . J_i^{Microp} = \sum_i v_i$$

$$v_i = k_i \prod_j C_j^{Microp} - k_{-i} \prod_{j'} C_{j'}^{Microp}$$

Multiscale Modeling, Fig. 5 Zeolite-based chemical reactor (diffusion in multiple scales and volume reactions within the crystal) (The catalytic reactor figure is courtesy of Prof. Christian Jallut, Université Claude Bernard Lyon 1)

Multiscale models can be based only on continuum mathematical descriptions, e.g., as in the zeolite-based chemical reactor illustrated in Fig. 5, where the conservation equation describing the reactant transport along the reactor has a source term being calculated from the boundary condition of a macropore diffusion model with a source term calculated in turn from the boundary condition of a micropore diffusion model with a reaction kinetics source term [44].

Multiparadigm Modeling Approaches

The mathematical descriptions in a multiscale model can be part of a single simulation paradigm (e.g., only continuum) or of a combination of different simulation paradigms (e.g., stochastic model describing a surface reaction coupled with a continuum description of reactant transport phenomena). In the latter, one speaks about *multiparadigm models*. Multiparadigm models can be classified in two classes: *direct* or *indirect*.

Direct Multiparadigm Models

Direct multiparadigm models are multiparadigm models which include "on-the-fly" mathematical couplings between descriptions of mechanisms realized with different paradigms, for example, coupling continuum equations describing transport phenomena of multiple reactants in a porous electrode with kinetic Monte Carlo (KMC) simulations describing electrochemical reactions among these reactants. Several numerical techniques are well established to develop such a type of models applied in the simulation of physicochemical processes, e.g., catalytic and electro-deposition processes [9, 10]. In the field of catalysis, KMC simulations have been used to calculate instantaneous kinetic reaction rates on a catalyst calculated iteratively from concentrations in turn calculated from computational fluid dynamics (CFD)-like continuum transport models [11]: the calculated reaction rates are in fact sink/source terms for the transport models (Fig. 6).

Indirect or "Hierarchical" Multiparadigm Models

Even if very precise, direct multiparadigm methods are computationally expensive. For this reason, *indirect (or hierarchical) multiparadigm models* consisting of injecting data extracted from a single-scale model into upper-scale models via their parameters constitute an elegant alternative (Fig. 7).

For example, the prediction of the efficiency of an electrochemical power generator as a function of the chemical and structural properties of the used electrode materials needs the calculation of the electrochemical activity and the transport of charges, reactants, and products as a function of these properties.

Electrochemical Activity Prediction A hypothetical indirect multiparadigm model allowing to calculate the electrochemical activity of the

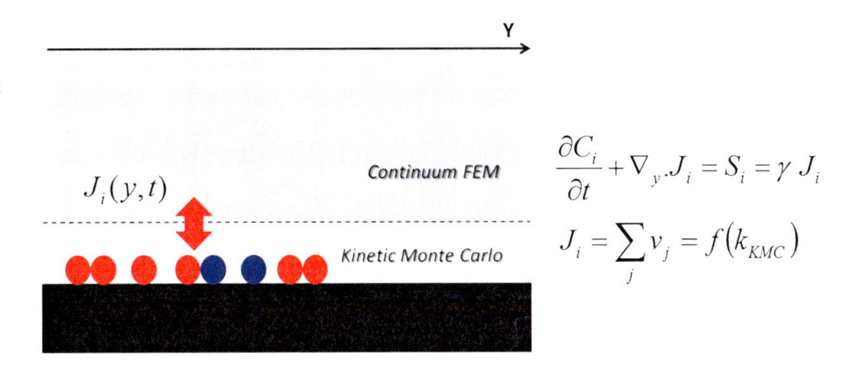

Multiscale Modeling, Fig. 6 Surface chemical reactor (volume transport coupled to surface reactions)

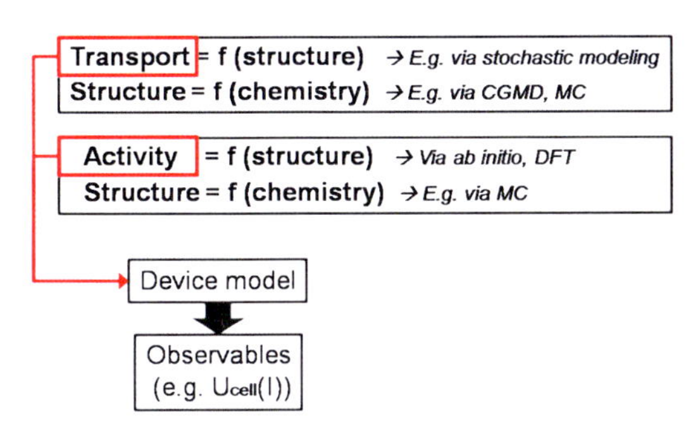

Multiscale Modeling, Fig. 7 The indirect multiparadigm simulation method of electrochemical power generators

cell could be built up by the following hierarchical procedure:

1. Calculating the structural properties of the electrochemical active material from its chemical composition (e.g., for the case of catalyst nanoparticles, by using MC or molecular dynamics)
2. Calculating the electrochemical activity of the calculated structure (e.g., still in the case of catalyst nanoparticles, by using ab initio Density Functional Theory (DFT)).

For example, in the field of catalysis, one can use Nudged Elastic Band (NEB) calculations [12] to estimate the values of the activation energies of single elementary reaction kinetic steps and then inject them into Eyring's expressions:

$$k = \kappa \frac{k_B T}{h} \exp\left(-\frac{E_{act}}{RT}\right)$$

for the calculation of the individual reaction rates at the continuum level [13],

$$v_i = k_i \prod_y a_y^{\nu} - k_{-i} \prod_{y\prime} a_{y\prime}^{\nu\prime}$$

which can in turn be used for the calculation of the evolution of the surface or volume concentrations of the reaction intermediates, reactants, and products, following

$$K_n \frac{da_y}{dt} = \sum_i v_i - \sum_j v_j$$

These kinetic rates and concentrations can then be used as on-the-fly inputs of coupled equations describing transport processes.

The activation energy in Eyring's equation has to be corrected by the effect of the electrochemical potential jump through the adlayer (compact layer). This potential can be, for example, estimated from [14, 15]

$$E_{act} = E_{act}^{NEB} + f(\sigma, \theta_i)$$

where the activation energy calculated by NEB is corrected by a function of the electronic charge density and the (time-dependent) coverage of the reaction intermediates. This function has to be calculated from a model describing the electrochemical double layer (EDL) structure at the interface between the electrolyte and the active material. Franco pionnered the development of such models describing the impact, in a PEMFC environment, of REDOX reactions onto the EDL structure and conversely, the impact of the EDL structure on the REDOX kinetics [15]. The theory has been recently extended by Franco et al. to describe the electric field screening by the solvent spatial heterogeneities within the diffuse layer region (K.H. Xue, M. Quiroga, A.A. Franco, in prepration 2013). This new EDL model also accounts by the presence of charged polymers (e.g., Nafion in PEMFCs), their dynamics of adsorption/desorption on/from the catalyst surface and their impact on the REDOX reactions kinetics.

Moreover so-called phase-field models start to become popular in describing phase formation and evolution in LIB electrodes. The phase-field modeling approach, initially developed for describing phase-separation and coarsening phenomena in a solid [16] and later for electrochemistry applications [17, 18], consists first on considering the total free energy of the intercalation (or conversion material) as follows:

$$\mathrm{F} = \oint_V (f_{bulk} + f_{grad} + f_{app})dV + \oint_V \oint_{V\prime} [f_{non\,local}]dV'dV$$

where f_{bulk} is the local chemical free-energy density (function of the composition), f_{grad} is the gradient energy density (accounting for the heterogeneities penalties), f_{app} is the coupling potential energy between the applied fields and order parameters, and the second integral accounts for the long-range interactions [2]. The chemical potential of each phase is given by

$$\mu_j = \frac{\partial F}{\partial c_j(\vec{r}, t)}$$

and the conservation equation governing the phases formation and displacement is given by

$$\frac{\partial c_i}{\partial t} = -\nabla.J = \nabla\left(M_{ij}\nabla\mu_j\right)$$

where M_{ij} refers to the mobility of each phase (could depend on the phases concentrations). This equation is known as the *Cahn-Hilliard* equation. This is a fourth-order equation, extremely sensitive to initial conditions and parameters values, which thus needs appropriate numerical schemes to be solved. This motivated a stronger interest for applied mathematics which brings onto the development of highly accurate but fast numerical methods such as the Chebyshev-spectral method, the generalized Newton's method, fast Fourier transform methods, and multigrid methods, each method having pros and cons depending on the application problem [19, 20]. An example of a result for the simulation of a two-phase system is reported in Fig. 8.

Furthermore, it should be noticed that phase-field modeling is an elegant approach in which parameters can in principle be estimated from first principles calculations, such as the interphase energies in conversion reactions (e.g., CoO/Co°, Co°/Li_2O) (Fig. 9) [22, 23]. It has to be underlined that until now phase-field modeling has been mainly used for the simulation of intercalation reactions in LIBs (e.g., Bazant et al. work [24, 25]) and of the reactions of the solid electrolyte interphase formation [26].

Transport Properties Prediction A hypothetical indirect multiparadigm model allowing to calculate the electrochemical transport properties of the cell could be built up by the following hierarchical procedure:

1. Calculating the structural properties of the component from its chemical composition (e.g., for the case of polymer electrolyte membrane fuel cell (PEMFC) electrodes, by using coarse-grain molecular dynamics (CGMD)) (Fig. 10)
2. Calculating the transport properties of the calculated structure (e.g., still in the case of PEMFC electrodes, by using pore network modeling of liquid water propagation in the porous electrode and its impact on the effective oxygen diffusion properties).

More precisely, for the first point, CGMD allows calculating the tortuosity and porosity of the electrode as a function of the materials chemistry, which are used in turn for the estimation of the effective diffusion parameters

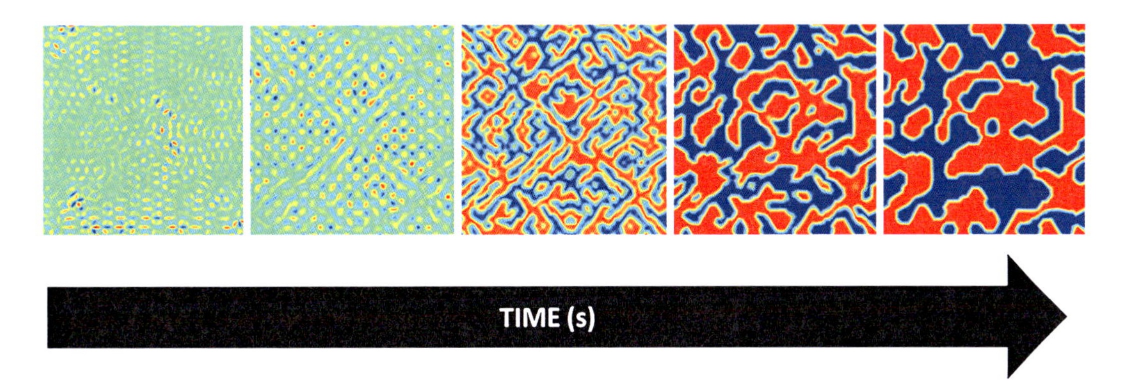

Multiscale Modeling, Fig. 8 Example of phase-separation dynamics calculation for a two-phase system [21]

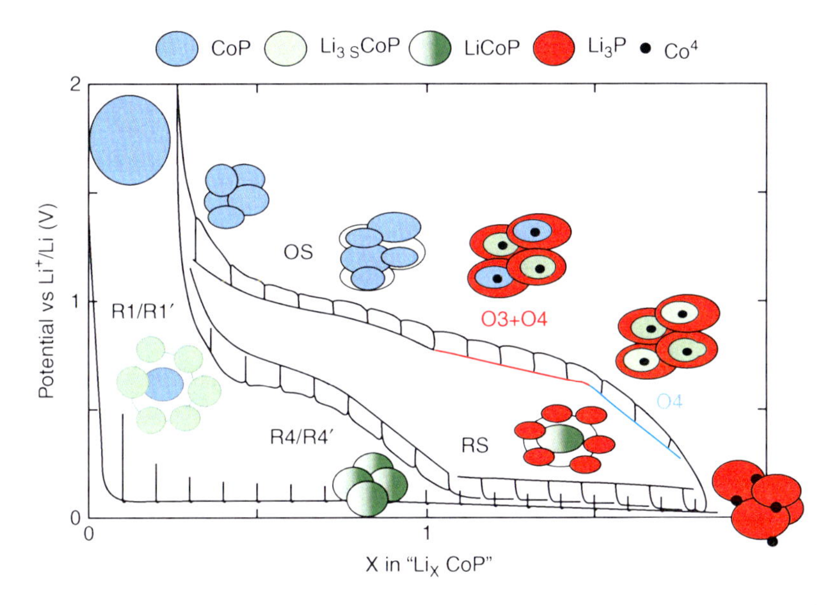

Multiscale Modeling, Fig. 9 GITT measurements carried out on the CoP/Li half-cell with steps of 1 h at C 10 (charge or discharge) and rest to open-circuit voltage until the potential slope is less than 30 mV h^{-1}. The experimental values are compared with the particles morphologies as estimated from DFT calculations [23]

used in continuum reactant transport models following, for example, Bruggeman-like relationships:

$$D_{eff} = \frac{\varepsilon}{\tau} D_0$$

CGMD has been recently used to predict structural properties of Nafion thin films inside the PEMFC electrodes (Fig. 11): the structural properties of such films are expected to be of crucial importance on the effectiveness of the electrochemical reactions and interfacial transport properties of protons and water [14, 15, 28, 29].

Material Degradation Prediction In 2002, Franco has invented a multiscale modeling approach (and an associated simulation package originally called MEMEPhys) dedicated to the simulation of PEM fuel cells (PEMFCs) (Fig. 12) but, later, extended for the simulation of other electrochemical systems, such as PEM

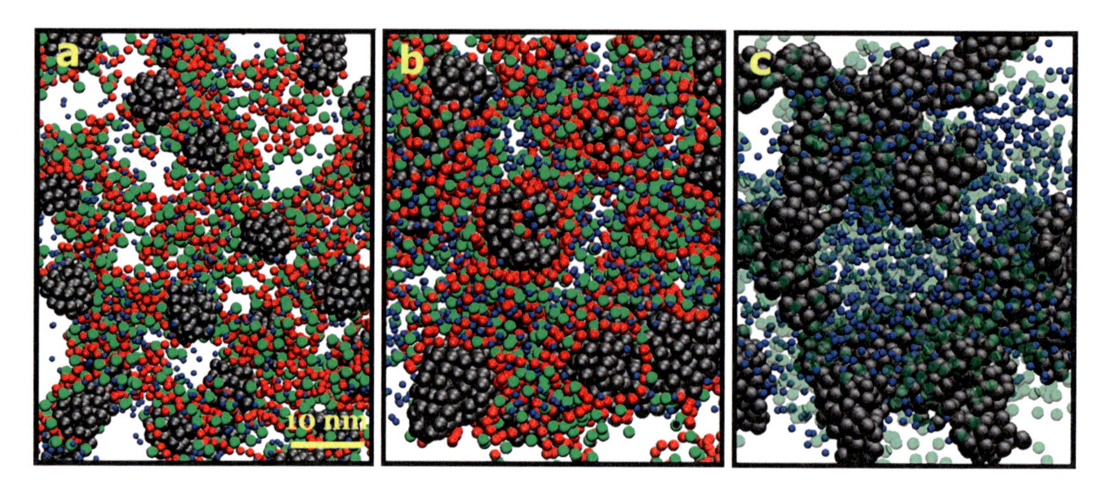

Multiscale Modeling, Fig. 10 Spontaneous microstructure formation: snapshots along the trajectory and during microstructure formation at (**a**) t) 100 ps, annealed configuration, (**b**) t) 40 ns, and (**c**) t) 320 ns, final equilibrated configuration. The *gray* spheres show the carbon particles. Nafion backbones correspond to *red* beads, while *green* beads show the side chains. Small *blue* beads represent hydronium ions inside the pore network. Water beads are not shown for better clarity. The equilibrated structure reveals separate hydrophilic (water, hydronium ions, and side chains) and hydrophobic (carbon particles and ionomer backbones) domains. In (**c**), the beads of Nafion backbones are removed, and Nafion side chains are represented by green transparent beads for better visualization of carbon clusters [27]

water electrolyzers (PEMWEs) [30] and lithium-air batteries (LABs) and LIBs [2, 31] (Fig. 13).

The model is an indirect multiparadigm approach: a continuum model describing electrochemical and transport mechanisms with parameters extracted from ab initio databases (for the reactions kinetic parameters as function of the catalyst chemistry and morphology) and CGMD calculations (for the materials' structural properties as function of their chemistry).

Franco has designed this model to connect within a nonequilibrium thermodynamics framework atomistic phenomena (elementary kinetic processes) with macroscopic electrochemical observables (e.g., I-V curves, EIS, $U_{cell}(t)$) with reasonable computational efforts. The model is a transient, multiscale, and multiphysics single electrochemical cell model accounting for the coupling between physical mechanistic descriptions of the phenomena taking place in the different component and material scales. For the case of PEMFCs, the modeling approach can account for detailed descriptions of the electrochemical and transport mechanisms in the electrodes, the membrane, the gas diffusion layers and the channels: H_2, O_2, N_2, and vapor H_2O transport macroscale description along the channels, H_2, O_2, N_2, and biphasic H_2O transport mesoscale description across the GDLs and the electrodes (including H_2O condensation between the C agglomerates), proton transport mesoscale description across the MEA, electron transport mesoscale description across the electrodes and GDLs, H_2, and O_2 diffusion microscale description across the on-catalyst ionomer film inside the electrodes, and the interfacial nanoscale mechanisms at the vicinity of the catalyst nanoparticles including both elementary kinetics and electrochemical double-layer effects. The model includes descriptions of coupled electrochemical aging processes (e.g., Pt and Pt_xCo_y oxidation/dissolution/ripening, carbon catalyst support corrosion, PEM degradation) [32–37].

For the simulation of the degradation, the model accounts for the numerical feedback between the sub-models describing the non-aging mechanisms (e.g., water transport across the porous electrode) and the sub-models describing the aging mechanisms (e.g., carbon corrosion or catalyst dissolution in the case of PEMFCs) (Fig. 14). At each time step of the simulation, the performance part of the model calculates the

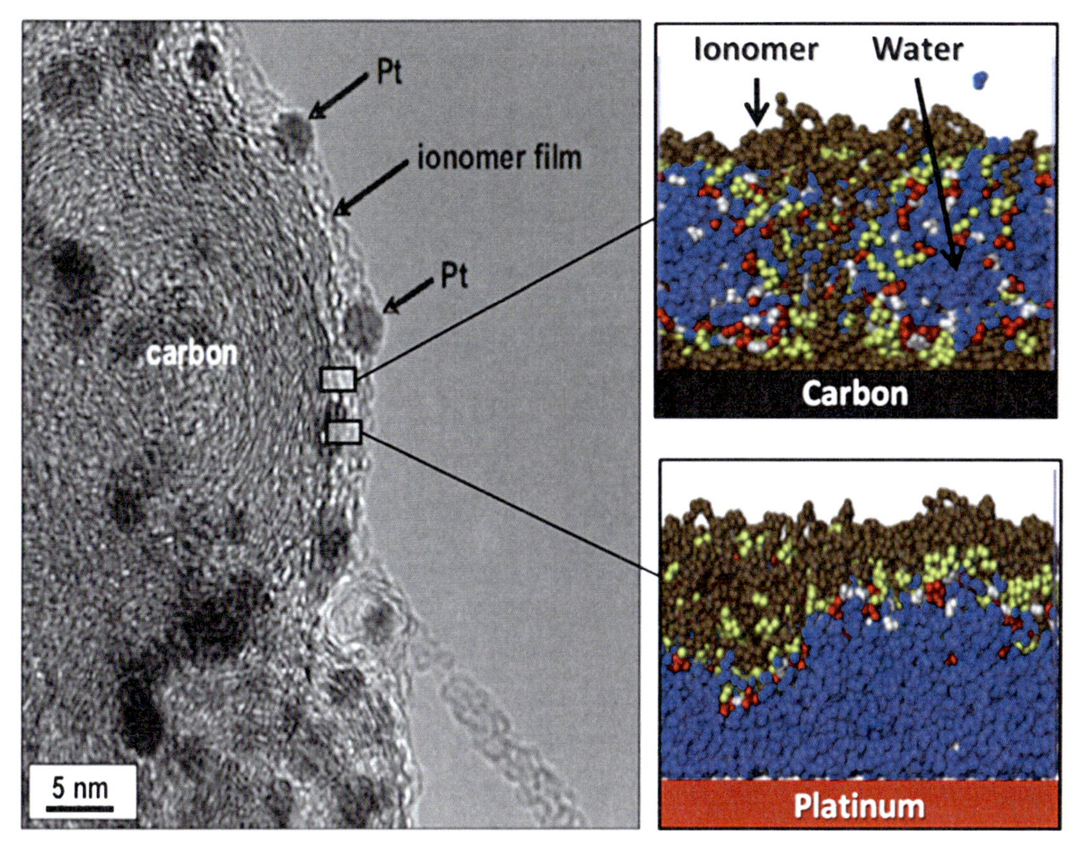

Multiscale Modeling, Fig. 11 Calculated Nafion thin-film structural properties at the vicinity of hydrophilic substrates (representative of Pt surfaces) and hydrophobic substrates (representative of C surfaces) for PEMFC electrodes (Figure built from the results presented in [28, 29])

local operation conditions (e.g., locally resolved liquid water transport inside the PEMFC electrode). These local operation conditions are used by the aging part of the model to calculate the material structural changes due to the material degradation, which will impact in the next time step the reactants transport properties. The calculation of the transport properties changes as function of the evolution of the material structural changes is carried out by numerically interpolating databases generated by CGMD calculations. This approach has been used, for example, for the simulation of the carbon corrosion (Fig. 15) [38] and of the membrane degradation [39] in PEMFCs. The approach provides then new insights on the interplaying between the different aging phenomena and analyzes the cell response sensitivity to operating conditions, initial catalyst/C/ionomer loadings, and temporal evolution of the electrocatalytic activity [40]. Generally speaking, there is still a lack of understanding on the interplaying between all the relevant transport processes, detailed electrochemistry and thermomechanical stresses, and on their relative impact on the global cell performance loss.

A recent extension of this simulation package has been recently reported by Franco [2, 41, 43], called MS LIBER-T *(Multiscale Simulator of Lithium Ion Batteries and Electrochemical Reactor Technologies)*. This model constitutes a breakthrough compared to the previously developed simulation package "MEMEPhys" penalized by its dependence on commercial software toolboxes and solvers such as Simulink. MS LIBER-T is coded on an independent C/Python language basis, highly flexible and portable (it can eventually be coupled to

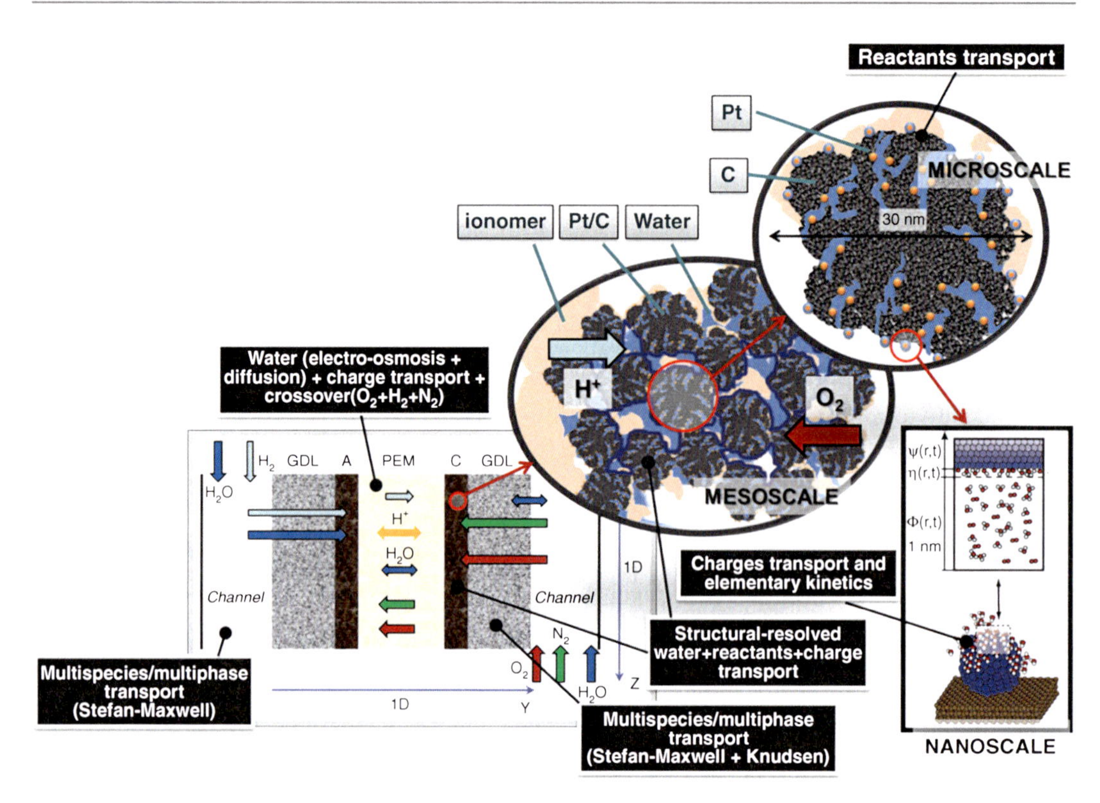

Multiscale Modeling, Fig. 12 Multiscale model of a PEMFC [2]

commercial software such as Matlab/Simulink), and is designed to support direct multiparadigm calculations, for instance, simulations coupling on the fly the numerical resolution of continuum models (e.g., describing reactant transport in a bulk) with the numerical resolution of discrete models (e.g., kinetic Monte Carlo codes resolving detailed electrocatalytic reactions). MS LIBER-T is fully modular meaning that the model represents explicitly the different physical phenomena as nonlinear sub-models in interaction. Such developed model is a multilevel one in the sense that it is made of a set of interconnected sub-models describing the phenomena occurring at different levels in the electrochemical reactor. However, this description remains macroscopic (suitable for engineering applications) in the sense that it is based on irreversible thermodynamic concepts as they are extensively used in chemical engineering: use of conservation laws coupled to closure equations (flux expressions, chemical rate models, thermodynamic models). Such an approach allows to easily modify the sub-models and to test new assumptions keeping the mathematical structure of the model and the couplings.

Future Directions

Numerical simulation and computer-aided engineering emerge nowadays as important tools to speed up the electrochemical power generators' R&D and to reduce their time-to-market for numerous applications.

In particular, integrative multiphysics, multiscale, and multiparadigm models spanning multiple scales and aiming to simulate competitions and synergies between electrochemical, transport, mechanical, and thermal mechanisms become now available. One of the major interests of such a class of models is its capability to

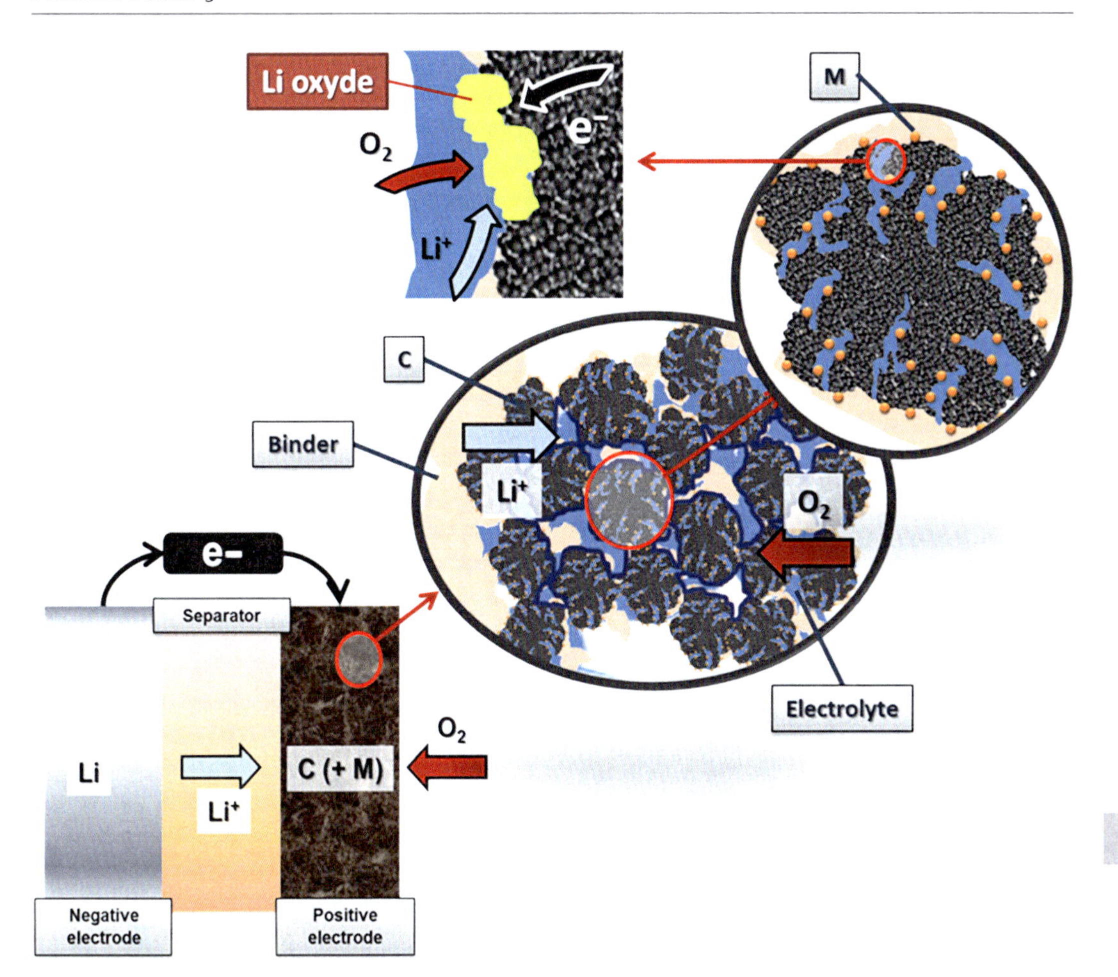

Multiscale Modeling, Fig. 13 Multiscale model of a lithium-air battery under the simulation package MS LIBER-T [2, 31]

analyze the impact of the materials' and components' structural properties on the global cell performance. Despite the tremendous progress achieved on developing multiscale models of electrochemical power generators with predictive capabilities, there are still major challenges to be overcome.

For the case of the electrocatalysis, more efforts should be devoted to perform ab initio calculations with solvent and electric field, in relation to the catalyst dissolution and oxidation (how the activity is affected by the catalyst degradation, and conversely, how the catalyst degradation kinetics is affected by the intrinsic catalytic activity?) [42].

For the case of batteries, the majority of the reported multiscale models focuses on the understanding of the operation and the impact of the structural properties of $LiFePO_4$ or graphite electrodes onto the global cell efficiency. And in the other hand, quantum mechanics and molecular dynamics models focalize on the understanding of the impact of the materials chemistry onto their storage or lithium transport properties at the nanoscale. It is now crucial to develop multiscale models that are able to incorporate both structure and chemical databases, in other words, that they are able to mimic the materials' behavior in realistic electrochemical environments. Within this sense, further intercalation and conversion

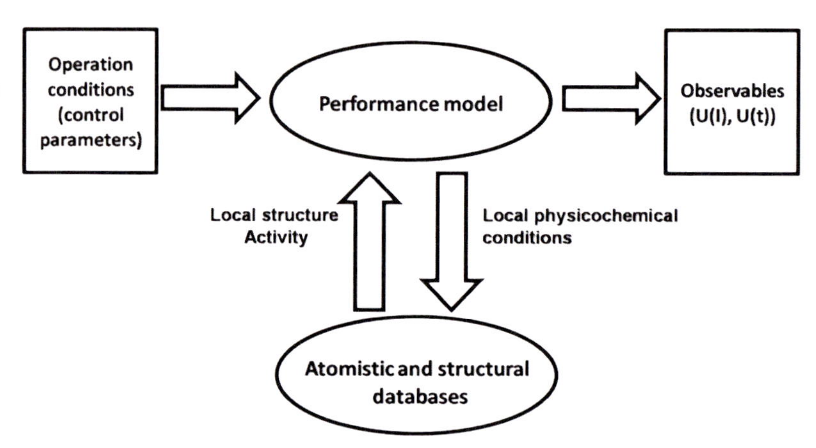

Multiscale Modeling, Fig. 14 Algorithmic approach for accounting the feedback between a performance model and a model describing the materials degradation mechanisms

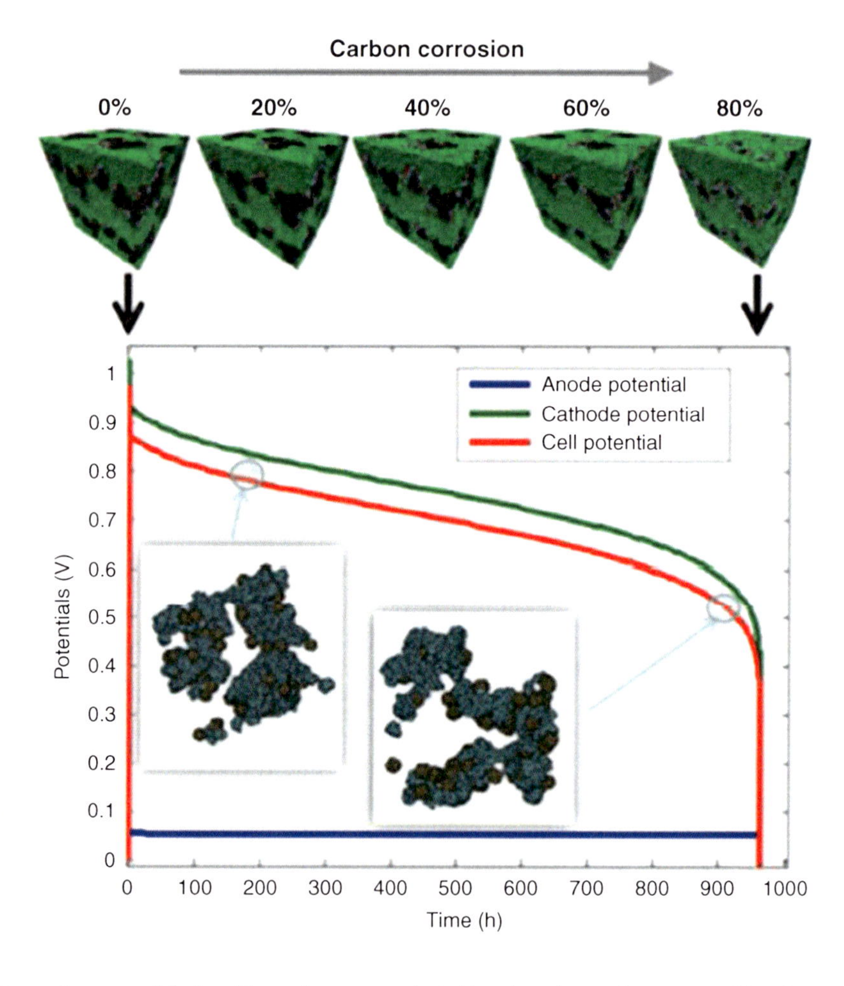

Multiscale Modeling, Fig. 15 Multiscale simulation of the PEMFC carbon corrosion by combining a continuum model with a CGMD structural data

materials have to be also modeled. The development of such a model tackles the issues related to how to couple discrete with continuum models and will need to set up rigorous methods to integrate ab initio data into elementary kinetics models of the lithiation/delithiation reactions.

Secondly, computational tools for the analysis of performance and degradation of fuel cells and batteries are currently fragmented and developed independently at different groups. Ideally, a multiscale model useful in engineering practice should have the following characteristics:

- It should be flexible, i.e., it should allow the developers to "virtually test" different cell designs to decide which cell satisfies their technical needs, as well as quickly studying new cell designs.
- It should be portable, i.e., the model should be a computer (or computing platform)-independent code.
- It should be scalable, i.e., the model code should allow developers to run it on single computers and multicore processors up to supercomputers.
- It should be easy to use, i.e., the model implementation details should be abstracted in a way that the developer interacts with a user-friendly interface.
- It should allow cloud computing and network development, i.e., simultaneous development by multiple researchers should be permitted as well as a performing exchange of information to benchmarking different model versions.
- Web platforms should be developed aiming to share physics and mathematical modules to facilitate different model calibration and benchmarking.

Approaches synergistically combining both top-down and bottom-up modeling viewpoints should be further developed. Macroscopic equations in top-down models should be written in terms of parameters with values calculated from lower-scale simulations. Implementation of such parameters into the macroscopic model should be done including empirical errors. Methodological evaluation of these parameters should be done systematically: for instance, coarse models should be developed first, with parameter sensitivity studies guiding further calculations at lower scales.

More from a materials engineering perspective, morphogenesis of the electrodes as a function of the ink properties and manufacture process (e.g., solvent used deposition time) should be further studied.

References

1. Franco AA (ed) (2013) Polymer electrolyte fuel cells: science, applications and challenges. Pan Stanford, Singapore
2. Franco AA (2013) Multiscale modeling and numerical simulation of rechargeable lithium ion batteries: concepts, methods and challenges. RSC Adv 3(32):13027–13058
3. Tarascon J-M, Armand M (2001) Issues and challenges facing rechargeable lithium batteries. Nature 414:359
4. Simon P, Gogotsi Y (2008) Materials for electrochemical capacitors. Nat Mater 7:845
5. Boudghene Stambouli A, Traversa E (2002) Renew Sustain Energy Rev 6:433
6. Christensen J et al (2012) J Electrochem Soc 159(2): R1
7. Holzapfel GA, Gasser TC (2001) Comput Method Appl Mech Eng 190:4379
8. Franco AA (2010) A multiscale modeling framework for the transient analysis of PEM fuel cells – from the fundamentals to the engineering practice. Habilitation Manuscript (H.D.R.), Université Claude Bernard Lyon 1. http://tel.archives-ouvertes.fr/docs/00/74/09/67/PDF/Alejandro_A_Franco_Manuscript_HDR_July_12_2010.pdf
9. Madec L, Falk L, Plasari E (2001) Chem Eng Sci 56:1731
10. Mandin P, Cense JM, Cesimiro F, Gbado C, Lincot D (2007) Comput Chem Eng 31(8):980
11. Reuter K, Deutschmann O (ed) (2009) Wiley-VCH, Weinberg
12. Sheppard D, Terrell R, Henkelman G (2008) Optimization methods for finding minimum energy paths. J Chem Phys 128:134106
13. Ferreira de Morais R, Loffreda D, Sautet P, Franco AA (2011) A multi-scale modeling methodology to predict electrochemical observables from ab initio data: application to the ORR in a Pt(111)-based PEMFC. Electrochim Acta 56(28):10842
14. Franco AA, Schott P, Jallut C, Maschke B (2007) Fuel Cells 7:99
15. Franco AA, Schott P, Jallut C, Maschke B (2006) J Electrochem Soc 153(6):A1053
16. Cahn J, Hilliard J (1958) J Chem Phys 28(2):258
17. Guyer JE, Boettinger WJ, Warren JA, McFadden GB (2004) Phys Rev E 69:021603
18. Guyer JE, Boettinger WJ, Warren JA, McFadden GB (2004) Phys Rev E 69:021604
19. Kim J (2007) Commun Nonlin Sci Numer Simulat 12:1560
20. Choo SM, Chung SK, Kim KI (2000) Comput Math Appl 39:229
21. Franco AA, Filhol JS, Doublet ML (2013) in preparation
22. Dalverny A-L, Filhol J-S, Doublet M-L (2011) J Mater Chem 21:10134

23. Khatib R, Dalverny AL, Saubanère M, Gaberscek M, Doublet ML (2013) J Phys Chem C 117:837
24. Singh GK, Ceder G, Bazant MZ (2008) Electrochim Acta 53(26):7599
25. Bai P, Cogswell DA, Bazant MZ (2011) Nano Lett 11:4890
26. Deng J, Wagner GJ, Muller RP (2013) J Electrochem Soc 160(3):A487
27. Malek K, Eikerling M, Wang Q, Navessin T, Liu Z (2007) J Phys Chem C 111:13627
28. Damasceno Borges D, Malek K, Mossa S, Gebel G, Franco AA (2012) ECS Trans (in press)
29. Damasceno Borges D, Franco AA, Malek K, Gebel G, Mossa S (2013) Nanoletters (submitted)
30. Lopes Oliveira LF, Laref S, Mayousse E, Jallut C, Franco AA (2012) Phys Chem Chem Phys 14:10215
31. Franco AA (2013) ECS Trans (in press)
32. Cheah SK, Sycardi O, Guetaz L, Lemaire O, Gelin P, Franco AA (2011) J Electrochem Soc 158(11):B1358
33. Franco AA, Passot S, Fugier P, Billy E, Guillet N, Guetaz L, De Vito E, Mailley S (2009) J Electrochem Soc 156:B410
34. Franco AA, Coulon R, Ferreira de Morais R, Cheah S-K, Kachmar A, Gabriel MA (2009) ECS Trans 25(1):65
35. Franco AA, Guinard M, Barthe B, Lemaire O (2009) Electrochim Acta 54(22):5267
36. Franco AA, Gerard M (2008) J Electrochem Soc 155(4):B367
37. Franco AA, Tembely M (2007) J Electrochem Soc 154(7):B712
38. Malek K, Franco AA (2011) J Phys Chem B 115(25):8088
39. Franco AA, Malek K (2012) A microstructural resolved model of the membrane degradation in PEMFCs, in preparation
40. Franco AA (2012) PEMFC degradation modeling and analysis. In: Hartnig C, Roth C (ed)polymer electrolyte membrane and direct methanol fuel cell technology (PEMFCs and DMFCs) – volume 1: fundamentals and performance. Woodhead, Cambridge
41. www.modeling-electrochemistry.com
42. Greeley J, Nørskov JK (2007) Electrochim Acta 52:5829
43. Franco AA, Xue KH (2013) Carbon-based electrodes for lithium air batteries: scientific and technological challenges from a modeling perspective. ECS Journal of Solid State Science and Technology 2(10): M3084
44. Baaiu A, Couenne F, Jallut C, Lefèvre L, Legorrec Y, Maschke B (2009) Port-based modelling of mass transport phenomena. Mathematical and Computer Modelling of Dynamical Systems 15(3): 233–254

Multivalence Cation Conductors

Nobuhito Imanaka, Shinji Tamura and Naoyoshi Nunotani
Department of Applied Chemistry, Faculty of Engineering, Osaka University, Osaka, Japan

Introduction

Since the discovery of the relationship between current passing through a solid and the resulting chemical changes that obey Faraday's law by Nernst et al., various types of ions have been reported to migrate in solids. However, until 1995, the ions were limited to 11 mono- and 20 divalent ions. Since the conduction of trivalent scandium cations (Sc^{3+}) in $Sc_2(WO_4)_3$ tungstate was reported, 18 trivalent cation species have been entered into the category of ionically conducting species in solids. The first discovery of a tetravalent cation solid conductor was for zirconium ion (Zr^{4+}) in $Zr_2O(PO_4)_2$ in 2000, and, until now, four such tetravalent cations have been discovered.

In this chapter, we briefly describe the tri- and tetravalent cation conductors, which are the newcomers in the field of solid state ionics.

Trivalent Cation Conductors

Until 1995, it was accepted that trivalent cations were poorly migrating species in solids due to strong electrostatic interaction between the conducting trivalent cation and the surrounding anions. Therefore, trivalent cations were commonly used as dopants in solid electrolytes for the adjustment of defect concentrations and lattice size. In order to realize trivalent cation conduction, such strong interactions between the mobile trivalent cations and the surrounding anion framework must, at least, be reduced. Although some solids such as Ln^{3+}-β/β"-alumina [1–9], β-alumina related materials [10–13], and

Ln^{3+}-β-$LaNb_3O_9$ [14, 15] have been reported to exhibit trivalent ion conduction, the conduction was not demonstrated, either directly or quantitatively. Solid electrolytes of which trivalent cation conduction has been directly demonstrated are $M_2(M'O_4)_3$ (M = Al, In, Sc, Y, Er–Lu; M' = W, Mo) with the $Sc_2(WO_4)_3$-type structure [6, 16–25], and NASICON-type $R_{1/3}Zr_2(PO_4)_3$ [26–28] and $(M_xZr_{1-x})_{4/(4-x)}Nb(PO_4)_3$ (M: rare earths or Al) [29–31].

$Sc_2(WO_4)_3$-Type Structure

$M_2(WO_4)_3$ tungstate with trivalent cations has two crystal structures, orthorhombic $Sc_2(WO_4)_3$-type and monoclinic $Eu_2(WO_4)_3$-type, depending on the trivalent cation size. The $Sc_2(WO_4)_3$-type structure is quasi-layered with sufficient space between the layers for the smooth migration of M^{3+} ions. In this structure, hexavalent tungsten (W^{6+}) bonds to four surrounding oxide anions to form a rigid WO_4^{2-} tetrahedron unit, which results in considerable reduction of the electrostatic interaction between M^{3+} and O^{2-}, and thereby allows M^{3+}-ion diffusion. While the $Eu_2(WO_4)_3$-type structure also contains W^{6+}, there is insufficient space for trivalent cation migration.

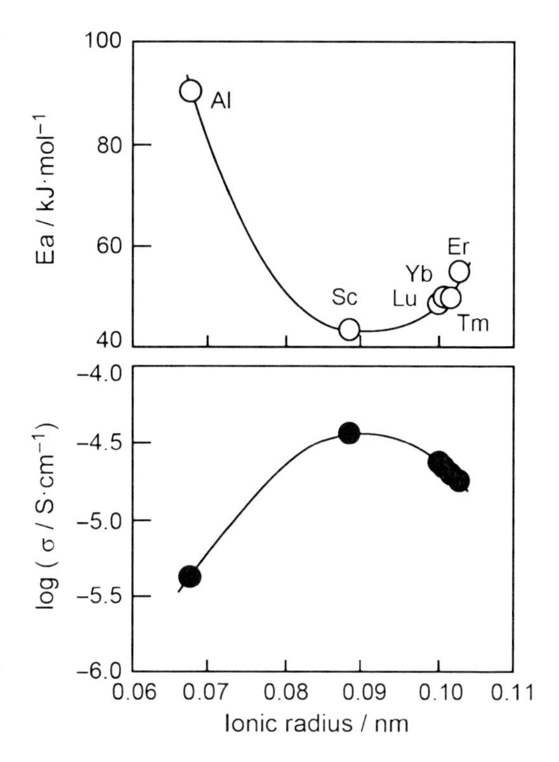

Multivalence Cation Conductors, Fig. 1 Electrical conductivity and activation energies for $M_2(WO_4)_3$ with the $Sc_2(WO_4)_3$-type structure at 600 °C in an air atmosphere

Figure 1 shows the electrical conductivities and activation energies of $M_2(WO_4)_3$ with the $Sc_2(WO_4)_3$-type structure at 600 °C. The conductivity of $M_2(WO_4)_3$ changes with variation of the ionic radius of M, and $Sc_2(WO_4)_3$ exhibits the highest conductivity among the $M_2(WO_4)_3$ series. The conductivity decreases in those solids with trivalent M^{3+} ions larger than Sc^{3+}. In contrast, for Al^{3+} with a much smaller ionic radius (0.0675 nm [32]) than Sc^{3+} (0.0885 nm [32]), the conductivity is also lower, presumably due to the larger decrease of the volume of the migrating trivalent ion (from Sc^{3+} to Al^{3+}) compared to that of the crystal lattice (from $Sc_2(WO_4)_3$ to $Al_2(WO_4)_3$).

Trivalent cation conduction in $M_2(MoO_4)_3$ molybdates with the $Sc_2(WO_4)_3$-type structure was also investigated. The transport of trivalent M^{3+} ions in molybdates was higher than that in tungstates, due to the slightly smaller radius of Mo^{6+} (0.055 nm [32]) than that of W^{6+} (0.056 nm [32]), which results in a smaller volume of the corresponding lattice elements suitable for trivalent ion conduction. However, molybdates tend to be more easily reduced than tungstates; $Sc_2(WO_4)_3$ is not reduced even at $Po_2 = 10^{-17}$ Pa, while $Sc_2(MoO_4)_3$ is readily reduced at $Po_2 = 10^{-13}$ Pa and 700 °C.

NASICON-Type Structure

NASICON is a family of high-conductivity Na^+ conducting solid electrolytes, described in the previous chapter. The NASICON structure has a three dimensional network, where PO_4 tetrahedra and ZrO_6 octahedra are linked by shared oxygens. The first report on trivalent cation conduction in a NASICON-type solid was for $Sc_{1/3}Zr_2(PO_4)_3$ obtained by a sol–gel

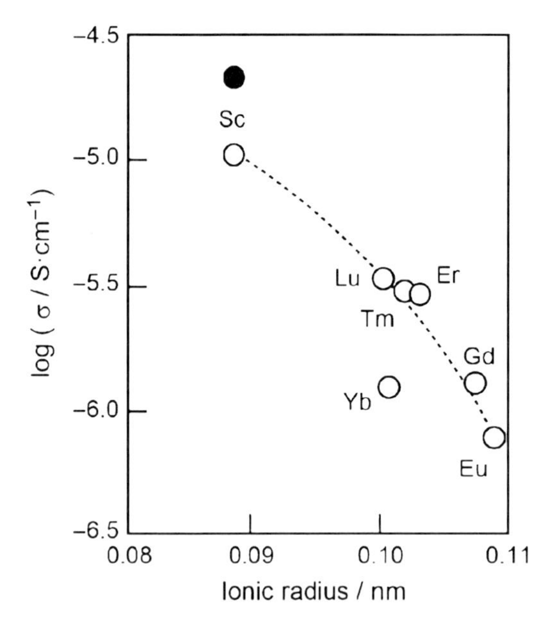

Multivalence Cation Conductors, Fig. 2 Relationship between the R^{3+} ionic radius and electrical conductivity at 600 °C in an air atmosphere for the $R_{1/3}Zr_2(PO_4)_3$ series prepared by a sol–gel method (○) and $Sc_{1/3}Zr_2(PO_4)_3$ obtained via ball milling (●)

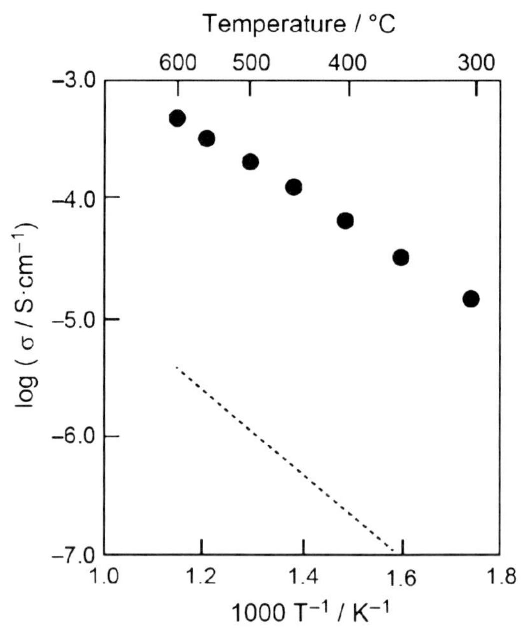

Multivalence Cation Conductors, Fig. 3 Temperature dependence of the Al^{3+} ion conductivity for $(Al_{0.2}Zr_{0.8})_{20/19}Nb(PO_4)_3$ (●) and corresponding data for $Al_2(WO_4)_3$ (– –)

method, where the Sc^{3+} ion conductivity was 1.07×10^{-5} S · cm^{-1} at 600 °C [26]. Similar to the $Sc_2(WO_4)_3$-type solid, tetravalent Zr^{4+} and pentavalent P^{5+} cations are contained as high valency cations that can effectively reduce the electrostatic interaction between Sc^{3+} cations and surrounding O^{2-} ions in the NASICON-type structure. Enhancement in the Sc^{3+} conductivity of $Sc_{1/3}Zr_2(PO_4)_3$ was achieved by applying the mechanical activation method for the synthesis. Ball milling can mix the starting powders and form an amorphous phase with sufficient high mechanical energy, which is effective for the realization of high crystallinity. As a result of improving the crystallinity of $Sc_{1/3}Zr_2(PO_4)_3$, the ionic conductivity was increased up to 2.9×10^{-5} S · cm^{-1} at 600 °C [27], as shown in Fig. 2 (closed circle).

Figure 2 also displays the relationship between the R^{3+} ionic radius and the R^{3+} ion conductivity at 600 °C for $R_{1/3}Zr_2(PO_4)_3$ (R = Sc, Eu, Gd, Er, Tm, Yb, Lu) solids. R^{3+} ion conductivity is monotonically increased with decreasing R^{3+} radius, and the Al^{3+} ion is considered to be a more suitable trivalent cation species in the NASICON-type solids, because the ionic radius of Al^{3+} (0.0675 nm [32]) is smaller than Sc^{3+} (0.0885 nm [32]). However, the $Al_{1/3}Zr_2(PO_4)_3$ solid cannot be obtained due to stereological limitations. In order to realize the formation of the NASICON-type structure with mobile Al^{3+}, the $(Al_xZr_{1-x})_{4/(4-x)}Nb(PO_4)_3$ solid was developed in 2002 [29] by shrinking the NASICON-type lattice with the partial substitution of smaller pentavalent Nb^{5+} (0.078 nm [32]) onto the Zr^{4+} (0.086 nm [32]) sites. Among the $(Al_xZr_{1-x})_{4/(4-x)}Nb(PO_4)_3$ series, only the samples with aluminum content (x) smaller than or equal to 0.2 can form a single phase of the NASICON-type structure and the highest Al^{3+} ion conductivity is obtained for the solubility limit composition of $(Al_{0.2}Zr_{0.8})_{20/19}Nb(PO_4)_3$. Figure 3 presents the conductivity of $(Al_{0.2}Zr_{0.8})_{20/19}Nb(PO_4)_3$ and the corresponding data for $Al_2(WO_4)_3$ with the quasi-layered $Sc_2(WO_4)_3$-type structure. The ionic conductivity of $(Al_{0.2}Zr_{0.8})_{20/19}Nb(PO_4)_3$ (4.5×10^{-4} S · cm^{-1} at 600 °C) is two orders of magnitude higher than that of $Al_2(WO_4)_3$.

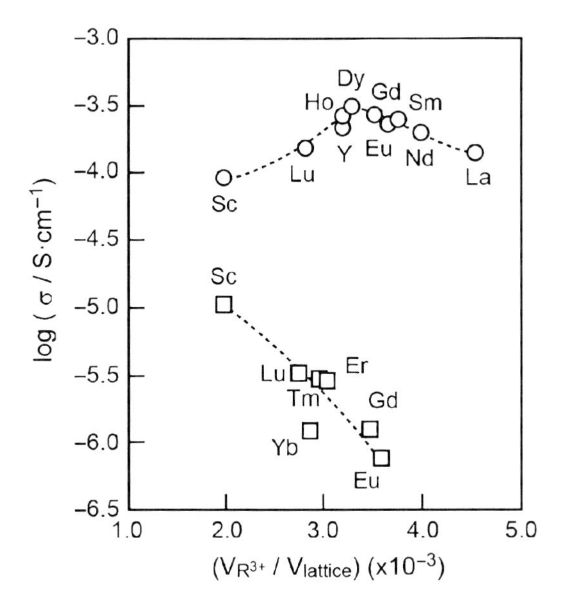

Multivalence Cation Conductors, Fig. 4 Relationship between the electrical conductivity at 600 °C in an air atmosphere and the R^{3+} size to lattice volume ratio of $(R_{0.05}Zr_{0.95})_{80/79}Nb(PO_4)_3$ (○) and $R_{1/3}Zr_2(PO_4)_3$ solids (□)

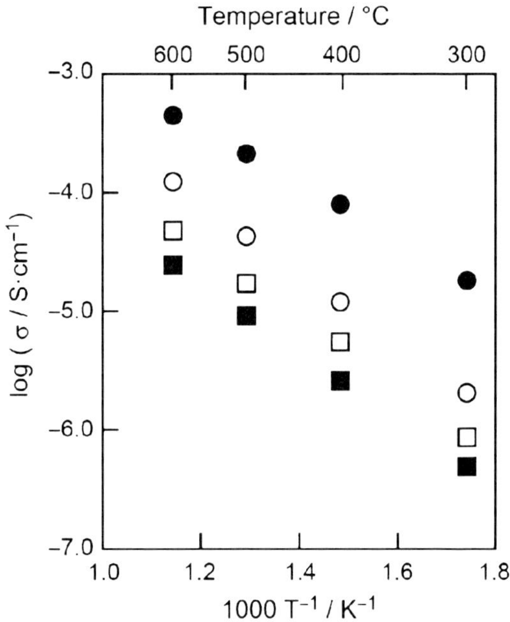

Multivalence Cation Conductors, Fig. 5 Temperature dependence of the M^{4+} cation conductivity for $MM'(PO_4)_3$ [(M, M') = (Zr, Ta) (■), (Zr, Nb) (□), (Hf, Nb) (○)] and $Zr_{39/40}TaP_{2.9}W_{0.1}O_{12}$ (●) solids in air atmosphere

These results imply that a suitable crystal lattice size for each trivalent cation species can be obtained by partial substitution of the constituent ions in the NASICON-type solid. The suitable sizes for rare earth ion conduction were also extensively investigated for $(R_{0.05}Zr_{0.95})_{80/79}Nb(PO_4)_3$ [31]. Figure 4 shows the relationship between the trivalent R^{3+} conductivity at 600 °C and the ratio of the R^{3+} size to the lattice volume for $(R_{0.05}Zr_{0.95})_{80/79}Nb(PO_4)_3$ or $R_{1/3}Zr_2(PO_4)_3$. Although the conductivity of $R_{1/3}Zr_2(PO_4)_3$ decreases monotonically with increasing R^{3+} size to lattice volume ratio, the highest conduction in the $(R_{0.05}Zr_{0.95})_{80/79}Nb(PO_4)_3$ series is obtained for R = Dy (the ratio is 3.29×10^{-3}), which suggests that the suitable lattice size for R^{3+} transport in the NASICON-type lattice is dependent on the constituent species.

Tetravalent Cation Conductors

The first report of tetravalent ion conduction was for Zr^{4+} ions in $Zr_2O(PO_4)_2$ in 2000 [33], although the conductivity was still lower that of the lower-valence ions. Zr^{4+} ion conduction in the $Zr_2O(PO_4)_2$ solid was demonstrated by DC electrolysis. Since 2001, high tetravalent cation conduction has been reported for $ZrM(PO_4)_3$ (M = Ta, Nb) solids with the NASICON-type structure [34–36]. Analogously, the NASICON-type $HfNb(PO_4)_3$ [35] has also been reported to be a Hf^{4+} ion conductor. The order of the conductivity variation, $HfNb(PO_4)_3 > ZrNb(PO_4)_3 > ZrTa(PO_4)_3$ (Fig. 5), may originate from the electronegativity of the constituent cations in the solids. High electronegativity of a cation leads to the formation of stronger covalent bonds with the surrounding anions, and therefore, by comparing the electronegativity of Zr^{4+} (1.33) with that of Hf^{4+} (1.3), the bonding of Hf^{4+} with O^{2-} anions should be weaker than that of Zr^{4+}; the slightly smaller ionic size of Hf^{4+} (0.085 nm [32]) compared with Zr^{4+} (0.086 nm [32]) may also facilitate migration. Recently, even higher Zr^{4+} conductivity was reported for one $ZrTa(PO_4)_3$ derivative, $Zr_{39/40}TaP_{2.9}W_{0.1}O_{12}$ [37], where the P^{5+} sites are partially substituted with larger ionic size and higher valency hexavalent W^{6+}. By doping W^{6+} into P^{5+} sites, further reduction of the electrostatic interaction between Zr^{4+} and O^{2-} is expected, in addition to expansion of the lattice volume. The highest Zr^{4+} ionic conductivity was obtained for the $Zr_{39/40}TaP_{2.9}W_{0.1}O_{12}$ solid and the conductivity was approximately one order of magnitude higher than that of the $ZrTa(PO_4)_3$ solid (Fig. 5).

Ions that can exhibit more than two valence states and/or have high electronegativity are generally considered to be inappropriate candidates for migration in solids, due to the appearance of electronic conduction by change in the valence state and/or by the formation of strong covalent bonds. Although titanium and germanium are such representative cation species, Ti^{4+} and Ge^{4+} ion migration has been demonstrated by selection of the NASICON-type structure. Although Ti can have two valence states of +3 and +4, only Ti^{4+} has been demonstrated to migrate in the $Ti(Nb_{0.8}\ W_{0.2})_{5/5.2}(PO_4)_3$ solid [38], in which the Nb^{5+} sites in NASICON-type $TiNb(PO_4)_3$ are partially substituted with higher valence W^{6+} ions to weaken the interaction between Ti^{4+} and O^{2-}. Similarly, tetravalent Ge^{4+} ion conduction in the $Ge_{41/40}NbP_{2.9}Si_{0.1}O_{12}$ solid [39] was demonstrated by the introduction of larger ionic size Si_{4+} (0.040 nm [32]) to the P^{5+} (0.031 nm [32]) site in $GeNb(PO_4)_3$ to expand the lattice size by lengthening the Ge-O bond. As a result of weakening the Ge-O bond, the Ge^{4+} ion can migrate in the NASICON-type structure.

Summary

Until recently, various ions have been discovered to migrate in crystal solids, but most of these conductive species were limited to mono- and divalent ions. After 1995, the migration of tri- and tetravalent cations has been successfully demonstrated in several structures by strict design of the crystal structure and its constituents. Furthermore, some solids in which tri- or tetravalent cations can be conducting ion species exhibit practically applicable ion conductivity, and there remains a possibility for new solids that exhibit high conductivity to appear in the field of solid state ionics and be applied as functional materials.

Cross-References

▶ Solid Electrolytes

References

1. Dunn B, Farrington GC (1983) Trivalent ion exchange in beta" alumina. Solid State Ionics 9–10:223–226
2. Sattar S, Ghosal B, Underwood ML, Mertwoy H, Salzberg MA, Frydrych WS, Farrington GC (1986) Synthesis of di- and trivalent β"-aluminas by ion exchange. J Solid State Chem 65:231–240
3. Köhler J, Urland W (1997) Strukturchemischer Einfluß auf die Ionenleitfähigkeit in Na^+/La^{3+}-β"-Al_2O_3-Kristallen (The influence of structural chemistry on the ionic conductivity in Na^+/La^{3+}-β"-alumina crystals). Z Anorg Allg Chem 623:231–238
4. Köhler J, Urland W (1996) Strukturchemische und impedanzspektroskopische Untersuchungen an Mg^{2+}-stabilisierten Na^+/Pr^{3+}-β"-Al_2O_3-Kristallen (Structural chemistry and impedance spectroscopy of Mg^{2+} stabilized Na^+/Pr^{3+}-β"-Aluminas). Z Anorg Allg Chem 622:191–196
5. Farrington GC, Dunn B, Thomas JO (1983) The lanthanide β" aluminas. Appl Phys A 32:159–161
6. Köhler J, Imanaka N, Urland W, Adachi G (2000) Direkter Nachweis dreiwertiger Kationenleitung in Nd^{3+}-β"-Al_2O_3 (Direct evidence for trivalent cationic conduction in Nd^{3+}-β"-Al_2O_3). Angew Chem Int Ed 39:904–907
7. Kumar RV (1997) Application of rare earth containing solid state ionic conductors in electrolytes. J Alloys Compd 250:501–509
8. Yang DL, Dunn B, Morgan PED (1991) Thermal decomposition of Nd^{3+}, Sr^{2+} and Pb^{2+} β"-aluminas. J Mater Sci Lett 10:485–490
9. Wen ZY, Lin ZX, Tian SB (1990) Preparation and ionic conduction property of polycrystalline Eu^{2+} β"-alumina. Solid State Ionics 40–41:91–94
10. Wang XH, Lejus AM, Vivien D, Collongues R (1988) Synthesis and characterization of lanthanide aluminum oxynitrides with magnetoplumbite like structure. Mater Res Bull 23:43–49
11. Sun WY, Yen TS, Tien TY (1991) Subsolidus phase relationships in the systems *Re*-Al-O-N (where *Re* = rare earth elements). J Solid State Chem 95:424–429
12. Kahn A, Lejus AM, Madsac M, Théry J, Vivien D, Bernier JC (1981) Preparation, structure, optical, and

magnetic properties of lanthanide aluminate single crystals ($LnMAl_{11}O_{19}$). J Appl Phys 52:6864–6869
13. Warner TE, Fray DJ, Davies A (1997) A high-temperature solid-state potentiometric sensor for nitrogen based on ceramic $LaAl_{12}O_{18}N$. J Mater Sci 32:279–282
14. Dyer AJ, White EAD (1964) A study of trivalent-pentavalent oxide system. Trans Br Ceram Soc 63:301–312
15. George AM, Virkar AN (1988) Mixed iono-electronic conduction in β-$LaNb_3O_9$. J Phys Chem Solids 49:743–751
16. Imanaka N, Kobayashi Y, Adachi G (1995) A direct evidence for trivalent ion conduction in solids. Chem Lett 24:433–434
17. Kobayashi Y, Egawa T, Tamura S, Imanaka N, Adachi G (1997) Trivalent Al^{3+} ion conduction in aluminum tungstate solid. Chem Mater 9:1649–1654
18. Tamura S, Egawa T, Okazaki Y, Kobayashi Y, Imanaka N, Adachi G (1998) Trivalent aluminum ionic conduction in the aluminum tungstate-scandium tungstate-lutetium tungstate solid solution system. Chem Mater 10:1958–1962
19. Imanaka N, Kobayashi Y, Fujiwara K, Asano T, Okazaki Y, Adachi G (1998) Trivalent rare earth ion conduction in the rare earth tungstates with the $Sc_2(WO_4)_3$-type structure. Chem Mater 10:2006–2012
20. Köhler J, Imanaka N, Adachi G (1998) Multivalent cationic conduction in solids. Chem Mater 10:3790–3812
21. Kobayashi Y, Tamura S, Imanaka N, Adachi G (1998) Quantitative demonstration of Al^{3+} ion conduction in $Al_2(WO_4)_3$ solids. Solid State Ionics 113–115:545–552
22. Köhler J, Imanaka N, Adachi G (1999) Indiumwolframat, $In_2(WO_4)_3$ - ein In^{3+} leitender Festkörperelektrolyt (Indium tungstate, $In_2(WO_4)_3$ - an In^{3+} conducting solid electrolyte). Z Anorg Allg Chem 625:1890–1896
23. Okazaki Y, Ueda T, Tamura S, Imanaka N, Adachi G (2000) Trivalent Sc^{3+} ion conduction in the $Sc_2(WO_4)_3$–$Sc_2(MoO_4)_3$ solid solution. Solid State Ionics 136–137:437–440
24. Imanaka N, Ueda T, Okazaki Y, Tamura S, Adachi G (2000) Trivalent ion conduction in molybdates having $Sc_2(WO_4)_3$-type structure. Chem Mater 12:1910–1913
25. Imanaka N, Hiraiwa M, Tamura S, Adachi G, Dabkowska H, Dabkowski A (2002) Anisotropic trivalent ion conducting behaviors in single crystals of aluminum tungstate–scandium tungstate solid solution. J Mater Sci 37:3483–3487
26. Tamura S, Imanaka N, Adachi G (1999) Trivalent Sc^{3+} ion conduction in $Sc_{1/3}Zr_2(PO_4)_3$ solids with the NASICON-type structure. Adv Mater 11:1521–1523
27. Tamura S, Imanaka N, Adachi G (2002) Trivalent ion conduction in NASICON type solid electrolyte prepared by ball milling. Solid State Ionics 154–155:767–771
28. Imanaka N, Adachi G (2002) Rare earth ion conduction in tungstate and phosphate solids. J Alloys Compd 344:137–140
29. Imanaka N, Hasegawa Y, Yamaguchi M, Itaya M, Tamura S, Adachi G (2002) Extraordinary high trivalent Al^{3+} ion conduction in solids. Chem Mater 14:4481–4483
30. Hasegawa Y, Tamura S, Imanaka N, Adachi G, Takano Y, Tsubaki T, Sekizawa K (2004) Trivalent praseodymium ion conducting solid electrolyte composite with NASICON type structure. J Alloys Compd 375:212–216
31. Hasegawa Y, Tamura S, Sato M, Imanaka N (2008) High-trivalent rare earth ion conduction in solids based on NASICON-type phosphate. Bull Chem Soc Jpn 81:521–524
32. Shannon RD (1976) Revised effective ionic radii and systematic studies of interatomic distances in halides and chalcogenides. Acta Crystallogr A 32:751–767
33. Imanaka N, Ueda T, Okazaki Y, Tamura S, Hiraiwa M, Adachi G (2000) Tetravalent ion (Zr^{4+}) conduction in solids. Chem Lett 29:452–453
34. Imanaka N, Ueda T, Adachi G (2001) Extraordinary high tetravalent cation conducting behaviors in solid. Chem Lett 30:446–447
35. Itaya M, Imanaka N, Adachi G (2002) Tetravalent Zr^{4+} or Hf^{4+} ion conduction in NASICON type solids. Solid State Ionics 154–155:319–323
36. Imanaka N, Ueda T, Adachi G (2003) Tetravalent Zr^{4+} ion conduction in the NASICON type phosphate solids. J Solid State Electrochem 7:239–243
37. Imanaka N, Tamura S, Itano T (2007) Extraordinarily High Zr^{4+} ion conducting solid. J Am Chem Soc 129:5338–5339
38. Nunotani N, Tamura S, Imanaka N (2009) First discovery of tetravalent Ti^{4+} ion conduction in a solid. Chem Mater 21:579–581
39. Nunotani N, Tamura S, Imanaka N (2009) A discovery of tetravalent Ge^{4+} ion conduction in solids. Chem Lett 38:658–659

N

Nanoionics at High Temperatures

J. Maier
Max Planck Institute for Solid State Research, Stuttgart, Germany

While in the bulk of materials local electroneutrality has to be fulfilled, at interfaces space charge zones occur which provide additional degrees of freedom defined by the contact chemistry. So it is the *individual* preference of a given charge carrier for a certain phase or the boundary core itself that is to the fore. The trade-off of this additional flexibility is the occurrence of an electrical field which confines the effects to the immediate vicinity of the interface.

In the case of crystalline solids, the calculation of the equilibrium charge carrier concentration as a function of the control parameters (typically partial pressures of the components, dopant concentration, and temperature) requires mass action laws of point defect generation or interaction, but in addition to these chemical constraints also an electrostatic constraint. In the bulk this is local electroneutrality, while at the boundaries excess charges can be tolerated but regulated by Poisson's equation and the respective boundary conditions. This excess charge decays to zero towards bulk within a characteristic length that typically is in the range of nanometers (maximum, about 100 nm for very pure materials; minimum, interatomic distance). Thus at interfaces, depending on the chemistry there, charge carriers can be accumulated or depleted by orders of magnitude compared to the bulk. Notwithstanding structural changes that might occur as well, this space charge effect occurs even and in particular for an abrupt junction. For surfaces of solids this has been addressed by Frenkel, Kliewer, and Blakeley [1–3]. A general treatment of space charge effects at interfaces as a function of the thermodynamic parameters and hence implementation in the defect chemical consideration was given by the author [4]. It is the latter procedure that paved the way for a systematic exploitation of space charge effects ("Heterogeneous Doping" or "Higher-dimensional Doping") with respect to materials design [5].

Figure 1 is a key figure in this context and highlights for a Frenkel-disordered solid the thermodynamic behavior of mobile electronic and ionic carriers at boundaries and their coupling [6]. Figure 2 is equally important as it shows the complete problem set to be solved using the example of surface space charges [7]. Based on this treatment it was possible to explain a great variety of anomalies as well as to predict and verify new systems of interest.

Figure 3 gives a collection of prototypes [8].

To emphasize the potential of this "higher-dimensional doping," let us consider a pure material in which cation vacancies and cation interstitials represent the relevant ionic carriers and in which the electronic carriers are in minority. Let us further imagine we are primarily interested in the cation vacancy owing to higher mobility. If the counter defect possesses a higher

G. Kreysa et al. (eds.), *Encyclopedia of Applied Electrochemistry*, DOI 10.1007/978-1-4419-6996-5,

defect formation energy than the cation vacancy, the cation vacancy concentration is very limited owing to electroneutrality irrespective of its quite favorable individual formation energy. A well-known remedy is (homogeneous) doping: in the special case considered, one would try to introduce a positively charged dopant. The benefits are very limited if the solubility for the dopant is small. If one, however, creates a situation where the metal ion is absorbed or adsorbed owing to the second phase or the core region of the contact (e.g., grain boundary core), metal vacancies are accumulated in the space charge zones of the material under concern. This accumulation effect can be huge but is restricted to the very surroundings of the contact. For properties that directly probe the interfacial situation (e.g., surface kinetics in fuel cells, electrochemical transfer reactions), this is immediately of great relevance. As far as overall transport properties are concerned, a high density of percolating interfaces is required for a substantial overall effect. Should the spacing of the interfaces be so small that the material is charged everywhere, then mesoscopic situations are realized.

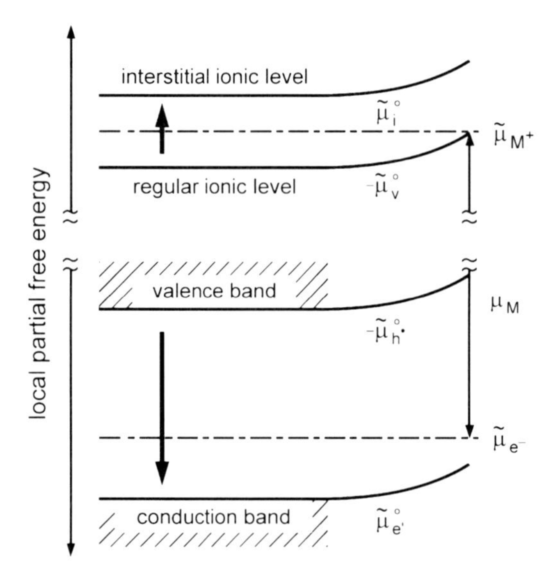

Nanoionics at High Temperatures, Fig. 1 The figure shows the energy level diagram for a mixed conductor MX with Frenkel disorder in the cation sublattice. *Top*: ionic disorder. *Bottom*: electronic disorder. The connection occurs via the chemical potential of the component (Reprinted from Ref. [6] by permission of Elsevier)

A thorough theoretical treatment together with various verified examples can be found in previous works of the author [4–9]. It is particularly important to note that all carriers that are mobile feel the field effect even if their contribution to it is marginal (fellow traveler effect) [4]. For that reason ionic space charge effects are also crucial for electronic conductors.

In the following, and referring to Fig. 3, a few characteristic situations are described that are of interest for high-temperature defect chemistry (see Ref. [10] and references therein):

1. A phenomenon of great practical relevance is the increase of ion conductivity by adding second surface active particles (a). The earliest example was $LiI:Al_2O_3$, [11] later examples are silver, copper, and thallium halides which showed similar effects [4, 5]. Similar effects can also be observed for anion conductors examples (e.g., PbF_2: SiO_2) (b) [4]. They all find their explanation in the concept of

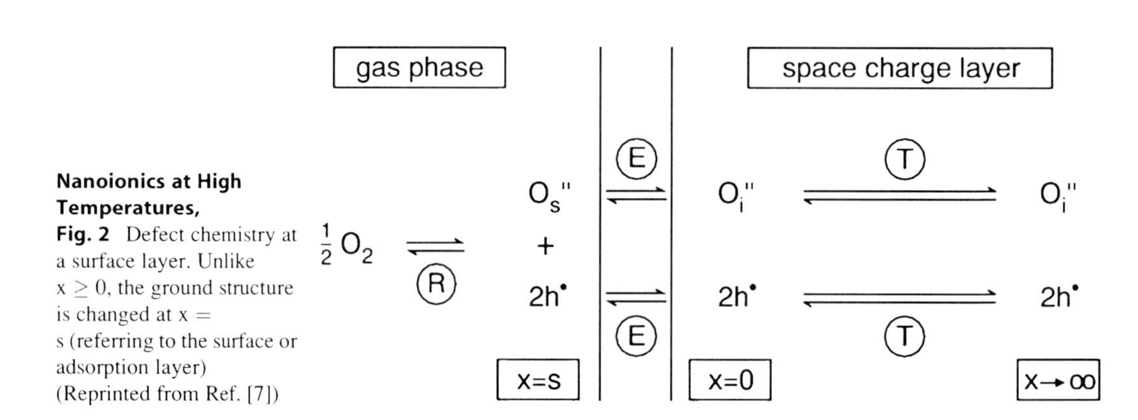

Nanoionics at High Temperatures, Fig. 2 Defect chemistry at a surface layer. Unlike $x \geq 0$, the ground structure is changed at x = s (referring to the surface or adsorption layer) (Reprinted from Ref. [7])

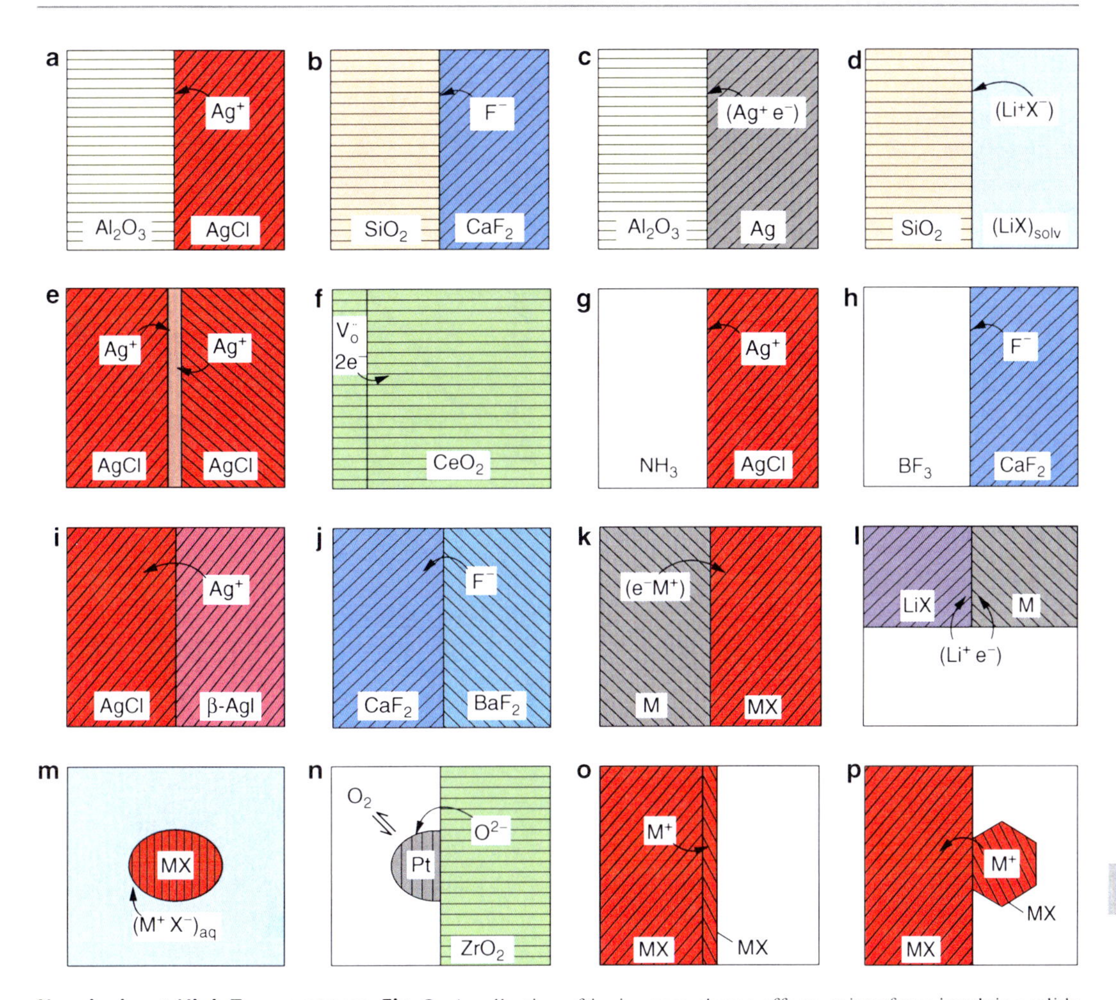

Nanoionics at High Temperatures, Fig. 3 A collection of ionic space charge effects at interfaces involving solids (Reprinted from Ref. [8] by permission of Elsevier)

heterogeneous doping, as quantitatively developed in Refs. [4–10,12].

A key parameter is the point defect concentration (c_0) directly adjacent to the interfacial core containing the information on the second phase. From a thermodynamic viewpoint, we better express c_0 in terms of surface charge density or even better in terms of the "energy levels" as the real invariants (see Fig. 1) [12].

2. Space charge zones also occur at surfaces. Exciting examples are Ag-halide surfaces the excess charge of which can be varied by acid–base active gases (g). This can be nicely exploited for chemical sensing of acid–base active gases such as NH_3. (The anionic analogue to NH_3: AgCl is BF_3: CaF_2 [5].) Other examples of interest are stoichiometry and lattice constant anomalies in CeO_2 [13, 14].
3. If the second phase is a Frenkel-disordered ionic conductor itself, Ag^+ can be absorbed and accommodated in its charge zones, rather than just adsorbed. Examples are AgI: AgBr, AgI: AgCl (i), or CaF_2: BaF_2 as anionic counterpart (j) where ions can redistribute in addition to possible neutral mixing demanded by the phase equilibrium. A particularly thorough study referred to CaF_2/BaF_2 heterolayers as here the morphology is simple and the layer thickness could be varied from nm to μm. The situation becomes exciting if the thickness of

the layers is so small that the space charge zones overlap, then not a single point within the layers remains electroneutral [15].

4. Excess charges also occur in polycrystalline materials owing to charging of the interfacial core. So in Ag-halides (i) or BaF_2, CaF_2 (j), we find Ag^+ or F^- accumulation in the core and vacancy accumulation in the space charge zones. This effect can be chemically augmented by contaminating the grain boundaries with Lewis acids or bases [5]:

   In oxide systems [16–22] the grain boundary core is often found positively charged owing to ex-corporation of O^{2-} in order to energetically stabilize the grain boundary. This positive core charge leads to a variety of significant phenomena (see Fig. 4): depletion of holes resulting in an increased resistance of p-type $SrTiO_3$, even greater depletion of oxygen vacancies. The increase of excess electrons is most clearly seen in CeO_2 where lightly doped CeO_2 switches from ion to electron conductivity if downsized. Depletion of protons in oxidic materials appears to be the reason for the enormous grain boundary resistances in $BaZrO_3$ preventing the material from being an excellent proton conductor in polycrystalline form. The joint effects on excess electrons; holes and oxygen vacancies can be nicely seen in the oxygen partial pressure dependence of the transport properties in nanocrystalline $SrTiO_3$. Also here the grains are in the mesoscopic regime. As far as the electronic contribution is concerned, downsizing to about 30 nm leads to a change in the electronic defect chemistry that is equivalent to a P_{O2} change of 12 orders of magnitude. The ionic contribution essentially vanishes (depression by 6 orders of magnitude) [23].

5. As point defects are also highly reactive centers, the space charge in particular at surfaces is very important for heterogeneous catalysis [24–26]. A few studies are available that highlight the significance of heterogeneous doping in that respect. Related effects that also rely on charging and the individual redistribution of ions are spillover and the variation of catalytic activity by polarizing a solid electrolyte (h) [24, 27].

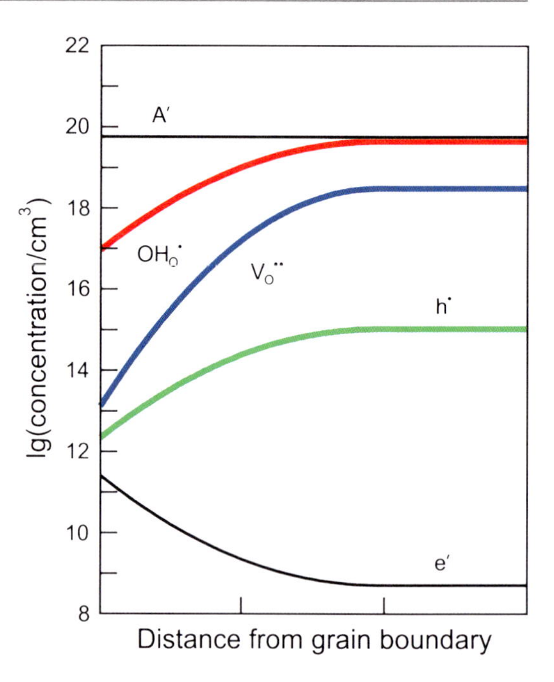

Nanoionics at High Temperatures, Fig. 4 Carrier concentrations near a positively charged grain boundary core in oxide systems (here: Mott-Schottky situation in an acceptor (A′) doped oxide)

6. A related area in which now experiments are underway is the significance of space charge zones for chemical or electrochemical reactions at surfaces or interfaces. Large effects are expected as the relevant interfacial carrier concentration is greatly affected (surface vacancies).

7. Space charge zones can also lead to storage anomalies, as shown recently in the context of Li-based batteries (l). As described in Ref. [28], early results [29, 30] on greatly varied miscibilities in the system (Ag-chalcogenide: Al_2O_3) may be explained in this way. Similarly one has to reckon with varied (charged) non-stoichiometries at oxide interfaces. This list can be prolonged, in particular as phenomena at lower temperature are concerned.

These selected master examples highlighted the significance of nanoionics for applied electrochemistry as to find better electrodes, electrolytes, catalysts, and chemical sensors. It is also a fundamentally exciting area, as it leads to (i) new mechanisms and (ii) to new adjusting screws. In addition to temperature component partial pressure and doping content, these new degrees of freedom are nature, density, and distribution of interfaces and hence include size effects.

A key point which is heavily investigated at the moment is how to influence and tune the core charge of given systems. Whether or not this will work out satisfactorily, it can be foreseen that nanoionics will play a similarly important role for energy research as nanoelectronics does for information technology.

So far the ionic space charge concepts have been mainly applied to transport and storage. An area that is not adequately addressed but expected to be greatly influenced by such effects is the area of interfacial transfer and reactions including reaction kinetics, catalysis, or in particular the electrochemical transfer reaction. In many of these processes, defect concentrations are involved in the rate determining steps and not only will space charge considerations help understand them but more importantly lead to a targeted improvement of electrochemical performance.

Cross-References

▶ Solid Electrolytes

References

1. Frenkel J (1926) Thermal movement in solid and liquid bodies. Z Physik 53:652–669
2. Kliewer KL, Koehler JS (1965) Space charge in ionic crystals. I. General approach with application to NaCl. Phys Rev A 140:1226–1240
3. Poeppel RB, Blakely JM (1969) Origin of equilibrium space charge potentials in ionic crystals. Surf Sci 15:507–523
4. Maier J (1987) Defect chemistry and conductivity effects in heterogeneous solid electrolytes. J Electrochem Soc 134:1524–1535
5. Maier J (1995) Ionic conduction in space charge regions. Prog Solid State Chem 23:171–263
6. Maier J (2003) Defect chemistry and ion transport in nanostructured materials. Part II. Aspects of nanoionics. Solid State Ion 157:327–334
7. Maier J (2004) Physical chemistry of ionic materials. Ions and electrons in solids. Wiley, Chichester
8. Maier J (2004) Ionic transport in nano-sized systems. Solid State Ion 175:7–12
9. Maier J (1987) Space charge regions in solid two phase systems and their conduction contribution-III: defect chemistry and ionic conductivity in thin films. Solid State Ion 23:59–67
10. Maier J (2009) Nanoionics: ionic charge carriers in small systems. Phys Chem Chem Phys 11:3011–3022
11. Liang CC (1973) Conduction characteristics of lithium iodide aluminium oxide solid electrolytes. J Electrochem Soc 120:1289–1292
12. Jamnik J, Maier J, Pejovnik S (1995) Interfaces in solid ionic conductors: Equilibrium and small signal picture. Solid State Ion 75:51–58
13. Kim S, Merkle R, Maier J (2004) Oxygen nonstoichiometry of nanosized ceria powder. Surf Sci 549:196–202
14. Kossoy A, Feldman Y, Wachtel E, Gartsman K, Lubomirsky I, Fleig J, Maier J (2006) On the origin of the lattice constant anomaly in nanocrystalline ceria. Phys Chem Chem Phys 8:1111–1115
15. Sata N, Eberman K, Eberl K, Maier J (2000) Mesoscopic fast ion conduction in nanometre-scale planar heterostructures. Nature 408:946–949
16. Denk I, Claus J, Maier J (1997) Electrochemical Investigations of SrTiO3 Boundaries. J Electrochem Soc 144:3526–3536
17. Balaya P, Jamnik J, Fleig J, Maier J (2006) Mesoscopic electrical conduction in nanocrystalline SrTiO3. Appl Phys Lett 88:062109
18. De Souza RA, Fleig J, Maier J, Kienzle O, Zhang Z, Sigle W, Rühle M (2003) Electrical and structural characterization of a low-angle tilt grain boundary in iron-doped strontium titanate. J Am Ceram Soc 86:922–928
19. Tschöpe A (2001) Grain size-dependent electrical conductivity of polycrystalline cerium oxide II: Space charge model. Solid State Ion 139:267–280
20. Chiang YM, Lavik EB, Kosacki I, Tuller HL, Ying JY (1996) Defect and transport properties of nanocrystalline CeO_{2-x}. Appl Phys Lett 69:185–187
21. Kim S, Maier J (2002) On the conductivity mechanism of nanocrystalline ceria. J Electrochem Soc 149 (10): J73–J83
22. Shirpour M, Lin CT, Merkle R, Maier J (2012) Nonlinear electrical grain boundary properties in proton conducting Y-$BaZrO_3$ supporting the space charge depletion model. Phys Chem Chem Phys 14:730–740
23. Lupetin P, Gregori G, Maier J (2010) Mesoscopic charge carriers chemistry in nanocrystalline $SrTiO_3$. Angew Chem Int Ed 49:10123–10126
24. Fleig J, Jamnik J (2005) Work function changes of polarized electrodes on solid electrolytes. J Electrochem Soc 152: E138–E145
25. Murugaraj P, Maier J (1989) Heterogeneous catalysis with composite electrolytes. Solid State Ion 32/33:993–999

26. Merkle R, Maier J (2006) The significance of defect chemistry for the rate of gas-solid reactions: three examples. Topics Catal 38:141–145
27. Riess I, Vayenas CG (2003) Fermi level and potential distribution in solid electrolyte cells with and without ion spillover. Solid State Ion 159:313–329
28. Maier J (2007) Mass storage in space charge regions of nano-sized systems (Nano-ionics. Part V). Faraday Discuss 134:51–56
29. Petuskey WT (1986) Interfacial effects on Ag: S nonstoichiometry in silver sulfide/alumina composites. Solid State Ion 21:117–129
30. Janek J, Mogwitz B, Beck G, Kreutzbruck M, Kienle L, Korte C (2004) The magnetoresistance of metal-rich Ag2+xSe - A prototype nanoscale metal/semiconductor dispersion? Prog Solid Sate Chem 32:179–205

Neural Stimulation Electrodes and Sensors

J. Thomas Mortimer
Case Western Reserve University,
Cleveland, OH, USA

Introduction

The electrical excitability phenomenon, inherent to neural tissues, provides an opportunity to effect external control over many body systems: paralyzed limbs can be made to move, the blind can experience visual sensations, the deaf people can experience the voice of others close by or by phone, pain can be alleviated, tremors suppressed, and mental disorders treated. Devices, often referred to as neuroprostheses, that perform these functions can be sold. Companies that make devices that provide the most function with the smallest, safest devices that also have long battery lifetimes usually have the market advantage. Function is directly related to electrode placement; improperly placed electrodes don't work or have undesirable side effects. The remedy has been reimplantation, but tunable electrodes, structures with multiple small contacts, are making it possible to avoid additional surgery by providing pathways to manipulate the shape of the excitatory fields. Device size is often dictated by the size of the battery, and the body usually accommodates smaller devices more easily. Safe devices don't cause tissue injury. Devices that have long battery life don't require replacement as often. On the down side, devices do fail and inject unintended charge through the electrode contacts causing changes local to the contact site that often lead to costly litigation. Life-changing experiences are opened with these devices, and money can be made and lost. Understanding how they work or made to work more efficiently is key to their growing use and success. Further, bringing new devices or improved devices to the market can involve costly animal trials because of safety concerns. Electrical charge, weather drawn from internal batteries or injected through metal contacts into the electrolytes found in the living system, is common to understanding and improving neural prostheses.

It should come as no surprise to an electrochemist that many of these issues reduce to injecting the least amount of charge with the least amount of electrical potential to get the job done. Minimizing the amount of charge injected lessens the drain on power sources and lessens the likelihood that bad things will happen at the electrode-electrolyte interface where the target cells live. Keeping the potential low, particularly the ohmic losses, usually means the battery is physically smaller for the implantable pulse generator.

Typically neural stimulating electrodes:

- Are platinum or stainless steel in commercially available devices.
- Subjected to current densities that can be in the range of 1 A/cm^2.
- Employ pulse configurations that are biphasic with the cathodic phase first.
- See pulse durations range from 50 to 200 μS, with lower and higher values also employed and are subjected to.
- Repetition rates range from 10 to 200 Hz with higher rates used in some applications.

The job to be done by an electrode, connected to a current or potential source, is to create a potential field in the region of neural tissues that is capable of initiating a propagated action potential. The propagated action potential is a signal traveling on a unity-gain transmission line, the axon or nerve fiber, from the signal source, usually the cell body of the nerve or neuron, to the terminal end where the invading potential change initiates the release of a neurotransmitter, transferring the signaling information to the target cell, e.g., muscle or another neuron. Basically, the axon is the means by which the brain receives information and gives commands to and receives commands from distant parts of the body. The unity-gain feature of the axon is the phenomenon we harness to take control of the nervous system.

Basics of the Stimulation Target, the Axon

Information is carried in the nervous system by pulsed signals carried on axons designated for specific functions, e.g., vision, hearing, pain, and limb movement. Usually a higher pulse repetition rate implies a more intense piece of information being transmitted, and often bursts are used to get the attention of certain cells. Repetition rates, sometimes called firing rates, run in the range of 10 Hz to a couple of 100 Hz, with some body systems operating above, for short periods, and below these values. The firing rate determines the nature of the transmitter released at the terminal end, which is the key information transferred.

The neural signal is a localized transient change in the transmembrane potential of the axon (each less than 20 μm in diameter). The unity-gain transmission is achieved by having repeaters spaced along the axon at distances below 2 mm over the entire length. The repeaters are clusters ($\sim$1,000 per μ^2) of voltage-gated sodium ion channels isolated to the nodes of Ranvier, the gap between cells acting to insulate the length of the axon. The resting membrane potential is maintained at or near the Nernst potential for potassium by voltage-gated potassium ion channels, allowing potassium to move from inside to outside the membrane. Depolarization, sodium moving from outside to inside the cell, of one node creates a potential difference between nodes and initiates the depolarization of adjacent nodes, which accounts for the propagation of the signal. Movement of potassium from the inside to the outside of the membrane reestablishes the resting membrane potential in preparation for the next signal to be transmitted.

Stimulation-Induced Action Potentials

When a current or voltage pulse is applied to an electrode in a living system, double-layer charging takes place and if the potential is sufficient electron transfer occurs across the electrode-tissue interface. Further away from the interface ions move, in response to the local changes, to the induced by the current injected into the electrode and create a spatial potential gradient, and if that area contains axons, action potentials can be initiated that propagate the entire length of the axon as if they were naturally initiated.

Strength-Duration Curve: Shorter Duration Pulses Require Larger Amplitudes

When the electrical excitability phenomenon of axons was discovered, investigators explored the relationship between pulse duration and pulse amplitude required to initiate an action potential on an axon. The results of their investigation looked something like the graph shown in Fig. 1a, which indicates that larger currents are required to initiate an action potential when shorter pulse durations were used. In order to account for the fact that the separation between the electrode and the excitable tissues determined the magnitude of the stimulus, the term rheobase current (I_r) was defined as the asymptote defining the minimum current and the chronaxie time (t_c) was the time defined by the point on the duration axis at twice the rheobase current. The chronaxie time is relatively stable from lab to lab and measurement to measurement for given tissues where as the rheobase is variable.

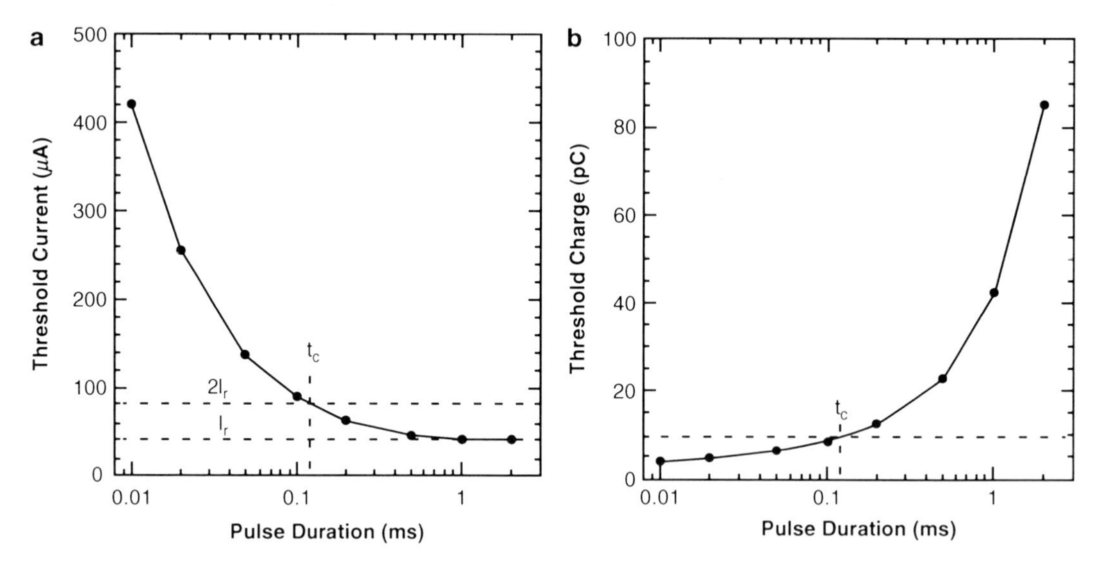

Neural Stimulation Electrodes and Sensors, Fig. 1 Strength-duration (**a**) and charge-duration (**b**) plots for nerve stimulation [1]. The current amplitude required to create a propagated action potential has been experimentally found to increase as the duration of the current pulse decreases. The actual magnitude depends on the separation between the electrode and nerve, which shifts the plot shown in **a** up or down. To accommodate for the shift and to make the result from one laboratory useful to another laboratory, the rheobase current is defined as the minimum current for a very long pulse, the asymptote. Doubling this value, $2I_r$, and measuring the intersection with the experimentally derived data defines the chronaxie time (t_c), a measurement easily transferable between laboratories. The chronaxie for neural tissue is in the range of 100 μS and the chronaxie for muscle tissue is in the range of 10 mS. Multiplying the current amplitude in **a** by the pulse duration yields the charge delivered to the nerve. Plotting the charge as a function of pulse width produces **b**. In **b** it is noted that the injected charge, required to effect a propagated action potentials, decreases with decreasing pulse width

Charge-Duration Curve: Short Duration Pulses Are More Efficient

The relationship between charge injection required to initiate a propagated action potential and the pulse duration can be found by reformulating Fig. 1a into a charge-duration curve (Fig.1b). We learn from this plot that charge required to initiate a propagated action potential decreases with decreasing pulse width and to minimize the amount of charge injected, one should use the shortest possible pulse width. Practical constraints are brought to bear when one actually builds a stimulator because of power supply considerations (Fig. 1a). Generally speaking, pulse durations in the range 50–200 μS are used in most electrical stimulation devices, right around the chronaxie.

Charge Injection Limits to Avoid Tissue and Electrode Damage

If target tissue is damaged or the electrode destroyed by a device with an intended purpose of restoring or providing a therapeutic effect the device fails, not good for many reasons. The mechanism for tissue damage is not now known. There are two schools of thought, (1) hyperactivity brought on by forcibly driving the target cells and (2) by toxic products of the charge transfer process. They are not mutually exclusive, but this author subscribes to the electrochemical aspect, which forces the focus on the metals used at the charge transfer site.

Platinum and stainless steel are two metals extensively used in commercially available implanted neurostimulating devices. Studies have

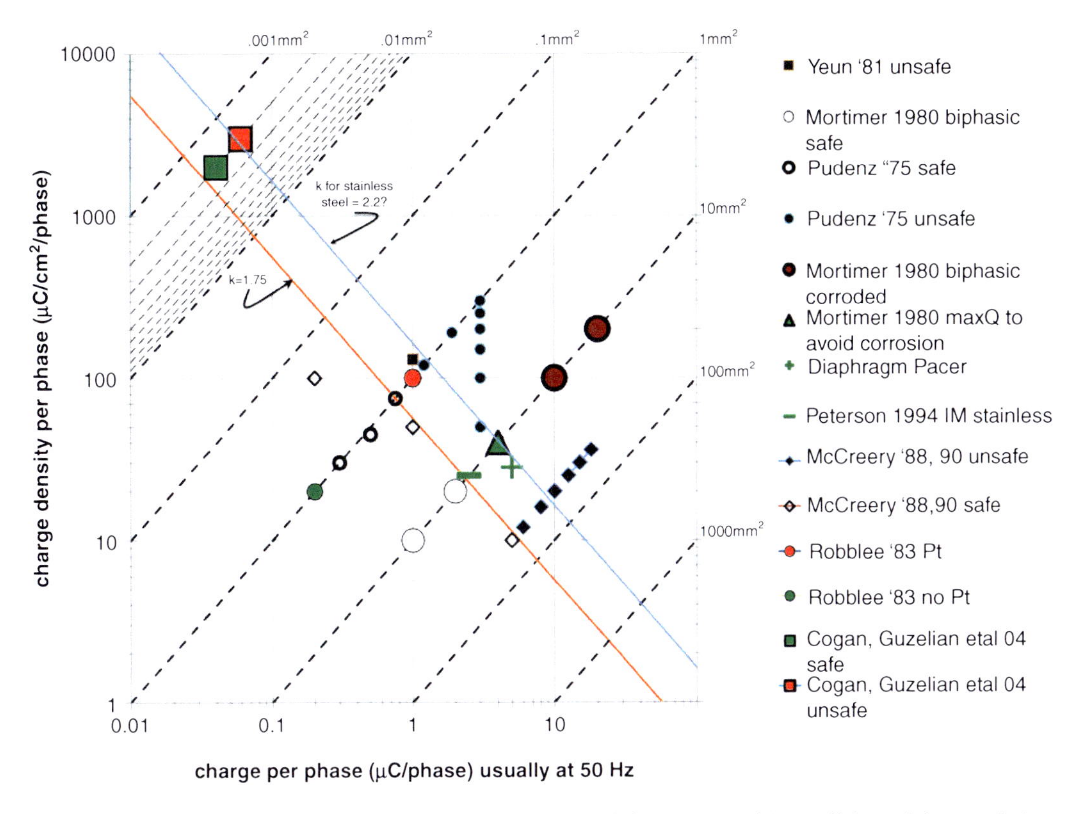

Neural Stimulation Electrodes and Sensors, Fig. 2 Shannon plot. This graph presents experimental data on safe and unsafe levels of stimulation. The safe and unsafe data are separable by the orange line defined by k = log (Q/A) + log (Q) and in this case k = 1.75. Charge and charge density values to the left of the line have been deemed safe from animal experiments using platinum electrodes applied to the brain, *small white filled symbols*, and values to the right have been deemed unsafe, *small black filled symbols*. The *small circles* indicate result where platinum was found in the brain tissue, *red filled*, and no platinum was found, *green filled*. The data appearing on the 10 mm^2 line (*dashed line* moving from *lower left* to *upper right*) are all for stainless steel electrodes in muscle. The safe values for stainless steel *white filled* or *green*. The *brown filled symbols* represent data from experiments in muscle where the electrodes were found to be corroded, but no significant tissue damage was observed. The *green filled triangle* depicts a point where the anodic corrosion potential limits maximum charge and charge density for a stainless steel electrode when charge balance is applied. The *green filled square* represents data collected from iridium oxide microelectrodes where the electrode appearance and tissue responses were deemed safe, and the *red filled* values represent electrode and tissue injury

been carried out with electrodes implanted in and on brain and muscle of animals and the tissues excised for microscopic analysis. The results of the early experiments using platinum electrodes were studied by Robert Shannon [2]. He found that plotting the results as log (charge per unit area) as a function of log (charge) yielded a plot where the "safe" and "unsafe" data could be separated by a line defined by:

$$k = logQ/A + logQ$$

There is no known explanation for these results, but they work and seem to apply to a range of electrode sizes.

Since Shannon's publication I have added experimental data to that plot; Shannon plot shown in Fig. 2 [3–10]. Consider first **platinum**: filled black symbols are data for levels deemed "unsafe" for platinum electrodes applied to brain tissues, and open symbols are deemed "safe." The line defined as k = 1.75 separates the "safe" (lower left) and "unsafe" (upper right) platinum data. The dashed lines are lines of constant electrode area (geometric).

A designer will use this plot to specify maximum limits for stimulation. For example, say that a device has been designed with a platinum electrode having a geometric area of 1 mm^2. From the Shannon plot, the maximum "safe" limit for charge injection is 0.75 μC. For this electrode and for a 100 μs pulse, the maximum "safe" amplitude would be 7.5 mA or 0.75 A/cm^2.

Now consider stainless steel, usually 316 LVM. These electrodes were all formed by winding stainless wire, stranded and single strand, into a helix and inserting them, with the aid of a hypodermic needle, into muscle of cat. The area of the conducting surface was estimated to 10 mm^2. These data are presented in the Shannon plot as filled symbols plotted on the 10 mm^2 line [6]. The two white symbols represent data collected for balanced charge biphasic, cathodic first, pulses at 1 and 2 μC/phase, which were deemed safe. Increasing the charge injection to 10 and 20 μC/phase for balanced charge biphasic, cathodic first, pulses, brown filled symbols, caused the electrodes to corrode but did not cause an increase in muscle tissue damage. It was determined that at charge density less than or equal to 40 $\mu C/cm^2$ would not cause corrosion of the stainless steel electrodes, green filled triangle symbol. If Q/A(Q) for stainless steel has a similar relationship to that found for platinum, then the k value for stainless steel would be 2.2, meaning that more charge can be "safely" injected with stainless steel than platinum! These results put the charge injection limits on stainless steel as electrode corrosion rather than tissue damage. Subsequent experiments using *imbalanced biphasic* pulses [11], less charge in the anodic phase than the cathodic phase, demonstrated that charge densities as great as 120 $\mu C/cm^2$, threefold increase, could be "safely" applied to muscle without causing corrosion or tissue damage.

Speculation on the Cause of Tissue Damage with Platinum Electrodes

The data presented on stainless steel point to corrosion as the limiting factor, and the results indicate that cells in the immediate vicinity of the electrode can tolerate the corrosion products of stainless steel. Now, consider the two filled circle symbols for Robblee'83 data. The red filled symbol represents data where platinum was measured in the capsule and in the first millimeter of brain tissue at charge density injection rates of 100 $\mu C/cm^2$/phase AND; this datum point lies to the right of the k = 1.75 line! No platinum was detected in the capsule or the brain tissue for charge injection rates of 20 $\mu C/cm^2$/phase, a datum point to the left of the k = 1.75 line. These observations suggest that the corrosion products of platinum may be a causative factor.

In the 1960s Rosenberg was studying the role of electric currents on cell division and serendipitously discovered that cisplatin, a reaction product of the electrochemistry, was toxic to the cells in the culture [12]. The mechanism of toxicity is believed to be cisplatin interacting with DNA of the cell to induce programmed cell death, apoptosis [13]. Cisplatin has subsequently become a widely used anticancer agent, particularly for treating testicular cancer. In 1977 Agnew and colleagues [14] reported the results of injecting platinum salts into the brain of cats. They reported that the ultrastructural changes bore a likeness to the ultrastructural changes induced by electrical stimulation. It is curious that neither Robblee and her colleagues nor Agnew and his colleagues, all very familiar with neural stimulation of the brain, never acknowledged a connection to cisplatin.

Drawing from the stainless steel experience, if cisplatin, a product of platinum corrosion, occurs during electrical stimulation of the brain with platinum electrodes, it will be possible to increase the charge injection through platinum electrodes to values beyond k = 1.75 by using imbalanced biphasic pulses, less charge in the anodic phase. When balanced charge biphasic pulses are employed, any charge going into irreversible reaction products during the cathodic phase will force the electrode potential positive to the potential prior to the initiation of the cathodic phase. When the electrode potential of platinum is forced positive, potentials may be reached to push the electrode potential into regions where platinum oxide is formed as well as platinum salts. Putting less charge in the secondary, anodic, phase forces the electrode less positive.

Why Biphasic Pulses and not Monophasic Pulses?

When a cathodic pulse is applied to an electrode, double-layer charging occurs along with any reduction reactions. At the termination of the current pulse, the electrode potential can decay by reactions driven by charge stored on the double layer (and pseudo capacitance), and in the case of electrodes in the living system, the reactions is oxygen reduction. Under resting conditions, the electrode potential of a platinum electrode can be found just positive where oxygen reduction occurs. The products of oxygen reductions are reactive oxygen species that, among other possibilities, react with nitrous oxide in the walls of blood vessels to induce vessel constriction and ischemia. Prolonged ischemia can lead to cell death. This reaction, occurring during the interpulse interval, can be eliminated by discharging the double layer (and pseudo capacitance) through a current pulse, anodic, supplied by the stimulator, thus a biphasic pulse. When Lilly [15] discovered biphasic pulses were less harmful than monophasic, he thought the harm was coming from displacement of the constituents of the cell wall and reasoned that a balance charge biphasic pulse would restore the displaced constituents. This is to origin of, what I call, the *dogma* in our field, *balanced charge biphasic pulses are the law*.

Adding a Delay Between the Cathodic Phase and the Anodic Phase of the Biphasic Pulse

In the preceding section, I have made the argument for biphasic stimulation. The first, usually cathodic, phase initiates the propagated action potential, and the second, anodic phase, is for the purpose of discharging the double layer (and pseudo capacitance) in order to terminate the electrochemistry driven by the cathodic phase. The down side to the second phase is that it can terminate/quench the development of the action potential [16]. This is because, when working with stimulus values slightly over threshold, the propagating action potential occurs after the termination of the cathodic phase. The quenching effect can be overcome by increasing the magnitude of the primary, cathodic, phase, but this violates the principal of minimizing charge injection. An alternative to increasing the cathodic charge injection is to add a delay between the primary and the secondary phases. The addition of the delay is illustrated in Fig. 3. The data presented were derived from an experiment where an intramuscular electrode was inserted into a muscle. The electrode was in the vicinity of the nerves innervating the muscle fibers, which means that some of the nerve fibers were closer to the electrode and some further from the current source, at the electrical threshold for excitation or slightly above threshold. In this experiment a monophasic pulse represents the control, dashed line. When the delay between the primary and the secondary pulse was zero, the evoked muscle response was $\sim$45 % of the monophasic response. As the delay between the two phases was increased, the evoked muscle response increased. Most devices, that are capable of adding a delay, use a 100 μS delay between the two phases. The results shown were for nerves innervating muscles, but the results apply to virtually all neurostimulation devices.

The Neural Stimulating Community Needs to Have a Better Understanding of the Processes that Takes Place at the Electrode-Electrolyte Interface

A better understanding will enable us to determine which mechanisms are at play in tissue damage and electrode damage. This knowledge will enable us to engineer technology to safely

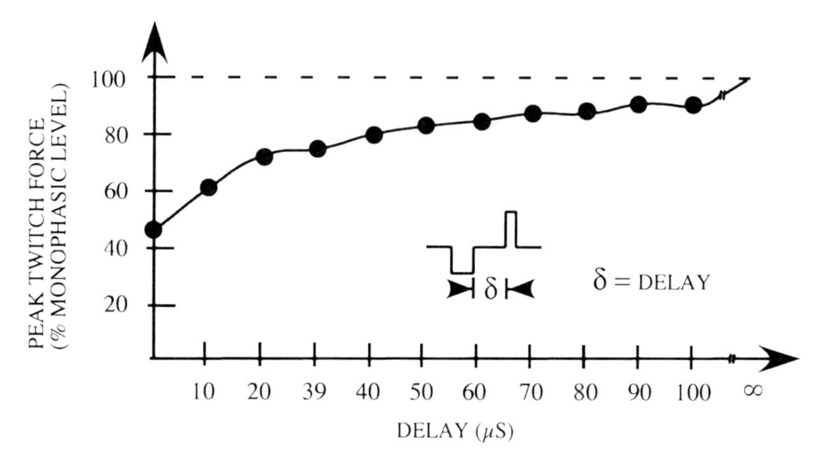

Neural Stimulation Electrodes and Sensors, Fig. 3 Evoked muscle force as a function of the delay between the primary, cathodic, and secondary, anodic, pulses of a biphasic stimulus. The *dashed line* represents the force recorded for a single cathodic stimulus pulse. The anodic phase is added to terminate the electrochemistry driven by charge accumulated on the double layer during the cathodic phase. If the anodic phase is applied immediately after the termination of the cathodic phase, the activation of nerve fibers not super maximally depolarized will be quenched, and no propagated action potential will occur. For most applications, a delay of 100 μS is deemed a reasonable period for the action potential to develop following a slightly depolarizing stimulus [16]

push more charge through the electrode than we now can. This knowledge may also be of use in litigation. At this time, too many neural stimulating devices must be operated at near the maximum safe limits before they begin to produce the desired result, be it a therapeutic effect or an evoked sensory percept, so a looming need is to find ways to inject more charge safely.

Future Directions

The neurostimulating community needs know the reactants that are generated during neural stimulation. If the reactions are known and the impact the reaction products have on cells in and around electrodes, life could be easier. For instance, workarounds could be reasoned rather than guessed, fewer animal experiments would be required, and litigation costs could be reduced. Too many devices in human use must use the maximum allowable stimulus to begin to attain the desired therapeutic effect; we need to be able to inject safely more charge than is currently possible. Further, it would be desirable to be able to do the job with batteries that are physically smaller than those currently used and preferable to not have to use larger batteries to get the job done. I can think of several ways to accomplish this increase:

1. Understand the charge transfer processes that **do** occur on the electrodes when current densities are applied in the range of neural stimulation applications.
2. Understand the charge transfer processes that occur on electrodes operating in living systems, lower order animals, and higher order animals and how these processes differ from studies in highly controlled electrolytes, e.g., sulfuric acid and phosphate buffered saline. As indicated in this document, narrow pulses are more charge efficient than are wider pulses. Therefore, studies must include pulses in the 10 μs to several hundred microsecond ranges.
3. Explore the imbalanced biphasic waveform, less anodic charge than cathodic, as a way to extend the excitation range.
4. Explore materials that will allow charge injections through reactions that are reversible within the time frame of the two phases of the stimulus pulse.

The cyclic voltammograms, when used in various electrolyte media, have been valuable tools to learn about potential reactions that **might** occur during neural stimulation in those media, but more information is needed to understand what actually occurs during the high current density, short pulse width, and conditions imposed during neural stimulation.

What is the potential range for the generation of corrosion products of platinum in different electrolytes? What are these products and do they interfere with cell function?

With stainless steel, we know that imbalanced biphasic stimulation of muscle cells is a safe way to avoid corrosion and more than triple charge injection limits imposed by balanced charge stimulation. There is no obvious reason to think imbalanced biphasic stimulation shouldn't be safe with electrodes in the brain, cells are cells, and it appears that the corrosion products are the culprits, which occur during the anodic phase. If a *reference electrode could be developed for implantable pulse generator*, IPG, closed-loop control could be used to eliminate electrode potentials that push the metal into regions where metal loss occurs.

The limits for the cathodic phase of an imbalanced biphasic pulse are not known because the mechanism for tissue injury is not known for this paradigm. The dogma in the neurostimulating community is that water reduction must be avoided; it surely was not with the imbalanced biphasic pulses that were safely applied to muscle.

Iridium oxide, belonging to a class of materials called super capacitors or variable valence materials, is an electrode material that has drawn considerable interest in the neural stimulating community [17] because the large pseudo capacitance, 100 times greater than that measured for platinum, based on slow cyclic voltammetry. The attractive feature of these materials is that the electron transfer processes are accommodated by valance changes within the oxide film, reversible reduction and oxidation between Ir^{+3} and Ir^{+4}, and movement of the counter ion in the hydrated oxide. This material, though used in research projects, is not yet used in commercially available devices. Though not a review article, Cogan [18] covers many of the findings for iridium oxide microelectrodes.

Concluding Remarks

The business end of a neural prosthetic device is the electrode, the metal-tissue interface, through which the device is to do a job, safely and efficiently. Understanding how these electrodes operate will provide insight into the mechanisms of tissue injury and ways to extend their charge injection capacities.

The most charge efficient way to activate neural tissues is to use the narrowest possible cathodic pulse followed by a delay of ~100 μS and then an anodic pulse. The function of the cathodic phase is to initiate a propagated action potential. The function of the delay phase is to avoid quenching the nascent action potential evoked in nerves driven to or slightly above threshold. The function of the anodic phase is to bring the interface potential back to the prepulse value. Bringing the electrode potential back to prepulse values (1) terminates reactions that can be driven by the potential across the electrode interface at the termination of the cathodic phase and (2) returns the interface potential to the prepulse value for the next succeeding stimulus pulse.

Tunable electrodes will be the way of the future. Using electrodes with multiple contacts through which electric fields in the tissue space can be shaped will enable purveyors of neural prostheses to electronically adjust or improve the efficacy without physically moving the electrode relative to the target tissue. These techniques, field steering, will likely utilize both anodic and cathodic first pulse paradigms.

The electrochemistry occurring on stimulating electrodes must be known much better than it is now. Cyclic voltammograms and classical ways of thinking of reactions do not tell the story for electrodes operating in the 1 A/cm^2 range for 100 μS. Porous surfaces may look good with slow cyclic voltammetry, but ions in the electrolyte may not move fast enough to utilize the full surface area.

Finally, knowledge related to electrode operation must be digestible by engineers and scientist who are not trained or familiar with electrochemistry: the electrochemistry community must do a better job of communicating knowledge about how electrodes work in living systems.

Cross-References

▶ Electrode
▶ Neurons, Coupling
▶ Sensors

References

1. Grill WM, Kirsch RF (2000) Neuroprosthetic applications of electrical stimulation. Assist Technol 12(1):6–20
2. Shannon RV (1992) A model of safe levels for electrical stimulation. IEEE Trans BME 39(4):424–426
3. Cogan SF, Guzelian AA et al (2004) Over-pulsing degrades activated iridium oxide films used for intracortical neural stimulation. J Neurosci Methods 137(2):141–150
4. McCreery DB, Agnew WF et al (1990) Charge density and charge per phase as cofactors in neural injury induced by electrical stimulation. IEEE Trans Biomed Eng 37(10):996–1001
5. McCreery DB, Agnew WF et al (1988) Comparison of neural damage induced by electrical stimulation with faradaic and capacitor electrodes. Ann Biomed Eng 16(5):463–481
6. Mortimer JT, Kaufman D et al (1980) Intramuscular electrical stimulation: tissue damage. Ann Biomed Eng 8(3):235–244
7. Peterson DK, Nochomovitz ML et al (1994) Long-term intramuscular electrical activation of the phrenic nerve: safety and reliability. IEEE Trans Biomed Eng 41(12):1115–1126
8. Pudenz RH, Bullara LA et al (1975) Electrical stimulation of the brain III. The neural damage model. Surg Neurol 4(4):389–400
9. Robblee LS, McHardy J et al (1983) Electrical stimulation with Pt electrodes. VII. Dissolution of Pt electrodes during electrical stimulation of the cat cerebral cortex. J Neurosci Methods 9(4):301–308
10. Yuen TG, Agnew WF et al (1981) Histological evaluation of neural damage from electrical stimulation: considerations for the selection of parameters for clinical application. Neurosurgery 9(3):292–299
11. Scheiner A, Mortimer JT et al (1990) Imbalanced biphasic electrical stimulation: muscle tissue damage. Ann Biomed Eng 18(4):407–425
12. Rosenberg B, Vancamp L et al (1965) Inhibition of cell division in *Escherichia coli* by electrolysis products from a platinum electrode. Nature 205:698–699
13. Alderden RA, Hall MD et al (2006) The discovery and development of cisplatin. J Chem Educ 83(5):728–735
14. Agnew WF, Yuen TGH et al (1977) Neuropathological effects of intracerebral platinum salt injections. Neuropathol Exp Neurol XXXVI(3):533–546
15. Lilly JC, Hughes JR et al (1955) Brief noninjurious electrical waveform for stimulation of the brain. Science 121:468–469
16. van den Honert C, Mortimer JT (1979) The response of the myelinated nerve fiber to short duration biphasic stimulating currents. Ann Biomed Eng 7(2):117–125
17. Robblee LS, Lefko JL et al (1983) Activated Ir: an electrode suitable for reversible charge injection in saline. J Electrochem Soc 130:731–732
18. Cogan SF, Ehrlich J et al (2009) Sputtered iridium oxide films for neural stimulation electrodes. J Biomed Mater Res B Appl Biomater 89B(2):353–361

Neurons, Coupling

Andreas Offenhäusser
Institute of Complex Systems, Peter Grünberg Institute: Bioelectronics, Jülich, Germany

Introduction

Optimization of methodologies toward engineering the life sciences and healthcare remains a grand challenge. In the field of neurotechnology, tremendous progress has been made in a fundamental understanding of the nervous system and in building technology to diagnose and treat some neurological diseases. However, our understanding of nervous system function and technological approaches to measuring and manipulating neuronal circuits needs to be improved.

Neural prosthetic devices are artificial extensions of body parts which allow a disabled

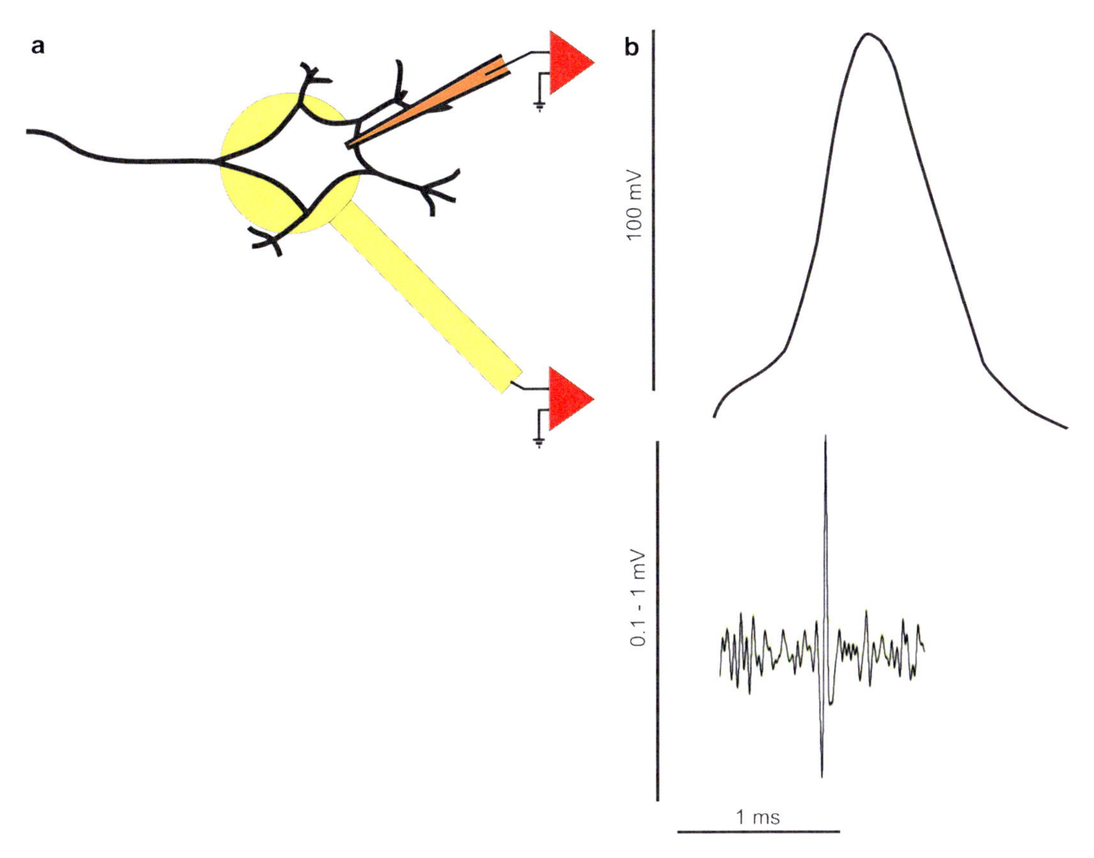

Neurons, Coupling, Fig. 1 (**a**) Schematic of a neuron on an extracellular electrode: intracellular (*upper orange* electrode) and extracellular (*lower yellow* electrode) signals can be recorded (**b**) Action potential of neuron (approx. 100 mV) recorded by an intracellular electrode (*upper trace*) and an extracellular electrode (*lower trace*)

individual to restore the body functions. Here, a neuroelectronic device which interfaces neuronal tissue with electronics is the key to restore the disabled body functions. Also, in vivo monitoring of the electrical signals from multiple cells during nerve excitation and cell-to-cell communication are important for design and development of novel materials and methods for laboratory analysis. In vitro biological applications such as drug screening and cell separation also require cell-based biosensors. Nowadays, the best approach to study the electrophysiological activity of neurons and cardiac cells in vitro and in vivo is based on planar microelectrode arrays or field-effect transistors which can be integrated with microfluidic devices. These methods allow the simultaneous monitoring and stimulation of large populations of excitable cells over many days and weeks and enable insights into long-term effects such as adaptivity in neuronal networks.

Basics

Silicon-based microstructures are gaining more and more importance in fundamental neuroscience and biomedical research. Precise and long-lasting neuroelectronic hybrid systems are in the center of research and development in this field. The interaction of a neuronal cell with an electronic device is schematically depicted in Fig. 1a. Sufficient electrical coupling between the cell and the (gate) electrode for extracellular signal recording is achieved only when a cell or a part of a cell is located directly on top of the (gate) electrode. Electrical signals recorded by these devices show lower signals and a higher noise level (owing to a weaker coupling to the (gate) electrode) compared to intracellular electrodes or patch pipettes (see Fig. 1b).

For extracellular signal recordings from electrically active cells in culture, two main concepts have been developed in the past: microelectrode arrays (MEAs) (see Fig. 2a) with metalized contacts on silicon or glass substrates have been used to monitor cardiac impulse propagation from dissociated embryonic myocytes [1–3], dissociated invertebrate neurons [4, 5] and mammalian neurons [6] spinal cord [7], and mouse dorsal root ganglia [8]. Alternatively, arrays of field-effect transistors (FETs) (see Fig. 2b) are used for extracellular recordings having either non-metalized transistor gates with cells growing directly on the gate oxide [9–11] or metalized gates. The latter were in direct contact with the electrolyte [12] or they were electrically insulated, so-called floating gates [13–15]. With these noninvasive methods, the electrical activity of single cells and networks of neurons can be observed over an extended period of time. Meanwhile both concepts are growing together

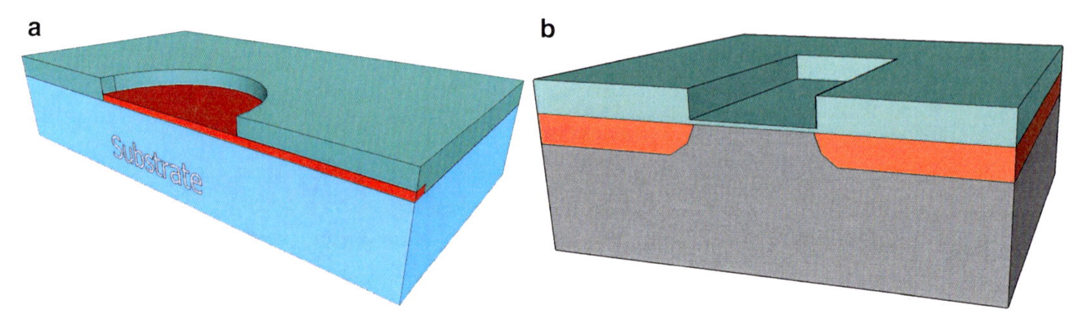

Neurons, Coupling, Fig. 2 (**a**) Substrate-embedded microelectrode: the metal electrode (*red*) is exposed to the electrolyte while the feed lines are covered with an isolation layer (*green-blue*) (**b**) Open-gate field-effect transistor for the recording of extracellular signals

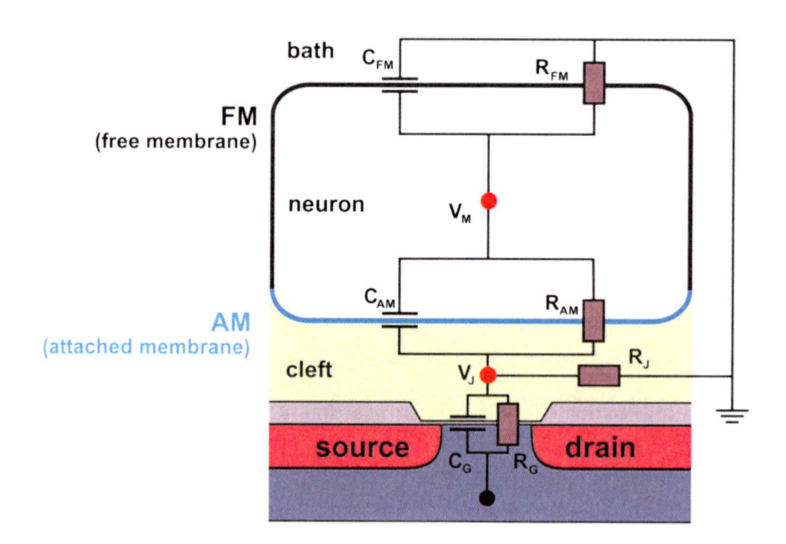

Neurons, Coupling, Fig. 3 Schematics of the neuroelectronic hybrid. The cell membrane is divided into free (*FM*) and attached membrane (*AM*) with the respective values of membrane area (A_{FM}, A_{JM}) and membrane capacitance (C_{FM}, C_{JM}) and resistance (R_{FM}, R_{JM}). C_G and R_G are the capacitance and the resistance of the (gate) electrode, respectively. The seal resistor R_J represents the electrical properties of the cleft between the membrane and the sensor surface. In case of patch-clamp experiments, the intracellular voltage V_M can be determined by designing MEAs inside a CMOS process with on-chip amplification and filtering [16, 17].

Neuron-Electrode Coupling

For a quantitative understanding of the extracellular signals recorded by electronic devices, it is necessary to explain the experimental situation in detail. A schematic picture of a typical experimental situation is depicted in Fig. 3. Here, the neuroelectronic hybrid is formed by the neuron, the cleft between neuron and the sensor surface, and the electronic device. Outside the neuron and inside the cleft, there is extracellular electrolyte solution. By electrical excitation, the ion channels in the cell's membrane open and ions can flow from across the cell membrane. While in the upper part of the cell (free membrane), these ions just enter the surrounding electrolyte bath directly; it is different at the attached membrane. Here, the ions have to pass the cleft before entering/leaving the bath. The cleft acts as a resistance typically called seal resistance R_J [10, 18]. The magnitude of R_J is typically in the order of several 100 kW up to MW corresponding to a typical cleft thickness of 40–150 nm [19, 20]. The voltage V_J, which determines the voltage at the (gate) electrode, is mainly determined by the seal resistance R_J, and the current the flows across it.

Future Directions

Although this noninvasive method of extracellular recordings allows monitoring the electrical activity of single cells and networks of neurons over an extended period of time with good time resolution, it does not allow detecting subthreshold signals of neuronal cells. In the last years, a number of research groups began to study the combination of planar (2D) electrodes with intracellular recordings. This includes the use of gold mushroom-shaped protrusions, nanopillar electrodes, and nanorods. These developments may improve our understanding of neuroscience in the future [21–23].

Cross-References

▶ Neural Stimulation Electrodes and Sensors

References

1. Thomas CA, Springer PA, Loeb GE, Berwald-Netter Y, Okun LM (1972) A miniature microelectrode array to monitor the bioelectric activity of cultured cells. Exp Cell Res 74:61–66
2. Israel DA, Barry WH, Edell DJ, Mark RG (1984) An array of microelectrodes to stimulate and record from cardiac cells in culture. Am J Physiol 247: H669–H674
3. Connolly P, Clark P, Curtis ASG, Dow JAT, Wilkinson CDW (1990) An extracellular microelectrode array for monitoring electrogenic cells in culture. Biosens Bioelectron 5:223–234
4. Breckenridge LJ, Wilson RJA, Connolly P, Curtis ASG, Dow JAT, Blackshaw SE, Wilkinson CDW (1995) Advantages of using microfabricated extracellular electrodes for in vitro neuronal recording. J Neurosci Res 42:266–267
5. Regehr WG, Pine J, Rutledge DB (1989) A long-term in vitro silicon-based microelectrode–neuron connection. IEEE Trans Biomed Eng 35:1023–1031
6. Pine J (1980) Recording action-potentials from cultured neurons with extracellular micro-circuit electrodes. J Neurosci Methods 2:19–31. doi:10.1016/0165-0270(80)90042-4
7. Gross GW, Williams AN, Lucas JH (1982) Recording of spontaneous activity with photoetched microelectrode surfaces from spinal cord neurons in culture. J Neurosci Methods 5:13–22
8. Jimbo Y, Kawana A (1992) Electrical stimulation and recording from cultured neurons using a planar electrode array. Biochem Bioenerg 29:193–204
9. Bergveld P, Wiersma J, Meertens H (1976) Extracellular potential recordings by means of a field-effect transistor without gate metal, called OSFET. IEEE Trans Biomed Eng 23:136–144
10. Fromherz P, Offenhäusser A, Vetter T, Weis J (1991) A neuron-silicon junction: a Retzius cell of the leech on an insulated-gate field-effect transistor. Science 252:1290–1293
11. Offenhäusser A, Sprössler C, Matsuzawa M, Knoll W (1997) Field-effect transistor array for monitoring electrical activity from mammalian neurons in culture. Biosens Bioelectr 12:819–826
12. Jobling DT, Smith JG, Wheal HV (1981) Active microelectrode array to record from the mammalian central nervous-system in vitro. Med Biol Eng Comput 19:553–560
13. Offenhäusser A, Rühe J, Knoll W (1995) Neuronal cells cultured on modified microelectronic device surfaces (1995). J Vac Sci Tech A 13:2606–2612

N

14. Cohen A, Spira ME, Yitshaik S, Borghs G, Shwartzglass O, Shappir J (2004) Depletion type floating gate p-channel MOS transistor for recording action potentials generated by cultured neurons. Biosens Bioelectron 19:1703–1709
15. Meyburg S, Goryll M, Moers J, Ingebrandt S, Böcker-Meffert S, Lüth H, Offenhäusser A (2006) N-channel field-effect transistors with floating gates for extracellular recordings. Biosens Bioelectr 21:1037–1044
16. Heer F, Hafizovic S, Franks W, Blau A, Ziegler C, Hierlemann A (2006) CMOS microelectrode array for bidirectional interaction with neuronal networks. IEEE J Sol State Circ 41:1620–1629
17. Imfeld K, Neukom S, Maccione A, Bornat Y, Martinoia S, Farine P-A, Koudelka-Hep M, Berdondini L (2008) Large-scale high-resolution data acquisition system for extracellular recording of electrophysiological activity. IEEE Trans Biomed Eng 55:2064–2073
18. Rutten W (2002) Selective electrical interfaces with the nervous system. Ann Rev Biomed Eng 4:407–452
19. Lambacher A, Fromherz P (1996) Fluorescence interference-contrast microscopy on oxidized silicon using a monomolecular dye layer. Appl Phys A 63:207–216
20. Wrobel G, Höller M, Ingebrandt S, Dieluweit S, Sommerhage F, Bochem HP, Offenhäusser A (2007) Cell-transistor coupling: transmission electron microscopy study of the cell-sensor interface. J R Soc Interface 5:213–222
21. Hai A, Shappir J, Spira ME (2010) In-cell recordings by extracellular microelectrodes. Nat Methods 7:200–202
22. Almquist BD, Melosh NA (2010) Fusion of biomimetic stealth probes into lipid bilayer cores. Proc Natl Acad Sci USA 107:5815–5820
23. Brüggemann D et al (2011) Nanostructured gold microelectrodes for extracellular recording from electrogenic cells. Nanotechnology 22:265104

Ni-Cadmium Batteries

Takashi Eguro
Frukawa Battery, Iwaki, Fukushima, Japan

Introduction

A storage battery has supported a recent rapid expansion of the portable electronic device market and has been developed to the market where a further development has been expected such as eco-friendly cars market such as EV and HEV or the power supply market of an electricity accumulation system of a renewable energy such as sunlight and wind power. A nickel–cadmium secondary battery plays a role as a pioneer making the importance of the storage battery recognized in these fields and has been used in many fields still now.

The nickel–cadmium secondary battery was invented in 1899 by Waldemar Jungner as a durable storage battery which endures severe conditions of use such as overcharge/overdischarge/long-term leaving to which a lead-acid storage battery has been unsuitable and has been used for a long time in various fields with the lead–acid storage battery, until a nickel–hydrogen battery and a lithium ion battery appeared in 1990s.

History

The nickel–cadmium secondary battery was invented in 1899 by Waldemar Jungner, and was sometimes referred to as a "Jungner battery." The practically used "Jungner battery" is a vented type battery using pocket-type electrodes. Then, a sintered-type electrode which is excellent in high-rate discharge performance and low-temperature performance was invented, and the nickel–cadmium secondary battery has come to be used for many uses, such as aircraft starting, railroads, vehicles for industry, miner lamps, and emergency lighting.

Then, A. B. Lange and others found how a cadmium electrode consumes the oxygen from a positive electrode during overcharging in 1938. Then, G. Neumann and others established the present sealing principle and basic structure of a sealed-type nickel–cadmium secondary battery in 1948. Then, the sealed-type nickel–cadmium secondary battery has come to rapidly spread mainly in the consumer electronic equipment market.

In Japan, the Furukawa Battery Co., Ltd. industrialized a sintered-type vented nickel–cadmium battery for airplane starting in 1955. The Furukawa Battery Co., Ltd. started mass production of the vented-type nickel–cadmium secondary battery and a sealed nickel–cadmium secondary battery for industrial use in 1962 and developed the same to the fields, such as aircrafts, railroads, backup power supply, and apparatus for emergency use.

On the other hand, Sanyo Electric Co., Ltd. started mass production of a consumer sealed type nickel–cadmium secondary battery. Then, other companies produced a consumer sealed-type nickel–cadmium secondary battery commercially one after another. Then, the batteries have been used as the main power supply of portable electronic devices such as cordless power tools and toys, video cameras, and notebook PCs in 1980s, and the production amount has enlarged drastically.

Although when a nickel–hydrogen battery and a lithium ion battery were produced commercially successively in 1990 and in 1991, respectively, the share of the portable electronic device market was taken by these batteries. However, the nickel–cadmium secondary battery is widely used currently in the fields, such as power tools, toys, aircrafts, railroads, backup power, and apparatus for emergency use.

Principle

The nickel–cadmium secondary battery contains NiOOH/nickel hydroxide as a positive active material, cadmium/cadmium hydroxide as a negative active material, and an aqueous solution containing potassium hydroxide as the main component as an electrolyte. Generally the charge-and-discharge reaction is shown in the following formulas 1, 2 and 3.

Positive active material:

$$NiOOH + H_2O + e^- \underset{\text{charge}}{\overset{\text{discharge}}{\rightleftarrows}} Ni(OH)_2 + OH^- \qquad E_0 = +0.52V \tag{1}$$

Negative active material:

$$Cd + 2OH^- \underset{\text{charge}}{\overset{\text{discharge}}{\rightleftarrows}} Cd(OH)_2 + 2e^- \qquad E_0 = -0.80V \tag{2}$$

Overall cell reaction

$$2NiOOH + Cd + 2H_2O \underset{\text{charge}}{\overset{\text{discharge}}{\rightleftarrows}} 2Ni(OH)_2 + Cd(OH)_2 \tag{3}$$

Electromotive force: 1.32V

The actual reaction mechanisms are more complicated and are considered that the charge-and-discharge reaction at a positive active material is a reaction accompanied by the diffusion in the solid phase of a proton, and reaction at a negative active material is a dissolution/deposit reaction accompanied by an intermediate product.

Type

The electrodes of the nickel–cadmium secondary battery are classified into pocket type, sintered type, and pasted type according to those manufacturing methods. Moreover, the batteries are classified into vented-type cell and sealed-type cell according to the existence of sealing structure. The batteries are classified into a prismatic cell, a cylindrical cell, and a button cell according to shape.

Manufacturing Methods of Electrode

Pocket type
Sintered type
Pasted type
Cell type
Vented-type cell
Sealed-type cell

Shape
Prismatic cell
Cylindrical cell
Button cell

Pocket Type

The pocket-type electrode is the electrode structure used as the foundations of the "Jungner battery" and is characterized by filling up active materials into a "pocket" formed with perforated iron sheets. As for the manufacture method of a pocket-type cell is as follows.

The positive active material of a pocket-type electrode consists of nickel hydroxide powder by which cobalt hydroxide was coprecipitated, and is obtained from those mixed sulfate by a neutralizing method. Graphite powder is mixed as a conductor. The addition of cobalt is performed for increasing the capacity of the positive active material. Cadmium hydroxide which is a negative active material is manufactured by a coprecipitation method or a dry-mixing method. Moreover, iron powder is usually mixed with negative active material of a pocket type electrode for the increase in capacity. Graphite powder and/or nickel flake is mixed as a conductor.

These active material powders are molded by pressurizing to be formed into long and slender tabular briquettes. Then, these briquettes are wrapped in perforated iron sheets of a ribbon shape to form a pocket of a long and slender plate shape. A plate frame is welded to a substance obtained by engaging a plurality of the pockets to form a plate. There is a projecting lug in the upper part of the plate frame, and a large number of positive and negative electrodes are put together alternately, and then each pole is fixed with a bolt and a nut to the lugs of each pole plate group. There is no sheet-like separator between positive and negative electrodes, and pin-like resin is inserted between the positive and negative electrodes to prevent contact thereof. Thus, an element referred to as a pole plate group is completed. The electrode group activated in the formation process is inserted in a battery container, and the container is filled with an electrolyte. Usually, lithium hydroxide is added to an electrolyte for the increase in capacity. A prismatic container made of resin or iron is mainly used for the battery container. The usual pocket-type nickel–cadmium battery is a vented type and is provided with a lid having a vent for escaping gases generated from the electrode group during charge operation. This vent is provided also with the role for filling a cell with water or electrolyte.

The pocket electrodes have a strong structure and are excellent in durability and the manufacturing cost thereof is low, and therefore, the pocket electrodes have been used for miner lamps, railway vehicles, emergency backup power, etc. for a long time.

However, the pocket-type nickel–cadmium battery has low energy density and low output performance, requires frequent water supply, and is difficult to seal, and therefore the production amount thereof has decreased.

Sintered Type

The sintered-type electrode was developed in Germany in 1932. This electrode is characterized by filling up active materials by an impregnation method into a porous sintered plaque in which carbonyl nickel powder having an average particle diameter of several microns is sintered.

The production method of the porous sintered plaque includes a dry method and a wet method. With respect to the dry method, carbonyl nickel powder is spread to a Ni wire grid with a sieve or the like, adjusted to a predetermined thickness, and then sintered at 800–1,000° [Celsius] in a reducible gas atmospheres, such as hydrogen gas or butane reformed gas. With respect to the wet method, carbonyl nickel powder, a binder such as CMC or MC, and water are mixed to prepare a slurry. The slurry is applied to a nickel-plated iron thin sheet (perforated sheet) which has an open area ratio of about 50 %, and the thickness is adjusted in a scratching portion. Then, after drying with a drying furnace, it is sintered at a temperature of 800–1,000° [Celsius] in a reducible gas atmosphere. The typical porosity of the porous sintered plaque is 80–87 %.

The impregnation method is a method including impregnating the sintered plaque with a hot impregnation liquid under normal pressure or decompression, neutralizing in a sodium hydroxide aqueous solution, and then filling up the inside of the porous sintered plaque with an active material. For a general impregnation liquid of a positive active material, an aqueous solution obtained by mixing nickel nitrate and a fixed quantity of cobalt nitrate is used. Then, a coprecipitated material of nickel hydroxide and cobalt hydroxide is obtained in a neutralization process. Usually, since a predetermined amount of an active material cannot be charged in one impregnation process, this process is repeated several times. This method is also referred to as a chemical impregnation method.

Moreover, there is a method referred to as an electrochemical impregnation method for depositing hydroxide in the plaque by carrying out the cathodic polarization of the plaque in the impregnation liquid. The electrochemical impregnation method has a merit such that an active material can be charged with a high density by a small number of times of impregnation. However, since the composition of the impregnation liquid changes with progress of electrochemical impregnation and by-products also tend to be formed, the method is used only for the limited use.

In the sintered electrode, a lug is formed at the end of the sheet to which the slurry is not applied or is attached by welding. Then, a large number of positive and negative electrodes are put together alternately inserting a separator. The pole is fixed to the lug with a bolt and a nut similarly as in the pocket type or is joined by resistance welding. Nylon or polypropylene cloth or nonwoven fabric is mainly used for a separator in the sintered-type nickel–cadmium secondary battery. Moreover, since the sintered-type plate is flexible compared with the pocket type, the sintered-type plate is able to be rolled in the shape of a bobbin combining positive electrodes, negative electrodes, and separators and can be stored in a cylindrical container. The cylindrical container is excellent in resistance to pressure and therefore is frequently used for the sealed-type nickel–cadmium secondary battery described later. The sintered-type electrode has high energy density, high-rate discharge characteristic, and high mass productionability by the wet sintering method. Therefore, currently, the sintered-type electrode is most widely produced in the nickel–cadmium secondary battery with the pasted-type electrode described later.

Pasted Type

The pasted-type electrode is developed in order to design a small and lightweight nickel–cadmium secondary battery. Main production methods of the pasted-type electrode include a method of applying to a nickel-coated steel sheet with perforations and a method of filling up a sponge-like nickel substrate.

The former is used for a cadmium negative electrode and the latter is mainly used for a nickel positive electrode.

A method for manufacturing a common pasted-type electrode for use in the cadmium negative electrode includes mixing a cadmium oxide activate material and PVA or the like with ethylene glycol to form a paste, applying the paste to a nickel-plated perforated iron sheet, and then drying the same to form a plate. The pasted-type electrode for use in the nickel positive electrode is obtained by filling up a slurry obtained by mixing a nickel hydroxide positive electrode active material, a binder, and water to a sponge-like nickel substrate having a relatively large pore diameter of 100 to several 100 μm, and then drying and pressing the same. There is also a method using a nickel substrate formed with nickel fiber in addition to the sponge-like nickel substrate. The assembly method of the electrode group is substantially similar to that of the sintered type.

Compared with the sintered-type electrode, the pasted-type electrode is able to achieve an increase in capacity of the electrode. Currently, the pasted-type electrode is used widely.

Vented Type Cell

In the completion or the end of charging, the storage battery containing an aqueous electrolyte causes electrolysis reaction of the water. Then oxygen generates from the positive electrode and hydrogen generates from the negative electrode in accordance with the following formulas 4 and 5:

The oxygen generating reaction in positive electrode:

$$4OH^- \longrightarrow 2H_2O+O_2\uparrow+4e^- \quad (4)$$

The hydrogen generating reaction in negative electrode:

$$2H_2O+2e^- \longrightarrow H_2\uparrow+2OH^- \quad (5)$$

The oxygen recombination reaction in negative electrode:

$$2Cd+O_2+2H_2O \longrightarrow 2Cd(OH)_2 \quad (6)$$

$$O_2+2H_2O+4e^- \longrightarrow 4OH^- \quad (7)$$

Thus, it is necessary to have a vent valve for releasing the gas. Moreover, since alkaline mist splashes from the electrolyte in the battery with gas venting, the vent valve has a splash proof structure which prevents diffusion of mist. The nickel–cadmium secondary battery with such a structure is referred to as the vented-type cell.

The vented-type battery needs to perform "water addition" in which water consumed by electrolysis is periodically added. However, in recent years, the vented-type nickel–cadmium secondary battery which has reduced the "water addition" frequency is produced commercially for trains and the like. The battery controls the electrolysis of the water under float charging by using the pasted-type cadmium electrode which has a high hydrogen overpotential characteristic for the negative electrode.

A nickel–cadmium secondary battery with the type of the gas recombination by catalyst in which oxygen and hydrogen gas generated in the end of charging are made to react and return to water by providing a catalyst to the vent valve is used partly.

The capacity range of the vented-type cell is broad from several Ah(s) to hundreds Ah(s), and a comparatively large cell occupies the mainstream. The shape of the vented-type cell is only prismatic.

There is also a mono-block-type battery using mono-block container with some independent cell rooms divided with partitions.

Sealed-Type Cell

A.B. Lange and others discover and G. Neumann and others established the sealed battery principles of the gas recombination on the negative electrode are as follows in general.

It is configured so that oxygen gas is previously generated from the positive electrode in the end of charging by adjusting the negative electrode capacity to an excessive degree rather than the positive electrode capacity (referred to as "charge reserve").

Oxygen generated from the positive electrode is easily diffused to the negative electrode by limiting the amount of the electrolyte and by using a separator with high gas permeability.

Thus, it is necessary to have a vent valve for releasing the gas. Moreover, since alkaline mist splashes from the electrolyte in the battery with gas venting, the vent valve has a splash proof structure which prevents diffusion of mist. The nickel–cadmium secondary battery with such a structure is referred to as the vented-type cell.

The vented-type battery needs to perform "water addition" in which water consumed by electrolysis is periodically added. However, in recent years, the vented-type nickel–cadmium secondary battery which has reduced the "water addition" frequency is produced commercially for trains and the like. The battery controls the electrolysis of the water under float charging by using the pasted-type cadmium electrode which has a high hydrogen overpotential characteristic for the negative electrode.

A nickel–cadmium secondary battery with the type of the gas recombination by catalyst in which oxygen and hydrogen gas generated in the end of charging are made to react and return to water by providing a catalyst to the vent valve is used partly.

The capacity range of the vented-type cell is broad from several Ah(s) to hundreds Ah(s), and a comparatively large cell occupies the mainstream. The shape of the vented-type cell is only prismatic.

There is also a mono-block-type battery using mono-block container with some independent cell rooms divided with partitions.

An oxygen gas recombination reaction occurs on the negative electrode.

This oxygen gas recombination reaction is considered that two kinds of reactions occur as follows.

Since the charging of the negative electrode is limited in both the cases, the negative electrode does not result in a full charge state, so that hydrogen gas does not generate. Moreover, since the charge–discharge performance of the negative electrode is inferior to that of the positive electrode, a part of the negative electrode is beforehand charged (referred to as "pre-charge"). By performing "charge reserve" and "pre-charge," the charge-and-discharge process of the sealed-type cell is always regulated by the positive capacity. The capacity range of the sealed type battery is from several mAh(s) to tens Ah(s), and a small cell occupies the mainstream. The sealed-type cell has high flexibility of shape, such as a square cell, a cylindrical cell, and a button cell.

This sealed-type nickel–cadmium secondary battery is easy to downsize, is excellent in discharging characteristic, and requires no maintenance at all. Therefore, the sealed-type nickel–cadmium secondary battery has spread as the main power supply of portable electronic devices, such as cordless power tools, toys, video cameras, and notebook PCs, and has supported the development thereof.

Future Directions

Although the market of the nickel–cadmium battery is decreasing gradually with a rapid development of a lithium ion battery in recent years, the nickel–cadmium battery is still used due to high reliability and achievements thereof for many uses. Moreover, the history of the nickel–cadmium battery is also the history of the development of a high-performance electrode. Various knowledge that many ancient people had is kept alive for a development of a next-generation high-performance battery.

Cross-References

▶ Ni-Metal Hydride Batteries

References

1. Uno Falk S, Salkind AJ (1969) Alkaline storage batteries. Wiley, New York
2. Kubokawa S, Ikari S, Ikeda K, Tagawa H, Shimizu K, Takagaki T, Takahashi H, Takehara Z (1975) Denchi handbook (Ed. Z. Takehara). Denki-syoin, Tokyo, Chap. 3. [in Japanese]
3. Awajitani T, Kaiya H (2001) Denchi Binran (Ed. Y. Matsuda, Z. Takehara). Maruzen, Tokyo, Chap. 3.4. [in Japanese]

Ni-Metal Hydride Batteries

Munehisa Ikoma
Panasonic, Moriguchi, Japan

Introduction

Nickel/metal hydride (Ni/MH) battery is a secondary battery using hydrogen storage alloy for the negative electrode, $Ni(OH)_2$ for the positive electrode, and alkaline solution for the electrolyte. Polypropylene nonwoven fabric is usually selected for the separator. The theoretical voltage is about 1.32 V, and the operating voltage is about 1.2 V which is almost the same as that of Ni/Cd battery [1]. The Ni/MH battery has been put to practical use for portable electric equipments in 1990 and for HEV (hybrid electric vehicle) in 1997 [2, 3].

Reaction Mechanism

The electrode reaction and battery reaction are shown in formulas (1), (2), and (3), at the positive electrode

$$NiOOH + H_2O + e^- \rightleftarrows Ni(OH)_2 + OH^- \quad (1)$$

at the negative electrode

$$MH + OH^- \rightleftarrows M + H_2O + e^- \quad (2)$$

total battery reaction

$$MH + NiOOH \rightleftarrows M + Ni(OH)_2 \quad (3)$$

M: hydrogen storage alloy

The reaction at the positive electrode is the same as that in Ni/Cd battery. At the negative electrode, hydrogen is absorbed in the alloy during the charge reaction, and the absorbed hydrogen is released and electrochemically consumed on the surface of the alloy during discharge. The battery reaction is relatively simple in that it basically involves hydrogen transportation between the positive and negative electrode.

During overcharge, oxygen is generated at the positive electrode as expressed formula (4). When the capacity of the negative electrode is sufficiently high, oxygen is absorbed by the negative electrode as shown in formula (5). This absorption reaction falls into equilibrium with the charging reaction as shown in formula (6), and so the state of charge of the negative electrode is kept:
at the positive electrode

$$OH^- \rightarrow 1/4O_2 + 1/2H_2O + e^- \quad (4)$$

at the negative electrode

$$MH + 1/4O_2 \rightarrow M + 1/2H_2O \quad (5)$$

$$M + H_2O + e^- \rightarrow MH + OH^- \quad (6)$$

Components

Positive Electrode

The sintered type and the pasted type are in popular in a positive electrode.

The sintered-type electrode is formed by filling of the nickel hydroxide active material into the sintered nickel porous layer on the punched steel metal. The pasted-type electrode is formed by filling of the nickel hydroxide active material into the formed nickel substrate with very high porosity. In the sintering layer of the sintered-type electrode, the pore size is around 10 μm and the porosity is approximately 75 %. On the contrary, in the formed nickel substrate of the pasted-type electrode, the pore size is around 500 μm and the porosity is approximately 95 %. Therefore, the sintered-type electrode is suitable for a high power use, and the pasted-type electrode is used for a high capacity battery [4].

Nickel hydroxide active material is provided by reacting nickel sulfate solution and an alkaline solution. A part of nickel of nickel hydroxide is substituted for Zn and Co for the improvement of the battery performance. The theory capacity of nickel hydroxide is 289 mAh/g by supposing one electron reaction. The utilization of nickel hydroxide of a sintered-type electrode is approximately 100 %, but that of a pasted-type electrode without a conductive additive is around 65 %. The improvement of the utilization is enabled by forming CoOOH conductive networks between nickel hydroxide particles. The cobalt compound (Co, CoO, $Co(OH)_2$) is filled into the formed nickel substrate with nickel hydroxide as an additive. The Co compounds form a conductive network as CoOOH by charging. To improve the conductivity of the cobalt conductive layer, $Co(OH)_2$ layer coating to the surface of nickel hydroxide particles and the oxidation treatment in an alkaline solution of this coated powder are suggested.

Negative Electrode

In the Ni/MH battery, hydrogen storage alloy, which contains no poison and is environment friendly compared with conventional Ni/Cd battery, is utilized as the negative electrode.

The negative electrode is obtained by coating the hydrogen storage alloy paste on the punched steel metal. The paste consists of the mixture of the hydrogen storage alloy powder, the conductive additives, and the binder.

Hydrogen storage alloy is able to store more than 1,000 times quantity of hydrogen compared with the liquid hydrogen. For application of hydrogen storage alloy to Ni/MH battery, the following conditions are required:

1. Electrochemically reversible absorption and desorption of hydrogen in large quantities
2. Minimal deterioration during repeated hydrogen absorption and desorption (superior in oxidation resistance)
3. Chemical stability over a wide humidity range
4. Easy to handle and of high safety
5. High environmental conformity
6. Abundance of resources and relatively inexpensive.

AB type (TiFe), AB_2 type ($ZrMn_2$,ZrV_2,$ZrNi_2$), AB_5-type ($CaNi_5$,$LaNi_5$,$MmNi_5$), A_2B type (Mg_2Ni,Mg_2Cu), bcc type (V based), and superlattice type (mixture of AB_5 type and AB_2 type) have been studied, but with consideration of the abovementioned condition (1), (2), (3), (4), (5), and (6), only the AB_5-type alloy and the superlattice-type alloy have been put to practical use.

The alloy composition suitable for a negative electrode with high capacity is selected from the pressure-composition isotherms (PCT curves) which are obtained from the absorption and desorption reactions of the alloy with hydrogen. The AB_5-type alloys which are most often adopted for battery consist of $LaNi_5$ or $MmNi_5$ in which La is displaced by inexpensive Mm (mischmetal, a mixture of rare earth elements). $MmNi_5$ alloy is promising for the low costs, but with a high hydrogen equilibrium pressure, charging and discharging are difficult under room temperature and nominal pressure. For achieving a high capacity alloy, substituting a part of Ni in a $MmNi_5$ alloy with other metal elements is attempted. In addition, thermal treatment is applied for homogenizing composition and enhancing crystallinity to flatten the plateau part of the PCT curve, thereby achieving a high capacity.

The cycle life of the Ni/MH battery strongly depends on the characteristics of the negative electrode. Hydrogen storage alloy is pulverized and corroded during charge–discharge cycles in an alkaline solution. The pulverization properties of alloy can be controlled by substitution a part of Ni with Co [5]. In the Ni/MH battery with AB_5-type alloy, there is the problem of the long-term storage performance caused by the dissolution of Co and/or Mn which are the constituent element of the alloy. This problem is solved by adopting the rare earth–Mg–Ni type superlattice alloy which does not include Co and Mn in a constituent element. This alloy has a structure that AB_5-type structure and AB_2-type structure arranged regularly [6].

Separator

The nonwoven polyolefin fabric is used. The thickness of the separator is 100–200 μm, and the basis weight of the fiber is 50–80 g/m^2. The hydrophilicity is provided to the fiber surface of the separator with the sulfonation treatment, plasma treatment or acrylic acid graft polymerization treatment.

Application

The Ni/MH battery has been put to practical use for the power supply for the camcorder and the cellular phone in 1990.

After practical use, the energy density of the battery is doubled from 180 to 350 Wh/L by the development of the various technologies such as active material composition, surface treatment, and additives. This battery has been put to practical use for the power supply for the power tool by the improvement of the high output performance. The AA size and AAA size batteries that are compatible with a dry cell have been also put to practical use. This battery system has been put to practical use for HEV (hybrid electric vehicle) by the establishment of the battery management technology and the improvement of the battery performance (the input/output performance and the durability) in 1997.

Future Directions

The energy density of the Ni/MH battery is smaller than that of the Lithium ion battery. But the Ni/MH battery is superior in the general performance balance such as input power density, output power density, cycle life, safety, reliability, recyclability, the cost. This superior general performance balance will bring up the market of the Ni/MH battery as the power supply for HEV and compatible with a dry cell in future.

The improvement of the hydrogen storage alloy is important to the performance enhancement of the Ni/MH battery. The detailed analysis and improvement of the chemical formula, structure and surface state of the hydrogen storage alloy shall lead the Ni/MH battery to the performance enhancement. In addition, the surface treatment technology, the other device and the battery management technology should be improved for the maximization of the battery performance.

Cross-References

► Ni-Cadmium Batteries

References

1. Ogawa H et al (1989) Proceedings of the 16th international power sources symposium, Bournemouth, pp. 393–410
2. Yuasa K et al (1991) Cylindrical type sealed nickel-metal hydride battery. Natl Tech Rep 37(1):44–51
3. Morishita N et al (1998) Nickel-Metal Hydride Battery for EV and HEV. Matsushita Tech J 44(4):426–433
4. Kaiya H et al (1986) New type high capacity nickel-cadmium battery (SM30). Natl Tech Rep 32(5):631–638
5. Ikoma M et al (1999) Effect of alkali-treatment of hydrogen storage alloy on the degradation of Ni/MH batteries. J Alloy Compd 284:92
6. Ochi M et al (2012) Nickel-metal hydride battery for hybrid electric vehicles (HEVs). Panasonic Tech J 57(4):278–283

Nickel Oxide Electrodes

Masanobu Chiku
Department of Applied Cemistry,
Graduate School of Engineering,
Osaka Prefecture University, Osaka, Japan

Introduction

A wide variety of metal oxides are investigated as electrode materials for electrochemical capacitors (super capacitors, ultracapacitors). Ruthenium oxide and iridium oxide are frequently used due to its high charging-discharging capacities. However, ruthenium and iridium are high cost and limited resources on the earth. Thus wide varieties of transition metals are studied as the electrode materials to alternate precious metals. Nickel oxide is one of the promising materials for high-performance electrode materials for electrochemical capacitors due to its low cost, high reservation on the earth, and low toxicity for environment.

Mechanisms

Nickel hydroxide exists in four different forms: α-$Ni(OH)_2$, β-$Ni(OH)_2$, β-NiOOH, and γ-NiOOH. β-$Ni(OH)_2$ and β-NiOOH are known as stable forms. The electrochemical reaction of NiO, $Ni(OH)_2$, and NiOOH is described as below:

$$Ni(OH)_2 + OH^- \leftrightarrow NiOOH + H_2O + e^- \quad (1)$$

$$NiO + OH^- \leftrightarrow NiOOH + e^- \quad (2)$$

Preparation of Nickel Oxide Electrodes

To prepare nickel oxide electrodes, nickel hydroxide precursors are frequently used, and nickel hydroxides are thermally oxidized to nickel oxide (Fig. 1). Industrially produced nickel hydroxide active materials are mainly used as

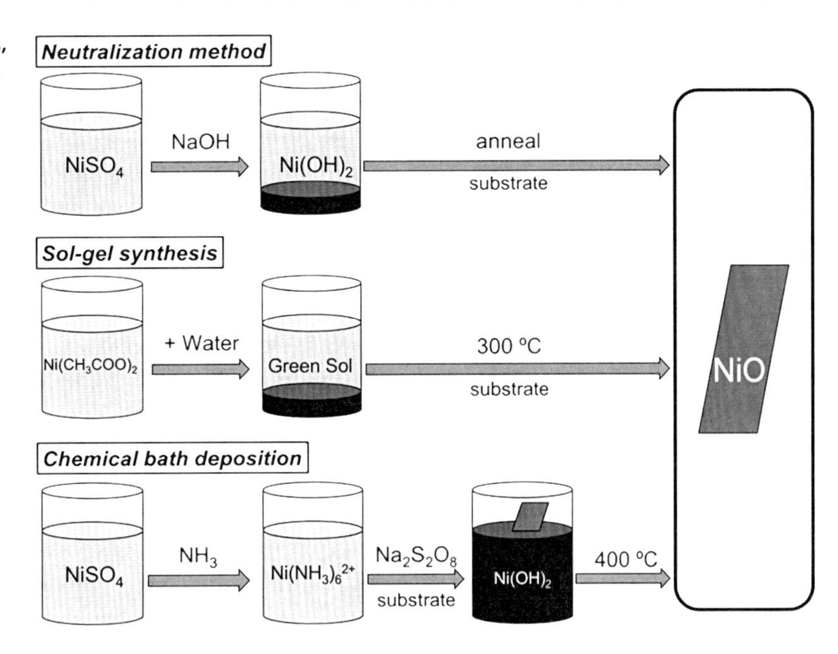

Nickel Oxide Electrodes, Fig. 1 Schematic images of the NiO electrode preparation

positive electrode for nickel-metal hydride secondary battery, and they are produced by using coprecipitation methods. Neutralization method is one of the most basic coprecipitation methods to prepare nickel hydroxide. The general neutralization method is described as follow: putting nickel sulfate solution by drops into alkali solution such as NaOH solution to neutralize and precipitate nickel hydroxide. After dried nickel hydroxide, it was washed with hot water to remove nickel sulfate and dried less than 150 °C to obtain nickel hydroxide powder.

Liu et al. applied sol–gel method to prepare nickel oxide active materials for electrochemical capacitors [1]. They dried nickel acetate tetrahydrate at 100 °C and stirred it in water for 2 days. The precipitants were separated by centrifugation and put it into water to get pale-green sol. Dipping coated the sol on nickel sheets and annealed in air at 300 °C for 1 h. The diameters of resultant nickel oxide were 3 ~ 8 nm and pore sizes were 2 ~ 3 nm. The electrochemical capacitors using resultant nickel oxide indicated 50 ~ 64 F g^{-1} capacitance with 1 M KOH aqueous solution.

Inoue et al. applied chemical bath deposition (CBD) method to prepare nickel oxide electrodes [2].

Ammonia water was put into nickel sulfate aqueous solution at 30 °C. Then potassium persulfate aqueous solution was added into the solution, followed by stirring for 3 min. The resultant solution is named a CBD bath. A Ni foam as a substrate was soaked in CBD bath. Then the nickel hydroxide was deposited on Ni foam and annealed in air at 400 °C. The diameter of resultant nickel oxide particle was ca. 5 μm and flowerlike form. The electrochemical capacitors using resultant nickel oxide indicated ca. 80 F g^{-1} capacitance with 10 M KOH aqueous solution.

Future Perspective

Several types of NiO or $Ni(OH)_2$ electrode such as flowerlike [3], mesoporous [4], nanotubes [5], and nanorod [6] were investigated as active materials for electrochemical capacitors. These Ni oxide electrodes are combined with carbon negative electrodes to construct the hybrid capacitors with high capacity and power density [2]. However, there still remaining serious problems with life cycle, reducibility, long-term stability, etc. It would need some more innovation for the practical use of these high-performance electrode materials.

Cross-References

▶ Electrode
▶ Super Capacitors

References

1. Liu KC, Anderson MA (1996) Porous nickel oxide/nickel films for electrochemical capacitors. J Electrochem Soc 143:124–130
2. Inoue H, Namba Y, Higuchi E (2010) Preparation and characterization of Ni-based positive electrodes for use in aqueous electrochemical capacitors. J Power Sources 195:6239–6244
3. Xu LP, Ding YS, Chen CH, Zhao LL, Rimkus C, Joesten R, Suib SL (2008) 3D flowerlike α-nickel hydroxide with enhanced electrochemical activity synthesized by microwave-assisted hydrothermal method. Chem Mater 20:308–316
4. Xing W, Li F, Yan ZF, Lu GQ (2004) Synthesis and electrochemical properties of mesoporous nickel oxide. J Power Sources 134:324–330
5. Kim JH, Zhu K, Yan Y, Perkins CL, Frank AJ (2010) Microstructure and pseudocapacitive properties of electrodes constructed of oriented nio-tio2 nanotube arrays. Nano Lett 10:4099–4104
6. Hasan M, Jamal M, Razeeb KM (2012) Coaxial NiO/Ni nanowire arrays for high performance pseudocapacitor applications. Electrochim Acta 60:193–200

Nitrogen Oxides (NOx) Removal

Jorge G. Ibanez[1] and Krishnan Rajeshwar[2]
[1]Department of Chemical Engineering and Sciences, Universidad Iberoamericana, México, Mexico
[2]The University of Texas at Arlington, Arlington, TX, USA

Introduction

Nitrogen oxides in the atmosphere participate in environmentally challenging processes such as the greenhouse effect, acid rain, and photochemical smog. Their chemical reduction involves the use of a reducing agent whose storage and leakage (or *slip*) may be problematic [1]. A plausible alternative involves their electrochemical treatment, although this necessitates that the gases be dissolved in aqueous solution, which may not be a straightforward task. In addition, complex chemical equilibria exist among the different nitrogen oxides (see Fig. 1), derived in part from the varied oxidation states of nitrogen (i.e., from −3 to +5) [2–4].

Specific aspects of their electrochemistry and electrochemical reduction aimed at their removal are discussed next.

Nitrogen (I) Oxide

Also called nitrous oxide, dinitrogen monoxide, sweet air or laughing gas, N_2O is the most stable nitrogen oxide which – coupled with its IR absorption properties – makes its participation inevitable in the greenhouse effect. It can be electroreduced in alkaline and acidic media. Current efficiencies up to 100 % have been reported. Pd and macrocyclic amine complexes catalyze this electroreduction to $N_{2(g)}$ [2, 5]. The presence of N_2O can be used to advantage by forming an N_2O–H_2 fuel cell [5].

Nitrogen (II) Oxide

Also called nitric oxide, nitrogen monoxide or oxidonitrogen, NO is a colorless, relatively unreactive radical that is essentially insoluble in aqueous solution. Simple absorption in alkaline solutions is not effective, since it is only physically absorbed [3, 4, 6]. Absorption in nitric acid decreases with acid concentration [4], and oxidation with ozone to produce acidic NO_2 facilitates its absorption in alkaline media [6]. Absorption of NO with simultaneous oxidation at a gas diffusion electrode in alkaline solution eliminates the need for an oxidizing agent [6]:

$$NO + OH^- = NO_2^- + H^+ + e^- \qquad (1)$$

Aqueous absorption of NO is also facilitated by aminopolycarboxylate chelates of Fe(II) and Co(II) [7–10]. The resulting compounds can

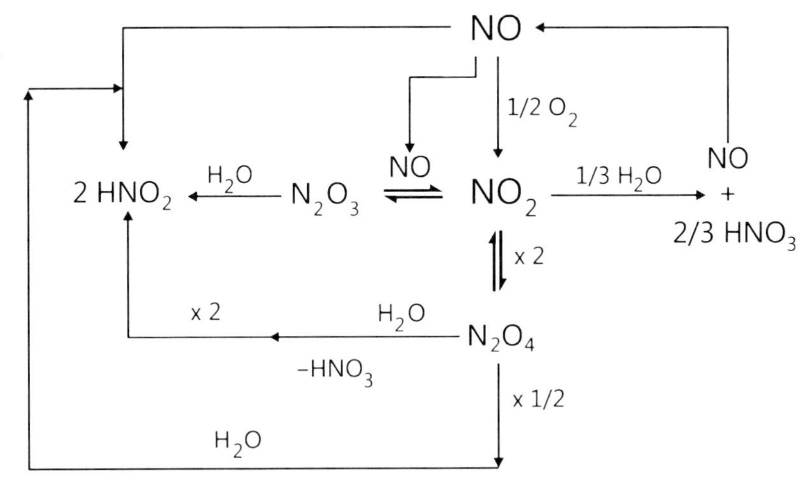

Nitrogen Oxides (NOx) Removal, Fig. 1 Nitrogen oxides chemical equilibria (Adapted from [2])

then be chemically or electrochemically reduced to yield hydroxylamine, hydrazine or ammonia, regenerating the chelate. Removal of NO from flue gases is facilitated by its preferential adsorption on noble metal catalysts. This allows for its reduction to occur at high yields even in dilute gas streams and in the presence of other competitive species (e.g., O_2, SO_2) [11]. In addition, by adjusting the electrode potential to place the valence electrons of the electrode at an energy level between that of the antibonding orbitals of NO and O_2, the competitive O_2 reduction can be inhibited [1]. Although electronically similar to CO, NO has an unpaired electron and a $2\pi^*$ orbital of lower energy that make it substantially more reactive towards oxidation and reduction [12].

The reduction of NO to produce ammonia, N_2O and hydroxylamine is thermodynamically allowed (see Table 1) and electrochemically viable [13, 14]. On this basis, an electrogenerative process has been proposed whereby NO reacts in a cell with protons and electrons at one electrode and H_2 at the other electrode to generate current [11]. Note that this configuration resembles that of a fuel cell (although the reactions are not necessarily the same), and the corresponding technology can be borrowed from that field. Electrogenerative processes can be defined as those in which favorable thermodynamic and kinetic factors are utilized for the production of coupled electrode reactions that take place in separate compartments in an electrochemical cell with simultaneous generation of electricity [11, 15–19].

Nitrogen Oxides (NOx) Removal, Table 1 Reduction of NO to environmentally friendlier products [2]

Reaction
$2NO + 2H^+ + 2e^- \rightarrow N_2O + H_2O$
$2NO + 4H^+ + 4e^- \rightarrow N_2 + 2H_2O$
$2NO + 6H^+ + 6e^- \rightarrow 2NH_2OH$
$2NO + 10H^+ + 10e^- \rightarrow 2NH_3 + 2H_2O$

Nitrogen (IV) Oxide

Also called nitrogen dioxide or dioxidonitrogen, NO_2 is a brownish acidic radical whose electrochemistry follows essentially that of NO, as they have a common electroactive precursor, namely NO^+ [20].

Complex Nitrogen Oxides

The complex nitrogen oxides N_2O_4 and N_2O_3 undergo chemical absorption in aqueous solutions more easily than NO and NO_2, forming nitrogenated species susceptible to reduction [3].

Other General Treatments for NO_x

The electroreduction of NO_x in industrial waste gas streams can simultaneously lead to useful

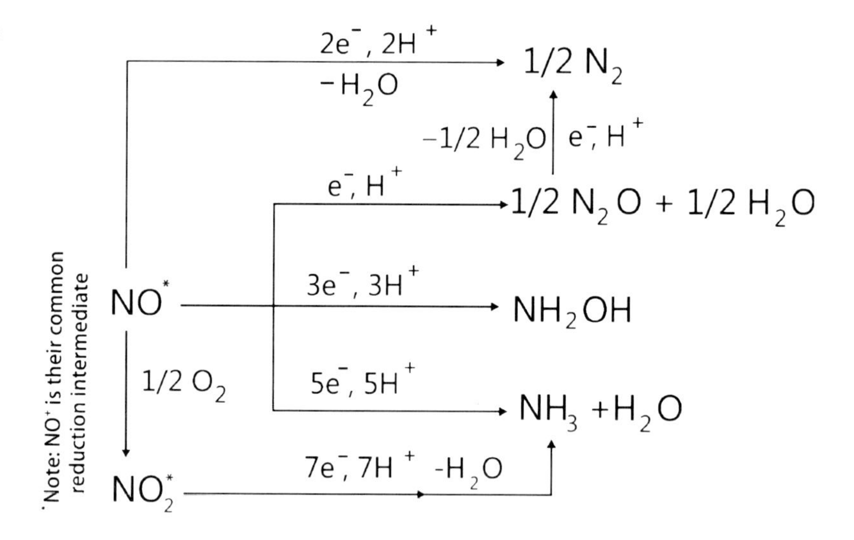

Nitrogen Oxides (NOx) Removal, Fig. 2 NOx reduction (Adapted from [2])

and/or inert compounds, as shown in Fig. 2 [8, 11, 15–18, 21, 22]. A major fraction of the current at low potentials may result in formation of ammonia and hydroxylamine [15–17]. Ammonia formation is strongly favored in N_2 diluent alone, whereas a trend towards hydroxylamine formation has been observed in the presence of CO or SO_2 [15–17]. A simultaneous desulfurization/denoxing process (called the Saarberg-Hölter-Lurgi, SHL process) is based on this principle [7]. Mixed ionic-electronic conducting catalytic membranes allow the reduction of NO_x by acting as short circuited devices, where the electrons reduce the oxides while the oxide ions thus produced move towards a low pressure end, completing the process [1, 2]. The catalytic activity and selectivity of porous metal films deposited on solid electrolytes for the reduction of NO_x with H_2 or CO can be substantially promoted by the application of an external potential or current [14, 23, 24]. This non-faradaic electrochemical modification of catalytic activity (NEMCA) is envisaged as a plausible technology in the near future [25].

Future Directions

Optimization of the conductivity of solid oxide ion conducting electrolytes is essential to lower the temperatures required for the reduction of NO_x [1]. In addition, the development of more selective layers can potentially increase the efficiency of electroreduction devices by avoiding the undesired reduction of O_2 [26–28].

References

1. Kammer Hansen K (2000) Electrochemical reduction of NO and O_2 on Cu/CuO. J Appl Electrochem 30:193–200
2. Rajeshwar K, Ibanez JG (1997) Environmental electrochemistry: fundamentals and applications in pollution abatement. Academic, San Diego
3. Weisweiler W, Deib K-H (1987) Influence of electrolytes on the absorption of nitrogen oxide components N_2O_4 and N_2O_3 in aqueous absorbents. Chem Eng Technol 10:131
4. Weisweiler W, Eidam K, Thiemann M, Scheibler E, Wiegand KW (1991) Absorption of nitric oxide in dilute nitric acid. Chem Eng Technol 14:270
5. Tomkiewics M, Yoneyama H, Haynes R, Hori Y (eds) (1993) Environmental aspects of electrochemistry and photoelectrochemistry, vol 93-18, The Electrochemical Society, proc. The Electrochemical Society, Pennington
6. Furuya N, Okada T (1991) Electrooxidation of NO on gas-diffusion electrode. In: 179th meeting of the Electrochemical Society, Washington, DC, 5–10 May 1991. Extended Abstract # 728
7. van Velzen D (1992) Electrochemical processes in the protection of the environment. In: Genders D, Weinberg N (eds) Electrochemistry for a cleaner environment. The Electrosynthesis Co. East Amherst, New York
8. Uchiyama S, Muto G (1981) Dependence of pH and chelate ligand on the electroreduction mechanism of

a nitrosyl-aminopolycarbonato-ferrous complex in weakly acidic media. J Electroanal Chem 127:275
9. Long X-L, Xiao W-D, Yuan W-K (2005) Kinetics of gas–liquid reaction between NO and $Co(en)_3^{3+}$. Ind Eng Chem Res 44:4200–4205
10. Ibanez JG, Hernandez-Esparza M, Doria-Serrano C, Fregoso-Infante A, Singh MM (2008) Environmental chemistry: microscale laboratory experiments. Springer, New York
11. Pate KT, Langer SH (1985) Electrogenerative reduction of nitric acid for pollution abatement. Environ Sci Technol 19:371–373
12. de Vooys ACA (2004) Mechanisms of electrochemical reduction and oxidation of nitric oxide. Electrochim Acta 49:1307–1314
13. de Vooys ACA (2001) Mechanistic study of the nitric oxide reduction on a polycrystalline platinum electrode. Electrochim Acta 46:923–930
14. Petrushina IM (2003) Electrochemical promotion of NO reduction by hydrogen on a platinum/polybenzimidazole catalyst. J Electrochem Soc 150: D87–D90
15. Langer SH, Foral MJ, Colucci JA, Pate KT (1986) Electrogeneration and related electrochemical methods for NO_x and SO_x control. Environ Prog 5:277
16. Foral MJ, Langer SH (1988) The effect of preadsorbed sulfur on nitric oxide reduction at porous platinum black electrodes. Electrochim Acta 33:257
17. Foral MJ, Langer SH (1991) Sulfur coverage effects on the reduction of dilute nitric oxide at platinum black gas diffusion electrodes. Electrochim Acta 36:299
18. Langer S (1994) Electrochemical processing without a power source: benefits and applications. In: Eighth international forum on electrol in the chemical industries. The Electrosynthesis Co., Orlando
19. Yap CY, Mohamed N (2008) Electrogenerative processes for environmental applications. Clean 36:443–452
20. Snider BG (1979) Reduction of nitric oxide, nitrous acid and nitrogen dioxide at platinum electrodes in acidic solutions: review and new voltammetric results. Anal Chim Acta 105:9–23
21. Langer SH, Pate KT (1980) Electrogenerative reduction of nitric oxide. Nature 284:434
22. Furuya N, Murase K (1994) Electroreduction of nitric oxide to nitrogen using a gas diffusion electrode loaded with noble metals. In: 186th meeting of the Electrochemical Society, Miami Beach, Abstract # 589
23. Pliangos C (2000) Electrochemical promotion of a classically promoted Rh catalyst for the reduction of NO. Electrochim Acta 46:331–339
24. Dorado F, de Lucas-Consuegra A, Vernoux P, Valverde JL (2007) Electrochemical promotion of platinum impregnated catalyst for the selective catalytic reduction of NO by propene in presence of oxygen. Appl Catal B Environ 73:42–50
25. Vernoux P, Gaillard F, Lopez C, Siebert E (2003) Coupling catalysis to electrochemistry: a solution to selective reduction of nitrogen oxides in lean-burn engine exhausts? J Catal 217:203–208
26. Hamamoto K (2006) Intermediate temperature electrochemical reactor for NO_x decomposition. J Electrochem Soc 153:D167–D170
27. Wang X (2004) Selective decomposition of NO in the presence of excess O_2 in electrochemical cells. J Appl Electrochem 34:945–952
28. Kammer K (2005) Electrochemical $deNO_x$ in solid electrolyte cells—an overview. Appl Catal B Environ 58:33–39

Non-Aqueous Electrolyte Solutions

Heiner Jakob Gores[1] and Hans-Georg Schweiger[2]
[1]Institute of Physical Chemistry, Münster Electrochemical Energy Technology (MEET), Westfälische Wilhelms-Universität Münster (WWU), Münster, Germany
[2]Faculty of Electrical Engineering and Computer Science, Ingolstadt University of Applied Sciences, Ingolstadt, Germany

Definition and Main Properties

Nonaqueous electrolyte solutions are ion conductors comprising a solvent or blends of solvents and a dissolved salt or several dissolved salts. They may also contain several additives, i.e., materials that improve a wanted property. Due to the huge number of possible solvents, salts, and additives, nonaqueous electrolyte solutions cover large ranges of selectable properties [1]. To give two examples, in comparison to aqueous electrolyte solutions, nonaqueous electrolyte solutions offer wider liquid ranges (down to −150 °C and up to about 300 °C) [2] and an appreciably larger voltage window, also called electro-inactivity range (up to about >4 V vs. about 1.2 V for aqueous systems). Both the large liquid range and the large voltage window not only extend the range of accessible investigations of dissolved materials in fundamental research but offer also applications for processes and in devices that would not be possible with electrolyte aqueous solutions [1].

Properties of nonaqueous electrolyte solutions have been widely studied in fundamental research due to the possibility to vary parameters such as the viscosity and dielectric permittivity of the solvent. The result of these studies mainly conducted in the last century was a better knowledge of spectroscopic and transport properties as well as the thermodynamics of electrolyte solutions [3–17]. The observed behavior was interpreted in terms of structure formation in solutions including solvation of ions, ion pair formation, formation of triple ions and clusters caused by the underlying interactions, the ion/solvent molecule interaction and the ion/ion interaction [2, 5, 6, 9, 14, 18–21].

In applied research the possibility to tailor the properties of the solutions have initiated many applications including processes such as [1, 2, 14, 15]:

- Electroplating of materials that cannot be electroplated from aqueous solutions because hydrogen formation would occur instead, including elements such as Al, Si, Ti, and Dy [22]
- Electrodeposition of nonconducting materials
- Electrosynthesis of organic and inorganic materials
- Deposition of conducting polymers
- Electromachining of metals and alloys.

Nonaqueous electrolyte solutions are also applied in each device that could not work without the large voltage window of these materials, including [23–29, 35–39]:

- Primary lithium cells
- Secondary lithium ion cells and batteries
- Dye sensitized solar cells
- Electrochemical double-layer capacitors.

The last three fields of application are currently under heavy research, development, and marketing worldwide. This statement is especially stressed for secondary lithium-ion cells and batteries that currently are booming. After their successful and increasing use in mobile equipments such as mobile phones, tablet computers, and laptops as well as in electric tools. lithium-ion batteries for battery-electric or hybrid-electric vehicles are beginning to reach the market in large numbers. Electrochemical double-layer capacitors may be useful in addition to lithium-ion batteries when very many cycles at high power densities are requested. Finally, solar cells and also in the future dye sensitized solar cells may be used to produce the electric energy for electric cars that are used for short distances (home to work and back). Finally, secondary lithium-ion batteries of cars can be used to buffer the unsteady delivery of electrical energy from so-called renewable resources.

Solvents

Solvents can be classified according to their properties, including physical properties such as dielectric permittivity, viscosity, and liquid range; chemical properties of the solvent including functional groups of solvent molecules; and physical properties of isolated solvent molecules, such as dipole moment and polarizability and last but not least empirical solvent parameters [1, 2]. Empirical solvent parameters can be obtained by measuring the interaction of the studied solvent molecule with other molecules. For example, Gutmann's donor number (DN) obtained by reaction enthalpy measurements of dissolved solvent molecules with the Lewis acid $SbCl_5$ is an excellent measure of the Lewis basicity of a solvent and hence its ability to solvate cations. The acceptor number (AN) scale reflects the Lewis acidity of the solvent and is measured by ^{31}P NMR chemical shift measurements of the strong base triethylphosphine oxide reflecting the ability of a solvent molecule to solvate anions. Other important empirical parameters include solvatochromic parameters including Kamlet and Taft's α-, β-, and π^* scales that are related to the ability to accept or donate hydrogen bonds and to the polarity and polarizability of a solvent. Dimroth and Reichardt's E_T (30) scale is a useful measure for the polarity (ionizing power) of a solvent as well. These solvatochromic parameters are spectroscopically determined by measuring the interaction of a solvent molecule with a selected dye. For details and references to original papers, see [1, 2, 18]. Based on those aforementioned criteria and solvent classes proposed by other workers, we [1] proposed to classify solvents according to eight classes, including:

1. Amphiprotic hydroxylic solvents, typical examples are the alcohols
2. Amphiprotic protogenic solvents, e.g., carboxylic acids
3. Protophilic H-bond donor solvents, e.g., amines
4. Dipolar aprotic protophilic solvents, e.g., pyridine
5. Dipolar aprotic protophobic solvents, e.g. esters
6. Low-permittivity electron donor solvents, e.g. ethers
7. Low polarity solvents of high polarizability, e.g. benzene
8. Inert solvents, e.g. alkanes and perfluoroalkanes.

For tables of solvents of these classes including physical properties (melting point, boiling point, dielectric permittivity, viscosity, density, dipole moment, and the mentioned empirical solvent parameters), see Ref. [1]. These and similar tables can also be found in Refs. [8, 14, 15, 18]. Suffice it to state here that selected solvents and solvent blends of classes 5 and 6 are often used in practical applications due to their superior electrochemical stability.

Salts

Generally, in nearly every publication devoted primarily to electrolyte solutions apart from solvents, the role of ions is taken into account as well, especially when solvent molecule–ion interaction (solvation) is considered. Marcus has published a review book [19] including several aspects of electrolyte solutions seen from the ion's side including topics such as ionic radii, partial molar quantities, ionic volumes, polarizability, and ionic transport. In contrast to the last century, where mainly salts consisting of atomic ions have been studied currently, molecular organic ions are gaining increasing interest. This statement is especially true for ionic liquids and their solutions in organic solvents consisting of both molecular organic cations and anions. It holds also for large molecular anions bearing many electron-withdrawing substituents, i.e., the so-called weakly coordinating anions that show apart from fascinating properties also many possible applications [23]. Some of the corresponding acids of these anions are very strong acids and may even reach the state of superacids. Whereas the role of specific ion effects is a mature field of research for aqueous solutions [30], it has rarely been studied for nonaqueous solutions. Another field of current interest is the supermolecular chemistry of anions [20] that focuses on anionic coordination chemistry and theoretical and practical aspects of anion complexation. An example from electrolyte research for lithium-ion cells may be appropriate here. It is well known that LiF is scarcely soluble in any solvent. However, it is possible to obtain highly conducting LiF solutions in nonaqueous solvents by means of anion receptors based on boron, such as tris(pentafluorophenyl) borane [31, 32], increasing the solubility of the nearly insoluble salt LiF by six orders of magnitude up to 1 $mol{\cdot}L^{-1}$ solutions.

Additives

Generally, every application of nonaqueous electrolyte solutions needs its specific additives to prevent problems in the intended application or to increase the shelf life of a device or a process. For example, in lithium-ion batteries there are several critical processes that decide on its shelf life. According to their role the following additives are used to reach this requirement: solid electrolyte interphase forming improvers, cathode protecting agents, overcharge protecting agents, wetting agents, flame retardant agents, and Al-current-collector corrosion inhibitors. For recent reviews see [33, 34].

Future Directions

The future of this research field is extremely bright. The meaning of the title of a review written nearly 20 years ago still holds: *Solution chemistry: A cutting edge in modern electrochemical technology* [1]. Driven by the economic interest research efforts will continue and will be expanded. Despite the fact that electrolyte solutions are under investigation since more than 100 years, several interesting topics related to fundamental research are not yet understood. To give an example, up to now we are waiting for a useful theory that is able to calculate transport properties of nonaqueous electrolyte solutions starting from quantum-mechanical levels. Also the role of ion pairs and higher clusters is not completely understood and invites workers for further research. To give an example, charge transport in electrolytes of low dielectric permittivity (including solid polymer electrolytes and gels) in lithium-ion cells may be based on transport of ion pairs instead of lithium-ion transport. A combined research program including spectroscopic (such as dielectric permittivity studies and infrared spectroscopy studies) and electrochemical studies is needed to resolve these open questions. A final cautionary statement is stressed here again: Many workers that are not familiar with nonaqueous systems are not aware of the fact that water affects the results of measurements directly (e.g., by hydrolysis of constituents of the solutions) and indirectly by changing the coupled equilibria governing the properties of electrolyte solutions, especially when the concentrations of the salts are low. We observed changes of $> 100\ \%$ for $0.1\ \%$ water content and repeated a diffusion measurement at thin layer cells ($\sim 30\ \mu m$ thickness) without protecting inert atmosphere and received doubled diffusion coefficients despite rather short times where the non-dried cells were in contact with the atmosphere. To sum up, working with nonaqueous electrolyte solutions means strictly controlling the water content be Karl–Fischer titration or other useful methods and working under controlled pure inert gas.

Cross-References

- ▶ Conductivity of Electrolytes
- ▶ Electrolytes for Electrochemical Double Layer Capacitors
- ▶ Electrolytes for Rechargeable Batteries
- ▶ Ion Mobilities
- ▶ Ion Properties
- ▶ Ionic Liquids

References

1. Barthel J, Gores HJ (1994) Chap 1. Solution chemistry: a cutting edge in modern electrochemical technology. In: Mamontov G, Popov AI (eds) Chemistry of nonaqueous electrolyte solutions. Current progress. VCH, New York, pp 1–148
2. Gores HJ, Barthel J, Zugmann S, Moosbauer D, Amereller M, Hartl R, Maurer A (2011) Chap. 17. Liquid nonaqueous electrolytes. In: Daniel C (ed) Handbook of battery materials, 2nd edn. Wiley-VCH, Weinheim, pp 525–626
3. Robinson RA, Stokes RH (1959) Electrolyte solutions, 2 revth edn. Butterworth, London (Reprint 2003, Dover Publications, New York)
4. Harned HS, Owen BB (1958) The physical chemistry of electrolyte solutions, 3rd edn. Reinhold, New York
5. Gurney RW (1953) Ionic processes in solution. McGraw-Hill, New York
6. Gurney RW (1962) Ions in solution. Dover Publications, New York (reprint)
7. Barthel J (1976) Ionen in nichtwäßrigen Elektrolytlösungen. Dr. D. Steinkopff, Darmstadt
8. Barthel J, Krienke H, Kunz W (1998) Physical chemistry of electrolyte solutions, modern aspects, vol 5. Dr. Dietrich Steinkopff Verlag GmbH & Co. KG, Darmstadt
9. Bockris JO'M, Reddy AKN (1998) Modern electrochemistry, vol. 1: ionics, vol 2. Plenum Press, New York
10. Luckas M, Krissmann J (2001) Thermodynamik der Elektrolytlösungen. Springer, Berlin
11. Lee LL (2008) Molecular thermodynamics of electrolyte solutions. World Scientific Publishing, Singapore
12. Falkenhagen H (1971) Theorie der Elektrolyte. S. Hirzel, Leipzig
13. Friedman HL (1962) Ionic solution theory based on cluster expansion methods, vol 3, Monographs in statistical physics. Wiley, New York
14. Izutsu K (2009) Electrochemistry in nonaqueous solutions, 2nd edn. Wiley-VCH, Weinheim
15. Aurbach D (1999) Nonaqueous electrochemistry. Marcel Dekker, New York
16. Mann CK (1969) Nonaqueous solvents for electrochemical use in electroanalytical chemistry. In: Bard AJ (ed) Electroanalytical chemistry, vol 3. Marcel Dekker, New York, p 57
17. Marcus Y (1977) Introduction to liquid state chemistry. Wiley, London
18. Marcus Y (1985) Ion solvation. Wiley, Chichester
19. Marcus Y (1997) Ion properties. Marcel Dekker, New York
20. Bianchi A, Bowman-James K, Enrique García-España E (1997) Supramolecular chemistry of anions. Wiley-VCH, Weinheim
21. Marcus Y, Hefter GT (2006) Ion pairing. Chem Rev 106(11):4585–4621
22. Lodermeyer J, Multerer M, Zistler M, Jordan S, Gores HJ, Kipferl W, Diaconu E, Sperl M,

Bayreuther G (2006) Electroplating of dysprosium, electrochemical investigations, and study of magnetic properties. J Electrochem Soc 153(4): C242–C248
23. Nakajima T, Groult H (eds) (2005) Fluorinated materials for energy conversion. Elsevier, Amsterdam
24. Xu K (2004) Nonaqueous liquid electrolytes for lithium-based rechargeable batteries. Chem Rev 104:4303–4418
25. Barthel J, Gores HJ, Schmeer G, Wachter R (1983) Nonaqueous electrolyte solutions in chemistry and technology. Top Curr Chem 111:33
26. von Schalkwijk W, Scrosati B (eds) (2002) Advances in lithium-ion batteries. Kluwer, New York
27. Yamamoto O, Wakihara M (1998) Lithium-ion batteries. Wiley-VCH, Weinheim
28. Yoshio M, Brodd RJ, Kozawa A (eds) (2003) Lithium ion batteries. Springer, New York
29. Nazri GA, Pistoia G (2003) Lithium batteries, science and technology. Springer, New York (first softcover printing 2009)
30. Kunz W (ed) (2010) Specific ion effects. World Scientific, New Jersey
31. McBreen J, Lee HS, Yang XQ, Sun X (2000) New approaches to the design of polymer and liquid electrolytes for lithium batteries. J Power Sources 89:163–167
32. Lee HS, Yang XQ, Sun X, McBreen J (2001) Synthesis of a new family of fluorinated boronate compounds as anion receptors and studies of their use as additives in lithium battery electrolytes. J Power Sources 97–98:566–569
33. Ue M (2003) Chap. 4. Role-assigned electrolytes: additives. In: Yoshio M, Brodd RJ, Kozawa A (eds) Lithium ion batteries. Springer, New York, pp 75–115
34. Zhang SS (2006) A review on electrolyte additives for lithium-ion batteries. J Power Sources 162:1379–1394
35. Linden DE, Reddy TB (2002) Handbook of batteries, 3rd edn. McGraw-Hill, New York
36. Winter M, Brodd RJ (2004) What are batteries, fuel cells, and supercapacitors? Chem Rev 104(10):4245–4270
37. Conway BE (1999) Electrochemical supercapacitors. Kluwer, New York
38. O'Reagan B, Grätzel M (1991) A low-cost, high-efficiency solar cell based on dye-sentizied colloidal TiO_2 films. Nature 353:737
39. Hinsch A, Behrens S, Berginc M, Boennemann H, Brandt H, Drewitz A, Einsele F, Fassler D, Gerhard D, Gores H, Haag R, Herzig T, Himmler S, Khelashvili G, Koch D, Nazmutdinova G, Opara-Krasovec U, Putyra P, Rau U, Sastrawan R, Schauer T, Schreiner C, Sensfuss S, Siegers C, Skupien K, Wachter P, Walter J, Wasserscheid P, Wuerfel U, Zistler M (2008) Material development for dye solar modules: results from an integrated approach. Prog Photovoltaics 16(6):489–501

Non-Faradaic Electrochemical Modification of Catalytic Activity (NEMCA)

Tatsumi Ishihara
Department of Applied Chemistry, Faculty of Engineering, International Institute for Carbon Neutral Energy Research(WPI-I2CNER), Kyushu University, Nishi-ku, Fukuoka, Japan

Introduction

One of the useful applications of a solid electrolyte is a convertor from chemical energy to electricity (fuel cell) or from chemicals to another useful compound (ceramic reactor). In particular, application of solid electrolyte for so-called electrochemical ceramic reactor is highly attracting because of a selective conversion achieved. In the ceramic reactor, ion species are supplied through electrolyte to the electrode catalyst electrochemically, and so, conversion of reactant to the objective products is performed by using the permeated ion species as a reactant. Therefore, the formation rate of reaction products is generally the same or of lower amount than that of pumped ion species. Namely, formation rate of products should obey the Faraday's law. However, in 1981, interesting phenomena were first reported by Vayenas et al. in the high-temperature electrochemical reactor, and they found that the catalytic activity and selectivity depended strongly on the catalyst potential and the electrochemically induced reversible change in the catalytic rate was found to exceed the rate of ion pumping to the catalyst by up to five orders of magnitude and thus the rate change does not obey the Faraday's law [1]: Since the improvement in catalytic activity does not obey the Faraday's law, this improvement in catalytic reaction rate under a condition of electrochemically pumping ions, typically, oxide ion or Na^+, is called "non-Faradaic electrochemical modification of catalytic activity process (NEMCA)" [2] or electrochemical promotion of catalysis

(EPOC) [3]. Here, unique effects of NEMCA or EPOC effects are explained briefly.

EMCA (or EPOC) Effects on Catalysis

The basic concept of this NEMCA or EPOC is shown in Fig. 1 where O^{2-} conducting solid electrolytes is used for C_2H_4 oxidation [4]. The porous metal catalyst electrode, typically 0.05–2 μm thickness, is deposited on the solid electrolyte and under open-circuit conditions, i.e., $I = 0$, no electrochemical pumping oxygen; no product forms in C_2H_4 oxidation as shown in Fig. 1. Application of an electrical current I of 1 mA or potential of 0.5 V between the Pt catalyst and a counter electrode causes significant influence. The catalytic rate increase (Δr) is 25 times larger than the rate r_0 before current application and 74,000 times larger than the rate I/2 F of O^{2-} supply to the catalyst electrode.

Similar improvement in non-Faradaic improvement in electrochemical reaction is also reported for various reactions not only for partial oxidation with oxide ionic conductor but also for hydrogenation reaction with proton conductor, or NO reduction with CO on Na^+ ion conductor [3]. Until recently [5], more than 70 different catalytic reactions, i.e., oxidations, hydrogenations, dehydrogenations, isomerizations, and decompositions, have been electrochemically promoted on Pt, Pd, Rh, Ag, Au, Ni, IrO_2, and RuO_2 catalysts deposited on ionic conductor of O^{2-} (Y_2O_3 stabilized ZrO_2, YSZ), Na^+ (β-Al_2O_3); H^+ ($CaZr_{0.9}In_{0.1}O_3$, Nafion); F^- (CaF_2); aqueous, molten salt; and mixed ionic-electronic (TiO_2, CeO_2) conductors. Table 1 summarizes the reaction and ionic conductor reported for NEMCA. Clearly NEMCA is not limited to any particular class of conductive catalyst, catalytic reaction, or ionic conducting support.

Two parameters Λ and ρ are commonly used to describe the magnitude of NEMCA effects:

1. The apparent Faradaic efficiency, Λ:

$$\Lambda = \Delta r_{catalytic}/(I/nF)$$

where $\Delta r_{catalytic}$ is the current or potential induced change in catalytic rate, I is the applied current, n is the charge of the applied ion, and F is the Faraday's constant.

2. The rate enhancement, ρ:

$$\rho = r/ro$$

where r is the electro-promoted catalytic rate and ro is the un-promoted (open-circuit) catalytic rate.

3. The promotion index, PI_j, where j is the promoting ion (e.g., O^{2-} for the case of YSZ or H^+ for the case of Nafion), defined from:

$$PI_j = (\Delta r/\Delta r_0)/\Delta\theta_j$$

where θ_j is a coverage of the promoting ion on the catalyst surface.

A reaction exhibits electrochemical promotion when $|\Lambda| > 1$, while electrocatalysis is limited to $|\Lambda| < 1$. A reaction is termed electrophobic when $\Lambda > 1$ and electrophilic when $\Lambda < -1$. In the former case, the rate increases with catalyst potential, U, while in the latter case the rate decreases with catalyst potential. Λ values up to 3×10^5 and ρ values up to 150 have been found for several systems. For example, ρ values between 300 and 1,400 have been observed for C_2H_4 and C_3H_8 oxidation on Pt/YSZ, respectively. In the experiment of Fig. 1, $\Lambda = 74{,}000$ and $\rho = 26$, i.e., the rate of C_2H_4 oxidation increases by a factor of 25, while the increase in the rate of O consumption is 74,000 times larger than the rate, I/2 F, of O^{2-} supply to the catalyst [4].

It is reported that not only productivity but also selectivity is affected by NEMCA effects. Figure 2 shows product distribution and current in hydration of electrochemical 1-butene supplied over a dispersed Pd/C catalyst electrode, deposited on Nafion, H^+ conductor at room temperature [6]. Obviously, the products are strongly dependent on the applied potential. The maximum ρ values for the production of *cis*-2-butene, *trans*-2-butene, and butane are of the order of 50, and the corresponding maximum Λ values are of the

Non-Faradaic Electrochemical Modification of Catalytic Activity (NEMCA), Fig. 1 Basic experimental setup (**a**, **b**). Catalytic rate r and turnover frequency TOF response of C_2H_4 oxidation on Pt deposited on YSZ, an O^{2-} conductor, upon step changes in applied current (T = 643 K, P_{O2} = 4.6 kPa, P_{C2H2} = 0.36 kPa)

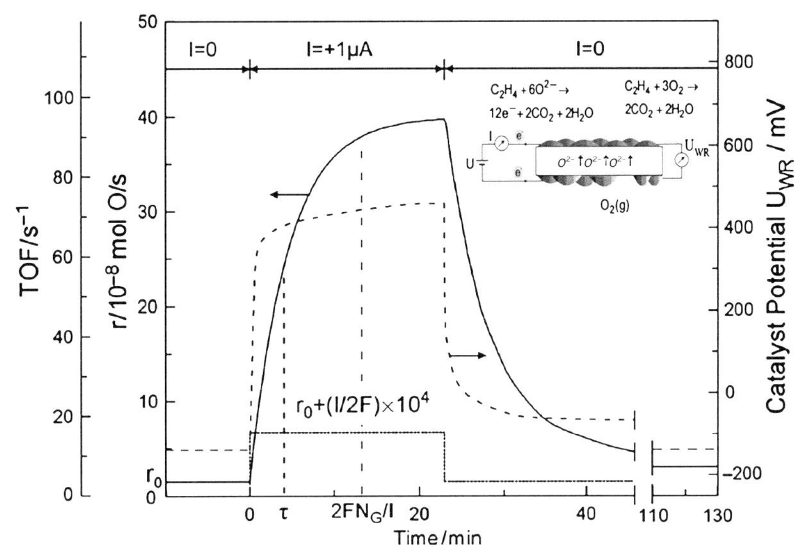

Non-Faradaic Electrochemical Modification of Catalytic Activity (NEMCA), Table 1 Typical case of NEMCA effects reported

Reaction	Cell configuration	Pumped oxygen	NEMCA effects	Refs.
C_2H_4, O_2	Pd/YSZ/Au	O^{2-}	Λ values up to 258 and ρ values up to 50	[13]
CO_2, H_2	Pd/β″-Al_2O_3/Au	Na^+	Water-gas shift reaction shows NEMCA effects	[14]
CO_2, H_2	Rh/YSZ/Au	O^{2-}	Formation of CH_4 is promoted by NEMCA effects	[15]
C_2H_4, O_2	Pt/BCN/Au	H^+	Λ values up to 1,000 and ρ values up to 12	[16]
C_2H_5OH, O_2	Pt/YSZ/Pt	O^{2-}	Acetaldehyde and CO_2 production rates are accelerated, Λ values for C_2H_5OH dehydrogenation up to 10^4	[17]
C_3H_6, O_2	LSCM/YSZ/Pt	O^{2-}	LSCM is restricted to oxidizing atmosphere	[18]
C_3H_6,O_2	Pt/β″-Al_2O_3/Au	K^+	ρ values up to 6	[19]
C_3H_6, O_2	Pt/NASICON/Au	Na^+	ρ values values up to 3 close to stoichiometric	[20]
C_3H_8, O_2,	Pt/β″-Al_2O_3/Au	Na^+	Changes in catalytic rate larger than 60 times than corresponding changes in sodium coverage	[21]
NO,O_2, C_3H_6	Ir/YSZ	O^{2-}	Increase in N_2 selectivity upon positive applied current	[22]
NH_3	Ru/SCY/Ag	H^+	75 % decrease in the reaction activation energy under positive polarization	[23]
C_3H_6, H_2O	Pd/SCY/Pd	H^+	4-fold increase in the rate of propane decomposition. During the electrochemical pumping of hydrogen, the rate of hydrogen pumped to the outer chamber over I/2F reaches values up to 0.9	[24]

order of 40 for *cis*-2-butene formation, 10 for *trans*-2-butene formation, and less than one for butane formation. Thus, each proton supplied to the Pd catalyst can cause the isomerization of up to 40 1-butene molecules to *cis*-2-butene and up to 10 1-butene molecules to *trans*-2-butene, while the hydrogenation of 1-butene to butane is electrocatalytic, i.e., Faradaic reaction. Therefore, this non-Faradaic activation of heterogeneous catalytic reactions is a novel

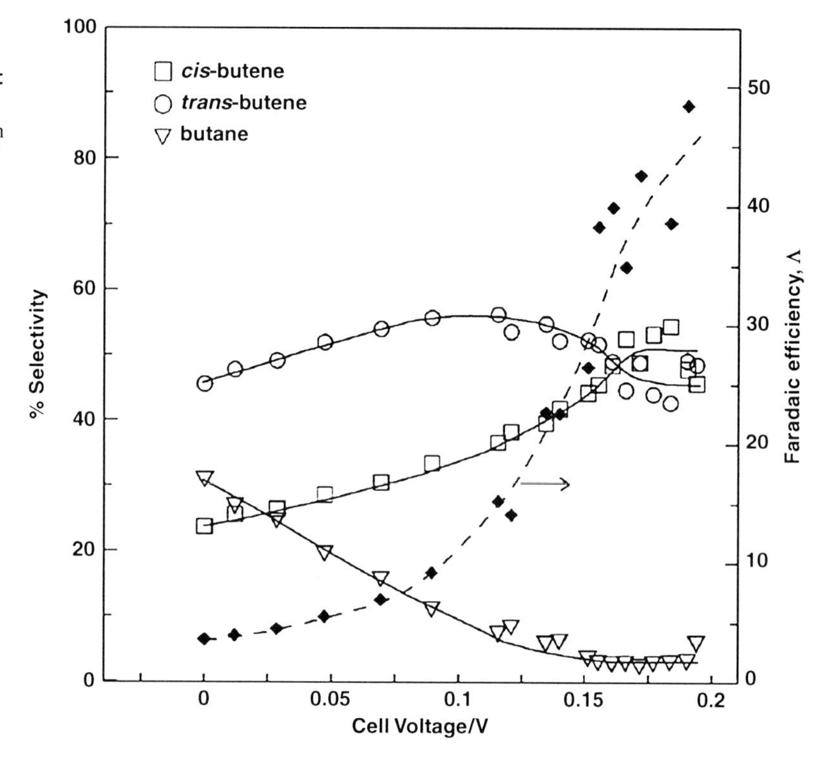

Non-Faradaic Electrochemical Modification of Catalytic Activity (NEMCA), Fig. 2 Product distribution and current in hydration of electrochemical 1-butene supplied over a dispersed Pd/C catalyst electrode, deposited on Nafion

application of electrochemistry with several technological possibilities, particularly in useful product selectivity modification and in exhaust gas treatment.

Mechanism for NEMCA Effects

The origin of NEMCA effects is also studied, and it is reported that the surface modification of reactant by ion back spillover of an effective double layer at the metal-gas interface is strongly related with unique improvement in reaction rate and selectivity [5, 7]. In situ work function measurements are performed by the Kelvin probe technique [7] or UPS [8] for explanation of NEMCA effects. These measurements show that over a wide range of temperatures, work function of metal catalyst linearly changes with increasing potential and the supplied ion species like oxide ion, proton, or Na^+ spill over [9] on the metal catalyst to form electric double layer. Schematic image of electrochemical modification is shown in Fig. 3 for the case of oxide ion conductor, in which δ value is not still determined yet [5]. Therefore, this electric double layer which forms at the catalyst-gas interface modifies the adsorption state of reactants like oxygen. Change in adsorption state under application of potential is also studied with temperature-programmed desorption [10], X-ray photoelectron spectroscopy [11], etc. Figure 4 shows the typical example of electrochemical modification of oxygen on Pt/YSZ catalyst [5]. In this experiment, desorption of oxygen from gas phase and electrochemically supplied was measured. Figure 4a presents the O_2-TPD spectra on Pt/YSZ when oxygen is supplied from the gas phase. A broad oxygen desorption peak (β2 state) was observed after catalyst exposure to oxygen atmosphere. In Fig. 4b, oxygen has been supplied electrochemically in the form of oxygen ions (O^{2-}) from the support, YSZ, using an external circuit and a constant positive current. Nonstoichiometric oxygen from the support migrates and gets adsorbed on the catalyst surface. O_2-TPD curve consists of one sharp oxygen peak (β3 state) desorbing at higher temperatures than the β2 state. This suggests that the back-spillover oxygen exists and is more strongly adsorbed on the

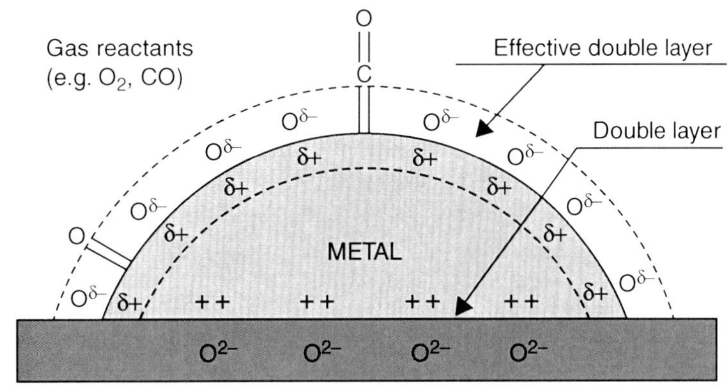

Non-Faradaic Electrochemical Modification of Catalytic Activity (NEMCA), Fig. 3 Schematic image of electrochemical modification

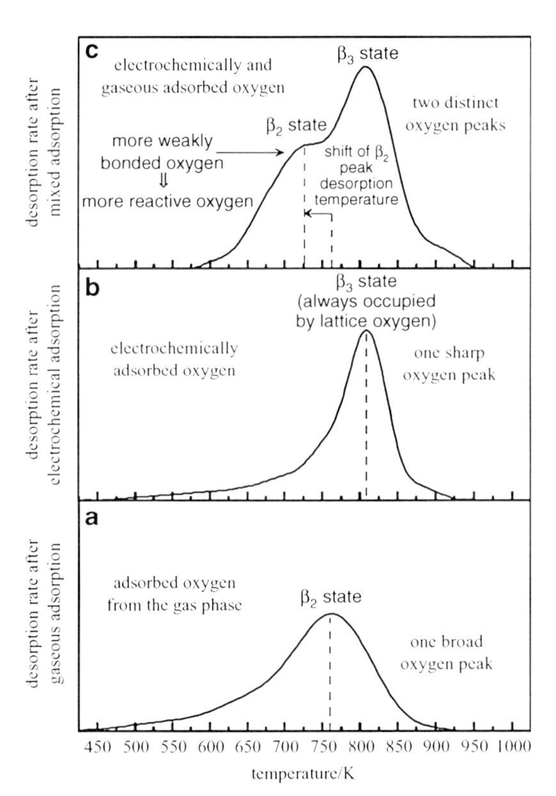

Non-Faradaic Electrochemical Modification of Catalytic Activity (NEMCA), Fig. 4 Typical example of electrochemical modification of oxygen on Pt/YSZ catalyst (**a**) gaseous adsorption, (**b**) electrochemical adsorption, and (**c**) mixed adsorption [5]

surface than the gaseous oxygen of β2 state. Figure 4c is the O_2-desorption under gas-phase adsorption and oxygen-pumping condition, namely, a mixed adsorption condition. Two oxygen peaks evidently appear, a broad one originating from the gas phase (β2 state) and a sharper one (β3 state) occupied by a more strongly bonded oxygen. It is worth noting that the β3 state which corresponds to the back-spillover oxygen species acts as a promoter for supplied oxygen since the last one moves to more weakly bonded states on the catalyst surface (lower desorption temperatures). Using ^{18}O tracer adsorption experiment, these weakly desorbed species are recombination ones of oxygen from gas phase and electrochemically pumped. Consequently, origin of NEMCA or EPOC effects could be assigned to such significant improvement in reactivity of adsorbed reactant by contributing the pumped ion species. This is more clearly observed on Na^+- pumping case on CO oxidation [5].

Summary

Many studies have been reported during the last almost 20 years regarding the effect of the electrochemical promotion of catalysis and its origin and application to several types of reactions of environmental and industrial interest. The effectiveness of NEMCA for catalytic oxidations, reductions, hydrogenerations, decompositions, and isomerizations using numerous types of solid electrolytes and catalysts underlines the importance of this phenomenon in both catalysis and electrochemistry. Application of these effects on the larger size reactor is also studied [12], and this seems to be more popular in the future. In particular, much increase in performance are required for fuel cells and electrocatalytic reactor. Therefore, new concepts based on NEMCA for these area will open a new reseach area.

References

1. Stoukides M, Vayenas CG (1981) The effect of electrochemical oxygen pumping on the rate and selectivity of ethylene oxidation on polycrystalline silver. J Catal 70:137
2. Vayenasa CG, Koutsodontis CG (2008) Non-Faradaic electrochemical activation of catalysis. J Chem Phys 128:182506
3. Katsaounis A (2010) Recent developments and trends in the electrochemical promotion of catalysis (EPOC). J Appl Electrochem 40:885
4. Bebelis S, Vayenas CG (1989) Non-faradaic electrochemical modification of catalytic activity: 1. The case of ethylene oxidation on Pt. J Catal 118:125
5. Vayenas CG, Bebelis S, Pliangos C, Brosda S, Tsiplakides D (2001) Electrochemical activation of catalysis: promotion, electrochemical promotion and metal-support interactions. Kluwer, New York
6. Ploense L, Salazar M, Gurau B, Smotkin ES (1997) Proton spillover promoted isomerization of n-Butylenes on Pd-Black cathodes/nafion 117. J Am Chem Soc 119:11550
7. Vayenas CG, Bebelis S, Ladas S (1990) Dependence of catalytic rates on catalyst work function. Nature (London) 343:625
8. Zipprich W, Wiemhöfer H-D, Vöhrer U, Göpel W (1995) In-situ photoelectron spectroscopy of oxygen electrodes on stabilized zirconia. Ber Bunsenges, Phys Chem 99:1406
9. Harkness I, Lambert RM (1995) Electrochemical promotion of the no + ethylene reaction over platinum. J Catal 152:211
10. Neophytides SG, Vayenas CG (1995) TPD and cyclic voltammetric investigation of the origin of electrochemical promotion in catalysis. J Phys Chem 99:17063
11. Ladas S, Kennou S, Bebelis S, Vayenas CG (1993) Origin of non-faradaic electrochemical modification of catalytic activity. J Phys Chem 97:8845
12. Balomenou S, Tsiplakides D, Katsaounis A, Thiemann-Handler S, Cramer B, Foti G, Comninellis C, Vayenas CG (2001) Novel monolithic electrochemically promoted catalytic reactor for environmentally important reactions. Appl Catal B Environ 52:181
13. Roche V, Karoum R, Billard A, Revel R, Vernoux P (2008) Electrochemical promotion of deep oxidation of methane on Pd/YSZ. J Appl Electrochem 38:1111
14. Bebelis S, Karasali H, Vayenas CG (2008) Electrochemical promotion of the CO2 hydrogenation on Pd/YSZ and Pd/β′′-Al2O3 catalyst-electrodes, Solid State Ionics 179:1391
15. Bebelis S, Karasali H, Vayenas CG (2008) Electrochemical promotion of CO2 hydrogenation on Rh/YSZ electrodes. J Appl Electrochem 38:1127
16. Thursfield A, Brosda S, Pliangos C, Schober T, Vayenas CG (2003) Electrochemical promotion of an oxidation reaction using a proton conductor. Electrochim Acta 48:3779
17. Tsiakaras PE, Douvartzides SL, Demin AK, Sobyanin VA (2002) The oxidation of ethanol over Pt catalyst-electrodes deposited on ZrO2 (8 mol% Y2O3). Solid State Ionics 152:721
18. Gaillard F, Li XG, Uray M, Vernoux P (2004) Electrochemical promotion of propene combustion in air excess on perovskite catalyst. Catal Lett 96:177
19. de Lucas-Consuegra A, Dorado F, Valverde JL, Karoum R, Vernoux P (2007) Low-temperature propene combustion over Pt/K-βAl2O3 electrochemical catalyst: Characterization, catalytic activity measurements, and investigation of the NEMCA effect. J Catal 251:474
20. Vernoux P, Gaillard F, Lopez C, Siebert E (2004) In-situ electrochemical control of the catalytic activity of platinum for the propene oxidation. Solid State Ionics 175:609
21. Kotsionopoulos N, Bebelis S (2007) Top Catal 44:379
22. Vernoux P, Gaillard F, Karoum R, Billard A (2007) Reduction of nitrogen oxides over Ir/YSZ electrochemical catalysts. Appl Catal B Environ 73:73
23. Skodra A, Ouzounidou M, Stoukides M (2006) NH3 decomposition in a single-chamber proton conducting cell. Solid State Ionics 177:2217
24. Karagiannakis G, Kokkofitis C, Zisekas S, Stoukides M (2005) Catalytic and electrocatalytic production of H2 from propane decomposition over Pt and Pd in a proton-conducting membrane-reactor. Catal Today 104:219

Numerical Simulations in Electrochemistry

Bernd Speiser
Institut für Organische Chemie, Universität Tübingen, Tübingen, Germany

Basic Aspects

Simulation is defined as the solution of mathematical equations (mathematical model) describing a particular physicochemical phenomenon (physical model), here in the electrochemical context. Numerical techniques are commonly involved, because in most cases the mathematical model cannot be solved in a closed form. Since, at present, numerical calculations are almost exclusively performed on digital computers, the term "digital simulation" is also common. Besides experimentation and theoretical description, numerical simulation has been qualified as a third means ("computer experiments") of knowledge generation in science [1].

Electrochemical simulations typically concern phenomena on the levels of (A) complete systems (electrochemical cell including electrodes and electrolyte as well as possibly environment; see, e.g., [2, 3]), (B) transport and reaction close to the electrode [4], or (C) atoms, their bonds and interactions [5]. This essay will be concerned with simulations on intermediate level B.

Owing to the extensive use of computation, numerical simulation is regarded as part of "computational electrochemistry" [6–8], although this term is sometimes restricted to calculations on level (C).

Numerical simulations in electrochemistry are used either in (1) a predictive way, i.e., the results describe a phenomenon which has not (yet) been or even cannot be studied experimentally, or (2) in qualitative (to determine a reaction mechanism) or quantitative (to determine characteristic properties) comparison to experimental data. Often simulation and "modelling" are used as synonyms [9].

Cyclic voltammetry is probably the electrochemical technique that is simulated most often, aiming at the analysis of electrode processes with respect to mechanism, kinetics, and thermodynamics of the reaction steps as well as transport properties of the molecules involved. The simulation of processes at (ultra)microelectrodes is also popular and highly important for the analysis of scanning electrochemical microscopy experiments [10].

For additional information, the reader is referred to earlier reviews on the topic [4, 6, 11–13].

Models

The formulation of physical and subsequently mathematical models is a simplification and abstraction of the complex real processes at the electrode. Thus, for the numerical simulation of electrode reactions, only the most important steps are considered:

- Transport, often restricted to diffusion (Fick's laws), but in principle also including convection (Navier-Stokes equation) and migration (Poisson-Boltzmann equation)
- Electron transfer, most often assumed to follow Butler-Volmer kinetics, but also in Nernstian equilibrium or according to Marcus kinetics
- Adsorption of molecules at the electrode surface, described by isotherms (e.g., Langmuir or Frumkin)
- Preceding or follow-up chemical reactions in the electrolyte phase or on the surface with appropriate kinetics, resulting in the great variety of electrode reaction mechanisms in electroanalytical and electrosynthetic contexts.

The combination of these steps defines the temporal (t) and spatial (for typical one-dimensional models: x; two- and three-dimensional simulations are also performed [10]) variation of the concentrations c of all educts, products, and intermediates involved. The concentrations at the electrode surface and in the bulk of the solution are defined through boundary conditions. Electrode boundary conditions also characterize the type of experiment performed, e.g., a triangular potential variation for cyclic voltammetry or a potential or current step function for chronoampero- or chronopotentiometry, respectively. The equations of the resulting mathematical model form a system of partial differential equations (PDEs) with the c as the unknowns. This system is solved starting from initial values of the unknowns (initial conditions) and subject to the boundary conditions. The primary solution provides concentration profiles $c = f(x)$ as a function of t. The experimental observable (in the case of a potential controlled experiment, the current i through the electrode;

for a current controlled experiment, the potential of the electrode E) is often calculated from these profiles. It may, however, also be considered as an additional unknown [14].

Algorithms

Numerically, the solution of the model equations (PDEs subject to initial and boundary conditions) corresponds to an integration with respect to the space and time coordinates. In general, this is an approximation to the mathematical model's exact solution. In simple cases, often restricted models, *analytical solutions* given by some, even complex, mathematical function are available. Additional work, e.g., Laplace transformation of the original mathematical model, may be required. Generating an analytical solution is commonly not termed simulation ("modelling... without... simulation" [15]). If such solutions are not practical, several techniques are applied, among these:

- *Finite difference approximation* of the concentration profiles. The space and time axes are discretized in the form of points or boxes, and the differential equations are changed into difference equations. The solution reduces to that of a system of ordinary equations.
- *Orthogonal collocation* – the substitution of the concentration profiles by polynomials that are forced to locally fulfil the differential equations exactly at certain points (discretization based on the zeroes of the polynomials). The partial differential equations are reduced to ordinary differential equations to be solved by standard methods.
- *Finite element methods* are based on global constraints imposed on the solution in certain, finite domains ("elements" defining the discretization) along the space coordinate. Thus, the integrated residual between the true and the approximated solutions is forced to zero subject to a weighting function.

Certain electrochemical conditions present problem cases for these algorithms, in particular [11]:

- The development of narrow diffusion layers close to the electrode or reaction fronts inside the electrolyte requires extremely high accuracy in these space regions.
- Abrupt changes in the controlling potential or current cause discontinuous changes of the boundary conditions or concentration profiles with t.
- Fast chemical reaction steps result in "stiff" PDE systems calling for specialized solvers.
- Nonlinearity of mathematical model equations results from reactions with orders higher than unity or particular boundary terms (e.g., adsorption isotherms).

To accommodate such situations, adaptive algorithms have been proposed that dynamically select the discretization step size in t and x according to the situation [16].

Simulators

Algorithms to solve mathematical models in electrochemistry are implemented as computer programs and used in the specific context of a particular experiment, problem, electrode reaction mechanism, etc., to be studied. However, for practical use in the electrochemical community, they have also been generalized in extended program packages (simulators) to be applied to a great variety of situations. Moreover, some electrochemical simulators provide means to easily define reaction mechanisms or experimental details and translate corresponding user input dynamically into the PDE system and its initial and boundary conditions. Finally, tools for data analysis and further tasks may complement a simulator to give a "problem-solving environment" [6].

A selected list of such simulators for electrochemical experiments is given below:

- `cvsim` [17] was an early attempt of a cyclic voltammetry simulator with a fixed selection of common electrode reaction mechanisms. `casim` (for chronoamperometry) and `gesim` (for galvanostatic electrolyses) were companion programs. Their development (based on orthogonal collocation) has been discontinued.

- `DigiSim` [18] was the first commercialized [19] cyclic voltammetry simulator and remains very popular with a window-based user interface and a rather general input tool for electrode reaction mechanisms, allowing nonprogrammers to formulate the reaction steps in a way intuitive to an electrochemist. It is based on finite difference algorithms.
- `DigiElch` [20] is a rewrite and extension of `DigiSim` with additional features by one of the original authors.
- `ELSIM` [6, 8] uses a somewhat different approach by providing an input tool on the level of the mathematical model which is solved by finite difference techniques.
- `EChem++` [9, 14] is based on adaptive finite elements. Both reaction mechanisms and experimental conditions (such as the potential or current program applied to the electrode as an "excitation" of the electron transfer processes) can flexibly be formulated. It is an open-source program maintaining principles of object-oriented software design [21] and is under continuous development.
- Online simulators have been described [13], but only one (for the specific problem of a monolayer covered electrode order cyclic voltammetric conditions [22]) seems to be working to date [23].

Applications

Simulations are generated in practice in both dimensioned and dimensionless forms. Running a dimensioned simulation requires to know (or assume) real values of parameters describing the modelled system, e.g., rate constants and diffusion coefficients, and leads to results that can directly be compared to experimentally observed currents or potentials.

On the other hand, in dimensionless simulations, all parameters values are normalized. Dimensionless results are particularly valuable because they can be transformed into many real contexts, depending on the normalization equations (usually linear).

It must be noted that, owing to the approximate character of simulations (see above) and the facts that (a) the solutions are not unique (several models may give the same results) and (b) possibly errors in calculation or inadequate model, mathematical, or algorithmic formulations will cancel each other (and consequently go undetected) [1], great care must be taken when drawing conclusions from comparing experiments and simulations [4, 11].

Typical example applications include:

- Simulations are used to explore the behavior of certain mechanisms; see as an example the influence of the attractive or repulsive interaction between molecules (characterized by a_{AB}) upon Frumkin-type adsorption on an electrode surface (Fig. 1) [24].
- A large number of dimensionless simulations can be condensed into working curves or zone diagrams [25].
- (Complex) mechanistic schemes are simulated, and the results are compared to experiments, helping to support or discard mechanistic hypotheses [26]. The example (Fig. 2) shows cyclic voltammograms for a mechanism including 3 electron transfers and 4 chemical steps [27].
- Quantitative comparison between simulation and experiment enables to determine rate constants, redox potentials, and other characteristics of the investigated electrochemical system (see also Fig. 2). In this case, it is strongly advised not to rely on comparison of a single curve but rather to fit simulations to experimental data recorded under widely varying conditions [26].

Conclusions

Numerical simulation is an indispensable tool in electrochemistry to model complex real systems, in particular at the level of chemical reaction and transport processes. The approximation and solution

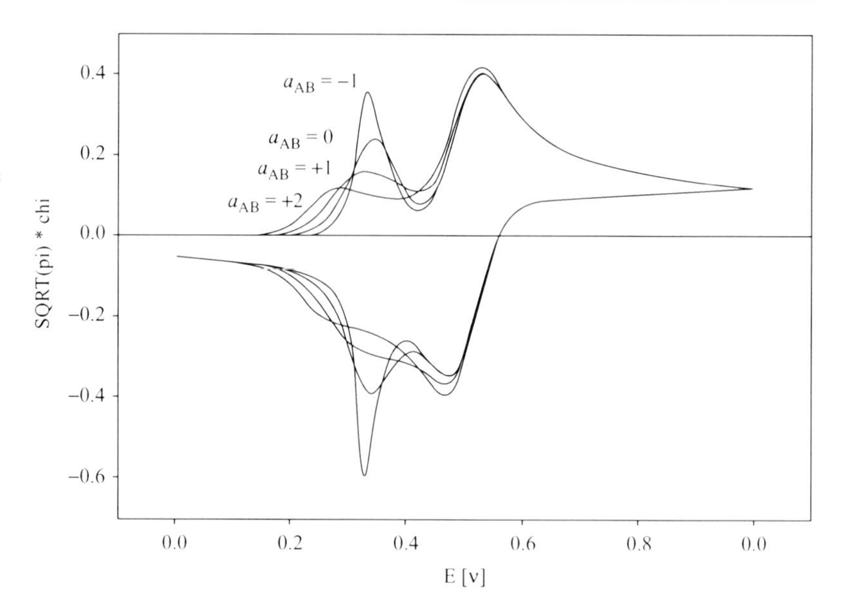

Numerical Simulations in Electrochemistry, Fig. 1 Simulated cyclic voltammograms with dimensionless current ($\pi^{1/2}$ ψ) for a redox couple with Frumkin adsorption, a_{AB} is the interaction parameter (Adapted from [24])

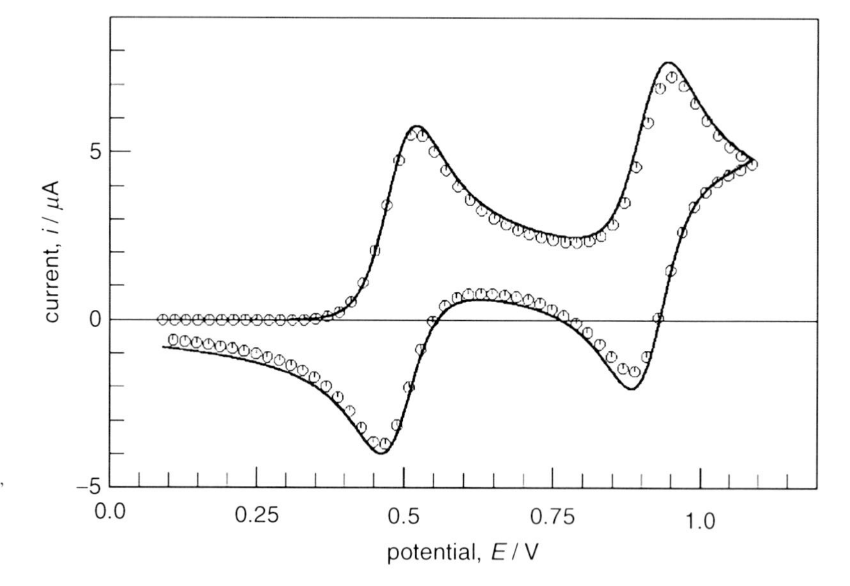

Numerical Simulations in Electrochemistry, Fig. 2 Comparison of simulated (*line*) and experimental (*symbols*) cyclic voltammograms of octamethyl-1,1′-bipyrrole, details, and simulation parameters; see [27] (Adapted from [27])

of partial differential equations that describe the real phenomena lead to an improved understanding of the pertinent reaction mechanisms and their kinetics and thermodynamics. It also allows to determine physicochemical constants and parameters related to the electrochemical process.

Cross-References

▶ Cyclic Voltammetry

References

1. Heymann M (2006) Understanding and misunderstanding computer simulation: the case of atmospheric and climate science – an introduction. Stud Hist Philos Mod Phys 41:193–200
2. Sorrentino M, Pianese C, Guezennec YG (2008) A hierarchical modeling approach to the simulation and control of planar solid oxide fuel cells. J Power Sources 180:380–392
3. Ramadesigan V, Northrop PWC, De S, Santhanagopalan S, Braatz RD, Subramanian VR (2012) Modeling and simulation of lithium-ion

batteries from a systems engineering perspective. J Electrochem Soc 159:R31–R45

4. Britz D (2003) Digital simulation in electroanalytical chemistry. In: Bard AJ, Stratmann M, Unwin P (eds) Encyclopedia of electrochemistry, vol 3, Instrumentation and electroanalytical chemistry. Weinheim, Wiley-VCH, pp 51–71
5. Jaque P, Marenich AV, Cramer CJ, Truhlar DG (2007) Computational electrochemistry: the aqueous $Ru^{3+}|Ru^{2+}$ reduction potential. J Phys Chem C 111:5783–5799
6. Bieniasz LK (2002) Towards computational electrochemistry – a kineticist's perspective. In: Conway BE, White RE (eds) Mod asp electrochem, vol 19. Marcel Dekker, New York, pp 135–195
7. Gooch KA, Fisher AC (2002) Computational electrochemistry: the simulation of voltammetry under hydrodynamic modulation control. J Phys Chem B 106:10668–10673
8. Bieniasz LK (2007) A unifying view of computational electrochemistry. In: Simons TE, Maroulis G (eds) Computational methods in science and engineering, theory and computation: old problems and new challenges. American Institute of Physics, Melville, pp 481–486
9. Ludwig K, Morales I, Speiser B (2007) EChem++ – an object-oriented problem solving environment for electrochemistry. Part 6. Adaptive finite element simulations of controlled-current electrochemical experiments. J Electroanal Chem 608:102–110
10. Combellas C, Fuchs A, Kanoufi F (2004) Scanning electrochemical microscopy with a band microelectrode: theory and application. Anal Chem 76:3612–3618
11. Speiser B (1996) Numerical simulation of electroanalytical experiments: recent advances in methodology. In: Bard AJ, Rubinstein I (eds) Electroanalytical chemistry, vol 19. Marcel Dekker, New York, pp 1–108
12. Bieniasz LK, Britz D (2004) Recent developments in digital simulation of electroanalytical experiments. Pol J Chem 78:1195–1219
13. Britz D (2005) Digital simulation in electrochemistry. Springer, Heidelberg
14. Ludwig K, Speiser B (2007) EChem++ – an object-oriented problem solving environment for electrochemistry: part 5. A differential-algebraic approach to the error control of adaptive algorithms. J Electroanal Chem 608:91–101
15. Oldham KB, Myland JC (2011) Modelling cyclic voltammetry without digital simulation. Electrochim Acta 56:10612–10625
16. Britz D (2011) The true history of adaptive grids in electrochemical simulation. Electrochim Acta 56:4420–4421
17. Speiser B (1990) EASIEST – a program system for electroanalytical simulation and parameter estimation – I. Simulation of cyclic voltammetric and chronoamperometric experiments. Comput Chem 14:127–140
18. Rudolph M (1995) Digital simulations with the fast implicit finite difference algorithm – the development of a general simulator for electrochemical processes. In: Rubinstein I (ed) Physical electrochemistry. Principles, methods, and applications. Marcel Dekker, New York, pp 81–129
19. Rudolph M, Reddy DP, Feldberg SW (1994) A simulator for cyclic voltammetric responses. Anal Chem 66:589A–600A
20. http://www.elchsoft.com/Default.aspx. Accessed 29 Jun 2013
21. Ludwig K, Rajendran L, Speiser B (2004) EChem++ – an object oriented problem solving environment for electrochemistry. Part 1. A C++ class collection for electrochemical excitation functions. J Electroanal Chem 568:203–214
22. Ohtani M (1999) Quasi-reversible voltammetric response of electrodes coated with electroactive monolayer films. Electrochem Commun 1:488–492
23. Ohtani M. http://www.kanazawa-bidai.ac.jp/~momo/qrcv/QRCV.html. Accessed 29 Jun 2013
24. Schulz C, Speiser B (1993) Electroanalytical simulations. Part 14. Simulation of frumkin-type adsorption processes by orthogonal collocation under cyclic voltammetric conditions. J Electroanal Chem 354:255–271
25. Savéant J-M (2006) Elements of molecular and biomolecular electrochemistry. Wiley, Hoboken
26. Speiser B (2004) Methods to investigate mechanisms of electroorganic reactions. In: Bard AJ, Stratmann M, Schäfer HJ (eds) Encyclopedia of electrochemistry, vol 8, Organic electrochemistry. Wiley-VCH, Weinheim, pp 1–23
27. Kuhn N, Kotowski H, Steimann M, Speiser B, Würde M, Henkel G (2000) Synthesis, oxidation and protonation of octamethyl-1,1′-bipyrrole. J Chem Soc Perkin Trans 2:353–363

O

Optimization of Electrolyte Properties by Simplex Exemplified for Conductivity of Lithium Battery Electrolytes

Heiner Jakob Gores[1], Hans-Georg Schweiger[2] and Woong-Ki Kim[2]
[1]Institute of Physical Chemistry, Münster Electrochemical Energy Technology (MEET), Westfälische Wilhelms-Universität Münster (WWU), Münster, Germany
[2]Faculty of Electrical Engineering and Computer Science, Ingolstadt University of Applied Sciences, Ingolstadt, Germany

Definition and Main Properties

Optimization of physical properties of a component of a device such as the conductivity of an electrolyte of a battery for storage of electrical energy via transformation to chemical energy is a key issue in developing new devices or improving existing ones.

Typical electrolytes of lithium-ion batteries are composed of a blend of several different solvents, at least one salt and several additives. Therefore, it is an expensive and time-consuming task to optimize the electrolyte if a single-step variation approach is used to increase the conductivity of the resulting electrolyte. To reduce the effort for this task, experimental design methods (DoE, Design of Experiments) can be used. Depending on knowledge about influencing factors, these methods can be divided into two groups. If already sufficient knowledge exists, sequential optimization procedures can be used for finding the optimum. If little or nothing is known about the influencing factors, factorial design plans are used for general optimization problems [1, 2]. The basic simplex method was introduced by G.B. Danzig, according to Weisstein [3], was presented by Spendley et al. [4], and was modified and extended by several groups [5–8]. Details of this mathematical procedure and its development are given in a number of publications [8–11]. The principles of this method are explained in Fig. 1. In this example, a peak of a hill is searched by the basic simplex method. Imagine a starting simplex is known represented by the Cartesian coordinates of three places X_n, Y_n, and Z_n, $n = 1$ to 3. The place with the lowest Z-value is dropped and a new place (X_4,Y_4) is proposed by reflection; there the height Z_4 is measured and the procedure is repeated until outcome Z_n oscillates around the optimal value. Of course, the method is not useful for only three variables. However, it can be generalized to multidimensional optimization problems and is advantageously applied when more than five variables have to be taken into account.

The extended simplex method gives a faster approach to the optimum in combination with better approximation, by adding expansion and contraction operations. The principle of this method is explained in Fig. 2, again by a 3D example, searching a peak of a hill.

The extended simplex method was applied to optimize conductivity of electrolytes for lithium-ion

G. Kreysa et al. (eds.), *Encyclopedia of Applied Electrochemistry*, DOI 10.1007/978-1-4419-6996-5,

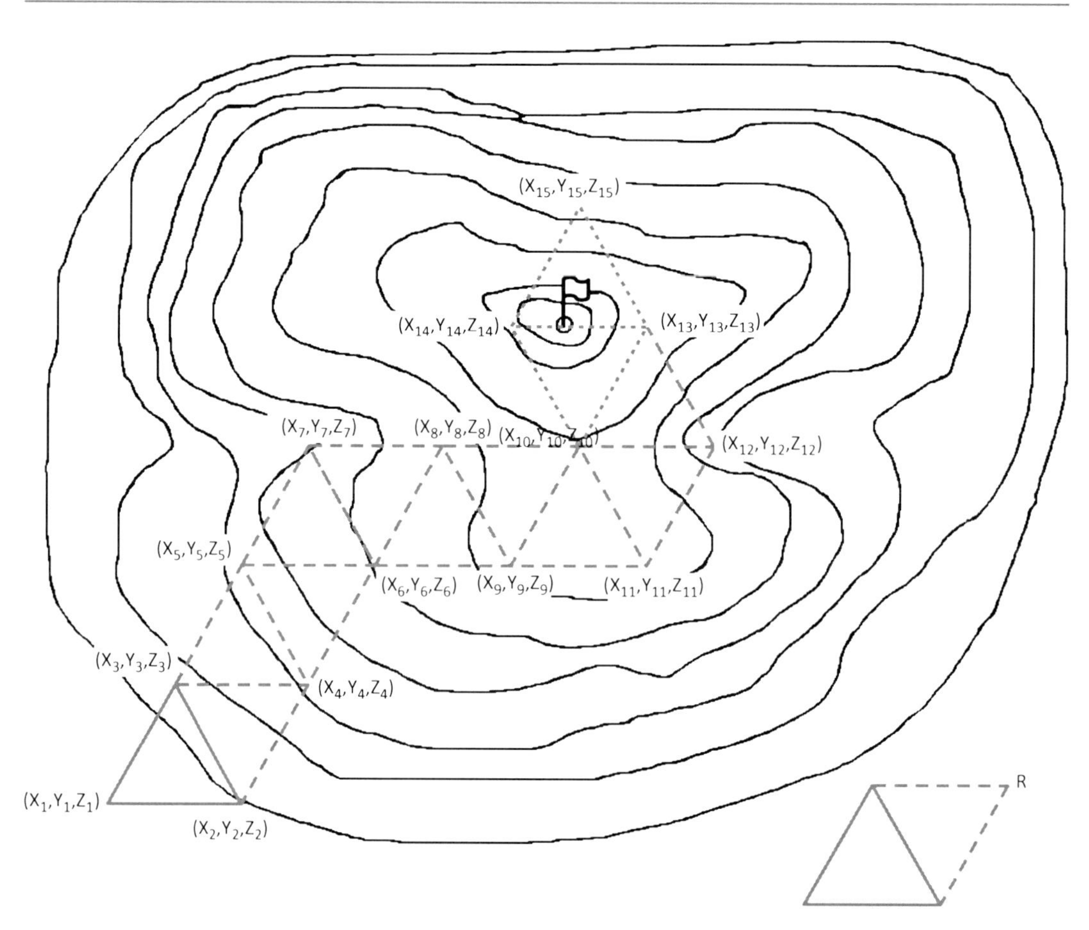

Optimization of Electrolyte Properties by Simplex Exemplified for Conductivity of Lithium Battery Electrolytes, Fig. 1 The basic simplex method is based on simple reflections (*R*) of the simplex (shown in the *lower right*). The optimization begins at starting simplex ($X_1,Y_1,Z_1|X_2,Y_2,Z_2|X_3,Y_3,Z_3$), solid line. A new set of control variables (X_4,Y_4) is generated by reflecting the simplex at line opposite to the worst corner here (X_1,Y_1,Z_1) and the outcome variable Z_4 is determined (measured). Repeating this procedure, the optimum is approximated. At the end of the optimization procedure, the simplex oscillates around the peak

batteries by these authors with the help of a program [12]. With this method it took about 20–30 optimization steps only to find the composition of an electrolyte based on a salt and five solvents with the optimum conductivity. It would have taken several 100 optimization steps (measurements) if a conventional step-by-step approach would have been used. It is interesting to stress that the so obtained conductivities are much better when compared to compositions selected on previous soft rules.

General mathematical software like Matlab®, Maple®, Mathematica®, or Mathcad® provides routines for optimization. But also specialized software packages can be used for optimization tasks. For sequential optimization ULTRAMAX®, Ultramax Corp. (Cincinnati, USA), can be applied. The basic principle of its operation is described by Moreno et al. [13].

ASSISTANT®, Design-Expert®, JMP®, MINITAB®, MODDE®, STATGRAPHICS®, STAVEX® or Visual-Xsel® may be used for designing the test plans, if the DoE method is used instead of the simplex method.

With featuring both classical and advanced simplex approaches, MultiSimplex® from Grabitech Solutions AB (Sundsvall, Sweden) is a program that is especially suitable for simplex optimization.

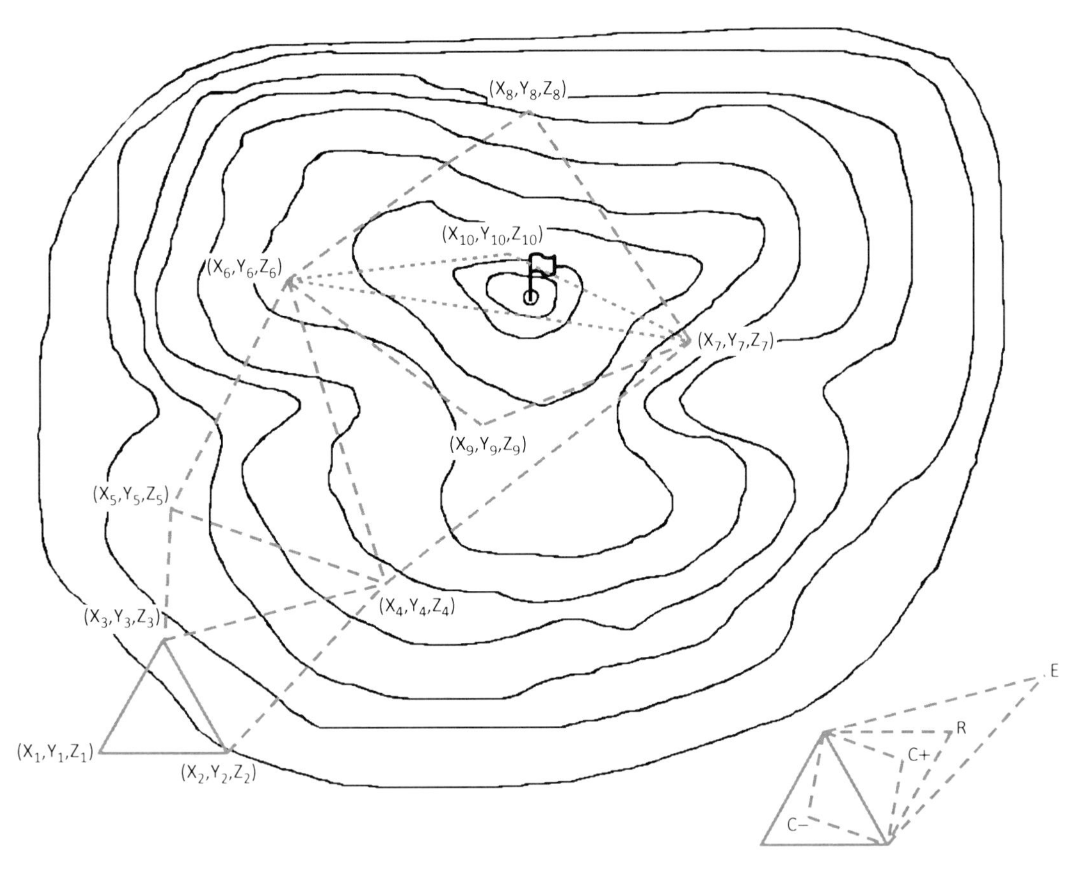

Optimization of Electrolyte Properties by Simplex Exemplified for Conductivity of Lithium Battery Electrolytes, Fig. 2 The extended simplex method includes other operations in addition to reflection (R) (basic method) expansions (*E*) and contractions (C+) and (C−). The operation is selected in dependence on change of the outcome variable. The optimization begins again at the starting simplex (X_1,Y_1,Z_1|X_2,Y_2,Z_2|X_3,Y_3,Z_3), solid line. A new set of control variables (X_4,Y_4) is generated by expanding the simplex at line opposite to the worst corner, here (X_1,Y_1,Z_1). At the beginning of optimization expansion steps provide a fast approach to the peak, while contractions give good approximation to the peak at the end of the optimization. Optimization is aborted if the variations of control variables fall below a preset value

Optimization of the Conductivity of an Electrolyte

A major objective in improving battery performance is the optimization of the power of a cell. The power of the cell is governed by thermodynamic, kinetic, transport, and geometric parameters [14–25]. The conductivity of electrolyte solutions is mainly determined by following parameters [26]:

- Concentration of the salt and temperature of the solution
- Dynamic viscosity of the electrolyte solution
- The radii of the solvated ions
- The ion-ion interaction entailing association of ions and ion solvent interaction entailing solvation
- Selective solvation in mixtures of solvents
- Competition of solvation and association

Acetonitrile (AN) and other nitriles combine low viscosities with reasonable high dielectric permittivities. So this solvent would be the ideal solvent for lithium-ion batteries. However, these solvents polymerize at lithiated carbon electrodes and are not suitable for secondary lithium-ion batteries. The class of organic carbonates shows good electrochemical stability in lithium-ion batteries by forming a stabile solid electrolyte interface (SEI). Unfortunately no member of this class combines low viscosity with high dielectric permittivity. On the one hand there are solvents such as propylene carbonate (PC) showing high dielectric permittivity but unfortunately in combination with high viscosity. On the other hand open-chain carbonates, such as diethyl carbonate (DEC), show low viscosity but coming in hand with low dielectric permittivity. Low conductivity results if an electrolyte is composed of a lithium salt and such a single solvent. It does not matter which type of solvent is chosen. Conductivity is limited either by high viscosity limiting the mobility of charge carriers or strong ion pairing due to low dielectric permittivity causing the formation of nonconducting ion pairs.

To overcome this problem the mixed solvent approach is used. By blending at least one solvent with high permittivity (but high viscosity) and at least one solvent of low viscosity (but low permittivity), the conductivity of the electrolyte can be improved [27–29]. But it is a time-consuming task to find the best composition, if more than two solvents are used. Then, the composition of the blend and the concentration of the salt have to be optimized by methods like simplex to obtain an optimal conductivity.

Of course, a theoretical approach predicting the conductivity of such electrolyte solutions would be the best way for finding the best electrolyte composition. By combining a theoretical approach with a mathematical optimization method, it would be possible to determine the electrolyte composition with the highest conductivity.

However, in contrast to dielectric permittivity of solvents which can be estimated with an acceptable degree of precision [30], there is up to now no valid approach to estimate the viscosity of solvents [31] and its temperature dependence and the conductivity of concentrated solutions and its temperature dependence, especially not when the solvent blend shows a low dielectric permittivity. A remarkable approach for the calculation of conductivity of electrolyte solutions (lithium salts in organic carbonates) was published in 2005 [31]. With this approach it is possible to obtain conductivities without any adjustable parameter, but only for low concentrations up to 10^{-2} M [31], far away from concentrations applied in batteries. Fitting of conductivity data and using mean spherical approximation including association (AMSA) can be used to describe the conductivity and its dependence from composition of the electrolyte [33] even for concentration ranges used in batteries (about 1 M) and are helpful to generate soft rules. More information about these and other concepts is given in literature [30, 32, 33].

Future Directions

Because the specific conductivity κ (S/m) of an electrolyte is determined readily and easy, this property is widely used for optimizing the battery performance. In contrast, other parameters which are more difficult to obtain, e.g., diffusion coefficients of ions near to or in the electrode materials or transference numbers of ions, are seldom studied and not yet included in optimization. We expect that automated measurement systems will be used in the future to optimize this and other critical parameters of solutions as long as no valid theoretical approach is available. These systems should be able to measure selected quantities automatically as a function of temperature and composition of solutions according to proposals made by optimization methods such as simplex. First steps on this way were undertaken by Schweiger et al., who presented an equipment that is able to measure $\kappa(T(t))$ and $T(t)$ automatically in up to 32 cells [34–38].

Simplex optimization was used for conductivity optimization but, of course, will be useful for the optimization of all battery parameters, e.g., aging of the battery due to unstable composition of the additives and electrolyte. The method can also be used for improving the power or energy density of the cell, if it is applied to the electrode composition. Because the simplex method is not only suitable for optimizing problems of one single outcome variable, this method can be used for the optimization of the total battery. By defining the desired parameters and their importance for the performance of the battery, a new outcome variable results that can be optimized. This outcome variable can be understood as the best battery for a specific application. If the simplex approach is applied for this outcome variable, the battery can be tuned into the desired direction. Because if every relevant parameter of the battery is taken into account, failing optimizations such as high-power batteries with a too short lifetime can be avoided.

Cross-References

- ► Conductivity of Electrolytes
- ► Electrolytes for Rechargeable Batteries
- ► Ion Mobilities
- ► Ion Properties
- ► Non-Aqueous Electrolyte Solutions

References

1. Verband der Automobilindustrie (2003) Qualitätsmanagement in der Automobilindustrie, Sicherung der Qualität während der Produktrealisierung Methoden und Verfahre. VDA, Oberursel
2. Kleppmann W (2006) Taschenbuch versuchsplanung. Hanser, München
3. Weisstein EW (1999) CRC concise encyclopedia of mathematics. CRC Press, Boca Raton
4. Spedley W, Hext GR, Himsworth FR (1962) Sequential application of simplex designs in optimisation and evolutionary operation. Technometrics 4:441
5. Nelder JA, Mead R (1965) A simplex method for function minimization. Comput J 7:308
6. Aberg RA, Gustavson AGT (1982) Design and evaluation of modified simplex methods. Anal Chim Acta 144:39
7. User's Guide to Multisimplex (Version 2.1). Grabitech Solutions AB Sundsvall, Sweden
8. Zadeh LA (1965) Fuzzy sets. Info Contr 8:338
9. Krabs W (1983) Einführung in die lineare und nichtlineare optimierung für ingenieure. B.G. Teubner, Stuttgart
10. Richter C (1988) Optimierungsverfahren und BASIC programme. Akademie Verlag, Berlin
11. Rao SS (1978) Optimization, theory and application. Wiley Eastern Ltd, New Dehli
12. Schweiger HG, Multerer M, Schweizer-Berberich M, Gores HJ (2005) Finding conductivity optima of battery electrolytes by conductivity measurements guided by a simplex algorithm. Electrochem Soc 152:A577
13. Moreno CW, Eilebrecht B (1992) Qualität und Zuverlässigkeit 37:53
14. Thomas KE, Newman J, Darling RM (2002) Mathematical modeling of lithium batteries. In: Van Schalkwijk W, Scrosati B (eds) Advances in lithium-ion batteries. Kluwer, New York
15. García RE, Chiang YM, Carter WC, Limthongkul P, Bishop CM (2005) Microstructural modeling and design of rechargeable lithium-ion batteries. J Electrochem Soc 152:A255
16. Song L, Evand JW (2000) Electrochemical-thermal model of lithium polymer batteries. J Electrochem Soc 147:2086
17. Georén P, Lindbergh G (2004) Characterisation and modelling of the transport properties in lithium battery gel electrolytes: Part I. The binary electrolyte PC/$LiClO_4$. Electrochim Acta 49:3497
18. Georén P, Lindbergh G (2001) Characterisation and modelling of the transport properties in lithium battery polymer electrolytes. Electrochim Acta 47:577
19. Bottle G, Subramanian VR, With RE (2000) Mathematical modeling of secondary lithium batteries. Electrochim Acta 45:2595
20. Nazri GA, Pistoia G (eds) (2004) Lithium batteries: science and technology. Kluwer, New York
21. Van Schalkwijk W, Scrosati B (eds) (2002) Advances in lithium-ion batteries. Kluwer, New York
22. Balbuena PB, Wang Y (eds) (2004) Lithium-Ion batteries: solid-electrolyte interphase. Imperial College Press, London
23. Yamamoto O, Wakihara M (1998) Lithium-Ion batteries. Wiley-VCH, New York/Weinheim
24. Besenhard JO (ed) (1999) Handbook of battery materials. VCH, New York
25. Linden D, Reddy TB (eds) (2002) Handbook of batteries, 3rd edn. New York, McGraw-Hill
26. Barthel J, Gores HJ (1994) Solution chemistry: a cutting edge in modern electrochemical technology. In: Mamontov G, Popov AI (eds) Chemistry of nonaqueous electrolyte solutions: current progress. VCH, New York, pp 1–147, Ch. 1
27. Gores HJ, Barthel J, Zugmann S, Moosbauer D, Amereller M, Hartl R, Maurer A (2011) In: Daniel C, Besenhard JO (eds) Handbook of battery materials, 2nd edn. Wiley-VCH, Weinheim, pp 525–626, Ch. 17
28. Barthel J, Gores HJ, Neueder R, Schmid A (1999) Electrolyte solutions for technology – new aspects and approaches. Pure Appl Chem 71:1705
29. Gores HJ, Barthel J (1980) Conductance of salts at moderate and high concentrations in propylene carbonate-dimethoxyethane mixtures at temperatures from −45 °C to 25 °C. J Solut Chem 9:939
30. Krienke H, Barthel J (2000) Ionic fluids. In: Sengers JV et al (eds) Equations of state for fluids and fluid mixtures. Elsevier, Amsterdam
31. Pu W, He X, Lu J, Jiang C, Wan C (2005) Molar conductivity calculation of Li-ion battery electrolyte based on mode coupling theory. J Chem Phys 123:231105

32. Krienke H, Barthel J, Holovko MF, Protsykevich I, Kalyushnyi Y (2000) Osmotic and activity coefficients of strongly associated electrolytes over large concentration ranges from chemical model calculations. J Mol Liquids 87:191
33. Barthel J, Krienke H, Holovko M, Kapko VI, Protsykevich I (2000) The application of the associative mean spherical approximation in the theory of nonaqueous electrolyte solutions. Cond Mat Phys 3(23):657
34. Wudy FE, Moosbauer DJ, Multerer M, Schmeer G, Schweiger HG, Stock C, Hauner FP, Suppan HG, Gores HJ (2011) Fast micro-Kelvin resolution thermometer based on NTC thermistors. J Chem Eng Data 56:4823
35. Wachter P, Schreiner C, Schweiger HG, Gores HJ (2010) Determination of phase transition points of ionic liquids by combination of thermal analysis and conductivity measurements at very low heating and cooling rates. J Chem Thermodyn 42:900
36. Schweiger HG, Wachter P, Simbeck T, Wudy FE, Zugmann S, Gores HG (2010) Multichannel conductivity measurement equipment for efficient thermal and conductive characterization of nonaqueous electrolytes and ionic liquids for lithium ion batteries. J Chem Eng Data 55:178
37. Wachter P, Schweiger HG, Wudy FE, Gores HJ (2008) Efficient determination of crystallisation and melting points at low cooling and heating rates with novel computer controlled equipment. J Chem Thermodyn 40:1542
38. Schweiger HG, Multerer M, Gores HJ (2007) Fast multi channel precision thermometer. IEEE Trans Instr Meas 56(N5):2002

Organic Electrochemistry, Industrial Aspects

Nicola Aust
BASF SE, Ludwigshafen, Germany

Introduction

One of the strengths of organic electrochemistry is its variety. Already in a simple beaker-type cell connected to a rectifier, one can conduct countless different electrolyses: anodic oxidations as well as cathodic reductions. The reactions are as manifold as chemistry itself is manifold thinking only about one of the oldest organic syntheses the *Kolbe* electrolysis [1] that is utilized till today [2, 3] or the versatile anodic substitution [4] or the electroreductive coupling as an important electrochemical form of C-C coupling [5]. In innumerable articles, reviews, and books, scientists have published, summarized, and illustrated this field of electrochemistry [6–9].

The application of organic electrochemistry in industry is not equally broad compared to the number of publications [10]. This is an obvious statement and true for synthesis methods in general. Not every chemical method or reaction gets applied in industry. In addition, it has to be emphasized that it is more than difficult to judge from publications and patents if an electrolysis or generally speaking a reaction is actually applied in industry because of the protection of intellectual property.

But what is the key to an application of organic electrochemistry in industry? The first and easy answer is that an electrosynthesis is applied in industry if the electrochemically synthesized product is economically successful. Applying organic electrochemistry is no end in itself regardless how novel or elegant it is. The process has to generate a competitive edge or a unique selling proposition. Besides the profitability today sustainability is important in industry, and the importance of this factor is rising. To this aspect of sustainability, organic electrochemistry renders an inherent advantage; it uses the purest possible reagent for synthesis: the electron.

Industrial Examples of Organic Electrochemistry

Three examples shall illustrate important aspects of a successful industrial electrosynthesis process. At the beginning often stands an idea to utilize a unique advantage rendered by organic electrochemistry.

Aromatic Aldehydes and Their Acetals by Anodic Substitution

The first example describes the electrochemical production of acetals of aromatic aldehydes established by BASF. This electrolysis is now successfully running for three decades at BASF.

Organic Electrochemistry, Industrial Aspects, Fig. 1 Electrosynthesis of aromatic aldehydes

The oxidation of the methyl group on aromatic rings is an important synthesis route to aromatic aldehydes. Often transition metals are proposed as oxidants [11]. Chlorination is another oxidation method that is, e.g., applied for the industrial synthesis of benzaldehyde from toluene. The treatment with chlorine leads to benzal chloride and hydrochloric acid together with some chlorinated side products. By hydrolysis the aldehyde is formed [12]. A general problem in the oxidation to the aldehyde stage is the overoxidation to the carboxylic acid. There are special reagents described in the literature like benzeneseleninic anhydride [13] or the combination of laccase and a mediator [11] that renders a good selectivity towards the aromatic aldehyde. Also the trapping of the aldehyde with acetic acid may omit overoxidation [11]. Organic electrochemistry renders an elegant way to overcome the problem of overoxidation. By anodic oxidation of the toluene derivate in the presence of methanol as nucleophile, first the intermediate ether and then the corresponding acetal are formed [14, 15]. The generated acetals are quite stable against overoxidation [16, 17], and the process can be controlled through the charge applied. As cathodic process, hydrogen is evolved (Fig. 1).

After the electrolysis, the acetals can be hydrolyzed to their aldehydes and methanol is recovered and can be recycled for the acetalization step. The recycling of methanol is an advantage also in regard to sustainability compared, for example, with the chlorination process where the resulting hydrochloric acid cannot be recycled like the recovered methanol for the acetalization [18]. The electrosynthesis takes place in good yields for toluene derivatives with electron pushing para-substituents like the *tert*-butyl or methoxy group [15, 18].

It is carried out in an undivided cell developed by BASF: the capillary gap cell [19] (Fig. 2). This cell contains a stack of bipolar round graphite electrodes. The electrodes with a central hole are separated by spacers and connected in series. The electrolyte flows through the middle channel which is generated by the stacking and outwards between the electrodes. By this stacking the capillary gap cell is one answer to the permanent question in electrochemistry on how to realize sufficient electrode area in as little space as possible, respectively, in one cell.

The combination of an elegant electrosynthesis and an elaborate cell leads to an optimal process with yields over 80 % [18], e.g., for 4-*tert*-butylbenzaldehyde dimethylacetal [20] which is an intermediate for aroma chemicals and fungicides.

This electrosynthesis has been developed further with regard to efficiency and sustainability by combining the anodic oxidation process to the acetals with the cathodic reduction of phthalic acid dimethylester to phthalide. Replacing the cathodic reduction of protons by reduction of phthalic acid dimethylester leads to a paired

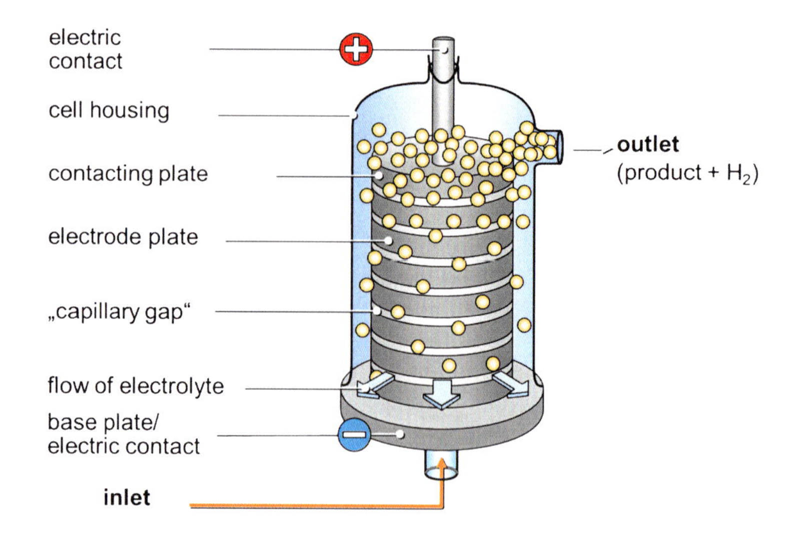

Organic Electrochemistry, Industrial Aspects, Fig. 2 BASF capillary gap cell

anode reaction

t-Bu

carbon anode

– 4e⁻, – 4H⁺
+ 2MeOH

cathode reaction

MeO2C

carbon cathode

+ 4e⁻, + 4H⁺
– 2MeOH

Organic Electrochemistry, Industrial Aspects, Fig. 3 Paired electrosynthesis of 4-*tert*-butylbenzaldehyde dimethylacetal and phthalide

electrosynthesis of 4-*tert*-butylbenzaldehyde dimethylacetal and phthalide [21, 22] (Fig. 3). This paired electrosynthesis is carried out in the same capillary gap cell as the acetal synthesis, and it has no increase in electric energy consumption compared to the non-paired process. In addition, the hydrogen that was required to produce phthalide in a catalytic hydrogenation is avoided completely. Therefore, the energy/fossil fuel to generate the hydrogen for the reduction step is economized [21].

Adiponitrile by Electrohydrodimerization

In the second example for an industrial application of organic electrochemistry, adiponitrile is produced by electrohydrodimerization of acrylonitrile (Fig. 4).

Electrohydrodimerization is a form of electrochemical reductive C-C coupling where the substrate dimerizes at the cathode in the presence of a proton source [5, 23]. At the beginning of the process development stands the idea to use this C-C coupling to hydrodimerize acrylonitrile to adiponitrile.

Like in the first example of the BASF acetal process, a unique feature of organic electrochemistry is employed. In this case the desired product is elegantly generated at the cathode in one step with only one substrate and water as proton source. The first main developments have been conducted at Monsanto by M. M. Baizer and his colleagues [24, 25]. Therefore, this adiponitrile process is often referred to as Monsanto process. As the electrohydrodimerization appeals by its atom efficiency, the realization of this efficiency has not been easy. The electrohydrodimerization takes place at quite negative potential [26]. In addition the needed proton source delivers at the

cathode reaction

$$2\ \text{CH}_2{=}\text{CH–CN} \xrightarrow[+\,2e^-,\ +\,2H^+]{} \text{NC–(CH}_2)_4\text{–CN}$$

Organic Electrochemistry, Industrial Aspects, Fig. 4 Electrohydrodimerization of acrylonitrile to adiponitrile

same time the substrate for the main cathodic side reaction the evolution of hydrogen. Therefore, one needs a cathode with a high overpotential of hydrogen like lead or cadmium [18], and all effects that suppress the hydrogen forming at the electrode surface improve the electrolysis. In this regard, the development of an electrolyte that contains special additives besides acrylonitrile and water has been essential to improve the yield. Starting from considerations on how to resolve more than 7 % acrylonitrile in water quaternary ammonium salts and later on bisquaternary ammonium salts as supporting electrolyte has been important in the developments towards an efficient process [18, 24, 25]. These salts not only improve the solubility of acrylonitrile but improve the conductivity and reduce the hydrogen formation and the hydrogenation of acrylonitrile to propionitrile by hydrophobization of the cathode [27]. The yield was improved to over 90 % [24].

Many aspects like the optimal current density or the optimization of the anode process by lowering of the anode overpotential have been explored [25]. Together with the development of a special adapted cell, an optimized process in a large tonnage could be achieved. First, a divided cell with a membrane was used. Later on an undivided cell with a stack of vertical bipolar electrodes was employed [18, 25]. As in the BASF capillary gap cell, one achieves a large electrode area by electrode stacking. The vertical assembly has turned out to be advantageous for the reductive process.

Also other companies like Asahi and BASF have applied this electrohydrodimerization of acrylonitrile to adiponitrile and have searched for the optimum process [18, 25]. Over many years, the electrosynthesis of adiponitrile has been the electroorganic process with the highest tonnage [10, 18, 25]. Based on the raw material, propylene for the substrate acrylonitrile adiponitrile is synthesized as intermediate for hexamethylenediamine in the production of nylon 66.

Though how optimized and elaborate the process has become, it competes with other routes. For adiponitrile, the nickel-catalyzed hydrocyanation of butadiene is today regarded as the most cost-effective route [28]. This route has originally been developed at DuPont [27] and it is sensitive to the natural gas price, while for the electrohydrodimerization, the propylene price is of interest. Today adiponitrile is produced for the most part by hydrocyanation [28]. This makes an industrial aspect of organic electrochemistry very clear: the electrolysis has to generate a profitable product. The efficiency of a process with high yields is one step to this profitable product. In the end factors like the raw material basis may make the difference between being competitive and not especially for a commodity like adiponitrile.

Perfluorinated Compounds by Electrochemical Fluorination

The perfluorination of hydrocarbons is the third example where the utilization of a unique electrochemical reaction leads to an industrial successful electrosynthesis.

J. H. Simons at 3M Co. was a pioneer in the field of electrochemical perfluorination. Beginning in 1941, he has electrolyzed organic molecules in liquid anhydrous HF below the fluorine evolution potential using nickel electrodes [29, 30]. Of course a high safety standard is required to handle anhydrous HF. But with a professional HF handling, the striking advantage of this electrolysis can be utilized: under tolerance of a variety of functional groups, perfluorination of hydrocarbons can be performed. Good yields in the electrochemical fluorination are achieved for substance classes like carboxylic acid, sulfonic acids, or trialkyl amines [31]. However, in many other cases, the yield of the electrochemical perfluorination is low [31].

The mechanism has been investigated and discussed for decades [31–34], but the decisive

Organic Electrochemistry, Industrial Aspects, Fig. 5 Electrochemical perfluorination

anode reaction

$$F-S(=O)_2-[CH_2]_n-CH_3 \xrightarrow[\text{anhydrous HF}]{\text{Ni anode}} F-S(=O)_2-[CF_2]_n-CF_3$$

$$F-C(=O)-[CH_2]_n-CH_3 \xrightarrow[\text{anhydrous HF}]{\text{Ni anode}} F-C(=O)-[CF_2]_n-CF_3$$

cathode reaction

$$2H^+ + 2e^- \longrightarrow H_2$$

use of nickel electrodes has been very clear throughout [32, 34]. An involvement of nickel fluorides NiF_x ($x = 2$ and higher) in the fluorination mechanism is proposed in many studies [31–34]. The importance of nickel electrodes is also underlined by the fact that by replacement of the nickel electrode material, e.g., with platinum and modification of the fluorinating agent to HF amine complexes, the fluorination is directed in a different pathway towards a partial fluorination [32, 35]. Successful industrial electrolyses are, e.g., the synthesis of perfluorinated alkyl carbonic acids or perfluorinated alkyl sulfonic acids [10, 31, 32] (Fig. 5).

In these cases the success of electrochemical perfluorination in anhydrous HF is based on the fact that it is a one-step reaction with a cheap fluorinating agent under toleration of the functional group. The resulting compounds supply interesting properties for their application through their combination of functional group and perfluorinated alkyl chain [32].

Summary and Future Direction

The chosen examples of applied organic electrochemistry are only a selection of successful industrial electrosyntheses. Other remarkable examples are shown in patents and reviews [10, 18, 36, 37]. Starting with an electrosynthesis that renders a unique feature, the industrial electrochemist explores the electrolysis to understand it thoroughly. Industrial research in these cases can be quite fundamental and the depth is often necessary to optimize the electrolysis. The combination with a product that has market potential triggers the question of a profitable process. The examples in their diversity illustrate that in the development of an industrial electrosynthesis, there is no such thing as a general optimal industrial cell or perfect electrolysis parameters. The process is developed to fit the special features of each electrolysis and is optimized in regard to its yield as it is necessary for every chemical process. And of course the process development includes optimizing electrochemical parameters like current density with regard to costs. Electrochemists with knowledge and passion drive the process development of their electrolysis to the point where an efficient if possible sustainable process renders a profitable product with the competitive edge.

Cross-References

- ▶ Anodic Substitutions
- ▶ Cathodic Hydrocoupling of Acrylonitrile (Electrosynthesis of Adiponitrile)
- ▶ Electrosynthesis of Fine Chemicals
- ▶ Kolbe and Related Reactions
- ▶ Paired Electrosynthesis
- ▶ Selective Electrochemical Fluorination

References

1. Kolbe H (1849) Untersuchung über die Elektrolyse organischer Verbindungen. Liebigs Ann Chem 69:257–294
2. Renaud P, Seebach D (1986) Electrochemical decarboxylation of hydroxyproline: a simple three-step conversion of (2S,4R)-4-hydroxyproline to (R)-γ-amino-β-hydroxybutanoic acid (GABOB). Synthesis 424–426
3. Klotz-Berendes B, Schäfer HJ, Grehl M, Fröhlich R (1995) Diastereoselektive Kupplung anodisch erzeugter Radikale mit chiralen Amidgruppen. Angew Chem 107:218–220
4. Eberson L, Nyberg K (1976) Synthetic uses of anodic substitution reactions. Tetrahedron 32:2185–2206
5. Schäfer HJ (1981) Anodic and cathodic CC-bond formation. Angew Chem Int Ed 20:911–934
6. Lund H, Hammerich O (2001) Organic electrochemistry. Marcel Dekker, New York
7. Schäfer HJ, Bard AJ, Stratmann M (2004) Organic electrochemistry. In: Encyclopedia of electrochemistry, vol 8. Wiley-VCH, Weinheim
8. Lund H (2002) A century of organic electrochemistry. J Electrochem Soc 149:S21–S33
9. Wendt H (1985) Organische Elektrochemie. Chemie in unserer Zeit 19:145–155
10. Steckhan E (2011) Electrochemistry, 3. Organic electrochemistry. In: Ullmann's encyclopedia of industrial chemistry, vol 12. Wiley-VCH, Weinheim, pp 315–349
11. Potthast A, Rosenau T, Chen CL, Gratzl JS (1995) Selective enzymatic oxidation of aromatic methyl groups to aldehydes. J Org Chem 60:4320–4321
12. Brühne F, Wright E (2011) Benzaldehyde. In: Ullmann's encyclopedia of industrial chemistry, vol 5. Wiley-VCH, Weinheim, pp 224–235
13. Barton DHR, Hui RAHF, Lester DJ, Ley SV (1979) Preparation of aldehydes and ketones by oxidation of benzylic hydrocarbons with benzeneseleninic anhydride. Tetrahedron Lett 20:3331–3334
14. Nilsson A, Palmquist U, Pettersson T, Ronlán A (1978) Anodic functionalisation in synthesis. Part 1. Methoxylation of methyl-substituted benzene and anisole derivatives and the synthesis of aromatic aldehydes by anodic oxidation. J Chem Soc Perkin Trans I:708–715
15. Degner D (BASF) DE2848397. Degner D, Barl M, Siegel H (BASF) DE2848397
16. Wendt H, Bitterlich S (1992) Anodic synthesis of benzaldehydes – 1. Voltammetry of the anodic oxidation of toluene in non-aqueous solutions. Electrochim Acta 37:1951–1958
17. Wendt H, Bitterlich S, Lodowicks E, Liu Z (1992) Anodic synthesis of benzaldehydes – 2. Optimization of the direct anodic oxidation of toluenes in methanol and ethanol. Electrochim Acta 37:1959–1969
18. Pütter H (2001) Industrial electroorganic chemistry. In: Lund H, Hammerich O (eds) Organic electrochemistry. Marcel Dekker, New York, pp 1259–1307
19. Beck F, Guthke H (1969) Entwicklung neuer Zellen für elektro-organische Synthesen. Chem-Ing-Tech 41:943–950
20. Degner D, Siegel H, Hannenaum H (BASF) EP 0029995
21. Hannebaum H, Pütter H (1999) Elektrosynthesen. Strom doppelt genutzt: Erste technische "Paired Electrosynthesis". Chemie in unserer Zeit 33:373–374
22. Hannebaum H, Pütter H (BASF) DE19618854
23. Li CJ (2005) Organic reactions in aqueous media with a focus on carbon-carbon bond formations: a decade update. Chem Rev 105:3095–3165
24. Baizer MM (1980) The electrochemical route to adiponitrile – 1 discovery. Chemtech 10: 161–164
25. Danly DE (1981) Processes for electrohydrodimerization of acrylonitrile to adiponitrile. AIChE sym ser 77:39–44
26. Beck F (1965) Elektro-organische Dimersierung von Adipinsäuremonomethylester und Acrylnitril. Chem-Ing-Tech 37:607–616
27. Weissermel K, Arpe HJ (1998) Industrielle organische chemie. Wiley-VCH, Weinheim
28. Market Publishers Ltd. (2012) News & press release on adiponitrile market research report 26 April 2012. http://marketpublishers.com/lists/13624/news.html. Accessed May 2012
29. Simons JH (1949) Production of fluorocarbons: I. The generalized procedure and its use with nitrogen compounds. J Electrochem Soc 95:47–52
30. Simons JH (1986) The seven ages of fluorine chemistry. J Fluor Chem 32:7–24
31. Suriyanarayanan N, Noel M (2008) Interaction between organic molecules and electrogenerated nickel fluoride films: the choice of organic reactants for electrochemical perfluorination. J Solid State Electrochem 12:1453–1460
32. Hollitzer E, Satori P (1986) Die elektrochemische Fluorierung – ein Überblick. Chem-Ing-Tech 58:31–38
33. Dimitrov A, Rüdiger S, Ignatyev NV, Datcenko S (1990) Investigations on the electrochemical fluorination of amines. J Fluor Chem 50:197–205
34. Sartori P, Ignat'ev N, Jünger C, Jüschke R, Rieland P (1998) Electrochemical synthesis of polyfluorinated compounds with functional groups. J Solid State Electrochem 2:110–116
35. Fuchigami T, Tajima T (2005) Highly selective electrochemical fluorination of organic compounds in ionic liquids. J Fluor Chem 126:181–187
36. Degner D (1988) Organic electrosyntheses in industry. Top Curr Chem 148:1–95
37. Sequeira CAC, Santos DMF (2009) Electrochemical routes for industrial synthesis, J Braz Soc 20:387–406

Organic Pollutants for Wastewater Treatment, Reductive Dechlorination

Sandra Rondinini, Alessandro Minguzzi and Alberto Vertova
Dipartimento di Chimica, Università degli Studi di Milano, Milan, Italy

Introduction: Motivation and Theoretical Background

Chlorinated organic compounds (COCs) undoubtedly represent one of the main categories of water pollutants. This is due, on one hand, to their wide use in many industrial fields: from solvents to chemical industry intermediates, pesticides (chlordane, 57-74-9 [1]), fungicides (hexachlorobenzene, 118-74-1), insecticides (aldrin, 309-00-2; dieldrin, 60-57-1; endrin, 72-20-8; heptachlor, 76-44-8), dielectrics and coolants (polychlorinated biphenyls – PCBs), plasticizers (PCBs), and drugs (chloral, 75-87-6, representing the first synthetic COC, by Liebig in 1832).

In addition, they may be generated as undesired by-products in oxidative waste treatments (e.g., in waste incineration, where dioxins can be produced) and in natural waters under the effect of UV radiation.

On the other hand, organic halides are hardly and slowly degraded by the environment. For this reason, most of the compounds listed among the persistent organic pollutants (POPs) by the Stockholm Convention [2] are COCs.

Moreover, COCs present high toxicity (often being carcinogenic) and high environmental impact. Particularly insidious are volatile organic halides, often used as solvents, whose emissions were limited by the European Community in 1999 [3]. In the past, they were the cause of significant accidents: hexachlorobenzene fatally poisoned 500 people in Turkey between 1955 and 1959; chlorofluorocarbons (CFCs) represented the main compounds responsible for the ozone layer depletion; dioxins caused several large-scale environmental contaminations (the United States, Italy, and Germany). A more recent case, although of lower level of concentration, involved the detection of dioxin in poultry.

All these issues clearly explain why many COCs have been banned during the last decades (mirex, 2385-85-5, kepone, 143-50-0, CFCs, and many of the molecules listed above) and why many methods for treating COC-containing wastes have been proposed thus far. In parallel, organohalogen compounds are placed at the beginning of the indicative list of main pollutants in Annex VIII of the Directive 2000/60/EC of the European Parliament and of the Council [4].

Many methods can be applied for cleaning wastewaters from organic chlorides. However, while oxidative treatments (chemical, photochemical, or electrochemical) may lead to products that are even more harmful than the original ones (e.g., dioxins), reductive methods may in principle lead to value-added compounds. For example, the reduction of polychloromethane may lead to methane, while the reduction of polychloroethanes and polychloroethenes may lead to ethane, ethylene, or acetylene. While the most adopted and studied methods are represented by reduction with molecular H_2 [5] or by zerovalent metals and organometallic compounds [6, 7], other methods involving photochemical [8] and electrochemical approaches have been intensively developed.

In particular, electrochemical methods guarantee mild operative conditions and avoid any secondary pollution effect due to any reagent excess. Moreover, they adapt well to both large- and small-scale processes, can be applied on stream, and are effective for treating both concentrated and dilute solutions. It is worth noting that electrochemical methods have also been successfully adopted to treat gas phases [9].

The first studies of electrochemical reduction of organic halides were published in 1936 by Winkel and Proske [10] and in 1949 by von Stackelberg and Stracke [11], to become a recurrent issue in both fundamental (development of dissociative electron transfer theory, see below) and applied electrochemistry, as well as for synthetic purposes.

The importance of carbon–halogen bond cleavage for synthetic and environmental applications prompted several studies to understand the reaction mechanism, a work mainly done by Savéant's group, adopting the general concepts introduced by the Marcus-Hush model [12]. The key step of halogenated compound reduction is the so-called dissociative electron transfer (DET):

$$R{-}Cl + e^- \rightarrow R\text{-}Cl^{\bullet -} \quad (SW1)$$

$$R{-}Cl^{\bullet -} \rightarrow R^{\bullet} + Cl^- \quad (SW2)$$

$$R{-}Cl + e^- \rightarrow R^{\bullet} + Cl^- \quad (C)$$

$$R^{\bullet} + e^- \rightarrow R^- \quad (1)$$

which involves the insertion of an electron and the breaking of the C–Cl bond. These two events can occur simultaneously (concerted mechanism, C) or separately (stepwise mechanism, SW). According to the specific conditions, the radical may undergo a second electron transfer that leads to the corresponding anion (1).

On inert electrodes, determining whether the mechanism is concerted or stepwise is relatively straightforward: the stepwise mechanism is typically associated with a smaller energy barrier in comparison to the one relevant to the concerted mechanism. In fact, the barrier relevant to the concerted mechanism includes the bond dissociation energy and requires a greater driving force, thus leading to peak potentials much more negative than the standard ones. Hence, the two mechanisms can be distinguished by evaluating the effect of the driving force on the energy barrier. This can be accomplished by determining α, the symmetry factor, expressed by the following relation:

$$\alpha = \frac{\partial \Delta G^{\ddagger}}{\partial \Delta G^0} = 0.5\left(1 + \frac{\Delta G^0}{4\Delta G_0^{\ddagger}}\right)$$

(where

ΔG^0 = standard Gibbs free energy, representing the reaction driving force

$\Delta G_0^{\ddagger}$ = intrinsic activation energy barrier) which assumes values lower than 0.5 for concerted pathways, while for stepwise mechanisms, α results in values higher than 0.5.

Note that this theoretical approach may fail in the case of electrode materials showing specific interactions with the reactants/intermediates/products of the dehalogenation process, as proved, for the first time, in a recent paper [13] and discussed in more detail in the next paragraph.

On inert electrodes like glassy carbon, many R–Cl compounds have been studied, both in terms of mechanism and in view of operative systems, for the dechlorination of the different classes of organic chlorides: geminals, aliphatics, aromatics, CFCs, etc. In this context, the intensive use of preparative electrolyses allowed the different products that can be obtained from $R^{\bullet}$ or R^- reacting with other molecules (solvent, other R–Cl molecules, etc.). Depending on the specific electrolysis conditions, many different reaction products can be obtained, some of which are listed below:

$$R^- + RX \rightarrow R{-}R + X^- \quad (2)$$

$$R^- + HA \rightarrow R{-}H + A^- \quad (3)$$

$$R^{\bullet} + HA \rightarrow R{-}H + A^{\bullet} \quad (4)$$

$$R^{\bullet} + R^{\bullet} \rightarrow R{-}R \quad (5)$$

Most of the studies have been carried out in nonaqueous solvents, mainly because of the highest potential window available and because of the very low solubility of most organic halides in water. Even if, as a first consideration, the use of organic solvents is not encouraged at industrial scale, one of the possible strategies for the effective dechlorination of wastewaters is represented by reductive treatments in organic media following extraction of the pollutants from the refluence (by solvent extraction or by adsorption on a specific solid filter) [14].

The choice of the electrolyte strongly influences the reaction product distribution. In a nonaqueous environment, the radical formation easily leads to the growth of polymeric chains that, in turn, can form an insulating layer on the cathode itself [15]. At the same time, free radicals are known to be responsible for the cathode (especially carbon-based) ablation [16]. In a nonaqueous environment in the complete absence of a proton donor, the formation of carbenes is possible in the case of polychloromethanes [14]. On the other hand, in the presence of proton donors, i.e., in an aqueous environment, the lifetime of radicals and of carbenes is very short, and the main products derive from chloro compound hydrodehalogenation:

$$R{-}Cl + 2e^- + H^+ \rightarrow R{-}H + Cl^- \quad (6)$$

Electrocatalysts for Reductive Dechlorination

One of the main issues in the electrochemical reductive dechlorination of organic chlorides is the energy consumption associated with the process. The cost of the electricity needed for driving the electrolysis can easily become decisive for the commercial success of this process, and this leads to the necessity of finding suitable electrocatalysts for lowering the cell potential under operative conditions.

In the beginning of this century, Ag was found to be an effective electrocatalyst for dehalogenation of organic halides [17], showing dechlorination reduction peaks at less negative potentials than on glassy carbon, leading to up to 1 V gains, generally considered an inert electrode material. However, it must be noted that the real inertness of carbon was recently brought into question [18]. After evidencing the silver activity for hydrodehalogenation reactions, several groups speculated on the reason for this impressive electrocatalytic power, and many substrates have been considered and studied, mainly in nonaqueous environments. Already during the first studies, the reason for the exceptional activity of Ag was ascribed to the existence of specific interaction between the reactants/intermediates/products and the silver surface [17, 19]. This hypothesis was initially based on the well-known interaction between silver and the halide atoms [17]. Another contribution evidenced that the catalytic power of silver manifested only in the case of those systems that, according to the Savéant model, follow a concerted path on inert electrodes [19].

The reasons behind the extraordinary activity of Ag were experimentally and theoretically investigated in a more recent contribution following collaboration of different groups, involving the combination of voltammetry, surface-enhanced Raman spectroscopy (SERS), and density functional theory (DFT) [20]. Quite interestingly, this study demonstrated, at least in the case of benzyl chloride hydrodehalogenation, that the exceptional activity of Ag is due to specific interactions between the silver surface and substrate/intermediates. This, in turn, is prompting the further extension of this study at higher voltammetric scanning rates, on the basis of the outcomes of the most recent treatment of the possible adsorption phenomena as reported by Klymenko et al. [21] in their unifying work of the all the limiting Laviron's cases.

Toward Operative Systems

Notwithstanding the relevant number of papers dealing with the mechanism of organic chloride dechlorination in organic solvents, the literature dealing with studies in mixed solvents [15] or in water [9, 22] is poor. Also, in the case of usage of electrochemistry for civil or industrial wastewater treatment, only a few studies involving laboratory-scale or pilot plants are available in the literature [23, 24]. As mentioned before, nonaqueous solvents might be used for wastewater treatment after the extraction of the chloro compounds. However, this procedure introduces a new potential pollutant (the solvent) and can be particularly expensive for all reasons associated with the use of flammable, volatile, and toxic solvents (dimethylformamide and acetonitrile are the solvents typically considered in these works) and with the adoption of low-conductivity electrolytes (as are organic solvent-based ones). Moreover, it was already demonstrated that the presence of a proton donor (water, acids) dramatically reduces the energy costs by shifting the reduction peak potential toward less negative values [14, 15, 22]. In this scenario, the main disadvantage of the presence of a proton donor in solution is the concomitant evolution of molecular hydrogen that can drastically reduce the overall current efficiency. An intermediate approach is possible in the case of divided cells, thanks to the use of membranes; in this case, the catholyte can be based on a nonaqueous solvent, while the anolyte can consist of an aqueous electrolyte. This approach has already been adopted in laboratory-scale preparative electrolyses [15].

From another point of view, efforts are devoted to optimize the silver catalyst in terms of its shape, size, and morphology and in view of its support on a low-cost matrix. In fact, though the most fundamental studies involve the use of a silver rod for common voltammetric measurements, electrodeposited silver and silver micro- and nanoparticles have been considered as well [9].

Future Directions

The research work on organic pollutant dechlorination and wastewater treatment will likely involve studies aimed at further disclosing the hows and whys for the impressive electrocatalytic activity of silver, as well as the discovery of new methods for obtaining active materials with a low silver content. In order to reduce the catalyst costs, researchers are looking for new metal catalysts [25], metals alloys [15], and advanced materials, taking into account the role of the supporting material that can show synergistic effects with the active material [26]. On the other hand, these processes are still not optimized in view of an industrial application, and many technical studies are needed to improve the plant design and operation.

Cross-References

- Activated Carbons
- Electrocatalysis - Basic Concepts, Theoretical Treatments in Electrocatalysis via DFT-Based Simulations
- Electrocatalysis, Fundamentals - Electron Transfer Process; Current-Potential Relationship; Volcano Plots
- Electrochemical Cell Design for Water Treatment
- Electrochemical Reactor Design and Configurations
- Green Electrochemistry
- Hydrogen Evolution Reaction
- Membrane Technology
- Organic Electrochemistry, Industrial Aspects
- Organic Reactions and Synthesis
- Permanent Electrochemical Promotion for Environmental Applications
- Wastewater Treatment, Electrochemical Design Concepts
- Water Treatment by Adsorption on Carbon and Electrochemcial Regeneration

References

1. All numbers in the parenthesis represent the relevant CAS number
2. Official Journal of the European Communities, COUNCIL DECISION of 14 October 2004 concerning the conclusion, on behalf of the European Community, of of the Stockholm Convention on Persistent Organic Pollutants (2006/507/EC)
3. Official Journal of the European Communities, COUNCIL DIRECTIVE 1999/13/EC of 11 March 1999 on the limitation of emissions of volatile organic compounds due to the use of organic solvents in certain activities and installations
4. Official Journal of the European Communities, DIRECTIVE 2000/60/EC OF THE EUROPEAN PARLIAMENT AND OF THE COUNCIL of 23 October 2000 establishing a framework for Community action in the field of water policy
5. Lien HL, Zhang WX (2007) Nanoscale Pd/Fe bimetallic particles: catalytic effects of palladium on hydrodechlorination. Appl Catal B-Environ 77:110–116
6. Hara T, Kaneta T, Mori K, Mitsudome T, Mizugaki T, Ebitani K, Kaneda K (2007) Magnetically recoverable heterogeneous catalyst: palladium nanocluster supported on hydroxyapatite-encapsulated c-Fe2O3

nanocrystallites for highly efficient dehalogenation with molecular hydrogen. Green Chem 9:1246–1251
7. Alonso F, Beletskaya IP, Yus M (2002) Metal-mediated reductive hydrodehalogenation of organic halides. Chem Rev 102:4009–4091
8. Shiraishi Y, Takeda Y, Sugano Y, Ichikawa S, Tanaka S, Hirai T (2011) Highly efficient photocatalytic dehalogenation of organic halides on TiO2 loaded with bimetallic Pd–Pt alloy nanoparticles. Chem Commun 47:7863–7865
9. Rondinini S, Aricci G, Krpetić Z, Locatelli C, Minguzzi A, Porta F, Vertova A (2008) Electroreductions on silver-based electrocatalysts: the use of Ag nanoparticles for CHCl3 to CH4 conversion. Fuel Cells 2:253–263
10. Winkel A, Proske G (1936) Über die elektrolytische Reduktion organischer Verbindungen an den Quecksilber-tropfelektrode. Ber. Dtsch. Chem. Ges. B, 69: Mitteil I. 693–706, Mitteil II. 1917–1929
11. von Stackelberg M, Stracke W (1949) Das Polarographische Verhalten ungesättiger und halogenierte Kohlenwasserstoffe. Z Elektrochem Angew Phys Chem 53:118–125
12. Savéant J-M (2006) Elements of molecular and biomolecular electrochemistry. An electrochemical approach to electron transfer chemistry. Wiley, Hoboken
13. Wang A, Huang Y-F, Sur U-K, Wu D-Y, Ren B, Rondinini S, Amatore C, Tian Z-Q (2010) In Situ identification of intermediates of benzyl chloride reduction at a silver electrode by SERS coupled with DFT calculations. J Am Chem Soc 132:9534–9536
14. Durante C, Isse AA, Sandoná G, Gennaro A (2009) Electrochemical hydrodehalogenation of polychloromethanes at silver and carbon electrodes. Appl Cat B 88:479–489
15. Rondinini S, Vertova A (2004) Electrocatalysis on silver and silver alloys for dichloromethane and trichloromethane dehalogenation. Electrochim Acta 49:4035–4046
16. Poizot P, Durand-Drouhin O, Lejeune M, Simonet J (2012) Changes in a glassy carbon surface by the cathodic generation of free alkyl radicals mediated by a silver–palladium catalyst. Carbon 50:73–83
17. Rondinini S, Mussini PR, Crippa F, Sello G (2000) Electrocatalytic potentialities of silver as a cathode for organic halide reductions. Electrochem Commun 2:491–496
18. Gennaro A, Isse AA, Bianchi CL, Mussini PR, Rossi M (2009) Is glassy carbon a really inert electrode material for the reduction of carbon–halogen bonds? Electrochem Commun 11:1932–1935
19. Isse AA, Falciola L, Mussini PR, Gennaro A (2006) Relevance of electron transfer mechanism in electrocatalysis: the reduction of organic halides at silver electrodes. Chem Comunn 344–346
20. Huang Y-F, Wu D-Y, Wang A, Ren B, Rondinini S, Tian Z-Q, Amatore C (2010) Bridging the gap between electrochemical and organometallic activation: benzyl chloride reduction at silver cathodes. J Am Chem Soc 132:17199–17210
21. Klymenko OV, Svir I, Amatore Ch (2013) New theoretical insights into the competitive roles of electron transfers involving adsorbed and homogeneous phases J Electroanal Chem 688:320–327
22. Scialdone O, Guarisco C, Galia A, Herbois R (2010) Electroreduction of aliphatic chlorides at silver cathodes in water. J Electroanal Chem 641:14–22
23. He J, Saez AE, Ela WP, Betterton EA, Arnold RG (2004) Destruction of aqueous-phase carbon tetrachloride in an electrochemical reactor with a porous cathode. Ind Eng Chem Res 43:913–923
24. He J, Ela WP, Betterton EA, Arnold RG, Saez AE (2004) Reductive dehalogenation of aqueous-phase chlorinated hydrocarbons in an electrochemical reactor. Ind Eng Chem Res 43:7965–7974
25. Huang B, Isse AA, Durante C, Wei C, Gennaro A (2012) Electrocatalytic properties of transition metals toward reductive dechlorination of polychloroethanes. Electrochim Acta 70:50–61
26. Minguzzi A, Lugaresi O, Aricci G, Rondinini S, Vertova A (2012) Silver nanoparticles for hydrodehalogenation reduction: evidence of a synergistic effect between catalyst and support. Electrochem Commun 22:25–28. doi:10.1016/j.elecom.2012.05.014

Organic Pollutants in Water Using BDD, Direct and Indirect Electrochemical Oxidation

Carlos Alberto Martinez-Huitle
Institute of Chemistry, Federal University of Rio Grande do Norte, Lagoa Nova, Natal, RN - CEP, Brazil

Electrochemical Oxidation of Organic Pollutants

Electrochemistry can offer much for solving or alleviating environmental problems. Over the past 15 years, electrochemical technology has been largely developed for its alternative use in wastewater remediation. The strategies include both the treatment of effluents and waste and the development of new processes or products with less harmful effects, often denoted as process-integrated environmental protection. Electrochemistry offers two options for the treatment of these pollutants with the aim of oxidizing them, not only to CO_2 and water (known as electrochemical combustion or mineralization) but also to biodegradable products [1–7]:

(i) Direct anodic oxidation (or direct electron transfer to the anode), which yields very poor decontamination.

(ii) Chemical reaction with electrogenerated species from water discharge at the anode, such as physically adsorbed "active oxygen" (physisorbed hydroxyl radical (•OH)) or chemisorbed "active oxygen" (oxygen in the lattice of a metal oxide (MO) anode). The action of these oxidizing species leads to total or partial decontamination, respectively [6, 7].

The existence of indirect or mediated electrochemical oxidation (MEO) with different heterogeneous species formed from water discharge has allowed the proposal of two main approaches for the pollution abatement in wastewaters by direct electrochemical oxidation (DEO), depending on the nature of the electrode material.

Results reported by several authors in the world, during the last two decades, corroborate the great mineralization attained for several pollutants in DEO with different anode materials (see Table 1). Also, in order to demonstrate the potential application of this process for treating wastewaters, as well as the influence the reactivity toward organic oxidation caused by the use of different electrode materials, other results obtained by DEO also have been recently reviewed and described in detail by other authors [1–5].

Nevertheless, the overall performance of electrochemical processes is established by the complex interaction of different parameters that may be optimized to obtain an effective and economical mineralization of pollutants. The principal parameters that determine an electrolysis performance are (i) electrode potential and current density, (ii) current distribution, (iii) mass transport regime, (iv) cell design, (v) electrolysis medium, and (vi) electrode materials. Even if we still remain far from meeting all the requirements needed for an ideal anode, significant steps have been made toward the production of better electrode materials [3].

Anodic Classification and Activity

According to the model proposed by Comninellis [6], anode materials are divided for simplicity into two classes: (i) active and (ii) non-active electrodes.

Active anodes, which present low oxygen evolution overpotential, are good electrocatalysts for the oxygen evolution reaction (OER) and, consequently, lead to selective oxidation of the organic pollutants. In this way, some electrode materials, such as carbon and graphite and platinum-based, iridium-based, and ruthenium-based oxides, can be considered in this classification. On the other hand, non-active anodes, which present high oxygen evolution overpotential, are poor electrocatalysts for the OER, and direct electrochemical oxidation is expected to occur for these electrodes [6]. Also, they present no higher oxidation state available, and the organic species are directly oxidized by an adsorbed hydroxyl radical, giving complete combustion. So, antimony-doped tin oxide, lead dioxide (PbO_2), and boron-doped diamond (BDD) are considered non-active anodes and, therefore, the most suitable for electrochemical combustion reactions.

Synthetic Boron-Doped Diamond (BDD) Thin Films

BDD thin film is a new electrode material that has recently received great attention, thanks to the development of technologies for synthesizing high-quality conducting diamond films at a commercially feasible deposition rate. Diamond films are grown on non-diamond materials, usually silicon, tungsten, molybdenum, titanium, niobium, tantalum, or glassy carbon, by energy-assisted (plasma or hot-filament) chemical vapor deposition. In order to make diamond films conducting, they are doped with different concentrations of boron atoms. The doping level of boron in the diamond layer, expressed as B/C ratio, is about 1,000–10,000 ppm. High-quality BDD electrodes possess several technologically important properties that distinguish them from conventional electrodes, such as the following [6, 7]:

Organic Pollutants in Water Using BDD, Direct and Indirect Electrochemical Oxidation, Table 1 Percentage of elimination, experimental conditions and energy consumption for the electrochemical oxidation with metal oxides, Pt, carbonaceous and BDD anodes of selected solutions. Adapted from ref. [2-5, 7]

Organic pollutant[a]	C_0/mg dm^{-3}	j^b/mA cm^{-2}	Electrolysis time/h	Color removal %	COD decay %	Energy consumption[c]
PbO_2 anode						
Acid red 2(methyl red)	235[f]	31.2	11	100	97	[e]
Basic brown 4	100	30	0.5	100	73	[e]
Ti/Sb_2O_5-SnO_2 anode						
Acid orange 7 (orange II)	750	20	6.25[h]	98	27	[e]
Reactive red 120 (Reactive red HE-3B)	1,500	20	6.25[h]	95	13	[e]
$Ti/Ru_{0.3}Ti_{0.7}O_2$ anode						
Reactive red 198	30	50	3	80	18[d]	95×10^6 kWh kg dye^{-1}
Direct red 81	0.1–1	25	3	100	57–61[d]	0.555 kWh (g TOC)$^{-1}$
Pt anode						
Acid red 27 (amaranth)	100	10–20	3	100	10	[e]
Reactive Orange 4	100	40	1	91	[e]	44.1 kWh m^{-3}
Activated carbon anode fiber						
Acid red 27 (amaranth)		80	0.5	8	99	52
Graphite anode						
Vat blue 1 (indigo)	200	0.43	0.5	14	[e]	6.93 kWh m^{-3}
Ti/BDD anode						
Acid orange 7 (orange II)	750	20	6.25[d]	90	92	[e]
15 reactive dyes	1,000	10	2–4[d]	>95	89–95	8.9–17.9 kWh m^{-3}
Nb/BDD anode						
Blue reactive 19	25	50	0.3[d]	95	82[f]	[e]
Basic red 29	40	1	0.03	98	[e]	[e]
Textile wastewater	566[g]	1	8.5	97	92	1.4 kWh (g COD)$^{-1}$
Si/BDD anode						
Acid red 27 (amaranth)	100	80	6.5	100	100, 94[f]	[e]
Acid orange 7 (orange II)	0.1[h]	20	2.5	100	100	[e]
Acid red 2 (methyl red)	200	31.2	100	6	98	[e]
Vat blue 1 (indigo)	0.29[h]	0.36–80	9.5	100	[e]	0.47–35 kWh m^{-3}
Mordant black 11 (Eriochrome Black T)	100[g]	30	4	100	100,100[f]	800 kWh m^{-3}
Acid blue 22	0.3[h]	20	12	100	97	70 kWh m^{-3}
Acid black 210	500	5	25	100	100	53 kWh m^{-3}
BDD with supports unspecified						
Blue reactive 19	50	50	2	100	100	[e]
Direct red 80	350	1.5	24	100	87	6.65 kWh m^{-3}
5 textile dyes	500	2.5	2	50–80	60–80	[e]
Acid orange 7	360	10	8	100	90	[e]

[a]Color Index (common) name
[b]Applied current density
[c]Different units reported
[d]Percentage of TOC decay
[e]Not determined
[f]Initial COD
[g]mM concentration
[h]Specific charge passed (Ah dm^{-3})

- An extremely wide potential window in aqueous and nonaqueous electrolytes
- Corrosion stability in very aggressive media
- Inert surface with low adsorption properties and strong tendency to resist deactivation
- Very low double-layer capacitance and background current.

Thanks to these properties, during electrolysis in the region of water discharge, BDD anodes promote the production of weakly adsorbed hydroxyl radicals, which unselectively and completely mineralize organic pollutants with a high current efficiency [3–5]:

$$BDD + H_2O \rightarrow BDD(\bullet OH) + H^+ + e^-$$

$$BDD(\bullet OH) + R \rightarrow BDD + CO_2 + H_2O$$

Experiments conducted to confirm this mechanism, using 5,5-dimethyl-1-pyrroline-*N*-oxide (DMPO) and salicylic acid as spin trapping, demonstrated that the oxidation process on BDD electrodes involves hydroxyl radicals as electrogenerated intermediates [3, 4].

Organic Oxidation Using BDD

A considerable number of laboratories investigating this material for wastewater treatment and the number of related publications have rapidly increased during the last two decades. Many papers have demonstrated that BDD anodes allow complete mineralizations up to near 100 % current efficiency of a large number of organic pollutants, such as carboxylic acids, benzoic acid, cyanides, cresols, herbicides, drugs, naphthol, phenolic compounds, polyhydroxybenzenes, polyacrylates, surfactants, and real wastewaters. The experimental conditions and some results of the most relevant research on this electrode material are summarized in Table 1 (results adapted from references [2–5, 7]).

In this context, several groups have proposed the use of BDD as anode material. The electrochemical behavior of such BDD thin films deposited on different substrates (Si, Ti, Nb, or Pt) has also been studied with the aim of developing applications for the electrochemical oxidation of different organic compounds for wastewater treatment.

During the last two decades, Comninellis and co-workers investigated the electrochemical oxidation of a wide range of pollutants using diamond electrodes [6], determining that, in particular, for high organic concentrations or low current densities, COD decreased linearly, forming a large amount of intermediates, while ICE remained about 100 %, indicating a kinetically controlled process [3, 4]. Conversely, for low organic concentrations or high current densities, pollutants were directly mineralized to CO_2, but ICE was below 100 %, due to mass transport limitation and side reactions of oxygen evolution. In order to describe these results, a comprehensive kinetic model to predict COD trends and current efficiency for the electrochemical combustion of the organic with BDD electrodes was proposed by them. The model, developed for an electrochemical reactor operating in a batch recirculation mode under galvanostatic conditions, is based on the following assumptions: (i) adsorption of the organic compounds at the electrode surface is negligible, (ii) all the organics have the same diffusion coefficient (D), (iii) the global rate of the electrochemical mineralization of organics is a fast reaction, and it is controlled by mass transport of organics to the anode surface [4, 7]. Based on these assumptions, anodic oxidation of different organic pollutants was performed in order to determine the influence of additional parameters such as nature of organic pollutant, concentration of organic compound, applied current density, flow rate, and temperature [4–7].

For example, ICE and COD evolution with the specific electrical charge passed during the anodic oxidation of different classes of organic compounds (acetic acid, isopropanol, phenol, 4-chlorophenol, and 2-naphthol) [4] demonstrated that for all these compounds, the electrochemical treatment is independent of the chemical nature of the organic pollutants, and the complete mineralization is obtained after the passage of 15 Ah dm^{-3}.

The effect of the initial concentration of the pollutants on the COD and ICE evolution during electrolysis of 4-chlorophenol (4-CP) when using a current density of 30 mA cm^{-2} indicated that, overall, COD removal is achieved in all cases, and the time for total mineralization increased with the 4-CP concentration as expected from the presence of a greater amount of organic matter in the solution. However, at a high initial concentration and at the beginning of the electrolysis, the COD decreases linearly with specific charge, and ICE remains at about 100 % [4]. This indicates that the electrolysis is performed at a current below the limiting one, and under these conditions the oxidation of 4-CP is controlled by the rate at which electrons are delivered at the anode [6, 7]. On the contrary, at a low concentration, the ICE decreases linearly to zero, meaning that the oxidation is carried out at a current density higher than the limiting one, and the process is under mass transport control. Under the latter conditions, which are characteristic of electrolyses with low COD values, the process is controlled by the rate at which organic molecules are transported from the bulk liquid to the electrode surface [4].

In the case of the influence of applied current density, the electrolyses of 1 g dm^{-3} 3,4,5-trihydroxybenzoic acid was performed at 10 and 60 mA cm^{-2}; the results demonstrated that at a low current density, the current efficiency remains at 100 % during almost all the oxidation, and the mineralization requires a low electrical charge but a long electrolysis time because some of the reactor capacity is under used. On the contrary, when operating current exceeds the limiting one, the electrolysis is fast, but the current efficiency decreases and therefore the specific charge consumed increases because a portion of the current is wasted on the secondary reaction of oxygen evolution [3–7]:

$$2 \cdot OH \rightarrow O_2 + 2H^+ + 2e^-$$

Concerning the effect of the flow rate, a specific case can be commented: when the synthetic dye methylene blue was electrochemically treated at 20 mA cm^{-2}, the COD and dye removals were faster at higher flow rates, meaning that the oxidation is a mass-controlled process [1, 3, 4]. In fact, the increase in the flow rate produces a higher concentration of organics that can react with electrogenerated hydroxyl radicals near the electrode surface, avoiding their decomposition to oxygen. Furthermore, these results demonstrate that there is an excellent agreement between the experimental data and the values predicted from the proposed model [6, 7].

On the other hand, with respect to the effect of temperature on the oxidation of 1 g dm^{-3} of the nitro dye acid yellow 1 in a medium that cannot generate oxidizing species, such as $HClO_4$, higher temperatures do not yield a significant increase of the oxidation rate during the electrochemical incineration process [1, 4, 6]. The small difference between the results obtained at the two temperatures is only due to an increase of the diffusion rate with rising temperature due to the decrease of the medium viscosity. However, several studies report that an increase of temperature favors organic oxidation; this behavior is not caused by the increase of the activity of the BDD anode but by the increase of the indirect reaction of organics with electrogenerated oxidizing agents from the electrolyte oxidation. In fact, electrolysis with BDD anodes in media containing chloride, sulfate, or phosphate ions generates chlorine, peroxodisulfate, and peroxodiphosphate [6]. These powerful oxidizing agents can oxidize the organic matter by a chemical reaction whose rate increases with temperature [6, 7].

Future Directions

The application of BDD electrodes for wastewater treatment has been mostly studied at Si-supported devices, in spite of the difficulties related to their industrial transposition due to the fragility and the relatively low conductivity of the Si substrate. BDD films synthesized on Nb, Ta, and W are promising, but their large-scale utilization is impossible due to the unacceptably high costs of these metal substrates. On the contrary, titanium would possess all required features to be a good substrate material; it is a preferable choice as a substrate due to its good conductivity, high strength, low price, high anticorrosion, and quick repassivation behavior [3, 4]. However, service life of a Ti/BDD electrode is relatively short because of large thermal residual stress due to substrate cooling from about 850 °C to ambient temperature as well as due to the formation of a TiC interlayer reducing diamond film adhesion to the substrate [4]. For this reason, future developments will rely upon the close collaboration of chemists, engineers, and electrochemists to ensure effective application and exploitation of new metallic supports, dimensions, and reactor design to answer important questions regarding the scale up of this process [4].

It is important to note that the viability of the electrooxidation technology provided with BDD electrodes for the treatment and reuse of the water used in different domestic and industrial processes is being evaluated in several parts of the world [7]. However, important advances are currently being achieved in Spain, Italy, and Switzerland, where industrial dimensions of BDD anodes are being tested for industrial application of this technology; these directions are the main goals of the electrochemical technologies nowadays [7].

Cross-References

► Wastewater Treatment, Electrochemical Design Concepts

References

1. Martínez-Huitle CA, Brillas E (2009) Decontamination of wastewaters containing synthetic organic dyes by electrochemical methods: A general review. Appl Catal B Environ 87:105
2. Chen G (2004) Sep Purif Technol 38:11
3. Martínez-Huitle CA, Ferro S (2006) Electrochemical oxidation of organic pollutants for the wastewater treatment: Direct and indirect processes. Chem Soc Rev 35:1324
4. Panizza M, Cerisola G (2009) Direct and mediated anodic oxidation of organic pollutants. Chem Rev 109:6541
5. Brillas E, Sirés I, Oturan MA (2009) Electro-fenton process and related electrochemical technologies based on fenton's reaction chemistry. Chem Rev 109:6570
6. Comninellis C, Chen G (2010) Electrochemistry for the environment. Springer, New York, NY-USA
7. Brillas E, Martínez-Huitle CA (2011) Synthetic diamond films preparation, electrochemistry, characterization and applications. John Wiley & Sons Inc., Hoboken, New Jersey-USA.

Organic Pollutants in Water Using DSA Electrodes, In-Cell Mediated (via Active Chlorine) Electrochemical Oxidation

Alexandros Katsaounis and Stamatios Souentie
Department of Chemical Engineering, University of Patras, Patras, Greece

Introduction

Toxic and nonbiodegradable organic pollutants, such as phenolic substances, are typically found in various industrial wastewaters, including pulp and paper mill industries, petrochemical refineries, plastics and glue manufacturing, and coke plants [1]. Such wastewaters usually have to be pretreated in order to minimize their organic charge, which is frequently quite toxic and biorecalcitrant. There are several methods of wastewater treatment, including incineration, wet air oxidation, adsorption, biological treatment, oxidation with strong oxidants (H_2O_2, O_3, active chlorine, etc.), or electrochemical oxidation. The choice of the treatment depends, among other things, on cost efficiency, as well as reliability and treatment efficiency.

Electrochemical oxidation processes have proved to be an efficient and versatile technology capable of handling a wide variety of wastewaters [2–4]. Enhanced efficiencies can be achieved through the use of compact bipolar electrochemical reactors utilizing suitable electrodes with large surface area. The role of the latter could be undertaken by the dimensionally stable anodes (DSA) developed in the mid-1960s, initially for the chlor-alkali industry in place of carbon anodes [5, 6]. DSA-type electrodes consist of electrocatalytic layers (such as RuO_2 and IrO_2)

on titanium substrates. Their advantages include low overpotential for O_2 or Cl_2 evolution, high electrocatalytic activity and electrical conductivity, shape and structure stability, corrosion resistance, and long service life. During the last decades, a lot of studies have been carried out focusing both on the DSA preparation process and resulting morphology and on fundamental studies of single- and mixed-oxide films [7–13].

The electrochemical mineralization of organic compounds can take place either directly at the anode by electron exchange between the adsorbed organics and the surface ($R_{ads} \rightarrow P_{ads} + ne^-$) or through the mediation of electrochemically generated redox reagent s, which act as intermediates for transferring electrons between the electrode and the organics. The latter, depending on the reaction kinetics, is carried out either in the vicinity of the electrode surface or in the bulk electrolyte. It has been found that direct oxidation of organics on DSA electrodes usually results in electrode fouling due to formation of polymeric layers on their surfaces [14, 15]. To avoid the above drawback, indirect oxidation in the oxygen evolution region is used, where the mineralization of the organic loading is carried out without deactivation of the anode. Another approach of indirect oxidation is the electrochemical generation of chemical reactants/mediators, which can interact and oxidize the organics of the effluent. Typical mediators used in indirect electrolysis are metallic redox couples, such as Ag(I/II), Ce(III/IV), Co(II/III), Fe(II/III), and Mn(II/III), and/or strong oxidizing chemicals, such as ozone, hydrogen peroxide, percarbonate, persulfate, perphosphate, and active chlorine. The final choice of the intermediates should be taken having in mind that their generation rate must be high and, at the same time, their interaction with the organics faster than any possible undesirable side reaction. One of the disadvantages of the mediators (at least in the case of the strong oxidants) is the nonselective oxidation process. The mediators interact with all types of organics, producing fully oxidized products (e.g., CO_2). This chapter will focus on the electrogeneration of active chlorine on DSA-type electrodes and its interaction with organics in aqueous media.

Electrogeneration of Active Chlorine

The terms "active chlorine" or "free available chlorine" are used to describe species such as Cl_2, HOCl, and OCl^-, which are recognized as key oxidants responsible for organic mineralization during wastewater treatment. Their production in electrolytic cells can be described by the following reaction mechanisms:

(i) Chlorine is electrochemically generated at the anode from chloride ions that are dissolved in the solution (Eq. 1):

$$2Cl^- \rightarrow Cl_2 + 2e^- \qquad (1)$$

(ii) Chlorine is hydrolyzed in water, forming hypochlorous acid (HClO) (Eq. 2):

$$Cl_2 + H_2O \rightarrow HClO + H^+ + Cl^- \qquad (2)$$

(iii) Depending on the solution pH, hypochlorous acid can be further dissociated to form hypochlorite anions (Eq. 3):

$$HClO \leftrightarrow ClO^- + H^+ \qquad (3)$$

The sum of hypochlorous acid (HClO), hypochlorite anion (ClO^-), and dissolved chlorine (Cl_2) concentration is usually termed "active chlorine" or "free available chlorine." As shown in Fig. 1, at pH between 5 and 6, active chlorine exists as hypochlorous acid (HClO). Above pH 6, following the forward reaction of Eq. 3, hypochlorous anions (OCl^-) are present too, while at higher pHs (>7.5), the hypochlorous anions (OCl^-) are the predominant most effective species. In terms of oxidation power, hypochlorous acid (HClO) is the most powerful oxidant among the active chlorine species. Therefore, oxidation of organics though active chlorine mediation seems to be favored in acidic and neutral environments. Taking into account that the anodic reaction of chlorine formation competes with the side reaction of oxygen evolution (Eq. 4), the local concentration of protons near the anode would be enhanced, and therefore the

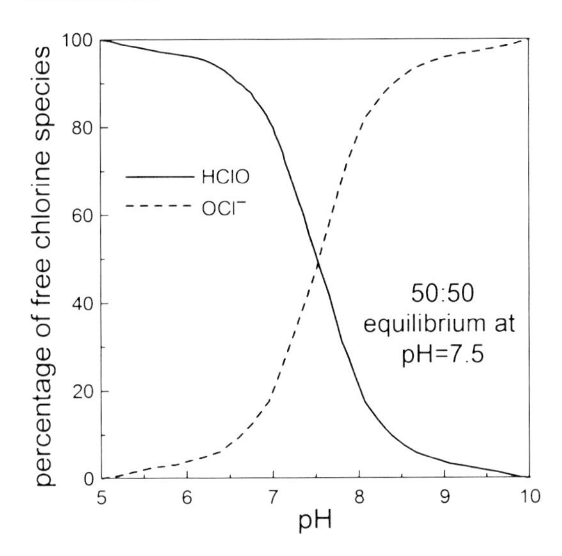

Organic Pollutants in Water Using DSA Electrodes, In-Cell Mediated (via Active Chlorine) Electrochemical Oxidation, Fig. 1 Effect of pH on the HClO and OCl^- concentration

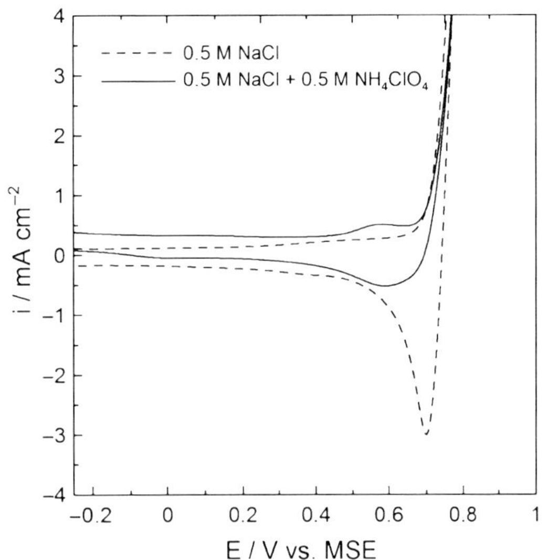

Organic Pollutants in Water Using DSA Electrodes, In-Cell Mediated (via Active Chlorine) Electrochemical Oxidation, Fig. 2 Cyclic voltammograms recorded in 0.5 M NaCl solution in the presence/absence of ammonia on a $Ti/PtO_x–IrO_2$ electrode (scan rate = 50 mVs^{-1}). Conditions: pH = 5.5, T = 25 °C (Adapted from [21]. Reprinted with permission from Elsevier)

local pH in this region can significantly differ from that in the bulk, leading to more acidic conditions.

$$2H_2O \rightarrow 4H^+ + O_2 + 4e^- \quad (4)$$

On the other hand, low pH values can affect the stability of the electrode due to corrosion phenomena.

Oxidation power of active chlorine is reduced when HOCl and/or OCl^- react with ammonia (according to Eqs. 5, 6, and 7) or organic nitrogen to form chloramines ("combined chlorine").

$$NH_3 + HOCl \rightarrow NH_2Cl \text{ (monochloramine)} + H_2O \quad (5)$$

$$NH_2Cl + HOCl \rightarrow NHCl_2 \text{(dichloramine)} + H_2O \quad (6)$$

$$NHCl_2 + HOCl \rightarrow NCl_3 \text{(trichloramine)} + H_2O \quad (7)$$

The fast interaction of active chlorine with ammonia [16–20] has been studied recently during ammonia oxidation on a DSA-type electrode [21]. Figure 2 shows voltammograms obtained in 0.5 M NaCl solution in the presence/absence of ammonia. In the absence of ammonia, a well-defined reduction peak is observed at 0.7 V versus MSE. This peak can be attributed to the reduction of active chlorine produced during the anodic scan. When ammonia is present, this reduction peak decreases significantly, indicating that the electrogenerated active chlorine species (mainly HClO at pH = 5.5) reacts with ammonia, according to Eqs. 5, 6, and 7, resulting in chloramines formation [21]. Although chloramines may also act as oxidants [22], their sanitizing power is less than 1 % of that of active chlorine, while at the same time they are responsible for unpleasant odors.

The production of active chlorine should be proportional to the chloride concentration and to the applied current density, as well. The first is shown in Fig. 3, where active chlorine concentration was measured during electrolysis of synthetic waters containing sulfate, nitrate,

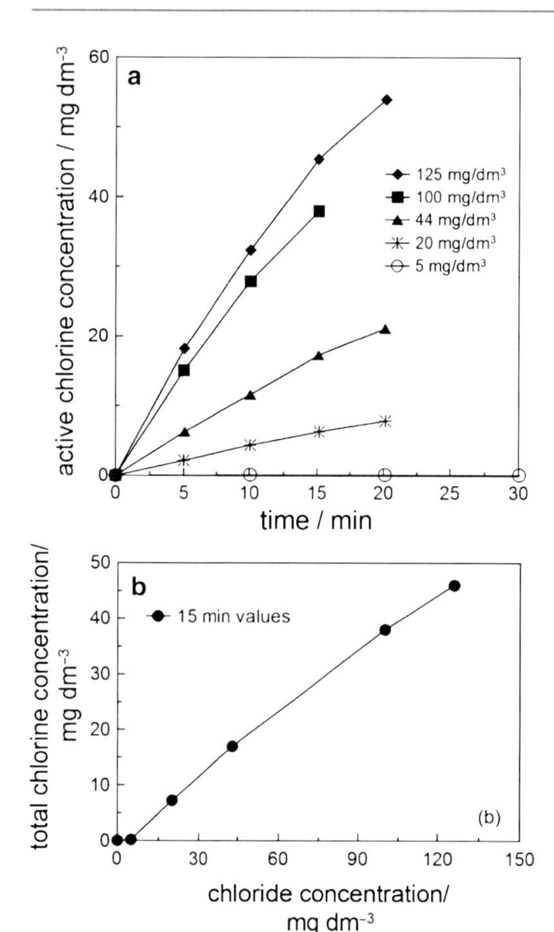

Organic Pollutants in Water Using DSA Electrodes, In-Cell Mediated (via Active Chlorine) Electrochemical Oxidation, Fig. 3 (**a**) Active chlorine concentration versus time in discontinuous experiments with varying chloride concentrations, (**b**) 15 min values of total chlorine concentration (current 300 mA; rotation rate 300 rpm; temperature 20–22 °C; V = 0.150 dm^3; Ti cathode, sulfate 240 mg dm^{-3}; nitrate 10 mg dm^{-3}; and carbonate 50 mg dm^{-3}, as sodium salts) (Adapted from [23]. Reprinted with permission from Springer)

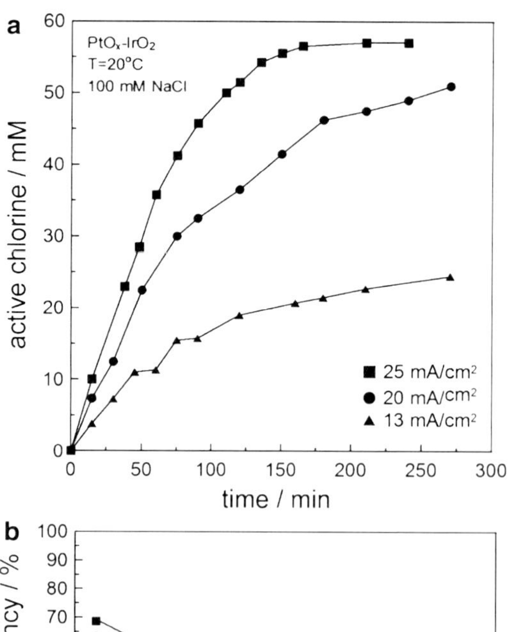

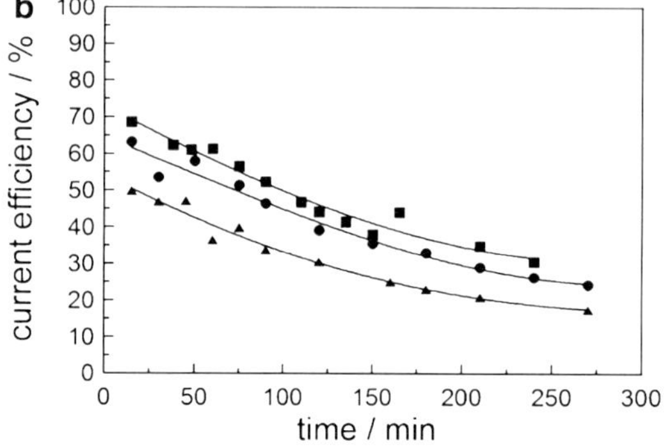

Organic Pollutants in Water Using DSA Electrodes, In-Cell Mediated (via Active Chlorine) Electrochemical Oxidation, Fig. 4 Effect of applied current density on (**a**) active chlorine concentration and (**b**) current efficiency. Conditions: T = 25 °C, pH 5.5, initial concentration of chloride: 0.1 M

carbonate/bicarbonate, and chloride ions as sodium salts [23]. According to Bergmann and Koparal [23], side effects such as:

- Stripping of chlorine by evolving oxygen from the electrode surface
- Electrochemical reaction of active chlorine in consecutive reactions at the anode [24]
- Chemical reaction of active chlorine in the bulk electrolyte and/or
- Cathodic reduction of active chlorine

can prevent a continuous increase in chlorine concentration. In a previous review paper [25], Trasatti reported that the reaction order of chlorine evolution with respect to chloride concentration is strictly one at constant pH over various oxide electrodes.

The effect of current on the production of active chlorine is shown in Fig. 4 during electrolysis of 0.1 M NaCl solution over a Ti/PtO_x-IrO_2 anode. Higher currents result in higher concentrations of active chlorine (Fig. 4a). It is worth noting that the current efficiencies are higher in the beginning of the process and decrease with time (Fig. 4b). In general, current efficiency can

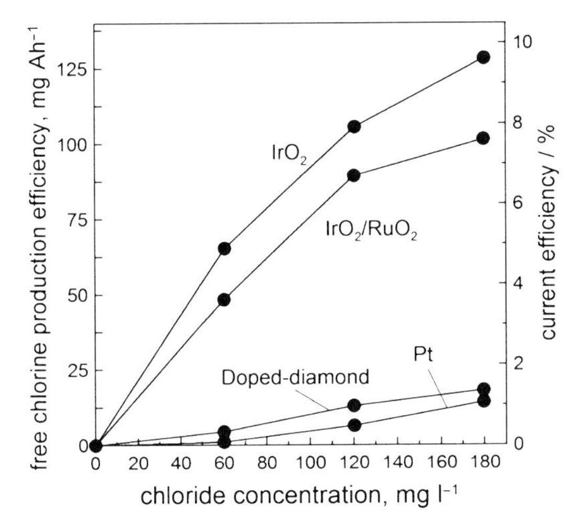

Organic Pollutants in Water Using DSA Electrodes, In-Cell Mediated (via Active Chlorine) Electrochemical Oxidation, Fig. 5 Dependence of the electrochemical free chlorine production efficiency on the chloride content of the electrolyzed water under standard conditions using four different anode materials (iridium oxide, mixed iridium/ruthenium oxides, platinum, doped diamond) (Adapted from [26]. Reprinted with permission from Johnson Matthey Plc)

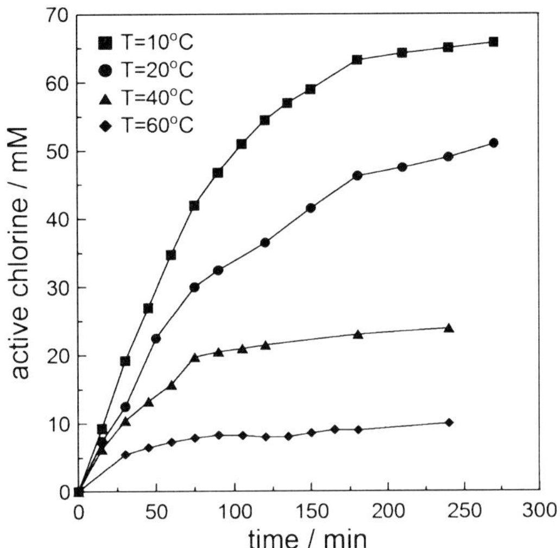

Organic Pollutants in Water Using DSA Electrodes, In-Cell Mediated (via Active Chlorine) Electrochemical Oxidation, Fig. 6 Effect of temperature on active chlorine concentration. Conditions: $i = 20$ mA cm^{-2}, pH 5.5, initial concentration of chloride: 0.1 M

be directly correlated with the type of the anode and its electrochemical properties. Kraft et al. [26] reported that DSA-type electrodes (IrO_2 and IrO_2-RuO_2) give higher current efficiencies during electrochemical chlorine production compared with boron-doped diamond (BDD) and platinum (Pt) anodes (Fig. 5). Based on the electrochemical production of active chlorine, DSA-type electrode materials clearly outperform BDD and Pt electrodes and could be perfect candidates as anodes for water disinfection. In the literature, the most widely used anode materials for electrogeneration of active chlorine are based on a mixture of metal oxides (such as IrO_2, TiO_2, RuO_2, and PbO_2) [15, 23, 26–39]. These electrodes are characterized by structure stability and excellent electrocatalytic activity.

Another important advantage of DSA-type electrodes during electrogeneration of active chlorine is the negligible production of chlorate and perchlorate species. Hypochlorite may further be oxidized to chlorate and perchlorate, especially on anodes characterized by high overpotential for oxygen and chlorine evolution (such as BDD) [40–44]. On the contrary, chlorate and perchlorate formation is not favored on DSA-type and Pt electrodes [45, 46].

Temperature is very important in active chlorine production. Figure 6 shows the concentration of active chlorine during electrolysis of a 0.1 M NaCl solution at four different temperatures using a DSA-type electrode (Ti/PtO_x-IrO_2). Higher concentrations are obtained at the lowest temperature. This can be attributed to the enhanced evolution of gaseous chlorine ($Cl_{2(g)}$) from the solution at high temperatures (according to Eq. 1 and the backward reaction of Eq. 2).

Interaction of Active Chlorine with Organics

One of the first studies on organic electrooxidation in the presence of NaCl was reported by Mieluch et al. [47] using phenol as model organic compound. According to their results, the oxidation rate was enhanced in NaCl solution through the mediation of Cl^- or ClO^- ions, which were formed at the anode. Since then, many studies have been reported emphasizing the

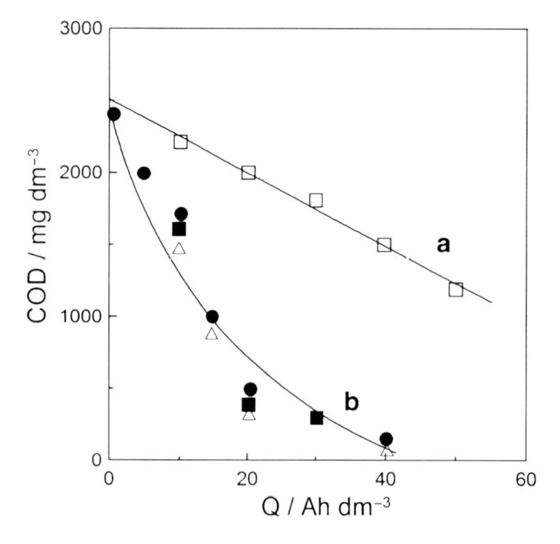

Organic Pollutants in Water Using DSA Electrodes, In-Cell Mediated (via Active Chlorine) Electrochemical Oxidation, Fig. 7 Influence of NaCl concentration on the rate of chemical oxygen demand (COD) elimination. *(a)* 150 g dm^{-3} Na_2SO_4; *(b)* 150 g dm^{-3} Na_2SO_4 + NaCl. NaCl concentration: (Δ) 433, (●) 85, and (■) 17 mM. Conditions: i = 0.1 A cm^{-2}, T = 50 °C, pH 12.2, and initial phenol concentration: 10 mM (Adapted from [48]. Reprinted with permission from Springer)

effective role of active chlorine during the electrochemical degradation of various organics. The study of Comninellis and Nerini on the anodic oxidation of phenol in the presence of NaCl [48] was one of the first reports where DSA-type electrodes were used as anodes. The presence of NaCl catalyzed the anodic oxidation of phenol at the IrO_2 anode due to the participation of the electrogenerated ClO^- in the oxidation. The rate of phenol oxidation and COD elimination in the absence and presence of NaCl is shown in Fig. 7 [48]. Interestingly, the enhancement is independent of NaCl concentration and occurs even in the presence of small amounts of NaCl (17 mM), indicating the high oxidation power of active chlorine. Similar results were obtained recently by Chatzisymeon et al. [39] during anodic oxidation of phenol on IrO_2 electrodes in the presence of various supporting electrolytes.

It should be mentioned that analysis of the reaction products during the process (in the case of phenol [48]) showed that organochlorinated compounds can be formed within the first steps of the oxidation, being further oxidized to volatile organics, such as $CHCl_3$. Similar behavior was reported by Panizza et al. during 2-napthol electrolysis [30]. However, the stability of some organochlorinated compounds and their resistance to oxidation is sometimes so high that even after long-term oxidation, the toxicity level of the solution remains above the desirable limits. Therefore, the formation of organochlorinated compounds during electrochemical indirect oxidation via active chlorine mediation must be followed. A characteristic case was reported by Gotsi et al. [49] during the electrochemical oxidation of olive mill wastewaters over a titanium-tantalum-platinum-iridium anode in the presence of high NaCl concentrations. Although acute toxicity to marine bacteria *V. fischeri* decreased slightly during the early stages of the reaction, it progressively increased thereafter due to the formation of organochlorinated byproducts, as confirmed by GC/MS analysis.

The case of olive mill wastewater is one of the cases where active chlorine and DSA-type electrodes were used for the electrochemical treatment of real wastewater. Another case is that of landfill leachates. Biological degradation is very effective for the treatment of landfill leachates with high values of biological oxygen demand (BOD), but it is not entirely efficient if recalcitrant compounds are present; therefore, alternative processes have been pursued, such as chemical precipitation, Fenton reaction, ozonation, photocatalysis, ultrasound irradiation, and combinations of the above. There are also reports studying the effectiveness of DSA-type electrodes on the electrochemical oxidation of landfill wastewaters [50–54]. This type of wastewater is perfectly treated via the mediation of active chlorine, since the effluent contains already significant amounts of chlorides [52], and there is no need for further addition of chloride salts. Turro et al. [52] reported a 90 % COD and 65 % total carbon decrease, together with complete color and total phenol removal, after 4 h of electrooxidation at 32 mA cm^{-2} current density at 80 °C and under acidic conditions using a DSA-type anode (Ti/IrO_2-RuO_2).

Indirect oxidation through active chlorine mediation has also been applied in wastewaters

Organic Pollutants in Water Using DSA Electrodes, In-Cell Mediated (via Active Chlorine) Electrochemical Oxidation,

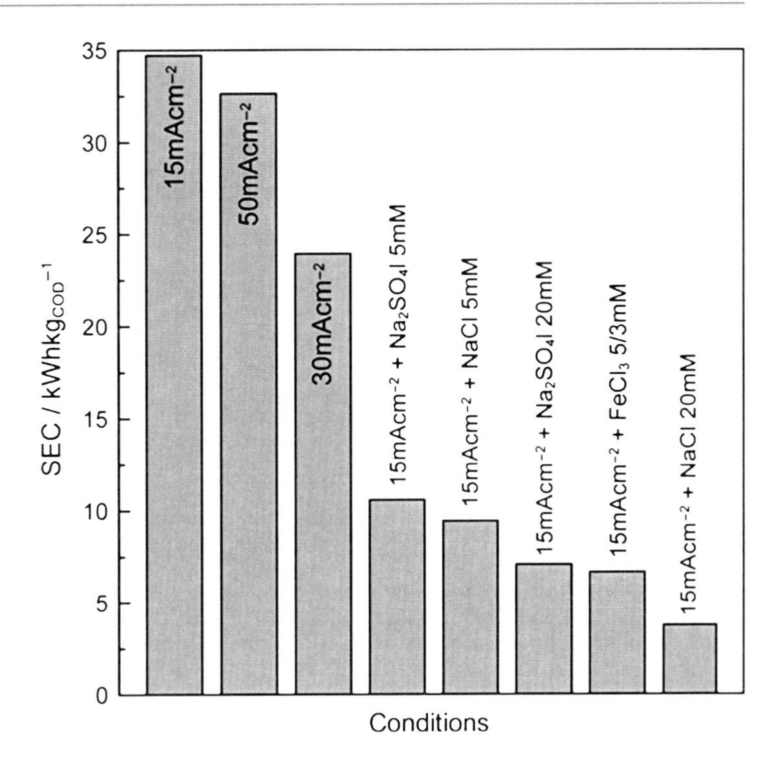

Fig. 8 Changes in 5 h-SEC during the electrochemical oxidation of olive mill wastewater (OMW) at 1,220 mg L^{-1} COD_0 and various conditions (Adapted from [65]. Reprinted with permission from Springer)

with high loadings of dyes. The development and expansion of various industries (textile, leather, agricultural, hair coloring, etc.) that use synthetic organic dyes has been remarkable during the last decades and leads to a continuously increased production of dye wastewaters. Recently, Martinez-Huitle and Brillas [55] reviewed several types of anodes that have been used for electrochemical oxidation of dyes. DSA-type electrodes composed of Ti, Ir, Ru, Sn, Pb, and Sb oxide mixtures were used in the above studies [32, 55–64].

An important advantage of the active chlorine mediation process is the reduction of the total energy that is required for treatment. Both treatment time and applied current densities are reduced, resulting in lower specific energy consumption. The latter can be defined as the amount of energy consumed per unit mass of organic loading (e.g., COD) removed. Figure 8 shows a characteristic example during DSA electrochemical treatment of olive mill wastewater on a RuO_2-based anode [65]. Addition of 20 mM NaCl in the solution resulted in a 90 % decrease in power demand, without any effect on the toxicity of the effluent.

Interpretation of the Active Chlorine: Organics Interaction on DSA Electrodes

In order to interpret the mechanism of organic oxidation via active chlorine mediation, the group of De-Battisti suggested the reaction scheme shown in Fig. 9 [27]. This scheme is based on the reaction pattern proposed by Comninellis for the direct electrochemical incineration of organics on "active" and "non-active" anodes [4]. In that case (absence of chlorine), electrochemical oxidation electron transfer occurs through the adsorbed hydroxyl radicals and their interaction with the anode. In the case of active chlorine mediation, oxygen transfer can be carried out by adsorbed oxychloro species (Fig. 9), which are considered intermediates of the chlorine evolution reaction.

Future Directions

The active chlorine-mediated electrochemical treatment of organics can be considered a powerful and effective tool in wastewater treatment. DSA-type electrodes can be excellent anodes for

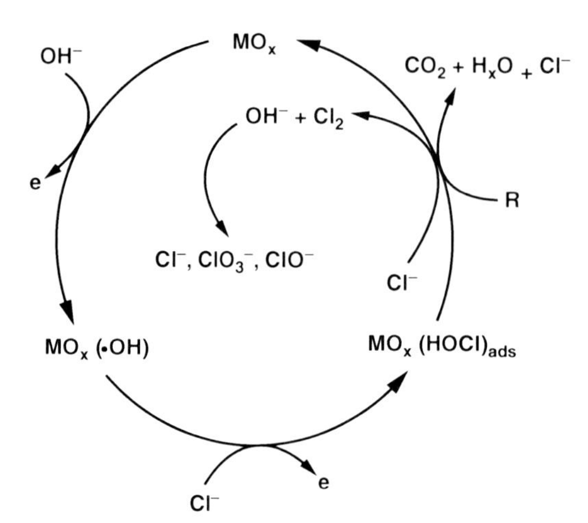

Organic Pollutants in Water Using DSA Electrodes, In-Cell Mediated (via Active Chlorine) Electrochemical Oxidation, Fig. 9 Reaction scheme of chlorine-mediated electrochemical oxidation of organics (Adapted from [27]. Reprinted with permission from The Electrochemical Society)

electrogeneration of active chlorine and therefore for the indirect electrochemical oxidation of organics. They have been effectively used for organics oxidation, both in model solutions and in actual wastewaters. The strong oxidation power of active chlorine allows for more effective and low-cost processes, while the side formation of organochlorinated intermediates (a drawback for real applications) can be suppressed by the development of novel DSA-type anodes composed of different metal oxides. This process can be effectively applied for water treatment, where the removal of emerging micropollutants and pathogens can be achieved under the mediation of electrogenerated active chlorine.

Cross-References

- ▶ Organic Pollutants, Direct Electrochemical Oxidation
- ▶ Organic Pollutants in Water, Direct Electrochemical Oxidation Using PbO_2
- ▶ Organic Pollutants in Water Using SnO_2, Direct Electrochemical Oxidation
- ▶ Wastewater Treatment, Electrochemical Design Concepts

References

1. Gattrell M, Kirk DW (1990) The electrochemical oxidation of aqueous phenol carbon electrode. Can J Chem Eng 68(6):997–1003
2. Anglada Á, Urtiaga A, Ortiz I (2009) Contributions of electrochemical oxidation to waste-water treatment: fundamentals and review of applications. J Chem Technol Biotechnol 84(12):1747–1755
3. Rajkumar D, Palanivelu K (2004) Electrochemical treatment of industrial wastewater. J Hazard Mater 113:123–129
4. Comninellis C (1994) Electrocatalysis in the electrochemical conversion/combustion of organic pollutants for waste water treatment. Electrochim Acta 39(11–12):1857–1862
5. Beer HB (1980) The invention and industrial development of metal anodes. J Electrochem Soc 127(8):303C–307C
6. Duby P (1993) The history of progress in dimensionally stable anodes. J Miner Met Mater Soc 45(3):41–43
7. Holden HS, Kolb JM, Holden HS, Kolb JM (1981) Metal anodes. In: Encyclopedia of chemical technology, vol 15. Wiley, New York
8. O'Leary KJ, Navin TJ (1974) Morphology of dimensionally stable anodes. In: Chlorine bicentennial symposium, San Francisco. p 174
9. Trasatti S, O'Grady WE (1981) Properties and applications of ruthenium oxide based electrodes. In: Gerisher H, Tobias CW (eds) Advances in electrochemistry and electrochemical engineering, vol 12. Wiley, New York, pp 177–261
10. Trasatti S (1980) Electrodes of conductive metal oxides, part A. Elsevier, Amsterdam
11. Trasatti S (1981) Electrodes of conductive metal oxides, part B. Elsevier, Amsterdam
12. Rolewicz J, Comninellis C, Plattner E, Hinden J (1988) Charactérisation des électrodes de type DSA pour le dégagement de O_2-I. L'électrode $Ti/IrO_2Ta_2O_5$. Electrochim Acta 33(4):573–580
13. Comninellis C (1989) Characterization of DSA-type oxygen evolving anodes. In: Hine F (ed) Performance of electrodes for industrial processes. The Electrochemical Society, Princeton
14. Rodrigo MA, Michaud PA, Duo I, Panizza M, Cerisola G, Comninellis C (2001) Oxidation of 4-chlorophenol at boron-doped diamond electrode for wastewater treatment. J Electrochem Soc 148(5):D60–D64
15. Chatzisymeon E, Dimou A, Mantzavinos D, Katsaounis A (2009) Electrochemical oxidation of model compounds and olive mill wastewater over DSA electrodes: 1. The case of Ti/IrO_2 anode. J Hazard Mater 167(1–3):268–274
16. Randtke S (2010) White's handbook of chlorination and alternative disinfectants. Wiley, Hoboken
17. Weil I, Morris JC (1949) Kinetic studies on the chloramines. I. The rates of formation of monochloramine, N-chlormethylamine and N-chlordimethylamine. J Am Chem Soc 71(5):1664–1671

18. Qiang Z, Adams CD (2004) Determination of monochloramine formation rate constants with stopped-flow spectrophotometry. Environ Sci Technol 38(5):1435–1444
19. Fair G, Morris J, Chang S, Weil I, Burden RP (1948) The behavior of chlorine as a water disinfectant. J Am Water Work Assoc 40:1051
20. Saguinsin L, Morris J (1975) Disinfection – water and wastewater. Ann Arbor Science, Ann Arbor
21. Kapałka A, Katsaounis A, Michels NL, Leonidova A, Souentie S, Comninellis C, Udert KM (2010) Ammonia oxidation to nitrogen mediated by electrogenerated active chlorine on Ti/PtO_x-IrO_2. Electrochem Commun 12(9):1203–1205
22. Lee W, Westerhoff P (2009) Formation of organic chloramines during water disinfection - chlorination versus chloramination. Water Res 43(8):2233–2239
23. Bergmann MEH, Koparal AS (2005) Studies on electrochemical disinfectant production using anodes containing RuO_2. J Appl Electrochem 35(12):1321–1329
24. Tasaka A, Tojo T (1985) Anodic oxidation mechanism of hypochlorite ion on platinum electrode in alkaline solution. J Electrochem Soc 132(8):1855–1859
25. Trasatti S (1987) Progress in the understanding of the mechanism of chlorine evolution at oxide electrodes. Electrochim Acta 32(3):369–382
26. Kraft A (2008) Electrochemical water disinfection: a short review. Platin Met Rev 52(3):177–185
27. Bonfatti F, Ferro S, Lavezzo F, Malacarne M, Lodi G, De Battisti A (2000) Electrochemical incineration of glucose as a model organic substrate II. Role of active chlorine mediation. J Electrochem Soc 147(2):592–596
28. Zhou M, Wu Z, Wang D (2002) Electrocatalytic degradation of phenol in acidic and saline wastewater. J Environ Sci Health A Tox Hazard Subst Environ Eng 37(7):1263–1275
29. Iniesta J, González-García J, Expósito E, Montiel V, Aldaz A (2001) Influence of chloride ion on electrochemical degradation of phenol in alkaline medium using bismuth doped and pure PbO_2 anodes. Water Res 35(14):3291–3300
30. Panizza M, Cerisola G (2003) Influence of anode material on the electrochemical oxidation of 2-naphthol: part 1. Cyclic voltammetry and potential step experiments. Electrochim Acta 48(23):3491–3497
31. Panizza M, Cerisola G (2003) Electrochemical oxidation of 2-naphthol with in situ electrogenerated active chlorine. Electrochim Acta 48(11):1515–1519
32. Panizza M, Cerisola G (2008) Electrochemical degradation of methyl red using BDD and PbO_2 anodes. Ind Eng Chem Res 47(18):6816–6820
33. Panizza M, Cerisola G (2009) Direct and mediated anodic oxidation of organic pollutants. Chem Rev 109(12):6541–6569
34. Martínez-Huitle CA, Andrade LS (2011) Electrocatalysis in wastewater treatment: recent mechanism advances. Quimica Nova 34(5):850–858
35. Martínez-Huitle CA, Ferro S (2006) Electrochemical oxidation of organic pollutants for the wastewater treatment: direct and indirect processes. Chem Soc Rev 35(12):1324–1340
36. Martínez-Huitle CA, Ferro S, De Battisti A (2004) Electrochemical incineration of oxalic acid: role of electrode material. Electrochim Acta 49(22–23 SPEC. ISS.):4027–4034
37. Martínez-Huitle CA, Quiroz MA, Comninellis C, Ferro S, De Battisti A (2004) Electrochemical incineration of chloranilic acid using Ti/IrO_2, Pb/PbO_2 and Si/BDD electrodes. Electrochim Acta 50(4):949–956
38. Quiroz MA, Reyna S, Martínez-Huitle CA, Ferro S, De Battisti A (2005) Electrocatalytic oxidation of p-nitrophenol from aqueous solutions at Pb/PbO_2 anodes. Appl Catal Environ 59(3–4):259–266
39. Chatzisymeon E, Fierro S, Karafyllis I, Mantzavinos D, Kalogerakis N, Katsaounis A (2010) Anodic oxidation of phenol on Ti/IrO_2 electrode: experimental studies. Catal Today 151(1–2):185–189
40. Landolt D, Ibl N (1970) On the mechanism of anodic chlorate formation in concentrated NaCl solutions. Electrochim Acta 15(7):1165–1183
41. Landolt D, Ibl N (1972) Anodic chlorate formation on platinized titanium. J Appl Electrochem 2(3):201–210
42. Landolt D, Ibl N (1968) On the mechanism of anodic chlorate formation in dilute NaCl solutions. J Electrochem Soc 115(7):713–720
43. Foerster F (1924) Trans Am Electrochem Soc 46:23
44. Palmas S, Polcaro AM, Vacca A, Mascia M, Ferrara F (2007) Characterization of boron doped diamond during oxidation processes: relationship between electronic structure and electrochemical activity. J Appl Electrochem 37(1):63–70
45. Kraft A, Stadelmann M, Blaschke M, Kreysig D, Sandt B, Schröder F, Rennau J (1999) Electrochemical water disinfection. Part I: hypochlorite production from very dilute chloride solutions. J Appl Electrochem 29(7):861–868
46. Kraft A, Wünsche M, Stadelmann M, Blaschke M (2003) Electrochemical water disinfection. Recent Res Dev Electrochem 6:27–55
47. Mieluch J, Sadkowski A, Wild J, Zoltowski P (1975) Electrochemical oxidation of phenolic compounds in aqueous solutions. Przemysl Chemiczny 54(9):513–516
48. Comninellis C, Nerini A (1995) Anodic oxidation of phenol in the presence of NaCl for wastewater treatment. J Appl Electrochem 25(1):23–28
49. Gotsi M, Kalogerakis N, Psillakis E, Samaras P, Mantzavinos D (2005) Electrochemical oxidation of olive oil mill wastewaters. Water Res 39(17):4177–4187
50. Cossu R, Polcaro AM, Lavagnolo MC, Mascia M, Palmas S, Renoldi F (1998) Electrochemical treatment of landfill leachate: oxidation at Ti/PbO_2 and Ti/SnO_2 anodes. Environ Sci Technol 32(22):3570–3573
51. Panizza M, Delucchi M, Sirés I (2010) Electrochemical process for the treatment of landfill leachate. J Appl Electrochem 40(10):1721–1727

52. Turro E, Giannis A, Cossu R, Gidarakos E, Mantzavinos D, Katsaounis A (2011) Electrochemical oxidation of stabilized landfill leachate on DSA electrodes. J Hazard Mater 190(1–3):460–465
53. Chiang LC, Chang JE, Wen TC (1995) Indirect oxidation effect in electrochemical oxidation treatment of landfill leachate. Water Res 29(2):671–678
54. Shi Y, Yu H, Xu D, Zheng X (2012) Degradation of landfill leachate by combined three-dimensional electrode and electro-Fenton. Adv Mater Res 347–353:440–443
55. Martínez-Huitle CA, Brillas E (2009) Decontamination of wastewaters containing synthetic organic dyes by electrochemical methods: a general review. Appl Catal Environ 87(3–4):105–145
56. Catanho M, Malpass GRP, Motheo AJ (2006) Photoelectrochemical treatment of the dye reactive red 198 using DSA® electrodes. Appl Catal Environ 62(3–4):193–200
57. del Río AI, Molina J, Bonastre J, Cases F (2009) Influence of electrochemical reduction and oxidation processes on the decolourisation and degradation of C.I. Reactive Orange 4 solutions. Chemosphere 75(10):1329–1337
58. Panakoulias T, Kalatzis P, Kalderis D, Katsaounis A (2010) Electrochemical degradation of Reactive Red 120 using DSA and BDD anodes. J Appl Electrochem 40(10):1759–1765
59. Aquino JM, Rocha-Filho RC, Bocchi N, Biaggio SR (2010) Electrochemical degradation of the Acid Blue 62 dye on a β-PbO_2 anode assessed by the response surface methodology. J Appl Electrochem 40(10):1751–1757
60. Aquino JM, Rocha-Filho RC, Bocchi N, Biaggio SR (2010) Electrochemical degradation of the reactive red 141 dye on a β-PbO_2 anode assessed by the response surface methodology. J Braz Chem Soc 21(2):324–330
61. Rajkumar D, Song BJ, Kim JG (2007) Electrochemical degradation of Reactive Blue 19 in chloride medium for the treatment of textile dyeing wastewater with identification of intermediate compounds. Dye Pigment 72(1):1–7
62. Vaghela SS, Jethva AD, Mehta BB, Dave SP, Adimurthy S, Ramachandraiah G (2005) Laboratory studies of electrochemical treatment of industrial azo dye effluent. Environ Sci Technol 39(8):2848–2855
63. De Oliveira GR, Fernandes NS, Melo JVD, Da Silva DR, Urgeghe C, Martínez-Huitle CA (2011) Electrocatalytic properties of Ti-supported Pt for decolorizing and removing dye from synthetic textile wastewaters. Chem Eng J 168(1):208–214
64. Solano AMS, Rocha JHB, Fernandes NS, Da Silva DR, Martinez-Huitle CA (2011) Direct and indirect electrochemical oxidation process for decolourisation treatment of synthetic wastewaters containing dye. Oxid Commun 34(1):218–229
65. Papastefanakis N, Mantzavinos D, Katsaounis A (2010) DSA electrochemical treatment of olive mill wastewater on Ti/RuO_2 anode. J Appl Electrochem 40(4):729–737

Organic Pollutants in Water Using SnO_2, Direct Electrochemical Oxidation

Yujie Feng
Harbin Institute of Technology, Harbin, China

Introduction

Organic pollutants in water, especially aromatic chemicals, are common pollutants in the wastewater from many industrial sectors, such as synthetic chemical plants, petroleum refineries, and paper, textile, and pesticide factories. Aromatic compounds are refractory and often toxic to biological treatment processes. Electrochemical oxidation is another alternative which has attracted considerable research attention. Electrochemical oxidation has the potential to be developed as a cost-effective technology for the treatment of toxic organic pollutants due to its unique advantages, such as simplicity, robustness in structure, and ease in operation [1–4].

Although organic pollutants in water usually can be oxidized at numerous anode materials, the electrochemical performance depends on the kinds and structure of anode materials [5]. The oxidation of organic pollutants at traditional anodes, such as Pt and graphite, was found to be slow. Since the discovery of dimensionally stable anodes (DSA) in 1976 by Beer [6], great development has been obtained. DSA anodes are composed of a thin layer (a few micrometers) of metal oxides (RuO_2, IrO_2, SnO_2, PbO_2, etc.) on a base metal (Ti, Zr, Ta, etc.). Many research results showed that anodes either composed of SnO_2 or with a component of SnO_2 in the catalytic layers can possess better oxidation ability for organic pollutants [5, 7–9].

Crystal Structure of SnO_2

The crystal structure of a unit cell of SnO_2 is shown in Fig. 1. An Sn atom is in the center of an octahedron, with six oxygen atoms around it. The crystal parameters of SnO_2 are a = b = 0.4737 nm and c = 0.3185 nm. The radii of O^{2-} and Sn^{4+} are 0.14 nm and 0.071 nm, respectively [10].

Pure SnO_2 crystal is a kind of n-type semiconductor, which has a wide Eg value (3.5 ~ 4.3 eV) [11], and its conductivity is too poor to be used. The conductivity of SnO_2 can be enhanced by adding some doping elements. In most cases, Sb is used as a doping element, and new energy bands can be induced.

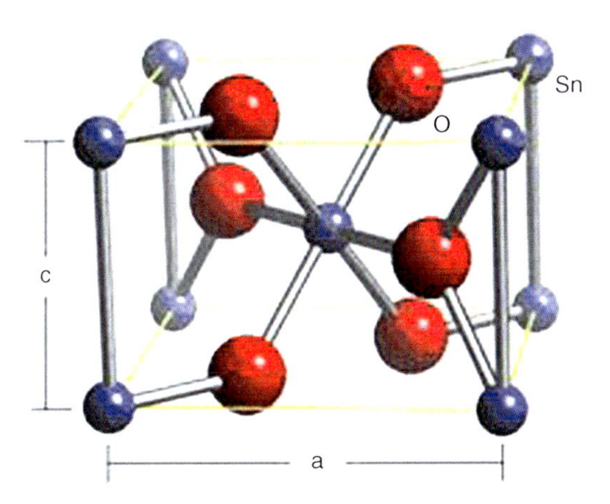

Organic Pollutants in Water Using SnO_2, Direct Electrochemical Oxidation, Fig. 1 Crystal structure of SnO_2

Preparation Methods of SnO_2 Anodes

Ti metal is often used as a base material because of its low cost. The key process for SnO_2 anode preparation is how to achieve catalytic coating on a Ti base. To ensure the catalytic coating is stable, it is important to try to make the coating with better adhesion to the base surface. There are several methods in the electrode preparation, such as dip coating, sputtering, chemical vapor deposition, electrodeposition, and sol–gel [9, 12, 13]. Almost all preparation methods follow the same basic preparation procedure. The basic preparation procedure includes three steps: pretreatment of the Ti base, coating solution preparation, and a pyrolysis process. Research results revealed that the electrochemical characteristics and service life of SnO_2 anodes have been influenced by the preparation method and technological factors. The electrochemical characteristics of SnO_2 anodes prepared by different methods are shown in Fig. 2.

In fact, the particle size of the SnO_2 crystal has a great influence on the electrochemical characteristics of SnO_2 anodes. Smaller particle size means larger surface area, and much more electrocatalytic reaction can take place at the same time. The particle size of SnO_2 crystals prepared by electrodeposition and sol–gel methods is on a nanometer scale. This is the

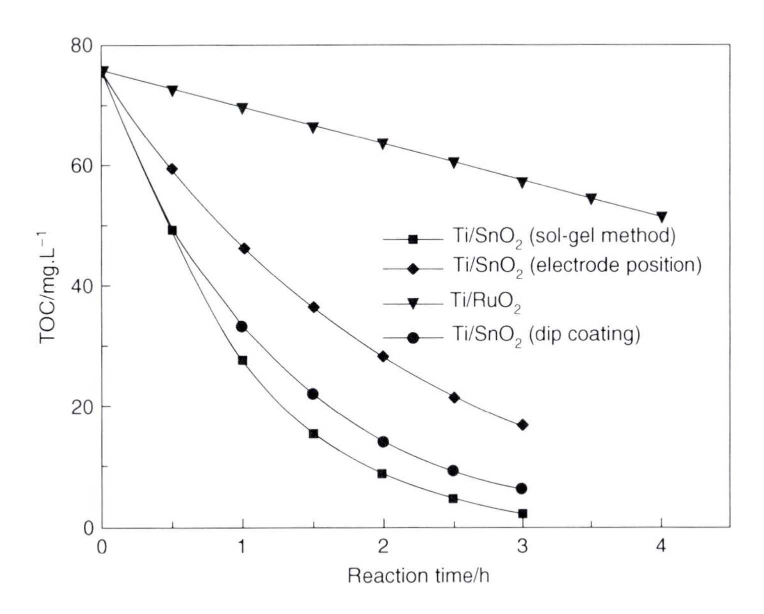

Organic Pollutants in Water Using SnO_2, Direct Electrochemical Oxidation, Fig. 2 Variation of TOC in different SnO_2 anodes

reason they have better electrochemical characteristics. In all preparation methods, the sol–gel method is a relatively simple, effective, and convenient way to realize SnO_2 nanocoatings [9].

Electrochemical Oxidation of Organic Pollutants

The anode material is certainly an important parameter in the electrochemical oxidation of organic pollutants because the oxidation mechanism and products of anodic reactions vary with the anode material. For example, the anodic oxidation of phenol in a Ti/RuO_2 anode can only yield hydroquinone and benzoquinone *while* CO_2 *in* Ti/SnO_2 *anode*. The electrochemical oxidation of phenol in a Ti/SnO_2 anode is complete.

In general, two different mechanisms can be distinguished for the oxidation of organic pollutants in SnO_2 anodes. One is direct oxidation, and another is indirect oxidation. Direct oxidation can only take place on the surface of SnO_2 anodes. Indirect oxidation can occur via hydroxyl radicals, which are generated by oxygen vacancies [14] in SnO_2 anodes continuously. The oxidizability of hydroxyl radicals is very strong. Electrochemical oxidation of organic pollutants in SnO_2 anodes mainly depends on indirect oxidation.

Cross-References

- ▸ Organic Pollutants, Direct Electrochemical Oxidation
- ▸ Organic Pollutants in Water, Direct Electrochemical Oxidation Using PbO_2
- ▸ Organic Pollutants in Water Using DSA Electrodes, In-Cell Mediated (via Active Chlorine) Electrochemical Oxidation
- ▸ Wastewater Treatment, Electrochemical Design Concepts

References

1. Brillas E, Calpe JC, Casado J (2000) Mineralization of 2, 4-D by advanced electrochemical oxidation processes. Water Res 34:2253–2262
2. Santiago E, Jaime G, Sandra C (2002) Comparison of different advanced oxidation processes for phenol degradation. Wat Res 36:1034–1042
3. Ricardo AT, Victor S, Walter T (2003) Electrochemical treatment of industrial wastewater containing 5-amino-6-methyl-2-benzimidazolone. Toward an electrochemical–biological coupling. Wat Res 37:3118–3124
4. Rajkumar D, Palanivelu K (2004) Electrochemical treatment of industrial wastewater. J Hazard Mater B113:123–129
5. Feng YJ, Li XY (2003) Electro-catalytic oxidation of phenol on several metal-oxide electrodes in aqueous solution. Water Res 37(10):2399–2407
6. Beer HB (1976) Electrodes and coating thereof. US Patent 3632498
7. Comninellis C, Pulgarin C (1993) Electrochemical oxidation of phenol for wastewater treatment using SnO_2 anodes. J Appl Electrochem 23:108–112
8. Correa-lozano B, Comninellis C, Battisti AD (1996) Preparation of SnO_2-Sb_2O_5 films by the spray pyrolysis technique. J Appl Electrochem 26:83–89
9. Liu JF, Feng YJ, Sun LX, Qian ZG (2006) Investigation on preparation and electrocatalytic characteristics of Ti-base SnO_2 electrode with nano-coating. Mater Sci Technol 14(2):200–203
10. Jarzebski ZM, Marton JP (1976) Physical properties of SnO_2 materials. Preparation and defect structure. J Electrochem Soc 123(7):199C–205C
11. Supothina S, Guire MRD (2000) Characterization of SnO_2 thin films grown from aqueous solutions. Thin Solid Films 371:1–9
12. Ding HY, Feng YJ, Liu JF (2007) Preparation and properties of Ti\\SnO_2-Sb_2O_5 electrodes by electrodeposition. Mater Lett 61(7):4920–4923
13. Liu J, Feng Y (2009) Investigation on the electrocatalytic characteristics of SnO_2 electrodes with nanocoating preparation by electrodeposition method. Sci China Ser E-Technol Sci 52(6):1799–1803
14. Cui YH, Feng YJ (2005) EPR study on Sb doped Ti-base SnO_2 electrodes. J Mater Sci 40(17):4695–4697

Organic Pollutants in Water, Direct Electrochemical Oxidation Using PbO_2

Romeu C. Rocha-Filho, Leonardo S. Andrade, Nerilso Bocchi and Sonia R. Biaggio
Universidade Federal de Goiás, Catalão, Brazil

Introduction

The growing need to avoid the contamination of the environment has increased the interest in the elimination of toxic organic compounds present in several types of industrial effluents or wastewater. The use of conventional methods, such as biological and chemical or incineration, for treatment of these effluents can often solve this problem. However, due to different inconveniences associated with these methods, electrochemical techniques have been proposed for treating various effluents [1–4]. The use of electrode materials such as PbO_2, SnO_2, and boron-doped diamond (BDD) allows organic compounds contained in effluents to be transformed into biodegradable compounds and, eventually, into CO_2 and H_2O, then referred as electrochemical combustion or mineralization [5].

The most critical aspect underlying the electrochemical treatment of effluents is the possible high energy consumption due mainly to parallel reactions, especially those related to the solvent; if electrooxidation processes are carried out in aqueous media, the oxygen evolution reaction (OER) takes place, decreasing the current efficiency. Consequently, the use of electrodes that present a high OER overpotential becomes necessary to attain a better faradaic efficiency. SnO_2 [6, 7], BDD [8], or PbO_2 electrodes [9–16] meet this criterion.

PbO_2 Electrodes

The rapid and easy preparation of PbO_2 electrodes, besides their low cost and good stability at high applied potentials in media of different pHs, have been considered the main qualities that justify their use in the electrochemical treatment of wastewater. On the other hand, problems such as lifetime and contamination of the effluent by Pb^{2+} could be a serious inconveniency in practical applications. However, the incorporation of dopants such as Fe^{2+} [17–19], Bi^{3+} [10, 12, 13, 19, 20], Co^{2+} [14, 15, 21], and F^- [15, 22, 23] into the PbO_2 crystalline matrix has led to significantly improved stability and electrocatalytic activity. For example, Andrade et al. [15] showed that the incorporation of Co and F ions into the PbO_2 film (Co, F-doped PbO_2) caused a significant improvement in its chemical stability compared to that of pure PbO_2. In accelerated service life tests (50 h at 0.1 A/cm^2 in 1 mol/L H_2SO_4, 40°C), the Co, F-doped PbO_2 electrode lost only 0.13 % of its mass (a mass loss rate of about 0.06 % per day), while the relative mass loss for the pure PbO_2 electrode was ~4.5 times greater.

PbO_2 electrodes have been obtained by different methods, e.g., anodization [24] and electrodeposition [9–16, 24, 25], but also by a thermo-electrochemical method [26, 27]. PbO_2 films were electrodeposited on platinized Ti [6, 14–16], graphite [24], Ta [9], Ti/(IrO_2-Ta_2O_5) [10], Au [19, 28–33], or Pt [18, 22, 34–37]. For practical use of these electrodes, the Ti substrate is the best choice. However, in some cases the electrode performance was not good [26]; thus, several authors have used platinized Ti as a substrate, e.g., [6, 14–16].The presence of Pt between the deposited film and Ti avoids the formation of a passivation film (n-TiO_2), which acts as a barrier for electron transfer.

Lead dioxide exists as three different polymorphs: orthorhombic, α-PbO_2; tetragonal, β-PbO_2; and cubic, γ-PbO_2, which is stable only at high pressures [38]. The α- and β-PbO_2 forms are highly conducting, and a mixture of them is used as the anode in lead-acid batteries; however, since in sulfuric acid α-PbO_2 presents a lower OER overvoltage than β-PbO_2 [39], the latter is preferred for electrooxidation purposes. According to Yeo et al. [40], β-PbO_2 is obtained from acidic solutions containing Pb^{2+} and α-PbO_2 from alkaline solutions.

Electrochemical Oxidation Using PbO_2

The electrochemical oxidation of an organic pollutant can occur by direct electron transfer to the anode, indirect electron transfer mediated by hydroxyl radicals ($^\bullet$OH) or other species electrogenerated from inorganic salts. In the direct electrochemical oxidation, which yields very poor decontamination, the organic compounds are adsorbed and oxidized on the electrode surface, i.e.,

$$R_{ads} \rightarrow P_{ads} + e^- \quad (1)$$

where R_{ads} and P_{ads} represent an adsorbed organic molecule and its adsorbed oxidation product, respectively.

In electrochemical oxidation processes mediated by hydroxyl radical on metal oxides (MO_x), the $^{\bullet}OH$ species generated from water discharge can become physically (Eqs. 2 and 3) or chemically (Eq. 4) adsorbed at the anode surface:

$$MO_x + H_2O \rightarrow MO_x[^{\bullet}OH] + H^+ + e^- \quad (2)$$

$$MO_x[\ ] + H_2O \rightarrow MO_x[^{\bullet}OH] + H^+ + e^- \quad (3)$$

$$MO_x[^{\bullet}OH] \rightarrow MO_{x+1} + H^+ + e^- \quad (4)$$

where $MO_x[]$ represents the surface sites at which $^{\bullet}OH$ species can be adsorbed and MO_{x+1} a higher oxide or superoxide. It should be noted that higher oxides cannot be formed in materials where the metal is already present in its maximum oxidation state, e.g., PbO_2 films. Anodes that present $^{\bullet}OH$ radicals physically or chemically adsorbed were later referred to as "non-active" or "active" anodes [41], respectively. More recently, these anodes were simply classified according to their oxidation power in acid media, taking into account the value of the adsorption enthalpy of hydroxyl radicals on the anode surface [42]; thus, non-active anodes (low $^{\bullet}OH$ adsorption enthalpy) present higher oxidation power concomitantly to a higher OER overpotential, whereas active anodes present lower oxidation power and OER overpotential.

Only partial oxidation of organics will occur at active anodes, because $^{\bullet}OH$ species interact strongly with their surface; thus, selective organic oxidation (Eq. 5) occurs, simply transforming R to by-products (RO):

$$MO_{x+1} + R \rightarrow MO_x + RO \quad (5)$$

On the other hand, at non-active anodes (such as PbO_2, BDD, or SnO_2), the electrooxidation process occurs at or near the electrode surface, mediated by hydroxyl radicals. Hence, total decontamination may occur at these anodes because R can be fully oxidized:

$$aMO_x[^{\bullet}OH] + R \rightarrow aMO_x + mCO_2 + nH_2O + xH^+ + xe^- \quad (6)$$

where R represents an organic compound with m carbon atoms and without any heteroatom, thus needing a ($= 2m + n$) oxygen atoms to be totally mineralized to CO_2.

It should be emphasized that reaction 6 in fact represents an indirect electrochemical process that is mediated by hydroxyl radicals. However, as in this case the surface interaction with $^{\bullet}OH$ is quite weak, this process is usually classified as direct oxidation [5]. Clearly, in the electrooxidation of organics, a non-active anode acts only as an inert substrate and a sink for the removal of electrons [43].

Inevitably and undesirably, OER takes place in parallel to reactions 2 and 4, i.e.,

$$MO_x[^{\bullet}OH] + H_2O \rightarrow MO_x[\ ] + O_2 + 3H^+ + 3e^- \quad (7)$$

$$MO_{x+1} \rightarrow MO_x + \frac{1}{2}O_2 \quad (8)$$

Since these reactions compete with reactions 6 and 5, respectively, the use of electrode materials that present a high OER overpotential is desirable, as previously pointed out.

Despite being non-active anodes, the performance of PbO_2 electrodes in the electrooxidation of organic compounds is worse than that of BDD electrodes. According to Zhu et al. [45], although the $^{\bullet}OH$ concentration on the PbO_2 surface is higher than that on BDD, the PbO_2 performance in the removal of organics is worse. The possible reason for that is the strength of the $^{\bullet}OH$ adsorption on the PbO_2 surface, i.e., its OER overpotential is lower.

As mentioned in the overall discussion above, all models assume that the performance in the

Organic Pollutants in Water, Direct Electrochemical Oxidation Using PbO_2, Table 1 Summarized results related to recent applications of PbO_2 electrodes in the electrochemical treatment of several types of organic pollutants

Organic pollutant (R)	$[R]_0$ (mg/L)	i(mA/cm^2)	Electrolysis charge (Ah/L)	COD or TOC decay (%)	Ref.
Reactive blue 19 (anthraquinone dye)	25	50	8.0	95	[14]
Phenol	1,000	100	32	75	[15]
	500	20	7.0	86	[48]
	500	20	4.3	70	[49]
Reactive orange 16 (azo dye)	100	50	2.0	85	[16]
Red X-GRL (azo dye)	500	4.8	1.2	38	[50]
4-Chlorophenol	129	20	7,0	86	[48]
	1,000	80	19	60	[51]
4-Chloro-3-ethyl phenol	300	10	6.0	60	[52]
Reactive red 141	100	75	3.0	100	[53]
Acid blue 62	100	125	3.0	100	[54]
Direct yellow 86	200	50	2.7	100	[55]
Real textile effluent	Unknown	15	9.0	40	[56]
Methyl red	100	10	13	90	[57]
Methyl orange	82	–	3.6	45	[58]

electrooxidation of organics is strongly related to the type of interaction between the anode surface and hydroxyl radicals. More details about recent mechanistic advances related to electrocatalysis in wastewater treatment can be found in [44].

Electrooxidation of Specific Organic Pollutants on PbO_2

The potential application of PbO_2 electrodes for the electrochemical treatment of wastewater containing different types of organic pollutants is indicated by many instances reported since 2007, as summarized in Table 1. Other results related to PbO_2 anodes, as well as to other electrode materials, can be found in some reviews [4, 46, 47].

Future Directions

As extensively discussed, the OER overpotential is a very important parameter that determines an anode oxidation power. Hence, modifications of PbO_2 films that lead to a higher OER overpotential are desirable. On the other hand, all the PbO_2 electrodes used in the electrooxidation of organics were planar, with one recent exception: films electrodeposited on reticulated vitreous carbon [58]. The obtention of stable tridimensional electrodes that present a good performance in the electrooxidation of organics is still a challenge; in flow reactors, such materials could certainly be very advantageous.

Cross-References

- Organic Pollutants, Direct Electrochemical Oxidation
- Organic Pollutants in Water Using DSA Electrodes, In-Cell Mediated (via Active Chlorine) Electrochemical Oxidation
- Organic Pollutants in Water Using SnO_2, Direct Electrochemical Oxidation
- Wastewater Treatment, Electrochemical Design Concepts

References

1. Rajeshwar K, Ibanez JG (1997) Environmental electrochemistry: fundamentals and applications in pollution abatement. Academic, San Diego

2. Walsh FC (2001) Electrochemical technology for environmental treatment and clean energy conversion. Pure Appl Chem 73:1819–1837
3. Chen G (2004) Electrochemical technologies in wastewater treatment. Sep Purif Technol 38:11–41
4. Martinez-Huitle CA, Ferro S (2006) Electrochemical oxidation of organic pollutants for the wastewater treatment: direct and indirect processes. Chem Soc Rev 35:1324–1340
5. Comninellis C (1994) Electrocatalysis in the electrochemical conversion/combustion of organic pollutants for waste water treatment. Electrochim Acta 39:1857–1862
6. Kötz R, Stucki S, Carcer B (1991) Electrochemical waste water treatment using high overvoltage anodes. Part I: physical and electrochemical properties of SnO_2 anodes. J Appl Electrochem 21:14–20
7. Comninellis C, Pulgarin C (1993) Electrochemical oxidation of phenol for waste water treatment using SnO_2 anodes. J Appl Electrochem 23:108–112
8. Panizza M, Cerisola G (2005) Application of diamond electrodes to electrochemical processes. Electrochim Acta 51:191–199
9. Tahar NB, Savall A (1998) Mechanistic aspects of phenol electrochemical degradation by oxidation on a Ta/PbO_2 anode. J Electrochem Soc 145: 3427–3434
10. Tahar NB, Savall A (1999) A comparison of different lead dioxide coated electrodes for the electrochemical destruction of phenol. J New Mater Electrochem Syst 2:19–26
11. Johnson DC, Feng J, Houk LL (2000) Direct electrochemical degradation of organic wastes in aqueous media. Electrochim Acta 46:323–330
12. Iniesta J, Gonzales-García J, Expósito E, Montiel V, Aldaz A (2001) Influence of chloride ion on electrochemical degradation of phenol in alkaline medium using bismuth doped and pure PbO_2 anodes. Water Res 35:3291–3300
13. Iniesta J, Expósito E, Gonzalez-García J, Montiel V, Aldaz A (2002) Electrochemical treatment of industrial wastewater containing phenols. J Electrochem Soc 149:D57–D62
14. Andrade LS, Ruotolo LAM, Rocha-Filho RC, Bocchi N, Biaggio SR, Iniesta J, García-García V, Montiel V (2007) On the performance of Fe and Fe, F doped Ti Pt/PbO2 electrodes in the electrooxidation of the blue reactive 19 dye in simulated textile wastewater. Chemosphere 66:2035–2043
15. Andrade LS, Rocha-Filho RC, Bocchi N, Biaggio SR, Iniesta J, García-García V, Montiel V (2008) Degradation of phenol using Co- and Co, F-doped PbO_2 anodes in electrochemical filter-press cells. J Hazard Mater 153:252–260
16. Andrade LS, Tasso TT, Silva DL, Rocha-Filho RC, Bocchi N, Biaggio SR (2009) On the performances of lead dioxide and boron-doped diamond electrodes in the anodic oxidation of simulated wastewater containing the Reactive Orange 16 dye. Electrochim Acta 54:2024–2030
17. Feng J, Johnson DC, Lowery SN, Carey JJ (1994) Electrocatalysis of anodic oxygen-transfer reaction: evolution of ozone. J Electrochem Soc 141:2708–2711
18. Velichenko AB, Amadelli R, Zucchini GL, Girenko DV, Danilov FI (2000) Electrosynthesis and physicochemical properties of Fe-doped lead dioxide electrocatalysis. Electrochim Acta 45:4341–4350
19. Treimer SE, Feng JR, Scholten MD, Johnson DC, Davenport AJ (2001) Comparison of voltammetric responses of toluene and xylenes at iron(III)-doped, bismuth(V)-doped, and undoped b-lead dioxide film electrodes in 050 M H_2SO_4. J Electrochem Soc 148: E459–E463
20. Kawagoe KT, Johnson DC (1994) Oxidation of phenol and benzene at Bi-doped lead dioxide electrodes in acid solutions. J Electrochem Soc 141:3404–3409
21. Velichenko AB, Amadelli R, Baranova EA, Girenko DV, Danilov FI (2002) Electrodeposition of Co-doped lead dioxide and its physicochemical properties. J Electroanal Chem 527:56–64
22. Amadelli R, Armelao L, Velichenko AB, Nikolenko NV, Girenko DV, Kovalyov SV, Danilov FI (1999) Oxygen and ozone evolution at fluoride modified lead dioxide electrodes. Electrochim Acta 45:713–720
23. Ai SY, Gao MN, Zhang W, Sun ZD, Jin LT (2003) Preparation of fluorine-doped lead dioxide modified electrodes for electroanalytical applications. Electroanalysis 15:1403–1409
24. De Sucre VS, Watkinson AP (1981) Anodic oxidation of phenol for waste water treatment. Can J Chem Eng 59:52–59
25. Idbelkas B, Takky D (2001) Electrochemical treatment of waste water containing phenol: a comparative study on lead dioxide and platinum electrodes. Ann Chim – Sci Mater 26:33–44
26. Laurindo EA, Bocchi N, Rocha-Filho RC (2000) Production and characterization of Ti/PbO_2 electrodes by a thermal-electrochemical method. J Braz Chem Soc 11:429–433
27. Andrade LS, Laurindo EA, Oliveira RV, Rocha-Filho RC, Cass QB (2006) Development of a HPLC method to follow the degradation of phenol by electrochemical or photoelectrochemical treatment. J Braz Chem Soc 17:369–373
28. Caldara F, Delmastro A, Maja M (1980) Properties of lead dioxide doped with antimony. J Electrochem Soc 127:1869–1876
29. Yeo I, Johnson DC (1987) Electrocatalysis of anodic oxygen-transfer reaction: effect of groups IIIA and VA metal oxides in electrodeposited b-lead dioxide electrodes in acidic media. J Electrochem Soc 134:1973–1977
30. LaCourse WR, Hsiao Y, Johnson DC, Weber WH (1989) Electrocatalytic oxidations at electrodeposited

bismuth(III)-doped beta-lead dioxide film electrodes. J Electrochem Soc 136:3714–3719
31. Feng J, Johnson DC (1990) Electrocatalysis of anodic oxygen-transfer reaction: Fe-doped beta-lead dioxide electrodeposited on noble metals. J Electrochem Soc 137:507–510
32. Larew LA, Gordon JS, Hsiao Y, Johnson DC, Buttry DA (1990) Electrocatalysis of anodic oxygen-transfer reaction: Application of an electrochemical quartz crystal microbalance to a study of pure and bismuth-doped beta-lead dioxide film electrodes. J Electrochem Soc 137:3071–3078
33. Pamplin KL, Johnson DC (1996) Electrocatalysis of anodic oxygen-transfer reaction: oxidation of Cr(III) to Cr(VI) at Bi(V)-doped PbO_2-film electrodes. J Electrochem Soc 143:2119–2115
34. Delmastro A, Maja M (1984) Some characteristics of PbO_2 doped with various elements. J Electrochem Soc 131:2756–2760
35. Thiagarajan N, Nagalingam N (1990) Electrodeposition of lead dioxide on titanium substrates. Bull Electrochem 6:604–605
36. Velichenko AB, Girenko DV, Kovalyov SV, Gnatenko AN, Amadelli R, Danilov FI (1998) Lead dioxide electrodeposition and its application: influence of fluoride and iron ions. J Electroanal Chem 454:203–208
37. Amadelli R, Velichenko AB (2001) Lead dioxide electrodes for high potential anodic processes. J Serb Chem Soc 66:835–845
38. Hill RJ (1982) The crystal-structures of lead dioxides from the positive plate of the lead acid battery. Mater Res Bull 17:769–784
39. Ruetschi P, Angstadt RT, Cahan BD (1959) Oxygen overvoltage and electrode potentials of alpha-PbO_2 and beta-PbO_2. J Electrochem Soc 106:547–551
40. Yeo I, Kim S, Jacobson R, Johnson DC (1989) Electrocatalysis of anodic oxygen-transfer reaction: comparison of structural data with electrocatalytic phenomena for bismuth-doped lead dioxide. J Electrochem Soc 136:1395–1401
41. Simond O, Schaller V, Comninellis C (1997) Theoretical model for the anodic oxidation of organics on metal oxide electrodes. Electrochim Acta 42:2009–2012
42. Kapalka A, Fóti G, Comninellis C (2008) Kinetic modelling of the electrochemical mineralization of organic pollutants for wastewater treatment. J Appl Electrochem 38:7–16
43. Marselli B, García-Gomez J, Michaud PA, Rodrigo MA, Comninellis C (2003) Electrogeneration of hydroxyl radicals on boron-doped diamond electrodes. J Electrochem Soc 150:D79–D83
44. Martínez-Huitle CA, Andrade LS (2011) Electrocatalysis in wastewater treatment: recent mechanism advances. Quim Nova 34:850–858
45. Zhu X, Tong M, SHi S, Zhao H, Ni J (2008) Essential explanation of the strong mineralization performance of boron-doped diamond electrodes. Environ Sci Technol 42:4914–4920
46. Panizza M, Cerisola G (2009) Direct and mediated anodic oxidation of organic pollutants. Chem Rev 109:6541–6569
47. Martinez-Huitle CA, Brillas E (2009) Decontamination of wastewaters containing synthetic organic dyes by electrochemical methods: a general review. Appl Catal B 87:105–145
48. Zheng Y, Su W, Chen S, Wu X, Chen X (2011) Ti/SnO_2-Sb_2O_5-RuO_2/a-PbO_2/b-PbO_2 electrodes for pollutants degradation. Chem Eng J 174:304–309
49. Yang X, Zou R, Huo F, Caia D, Xia D (2009) Preparation and characterization of Ti/SnO_2-Sb_2O_3-Nb_2O_5/PbO_2 thin film as electrode material for the degradation of phenol. J Hazard Mater 164:367–373
50. Zhou M, He J (2008) Degradation of cationic red X-GRL by electrochemical oxidation on modified PbO_2 electrode. J Hazard Mater 153:357–363
51. Tan C, Xiang B, Li Y, Fang J, Huang M (2011) Preparation and characteristics of a nano PbO_2 anode for organic wastewater treatment. Chem Eng J 166:15–21
52. Song S, Zhan L, He Z, Lin L, Tu J, Zhang Z, Chen J, Xu L (2010) Mechanism of the anodic oxidation of 4-chloro-3-methyl phenol in aqueous solution using Ti/SnO_2-Sb/PbO_2 electrodes. J Hazard Mater 175:614–621
53. Aquino JM, Rocha-Filho RC, Bocchi N, Biaggio SR (2010) Electrochemical degradation of the Reactive Red 141 dye on a β-PbO_2 anode assessed by the response surface methodology. J Braz Chem Soc 21:324–330
54. Aquino JM, Rocha-Filho RC, Bocchi N, Biaggio SR (2010) Electrochemical degradation of the Acid Blue 62 dye on a β-PbO_2 anode assessed by the response surface methodology. J Appl Electrochem 40:1751–1757
55. Aquino JM, Irikura K, Rocha-Filho RC, Bocchi N, Biaggio SR (2010) A comparison of electrodeposited Ti/β-$PbO_2$and Ti-Pt/β-PbO_2anodes in the electrochemical degradation of the Direct Yellow 86 Dye. Quim Nova 33:2124–2129
56. Aquino JM, Pereira GF, Rocha-Filho RC, Bocchi N, Biaggio SR (2011) Electrochemical degradation of a real textile effluent using boron-doped diamond or β-PbO_2 as anode. J Hazard Mater 192:1275–1282
57. Panizza M, Cerisola G (2008) Electrochemical degradation of methyl red using BDD and PbO_2 anodes. Ind Eng Chem Res 47:6816–6820
58. Recio FJ, Herrasti P, Sirés I, Kulak AN, Bavykin DV, Ponce-de-León C, Walsh FC (2011) The preparation of PbO_2 coatings on reticulated vitreous carbon for the electro-oxidation of organic pollutants. Electrochim Acta 56:5158–5165

Organic Pollutants, Direct and Mediated Anodic Oxidation

Marco Panizza
University of Genoa, Genoa, Italy

Introduction

Oxidative electrochemical technologies offer an alternative solution to many environmental problems in the process industry because electrons provide a versatile, efficient, cost effective, easily automatizable and clean reagent. Thanks to intensive investigations that have improved the electrocatalytic activity and stability of electrode materials and optimized reactor geometry, electrochemical technologies have reached a promising state of development and can be effectively used for disinfection and purification of wastewater polluted with organic compounds [1, 2].

The overall performance of the electrochemical processes is determined by the complex interplay of parameters that may be optimized to obtain an effective and economical incineration of pollutants. The principal factors determining the electrolysis performance will be:

(I) Electrode potential and current density: these control which reaction occurs and its rate and commonly determine the efficiency of the process;

(II) Current distribution: determines the spatial distribution of the consumption of reactants and hence must be as homogeneous as possible;

(III) Mass transport regime: a high mass transport coefficient leads to a greater uniformity of pollutant concentration in the reaction layer near the electrode surface and to a generally higher efficiency;

(IV) Cell design: the cell dimension, the presence or the absence of a separator, the design of the electrode, etc. affect the figures of merit of the electrochemical process;

(V) Electrolysis medium: the choice of electrolyte and its concentration, pH, and temperature influence the reaction mechanism and electrogeneration of chemical oxidants;

(VI) Electrode materials: the ideal electrode material for the degradation of organic pollutants should be totally stable in the electrolysis medium, cheap and exhibit high activity towards organic oxidation and low activity towards secondary reactions (e.g. oxygen evolution).

The electrochemical oxidation of organics for wastewater treatment can be obtained in different ways that are schematized in Fig. 1. It has been generally observed that the nature of the electrode material, the experimental conditions and the electrolyte composition strongly influence the oxidation mechanism.

Direct Oxidation

In direct electrolysis, the pollutants are oxidized after adsorption on the anode surface without the involvement of any substance other than the electron, which is a "clean reagent":

$$R_{ads} - ze^{-} \rightarrow P_{ads} \quad (1)$$

Direct electro-oxidation is theoretically possible at low potentials, before oxygen evolution, but the reaction rate usually has low kinetics that depend on the electrocatalytic activity of the anode. High electrochemical rates have been observed using noble metals such as Pt and Pd and metal-oxide anodes such as iridium dioxide, ruthenium-titanium dioxide and iridium-titanium dioxide.

However, the main problem of electro-oxidation at a fixed anodic potential before oxygen evolution is a decrease in the catalytic activity, commonly called the poisoning effect, due to the formation of a polymer layer on the anode surface. This deactivation, which depends on the adsorption properties of the anode surface and the concentration and the nature of the organic compounds, is more accentuated in the

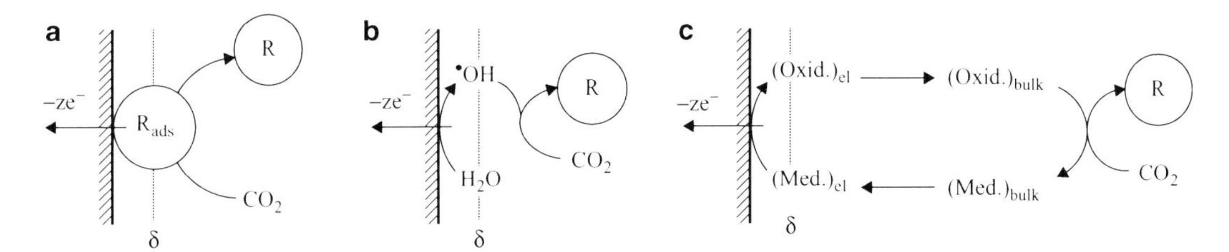

Organic Pollutants, Direct and Mediated Anodic Oxidation, Fig. 1 Scheme of the electrochemical processes for the removal of organic compounds (R): (**a**) direct electrolysis; (**b**) via hydroxyl radicals produced by the discharge of the water; (**c**) indirect electrolysis, where *Med* is the mediator and *Oxid* is the electrogenerated oxidant

presence of aromatic organic substrates such as phenol, chlorophenols, naphthol, etc. [3].

The poisoning effect can be avoided by performing the oxidation in the presence of inorganic oxidation mediators or in the potential region of water discharge with simultaneous oxygen evolution.

Indirect Oxidation

In indirect oxidation, organic pollutants do not exchange electrons directly with the anode surface but through the mediation of some electroactive species regenerated there, which acts as an intermediary for shuttling electrons between the electrode and the organics.

The principal requirements for obtaining high efficiencies in the indirect electrolytic processes are that (i) the potential at which the intermediate species is produced must not be near the potential of oxygen evolution; (ii) the generation rate of the intermediate must be large; (iii) the rate of the reaction of the intermediate species and the pollutant must be higher than the rate of any competing reactions and (iv) adsorption of the pollutant must be minimized.

The mediators of oxidation can be a metallic redox couple or strong oxidizing chemicals.

The indirect electrolysis that uses metallic couples such as Ag(II), Ce (IV), Co(III), and Fe (III) as redox reagents is called Mediated Electrochemical Oxidation (MEO) [4]. In this process, metal ions in acidic solutions are oxidized anodically from their stable oxidation state (M^{n+}) to the higher reactive oxidation state ($M^{(n+1)+}$) in which they attack the other organic feed, breaking it down into carbon dioxide, insoluble inorganic salts and water:

$$M^{n+} \rightarrow M^{(n+1)+} + e^- \quad (2)$$

$$xM^{(n+1)+} + \text{organics} \rightarrow xM^{n+} + y\ CO_2 \quad (3)$$

This reaction brings the couple back to the stable state (M^{n+}), and it is then recycled through the cell to continuously regenerate further reactions. This process has the advantages of operating at low temperatures (up to 90 °C) and near-atmospheric pressure, it is easily controllable by switching off the power supply and it does not produce dioxins. The main drawback of the use of a solution redox couple is the need to subsequently separate the oxidation products from the mediator.

Beside MEO, organic pollutants can be destroyed by indirect electrolysis by generating strong oxidizing chemicals in situ. Chlorine and hypochlorite, whose use is well established in the disinfection of potable and swimming pool waters and paper and pulp bleaching, have also found wide applications as electro-oxidation mediators for wastewater treatment [5–7]. Gaseous chlorine is anodically produced by the oxidation of chloride ions according to the following reaction:

$$2Cl^- \rightarrow Cl_2 + 2e^- \quad (4)$$

As a function of pH, chlorine remains in solution as aqueous chlorine (pH < 3.3) or disproportionates to hypochloric acid (pH < 7.5) or hypochlorite ions (pH > 7.5):

$$Cl_2 + H_2O \rightarrow HOCl + H^+ + Cl^- \quad (5)$$

$$HOCl \rightarrow H^+ + OCl^- \quad (6)$$

The more suitable electrode materials for in situ generation of active chlorine are based on platinum or on a mixture of metal oxides (e.g. RuO_2, TiO_2, IrO_2) that have good electrocatalytic properties for chlorine evolution and long-term mechanical and chemical stability. This process can effectively oxidize many pollutants; however it has the drawback of permitting the formation of chlorinated organic compounds during the electrolysis.

Organic pollutants can also be treated by electrochemically assisted Fenton's reaction (reaction 8), where hydrogen peroxide is generated in situ from the two-electron reduction of O_2 on cathodes such as gas diffusion electrodes (GDE), reticulated vitreous carbon (RVC) or graphite-felt following the reaction 7 [8, 9]:

$$O_2 + 2H^+ + 2e^- \rightarrow H_2O_2 \quad (7)$$

$$Fe^{2+} + H_2O_2 \rightarrow Fe^{3+} + OH^- + OH \quad (8)$$

Reaction 8 is propagated through the continuous regeneration of ferrous iron at the cathode (reaction 9), thus avoiding Fe^{3+} accumulation in the medium and consequently eliminating the production of iron sludge:

$$Fe^{3+} + e^- \rightarrow Fe^{2+} \quad (9)$$

Oxidation via Intermediates of Oxygen Evolution

Organic compounds (R) can be also oxidized performing the electrolysis at high anodic potentials in the region of water discharge due to the participation of intermediates of oxygen evolution:

$$M + H_2O \rightarrow M(OH) + H^+ + e^- \quad (10)$$

$$R + M\ (OH) \rightarrow M + CO_2 + H_2O \quad (11)$$

This process has an advantage over indirect electrolysis in that it does not need to add oxidation catalysts to the solution and does not produce byproducts; however the current efficiency is diminished by the secondary reaction of oxygen evolution occurring during the oxidation:

$$M(OH) \rightarrow M + {}^1/_2O_2 + H^+ + e^- \quad (12)$$

The nature of the electrode material strongly influences both the selectivity and the efficiency of the process; in particular anodes with low oxygen evolution overpotential, such as IrO_2, RuO_2 or Pt, have "active" behavior and favor the partial and selective oxidation of pollutants (i.e. conversion), while anodes with high oxygen evolution overpotential, such as SnO_2, PbO_2 or boron-doped diamond (BDD), have "non-active" behavior and so are ideal electrodes for the complete oxidation of organics to CO_2 in wastewater treatment [3, 10–12].

In particular, BDD electrodes have received great attention because they exhibit good chemical and electrochemical stability, a long life and a wide potential window for water discharge and are thus promising anodes for industrial-scale wastewater treatment. Using BDD anodes, the current efficiency and the amount of intermediates are strongly affected by the experimental conditions. In particular, for high concentrations of organic compounds or low current densities, chemical oxygen demand (COD) decreases linearly, forming a large amount of intermediates, whereas current efficiency remains at about 100 %, indicating a kinetically controlled process. Conversely, for low concentrations of organic compounds or high current densities,

Organic Pollutants, Direct and Mediated Anodic Oxidation, Table 1 Some examples of organic compounds oxidized on BDD anodes

Pollutants	Experimental conditions	Ref.
Phenol	$i = 5–60$ mA cm^{-2}; concentration: 20 mM	[11]
4-chlorophenol	$i = 15–60$ mA cm^{-2}; concentration: 3.9–15.6 mM; T = 25–70 °C	[3, 13]
Naphthol	$i = 15–60$ mA cm^{-2}; concentration: 2–9 mM; T = 30–60 °C	[14]
Anionic surfactants	$i = 25–75$ mA cm^{-2}; flow-rate: 60–180 dm^{-3} h^{-1}; concentration: 750 mg dm^{-3}	[10]
Landfill leachate	Initial COD = 350 mg dm^{-3}; $i = 10–60$ mA cm^{-2}	[15–17]
Mecoprop herbicide	$i = 6–40$ mA cm^{-2}; flow-rate: 75–300 dm^{-3} h^{-1}; concentration: 180–700 mg dm^{-3}	[18]
3-methylpyridine	$i = 2.5–60$ mA cm^{-2}; concentration: 5 mM	[19]
Synthetic dyes	Methylene Blue, Alizarin Red, Eriochrome Black T, Methyl Red, Acid Yellow 1, Acid Blue 22	[20–25]

pollutants are directly mineralized to CO_2, but current efficiency is below 100 %, due to mass transport limitation and side reactions of oxygen evolution. Some examples of organic compounds treated using a BDD anode are shown in Table 1.

Future Directions

Although laboratory and pilot tests have been successful, industrial applications of these methods are still limited, due to the relatively high energy consumption of the electrochemical methods. However, thanks to the development of new electrode materials, electrochemical oxidation could be increasingly applied in the future, due to specific advantages for certain applications over other technologies. Moreover, energy consumption could be reduced using so-called "advanced electrochemical oxidation processes", based on the combination of anodic and cathodic electrogeneration of highly oxidizing hydroxyl radicals.

Cross-References

- ▶ Boron-Doped Diamond for Green Electro-Organic Synthesis
- ▶ Electro-Fenton Process for the Degradation of Organic Pollutants in Water

References

1. Panizza M, Cerisola G (2009) Direct and mediated anodic oxidation of organic pollutants. Chem Rev 109:6541–6569
2. Martínez-Huitle CA, Ferro S (2006) Electrochemical oxidation of organic pollutants for the wastewater treatment: direct and indirect processes. Chem Soc Rev 12:1324–1340
3. Rodrigo MA, Michaud PA, Duo I, Panizza M, Cerisola G, Comninellis C (2001) Oxidation of 4-chlorophenol at boron-doped diamond electrodes for wastewater treatment. J Electrochem Soc 148:D60–D64
4. Steele DF (1990) Electrochemical destruction of toxic organic industrial waste. Platin Met Rev 34:10–14
5. Bonfatti F, Ferro S, Lavezzo F, Malacarne M, Lodi G, De Battisti A (2000) Electrochemical incineration of glucose as a model organic substrate. II. Role of active chlorine mediation. J Electrochem Soc 147:592–596
6. Panizza M, Cerisola G (2003) Electrochemical oxidation of 2-naphthol with in situ electrogenerated active chlorine. Electrochim Acta 48:1515–1519
7. Martinez-Huitle CA, Ferro S, De Battisti A (2005) Electrochemical incineration in the presence of halides. Electrochem Solid-State 8:D35–D39
8. Brillas E, Sirés I, Oturan MA (2009) Electro-Fenton process and related electrochemical technologies based on Fenton's reaction chemistry. Chem Rev 109:6570–6631
9. Ponce de Leon C, Pletcher D (1995) Removal of formaldehyde from aqueous solutions via oxygen reduction using a reticulated vitreous carbon cathode cell. J Appl Electrochem 25:307–314
10. Panizza M, Cerisola G (2005) Application of diamond electrodes to electrochemical processes. Electrochim Acta 51:191–199

11. Iniesta J, Michaud PA, Panizza M, Cerisola G, Aldaz A, Comninellis C (2001) Electrochemical oxidation of phenol at boron-doped diamond electrode. Electrochim Acta 46:3573–3578
12. Cossu R, Polcaro AM, Lavagnolo MC, Mascia M, Palmas S, Renoldi F (1998) Electrochemical treatment of landfill leachate: oxidation at Ti/PbO_2 and Ti/SnO_2 anodes. Environ Sci Technol 32:3570–3573
13. Gherardini L, Michaud PA, Panizza M, Comninellis C, Vatistas N (2001) Electrochemical oxidation of 4-chlorophenol for wastewater treatment. Definition of normalized current efficiency. J Electrochem Soc 148:D78
14. Panizza M, Michaud PA, Cerisola G, Comninellis C (2001) Anodic oxidation of 2-naphthol at boron-doped diamond electrodes. J Electroanal Chem 507:206
15. Panizza M, Zolezzi M, Nicolella C (2006) Biological and electrochemical oxidation of naphthalene sulfonates in a contaminated site leachate. J Chem Technol Biotechnol 81:225–232
16. Anglada A, Urtiaga A, Ortiz I, Mantzavinos D, Diamadopoulos E (2011) Boron-doped diamond anodic treatment of landfill leachate: evaluation of operating variables and formation of oxidation by-products. Water Res 45:828–838
17. Anglada A, Urtiaga AM, Ortiz I (2010) Laboratory and pilot plant scale study on the electrochemical oxidation of landfill leachate. J Hazard Mater 181:729–735
18. Sirés I, Brillas E, Cerisola G, Panizza M (2008) Comparative depollution of mecoprop aqueous solutions by electrochemical incineration using BDD and PbO_2 as high oxidation power anodes. J Electroanal Chem 613:151–159
19. Iniesta J, Michaud PA, Panizza M, Comninellis C (2001) Electrochemical oxidation of 3-methylpyridine at a boron-doped diamond electrode: application to electroorganic synthesis and wastewater treatment. Electrochem Commun 3:346–351
20. Panizza M, Cerisola G (2008) Removal of colour and COD from wastewater containing acid blue 22 by electrochemical oxidation. J Hazard Mater 153:83–88
21. Panizza M, Barbucci A, Ricotti R, Cerisola G (2007) Electrochemical degradation of methylene blue. Sep Purif Technol 54:382–387
22. Sáez C, Panizza M, Rodrigo MA, Cerisola G (2007) Electrochemical incineration of dyes using a boron-doped diamond anode. J Chem Technol Biotechnol 82:575–581
23. Rodriguez J, Rodrigo MA, Panizza M, Cerisola G (2009) Electrochemical oxidation of Acid Yellow 1 using diamond anode. J Appl Electrochem 39:2285–2289
24. Panizza M, Cerisola G (2008) Electrochemical degradation of methyl red using BDD and PbO_2 anodes. Ind Eng Chem Res 47:6816–6820
25. Panizza M, Cerisola G (2007) Electrocatalytic materials for the electrochemical oxidation of synthetic dyes. Appl Catal B-Environ 75:95–101

Organic Pollutants, Direct Electrochemical Oxidation

Christos Comninellis[1], Agnieszka Kapałka[1], Stéphane Fierro[1], György Fóti[1], Pierre-Alain Michaud[1] and Petros Dimitriou-Christidis[2]
[1]Institute of Chemical Sciences and Engineering, Ecole Polytechnique Fédérale de Lausanne (EPFL), Lausanne, Switzerland
[2]Environmental Chemistry Modeling Laboratory, Ecole Polytechnique Fédérale de Lausanne (EPFL), Lausanne, Switzerland

Introduction

Biological treatment of polluted water is the most economical process used for the elimination of "readily degradable" organic pollutants present in wastewater. The situation is completely different when the wastewater contains toxic and refractory (i.e., resistant to biological treatment) organic pollutants. One interesting possibility is to use a coupled process: partial oxidation–biological treatment. The goal is to decrease the toxicity and to increase the biodegradability of the wastewater before biological treatment. The optimization of this coupled process is complex, however, and complete mineralization of the organic pollutants is usually preferred.

The electrochemical method for mineralization of organic pollutants is a new technology and has attracted a great deal of attention recently. The technology is interesting for the treatment of dilute wastewater (COD < 5 g/l), and it is in competition with the process of chemical oxidation using strong oxidants (O_3, H_2O_2/Fe^{2+}, etc). The main advantage of this technology is that chemicals are not used. In fact, only electrical energy is consumed for the mineralization of organic pollutants.

The aim of the present work is to elucidate the fundamentals of direct electrochemical oxidation (mineralization), including thermodynamics, reaction mechanisms and reaction kinetics.

Thermodynamics of the Electrochemical Mineralization (EM) of Organics

Thermodynamically, the electrochemical mineralization (EM) of any soluble organic compound in water should be achieved at low potentials, widely before the thermodynamic potential of water oxidation to molecular oxygen (1.23V/SHE under standard conditions), as given by Eq. 1:

$$2H_2O \rightarrow O_2 + 4H^+ + 4e^- \quad (1)$$

A typical example of EM is the anodic oxidation of acetic acid to CO_2 (Eq. 2).

$$CH_3COOH(aq) + 2H_2O(l) \rightarrow 2CO_2(g) + 8H^+(aq) + 8e^- \quad (2)$$

The thermodynamic potential of this reaction (E°) under standard conditions (1 mol/L CH_3COOH, P = 1 atm, T = 298 K) can be calculated using Eq. 3:

$$E^\circ = \frac{-\Delta_r G^\circ}{n \cdot F} = \frac{-(-81.89 \times 10^3)}{8 \times 96485} = 0.10V/SHE \quad (3)$$

Where:
$\Delta_r G^\circ$ = standard free energy of the reaction ($\Delta_r G^\circ = -81.89$ kJ/mol)
n = number of electrons exchanged (n = 8)
F = Faraday's constant.

Taking this result into consideration, it will be theoretically possible to treat an aqueous organic pollutant stream (in this case acetic acid) in a fuel cell with the cogeneration of electrical energy. In this device, the organic pollutant is oxidized at the anode (Eq. 2 in the case of acetic acid), and oxygen is reduced at the cathode (Eq. 4):

$$O_2 + 4H^+ + 4e^- \rightarrow 2H_2O \quad (4)$$

The total reaction of this fuel cell, considering acetic acid as a fuel, is given by Eq. 5:

$$CH_3COOH + 2O_2 \rightarrow 2CO_2 + 2H_2O \quad (5)$$

The standard thermodynamic potential of the cell based on Eq. 5, i.e. incineration of the considered pollutant (acetic acid) with cogeneration of electrical energy, can be calculated using the relationship (Eq. 6):

$$E^\circ_{cell} = \frac{-\Delta_r G^\circ}{n \cdot F} = \frac{-(-866.6 \times 10^3)}{8 \times 96485} = 1.12 \text{ V} \quad (6)$$

where $\Delta_r G^\circ$ is the standard free energy change of reaction 5.

Similar values are obtained for the EM of other organic compounds [1].

In contrast to such promising thermodynamic data, the kinetics of electrochemical incineration is very slow and, in practice, can be achieved close to the thermodynamic potentials only in very limited cases. In fact, only platinum-based electrodes can allow EM of simple C_1 organic compounds. A typical example is the use of Pt–Ru catalyst in the electrochemical mineralization of methanol. This system is largely studied for fuel cell applications.

In conclusion, in the actual state of the art, electrochemical mineralization of organic pollutants with cogeneration of electrical energy is not feasible, due to the lack of active electrocatalytic anode material. Bioelectrocatalysis is a new and active field and can overcome this problem as has been demonstrated recently in the development of biofuel cells. Nevertheless, this technology is yet in its infancy.

Recently, however [1–19], we have demonstrated that the electrochemical oxidation of organics can be achieved via the intermediates involved in the oxygen evolution reaction at potentials largely above the thermodynamic potential of oxygen evolution (1.23V/SHE under standard conditions). Even if in this process electrical energy should be consumed, this system opens new possibilities for treatment at room temperature of very toxic organic pollutants present in industrial wastewater.

Mechanism of the Electrochemical Oxidation (Mineralization) of Organics via the Intermediates of O_2 Evolution Reactions

According to this mechanism in acid media, water is discharged at the anode, producing hydroxyl radicals (Eq. 7). These hydroxyl radicals compete with the reaction of oxygen evolution (Eq. 8) and the reaction of organics (RH) oxidation (Eq. 9):

$$H_2O + M \rightarrow M\left({}^{\bullet}OH\right) + H^+ + e^- \quad (7)$$

$$M\left({}^{\bullet}OH\right) \rightarrow M + H^+ + e^- + \frac{1}{2}O_2 \quad (8)$$

$$RH(aq) + M\left({}^{\bullet}OH\right)_n \rightarrow M + \text{ oxidation products} \quad (9)$$

We consider here that RH is not adsorbed on the anode surface and that the hydroxyl radicals are chemisorbed or physisorbed (depending on anode material) on the anode surface.

The electrochemical activity (rate of reaction 8) and the chemical reactivity (rate of reaction 9) of these electrogenerated hydroxyl radicals are strongly linked to their interaction with the electrode surface M. As a general rule, the weaker the interaction, the lower the electrochemical activity (reaction 8 is slow) toward oxygen evolution (high O_2 overvoltage anodes) and the higher the chemical reactivity toward oxidation of organics (reaction 9 is fast). On the basis of this approach, we can classify the different anode materials according to their oxidation power in acid media [1, 15].

A low oxidation power anode is characterized by a strong electrode–hydroxyl radical interaction, resulting in a high electrochemical activity for the oxygen evolution reaction (low overvoltage anode) and in a low chemical reactivity for oxidation of organics (low current efficiency for oxidation of organics). A typical low oxidation power anode is the IrO_2-based electrode. In regard to this anode, it has been reported that the IrO_2–hydroxyl radical interaction is so strong (chemisorbed hydroxyl radicals) that a higher oxidation state oxide IrO_3 can be formed [18, 19]. This higher oxide can act as a mediator for both oxidation of organics and oxygen evolution [19].

In contrast to this low oxidation power anode, a high oxidation power anode is characterized by a weak electrode–hydroxyl radical interaction, resulting in a low electrochemical activity for the oxygen evolution reaction (high overvoltage anode) and a high chemical reactivity for organic oxidation (high current efficiency for organic oxidation).

A boron-doped diamond-based anode (BDD) is a typical high oxidation power anode. It has been reported that the BDD–hydroxyl radical interaction is so weak (no free p or d orbitals on BDD) that the hydroxyl radicals can even be considered as quasi-free physisorbed hydroxyl radicals.

These quasi-free hydroxyl radicals are very reactive and can result in mineralization of the organic compounds [1, 4–9, 11–15, 17].

Electrogeneration of Free Hydroxyl Radicals on BDD Electrodes

The evidence for the formation of free hydroxyl radicals on BDD in acid solutions was found by Marselli et. al. [13] using 5.5-dimethyl-1-pyrroline-N-oxide (DMPO) as a spin trap (Eq. 10):

H_3C, H_3C, N^+, O^- + HO• ⟶ H_3C, H_3C, N, OH, O• (10)

DMPO spin trap — DMPO hydroxyl radical spin adduct

In the absence of reactive compounds, one possible reaction of these electrogenerated highly reactive hydroxyl radicals (Eq. 7) is their recombination into hydrogen peroxide (Eq. 11). Indeed, H_2O_2 has been formed during electrolysis of $HClO_4$ solution on BDD anodes.

$$HO^{\bullet} + HO^{\bullet} \rightarrow H_2O_2 \quad (11)$$

Hydrogen peroxide can be further oxidized to oxygen, either by its direct discharge on the electrode surface (Eq. 12) or mediated by hydroxyl radicals (Eq. 13).

$$H_2O_2 \rightarrow O_2 + 2H^+ + 2e^- \quad (12)$$

$$H_2O_2 + 2HO^{\bullet} \rightarrow O_2 + 2H_2O \quad (13)$$

Assuming that the recombination reaction (Eq. 11) is the rate-determining step, the concentration profile of hydroxyl radicals as a function of distance from the electrode surface during oxygen evolution can be calculated as (Eq. 14 and 15) [17]:

$$c^s_{HO^{\bullet}} = \frac{3D_{HO^{\bullet}}}{2k_{HO^{\bullet}}\left(x + \sqrt{\frac{3D_{HO^{\bullet}}}{2k_{HO^{\bullet}}c^s_{HO^{\bullet}}}}\right)^2} \quad (14)$$

$$c^s_{HO^{\bullet}} = \sqrt[3]{\frac{j^2}{2.67F^2k_{HO^{\bullet}}D_{HO^{\bullet}}}} \quad (15)$$

where $D_{HO^{\bullet}}$ ($m^2\ s^{-1}$) is the diffusion coefficient of $HO^{\bullet}$, $k_{HO^{\bullet}}$ ($m^3\ mol^{-1}\ s^{-1}$) is the rate constant of the recombination reaction (Eq. 11), $c_{HO^{\bullet}}$($mol\ m^{-3}$) is the concentration of $HO^{\bullet}$, $c_{HO^{\bullet}}$ is the electrode surface $HO^{\bullet}$ concentration, and j is the applied current density.

Figure 1 shows the simulated concentration profile of hydroxyl radicals (Eq. 14 and 15) during oxygen evolution as a function of the distance from the electrode surface. It can be seen that, for a current density of $j = 300\ A\ m^{-2}$, the thickness of the reaction layer is about 1 μm, whereas the maximum (surface) concentration of hydroxyl radicals reaches values in the range of several tenths of μM.

Using a similar approach, it is possible to determine the concentration profile of hydroxyl radicals during oxidation of organic compounds (Eq. 9), assuming that (i) the concentration of organic compounds is high enough to be considered as a constant within the reaction layer and (ii) oxygen evolution, via H_2O_2 oxidation, is negligible.

The surface concentration $c_{HO^{\bullet}_R}{}^s$ and the concentration profile $c_{HO^{\bullet}_R}$ of hydroxyl radicals during oxidation of organics can be given by the relationships (Eq. 16 and 17, respectively) [17]:

$$c^s_{HO^{\bullet}_R} = \frac{j}{F\sqrt{zk_Rc_RD_{HO^{\bullet}}}} \quad (16)$$

$$c_{HO^{\bullet}_R} = c^s_{HO^{\bullet}_R}\ \exp\left(-\sqrt{\frac{zk_Rc_R}{D_{HO^{\bullet}}}}x\right) \quad (17)$$

where k_R ($m^3\ mol^{-1}\ s^{-1}$) is the rate constant of oxidation of the target organic compound R with $HO^{\bullet}$, and c_R ($mol\ m^{-3}$) is the concentration of R.

The thickness of the reaction layer (reaction cage) depends on the concentration of organic compounds, the rate constant of organic oxidation (via hydroxyl radicals), and the applied current density. As a typical example, Fig. 2 shows the simulated concentration profile during the oxidation of formic acid (0.25–1 M) at $300\ A\ m^{-2}$.

It can be seen that the higher the formic acid concentration, the lower the surface concentration and the smaller the thickness of the reaction layer. For the range of formic acid concentrations investigated in this work, the thickness of the reaction layer (reaction cage) drops to barely tenths of nanometers (Fig. 2), which is significantly lower when compared with that in the absence of organics (Fig. 1).

Kinetic Model of Oxidation of Organics (Mineralization) on BDD Anodes

In this section, a kinetic model for the electrochemical oxidation (mineralization) of organics

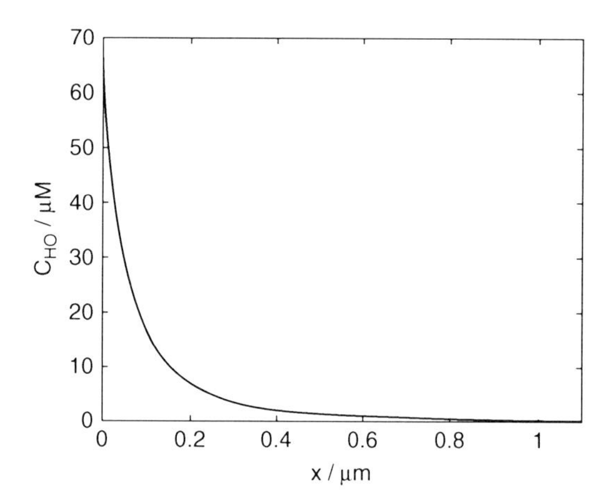

Organic Pollutants, Direct Electrochemical Oxidation, Fig. 1 Simulated concentration profile of hydroxyl radicals during oxygen evolution according to Eq. 14 and 15. $j = 300\ A\ m^{-2}$; $D_{HO\bullet} = 2.2 \cdot 10^{-9}\ m^2\ s^{-1}$; $k_{HO\bullet} = 5.5 \cdot 10^9\ M^{-1}\ s^{-1}$

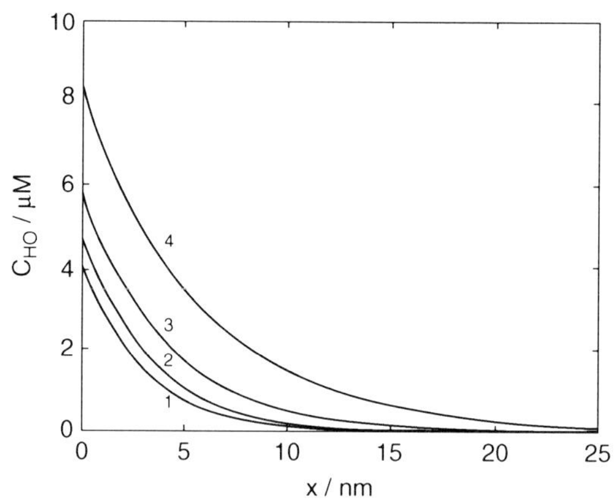

Organic Pollutants, Direct Electrochemical Oxidation, Fig. 2 Simulated concentration profiles of hydroxyl radicals during oxidation of (*1*) 1 M, (*2*) 0.75 M, (*3*) 0.5 M, and (*4*) 0.25 M HCOOH, according to Eq. 16 and 17. $j = 300\ A\ m^{-2}$; $D_{HO\bullet} = 2.2 \cdot 10^{-9}\ m^2\ s^{-1}$; $k_{HCOOH} = 1.3 \cdot 10^8\ M^{-1}\ s^{-1}$; $z = 2$

on BDD anodes, a recirculated batch system is presented [1, 7, 15]. In this model, electrogenerated hydroxyl radicals (Eq. 7) have been considered to be the intermediates for both the main reaction of oxidation of organics (Eq. 9) and the side reaction of oxygen evolution (Eq. 8). The main assumption of this model is that the global rate of the electrochemical oxidation (mineralization) of organic is a fast reaction and can be controlled by mass transport of organics to the anode surface. Furthermore, it has been considered that the rate of oxidation of organics is independent of the chemical nature of the organic compound present in the electrolyte. Under these conditions, the limiting current density for electrochemical oxidation of an organic compound under given hydrodynamic conditions can be written as (Eq. 18):

$$i_{lim} = 4 \cdot F \cdot k_m \cdot COD \qquad (18)$$

where i_{lim} is the limiting current density for oxidation of organics ($A\ m^{-2}$), F is Faraday's constant ($C\ mol^{-1}$), k_m is the average mass transport coefficient ($m\ s^{-1}$), and COD is the initial chemical oxygen demand ($mol\ O_2\ L^{-1}$).

At the beginning of electrolysis ($t = 0$), the initial limiting current density i^o_{lim} is:

$$i^o_{lim} = 4 \cdot F \cdot k_m \cdot COD^\circ \qquad (19)$$

where COD° is the initial chemical oxygen demand.

Furthermore, we have defined the characteristic parameter α of the electrolysis process as (Eq. 20):

$$\alpha = i / i^o_{lim} \qquad (20)$$

where i is the applied current density and i^o_{lim} is the initial limiting current density.

Working under galvanostatic conditions (constant current density), it is possible to identify two different operating regimes: at $\alpha < 1$, electrolysis is controlled by the applied current, while at $\alpha > 1$, it is controlled by mass transport [1, 7, 15].

Electrolysis Under Current Control ($\alpha < 1$)

In this operating regime, $i < i^o_{lim}$ the current efficiency is 100%, and the temporal evolution of COD(t) ($mol\ O_2\ L^{-1}$) is given by:

$$COD(t) = COD^\circ \left(1 - \alpha \frac{A k_m}{V_R} t\right) \qquad (21)$$

where V_R is the electrolyte volume and A is the anode surface area.

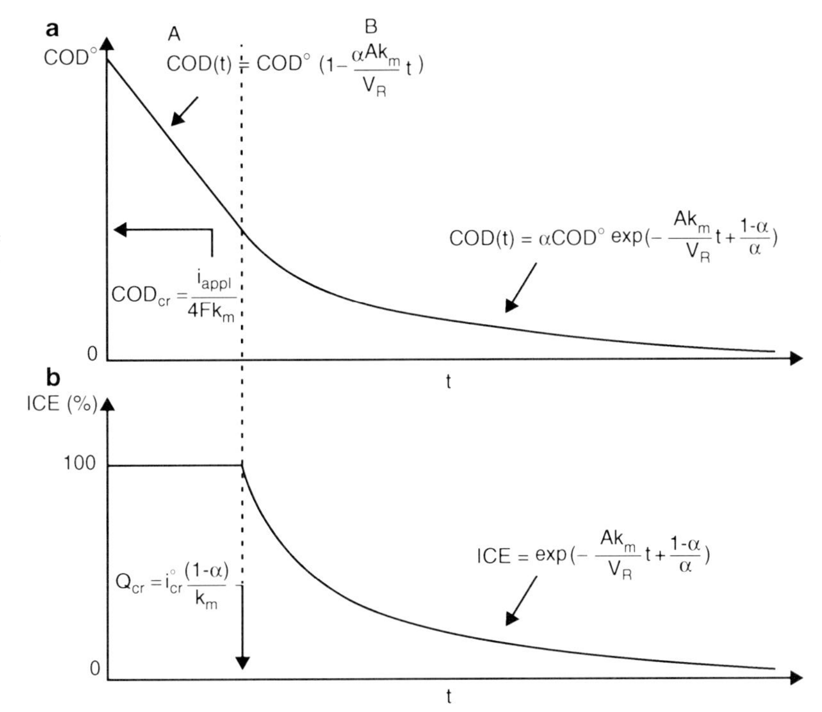

Organic Pollutants, Direct Electrochemical Oxidation, Fig. 3 Simulated profiles for the evolution of (**a**) COD and (**b**) ICE as a function of time; (*A*) represents the charge transport controlled regime (Eq. 21) and (*B*) represents the mass transport controlled regime (Eq. 23 and 24)

This behavior persists until a critical time (t_{cr}, given by Eq. 22), at which the applied current density is equal to the limiting current density:

$$t_{cr} = \frac{1-\alpha}{\alpha}\frac{V_R}{Ak_m} \tag{22}$$

Electrolysis Under Mass Transfer Control ($\alpha > 1$)

When the applied current exceeds the limiting one ($i > i_{lim}$), secondary reactions (such as oxygen evolution) commence, resulting in a decrease of the instantaneous current efficiency (ICE). In this case, the temporal evolution of COD(t) is given by:

$$COD(t) = \alpha COD^{\circ} \quad \exp\left(-\frac{Ak_m}{V_R}t + \frac{1-\alpha}{\alpha}\right) \tag{23}$$

and the instantaneous current efficiency (ICE) by the relationship (Eq. 24):

$$ICE = \exp\left(-\frac{Ak_m}{V_R}t + \frac{1-\alpha}{\alpha}\right) \tag{24}$$

In Fig. 3, simulated profiles for the evolution of COD and ICE as a function of time are given (Eq. 21, 22, 23, and 24).

Using this model, almost 100 % current efficiency can be achieved on BDD anodes using modulated current electrolysis [16].

Future Directions

Future developments will require research and development activities in the following areas:

1. Low-cost and efficient electrodes with high service life. The most efficient electrodes actually used are BDD. However, these electrodes are expensive, and new developments are necessary in order to optimize the preparation technique to decrease cost. Furthermore, new electrode materials should be developed. Possible candidates are doped valve metal oxides (Ti, Ta, Nb, Zr oxides).

2. Development of new accelerated tests for the estimation of electrode service life for the wastewater under investigation.
3. New three-dimensional electrodes for the treatment of very dilute wastewaters.
4. Avoidance of the formation of halogenated organics if chloride ions are present in the wastewater.
5. Investigation of the possibility of the treatment of pharmaceutical micropollutants.

Cross-References

- ▶ Electrochemical Cell Design for Water Treatment
- ▶ Ex-Cell-Mediated Oxidation (via Persulfate) of Organic Pollutants
- ▶ General Concepts and Global Parameters (EOD, COD, O_x)
- ▶ Organic Pollutants, Direct and Mediated Anodic Oxidation
- ▶ Organic Pollutants in Water, Direct Electrochemical Oxidation Using PbO_2
- ▶ Organic Pollutants in Water Using DSA Electrodes, In-Cell Mediated (via Active Chlorine) Electrochemical Oxidation
- ▶ Organic Pollutants in Water Using SnO_2, Direct Electrochemical Oxidation
- ▶ Organic Pollutants, Oxidation on Active and Non-Active Anodes
- ▶ Water Treatment with Electrogenerated Fe(VI)

References

1. Kapalka A, Foti G, Ch C (2010) Basic principles on the electrochemical mineralization of organic pollutants for wastewater treatment. In: Ch C, Chen G (eds) Electrochemistry for the environment. Springer, New York, pp 1–23
2. Comninellis C, Plattner E (1988) Electrochemical wastewater treatment. Chimia 42:250–252
3. G Foti, D Gandini, Ch Comninellis (1997) Anodic oxidation of organics on thermally prepared oxide electrodes. In Current topics in electrochemistry. Research Trend, Trivandrum, 5: 71–91
4. Fóti G, Ch C (2004) Electrochemical oxidation of organics on IrO_2 and BDD based electrodes. In: White RE, Conway BE, Vayenas CG, Gamboa-Adelco ME (eds) Modern aspects of electrochemistry, vol 37. Kluwer/Plenum, New York, pp 87–130
5. Comninellis C (1994) Electrochemical conversion/combustion of organic pollutants for waste water treatment. Electrochim Acta 39:1857–1862
6. Foti G, Gandini D, Comninellis C, Perret A, Haenni W (1999) Oxidation of organics by intermediates of water discharge on IrO_2 and synthetic diamond anodes. Electrochem Solid-State Lett 2(5):228–230
7. Panizza M, Michaud PA, Cerisola G, Comninellis C (2001) Anodic oxidation of 2-naphthol at boron-doped diamond electrodes. J Electroanal Chem 507:206–214
8. Iniesta J, Michaud PA, Panizza M, Comninellis C (2001) Electrochemical oxidation of 3-methylpyridine at a boron-doped diamond electrode: application to electroorganic synthesis and wastewater treatment. Electrochem Commun 3(7):346–351
9. Iniesta J, Michaud PA, Panizza M, Cerisola G, Aldaz A, Comninellis C (2001) Electrochemical oxidation of phenol at boron-doped diamond electrode. Electrochim Acta 46(23):3573–3578
10. Panizza M, Michaud PA, Cerisola G, Comninellis C (2001) Electrochemical treatment of wastewaters containing organic pollutants on boron-doped diamond electrodes: prediction of specific energy consumption and required electrode area. Electrochem Commun 3(7):336–339
11. Rodrigo MA, Michaud PA, Duo I, Panizza M, Cerisola G, Comninellis C (2001) Oxidation of 4-chlorophenol at boron-doped diamond electrode for wastewater treatment. J Electrochem Soc 148(5): D60–D64
12. Montilla F, Michaud PA, Morallon E, Varquez JL, Comninellis C (2002) Electrochemical oxidation of benzoic acid at boron-doped diamond electrodes. Electrochim Acta 47(21):3509–3513
13. Marselli B, Garcia-Gomez J, Michaud PA, Rodrigo MA, Comninellis C (2003) Electrogeneration of hydroxyl radicals on boron doped diamond electrodes. J Electrochem Soc 150:D79–D83
14. Kapalka A, Foti G, Comninellis C (2007) Investigations of electrochemical oxygen transfer reaction on boron-doped diamond electrodes. Electrochim Acta 53:1954–1961
15. Kapalka A, Foti G, Ch C (2008) Kinetic modeling of the electrochemical mineralization of organic pollutants for wastewater treatment. J Appl Electrochem 38:7–16
16. Panizza M, Kapalka A, Comninellis C (2008) Oxidation of organic pollutants on BDD anodes using modulated current electrolysis. Electrochim Acta 53:2289–2295
17. Kapalka A, Foti G, Comninellis C (2009) The importance of electrode material in environmental electrochemistry: formation and reactivity of free hydroxyl radicals on boron-doped diamond electrodes. Electrochim Acta 54:2018–2023

18. Fierro S, Nagel T, Baltruschat H, Comninellis C (2007) Investigation of the oxygen evolution reaction on Ti/IrO$_2$ electrodes using isotope labeling and on-line mass spectrometry. Electrochem Commun 9(8):1969–1974
19. Fierro S, Nagel T, Baltruschat H, Comninellis C (2008) Investigation of formic acid oxidation on Ti/IrO$_2$ electrodes using isotope labeling and online mass spectrometry. Electrochem Solid-State Lett 11:E20–E23

Organic Pollutants, Oxidation on Active and Non-Active Anodes

Nigel J. Bunce and Dorin Bejan
Department of Chemistry, Electrochemical Technology Centre, University of Guelph, Guelph, ON, Canada

Introduction

Electrochemical oxidation has attracted research interest for over two decades as a possible technology for remediation of water pollutants that resist conventional biological and chemical treatment. Several reviews have appeared in recent years [1–10]. Advantages commonly put forward include simple equipment, versatility, operation at ordinary temperature and pressure, amenability to automation [7], low cost per mol of electrons, and environmental friendliness as a reagent, as compared with chemical oxidants. Electrochemical oxidations may be categorized as follows, with minor differences in terminology among investigators:

- Direct oxidation: the substrate transfers electron(s) directly to the anode to form, initially, the substrate radical cation.
- Indirect oxidation: a substance other than the substrate is oxidized at the anode and in turn oxidizes the substrate.
- Mediated oxidation: a subset of indirect oxidation in which the substance that is formed electrochemically is reduced back to its original form upon oxidation of the substrate and recycled (e.g., Co(II)/Co(III); Cl^-/HOCl).

The processes considered in this entry are indirect oxidations in which water, rather than the substrate, is the substance being oxidized at the anode. This research area developed in the early 1990s, following observations that the course of oxidation often depended on the anode material. Some anodes promote the formation of well-defined oxidation products (electrochemical conversion), while others initiate an unselective oxidation that may result in complete mineralization of the substrate (electrochemical combustion).

Mechanisms of Oxidation at Active Versus Non-active Anodes

In a seminal paper, Comninellis [11] recognized two subgroups of indirectly acting anodes. One type, later called "active" anodes, is based on noble metal oxides (Ti/RuO$_2$ and Ti/IrO$_2$ and also platinum, which forms a surface oxide PtO$_x$ under anodic polarization). "Non-active" anodes are based on main group elements: Ti/SnO$_2$ and subsequently β-Pb/PbO$_2$ and the then-novel anode material, boron-doped diamond (BDD), which has exceptional anodic and cathodic stability in aqueous solution and is known for long service life.

At both groups of anode (**A**), water is initially oxidized to anode-sorbed hydroxyl radicals **A**~$^\bullet$OH:

$$\mathbf{A} + H_2O \rightarrow \mathbf{A}\sim{}^\bullet OH + H^+ + e^- \qquad (1)$$

Non-active anodes do not bind the oxygen atom of **A**~$^\bullet$OH covalently. The sorbed hydroxyl radicals (physisorbed active oxygen) have high and nonspecific reactivity towards organic substrates through free radical addition and/or hydrogen abstraction reactions, which can lead to full or partial mineralization of the substrate. This mode of oxidation is variously known as electrochemical combustion, electrochemical incineration, or an electrochemical advanced oxidation process (EAOP), the latter by analogy with conventional AOPs that also involve hydroxyl radicals:

$$x\mathbf{A}\sim{}^\bullet OH + SH \rightarrow x\mathbf{A} + mCO_2 + nH_2O \qquad (2)$$

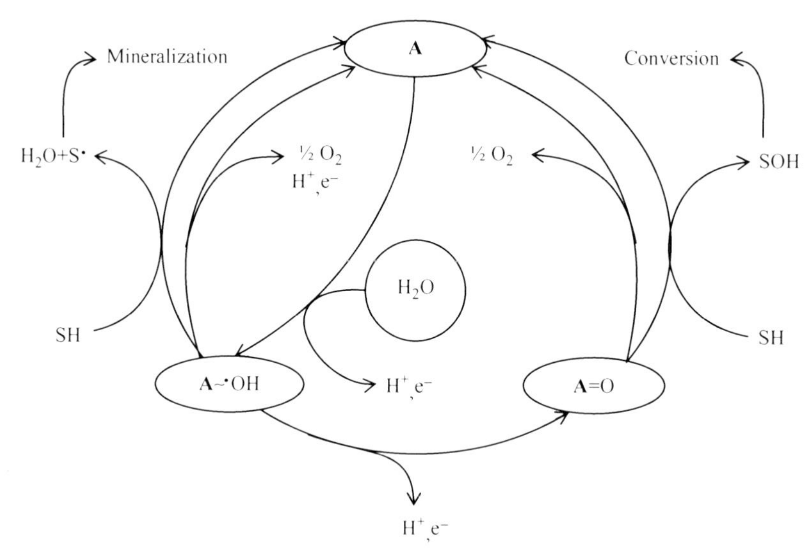

Organic Pollutants, Oxidation on Active and Non-Active Anodes, Fig. 1 Outline of the processes of oxidation at active and non-active anodes (Adapted from Ref. [11])

The kinetic stability of **A**~•OH at these anodes is attributed to their high overpotential for oxygen evolution:

$$\mathbf{A}\sim{}^{\bullet}\mathrm{OH} \rightarrow \mathbf{A} + {}^{1}\!/_{2}\mathrm{O}_2 + \mathrm{H}^+ + \mathrm{e}^- \quad (3)$$

Substrate oxidation can compete because hydroxyl radicals react much faster with organic substrates than they dimerize to H_2O_2 en route to O_2. The O_2 formed at the anode may also participate in substrate oxidation by intercepting some of the radical intermediates. This is implied in the oxidation of acetic acid at BDD in the presence of $^{18}O_2$, where differential electrochemical mass spectrometry revealed partial incorporation of the label into the carbon dioxide product [1].

At active anodes based on noble metal oxides, **A**~•OH is rapidly oxidized further to a covalently bound species **A**=O (chemisorbed active oxygen). For example, at a Ti/IrO_2-based anode, **A**=O would be a surface layer of IrO_3. This reaction is evidenced by incorporation of ^{18}O from ^{18}O-enriched water into the metal oxide. Substrate oxidation, again in competition with oxygen evolution, involves oxygen transfer from **A**=O, as shown by incorporation of ^{18}O into CO_2 when formic acid is electrooxidized at an ^{18}O-enriched metal oxide in unlabelled water [4]:

$$\mathbf{A}\sim{}^{\bullet}\mathrm{OH} \rightarrow \mathbf{A} = \mathrm{O} + \mathrm{H}^+ + \mathrm{e}^- \quad (4)$$

$$\mathbf{A}=\mathrm{O} \rightarrow \mathbf{A} + {}^{1}\!/_{2}\mathrm{O}_2 \quad (5)$$

$$\mathbf{A}=\mathrm{O} + \mathrm{SH} \rightarrow \mathbf{A} + \mathrm{SOH} \quad (6)$$

Oxygen transfer from **A**=O to the substrate does not initiate free radical chemistry, and so significant mineralization of more complex substrates rarely occurs at active anodes. The surface of an active anode acts as a mediator, and the outcome is selective oxidation. As an example, acetaminophen undergoes efficient conversion to benzoquinone at Ti/IrO_2-Ta_2O_5 but is mineralized at BDD.

Figure 1, adapted from reference [11], summarizes these processes.

Kapalka et al. [3] have distinguished active and non-active anodes in terms of the oxidation potential of water and the overpotential for oxygen evolution (Table 1). Active anodes based on IrO_2 and RuO_2 exhibit water oxidation near +1.6 V vs. SHE. They have a low overpotential for oxygen evolution and relatively weak oxidizing power towards external substrates. Non-active anodes based on PbO_2 and SnO_2 have intermediate properties, while BDD represents the other extreme (water oxidation at >2.2 V vs. SHE, and the overpotential for oxygen formation is 1.3 V). In practice, active and non-active behavior cannot be cleanly separated [6, 10], especially for anodes in the middle of Table 1.

Organic Pollutants, Oxidation on Active and Non-Active Anodes, Table 1 Oxidation potentials for water oxidation and overpotentials for oxygen evolution at different anodes

Electrode	Oxidation potential for organics, V	Overpotential for O_2 evolution, V
Ti/RuO_2-TiO_2	1.4–1.7	0.18
Ti/IrO_2-Ta_2O_5	1.5–1.8	0.25
Ti/Pt	1.7–1.9	0.3
Ti/PbO_2	1.8–2.0	0.5
Ti/SnO_2-Sb_2O_5	1.9–2.2	0.7
p-Si/BDD	2.2–2.6	1.3

Very easily oxidized organic substrates (oxidation potential <1.4 V) undergo direct oxidation, not catalyzed by-products of water oxidation, but frequently cause anode fouling by oligomeric products of partial oxidation [2, 6, 9]. At active anodes, the intervention of Eqs. 4 and 6 has been described as a potential buffer that limits oxidation to substrates that are moderately easily oxidized [1, 4, 8]. The resistance of **A**~$^{\bullet}$OH to further oxidation at non-active anodes allows the most recalcitrant organic substrates to undergo hydroxyl radical catalyzed oxidation (E° for formation of $^{\bullet}$OH at pH 7 = 2.3 V vs. SHE) and also allows self-cleaning of the anode by oxidation of any oligomeric deposits.

Non-active anodes have attracted the most interest for contaminant remediation because of their ability to promote mineralization, as shown by production of inorganic carbon or by loss of total organic carbon, but none of them satisfy all requirements. Pb/PbO_2 electrodes and inexpensive but may release toxic Pb^{2+} ions into the treated solution; Ti/SnO_2-based anodes suffer from short service life; BDD is presently expensive [6, 8].

Mineralization at BDD can be observed even at the lowest conversion of substrate, indicating the following mechanistic paradigm [1]: (a) Organic substrates (including intermediate oxidation products) have negligible adsorption to the anode surface. (b) All small molecule organics have similar rate constants for diffusion. (c) Mineralization under mass transport control is characterized by multiple acts of oxidation once the substrate arrives at the anode surface; this inhibits significant diffusion of intermediates into the bulk solution, and significant yields of intermediate oxidation products are not observed. This is a crucial difference between EAOP and AOP; in AOP, the reactions between $^{\bullet}$OH and the substrate occur in the bulk solution, and mineralization is characterized by the successive formation and oxidation of intermediate oxidation products.

Detection of Hydroxyl Radicals at Non-active Anodes

Much effort has been devoted to demonstrating the involvement of hydroxyl radicals in these reactions, most often at BDD anodes, at which voltammetric evidence for Eq. 1 has recently been provided [2]. The high reactivity of $^{\bullet}$OH impedes direct observation, and the hydroxyl species are usually identified indirectly by trapping with an appropriate reagent. Analytical methods include bleaching of *p*-nitrosodimethylaniline (absorption spectroscopy), spin trapping with nitrones (electron spin resonance), formation of hydrogen peroxide (titration), formation of hydroxylated products from salicylic acid and similar compounds (HPLC), and competitive oxidation (in the case of formic and oxalic acid: HPLC) [1, 11]. Most of these methods (although useful for analyzing chemically produced hydroxyl radicals) suffer from the ambiguity that the trapping reagents are oxidized at similar potentials as the oxidation of water to $^{\bullet}$OH. This makes it difficult to distinguish oxidation of water and reaction of $^{\bullet}$OH with the trapping agent from oxidation of the trapping agent followed by nucleophilic attack of water. The hydroxylation of coumarin (analysis by fluorescence spectroscopy) is useful qualitatively because coumarin is not itself oxidized readily but cannot afford quantitative information because the fluorescent product is formed in very low chemical yield [12]. It can also be argued that substrate mineralization is in itself indirect evidence for the intermediacy of hydroxyl species in these reactions.

Recent work has examined the question of whether A~˙OH is exclusively anode-sorbed or whether there is a continuum of species between the extremes of physisorbed and bulk hydroxyl radicals. Vatistas [13] has considered the dissociation of hydroxyl radicals from an adsorption layer ("true" A~˙OH) into a three-dimensional "reactive" layer close to the anode surface. A~˙OH is thus a combination of surface and reactive layer species, both of which experience the anode electrostatically and, in principle, have a reactivity different from that of bulk ˙OH(aq). Kapalka et al. [1] have estimated the profile of hydroxyl species adjacent to a BDD anode concurrent with the evolution of O_2 in the absence or presence of an organic substrate. Their model describes the hydroxyl species as "quasi-free." In the absence of substrate, the hydroxyl radicals form H_2O_2 within a stagnant layer of solution close to the anode. It is concluded that their concentration falls to <10 % of the value at the anode surface within 0.2 μm and almost to zero by 1 μm; almost no hydroxyl radicals escape the anode surface completely and become bulk ˙OH (aq). When a reactive organic substrate is also present, the hydroxyl radicals are trapped much closer to the anode because of the higher rate constant for reaction of ˙OH with a substrate as compared with that for dimerization, and their concentration falls to almost zero within tens of nm. No data have yet appeared in which the reactivities of anode-sorbed and bulk hydroxyl radicals have been compared.

Kinetic Models

The extremes of kinetic behavior in an electrochemical reaction are known as mass transport controlled oxidation and current controlled oxidation (also called "reaction oxidation control"). Mass transport control is promoted at low substrate concentration and/or high current density, and the rate of oxidation is limited by the rate of arrival of substrate molecules into the reactive zone at or near the anode. The kinetic rate law shows first-order dependence on substrate concentration but is independent of the applied current. There is a plentiful supply of reactive intermediates (A~˙OH or A=O) for every substrate molecule that arrives in the reaction zone. At non-active anodes, the abundance of (sorbed) hydroxyl radicals under mass transport control explains why multiple oxidations can take place without significant loss of intermediates into the bulk solution, and the rate of mineralization often tracks the rate of substrate disappearance closely. At both types of anode, the rate of water oxidation exceeds the rate of arrival of substrate, and so evolution of O_2 at the anode always competes with substrate oxidation, leading to a current efficiency below 100 %.

Under current control, typified by high substrate concentration and/or low current density, there is a plentiful supply of substrate molecules available to trap A~˙OH or A=O. In the limit of current control, the rate of substrate oxidation is directly proportional to the applied current and independent of the substrate concentration, whose rate of disappearance is linear with time. No oxygen is evolved, and the current efficiency is 100 %. Current efficiency is defined as the fraction (or %) of all charges passed through the solution that carry out the electrochemical process of interest – for example, mineralization. Under current control, outward diffusion of partly oxidized intermediates is likely, and these are often observed even at non-active anodes, e.g., formic, oxalic, and maleic acids from phenolic precursors. Current controlled kinetics are generally seen for substrate concentrations >50 mM at moderate current densities [3], and at non-active anodes, mineralization is slower than the initial loss of substrate.

Several research groups have shown that under batch conditions and constant current, the current efficiency declines with substrate conversion as the contaminant concentration falls and the kinetics shift away from current control towards mass transport control [1, 10]. Between the extremes of mass transport and current control, a general rate equation can be written with fractional kinetic order, the values of m and n changing with the experimental conditions [14]:

$$\text{Rate} = [\text{Substrate}]^{m} \times \text{Current}^{n} \qquad (7)$$

A useful model has been developed for BDD anodes, in which a current i_{lim} represents the high current limit of current controlled behavior.

$$i_{lim} = n\ Fk_m[SH] \quad (8)$$

In Eq. 8, i_{lim} is the limiting current for contaminant oxidation (A m^{-2}). SH is the organic contaminant, and n is the number of electrons required to mineralize SH (concentration in mol m^{-3}). F is the Faraday constant (96,485 C mol^{-1}), and k_m is the mass transport coefficient (m s^{-1}), which is similar in magnitude for all small molecule organics [6]. Because i_{lim} depends on (n × [SH]), the same i_{lim} can be found for a large concentration of a simple organic or a small concentration of a complex contaminant. The simplification is often made that oxidation occurs with 100 % current efficiency for $i_{app} < i_{lim}$ and that the system operates under mass transport control for $i_{app} > i_{lim}$, although it is recognized that in reality, the transition is gradual [4]. In addition, current control is usually associated with the escape of partly oxidized intermediates from the anode; complete mineralization involves such a plentiful supply of A~•OH that O_2 is evolved competitively, and the current efficiency is less than 100 %. Distinction must also be made with respect to the kinetics for loss of substrate versus kinetics of mineralization: at low values of the applied current, the oxidation of substrate may be mass transport controlled (first-order loss of substrate), while mineralization follows current control (linear loss of total organic carbon, TOC) [5].

Practicality of EAOP Technology

The kinetics of mineralization bears importantly on the practicality of EAOP technology. Most contaminant streams are dilute (otherwise the contaminant would be worth recovering). At non-active anodes, excellent conversion of substrate to mineralized products is possible under mass transport control, at the expense of low current efficiency. When $i_{app} < i_{lim}$, the need for low current densities means that high current efficiency is attained at the expense of long reaction times and incomplete mineralization. The compromise is a kinetic regime in between current and mass transport control, in order to speed up the process, at the expense of maximal current efficiency. This represents a compromise between efficient utilization of charge and long reaction times, which require a reactor of large volume to treat a given flow rate of contaminated water. From the engineering perspective, mass transport is optimized by the use of electrodes of large area and close separation in order to minimize ohmic drop; fortunately, the irreversibility of mineralization means that undivided cells can be used.

The instantaneous current efficiency (ICE) and average current efficiency (ACE) are used to assess the efficiency of the use of electrical charge. Most authors follow their electrolyses by analysis of chemical oxygen demand (COD). COD is monitored most directly by following the evolution of molecular oxygen in the absence and presence of the organic contaminants (oxygen flow rate technique, OFR). V′ is the rate of oxygen evolution in the absence (V'_0) and presence of the organic contaminants over the time increment Δt (ICE) or the complete time t of electrolysis (ACE):

$$ICE_{OFR} = (V'_0 - V'_{\Delta t})/V'_0 \quad (9)$$

$$ACE_{OFR} = (V'_0 - V'_{\Delta t})/V'_0 \quad (10)$$

COD is more conveniently determined by the acid dichromate method, whose results are usually reported in the units (mol O_2 m^{-3}). In this case, the ICE and ACE are given by Eqs. 11 and 12, in which F is Faraday's constant (C mol^{-1}), V is the electrolyte volume (m^3), i is the applied current (A), 4 is the conversion factor between mol e^- and mol O_2, and t is in seconds:

$$ICE_{COD} = 4FV(COD_t - COD_{t+\Delta t})/i\Delta t \quad (11)$$

$$ACE_{COD} = 4FV(COD_0 - COD_t)/it \quad (12)$$

Stripping of volatile substances, evolution of chlorine from solutions containing chloride ion, deposition of insoluble products, and failure to account for incompletely oxidized substances can interfere with COD analyses [3].

Mineralization may also be followed by analyzing for TOC, in which the carbon atoms of all organic compounds in the solution are converted to inorganic carbon (CO_2). TOC analysis cannot account accurately for partial oxidation – e.g., ethanol and acetic acid have the same carbon content.

Many research papers have modeled the change in current efficiency based on COD (more rarely TOC) with conversion, but quantitative correlation is only possible for synthetic wastewaters, whose compositions are known. These limitations are not an obstacle in the industrial setting, where the total carbon load (kg m^{-3}) is more important than the identities of specific contaminants, and the practical objective is to achieve some final objective in terms of the TOC or the COD of the treated solution. The efficiency of the process can then be expressed indirectly, for example, as specific energy consumption (kWh per kg C or per kg COD removed) [6].

A significant concern for electrochemical oxidation is the presence of chloride ion in the contaminated solution. Under these conditions, anodic oxidation of chloride to hypochlorite competes with oxidation of water to hydroxyl species. The result is a hypochlorite-mediated process of oxidation/chlorination and the formation of persistent organochlorine by-products. Although these can be mineralized in turn, their destruction is much more demanding of energy (charge) [2, 6, 8, 9].

Future Directions

The focus of research in this area has trended increasingly towards mineralization as an objective, with BDD as the anode material of greatest interest. Small-scale commercial cells incorporating BDD electrodes have recently become available. As a competitor to biological oxidation, electrochemical oxidation offers the prospective advantages of treatment in a one-pass flow-through reactor, in which even modest current efficiency can be an acceptable trade-off with the large reactors and long residence times typically required for biological oxidation.

As always, displacement of an existing technology will depend on cost. One factor is the need to decrease the cost of BDD, the preferred anode material. Another, signaled by the introduction of the limiting current concept, is to optimize the use of electrical charge. Possibilities include cascades of electrochemical reactors in which the applied current is progressively reduced, corresponding to the decrease of i_{lim} with contaminant concentration. On a large enough scale, even the recovery of the hydrogen formed by cathodic reduction of water might constitute an offset to the cost of electrical energy.

Cross-References

▶ Boron-Doped Diamond for Green Electro-Organic Synthesis
▶ Organic Pollutants in Water Using DSA Electrodes, In-Cell Mediated (via Active Chlorine) Electrochemical Oxidation

References

1. Kapalka A, Baltruschat H, Comninellis C (2011) Electrochemical oxidation of organic compounds induced by electro-generated free hydroxyl radicals on BDD electrodes. In: Brillas E, Martinez-Huitle CA (eds) Synthetic diamond films: preparation, electrochemistry, characterization and applications. Wiley, Hoboken/New Jersey
2. Scialdone O, Galia A (2011) Modeling of electrochemical process for water treatment using diamond films. In: Brillas E, Martinez-Huitle CA (eds) Synthetic diamond films: preparation, electrochemistry, characterization and applications. Wiley, Hoboken/New Jersey
3. Kapalka A, Foti G, Comninellis C (2010) Basic principles of the electrochemical mineralization of organic pollutants for wastewater treatment. In: Comninellis C, Chen G (eds) Electrochemistry for the Environment. Springer Science + Business Media, LLC, New York
4. Fierro S (2010) Electrochemical oxidation of organic compounds in aqueous acidic media on 'active' and 'non-active' type electrodes. In: Kuai S, Meng J (eds) Electrolysis. Nova Science Publishers, Inc, Hauppauge/New York

5. Polcaro AM, Mascia M, Palmas S, Vacca A (2010) Case studies in the electrochemical treatment of wastewater containing organic pollutants using BDD. In: Comninellis C, Chen G (eds) Electrochemistry for the environment. Springer Science + Business Media, LLC, New York
6. Panizza M, Cerisola G (2009) Direct and mediated anodic oxidation of organic pollutants. Chem Rev 109:6541–6569
7. Anglada A, Urtiaga A, Ortiz I (2009) Contributions of electrochemical oxidation to wastewater treatment: fundamentals and review of applications. J Chem Technol Biotechnol 84:1747–1755
8. Ferro S, Martinez-Huitle CA (2006) Electrochemical oxidation of organic pollutants for the wastewater treatment: direct and indirect processes. Chem Soc Rev 35:1324–1340
9. Kraft A (2007) Doped diamond: a compact review on a new, versatile electrode material. Int J Electrochem Sci 2:355–385
10. Foti G, Comninellis C (2004) Electrochemical oxidation of organics on iridium oxide and synthetic diamond based electrodes. In: White RE, Conway BE, Vayenas CG, Gamboa-Adelco ME (eds) Modern aspects of electrochemistry 37. Kluwer Academic/Plenum Publishers, New York
11. Comninellis C (1994) Electrocatalysis in the electrochemical conversion/combustion of organic pollutants for waste water treatment. Electrochim Acta 39:1857–1862
12. Nosaka Y, Ohtaka K, Ohguri N, Nosaka AY (2011) Detection of OH radicals generated in polymer electrolyte membranes of fuel cells. J Electrochem Soc 158:B430–B433
13. Vatistas N (2012) Electrocatalytic properties of BDD anodes: its loosely adsorbed hydroxyl radicals. Int J Electrochem. doi:10.1155/2012/507516
14. Li S, Bejan D, McDowell MS, Bunce NJ (2008) Mixed first and zero order kinetics in the electrooxidation of sulfamethoxazole at a boron-doped diamond (BDD) anode. J Appl Electrochem 38:151–159

Organic Reactions and Synthesis

Atusko Nosaka
Department Materials Science and Technology, Nagaoka University of Technology, Nagaoka, Niigata, Japan

Photocatalysts

Selective photocatalytic reactions have a potential to provide alternative green routes by replacing environmentally hazardous processes with safe and energy-efficient routes [1–5]. Among the various photocatalysts, semiconductors such as TiO_2, ZnO, WO_3, CdS, and NiO are commonly anticipated. These catalysts are often metal-doped to shift the radiation absorption towards larger wavelengths and supported over various materials such as silica and zeolites to increase the surface area and hence the reaction rate. Among these photocatalysts it is generally recognized that TiO_2 and its related materials are the most reliable materials for photocatalytic reactions, because it is low-cost and exhibits high catalytic activity and high stability under photoirradiation. The yield and selectivity of the organic products are significantly influenced by the crystal form and size of the TiO_2 particles [3].

The Photocatalytic Reaction Systems

The reaction systems can be designed by selecting the photocatalysts and reaction conditions, such as the type of the photocatalyst (bare TiO_2, platinum-loaded TiO_2, mesoporous TiO_2, supported TiO_2), the phase (gas–solid or liquid–solid system), the atmosphere (the presence or absence of molecular oxygen), the solvent (water or organic), and the wavelength of irradiation light (in UV light or visible light lamps, solar radiation simulators are often utilized) [4].

Photocatalytic Reaction Mechanism

TiO_2 and related photocatalysts can promote productive reactions for organic syntheses. Most of the studies report on oxidations or reductions. Photocatalytic reactions occurring on TiO_2 particle can be summarized as Scheme 1. When TiO_2 particle absorbs bandgap photons, photoexcited electrons (e^-) and holes (h^+) are produced in the conduction band (cb) and the valence band (vb), respectively. An efficient charge separation of the electron–hole pairs enables the respective oxidation and reduction reactions on the particle surface. In the presence of molecular oxygen (O_2), it acts as an electron acceptor to be

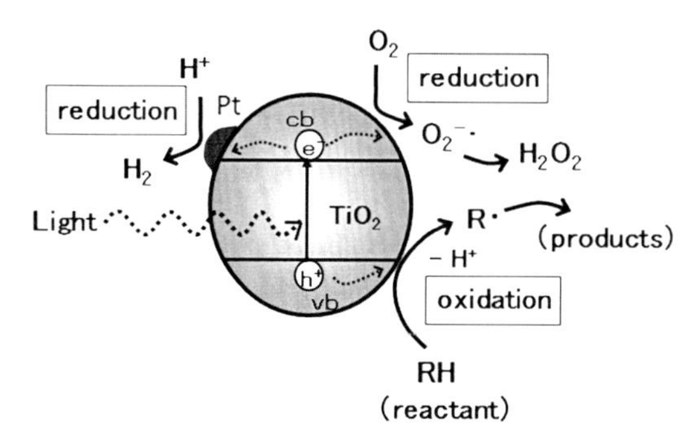

Organic Reactions and Synthesis, Scheme 1 Mechanism for photocatalytic reaction of organic substrates RH

converted to a superoxide anion ($O_2\cdot^-$) and H_2O_2. On the other hand, the hole oxidizes the substrates directly. Thus, in a photocatalytic system, photoenergy produces both excited electrons and holes simultaneously, which could promote reductive and oxidative elementary reactions, respectively. The characteristics of photocatalysis enable a wide variety of unique reactions. By the use of appropriate substrates under specific reaction conditions, selective conversion can be promoted [2].

Some of the recent representative researches on selective organic transformations carried out by TiO_2 photocatalysis are summarized as follows.

Oxidation

Most of the photocatalytic oxidations concern aliphatic and aromatic alkanes and alkenes derivatives, and the selective oxidation of alcohols to carbonyls.

Hydrocarbons Oxidation in the Presence of O_2

Hydrocarbons, such as toluene, cyclohexane, methylcyclohexane, and ethylbenzene, can be oxidized with good yields and high selectivity to alcohols, aldehydes, and ketones [3].

Oxidation of Cyclohexane Liquid Phase: Cyclohexanone is obtained with an almost complete selectivity in the TiO_2 suspension irradiated with near-UV light (≃280 nm) [5]. For the formation of cyclohexanol and cyclohexanone, dichloromethane is the best solvent [3].

Gas Phase: Cyclohexanol and cyclohexanone are obtained with TiO_2 film reactors produced by flame aerosol, with selectivity for the two products of 98 % [3].

Selective Oxidation of Propene Acetone is the main product with TiO_2/SiO_2 with the selectivity of 95 % [4].

Olefins Oxidative Cleavage Photoirradiation (>280 nm) to an acetonitrile solution containing various cyclic and linear olefins with mesoporous silica containing isolated Ti-oxide species produces the corresponding epoxide with high selectivity (>98 %) [2].

Alcohol Oxidation

The oxidation of alcohols by using irradiated TiO_2 in O_2-saturated acetonitrile yields aldehydes or ketones and traces of acids as products.

Liquid Phase

Benzyl Alcohol Benzyl alcohol and some of its derivatives (para-structures) can be transformed into the corresponding aldehydes with a conversion and selectivity of ca. 99 % both under UV and visible light irradiation [5].

1-Phenylethanol, Benzhydrol, Hydrobenzoin, 4,4′-Dimethoxyhydrobenzoin, 4,4′-Dichlorohydrobenzoin, and 4-Chloro-benzhydrol These alcohols can be oxidized with conversions higher than 90 % and high yields into the desired products [3].

4-Methoxybenzyl Alcohol 4-Methoxybenzaldehyde (or p-anisaldehyde) is obtained by the selective oxidation in organic-free water suspensions of home-prepared TiO_2, at room temperature with a yield of 41.5 %mol. The by-products are traces of 4-methoxybenzoic acid, open ring products, and CO_2 [3].

2-Propanol Only propanone is obtained by suspending in TiO_2-zeolite composites in pure 2-propanol saturated with O_2 or mixtures of N_2 and O_2 [3].

Phenanthrene Coumarin derivative can be obtained by UV irradiation (≃365 nm) to an acetonitrile solution containing 8 % water and phenanthrene with commercial P25 TiO_2 (Japan Aerosil) with good yield (45 %). The quantum efficiency of the product formation is relatively high (0.17) [1].

Gas Phase

1-Pentanol, Cyclohexanol, Benzyl Alcohol, 1-Phenylethanol, and Methanol 1-Pentanol, cyclohexanol, benzyl alcohol, and 1-phenylethanol can be oxidized to the corresponding carbonyls in a continuous gas–solid reactor with commercial anatase TiO_2 particles at about 463 K in moderate yield (37 %) but with high selectivity (>95 %) [2]. *Methanol* gives methyl formate at room temperature with a selectivity of 87–91 %, whereas formaldehyde, CO, and CO_2 are detected as minor products [5].

In the Absence of Oxygen

Methanol, Ethanol, 2-Propanol, and t-Butyl Alcohol Alcohols such as *methanol, ethanol, and 2-propanol* are converted with Pt/TiO_2 to the corresponding dehydrogenation products such as formaldehyde, acetaldehyde, and acetone, respectively, with the formation of equimolar H_2. *t-Butyl alcohol* is converted into 2,5-dimethyl-2,5-hexanediol with the formation of H_2 through intermolecular dimerization since t-butyl alcohol has no alpha-hydrogen. In these oxidative reactions, alcohols are oxidized by photoformed holes, while photoexcited electrons reduce water (or H^+) to form H_2 [4].

Hydroxylation of Aromatics The aromatic compounds photoadsorbed on the catalyst surface undergo two competing reaction pathways: (a) hydroxylation of the aromatic ring or (b) multi-step oxidation reactions to complete mineralization. In the first case, the OH radical attack follows the selectivity rules known for homogeneous electrophilic aromatic substitution when the oxidized compound contains an electron donor group. Hence the only ortho- and *para*-isomers are obtained. In the presence of an electron-withdrawing group, all three hydroxylated isomers are formed [3].

Benzene TiO_2-pillared clays are effective for benzene because of their pore structure. The distribution among the different oxidation products depends on the type of clay host [3].

The mesoporous TiO_2 system has prominent advantages: (1) additive-free, (2) low-cost source of oxidant (H_2O), and (3) mild reaction condition (room temperature). The high selectivity over the additive-free TiO_2 can be obtained owing to the adsorption property of mesoporous TiO_2 because this TiO_2 adsorbs benzene but scarcely adsorbs phenol and phenol can be produced under nitrogen with 81 % selectivity (34 % yield) in 42 % benzene conversion [2].

Toluene, tert-Butyl Benzene Selective aromatic-ring hydroxylation of benzene and alkylbenzenes to produce corresponding phenols over Pt-loaded TiO_2 photocatalyst can be conducted by using water molecule as an oxidant. The hydroxylation selectivity is 97 % for *toluene* and more than 99 % for *tert-butyl benzene* after photoirradiation for 4 h. The phenols are hardly converted further, resulting in the high selectivity [4].

Reductions

Photocatalytic reductions of organic molecules are usually conducted in the absence of O_2 with employing sacrificial electron donors such as methanol, which can scavenge valence band holes to reduce the degree of recombination of photoexcited electron–hole pairs within the TiO_2 particle. To improve the reduction efficiency, the deoxygenation of the system is necessary because oxygen acts as a competitive electron scavenger.

Reduction of Nitro Compounds

***p*-Nitroacetophenone** A hydroxylamine intermediate is detected by the photo-induced reduction of *p*-nitroacetophenone with high yields by the irradiation of suspensions of P25 TiO_2 in ethanol. The photogenerated electrons reduce the nitro compound, whereas the holes oxidize ethanol to acetaldehyde with the formation of hydrogen.

The hydrogen produced during the process can also cause the reduction of the nitro group [3].

Nitroarenes The reduction of nitroarenes gives corresponding amines in excellent yield (>90 %) with N-doped TiO_2 catalyst. N-doped TiO_2 catalyst shows good activity under visible light irradiation for the synthesis of amines from nitro compounds. The electronic properties of substituent attached to the aromatic ring do not have any effect on the reduction of nitro groups. Nitro compounds bearing electron-withdrawing groups produce the corresponding amines in better yields and in shorter reaction time with respect to the aromatic nitro compounds substituted with strong electron-releasing groups [5].

Nitrobenzene and Nitrotoluene The high yields of aniline products are obtained in TiO_2-suspended aqueous solution, with the *addition of alcohols, such as methanol and 2-propanol, and with the removal of* O_2 [3].

4-Nitrophenol 4-Aminophenol is obtained with almost complete selectivity in various alcohols (methanol, ethanol, 1-propanol, 2-propanol, 1-butanol, 2-butanol) in P25 TiO_2 suspensions under UV irradiation. Viscosity and polarity/polarisability parameter, i.e., the ability of a solvent to stabilize a charge or dipole by means of its dielectric effect, plays an important role in the conversion rate. Increase in the polarity parameter leads to better stabilization of the charged produced intermediate and hence, accelerates the photocatalytic reduction [3].

***p*-Chloronitrobenzene** *p*-Chloroaniline is obtained for UV-irradiated P25 TiO_2 slurries with the yield of almost 99 % in the presence of electron donor of a 2-propanol/formic acid (ratio equal to nine on volume base) [3].

Reduction of CO_2 to Organic Molecules

Photocatalytic CO_2 reduction is gathering attention due to the environmental problems caused by the continuous increase in CO_2 concentration in the atmosphere. Since CO_2 is inexpensive, nonflammable, and nontoxic, the conversion of the high amount of the excessive CO_2 in the atmosphere to useful products by photocatalytic reduction is highly desirable.

HCOOH, HCHO, and CH_3OH Formation HCOOH, HCHO, and CH_3OH can be obtained in alkaline aqueous suspension of TiO_2 doped with Nb, Cr, or RuO_2. HCOOH is the main product in all runs with a very modest amount of methanol [3].

CH_3OH Formation Addition of copper (II) into the TiO_2 matrix improves the efficiency and selectivity for the CH_3OH formation. The highest yield is achieved with a 3 % CuO/TiO_2 catalyst. Cu (I) acts as an electron trap and avoids the recombination of electrons and holes, which increases the photoefficiency of the process [3].

CH_4 Formation Liquid Phase: CH_4 is the main reaction product when CO_2 is reduced in aqueous 2-propanol solutions, with P25 TiO_2 suspension [3].

Gas Phase: The photocatalytic process with the pellet forms has the advantage because it is free from filtration to recover the catalysts from the gaseous product mixture. With a fixed-bed gas–solid photocatalytic reactor provided with commercial TiO_2 pellets, irradiated by UV light, the yield of CH_4 is higher than that obtained from processes using thin films or anchored TiO_2 catalysts and comparable to that of the powders [3].

CH_4 and CH_3OH Formation Silica containing highly dispersed Ti-oxide species promotes highly selective and efficient photoreduction of CO_2 [2].

(a) *Zeolites and mesoporous molecular sieves* promote selective production of CH_3OH and CH_4 with minor amounts of CO and O_2.
(b) *Isolated and tetrahedrally coordinated Ti-oxide species within the zeolite framework* exhibits high selectivity for CH_3OH formation.
(c) *Hydrophobic treatment of the catalysts* promotes selective CH_3OH formation.
(d) *Immobilized photocatalysts* are employed since they do not require catalyst separation step and facilitate catalyst reusability.

(e) *TiO_2/SiO_2-mixed mesoporous thin films* show higher reduction yields and quantum yields than the powdered catalysts.
(f) *The transparent self-standing thin films prepared* by a surfactant-templating sol–gel method promote photocatalytic conversion of CO_2 by H_2O to CH_4 and CH_3OH with a total quantum yield of almost 0.3. The increased catalytic activity is considered to be due to the improved light absorption by the shaped thin film.

Other Reactions

Besides the oxidative and reductive reactions, photocatalytic reactions can be applied to the formation of perfluorinated aromatic compounds and aromatic carbamate [3] and one-pot cyclizations [1] and coupling reactions (C–N coupling and C–C coupling) [2].

Future Directions

The reacting systems with the TiO_2-based photocatalysts are capable of organic synthesis by selective oxidation and reduction. Under appropriate reaction conditions, photocatalytic reactions proceed highly efficiently and selectively. The photocatalytic reactions allow us to avoid both the use of harmful and dangerous chemical reagents and solvents and the release of harmful wastes into the environment. In addition, the metal oxides catalysts are easily recyclable, since they can be simply separated and usually are not readily deactivated. These prominent advantages imply that photocatalysis can be an alternative "green" synthetic route. However, for practical applications, photocatalytic systems need to be designed more economically and simply. From this point of view, the employment of sunlight would be desired option especially in scaled up processes. For wider applications, photocatalytic processes need to be accelerated. Hence, the elucidation of the detailed mechanism of photocatalytic reactions would be prerequisite. If these issues are circumvented, photocatalytic transformations would play an important role in organic synthesis in economically and environmentally preferable ways.

Cross-References

▶ Photocatalyst
▶ Photoelectrochemical CO_2 Reduction
▶ Photoelectrochemistry, Fundamentals and Applications
▶ Semiconductors, Principles
▶ Solvents and Solutions
▶ Specific Ion Effects, Theory
▶ TiO_2 Photocatalyst

References

1. Palmisano G, Augugliaro V, Pagliaro M, Palmisano L (2007) Photocatalysis: a promising route for 21st century organic chemistry. Chem Commun 33:3425–3437. doi:10. 1039/B700395C
2. Shiraishi Y, Hirai T (2008) Selective organic transformations on titanium oxide-based photocatalysts. J Photochem Photobiol C Photochem Rev 9:157–170. doi:10.1016/j.jphotochemrev.2008.05.001
3. Augugliaro V, Di Paola A, Marcí G, Pagliaro M, Palmisano G, Palmisano L (2010) Photocatalytic organic syntheses. In: Anpo M, Kamat PV (eds) Environmentally benign photocatalysts, 1st edn. Springer, New York
4. Yoshida S (2010) TiO_2-based photocatalysis for organic synthesis. In: Anpo M, Kamat PV (eds) Environmentally benign photocatalysts. Nanostructure science and technology, 1st edn. Springer, New York
5. Palmisano G, García-López E, Marcí G, Loddo V, Yurdakal S, Augugliaro V, Palmisano L (2010) Advances in selective conversions by heterogeneous photocatalysis. Chem Commun 46:7074–7089. doi:10. 1039/C0CC02087G

Overpotentials in Electrochemical Cells

Eric M. Stuve
Department of Chemical Engineering, University of Washington, Seattle, WA, USA

Definition

Electrochemical devices – whether a reaction cell, battery, or fuel cell – operate at potentials

Overpotentials in Electrochemical Cells, Table 1 Example reactions for a galvanic cell, exemplified by a H_2/O_2 fuel cell, and an electrolysis cell operating in alkaline media

Reaction	Galvanic cell	Electrolysis cell
Overall	$H_2 + 1/2O_2 \rightleftharpoons H_2O$	$H_2O \rightleftharpoons H_2 + 1/2O_2$
Cathode	$1/2O_2 + H_2O + 2e^- \rightleftharpoons 2OH^-$	$2H_2O + 2e^- \rightleftharpoons H_2 + 2OH^-$
Anode	$H_2 + 2OH^- \rightleftharpoons 2H_2O + 2e^-$	$2OH^- \rightleftharpoons 1/2O_2 + H_2O + 2e^-$

substantially different than their reversible potentials. The difference between actual potential and reversible potential is called the overpotential.

> An overpotential is the potential beyond a reversible potential that produces an increased thermodynamic driving force for the process.

Overpotentials may be positive or negative. The sign is determined by the nature of the process: overpotentials for reduction are negative, since a more negative potential is more reducing. Conversely, overpotentials for oxidation are positive, since a more positive potential is more oxidizing. Overpotentials are not limited to reactions. They can be applied to electrode and electrolyte resistance, and to more complex processes, such as mass transfer limitations. Thus, overpotentials for cathodes are negative, whereas those for anodes are positive. In all cases, overpotentials express inefficiencies of the process.

The overpotential η of a cell is defined as the difference between the actual E and reversible E_r cell potentials,

$$\eta = E - E_r. \tag{1}$$

In the following treatment, the nature of overpotentials is presented for two cases: (1) a galvanic cell, such as a fuel cell or battery in discharge, that produces electrical work and (2) an electrolytic cell, such as an electrochemical reactor or battery in charge, that consumes electrical work. Table 1 lists the reactions of a H_2/O_2 fuel cell and of a water electrolysis cell operating in alkaline media (An alkaline medium is used for these examples to avoid coincidental values of zero that arise for H_2/O_2 cells in acidic media). Each process is the reverse of the other. Table 2 lists characteristic potentials and other quantities for the fuel cell and electrolysis cell operating in alkaline media at assumed practical potentials of 0.7 and 1.9 V, respectively. The other values are for standard state conditions of 25 °C and 1 atm. As each quantity is covered below, reference will be made to the corresponding line in Table 2 (e.g., L1 for line 1) to serve as an example.

Cell Potentials

The Gibbs free energy expresses the maximum work of an electrochemical cell operating at constant temperature and pressure. The Gibbs free energy of reaction ΔG_{rxn} is evaluated for the reaction as written and is given by the general formula

$$\Delta G_{rxn} = \sum_i v_i \Delta G_{f,i}, \tag{2}$$

where v_i is the stoichiometric coefficient of species i (positive for products, negative for reactants) and $\Delta G_{f,i}$ is the Gibbs free energy of formation of species i. For the fuel cell reaction in L1, the Gibbs free energy of reaction is

$$\Delta G_{rxn} = \Delta G_{f,H_2O} - \Delta G_{f,H_2} - 1/2\Delta G_{f,O_2}. \tag{3}$$

The electrolysis reaction in L1 is the reverse of the fuel cell reaction, so the Gibbs free energy has the opposite sign, as indicated in L2. The reversible cell potential is given by

$$E_r = \frac{-\Delta G_{rxn}}{nF} \tag{4}$$

where n is the number of electrons transferred in the reaction and F is Faraday's constant. As the Gibbs free energies are of opposite signs, so too are the reversible cell potentials (L3).

Overpotentials in Electrochemical Cells, Table 2 Example of potentials, overpotentials, and work of a H_2/O_2 fuel cell and electrolysis cell operating in alkaline media at standard state conditions (25 °C and 1 bar)

Line	Thermodynamic convention			
Overall reactions				
L1	Fuel cell		Electrolysis cell	
	$H_2 + 1/2O_2 \rightleftharpoons H_2O$		$H_2O \rightleftharpoons H_2 + 1/2O_2$	
Cell potentials				
	Quantity	Units	Fuel cell	Electrolysis cell
L2	(ΔG^o_{rxn}) [a]	kJ/mol	−237.141	237.141
L3	$E^o_r = -\Delta G^o_{rxn}/(nF)$	V	1.229	−1.229
L4	$E = E_c - E_a$	V	0.7[b]	−1.9[b]
L5	$\eta = E - E_r$	V	−0.529	−0.671
Electrode reactions				
	Fuel cell		Electrolysis cell	
L6[c]	$1/2O_2 + H_2O + 2e^- \rightleftharpoons 2OH^-$		$2H_2O + 2e^- \rightleftharpoons H_2 + 2OH^-$	
L7[d]	$H_2 + 2OH^- \rightleftharpoons 2H_2O + 2e^-$		$2OH^- \rightleftharpoons 1/2O_2 + H_2O + 2e^-$	
Electrode potentials				
	Quantity	Units	Fuel cell	Electrolysis cell
L8	$E_{c,r}$	V_{she}	0.401	−0.828
L9	$E_{a,r}$	V_{she}	−0.828	0.401
L10	E_c	V_{she}	0.100[b]	−1.100
L11	E_a	V_{she}	−0.600	0.800[b]
L12	$\eta_c = E_c - E_{c,r}$	V	−0.301	−0.272
L13	$\eta_a = E_a - E_{a,r}$	V	0.228	0.399
Electrical work				
	Quantity	Units	Fuel cell	Electrolysis cell
L14	$W_r = -nFE_r$	kJ/mol	−237.141	237.141
L15	$W = -nFE$	kJ/mol	−135.079	366.643
L16	$W_{lost} = -nF\eta$	kJ/mol	102.081	129.483

[a]Water in liquid phase
[b]Typical value
[c]Cathode
[d]Anode

The signs follow from the work aspects of the cell; galvanic cells are spontaneous and thus have $\Delta G_{rxn} < 0$ and $E_r > 0$, whereas electrolytic cells consume electrical work and thus have $\Delta G_{rxn} > 0$ and $E_r < 0$. Equation 4 establishes the thermodynamic convention for cell potentials used in Table 2.

The entries in Table 2 are for representative cell potentials of 0.7 V for the fuel cell and 1.9 V for the electrolysis cell (L4). Determination of the cell overpotential (L5) follows immediately by Eq. 1 with values from L3 and L4. Note that both cell overpotentials are negative, which is significant with respect to electric work (L14–L16).

Electrode Potentials

According to the thermodynamic convention, cell potentials are given by the difference of cathode and anode potentials,

$$E = E_c - E_a \tag{5}$$

where E_c is the cathode potential and E_a is the anode potential, both based on standard reduction potentials. Lines L8 and L9 show that standard reduction potentials are used for the reversible potentials of *both* cathode and anode. The minus sign in Eq. 5 distinguishes between reduction and oxidation. Equation 5 indicates that cell potential is measured by placing the positive terminal of the voltmeter on the cathode and the negative terminal on the anode. This convention follows from the fact that positive current always flows from the cathode.

While it is not possible to measure the potential of an individual electrode, cathode and anode potentials can be measured with respect to a reference electrode. There are many possible reference electrodes; a good choice for a fuel cell is a hydrogen electrode (rhe),

$$2\,H^+_{soln} + 2e^- \rightleftharpoons H_{2gas}, \tag{6}$$

which can be considered either at standard state or the conditions of measurement. A standard hydrogen electrode (SHE) is the reference for standard reduction potentials. It is based on unit activity of H^+_{soln} and $H_{2,gas}$: 1 mol/l and 1 atm in ideal (non-interacting) states. The potential of the SHE is 0.000 V at all temperatures. The SHE is not a practical electrode, however, because H^+_{soln} is not ideal at 1 mol/l. A practical choice is the reversible hydrogen electrode (RHE), which is reversible under the conditions of the measurement and is thus defined as 0.000 V for those conditions. In the example of Table 2, electrode potentials are reported with respect to the SHE.

The reversible oxygen reaction is oxygen reduction in the fuel cell (L8) and oxygen evolution in the electrolysis cell (L9). Its potential of 0.401 V_{she} is easily determined from the reversible cell potential (L3) and the reversible hydrogen potential, L8 (electrolysis cell) or L9 (fuel cell).

Actual cathode and anode potentials can be measured by a number of means relative to a reference electrode. For the fuel cell, the oxygen electrode, when measured against a standard hydrogen electrode, is assumed for this example to have a potential of 0.100 V_{she} (L10). For the electrolysis cell, the oxygen electrode potential is assumed to be 0.800 V_{she} (L11). The fuel cell anode potential and electrolysis cell cathode potentials are determined by Eq. 5 and have the values in L11 and L10, respectively.

Since the cell potential can be separated into cathodic and anodic components, it follows that the cell overpotential can be, too. Beginning with the definition of cell overpotential, and incorporating the definitions of cathode and anode potentials, the relationship of cell overpotential to electrode overpotentials is analogous to Eq. 5,

$$\begin{aligned} \eta &= E - E_{rev} \\ &= (E_c - E_a) - (E_{c,rev} - E_{a,rev}) \\ &= (E_c - E_{c,rev}) - (E_a - E_{a,rev}) \\ &= \eta_c - \eta_a. \end{aligned} \tag{7}$$

With the values of reversible and actual electrode potentials in L8–L11, the respective cathode and anode overpotentials are given in lines L12 and L13. Note that overpotential has units of Volts, since the effect of a reference electrode cancels in Eqs. 1 and 7. As mentioned above, overpotential is defined with respect to an increase in thermodynamic driving force. Thus, cathode (reduction) overpotentials are negative and anode (oxidation) overpotentials are positive. The cell overpotential will therefore be negative, according to Eq. 7.

Electrical Work

The general equation for electrical work is

$$W_i = -nFE_i \tag{8}$$

where W_i is the electrical work due to some type of potential E_i. Electrical work is positive when work is applied to the cell (electrolytic cell) and negative when the cell supplies work (galvanic cell). Table 2 lists the reversible work (L14) and actual work (L15) of the two cells.

The direct link between electrical work and overpotential is given by Eq. 8 written for overpotentials. Since overpotential is the deviation from reversibility, it follows that the work due to overpotential is work lost in the system. This is expressed explicitly by

$$W_{lost} = -nF\eta, \tag{9}$$

where W_{lost} is the lost work. Since cell overpotential is always negative (L5), it follows that W_{lost} is always positive. That is, work must be added to the electrochemical cell to make up for the lost work. Values for lost work are listed in L16.

As overpotential is the difference of actual and reversible potentials, Eq. 1, it follows that the lost work is the difference of actual and reversible work, as indicated in L14–L16. The fuel cell produces less work than its reversible work, and the electrolysis cell requires more work than its reversible work. Both cases arise from lost work: the capability of harvesting the maximum (reversible) work is diminished by the lost work of the fuel cell, and the magnitude of actual work is increased by the lost work of the electrolysis cell.

Practical Potentials

The thermodynamic convention of cell potential defined in Eq. 5 is not practical for most electrochemical work. For a battery switching from discharge to charge, cathode and anode interchange, but the electrode polarities remain the same. It is not practical to switch the leads of the voltmeter when the battery switches to the charge mode. Rather, the voltmeter stays connected during the switch with the positive terminal attached to the positive terminal of the battery and likewise for the negative terminal. The practical convention for cell potential is

$$|E| = E_+ - E_-, \tag{10}$$

where E is the cell potential according to the thermodynamic convention, defined in Eq. 5, and E_+ and E_- are the potentials of the positive and negative electrodes, respectively.

The distinguishing factor between the thermodynamic and practical conventions is the direction of current flow. In the thermodynamic convention, positive current always flows *from* the cathode, whether the cell is galvanic or electrolytic. In the practical convention, positive current flows *from* the positive electrode of a galvanic cell, yet *to* the positive electrode of an electrolytic cell. Interconversion between the two conventions can be accomplished by defining a current direction factor κ by

$$\kappa = \frac{j}{|j|} \tag{11}$$

where j is the current, defined as positive from the cathode. For the galvanic cell, $\kappa = 1$, while for the electrolysis cell, $\kappa = -1$. The relationship between the two conventions is given by

$$E = \kappa(E_+ - E_-). \tag{12}$$

The calculations of Table 2 can also be done in the practical convention by using this conversion. The interested reader should repeat the calculations of the electrolytic cell and verify that the work terms in L14–L16 have the same sign and values as with the thermodynamic convention.

For a cell that can have both galvanic and electrolytic modes (such as a battery), the thermodynamic convention requires an interchange of voltmeter electrodes for practicality, while the practical convention requires a sign reversal in computational and theoretical treatments of the two modes. In practice, most electrochemists switch between the two conventions automatically, but for computer calculations, the thermodynamic convention is preferred, with careful attention paid to converting laboratory data to the quantities needed for calculation.

Overview of Overpotentials

Figure 1 illustrates the effect of cell overpotential on cell performance. Overpotentials and potentials of a fuel cell are shown in the upper portion and of an electrolytic cell in the lower portion. Both refer to common potential scales of V_{rhe} and V_{she} shown in the scales at mid-level. The length of each arrow gives the magnitude of the quantity and the direction gives the sign. For example, the arrows for cathode overpotential η_c point to negative potential (left) since cathode overpotential is negative.

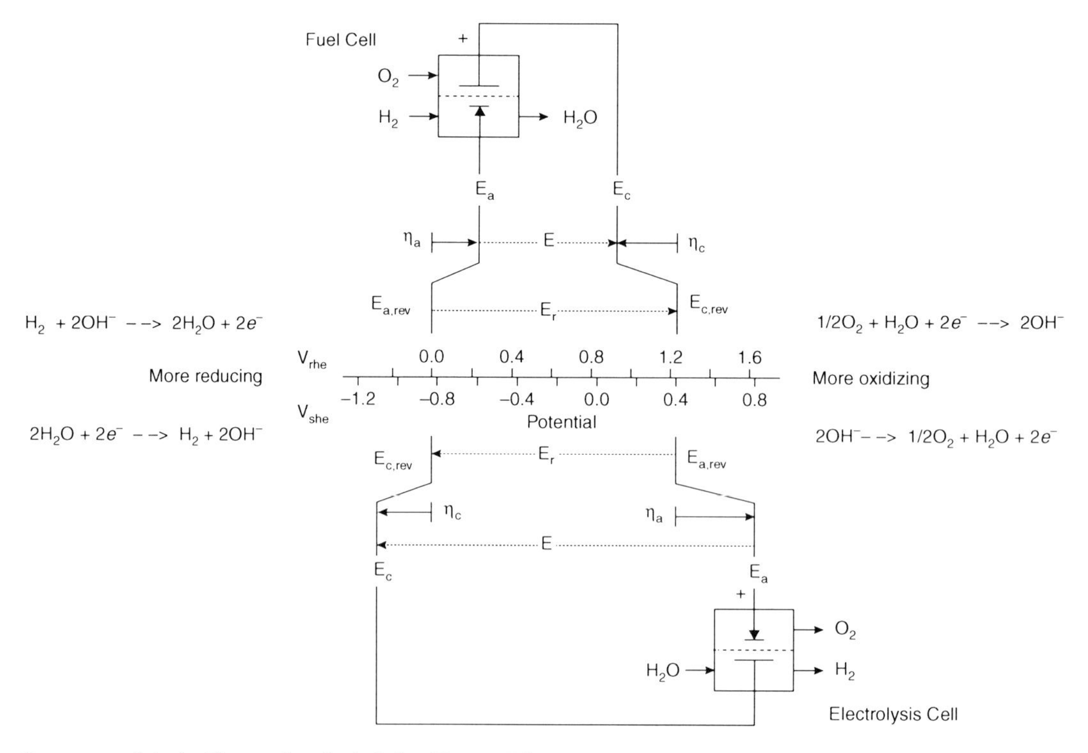

Overpotentials in Electrochemical Cells, Fig. 1 Effect of overpotential on a H_2/O_2 fuel cell (*above*) and an electrolysis cell (*below*), operating in alkaline medium

The symbol for the fuel cell and electrolysis cell is derived from the battery symbol; the longer and shorter lines represent, respectively, the cathode and anode, and the dashed line represents the electrolyte. An arrow drawn in the direction of positive current flow points toward an electrolyte with negative charge carriers, as in the manner of the transistor symbol. Galvanic and electrolytic cells are distinguished by the location of the positive terminal: a positive terminal at the cathode indicates a galvanic cell, while a positive terminal at the anode indicates an electrolytic cell. The outer box represents the system enclosure, which may or may not be open. The values of potential and overpotential are consistent with Table 2.

The effect of overpotential is immediately evident for both processes. The potential decreases substantially from E_r to E for the fuel cell and increases for the electrolysis cell. Cathode overpotentials are negative (to the left) and located at the positive terminal of the fuel cell (top) and negative terminal of the electrolysis cell (bottom). Anode overpotentials are positive (to the right) and located at the negative terminal of the fuel cell and positive terminal of the electrolysis cell. Both cells exhibit marked limitations at both anode and cathode, indicating, at a minimum, that slow reaction kinetics affect both electrodes. Also note the direction of current flow. Positive current of the fuel cell flows from high potential to low potential (to the left of the potential scale), meaning that it can do work. Conversely, positive current of the electrolysis cell flows in the opposite direction, meaning that it requires work.

Component Overpotentials

Overpotentials can be applied to an entire cell (η), electrodes (η_c, η_a), and to specific processes within the cell. Just as the cell overpotential can be divided into individual electrode overpotentials, Eq. 7, so too can the electrode overpotential

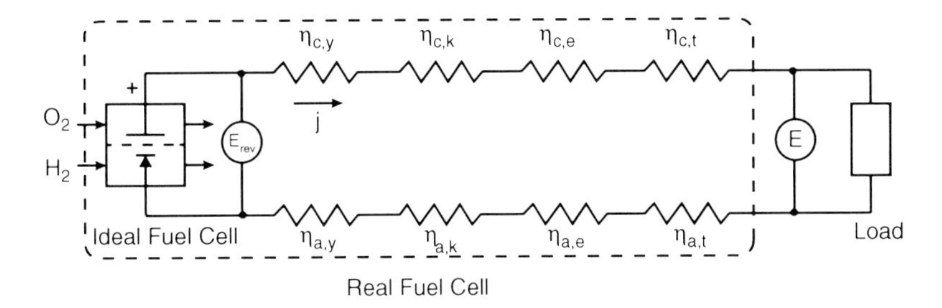

Overpotentials in Electrochemical Cells, Fig. 2 Equivalent circuit showing an ideal fuel cell at reversible potential E_{rev} in series with overpotentials representing electrolyte resistance, activation losses, electronic resistance, and concentration overpotential. Each component overpotential has cathodic and anodic components

be divided into its component overpotentials. The relationship of cell overpotential to its component overpotentials η_i is

$$\eta = \sum_i \eta_i. \tag{13}$$

In specific terms, the overpotential can be written as

$$\eta = \eta_y + \eta_k + \eta_e + \eta_t \tag{14}$$

where the component overpotentials are: η_y, electrolyte resistance; η_k, activation or reaction kinetics limitation; η_e, electronic resistance of electrodes; and η_t, concentration loss or mass transfer limitation. Each component relates to a specific property of the electrochemical cell and is composed of a cathode and an anode component. Thus, all of the component overpotentials can be expressed as

$$\eta = \eta_c - \eta_a = \sum_i \eta_{c,i} - \sum_i \eta_{a,i} \tag{15}$$

in which Eq. 14 has been written for the cathode and for the anode.

Figure 2 shows the equivalent circuit with component overpotentials represented by resistors. The circuit consists of an ideal fuel cell operating at its reversible potential, but non-idealities in the cell reduce the potential from E_r to E. The resistors symbolize the loss in potential for each component. Each resistor has a voltage drop, which establishes the overpotential for that resistor. Depiction of overpotentials as resistive elements is not entirely accurate, however. Electrolyte and electrode resistances are ohmic in nature, meaning that they behave Ohm's law with a constant resistance. Activation and concentration losses are not ohmic, so the use of resistors for these components is only for illustration.

Overpotentials in Electrochemical Cells, Table 3 Component overpotentials of electrochemical cells

Overpotential	Equation
Cell	$\eta = \eta_y + \eta_k + \eta_e + \eta_t$
Electrolyte resistance	$\eta_y = -j(R_{c,y} + R_{a,y})$
Activation (reaction kinetics)	See Eqs. 21, 25, 26, and discussion
Electronic resistance	$\eta_e = -j(R_{c,e} + R_{a,e})$
Concentration losses (transport)	$\eta_t = \frac{RT}{nF}\left[\ln(1 - j/j_{L,c}) + \ln(1 - j/j_{L,a})\right]$

Table 3 lists component overpotentials and corresponding equations, and Table 4 defines the terms used in Table 3. The components of electrolyte and electronic resistance are represented as ionic ($R_{c,y}, R_{a,y}$) and electronic ($R_{c,e}, R_{a,e}$) resistances, respectively. The electrolyte resistance can usually be assumed equal for the anode and cathode. The electronic resistance relates to specific properties of the electrode and is not necessarily equal for the anode and cathode.

In Table 3, it is worth noting that the entries for electrolyte resistance, electrode resistance, and concentration losses appear to have the same

Overpotentials in Electrochemical Cells, Table 4 Terms in component overpotential equations. Subscript "c" refers to the cathode, and subscript "a" refers to the anode

Symbol	Description	Remarks
$R_{c,y}, R_{a,y}$	Ionic resistance of electrolyte	$R_{c,y} = R_{a,y}$
$j_{o,c}, j_{o,a}$	Exchange current density	$j_{o,c} \neq j_{o,a}$
β_c, β_a	Symmetry factor	$\beta = 1/2$ (typ.)
$R_{c,e}, R_{c,a}$	Electronic resistance	$R_{c,e} \neq R_{a,e}$
$j_{L,c}, j_{L,a}$	Mass transfer limited current density	$j_L = nFDC_b/\delta$
D	Diffusion coefficient of reacting species	
C_b	Bulk concentration of reacting species	
δ	Thickness of diffusion layer	

sign of cathode and anode components of overpotential. Since every overpotential leads to lost work, according to Eq. 9, each component of overpotential must be negative when considered for the cathode and positive when considered for the anode. For the case of electrolyte resistance

$$\eta_{c,y} = -jR_{c,y} \tag{16}$$

$$\eta_{a,y} = jR_{a,y} \tag{17}$$

$$\eta_y = \eta_{c,y} - \eta_{a,y} \tag{18}$$

$$= -j(R_{c,y} + R_{a,y}) \tag{19}$$

which is the equation shown in Table 3. The expressions for electrode resistance and concentration losses in Table 3 are likewise derived. A general way to handle overpotentials as unsigned quantities is to rewrite Eq. 15 as

$$\eta = -|\eta_c| - |\eta_a| = -\sum_i |\eta_{c,i}| - \sum_i |\eta_{a,i}| \tag{20}$$

Equations 15 and 20 are fully equivalent, but a possible source of confusion can arise in evaluating the activation overpotential, as the Butler-Volmer equation (discussed next) requires *signed* values of cathode and anode overpotential.

The equation for the activation overpotential is derived from the Butler-Volmer equation, the general form of which is [1]

$$j = j_o\{\exp[(1-\beta)X] - \exp(-\beta X)\}, \tag{21}$$

where X is the dimensionless kinetic overpotential

$$X = \frac{nF\eta_k}{RT} \tag{22}$$

and the other terms are defined in Table 4. Equation 21 is written separately for the cathode and for the anode. By convention, the Butler-Volmer equation is written with reduction current as negative, though the thermodynamic convention treats current as unsigned. Thus, to make reduction current positive, the absolute value of j is used. A single-term approximation to the Butler-Volmer equation can be used for a sufficiently large overpotential, namely,

$$|X| \gtrsim 3.2, \tag{23}$$

which corresponds to anodic $j_{a,lim}$ and cathodic $j_{c,lim}$ current limits of

$$\left|\frac{j_{a,lim}}{j_{o,a}}\right| \gtrsim \exp[3.2(1-\beta_a)]; \quad \left|\frac{j_{c,lim}}{j_{o,c}}\right| \gtrsim \exp(3.2\beta_c). \tag{24}$$

For $\beta = 1/2$, the ratio j/j_o must be greater than 4.95 for the single-term Butler-Volmer equation to be valid. The single-term cathode and anode kinetic overpotentials are

$$\eta_{a,k} = b_a \log_{10}\left|\frac{j_a}{j_{a,k}}\right| \tag{25}$$

$$\eta_{c,k} = b_c \log_{10}\left|\frac{j_c}{j_{o,c}}\right|, \tag{26}$$

where b is the Tafel slope, defined by

$$b_a = 2.303\frac{RT}{(1-\beta_a)n_aF}; \qquad b_c = 2.303\frac{RT}{\beta_c n_c F}. \tag{27}$$

Equations 25 and 26 are used to measure Tafel slopes. The equations apply only when the criteria of Eq. 23 or 24 are met. Otherwise, the full Butler-Volmer equation, Eq. 21, must be solved iteratively to obtain kinetic overpotential as a function of current.

The concentration and transport overpotential can be estimated by solution of Fick's law of diffusion across a small layer of thickness 10–100 μm at the electrode surface. The result is [2]

$$\eta_{c,t} = \frac{RT}{nF} \ln\left(1 - \frac{j}{j_{c,L}}\right), \tag{28}$$

and the terms are defined in Table 4. A similar equation applies to the anode. Typical values of limiting current densities range from 1 to 1.5 A/cm^2. In some cases, higher values are obtained.

Summary

The concept of overpotential is a powerful means for quantifying the performance of an electrochemical system. It applies at multiple levels: the overall cell, the individual electrodes, and the processes occurring at those electrodes. All components of overpotential can be measured, so the method is of value in any electrochemical application. In most cases, component overpotentials may be considered independently. This simplification greatly aids the development of an electrochemical system as it enables isolated study of a part of the overall process. Since overpotentials are a deviation from ideality, which is the condition of maximum work, they quantitatively express the work lost in a system that results from a particular electrode process. In so doing, they express the inefficiency of electrochemical systems.

Cross-References

▶ Electrocatalysis - Basic Concepts, Theoretical Treatments in Electrocatalysis via DFT-Based Simulations
▶ Electrode

References

1. Newman J, Thomas-Alyea KE (2004) Electrochemical systems, 3rd edn, Electrochemical society series. Wiley, Hoboken
2. O'Hayre R, Cha S-W, Colella W, Prinz FB (2006) Fuel cell fundamentals. Wiley, Hoboken

Oxide Ion Conductor

Tatsumi Ishihara
Department of Applied Chemistry, Faculty of Engineering, International Institute for Carbon Neutral Energy Research (WPI-I2CNER), Kyushu University, Nishi ku, Fukuoka, Japan

Oxide ion conductivity generally occurs through oxygen vacancy which could be introduced by doping aliovalent cation. In case of perfect crystal which means no defects, no ion can diffuse excepting for interstitial position. However, after vacancy once introduces in lattice position, vacancy can diffuse and so ion conductivity is appeared. The typical theory for ion diffusion is treated with random work theory, [1, 2] in which diffusion of vacancy is related with jump frequency and number of jump site. Therefore, diffusivity of oxygen vacancy is strongly related with crystal structure and kind of dopant. Typical case for oxide ion conductor is stabilized ZrO_2 and the oxygen vacancy is introduced as the following equation for the case of Y doping.

$$Y_2O_3 = 2Y_{Zr}{}' + 3O_O{}^X + V_O{}^{\bullet\bullet} \tag{1}$$

Oxide ion conductivity was first found in ZrO_2 with 15 wt% Y_2O_3 (stabilized zirconia denoted as YSZ) by Nernst [3] as early as 1899 and so, the history of oxide ion conductor is longer than a century. Figure 1 shows the comparison of oxide ion conductivity in typical materials. In the history of oxide ion conductor, fluoride structure oxides consisting of tetravalent cation (ZrO_2, CeO_2, ThO_2 etc.) have been widely studied.

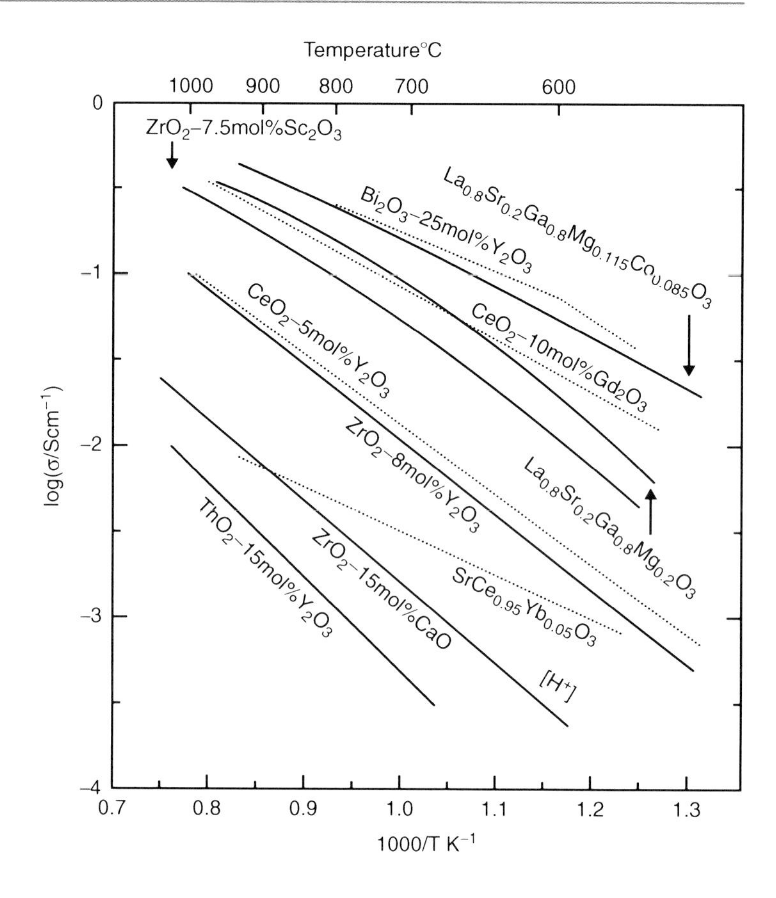

Oxide Ion Conductor, Fig. 1 Comparison of oxide ion conductivity in typical materials

Fluorite-type structure is a face-centered cubic arrangement of cations with anions occupying all tetrahedral sites. It has a large number of octahedral interstitial free volume sites. Thus this structure is a rather open structure and rapid ion diffusivity is achieved. Oxide ion conductivity has a strong relationship between oxygen vacancies formation and its diffusivity. Therefore, in the previous open literature, there are many reports studied on fluorite oxide and suggest that ionic size of dopant is highly important for determining the oxide ion conductivity.

ZrO_2 has typically three crystal structures, Monoclinic, Tetragonal (>1,443 K), and Cubic (>2,643 K) [4]. In pure ZrO_2, cubic phase is stable only at high temperature and at room temperature, monoclinic phase is the stable one from chemical equilibrium. ZrO_2 is principally classified as an electrical insulator. As discussed, in case of Y doped ZrO_2, formation of oxygen vacancy shown by Eq. 1 is essential. Here, high temperature phase of cubic structure is stabilized down to room temperature by substituting the lattice position with lower valence cation, and so oxygen vacancy introduced ZrO_2 is called "stabilized ZrO_2," which includes tetragonal (partially stabilized) and cubic (fully stabilized) phase. Addition of Y_2O_3 to ZrO_2 reduces the tetragonal/monoclinic transformation temperature. The minimum amount required to fully stabilize the cubic phase of ZrO_2 is about 8–10 mol% at 1,273 K. Other ZrO_2-M_2O_3 system where M is Y, Sc, Nd, Sm, or Gd have also shown stabilized solution in a certain amount range. The minimum amount of dopant necessary to stabilize ZrO_2 in cubic structure is close to the composition which gives the highest conductivity (8 mol% Y_2O_3, 10 mol% Sc_2O_3, 15 mol% Nd_2O_3, 10 mol% Sm_2O_3, and 10 mol% Gd_2O_3).

The concentration of the vacancies is given simply by the electron neutrality condition. In this case, therefore, $2[Y_{Zr}'] = [V_o^{\bullet\bullet}]$ for Y_2O_3

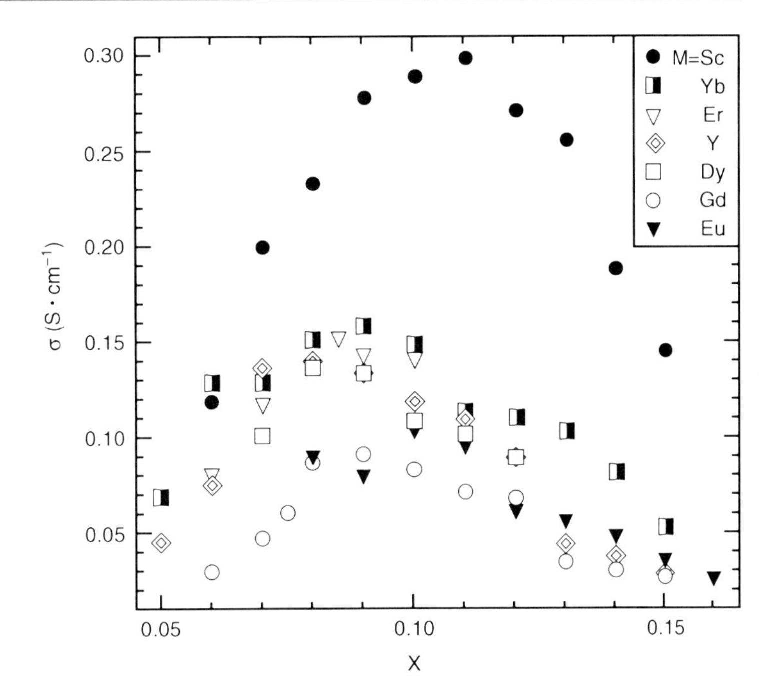

Oxide Ion Conductor, Fig. 2 Oxide ionic conductivity in ZrO_2 as a function of dopant amount

stabilized ZrO_2. On the other hand, it is well known that the ionic conductivity, σ, can be expressed by Eq. 2,

$$\sigma = en\mu \quad (2)$$

where n is the number of mobile oxide ion vacancies, μ its mobility, and e the charge (in case of oxide ion, 2). By introduction of oxygen vacancy, oxide ion conductivity is appeared in cubic phase ZrO_2. Therefore, the vacancy concentration is linearly dependent upon the dopant level. However, this is not true and the higher dopant concentration just leads to the formation of vacancy and dopant cluster resulting in the decreased oxide ion conductivity.

The conductivity of stabilized ZrO_2 varies with dopant concentration. As shown in Fig. 2, the conductivity of doped zirconia shows a maximum at a specific concentration of dopant [5]. It is obvious that the electrical conductivity in ZrO_2 is strongly dependent on the dopant element and its concentration. In the small doping amount, conductivity monotonically increases with increasing the dopant amount, that is expected from the theory and evidently, introduced defects behave as a point defect. Therefore, conductivity is mainly determined by the amount of oxygen vacancy, namely, the amount of dopant. On the other hand, conductivity as well as activation energy for conduction are strongly affected by dopant ionic size, which is reported by Arachi et al. [6] for the ZrO_2-Ln_2O_3 (Ln = lanthanide) system. Figure 3 shows the maximum oxide ionic conductivity and apparent enthalpy of oxide ion transport as a function of ionic size of dopant. It is evident that the conductivity increased with decreasing ionic size of doped cations. Explanations for this conductivity behavior are tried based on structural effects. The content of dopant with the highest conductivity in the ZrO_2-Ln_2O_3 system decreases with increasing dopant ionic radius. The dopant, Dy^{3+} and Gd^{+3}, with larger ionic radii shows a limiting value of 8 mol%. The dopant Sc^{3+}, which has the closest ionic radius to the host ion, Zr^{4+}, shows the highest conductivity and the highest dopant content at which cluster formation starts. Scandia doped zirconia is quite attractive as the electrolyte for solid oxide fuel cells (SOFCs), especially for the intermediate temperature (873–1,073 K) SOFCs and as-sintered sample shows the conductivity of 0.3 S/cm at 1,273 K. However, the degradation in conductivity is more

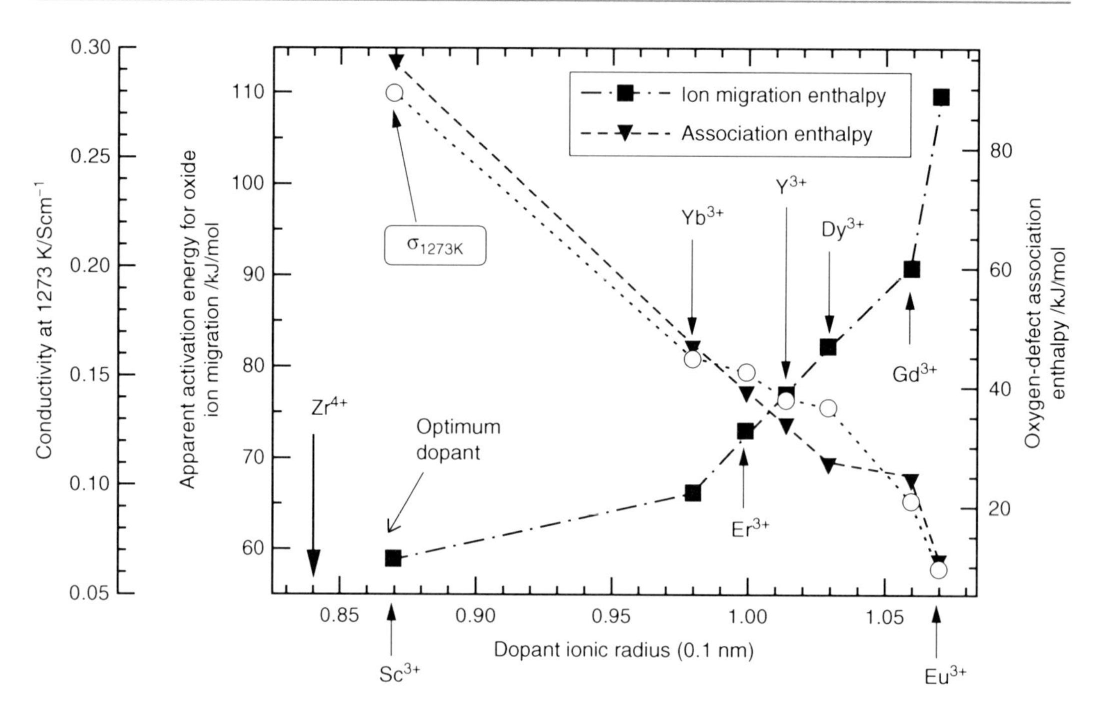

Oxide Ion Conductor, Fig. 3 Maximum oxide ionic conductivity and apparent enthalpy of oxide ion transport as a function of ionic size of dopant

clearly observed for Sc stabilized ZrO_2 due to annealing effects, which is a result of phase change from cubic to tetragonal phase.

Doped ceria is also an alternative oxide ion conductor which shows higher ionic conductivity than ZrO_2 [7]. Ceria possesses the same fluorite structure as that of stabilized zirconia. In a similar manner with ZrO_2, mobile oxygen vacancies are introduced by substituting Ce^{4+} with trivalent rare earth ions, typically, Gd^{3+} or Sm^{3+}. The conductivity of doped ceria systems also depends on the kind of dopant and its concentration, which is similar with the case of stabilized ZrO_2. Since cubic phase is originally stable phase for ceria, oxide ion conducting ceria is called "doped CeO_2."

Ceria based oxide ion conductors have purely ionic conductivity at high oxygen partial pressures. At lower oxygen partial pressures, the materials become partially reduced. This leads to an electronic conductivity in a large volume fraction of the electrolyte extending from the anode side [7].

Another promising group for oxide ion conducting materials is Perovskite. Although the oxide with perovskite structure is anticipated to be a superior oxide ion conductor, typical perovskite oxides such as $LaCoO_3$ and $LaFeO_3$ are known as a famous mixed electronic and oxide ionic conductors so called "mixed conductor." On the other hand, as shown in Fig. 1, the high oxide ion conductivity in $LaGaO_3$ based perovskite, which is the first pure oxide ion conductor in Perovskite group is reported in 1994 [8]. The high oxide ion conductivity in this oxide is achieved by double doping of lower valence cation into A and B site of perovskite oxide, ABO_3. It is obvious that the oxide ion conductivity strongly depends on the cations for A site and the highest conductivity is achieved on $LaGaO_3$, which is also the largest unit lattice volume in Ga based perovskite among $LnGaO_3$ (Ln; rare earth cation). The electrical conductivity of Ga-based perovskite oxides is almost independent of the oxygen partial pressure in P_{O2} range from 1 to 10^{-21} atm. Therefore, it is expected that the oxide ion conduction will be dominant in doped $LaGaO_3$ perovskite oxides.

Recently, there are some oxides with new crystal structure reported for promising oxide

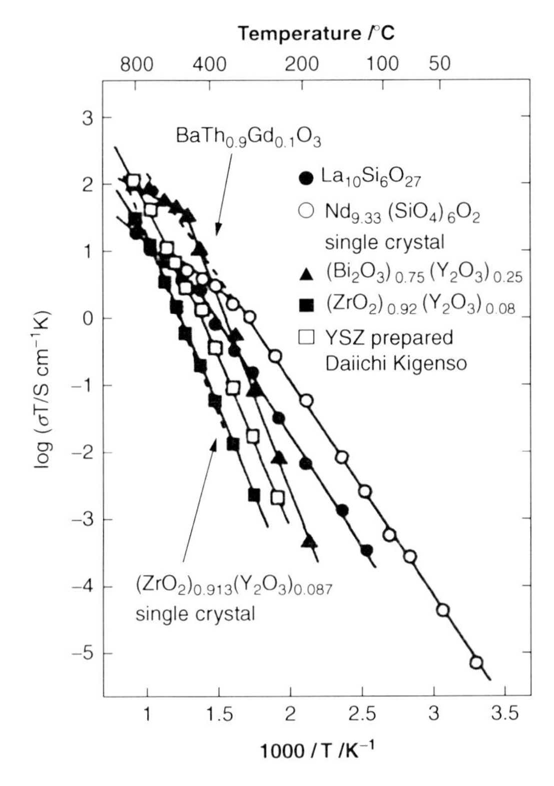

Oxide Ion Conductor, Fig. 4 Comparison of oxide ion conductivity in $La_{10}Si_6O_{27}$ oxide with those of doped bismuth oxides and zirconia

ion conductor. So far there have been no reports in the literature for a notable oxide ion conductivity in non-cubic structured oxides. Among few exceptions, the oxide ion conductivity in hexagonal apatite oxide of $La_{10}Si_6O_{27}$ and $Nd_{10}Si_6O_{27}$ is reported by Nakayama et al. [9]. Figure 4 shows the comparison of oxide ion conductivity in $La_{10}Si_6O_{27}$ oxide with those of doped bismuth oxides and zirconia. The electrical conductivity values of this oxide at temperatures higher than 873 K are not high enough compared with that of conventional fast oxide ion conductors such as 8 mol% Y_2O_3 stabilized ZrO_2. However, at lower temperatures, $La_{10}Si_6O_{27}$ exhibits higher oxide ion conductivity than those of the conventional oxide ion conductors. The refinement structure suggests that the $La_{10}Si_6O_{26}$ oxide has a unique structure in channel oxygen sites and the high oxygen ion conductivity could be assigned to this disorder in the channel site. In an analogue of $La_{10}Si_6O_{10}$ hexagonal apatite, La_2GeO_5 is also reported as a high oxide ion conductor [10].

On the other hand, bismuth based oxide, so called BIMEVOX, is also reported as high oxide ion conductor with non-cubic structure, however, this oxide exhibits wholly oxide ionic conductivity only in a limited P_{O2} range [11] and is not interesting for the electrolyte because of high electronic conductivity in reducing atmosphere. Oxide ion conductivity in β-phase $La_2Mo_2O_9$ is reported at 2000 by Lacorre et al. [12]. This oxide also shows large jump in conductivity around 973 K, at which temperature, α-phase is transferred to β-phase, more disordered structure. In this oxide, much higher electronic conduction was reported like Bi_2O_3, however, this oxide is easily reduced to form electronic conduction. Therefore, application to SOFC electrolyte is highly difficult. On the other hand, by doping lower valence cation, in particular, W for Mo site or Ca, Sr for La site, stability against reduction becomes little better [13]. It is reported that the highest conductivity in this series is achieved at $La_2Mo_{1.7}W_{0.3}O_9$ and the conductivity is around $\log(\sigma/\mathrm{Scm}^{-1}) = -0.8$ at 1048 K. Although chemical stability in reducing atmosphere is improved by dopant and also conductivity increases, higher electrical conductivity in reducing atmosphere is still not enough for electrolyte of SOFCs.

Transport number t_i, is a ratio of coulomb number carried by ion and electronic charge, i.e., electron and hole and t_i is an important when the materials are considered for electrolyte. For application to electrolyte of fuel cell or battery, at least, t_i should be higher than 0.99, which mean 99 % of oxide ion and 1 % of electric charge conducting. Figure 5 shows the evaluated boundaries of the electrolytic domain ($t_i > 0.99$) of various oxide ion conductor plotted in the axis of log (P_{O2}/atm) versus reciprocal temperature. The lower boundary of the electrolytic domain (defined as $t_{ion} > 0.99$) for LSGM is 10^{-23} atm at 1273 K. This pressure is even lower than that of CaO-stabilized ZrO_2 and that of YSZ. Evidently, electrolyte domain expanded with Consequently, it is clear that the electrolytic domain of LSGM or YSZ covers the P_{O2} range required for the operation of SOFCs.

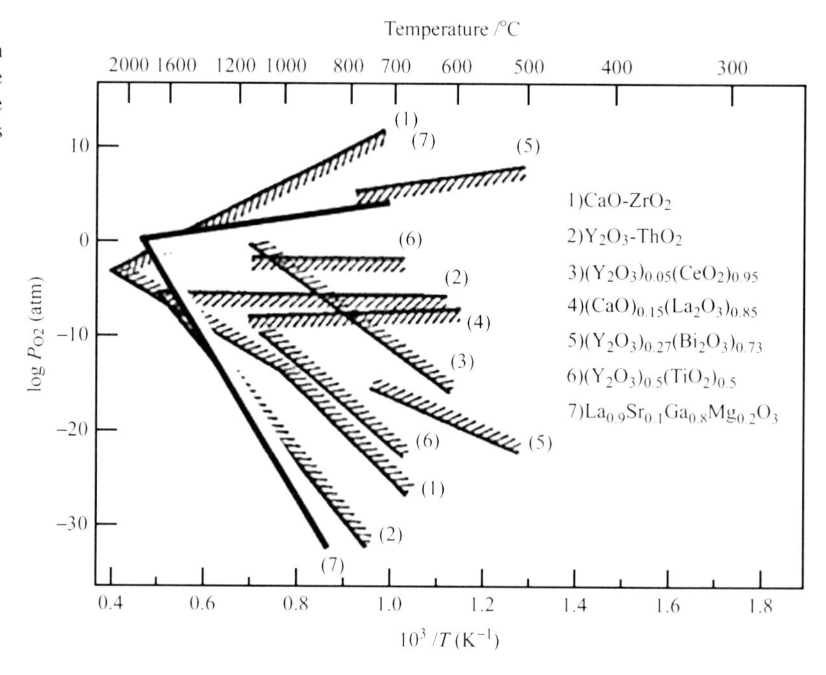

Oxide Ion Conductor, Fig. 5 Electrolytic domain ($ti > 0.99$) of various oxide ion conductor plotted in the axis of log (P_{O2}/atm) versus reciprocal temperature

As discussed, oxide ion conductivity in the conventional materials is mainly appeared through oxygen vacancy in lattice, and so oxide with cubic lattice is preferable for fast oxide ion conductivity. However, in the newly developed materials, interstitial oxygen contributes the fas oxide ion conductivity. Therefore, new oxide ion conductor family based on interstitial oxygen will be more popular in future. For this future directions, study on oxide ion conductivity in defect perovskite in which interstitial oxygen could be easily introduced, will be studied intensively.

Cross-References

▶ Solid Electrolytes
▶ Solid Oxide Fuel Cells, Introduction

References

1. Glicksman ME (2000) Diffusion in solids. Wiley, New York, p 191
2. Montroll EW, Weiss GH (1967) Random walks on lattices. II. J Math Phys 6:167
3. Nernst W (1899) Uber die elektrolytische Leitung fester Korper bei sehr hohen Temperaturen. Z Elektrochem 6:41
4. Scott HG (1975) Phase relationships in the zirconia-yttria system. J Mater Sci 10:1527
5. Baumard JF, Abelard P (1984) Science and technology of Zirconia II. In: Claussen N, Ruhle M, Heuer AH (eds) American ceramic Society, Ohio, p 555
6. Arachi Y, Sakai H, Yamamoto O, Takeda Y, Imanishi N (1999) Electrical conductivity of the ZrO2–Ln2O3 (Ln=lanthanides) system. Solid State Ionics. 121:133
7. Tuller HL, Nowick AS (1975) Doped ceria as a solid oxide electrolyte. J Electrochem Soc 122:255
8. Ishihara T, Matsuda H, Takita Y (1994) Doped LaGaO3 perovskite type oxide as a new oxide ionic conductor. J Am Chem Soc 116:3801
9. Nakayama S, Sakamoto M (1998) Electrical properties of new type high oxide ionic conductor RE10Si6O27 (RE = La, Pr, Nd, Sm, Gd, Dy). J Eur Ceram Soc 18:1413
10. Ishihara T, Arikawa H, Akbay T, Nishiguchi H, Takita Y (2001) Nonstoichiometric La2-xGeO5-δ monoclinic oxide as a new fast oxide ion conductor. J Am Chem Soc 123:203
11. Abraham F, Boivin JC, Mairesse G, Nowogrocki G (1990) The bimevox series: A new family of high performances oxide ion conductors. Solid State Ionics 40–41:934
12. Lacorre P, Goutenoire F, Bohnke O, Retoux R, Laligant Y (2000) Designing fast oxide-ion conductors based on La2Mo2O 9. Nature 404:856
13. Tealdi C, Malavasi L, Ritter C, Flor G, Costa G (2008) Lattice effects in cubic La2Mo2O9: Effect of vacuum and correlation with transport properties. J Solid State Chem 181:603

Oxide Ion Conductor Steam Electrolysis

Hiroshige Matsumoto
Kyushu University, Fukuoka, Japan

Introduction

Steam electrolysis splits water in the form of steam into hydrogen and oxygen by use of electricity and thus can be used as a method to produce hydrogen. Unlike other types of electrolysis, e.g., alkaline water electrolysis and polymer electrolyte water electrolysis, the steam electrolysis operates typically at 700–1,000 °C since the electrolysis uses solid electrolyte that works at the high temperatures. Such high operation temperature leads to fast kinetics for the electrode reactions, so that precious metals are not necessary for the electrocatalysis.

Working Principle

Schematic illustration of steam electrolysis is shown in Fig. 1(a). Oxide ion conductors are used as the solid electrolyte. Steam is introduced to the cathode compartment, and hydrogen is generated at the cathode on sending a direct current to the electrolysis cell.

The cathode reduces a gaseous water molecule to generate a hydrogen molecule and an oxide ion by the following electrode reaction:

$$H_2O\ (g) + 2e^- \rightarrow H_2(g) + O^{2-}(\text{cathode}) \quad (1)$$

The anode oxidizes the oxide ions to generate oxygen by the following electrode reaction:

$$O^{2-} \rightarrow 1/2O_2(g) + 2e^-(\text{anode}) \quad (2)$$

The electrode reactions involve hydrogen, oxygen, and/or water vapor in the gaseous phase, oxide ion in the electrolyte phase, and electron in the electrode phase so that the reactions take place at the boundary of the three phases; an oxide-ion-electron mixed conducting electrode can work at the two-phase boundary with the gaseous phase.

Other types of electrolysis, alkaline water electrolysis and polymer electrolyte water electrolysis, typically work in the temperature range from room temperature to 80–150 °C. In contrast, the steam electrolysis operates at much higher temperature, and thus electrode overpotential is low, and precious metals are not necessary. Steam electrolysis can be regarded as a reverse operation of a hydrogen-oxygen solid oxide fuel cell, and thus, the materials for electrodes and electrolytes are similar to each other. Since the operation temperature is rather high, materials not only for electrode and electrolyte but also for the housing and gas tubing should be durable to the temperature.

Materials

Stabilized zirconia, e.g., $(ZrO_2)_{0.92}(Y_2O_3)_{0.08}$, is most typically used as the electrolyte. $LaGaO_3$-based oxide ion conductors, which are characterized by high ionic conductivity, are also useful. Proton-conducting oxide, e.g., $BaZr_{1-x}Y_xO_{3-\delta}$, ($x = 0.05$–$0.2$), is available as an electrolyte for the steam electrolysis. As shown in Fig. 1(b), steam is introduced to the anode, and hydrogen is generated at the counter electrode. This is a merit of the use of proton-conducting electrolyte, due to no need of the separation of hydrogen from steam that is required in the case of oxide-ion conducting steam electrolyzers (Fig. 1a).

Porous nickel or the Ni/YSZ cermet is used as the cathode (hydrogen-evolving electrode). The latter is a composite of nickel and electrolyte and is effective in increasing the amount of the three-phase boundary to enhance the cathode activity. The anode (oxygen-evolving electrode) consists of complex oxide containing transition metals as a component cation and thus has high electronic conductivity. $La_{1-x}Sr_xMnO_{3-\delta}$, $La_{1-x}Sr_xCo_{1-y}Fe_yO_{3-\delta}$, and $Sm_{0.5}Sr_{0.5}CoO_{3-\delta}$ are the example of the anode materials.

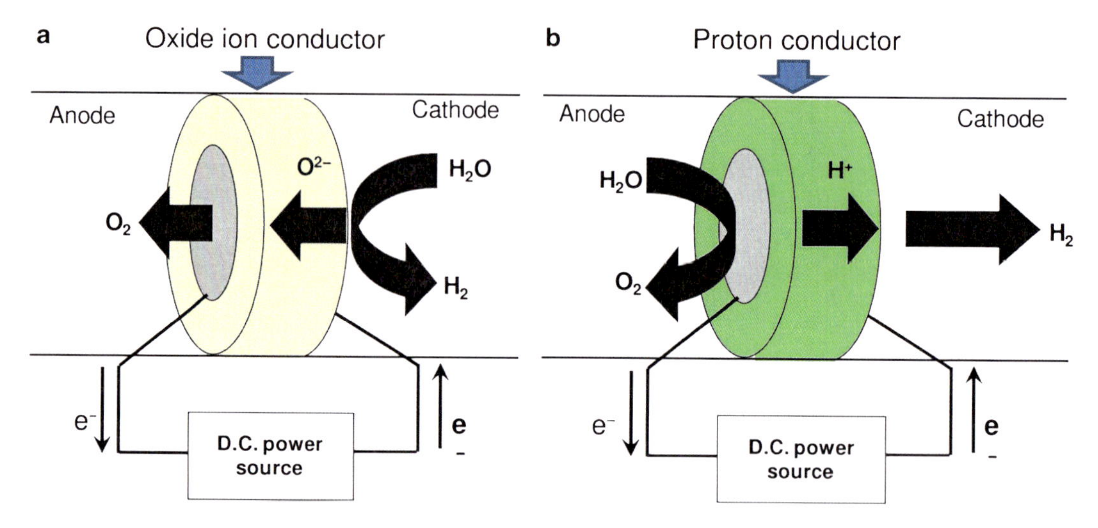

Oxide Ion Conductor Steam Electrolysis, Fig. 1 Schematic illustrations of steam electrolysis using (**a**) oxide ion conductor and (**b**) proton conductor

Thermodynamics

The sum of the electrode reactions in (Eq. 1) and (Eq. 2) is the total water splitting reaction.

$$H_2O\ (g)\ \rightarrow\ H_2(g) + 1/2O_2(g) \qquad (3)$$

Gibbs free energy change of the reaction in (Eq. 3), $\Delta_r G$ that is positive, is equivalent to the minimum energy required as work. Thus, the energy divided by $2F$ (i.e., $\Delta_r G/2F$) is the theoretical electrolysis voltage that is the minimum voltage for the steam electrolysis, V_{th}.

$$V_{th} = \frac{\Delta_r G}{2F} = \frac{\Delta_r G^o}{2F} + \frac{RT}{2F}\ln\frac{pH_2, C \cdot p_{O_2}^{1/2}, A}{pH_2O, C} \qquad (4)$$

F is the Faraday constant, $\Delta_r G^o$ is the standard Gibbs free energy of the reaction in (Eq. 3), and p is the partial pressure of the subscript gas species in either anode or cathode compartment distinguished by the subscript of A or C. In actuality, a voltage higher than the theoretical one must be applied to the steam electrolysis cell in order to precede the electrode reactions. The difference from the theoretical voltage consists of the anode and cathode overvoltages (η_A,η_C) and the ohmic loss (Ri).

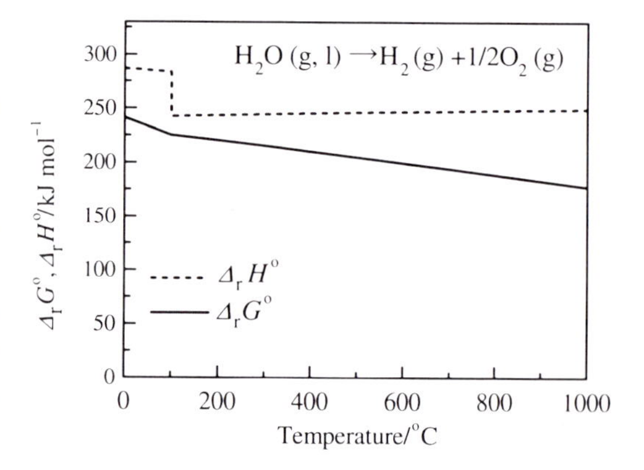

Oxide Ion Conductor Steam Electrolysis, Fig. 2 Standard free energy and enthalpy of water split reaction (Eq. 3)

$$V_t = V_{th} + \eta_A + \eta_C + Ri \qquad (5)$$

The standard Gibbs energy and enthalpy of water splitting reaction ($\Delta_r G^o$, $\Delta_r H^o$) are shown in Fig. 2 as a function of temperature. $\Delta_r G^o$ decreases with temperature. According to (Eq. 4), the theoretical steam electrolysis voltage can be expressed as $\Delta_r G^o/2F$ when all the gas species taking part in the water splitting reaction are assumed to be in the standard state. Therefore, the electrolysis voltage decreases with temperature. For example, the theoretical voltage is 1.23 V at room temperature (25 °C) and is 0.92 V at 1,000 °C. On the other hand, $\Delta_r H^o$-$\Delta_r G^o$ stands for heat additionally needed for electrolysis under reversible condition. $\Delta_r H^o/2F$ is a voltage for thermal neutrality. When the voltage is lower than $\Delta_r H^o/2F$, steam electrolysis is endothermic, that is, heat should be supplied from outside to maintain the cell temperature. The quantity of $\Delta_r H^o/2F$ is not so dependent on temperature, 1.28–1.29 V in the temperature range of 700–1,000 °C. Hence, the difference between $\Delta_r H^o/2F$ and $\Delta_r G^o/2F$ becomes larger with increasing temperature.

In Practice

As a practical application, steam electrolysis has been studied as a hydrogen production method [1–7]. Since hydrogen can be produced much cheaply by reforming of fossil fuels, steam electrolysis cannot pay economically and is not widely commercialized at present. However, increasing concern about the climate change stimulates the shift of the fossil energy to the renewable ones. As a highly efficient hydrogen method from the renewable energies, steam electrolysis is considered to be an important technique.

Cross-References

▶ Oxide Ion Conductor
▶ Solid Electrolytes
▶ Solid Electrolytes Cells, Electrochemical Cells with Solid Electrolytes in Equilibrium

References

1. Doentz W et al. (1980) Hydrogen production by high temperature electrolysis of water vapour. Int J Hydrogen Energy 5:55–63
2. Isenberg AO (1981) Energy conversion via solid oxide electrolyte electrochemical cells at high temperatures. Solid State Ionics 3(4):431–437
3. Maskalick NJ (1986) High temperature electrolysis cell performance characterization. Int J Hydrogen Energy 9:563–570
4. Kusunoki D et al. (1995) Development of Mitsubishi-planar reversible cell–Fundamental test on hydrogen-utilized electric power storage system. Int J Hydrogen Energy 20:831–834
5. Hino R et al. (2004) R&D on hydrogen production by high-temperature electrolysis of steam. Nuc Eng Des 233:363–375
6. O'Brien JE (2005) Performance Measurements of Solid-Oxide Electrolysis Cells for Hydrogen Production. J Fuel Cell Sci Technol 2:156–163
7. Hauch A et al. (2008) Highly efficient high temperature electrolysis. J Mater Chem 18:2331–2340

Oxygen Anion Transport in Solid Oxides

Steven McIntosh
Department of Chemical Engineering, Lehigh University, Bethlehem, PA, USA

Introduction

Crystalline solid oxides with facile oxygen anion transport find application in solid oxide fuel cells (SOFCs), solid oxide electrolysis cells (SOECs), solid oxide electrochemical reactors (SOERs), and oxygen ion transport membranes (ITMs or OTMs). These technologies efficiently convert between chemical and electrical energy (SOFCs and SOECs), generate value-added chemicals (SOERs), and separate oxygen from air (ITMs and OTMs). The operating temperature of current solid oxide electrochemical systems (typically > 750 °C) is primarily dictated by the high activation energy for oxygen anion transport in solid oxide materials. This high operating temperature increases the cost of balance of plant components and decreases the cell lifetime; thus there is great interest in developing electrolyte and electrode materials that provide high conductivity at reduced temperatures. The materials of interest are crystalline metal oxides with typically more than one metal cation in the lattice. Understanding how material composition and crystal structure dictate ion transport is essential to guide

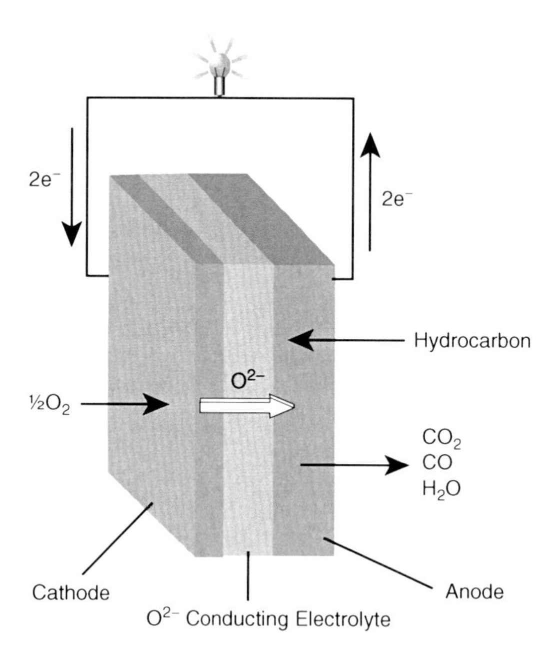

Oxygen Anion Transport in Solid Oxides, Fig. 1 Schematic of Solid Oxide Fuel Cell (SOFC) operation with a hydrocarbon fuel. Note that the reaction stoichiometry is unbalanced in this general example [8]

researchers in this area. The material compositions developed by researchers in this field can appear confusing to the novice researcher; for example, the materials $La_{0.6}Sr_{0.4}Co_{0.8}Fe_{0.2}O_{3-\delta}$ is a commonly utilized air electrode material for SOFCs. The goal of this chapter is to explain the origin of such material compositions and provide the reader with a deeper understanding of oxygen anion transport in solids. This brief overview uses illustrative examples of real materials to introduce the reader to some of the numerous complexities of this topic. The reader is directed to some of the more in-depth literature reviews for a deeper understanding of these topics [1–7].

SOFCs, SOECs, and SOERs all consist of three parts: an electrolyte, an anode, and a cathode. The electrolyte is an oxygen anion conducting but electrically insulating solid oxide. This dense ceramic layer also acts as a barrier to prevent gas mixing between the two gas chambers. If we take the SOFC as an example, Fig. 1, the cathode is in contact with air, while the anode is in contact with fuel. The difference in oxygen partial pressure (pO_2) between these two electrodes provides a driving force for oxygen anion reduction at the cathode, anion transport through the electrolyte, oxidation of fuel at the anode, and electron transport through the external circuit. This series of processes defines the operating mechanism of the cell to provide useful work. The theoretical, equilibrium, maximum driving force is given by the Nernst potential and is ~1.1 V under typical operating conditions with air at the cathode and humidified hydrogen fuel at the anode. Assuming no electronic leakage through the electrolyte and good sealing of the cell, a typical SOFC will provide this cell potential at open circuit (zero current). Under operation, this driving force is consumed by kinetic barriers associated with all of the operating processes in the cell. To maximize power density we wish to maximize the current generated for a given potential drop from the open circuit value. That is, we wish to minimize the cell resistance by minimizing the kinetic barriers to ion and electron transport, electrochemical reaction, and gas transport. Here we will focus on oxygen anion transport.

SOECs operate by a similar principle but in the reverse direction; instead of consuming fuel to generate electrical power, they consume electrical power to generate a fuel through the reduction of an oxidized species. Typical targets are the generation of H_2 and/or CO by reduction of H_2O and CO_2 [9, 10]. One potential application is to design a reversible SOFC/SOEC system for use in large scale storage of renewable energy. SOERs are typically operated in the same 'direction' as SOFCs but instead to total oxidation of the fuel, which generates the maximum current per unit fuel, the goal of the SOER is partial oxidation to generate value added chemicals [11]. ITMS and OTMs typically operate with a single mixed ionic and electronically conducting dense membrane. While this membrane acts as a barrier to gas mixing, the 'internal' electrical short removes the need for an external electrical circuit. Thus the membrane operates to generate maximum ion flux, but does not generate electrical power. These membranes are of interest primarily for air separation [12, 13], but can also find application in partial oxidation and oxidative coupling reactions [14].

Oxygen anion transport in crystalline solid oxides primarily occurs via a thermally activated hopping mechanism from occupied to vacant oxygen sites on the crystal lattice. Thus the anion conductivity is dictated by the density of vacant sites, energy barrier to hopping, and site-to-site distance and curvature along the crystallographic pathway. Oxygen anion vacancies are generated either through doping with cations of lower oxidation state or by reduction of one of the cations within the lattice. The reduction in total positive charge induced by these mechanisms is charge compensated either by the removal of negatively charged oxygen anions from the lattice, or by the increase in charge of another cation. This first mechanism generates the vacancies on the oxygen sublattice required for anion transport, while the second mechanism typically leads to an increase in *p*-type (hole mediated) or *n*-type (electron mediated) electronic conductivity due to changes in cation oxidation state. The stable cation oxidation states are also dictated by the temperature and oxygen partial pressure of interest. Thus the relative contributions of ionic and electronic carriers and the total conductivity of materials are functions of these parameters.

Whether doping is charge compensated by vacancy formation or the generation of electronic carriers depends on whether there is a redox state in the band gap between the O-2p valence band and the conduction band of crystal. If the band gap is sufficiently large, and there are no redox energy states in the band gap, aliovalent doping is typically charge compensated by the formation of ionic defects. In our case, we are primarily interested in doping with lower valence cations to generate oxygen anion vacancies. If there is a redox state in the band gap, aliovalent doping can be charge compensated by filling of this redox state with associated electronic conductivity. In our case, doping with a lower valence cation typically leads to charge compensation through the generation of polarons with associated *p*-type conductivity.

The oxygen stoichiometry of these materials is typically determined through thermogravimetry where the weight loss is measured as a function of temperature and pO_2. This relative measurement requires the determination of a reference oxygen stoichiometry. This is typically found by recording the mass change following reduction in H_2 at high temperature with the assumption that the material is fully reduced. This oxygen stoichiometry information can then be combined with X-ray diffraction (XRD) studies to understand the interaction between crystal structure and oxygen stoichiometry. An alternative technique is to utilize *in-situ* powder neutron diffraction that enables simultaneous measurement of both crystal structure and oxygen stoichiometry. In either case, these measurements can further be combined with measurements of total electrical conductivity as a function of temperature and pO_2 to determine the charge compensation mechanism.

The total electrical conductivity of a materials is given by [15]:

$$\sigma = \sum q_i \mu_i c_i$$

Where the summation is over all i charge carriers (ions, electrons, electron holes), and q_i is the carrier charge, μ_i is the carrier mobility, and c_i is the carrier concentration. Based on this, one simple approach to increasing the conductivity is to increase the carrier concentration. There are limitations to this in the case of oxygen vacancies, as vacancy clustering or ordering within the crystal can lead to a significant decrease in the mobility term. It is the high thermal activation energy (>0.5 eV) of this mobility term for ion hopping that leads to the high operating temperature of current solid oxide devices. The ionic mobility is related to the diffusivity, D_i, through the Nernst-Einstein equation [15]:

$$\mu_i = \frac{D_i q_i}{k_B T} \quad (1)$$

Where k_B is the Boltzmann constant. The diffusion coefficient can be estimated as

$$D_i = z v_0 a^2 \exp\left(\frac{-\Delta G_m}{k_{BT}}\right) \quad (2)$$

Where z is a geometric term, v_0 is the attempt frequency, a is the hop distance, and ΔG_m is the

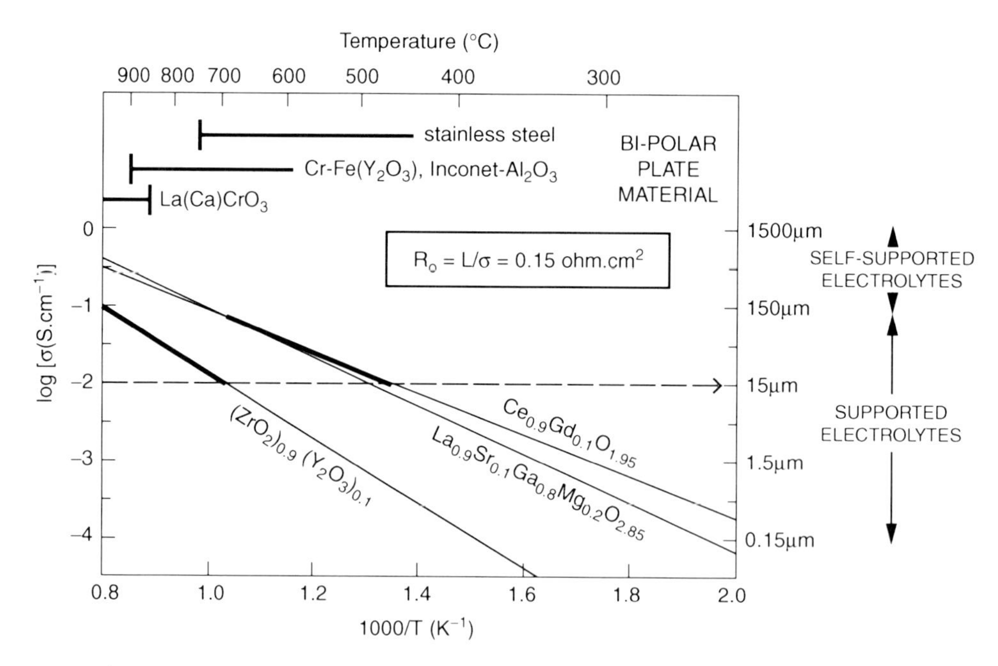

Oxygen Anion Transport in Solid Oxides, Fig. 2 Conductivity versus temperature for the three most common SOFC electrolyte materials. The required electrolyte thickness to meet the $0.15\Omega.cm^2$ performance target, and the temperature range for common materials used as interconnects in SOFC stacks are also noted [2]

free energy of migration. If we also assume that the formation of vacancies is thermally activated with a known heat of formation, ΔH_f, we can write:

$$C_i = C_{i0}\exp\left(\frac{-\Delta H_f}{2k_BT}\right) \quad (3)$$

If we combine what we have so far we get:

$$\sigma_{ion} = \frac{q_i^2}{k_BT}C_{i0}z\nu_0a^2\exp\left(\frac{-\Delta G_m}{k_{BT}}\right) \times \exp\left(\frac{-\Delta H_f}{2k_BT}\right) \quad (4)$$

If we assume that the entropy of vacancy migration, crystal geometry, attempt frequency, and hop distance are constant, we can write a general ionic conductivity as:

$$\sigma_{ion} = \frac{\sigma_0}{T}\exp\left(\frac{-\Delta H_A}{k_BT}\right) \quad (5)$$

Where ΔH_f is the sum of the migration and one half of the formation enthalpy. These assumed constant parameters are important to consider when we study these materials. In particular, changes in crysal morphology can lead to significant deviation from the idealized behavior presented above.

Electrolyte Materials

Electrolyte (pure ion conducting) materials form the separating membrane between air and fuel components in SOFCs, SOECs, and SOERs. A commonly quoted performance criterion for SOFC electrolytes is that the electrolyte resistance, the ratio of electrolyte thickness to conductivity, should be less than 0.15 $\Omega.cm^2$ at the operating temperature, Fig. 2 [2]. Numerous constraints make development of high conductivity electrolyte materials extremely challenging for materials chemists. Materials are suitable electrolytes only if they possess negligible electronic conductivity and are thermodynamically stable under both the oxidizing and reducing conditions at the cell air and fuel. Further constraints include ease of manufacture, mechanical stability, and cost. Negligible electronic conductivity is essential to prevent the creation of an 'internal short' between the electrodes. Any electronic current going through the electrolyte instead of the external circuit represents decreased cell efficiency. The thermodynamic efficiency is reduced as the internal current flow reduces the electrical potential between the electrodes. This is accompanied by an operating inefficiency of wasted fuel as current passing through the electrolyte instead of around the external circuit consumes fuel but does not provide useful work. Electronic conductivity in solid oxide fuel cells is typically observed as an open circuit potential (OCP) below the theoretical limit.

Doped Zirconia

The most commonly utilized solid oxide electrolyte, yttria-stabilized zirconia, utilizes acceptor doping of Y_2O_3 into a ZrO_2 lattice to generate mobile oxygen vacancies. Y^{3+} is substituted for the Zr^{4+} in the lattice, with one oxygen vacancy generated by charge compensation for every two Y^{3+} substituted. This vacancy concentration is stable under both the oxidizing and reducing atmospheres of interest to application, due to the stability of both cations. Y^{3+} doping, typically at 8 mol% Y_2O_3, leads to sufficient oxygen vacancies to provide a relatively facile transport pathway with sufficient ionic conductivity at ~800 °C. Yttria doping also serves to stabilize the cubic fluorite structure of zirconia: yttria-stabilized zirconia (YSZ) is the commonly utilized term to refer to materials with 4–10 mol% Y_2O_3. YSZ is strong enough to enable fabrication of thin films, commonly less than 50 μm thick, with sintering temperatures in the 1,500 °C range.

Undoped zirconia does not show significant ionic conductivity under typical operating conditions as the Zr^{4+} oxidation state is very stable, such that ZrO_2 does not show a significant oxygen vacancy concentration. Undoped zirconia is cubic at high temperature but transforms to tetragonal and monoclinic polymorphs upon cooling. A cubic structure is often considered desirable as the transport pathway is homogenous in all directions through the crystal. This is particularly important when we consider that manufactured components are typically polycrystalline, with uncontrolled orientation of individual grains. Other dopants may be introduced into zirconia, with most trivalent rare earth oxides and divalent alkaline earth oxides leading to the formation of oxygen vacancies. However, most of these lead to lower oxygen anion conductivity. The highest reported conductivities are found for Sc_2O_3 doping but the increased cost of Sc over Y relative to the improvement in performance has limited wide adoption. If all of the charge difference upon doping is balanced by the formation of oxygen vacancies, the carrier density between will be independent of the dopant. Thus the observed differences in conductivity are due to secondary effects, primarily suggested to be the generation of local strain within the lattice that alters the local structure. The highest conductivities occur when the size of the dopant cation most closely matches the host, a condition which coincides with a minimum in the activation enthalpy for the vacancy-cation pairs [16].

Doped Ceria

Oxygen vacancies can be formed in fluorite structures CeO_2 by charge compensation upon reduction of the Ce^{4+} to Ce^{3+} to create $CeO_{2-\delta}$. However, a localized $4f^1$ electronic level exists within the band gap of these materials, trapping electrons and enabling small polaron (*p*-type) electronic transport in these materials. Thus, while CeO_2 can shows significant oxygen nonstoichiometry and associated ionic conductivity under reducing conditions, it is not a pure electrolyte. It also shows only a small oxygen vacancy concentration, and is thus a poor oxygen anion conductor under oxidizing conditions. Pure CeO_2 is not a suitable electrolyte as it does not meet the condition of significant, purely ionic, conductivity in both oxidizing and reducing conditions.

CeO_2 is most commonly doped with Gd_2O_3 to generate oxygen vacancies under oxidizing conditions: one oxygen vacancy is formed for every two Gd^{3+} atoms doped into the Ce^{4+} lattice. These GD-doped CeO_2materials are commonly referred to as Gadolinia-doped Ceria (GDC) or Ceria Gadolinium Oxide (CGO). Unlike reduction of Ce^{4+}, doping with Gd does not introduce an electronic energy level into the band gap, and thus these materials are pure ionic conductors under oxidizing conditions. However, as with pure CeO_2, reduction of Ce^{4+} to Ce^{3+} under reducing conditions does lead to *p*-type electronic conductivity. Nevertheless, the two most common compositions $Gd_{0.1}Ce_{0.9}O_{1.95-\delta}$ and $Gd_{0.2}Ce_{0.8}O_{1.9-\delta}$ show significantly greater ionic conductivity that YSZ in the temperature range of interest to application; the same conductivity is achieved at ~ 200 °C lower temperature with CGO compared with YSZ. This comes at the expense of a reduced open circuit cell potential (OCP) due to the internal electronic transport in the electrolyte.

Doped Lanthanum Gallates

Unlike fluorite structured YSZ and CGO, the electrolyte material $La_{1-x}Sr_xGa_{1-y}Mg_yO_3$ (LSGM) [6] is a perovskite structured oxide. The perovskite oxide structure has general composition ABO_3. While a wide range of materials can form the perovskite structure, A is typically occupied by lanthanide or alkaline earth elements and B is typically occupied by transition metals for oxygen anion transport applications. The perovskite structure forms when the radii of A (r_A) and B (r_B) cations and oxygen (r_O) anions meets, or is close to meeting, the Goldschmidt tolerance factor:

$$t_G = \frac{r_A + r_o}{\sqrt{2}(r_B + r_o)}$$

This tolerance factor, a geometric packing argument, is equal to unity for a cubic perovskite structure with space group *Pm-3m*. Deviations from this value lead to distortions in the structure with deviations to other crystal symmetries. The perovskite structure leads to a relatively high free volume within the lattice that is often considered to favor the migration of oxygen anions. The oxygen sublattice is fully occupied ($\delta = 0$) when the sum of the positive charges on the A and B site cations is six. As with the fluorites, ionic and electronic conductivity can be introduced through aliovalent doping on either site or by changes in the oxidation state of the B site cations.

Initial efforts towards introducing oxygen vacancies, and thus anion conductivity, into $LaGaO_3$ focused on substitution of two positive alkaline earths onto the La^{3+} site. The conductivity was found to follow $Sr^{2+} > Ba^{2+} > Ca^{2+}$ with conductivity increasing with the amount of dopant used. Unfortunately, the solid solubility of Sr in the lattice was found to be ~10 %, limiting the total doping, and hence vacancy concentration that could be generated by single site doping. Additional vacancies can be generated by additional doping onto the B-site, with Mg^{2+} proving particularly effective. An unexpected side-effect of this doping is that there is a commensurate increase in the Sr solubility to ~ 20 %, possibly due to the increase lattice volume due to Mg doping. Figure 3 shows a contour plot of conductivity as a function of x and y in $La_{1-x}Sr_xGa_{1-y}Mg_yO_{3-\delta}$ at 800 °C, indicating an optimum close to the commonly utilized composition $La_{0.8}Sr_{0.2}Ga_{0.8}Mg_{0.2}O_{3-\delta}$.

LSGM shows some *p*-type conductivity under very reducing conditions and some *n*-type conductivity under very oxidizing conditions [18, 19]. However, LSGM shows almost purely ionic conductivity between $10^{-25} < pO_2 < 10^5$ atm. This electrolyte domain has been fully characterized by Kim and Yoo [18], for $La_{0.9}Sr_{0.1}Ga_{0.8}Mg_{0.2}O_{3-\delta}$, Fig. 4. As shown in Fig. 2, LSGM provides a total conductivity similar to CGO at intermediate temperatures; combining this with the insignificant electronic conductivity makes LSGM a promising electrolyte material for future development.

Mixed Conducting Oxides

Mixed oxygen ionic and electronic conducting (MIEC) oxides find application in the electrodes of SOFCs, SOECs, SOERs, and as the membrane materials in ITMs. There are many materials

Oxygen Anion Transport in Solid Oxides, Fig. 3 Contour plot of total conductivity at 800 °C as a function of x and y in $La_{1-x}Sr_xGa_{1-y}Mg_yO_{3-\delta}$ [17]

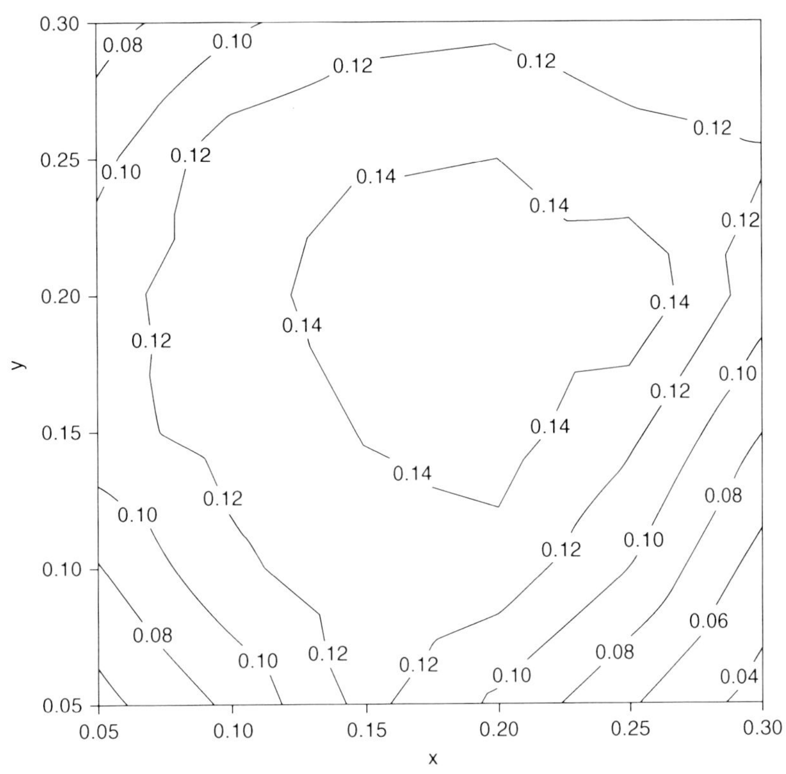

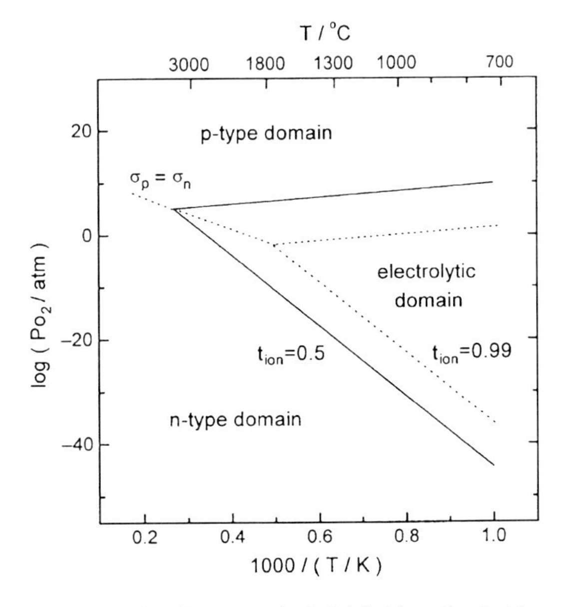

Oxygen Anion Transport in Solid Oxides, Fig. 4 The electrolytic domain for $La_{0.9}Sr_{0.1}Ga_{0.8}Mg_{0.2}O_{3-\delta}$ as a function of temperature and pO_2 [18]

suggested for these applications, and numerous thorough reviews on these topics. Here we will limit the discussion to some illustrative examples, focusing on materials utilized for SOFC cathodes as these have received the most attention in the literature [20]. The majority of these materials fall into the class of single perovskites where the crystallographic unit cell is that shown in Fig. 5 and all of the cations are randomly distributed across the relevant sites. While many of these materials show significant distortions away from a cubic structure at room temperature, the majority are cubic perovskites (space group *Pm-3 m*) under operating conditions. There is increasing interest in the use of more complex structures that typically lead to localization of oxygen vacancies within specific crystal planes. These include double perovskites where the cations are ordered in alternating layers with associated localization of oxygen vacancies, and A_2BO_4-structured materials [4, 21].

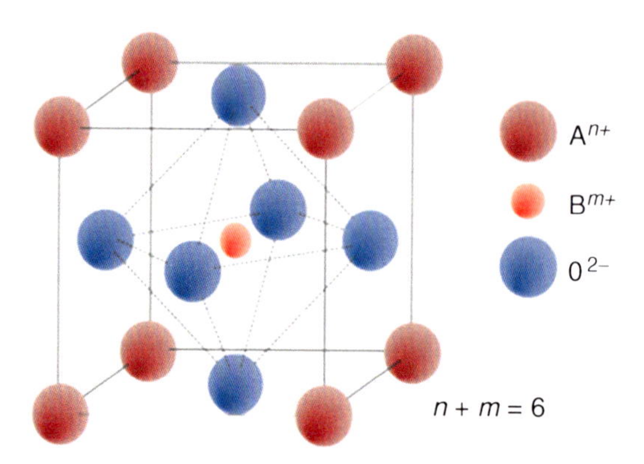

Oxygen Anion Transport in Solid Oxides, Fig. 5 Cubic perovskite unit cell

Single Perovskites

The first widely used material in the SOFC cathode was $La_{1-x}Sr_xMnO_3$ (LSM), typically with $x = 0.2$. This is primarily an electronic conductor under SOFC operating conditions, $\delta \approx 0$ as charge compensation for the Sr^{2+} doping occurs through the generation of electron holes (Mn^{4+}) to provide *p*-type conductivity. LSM is thus used as a two phase composite electrode with the electrolyte material with the required functionality of the electrode being met at the contact points between the LSM and electrolyte phases – the commonly discussed triple-phase boundary (TPB) between the two solid phases and the gas phase. The electrolyte provides ionic conductivity while LSM provides electronic conductivity and catalytic activity. Even with only limited ionic conductivity, experiments on thin film electrodes indicate that bulk ionic transport can occur, especially under reducing operating potentials found in the SOFC cathode [22, 23]. The reducing electrochemical potential leads to reduction of Mn^{4+} to Mn^{3+} and concomitant formation of oxygen vacancies and associated ionic conductivity.

There is an increasing move from LSM to mixed ionic-electronic conductors for use in the SOFC cathode. These mixed conductors can remove theoretically provide all of the electrode material requirements in a single phase. This can open the entire electrode surface for reaction by removing the restriction that reaction only occurs at the TPB. The primary challenge for materials chemists has been to design materials that provide high levels of functionality at reduced temperature, while maintaining stability and chemical compatibility with the electrolyte materials. A large number of materials have been suggested, primarily with Co and/or Fe on the B-site as these transition metals are partially reduced under typical SOFC operating conditions, providing significant concentrations of both oxygen vacancies and electron holes [24]. Figure 6 shows the oxygen stoichiometry as a function of La/Sr and Fe/Co ration in $La_{0.w}Sr_{0.x}Co_{0.y}Fe_{0.z}O_{3-\delta}$ materials. Both sources of oxygen vacancies are observed; the oxygen stoichiometry decreases (increasing vacancy concentration) with increasing Sr^{2+} substitution for La^{3+}, and increasing Co content (Co is more easily reduced than Fe).

The oxygen migration pathway in single perovskites has been modeled by the Islam group [25]; who provide evidence for a curved pathway from one site to the next, Fig. 7. This is a logical pathway for oxygen anion migration between the positively charged A- and B-site cations. This migration pathway is generally true for all single perovskites, including the electrolyte LSGM.

Experimental visualization of the oxygen migration is possible through *in-situ* powder neutron diffraction. The most common experimental technique to examine oxide materials is X-ray diffraction (XRD); however, X-rays interact with the electron cloud of the atoms and the low electron density of oxygen renders it a weak scatterer. XRD provides useful information about cation structure but does not yield the desired information about the oxygen anions. In contrast, neutrons interact with the atomic nucleus such that refinement of powder neutron diffraction data provides information regarding both cation and anion structure. This includes direct measurement of the oxygen stoichiometry by Rietveld refinement [26, 27] of the oxygen site occupancies, and indication of preferred oxygen migration pathways by refinement of the anisotropic atomic displacement parameters and

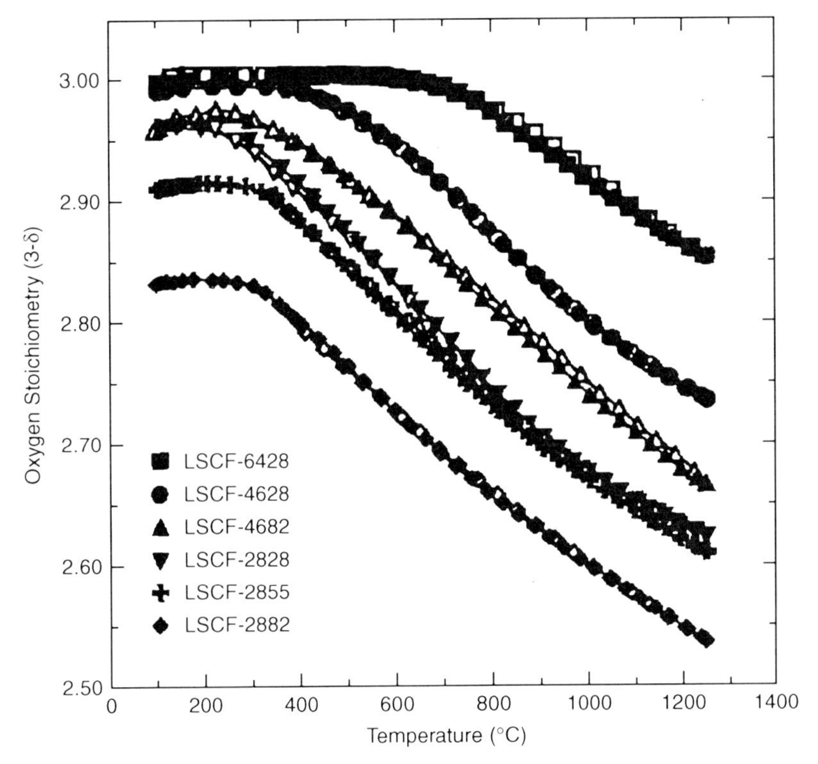

Oxygen Anion Transport in Solid Oxides, Fig. 6 Oxygen stoichiometry of $La_{0.w}Sr_{0.x}Co_{0.y}Fe_{0.z}O_{3-\delta}$ (denoted LSCFwxyz) as a function of temperature in air. Open symbols denote data collected on heating, closed symbols denote data collected on cooling

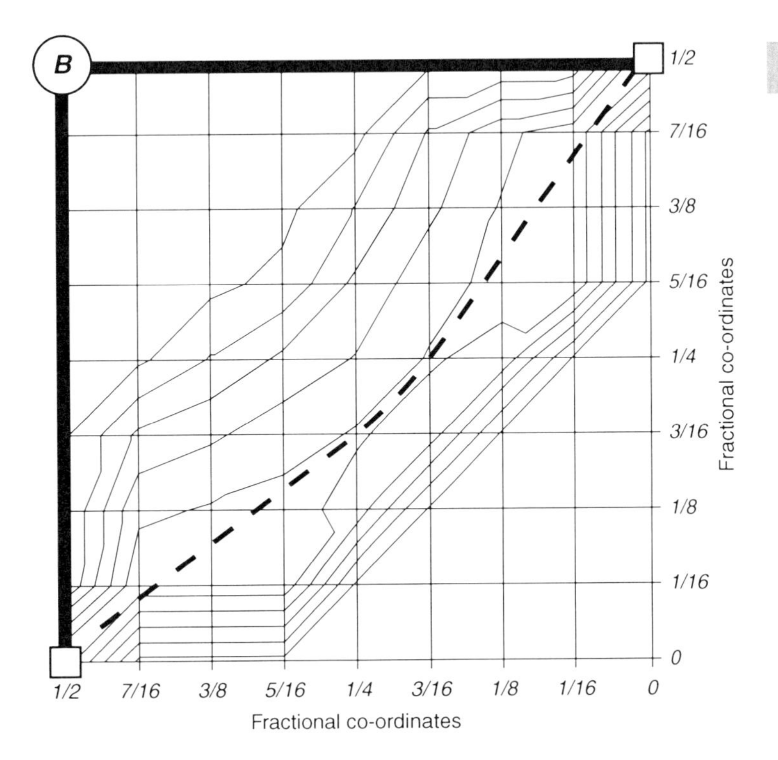

Oxygen Anion Transport in Solid Oxides, Fig. 7 Calculated curved oxygen migration pathway in a cubic perovskite [25]

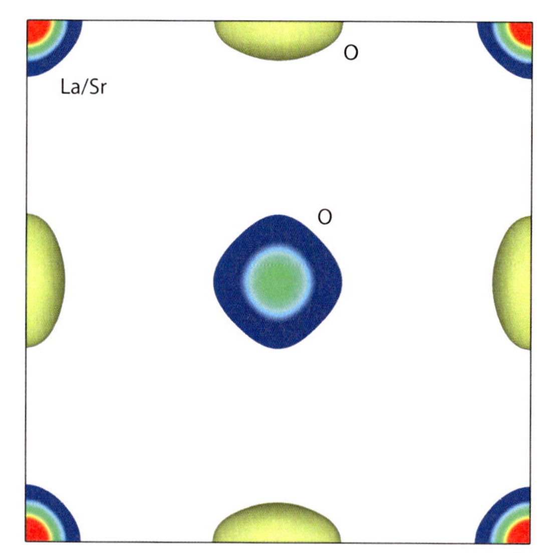

Oxygen Anion Transport in Solid Oxides, Fig. 8 Projection of the observed nuclear density map of $La_{0.5}Sr_{0.5}Co_{0.8}Fe_{0.2}O_{3-\delta}$ measured at 800 °C, $pO_2 = 10^{-2}$ atm. Note the cross-shaped displacement of the O-atoms [28]

examination of the observed nuclear density maps [28–33]. Figure 8 is such an observed nuclear density map for $La_{0.5}Sr_{0.5}Co_{0.8}Fe_{0.2}O_{3-\delta}$ measured at 800 °C, $pO_2 = 10^{-2}$ atm [28]. The cross-shaped pattern of the O-site nuclear density indicates preferred oxygen anion motion towards the surrounding oxygen sites and away from the nearest A-site cations. The B-site cation sits behind the central oxygen in this projection, with the observed nuclear density 'flattened' away from this central cation. Thus we can directly visualize the motion of the oxygen anions in the structure.

$Ba_{0.5}Sr_{0.5}Co_{0.8}Fe_{0.2}O_{3-\delta}$ (BSCF) is a particularly curious single perovskite that has received significant attention in the literature. Initial reports indicated that this material provides one of the highest reported performances as an SOFC cathode at intermediate temperature [34]. Neutron diffraction studies by McIntosh et al. indicated that this was due to an exceptionally low oxygen stoichiometry (i.e. a very high vacancy concentration) in this material [33], Fig. 9a. A number of groups have since attempted to verify this low oxygen stoichiometry via thermogravimetric measurements [35–39] Fig. 9b; however, these alternative measurements show significant scatter in the determined values and none have reproduced the low oxygen stoichiometry reported by McIntosh et al. Švarcová et al. [35] reported that BSCF can undergo a phase transition from cubic to hexagonal symmetry. While this was not observed in the neutron diffraction studies of McIntosh et al., it has been suggested, with additional first principle calculations, that this may explain the scatter in the thermogravimetric data [40]. The debate surrounding BSCF illustrates the complexity of accurately determining oxygen stoichiometry.

The oxygen vacancies in these materials only contribute to ion transport if they are free to move within the lattice. Oxygen vacancy ordering, where the vacancies are fixed at specific crystallographic positions dramatically lowers the ionic conductivity by reducing the mobility term in the conductivity: a high concentration of fixed vacancies leads to low conductivity. Vacancy ordering can be observed at both low and high temperature. One example is the system $SrCo_{0.8}Fe_{0.2}O_{3-\delta}$ (SCF). SCF is a cubic perovskite between 600 and 900 °C at $pO_2 = 1$ atm with oxygen stoichiometry between 2.37 and 2.57. However, as the pO_2 is lowered, SCF undergoes oxygen vacancy ordering to a browmillerite structure with fixed oxygen stoichiometry of $\delta = 0.5$, $Sr_2Co_{1.6}Fe_{0.4}O_5$. At $pO_2 = 10^{-2}$ atm, this vacancy ordered structure exists below 750 °C, but disorders above this temperature to again form a vacancy disordered cubic perovskite as the enthalpic term overcomes entropy driven ordering. The brownmillerite structure, space group *Icmm*, determined from Rietvled refinement of *in-situ* powder neutron diffraction data is shown in Fig. 10. The brownmillerite structure consists of alternating layers of Co/Fe in 6 coordinated octahedra and 4 coordinated tetrahedra, that is, the oxygen vacancies are fixed in the tetrahedral coordinate layers as opposed to randomly distributed as in the cubic perovskite. This structure shows significant tilting of the octahedra and tetrahedra with associated anisotropic atomic displacement of the oxygen anions.

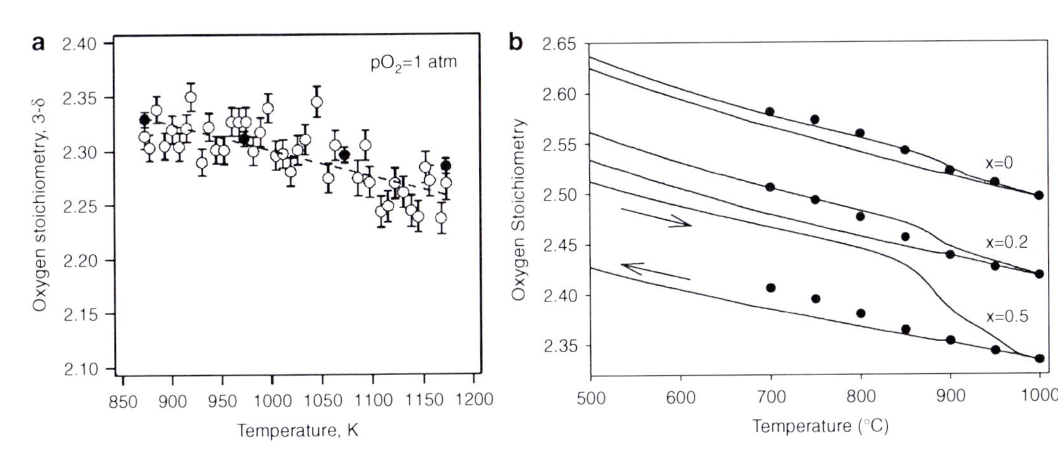

Oxygen Anion Transport in Solid Oxides, Fig. 9 Oxygen stoichiometry of (**a**) $Ba_{0.5}Sr_{0.5}Co_{0.8}Fe_{0.2}O_{3-\delta}$ at $pO_2 = 10^{-1}$ atm determined by *in-situ* neutron diffraction (open symbols) and thermogravimetry (closed symbols) [33] and (**b**) $Ba_xSr_{1-x}Co_{0.8}Fe_{0.2}O_{3-\delta}$ determined by thermogravimetry in air [35]. Circles in b) denote data for the cubic polymorph

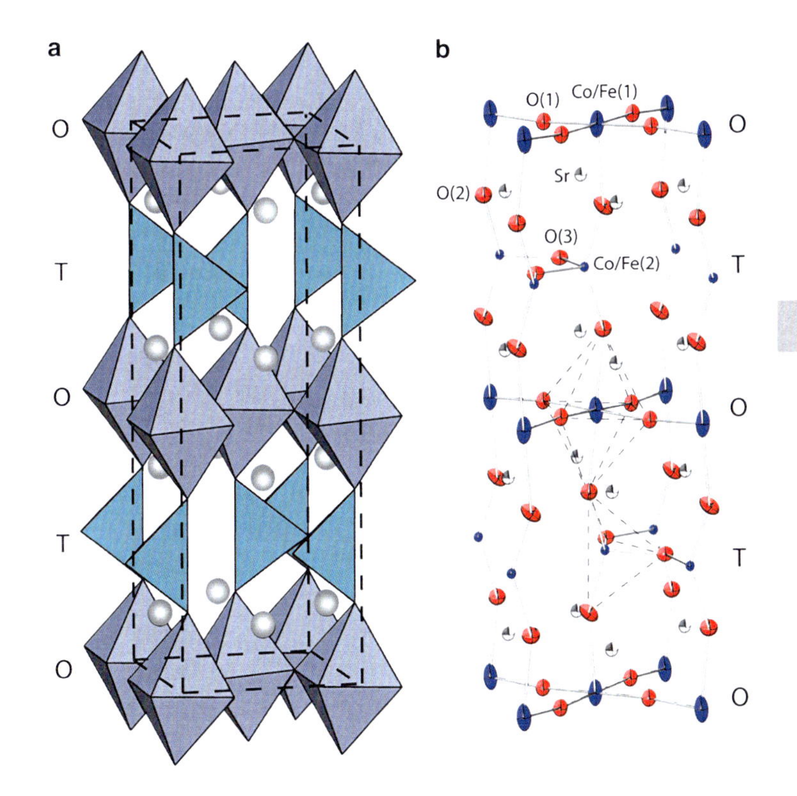

Oxygen Anion Transport in Solid Oxides, Fig. 10 The vacancy ordered brownmillerite $Sr_2Co_{1.6}Fe_{0.4}O_5$ determined by *in-situ* powder neutron diffraction at 650 °C, $pO_2 = 5x10^{-4}$ atm represented as (**a**) polyhedral and (**b**) refined atomic displacement ellipsoids

Double Perovskites

In addition to oxygen vacancy ordering, as in the brownmillerite case, the A and B-site cations can also undergo ordering. One class of materials gaining significant traction as potential SOFC anodes are the A-site ordered double perovskites, in particular those in the series $LaBaCo_2O_{5+\delta}$ where La is a lanthanide. The A-site ordering in these materials is described by doubling of the

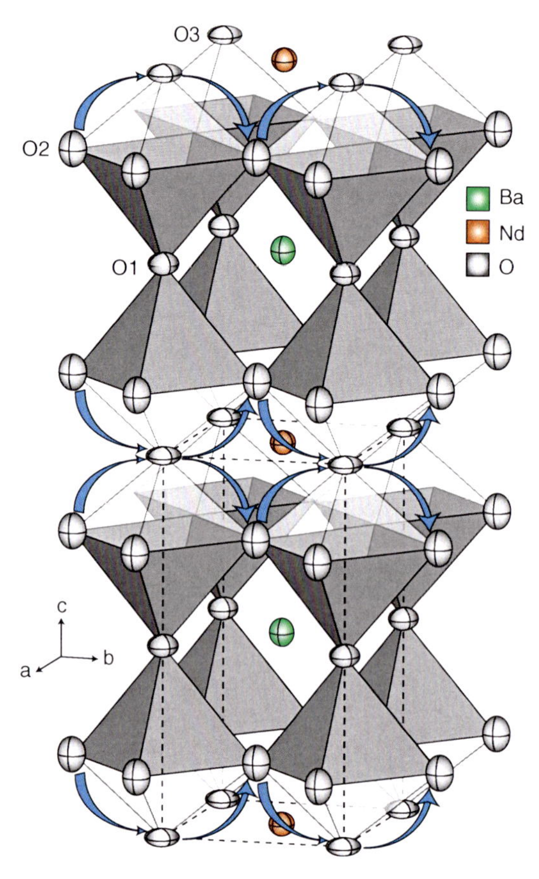

Oxygen Anion Transport in Solid Oxides, Fig. 11 Refined structure of $NdBaCo_2O_{5+\delta}$ at 573 °C and $pO_2 = 10^{-1}$ atm [29]. Atoms are represented by refined atomic displacement ellipsoids, partial shading indicates location of oxygen vacancies at the O3 site, and arrows are illustrative indications of the ion transport pathway

perovskite unit cell in one direction to form alternating layers of A-site cations. Figure 11 shows a refined structure of $NdBaCo_2O_{5+\delta}$ determined from Rietveld refinement of *in-situ* powder neutron diffraction data. The doubling of the perovskite unit cell is shown in the c-direction in space group *P4/mmm* There are three distinct oxygen sites within this structure; O1 is the axial oxygen of the Co octahedron that sits in the Ba layer, O2 is the equatorial oxygen of the Co octahedra, and O3 is the axial oxygen of the Co octahedra sitting in the Nd layer. Refining the oxygen occupancies of all of these sites leads to the conclusion the oxygen vacancies are almost exclusively located at the O3 site in the Nd layer. The fractional occupancies of the O3 sites under the conditions depicted at 0.99(1), 1.00(5), and 0.50(1) for the O1, O2 and O3 sites respectively. The Co and O2 sites both also show a significant shift from their ideal positions towards the vacancy rich layer.

Combining these positions and occupancies with the refined anisotropic atomic displacement parameters of the oxygen sites points to the preferred oxygen migration pathway, indicated by arrows in Fig. 11. The O3 site in the vacancy-rich layer shows strong anisotropic motion within the layer (the a-b plane), while the O2 site shows strong anisotropy along the c-axis towards the vacancy rich layer. The O1 site also shows anisotropic motion in the a-b plane. Thus the vacancy rich oxygen sites and their nearest neighbor show displacements consistent with a curved hopping path between these sites. This is further mediated by the shift in position of the O2 site towards the vacancy rich layer. Thus it is concluded that oxygen transport is confined to hops between the O3 site in the Nd plane and its nearest neighbor. A similar conclusion has been reported for $PrBaCo_2O_{5+\delta}$ [41].

Further experimental evidence for this preferred oxygen migration path is provided by oxygen tracer diffusion measurements of Burriel et al. [42] on the related material $PrBaCo_2O_{5+\delta}$. They report at least a factor of four increased oxygen isotope tracer diffusion coefficient through grains oriented in the a-b plane versus those oriented in the c-axis direction. Molecular dynamics studies on $LnBaCo_2O_{5+\delta}$ materials by Seymour et al. corroborate this suggested transport mechanism [43, 44], Fig. 12. This MD simulation was able to directly calculate an oxygen diffusion pathway consisting of oxygen anion transport between the oxygen sites equivalent to the O3 site (Ln layer in the general form $LnBaCo_2O_{5+\delta}$) and O2 site (Co layer) in these double perovskites.

A_2BO_4 Structure

The A_2BO_4 structure, often referred to as the K_2NiF_4 structure or the n = 1 member of the Ruddlesden-Popper structure family, is another

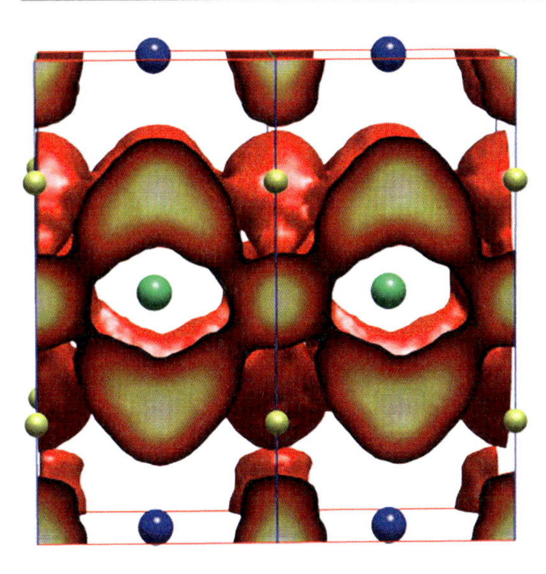

Oxygen Anion Transport in Solid Oxides, Fig. 12 Oxygen density plot for $PrBaCo_2O_{5.5}$ lattice at 927 °C. Pr^{3+} ions are shown in green, Ba^{2+} are shown in blue and Co^{3+} are shown in yellow [41]

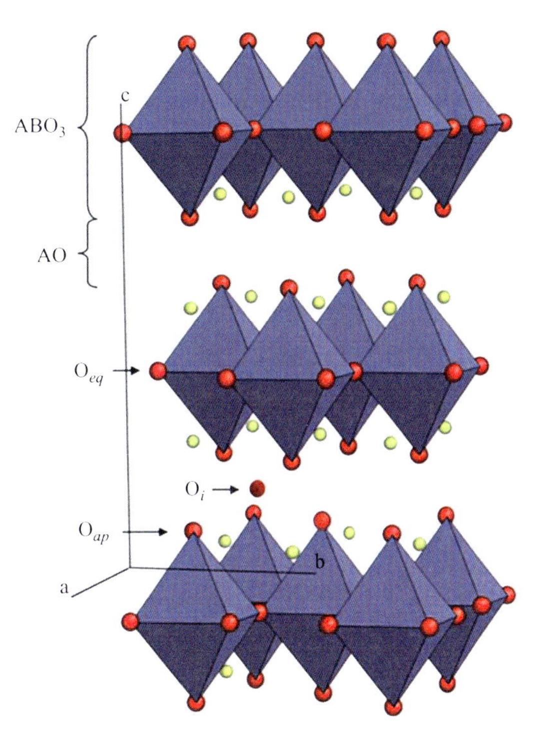

Oxygen Anion Transport in Solid Oxides, Fig. 13 A polyhedral representation of the A_2BO_4 structure with equatorial, apical and interstitial oxygen locations labeled [4]

layered system that can be described as consisting of alternating rock salt and perovskite structured layers, Fig. 13 [4, 45]. The most extensively studied oxygen anion conducting material with this structure is La_2NiO_4 with alternating LaO rock salt and $LaNiO_3$ perovskite layers [4, 21]. As with the double perovskites, these materials show strongly anisotropic oxygen anion transport, occurring primarily in the a-b plane; that is, in the direction between the layers; Burriel et al. [46] report a factor of three higher oxygen diffusivity in the a-b direction when compared to the c-direction (across the layers) for epitaxial films.

The oxygen anion transport mechanism in these materials is dependent on the oxygen stoichiometry, and has been confirmed by a number of groups through simulation [47–52]. These materials can incorporate oxygen excess (positive δ) through incorporation of oxygen interstitials upon oxidation of the lattice. These interstitials exist and move only in the rock salt layer, leading to highly anisotropic oxygen diffusivity within these layers in the a-b plane. Oxygen non-stoichiometry can be introduced through doping with Sr on the La site with associated charge compensation. Calculations indicate that these vacancies preferentially form at the equatorial positions in the perovskite layer. Thus a vacancy hopping mechanism again favors transport in the a-b plane between these equatorial sites.

Summary

As discussed above, oxygen anion transport in oxides is a complex topic. While the underlying general principles of increasing ion transport rates with increased vacancy concentration and ion mobility are clear, the details for each material system are often not. The influence of crystal structure and composition on the location, and clustering, of oxygen vacancies, and the possible presence of interstitial species, are not always obvious nor are they easily understood. Thus a series of experimental and theoretical computational studies are required to truly understand oxygen transport in any given material system.

Cross-References

- ► High-Temperature Polymer Electrolyte Fuel Cells
- ► Oxide Ion Conductor
- ► Solid Electrolytes

References

1. Goodenough JB (2003) Oxide-ion electrolytes. Ann Rev Mater Res 33(1):91–128
2. Brandon NP, Skinner S, Steele BCH (2003) Recent advances in materials for fuel cells. Ann Rev Mater Res 33(1):183–213
3. Carter S, Selcuk A, Chater RJ, Kajda J, Kilner JA, Steele BCH (1992) Oxygen transport in selected nonstoichiometric perovskite-structure oxides. Solid State Ion 53–56 (Part 1):597
4. Tarancon A, Burriel M, Santiso J, Skinner SJ, Kilner JA (2010) Advances in layered oxide cathodes for intermediate temperature solid oxide fuel cells. J Mater Chem 20(19):3799–3813
5. Inaba H, Tagawa H (1996) Ceria-based solid electrolytes. Solid State Ion 83(1–2):1–16
6. Ishihara T (2006) Development of new fast oxide ion conductor and application for intermediate temperature solid oxide fuel cells. Bull Chem Soc Jpn 79(8):1155–1166
7. Yokokawa H, Sakai N, Horita T, Yamaji K, Brito ME (2005) Electrolytes for solid-oxide fuel cells. MRS Bull 30(8):591–595
8. van den Bossche M, McIntosh S (2008) Rate and selectivity of methane oxidation over $La_{0.75}Sr_{0.25}Cr_xMn_{1-x}O_{3-d}$ as a function of lattice oxygen stoichiometry under solid oxide fuel cell anode conditions. J Catal 255(2):313–323
9. Ebbesen SD, Mogensen M (2009) Electrolysis of carbon dioxide in solid oxide electrolysis cells. J Power Sources 193(1):349–358
10. Hauch A, Ebbesen SD, Jensen SH, Mogensen M (2008) Solid oxide electrolysis cells: microstructure and degradation of the Ni/yttria-stabilized zirconia electrode. J Electrochem Soc 155(11):B1184–B1193
11. Zhan Z, Lin Y, Pillai M, Kim I, Barnett SA (2006) High-rate electrochemical partial oxidation of methane in solid oxide fuel cells. J Power Sources 161(1):460–465
12. Smith AR, Klosek J (2001) A review of air separation technologies and their integration with energy conversion processes. Fuel Processing Technol 70(2):115–134
13. Bouwmeester HJM, Burggraaf AJ (1997) Dense ceramic membranes for oxygen separation. In: Gellings PJ, Bouwmeester HJM (eds) CRC handbook of solid state electrochemistry; CRC Press, Boca Raton, pp 481–553
14. Bouwmeester HJM (2003) Dense ceramic membranes for methane conversion. Catal Today 82(1–4):141
15. Knauth P, Tuller HL (2002) Solid-state Ionics: roots, status, and future prospects. J Am Ceram Soc 85(7):1654–1680
16. Kilner JA (2000) Fast oxygen transport in acceptor doped oxides. Solid State Ion 129(1–4):13
17. Huang K, Tichy RS, Goodenough JB (1998) Superior perovskite oxide-Ion conductor; strontium- and magnesium-doped $LaGaO_3$: I, phase relationships and electrical properties. J Am Ceram Soc 81(10):2565–2575
18. Kim J, Yoo H (2001) Partial electronic conductivity and electrolytic domain of $La_{0.9}Sr_{0.1}Ga_{0.8}Mg_{0.2}O_{3-\delta}$. Solid State Ion 140(1–2):105–113
19. Baker RT, Gharbage B, Marques FMB (1997) Ionic and electronic conduction in Fe and Cr Doped (La,Sr) $GaO_{3-\delta}$. J Electrochem Soc 144(9):3130–3135
20. Adler SB (2004) Factors governing oxygen reduction in solid oxide fuel cell cathodes. Chem Rev 104 (10):4791
21. Tarancon A, Skinner SJ, Chater RJ, Hernandez-Ramirez F, Kilner JA (2007) Layered perovskites as promising cathodes for intermediate temperature solid oxide fuel cells. J Mater Chem 17(30): 3175–3181
22. Yasumoto K, Shiono M, Tagawa H, Dokiya M, Hirano K, Mizusaki J (2002) Effect of oxygen nonstoichiometry on a $La_{1-x}A_xMnO_{3+d}$ cathode under a polarized state. J Electrochem Soc 149(5): A531
23. Mizusaki J, Saito T, Tagawa H (1996) A chemical diffusion-controlled electrode reaction at the compact $La_{1-x}Sr_xMnO_3$/stabilized zirconia interface in oxygen atmospheres. J Electrochem Soc 143(10): 3065–3073
24. Mizusaki J (1992) Nonstoichiometry, diffusion, and electrical properties of perovskite-type oxide electrode materials. Solid State Ion 52(1–3):79
25. Islam MS (2000) Ionic transport in ABO_3 perovskite oxides: a computer modelling tour. J Mater Chem 10(4):1027–1038
26. Von Dreele RB, Jorgensen JD, Windsor CG (1982) Rietveld refinement with spallation neutron powder diffraction data. J Appl Crystallog 15(6):581–589
27. Rietveld HM (1969) A profile refinement method for nuclear and magnetic structures. J Appl Crystallogr 2:65–71
28. Tamimi MA, Tomkiewicz AC, Huq A, McIntosh S To be published.
29. Cox-Galhotra R, Huq A, Hodges JP, Kim J, Yu C, Wang X, Jacobson AJ, McIntosh S (2013) Visualizing oxygen anion transport pathways in $NdBaCo_2O_{5+\delta}$ by in situ neutron diffraction. J Mater Chem A 1(9):3091–3100
30. Yashima M, Enoki M, Wakita T, Ali R, Matsushita Y, Izumi F, Ishihara T (2008) Structural disorder and diffusional pathway of oxide ions in a doped

Pr_2NiO_4-based mixed conductor. J Am Chem Soc 130(9):2762–2763
31. Chen Y, Yashima M, Ohta T, Ohoyama K, Yamamoto S (2012) Crystal structure, oxygen deficiency, and oxygen diffusion path of perovskite-type lanthanum cobaltites $La_{0.4}Ba_{0.6}CoO_{3-d}$ and $La_{0.6}Sr_{0.4}CoO_{3-d}$. J Phys Chem C 116(8):5246–5254
32. McIntosh S, Vente JF, Haije WG, Blank DHA, Bouwmeester HJM (2006) Phase stability and oxygen non-stoichiometry of $SrCo_{0.8}Fe_{0.2}O_{3-d}$ measured by in situ neutron diffraction. Solid State Ion 177(9–10):833–842
33. McIntosh S, Vente JF, Haije WG, Blank DHA, Bouwmeester HJM (2006) Oxygen stoichiometry and chemical expansion of $Ba_{0.5}Sr_{0.5}Co_{0.8}Fe_{0.2}O_{3-d}$ measured by in situ neutron diffraction. Chem Mater 18(8):2187–2193
34. Shao ZP, Haile SM (2004) A high-performance cathode for the next generation of solid-oxide fuel cells. Nature 431(7005):170–173
35. Švarcová S, Wiik K, Tolchard J, Bouwmeester HJM, Grande T (2008) Structural instability of cubic perovskite BaxSr1-xCo1 − yFeyO3−δ. Solid State Ion 178(35–36):1787–1791
36. Mueller DN, De Souza RA, Yoo H, Martin M (2012) Phase stability and oxygen nonstoichiometry of highly oxygen-deficient Perovskite-type oxides: a case study of $(Ba,Sr)(Co,Fe)O_{3-\delta}$. Chem Mater 24 (2):269–274
37. Bucher E, Egger A, Ried P, Sitte W, Holtappels P (2008) Oxygen nonstoichiometry and exchange kinetics of $Ba_{0.5}Sr_{0.5}Co_{0.8}Fe_{0.2}O_{3-\delta}$. Solid State Ion 179(21–26):1032–1035
38. Kriegel R, Kircheisen R, Töpfer J (2010) Oxygen stoichiometry and expansion behavior of $Ba_{0.5}Sr_{0.5}Co_{0.8}Fe_{0.2}O_{3-\delta}$. Solid State Ion 181(1–2):64–70
39. Jung J, Misture ST, Edwards DD (2010) Oxygen stoichiometry, electrical conductivity, and thermopower measurements of BSCF ($Ba_{0.5}Sr_{0.5}Co_xFe_{1-x}O_{3-\delta}$, $0 \leq x \leq 0.8$) in air. Solid State Ion 181(27–28):1287–1293
40. Kuklja MM, Mastrikov YA, Jansang B, Kotomin EA (2013) First principles calculations of (Ba, Sr)(Co, Fe) O3−δ structural stability. Solid State Ion 230:21–26
41. Seymour ID, Tarancón A, Chroneos A, Parfitt D, Kilner JA, Grimes RW (2012) Anisotropic oxygen diffusion in $PrBaCo_2O_{5.5}$ double perovskites. Solid State Ion 216:41–43
42. Burriel M, Pena-Martinez J, Chater RJ, Fearn S, Berenov AV, Skinner SJ, Kilner JA (2012) Anisotropic oxygen Ion diffusion in layered $PrBaCo_2O_{5+\delta}$. Chem Mater 24(3):613–621
43. Seymour ID, Chroneos A, Kilner JA, Grimes RW (2011) Defect processes in orthorhombic LnBaCo2O5.5 double perovskites. Phys Chem Chem Phys 13(33):15305–15310
44. Seymour ID, Tarancón A, Chroneos A, Parfitt D, Kilner JA, Grimes RW (2012) Anisotropic oxygen diffusion in $PrBaCo_2O_{5.5}$ double perovskites. Solid State Ion 216:41–43
45. Tarancón A, Skinner SJ, Chater RJ, Hernandez-Ramirez F, Kilner JA (2007) Layered perovskites as promising cathodes for intermediate temperature solid oxide fuel cells. J Mater Chem 17(30): 3175–3181
46. Burriel M, Garcia G, Santiso J, Kilner JA, Chater RJ, Skinner SJ (2008) Anisotropic oxygen diffusion properties in epitaxial thin films of La_2NiO_{4+d}. J Mater Chem 18(4):416–422
47. Allan NL, Lawton JM, Mackrodt WC (1989) A comparison of the calculated lattice and defect structures of La_2CuO_4, La_2NiO_4, Nd_2CuO_4, Pr_2CuO_4, Y_2CuO_4, Al_2CuO_4: relationship to high-Tc superconductivity. Philos Mag Part B 59(2):191–206
48. Allan NL, Mackrodt WC (1991) Oxygen ion migration in La_2CuO_4. Philos Mag A 64(5):1129–1132
49. Mazo GN, Savvin SN (2004) The molecular dynamics study of oxygen mobility in $La_{2-x}Sr_xCuO_{4-\delta}$. Solid State Ion 175(1–4):371–374
50. Read MSD, Islam MS, King F, Hancock FE (1999) Defect chemistry of $La_2Ni_{1-x}MxO_4$ (M = Mn, Fe, Co, Cu): relevance to catalytic behavior. J Phys Chem B 103(9):1558–1562
51. Minervini L, Grimes RW, Kilner JA, Sickafus KE (2000) Oxygen migration in LaNiO. J Mater Chem 10(10):2349–2354
52. Cleave AR, Kilner JA, Skinner SJ, Murphy ST, Grimes RW (2008) Atomistic computer simulation of oxygen ion conduction mechanisms in La_2NiO_4. Solid State Ion 179(21–26):823–826

Oxygen Evolution Reaction

Marcel Risch[1], Jin Suntivich[2] and Yang Shao-Horn[1]
[1]MIT, Cambridge, MA, USA
[2]Cornell University, Ithaca, NY, USA

Keywords

Oxygen evolution; Water oxidation; Water splitting

Introduction

The oxygen evolution reaction (OER) is an enabler of several technological applications. In energy storage, these include regenerative fuel cells (see entry ▶ Fuel Cells, Principles and Thermodynamics), electrolyzers, ▶ metal-air batteries, and solar-driven water-splitting devices. Traditional metal refinery techniques such as electro-deposition (see entry ▶ Electro-deposition of Electronic Materials for Applications in Macroelectronic- and Nanotechnology-Based Devices), electro-synthesis, and electro-refining also generally involve the OER on the anode. In addition, the OER is central to natural photosynthesis in plants, algae, and cyanobacteria as well as ▶ artificial photosynthesis. Due to the ubiquitous nature of the OER, numerous efforts from chemists, physicists, biologists, and engineers have been made to understand the OER mechanism, as mastery of it holds the key to unlocking fundamental mechanism of oxygen chemistry and strategy to develop more cost-effective OER-based technology. However, to date, the exact mechanism of the OER is not known. Furthermore, the high overpotential of the OER (Fig. 1) due to the slow kinetics of the reaction poses significant inefficiency.

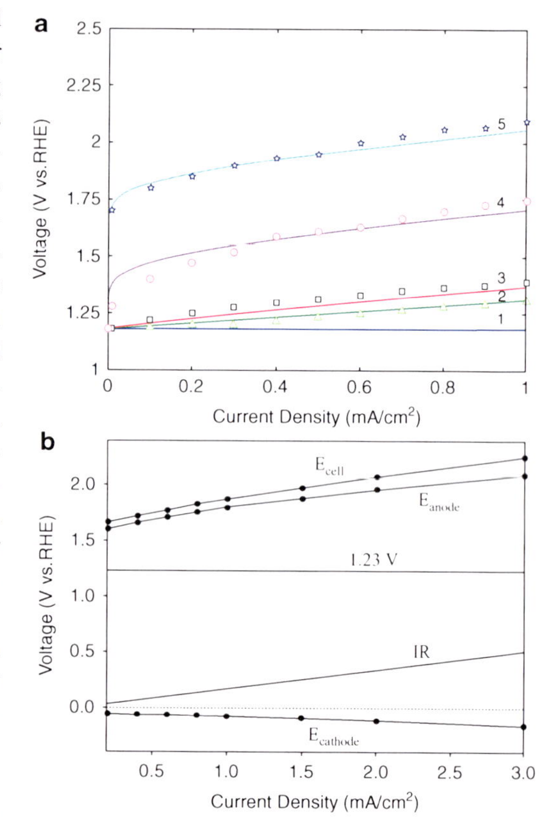

Oxygen Evolution Reaction, Fig. 1 Activity losses during the evolution reaction (OER) in acid (**a**) at 80 °C (*1* equilibrium voltage, *2* ohmic drop, *3* cathode overpotential, *4* anode overpotential on Pr-IrO_2, *5* anode overpotential on Pt) [19] and (**b**) at 25 °C on IrO_2/N-105/40 % Pt/C [20] (Both figures are reprinted with permission from Elsevier)

From a technology viewpoint, the OER is facilitated commonly by an electrocatalyst to minimize the energy loss due to overpotential. The influence of an electrocatalytic surface on the OER has been intensively studied for many decades in both nonaqueous and aqueous electrolytes. The latter cover a wide range of pH from alkaline, to neutral and acidic conditions. Most commonly, the reaction is catalyzed by heterogeneous transition metal catalysts such as Pt, Ru, and Ir [1–3]. However, it was later found that these metals undergo oxidation to form oxides during the OER condition. Since then, research efforts have focused on studying a variety of oxides including rutile oxides (e.g., RuO_2, IrO_2, MnO_2) [4, 5], spinel oxides (e.g., Co_3O_4, $NiCo_2O_4$) [5–8], perovskite oxides (e.g. $LaCoO_3$, $LaNiO_3$, $Ba_{0.5}Sr_{0.5}Co_{0.8}Fe_{0.2}O_3$) [9–11], and layered hydroxides [12–15]. Thus far, the most active OER surfaces (as defined by the highest OER current per unit of estimated active sites) have been RuO_2 and IrO_2 based oxides [16–18]. Recent fundamental works on 3d transition metal-based oxides have begun to bridge the activity difference to these precious metal oxide catalysts, although further improvement is still needed for more efficient OER applications. Finding electrocatalytic surfaces that can further minimize this inefficiency represents one of the grand challenges in electrocatalysis and is a goal for many laboratories around the world.

History

The earliest discussion of electrocatalytic oxygen evolution was reported in the context of the electrolysis of water by van Troostwijk and Deiman in 1789 [21, 22]. In 1800, Nicolson and Carlisle reproduced Volta's pile (a primitive battery) to study water electrolysis [21, 23]. Volta may also

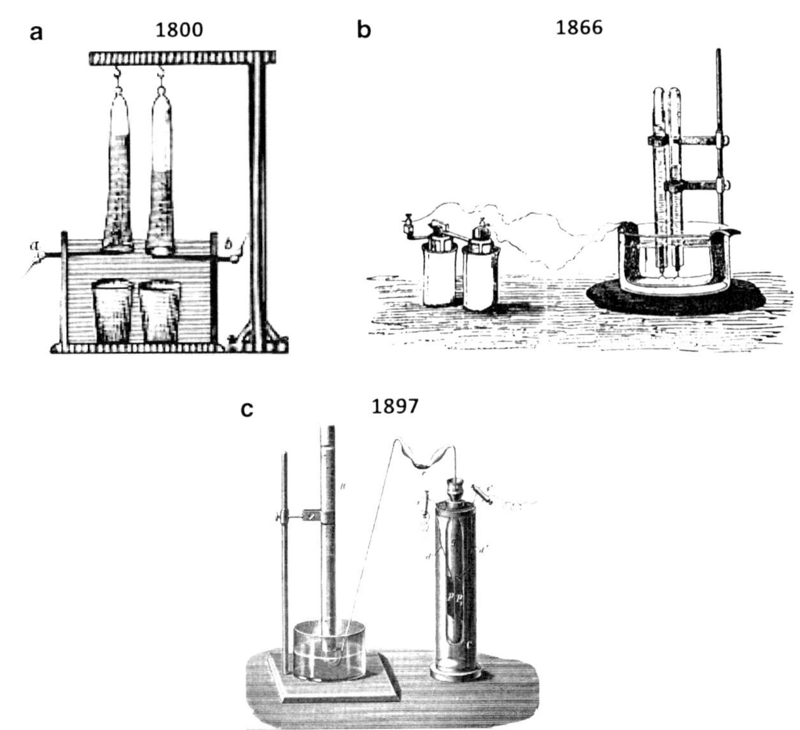

Oxygen Evolution Reaction, Fig. 2 Historical apparatuses used to study the oxygen evolution reaction (OER) by Johann Wilhelm Ritter (**a**) [27] and August Wilhelm von Hoffmann (**b**) [28] as well as a *Knallgasvoltameter* (oxyhydrogen gas voltameter; **c**) [29], in which the oxygen evolution reaction is employed to measure electric current. In all drawings, the ratio of 2:1 for hydrogen gas to oxygen gas can be seen clearly

have used his primitive battery to study water electrolysis [23]. Independently, Ritter also performed experiments on water electrolysis in these early years (Fig. 2) [24].

Systematic studies of the electrochemical oxygen evolution however would have to wait until the work of Hickling and Hill in 1947 [25]. These authors determined the activity of various metallic electrodes and graphite with applied overpotential, but could not identify any conclusive trend. Rüetschi and Delahay revisited the Hickling and Hill's dataset several years later and rationalized that the activity could depend on the energy of the metal-hydroxide bond [26]. This was the first breakthrough in rationalizing property of materials that can be useful for predicting the activity. This is now commonly called the "descriptor" approach to electrocatalysis as it allows the researchers to describe why certain surfaces are more active than others.

The descriptor approach to the OER electrocatalysis continues in the latter half of the twentieth century. In 1976, Tseung and Jasem suggested the redox potential of the active metal site ("higher to lower oxide transition") as a descriptor for the catalytic activity of the spinel $NiCo_2O_4$ [6]. Subsequently, Trasatti proposed the enthalpy of such redox transitions as a descriptor for the catalytic activity of transition metal oxides [16]. Most critically, Trasatti demonstrated that the descriptor for the OER has an optimum value (Sabatier's principle [30]) – a catalyst with too strong or too weak of the enthalpy is not a good catalyst. This is one of the earliest examples of a "volcano" plot in electrocatalysis, a concept that is central to the area of catalyst engineering in the present day.

To add predictive power to the descriptor approach, researchers have attempted to develop activity descriptors based on material properties that can be intuitively determined prior to material synthesis. One of the popular approaches uses parameters based on electronic structure. For example, Bockris and Otagawa studied the correlation between catalytic activity and the electronic structure of perovskites [9]. They concluded that the d-electron count, specifically high occupancy of the anti-bonding orbitals of the metal hydroxide on the surface, describes the catalytic activity. Nørskov and coworkers performed extensive calculations d-band center for transition metals [31]. For oxide surfaces, these authors proposed the binding energy of surface oxides as a descriptor for the activity [32], among other descriptors [33].

Contemporary with the proposal of the enthalpy of the redox transition, Matsumoto and coworkers suggested the overlap between the e_g orbital (or σ* band) and the orbital of the hydroxide ion could explain the trends in the oxygen evolution of perovskites [34, 35]. Recently, researchers predicted and verified a volcano relationship with an e_g occupancy close to unity for optimum oxygen evolution activity for perovskites [10]. In addition, they suggested that the covalency of the metal-oxygen bond could serve as a secondary predictor for the catalytic activity.

Reaction Mechanism of the Aqueous OER

In acidic media, the half-cell reaction of the OER is expressed as

$$2H_2O \rightarrow 4e^- + 4H^+ + O_2,$$

whereas in alkaline media, the equation reads

$$4OH^- \rightarrow 4e^- + 4H_2O + O_2.$$

In the context of heterogeneous catalysis, it is widely believed that such reactions occur in four discrete electron transfer steps. In general, several transient intermediate steps are assumed to occur [4, 9, 36–38]. A minimalistic set consists of:

1. Water adsorption
2. Oxygen-oxygen bond formation
3. Release of molecular dioxygen.

Mechanistic details are discussed commonly based on the Tafel slope in electrochemical experiments. However, the spread of reported Tafel slope values is considerable for most catalysts [36]. The current most widely accepted molecular model for each of the four steps originates from the computational work of Nørskov and coworkers [31]. Specifically, the reaction proceeds as follows, which is written for elementary reactions in acid:

$$H_2O \rightarrow e^- + H^+ + OH_{ad}. \quad (1)$$

$$OH_{ad} \rightarrow e^- + H^+ + O_{ad}. \quad (2)$$

$$O_{ad} + H_2O \rightarrow e^- + H^+ + OOH_{ad}. \quad (3)$$

$$OOH_{ad} \rightarrow e^- + H^+ + O_2, \quad (4)$$

To explain the volcano relationship that was observed experimentally by Trasatti, Nørskov proposed that the rate-limiting step is (2), when the surface interacts too weakly with the oxygenated species, or (3), when the surface interacts too strongly. This allowed Nørskov to draw the conclusion that the origin of the volcano plot is driven by the need to balance the kinetics between different elementary steps. Recently, it has been suggested that a perfect OER catalyst must catalyze four electron transfers with equal efficiency (Fig. 3). Ongoing efforts in the OER research community are examining ways to more efficiently balance the kinetics between each of these steps. An alternative, more radical strategy based on developing new catalysts with entirely different OER mechanisms has also been proposed [39], although experimental demonstration remains to be done. In these analyses, it should be noted that a key hypothesis is that each step of the electron transfer is accompanied by a concerted neutralization of the electric charge by the ions (H^+ or OH^-), although direct experimental evidence of such proton-coupled electron transfer reactions in OER still remains elusive to date. Proton-decoupled electron transfer may explain the pH-dependent OER activities reported in the literature, which requires further investigation.

Outlook

Despite tremendous progress in understanding the oxygen evolution reaction, many important

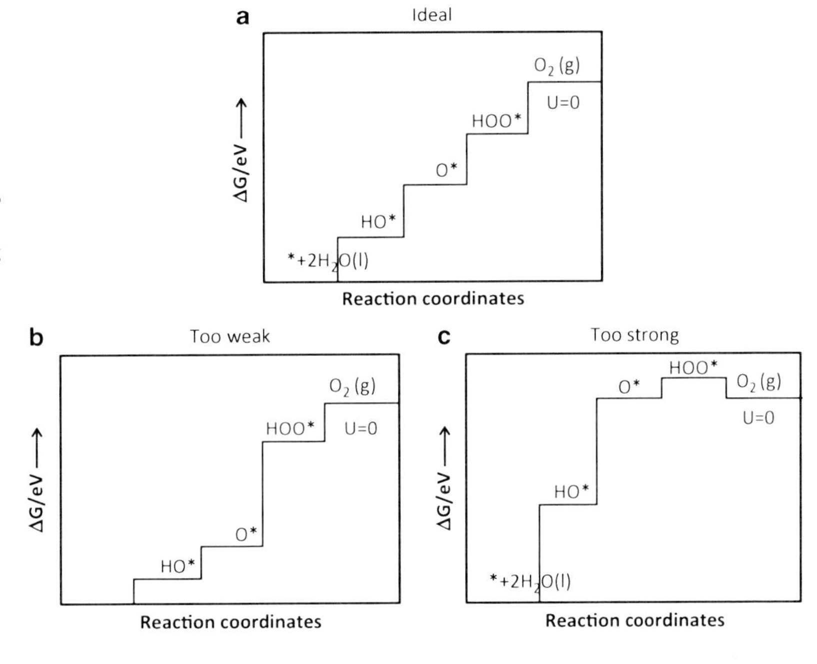

Oxygen Evolution Reaction, Fig. 3 Standard free energy diagram for the OER at zero overpotential (1.23 V vs. RHE) for catalysts with (**a**) ideal interaction of oxygen with the catalytic surface, (**b**) too weak interaction as in $LaMnO_3$, and (**c**) too strong interaction as in $LaCuO_3$ (Picture adapted from ref [32] with permission. Copyright 2011 Wiley and Sons)

questions on the reaction mechanism and the influence of the electrocatalytic surfaces on the reaction mechanism still remain open. This represents a great deal of opportunities for researchers to apply new ideas and insights into this long-standing problem in electrocatalysis science. We have discussed here the current understanding of the reaction mechanism on the atomic scale, which to date has relied heavily on computational studies. Progress in the instrumentation of in situ spectroscopy and the synergistic effects of combining experimental studies with a very mature theoretical framework is essential to unravel the mechanism of the oxygen evolution reaction. Such understanding is key to designing future generation OER electrocatalysts.

Cross-References

▶ Artificial Photosynthesis
▶ Electrocatalysis of Anodic Reactions
▶ Electrodeposition of Electronic Materials for Applications in Macroelectronic- and Nanotechnology-Based Devices
▶ Electrolytes, Thermodynamics
▶ Metal-Air Batteries
▶ Photolysis of Water

References

1. Damjanovic A, Dey A, Bockris JOM (1966) Kinetics of oxygen evolution and dissolution on platinum electrodes. Electrochim Acta 11:791–814
2. Miles MH, Klaus EA, Gunn BP, Locker JR, Serafin WE, Srinivasan S (1978) The oxygen evolution reaction on platinum, iridium, ruthenium and their alloys at 80 °C in acid solutions. Electrochim Acta 23:521–526
3. Bockris JO, Huq AKMS (1956) The mechanism of the electrolytic evolution of oxygen on platinum. Proc R Soc London Ser-A 237:277–296
4. Rossmeisl J, Qu ZW, Zhu H, Kroes GJ, Nørskov JK (2007) Electrolysis of water on oxide surfaces. J Electroanal Chem 607:83–89
5. Trasatti S (1984) Electrocatalysis in the anodic evolution of oxygen and chlorine. Electrochim Acta 29:1503–1512
6. Tseung ACC, Jasem S (1977) Oxygen evolution on semiconducting oxides. Electrochim Acta 22:31–34
7. Davidson CR, Kissel G, Srinivasan S (1982) Electrode-kinetics of the oxygen evolution reaction at $NiCo_2O_4$ from 30-percent KOH – dependence on temperature. J Electroanal Chem 132:129–135

8. Singh RN, Hamdani M, Koenig JF, Poillerat G, Gautier JL, Chartier P (1990) Thin films of Co_3O_4 and $NiCo_2O_4$ obtained by the method of chemical spray pyrolysis for electrocatalysis III. The electrocatalysis of oxygen evolution. J Appl Electrochem 20:442–446
9. Bockris JO, Otagawa T (1984) The electrocatalysis of oxygen evolution on perovskites. J Electrochem Soc 131:290–302
10. Suntivich J, May KJ, Gasteiger HA, Goodenough JB, Shao-Horn Y (2011) A perovskite oxide optimized for oxygen evolution catalysis from molecular orbital principles. Science 334:1383–1385
11. Jain AN, Tiwari SK, Singh RN, Chartier P (1995) Low-temperature synthesis of perovskite-type oxides of Lanthanum and Cobalt and their electrocatalytic properties for oxygen evolution in alkaline-solutions. J Chem Soc Faraday Trans 91:1871–1875
12. Kanan MW, Nocera DG (2008) In situ formation of an oxygen-evolving catalyst in neutral water containing phosphate and Co^{2+}. Science 321:1072–1075
13. Zaharieva I, Chernev P, Risch M, Klingan K, Kohlhoff M, Fischer A, Dau H (2012) Electrosynthesis, functional, and structural characterization of a water-oxidizing manganese oxide. Energ Environ Sci 5:7081–7089
14. Dinca M, Surendranath Y, Nocera DG (2010) Nickel-borate oxygen-evolving catalyst that functions under benign conditions. Proc Natl Acad Sci U S A 107:10337–10341
15. Hocking RK, Brimblecombe R, Chang LY, Singh A, Cheah MH, Glover C, Casey WH, Spiccia L (2011) Water-oxidation catalysis by manganese in a geochemical-like cycle. Nat Chem 3:461–466
16. Trasatti S (1980) Electrocatalysis by oxides – Attempt at a unifying approach. J Electroanal Chem 111:125–131
17. Nakagawa T, Beasley CA, Murray RW (2009) Efficient electro-oxidation of water near its reversible potential by a mesoporous IrO_x nanoparticle film. J Phys Chem C 113:12958–12961
18. Lee Y, Suntivich J, May KJ, Perry EE, Shao-Horn Y (2012) Synthesis and activities of Rutile IrO_2 and RuO_2 nanoparticles for oxygen evolution in acid and alkaline solutions. J Phys Chem Lett 3:399–404
19. Choi P, Bessarabov DG, Datta R (2004) A simple model for solid polymer electrolyte (SPE) water electrolysis. Solid State Ion 175:535–539
20. Marshall A, Borresen B, Hagen G, Tsypkin M, Tunold R (2007) Hydrogen production by advanced proton exchange membrane (PEM) water electrolysers - Reduced energy consumption by improved electrocatalysis. Energy 32:431–436
21. de Levie R (1999) The electrolysis of water. J Electroanal Chem 476:92–93
22. van Troostwijk AP, Deiman JR (1789) Lettre à M. de la Mètherie, sur une manière de dècompose l'eau en air inflammable et en air vital. Journal de physique, de chimie et de l'histoire naturelle 35:369–378
23. Trasatti S (1999) 1799–1999: Alessandro Volta's 'electric pile' – Two hundred years, but it doesn't seem like it. J Electroanal Chem 460:1–4
24. Berg H (2008) Johann Wilhelm Ritter – the founder of scientific electrochemistry. Rev Polagraphy 54:99–103
25. Hickling A, Hill S (1947) Oxygen overvoltage. Part I. The influence of electrode material, current density, and time in aqueous solution. Discuss Faraday Soc 1:236–246
26. Ruetschi P, Delahay P (1955) Influence of electrode material on oxygen overvoltage: a theoretical analysis. J Chem Phys 23:556–560
27. Ritter JW (1800) Volta's galvanische batterie: selbst Versuche mit derselben angestellt. Voigts Magazin für den neuesten Zustand der Naturkunde 2:356–400
28. von Hofmann AW (1866) Introduction to modern chemistry. Walton and Maberly, London
29. Classen A, Danneel H (1897) Quantitative analyse durch elektrolyse. Springer, Berlin
30. Sabatier P (1911) Announcement. Hydrogenation and dehydrogenation for catalysis. Ber Dtsch Chem Ges 44:1984–2001
31. Hammer B, Nørskov JK (2000) Theoretical surface science and catalysis – calculations and concepts. Adv Catal 45:71–129
32. Man IC, Su H-Y, Calle-Vallejo F, Hansen HA, Martínez JI, Inoglu NG, Kitchin J, Jaramillo TF, Nørskov JK, Rossmeisl J (2011) Universality in oxygen evolution electrocatalysis on oxide surfaces. Chemcatchem 3:1159–1165
33. Vojvodic A, Nørskov JK (2011) Optimizing perovskites for the water-splitting reaction. Science 334:1355–1356
34. Matsumoto Y, Yoneyama H, Tamura H (1977) Influence of nature of conduction-band of transition-metal oxides on catalytic activity foroxygen reduction. J Electroanal Chem 83:237–243
35. Matsumoto Y, Sato E (1979) Oxygen evolution on $La_{1-X}Sr_xMnO_3$ electrodes in alkaline-solutions. Electrochim Acta 24:421–423
36. Dau H, Limberg C, Reier T, Risch M, Roggan S, Strasser P (2010) The mechanism of water oxidation: from electrolysis via homogeneous to biological catalysis. Chemcatchem 2:724–761
37. Gerken JB, McAlpin JG, Chen JYC, Rigsby ML, Casey WH, Britt RD, Stahl SS (2011) Electrochemical water oxidation with cobalt-based electrocatalysts from pH 0–14: the thermodynamic basis for catalyst structure, stability, and activity. J Am Chem Soc 133:14431–14442
38. Surendranath Y, Kanan MW, Nocera DG (2010) Mechanistic studies of the oxygen evolution reaction by a cobalt-phosphate catalyst at neutral pH. J Am Chem Soc 132:16501–16509
39. Nørskov JK, Bligaard T, Rossmeisl J, Christensen CH (2009) Towards the computational design of solid catalysts. Nat Chem 1:37–46

Oxygen Nonstoichiometry of Oxide

Junichiro Mizusaki
Tohoku University, Funabashi, Chiba, Japan

What Is Nonstoichiometry?

Molecules consist of elements. The number of atoms of the elements of a certain molecule is uniquely determined for each chemical species. If one atom is added or missing from the molecule, the molecule is no more the original chemical species but becomes a different one. Let us consider, for example, a water molecule, H_2O. When we add one oxygen atom to this molecule, it becomes H_2O_2 and is called hydrogen peroxide. When we take out an oxygen atom from a water molecule, it becomes a hydrogen molecule. It is no more a water molecule. Molecules are always stoichiometric, that is, being consisted of the whole number of atoms of the component elements.

However, when atoms or molecules are accumulated in a huge amount to be a visual size or to be seen under optical micrographs and becomes suitable size to be treated by statistical dynamics or thermodynamics, in other words, when atoms or molecules are accumulated to form phase, things becomes very different.

You may know that perfect crystal can exists only at 0 K. That is, in the real world, thermally activated defects always exist in the crystal. Defects increase with temperature.

You may easily recognize that it is impossible to obtain a pure element. Semiconductor technology requires "pure" silicon or germanium. By repeated zone refining or some other techniques, 15 N (eleven-nine, 99.9999999999999 %) silicon is obtained. This purity is one of the highest so far obtained. However, even in such high purity bulk, impurity density is more than the order of 1/cubic micrometer, or the average impurity atom distance is several hundred nanometers.

In solid ionic crystalline compounds such as metallic oxides, the crystal lattice consists of cation sublattice and oxide-ion sublattice. Then, each sublattice contains defects and impurities and thermodynamic equilibrium governs the defects in the same sublattice and also among different sublattices. The details of such equilibriums are treated in the field of defect chemistry.

Impurity elements are incorporated in either of the appropriate sublattices, that is, impurity cation is incorporated in cation sublattice and impurity anions are in anion sublattice, both to the amount of their solubility limits. In general, the solubility of impurity ions for different sublattice is different. The number of atoms in respective sublattices is different from each other. That is, the number of total cations and total anions deviates from the ratio of simple whole number. Such deviation of composition from the nominal simple chemical formula of the ionic compound is called nonstoichiometry.

Similar to the dissolution of impurity elements, the dissolution of the component element to each sublattice may also take place. The total number of atoms in one sublattice may differ from that in another, resulting in nonstoichiometric deviation of the composition.

Every condensed phase of compounds shows some extent of nonstoichiometry, the deviation from its nominal stoichiometric composition. It comes from the thermodynamic nature of condensed phase and thus comprehensive. That is, there is no compound phase without nonstoichiometric compositional deviations [1].

Thermodynamic Base of Nonstoichiometry

The free energy diagram for metal (M)–oxygen system is shown in Fig. 1. We can draw a cotangential line to the free energy curve of M and to that of $M_{1-x}O_x$. This cotangential line gives the equilibrium of metal M phase and oxide phase $M_{1-x}O_x$. The tangential point to the $M_{1-x}O_x$ curve, x_2, gives the lowest oxygen content of $M_{1-x}O_x$. When we draw a tangential line from the point of $\mu_O{}^0(P(O_2) = 10^5$ Pa) to $M_{1-x}O_x$, the tangential point, x_3, to the free energy curve of $M_{1-x}O_x$ gives the oxygen content of $M_{1-x}O_x$ at

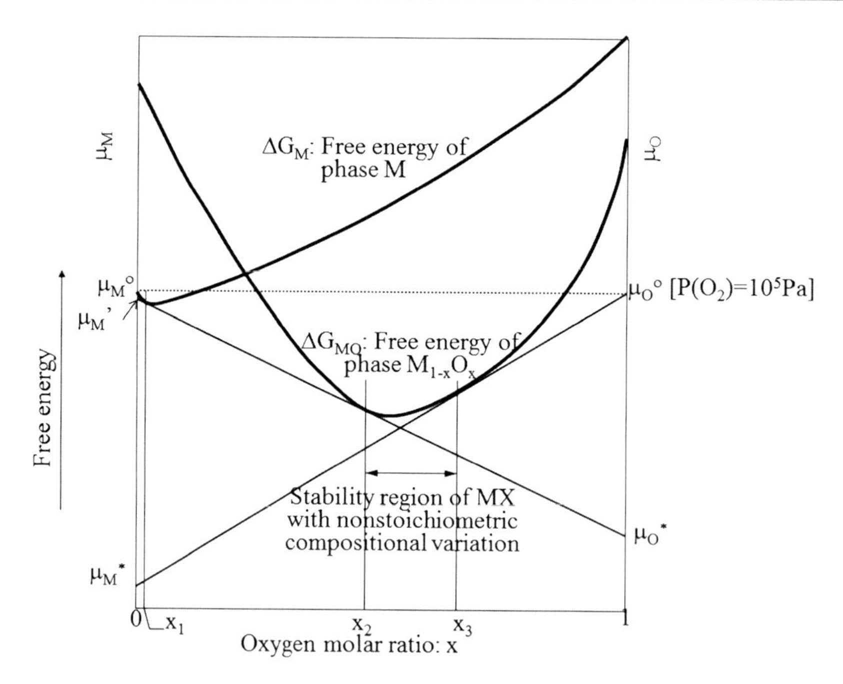

Oxygen Nonstoichiometry of Oxide, Fig. 1 Isothermal free energy diagram for the metal (M)–oxygen(O) system. The region $0 < x < x_1$ is the oxygen dissolution region into M. The region $x_1 < x < x_2$ is the two-phase region of M and oxide $M_{1\text{-}x}O_x$. The region $x_2 < x < x_3$ is the stability region of the oxide $M_{1\text{-}x}O_x$. Then, x_2 is the composition of minimum oxygen content in $M_{1\text{-}x}O_x$, and x_3 is the maximum of that. The line μ_M' and μ_O* is the cotangential line of the free energy curve for M and that for $M_{1\text{-}x}O_x$. The line μ_O^0 and μ_M* is the tangent of the free energy curve for $M_{1\text{-}x}O_x$

$P(O_2) = 10^5$ Pa. If we limit the maximum oxygen pressure to 10^5 Pa, the nonstoichiometric composition of $M_{1\text{-}x}O_x$ is between x_2 and x_3. When the $P(O_2)$ is higher than 10^5 Pa, we obtain larger oxygen content, x.

Extent of Nonstoichiometry from Thermodynamical Standpoint

When the bottom of the free energy curve for $M_{1\text{-}x}O_x$ shows small curvature as the red broken curve of Fig. 2, the distance between the two tangential points, $x_3\text{-}x_2$, becomes large, and hence, the nonstoichiometric compositional region becomes wide. On the other hand, if the bottom of the free energy curve for $M_{1\text{-}x}O_x$ is sharp and the curvature of the tip is large, like the blue dashed curve, the nonstoichiometric compositional region becomes narrow. As shown in Figs. 1 and 2, when the compound $M_{1\text{-}x}O_x$ exists, there always exist two tangential lines to show the maximum and minimum oxygen content in $M_{1\text{-}x}O_x$. That is, there always exists nonstoichiometric compositional deviation in any metallic oxides.

Extent of Nonstoichiometry from the Chemical Nature of Cationic Element

The extent of nonstoichiometry differs among variety of oxides. For example, the oxide superconductor $YBa_2Cu_3O_{7\text{-}d}$ is known to show large nonstoichiometric composition variation of $0 < d < 1$ [2]. On the other hand, silver chloride, AgCl, is known as a typical "stoichiometric compound." As described above, any metallic oxide should theoretically have some extent of nonstoichiometry. It is of interest how large is the nonstoichiometry in AgCl. It is known [3] that AgCl in equilibrium with silver metal has silver

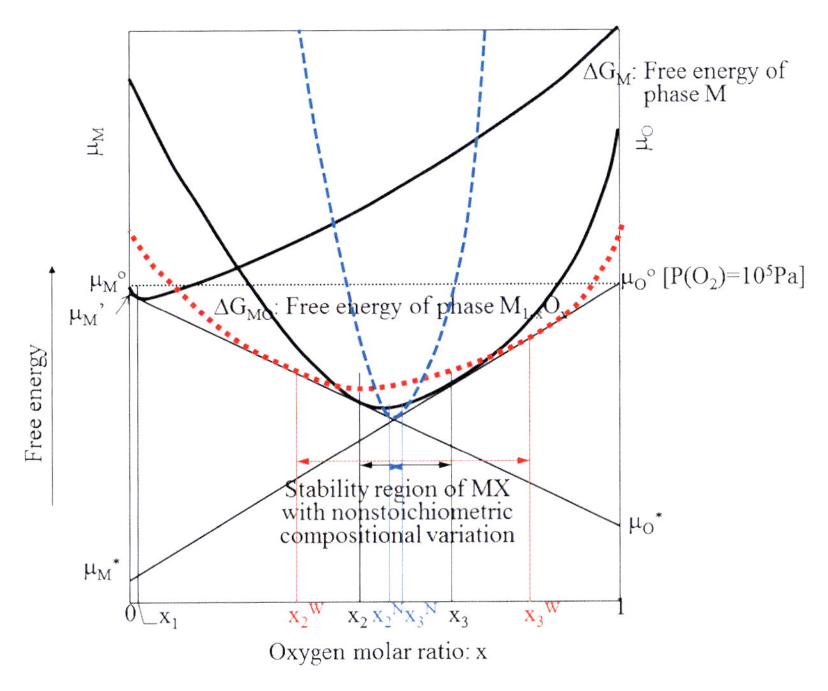

Oxygen Nonstoichiometry of Oxide, Fig. 2 Relationship between the width of nonstoichiometric compositional variation, $x_3 - x_2$ and the shape of the bottom of free energy curve for $M_{1-x}O_x$. The three free energy curves, *red*, *black*, and *blue*, for $M_{1-x}O_x$, possess the same tangential lines and the chemical stability of $M_{1-x}O_x$ represented by these curves are the same. However, the nonstoichiometric compositional ranges are different in these three types of $M_{1-x}O_x$. The *red* curve shows a large curvature and large nonstoichiometric variation of composition, $x_3^W - x_2^W$, while the *blue* curve shows a sharp bottom curvature and small value of $x_3^N - x_2^N$

excess composition, and when we express the nonstoichiometric composition by $Ag_{1+d}Cl$, d at 440^{0}C is roughly 10^{-6} while AgCl is in equilibrium with Cl_2 gas of 10^5 Pa, d at 440^{0}C is roughly -10^{-4}.

The range of nonstoichiometric compositional variation is in variety. The metallic components of the oxides with very small nonstoichiometric compositional variation, conventionally assigned as stoichiometric compounds such as AgCl, are characterized to have fixed valence state. Typical examples are Li_2O(+), Na_2O(+), MgO(2+), SrO(2+), Al_2O_3(3+), Bi_2O_3(3+), ZrO_2(4+), SnO_2(4+), and so on. Here, the valence states of cations are shown in parentheses. The oxides with large nonstoichiometric composition variation are often called nonstoichiometric oxides. They are characterized to have the metallic component with mixed valence state. Typical crystal structures for these nonstoichiometric oxides are tolerant toward oxide-ion vacancy formation and valence change of ionic state. Typical examples are perovskite type, such as $La_{1-x}Sr_xCoO_{3-d}$, $La_{1-x}Sr_xFeO_{3-d}$, $La_{1-x}Sr_xMnO_{3-d}$, and $SrCe_{1-x}Y_xO_{3-d}$, or perovskite-related type, such as $La_{2-x}Sr_xNiO_{4+d}$, $La_{2-x}Sr_xCuO_{4-d}$, $SrFeO_{2.5+d}$, and $YBa_2Cu_3O_{7-d}$; fluorite type or bixbyite type, such as UO_{2+d}, CeO_{2-d}, and Tb_2O_{3+d}; and some layered intercalation compounds typically Li_xCoO2.

Future Directions

As described above, the variation in oxygen nonstoichiometry of oxide is followed by the formation of defects in the oxide. The oxide-ion vacancies may be accompanied by the oxygen deficient type nonstoichiometry. The interstitial oxide ions or the cation vacancies

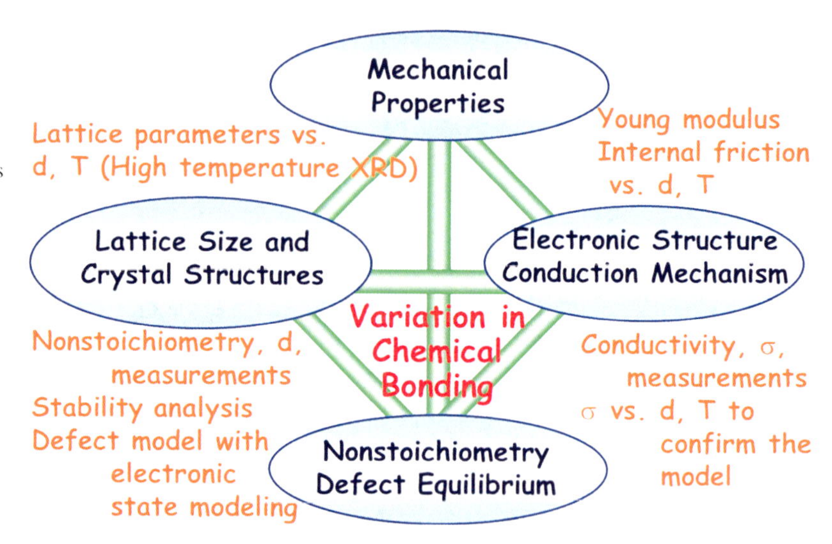

Oxygen Nonstoichiometry of Oxide, Fig. 3 Relationship between science and engineering of solid oxides and nonstoichiometry

may be formed in the oxide with the increase in the oxygen excess or the metal deficient type nonstoichiometry. As "defect chemistry" shows, in order to maintain electroneutrality in the oxide and due to the quasi-chemical equilibrium among defects, when the concentration of a certain type of defects varies, the concentrations of all the other defects also vary

When the concentrations of ionic defects such as oxide ion vacancies, interstitial oxide-ions or cation vacancies increase, the oxide-ion conductivity or the cation conductivity increases. When the concentrations of electronic defects such as electrons or holes both either localized or itinerant increase, electronic conductivity may increase.

With the variation of average valence state caused by the variation of the concentration of localized electrons, the ionic radii vary and with the variation of the concentrations of vacancies or interstitial cation formation, the binding energy of the crystal may vary. These variations result in the variation of the lattice constants of the crystal. The binding energy and the lattice constants are closely related to the mechanical properties. Then, with the nonstoichiometric compositional variation, the mechanical properties of the oxide may vary. The relationship of materials properties and nonstoichiometry is schematically shown in Fig. 3. In the near future, nonstoichiometry and defect chemistry become very important base for the science and engineering of mechanical properties of oxides and ceramics.

Cross-References

- ▶ Defect Chemistry in Solid State Ionic Materials
- ▶ Defects in Solids
- ▶ Electrolytes, Thermodynamics
- ▶ Kröger-Vinks Notation of Point Defects
- ▶ Mixed Conductors, Determination of Electronic and Ionic Conductivity (Transport Numbers)
- ▶ Nanoionics at High Temperatures
- ▶ Oxide Ion Conductor
- ▶ Oxygen Anion Transport in Solid Oxides
- ▶ Perovskite Proton Conductor
- ▶ Solid Electrolytes
- ▶ Solid State Electrochemistry, Electrochemistry Using Solid Electrolytes

References

1. Kofstad P (1972) Nonstoichiometry, diffusion and electrical conductivity in binary metal oxides. Wiley, New York
2. Strobel P, Capponi JJ, Marezio M, Monod P (1987) High-temperature oxygen defect equilibrium in superconducting oxide $YBa_2Cu_3O_7$-x, Solid State Commun 64(4):513–15
3. Mizusaki J, Fueki K (1982) Electrochemical Determinations of the Chemical Diffusion Coefficient and Nonstoichiometry in AgCl, Solid State Ionics 6:85–91

Oxygen Reduction Reaction in Acid Solution

Lj Vracar
Faculty of Technology and Metallurgy
University of Belgrade, Belgrade, Serbia

Introduction

A widespread interest for the electrochemical oxygen reduction reaction (ORR) has two aspects. The reaction attracts considerable attention from fundamental point of view, as well as it is the most important reaction for application in electrochemical energy conversion devices. It has been in the focus of theoretical considerations as four-electron reaction, very sensitive to the electrode surface structural and electronic properties. It may include a number of elementary reactions, involving electron transfer steps and chemical steps that can form various parallel-consecutive pathways [1–3].

In addition to electrochemical energy conversion in fuel cells, the reaction has applications in energy storage in metal-air batteries, in several industrial processes as the chloralkali electrolysis, and it causes corrosion of metals and alloys in the presence of air. That is why the efforts have been focused on elucidating the mechanism of this reaction and developing proper catalysts.

Thermodynamic Aspects

The overall electrochemical reduction of oxygen in acid aqueous solution is

$$O_2 + 4H^+ + 4e \Leftrightarrow 2H_2O \qquad \left[E^0_{SHE}\right]_{298\,K} = 1.229\,V \tag{1}$$

where the standard reaction potential, relative to the standard hydrogen electrode, has been calculated from the standard Gibbs energy formation of H_2O that is $\Delta G^0_{298K}(H_2O) = -23,18\,kJ\,mol^{-1}$.

The oxygen reduction is highly irreversible reaction even at very high temperatures, so it was experimentally difficult to verify its thermodynamic (reversible) potential in aqueous solutions. The earliest referred experimentally obtained reversible potential of 1.23 V (SHE), established on Pt electrode in oxygen-saturated 0.05 mol dm^{-3} H_2SO_4, previously carefully purified by pre-electrolysis, [4] was stable only for 1 h. More recently, the potential of 1.23 V was obtained and maintained a few hours, by using special treatment of the Pt electrode and different very long and careful solution purification procedure [5–7].

The oxygen reversible potential of 1.23 V was confirmed indirectly by extrapolation of anodic and cathodic polarization curves on various noble metal electrodes. The interception of the polarization curves at the potential close to 1.23 V was observed with highly oxidized Pt electrode [1, 8]. These results were unusual as the extrapolations were made from high cathodic and high anodic overpotentials where the surface conditions were different and it was expected that the mechanisms for reduction and oxidation of oxygen might be different.

Nature of Open-Circuit Potential

The irreversibility of the reaction is the source of energy losses in processes involving oxygen electrodes. It is desirable to have the ORR occurring at potentials, as close as possible to the reversible potential, with a satisfactory reaction rate. The overpotential at which the oxygen cathodes operate is influenced by adsorbed species at the metal surface that block the approach of oxygen molecules to the surface sites where they are going to be reduced.

Irreversibly established rest potential, also called open-circuit potential, experimentally obtained in O_2 saturated solutions varies for different electrode materials. On the most active platinum in pure acid solution saturated with O_2 at $p = p^0$ and $t = 25\,°C$, the open-circuit potential usually has value close to 1.0 V versus RHE, not exceeding 1.1 V. The value of open-circuit potential is an important parameter being the starting point for the oxygen reduction as cathodic reaction in electrochemical processes.

For oxygen cathode in a fuel cell, the loss of potential is ca. 0.4 V since its working potential is close to 0.8 V.

To explain the nature of open-circuit potential, the concept of a mixed potential has been usually suggested as a possible mechanism. According to this mechanism, as a result of simultaneous occurrence of cathodic (O_2 reduction) and some side anodic reaction that compete with it, the rest potential is going to be established in between the equilibrium potentials of these competing reactions. It has been proposed that relevant anodic reactions could be either oxidation of impurities from the solution [4, 9] or chemisorbed oxygen-containing species, as OH, or anodic dissolution of the electrode [10, 11].

Reaction Pathways of ORR in Acid Solution

Electrochemical oxygen reduction in acidic aqueous solution can proceed by two overall partways:

(a) The direct four-electron partway (Eq. 1 in reduction direction)
(b) The peroxide partway through peroxide formation

$$O_2 + 2H^+ + 2e \rightarrow 2H_2O_2 \quad \left[E^{0}_{SHE}\right]_{298\ K} = 0.67\ V \tag{2}$$

followed by its reduction

$$H_2O_2 + 2H^+ + 2e \rightarrow 2H_2O \left[E^{0}_{SHE}\right]_{298\ K} = 1.77V \tag{3}$$

or decomposition

$$2H_2O_2 \rightarrow 2H_2O + O_2 \tag{4}$$

According to the literature data for the ORR, all metallic electrode materials can be divided into two groups: (a) the first one that is represented by platinum, palladium [12–15], platinum alloys [16, 17], silver [18], some metal oxides, and some transition metals on which oxygen reduction proceeds to water both directly or in parallel with hydrogen peroxide formation, which is decomposed by electrochemical or chemical reactions, and (b) the second one that is represented by gold [19, 20], carbon materials [21, 22], mercury [23], and some oxide-covered metals (Ni, Co) with reduction that predominantly results in hydrogen peroxide formation.

Based on the use of the rotating ring-disk methodology, it was possible to follow reduction of O_2 on the disk and to monitor the H_2O_2 production on the ring to distinguish whether oxygen reduction proceeds through hydrogen peroxide as an intermediate or through direct four-electron multiple step reduction without H_2O_2 as an intermediate. One reaction scheme, based on the ring-disk experimental results, was proposed by Bagotskii et al. [24] (Fig. 1).

According to this scheme the solution phase molecular oxygen first adsorbs at the electrode surface. From the adsorbed state, its electrochemical reduction goes either directly to water with constant k_1 or through adsorbed hydrogen peroxide as intermediate that could be further reduced to H_2O, with the constant k_3, in which case the reduction still proceed by direct $4e^-$ pathway, or peroxide can desorbed and, as a final product of reaction, diffuse into the solution following $2e^-$ pathway. The scheme also includes chemical decomposition of two hydrogen peroxide molecules to H_2O and O_2 [25]. The essential problem in determining the correct mechanism according to the Bagotskii scheme is unsatisfactory number of experimentally obtained parameters from the ring-disk measurements meaning that many rate constants cannot be determined. Some modifications of the Bagotskii scheme are made by Wroblowa et al. [20] proposing mechanism with two rate constants missing, by Damjanovic et al. [26] that omitted $(H_2O_2)_{ads}$ state and did not consider slow desorption of (H_2O_2), or by Anastasijevic et al. [27] whose scheme includes almost all the possible intermediates. None of them did fully succeed in solving problem since the number of variables overpasses the number of calculate parameters that are obtained by measuring.

Despite the problems in determining the exact values of all rate constants, the careful

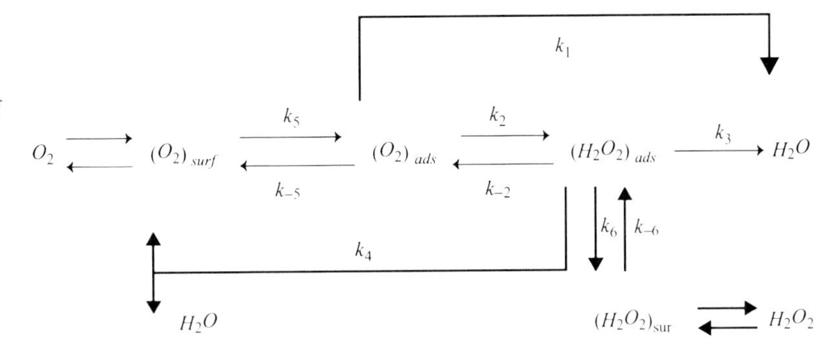

Oxygen Reduction Reaction in Acid Solution, Fig. 1 General scheme of ORR in acid solution

analysis of the experimental data collected by frequently used electrochemical techniques as steady-state polarization, cyclic voltammetry, rotating disk electrode, and rotating ring-disk electrode, taking into consideration the surface properties, allowed one to propose the course of the reaction.

ORR Mechanism on Platinum Family Metals and Their Alloys

The generally accepted statement for oxygen reduction on Pt family metals and their alloys is that predominantly direct $4e^-$ mechanism is operative in acid solution, without hydrogen peroxide being detected on the electrode ring (Fig. 2b). The polarization curves are characterized by a single wave with a limiting current close to the diffusion current for the $4e^-$ process (Fig. 2a).

It is also accepted that, in the range of high current densities, the slow charge transfer is rate-determining step characterized with experimentally determined Tafel slope of -0.120 V dec^{-1} ($-2\cdot2.3$ RT/F) and reaction order with respect to H^+ being 1.

Sepa et al. [12] suggested that the proton transfer occurs simultaneously with the charge transfer and proposed the mechanism:

$$O_2 \Leftrightarrow O_{2,ads} \tag{5}$$

$$O_{2,ads} + H^+ + e^- \rightarrow \text{product(s)} \tag{6}$$

According to them, the experimentally determined Tafel slope of -0.060 V dec^{-1} (-2.3 RT/F), in the range of low current densities, arose from oxygen-containing species (PtOH) formed in reaction of Pt with H_2O and adsorbed under Temkin conditions, meaning that their coverage is potentially dependent. The reaction order with respect to H^+ that is found to be 3/2 supports this mechanism.

Clouser et al. [29] have the same experimentally obtained parameters explained differently suggesting that charge transfer and dissociative chemisorptions of the O_2 on the electrode surface occurred simultaneously as a rate-determining step.

The exchange current density is mainly reported on Pt and Pt alloys catalysts. As there are two Tafel regions, two exchange current densities have been reported. The extrapolation from the lower slope gave the value close to 10^{-10} A cm^{-2}, and from the higher, the value close to 10^{-6} A cm^{-2}. On the other catalysts the exchange current density has occasionally been determined and their values vary on the catalysts, as well as, on the research method.

ORR on Carbon Materials, Gold, and Mercury

Two different reaction mechanisms following a two-electron reduction were suggested on the basis of detailed mechanistic studies of the reduction. The polarization curves are characterized by two waves: the first with a limiting current close to the current that corresponds to the oxygen

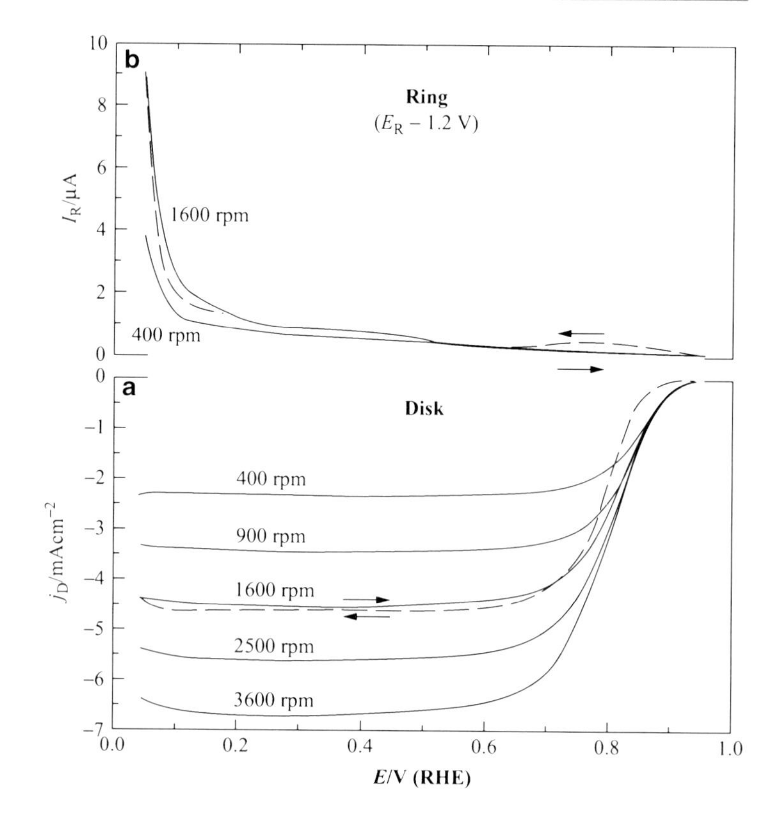

Oxygen Reduction Reaction in Acid Solution, Fig. 2 (a) Potentiodynamic ($5\ mV\ s^{-1}$, positive sweep) O_2 reduction current densities (j_D) on the thin film RRDE with Pt/Vulcan catalyst ($14\ \mu g_{Pt}\ cm^{-2}$) at 60 °C in 0.5 M H_2SO_4 saturated with 1 bar O_2. (**b**) simultaneously recorded ring currents (I_R) at 400 and 1,600 rpm for a ring potential of $E_R = 1.2$ V. The dashed lines show the negative going sweep at 1,600 rpm (Reproduced from [28] with permission)

reduction to hydrogen peroxide and the second one that is difficult to distinguish as it represent very irreversible reduction of peroxide. The referred values of Tafel slope for the ORR in acid solutions of different pH varies from $-0.120\ V\ dec^{-1}$ to $-0.160\ V\ dec^{-1}$, depending on the electrode materials.

Androuseva et al. [22] proposed oxygen adsorption as rate-determining step that should be followed by a charge transfer and peroxide formation:

$$O_2 \rightarrow O_{2,ads} \quad (7)$$

$$O_{2,ads} + e^- \rightarrow O^-_{2,ads} \quad (8)$$

$$O^-_{2,ads} + H^+ + e \rightarrow HO^-_2 \quad (9)$$

while Zurilla et al. [19] suggested combination of reactions (Eqs. 8 and 9) as a single rate-determining step.

ORR as a Structure-Sensitive Reaction

Oxygen reduction is a structure-sensitive reaction. According to the results on Pt single crystals, the effect of crystallographic orientation varies depending on the solution where strong effect was reported in H_3PO_4 solution, considerable in H_2SO_4 and HCl solutions, while there is no generally accepted agreement for $HClO_4$ solution. The rate of oxygen reduction generally increases as the surface area decreases. This effect is, in case of Pt, explained by variation of the different crystal planes exposed to the electrolyte as a function of the particle size (Table 1).

It has been demonstrated that the decrease of Pt particle size reduces the true d-electron number through the more favorable hybridization of the 5d state with the empty states above the Fermi level, increasing, as a consequence, the energy of adsorbed oxygen species.

Oxygen Reduction Reaction in Acid Solution, Table 1 Comparison of ORR activity on Pt single crystals in 0.05 mol dm^{-3} H_2SO_4, at $t = 60$ °C [30]

Electrode	j_K at 0.90 V (μA cm$^{-2}_{real}$)	j_K at 0.85 V (μA cm$^{-2}_{real}$)
Pt(111)	80	650
Pt(100)	480	1,450
Pt(110)	1,900	5,400

Requirements for ORR Catalysts

Platinum and palladium are single metals known to possess the best catalytic activity for oxygen reduction in acid solutions in processes of interest, among which the most attention is focused to their use in fuel cell technology. Due to Pt and Pd cost and supply limitations, it is required to reduce their loading in the catalyst. Generally, there are two possibilities to reduce the amount of single metal and even enhance the catalyst activity. The first widely adopted approach to do it is to increase the surface area of the catalyst by dispersing nano-sized Pt or Pd particles on the proper support and the second one is to use bi- or tri-metallic alloys containing Pt or Pd with other metals such as Ni, Co, Fe, and Cr. It is now generally accepted that the size and distribution of the catalyst particles are affected by the physical structure of the support (porosity and surface area) and the nature of the catalyst precursor.

Catalyst Supports

Catalyst support has to offer highly developed surface combined with excellent electronic conductivity and corrosion resistance.

Generally, carbon has been used as a support in fuel cell systems since it allows one to decrease Pt loading from ca. 4 to 0.2 mg cm^{-2} [31, 32]. However, thermodynamically favorable and kinetically slow electrochemical oxidation of carbon support during start/stop of a simple fuel cell (Eq. 10) or carbon corrosion that could proceed during recharge of the regenerative fuel cell at the air electrode leads to degradation in the fuel cell performance.

$$\begin{aligned} &C + H_2O \rightarrow CO_2 + 4H^+ + 4e^- \\ &E = 0.118\ \text{V (RHE)} \end{aligned} \tag{10}$$

The other catalyst supports as a nonstoichiometric mixture of several titanium oxide phases that have hypo-d-electron character have an ability to interact with inert noble metals, like platinum, changing additionally the catalytic activity of the platinum as catalyst. However, very low specific surface area (the maximum value referred is 15 m^{2}g^{-1}) prevents high dispersion and good compositional homogeneity of the catalyst clusters which is an important request for the activity of the catalyst.

Tungsten-based materials as n-type semiconductor with nonstoichiometric compositions are extremely stable under electrochemical oxidation conditions and could be used as a non-carbon support for catalyst. The interest in their use as catalyst support is due to the possible synergetic effect between metal catalyst and support.

Catalyst deposition using conventional methods, such as impregnation, coprecipitation, electrodeposition, and ion exchange, is not adequate from the viewpoint of offering high metal surface area or good dispersion of metal particles and that is why the alternative method of prepared metal colloids and subsequent deposition is frequently in use.

Pt Alloys Catalysts

However, numerous studies reported that Pt and Pd alloys, with transition metals as Co, Ni, Cr, and Fe, showed the enhancement factor up to 10 for ORR if compared with pure metal (Table 2) [33].

Some of the studies investigated carbon-supported alloys while the others sputtered alloy films or metallurgically prepared bulk alloys. The enhancement of the oxygen reduction rate was ascribed to the several effects and their interplay: the inhibition of the OH adsorption on Pt surface by the alloying metal [34], the geometric

Oxygen Reduction Reaction in Acid Solution, Table 2 Structural and electrochemical characteristics of the Pd/C and Pd–Fe/C electrocatalysts for ORR in 0.1 mol dm^{-3} $HClO_4$ (From [33])

Sample	Particle size (nm)	Pd–Pd bond distance (nm)	Electrochemical area ($cm^2 mg^{-1}_{Pt}$)	j_k at 0.85 V (mA cm^{-2})
Pd/C	10.2	0.2753	446	0.131
Pd_4Fe/C	9.6	0.2735	501	0.500
Pd_3Fe/C	9.7	0.2730	726	0.791
Pd_2Fe/C	7.6	0.2742	874	0.346
PdFe/C	7.7	0.2743	721	0.420

modification of Pt [35], extensive surface roughening [36], as well as through the lowering of the strength of adsorbed oxygen species because of the electronic interactions [16]. The particle size effect on oxygen reduction was not unambiguously determined as far as the different Pt alloys are concerned. So, in case of Pt–Co alloy, it is found that the decrease in particle size decreases the specific activity [37], while in case of Pt-Fe alloy, the specific activity is found to be independent of the possible particle size [38].

The electrochemical results correlated with in situ X-ray adsorption spectroscopy that provides information on the electronic and geometric factors show that the alloys have higher Pt 5d-orbital vacancies and shorter Pt–Pt distances and that inhibit the chemisorptions of OH on the platinum facilitating faster oxygen reduction.

Cross-References

- ► Electrocatalysis - Basic Concepts, Theoretical Treatments in Electrocatalysis via DFT-Based Simulations
- ► Oxygen Reduction Reaction in Alkaline Solution
- ► Platinum-Based Cathode Catalysts for Polymer Electrolyte Fuel Cells
- ► Polymer Electrolyte Fuel Cells (PEFCs), Introduction

References

1. Tarasevich MR, Sadkowskiand A, Yeager E (1983) Comprehensive treatise of electrochemistry. In: Conway B, Bockris JOM, Yeager E, Khan SUM, White RE (eds) Vol 7. Plenum, New York
2. Kinoshita K (1992) Electrochemical oxygen technology. Wiley, New York
3. Adzic R (1998) Recent advances in the kinetics of oxygen reduction. In: Lipkowski J, Ross PN (eds) Electrocatalysis. Wiley, New York
4. Bockris JOM, Huq AKMS (1956) The mechanism of the electrolytic evolution of oxygen on platinum. Proc R Soc Lon Ser A 237:277–296
5. Watanabe N, Devenathan MAV (1964) Reversible oxygen electrode. J Electrochem Soc 111:615–619
6. Hoare JP (1979) Some aspects of the reduction of oxygen at a platinum-oxygen alloy diaphragm. J Electrochem Soc 126:1502–1504
7. Hoare JP (1975) On the reduction of oxygen platinum-oxygen alloy diaphragm electrodes. Electrochim Acta 20:267–272
8. Damjanovic A (1969) Modern aspects of electrochemistry. In: Bockris JOM, Conway B (eds) Vol 5. Plenum, New York
9. Wroblowa HS, Rao MLB, Damjanovic A, Bockris JOM (1967) Adsorption and kinetics at platinum electrodes in the presence of oxygen at zero net current. J Electroanal Chem 15:139–150
10. Rand DAJ, Woods R (1972) A study of the dissolution of platinum, palladium, rhodium and gold electrodes in 1 M sulphuric acid by cyclic voltammetry. J Electroanal Chem 35:209–218
11. Bindra P, Clouser S, Yeager E (1979) Platinum dissolution in concentrated phosphoric acid. J Electrochem Soc 126:1631–1632
12. Sepa DB, Vojnovic MV, Damjanovic A (1981) Reaction intermediates as a controlling factor in the kinetics and mechanisms of oxygen reduction at platinum electrodes. Electrochim Acta 26:781–793
13. Sepa DB, Vojnovic MV, LjM V, Damjanovic A (1987) Different views regarding the kinetics and mechanisms of oxygen reduction at Pt and Pd electrodes. Electrochim Acta 32:129–134
14. Vracar LJM, Sepa DB, Damjanovic A (1986) Palladium electrode in oxygen-saturated aqueous solutions, reduction of oxygen in activation-controlled region. J Electrochem Soc 133:1835–1839
15. Antropov LI, Vrzhosek CG, Tarasevich MR, Marinich MA (1972) Issledovanie vliyaniya kationov tetrabutilamminiisulfata na katodnoe vosstavlenie kisloroda na platine v kislikh rastvorakh. Elektrokhimiya 8:149–151

16. Mukerjee S, Srinivasan S, Soriaga M, Breen JMC (1995) Effect of preparation conditions of Pt alloys on their electronic, structural, and electrocatalytic activities for oxygen reduction, XRD, XAS and electrochemical studies. J Phys Chem 99:4577–4589
17. Mukerjee S (1990) Particle size and structural effects in platinum electrocatalysis. J Appl Electrochem 20:537–548
18. Sawyer DT, Day RJ (1963) Kinetics for oxygen reduction at platinum, palladium and silver electrodes. Electrochim Acta 8:589–594
19. Zurilla RW, Sen RK, Yeager E (1978) The kinetics of the oxygen reduction reaction on gold in alkaline solution. J Electrochem Soc 125:1103–1109
20. Wroblowa HS, Yen-Chi-Pan RG (1976) Electroreduction of oxygen: a new mechanistic criterion. J Electroanal Chem 69:195–201
21. Taylor RJ, Humffray AA (1975) J electrochemical studies on glassy carbon electrodes: oxygen reduction in solution of high pH. J Electroanal Chem 64:63–84
22. Andruseva SI, Tarasevich MR, Radyushkina KA (1977) K vokhprosu ob elektrovosstanovlenii kisloroda na uglerodistikh materialakh. Elektrokhimiya 13:253–255
23. Van Velzen CJ, Remijuse AG, Sluyters-Rehibach M, Sluyters JH (1982) The electrochemical reduction of oxygen to hydrogen peroxide at the dropping mercury electrode part II. Its kinetics $0.4 \leq pH \leq 5.9$. J Electroanal Chem 142:229–242
24. Bagotskii VS, Tarasevich MR, Filinovskii VY (1972) Ichet adsorbtsionnoi stadii pri raschete kineticheskiikh parametrovreakcii kisloroda I perekisi vodoroda. Elektrokhimiya 8:84–87
25. Tarasevich MR, Zakharkin GI, Smirnova RM (1973) Oxygen and hydrogen peroxide reactions using oxygen – 18, hydrogen peroxide decomposition on platinum in the presence of various cations and anions. Elektrokhimiya 9:645–648
26. Damjanovic A, Genshaw MA, Bockris JOM (1966) Distinction between intermediates produced in main and side electrodic reactions. J Chem Phys 45:4057–4059
27. Anastasijevic NA, Vesovic V, Adzic RR (1987) Determination of the kinetic parameters of the oxygen reduction reaction using the rotating ring-disk electrode part I. Theory, part II. Applications. J Electroanal Chem 229:305–316, 229:317–325
28. Paulus UA, Schmidt TJ, Gasteiger HA, Behm RJ (2001) Oxygen reduction on a high-surface area Pt/Vulcan carbon catalyst: a thin-film rotating ring-disk electrode study. J Electroanal Chem 495:134–145
29. Clouser SJ, Huang JC, Yeager E (1993) Temperature dependence of the Tafel slope for oxygen reduction on platinum in concentrated phosphoric acid. J Appl Electrochem 23:597–605
30. Grgur B, Markovic NM, Ross PN Jr (1997) Temperature-dependent oxygen electrochemistry on platinum low-index single crystal surfaces in acid solutions. Can J Chem 75:1465–1471
31. Wilson MS, Valerio JA, Gottesfeld S (1995) Low platinum loading electrodes for polymer electrolyte fuel cells fabricated using thermoplastic ionomers. Electrochim Acta 40:355–363
32. Kumar GS, Raja M, Parthasarathy S (1995) High performance electrodes with very low platinum loading for polymer electrolyte fuel cells. Electrochim Acta 40:285–290
33. Shao MH, Sasaki K, Adzic RR (2006) Pt-Fe nanoparticles as electrocatalysts for oxygen reduction. J Am Chem Soc 128:3526–3527
34. Paulus UA, Wokaun A, Sherer GG, Schmidt TJ, Stamenkovic V, Radmilovic V, Markovic NM, Ross PN (2002) Oxygen reduction on carbon-supported Pt-Ni and Pt-Co alloy catalysts. J Phys Chem B 106:4181–4191
35. Jalan V, Tayler EJ (1983) Importance of interatomic spacing in catalytic reduction of oxygen in phosphoric acid. J Electrochem Soc 130:2299–2302
36. Kim KT, Kim YG, Chung JS (1995) Effect of surface roughening on the catalytic actvity of Pt-Cr electrocatalysts for oxygen reduction in phosphoric fuel cell. J Electrochem Soc 142:1531–1538
37. Lima FHB, Lizcano-Valbuena WH, Teiheira-Neto E, Fc N, Gonzalez ER, Ticianelli EA (2006) Pt-Co/C nanoparticles as electrocatalysts for oxygen reduction in H_2SO_4 and H_2SO_4/CH_3OH electrolytes. Electrochim Acta 52:385–393
38. Hwang JT, Chung JS (1993) The morphological and surface properties and their relationship with oxygen reduction activity for platinum-iron electrocatalysts. Electrochim Acta 38:2715–2723

Oxygen Reduction Reaction in Alkaline Solution

Andrzej Wieckowski[1] and Jacob Spendelow[2]
[1]University of Illinois, Urbana, IL, USA
[2]Los Alamos National Laboratory, Los Alamos, NM, USA

Introduction

The oxygen reduction reaction (ORR) has been, and continues to be, the focus of widespread R&D efforts, in part due to the importance on this reaction in fuel cells. Fuel cells typically operate with the ORR as the cathode half reaction, but slow ORR kinetics and the use of expensive ORR catalysts have hindered efforts to commercialize fuel cell technology on a wide scale. The problem of ORR catalysis has been particularly acute in low temperature fuel cells, including acidic polymer electrolyte membrane fuel cells (PEMFCs), liquid-electrolyte alkaline fuel cells (AFCs), and alkaline membrane fuel cells (AMFCs). The ORR is also an important reaction in diverse fields such as chlor-alkali electrolysis, metal-air batteries, electrochemical sensing, and corrosion.

The development and increasing maturation of alkaline ionomer membranes in recent years has prompted increased attention to the catalysis of the ORR in alkaline media. As compared with PEMFCs, the lower corrosivity of the high pH electrolyte in AFCs and AMFCs results in an increased variety of potential catalysts, including non-noble metals, that may be used as cathode materials. Along with the platinum-group metals (PGMs) that have been widely explored for applications in PEMFC cathodes, additional materials under development for cathode applications in AFCs and AMFCs include Ag, transition metals, and various materials containing carbon and nitrogen. With all these catalysts, cathode kinetic overpotentials greater than 0.3 V exist during fuel cell operation. Therefore, accelerated ORR kinetics would provide significant improvements in fuel cell performance and efficiency.

The ORR is a multistep reaction capable of proceeding through multiple pathways. Final products depend on whether the reaction is complete (four electron reduction, Eq. 1) or incomplete (two electron reduction, Eq. 2).

$$O_2 + 2H_2O + 4e^- \rightarrow 4OH^- \quad (1)$$

$$O_2 + H_2O + 2e^- \rightarrow HO_2^- + HO^- \quad (2)$$

In addition to being a final product of the two electron ORR, hydroperoxide ion can also be an intermediate in the four electron ORR, in an "indirect pathway" (Eq. 2, followed by Eq. 3).

$$HO_2^- + H_2O + 2e^- \rightarrow 3OH^- \quad (3)$$

Platinum Group Metals

The four electron ORR has been found to predominate on Pt [1]. Selectivity depends on multiple factors, including surface composition and structure, pH, potential, presence of adsorbates or contaminants, and mass transport characteristics [2]. The reported dependence on mass transport characteristics suggests a role of the indirect pathway [3]. The indirect pathway may also occur on catalysts with multiple types of active sites; for instance, on carbon-supported Pt catalysts, HO_2^- generated through the two electron ORR on C sites can be further reduced to OH^- on Pt sites [4].

The ORR on PGMs is widely accepted to occur through a mechanism in which electron and proton transfer to adsorbed O_2 precedes O—O bond scission. Kinetic [5] and spectroscopic [6] studies have indicated a likely rate-limiting step of superoxide formation (Eq. 4).

$$O_{2,ads} + e^- \rightarrow O_{2,ads}^- \quad (4)$$

Following superoxide formation, subsequent proton and electron transfer steps continue in the complete ORR (Eq. 1) or incomplete ORR (Eq. 2), though the pathway through which this occurs remains a topic of debate.

The rate of the ORR on PGMs, Au, and Ag has been found to depend on the electronic properties of the surface, with a "volcano curve" observed in plots of ORR activity vs. d-band center [7]. The peaked nature of this curve is attributed to the conflicting effects of adsorption strengths of reaction intermediates and products on the ORR. In metals with weaker adsorption of O-containing species, relatively slow O—O bond scission results in lower ORR rates, while in metals with stronger adsorption of O-containing species, surface sites become blocked by a high coverage of adsorbed O and OH, decreasing the number of sites available for the ORR. The position of Pt and Pd near the peak of this curve indicates a favorable balance between the rate of bond scission on available sites and the number of sites available for the ORR.

Kinetic studies of the ORR on Pt and related metals have revealed the presence of two distinct potential regions: at low overpotentials, the ORR is typically found to occur with a Tafel slope on the order of -60 mV per decade and a pH dependence of -30 mV per pH unit, while at larger overpotentials, the Tafel slope is on the order of -120 mV per decade, and the pH dependence is 0 mV per pH unit. These observations, along with the first-order dependence on O_2 partial pressure, are consistent with the proposed rate-limiting first electron transfer to adsorbed O_2 to form superoxide (Eq. 4). Under this interpretation, the distinct potential regimes are due to a transition from low surface coverage of O and OH at high overpotentials to higher surface coverage at low overpotentials. The role of adsorbed O and OH is also apparent in the hysteresis observed during potential cycling experiments, in which O and OH adsorption at higher potentials causes significantly lower ORR currents on negative-going sweeps than on positive-going sweeps.

Silver

Silver has been widely used as a cathode material in AFCs, in part due to its relatively low cost (~1 % that of Pt). The performance of Ag-based electrodes is typically reported to be slightly lower than that of Pt, but under high pH conditions, the ORR activity of Ag compares favorably with that of Pt [8]. The application of Ag as a catalyst in low temperature fuel cells is favorable only in alkaline media, as Ag is unstable in acidic electrolytes in the potential range of interest for the ORR.

The ORR mechanism on Ag in alkaline media shows many similarities with that on Pt, with similar Tafel behavior and reaction order in O_2, with both two-electron and four-electron pathways occurring, and with the reaction most likely proceeding via a superoxide intermediate [2]. On clean Ag surfaces, the four-electron ORR pathway predominates, but this selectivity is sensitive to surface contamination. On iodine-poisoned Ag surfaces, the two-electron ORR pathway predominates.

Unlike Pt, ORR currents on Ag electrodes are similar on positive-going and negative-going potential sweeps, i.e., they do not show the strong hysteresis exhibited by Pt electrodes [9]. This is due to the weaker affinity of O-containing species for Ag and the higher reversibility of OH adsorption on Ag than on Pt. An additional difference has been reported in the pH dependence of the ORR reaction, which is near zero for Ag electrodes. This difference may reflect ORR mechanistic differences between Ag and Pt; Sepa et al., for instance, interpreted the lack of pH dependence at low overpotential to indicate a different rate-limiting step on Ag surface [10].

N-M-C Catalysts

A variety of materials containing carbon and nitrogen, and in some cases transition metals, have been synthesized and tested for the ORR in alkaline media, and some of these non-precious catalysts have proven to be promising candidates to replace Pt and Ag in AFCs. One class of non-precious N-M-C ORR catalyst is based on metal macrocycles. Some macrocycles, such as Fe phthalocyanine, exhibit high ORR activity in alkaline media, but such catalysts are subject to rapid deactivation [11]. In AMFC testing, Co phthalocyanine cathode catalysts have yielded performance as high as nearly 100 mW/cm^2 in H_2/O_2, vs. around 130 mW/cm^2 for Pt/C [12].

Pyrolysis of macrocyclic precursors has been shown in some cases to yield more active and stable catalysts; for instance, pyrolysis of a mixture of Fe and Cu phthalocyanine-based complexes on carbon black produces a catalyst with 4-electron ORR activity higher than Pt/C at potentials lower than 0.84 V (RHE) in RDE experiments [13]. Such pyrolyzed macrocycle catalysts contain nitrogen coordinated to metal atoms, as well as nitrogen in C_xN_y compounds in which nitrogen is not coordinated to any metal. While the central importance of nitrogen to such catalysts is widely attested, the complex heterogeneous nature of these catalysts has thus far prevented any consensus on the identity of the ORR active site(s). Similarly, pyrolysis of Fe phthalocyanine to produce N-doped carbon nanotubes, followed by a chemical and electrochemical leaching to remove Fe, produced a highly-active ORR catalyst in which the activity was attributed to C-N sites, due to the lack of remaining Fe [14]; however, the evidence cited in this work did not rule out the presence of low levels of residual Fe. Furthermore, the possibility of Pt contamination must be considered in all experiments that use a Pt counter electrode. Therefore, the attribution of the high activity to C-N sites remains uncertain.

A related approach to producing N-M-C ORR catalysts makes use of mixtures of inorganic metal precursors with organic nitrogen compounds. For instance, ball milling of carbon black with phenanthroline and Fe(II) acetate, followed by pyrolysis in Ar and a subsequent pyrolysis in NH_3, produces an ORR catalyst with RDE-measured mass activity and volumetric activity 1/3 and 1/5 that of 46 wt% Pt/C, respectively [15]. The ORR on these materials occurs primarily via the 4-electron pathway. Similarly, pyrolysis of mixtures of carbon black with ethylenediamine, Co (II) nitrate, and Fe (II) sulfate, followed by acid leaching and a second pyrolysis, produces ORR catalysts that are competitive with Pt/C [16]. Fuel cell testing of such catalysts in an AMFC has yielded performance only slightly lower than that of an AMFC with 0.4 mg/cm^2 Pt at the cathode; notably, the performance is about the same in the 0.55–0.75 V window in which most fuel cell operation would occur, with the performance decline at lower voltage attributed to mass transport limitations arising from the thicker non-precious electrode layer (loading 4 mg/cm^2). RDE testing of these catalysts produced ORR performance apparently much lower than would be expected based on the MEA testing, but given the similarly poor performance of a Pt/C baseline, the veracity of these RDE results is questionable.

Along with nitrogen, inclusion of other heteroatoms in the catalyst synthesis, including sulfur, may play a role in enhancing ORR activity of non-PGM catalysts. Extremely high ORR activity has been achieved with a catalyst consisting of Co_9S_8 nanoparticles surrounded with nitrogen-doped graphene-like carbon (Co_9S_8-N-C) [17]. The Co_9S_8-N-C catalyst exhibits ORR activity in RDE testing 30–40 mV higher than that of Pt/C at low overpotential; notably, the improvement is even larger at high overpotential, due to the constant ~60 mV/decade Tafel slope throughout the experimental range, in contrast to the shift of Pt/C to a ~120 mV/decade Tafel slope at increased overpotential.

A related nanomaterial involves Co_3O_4 nanoparticles on N-doped reduced graphene oxide (Co_3O_4/N-rmGO), which was recently demonstrated as both an ORR and oxygen evolution catalyst. Neither the Co_3O_4, the reduced graphene oxide, nor the N-doped reduced graphene oxide exhibited acceptable ORR activity when tested separately, but the combined Co_3O_4/N-rmGO exhibited excellent ORR activity, highlighting an apparent synergy between the metal oxide and the N-doped carbon. In RDE testing, the ORR half-wave potential of 0.83 V for Co_3O_4/N-rmGO was only slightly lower than the 0.86 V measured for a Pt/C catalyst, though it should be noted that the Pt/C benchmark in these experiments was somewhat lower than that reported elsewhere [13, 15, 18]. Higher durability than Pt/C was also reported in this work, though the reported decay in Pt/C performance during a potential hold at 0.7 V is not representative of typical Pt behavior and may reflect the presence of contaminants. Despite the noted limitations of the Pt benchmark, the promising activity of the Co_3O_4/N-rmGO catalyst warrants further investigation.

While the cost advantage of using non-precious catalysts has long been apparent, the poor durability of most N-M-C catalysts has presented an obstacle to further development of these otherwise-promising materials. Despite the low durability evinced by metal macrocycles and other earlier formulations of N-M-C catalysts, recently developed catalysts based on an Fe-doped, N-containing carbon nanotube/nanoparticle composite have proved remarkably stable in RDE testing [18]. Indeed, the ORR half-wave potential of this N-Fe-CNT/CNP catalyst, which is initially 0.93 V (cf. 0.91 V for a Pt/C benchmark), increases to 0.95 V over the course of 5,000 potential cycles from 0.6 to 1.0 V (cf. a decrease to 0.90 V for the Pt/C under the same conditions). This remarkable net ~50 mV advantage over Pt/C after potential cycling suggests the extraordinary potential of these non-precious materials to replace precious metal catalysts in AMFCs.

Future Directions

Pt-containing catalysts have historically been the material of choice for the ORR in alkaline media, and a major role for these catalysts is anticipated in the future in the most demanding applications, such as in AFCs for space flight applications, where cost is relatively unimportant and extremely high catalyst loadings are acceptable. Opportunities for improvement of this catalyst technology are limited, as extensive optimization efforts have already been underway for more than five decades, and the technology has already reached a high degree of maturity.

In contrast, in more cost-sensitive applications where AMFCs may find their first commercial employment, non-precious N-M-C catalysts are expected to become the material of choice, though considerable maturation of these as of yet unproved materials is required before such a scenario could become a reality. The majority of recently described catalysts are either less active than Pt, less durable than Pt, or both. The very best N-M-C catalysts appear thus far to have advantages over Pt in both activity and durability, but the preliminary results on these materials come from half-cell studies in liquid electrolytes; considerable effort will be required to adapt these materials to AMFC electrodes. Such efforts are complicated by the immature state of AMFC membrane and electrode ionomer technology, both of which lag behind catalyst development. The lack of widespread availability of high-performance membranes and ionomers to catalyst developers has limited MEA testing to date, but such testing is required to confirm the activity and durability of alkaline ORR catalysts in the AMFC environment. Mass transport limitations are expected to be a major obstacle for many N-M-C catalysts, due to the relatively thick catalyst layers required. Furthermore, while high durability has been demonstrated for a few catalysts during potential cycling in liquid electrolytes, the application of these catalysts in AMFCs will require tolerance to much more destructive conditions, including large potential transients associated with startup/shutdown, fuel starvation, and cell reversal; temperature and humidity cycling; freeze/thaw cycling and operation in freezing environments; and exposure to fuel, air, and system-derived contaminants.

While the largest challenges for AMFC technology remain in the membrane and the electrode ionomer, further catalyst improvement is still required. The inherent limitations of hydroxide ion transport vs. proton transport suggest that AMFCs may always suffer an ohmic penalty when compared to their acidic counterparts; such a penalty must be counterbalanced by an equivalent advantage, and this advantage seems most likely to come in the form of low electrode overpotentials. Therefore, for AMFC technology to outcompete PEMFC technology in certain applications, AMFC catalysts must *significantly* outperform PEMFC catalysts. The rapid pace of improvement in N-M-C catalysts in the last few years, which has occurred despite continuing uncertainty as to what characteristics are required for high activity (including imprecise knowledge of the nature of the active site(s), and debate over the role of metals vs. nitrogen and other heteroatoms), suggests that major further improvements in non-precious alkaline ORR catalysts are to be expected in the next few years.

Cross-References

▶ Alkaline Membrane Fuel Cells
▶ Anion-Exchange Membrane Fuel Cells, Oxide-Based Catalysts
▶ Electrocatalysis - Basic Concepts, Theoretical Treatments in Electrocatalysis via DFT-Based Simulations
▶ Oxygen Reduction Reaction in Acid Solution

References

1. Yeager E (1984) Electrocatalysts for O2 reduction. Electrochim Acta 29(11):1527–1537
2. Spendelow JS, Wieckowski A (2007) Electrocatalysis of oxygen reduction and small alcohol oxidation in alkaline media. Phys Chem Chem Phys 9(21):2654–2675
3. Pletcher D, Sotiropoulos S (1993) A study of cathodic oxygen reduction at platinum using microelectrodes. J Electroanal Chem 356(1–2):109–119
4. Striebel K, McLarnon F, Cairns E (1990) Fuel-cell cathode studies in aqueous K2CO3 and KOH. J Electrochem Soc 137(11):3360–3367
5. Damjanovic A, Brusic V (1967) Electrode kinetics of oxygen reduction on oxide-free platinum electrodes. Electrochim Acta 12(6):615–628
6. Shao M, Liu P, Adzic R (2006) Superoxide anion is the intermediate in the oxygen reduction reaction on platinum electrodes. J Am Chem Soc 128(23):7408–7409
7. Lima FHB, Zhang J, Shao MH, Sasaki K, Vukmirovic MB, Ticianelli EA, Adzic RR (2007) Catalytic activity-d-band center correlation for the O-2 reduction reaction on platinum in alkaline solutions. J Phys Chem C 111(1):404–410
8. Chatenet M, Genies-Bultel L, Aurousseau M, Durand R, Andolfatto F (2002) Oxygen reduction on silver catalysts in solutions containing various concentrations of sodium hydroxide – comparison with platinum. J Appl Electrochem 32(10):1131–1140
9. Blizanac B, Ross P, Markovic N (2006) Oxygen reduction on silver low-index single-crystal surfaces in alkaline solution: rotating ring disk(Ag(hkl)) studies. J Phys Chem B 110(10):4735–4741
10. Šepa D, Vojnovíc M, Damjanovic A (1970) Oxygen reduction at silver electrodes in alkaline solutions. Electrochim Acta 15(8):1355–1366
11. Chen R, Li H, Chu D, Wang G (2009) Unraveling oxygen reduction reaction mechanisms on carbon-supported Fe-Phthalocyanine and Co-Phthalocyanine catalysts in alkaline solutions RID B-9937-2012. J Phys Chem C 113(48):20689–20697
12. Kruusenberg I, Matisen L, Shah Q, Kannan AM, Tammeveski K (2012) Non-platinum cathode catalysts for alkaline membrane fuel cells. Int J Hydrog Energy 37(5):4406–4412
13. He Q, Yang X, Ren X, Koel BE, Ramaswamy N, Mukerjee S, Kostecki R (2011) A novel CuFe-based catalyst for the oxygen reduction reaction in alkaline media. J Power Sources 196(18):7404–7410
14. Gong K, Du F, Xia Z, Durstock M, Dai L (2009) Nitrogen-doped carbon nanotube arrays with high electrocatalytic activity for oxygen reduction. Science 323(5915):760–764
15. Meng H, Jaouen F, Proietti E, Lefèvre M, Dodelet J-P (2009) pH-effect on oxygen reduction activity of Fe-based electro-catalysts. Electrochem Commun 11(10):1986–1989
16. Li X, Popov BN, Kawahara T, Yanagi H (2011) Non-precious metal catalysts synthesized from precursors of carbon, nitrogen, and transition metal for oxygen reduction in alkaline fuel cells. J Power Sources 196(4):1717–1722
17. Wu G, Chung HT, Nelson M, Artyushkova K, More KL, Johnston CM, Zelenay P (2011) Graphene-enriched Co9S8-N-C non-precious metal catalyst for oxygen reduction in alkaline media. ECS Trans 41(1):1709–1717
18. Chung HT, Won JH, Zelenay P (1922) Active and stable carbon nanotube/nanoparticle composite electrocatalyst for oxygen reduction. Nature Communications 4, doi:10.1038/ncomms2944

Oxygen Separation

Hitoshi Takamura
Department of Materials Science, Graduate School of Engineering, Tohoku University, Sendai, Japan

Description

Oxygen separation is expected to be a real possibility thanks to developments in mixed ionic and electronic conductors (MIECs). With mixed ionic and electronic conduction, oxide-ion conductors selectively permeate oxygen as a form of oxide ion. The mixed oxide-ion and electronic conductors used for this purpose are referred to as oxygen permeable membranes. An oxygen permeable membrane subjected to an oxygen potential gradient at elevated temperatures of around 700–1,000 °C leads to the ambipolar conduction of oxide ions and electrons, as shown in Fig. 1; the oxide ions transfer from the high oxygen pressure side to the lower side, while the electrons transfer from the lower to the higher oxygen pressure side. In this case, no electric field is required; the only driving force for the oxygen separation is the oxygen potential gradient, i.e., the difference in oxygen partial pressure.

For an oxygen permeable membrane with oxide-ion and electronic conductivities of σ_i(S/cm) and σ_e(S/cm), respectively, its theoretical oxygen flux density,

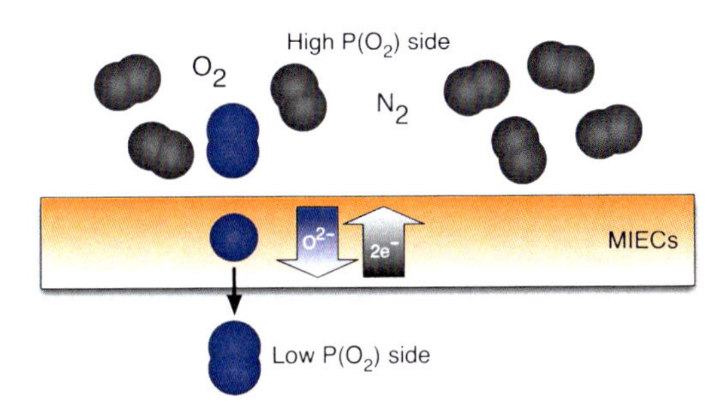

Oxygen Separation, Fig. 1 Oxygen permeable membrane based on mixed ionic and electronic conductor

$jO_2(\mathrm{mol} \cdot \mathrm{cm}^{-2} \cdot \mathrm{s}^{-1})$, under a given oxygen partial pressure difference between $P(O_2)'$ and $P(O_2)''$ can be calculated by Wagner's equation:

$$jO_2 = -\frac{RT}{16F^2L}\int_{lnP(O_2)'}^{lnP(O_2)''}\frac{\sigma_i\sigma_e}{\sigma_i+\sigma_e}dlnP(O_2) \quad (1)$$

where R and F are the gas constant ($\mathrm{J{\bullet}mol^{-1}{\bullet}K^{-1}}$) and the Faraday constant (C/mol), respectively, and T and L denote the temperature (K) and the membrane thickness (cm), respectively. Since $\sigma_i << \sigma_e$ and $\sigma_i(P(O_2)) \approx$ const, which is the case for most mixed conductors, (Eq. 1) can be simplified as follows:

$$jO_2 = -\frac{RT}{16F^2L}\sigma_i ln\frac{P(O_2)''}{P(O_2)'} \quad (2)$$

Based on (Eqs. 1) and (2), jO_2 depends on the conductivities, temperature, pressure gradients, and membrane thickness; the actual jO_2 reported are in the order of 10^{-6} $\mathrm{mol{\bullet}cm^{-2}{\bullet}s^{-1}}$. The unit of oxygen flux density can be converted to other units as follows:

$$1\,\mu\mathrm{mol{\bullet}cm^{-2}{\bullet}s^{-1}} = 1.34\,\mathrm{cm^3[STP]{\bullet}cm^{-2}{\bullet}min^{-1}} = 386\,\mathrm{mA{\bullet}cm^{-2}} \quad (3)$$

It should be noted that (1) jO_2 is controlled by the ambipolar conductivity of oxide ions and electrons, which is governed by a minor carrier with lower conductivity, and (2) in principle, a thinner membrane should show higher jO_2 proportional to the inverse of membrane thickness. However, as surface exchange kinetics is involved, actual jO_2 tends to be constant at a certain thickness, i.e., the characteristic thickness of L_c can be determined by the ratio of the diffusion coefficient, D, and the surface exchange coefficient, k.

To date, a variety of oxygen permeable membranes have been reported. In 1985, Teraoka et al. reported that La-Sr-Co-Fe-based perovskite-type oxides (LSCF), especially with high Sr and Co content, have a high oxygen permeation rate [1]. LSCF has been widely used not only for basic research but also for cathode of solid oxide fuel cells and oxygen separation from air. Because the transition metals used, like cobalt and iron, reduce easily in a reducing atmosphere, LSCF can be used only for high oxygen partial pressure. Balachandran et al. reported in 1995 that $SrFeCo_{0.5}O_x$ can be used even in a reducing atmosphere and can be used for syngas production, which is a mixture of CO and H_2, from methane [2]. For syngas production, the La-Sr-Ga-Fe-based perovskite-type oxides reported by Ishihara et al. in 2000 can also be utilized [3]. Among the many perovskite-type oxides, it is the Ba-Sr-Co-Fe-based ones (BSCF), and especially the composition of $Ba_{0.5}Sr_{0.5}Co_{0.8}\ Fe_{0.2}O_{3\text{-}\delta}$ reported by Shao et al. in 2000, which have a superior oxygen permeation rate [4].

The other type of oxygen permeable membranes is a composite type comprised of dual phases which have oxide-ion and electronic conductivities. Typical composites are comprised of a mixture of oxide-ion conductors, such as yttria-stabilized zirconia and doped ceria, and electronic conductors, such as precious metals and perovsikte-type oxides [5–7]. The advantage of the composite-type oxygen permeable membranes is their high chemical stability in wide oxygen partial pressure regions. They do, however, have lower oxygen permeation flux than the perovskite-type oxides mentioned above, so this needs to be improved. The lower oxygen flux density comes from their lower oxide-ion conductivity and/or limited reaction area; that is to say, the surface exchange reaction takes place only at the triple phase boundary where gaseous oxygen, oxide-ions, and electrons meet.

The typical application of oxygen permeable membrane is pure oxygen production from air. Cryogenic separation is currently used for large-scale pure oxygen production in industry. The energy consumption required for cryogenic separation is around 0.35 – 0.4 kWh/m^3-O_2. The oxygen separation process with the oxygen permeable membranes is believed to reduce energy consumption and make it possible to downsize oxygen production.

The other promising application is a so-called membrane reactor, which enables hydrocarbons such as methane to be converted into syngas or hydrogen. The technique currently used for syngas production from methane is steam reforming:

$$\begin{aligned} CH_4 + H_2O &= CO + 3H_2 \\ \Delta H_{298} &= 206.1\,kJ/mol \end{aligned} \tag{4}$$

This endothermic reaction is efficient; however, the reactor requires both a large volume to absorb a large amount of heat and a long wake-up time. Meanwhile, the following partial oxidation process, which utilizes pure oxygen, has been attracting a great deal of attention:

$$\begin{aligned} CH_4 + 1/2O_2 &= CO + 2H_2 \\ \Delta H_{298} &= -35.7\,kJ/mol \end{aligned} \tag{5}$$

This reaction is exothermic; the reactor can be downsized and woken up quickly. In this system, the oxygen permeable membrane can be used to separate air and methane; pure oxygen can be extracted from the air side to the methane side (low$P(O_2)$ side in Fig. 1) without the need for any additional energy. The driving force for oxygen separation is the large oxygen partial pressure gradients caused by air and the reducing gases, such as methane. By appropriately managing heat, including the recovery of oxygen-deficient air, the theoretical efficiency of the partial oxidation reactor reaches that for steam reforming.

Future Directions

In the future, the long-term stability of the oxygen permeable membrane, especially during oxygen separation and syngas production, has to be clarified. In principle, cation migration takes place in the opposite direction to oxygen migration, which results in the degradation of membrane materials. In addition, the surface exchange kinetics that limit the overall oxygen flux needs to be improved.

References

1. Teraoka Y, Zhang HM, Furukawa S, Yamazoe N (1985) Chem Lett 1743
2. Balachandran U, Dusek JT, Sweeney SM, Poeppel RB, Mieville RL, Maiya PS, Kleefisch MS, Pei S, Kobylinski TP, Udovich CA, Bose AC (1995) Methane to syngas via ceramic membranes. Am Ceram Soc Bull 74:71
3. Ishihara T, Yamada T, Arikawa H, Nishiguchi H, Takita Y (2000) Mixed electronic- oxide ionic conductivity and oxygen permeating property of Fe-, Co- or Ni-doped $LaGaO_3$ perovskite oxide. Solid State Ionics 135:631
4. Shao Z, Yang W, Cong Y, Dong H, Tong J, Xiong G (2000) Investigation of the permeation behavior and stability of a $Ba_{0.5}Sr_{0.5}Co_{0.8}Fe_{0.2}O_{3-d}$ oxygen membrane. J Membr Sci 172:177
5. Nigge U, Wiemhofer H-D, Romer EWJ, Bouwmeester HJM, Schulte TR (2002) Composites of $Ce_{0.8}Gd_{0.2}O_{1.9}$ and $Gd_{0.7}Ca_{0.3}CoO_{3-d}$ as oxygen permeable membranes for exhaust gas sensors. Solid State Ionics 146:163
6. Kharton VV, Kovalevsky AV, Viskup AP, Figueiredo FM, Yaremchenko AA, Naumovich EN, Marques FMB (2001) Oxygen permeability and faradaic efficiency of $Ce_{0.8}Gd_{0.2}O_{2-d}$ -$La_{0.7}Sr_{0.3}MnO_{3-d}$ composites. J Eur Ceram Soc 21:1763
7. Takamura H, Okumura K, Koshino Y, Kamegawa A, Okada M (2004) Oxygen permeation properties of ceria-ferrite-based composites. J Electroceramics 13:613

Oxygen Solid Electrolyte Coulometry

Ulrich Guth and Vladimir Vashook
Kurt-Schwabe-Institut für Mess- und Sensortechnik e.V. Meinsberg, Waldheim, Germany
FB Chemie und Lebensmittelchemie, Technische Universität Dresden, Dresden, Germany

Introduction

Coulometry is a quantitative electrochemical analysis based on counting the total electrical

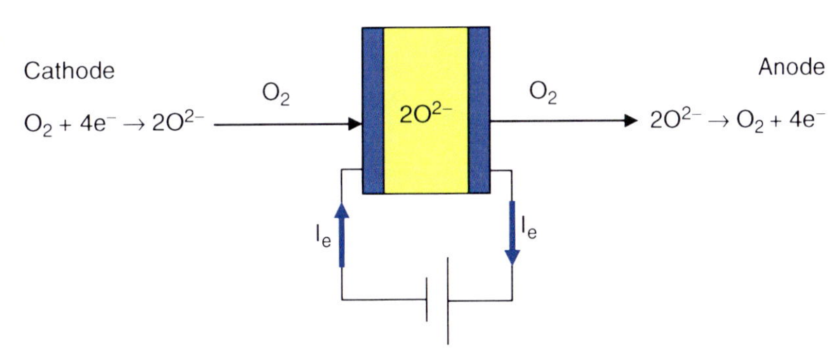

Oxygen Solid Electrolyte Coulometry, Fig. 1 Schematic presentation of oxygen solid electrolyte coulometry method (OSEC)

charge which flows through the electrochemical cell due to the electrochemical reaction. The solid state coulometry (SEC) can be performed with *solid electrolyte cells* [1] or *gas cells in equilibrium* [2]. Most relevant for analytical purposes are cells in which oxygen exchange is measured (OSEC). For this reason solid oxide electrolytes based on *stabilized zirconia* (ZrO_2, e.g., yttria-stabilized zirconia, YSZ), ceria (CeO_2, e.g., $Ce_{0.8}Gd_{0.2}O_{2-\delta}$, CGO), or gallates ($LaGaO_3$, e.g., $La_{1-x}Sr_xMg_yGa_{1-y}O_{3-\delta}$, LSGM) are used. Oxygen transport through gastight solid electrolyte membranes of these cells in form of O^{2-} ions is connected with equivalent electron transport in extrinsic circuit. Catalytically active electrode material (Pt) and high temperature of electrochemical cells are preconditions for the fast transport of oxide ions and sufficient high reaction rate of electrode reactions.

Easily presented in Fig. 1, electrode reactions are much more difficult (see also ▶ Solid Electrolytes Cells, Electrochemical Cells with Solid Electrolytes in Equilibrium). The mass of oxygen transported through gas-dense solid electrolyte membranes (m_{O_2}) can be exactly determined by the measurement of electrical current (I) following due to the Faraday's law:

$$m_{O_2} = \frac{M_{O_2}}{F * z} * \int_{t=\text{start}}^{t=\text{end}} I_t * dt = \frac{32}{96,500 * 4} * \int_{t=\text{start}}^{t=\text{end}} I_t * dt \quad (1)$$

where M_{O_2} is the molar mass of oxygen, F is the Faraday constant, z is the number of electrons needed for transport of one oxygen molecule, and I_t is the current in the cell at the time t.

Oxygen solid electrolyte coulometry can be successfully used for the following:

- Preparation of the gas mixtures having a defined oxygen concentration from 10^{-20} to 10^6 Pa (10^{-25} – 10 atm).
- Quantitative measurements of oxygen exchange between solid or liquid substances and the gas phase for scientific investigations, in different processes of the chemical and biochemical industry, in metallurgy, at semiconductor fabrication, and production of ceramics, etc. This method can be applied alternatively to thermogravimetry and can support the results of chemical analysis, spectroscopy, X-ray, or neutron diffractometry.
- Measurements of oxygen permeability (diffusivity) of ceramic, metal, or polymer membranes.
- Determination of humidity of gases.

Differently to many other analytical methods such as spectroscopy, X-ray, or neutron diffractometry, the method of solid electrolyte coulometry does not need calibration for the determination of oxygen content. Therefore, the measurement of oxygen content can be performed easy under in situ conditions. Unlike thermogravimetry, the solid electrolyte coulometry allows to determine a real oxygen exchange of materials, if their treatment is complicated by the other mass changes like gas or water desorption, fugacity, or decomposition. Oxygen exchange measurement can be realized in closed (Fig. 2) as well as in open (Fig. 3) systems.

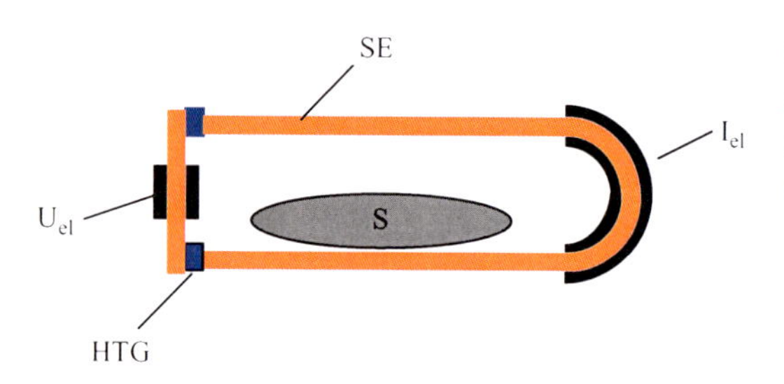

Oxygen Solid Electrolyte Coulometry, Fig. 2 Schematic representation of the cells for coulometric experiments: *SE* solid electrolyte tube, U_{el} potential electrodes, I_{el} current electrodes, *S* sample, *HTG* high temperature glass

Oxygen Solid Electrolyte Coulometry in Quasi-Closed Systems

Oxygen solid electrolyte coulometry in quasi-closed systems is understood at first the complete isolation of the material to be investigated in a minimal chamber with gas-dense solid electrolyte wall attached out- and inside with electrochemical active electrodes. Then oxygen can only be admitted to the chamber or removed from it quantitatively by coulometric titration. The chamber can be made completely (Fig. 2) or partly of solid electrolyte material. Current (I_{el}) and potential (U_{el}) electrodes are insulated from each other with gas-dense high temperature glass (HTG) layer, so titration current and voltage are simultaneously measurable at this construction of the cell.

The exchange of oxygen between solid sample S and gas environment could be determined as a function of temperature and oxygen partial pressure in known chamber volume. For example, if some solid oxide sample AO_x is placed and sealed at room temperature in vacuum into solid electrolyte cell (Fig. 2), the next reaction could be observed after the cell is heated to enough high temperature:

$$AO_x \rightarrow AO_{x-\delta} + \delta/2\ O_2 \tag{2}$$

If a defined equilibrium oxygen partial pressure (pO'_2) in the chamber at temperature T and oxygen titration current $I = 0$ was reached, oxygen content $(x - \delta_1)$ of oxide could be calculated according the potentiometric method following the next equation:

$$\begin{aligned} x - \delta_1 &= x - \frac{V * T_0 * M}{11.2 * T * m} * pO'_2 \\ &= x - \frac{V * T_0 * M}{11.2 * T * m} * \exp\left(-\frac{4FU_1}{RT}\right) \end{aligned} \tag{3}$$

where V is the volume of the chamber, T_0 is the normal temperature (273 K), M is the molar mass of oxide AO_x, m is the mass of the sample, pO_2' is the oxygen partial pressure, U_1 is the voltage in the Nernst's cell, F is the Faraday's constant, and R is the gas constant.

By coulometric titration of oxygen at constant temperature inside or outside of chamber through the current electrodes, the new oxygen partial pressure in chamber can be established and determined by voltage (U_2) measured on potential electrodes (U_{el}). Then the changed oxygen content of solid oxide material ($\Delta\delta_s$) can be calculated for constant temperature as follows:

$$\begin{aligned} \Delta\delta_s &= \Delta\delta - \Delta\delta_g \\ &= \frac{\int_{t=start}^{t=end} I_t * dt}{z * F} - \frac{V}{RT} * \Delta pO_2 \\ &= \frac{\int_{t=start}^{t=end} I_t * dt}{z * F} - \frac{V}{RT} \\ &\quad * \exp\left(\frac{-4F * \Delta U}{RT}\right) \end{aligned} \tag{4}$$

where $\Delta\delta$ is the total change of oxygen content in chamber, $\Delta\delta_g$ is the changed oxygen content of gas phase for new oxygen partial pressure pO''_2, and ΔpO_2 is the changed oxygen partial pressure in the chamber corresponding to changed voltage on the potential electrodes U_{el} ($\Delta U = U_2 - U_1$).

The known p-T-x diagrams can be constructed for some simple and mixed oxides using this gas coulometric method. Closed systems

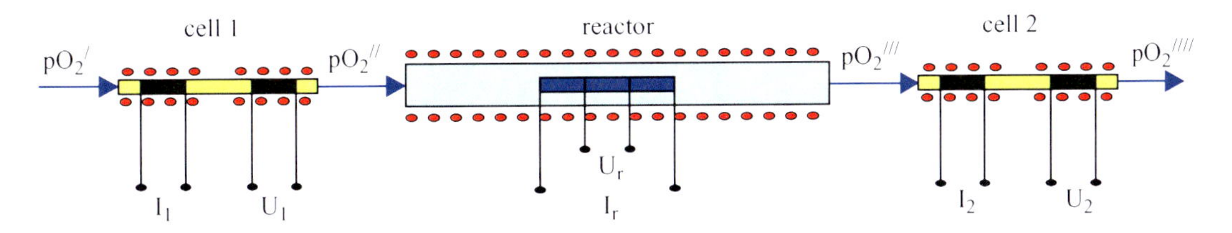

Oxygen Solid Electrolyte Coulometry, Fig. 3 Scheme of device for simultaneous in situ solid electrolyte coulometry and electrical conductivity measurement

allow correct investigations of oxygen nonstoichiometry when all construction elements of the cell are stable and gastight. The sealing consisting of a high-temperature glass is normally the restriction elements of such cells. Utilization of these systems is normally possible at temperatures 600–800°C because of low oxygen conductivity of solid electrolytes under 600° C and high oxygen permeation and reactivity of glasses over 800°C. The range of possible oxygen partial pressures is also restricted by oxygen pressures over 10 Pa.

Oxygen Solid Electrolyte Coulometry in Open Systems

Oxygen solid electrolyte coulometry in open systems (in carrier gas mode) has no restrictions of measurement conditions of samples like temperature or oxygen concentration and is more flexible in comparison with closed systems using equal temperature of both solid electrolyte cell and sample under investigation [3]. Materials can be investigated at every temperature and gas atmospheres accept of combustaible gases and such which are chemically aggressive to electrodes and solid electrolyte gases. Gas mixtures with small concentration of burning gases equilibrated with oxygen at experimental conditions (H_2O,H_2,O_2; CO_2,CO,O_2) can be investigated by this method as well. The other properties of materials like electrical conductivity, magnetization, and crystal structure could be easy determined online simultaneously with oxygen exchange.

Figure 3 illustrates operating principle of the open solid electrolyte coulometric-potentiometric system combined by a four-electrode conductivity measurement.

Two identical solid electrolyte cells are utilized in this measurement system. Every cell has both a pair of oxygen pumping (I) and potential (U) electrodes. The cells let to prepare steady-state gas flows at given oxygen concentration by preset of electrode voltage U at known temperature. The given voltage is controlled by feedback adjustment of coulometric titration current I between oxygen pumping electrodes. If an experimental reactor with a sample is placed between two such cells, an oxygen partial pressure in environment of sample can vary in a range of 10^{-15}–10^{5} Pa. Temperature of reactor is restricted only by material of reactor, furnace, and sample properties, because the cells are working at optimal for electrodes and solid electrolyte conditions. As carrier gas, inert gases like Ar, N_2 mixed with O_2, H_2O, H_2, O_2, and CO_2, CO, O_2, can be used.

A stable oxygen partial pressure pO_2' of inflowing gas can be modified in the cell 1 by coulometric oxygen dosing for the needed pO_2'' values. No change of oxygen concentration after reactor is observed if no oxygen exchange processes take place in reactor ($pO_2'' = pO_2'''$). Constant basic oxygen dosing current $I_{2,base}$ could be caused inside of cell 2, if the given constant voltage U_2 is lower as U_1 ($pO_2'' < pO_2''''$). After enough long sweeping time, all controlled electrochemical parameters I_1, I_2, U_1, and U_2 reach constant values; if gas fittings are dense, carrier gas flow is constant and no oxygen exchange processes take place in reactor. Every oxygen exchange in reactor should be then accompanied by deviation of coulometric

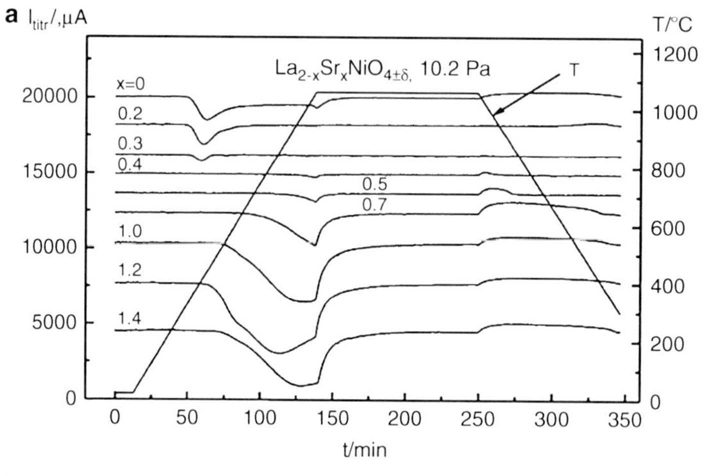

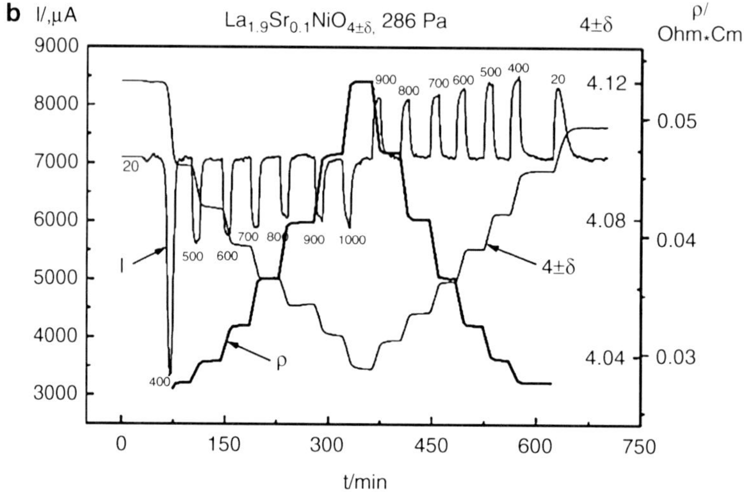

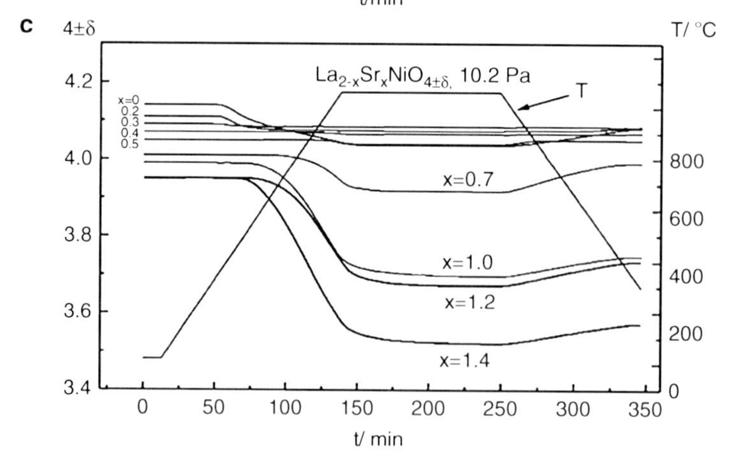

Oxygen Solid Electrolyte Coulometry, Fig. 4 (continued)

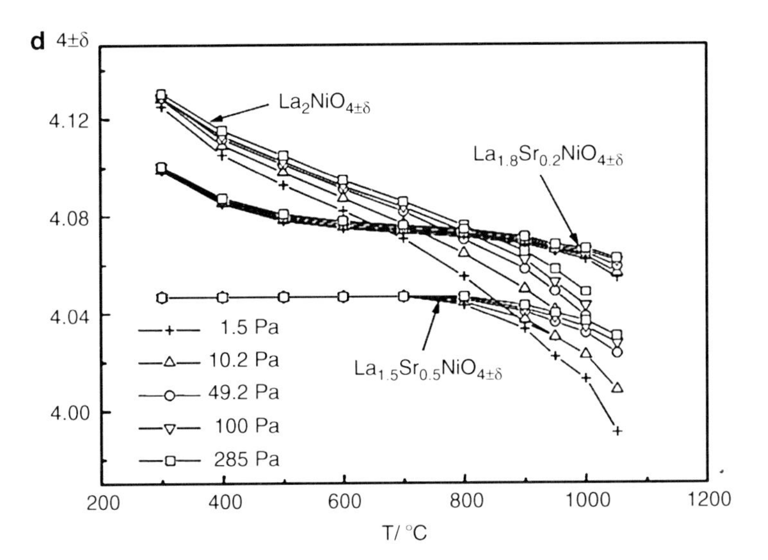

Oxygen Solid Electrolyte Coulometry, Fig. 4 (**a**) Titration current (I_2) during programmed treatment of $La_{2-x}Sr_xNiO_{4\pm\delta}$ powders for different compositions. (**b**) Titration current (I_2), oxygen content ($4\pm\delta$), and resistance of $La_{2-x}Sr_xNiO_{4\pm\delta}$ at defined oxygen partial pressure during stepwise heating and cooling. (**c**) Oxygen content of $La_{2-x}Sr_xNiO_{4\pm\delta}$ powder depending on the composition. (**d**) Oxygen nonstoichiometry some of $La_{2-x}Sr_xNiO_{4\pm\delta}$ compositions depending on oxygen partial pressure in ambient gas

titration current I_2 from $I_{2,\text{base}}$ value, if U_2 is kept constant. The mass of exchanged oxygen (Δm_{O_2}) is calculated according to the Faraday's law:

$$\Delta m_{O_2} = \frac{M_{O_2}}{F * z} * \int_{t=\text{start}}^{t=\text{end}} (I_{2,\text{base}} - I_t)\, dt = \frac{32}{96,500 * 4} * \int_{t=\text{start}}^{t=\text{end}} (I_{2,\text{base}} - I_t)\, dt \quad (5)$$

Knowing mass and chemical composition of the investigated sample, so named p-T-x and p-T-σ diagrams of oxides are easy to construct by jumping of oxygen partial pressure and temperature in reactor [4]. As an example, Fig. 4a–d illustrates the possibilities of that method by investigation of $La_{2-x}Sr_xNiO_{4\pm\delta}$ mixed oxides [5].

Beside the gas coulometry, solid state coulometry is used to investigate transport processes in solid conductors. In such a way, the transference number of solid conductors (e.g., Ag^+ in $\alpha-AgI$, at T > 149 °C) could be determined according to Tubandt's pellet cell arrangement [6]:

$$+ Ag|AgI|AgI|AgI|Ag- \quad (6)$$

Two Ag pellets are electrodes. The electrolyte to be investigated is divided into three disks. After a current flow, the changes in weight in the three pellets are measured and the transference number of Ag^+ ions can be calculated [7].

Cross-References

- Oxide Ion Conductor
- Solid Electrolytes
- Solid Electrolytes Cells, Electrochemical Cells with Solid Electrolytes in Equilibrium

References

1. Wagner C (1953) Investigations on silver sulfide. J Chem Phys 21:1819–1827

2. Tretyakov Yu D, Rapp R (1969) Nonstoichiometries und defect structures in pure nickel oxide and lithium ferrite. Trans Met Soc AIME 245:1235
3. Teske K, Ullmann H, Rettig D (1983) Investigation of the oxygen activity of oxide fuels and fuel-fission product systems by solid electrolyte techniques. Part I: qualification and limitations of the method. J Nucl Mater 116:260–266
4. Vashook V, Vasylechko L, Zosel J, Gruner W, Ullmann H, Guth U (2004) Crystal structure and electrical conductivity of lanthanum–calcium chromites–titanates $La_{1-x}Ca_xCr_{1-y}Ti_yO_{3-\delta}$ (x = 0–1, y = 0–1). J Solid State Chem 177:3784–3794
5. Vashook VV, Tolochko SP, Yushkevich II, Makhnach LV, Kononyuk IF, Altenburg H, Hauck J, Ullmann H (1998) Oxygen nonstoichiometry and electrical conductivity of the solid solutions $La_{2-x}Sr_xNiO_y$ ($0<x<0.5$). Solid State Ion 110:245–253
6. Tubandt C (1932) Conductivity and transference numbers in solid electrolytes (in German). In: Wien W, Harms F (eds) Hdb. Exp. Physik XII, part 1 381 et sqq. Leipzig, Germany
7. Rickert H (1982) Electrochemistry of solids Springer Berlin. New York, Heidelberg

Printed by Publishers' Graphics LLC